# HANDBOOK OF
# NETWORK AND SYSTEM
# ADMINISTRATION

# HANDBOOK OF NETWORK AND SYSTEM ADMINISTRATION

*Edited by*

Jan Bergstra
*Informatics Institute, University of Amsterdam*
*Amsterdam, The Netherlands*

Mark Burgess
*Faculty of Engineering, University College Oslo*
*Oslo, Norway*

ELSEVIER

Amsterdam – Boston – Heidelberg – London – New York – Oxford – Paris
San Diego – San Francisco – Singapore – Sydney – Tokyo

Elsevier
Radarweg 29, PO Box 211, 1000 AE Amsterdam, The Netherlands
Linacre House, Jordan Hill, Oxford OX2 8DP, UK

First edition 2007

*Library of Congress Cataloging-in-Publication Data*
A catalog record for this book is available from the Library of Congress

*British Library Cataloguing in Publication Data*
A catalogue record for this book is available from the British Library

ISBN:     978-0-444-52198-9

For information on all Elsevier publications
visit our website at books.elsevier.com

Printed and bound in the United Kingdom

Transferred to Digital Printing

# Preface

The volume, of which you are in some doubt could reasonably be called a book to be held in human hands, is a collection of chapters sampling the current State of the Art of network and system administration or management. It is written by researchers, for researchers, educators and advanced practitioners.

The field of Network and System Administration is (some would claim) still an emerging field, one that bridges disciplines, and challenges the traditional boundaries of computer science. It forces us to confront issues like partial predictability and partial reliability that are more commonly associated with engineering disciplines in an uncertain environment. It does so because human–computer networks *are* complex and uncertain environments. In other words, it rediscovers the lesson that other sciences have already made peace with: that *approximation* is the only realistic approach to its difficulties.

Whatever the novelty of the field, there is already a large body of knowledge about it, contained in a diverse array of sources. This has only recently begun to be formalized however, in the way we conventionally expect of an academic discipline. Much of it has been only sketchily formulated and does not lend itself to future development by researchers and designers. We hope to partially rectify that situation with this volume.

So what is administration or management? This is a difficult question to answer in full, but it clearly covers few themes:

- Making choices and decisions about design, implementation, practice and operations.
- Provision of resources.
- Infrastructure and service design.
- Maintenance and change management.
- Verification and validation of specifications.
- Economic usage and business integration.

Within the list above, we also find the foundations for more abstract concerns like security and commerce. We have not been able to cover every aspect of the field in this book, but we have hopefully covered enough to take us forward.

Our *Handbook of Network and System Administration* is not a textbook for day-to-day practitioners. It is aimed at those who are seeking to move beyond recipies and principles, on to creative innovation in application, development or research; it offers a solid scientific and engineering foundation for the field. We hope that it will also inspire less experienced readers. For them we recommend a background for this volume in a number of recent textbooks:

- For a technical introduction to the principles and techniques of system administration one has *Principles of Network and System Administration*, Mark Burgess. Wiley (2000,2004).

- An introduction to the heuristics and day to day practices of system administration is found in *The Practice of System and Network Administration*, T.A. Limoncelli and C. Hogan. Addison Wesley (2002).
- Advanced matters of modelling and in-depth understanding of systems can be found in *Analytical Network and System Administration*, Mark Burgess. Wiley (2004).

In addition to this educationally honed sequence, there are numerous resources on specific technologies, for instance in the 'animal series' of books by the publisher O'Reilly.

If we add to the above a book of collected papers, which documents the early work in system administration,

- *Selected Papers in Network and System Administration*, Erik Anderson, Mark Burgess and Alva Couch, Eds. Wiley (2001).

and a book of chapters on the early parts of Network Management

- *Network and Distributed Systems Management*, Morris Sloman, Ed. Addison Wesley (1994).

then we could present this volume as the next piece of a series covering the academic spectrum in the field.

An editor quickly discovers that soliciting contributions is harder than one might think. Shamelessly appealing to the vanity of contributors only goes so far in coaxing words onto pages. Those who have practiced network and system management, in particular, rarely like to write down their knowledge. This has made our volume end up being more forward-looking than historical, although we have succeeded in covering a few examples of classic materials, such as E-mail and backup. It has left us to try to mention the omissions in our commentary, with obvious limitations. Fortunately many traditional topics are already discussed in the books above.

Our book is already quite thick, so let us not prolong the introduction unnecessarily. We simply invite you to read the chapters and we offer some brief commentary along the way.

## Acknowledgements

Jan Bergstra acknowledges support from the NWO-Jacquard grant Symbiosis. The Symbiosis project focuses on IT sourcing in general and software asset sourcing more specifically.

Mark Burgess acknowledges support from the EC IST-EMANICS Network of Excellence (#26854).

Jan Bergstra (Amsterdam),
Mark Burgess (Oslo)

# List of Contributors

Badonnel, R., *LORIA, Lorraine, France* (Ch. 3.2)

Bandara, A., *Imperial College London, London, UK* (Ch. 4.3)

Bergstra, J.A., *University of Amsterdam, Amsterdam, and Utrecht University, Utrecht, The Netherlands* (Ch. 5.4)

Bethke, I., *University of Amsterdam, Amsterdam, The Netherlands* (Ch. 5.4)

Bishop, M., *University of California at Davis, Davis, CA, USA* (Ch. 4.2)

Brenner, M., *Ludwig Maximilians University, Munich, Germany* (Ch. 7.3)

Brown, A.B., *IBM Software Group, Somers, NY, USA* (Ch. 1.3)

Burgess, M., *University College Oslo, Oslo, Norway* (Chs. 1.2, 5.2, 6.2, 6.3)

Canright, G.S., *Telenor R&D, Fornebu, Norway* (Ch. 3.3)

Chalup, S.R. (Ch. 8.2)

Choi, M.-J., *POSTECH, Pohang, South Korea* (Ch. 2.4)

Couch, A.L., *Tufts University, Medford, MA, USA* (Chs. 1.4, 5.3)

Damianou, N., *Imperial College London, London, UK* (Ch. 4.3)

Delen, G., *Amsterdam School of ICT, Amsterdam, The Netherlands* (Ch. 7.5)

Dreo Rodosek, G., *University of Federal Armed Forces Munich, Munich, Germany* (Ch. 7.3)

Dulay, N., *Imperial College London, London, UK* (Ch. 4.3)

Engø-Monsen, K., *Telenor R&D, Fornebu, Norway* (Ch. 3.3)

Fagernes, S., *University College Oslo, Oslo, Norway* (Ch. 8.3)

Festor, O., *LORIA, Lorraine, France* (Ch. 3.2)

Frisch, Æ., *Exponential Consulting, Wallingford, CT, USA* (Ch. 2.6)

Hanemann, A., *Leibniz Supercomputing Center Munich, Munich, Germany* (Ch. 7.3)

Hausheer, D., *ETH Zürich, Computer Engineering and Networks Laboratory, TIK, Switzerland* (Ch. 7.2)

Hegering, H.-G., *Leibniz Supercomputing Center Munich, Munich, Germany* (Ch. 7.3)

Hellerstein, J.L., *Microsoft Corporation, Redmond, WA, USA* (Ch. 1.3)

Hong, J.W., *POSTECH, Pohang, South Korea* (Ch. 2.4)

Hoogeveen, D., *Verdonck, Klooster & Associates B.V., Zoetermeer, The Netherlands* (Ch. 7.6)

Keller, A., *IBM Global Technology Services, New York, NY, USA* (Ch. 1.3)

Koenig, R., *Ludwig Maximilians University, Munich, Germany* (Ch. 7.3)

Koymans, C.P.J., *University of Amsterdam, Amsterdam, The Netherlands* (Chs. 2.3, 2.5)

Koymans, K., *University of Amsterdam, Amsterdam, The Netherlands* (Ch. 3.5)

Kristiansen, L., *University of Oslo, Oslo, Norway* (Ch. 5.2)

Lupu, E., *Imperial College London, London, UK* (Ch. 4.3)

Martin-Flatin, J.-P., *NetExpert, Gland, Switzerland* (Ch. 2.7)

Pras, A., *University of Twente, Enschede, The Netherlands* (Ch. 2.7)

Reitan, T., *University of Oslo, Oslo, Norway* (Ch. 6.4)

Ribu, K., *University College Oslo, Oslo, Norway* (Ch. 8.3)

Scheerder, J. (Chs. 2.3, 2.5)

Scheffel, P.F.L., *Waterford Institute of Technology, Waterford, Ireland* (Ch. 7.4)

Schönwälder, J., *Jacobs University Bremen, Bremen, Germany* (Ch. 2.8)

Sloman, M., *Imperial College London, London, UK* (Ch. 4.3)

State, R., *LORIA, Lorraine, France* (Ch. 3.2)

Stav, K., *IBM Corporation, USA* (Ch. 2.2)

Stiller, B., *University of Zürich and ETH Zürich, Switzerland* (Ch. 7.2)

Strassner, J., *Motorola Labs, Schaumburg, IL, USA* (Chs. 3.4, 7.4)

Sun, Y., *Tufts University, Medford, MA, USA* (Ch. 5.3)

Vrancken, J., *Delft University of Technology, Delft, The Netherlands* (Ch. 3.5)

# Contents

# 1. The Arena

## 1.1. *Commentary*

The arena of system management is as broad as it is long. What began in consolidated computer centres has spread to the domestic arena of households and even to the outdoors, through a variety of personal consumer electronic items. It is an open question how much management such computing devices need in such different milieux.

In this part the authors describe some of the issues involved in the administration of systems in a more traditional organizational settings. Many matters have been left out; others are covered in later chapters, such as Chapter 3.2 about ad hoc networking in Part 3.

During the 1970s one spoke of computer operations, during the 1980s and 1990s 'system administration' or sometimes 'network administration' was the preferred appellation. Now in the 2000s we have become *service oriented* and departments and businesses are service providers. The organization is required to provide a service to customers within a 'business' framework.

Efficiency and productivity are key phrases now, and most computers in a business or university environment are located within some kind of data centre. Since the limits of performance are constantly being pushed Automation is an important approach for scalability. From an economic perspective, automation is a good investment for tasks that are frequently repeated and relatively easy to perform. As they become infrequent or difficult, the economics of automation become less attractive.

Configuration management has been one of the main areas of research in the field over the past twenty years. In other management arenas, as well as in software engineering, configuration management has a high level meaning and concerns the arrangement of resources in an enterprise. Do not be confused by the low-level interpretation of configuration management used by system administrators. This concerns the bits and bytes of files and databases that govern the properties of networks of computing devices.

# – 1.2 –

# Scaling Data Centre Services

## Mark Burgess

*Faculty of Engineering, University College Oslo, Room 707, Cort Adelers Gate 30, 0254 Oslo, Norway*
*E-mail: mark.burgess@iu.hio.no*

## 1. Introduction

The management of computer resources is a key part of the operation of IT services. Its trends tend to proceed in cycles of distribution and consolidation. Today, the data centre plays an increasingly important role in services: rather than managing computer resources in small private rooms, many companies and organizations are once again consolidating the management of systems into specialized data centres. Such organized management is linked to the rise of de facto standards like ITIL [4,25,26] and to the increased use of outsourcing and service oriented computing which are spreading through the industry. Management standards like ITIL are rather abstract however; this chapter offers a brief technical overview of the data centre environment.

## 2. Computing services

Computing services are increasingly required to be scalable and efficient to meet levels of demand. The delivery of high levels of computing service is associated with three common phrases:
- High Performance Computing (HPC).
- High Volume Services (HVS).
- High Availability Services (HAS).

These phrases have subtly different meanings.

HANDBOOK OF NETWORK AND SYSTEM ADMINISTRATION
Edited by Jan Bergstra and Mark Burgess

## 2.1. *High performance computing*

HPC is the term used to describe a variety of CPU intensive problems. This often involves supercomputers and/or networked clusters of computers to take a large computational problem, divide it into to smaller chunks and solve those chunks in parallel. HPC requires very high-capacity and/or low-latency connections between processors. The kinds of problems solved in this way are typically of a scientific nature, such as massive numerical problems (e.g., weather and climate modelling), or search problems in bioinformatics.

For problems that require massive computational power but which are less time-critical, another strategy is *Grid computing*, an extension of the idea in which the sharing takes place over anything from a short cable to a wide area network.

## 2.2. *High volume services*

This refers to applications or services, normally based on content provision, that are designed to handle large numbers of transactions. HVS are a generalization of HPC to non-specifically CPU-oriented tasks. High volume services are achieved by implementing strategies for workload sharing between separate server-hosts (often called load balancing).

HVS focuses on serving large volumes of requests, and must therefore address scalability. System throughput demands a careful analysis of system bottlenecks. Load-sharing is often required to attain this goal. HVS includes the economics and the performance tuning of services. The ability to deliver reliable service levels depends on both the resources that are available in the data centre and the pattern of demand driven by the users.

## 2.3. *High availability services*

These are applications designed for *mission-critical* situations, where it is essential that the service is available for the clients with a *low latency*, or short response time. To achieve this, a mixture of redundancy and fault detection is implemented in the system. HAS is about ensuring that a service is available with an assured quality of response time, with target service level. One must decide what 'available' means in terms of metric goals (e.g., upper bound response time).

## 3. Resources

To achieve a high performance (and hence high volume) computing we need resource management. The resources a computer system has to utilize and share between different tasks, users and processes are:
- CPU.
- Disk.

- Memory.
- Network capacity.[1]

Many technologies are available to do optimization and perform this sharing, as noted in the forthcoming sections.

Computer processing is limited by *bottlenecks* or throughput limitations in a system. Sometimes these performance issues can be separated and conquered one by one, other times interdependencies make it impossible to increase performance simply by making changes to components. For instance, some properties of systems are *emergent*, i.e., they are properties of the whole system of all components, not of any individual part.

### 3.1. *CPU*

For CPU intensive jobs, the basic sharing tools are:
- Multi-tasking and threading at the operating system level.
- Multi-processor computers, vector machines and Beowulf clusters which allow parallelism.
- Mainframe computing.
- Grid Engine job distribution systems for parallel computing.
- MPI (Message Passing Interface) programming tools.

Flynn's taxonomy is a classification scheme for parallel computers based on whether the parallelism is exhibited in the instruction stream and/or in the data stream. This classification results in four main categories, SISD (single instruction single data), SIMD (single instruction multiple data), MISD (multiple instruction single data) and MIMD (multiple instruction multiple data). Different strategies for building computers lead to designs suited to particular classes of problems.

Load-balancing clusters operate by having all workload come through one or more load-balancing front ends, which then distribute work to a collection of back end servers. Although they are primarily implemented for improved performance, they commonly include high availability features as well. Such a cluster of computers is sometimes referred to as a server farm.

Traditionally computers are built over short distance buses with high speed communications. Dramatic improvements in network speeds have allowed wide area connection of computer components. Grid computing is the use of multiple computing nodes connecting by such wide area networks. Many grid projects exist today; many of these are designed to serve the needs of scientific projects requiring huge computational resources, e.g., bioinformatics, climate and weather modelling, and the famous Search for Extra-terrestrial Intelligence (SETI) analysis project.

### 3.2. *Disk*

Fast, high volume storage is a key component of delivery in content based services. Because disk-storage requires mechanical motion of armatures and disk heads, reading from

---

[1]The use of the term 'bandwidth' is commonly, if incorrectly, used in everyday speech. Bandwidth refers to frequency ranges. Channel capacity (or *maximum throughput*) is proportional to bandwidth – see chapter System Administration and the Scientific Method in this volume.

and writing to disk is a time-consuming business, and a potential bottleneck. Disk services that cache pages in RAM can provide a large speed-up for read/write operations.

RAID includes a number of strategies for disk performance and reliability. Redundant Arrays of Independent Disks[2] are disk arrays that have special controllers designed to optimize speed and reliability. RAID deals with two separate issues: fault tolerance and parallelism. Fault tolerant features attempt to reduce the risk of disk failure and data-loss, without down-time. They do this using redundant data encoding (parity correction codes) or mirroring. Parallism (mirroring and striping of disks) is used to increase data availability. These are often conflicting goals [11,15,16,31].

*Disk striping*, or parallelization of data access across parallel disks can result in an up-to-$N$-fold increase in disk throughput for $N$ disks in the best case.

*Error correction*, sometimes referred to as *parity*, is about reliability. Error correction technology adds additional search and verification overheads, sometimes quoted at as much as 20% to the disk search response times. It can be a strategy for increasing up-time however, since RAID disks can be made hot-swappable without the need for backup.

Ten years ago networks were clearly slower than internal databuses. Today network speeds are often faster than internal databuses and there is a potential gain in both performance and administration in making disk access a network service. This has led to two architectures for disk service:

- Storage Area Networks (SAN).
- Network Attached Storage (NAS).

A Storage Area Network is an independent network of storage devices that works in place of disks connected on a local databus. This network nevertheless appears as a low level disk interface, usually SCSI. Disk storage appears as locally attached device in a computer operating system's device list. The device can be formatted for any high level filesystem.

SAN uses ISCSI (SCSI over Internet) using a dedicated network interface to connect to a storage array. This gives access to an almost unlimited amount of storage, compared to the limited numbers of disks that can be attached by the various SCSI versions. SAN can also use Fibre channel, which is another fibre-based protocol for SCSI connection.

NAS works more like a traditional network service. It appears to the computer as a filesystem service like NFS, Samba, AFS, DFS, etc. This service is mounted in Unix or attached as a logical 'drive' in Windows. It runs over a normal network protocol, like IP. Normally this would run over the same network interface as normal traffic.

### 3.3. *Network*

Network service provision is a huge topic that evolves faster than the timescale of publishing. Many technologies compete for network service delivery. Two main strategies exist for data delivery:

- Circuit switching.
- Packet switching.

---

[2]Originally the 'I' meant Inexpensive, but that now seems outdated.

These are roughly analogous to rail travel and automobile travel respectively. In circuit switching, one arranges a high-speed connection between certain fixed locations with a few main exchanges. In the packet switching, packets can make customized, individually routed journeys at the expense of some loss of efficiency.

Network throughput can be improved by adapting or tuning:

- Choice of transport technology, e.g., Fibre, Ethernet, Infiniband, etc.
- Routing policies, including the strategies above.
- Quality of service parameters in the underlying transport.

Most technological variations are based on alternative trade-offs, e.g. high speed over short distances, or stable transport over wide-areas. The appropriate technology is often a matter of optimization of performance and cost.

## 4. Servers and workstations

Computer manufacturers or 'vendors' distinguish between computers designed to be 'workstations' and computers that are meant to be 'servers'. Strictly speaking, a server is a *process* that provides a service not necessarily a whole computer; nevertheless, this appellation has stuck in the industry.

In an ideal world one might be able to buy a computer tailored to specific needs, but mass production constraints lead manufacturers to limit product lines for home, businesses and server rooms. Home computers are designed to be cheap and are tailored for gaming and multi-media. Business computers are more expensive versions of these with different design look. Servers are designed for reliability and have their own 'looks' to appeal to data centre customers.

A server computer is often designed to make better use of miniaturization, has a more robust power supply and better chassis space and expansion possibilities. High-quality hardware is also often used in server-class systems, e.g. fast, fault tolerant RAM, and components that have been tested to higher specifications. Better power supply technology can lead to power savings in the data centre, especially when many computers are involved in a large farm of machines.

We sometimes think of servers as 'large computers', but this is an old fashioned view. Any computer can now act as a server. Basic architectural low-level redundancy is a selling point for mainframes designs, as the kind of low-level redundancy they offer cannot easily be reproduced by clustering several computers.

### 4.1. *Populating a data centre*

How many computers does the data centre need, and what kind? This decision cannot be made without an appraisal of cost and performance. In most cases companies and organizations do not have the necessary planning or experience to make an informed judgment at the time they need it, and they have to undergo a process of trial-and-error learning. Industry choices are often motivated by the deals offered by sellers rather than by an engineering appraisal of system requirements, since fully rational decisions are too complicated for most buyers to make.

One decision to be made in choosing hardware for a server function is the following. Should one:

- Buy lots of cheap computers and use high-level load sharing to make it perform?
- Buy expensive computers designed for low-level redundancy and performance?

There is no simple answer as to which of these two approaches is the best strategy. The answer has fluctuated over the years, as technological development and mass production have advanced. Total cost of ownership is usually greater with custom-grown systems and cheap-to-buy computers. The cost decision becomes a question of what resources are most available in a given situation. For a further discussion of this see the chapter on System Administration and Business in this volume.

Computer 'blades' are a recent strategic design aimed at data centres, to replace normal chassis computers.

- Blades are small – many of these computers can be fitted into a small spaces specially designed for racks.
- Power distribution to blades is simplified by a common chassis and can be made redundant (though one should often read the small-print).
- A common switch in each chassis allows efficient internal networking.
- VLANs can be set up internally in a chassis, or linked to outside network.
- Blade chassis are increasingly designed for the kind of redundancy that is needed in a large scale computing operation, with two of everything provided as options (power, switch, etc.).
- They often have a separate network connection for an out-of-band management console.

The blade servers are a design convenience, well suited to life in data centre where space, power and cooling are issues.

### 4.2. *Bottlenecks – why won't it go faster?*

Service managers are sometimes disappointed that expensive upgrades to not lead to improved performance. This is often because computing power is associated with the CPU frequency statistics. Throwing faster CPUs at service tasks is a naive way of increasing processing. This will only help if the service tasks are primarily CPU bound.

So where are the main performance bottlenecks in computers? This is probably the wrong question to ask. Rather we have to ask, where is the bottleneck in an *application*? The resources needed by different applications are naturally different. We must decide whether a process spends most of its time utilizing:

- CPU?
- Network?
- Disk?
- Memory?

We also need to understand what *dependencies* are present in the system which could be a limiting factor [6].

Only when we know how an application interacts with the hardware and operating system can we begin to tune the performance of the system. As always, we have to look at the resources being used and choose a technology which solves the problem we are facing. Technologies have design trade-offs which we have to understand. For instance, we might choose Infiniband for short connection, high-speed links in a cluster, but this would be no good for connecting up a building. Distance and speed are often contrary to one another.

Read versus write performance is a key factor in limiting throughput. Write-bound operations are generally slower than read-only operations because data have to be altered and caching is harder to optimize for writing. Redundancy in subsystems has a performance cost, e.g., in RAID 5 data are written to several places at once, whereas only one of the copies has to be read. This explains why RAID level 5 redundancy slows down disk writing, for example. The combination of disk striping with each disk mirrored is often chosen as the optimal balance between redundancy and efficiency.

Applications that maintain *internal state* over long-term *sessions* versus one-off transactions often have complications that make it harder to share load using high-level redundancy. In such cases low-level redundancy, such as that used in mainframe design is likely to lead to more efficient processing because the cost of sharing state between separate computers is relatively high.

## 5. Designing an application architecture

### 5.1. *Software architecture*

In an ideal world, software engineers would be aware of data centre issues when writing application software. Service design involves concerns from the low-level protocols of the network to the high-level web experiences of the commercial Internet.

While low-level processor design aims for greater and greater integration, the dominant principle in soft-system architecture is the *separation of concerns*. Today's architipycal design for network application services is the so-called *three tier architecture*: webserver, application, database (see Figure 1).

Service design based entirely on the program code would be a meaningless approach in today's networked environments. A properly functioning system has to take into account everything from the user experience to the basic resources that deliver on the service promises. Design considerations for services include:

- Correctness.
- Throughput and scalability.
- Latency (response time).
- Reliability.
- Fault tolerance.
- Disaster recovery.

High volume services require efficiency of resource use. High availability requires, on the other hand, both efficiency and a reliability strategy that is fully integrated with the software deployment strategy. Typically one uses load sharing and redundancy for this.

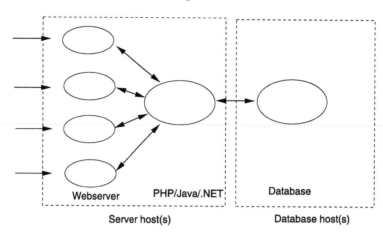

Fig. 1. A three tier architecture.

## 5.2. *Scalability*

Like security, scalability is a property of systems that researchers and designers like to claim or even boast, backed up with at best spurious evidence, as if the mere mention of the word could bring fortune. Like security, it has become a word to mistrust. Part of the problem is that very little attention has been given to defining what scalability means. This chapter is not the place for a comprehensive discussion of scalability, but we can summarize some simple issues.

Scalability is about characterizing the output of a system as a function of input (usually a task). The ability of a system to complete tasks depends on the extent to which the task can be broken up into independent streams that can be completed in parallel, and thus the topology of the system and its communication channels. Some simple ideas about scalability can be understood using the flow approximation of queueing systems (see chapter of System Administration and the Scientific Method in this volume) to see how the output changes as we vary the input.

We shall think of scalability as the ability of a system to deal with large amounts of input, or more correctly, we consider how increasing the input affects the efficiency with which tasks are discharged at the output. This sounds somewhat like the problem posed in queueing theory. However, whereas queueing theory deals with the subtleties of random processes, scalability is usually discussed in a pseudo-deterministic flow approximation.

If we think of rain falling (a random process of requests) then all we are interested in over time is how much rain falls and whether we can drain it away quickly enough by introducing a sufficient number of drains (processors or servers). Thus scalability is usually discussed as throughput as a function of load. Sometimes the input is a function of a number of clients, and sometimes the output is a function of the number of servers (see Figure 2). There are many ways to talk about scaling behavior.

Amdahl's law, named after computer designer Gene Amdahl, was one of the first attempts to characterize scalability of tasks in High Performance Computing, in relation to the number of processors [1]. It calculates the expected 'speed-up', or fractional increase

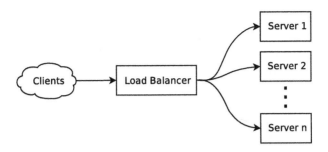

Fig. 2. The topology for low-level load sharing using a commercial dispatcher.

in performance, as a result of parallelizing part of the processing (load balancing) between $N$ processors as:

$$S = \frac{t_{\text{serial}}}{t_{\text{parallel}}} = \frac{t(1)}{t(N)}. \tag{1}$$

Suppose a task of total size $\sigma + \pi$, which is a sum of a serial part $\sigma$ and a parallelizable part $\pi$, is to be shared amongst $N$ servers or processors and suppose that there is an overhead $o$ associated with the parallelization (see Figure 3). Amdahl's law says that:

$$S = \frac{\sigma + \pi}{\sigma + \pi/N + o}. \tag{2}$$

In general, one does not know much about the overhead $o$ except to say that it is positive ($o \geqslant 0$), thus we write

$$S \leqslant \frac{\sigma + \pi}{\sigma + \pi/N}. \tag{3}$$

This is usually rewritten by introducing units of the *serial fraction*, $f \equiv \sigma/(\sigma + \pi)$, so that we may write:

$$S \leqslant \frac{1}{f + (1 - f)/N}. \tag{4}$$

This fraction is never greater than $1/f$, thus even with an infinite number of processors the task cannot be made faster than the processing of the serial part. The aim of any application designer must therefore be to make the serial fraction of an application as small as possible. This is called the bottleneck of the system.

Amdahl's law is, of course, an idealization that makes a number of unrealistic assumptions, most notably that the overhead is zero. It is not clear that a given task can, in fact, be divided equally between a given number of processors. This assumes some perfect knowledge of a task with very fine granularity. Thus the speedup is strictly limited by the largest chunk (see Figure 3) not the smallest. The value of the model is that it predicts two issues that limit the performance of a server or high performance application:

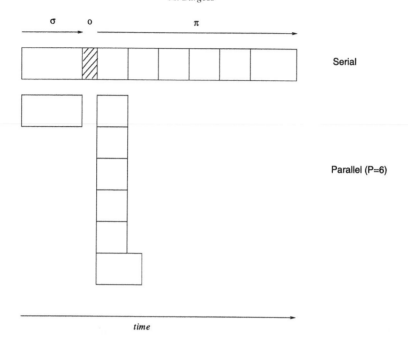

Fig. 3. Representation of Amdahl's law. A typical task has only a fraction that is parallelizable. The serial part cannot be sped up using parallel processing or load balancing. Note that one is not always able to balance tasks into chunks of equal size, moreover there is an overhead (shaded) involved in the parallelization.

- The serial fraction of the task.[3]
- The processor entropy or even-ness of the load balancing.

Amdahl's law was written with High Performance Computing in mind, but it applies also to server load balancing for network services. If one thinks of an entire job as the sum of all requests over a period of time (just as drain-water is a sum of all the individual rain-drops) then the law can also be applied to the speed up obtained in a load-balancing architecture such as that shown in Figure 2. The serial part of this task is then related to the processing of headers by the dispatcher, i.e., the dispatcher is the fundamental limitation of the system.

Determining the serial fraction of a task is not always as straightforward as one thinks. Contention for dependent resources and synchronization of tasks (waiting for resources to become available) often complicates the simple picture offered by the Amdahl law.

Karp and Flatt [21] noted that the serial part of a program must often be determined empirically. The Karp–Flatt metric is defined as the empirically determined serial fraction, found be rearranging the Amdahl formula (4) to solve for the serial fraction.

$$f = \frac{S^{-1} - N^{-1}}{1 - 1/N}. \tag{5}$$

[3]It is often quoted that a single woman can have a baby in nine months, but nine women cannot make this happen in a month.

Here we assume no overhead (it will be absorbed into the final effective serial fraction). Note that, as the number of processors becomes large, this becomes simply the reciprocal of the measured speed up. This formula has no predictive power, but it can be used to measure the performance of applications with different numbers of processors. If $S$ is small compared to $N$ then we know that the overhead is large and we are looking for performance issues in the task scheduler.

Another way of looking at Amdahl's law in networks has been examined in refs. [7,8] to discuss centralization versus distribution of processing. In network topology, serialization corresponds to centralization of processing, i.e. the introduction of a bottleneck by design. Sometimes this is necessary to collate data in a single location, other times designers centralize workflows unnecessarily from lack of imagination. If a centralized server is a common dependency of $N$ clients, then it is clear that the capacity of the server $C$ has to be shared between the $N$ clients, so the workflow per client is

$$W \simeq \frac{C}{N}. \tag{6}$$

We say then that this architecture scales like $1/N$, assuming $C$ to be constant. As $N$ becomes large, the workflow per client goes to zero which is a poor scaling property. We would prefer that it were constant, which means that we must either scale the capacity $C$ in step with $N$ or look for a different architecture. There are two alternatives:

- Server scaling by load balancing (in-site or cross-site) $C \rightarrow CN$.
- Peer-to-peer architecture (client-based voluntary load balancing) $N \rightarrow 1$.

There is still much mistrust of non-centralized systems although the success of peer to peer systems is now hard to ignore. Scaling in the data centre cannot benefit from peer to peer scaling unless applications are designed with it in mind. It is a clear case where application design is crucial to the scaling.

### 5.3. *Failure modes and redundancy*

To gauge reliability system and software engineers should map out the failure modes of (or potential threats to) the system [6,18] (see also the chapter on System Reliability in this volume). Basic failure modes include:

- Power supply.
- Component failure.
- Software failure (programming error).
- Resource exhaustion and thrashing.
- Topological bottlenecks.
- Human error.

So-called 'single points of failure' in a system are warning signs of potential failure modes (see Figure 4). The principle of separation of concerns tends to lead to a tree-like structure which is all about *not* repeating functional elements and is therefore in basic conflict with the idea of redundancy (Figure 4(a)). To counter this, one can use dispatchers, load balancers and external parallelism (i.e. not built into the software, but implemented

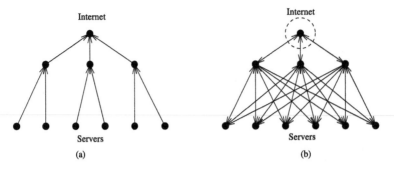

Fig. 4. Load balancing is a tricky matter if one is looking for redundancy. A load balancer is a tree – which is a structure with many intrinsic points of failure and bottlenecks.

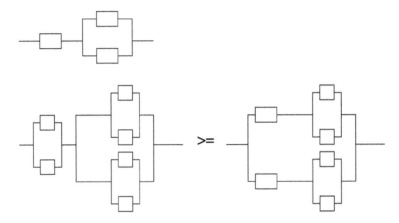

Fig. 5. Redundancy folk theorem.

afterwards). This can cure some problems, but there might still be points of failure left over (Figure 4(b)). Ideally one tries to eliminate these through further redundancy, e.g., redundant Internet service provision, redundant power supplies, etc. One should not forget the need for human redundancy in this reckoning: human resources and competence are also points of failure. Redundancy is the key issue in handling quality of service: it answers the issues of parallelism for efficiency and for fault tolerance.

When planning redundancy, there are certain thumb rules and theorems concerning system reliability. Figure 5 illustrates the folk theorem about parallisms which says that redundancy at lower system levels is always better than redundancy at higher levels. This follows from the fact that a single failure at a low level could stop an entire computing component from working. With low-level redundancy, the weakest link only brings down a low-level component, with high-level redundancy, a weakest link could bring down an entire unit of dependent components.

A minimum dependency and full redundancy strategy is shown in Figure 6. Each of the doubled components can be scaled up to $n$ to increase thoughput.

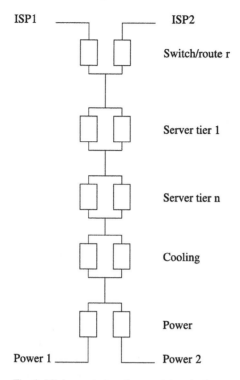

Fig. 6. Minimum strategy for complete redundancy.

Security is another concern that can be addressed in terms of failure modes. It is often distinguished from system failure due to 'natural causes' by being a 'soft failure', i.e., a failure relative to a policy rather than continuation. The distinction is probably unnecessary, as we can simply consider all failures relative to policy.

A secure system can be defined as follows [6]: *A secure system is one in which all the risks have been identified and accepted as a matter of policy.* See also the chapter by Bishop in this volume. This definition underlines the point that there is always an arbitrary threshold (or policy) in making decisions, about when faults are sufficiently serious to warrant a change of attitude, or a response.

An important enemy in system reliability is *human error*, either by mistake or incompetence. Human fault paths deal with many issues. Hostile parties expose us to risk through:

- Greed.
- Vanity.
- Bribery and blackmail.
- Revenge.
- Vandalism.
- Warfare.

Friends, on the other hand, expose us to risk by:

- Forgetfulness.
- Misunderstanding/miscommunication.

- Confusion/stress/intoxication.
- Arrogance.
- Ignorance/carelessness.
- Slowness of response.
- Procedural errors.
- Inability to deal with complexity.
- Inability to cooperate with others.

The lists are long. It is also worth asking why such incompetence could be allowed to surface. Systems themselves are clearly at fault in many ways through:

- Design faults – inadequate specification.
- Run-time fault – system does not perform within its specification.
- Emergent faults – behavior that was not planned for or explicitly designed for, often provoked by the environment.
- Changing assumptions – about technology or the environment of the system, e.g., a system is designed for one scenario that becomes replaced by another.

Documentation of possible fault modes should be prioritized by their relative likelihood.

### 5.4. *Safe and reliable systems*

The construction of safe systems [5,6] is both controversial and expensive. Security has many interpretations, e.g., see ISO 17799 [3] and RFC 2196 (replacing RFC 1244), or The Orange Book Trusted Computer Security Evaluation Criteria (TSEC) (now somewhat dated).

Security, like fault analysis, begins with a threat evaluation. It includes issues from human interface design to resistance to hostile actions. In all cases a preventative response is a *risk reducing strategy*.

There are many unhealthy attitudes to security amongst system designers and even administrators, e.g. 'Encryption solves 90% of the world's security problems' as stated by a naive software analyst known to the author. Security is a property of complete systems, not about installable features. Encryption deals with only a risk of information capture, which is one part of a large problem that is easily dealt with. Encryption key management, on the other hand, is almost never addressed seriously by organizations, but is clearly the basis for the reliability of encryption as a security mechanism.

Just as one must measure fault issues relative to a policy for severity, security requires us to place *trust boundaries*. The basis of everything in security is what we consider to be trustworthy (see the definition in the previous section). Technology can shift the focus and boundaries of our trust, but not eliminate the assumption of it. A system policy should make clear where the boundaries of trust lie. Some service customers insist on source code-review of applications as part of their Service Agreement. Threats include:

- Accidents.
- Human error, spilled coffee, etc.
- Attacks.
- Theft, spying, sabotage.
- Pathogenic threats like virus, Trojan horse, bugs.

- Human–computer interfaces.
- Software functionality.
- Algorithms.
- Hardware.
- Trust relationships (the line of defense).

Restriction of privilege is one defense against these matters. Access is placed on a 'need to know/do' basis and helps one to limit the scope of damage. System modelling is an important tool in predicting possible failure modes (see chapters on System Administration and the Scientific Method by the author, and the chapter on Security Management and Policies by Bishop).

In short, a well-designed system fails in a predictable way, and allows swift recovery.

## 6. Data centre design

The data centre is a dry and noisy environment. Humans are not well suited to this environment and should try to organize work so that they do not spend more time there than necessary. Shoes and appropriate clothing shall be worn in data centres for personal safety and to avoid contaminating the environment with skin, hair and lost personal items. Unauthorized persons should never be admitted to the data centre for both safety and security reasons.

Data centres are not something that can be sold from a shelf; they are designed within the bounds of a specific site to cope with a specific tasks. Data centres are expensive to build, especially as performance requirements increase. One must consider power supply, cooling requirements, disasters (flooding, collision damage, etc.) and redundancy due to routine failure. This requires a design and requirement overview that companies rarely have in advance. The design process is therefore of crucial importance.

It is normal for inexperienced system administrators and engineers to underestimate the power and cooling requirements for a data centre. Today, with blade chassis computers and dense rack mounting solutions it is possible to pack even more computers into a small space than ever before, thus the density of heat and power far exceeds what would be required in a normal building. It is important to be able to deliver peak power during sudden bursts of activity without tripping a fuse or circuit-breaker.

With so much heat being generated and indeed wasted in such a small area, we have to think seriously in the future about how buildings are designed. How do we re-use the heat? How can cooling be achieved without power-driven heat-exchangers?

### 6.1. *Power in the data centre*

Power infrastructure is the first item on the agenda when building a data centre. Having sufficient access to electrical power (with circuit redundancy) is a prerequisite for stable operation. Data centres are rated in a tier system (see Section 6.10) based on how much redundancy they can provide.

Every electrical device generates heat, including cooling equipment. Electrical power is needed to make computers run. It is also needed for lighting and cooling. Once power needs have been estimated for computers, we might have to add up to 70% again for the cost of cooling, depending on the building. In the future buildings could be designed to avoid this kind of gratuitous compounding of the heat/energy problem with artificial air-conditioning by cooling naturally with underground air intakes.

Power consumption is defined as the flow of energy, per unit time, into or out of a system. It is measured in Joules per second, or Watts or Volt-Amperes, all of which are equivalent in terms of engineering dimensions. However, these have qualitatively different interpretations for alternating current sources.

Computers use different amounts of electrical current and power depending on where they are located. Moreover, since electricity companies charge by the current that is drawn rather than the power used, the current characteristics of a device are important as a potential source of cost saving.

Electronic circuitry has two kinds of components:

- *Resistive components* absorb and dissipate power as heat. They obey Ohm's law.
- *Reactive components* absorb and return energy (like a battery), by storing it in electro-magnetic fields. They include capacitors (condensers) and inductors (electro-magnetic coils).

Reactive components (inductors and capacitors) store and return most of the power they use. This means that if we wait for a full cycle of the alternating wave, we will get back most of what we put in again (however, we will still have to pay for the current). Resistive components, on the other hand, convert most of the energy going into them into heat.

In direct current (DC) circuits, the power (energy released per unit time) is simply given by $P = IV$, where $I$ is the constant current and $V$ is the constant voltage. In a device where a DC current can flow, reactive components to not have an effect and Ohm's law applies: $V = IR$, where $R$ is the electrical resistance. In this case, we can write

$$P = IV = I^2 R = \frac{V^2}{R}. \tag{7}$$

For an alternating current (wave) source, there are several cases to consider however. Ohm's law is no longer true in AC circuits, because capacitors and inductors can borrow power for a short time and then return it, like a wave shifting pebbles on a beach. The voltage and current are now functions of time: $I(t)$, $V(t)$. The instantaneous power consumption (the total energy absorbed per unit time) is still $P = IV = I(t)V(t)$. However, mains power is typically varying at 50–60 Hz, so this is not an true reflection of the long term behavior, only what happens on the scale of tenths of a second.

For a system driven by a single frequency (clean) wave, the short term borrowing of current is reflected by a phase shift between the voltage (wave) and the current, which is the response of the system to that driving force (pebbles). Let us call this phase shift $\phi$.

$$V(t) = V_0 \sin(2\pi f t),$$

$$I(t) = I_0 \sin(2\pi f t + \phi), \tag{8}$$

where $f$ is frequency and $T = 1/f$ is the period of the wave. We can compute the average power over a number of cycles $nT$ by computing

$$\langle P \rangle_\phi = \frac{1}{nT} \int_0^{nT} I(t)\, V(t)\, \mathrm{d}t. \tag{9}$$

We evaluate this using two trigonometric identities:

$$\sin(A + B) = \sin A \cos B + \cos A \sin B, \tag{10}$$

$$\sin^2 X = \frac{1}{2}(1 - \cos 2X) \tag{11}$$

Using the first of these to rewrite $I(t)$, we have

$$\langle P \rangle = \frac{I_0 V_0}{nT} \int_0^{nT} \left[ \sin^2\left(\frac{2\pi t}{T}\right) \cos \phi + \sin\left(\frac{2\pi t}{T}\right) \cos\left(\frac{2\pi t}{T}\right) \sin \phi \right] \mathrm{d}t. \tag{12}$$

Rewriting the first term with the help of the second identity allows us to integrate the expression. Most terms vanish showing the power that is returned over a whole cycle. We are left with:

$$\langle P \rangle_\phi = \frac{1}{2} I_0 V_0 \cos \phi. \tag{13}$$

In terms of the normally quoted root-mean-square (RMS) values:

$$V_{\mathrm{rms}} = \sqrt{\frac{1}{nT} \int_0^{nT} \left( V(t) \right)^2}. \tag{14}$$

For a single frequency one has simply:

$$V_{\mathrm{rms}} = V_0/\sqrt{2},$$
$$I_{\mathrm{rms}} = I_0/\sqrt{2}, \tag{15}$$

$$\langle P \rangle_\phi = I_{\mathrm{rms}} V_{\mathrm{rms}} \cos \phi. \tag{16}$$

The RMS values are the values returned by most measuring devices and they are the values quoted on power supplies. The cosine factor is sometimes called the 'power factor' in electrical engineering literature. It is generalized below. Clearly

$$\langle P \rangle_{\phi > 0} \leqslant \langle P \rangle_{\phi = 0}. \tag{17}$$

In other words, the actual power consumption is not necessarily as bad as the values one would measure with volt and ammeters. This is a pity when we pay our bill, because the power companies measure current, not work-done. That means we typically pay more than we should for electrical power. Large factories can sometimes negotiate discounts based on the reactive power factor.

**6.2.** *Clipped and dirty power*

A direct calculation of the reduction in power transfer is impractical in most cases, and one simply defines a power factor by

$$PF = \cos\phi \equiv \frac{\langle P \rangle}{I_{\mathrm{rms}} V_{\mathrm{rms}}}. \tag{18}$$

More expensive power supply equipment is designed with 'power factor corrected' electronics that attempt to reduce the reactance of the load and lead to a simple $\phi = 0$ behavior. The more equipment we put onto a power circuit, the more erratic the power factor is likely to be. This is one reason to have specialized power supplies (incorporated with Uninterruptible Power Supplies (UPS)).

Actual data concerning power in the computer centres is hard to find, so the following is based on hearsay. Some authors claim that power factors of $\cos\phi = 0.6$ have been observed in some PC hardware. Some authors suggest that a mean value of $\cos\phi = 0.8$ is a reasonable guess. Others claim that in data centres one should assume that $\cos\phi = 1$ to correctly allow for enough headroom to deal with power demand.

Another side effect of multiple AC harmonics is that the RMS value of current is not simply $1/\sqrt{2} = 1/1.4$ of the peak value (amplitude). This leads to a new ratio called the 'crest factor', which is simply:

$$\mathrm{Crest} = \frac{I_0}{I_{\mathrm{rms}}}. \tag{19}$$

This tells us about current peaking during high load. For a clean sinusoidal wave, this ratio is simply $\sqrt{2} = 1.4$. Some authors claim that the crest factor can be as high as 2–3 for cheap computing equipment. Some authors claim that a value of 1.4–1.9 is appropriate for power delivered by a UPS. The crest factor is also load dependent.

**6.3.** *Generators and batteries*

Uninterruptible Power Supplies (UPS) serve two functions: first to clean up the electrical power factor, and second to smooth out irregularities including power outages. Batteries can take over almost instantaneously from supply current, and automatic circuit breakers can start a generator to take over the main load within seconds. If the generator fails, then battery capacity will be drained.

UPS devices require huge amounts of battery capacity and even the smallest of these weighs more than a single person can normally lift. They must be installed by a competent electrician who knows the electrical installation details for the building.

Diesel powered generators should not be left unused for long periods of time, as bacteria, molds and fungi can live in the diesel and transform it into jelly over time.

**6.4.** *Cooling and airflow*

Environmental conditions are the second design priority in the data centre. This includes cooling and humidity regulation. Modern electronics work by dumping current to ground through semi-conducting (resistive) materials. This is the basic transistor mode of operation. This generates large amounts of heat. Essentially all of the electrical power consumed by a device ultimately becomes heat.

The aim of cooling is to prevent the temperature of devices becoming too great, or changing too quickly. If the heat increases, the electrical and mechanical resistance of all devices increases and even more energy is wasted as heat. Eventually, the electrical properties of the materials will become unsuitable for their original purpose and the device will stop working. In the worst case it could even melt.

Strictly speaking, heat is not dangerous to equipment, but temperature is. Temperature is related to the density of work done. The more concentrated heat is, the higher the temperature. Thus we want to spread heat out as much as possible.

In addition to the peak temperature, changes in temperature can be dangerous in the data centre. If temperature rises or falls too quickly it can result in mechanical stress (expansion and contraction of metallic components), or condensation in cases of high humidity.

- Polymer casings and solder-points can melt if they become too hot.
- Sudden temperature changes can lead to mechanical stresses on circuits, resulting in cracks in components and circuit failures.
- Heat increases electrical and mechanical resistance, which in turn increases heat production since heat power conversion goes like $\sim I^2 R$.

Air, water and liquefied gases can be used to cool computing equipment. Which of these is best depends on budget and local environment. Good environmental design is about the constancy of the environment. Temperature, humidity, power consumption and all variables should be evenly regulated in space and time.

**6.5.** *Heat design*

Cooling is perhaps the most difficult aspect of data centre design to get right. The flow of air is a complicated science and our intuitions are often incorrect. It is not necessarily true that increasing the amount of cooling in the data centre will lead to better cooling of servers.

Cooling equipment is usually rated in BTUs (British Thermal Units), an old fashioned measurement of heat. 1 Watt = 3413 BTUs. A design must provide cooling for every heat generating piece of equipment in the data centre, including the humans and the cooling equipment itself. Temperature must be regulated over both space and time, to avoid gradient effects like condensation, disruptive turbulence and heat-expansion or contraction of components.

The key to temperature control is to achieve a constant ambient temperature throughout a room. The larger a room is, the harder this problem becomes. If hot spots or cold spots develop, these can lead to problems of condensation of moisture, which increases in like-

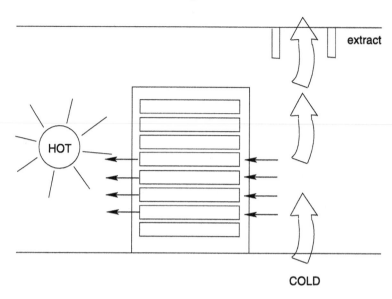

Fig. 7. A bad cooling design. Too much cold air comes through the floor, driving all the cold air past the racks into the extractor instead of passing through them. This leaves hot spots and wastes most of the cooling capacity.

lihood with increasing humidity and is both harmful to the computing equipment and can even lead to bacterial growth which is a health hazard (e.g., Legionnaires disease).

If air moves too quickly past hot devices, it will not have time to warm up and carry away heat from hot spots; in this case, most of the air volume will simply cycle from and back to the cooling units without carrying away the heat. If air moves too slowly, the temperature will rise. (See Figures 7 and 8.)

As everyone knows, hot air rises if it is able to do so. This can be both a blessing and a curse. Raised flooring is a standard design feature in more advanced data centres which allows cool air to come up through holes in the floor. This is a useful way of keeping air moving in the data centre so that no hot spots can develop. However, one should be careful not to have cold air entering too quickly from below, as this will simply cause the hot air to rise into the roof of the centre where it will be trapped. This will make some servers too hot and some too cold. It can also cause temperature sensors to be fooled into misreading the ambient temperature of the data centre, causing energy wastage from over cooling, or over regulation of cooling.

Over-cooling can, again, lead to humidity and condensation problems. Large temperature gradients can cause droplets of water to form (as in cloud formation). Hot computer equipment should not be placed too close to cooling panels as this will cause the coolers to work overtime in order to cool the apparent imbalance, and condensation-ice can form on the compressors. Eventually they will give up and water can then flood into the data centre under the flooring. The solution, unfortunately, is not merely to dry out the air, as this can lead to danger of static electrical discharges and a host of related consequences.

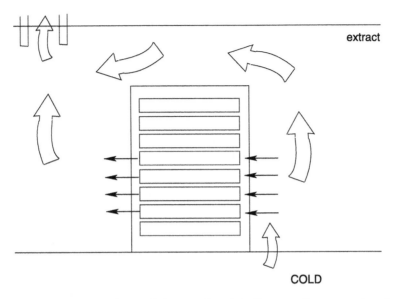

Fig. 8. A good cooling design. A trickle cold air comes through the floor, spreading out near the racks into the room where has time to collect heat. The air circulates from the cold aisle into the warm aisle where it is extracted and recycled.

### 6.6. *Heat, temperature and humidity*

Heat and temperature are related through the specific heat capacity (symbol $c_s$, also called specific heat) of a substance is defined as heat capacity per unit mass. The SI unit for specific heat capacity is the Joule per kilogramme Kelvin, $J\,kg^{-1}\,K^{-1}$ or $J/(kg\,K)$, which is the amount of energy required to raise the temperature of one kilogram of the substance by one Kelvin. Heat capacity can be measured by using calorimetry.

For small temperature variations one can think of heat capacity as being a constant property of a substance, and a linear formula for the relates temperature and heat energy.

$$\Delta Q = mc_s \Delta T, \tag{20}$$

where $\Delta Q$ is a change of heat energy (Watts $\times$ time). In other words temperature increase $\Delta T$, in a fixed mass $m$ of any substance, is proportional to the total power output and the time devices are left running.

If air humidity is too high, it can lead to condensation on relatively cool surfaces. Although a small amount of clean, distilled water is not a serious danger to electrical equipment, water mixed with dust, airborne dirt or salts will conduct electricity and can be hazardous.

Computer equipment can work at quite high temperatures. In ref. [24], experienced engineers suggest that the maximum data centre temperature should be 25 degrees Celcius (77 Fahrenheit), with a relative humidity of at least 40% (but no more than 80%). At less than 30% humidity, there is a risk of static electrical discharge, causing read errors or even unrecoverable runtime failures.

Heat is transmitted to and from a system by three mechanisms:

- *Conduction* is the direct transfer of heat by material contact (molecular vibration). For this, we use the heat capacity formula above.
- *Convection* is a transport of gas around a room due to the fact that, as a gas is heated its density decreases under constant pressure, so it gets lighter and rises. If the hot air rises, cooler air must fall underneath it. This results in a circulation of air called convection. If there is not enough space to form convection cells, hot air will simply sit on top of cold air and build up.
- *Radiation* is the direct transfer of heat without material contact. Electromagnetic radiation only passes through transparent substances. However, a substance that is transparent to visible light is not necessarily transparent to heat (infra-red), which has a much longer wavelength. This is the essence of the Greenhouse effect. Visible light can enter a Greenhouse through the glass walls, this energy is absorbed and some of the energy is radiated back as heat (some of it makes plants grow). However not all of the long wavelength heat energy passes through the glass. Some is absorbed by the air and heats it up. Eventually, it will cool off by radiation of very long wavelengths, but this process is much slower than the rate at which new energy is coming in, hence the energy density increases and temperature goes up.

If a material is brought into contact with a hotter substance, heat will flow into it by *conduction*. Then, if we can remove the hotter substance e.g. by flow convection, it will carry the heat away with it. Some values of $c_s$ for common cooling materials are shown in Table 1.

Water is about four times as efficient at carrying heat per degree rise in temperature than is air. It is therefore correspondingly better at cooling than air.

Melting temperatures in Table 2 show that it is not likely that data centre heat will cause anything in a computer to melt. The first parts to melt would be the plastics and polymers and solder connections (see Table 2).

Today's motherboards come with temperature monitoring software and hardware which shuts the computer off the CPU temperature gets too hot. CPU maximum operating temperatures lie between 60 and 90°C.

Table 1

| Substance | Phase | Specific heat capacity (J/(kg K)) |
|---|---|---|
| Air (dry) | gas | 1005 |
| Air (100% humidity) | gas | 1030 |
| Air (different source) | gas | 1158 |
| Water | liquid | 4186 |
| Copper (room temp) | solid | 385 |
| Gold (room temp) | solid | 129 |
| Silver (room temp) | solid | 235 |

Table 2

| Material | Melting point (°C) |
| --- | --- |
| Silicon | 1410 |
| Copper | 1083 |
| Gold | 1063 |
| Silver | 879 |
| Polymers | 50–100 |

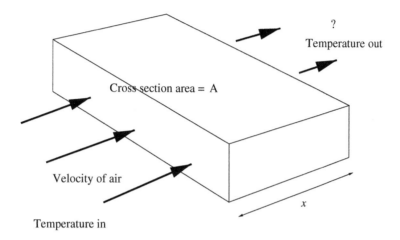

Fig. 9. Schematic illustration of air flowing through and around a rack device.

### 6.7. *Cooling fans*

Fans are place in computers to drive air through them. This air movement can also help to keep air circulating in the data centre. Rack doors and poorly placed fans can hinder this circulation by generating turbulence.

Cooling fans exploit the same idea as convection, to transport energy away by the actual movement of air (or sometimes fluid). The rate of flow of volume is proportional to the fixed area of the channel cross-section and the length per unit time (velocity) of the coolant passing through it.

$$\frac{dV}{dt} \propto A\frac{dx}{dt}, \tag{21}$$

$$F \propto Av. \tag{22}$$

You should bear in mind that:
- It is the temperature of components inside the computer that is the main worry in cooling design.

- The temperature of the air coming out depends on many factors, including the ambient temperature around the box, the rate of air flow through the device and around it, the amount of convection in and around the device, humidity etc.
- Air is a poor carrier of heat, so if air is getting hot, you know that the metallic components are sizzling! Also, air is a good insulator so it takes some time to warm up. If air moves too fast, it will not carry away the heat.

In very high speed computers that generate large amounts of heat quickly, liquids such as water and nitrogen are use for cooling, piped in special heat exchangers. The density of a liquid means that it is able to carry much more heat, and hence a smaller flow is needed to carry a larger amount of waste power. Water-based cooling is many times more expensive than air-cooling.

### 6.8. *Placement of cooling flow ducts*

The use of raised flooring is common in datacentres. Cold air usually passes underneath raised flooring from compressors and escapes into the data centre through holes in the floor.

If cold air rose from everywhere quickly and evenly, hot air would simply rise to the roof and there would be no circulation (see Figure 7). A standard strategy to maintain convectional circulation is to limit the number of holes in flooring to prevent air from circulating too quickly, and the have alternating hot and cold aisles. In a cold aisle, there are holes in the floor. In a hot aisle there are no holes in the floor, but there are cooling intakes (see Figure 8).

To ensure sufficient circulation, there should be at least ten centimetres of clearence between racks in the roof and computer racks should not be within a metre or more of cooling units.

### 6.9. *Fire suppression*

A data center fire suppression system uses gas rather than water or foam. There are many gaseous suppression systems that use gases hazardous to humans and environment and are therefore unsuitable. The gas used to suppress the fire should not be toxic to people nor damage the equipment causing data to be lost. Inergen and Argonite are both gasses commonly used in data centers. Inergen is a gaseous mixture composed of nitrogen, argon and carbon dioxide. Argonite is a gas composed of argon and nitrogen. Since the fire suppression systems is based on gas one should make sure that in case of gas release there is no leakage to the outside.

### 6.10. *Tier performance standards*

The Uptime Institute has created a 4 Tier rating system of a data centre [33]. Tier 1 level is the most basic where no redundancy is required, while Tier 4 level is a 100% redundant data center. A system is as good as the weakest link and hence terms like 'near 4 Tier' or 'Tier 3 plus' do not exist.

The ratings use the phrases *capacity components* and *distribution paths*. Capacity components are active components like servers which deliver the service provided by the data centre. Distribution paths refer to communication infrastructure, such as network.

The Tier 1 describes a basic site, i.e., a non-redundant data center. Communication paths, cabling, power and cooling are non-redundant, the computer system is affected by a component failure and the system has to be shut down to do maintaince work.

Tier 2 describes a data center that has redundant capacity components, but non-redundant distribution paths. The computer system might be affected by a capacity component failure, and will certainly be affected by a path failure. The system might be inoperative during maintenance work.

Tier 3 describes a concurrently maintainable data center. In other words, the data center is equipped with redundant capacity components and multiple distribution paths (of which one is always active). The computer system will not be affected either in case of a capacity component failure or path failure and does not require shutdown to do maintaince work, but interruptions could occur.

Tier 4 describes a fault tolerant data center. In other words, the data center is equipped with redundant capacity component sand multiple distribution active paths. The computer system will not be affected of any worst-case failure of any capacity system or distribution element and does not require to be shut down to do maintaince work, no interruptions may occur.

## 7. Capacity planning

Simple results from queueing theory can be used as estimators of the capacity needs for a system [2,9]. Most service delivery systems are I/O bound, i.e. the chief bottleneck is input/output. However, in a data centre environment we have another problem that we do not experience on a small scale: power limitations.

Before designing a data center we need to work out some basic numbers, the scales and sizes of numbers involved.

- How fast can we process jobs? (CPU rate and number of parallel servers.)
- How much memory do we need: how many jobs will be in the system (queue length) and what is the average size of a job in memory?
- How many bytes per second of network arrivals and returns do we expect?
- How many bytes per second of disk activity do we expect?
- What is the power cost (including cooling) of running out servers?

### 7.1. *Basic queueing theory*

Service provision is usually modelled as a relationship between a client and a server (or service provider). A stream of requests arrives randomly at the input of a system, at a mean rate of $\lambda$ transactions per second, and they are added to a queue, whereupon they are serviced at a mean rate $\mu$ transactions per second by a processor.

Generalizations of this model include the possibility of multiple (parallel) queues and multiple servers. In each case one considers a single incoming stream of transaction requests; one then studies how to cope with this stream of work.

As transaction demand is a random process, queues are classified according to the type of arrival process for requests, the type of completion process and the number of servers. In the simplest form, Kendall notation of the form $A/S/n$ is used to classify different queueing models.

- $A$: the arrival process distribution, e.g., Poisson arrival times, deterministic arrivals, or general spectrum arrivals.
- $S$: the service completion distribution for job processing, e.g., Poisson renewals, etc.
- $n$: the number of servers processing the incoming queue.

The basic ideas about queueing can be surmised from the simplest model of a single server, memoryless queue: $M/M/1$ (see chapter by Burgess, System Administration and the Scientific Method in this volume).

Queues are described in terms of the statistical pattern of job arrivals. These processes are often approximated by so-called memoryless arrival processes with mean arrival rate $\lambda$ jobs per second and mean processing rate $\mu$ jobs per second. The quantity $\rho = \lambda/\mu < 1$ is called the traffic intensity [6,19]. The expected length of the simplest $M/M/1$ queue is

$$\langle n \rangle = \sum_{n=0}^{\infty} p_n n = \frac{\rho}{1 - \rho}. \tag{23}$$

Clearly as the traffic intensity $\rho$ approaches unity, or $\lambda \to \mu$, the queue length grows out of control, as the server loses the ability to cope.

The job handling improves somewhat for $s$ servers (here represented by the $M/M/s$ queue), where one finds a much more complicated expression which can be represented by:

$$\langle n \rangle = s\rho + P(n \geqslant s)\frac{\rho}{1 - \rho}. \tag{24}$$

The probability that the queue length exceeds the number of servers $P(n \geqslant s)$ is of order $\rho^2$ for small load $\rho$, which naturally leads to smaller queues.

It is possible to show that a single queue with $s$ servers is always at least as efficient as $s$ separate queues with their own server. This satisfies the intuition that a single queue can be kept moving by any spare capacity in any of its $s$ servers, whereas an empty queue that is separated from the rest will simply be wasted capacity, while the others struggle with the load.

## 7.2. *Flow laws in the continuum approximation*

As we noted, talking about scalability, in the memoryless approximation we can think of job queues as fluid flows. Dimensional analysis provides us with some simple continuum relationships for the 'workflows'. Consider the following definitions:

$$\text{Arrival rate } \lambda = \frac{\text{No. of arrivals}}{\text{Time}} = \frac{A}{T}, \tag{25}$$

$$\text{Throughput } \mu = \frac{\text{No. of completions}}{\text{Time}} = \frac{C}{T}, \tag{26}$$

$$\text{Utilization } U = \frac{\text{Busy time}}{\text{Total time}} = \frac{B}{T}, \tag{27}$$

$$\text{Mean service time } S = \frac{\text{Busy time}}{\text{No. of completions}} = \frac{B}{C}. \tag{28}$$

The utilization $U$ tells us the mean level at which resources are being scheduled in the system. The 'utilization law' notes simply that:

$$U = \frac{B}{T} = \frac{C}{T} \times \frac{B}{C} \tag{29}$$

or

$$U = \mu S. \tag{30}$$

So utilization is proportional to the rate at which jobs are completed and the mean time to complete a job. Thus it can be interpreted as the probability that there is at least one job in the system. Often this law is written $U = \lambda S$, on the assumption that in a constant flow the flow rate in is equal to the flow rate out of the system $C/T \rightarrow \lambda$. This is reasonable at low utilization because $\mu$ is sufficiently greater than $\lambda$ that the process can be thought of as deterministic, thus:

$$U = \lambda S. \tag{31}$$

This handwaving rule is only applicable when the queue is lightly loaded and all jobs are essentially completed without delay; e.g., uppose a webserver receives hits at a mean rate of 1.25 hits per second, and the server takes an average of 2 milliseconds to reply. The law tells us that the utilization of the server is

$$U = 1.25 \times 0.002 = 0.0025 = 0.25\%. \tag{32}$$

This indicates to us that the system could probably work at four hundred times this rate before saturation occurs, since $400 \times 0.25 = 100\%$. This conclusion is highly simplistic, but gives a rough impression of resource consumption.

Another dimensional truism is known as Little's law of queue size. It says that the mean number of jobs in a queue $\langle n \rangle$, is equal to the product of the mean arrival rate $\lambda$ (jobs per second) and the mean response time $R$ (seconds) incurred by the queue:

$$\langle n \rangle = \lambda R. \tag{33}$$

Note that this equation has the generic form $V = IR$, similar to Ohm's law in electricity. This an analogy is a direct consequence of a simple balanced flow model. Note that $R$ differs from the mean service time. $R$ includes the waiting time in the queue, whereas the mean service time assumes that the job is ready to be processed.

In the $M/M/1$ queue, it is useful to characterize the expected response time of the service centre. In other words, what is the likely time a client will have to wait in order to have a task completed? From Little's law, we know that the average number of tasks in a queue is the product of the average response time and the average arrival rate, so

$$R = \frac{Q_n}{\lambda} = \frac{\langle n \rangle}{\lambda} = \frac{1}{\mu(1 - \rho)} = \frac{1}{(\mu - \lambda)}. \tag{34}$$

Notice that this is finite as long as $\lambda \ll \mu$, but as $\lambda \to \mu$, the response time becomes unbounded.

Load balancing over queues is a topic for more advanced queueing theory. Weighted fair queueing and algorithms like round-robin etc., can be used to spread the load of an incoming queue amongst multiple servers, to best exploit their availability. Readers are referred to refs. [6,13,19,35] for more discussion on this.

### 7.3. *Simple queue estimates*

The $M/M/1$ queue is simplistic, but useful for its simplicity. Moreover, there is a point of view amongst system engineers that says: most systems will have a bottleneck (a point at which there is a single server queue) somewhere, and so we should model weakest links as $M/M/1$. Let us use the basic $M/M/1$ formulae to make some order-of-magnitude calculations for capacity planning.

For example, using the formulae for the $M/M/1$ queue, let us calculate the amount of RAM, disk rate and CPU speed to handle the average loads:

#### 7.3.1. *Problem*
1. 10 downloads per second of image files 20 MB each.
2. 100 downloads per second of image files 20 MB each.
3. 20 downloads per second of resized images to a mobile phone, each image is 20 MB and takes 100 CPU cycles per byte to render and convert in RAM.
Given that the size of the download request is of the order 100 bytes.

#### 7.3.2. *Estimates*
1. If we have 10 downloads per second, each of 20 MB/s, then we need to process

$$10 \times 20 = 200 \text{ MB/s}. \tag{35}$$

Ideally, all of the components in the system will be able to handle data at this rate or faster. (Note that when the averages rates are equal ($\lambda = \mu$), the queue length is not zero but infinite as this means that there will be a significant probability that the stochastic arrival rate will be greater than the stochastic processing rate.) Let us suppose that the system disk has an average performance of 500 MB/s, then we have:

$$\lambda = \frac{200}{20} = 10 \text{ j/s},$$

$$\mu = \frac{500}{20} = 25 \text{ j/s}. \tag{36}$$

Note that we cannot use MB/s for $\lambda$, or the formulae will be wrong. In the worst case, each component of the system must be able to handle this data rate. The average queue length is:

$$\langle n \rangle = \frac{\rho}{1 - \rho} = \frac{\lambda}{\mu - \lambda} = 2/3 \simeq 0.7. \tag{37}$$

This tells us that there will be about 1 job in the system queue waiting to be processed at any given moment. That means the RAM we need to service the queue will be $1 \times 100$ bytes, since each job request was 100 bytes long. Clearly RAM is not much of an issue for a lightly loaded server.

The mean response time for the service is

$$R = \frac{1}{\mu - \lambda} = \frac{1}{25 - 10} = \frac{1}{15} \text{ s}. \tag{38}$$

2. With the numbers above the server would fail long before this limit, since at queue size blows up to infinity as $\lambda \to 25$ and the server starts thrashing, i.e. it spends all its time managing the queue and not completing jobs.

3. Now we have an added processing time. Suppose we assume that the CPU has a clock speed of $10^9$ Hz, and an average of 4 cycles per instruction. Then we process at a rate of 100 CPU cycles per byte, for all 20 MB.

$$\frac{20 \times 10^6 \times 100}{10^9} = 2 \text{ s}. \tag{39}$$

Thus we have a processing overhead of 2 seconds per download, which must be added to the normal queue response time.

$$\lambda = \frac{400}{20} = 20 \text{ j/s},$$

$$\mu = \frac{500}{20} = 25 \text{ j/s},$$

$$\langle n \rangle = \frac{\rho}{1 - \rho} = \frac{\lambda}{\mu - \lambda} = 4, \tag{40}$$

$$R = \frac{1}{\mu - \lambda} + \text{CPU} = \frac{1}{25 - 20} + \text{CPU} = \frac{1}{5} + 2 \text{ s}.$$

Notice that the CPU processing now dominates the service time, so we should look at increasing the speed of the CPU before worrying about anything else.

Developing simple order of magnitude estimates of this type forces system engineers to confront resource planning in a way that they would not normally do.

### 7.4. *Peak load handling – the reason for monitoring*

Planning service delivery based on average load levels is a common mistake of inexperienced service providers. There is always a finite risk that a level of demand will occur that is more than we can the system can provide for. Clearly we would like to avoid such a time, but this would require expensive use of redundant capacity margins (so-called 'over-provisioning').

A service centre must be prepared to either provide a constant margin of headroom, or be prepared to deploy extra server power on demand. Virtualization is one strategy for the latter, as this can allow new systems to be brought quickly on line using spurious additional hardware. But such extended service capacity has to be planned for. This is impossible if one does not understand the basic behavior of the system in advance (see the chapter on System Administration and the Scientific Method).

Monitoring is essential for understanding normal demand and performance levels in a system. Some administrators think that the purpose of monitoring is to receive alarms when faults occur, or for reviewing what happened after an incident. This is a dangerous point of view. First of all, by the time an incident has occurred, it is too late to prevent it. The point of monitoring ought to be to predict possible future occurrences. This requires a study of both average and deviant ('outlier') behavior over many weeks (see the chapter on System Administration and the Scientific Method).

The steps in planning for reliable provision include:

- Equipping the system for measurement instrumentation.
- Collecting data constantly.
- Characterizing and understand workload patterns over the short and the long term.
- Modelling and predicting performance including the likelihood and magnitude of peak activity.

Past data can indicate patterns of regular behavior and even illuminate trends of change, but they cannot truly predict the future.

In a classic failure of foresight in their first year of Internet service, the Norwegian tax directorate failed to foresee that there would be peak activity on the evening of the deadline for delivering tax returns. The system crashed and extra time had to be allocated to cope with the down-time. Such mistakes can be highly embarrassing (the tax authorities are perhaps the only institution that can guarantee that they will not lose money on this kind of outage).

Common mistakes of prediction include:

- Measuring only average workload without outlier stress points.
- Not instrumenting enough of the system to find the source of I/O bottlenecks.
- Not measuring uncertainties an data variability.
- Ignoring measurement overhead from the monitoring itself.

- Not repeating or validating measurements many times to eliminate random uncertainties.
- Not ensuring same conditions when making measurements.
- Not measuring performance under transient behavior (rapid bursts).
- Collecting data but not analyzing properly.

## 8. Service level estimators

Service level agreements are increasingly important as the paradigm of service provision takes over the industry. These are legal documents, transforming offers of service into obligations. Promises that are made in service agreements should thus be based on a realistic view of service delivery; for that we need to relate engineering theory and practice.

### 8.1. *Network capacity estimation*

The data centre engineer would often like to know what supply-margin ('over-provision') of service is required to be within a predictable margin of an Service Level Agreement (SLA) target? For example, when can we say that we are 95% certain of being able to deliver an SLA target level 80% of the time? Or better still: what is the full probability distribution for service within a given time threshold [2]?

Such estimates have to be based either on empirical data that are difficult to obtain, or simplified models that describe traffic statistics. In the late 1980s and early 1990s it became clear that the classical Poisson arrivals assumptions which worked so well for telephone traffic were inadequate to describe Internet behavior. Leland et al. [22] showed that Ethernet traffic exhibited self similar characteristics, later followed up in ref. [29]. Paxson and Floyd [28] found that application layer protocols like TELNET and FTP were modelled quite well by a Poisson model. However, the nature of data transfer proved to be bursty with long-range dependent behavior.

Web traffic is particularly interesting. Web documents can contain a variety of in-line objects such as images and multimedia. They can consist of frames and client side script. Different versions of HTTP (i.e., version 1.0 and 1.1) coexist and interact. Implementations of the TCP/IP stack might behave slightly different, etc. depending on the operating system. This leads to great variability at the packet level [12]. Refs. [14,27] have suggested several reasons for the traffic characteristics.

In spite of the evidence that for self-similarity of network traffic, it is known that if one only waits for long enough, one should see Poisson behavior [20]. This is substantiated by research done in [10] and elsewhere, and is assumed true in measurements approaching the order of $10^{12}$ points. However this amounts to many years of traffic [17].

### 8.2. *Server performance modelling*

To achieve a desired level of quality of service under any traffic conditions, extensive knowledge about how web server hardware and software interacts and affects each other is required. Knowing the expected nature of traffic and queueing system only tells us how we might expect a system to perform; knowledge about software and hardware interaction enables performance tuning.

Slothouber [32] derived a simple model for considering web server systems founded on the notion of serial queues. Several components interact during a web conversation and a goes through different stages, with different queues at each stage. The response time is therefore an aggregate of the times spent at different levels within the web server.

Mei et al. [34] followed this line of reasoning in a subsequent paper. One of the focal points in their work was to investigate the effects of for response time and server blocking probability due to congestion in networks. High end web servers usually sit on a network with excess bandwidth to be able to shuffle load at peak rates. However, customers inherently has poorly performing networks with possible asynchronous transfer mode, such as ADSL lines. Therefore, returning ACKs in from client to server can be severely delayed. In turn this causes the request to linger for a longer time in the system than would be the case if the connecting network was of high quality. Aggregation of such requests could eventually cause the TCP and HTTP listen queue to fill up, making the server refuse new connection requests even if it is subject to fairly light load.

These models are simple and somewhat unrealistic, but possess the kind of simplicity we are looking for our comparisons.

### 8.3. *Service level distributions*

The fact that both arrival and service processes are stochastic processes without a well-defined mean, combined with the effect of processor sharing has on response time tails, leads to the conclusion that there is an intrinsic problem in defining average levels for service delivery. Instead, it is more fruitful to consider the likelihood of being within a certain target. For instance, one could consider 80% or 90% compliance of a service level objective. Depending on the nature of traffic the differences between these two levels can be quite significant.

This becomes quite apparent if we plot the Cumulative Distribution Functions (CDF) for the experimental and simulated response time distributions (see Figure 10) [2]. Because of the large deviations of values, going from 80% to 90% is not trivial from a service providers point of view. The CDF plots show that going from 80% to 90% in service level confidence means, more or less, a double of the service level target.

The proper way to quote service levels is thus distributions. Mean response times and rates are useless, even on the time scale of hours. A cumulative probability plot of service times is straightforwardly derived from the experimental results and gives the most realistic appraisal of the system performance. The thick cut lines in the figure show the 80% and 90% levels, and the corresponding response times associated with them. From this one can make a service agreement based on probably compliance with a service time, i.e. not merely expectation value but histogram.

For instance, one could consider 80% or 90% compliance of a maximum service level target. Depending on the nature of traffic, the differences between these two levels can be quite significant. Since customers are interested in targets, this could be the best strategy available to the provider.

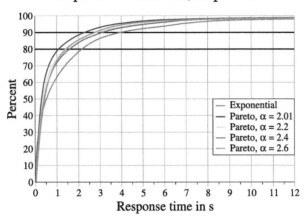

Fig. 10. The cumulative percentage distribution for response processing for sample exponential and Pareto arrivals. The thick cut lines show the 80% and 90% levels, and the corresponding response times associated with them. The theoretical estimators are too pessimistic here, which is good in the sense that it would indicate a recommended over-capacity, but the value if surely too much.

**8.4.** *Over-capacity planning*

The basic results for $M/M/1$ queues can be used to make over-estimates for capacity planning. The best predictive prescription (with noted flaws) is provided in ref. [2]:

1. Calculate the server rate $\mu$ for the weakest server that might handle a request. E.g. for a CPU bound process

$$\mu = \text{av. job size} \times \frac{\text{av. instructions}}{\text{job size}} \times \frac{\text{CPU cycles}}{\text{instruction}} \times \frac{\text{time}}{\text{cycles}}$$

$$= \frac{\text{av. instructions} \times \text{RISC/CISC ratio}}{\text{MHz}}. \tag{41}$$

2. Estimate traffic type at maximum load.
3. Construct the Cumulative Probability Distribution as a function of response times using FCFS.
4. Make SLA promises about probability of transaction rate based on where the curves cross basic thresholds.

With this recipe, one overestimates service promise times to achieve a given level of certainty (or equivalently overestimates required capacity for a given time) in a non-linear way – the amount can be anything up to four hundred percent! This explains perhaps why most sites have seemingly greatly over-dimensioned webservers. Given the problems that can occur if a queue begins to grow, and the low cost of CPU power, this over-capacity might be worth the apparent excess.

Readers are also referred to the cost calculations described in the chapter System Administration and the Business Process by Burgess, in this volume.

## 9. Network load sharing

Load sharing in the data centre is an essential strategy for meeting service levels in high volume and high availability services. Figure 2 shows the schematic arrangement of servers in a load balancing scenario. Classical queueing models can also be used to predict the scaling behavior of server capacity in a data centre.

Since service provision deals with random processes, exact prediction is not an option. A realistic picture is to end up with a probability or confidence measure, e.g., what is the likelihood of being able to be within 80% or 90% of an SLA target value? The shape of this probability distribution will also answer the question: what is the correct capacity margin for a service in order to meet the SLA requirements.

Once again, in spite of the limitations of simple queueing models, we can use these as simple estimators for service capacity [9]. The total quality of service in a system must be viewed as the combined qualities of the component parts [30]. It is a known result of reliability theory [18] that low level parallelism is, in general, more efficient than high-level parallelism, in a component-based system. Thus a single $M/M/n$ queue is generally superior in performance to $n$ separate $M/M/1$ queues (denoted $(M/M/1)^n$).

To understand load, we first have to see its effect on a single machine (Figure 11). We see that a single computer behaves quite linearly up to a threshold at which it no longer is able to cope with the resource demands placed upon it. At that point the average service delivery falls to about 50% of the maximum with huge variations (thrashing). At this point one loses essentially all predictability.

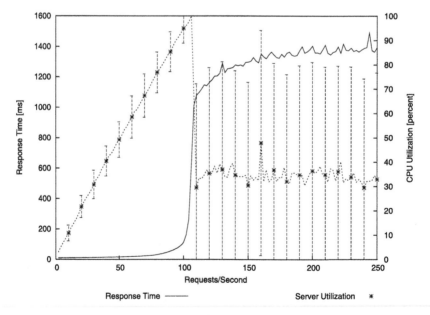

Fig. 11. Response time and CPU utilization of a single server as a function of requests per second, using Poisson distribution with exponential inter-arrival times.

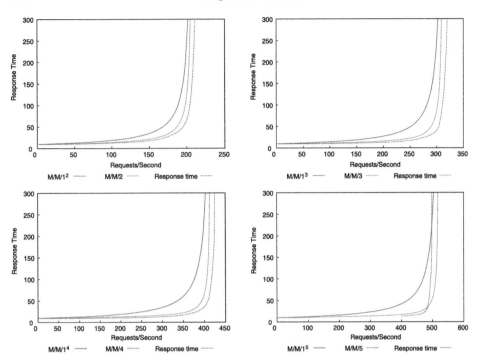

Fig. 12.  A comparison of $M/M/n$ and $(M/M/1)^n$ queue estimators with response times from a Poisson traffic on real web servers.

Experiments show that, as long as one keeps *all* servers in a server farm underneath this critical threshold, performance scales approximately linearly when adding identical servers hence we can easily predict the effect of adding more servers as long as load balancing dispatchers themselves behave linearly. Figure 13 shows the addition of servers in four traffic saturation experiments. In the first graph we see that the response time is kept low in all scenarios. This fits with the assumption that even a single server should be capable of handling requests. The next graph shows double this, and all scenarios except the one with a single server are capable of handling the load. This fits the scaling predictions. With four times as much load, in the third graph, four servers just cut it. In the fourth graph we request at six times the load we see that the response rate is high for all scenarios. These results make approximate sense, yet the scaling is not exactly linear; there is some overhead when going from having 1 server to start load balancing between 2 or more servers.

As soon as a server crosses the maximum threshold, response times fall off exponentially. The challenge here is how to prevent this from happening in a data centres of inhomogeneous servers with different capacities. Dispatchers use one of a number of different sharing algorithms:

- Round robin: the classic load sharing method of taking each server in turn, without consideration of their current queue length or latency.

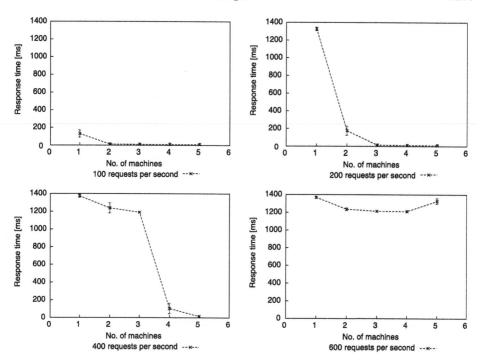

Fig. 13. How response rates scale when adding extra servers. For low traffic, adding a new server results in a discontinuous improvement in response time. At very high load, adding a server seems to reduce performance, leading to the upward curve in the last figure as the limitations of the bottleneck dispatcher become apparent.

- Least connections: the dispatcher maintains state over which back end server currently has fewest on-going TCP connections and channels new arrivals to the least connected host.
- Response time: the dispatcher measures the response time of each server in the back end by regularly testing the time it takes to establish a connection. This has many tunable parameters.

Tests show that, as long as server load is homogeneous, a round robin algorithm is most efficient. However the Least Connections algorithm holds out best against inhomogeneities [9].

## 10. Configuration and change management

A flexible organization is ready to adapt and make changes at any time. Some changes are in response to need and some are preemptive. All change is fraught with risk however. It is therefore considered beneficial to monitor and control the process of change in an organization in such a way that changes are documented and undergo a process of review. Two phrases are used, with some ambiguity, in this connection:

- Change management.
- Configuration management.

Change management is the planning, documentation and implementation of changes in system practices, resources and their configurations. The phrase configuration management invites some confusion, since it is used with two distinct meanings that arise from different cultures. In system administration arising from the world of Unix, configuration management is used to refer to the setup, tuning and maintenance of data that affect the behavior of computer systems. In software engineering and management (e.g., ITIL and related practices) it refers to the management of a database (CMDB) of software, hardware and components in a system, including their relationships within an organization. This is a much higher level view of 'configuration' than is taken in Unix management.

## 10.1. *Revision control*

One of the main recommendations about change management is the use of revision control to document changes. We want to track changes in order to:

- Be able to reverse changes that had a negative impact.
- Analyze the consequences of changes either successful or unsuccessful.

Some form of documentation is therefore necessary. This is normally done with the aid of a revision control system.

Revision control is most common in software release management, but it can also be used to process documentation or specifications for any kind of change (Figure 14). A revision control system is only a versioning tool for the actual change specification; the actual decisions should also be based on a process. There are many process models for change, including spiral development models, agile development, extreme change, followed by testing and perhaps and regret and reversal, etc. In mission critical systems more care in needed than 'try and see'. See for instance the discussions on change management in ITIL [25,26].

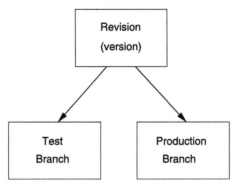

Fig. 14. Revision control – changes are tested and only later committed to production.

## 10.2. *'Rollback'*

In software engineering one has the notion of going back to an earlier version in the revision control system is a change is evaluated to have negative consequences. Many service managers and data centre engineers would like to have a similar view of live production systems, however there are problems associated with this.

The difference between software code and a live system is that code does not have *run-time operational state*. The behavior of a system depends in general on both system configuration and learned state that has been acquired through the running of the system. Even if one undoes changes to the configuration of a system, one cannot easily undo operational state (e.g. user sessions in an extended transaction). Thus simply undoing a configuration change will not necessarily return the behavior of a system to its previous condition. An obvious example of this is: suppose one removes all access control from a system so that it becomes overrun by hackers and viruses etc, simply restoring the access control will not remove these users or viruses from the system.

## 11. Human resources

In spite of the density of technology in a data centre, humans are key pieces of the service delivery puzzle. Services are increasingly about user-experience. Humans cannot therefore be removed from the system as a whole – one deals with human–computer systems [6]. Several issues can be mentioned:

- *Redundancy of workforce* is an important security in an organization. A single expert who has knowledge about an organization is a *single point of failure* just as much as any piece of equipment. Sickness or even vacation time places the organization at risk if the human services are not available to the organization.
- *Time management* is something that few people do well in any organization. One must resist the temptation to spend all day feeling busy without clearing time to work on something uninterrupted. The state of being busy is a mental state rather than a real affliction. Coping with the demands on a person's time is always a challenge. One has to find the right balance between fire-fighting and building.
- *Incident response procedures* are about making a formal response to a system event that is normally viewed to be detrimental to the system. This is a human issue because it usually involves human judgment. Some responses can be automated, e.g. using a tool like cfengine, but in general more cognition is required. This might involve a fault, a security breach or simply a failure to meet a specification, e.g. a Service Level Agreement (SLA).
- *Procedural documentation* of experience-informed procedures is an essential asset to an organization, because:
  - Past experience is not necessarily available to every data centre engineer.
  - People do not easily remember agreed procedures in a stressful situation or emergency.
  - Consistent behavior is a comfort to customers and management.

These matters are all a part of managing system predictability.

## 12. Concluding remarks

Data centres are complex systems that are essential components in a holistic view of IT service delivery. They require both intelligence and experience to perfect. There are many issues in data centres that system administrators have to deal with that are not covered in this overview. An excellent introduction to heuristic methods and advice can be found in ref. [23].

The main slogan for this review is simple: well designed systems fail rarely to meet their design targets and when they do so, they do so predictably. They are the result of exceptional planning and long hard experience.

## Acknowledgements

I have benefited from the knowledge and work of several former students and colleagues in writing this chapter: Claudia Eriksen, Gard Undheim and Sven Ingebrigt Ulland (all now of Opera Software) and Tore Møller Jonassen. This work was supported in part by the EC IST-EMANICS Network of Excellence (#26854).

## References

[1] G.M. Amdahl, *Validity of a the single processor approach to achieving large scale computer capabilities*, Proceedings of the AFTPS Spring Joint Computer Conference (1967).

[2] J.H. Bjørnstad and M. Burgess, *On the reliability of service level estimators in the data centre*, Proc. 17th IFIP/IEEE Integrated Management, volume submitted, Springer-Verlag (2006).

[3] British Standard/International Standard Organization, BS/ISO 17799 Information Technology – Code of Practice for Information Security Management (2000).

[4] British Standards Institute, *BS15000 IT Service Management* (2002).

[5] M. Burgess, *Principles of Network and System Administration*, Wiley, Chichester (2000).

[6] M. Burgess, *Analytical Network and System Administration – Managing Human–Computer Systems*, Wiley, Chichester (2004).

[7] M. Burgess and G. Canright, *Scalability of peer configuration management in partially reliable and ad hoc networks*, Proceedings of the VIII IFIP/IEEE IM Conference on Network Management (2003), 293.

[8] M. Burgess and G. Canright, *Scaling behavior of peer configuration in logically ad hoc networks*, IEEE eTransactions on Network and Service Management **1** (2004), 1.

[9] M. Burgess and G. Undheim, *Predictable scaling behavior in the data centre with multiple application servers*, Proc. 17th IFIP/IEEE Distributed Systems: Operations and Management (DSOM 2006), volume submitted, Springer (2006).

[10] J. Cao, W.S. Cleveland, D. Lin and D.X. Sun, *On the nonstationarity of Internet traffic*, SIGMETRICS '01: Proceedings of the 2001 ACM SIGMETRICS International Conference on Measurement and Modeling of Computer Systems, ACM Press (2001), 102–112.

[11] P.M. Chen, E.K. Lee, G.A. Gibson, R.H. Katz and D.A. Patterson, *RAID: High-performance, reliable secondary storage*, ACM Comput. Surv. **26** (2) (1994), 145–185.

[12] H.-K. Choi and J.O. Limb, *A behavioral model of web traffic*, ICNP'99: Proceedings of the Seventh Annual International Conference on Network Protocols, IEEE Computer Society, Washington, DC (1999), 327.

[13] R.B. Cooper, *Stochastic Models*, Handbooks in Operations Research and Management Science, Vol. 2, Elsevier (1990), Chapter: Queueing Theory.

[14] M.E. Crovella and A. Bestavros, *Self-similarity in world wide web traffic: Evidence and possible causes*, IEEE/ACM Trans. Netw. **5** (6) (1997), 835–846.

[15] G.R. Ganger, B.L. Worthington, R.Y. Hou and Y.N. Patt, *Disk subsystem load balancing: Disk striping vs. conventional data placement*, Proceeding of the Twenty-Sixth Hawaii International Conference on System Sciences, Vol. 1 (1993), 40–49.

[16] G.R. Ganger, B.L. Worthington, R.Y. Hou and Y.N. Patt, *Disk arrays: High-performance, high-reliability storage subsystems*, Computer **27** (3) (1994), 30–36.

[17] K.I. Hopcroft, E. Jakeman and J.O. Matthews, *Discrete scale-free distributions and associated limit theorems*, J. Math. Phys. **A37** (2004), L635–L642.

[18] A. Høyland and M. Rausand, *System Reliability Theory: Models and Statistical Methods*, Wiley, New York (1994).

[19] R. Jain, *The Art of Computer Systems Performance Analysis*, Wiley Interscience, New York (1991).

[20] T. Karagiannis, M. Molle and M. Faloutsos, *Long-range dependence: Ten years of Internet traffic modeling*, IEEE Internet Computing **8** (5) (2004), 57–64.

[21] A.H. Karp and H.P. Flatt, *Measuring parallel processor performance*, Comm. ACM **33** (5) (1990), 539–543.

[22] W.E. Leland, M.S. Taqqu, W. Willinger and D.V. Wilson, *On the self-similar nature of ethernet traffic (extended version)*, IEEE/ACM Trans. Netw. **2** (1) (1994), 1–15.

[23] T. Limoncelli and C. Hogan, *The Practice of System and Network Administration*, Addison–Wesley (2003).

[24] R. Menuet and W.P. Turner, *Continuous cooling is required for continuous availability*, Technical report, Uptime Institute (2006).

[25] Office of Government Commerce, ed., *Best Practice for Service Delivery*, ITIL: The Key to Managing IT Services, The Stationary Office, London (2000).

[26] Office of Government Commerce, ed., *Best Practice for Service Support*, ITIL: The Key to Managing IT Services, The Stationary Office, London (2000).

[27] K. Park, G. Kim and M. Crovella, *On the relationship between file sizes, transport protocols, and self-similar network traffic*, ICNP'96: Proceedings of the 1996 International Conference on Network Protocols (ICNP'96), IEEE Computer Society, Washington, DC (1996), 171.

[28] V. Paxson and S. Floyd, *Wide area traffic: The failure of Poisson modeling*, IEEE/ACM Trans. on Netw. **3** (3) (1995), 226–244.

[29] K. Raatikainen, *Symptoms of self-similarity in measured arrival process of ethernet packets to a file server*, Preprint, University of Helsinki (1994).

[30] G.B. Rodosek, *Quality aspects in it service management*, IFIP/IEEE 13th International Workshop on Distributed Systems: Operations and Management (DSOM 2002) (2002), 82.

[31] H. Simitci and D.A. Reed, *Adaptive disk striping for parallel input/output*, 16th IEEE Symposium on Mass Storage Systems (1999), 88–102.

[32] L.P. Slothouber, *A model of web server performance*, The 5th International World Wide Web Conference, Paris, France (1996).

[33] W.P. Turner, J.H. Seader and K.G. Brill, *Tier classifications define siet infrastructure performance*, Technical report, Uptime Institute (1996), 2001–2006.

[34] R.D. van der Mei, R. Hariharan and P.K. Reeser, *Web server performance modeling*, Telecommunication Systems **16** (March 2001), 361–378.

[35] J. Walrand, *Stochastic Models*, Handbooks in Operations Research and Management Science, Vol. 2, Elsevier (1990), Chapter: Queueing Networks.

# – 1.3 –

# Automating System Administration: Landscape, Approaches and Costs

Aaron B. Brown[1], Joseph L. Hellerstein[2], Alexander Keller[3]

[1]*IBM Software Group, Route 100, Somers, NY 10589, USA*
*E-mail: abbrown@us.ibm.com*

[2]*Microsoft Corporation, One Microsoft Way, Redmond, WA 98052, USA*
*E-mail: Joehe@microsoft.com*

[3]*IBM Global Technology Services, 11 Madison Avenue, New York, NY 10010, USA*
*E-mail: alexk@us.ibm.com*

## 1. Introduction

The cost of systems administration in Information Technology (IT) systems often exceeds the cost of hardware and software. Our belief is that automating system administration can reduce these costs and increase the business value provided by IT.

Making progress with automating system administration requires addressing three questions. What to automate? How should this automation be done? When does the automation provide business value?

*What to automate* is often answered in a simplistic way – everything! The problem here is that automation requires investment and inevitably causes some disruption in the 'as-is' IT environment. As a result, it is important to target aspects of System Administration where automation provides the most value. We believe that a process-based perspective provides the kind of broad context in which such judgments can be made.

*How to automate* system administration can be approached in many ways. Among these are rule-based techniques, control theory and automated workflow construction. These approaches provide different kinds of benefits, and, in many ways, address different aspects of the automation challenge.

HANDBOOK OF NETWORK AND SYSTEM ADMINISTRATION
Edited by Jan Bergstra and Mark Burgess

*When to automate* is ultimately a business decision that should be based on a full understanding of the costs and benefits. The traditional perspective has been that automation is always advantageous. However, it is important to look at the full costs imposed by automation. For example, automating software distribution requires that: (a) the distribution infrastructure be installed and maintained; (b) software packages be prepared in the format required by the distribution infrastructure; and (c) additional tools be provided to handle problems with packages that are deployed because of the large scale of the impact of these problems. While automation often provides a net benefit despite these costs, we have observed cases in which these costs exceed the benefits.

There is a variety of existing literature on systems administration. One of the earliest efforts in automating systems administration is an expert systems for managing mainframe systems [39]. More recently, [9] addresses distributed resource administration using cfengine. Historically, authors have addressed various aspects of systems administration [21,23,24,47,49] and [36]. There has been theoretical work as well such as [46], who discusses how to structure actions for systems administration in configuration management, and [50], who addresses automated responses to anomalies using immunological algorithms.

The remainder of this chapter provides more detail on the questions of what, how and when to automate.

## 2. What system administrators do

We start our consideration of automating system administration by examining what system administrators do, as these tasks are the starting points for building automation.

System administrators are a central part of IT operations. Anyone who has owned a personal computer knows some elements of system administration. Software must be installed, patched and upgraded. Important data must be backed up, and occasionally restored. And, when a problem occurs, time must be spent on problem isolation, identification, and resolution.

In many ways, system administrators perform the same activities as individuals. There are, however, some important differences. The first is scale. Data centers consist of hundreds to tens of thousands of servers, and administrators are typically responsible for tens to hundreds of machines. This scale means that management tools are essential to deal with repetitive tasks such as software installation.

A second important difference is that system administrators are usually responsible for mission critical machines and applications. Seconds of downtime at a financial institution can mean millions of dollars of lost revenue. Failures in control systems for trains and aviation can cost lives. These considerations place tremendous emphasis on the speed and accuracy with which administrators perform their jobs.

Large data centers structure system administration by technology.

Examples of these technology areas are servers, databases, storage and networks. Administrators will typically train and develop specialized expertise in one technology area –

for example a database administrator (DBA) becomes an expert in the details of database installation, tuning, configuration, diagnosis and maintenance. However, the different technology areas often contain similar tasks, such as diagnosis and capacity planning, so the techniques developed in one area will often translate relatively easily to others. Even so, the tools used to accomplish these tasks in the separate areas remain different and specialized.

The remainder of this section is divided into four parts. The first two parts describe system administration in the words of experts. We focus on two technology areas – database administration and network administration. The third part of the section is a broader look at system administration based on a USENIX/SAGE survey. The section concludes with a discussion of best practices.

### 2.1. *Database administration*

We begin our look at expert system administration in the area of database administration. The material in this section is based on a database administration course [10].

The job of a database administrator (DBA) is largely driven by the needs of the organization. But at the heart of this job is a focus on managing data integrity, access, performance and security/privacy. Typical tasks performed by a DBA might include:
- designing and creating new databases;
- configuring existing databases to optimize performance, for example by manipulating index structures;
- managing database storage to ensure enough room for stored data;
- diagnosing problems such as 'stale' database data and deadlocks;
- ensuring data is replicated to on-line or off-line backup nodes;
- federating multiple independent database instances into a single virtual database;
- interfacing with other administrators to troubleshoot problems that extend beyond the database (e.g., storage or networking issues);
- generating reports based on database contents and rolling up those reports across multiple database systems.

The tools that a DBA uses to perform tasks like these vary greatly depending on the scale and needs of the organization. Small organizations and departments may use simple tools such as Paradox, Visual FoxPro, or Microsoft Access to maintain data. Larger organizations and governments with corporate-wide data requirements demand industrial-strength database tools such as IBM DB2, Oracle database, Microsoft SQL Server, Ingres, etc. And for organizations with multiple databases (often a situation that occurs after acquisitions, or when independent departments are transitioned to a centralized system), additional tools, often called 'middleware', will be necessary to federate and integrate the existing databases across the corporation. Middleware poses its own administration considerations for installation, configuration, optimization, and maintenance.

In addition to the core tools – the database engine(s) and middleware – DBAs use many specialized administration tools. These may be scripts developed by the DBA for specific situations, or they may be vendor-supplied tools for monitoring, administration and reporting.

As we look toward automation of a DBAs system administration responsibilities, we must recognize the challenges faced by a DBA. First is the integration of large numbers of tools. It should be clear from the discussion above that anything larger than a small department environment will involve multiple database engines, possibly middleware and a mix of home-grown and vendor-supplied tools. Second is the need to 'roll-up' data in large environments, e.g., integrating department reports into corporate-wide reports. Part of the challenge of the roll-up is the need for processes to scrub data to ensure correctness and consistency. The challenge of providing roll-ups creates a tension between the small scale systems with ease of entry and the large scale systems that provide robustness and scalability.

Finally, DBAs do not operate in a vacuum. There is considerable interaction with other administration 'towers', such as network and storage management. The following from [10] provides an illustrative example:

> Unfortunately, one situation that can occur more often than planned, or more accurately, more than it should, is when the database does not update despite the careful steps taken or the time involved in trying to accomplish a successful update [...]
>
> The first step is to investigate the problem to learn the true 'age' of the data. Next, the DBA would attempt to update a recent set of data to determine what may have occurred. It could have been a malfunction that day, or the evening the update was attempted. Assuming the situation does not improve by this upload, the next step is to contact the network administrator about possible server malfunctions, changes in standard record settings, or other changes that might affect the upload of data.
>
> Occasionally, the network server record lock settings were changed to 'disallow' any upload to a network server over a certain limit [...] If the DBA is lucky, or has positive karma due for collection, all that may be required is for the network administrator to reset the record lock setting at a higher level. Updates can then be repeated and will result in current data within the business unit database for purposes of management reporting.
>
> However, on a bad day, the DBA may learn from the network administrator that the server was or is malfunctioning, or worse, crashed at the precise time of the attempted update [...] In this situation the investigation must retrace the steps of data accumulation to determine the validity of the dataset for backup and experimental upload (network administrators at ringside) to watch for any type of malfunction.

In environments such as the foregoing, automation usually proceeds in an incremental manner, starting with special-purpose scripts that automate specific processes around the sets of existing tools, to generalizations of those scripts into new management tools, to rich automation frameworks that integrate tools to close the loop from detecting problems to reacting to them automatically.

### 2.2. *Network administration*

Next, we look at another technology area for system administration: network administration. The material in this section is based on a course on network administration [54]. We have included quotes as appropriate.

Network administrators are responsible for the performance, reliability, and scalability of corporate networks. Achieving these goals requires a substantial understanding of the business for which these services are being delivered as well as the nature and trends in network technologies. In particular, in today's network-connected, web-facing world, network administrators have gone from supporting the back-office to enabling the online front-office, ensuring the network can deliver the performance and reliability to provide customer information, sales, support, and even B2B commerce online via the Internet. The network administrator's job has become a critical function in the revenue stream for many businesses.

Typical tasks performed by a network administrator might include:

- designing, deploying, and redesigning new networks and network interconnections;
- deploying new devices onto a network, which requires a good understanding of the network configuration, the device requirements, and considerations for loads imposed by network traffic;
- setting and enforcing security policies for network-connected elements;
- monitoring network performance and reconfiguring network elements to improve performance;
- managing network-sourced events;
- tracking the configuration of a network and the inventory of devices attached to it;
- detecting failures, diagnosing their root causes, and taking corrective actions such as reconfiguring the network around a failed component.

Network administrators make use of a variety of tools to perform these tasks. Some are as simple as the TCP `ping`, `traceroute` and `netstat` commands, which provide simple diagnostics. Administrators responsible for large corporate networks frequently make use of network management systems with sophisticated capabilities for filtering, eventing, and visualization. Examples of such systems are Hewlett–Packard's OpenView product and IBM's NetView product. Administrators of all networks will occasionally have to use low-level debugging tools such as packet sniffers and protocol decoders to solve compatibility and performance problems. No matter the scale of the network, tools are critical given the sizes of modern networks and the volumes of data that move across them. To quote from the advice for network administrators in [54], 'A thorough inventory and knowledge of the tools at your disposal will make the difference between being productive and wasting time. A good network administrator will constantly seek out ways to perform daily tasks in a more productive way'.

As we look toward automation of a network administrator's activities, there are two key aspects to consider. First is to provide more productive tools and to integrate existing tools. Here, automation must help administrators become more productive by absorbing the burden of monitoring, event management, and network performance optimization. Again this kind of automation progresses incrementally, starting with better scripts to integrate existing tools into situation-specific problem solvers, moving to integrated tools that, for example, merge monitoring with policy-driven response for specific situations, and culminating in integrated frameworks that close the loop and take over management of entire portions of a corporate network from the human administrator's hands.

The second key aspect of automating a network administrator's activities is to improve the task of problem diagnosis. Here, we refer to both proactive maintenance and reactive problem-solving. In the words of [54]:

> Resolving network, computer, and user related issues require the bulk of an administrator's time.
> Eventually this burden can be lessened through finding permanent resolutions to common prob-

lems and by taking opportunities to educate the end user. Some reoccurring tasks may be automated to provide the administrator with more time [...] However, insuring the proper operation of the network will preempt many problems before the users notice them.

Finally, it is important to point out that for any system administration role, automation can only go so far. There always is a human aspect of system administration that automation will not replace; in the case of network administration, again in the words of [54], this comes in many forms.

> Apart from the software and hardware, often the most difficult challenge for a network administrator is juggling the human aspects of the job. The desired result is always a productive network user, not necessarily just a working network. To the extent possible, the administrator should attempt to address each user's individual needs and preferences. This also includes dealing with the issues that arise out of supporting user skill levels ranging from beginner to knowledgeable. People can be the hardest and yet most rewarding part of the job.
>
> As problems are addressed, the solutions will be documented and the users updated. Keeping an open dialogue between the administrator and users is critical to efficiently resolving issues.

### 2.3. *The broader view*

Now that we have looked at a few examples of system administration, next we move up from the details of individual perspectives to a survey of administrators conducted by SAGE, the Special Interest Group of the USENIX Association focusing on system administration. Other studies support these broad conclusions, such as [4].

The SAGE study [20] covered 128 respondents from 20 countries, with average experience of 7 years, 77% with college degrees, and another 20% having some college. The survey focused on server administration. Within this group, each administrator supported 500 users, 10–20 servers, and works in a group of 2 to 5 people. As reported in the survey, administrators have an 'atypical day' at least once a week.

With this background, we use the survey results to characterize how administrators spend their time, with an eye to assessing the opportunity for automation. This is done along two different dimensions. The first is *what* is being administered:

- 20% miscellaneous;
- 12% application software;
- 12% email;
- 9% operating systems;
- 7% hardware;
- 6% utilities;
- 5% user environment.

There is a large spread of administrative targets, and the reason for the large fraction of miscellaneous administration may well be due to the need to deal with several problems at once.

> Broad automation is needed to improve system administration.

Automation that targets a single domain such as hardware will be useful, but will not solve the end-to-end problems that system administrators typically face. Instead, we should consider automation that takes an end-to-end approach, cutting across domains as needed to solve the kinds of problems that occur together. We will address this topic in Section 4 in our discussion of process-based automation.

Another way to view administrators' time is to divide it by the kind of activity. The survey results report:

- 11% meetings;
- 11% communicating;
- 9% configuring;
- 8% installing;
- 8% 'doing';
- 7% answering questions;
- 7% debugging.

We see that at least 29% of the administrator's job involves working with others. Sometimes, this is working with colleagues to solve a problem. But there is also substantial time in reporting to management on progress in resolving an urgent matter and dealing with dissatisfied users. About 32% of the administrator's job involves direct operations on the IT environment. This is the most obvious target for automation, although it is clear that automation will need to address some of the communication issues as well (e.g., via better reporting).

The survey goes on to report some other facts of interest. For example, fire fighting only takes 5% of the time. And, there is little time spent on scheduling, planning and designing. The latter, it is noted, may well be reflected in the time spent on meetings and communicating. We can learn from these facts that automation must address the entire lifecycle of administration, not just the (admittedly intense) periods of high-pressure problem-solving.

The high pressure and broad demands of system administration may raise serious questions about job satisfaction. Yet, 99% of those surveyed said they would do it again.

### 2.4. *The promise of best practices and automation*

One insight from the foregoing discussion is that today systems administration is more of a craft than a well-disciplined profession. Part of the reason for this is that rapid changes in IT make it difficult to have re-usable best practices, at least for many technical details.

There is a body of best practices for service delivery that is gaining increasing acceptance, especially in Europe. Referred to as the Information Technology Infrastructure Library (ITIL), these best practices encompass configuration, incident, problem management, change management, and many other processes that are central to systems administration [27]. Unfortunately, ITIL provides only high level guidance. For example, the ITIL process for change management has activities for 'authorizing change', 'assign priority' and 'schedule change'. There are few specifics about the criteria used for making decisions, the information needed to apply these criteria, or the tools required to collect this information.

> The key element of the ITIL perspective is its focus on *process*.

ITIL describes end-to-end activities that cut across the traditional system administration disciplines, and suggests how different 'towers' like network, storage, and application/database management come together to design infrastructure, optimize behavior, or solve cross-cutting problems faced by users. ITIL thus provides a philosophy that can guide us to the ultimate goals of automation, where end-to-end, closed-loop activities are subsumed entirely by automation, the system administrators can step out, and thus the costs of delivering IT services are reduced significantly.

We believe that installations will move through several phases in their quest to reduce the burden on systems administrators and hence the cost of delivering IT services. In our view, these steps are:

1. *Environment-independent automation*: execution of repetitive tasks without system-dependent data within one tower, for example static scripts or response-file-driven installations.

2. *Environment-dependent automation*: taking actions based on configuration, events, and other factors that require collecting data from systems. This level of automation is often called 'closed-loop' automation, though here it is still restricted to one discipline or tower.

3. *Process-based automation*: automation of interrelated activities in best practices for IT service delivery. Initially, this tends to be open-loop automation that mimics activities done by administrators such as the steps taken in a software install. Later on, automation is extended to closed-loop control of systems such as problem detection and resolution.

4. *Business level automation*: automation of IT systems based on business policies, priorities, and processes. This is an extension of process-based automation where the automation is aware of the business-level impact of various IT activities and how those IT activities fit into higher-level business processes (for example, insurance claim processing). It extends the closed-loop automation described in (3) to incorporate business insights.

Business-level automation is a lofty goal, but the state of the art in 2006 is still far from reaching it outside very narrowly-constrained domains. And, at the other extreme, environment-independent automation is already well-established through ad-hoc mechanisms like scripts and response files. Thus in the remainder of our discussion we will focus on how to achieve environment-dependent and process-based automation. Section 3 describes a strategy for reaching process-based automation, and follows that with a description of some automation techniques that allow for environment-dependent automation and provide a stepping stone to higher levels of automation.

## 3. How to automate

This section addresses the question of how to automate system administration tasks.

### 3.1. *Automation strategies*

We start our discussion of automating IT service delivery by outlining a best-practice approach to process-based automation (cf. [7]). We have developed this approach based on our experiences with automating change management, along with insight we have distilled from interactions and engagements with service delivery personnel. It provides a roadmap for achieving process-level automation.

Our approach comprises six steps for transforming existing IT service delivery processes with automation or for introducing new automated process into an IT environment. The first step is to

(1)  identify best practice processes for the domain to be automated.

To identify these best practices we turn to the IT Infrastructure Library (ITIL), a widely used process-based approach to IT service management. ITIL comprises several disciplines such as infrastructure management, application management, service support and delivery. The ITIL Service Support discipline [27] defines the Service Desk as the focal point for interactions between a service provider and its customers. To be effective, the Service Desk needs to be closely tied into roughly a dozen IT support processes that address the lifecycle of a service. Some examples of IT service support processes for which ITIL provides best practices are: Configuration Management, Change Management, Release Management, Incident Management, Problem Management, Service Level Management and Capacity Management.

ITIL provides a set of process domains and best practices within each domain, but it does not provide guidance as to which domain should be an organization's initial target for automation. Typically this choice will be governed by the cost of existing activities.

After identifying the best practices, the next step is to

(2)  establish the scope of applicability for the automation.

Most ITIL best practices cover a broad range of activities. As such, they are difficult to automate completely. To bound the scope of automation, it is best to target a particular subdomain (e.g., technology area). For example, Change Management applies changes in database systems, storage systems and operating systems.

We expect that Change Management will be an early focus of automation. It is our further expectation that initial efforts to automate Change Management will concentrate on high frequency requests, such as security patches. Success here will provide a proof point and template for automating other areas of Change Management. And, once Change Management is automated to an acceptable degree, then other best practices can be automated as well, such as Configuration Management. As with Change Management, we expect that the automation of Configuration Management will be approached incrementally. This might be structured in terms of configuration of servers, software licenses, documents, networks, storage and various other components.

The next step in our roadmap follows the ITIL-based approach:

(3)  identify delegation opportunities.

Each ITIL best practice defines (explicitly or implicitly) a process flow consisting of multiple activities linked in a workflow. Some of these activities will be amenable to automation, such as deploying a change in Change Management. These activities can be *delegated* to an automated system or tool. Other activities will not be easy to automate, such as obtaining change approvals from Change Advisory Boards. Thus, an analysis is needed to determine what can be automated and at what cost. Sometimes, automation drives changes in IT processes. For example, if an automated Change Management system is trusted enough, change approvals might be handled by the Change Management System automatically.

The benefit of explicitly considering delegation is that it brings the rigor of the best-practice process framework to the task of scoping the automation effort. The best practice defines the set of needed functionality, and the delegation analysis explicitly surfaces the decision of whether each piece of functionality is better handled manually or by automation. Using this framework helps prevent situations like the one discussed later in Section 4 of this chapter, where the cost of an automation process outweighs its benefits in certain situations.

With the delegated activities identified, the fourth step in our automation approach is to

(4)  identify links between delegated activities and external activities, processes and data sources.

These links define the control and data interfaces to the automation. They may also induce new requirements on data types/formats and APIs for external tools. An example in Change Management is the use of configuration data. If Change Management is automated but Configuration Management remains manual, the Configuration Management Database (CMDB) may need to be enhanced with additional APIs to allow for programmatic access from the automated Change Management activities. Moreover, automation often creates *induced processes*, additional activities that are included in support of the automation. Examples of induced processes in automated software distribution are the processes for maintaining the software distribution infrastructure, error recovery and preparation of the software packages.

The latter point motivates the following step:

(5)  Identify, design, and document induced processes needed to interface with or maintain the automation.

This step surfaces many implications and costs of automation and provides a way to do cost/benefit tradeoffs for proposed automation.

The last step in our automation approach is to

(6)  implement automation for the process flow and the delegated activities.

Implementing the process flow is best done using a workflow system to automatically coordinate the best-practice process' activities and the flow of information between them. Using a workflow system brings the additional advantage that it can easily integrate automated and manual activities.

For the delegated activities, additional automation implementation considerations are required. This is a non-trivial task that draws on the information gleaned in the earlier steps. It uses the best practice identified in (1) and the scoping in (2) to define the activities' functionality. The delegation choices in (3) scope the implementation work, while the interfaces and links to induced process identified in (4) and (5) define needed APIs, connections with external tools and data sources, and user interfaces. Finally, the actual work of the activity needs to be implemented directly, either in new code or by using existing tools.

In some cases, step (6) may also involve a recursive application of the entire methodology described here. This is typically the case when a delegated ITIL activity involves a complex flow of work that amounts to a process in its own right. In these cases, that activity sub-process may need to be decomposed into a set of subactivities; scoped as in steps (2) and (3), linked with external entities as in (4), and may induce extra sub-process as in (5). The sub-process may in turn also need to be implemented in a workflow engine, albeit at a lower level than the top-level best practice process flow.

**3.2.** *Automation technologies*

This section introduces the most common technologies for automation. Considered here are rules, feedback control techniques, and workflows technology.

**3.2.1.** *Rule-based techniques*

Rules provide a condition–action representation for automation.

An example of a rule is
- Rule: If there is a `SlowResponse` event from system ?S1 at location ?L1 within 1 minute of another `SlowResponse` event from system ?S2 ≠ ?S1 at location ?L1 and there is no `SlowResponse` event from system S3 at location ?L2 ≠ ?L1, then alert the Network Manager for location ?L1.

In general, rules are if-then statements. The if-portion, or left-hand side, describes a situation. The then-portion, or right-hand side, specifies actions to take when the situation arises.

As noted in [52], there is a great deal of work in using rule-based techniques for root cause analysis (RCA). [39] describes a rule-based expert system for automating the operation of an IBM mainframe, including diagnosing various kinds of problems. [22] describes algorithms for identifying causes of event storms. [28] describe the application of an expert system shell for telecommunication network alarm correlation.

The simplicity of rule-based systems offers an excellent starting point for automation. However, there are many shortcomings. Among these are:

1. A pure rule-based system requires detailed descriptions of many situations. These descriptions are time-consuming to construct and expensive to maintain.
2. Rule-based systems scale poorly because of potentially complex interactions between rules. Such interactions make it difficult to debug large scale rule-based systems, and create great challenges when adding new capabilities to such systems.
3. Not all automation is easily represented as rules. Indeed, step-by-step procedures are more naturally expressed as workflows.

There are many approaches to circumventing these shortcomings. One of the most prominent examples is [32], which describes a code-book-based algorithm for RCA. The practical application of this approach relies on explicit representations of device behaviors and the use of configuration information. This has been quite effective for certain classes of problems.

**3.2.2.** *Control theoretic approaches*

Control theory provides a formal framework for optimizing systems.

Over the last sixty years, control theory has developed a fairly simple reference architecture. This architecture is about manipulating a target system to achieve a desired objective. The component that manipulates the target system is the controller.

As discussed in [15] and depicted in Figure 1, the essential elements of feedback control system are:

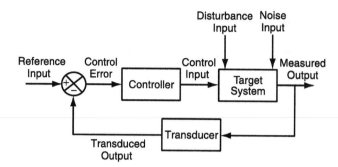

Fig. 1. Block diagram of a feedback control system. The reference input is the desired value of the system's measured output. The controller adjusts the setting of control input to the target system so that its measured output is equal to the reference input. The transducer represents effects such as units conversions and delays.

- target system, which is the computing system to be controlled;
- control input, which is a parameter that affects the behavior of the target system and can be adjusted dynamically (such as the `MaxClients` parameter in the Apache HTTP Server);
- measured output, which is a measurable characteristic of the target system such as CPU utilization and response time;
- disturbance input, which is any change that affects the way in which the control input influences the measured output of the target system (e.g., running a virus scan or a backup);
- noise input, which is any effect that changes the measured output produced by the target system. This is also called sensor noise or measurement noise;
- reference input, which is the desired value of the measured output (or transformations of them), such as CPU utilization should be 66%. Sometimes, the reference input is referred to as desired output or the setpoint;
- transducer, which transforms the measured output so that it can be compared with the reference input (e.g., smoothing stochastics of the output);
- control error, which is the difference between the reference input and the measured output (which may include noise and/or may pass through a transducer);
- controller, which determines the setting of the control input needed to achieve the reference input. The controller computes values of the control input based on current and past values of control error.

Reference inputs specify policies.

For example, a service level agreement may specify constraints on response times for classes of services such as "requests to browse the product catalogue should be satisfied within two seconds". In this case, the measured output is response time for a browse request, and the reference input is two seconds. More details can be found in [8].

The foregoing is best understood in the context of a specific system. Consider a cluster of three Apache Web Servers. The Administrator may want these systems to run at no greater than 66% utilization so that if any one of them fails, the other two can immediately absorb the entire load. Here, the measured output is CPU utilization. The control input is the maximum number of connections that the server permits as specified by the MaxClients parameter. This parameter can be manipulated to adjust CPU utilization. Examples of disturbances are changes in arrival rates and shifts in the type of requests (e.g., from static to dynamic pages).

To illustrate how control theory can be applied to computing systems, consider the IBM Lotus Domino Serveras described in [41]. To ensure efficient and reliable operation, Administrators of this system often regulate the number of remote procedure calls (RPCs) in the server, a quantity that we denote by *RIS*. *RIS* roughly corresponds to the number of *active users* (those with requests outstanding at the server). Regulation is accomplished by using the MaxUsers tuning parameter that controls the number of *connected users*. The correspondence between MaxUsers and *RIS* changes over time, which means that MaxUsers must be updated almost continuously to achieve the control objective. Clearly, it is desirable to have a controller that automatically determines the value of MaxUsers based on the objective for *RIS*.

Our starting point is to model how MaxUsers affects *RIS*. The input to this model is MaxUsers, and the output is *RIS*. We use $u(k)$ to denote the $k$th value of the former and $y(k)$ to denote the $k$th value of the latter. (Actually, $u(k)$ and $y(k)$ are offsets from a desired operating point.) A standard workload was applied to a IBM Lotus Domino Server running product level software in order to obtain training and test data. In all cases, values are averaged over a one minute interval. Based on these experiments, we constructed an empirical model (using least squares regression) that relates MaxUsers to *RIS*:

$$y(k+1) = 0.43y(k) + 0.47u(k). \tag{1}$$

To better facilitate control analysis, Equation (1) is put into the form of a transfer function, which is a Z-transform representation of how MaxUsers affects *RIS*. Z-transforms provide a compact representation for time varying functions, where $z$ represents a time shift operation. The transfer function of Equation (1) is

$$\frac{0.47}{z - 0.43}.$$

The poles of a transfer function are the values of $z$ for which the denominator is 0. It turns out that the poles determine the stability of the system represented by the transfer function, and they largely determine its settling time. This can be seen in Equation (1). Here, there is one pole, which is 0.43. The effect of this pole on settling time is clear if we solve the recurrence in Equation (1). The result has the factors $0.43^{k+1}, 0.43^{k}, \ldots$. Thus, if the absolute value of the pole is greater than one, the system is unstable. And the closer the pole is to 0, the shorter the settling time. A pole that is negative (or imaginary) indicates an oscillatory response.

Many researchers have applied control theory to computing systems. In data networks, there has been considerable interest in applying control theory to problems of flow control, such as [31] which develops the concept of a Rate Allocating Server that regulates the flow of packets through queues. Others have applied control theory to short-term rate variations in TCP (e.g., [35]) and some have considered stochastic control [2]. More recently, there have been detailed models of TCP developed in continuous time (using fluid flow approximations) that have produced interesting insights into the operation of buffer management schemes in routers (see [25,26]). Control theory has also been applied to middleware to provide service differentiation and regulation of resource utilizations as well as optimization of service level objectives. Examples of service differentiation include enforcing relative delays [1], preferential caching of data [37], and limiting the impact of administrative utilities on production work [40]. Examples of regulating resource utilizations include a mixture of queueing and control theory used to regulate the Apache HTTP Server [48], regulation of the IBM Lotus Domino Server [41], and multiple-input, multiple-output control of the Apache HTTP Server (e.g., simultaneous regulation of CPU and memory resources) [14]. Examples of optimizing service level objectives include minimizing response times of the Apache Web Server [16] and balancing the load to optimize database memory management [17].

All of these examples illustrate situations where control theory-based automation was able to replace or augment manual system administration activities, particularly in the performance and resource management domains.

**3.2.3.** *Automated workflow construction*    In Section 2.4, we have observed that each ITIL best practice defines (explicitly or implicitly) a process flow consisting of multiple activities linked in a workflow. Some of these activities will be amenable to automation – such as deploying a change in Change Management – whereas others will not (such as obtaining change approvals from Change Advisory Boards). On a technical level, recent efforts aim at introducing extensions for people facing activities into workflow languages [33]. The goal is to facilitate the seamless interaction between the automated activities of an IT service management process and the ones that are carried out by humans. Here, we summarize work in [5] and [29] on automating system administration using workflow.

In order to provide a foundation for process-based automation, it is important to identify the control and data interfaces between delegated activities and external activities, processes, and data sources. The automation is provided by lower-level automation workflows, which consist of atomic administration activities, such as installing a database management system, or configuring a data source in a Web Application Server. The key question is to what extent the automation workflows can be automatically generated from domain-specific knowledge, instead of being manually created and maintained. For example, many automation workflows consist of activities whose execution depends on the current state of a distributed system. These activities are often not specified in advance. Rather, they are merely implied. For example, applications must be recompiled if they use a database table whose schema is to change. Such implicit changes are a result of various kinds of relationships, such as service dependencies and the sharing of the service provider's resources among different customers. Dependencies express compatibility requirements between the various components of which a distributed system is composed. Such requirements typically comprise software pre-requisites (components that must be present on the system for an installation to succeed), co-requisites (components that must be jointly installed) as well as ex-requisites (components that must be removed prior to installing a new component). In addition, version compatibility constraints and memory/disk space requirements need to be taken into account. All of this information is typically captured during the development and build stages of components, either by the developer, or by appropriate tooling. Dependency models, which can be specified e.g., as *Installable Unit Deployment Descriptors* [53] or *System Description Models* [38], are a key mechanism to capture this knowledge and make it available to the tools that manage the lifecycle of a distributed system.

An example of a consumer of dependency information is the *Change Management System*, whose task is to orchestrate and coordinate the deployment, installation and configuration of a distributed system. Upon receiving a request for change (RFC) from an administrator, the change management system generates a *Change Plan*. A change plan describes the partial order in which tasks need to be carried out to transition a system from a workable state into another workable state. In order to achieve this, it contains information about:

- The type of change to be carried out, e.g., install, update, configure, uninstall;
- the roles and names of the components that are subject to a change (either directly specified in the RFC, or determined by the change management system);
- the precedence and location constraints that may exist between tasks, based on component dependency information;
- an estimate of how long every task is likely to take, based on the results of previous deployments. This is needed to estimate the impact of a change in terms of downtime and monetary losses.

The change management system exploits dependency information. The dependency information is used to determine whether tasks required for a change must be carried out sequentially, or whether some of them can be parallelized. The existence of a dependency between two components – each representing a managed resource – in a dependency graph indicates that a precedence/location constraint must be addressed. If a precedence constraint exists between two tasks in a workflow (e.g., X must be installed before Y), they need to be carried out sequentially. This is typically indicated by the presence of a link; any task may have zero or more incoming and outgoing links. If two tasks share the same predecessor and no precedence constraints exist between them, they can be executed concurrently. Typically, tasks and their constraints are grouped on a per-host basis. Grouping on a per-host basis is important because the actual deployment could happen either push-based (triggered by the provisioning system) or pull-based (deployment is initiated by the target systems). An additional advantage of grouping activities on a per-host basis is that one can carry out changes in parallel (by observing the presence of cross-system links) if they happen on different systems. Note that parallelism on a single host system is difficult to exploit with current operating systems as few multithreaded installers exist today. In addition, some operating systems require exclusive access to shared libraries during installation.

Different types of changes require different traversals through the dependency models: If a request for a new installation of a component is received, one needs to determine which components must already be present before a new component can be installed. On the other hand, a request for an update, or an uninstall of an component leads to a query to determine the components that will be impacted by the change.

Precedence constraints represent the order in which provisioning activities need to be carried out. Some of these constraints are implicit (e.g., by means of a containment hierarchy expressed as 'HasComponent' relationships between components), whereas others (typically resulting from communication dependencies such as 'uses') require an explicit representation (e.g., the fact that a database client needs to be present on a system whenever a database server located on a remote host needs to be accessed).

**3.2.3.1.** *Example for software Change Management*   The example is based on the scenario of installing and configuring a multi-machine deployment of a J2EE based enterprise application and its supporting middleware software (including IBM's HTTP Server, WebSphere Application Server, WebSphere MQ embedded messaging, DB2 UDB database and DB2 runtime client). The specific application we use is taken from the SPEC-jAppServer2004 enterprise application performance benchmark [51]. It is a complex, multi-tiered on-line e-Commerce application that emulates an automobile manufacturing company and its associated dealerships. SPECjAppServer2004 comprises typical manufacturing, supply chain and inventory applications that are implemented with web, EJB, messaging, and database tiers. We jointly refer to the SPECjAppServer2004 enterprise application, its data, and the underlying middleware as the SPECjAppServer2004 *solution*. The SPECjAppServer2004 solution spans an environment consisting of two systems: one system hosts the application server along with the SPECjAppServer2004 J2EE application, whereas the second system runs the DBMS that hosts the various types of SPEC-jAppServer2004 data (catalog, orders, pricing, user data, etc.). One of the many challenges in provisioning such a solution consists in determining the proper order in which its components need to be deployed, installed, started and configured. For example, 'HostedBy' dependencies between the components 'SPECjAppServer2004 J2EE Application' and 'Web-Sphere Application Server v5.1' (WAS) state that the latter acts as a hosting environment for the former. This indicates that all the WAS components need to be operating before one can deploy the J2EE application into them.

   A provisioning system supports the administrator in deploying, installing and configuring systems and applications. As mentioned above, a change plan is a procedural description of activities, each of which maps to an operation that the provisioning system exposes, preferably by means of a set of interfaces specified using the Web Services Description Language (WSDL). As these interfaces are known well in advance before a change plan is created, they can be referenced by a change plan. Every operation has a set of input parameters, for example, an operation to install a given software component on a target system requires references to the software and to the target system as input parameters. Some of the parameters may be input by a user when the change plan is executed (e.g., the hostname of the target system that will become a database server) and need to be forwarded to several activities in the change plan, whereas others are produced by prior activities.

**3.2.3.2.** *Executing the generated change plan*   Upon receiving a newly submitted change request, the change management system needs to determine on which resources and at what time the change will be carried out. The change management system first inspects the resource pools of the provisioning system to determine which target systems are best assigned to the change by taking into account which operating system they run, what their system architecture is, and what the cost of assigning them to a change request is. Based on this information, the change management system creates a change plan, which may be composed of already existing change plans that reside in a change plan repository. Once the change plan is created, it is submitted to the workflow engine of the provisioning system. The provisioning system maps the actions defined in the change plan to operations that are understood by the target systems. Its object-oriented data model is a hierarchy of *logical devices* that correspond to the various types of managed resources (e.g., software, storage, servers, clusters, routers or switches). The methods of these types correspond to *Logical Device Operations (LDOs)* that are exposed as WSDL interfaces, which allows their inclusion in the change plan. *Automation packages* are product-specific implementations of logical devices: e.g., an automation package for the DB2 DBMS would provide scripts that implement the software.install, software.start, software.stop, etc. LDOs. An automation package consists of a set of Jython scripts, each of which implements an LDO. Every script can further embed a combination of PERL, Expect, Windows scripting host or bash shell scripts that are executed on the remote target systems. We note that the composition pattern applies not only to workflows, but occurs in various places within the provisioning system itself to decouple a change plan from the specifics of the target systems.

The workflow engine inputs the change plan and starts each provisioning operation by directly invoking the LDOs of the provisioning system. These invocations are performed either in parallel or sequentially, according to the instructions in the change plan. A major advantage of using an embedded workflow engine is the fact that it automatically performs state-checking, i.e., it determines whether all conditions are met to move from one activity in a workflow to the next. Consequently, there is no need for additional program logic in the change plan to perform such checks. Status information is used by the workflow engine to check if the workflow constraints defined in the plan (such as deadlines) are met and to inform the change management system whether the roll-out of changes runs according to the schedule defined in the change plan.

**3.2.3.3.** *Recursive application of the automation pattern*    The automation described in this section essentially makes use of three levels of workflow engines: (1) top-level coordination of the Change Management workflow; (2) implementation of the generated change plan; and (3) the provisioning system where the LDOs are implemented by miniature-workflows within their defined scripts. This use of multiple levels of workflow engine illustrates a particular pattern of composition that we expect to be common in automation of best-practice IT service management processes, and recalls the discussion earlier in Section 3.1 of recursive application of the automation approach.

In particular, the delegated Change Management activity of Distribute and Install Changes involves a complex flow of work in its own right – documented in the change plan. We can see that the approach to automating the construction of the change plan follows the same pattern we used to automate change management itself, albeit at a lower level. For example, the creation of the change workflow is a lower-level analogue to using ITIL best practices to identify the Change Management process activities. The execution of change plan tasks by the provisioning system represents delegation of those tasks to that provisioning system. The provisioning system uses external interfaces and structured inputs and APIs to automate those tasks – drawing on information from the CMDB to determine available resources and invoking lower-level operations (automation packages) to effect changes in the actual IT environment. In this latter step, we again see the need to provide such resource information and control APIs in the structured, machine-readable formats needed to enable automation. The entire pattern repeats again at a lower level within the provisioning system itself, where the automation packages for detailed software and hardware products represent best practice operations with delegated functionality and external interfaces for data and control.

One of the key benefits of this type of recursive composition of our automation approach is that it generates reusable automation assets. Namely, at each level of automation, a set of automated delegated activities is created: automated ITIL activities at the top (such as Assess Change), automated change management activities in the middle (such as Install the DB2 Database), and automated software lifecycle activities at the bottom (such as Start DB2 Control Process). While created in the context of change management, it is possible that many of these activities (particularly lower-level ones) could be reused in other automation contexts. For example, many of the same lower-level activities created here could be used for performance management in an on-demand environment to enable creating, activating, and deactivating additional database or middleware instances. It is our hope that application of this automation pattern at multiple levels will reduce the long-term costs of creating system management automation, as repeated application will build up a library of reusable automation components that can be composed together to simplify future automation efforts.

**3.2.3.4.** *Challenges in workflow construction*   An important question addresses the problem whether the change management system or the provisioning system should perform error recovery for the complete change plan in case an activity fails during workflow execution or runs behind schedule. One possible strategy is not to perform error recovery or dealing with schedule overruns within the change plan itself and instead delegate this decision instead to the change management system. The reason for doing so is that in service provider environments, resources are often shared among different customers, and a change of a customer's hosted application may affect the quality of service another customer receives. Before rolling out a change for a given customer, a service provider needs to trade off the additional benefit he receives from this customer against a potential monetary loss due to the fact that an SLA with another customer may be violated because of the change. The scheduling of change plans, a core change management system function, is the result of solving an optimization problem that carefully balances the benefits it receives from servicing one customer's change request against the losses it incurs from not being able to service other customers. In an on-demand environment, the cost/profit situation may change very rapidly as many change plans are concurrently executed at any given point in time. In some cases, it may be more advantageous to carry on with a change despite its delay, whereas in other cases, aborting a change and instead servicing another newly submitted request that is more profitable may lead to a better global optimum. This big picture, however, is only available to the change management system.

Another important requirement for provisioning composed applications is the dynamic aggregation of already existing and tested change plans. For example, in the case of SPEC-jAppServer2004 and its underlying middleware, change plans for provisioning some of its components (such as WebSphere Application Server or the DB2 DBMS) may already exist. Those workflows need to be retrieved from a workflow repository and aggregated in a new workflow for which the activities for the remaining components have been generated. The challenge consists in automatically annotating the generated workflows with metadata so that they can be uniquely identified and evaluated with respect to their suitability for the specific request for change. The design of efficient query and retrieval mechanisms for workflows is an important prerequisite for the reuse of change plans.

## 4. When to automate

Now that we have identified the opportunities for automation and discussed various approaches, we now step back and ask: when is it appropriate to deploy systems management automation? The answer is, surprisingly, not as straightforward as one might expect.

Automation does not always reduce the cost of operations.

This is not a new conclusion, though it is rarely discussed in the context of automating system administration. Indeed, in other fields practitioners have long recognized that automation can be a double-edged sword. For example, early work with aircraft autopilots illustrated the dangers of imperfect automation, where pilots would lose situational awareness and fail to respond correctly when the primitive autopilot reached the edge of its envelope and disengaged [44], causing more dangerous situations than when the autopilot was not present. There are many other well-documented cases where either failures or unexpected behavior of industrial automation caused significant problems, including in such major incidents as Three Mile Island [43,45].

How are we to avoid, or at least minimize, these problems in automation of system administration? One solution is to make sure that the full consequences of automation are understood before deploying it, changing the decision of 'when to automate?' from a simple answer of 'always!' to a more considered analytic process.

In the next several subsections, we use techniques developed in [6] to consider what this analysis might look like, not to dissuade practitioners from designing and deploying automation, but to provide the full context needed to ensure that automation lives up to its promises.

### 4.1. *Cost-benefit analysis of automation*

We begin by exploring the potential hidden costs of automation. While automation obviously can reduce cost by removing the need for manual systems management, it can also induce additional costs that may offset or even negate the savings that arise from the transfer of manual work to automated mechanisms. If we take a process-based view, as we have throughout this paper, we see that automation transforms a manual process, largely to reduce manual work, but also to introduce new process steps and roles needed to maintain, monitor, and augment the automation itself.

Figure 2 illustrates the impact of automation on an example system management process, drawn from a real production data center environment. The process here is software distribution to server machines, a critical part of operating a data center. Software

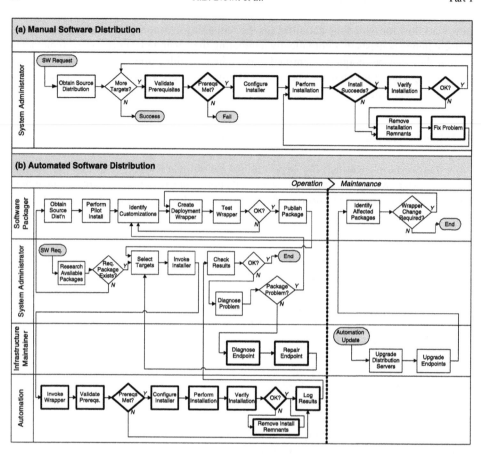

Fig. 2. Manual and automated processes for software distribution. Boxes with heavy lines indicate process steps that contribute to variable (per-target) costs, as described in Section 4.2.

distribution involves the selection of software components and their installation on target machines. We use the term 'package' to refer to the collection of software resources to install and the step-by-step procedure (process) by which this is done. We represent the process via 'swim-lane' diagrams – annotated flowcharts that allocate process activities across *roles* (represented as rows) and *phases* (represented as columns). Roles are typically performed by people (and can be shared or consolidated); we include automation as its own role to reflect activities that have been handed over to an automated system.

Figure 2(a) shows the 'swim-lane' representation for the manual version of our example software distribution process, and Figure 2(b) shows the same process once automated software distribution has been introduced.

Notice that, while key parts of the process have moved from a manual role to an automated role, there are additional implications. For one thing, the automation infrastructure is another software system that must itself be installed and maintained, creating initial transition costs as well as periodic costs for update and maintenance. (For simplicity, in the figure we have assumed that the automation infrastructure has already been installed, but we do consider the need for periodic updates and maintenance.) Next, using the automated infrastructure requires that information be provided in a structured form. We use the term *software package* to refer to these structured inputs. These inputs are typically expressed in a formal structure, which means that their creation requires extra effort for package design, implementation, and testing. Last, when errors occur in the automated case, they happen on a much larger scale than for a manual approach, and hence additional processes and tools are required to recover from them. These other impacts manifest as additional process changes, namely extra roles and extra operational processes to handle the additional tasks and activities induced by the automation.

We have to recognize as well that the effects of induced processes – for error recovery, maintenance, and input creation/structuring – are potentially mitigated by the probability and frequency of their execution. For example, if automation failures are extremely rare, then having complex and costly recovery procedures may not be a problem. Furthermore, the entire automation framework has a certain lifetime dictated by its flexibility and generalization capability. If this lifetime is long, the costs to create and transition to the automated infrastructure may not be an issue. But even taking this into account, it is apparent from inspection that the collection of processes in Figure 2(b) is much more complicated than the single process in Figure 2(a). Clearly, such additional complexity is unjustified if we are installing a single package on a single server. This raises the following question – at what point does automation stop adding cost and instead start reducing cost?

The answer comes through a cost-benefit analysis, where the costs of induced process are weighted by their frequency and balanced against the benefits of automation. Our argument in this section is that such analyzes are an essential part of a considered automation decision. That is, when considering a manual task for automation, the benefits of automating that task (reduced manual effort, skill level, and error probability) need to be weighed against the costs identified above (extra work from induced process, new roles and skills needed to maintain the automation, and potential consequences from automation failure).

In particular, this analysis needs to consider both the changes in tasks (tasks replaced by automation and tasks induced by it) and the changes in roles and required human skills. The latter consideration is important because automation tends to create induced tasks requiring more skill than the tasks it eliminates. For example, the tasks induced by automation failures tend to require considerable problem-solving ability as well as a deep understanding of the architecture and operation of the automation technology. If an IT installation deploys automation without having an experienced administrator who can perform these failure-recovery tasks, the installation is at risk if a failure occurs. The fact that automation often increases the skill level required for operation is an example of an *irony of automation* [3]. The irony is that while automation is intended to reduce costs overall, there are some kinds of cost that increase. Many ironies of automation have been identified in industrial contexts such as plant automation and infrastructure monitoring [45]. Clearly, the ironies of automation must be taken into account when doing a cost-benefit analysis.

### 4.2. *Example analysis: software distribution for a datacenter*

To illustrate the issues associated with automation, we return to the example of software distribution. To perform our cost-benefit analysis, we start by identifying the fixed and variable costs for the process activities in Figure 2. The activities represented as boxes with heavy outlines represent variable-cost activities, as they are performed once for each machine in the data center. The other activities are fixed-cost in that they are performed once per software package.

A key concept used in our analysis is that of the *lifetime of a software package*. By the lifetime of a package, we mean the time from when the package is created to when it is retired or made obsolete by a new package. We measure this in terms of the number of target machines to which the package is distributed.

For the benefits of automation to outweigh the cost of automation, the variable costs of automation must be lower than the variable costs of the manual process, *and* the fixed cost of building a package must be amortized across the number of targets to which it is distributed over its lifetime. Using data from a several computer installation, Figure 3 plots the cumulative fraction of packages in terms of the lifetimes (in units of number of targets to which the package is distributed). We see that a large fraction of the packages are distributed to a small number of targets, with 25% of the packages going to fewer than 15 targets over their lifetimes.

Next, we look at the possibility of automation failure. By considering the complete view of the automated processes in Figure 2(b), we see that more sophistication and people are required to address error recovery for automated software distribution than for the manual process. Using the same data from which Figure 3 is extracted, we determined that 19% of the requested installs result in failure. Furthermore, at least 7% of the installs fail due to issues related to configuration of the automation infrastructure, a consideration that does not exist if a manual process is used. This back-of-the envelope analysis underscores the

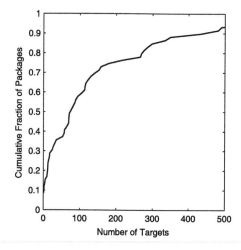

Fig. 3. Cumulative distribution of the number of targets (servers) on which a software package is installed over its lifetime in several data centers. A larger number of packages are installed on only a small number of targets.

importance of considering the entire set of process changes that occur when automation is deployed, particularly the extra operational processes created to handle automation failures. Drawing on additional data from IBM internal studies of software distribution and netting out the analysis (the details can be found in [6]), we find that for complex software packages, there should be approximately 5 to 20 targets for automated software distribution to be cost effective. In terms of the data in Figure 3, these numbers mean that from 15% to 30% of the installs in the data centers we examined should not have been automated from a cost perspective.

Automation can be a double-edged sword.

While in these datacenters automation of software distribution provided a net cost benefit for 70–85% of installs, it increased costs for the remaining 15–30%. In this case the choice was made to deploy the automation, with the assessment that the benefit to the large fraction of installs outweighed the cost to the smaller fraction. The insight of a cost-benefit analysis allows such decisions to be made with eyes open, and full awareness of the trade-offs being made.

We note in passing that cost models for system administration have been developed based on other frameworks. For example, [12] models direct costs (e.g., salaries of system administrators) and indirect costs (e.g., diminished productivity due to downtime). Clearly, different cost models can yield different results in terms of when to employ automation.

### 4.3. *Additional concerns: adoption and trust*

Automation is only useful if it is used. Even the best automation – carefully designed to maximize benefit and minimize induced process and cost – can lack traction in the field. Often these situations occur when the automation fails to gain the trust of the system management staff tasked with deploying and maintaining it.

So what can automation designers do to help their automation gain trust? First, they must recognize that automation is a disruptive force for IT system managers. Automation changes the way system administrators do their jobs. It also creates uncertainty in terms of the future of the job itself. And since no automation is perfect, there is concern about the extent to which automation can be trusted to operate correctly.

Thus adoption of automation is unlikely without a history of successful use by administrators. This observation has been proven in practice many times. Useful scripts, languages, and new open-source administration tools tend to gain adoption via grass-roots movements where a few brave early adopters build credibility for the automation technology. But, not all new automation can afford to generate a community-wide grass-roots movement behind it. And in these cases we are left with a kind of circular logic – we cannot gain adoption without successful adoptions!

In these situations, the process-based view can help provide a transition path. From a process perspective, the transition to automation can be seen as an incremental delegation of manual process steps to an automated system. And by casting the automation into the same terms as the previously-manual process steps (for example, by reusing the same work products and tracking metrics), the automation can provide visibility into its inner workings that assuages a system administrator's distrust. Thus the same work that it takes to do the cost/benefit analysis for the first cut at an automation decision can be leveraged to plan out an automation deployment strategy that will realize the automation's full benefits.

We believe that the process-based perspective described above can provide the basis for making a business case for automation. With a full understanding of the benefits of automation as well as all the cost factors described above, we finally have a framework to answer the question of when to automate. To determine the answer we compute or estimate the benefits of the proposed automation as well as the full spectrum of costs, and compute the net benefit. In essence, we boil the automation decision down to the bottom line, just as in a traditional business case.

The process for building an automation business case consists of 5 steps:

1. Identify the benefit of the automation in terms of reduction in 'core' manual activity, lower frequency of error (quantified in terms of the expected cost of errors, e.g., in terms of business lost during downtime), and increased accuracy of operations.
2. Identify and elucidate the induced processes that result from delegation of the manual activity to automation, as in the software distribution example above. 'Swim-lane' diagrams provide a useful construct for structuring this analysis.
3. Identify or estimate the probability or frequency of the induced processes. For example, in software distribution we estimate how often new packages will be deployed or new versions released, as those events require running the packaging process. Often estimates can be made based on existing runtime data.
4. Assess the possible impact of the 'automation irony'. Will the induced processes require additional skilled resources, or will they change the existing human roles enough that new training or hiring will be needed?
5. Collate the collected data to sum up the potential benefits and costs, and determine whether the automation results in net benefit or net cost.

These steps are not necessarily easy to follow, and in fact in most cases a rough, order-of-magnitude analysis will have to suffice. Alternately, a sensitivity analysis can be performed. For example, if the probability of automation failure is unknown, a sensitivity analysis can be used to determine the failure rate where costs break even. Say this is determined to be one failure per week. If the automation is thought to be stable enough to survive multiple weeks between failures, then the go-ahead decision can be given.

Furthermore, even a rough business-case analysis can provide significant insight into the impact of automation, and going through the exercise will reveal a lot about the ultimate utility of the automation. The level of needed precision will also be determined by the magnitude of the automation decision: a complete analysis is almost certainly unnecessary when considering automation in the form of a few scripts, but it is mandatory when deploying automation to a datacenter with thousands of machines.

### 4.4. *Measuring complexity*

In this chapter, we have provided a variety of reasons why IT departments of enterprises and service providers are increasingly looking to automation and IT process transformation as a means of containing and even reducing the labor costs of IT service management. However, Section 4.2 provides evidence of a *reduction* in efficiencies and productivity when automation is introduced into business operations. This raises a critical question from the point of view of both the business owners and CIOs as well as service and technology providers: *how can one quantify, measure and (ultimately) predict whether and how the introduction of a given technology can deliver promised efficiencies?*

Many automation strategies focus on the creation of standardized, reusable components that replace today's custom-built solutions and processes – sometimes at a system level, sometimes at a process level. While both service providers and IT management software vendors heavily rely on qualitative statements to justify investment in new technologies, a qualitative analysis does not provide direct guidance in terms of *quantitative* business-level performance metrics, such as labor cost, time, productivity and quality.

Our approach to measuring complexity of IT management tasks is inspired by the widely successful system performance benchmark suites defined by the Transaction Processing Council (TPC) and the Standard Performance Evaluation Corporation (SPEC). In addition, we have borrowed concepts from Robert C. Camp's pioneering work in the area of business benchmarking [11]. Furthermore, the system administration discipline has started to establish cost models: The most relevant work is [13], which generalizes an initial model for estimating the cost of downtime [42], based on the previously established System Administration Maturity Model [34].

In [5], we have introduced a framework for measuring IT system management complexity, which has been subsequently extended to address the specifics of IT service management processes [18]. The framework relies on categorical classification of individual complexities, which are briefly summarized as follows:

*Execution complexity* refers to the complexity involved in performing the tasks that make up the configuration process, typically characterized by the number of tasks, the context switches between tasks, the number of roles involved in an action and their degree of automation. *Decision complexity*, a sub-category of execution complexity, quantifies decision making according to: (1) the number of branches in the decision, (2) the degree of guidance, (3) the impact of the decision, and (4) the visibility of the impact.

*Business item complexity* addresses the complexity involved in passing data between the various tasks within a process. Business items can be nested; they range from simple scalar parameters to complex data items, such as build sheets and run books.

*Memory complexity* takes into account the number of business items that must be remembered, the length of time they must be retained in memory, and how many intervening items were stored in memory between uses of a remembered business item.

*Coordination complexity* represents the complexity resulting from coordinating between multiple roles. The per-task metrics for coordination complexity are computed based on the roles involved and whether or not business items are transferred.

While we have found it useful to analyze and compare a variety of configuration procedures and IT service management processes according to the aforementioned four complexity dimensions, a vector of complexity scores reflecting such a categorization does not directly reveal the actual labor reduction or cost savings. For both purchasers and vendors of IT service management software, it is this cost savings that is of eminent concern.

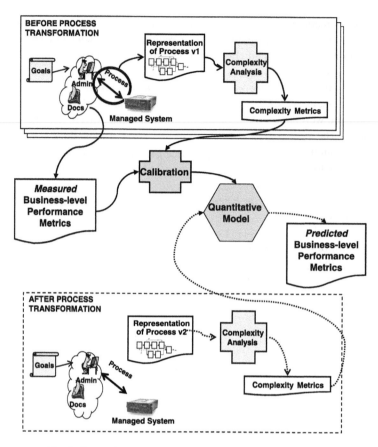

Fig. 4. Approach to constructing and applying a quantitative model.

In order to address this issue, one needs to relate the metrics of our IT management complexity framework to key business-level performance metrics (such as labor cost, effort and duration; cf. Section 4.4.2). In [19], we have developed a quantitative model based on a qualitative description of IT complexity parameters together with quantitative calibration data of the IT process. We found that such a hybrid approach is more practical than a pure qualitative or quantitative approach, since extensive quantitative IT process measurements are notoriously difficult to obtain in the field.

Our approach to predicting labor cost through IT management complexity metrics is depicted in Figure 4. It consists of four distinct steps, which can be summarized as follows:

**4.4.1.** *Step 1: collecting complexity metrics*   As depicted in the upper part of Figure 4, an administrator configures one or more managed systems, according to the goals he wants to accomplish. To do so, he follows a process that may either be described in authoritative documents, or as a workflow in an IT process modeling tool.

This process (called 'Process version 1' in the figure) reflects the current ('as-is') state of the process, which serves as a baseline for our measurements. While carrying out the configuration process, an administrator logs – by interacting with a web-based graphical user interface [18] – the various steps, context switches and parameters that need to be input or produced by the process. In addition, the administrator assesses the complexity for each of the parameters, as perceived by him. The complexity data that an administrator records by means of the web-based graphical user interface is recorded by our tooling on a step-by-step basis. The key concept of the process representation is that it is 'action-centric', i.e., the representation decomposes the overall process into individual, atomic actions that represent the configurations steps an administrator goes through.

Once the process has been carried out, the process capture is input to a complexity analysis tool that we implemented, which contains the scoring algorithms for a variety of complexity measures from the four complexity dimensions mentioned above. The tool also outputs aggregated complexity metrics, such as the maximum number of parameters that need to be kept in memory during the whole procedure, the overall number of actions in the process, or the context switches an administrator needs to perform. Both the individual and the aggregated complexity metrics are used to construct the quantitative model.

We note that by assigning a complexity score to each configuration action, it is possible to identify those actions in the process (or combinations of actions) that have the greatest contribution to complexity (i.e., they are complexity hotspots). As a result, the methodology facilitates the task of process designers to identify targets for process transformation and optimization: If, for example, a given step has a very high complexity relative to other steps in the process, a designer obtains valuable guidance on which action(s) his improvement efforts need to focus first in order to be most effective.

**4.4.2.** *Step 2: measuring business-level performance metrics*      The second step consists in measuring the business-level performance metrics. The most common example of a business-level metric, which we use for our analysis, is the time it takes to complete each action in the process, and the aggregated, overall process execution time. This is also commonly referred to as the *Full Time Equivalent (FTE)*. Labor Cost, another key business-level performance metric, is derived in a straightforward way by multiplying the measured FTEs with the administrator's billable hourly rate.

When assessing the time it takes to complete a task or the overall process, great care must be taken to distinguish between the *effort* (how long it actually takes to carry out a task) and the *duration*, the agreed upon time period for completing a task, which is often specified in a Service Level Agreement. While, for example, the effort for the installation of a patch for a database server may be only 30 minutes, the overall duration may be 2 days. Often, the difference is the time period that the request sits in the work queue of the administrator.

While our tooling allows an administrator to record the time while he captures the actions in a configuration process, the time measurements are kept separate from the complexity metrics.

**4.4.3.** *Step 3: model construction through calibration*      The step of constructing a quantitative model by means of calibration, depicted in the middle of Figure 4, is at the heart of the approach. Calibration relates the complexity metrics we computed in step 1 to the FTE measurements we performed in step 2. To do so, we have adapted techniques we initially developed in [30] (where we were able to predict the user-perceived response time of a web-based Internet storefront by examining the performance metrics obtained from the storefront's database server) to the problem domain of IT management complexity. The purpose of calibration is to set the time it takes to execute a configuration process in relation to the recorded complexity metrics, i.e., the former is being explained by the latter.

**4.4.4.** *Step 4: predict business-level performance metrics*   Once a quantitative model has been built based on the 'as-is' state of a process, it can be used to predict the FTEs and therefore the labor cost for an improved process that has been obtained through process transformation based on analyzing and mitigating the complexity hotspots that were identified in step 1. This 'to-be' process is referred to as 'Process version 2' in the bottom part of the figure as it accomplishes the very same goal(s) as the 'as-is' version 1 of the process. It is therefore possible to not only apply the quantitative model that has been developed for the 'as-is' state in order to estimate the FTEs from the complexity metrics of the 'to-be' state, but also to directly compare both the complexity metrics and the FTEs of the before/after transformation versions of the process. The difference between the 'as-is' and the 'to-be' FTEs and, thus, labor cost yields the savings that are obtained by process transformation. For a detailed example of applying the approach to an install scenario, the reader is referred to [19].

We envision a future scenario where our validated quantitative model is built directly into an IT process modeling tool, so that a process designer can simulate the execution of an IT management process during the design phase. The designer is then able to obtain the predicted FTEs by running the complexity metrics that were automatically collected from the process model – by means of a plugin – through the quantitative model. Such a quantitative model has many benefits:

First, the model can be used to assist in return-on-investment (ROI) determination, either a-priori or post-facto. Across multiple products (for example, IT offerings from different providers), a customer can use models involving each product for a comparison in terms of the impact on the bottom line. Conversely, IT sales personnel can make a substantiated, quantitative argument during a bidding process for a contract.

Second, using the cost prediction based on the model, the customer can further use the model output for budgeting purposes.

Third, a calibrated model for a particular process can reveal which are the important factors that contribute to the overall complexity of the process, along with a measure of their relative contributions. IT providers can use this data to improve their products and offerings, focusing on areas which yield the largest customer impact.

Fourth, a particular process can also be studied in terms of its sensitivity to skill levels of individual roles. Because labor cost of different skill levels varies, this sensitivity analysis can be used for hiring and employee scheduling purposes.

Finally, process transformation involves cost/benefit analysis. These decisions can be guided by the quantitative predictions from the model.

## 5. Conclusions

This chapter addresses the automation of system administration, a problem that is addressed as a set of three interrelated questions: what to automate, how to automate, and when to automate.

We address the 'what' question by studying what system administrators do and therefore where automation provides value. An observation here is that tasks such as configuration and installation consume a substantial fraction of the time of systems administrators, approximately 20%. This is fortunate since it seems likely that it is technically feasible to increase the degree of automation of these activities. Unfortunately, meetings and other forms of communication consume almost of third of the time of systems administrators. Here, there seems to be much less opportunity to realize benefits from automation.

The 'how' question is about technologies for automation. We discuss three approaches – rule-based systems, control theoretic approaches, and automated workflow construction. All three have been used in practice. Rules provide great flexibility in building automation, but the complexity of this approach becomes problematic as the scope of automation increases. Control theory provides a strong theoretical foundation for certain classes of automation, but it is not a universal solution. Workflow has appeal because its procedural structure is a natural way for humans to translate their activities into automation.

The 'when' question is ultimately a business question that should be based on a full understanding of the costs and benefits. The traditional perspective has been that automation is always advantageous. However, it is important to look at the full costs imposed by automation. For example, automating software distribution requires that: (a) the distribution infrastructure be installed and maintained; (b) software packages be prepared in the format required by the distribution infrastructure; and (c) additional tools be provided to handle problems with packages that are deployed because of the large scale of the impact of these problems. While automation often provides a net benefit despite these costs, we have observed cases in which these costs exceed the benefits.

As the scale of systems administration increases and new technologies for automation are developed, systems administrators will have even greater challenges in automating their activities.

## References

[1] T.F. Abdelzaher and N. Bhatti, *Adaptive content delivery for Web server QoS*, International Workshop on Quality of Service, London, UK (1999), 1563–1577.

[2] E. Altman, T. Basar and R. Srikant, *Congestion control as a stochastic control problem with action delays*, Automatica **35** (1999), 1936–1950.

[3] L. Bainbridge, *The ironies of automation*, New Technology and Human Error, J. Rasmussen, K. Duncan and J. Leplat, eds, Wiley (1987).

[4] R. Barrett, P.P. Maglio, E. Kandogan and J. Bailey, *Usable autonomic computing systems: The system administrators' perspective*, Advanced Engineering Informatics **19** (2006), 213–221.

[5] A. Brown, A. Keller and J.L. Hellerstein, *A model of configuration complexity and its application to a change management system*, 9th International IFIP/IEEE Symposium on Integrated Management (IM 2005), IEEE Press (2005), 531–644.

[6]  A.B. Brown and J.L. Hellerstein, *Reducing the cost of IT operations – Is automation always the answer?*, Tenth Workshop on Hot Topics in Operating Systems (HotOS-X), Santa Fe, NM (2005).

[7]  A.B. Brown and A. Keller, *A best practice approach for automating IT management processes*, 2006 IEEE/IFIP Network Operations and Management Symposium (NOMS 2006), Vancouver, BC, Canada (2006).

[8]  M. Burgess, *On the theory of system administration*, Science of Computer Programming **49** (2003).

[9]  M. Burgess and R. Ralston, *Distributed resource administration using cfengine*, Software Practice and Experience **27** (1997), 1083.

[10]  P. Byers, *Database Administrator: Day in the Life*, Thomson Course Technology, http://www.course.com/careers/dayinthelife/dba_jobdesc.cfm.

[11]  R.C. Camp, *Benchmarking – The Search for Industry Best Practices that Lead to Superior Performance*, ASQC Quality Press (1989).

[12]  A. Couch, N. Wu and H. Susanto, *Toward a cost model for system administration*, 19th Large Installation System Administration Conference (2005).

[13]  A.L. Couch, N. Wu and H. Susanto, *Toward a cost model for system administration*, Proc. 19th Large Installation System Administration Conference (LISA'05), D.N. Blank-Edelman, ed., USENIX Association, San Diego, CA, USA (2005), 125–141.

[14]  Y. Diao, N. Gandhi, J.L. Hellerstein, S. Parekh and D. Tilbury, *Using MIMO feedback control to enforce policies for interrelated metrics with application to the Apache Web server*, IEEE/IFIP Network Operations and Management Symposium (NOMS 2002) (2002), 219–234.

[15]  Y. Diao, R. Griffith, J.L. Hellerstein, G. Kaiser, S. Parekh and D. Phung, *A control theory foundation for self-managing systems*, Journal on Selected Areas of Communications **23** (2005).

[16]  Y. Diao, J.L. Hellerstein and S. Parekh, *Optimizing quality of service using fuzzy control*, IFIP/IEEE International Workshop on Distributed Systems: Operations and Management (DSOM 2002) (2002), 42–53.

[17]  Y. Diao, J.L. Hellerstein, A. Storm, M. Surendra, S. Lightstone, S. Parekh and C. Garcia-Arellano, *Using MIMO linear control for load balancing in computing systems*, American Control Conference (2004), 2045–2050.

[18]  Y. Diao and A. Keller, *Quantifying the complexity of IT service management processes*, Proceedings of 17th IFIP/IEEE International Workshop on Distributed Systems: Operations and Management (DSOM'06), Springer-Verlag, Dublin, Ireland (2006).

[19]  Y. Diao, A. Keller, V.V. Marinov and S. Parekh, *Predicting labor cost through IT management complexity metrics*, 10th International IFIP/IEEE Symposium on Integrated Management (IM 2007), IEEE Press (2007).

[20]  B. Dijker, *A day in the life of systems administrators*, http://www.sage.org/field/ditl.pdf.

[21]  J. Finke, *Automation of site configuration management*, Proceedings of the Eleventh Systems Administration Conference (LISA XI), USENIX Association, Berkeley, CA (1997), 155.

[22]  A. Finkel, K. Houck and A. Bouloutas, *An alarm correlation system for heterogeneous networks*, Network Management and Control **2** (1995).

[23]  M. Fisk, *Automating the administration of heterogeneous LANS*, Proceedings of the Tenth Systems Administration Conference (LISA X), USENIX Association, Berkeley, CA (1996), 181.

[24]  S.E. Hansen and E.T. Atkins, *Automated system monitoring and notification with swatch*, Proceedings of the Seventh Systems Administration Conference (LISA VII), USENIX Association, Berkeley, CA (1993), 145.

[25]  C.V. Hollot, V. Misra, D. Towsley and W.B. Gong, *A control theoretic analysis of RED*, Proceedings of IEEE INFOCOM'01, Anchorage, Alaska (2001), 1510–1519.

[26]  C.V. Hollot, V. Misra, D. Towsley and W.B. Gong, *On designing improved controllers for AQM routers supporting TCP flows*, Proceedings of IEEE INFOCOM'01, Anchorage, Alaska (2001), 1726–1734.

[27]  IT Infrastructure Library, *ITIL service support, version 2.3*, Office of Government Commerce (2000).

[28]  G. Jakobson, R. Weihmayer and M. Weissman, *A domain-oriented expert system shell for telecommunications network alarm correlation*, Network Management and Control, Plennum Press (1993), 277–288.

[29]  A. Keller and R. Badonnel, *Automating the provisioning of application services with the BPEL4WS work-flow language*, 15th IFIP/IEEE International Workshop on Distributed Systems: Operations and Management (DSOM 2004), Springer-Verlag, Davis, CA, USA (2004), 15–27.

[30]  A. Keller, Y. Diao, F.N. Eskesen, S.E. Froehlich, J.L. Hellerstein, L. Spainhower and M. Surendra, *Generic on-line discovery of quantitative models*, IEEE Eight Symposium on Network Management (2003), 157–170.

[31]  S. Keshav, *A control-theoretic approach to flow control*, Proceedings of ACM SIGCOMM'91 (1991), 3–15.

[32]  S. Kliger, S. Yemini, Y. Yemini, D. Ohlse and S. Stolfo, *A coding approach to event correlation*, Fourth International Symposium on Integrated Network Management, Santa Barbara, CA, USA (1995).

[33]  M. Kloppmann, D. Koenig, F. Leymann, G. Pfau and D. Roller, *Business process choreography in Web-Sphere: Combining the power of BPEL and J2EE*, IBM Systems Journal **43** (2004).

[34]  C. Kubicki, *The system administration maturity model – SAMM*, Proc. 7th Large Installation System Administration Conference (LISA'93), USENIX Association, Monterey, CA, USA (1993), 213–225.

[35]  K. Li, M.H. Shor, J. Walpole, C. Pu and D.C. Steere, *Modeling the effect of short-term rate variations on tcp-friendly congestion control behavior*, Proceedings of the American Control Conference (2001), 3006–3012.

[36]  D. Libes, *Using expect to automate system administration tasks*, Proceedings of the Fourth Large Installation System Administrator's Conference (LISA IV), USENIX Association, Berkeley, CA (1990), 107.

[37]  Y. Lu, A. Saxena and T.F. Abdelzaher, *Differentiated caching services: A control-theoretic approach*, International Conference on Distributed Computing Systems (2001), 615–624.

[38]  Microsoft, *Overview of the system description model*, http://msdn2.microsoft.com/en-us/library/ms181772.aspx.

[39]  K.R. Milliken, A.V. Cruise, R.L. Ennis, A.J. Finkel, J.L. Hellerstein, D.J. Loeb, D.A. Klein, M.J. Masullo, H.M. Van Woerkom and N.B. Waite, *YES/MVS and the automation of operations for large computer complexes*, IBM Systems Journal **25** (1986), 159–180.

[40]  S. Parekh, K. Rose, J.L. Hellerstein, S. Lightstone, M. Huras and V. Chang, *Managing the performance impact of administrative utilities*, 14th IFIP/IEEE International Workshop on Distributed Systems: Operations and Management (DSOM 2003) (2003), 130–142.

[41]  S. Parekh, N. Gandhi, J. Hellerstein, D. Tilbury, J. Bigus and T.S. Jayram, *Using control theory to achieve service level objectives in performance management*, Real-time Systems Journal **23** (2002), 127–141.

[42]  D. Patterson, *A simple way to estimate the cost of downtime*, Proc. 16th Large Installation System Administration Conference (LISA'05), A.L. Couch, ed., USENIX Association, Philadelphia, PA, USA (2002), 185–188.

[43]  C. Perrow, *Normal Accidents: Living with High-Risk Technologies*, Perseus Books (1990).

[44]  J. Rasmussen and W. Rouse, eds, *Human Detection and Diagnosis of System Failures: Proceedings of the NATO Symposium on Human Detection and Diagnosis of System Failures*, Plenum Press, New York (1981).

[45]  J. Reason, *Human Error*, Cambridge University Press (1990).

[46]  F.E. Sandnes, *Scheduling partially ordered events in a randomized framework – empirical results and implications for automatic configuration management*, Proceedings of the Fifteenth Systems Administration Conference (LISA XV), USENIX Association, Berkeley, CA (2001), 47.

[47]  P. Scott, *Automating 24 × 7 support response to telephone requests*, Proceedings of the Eleventh Systems Administration Conference (LISA XI), USENIX Association, Berkeley, CA (1997), 27.

[48]  L. Sha, X. Liu, Y. Lu and T. Abdelzaher, *Queueing model based network server performance control*, IEEE RealTime Systems Symposium (2002), 81–90.

[49]  E. Solana, V. Baggiolini, M. Ramlucken and J. Harms, *Automatic and reliable elimination of e-mail loops based on statistical analysis*, Proceedings of the Tenth Systems Administration Conference (LISA X), USENIX Association, Berkeley, CA (1996), 139.

[50]  A. Somayaji and S. Forrest, *Automated response using system-call delays*, Proceedings of the 9th USENIX Security Symposium, USENIX Association, Berkeley, CA (2000), 185.

[51]  SPEC Consortium, *SPECjAppServer2004 design document, version 1.01* (2005), http://www.specbench.org/osg/jAppServer2004/docs/DesignDocument.html.

[52]  D. Thoenen, J. Riosa and J.L. Hellerstein, *Event relationship networks: A framework for action oriented analysis in event management*, International Symposium on Integrated Network Management (2001).

[53] M. Vitaletti, ed., *Installable unit deployment descriptor specification, version 1.0*, W3C Member Submission, IBM Corp., ZeroG Software, InstallShield Corp., Novell (2004), http://www.w3.org/Submission/2004/SUBM-InstallableUnit-DD-20040712/.

[54] T. Woodall, *Network Administrator: Day in the Life*, Thomson Course Technology, http://www.course.com/careers/dayinthelife/networkadmin_jobdesc.cfm.

# – 1.4 –

# System Configuration Management

Alva L. Couch

*Computer Science, Tufts University, 161 College Avenue, Medford, MA 02155, USA*
*E-mail: couch@cs.tufts.edu*

## 1. Introduction

*System configuration management* is the process of maintaining the function of computer networks as holistic entities, in alignment with some previously determined policy [28]. A *policy* is a list of high-level goals for system or network behavior. This policy – that describes how systems should behave – is translated into a low-level *configuration*, informally defined as the contents of a number of specific files contained within each computer system. This policy translation typically involves installation of predefined hardware and software components, reference to component documentation and experience, and some mechanism for controlling the contents of remotely located machines (Figure 1).

For example, a high-level policy might describe classes of service to be provided; a 'web server' is a different kind of machine than a 'mail server' or a 'login server'. Another example of a high-level goal is that certain software must be either available or unavailable to interactive users (e.g., programming tools and compilers) [23].

The origin of the phrase 'system configuration management' is unclear, but the practice of modern system configuration management was, to our knowledge, first documented in Paul Anderson's paper [2] and implemented in the configuration tool *LCFG* [2,4]. This paper contains several ideas that shaped the discipline, including the idea that the configuration of a system occurs at multiple levels, including a high-level policy, intermediate code and physical configuration.

HANDBOOK OF NETWORK AND SYSTEM ADMINISTRATION
Edited by Jan Bergstra and Mark Burgess

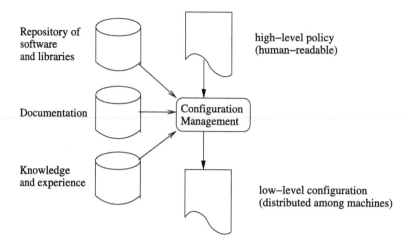

Fig. 1. An overview of the configuration management process.

### 1.1. *Overview of configuration management*

A typical system to be managed consists of a central-processing unit, hard disk and associated peripherals, including network card, video card, etc. Depending upon the system, many of these components may actually be contained on a single computer board. A system typically starts its life with no operating system installed, often called 'bare metal'. As software components are installed on the hard disk, the system obtains an operating system, system modules that extend the operating system (e.g., I/O drivers), and software packages that provide application-level functions such as web service or compilation. During this process, each operating system, module, or package is assigned a *default configuration* specifying how it will behave. This is best thought of as a set of variables whose values control behavior. For example, one variable determines the number of threads utilized by a web server, while another variable determines whether certain kinds of network tunneling are allowed. The configuration of each module or package may be contained in text files (for Unix and Linux) or in a system database (e.g., the Windows Registry). The configuration of a component may be a file of key-value associations (e.g., for the secure shell), an XML file (e.g., for the Apache web server), or a registry key-value hierarchy in Windows. Several crucial configuration files are shared by most Linux applications and reside in the special directory /etc. Configuration management is the process of defining and maintaining consistent configurations for each component so that they all work together to accomplish desired tasks and objectives.

### 1.2. *Configuration management tasks*

There are several tasks involved in typical configuration management (Figure 2). First, *planning* involves deciding what constitutes desirable behavior and describing an overall operating *policy*. The next step is to *code* this policy into machine-readable form. This step

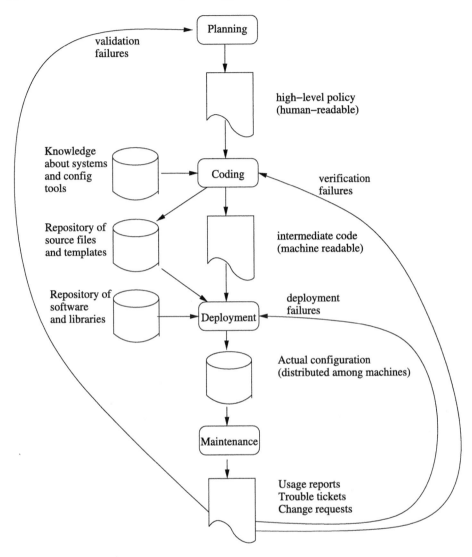

Fig. 2. The configuration management process includes creating high-level policies, translating them into machine-readable form, and deploying the configuration throughout a network.

utilizes knowledge about machines and configurations to create both the intermediate code and a repository of *source and template files* that can be used to create individual machine configurations. A source file is a verbatim copy of some configuration file, while an template file is a configuration file with some details left out, to be filled in later. The next step is to *deploy* the configuration on target machines, which involves changing the contents of each machine's disks. After deployment, the configuration must be *maintained* through an ongoing process of quality control, including data sources such as trouble tickets and new user requests.

At each step, there can be failures. A *verification failure* is a failure to create a system with a proper configuration, while a *validation failure* occurs when a correctly deployed configuration fails to meet user requirements.[1] For example, if a file fails to be copied to a machine to be configured, this is a verification failure, while a validation failure results if a machine that is intended to become a web server does not provide web services after being configured to do so. The former is a mistake in coding; the latter is a behavioral problem. A system fails to verify if the information that should be in configuration files on each machine does not ever get copied there; a system fails to validate if the configuration files are precisely as intended but the behavior fails to meet specifications.

### 1.3. *Objectives of configuration management*

It may seem to the reader at this point that configuration management is just about assuring the behavior of machines and nothing else. If this were the only goal of configuration management, it would be a simple and straightforward task. Configuration management is easy if one has access to unlimited hardware, and can encapsulate each distinct service on a separate machine. The complexities of configuration management arise from the need to combine or compose distinct and changing needs while utilizing limited amounts of hardware. For more details on complexity, see Section 9.5 and the chapter by Sun and Couch on Complexity of System Configuration Management.

In fact, the true goal of configuration management is to *minimize the lifecycle cost* and *total cost of ownership* for a computing system, while *maximizing value* of the computing system to the enterprise [36]. The cost of management has several components, including the cost of planning, deployment and maintenance. As in software engineering, the cost of maintenance seems to dominate the others for any lifecycle of reasonable length.

The relationships between configuration management and cost are many and varied. There is an intimate relationship between the methods utilized for configuration management and the ability to react to contingencies such as system outages quickly and effectively. One characterization of an effective configuration management strategy is that one should be able to pick a random server, unplug it and toss it out of a second story window, and *without backup*, be up and running again in an hour [73].

## 2. Relationship with other disciplines

### 2.1. *System administration tasks*

We define:

---

[1]The words 'verification' and 'validation' are used here in the precise software engineering sense, though in system administration, the word 'validation' has often been used to mean 'verification'. For more details, see [8].

1. *Configuration management* is the process of insuring that the operating system and software components work properly and according to externally specified guidelines.
2. *Package management* is the process of insuring that each system contains software components appropriate to its assigned task, and that components interact properly when installed.
3. *Resource management* is the process of insuring that appropriate resources (including memory, disk and network bandwidth) are available to accomplish each computing task.
4. *User management* is the process of assuring that users have access to computing resources, through appropriate authentication and authorization mechanisms.
5. *Monitoring* is the process of measuring whether computing systems are correctly performing their assigned missions.

These activities are not easily separable: one must install software through package management before one can configure it via configuration management, and monitor it via monitoring mechanisms. Users must be able to access the configured software, and have appropriate resources available from which to complete their tasks.

This is an inevitable convergence; part of monitoring is to ensure that configuration management is working properly, and an ideal configuration management tool would employ both static analysis and 'dynamic feedback' to understand whether the configuration is working properly and whether corrective action is needed.

## 2.2. *Configuration management and system administration*

System administration and system configuration management appear on the surface to be very similar activities. In both cases, the principal goal is to assure proper behavior of computing systems. The exact boundary between 'system configuration management' and regular 'system administration' is unclear, but there seems to be a solid distinction between the two at the point at which the overall configuration of a network of systems is being managed as an entity, rather than managing individual machines that contain distinct and unrelated configurations. At this point, we say that one is engaging in 'system configuration management' rather than merely acting as a system administrator for the same network.

## 3. Configuration management and systems architecture

There is an implicit assumption that configuration management begins with an overall 'architectural plan' of how services will be mapped to servers, and that changes in overall architecture are not allowed after that architecture has been mapped out. This is an unfortunate distinction for several reasons:

1. The choice of architecture often affects the cost of management. A classical example of this is that cramming many services onto a single server seems to save money but in fact, the cost of configuration management might be dramatically higher due to conflicts and constraints between services.

2. The exact impact of architectural choices may not be known until one considers how the architecture will be managed. For example, it might be unclear to someone not involved in system administration that certain pairs of web server modules cannot be loaded at the same time without behavior problems. Thus there could be a need for segmentation of web services that is only known to someone in configuration management, and not known to people who might be designing an architecture.

3. Several hard problems of configuration management can be mitigated or even eliminated via architectural changes. For example, one can eliminate subsystem dependencies by locating potentially conflicting subsystems on disjoint or virtual operating systems.

Separating architectural design from configuration management strategy can be a costly and significant mistake, especially because system administrators are often the only persons with enough knowledge to design an efficiently manageable solution architecture.

### 3.1. *Other kinds of configuration management*

The term *configuration management* has several other meanings with the potential to overlap with the meaning of that term in this chapter.

1. *Software configuration management* [52] (in software engineering) is the activity of insuring that the components of a software product interact correctly to accomplish the objectives of the product. The 'configuration' – in this case – is a set of software module revisions that are 'composed' to form a particular software product.

2. *Network configuration management* is the activity of insuring that routers, switches, load balancers, firewalls, and other networking hardware accomplish the objectives set by management.

System configuration management combines the problems inherent in software configuration management and network configuration management into a single problem. There is the problem of *component composition* [39] that arises in software configuration management, i.e., ensuring that a particular system contains sufficient components to accomplish a task, along with the problem of *component configuration*, of limiting the behavior of components according to some established norms.

Software configuration management in itself is only weakly related to system configuration management. It concentrates upon keeping versions of software components that comprise a specific software product synchronized, so that the resulting product works properly. The key difference between software and system configuration management is that a software product is not usually configured 'differently' for different customers. The goal of software configuration management is to produce a single product to which multiple individuals contribute, but the issue of providing distinct behavior in different environments does not arise except in very advanced cases where vendor customization is requested. In a way, system configuration management takes over where software configuration management leaves off, at the point where a product must be customized by choices for control parameters that are not made within the structure of the software package itself. Making these choices is typically distinct from software compilation.

Network configuration management is closely related to system configuration management, with almost the same objectives. The principal difference is that network devices – unlike computing systems – are seldom extensible in function by adding new programs to the ones they already execute. There is a single program running on each networking device that does not change. The key issues in network configuration management include:

1. Choosing an appropriate version for the single program executing on each device, so that this program interoperates well with programs running on other devices.
2. Determining values for parameters that affect this program and its ability to interoperate with programs on other devices.

Choices for these two parts of configuration are often determined more by device limitations rather than human desires; devices have limited interoperability and not all combinations of software versions will work. Once the component composition has been determined (usually by either reading documentation or by experience), programs that enforce parameter choices usually suffice to manage network configuration.

There are a few issues that arise in system configuration management and nowhere else. First, we must compose and configure software components to accomplish a large variety of tasks, where the success of composition is sometimes unsure. Second, the systems being configured are also mostly active participants in the configuration, and must be given the capability to configure themselves via a series of bootstrapping steps. This leads to problems with sequencing, and requirements that changes occur in a particular order, in ways that are less than intuitive.

## 4. Cost models

The basic goal of configuration management is to minimize cost and to maximize value (see chapter by Burgess on System Administration and Business), but many sources of cost have been intangible and difficult to quantify. A typical administrative staff has a fixed size and budget, so all that varies in terms of cost is how fast changes are made and how much time is spent waiting for problem resolution. Patterson [61] quantifies the 'cost of downtime' as

$$C_{\text{downtime}} = C_{\text{revenue lost}} + C_{\text{work lost}} \tag{1}$$

$$\approx (C_{\text{avg revenue lost}} + C_{\text{avg work lost}}) \tag{2}$$

$$\times T_{\text{downtime}},$$

where

- $C_{\text{revenue lost}}$ represents the total revenue lost due to downtime,
- $C_{\text{work lost}}$ represents the total revenue lost due to downtime,
- $C_{\text{avg revenue lost}}$ represents the average revenue lost per unit of downtime,
- $C_{\text{avg work lost}}$ represents the average work lost per unit of downtime,
- $T_{\text{downtime}}$ represents the downtime actually experienced.

Of course, this can be estimated at an even finer grain by noting that the revenue lost is a function of the number of customers inconvenienced, while the total work lost is a function of the number of workers left without services.

Patterson's simple idea sparked other work to understand the nature of cost and value. Configuration management was – at the time – completely concerned with choosing tools and strategies that make management 'easier'. Patterson defined the problem instead as minimizing cost and maximizing value. This led (in turn) to several observations about configuration management that may well transform the discipline in the coming years [36]:

1. Configuration management should minimize *lifecycle cost*, which is the integral of unit costs over the lifecycle of the equipment:

$$C_{\text{lifecycle}} = \int_{t_{\text{start}}}^{t_{\text{end}}} c_r(t)\,dt, \tag{3}$$

   where $c_r(t)$ is the cost of downtime at time $t$, and $t_{\text{start}}$ and $t_{\text{end}}$ represent the lifespan of the system being configured.

2. The instantaneous cost of downtime in the above equation is a sum of two factors:

$$c_r(t) = c_{rm}(t) + c_{ri}(t), \tag{4}$$

   where

   - $c_{rm}(t)$ represents constant costs, such as workers left idle (as in Patterson's model),
   - $c_{ri}(t)$ represents intangible costs, such as contingencies that occur *during* an outage.

3. We can safely approximate contingencies during an outage as multi-class with Poisson arrival rates, so that

$$c_r(t) = c_{rm}(t) + \sum_{d \in D_r} \lambda_d c_d, \tag{5}$$

   where $D_r$ is a set of classes of contingencies, and $d$ represents a contingency class with arrival rate $\lambda_d$ and potential average cost $c_d$.

4. To a first approximation, the cost of downtime is proportional to the time spent troubleshooting.

5. Troubleshooting time is controlled by the complexity of the troubleshooting process, as well as the sequence of choices one makes in understanding the problem.

6. The best sequence of troubleshooting tasks is site-specific, and varies with the likelihood of specific outages.

The key discovery in the above discussion is that *good practice is site-relative*.[2] The practices that succeed in minimizing downtime and increasing value depend upon the likelihood of each contingency.

One way to study potential costs is through simulation of the system administration process [36]. We can easily simulate contingencies requiring attention using classical queueing theory; contingencies are nothing more than multi-class queues with different average service times for each class. We can observe arrival frequencies of classes of tasks

---

[2] The term 'best practice' is often used in this context, albeit speculatively.

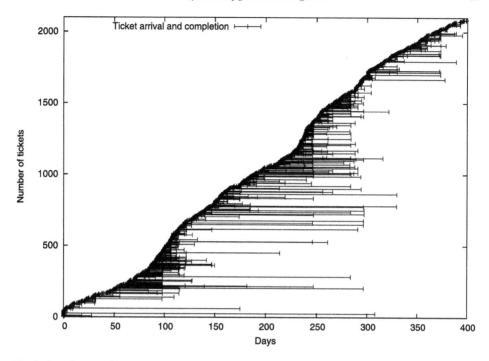

Fig. 3. Data from trouble ticket systems indicates arrival rates and service rates for common tasks. Reprinted from [36] by permission of the authors.

in practice by using real data from trouble ticketing systems (Figure 3), and can utilize this to form an approximate arrival and service models for requests.

Simple simulations show just how sensitive downtime costs are to the number of staff available. In an understaffed situation, downtime costs increase rapidly when request queues expand without bound; the same scenario with more system administrators available shows no such behavior (Figure 4).

Configuration management must balance the cost of management against the cost of downtime. Since sites differ in how much downtime costs, there are several distinct strategies for configuration management that depend upon site goals and mission.

### 4.1. *Configuration management lifecycles*

To understand the lifecycle cost of configuration management, it helps to first consider the lifecycle of the systems to be configured. Configuration management is an integral and inseparable part of the lifecycle of the computing systems being managed. Each system goes through a series of configuration states in progressing from 'bare metal' (with no information on disk) to 'configured', and finally to being 'retired' (Figure 5). We define a *bare*

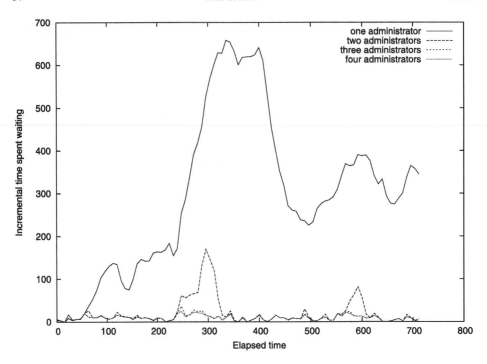

Fig. 4. Chaotic interactions ensue when there are too few system administrators to handle a trouble-ticket queue; the hidden cost is that of downtime due to insufficient staffing. Reprinted from [36] by permission of the authors.

*metal* system is one in which the operating system has not yet been installed; a *bare-metal rebuild* involves starting from scratch, with a clean disk, and re-installing the operating system, erasing all data on disk in the process. After system building, *initial configuration* gives the system desirable behavioral attributes, after which there is an ongoing *maintenance* phase in which configuration changes are made on an ongoing basis. Finally, the system is *retired* from service in some way.

In almost precise correspondence with the system lifecycle, configuration management itself has its own lifecycle, consisting of planning, deployment, maintenance, and retirement states. *Planning* consists of the work needed in order to begin managing the network as an entity. *Deployment* consists of making tools for configuration management available on every host to be managed. *Maintenance* consists of reacting to changes in policy and contingencies on an ongoing basis, and *retirement* is the process of uninstalling or de-commissioning configuration management software.

Part of what makes configuration management difficult is that the cost of beginning to utilize a particular method for configuration management can be substantial. To decide whether to adopt a method, one must analyze one's needs carefully to see whether the cost of a method is justified.

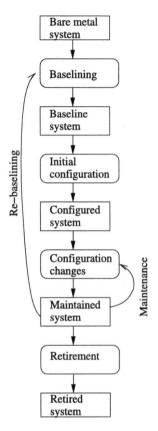

Fig. 5. Managed systems go through a series of states in their lifecycle, from 'bare metal' (unbuilt) to 'retired'.

## 5. Configuration management strategies

To minimize cost and maximize value, the configuration manager adopts an overall *configuration management strategy* to control the configuration of a network. The management strategy details and documents how configuration changes are to be made, tracked and tested. This strategy is best considered to be independent of the software tools used to enforce the changes. Key components of a typical configuration management strategy include:

1. A mechanism for making changes in the configuration, through tools or practices.
2. A mechanism for tracking, documenting and auditing configuration changes.
3. A mechanism for verifying and validating configurations.
4. A procedure for staging deployment of new configurations, including prototyping and propagating changes.

For example, in an academic research environment, staging may often be skipped; the prototype environment and the production environment may be the same and downtime may not be extremely important.

One should not confuse a choice of configuration management strategy with a particular choice of tools to perform configuration management. Tools can support and ease implementation of particular strategies, but are no substitute for a coherent strategic plan.

There are three main kinds of strategies for configuration management, that we shall call 'scripted', 'convergent' and 'generative' [35]. A *scripted* strategy uses some form of custom scripts to make and manage changes in configuration; the scripts mimic the way an administrator would make the same changes manually [7]. A *convergent* strategy utilizes a special kind of scripts that can be applied multiple times without risk of damage. These scripts are usually safer to employ than unconstrained scripts, but are somewhat more complex in character and more complex than the manual procedures they replace. A *generative* strategy involves generating all of the configuration from some form of database of requirements. Note that all convergent strategies are in some sense scripted, and all generative strategies are (trivially) convergent, in the sense that generating the whole configuration twice yields the same answer as generating it once.

These approaches differ in scope, capabilities and complexity; scripting has a low startup cost, convergent scripting is safer than unrestricted scripting but requires recasting manual procedures in convergent terms, and generative mechanisms require specification of a comprehensive site policy, and thus incur a high startup cost. One crucial choice the practicioner must make in beginning configuration management is to weigh cost and value, and choose between one of these three base strategies.

There are several reasons for adopting a comprehensive management strategy rather than administering machines as independent and isolated entities:

1. To reduce the cost of management by adopting standards and eliminating unnecessary variation and heterogeneity.
2. To reduce downtime due to troubleshooting.
3. To assure reproducibility of systems during a contingency.
4. To increase the interchangeability of administrative staff in dealing with changes in configuration.
5. To increase the organizational maturity of the system administration organization.

The last one of these deserves careful consideration. The basic idea of the Capabilities Maturity Model (CMM) of software engineering [20,21] is that the quality of a software product is affected by the 'maturity' of the organization that produces it. Maturity measures include size of programming staff, consistency and documentability of engineering process in creating software, interchangeability of function among programming staff, and ability to deal with contingencies. Applying this principle to system administration [51], we find that important factors in the maturity of a system administration organization include documentation of process, reproducibility of results, and interchangeability of staff. All of these maturity factors are improved by the choice of a reasonable configuration management strategy. The strategy defines how results are to be documented, provides tools for reproducing results and allows staff to easily take over configuration management when system administrators leave and enter the organization.

## 5.1. *The effect of site policy*

Configuration management is a discipline in which the nature of one's goals determines the best possible management strategy. Thus the choice of tools is dependent upon the choice of operating policies.

Let us consider two possible extremes for site policy, that we shall call 'loose' and 'tight'. At a 'tight' site, there is a need for operations to be auditable and verifiable, and for absolute compliance between policy and systems. Examples of 'tight' organizations include banks and hospitals. In these situations, mistakes in configuration cost real money, both in terms of legal liability and lost work. In some cases a configuration mistake could threaten the life of a person or the organization as a whole.

By contrast, at a 'loose' site, the consequences of misconfiguration are less severe. In a development lab, e.g., the risks due to a misconfiguration are much smaller. The same is true for many academic labs.

Of course, many sites choose to function between the two extremes. For example, within a startup firm, it is common practice to skimp on system administration and cope with lack of reliability, in order to save money and encourage a liquid development process. However, even in such an organization, *certain* systems – including accounts payable and receivable – must be managed to a 'tight' set of standards. This has led many startups to 'outsource' critical business functions rather than running them in-house. The choice of 'loose' or 'tight', however, comes with a number of other choices that greatly affect one's choice of management strategy.

First, we consider the 'tight' site. In this kind of site:

1. Security risks are high, leading to a lack of trust in binary packages not compiled from source code.
2. Thus many packages are compiled from source code rather than being installed from vendor-supplied binary packages.
3. Configuration is considered as a one-time act that occurs during system building. Changes are not routinely propagated.
4. Almost any deviation from desired configuration is handled by rebuilding the system from scratch. Thus 'immunizing' methods are not considered important, and may even obstruct forensic analysis of security break-ins.

Thus the ability to rebuild systems is most important. Day-to-day changes are not as important to handle.

The administrator of a 'tight' site is not expected to react to many changes. Instead, the configuration is specified as a series of steps toward rebuilding a system. 'Convergent' methods are considered irrelevant, and 'generative' methods are simply unnecessary. All that is needed for a 'tight' site is a script of commands to run in order to assure the function of the site, and no more. Thus 'managed scripting' is the mechanism of choice.

By contrast, consider the 'loose' site, e.g., a development laboratory or academic research site. In this kind of site:

1. Security risks are low.
2. It might be important to have up-to-date tools and software.

3. Manpower is often extremely short.
4. Changes are extremely frequent, thus, it is impractical to rebuild systems from scratch for every change.

Here some form of 'generative' or 'convergent' strategy could be advisable.

## 5.2. *Making changes*

With an understanding of the nature of one's site in hand, the task of the configuration manager is to choose a strategy that best fits that site's mission. We now turn to describing the components of an effective configuration management strategy in detail.

The most crucial component in any configuration management strategy is the ability to make changes in configuration. Change mechanisms usually involve a *configuration description* that specifies desired qualities of the resulting (possibly distributed) system in an intermediate language, as well as some *software tool or infrastructure* that interprets this and propagates changes to the network.

The process of actually making changes to configuration is so central to system configuration management that there is a tendency to ignore the other components of a management strategy and concentrate solely upon the change propagation mechanism. In a simple configuration management strategy, the configuration file suffices as documentation and as input for auditing and verification of the results, while a software tool provides mechanisms for propagating those changes. As of this writing, an overwhelming majority of sites use an extremely simple configuration management strategy of this kind, first noted by [43] and apparently still true today.

## 5.3. *Tracking and auditing changes*

A related problem in configuration management is that of tracking and auditing changes to machines after they are made. Today, many sites use only a simple form of tracking, by keeping a history of versions of the configuration description via file revision control [70]. A small number of sites, motivated by legal requirements, must use stricter change control mechanisms in which one documents a *sequence of changes* rather than a *history of configurations*.

Today, few sites track *why* changes are made in a formal or structured way, although one could argue that the changes themselves are meaningless without some kind of motivation or context. A change can often be made for rather subtle reasons that are soon forgotten, for example, to address a hidden dependency between two software packages. In the absence of any documentation, the system administrator is left to wonder why changes were made, and must often re-discover the constraint that motivated the change, all over again, the next time the issue arises. Thus, effective documentation – both of the systems being configured and the motives behind configuration features – is a cost-saving activity if done well.

**5.4.** *Verification and validation*

Verification and validation are technical terms in software engineering whose meaning in system configuration management is often identical.

1. *Verification* is the process of determining whether the software tools or procedures accomplish changes in accordance with the configuration policy.
2. *Validation* is the process of determining whether the intermediate policy code actually produces a system that satisfies human goals.

In the words of Fredrick Brooks [8], *verification* is the process of determining "whether one is making the product right", while *validation* is the process of determining "whether one is making the right product".

Failures of validation and verification bring 'loops' into the configuration management process (see Figure 2). A failure of verification requires recoding the intermediate representation of configuration and propagating changes again, while a failure of validation requires modifying the original operating policy.

Currently, many software tools for configuration management have strong mechanisms for verification: this is the root of 'convergent' management strategies as discussed in Section 13. In verification, the agreement between source descriptions and target files on each computer is constantly checked, on an ongoing basis, and lack of agreement is corrected.

Currently, the author is not aware of any tools for *validation* of system configurations.[3] While validation has been a central part of software engineering practice for the last 30 years, it has been almost ignored as part of configuration management. It is almost entirely performed by a human administrator, either alone or by interacting with users. Traditionally, validation has been mislabeled as other activities, including 'troubleshooting', 'help desk', etc. User feedback is often utilized as a substitute for quality assurance processes. For more details on structured testing and its role in configuration management, see [75].

Why do configurations fail to function properly? There are several causes, including:

(1) Human errors in coding the intermediate representation of configuration;
(2) Errors in propagating changes to configured nodes;
(3) Changes made to configured nodes without the administrator's knowledge;
(4) Errors or omissions in documentation for the systems being configured.

Human error plays a predominant role in configuration failures.

**5.5.** *Deployment*

Deployment is the process of enforcing configuration changes in a network of computing devices. It can take many forms, from rebuilding all workstations from scratch to invoking an intelligent agent on each host to make small changes.

Staging is the process of implementing a configuration change in small steps to minimize the effect of configuration errors upon the site (Figure 6). A typical staging strategy involves two phases, *prototyping* and *propagation*. In the prototyping phase, a small testbed of representative computers is assembled to test a new configuration.

---

[3]Note that some tools that perform verification are commonly referred to as 'validation' tools.

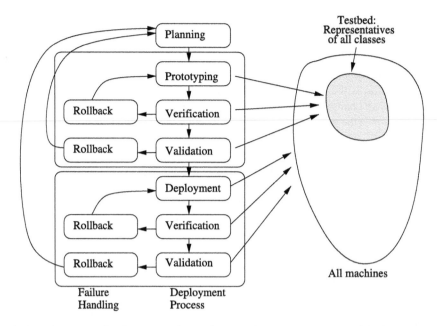

Fig. 6. A staging strategy involves prototyping, testing, actual deployment and potential rollback in case of difficulties.

A good testbed contains at least one computer of each type available in the network. The configuration is first tried on the prototype network, and debugging and troubleshooting irons out the problems. Once that is complete, propagation ensues and the configuration is enforced on all hosts. If this does not work properly, *rollback* may be utilized: this is the process of restoring a previous configuration that was known to function.

Problems due to deployment can cost money in terms of work and revenue lost; see [61] for some statistics. Testing is just one way to assure that deployment does not cause downtime. Another way is to wait a short time before installing a security patch, rather than installing it immediately [37].

## 6. Challenges of configuration management

Before discussing the traditions that led to current configuration management practices, it is appropriate to understand why configuration management is difficult and deserves further research. These brief remarks are intended to provide a non-practicioner with a background on the subtle challenges and rewards of a good configuration management strategy.

Configuration management would be more straightforward if all machines were identical; then a method for configuring one machine would be reusable for all the others. It would be more straightforward if there were no large-scale networks and no requirement for uptime and disaster recovery; one could simply configure all machines by hand. It would be more straightforward if policy never changed. But none of these is true.

The principal drivers of configuration management as a practice include changes in policy, scale of application, heterogeneity of purpose, and need for timely response to unforeseen contingencies. These have influenced current approaches to configuration management, both by defining the nature of desirable behavior and by identifying problematic conditions that should be avoided.

## 6.1. *Change*

Perhaps the most significant factor in configuration management cost is that the high-level policy that defines network function is not constant, but rather changes over time. One must be able to react effectively, efficiently, unintrusively and quickly to accomplish desired changes, without unacceptable downtime. Mechanisms for configuration management vary in how quickly changes are accomplished, as well as how efficiently one can avoid and correct configuration mistakes. One recurrent problem is that it is often not obvious whether it is less expensive to change an existing configuration, or to start over and configure the machine from scratch: what practicioners call a 'bare-metal rebuild'.

## 6.2. *Scale*

Another significant driver of configuration management cost and strategy is the scale of modern enterprise networks. Tasks that were formerly accomplished by hand for small numbers of stations become impractical when one must produce thousands of identical stations, e.g., for a bank or trading company. Scale also poses several unique problems for the configuration manager, no matter which scheme is utilized. In a large-enough network, it is not possible to assure that changes are actually made on hosts when requested; the host could be powered down, or external effects could counteract the change. This leads to strategies for insuring that changes are committed properly and are not changed inappropriately by other external means, e.g., by hand-editing a managed configuration file.

## 6.3. *Heterogeneity*

Heterogeneity refers to populations of multiple types of machines with differing hardware, software, pre-existing configuration, or mission constraints. Hardware heterogeneity often occurs naturally as part of the maintenance lifecycle; stations persist in the network as new stations are purchased and deployed. A key question in the choice of a configuration management strategy is how this heterogeneity will be managed; via a centrally maintained database of differences [2,4,40,41,44,45], or via a distributed strategy with little or no centralized auditing [9–11,18].

Another kind of heterogeneity is that imposed by utilizing differing solutions to the same problem; e.g., a group of administrators might all choose differing locations in which to install new software, for no particularly good reason. This 'unintentional heterogeneity'

[31] arises from lack of coordination among human managers (or management tools) and not from the site itself.

Unintentional heterogeneity can become a major stumbling block when one desires to transition from one configuration management scheme to another. It creates situations in which it is extremely difficult to construct a script or specification of what to do to accomplish a specific change, because the content of that script would potentially differ on each station to be changed.

For example, consider a network in which the active configuration file for a service is located at a different path on each host. To change this file, one must remember which file is actually being used on a host. One thus has to *remember* where the file is on all hosts, forever. This is common in networks managed by groups of system administrators, where each administrator is free to adopt personal standards. To manage such a network, one must remember *who* set up each host, a daunting and expensive task.

### 6.4. *Contingency*

A final driver of configuration management cost is contingency. A configuration can be modified by many sources, including package installation, manual overrides, or security breaches. These contingencies often violate assumptions required in order for a particular strategy to produce proper results. This leads to misconfigurations and potential downtime.

Contingencies and conflicts arise even between management strategies. If two tools (or administrators) try to manage the same network, conflicts can occur. Each management strategy or tool presumes – perhaps correctly – that it is the sole manager of the network. As the sole arbiter, each management strategy encourages making arbitrary configuration decisions in particular ways. Conflicts arise when arbiters disagree on how to make decisions. This also means that when an administrator considers adopting a new management strategy, he or she must evaluate the cost of changing strategies, which very much depends upon the prior strategy being used. More often than not, adopting a new strategy requires rebuilding the whole network from scratch.

## 7. High-level configuration concepts

The first task of the system administrator involved in configuration management is to describe how the configured network should behave. There are two widely used mechanisms for describing behavior at a high level. Machines are categorized into *classes* of machines that have differing behaviors, while *services* describe the relationships between machines in a network.

### 7.1. *Classes*

The need to specify similarities and differences between sets of hosts led to the concept of *classes*. Classes were introduced in *rdist* [24] as static lists of hosts that should agree on contents of particular files. Each host in the class receives the same exact file from a server; classes are defined and only meaningful on the *server* that stores master copies of configuration files.

The idea of classes comes into full fruition in *Cfengine* [9–11,18], where a class instead represents a set of hosts that *share a particular dynamic property*, as in some forms of object-oriented programming. In *Cfengine*, class membership is a *client* property, determined when *Cfengine* runs as a result of dynamic tests made during startup. There is little need to think of a class as a list of hosts; rather, a class is something that a client *is* or *is not*.[4]

In either case, a *class* of machines is a set of machines that should share some concept of state or behavior, and many modern tools employ a mix of *rdist*-like and *Cfengine*-like classes to specify those shared behaviors in succinct form. For example, the class 'web server' is often disjoint from the class 'workstation'; these are two *kinds* of hosts with different configurations based upon kind.

*Cfengine* also introduces two different kinds of classes:

(1) *hard* classes define attributes of machines that cannot change during the software configuration process, e.g., their hardware configurations;

(2) *soft* classes define attributes of machines that are determined by human administrators and the configuration process.

For example, a 'hard class' might consist of those machines having more than a 2.0 GB hard drive, while a 'soft class' might consist of those machines intended to be web servers. It is possible to characterize the difference between hard classes and soft classes by saying that the hard classes are part of the *inputs* to the configuration management process, while the soft classes describe *outputs* of that process.

Figure 7 shows the class diagram for a small computer network with both hard and soft classes of machines. In the figure there are two hard classes, 'x386' and 'PowerPC', that refer to machine architectures. A number of soft classes differentiate machines by

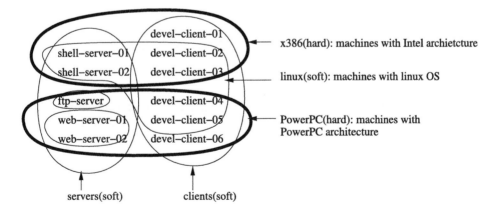

Fig. 7. Classes of machines include hard classes (bold boundaries) and soft classes (regular weight boundaries).

---

[4]This is the basis of 'client pull' configuration management, described in Section 13.1.

function: some machines are clients, while others are servers; among servers, there are different kinds of services provided.

Note that the concepts of 'hard' and 'soft' depend upon the capabilities of the tools being used and a particular machine's state within a configuration process. During system bootstrapping from bare metal, the operating system of the machine is a 'soft' class. If one is maintaining the configuration of a live system, the same class becomes 'hard', because the live maintenance process has no capacity to change the operating system itself without bringing the system down.

### 7.2. *Services*

Classes provide a way to describe the function of machines in isolation: two machines should exhibit common features or behavior if they are members of the same class. To describe how machines interoperate at the network level we use the concept of 'services'.

In informal terms, a *service* is a pattern of interaction between two hosts, a *client* requesting the service and a *server* that provides it. We say that the client *binds* to the server for particular services.

Traditionally, a *service* has been defined as a behavior provided by a single server for the benefit of perhaps multiple clients. In modern networks, this is too limiting a definition, and it is better to define a service as a set of tasks undertaken by a distinguished *set* of hosts for some client consumer.

Common services include:

1. Dynamic Host Configuration Protocol: MAC address to IP address mapping.
2. Domain Name Service: hostname to IP address mapping.
3. Directory Service: user identity.
4. File service: persistent remote storage for files.
5. Mail service: access to (IMAP, POP) and delivery of (SMTP) electronic mail.
6. Backup service: the ability to make offline copies of local disks.
7. Print service: the ability to make hard copies of files.
8. Login service: the ability to get shell access.
9. Window (or 'remote desktop') service: the ability to utilize a remote windowing environment such as X11 or Windows XP.
10. Web service: availability of web pages (HTTP, HTTPS),

etc.

A 'service architecture' is a map of where services are provided and consumed in an enterprise network (Figure 8). In typical practice, this architecture is labeled with names of specific hosts providing specific services.

### 8. Low-level configuration concepts

Our next step is to develop a model of the configuration itself. We describe 'what' system configuration should be without discussing 'how' that configuration will be assured or managed. This is a *declarative* model of system configuration, in the same way that a

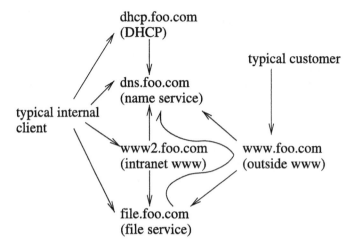

Fig. 8. A service architecture includes bindings between hosts and their servers. In the diagram $A \to B$ means that '$A$ is a client of $B$' or '$B$ serves $A$'.

'declarative language' in the theory of programming languages specifies 'what' goals to accomplish but not 'how' to accomplish them.

### 8.1. *Components*

A *component* of a computer network is any device that can be considered as an entity separate from other entities in the network. Likewise, a *component* of an individual computing system is a subsystem within the system that is in some sense independent from other subsystems. Hardware components are devices that may or may not be present, while software components are programs, libraries, drivers, kernel modules, or other code that are optional and/or independently controlled.

### 8.2. *Parameters*

Likewise, each *component* of a system may have *configuration parameters* that must be set in order for it to operate properly. Parameters can take many forms, from complex hierarchical files describing desired behaviors (e.g., the Apache web server's httpd.conf), to simple key/value pairs (such as kernel parameters). A parameter value describes or limits the *behavior* of a component.

### 8.3. *Configuration*

A configuration of a system usually has two parts:
1. A set of (hardware and software) components that should be present.

2. A collection of parameter values describing appropriate behavior for each component.

The way that components are installed, and the way their configuration is specified, varies
from component to component. There is in all cases some form of persistent storage containing the parameter values for each component, such as a file on disk (though certain
configuration parameters may be stored in unusual places, including flash memory).

In many cases, parameter values for a single component are spread out among several
files within a filesystem. For example, in implementing an FTP service, we might specify:

(1) the root directory for the FTP service;
(2) the security/access policy for the FTP service;
(3) the list of users allowed to access the FTP service

in *different* files. Files that specify parameters for components are called *configuration files*
and are considered to be *part of system configuration*.

## 8.4. *Behavior*

The configuration management problem is made more difficult by the lack of any efficient mechanism for mapping from low-level choices to the system behaviors that results
from these choices. Usually, the behaviors that arise from configuration choices are 'documented in the manual'. A manual is a map from behaviors to choices, and not the reverse.
This means that in practice, the meaning of a particular choice may be virtually impossible
to reverse-engineer from the list of choices.

## 8.5. *Constraints*

Configuration management is really a 'constraint satisfaction problem'. The input is a set
of constraints that arise from two sources:

1. The *hard constraints* that specify which configurations will work properly. Failing to
   satisfy a hard constraint leads to a network that does not behave properly.
2. The *soft constraints* (e.g., policies) that specify desirable options among the space
   of all possible options. Failing to satisfy a soft constraint does not meet the needs of
   users or enterprise.

The problem of configuration management is to choose one configuration that satisfies both
hard and soft constraints.

Constraints arise and can be described in several ways, including considering required
consistencies, dependencies, conflicts and aspects of the network to be configured.

## 8.6. *Consistency*

First, there are many *consistency* requirements for parameters, both within one machine
and within a group of cooperating machines. Parameters of cooperating components must
agree in value in order for the components to cooperate. As a typical example, machines

wishing to share a file service must agree upon a server that indeed contains the files they wish to share. But many simpler forms of coordination are required, e.g., a directory declared as containing web content must exist and contain web content, and must be protected so that the web server can access that content. But there are many other ways to express the same constraints.

### 8.7. *Dependencies*

A *dependency* refers to a situation in which one parameter setting or component depends upon the existence or prior installation of another component or a particular parameter setting in that component [33,74]. A program can depend upon the prior existence of another program, a library, or even a kernel module. For example, many internet services cannot be provided unless an internet service daemon `inetd` or `xinetd` is present.

### 8.8. *Conflicts*

It is sometimes impossible to select parameter settings that make a set of components work properly together, because components can make conflicting requirements upon parameters and/or other components. The classic example is that of two components that require different versions of a third in order to function properly. For example, two programs might require different versions of the same dynamic library; one might only work properly if the old version is installed, and the other might only work properly if a newer version is instead installed. We say in that case that there is a *conflict* between the two sets of component requirements. Equivalently, we could say that the two sets of requirements are *inconsistent*.

### 8.9. *Aspects*

Aspects provide a conceptual framework with which to understand and critically discuss configuration management tools and strategies [3,17]. One key problem in configuration management is that there are groups of parameters that only make sense when set together, as a unit, so that there is agreement between the parts. We informally call such groups *aspects*, in line with the current trends in 'aspect-oriented programming'.

Informally, an aspect is a set of parameters, together with a set of constraints that define the set of reasonable values for those parameters. The *value of an aspect* is a set of values, one per parameter in the aspect, where the individual parameter values conform to the constraints of the aspect.

The notion of aspects here is somewhat different than the notion of aspects in 'aspect-oriented programming' but there are some similarities. In aspect-oriented programming, an aspect is a snippet of code that changes the function of an existing procedure, perhaps by modifying its arguments and when it is invoked. It is most commonly utilized to describe side-effects of setting a variable to a particular value, or side-effects of reading that value. Here, an aspect is a set of parameters whose values must conform to some constraints.

**8.10.** *Local and distributed aspects*

There are many kinds of aspects. For example, in configuring Domain Name Service and Dynamic Host Configuration Protocol, the address of a hostname in DHCP should correspond with the same address in DNS. It makes no sense to set this address in one service and not another, so the host identity information (name, IP address, MAC address) form a single aspect even though the (name, IP address) pair is encoded in DNS and the (IP address, MAC address) pair is encoded in DHCP. Further, it is possible that the DNS and DHCP servers execute on different hosts, making this a *distributed aspect* that affects more than one machine at a time.

By contrast, a *local aspect* is a set of parameters that must agree in value for a single machine. For example, the host name of a specific host appears in multiple files in /etc, so that the parameters whose value should reflect that host name form a single aspect.

There are many cases in which configuration information must be replicated. For example, in the case of an apache web server, there are many places where parameters must agree in value (Figure 9). Each required agreement is an aspect.

Another example of a distributed aspect is that in order for web service to work properly, there must be a record for each virtual server in DNS. The parameter representing each virtual name in the server configuration file must appear in DNS and map to the server's address in DNS.

Another example of a distributed aspect is that for clients to utilize a file server, the clients must be configured to mount the appropriate directories. For the set of all clients and the server, the identity of the directory to be mounted is an aspect of the network; it makes no sense to configure a server without clients or a client without a corresponding server. The constraint in this aspect is that one choice – that of the identity of a server –

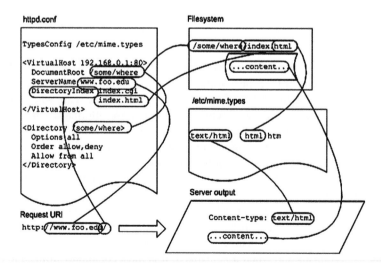

Fig. 9. Aspects of an apache web server include many local configuration parameters whose values must be coordinated. Reprinted from [68] with permission of the authors.

determines the choices for where clients will receive the service. A *binding* is a simple kind of aspect relationship between a client and a server, in which a single client takes service from a server known to provide that service.

### 8.11. *Aspect consistency*

Aspects can overlap and set differing constraints on the same parameters. There might be an aspect of web service that says that we need at least 2 GB of space in the web root directory, and another aspect that says the web root directory must be on a local disk. The complete configuration must satisfy the constraints of both aspects. As another example, the individual aspects for each host DHCP+DNS record overlap with any aspect determining the whole configuration of the DHCP or DNS server.

A set of several (potentially overlapping) aspects is *consistent* if there are no inherent conflicts between the constraints of the aspects. A set of (potentially overlapping) aspects is *inconsistent* if not. A set of inconsistent aspects represents a system that is not likely to function properly.

As another example, a particular web server could be described by specifying several aspects of the service:

(1)  where files for each site to be served are stored;
(2)  how much space must be available in that directory;
(3)  the URL that is used to access each site;
(4)  the way that users are authorized to view site materials,

etc.

Aspects need not be limited to a single machine, but can be properties of an entire network. The *scope* of an aspect is the number of components to which it applies. For example,

- the identity of the mail server,
- the identity of the server containing home directories,
- the identity of the gateway to the internet

are all aspects whose scope might be that of an entire network.

Anderson [3] refers to aspects as units of configuration managed by different administrators. This is very similar to our definition, because the administrators are typically responsible for different service layers requiring coordination of the kind we describe. We utilize a lower-level, mathematical definition for the purpose of separating the problem of configuration from that of mapping humans to configuration tasks. Our aspects exist as mathematical boundaries, without reference to human administrative structure. Thus,

PRINCIPLE 1. A configuration is a consistent composition of aspects.

This view of configuration management, as compositional in nature, motivates the chapter by Sun and Couch on the complexity of configuration management.

## 9. An operational model of configuration management

So far, we have considered configuration management as an abstract process of declaring specific state for a set of machines. In practice, however, this is not enough; we must also consider how to implement and manage that state on an on-going basis. Burgess formalized this discussion by introducing the notion of configuration operators with *Cfengine*, whose behavior can then be discussed in terms of certain properties [10,12,14], based on whether a system is in a known state or an unknown state. An operation would then bring the systems into conformance with specified requirements or a desired final state.

### 9.1. *Baselining*

The first step in an operational model of configuration management is to define a *baseline state* of each machine to which changes will be applied (see Figure 5). This might be a machine state, a state created by a differing management method, or a 'clean' system state created by a fresh install of the operating system onto a blank or erased disk. The choice of baseline state is one of many factors that influences which configuration management strategies may be least costly.

### 9.2. *Pre-conditions and post-conditions*

Operations utilized to configure a system are often dependent upon its current state; their outcome thus depends upon the order in which they are performed. This is particularly true during bootstrapping scripts, in which there is often only one order in which to complete the bootstrapping steps. Technically, we say that each change to a system has *pre-conditions* that must be true before the change can be accomplished properly, and *post-conditions* that will be true after the change, if the pre-conditions were true before the change.

PRINCIPLE 2. Every configuration operation can be associated with a set of pre-conditions that must be true for proper function, and a set of post-conditions that will be true after the operation if pre-conditions are met beforehand.

Preconditions and post-conditions of operations are often ignored but have an overall effect upon management. For example, many operations must be applied to previously configured systems; and almost all configuration management scripts presume that the system being configured has already been built by some other means.

Alternatively, we say that each operation *depends* upon the execution of its predecessors for success. For example, one must install and configure the network card driver and configuration before software can be downloaded from the network.

There are two kinds of dependencies: known and hidden (or latent). A known dependency is something that is documented as being required. A hidden dependency or 'latent precondition' [35,48] is a requirement not known to the system administrator. Many configuration changes can create latent pre-conditions. A typical example would be to set

security limits on a service and forget that these were set. As long as the service is never turned on, the latent precondition of security policy is never observed. But if the service is later turned on, mysteriously, it will obey the hidden security policy even though the system administrator might have no intent of limiting the service in that way.

Looking more carefully at this example, we see that there is another very helpful way to look at latent pre-conditions. Rather than simply being hidden state, they involve asserting behaviors that were not known to be asserted. Accordingly,

PRINCIPLE 3. Every latent precondition is a result of violating a (perhaps forgotten) aspect constraint.

In other words, since an aspect is (by definition) the functional unit of behavior, producing unexpected behavior is (by definition) equivalent with violating the atomicity of an aspect.

### 9.3. *Properties of configuration operations*

Suppose that we define a configuration operation to be a program that, when executed on a system or on the network, effects changes in configuration. There is some controversy amongst designers about how one should choose and define configuration operations.

One school of thought, introduced in the tool *Cfengine* [9–11,18], is that 'all operations should be declarative', which is to say that every operation should assert a particular system state or states, and leave systems alone if those states are already present. The more traditional school of thought claims that 'imperative operations are necessary' in order to deal with legacy code and systems, such as software that is installed and configured through use of make and similar tools.

The solution to this quandary is to look carefully at the definitions of what it means to be declarative or imperative. A set of declarative operations is simply a set in which the order of operation application does not matter [68]. Likewise, an imperative set of operations may have a different effect for each chosen order in which operations are applied. A set of operations where each one operates on a distinct subset parameters is trivially declarative, as the order of changing parameters does not affect the total result. Likewise, a set of operations each of which assigns a different value to one shared parameter is trivially imperative.

The impact of certain properties of configuration operations upon configuration management has been studied in some detail in [34]. Properties of configuration operations include:

1. *Idempotence:* an operation $p$ is *idempotent* if its repetition has the same effect as doing the operation once. If $p$ is an idempotent operation, then doing $pp$ in sequence has the same result as doing $p$.
2. *Sequence idempotence:* a set of operations $P$ is *sequence idempotent* if for any sequence $Q$ taken from $P$, applying the *sequence* $Q$ twice ($QQ$) has the same effect as applying it once. I.e., if one has a sequence of operations $pqr$, then the effect of $pqrpqr$ is the same as the effect of $pqr$ applied once. This is a generalization of individual operation idempotence.

3. *Statelessness:* if $p$ is an operation in a set $P$ and $Q$ is any sequence of operations from $P$, then $p$ is *stateless with respect to $P$* if $pQp$ has the same result as $Qp$, i.e., the first application of $p$ can be omitted with no change in effect. The set of operations $P$ is stateless if this is true for any $p \in P$. This is another generalization of idempotence.

4. *Finite convergence:* An operation $p$ is *finitely convergent* if there is a positive number $n$ such that the result of applying $p$ a total of $n + 1$ times has the same result as applying $p$ a total of $n$ times: $p^n p$ has the same result as $p^n$.

Note that this definition of convergence is not the same as the Cfengine notion of (fixed point) convergence – often called simply 'convergent' maintenance – which we discuss in Section 9.5.

Idempotent operations are those that do not change a system unless it 'needs' to be changed; systems that are already in an appropriate state remain untouched. The simplest idempotent operation is to set a parameter to a constant value. If repeated, the same value is set again.

Statelessness is a generalization of idempotence. Many common operations are stateless, i.e., one does not need to repeat them to accomplish an effect, and the last invocation 'always wins' in producing an appropriate behavior. For example, an operation that sets a particular single parameter to a constant value is stateless. Conversely, through the use of semigroup theory, [34] shows that any stateless collection of operations corresponds directly with a set of parameters.

The main result of the semigroup work so far is that [34]:

THEOREM 1. *If a set of operations is stateless, then $\mathcal{P}^*/\equiv$ is factorable into disjoint sets of anti-commutative sub-semigroups, each of which commute with one another.*

In the theorem, $\mathcal{P}^*$ represents the set of all sequences of operations taken from $\mathcal{P}$.

The meaning of this theorem is subtle. The anti-commutative sub-semigroups represent individual configuration parameters; each operation within a sub-semigroup represents a parameter value. The fact that any sub-semigroup commutes with any other indicates that each sub-semigroup operates on one facet of configuration, independently. The fact that operations within the sub-semigroup *never* commute indicates that they assert different values of configuration parameters.

In this chapter, we will discuss several theorems of the above kind, broadly characterized as *emergent property theorems*. The above theorem can be paraphrased in the following important way:

PRINCIPLE 4. Configuration parameters are an emergent property of the operations that manipulate them.

In other words, even if we do not assume that there is a concept of configuration parameter, that concept persists, especially when operators are stateless. This is very important in reducing the complexity of the configuration management task, because the parameters that emerge from a set of operators is often more simple in structure than the set of parameters one intends to manipulate with those operators. In particular, from the structure of operations, one can achieve a simpler *notion* of what parameters control a system than the one given by default! This is the key to limiting 'incidental complexity' [31] that arises from the way particular components are configured.

## 9.4. *Operations and complexity*

The most surprising result about operations, however, is the following recent result (paraphrased for clarity) [69]:

THEOREM 2. *Given a target behavior that one wishes to reproduce and a set of operations $\mathcal{P}$ that can be used to assure behaviors, the problem of choosing a smallest operation sequence $\bar{p}$ from $\mathcal{P}$ that assures that behavior is NP-hard with respect to the number of operations in $\mathcal{P}$. Further, the problem remains NP-hard if operations in $\mathcal{P}$ are idempotent, sequence idempotent, stateless, or finitely convergent.*

For details of the proof and further discussion, see the chapter by Sun and Couch on Complexity of System Configuration Management in this Handbook.

In non-mathematical terms, the hardness of the problem of choosing appropriate operations to achieve a given result is not made easier by limiting operations to have the seemingly desirable properties listed above and – in some fundamental sense – these operation properties do *not* make configuration management easier!

To understand why this theorem is true, we need to rephrase the problem in terms of *aspects* rather than in terms of pure operations. Recall that an aspect is a part of configuration that is subject to constraints that force it to be modified as a unit. Part of what is difficult about choosing appropriate operations is that their effects upon aspects are not necessarily known. The fact that an operation is stateless has no relation to whether it leaves a given aspect in a consistent and functional state! Conversely, if we limit our operations to those that leave aspects in a consistent state, and have some mechanism for choosing operations that are consistent with others, the above 'operation composition' problem becomes trivial rather than NP-hard. In other words,

PRINCIPLE 5. The problem of composing configuration operations toward desired behavioral goals is difficult if the state of all aspects is not guaranteed to be consistent after each operation, and trivial if this guarantee can be made.

The arguments about configuration language miss the point; it is consistency in constraint space that is difficult to achieve, and lack of aspect consistency is the real problem that configuration management must solve.

The impact of this result is subtle. The underlying problem of configuration management – choosing appropriate parameter settings – is fairly difficult in the general case in which one does not understand the couplings between parameter meanings. In the presence of a hard problem, humans simplify that problem by limiting the problem domain. For example, we do understand the couplings between most parameter values, and how they interact or do not interact. Thus we have a 'gray box' understanding of the problem rather than the 'black box' understanding [74] that proves to be intractable.

**9.5.** *Configuration operators*

The above arguments show that aspect consistency is an important property of configuration operations. There has been much study of how to make operations consistent, which has led to a theory of configuration operators. There is an important distinction to be made here:

- *Configuration operations* include any and all means of effecting changes in configuration parameters.
- *Configuration operators* (in the sense that *Cfengine* attempts to approximate) are operations that are fixed-point convergent in a strict mathematical sense, as defined below.

An alternative and more general view of convergence and configuration management turns the whole concept of classical configuration management upside down [12]. We view the system to be configured as a dynamical system, subject to the influences of multiple external operators that determine configuration. We specify configuration by specifying *operators to apply* under various circumstances, and cease to consider any parameters at all, except in the context of defining operators.

DEFINITION 1. A set of operators $\mathcal{P}$ is *fixed-point convergent* if there is a set of fixed points of all operators that are achieved when the operators are applied in random order with repeats, regardless of any outside influences that might be affecting the system. These fixed points are *equilibrium points* of the dynamical system comprising hosts and operators.

Whereas in the classical formulation, a policy is translated into 'parameter settings', in the dynamical systems characterization, a policy is translated into and embodied as a *set of operators to apply under various conditions*. We employ the equivalence between operation structure and parameters characterized in Section 9.3 *in reverse*, and substitute operators for parameters in designing a configuration management strategy.

The differences between classical configuration management and the dynamical systems characterization are profound. While most configuration management strategies assume that the configuration management problem lives in a closed world devoid of outside influences, the dynamical systems theory of configuration management views the system as open to multiple unpredictable influences, including 'configuration operators' that attempt to control the system and other operators and influences such as hacking, misbehaving users, etc. While the goal of traditional configuration management is to manage the configuration as a relatively static entity, the dynamical systems characterization views the configuration of each target system as dynamic, ever-changing, and open to unpredictable outside influences.

Traditional configuration management considers policy as a way of *setting parameters*, and the dynamical systems theory views the role of policy as *specifying bindings between events and operators to apply*. While the goal of traditional configuration management is deterministic *stability*, in which the configuration does not change except through changes in policy, the goal of dynamical system configuration management is *equilibrium*; a state in which the system is approximately behaving in desirable ways, with counter-measures to each force that might affect its behavior adversely. This is the key to computer immunology [10,14].

The ramifications of this idea are subtle and far-reaching. If one constitutes an appropriate set of (non-conflicting) operators as a policy, and applies them in any order and schedule whatsoever, the result is still equilibrium [14,66]. This has been exploited in several ways to create immunological configuration management systems that implement self-management functions without the feedback loops found in typical autonomic computing implementations. Because these solutions are feedback-free and self-equilibrating *without* a need for centralized planning, they are typically easier to implement than feedback-based self-management solutions, and inherently more scalable to large networks.

There are several applications of this theory already, with many potential applications as yet unexplored. By expressing the problem of managing disk space as a two-player game between users and administrator, one can solve the problem through random application of convergent 'tidying' operators that clean up after users who attempt to utilize more than their share of disk space [11]. By defining operators that react to abnormal conditions, a system can be 'nudged' toward a reasonable state in a statespace diagram. These applications go far beyond what is traditionally considered configuration management, and it is possible to model the entire task of system administration, including backup and recovery, via dynamical systems and operators [19].

### 9.6. *Fixed points of configuration operators*

The base result for configuration operators concerns the relationship between configuration operations and constraints. A common misconception is that we must predefine constraints before we utilize configuration management. If we consider the configuration operations as *operators upon a space*, we can formulate convergent operators as applying parameter changes only if parameters are different, by use of the Heaviside step function. This characterizes configuration operations as linear operators upon configuration space.

This relatively straightforward and obvious characterization has far-reaching consequences when one considers what happens when a *set* of such operators is applied to a system:

THEOREM 3. *Let $\mathcal{P}$ be a set of convergent operators represented as linear operators on parameter space, and let $\tilde{p}$ represent a randomly chosen sequence of operators from $\mathcal{P}$. As the length of $\tilde{p}$ increases, the system configured by $\tilde{p}$ approaches a set of states $S$ in which all operators in $\mathcal{P}$ are idempotent, provided that $S$ exists as a set of fixed points for the operators in $\mathcal{P}$.*

In other words, for any $\tilde{p}$ long enough, and for any $q \in \mathcal{P}$, $q\tilde{p} \equiv \tilde{p}$. Paraphrasing the theorem, provided that one selects one's operators upon configuration so that they are convergent and that there is a consistent state that is a fixed point, this state will be reached through any sufficiently long path of random applications of the operators. Thus the consistent state need not be planned in advance; *it is an emergent property of the system of operations being applied.*[5]

This is our second example of an emergent property theorem. Recall that previously, we observed that the concept of a parameter emerges from the structure of operations. Here we observe that the result of a set of operators over time is a fixed point that need not be known in advance.

This result challenges the concept that the configuration itself is a static aggregate state that does not change between configuration operations. Within the operator theory, a configuration can and does change due to 'outside influences' and the only way to bound it is via application of particular operators. Security violations are seen in traditional configuration management as being somewhat outside the scope of the management process; when formulating management in terms of operators, they are simply contingencies to be repaired. In the operator theory, security violations are just another operator process that occurs asynchronously with the operator process of ongoing management. A properly designed set of operators can automatically repair security problems and one can consider security violations as just one more aspect of *typical* operation.

The purport of this theorem is subtle and far-reaching. As network components increase in complexity, the act of planning their entire function becomes impractical and one must be content with planning parts of their function and leaving other parts to the components themselves. It says that if we control only parts of a thing, and the controls on parts are not conflicting, then the thing will eventually be configured in a consistent and compliant state, even though the exact nature of that state is *not known in advance*.

In our view, the controversy over operators is a problem with the definition of the problem, and not the operators. If configuration management consists just of controlling the contents of particular files, then operators are a difficult way of granting that kind of control, but if configuration management is instead a process instead of assuring particular *behaviors*, then operators can help with that process. The operators are behavioral controls that – instead of controlling the literal configuration – control the behavior that arise from that configuration.

The belief that one must control specific files seems to arise from a lack of trust in the software base being managed [72]. If minor undocumented differences in the configuration files can express latent effects, then one must indeed manage the fine structure of configuration files in order to assure particular behaviors. In that environment, operators upon the physical configuration do not seem useful, because there are parts of configuration that must be crafted in specific and complex ways so that applications will function properly.

This is again a problem with how we define configuration management as a practice. If we define the practice as controlling the entire configuration so that errors cannot occur, we then must do that and any operators that establish only partial control are unsuitable. If, conversely, we define the practice as assuring behaviors, then operators can be crafted that assure particular behaviors by direct feedback, and whose composition converges to a consistent network state in which all behaviors are assured, regardless of exactly how that state is accomplished, or even its nature.

---

[5]This is a similar result to that of the Maelstrom theorem for troubleshooting, which demonstrates that the optimal order for troubleshooting steps is an emergent property of the results of the steps [29].

**9.7.** *Observability*

In the previous section, we discussed the intimate relationship between configuration operators and behaviors, and how the operators – by assuring behaviors – both give rise to an implicit notion of parameter, as well as an implicit notion of consistency. But what is meant by behavior?

In [35], the concept of 'observability' is used to build a model of the effects of operations, using the idea that systems that behave equivalently are effectively indistinguishable. The configuration process is represented as a state machine, where operations cause transitions between behavioral states, represented as sets of tests that succeed. The configuration operators discussed above are then simply ways to accomplish state transitions, and the fixed points of Theorem 3 are simply a set of states achieved by repeated application of the operators.

The main result of [35] is that:

THEOREM 4. *One can determine whether the configuration state machine is deterministic by static analysis of the structure of its operations.*

The meaning of this result is that the state machine is deterministic if an operation applied to one observed state always leads to another observed state, which in turn means that there is no hidden state information that is (i) not observed and (ii) utilized during the application of an operation. This is a property of configuration operations that can be statically verified. Thus the theorem can be summarized:

PRINCIPLE 6. It is possible to assure freedom from latent configuration effects by careful design of convergent operators.

In other words, a good set of operators are all that one needs to accomplish configuration management.

**9.8.** *Promises*

Previous sections defined the idea of configuration operators and demonstrated how those operators can be constructed. But what are appropriate configuration operators? One issue of interest is whether it is possible to manage the configuration of a network without centralized authority or control. Can a network self-organize into a functional community when no component of the network has authority over another? Recently, Burgess et al. have suggested that this is possible, through a mathematical idea known as 'promise theory' [15,16].

The basic idea of promise theory is that stations in a network are autonomous competitors for resources who only collaborate for mutual gain, like animals in a food chain. Autonomous agents in the network are free to 'promise' facts or commitments to others. Thus there is a concept of 'service' from one station to another, in the form of 'promises kept'. This involves a kind of reciprocal trust agreement between computers who otherwise have no reason to trust one another. Preliminary results show that networks managed in this way do indeed converge to a stable state, even in the presence of harmful influences, simply because promises can be viewed as convergent operators in the sense of Theorem 3.

Many practicioners of configuration management are in opposition to the idea that computers will broker their own services; the definition of the job of configuration manager is to insure that there is no disruption of services. Politically, many people who practice configuration management do not yet accept the idea that systems could manage their own configurations and services. But as networks become larger, more heterogeneous and more complex, something like promise theory will be necessary to manage them.

### 9.9. *Myths of configuration management operations*

The above section lays to rest several ideas that have plagued the practice for years. One cannot sidestep the complexities of configuration management via a simple choice of language. Nor is the process of defining network state actually necessary in order to achieve appropriate behaviors. But there are other myths that also misdirect practicioners in creating effective and cost-effective configuration management strategies.

Let us consider these:

PRINCIPLE 7. Configuration management is not undecidable.

The conjecture that it is undecidable, proposed in [72], is easily proven false by a simple proof that no finite system can be undecidable, only intractable. The intractability of configuration management was partly refuted in [35], by showing that most configuration operations take a simple form whose completion is statically verifiable.

There has been some controversy about whether it is possible to 'roll back' systems from a new state into an older state. This stems partly from disagreement about what rollback means, i.e. whether it means a complete return to previous behavior. The argument has been that since it is impossible to roll back state using certain tools (including *ISConf* and *Cfengine*) that rollback is impossible in general. We assert that, as long as this refers to the managed state only, this is a limitation of the tools, not a theoretical limit:

PRINCIPLE 8. Rollback of configuration operations is in fact possible.

Any state that is not managed, like runtime or operational data, that affects the operation of a computer will, in general, prevent a computer from behaving identically after rollback.

### 10. Traditions of configuration management

In the above sections, we have discussed many theoretical issues that control the practice of configuration management. We now turn to the practice itself, and describe how practicioners actually accomplish the task of configuration management. Our first step is to study and analyze the traditions from which current practices arose.

System configuration management evolved from a number of distinct traditions in system administration, starting as an ad-hoc set of practices and driven by the overarching need for consistent and reproducible behavior among a population of machines. The current state of the art arises from and is influenced by several distinct historical traditions, including manual management, scripting, declarative specification, and other forms of system management with similar aims. In the following sections, we will explore the historical roots of modern configuration management, tracing several threads of development to their modern counterparts.

These roots of the practice are important even today, because many sites still employ relatively simple forms of configuration management. As well, the design of many currently utilized configuration management tools is based upon the historical practices listed below, rather than the theory explained above. This increases configuration management cost in perhaps subtle ways.

## 11. Scripting

Configuration management arises from the practice of manual system administration in Unix and VMS. The configuration of a machine is found in files within the machine's filesystem. To assure behaviors, one edits these files by hand, or with an interface tool. For particularly complex files, one can use a command-line interface to edit them or write a script to perform the edit. To scale this to a network, one learns to edit files on different machines in exactly the same way.

The first stage in the evolution of modern configuration management was to automate repetitive manual tasks via scripting [7]. Scripts can copy pre-edited 'template' files into place within a particular machine, or utilize inline editing commands such as a stream editor like `sed` to accomplish the same result as editing a file manually. The script can then be used to replace the manual configuration procedure for any machine in the initial state that the script expects.

A script's behavior is thus highly dependent upon assumptions about the machine and the environment in which it executes. An ideal script has no *pre-conditions*: there are no requirements for prior state without which it will fail to work properly. If the script has evolved from human practice, it is both important and perhaps impractical to encode all of the human steps into the script, including sanity checks as to whether the script is applicable to the current situation.

The most severe problem with pre-conditions occurs when the script embodies a 'transaction' consisting of multiple distinct operations that should occur together, for example, installing a particular network service. Running the script should have only one possible outcome: that of performing all of the changes. But a typical script – written without regard for potential system states before the script begins – may fail in the middle of a series of operations, leaving some done and some undone. In the worst case, this leads to security flaws in the system being configured.

Consider a network of five mobile or intermittently available workstations (Figure 10). The availability of each workstation is shown via a Gantt chart, where boxes indicate

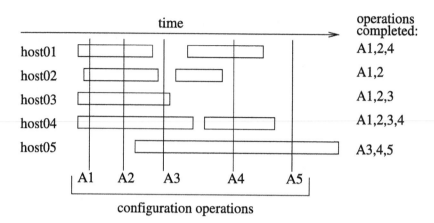

Fig. 10. Intermittent availability of workstations (especially in mobile environments) leads to inconsistency in applying configuration operations over time.

availability. This availability varies, leading to variation in which operations are applied to which workstations. The right-hand column shows the operations actually applied to each. This inconsistency leads to 'script management' mechanisms in which scripts are queued until systems become available.

The most desirable scripts are those that exhibit the following properties [29]:

1. *Idempotence:* applying the script twice has no harmful effect and the results are identical to applying it once.
2. *Minimal intrusiveness:* live services are not disrupted unless absolutely necessary. In particular, files not requiring changes are not touched.
3. *Awareness:* a script is aware of whether its pre-conditions are met, and fails otherwise.
4. *Atomicity:* a script either succeeds completely, or fails without making any changes to a system; the resulting outcome is one of only two system states.

Scripts without appropriate handling of error conditions are relatively harmless when applied once, but create undesirable effects when utilized repeatedly over a longer time period. When a script fails and no corrective action is taken, the target machine enters an unknown and perhaps unpredictable state that, in turn, can cause other scripts to fail. Unpredictable computer states can cause undesired behaviors as a result. Thus, much of configuration management revolves around avoiding unpredictable states for computers.

Since most system administrators who write scripts are not programmers (and do not consider themselves to be programmers), the scripts that mimic manual operations are not typically designed for reuse or maintenance, leading to maintenance problems [27]. This has led to many strategies for reducing the amount of required scripting, including supplying commonly scripted actions via libraries or agents that are not programmed by the administrator [25]. The most comprehensive of these libraries evolved into complete configuration tools such as *Psgconf* [64] and *Arusha* [47].

**11.1.** *File distribution*

The difficulties inherent in writing unconstrained scripts led to several alternative approaches that attempt to accomplish the same end results as scripts, while improving maintainability of the final result. Alternatives to scripting arose opportunistically, automating parts of the operations traditionally accomplished by scripts and leaving other operations to be done by scripts at a later time.

*File distribution* is the practice of configuring systems solely or primarily by copying files from a master repository to the computers being configured. This has its roots in the practice of using file copying commands in scripts. File distribution in its simplest form is accomplished by a script that copies files from a locally mounted template directory into system locations, using a local file copying command. More sophisticated versions of file distribution include use of remote copying commands ('*rcp*', '*scp*', '*rsync*', etc.), which naturally evolved into file distribution subsystems such as '*rdist*' [24], and related tools [26].

*Rdist* is a simple file distribution system that functions by executing on a machine with administrative privilege over managed machines. An input file, usually called `Distfile`, describes files to copy, the target hosts to which to copy them, and the locations for which they are destined. The input file can also direct *rdist* to execute appropriate remote commands on the hosts to be configured, after a file is distributed.

For example, the `Distfile` entries in Figure 11 direct *rdist* to ensure that `/usr/lib/sendmail` and all header files in `/usr/include` are the same for the source host as for the target hosts `host01` and `host02`. Additionally, whenever `/usr/lib/sendmail` has to be copied, *rdist* will run a command to recompile the sendmail ruleset (`/usr/lib/sendmail-bz`).

*Rdist* improved greatly upon the robustness of custom scripts, and exhibits several features that have become standard in modern configuration management tools. A *file repository* contains master copies of configuration files. This is copied to a target host via a *push strategy* in which the master copy of *rdist* (running on an *rdist* master server) invokes a *local agent* to modify the file on each client host. A *class mechanism* allows one to categorize hosts into equivalent classes and target specific file content at classes of hosts. Files are not copied unless the timestamp of the target is older than the timestamp of the source. This minimizes intrusiveness of the distribution process; 'gratuitous' copying that serves no purpose is avoided. Finally, after a file is copied, 'special' *post-copying commands* can be issued, e.g., to email a notification to a system administrator or signal a daemon to re-read its configuration.

```
HOSTS = ( host01 host02 )
FILES = ( /usr/lib/sendmail /usr/include/{*.h} )
${FILES} -> ${HOSTS}
        install ;
        special /usr/lib/sendmail "/usr/lib/sendmail -bz"
```

Fig. 11. A simple Distfile instructs *rdist* to copy some files and execute one remote command if appropriate.

File movement

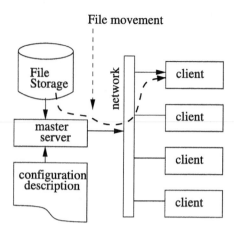

Fig. 12. A push strategy for file distribution stores all configuration information on a master server, along with files to be distributed.

*Rdist* implements what is often called a *server-push strategy* (also sometimes called 'master-slave') for configuration management. A push strategy has several advantages and disadvantages (Figure 12). First, all information about what is to be configured on clients must be contained on the master server in a configuration file. This means that the master server must somehow be informed of all deviations and differences that distinguish hosts on the network. This includes not only soft classes that designate behaviors, but also hard classes that indicate hardware limitations. One advantage of push strategies is that they are relatively easy to set up for small networks.

File distribution of this kind was useful on moderate-size networks but had many drawbacks on networks of large scale or high heterogeneity. First, in a highly heterogeneous network, the file repository can grow exponentially in size as new copies of configuration files must be stored; there is a *combinatorial explosion* as file repository size is an exponential function of heterogeneity. A more subtle limit of *rdist*'s push strategy is that (originally) only one host could be configured at a time; running *rdist* for a large network can take a very long time. Attempts at utilizing parallelism [24] address the problem somewhat, but utilizing push-based file distribution on large networks remains relatively time-consuming.

Server-push configuration management becomes impractical for large and complex networks for other reasons. In a *push* strategy, each host must receive content from a master host that has been configured specifically to contact the target. This requires configuring the master with a *list* of hosts to be contacted. If there are a very large number of hosts, keeping this list up to date can be expensive. For example, every new host to be managed must be 'registered' and data must be kept on its hardware configuration, etc.

File distribution used alone is impractical for large networks in two other ways:
1. Distribution to a large number of clients is a slow process, involving maintenance of client databases and inherently serial transactions.
2. The size of files that must be stored in the repository is proportional to the number of classes, including all combinations of classes actually employed to describe some host.

In other words, there is little economy that comes with scale. This has led to innovations that address each weakness.

## 11.2. *Client pull strategies*

The slowness of server push strategies can be eliminated via *client pull* strategies, in which clients instead contact a server regularly to check for changes. Cfengine has championed this approach [9–11,18]. In a pull strategy, each client has a list of *servers* (Figure 13). This list is smaller and easier to maintain. The clients make decisions about which servers to employ, allowing extremely heterogeneous networks to be managed easily.

Another related strategy developed earlier this year is that of *pull–push*. This is a hybrid strategy in which a centralized server first polls clients for their configurations, and then pushes a configuration to each one. This allows one to customize each client's configuration according to properties obtained by querying the client. This allows a centralized planning process to utilize data about client capabilities and current configuration. Pull–push management is a design feature of the *Puppet* toolset for configuration management [62], and also implemented to some extent in both *BCFG2* [40,41] and *LCFG* [2,4].

## 11.3. *Generating configuration files*

The second problem with file distribution is that the number of files to be stored increases exponentially with the heterogeneity of the network. One way of dealing with the second problem is to copy generic files to each host and then edit files on the client in order to customize them. This is particularly viable if the changes required are minor or can be accomplished by appending data to an existing master copy. This can be accomplished

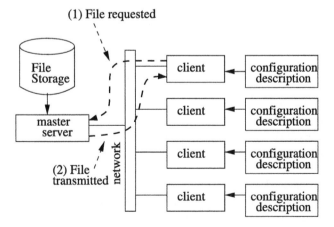

Fig. 13. A pull strategy for file distribution stores configuration information on clients, including the location of one or more master file servers.

via scripting, use of *Cfengine* [9–11,18], or post-copy scripts that are invoked during file distribution.

Another technique for avoiding an exponential explosion of files is to generate all of the configuration files for each host from a database [44,45] rather than storing each one explicitly in a repository. The database contains both requirements and hardware specifications (hard limitations) for each host. One can generate configuration files either on a centralized master distribution host or via an agent that runs on the target host. Generation scripts are easier to understand and maintain than scripts that modify pre-existing files. These rules form the core logic of many current-generation configuration management tools.

A third way to deal with the combinatorial explosion is via templating [58]. This is a compromise between generating client configuration files and editing them. In this strategy, one stores generic templates of configuration files in a file repository with some details left blank, and fills in these details from a master database or file. This strategy takes advantage of the fact that in very large configuration files, such as those for web servers, much of the configuration is fixed and need not be changed for different hosts in the network [68]. By templating the unchanging data, and highlighting changes, one limits the number of files being stored in a repository.

### 11.4. *Script portability management*

There remain many scripts that are not easily expressible in terms of templates, copying or generation. One common example is a script that compiles source files into binary code. The *script management problem* is to control script execution to achieve consistent results.

One form of script management evolved from mechanisms originally invented to help in maintaining portable software. When compiling a program via the *make* [59] program, there are a number of possible program parameters that change from machine to machine, but are constant on any particular machine, such as the full pathnames of specific programs, libraries, and include files. Imake is a program that codes these constant parameter values into a file for reusability, and generates Makefiles for *make* that have the values of these parameters pre-coded. Likewise, xmkmf is a version of Imake that utilizes pre-coded values for the X11 Window system, to allow building graphics applications. The idea of Imake and xmkmf is that *local dependencies can be isolated from portable code in crafting a portable application.*

It is not surprising that these tools for software configuration management have their counterparts in network configuration management. The Process-Informant Killer Tool (*PIKT*) [60] is a managed scripting environment that can be utilized for configuration management. *PIKT* scripts employ special variables to refer to system locations that may vary, e.g., the exact location of configuration files for various system functions. By recording these locations for each architecture or version of operating system, one can utilize one script on a heterogeneous network without modification. PIKT also contains a variety of programming aids, including a declarative mechanism for specifying loops that is remarkably similar to the much later XQUERY mechanism for XML databases.

## 11.5. *Managing script execution*

Scripts have several major drawbacks as configuration management tools: they are sensitive to their execution environments, difficult to maintain, and must be applied consistently to all hosts in a soft class. Particularly, if a sequence of scripts is to be applied to a homogeneous set of hosts, the effect can differ if the scripts are applied in different orders on each host. Consider two scripts, one of which edits a file (S1) and the other of which creates the file to be edited (S2). Unless S2 precedes S1, the result will not be what was intended. Again, one solution to this problem was first invented for use in software configuration management.

*ISConf* [72,73] utilizes a variant of the software compilation tool *make* [59] to manage system and software configuration at the same time. Since many software packages are delivered in source form, and many site security policies require use of source code, use of *make* is already required to compile and install them. The first version of *ISConf* utilizes *make* in a similar fashion, to sequence installation scripts into linear order and manage whether scripts have been executed on a particular host. Later versions of *ISConf* include Perl scripts that accomplish a similar function.

*ISConf* copes with latent effects by keeping records of which scripts have been executed on which hosts, and by insuring that every host is modified by the exact same sequence of scripts. *ISConf*'s input is a set of 'stanzas': a list of installation and configuration scripts that are to be executed in a particular order. Each stanza is an independent script in a language similar to that of the Bourne shell. A class mechanism determines which stanzas to execute on a particular host. *ISConf* uses time stamp files to remember which stanzas have been executed so far on each host. This assures that hosts that have missed a cycle of configuration due to downtime are eventually brought up to date by running the stanzas that were missed (Figure 14).

## 11.6. *Stateful and stateless management*

A configuration management tool is *stateful* if it maintains a concept of state on the local machine other than the configuration files themselves, and conditions its actions based upon that state. Otherwise, we call the tool *stateless*. Because it keeps timestamps of the scripts that have been executed on disk, *ISConf* supports a *stateful* configuration management strategy, as opposed to *Cfengine*, which is (in the absence of user customizations) a mainly *stateless* strategy.[6]

The stateful behavior of *ISConf* solves the problem of script pre-conditions in a very straightforward (but limited) way: each script utilizes the post-conditions of the previous one and assures the pre-conditions of the next, forming a 'pipeline' architecture. The advantage of this architecture is that – provided that one starts using *ISConf* on a cleanly installed host – scripts are (usually) only executed when their pre-conditions are present, so that repeatable results are guaranteed. There are rare exceptions to this, such as when the environment changes drastically due to external influences between invocations of stanzas.

---

[6]Some long term state is memorized in *Cfengine* through machine learning and associated policies, but this is not a part of the configuration operators.

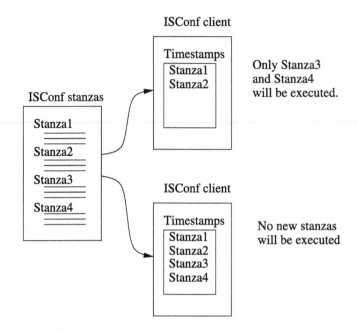

Fig. 14. Basic *ISConf* operation includes a timestamp file that keeps track of which stanzas have been executed.

The strength of *ISConf* is that when changes are few and the environment is relatively homogeneous, it produces repeatable results for each host in the network. This means in particular that if one host is configured with *ISConf* and exhibits correct behavior, it is likely that other hosts with the same hardware configuration will exhibit the same behavior as well. This means that one can test one node of a large homogeneous network and be reasonably sure that the rest of the network nodes in the same class function similarly.

*ISConf* does not cope well with frequent configuration changes. When using *ISConf*, it is impractical to add a stanza in the middle of the ISConf configuration, to delete a pre-existing stanza, or to change stanza order. Appending to the file always works properly, but inserting stanzas in the middle leads to executing stanzas out of order and risking a potentially inconsistent system state (Figure 15). *ISConf never* knows whether its stanzas have been applied to *all* hosts or not. If, e.g., a new host is added to the network after deletion of a stanza, the new host will never execute that stanza although all other hosts will have already executed it. This leads to a permanent divergence in configurations between hosts that have applied the stanza and those who have not.

Modifying the configuration file by any other mechanism than appending stanzas violates the 'pipeline' assumptions stated above. Thus, when there are many changes, the *ISConf* input file can increase linearly in size with each change [48]. Even erroneous stanzas that misconfigure a host cannot be safely deleted; they must instead be undone by later explicit stanzas. Thus the size and complexity of the input file grows in proportion to changes and states, and quickly becomes difficult to understand if changes and/or errors are frequent. In this case, one must start over (from 'bare metal'), re-engineering the entire script sequence from scratch and testing each stanza individually. A similar process of

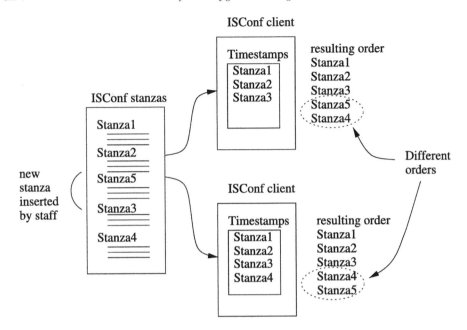

Fig. 15. *ISConf* order violations occur if stanzas are inserted out of order, leading to potential system inconsistencies.

starting over is required if one wishes to apply an *ISConf* input file for one architecture to a different architecture; *ISConf* input files are not always portable, because locations of system files and resources may be hard-coded into the stanzas.

## 12. Package management

Another historical tradition related to configuration management is that of package management. Package management refers to the act of managing software installation in (usually disjoint) modular units, called 'packages'. Depending upon the tool author, package management may or may not be considered as part of configuration management. For example, it is clearly within the scope of *ISConf*, *LCFG* and *Cfengine*, through interfaces to native mechanisms. Historically important package managers include *CICERO* [6], *Depot* [22,56,65,77], Solaris *pkgadd*, RedHat Package Manager (*RPM*) [5], Debian Package Manager, and many others.

The scope of package management as a practice has remained surprisingly unchanged since its inception. A package consists of files and scripts related to an application to be installed. Installing a package consists of running an optional pre-installation script, copying the files from the package into the filesystem, and then running an optional post-installation script. Some current package management systems, including *SUSEConfig*, also support an optional package configuration script that is run for all appropriate packages after any system configuration change. Package management is thus a combination of file copying and scripting strategies.

A key concept in package management is that of package *dependencies*. The function of a particular package often requires installation of others as prior events. This is accomplished by keeping track of both installed packages and the dependencies between packages. To illustrate this, consider the structure of the RedHat package manager (*RPM*). Each package internally declares the attributes that it *provides* and the attributes that it *requires*, as character strings. These are *abstract attributes*; there is no correspondence between requirements and files in the package. The package manager command, with access to data on which package requires which others, is able to pre-load prerequisites so that packages are always loaded in dependency order. Correctness of dependencies is left to the package creator. While experience with packages in major operating systems is good, incorrectness of dependencies is a major problem for contributed software packages maintained by individuals [46].

The intimate relationship between package management and configuration management arises from the fact that most packages have a configuration that must be managed and updated to control the function of the software in the package. Thus it is difficult to separate delivery and updating of the software package itself from the process of configuring that software for proper function. In a way, package management is doomed to be confused with configuration management, forever, because of the overlaps between package and configuration management in both mission and method.

The author is aware of at least a few sites that currently utilize a package management mechanism to manage configuration scripts as well as software packages; each script is encapsulated in a package and each such package depends upon the previous. Then the package manager itself manages the 'installation' of the configuration scripts, and insures that each script is run once and in the appropriate order, exactly as if ISConf were managing the scripts. This is exactly the mechanism by which enterprise management is accomplished in Microsoft environments, using the MSI package format [76].

## 13. Declarative specification

One alternative to the complexity of scripting is to recode script actions as declarative statements about a system. We no longer execute the declarations as a script; we instead interpret these declarations and assure that they are true about the system being configured (through some unspecified mechanism). This avoids the script complexity problems mentioned in the previous sections, and leaves programming to trained programmers rather than system administrators.

### 13.1. *Cfengine*

The most-used tool for interpreting declarative specifications is *Cfengine* [9–11,18]. Unlike rdist, whose input file is interpreted as a script, the input file for *Cfengine* is a set of declarations describing desirable state. These declarations subsume all functions of rdist and add considerably more functions, including editing of files and complete management of common subsystems such as NFS mounts. An autonomous agent (a software tool) running on the host to be configured implements these declarations through fixed-point convergent and 'idempotent' operations that have desirable behavioral properties.

A 'convergent' operation is one that – over time – moves the target system toward a desirable state. For a system in a desirable state, a 'convergent' operation is also 'idempotent'; it does not change anything that conforms to its model of 'system health'. Convergent and idempotent operations support an *immunological model of configuration management* [11, 14], in which repeated applications of an immunizing agent protect a system from harm. This approach is distinct from most others; most configuration management tools rely upon a single action at time of change to accomplish configuration management.

Constructing system operations that are idempotent and/or convergent is a challenging programming task – involving many branches – that is beyond the programming skills of a typical system administrator (Figure 16). A simple linear script becomes a multiple-state, complex process of checking for conditions before assuring them. Thus we can view the declarative input file for *Cfengine* as a form of information hiding; we specify 'what' is to be done while omitting a (relatively complex) concept of 'how' to accomplish that thing. The configuration manager need not deal with the complexity of 'how'. This makes the *Cfengine* input file considerably simpler in structure than a script that would accomplish the same effect. The same strategy makes the input file for the package manager *Slink* [25, 32] considerably less complex in structure than a script that accomplishes the same actions.

Unfortunately, there is a cost implication in use of this strategy. Many packages are provided to the system administrator with imperative scripts that install them, either via the

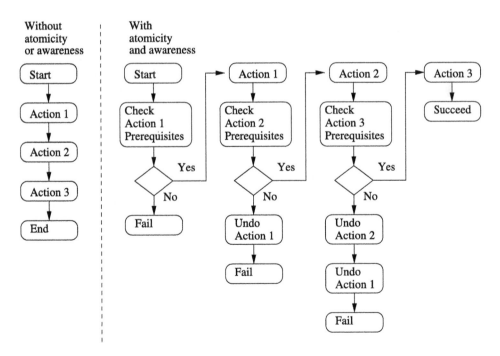

Fig. 16. Adding atomicity and awareness to a script causes greatly increased branch complexity.

*make* compilation control program or via a package manager post-install script written in a shell language. To make the process of configuration 'truly declarative', one must reverse-engineer the impact of these scripts and code their actions instead as declarative statements. This is a non-trivial task, and must be repeated each time a package is updated or patched. For example, many Redhat Package Manager packages contain post-installation scripts that make non-trivial changes to configuration files, that must be taken into account when installing the RPM as part of a configuration change.

*Cfengine* introduces several new strategies to configuration management. Unlike `rdist`, which utilizes a *server-push strategy* (master to slave) (Figure 12), *Cfengine* implements a *client-pull strategy* for file copying (Figure 13). The *Cfengine* agent runs on the host to be configured and can be invoked manually, under control of the *cron* timing daemon, or remotely. Once running, *Cfengine* can perform local edits to files and request copies of configuration files from perhaps multiple servers. Since the action of configuration is controlled from the client being configured, the server bottleneck that plagues `rdist` can be effectively load-balanced (also called 'splaying').

*Cfengine* employs classes to aid in building different kinds of hosts. *Cfengine*'s class variables define classes or groups of hosts differently from the way classes are defined in other configuration tools. In many of *Cfengine*'s predecessors, a class is simply a list of hosts. In *Cfengine*, a class is a *set of conditions upon a host*. This new meaning is more consistent with the meaning of the word 'class' in object-oriented programming than the former meaning. This definition, however, means that a particular instance of *Cfengine* only knows whether the host being configured is a member of each class, and cannot infer the identities of other members of a class from the class description.

Each class variable is a boolean value; either true or false. 'Hard classes' describe hardware and other system invariants, e.g., the operating system. For example, the hard class 'linux' is instantiated if linux is the host operating system. 'Soft classes' describe the desired state for the current configuration. These can be created by constructing simple scripts that probe system state. Each action in *Cfengine* is only applicable to a particular system state, which is expressed as a boolean expression of (both hard and soft) class variables.

A typical Cfengine declaration includes a stanza, guard clause, and action. For example, the stanza in Figure 17, copies the file `cf.server.edu:/repo/etc/fstab` into `/etc/fstab` on every host that is in the hard class `linux` and *not* (!) in the soft class `server`. The stanza identifier is `copy`, which determines what to do. The guard clause is `linux.!server`, which says to perform the following actions for linux machines that are not servers. The next line describes what to do. Note that this is a declarative, not an imperative, specification; the English version of this statement is that `/etc/fstab` *should* be a copy of `cf.server.edu:/repo/etc/fstab`; this is not interpreted as a command, but as a condition or state to assure.

```
copy:
    linux.!server::
        /repo/etc/fstab dest=/etc/fstab server=cf.server.edu
```

Fig. 17. A Cfengine copy stanza includes stanza identifier, guard clause and action clauses.

There is a striking similarity between *Cfengine*'s configuration and logic programming languages [30]. Each *Cfengine* class variable is analogous to a Prolog fact. Each declaration is analogous to a Prolog goal. This makes a link between the meaning of 'declarative' in programming languages and the meaning of 'declarative' in system administration: both mean roughly the same thing.

Another new idea in *Cfengine* is that of 'immunization' of an existing system to keep it healthy [11,14]. Most configuration management systems are executed only when changes are required. By contrast, *Cfengine* can be set to monitor a system for changes and correct configuration files as needed.

*Cfengine* has several other unique properties as a configuration management aid. Software subsystems, such as networking, remotely accessible disk, etc., are manageable through convergent declarations. *Cfengine* introduces the possibility of randomization of configuration actions. An action can be scheduled to run some of the time, or with a certain probability. This enables game-theoretic resource management strategies for limiting filesystem use without quotas [11,66]. Finally, it is also possible to set up *Cfengine* to react to changes in measured performance, so that it makes changes in configuration in accordance with system load and/or deviation from normal load conditions.

*Cfengine* breaks down the boundaries between configuration management, resource management, user management, and monitoring. *Cfengine* directives can clean up user home directories, kill processes, and react to abnormal load conditions. The traditional meaning of configuration management as acting on only system files is gone; all aspects of the filesystem and process space are managed via one mechanism.

### 13.2. *File editing*

One perhaps controversial aspect of *Cfengine* is the ability to edit configuration files to contain new contents. Users of *Cfengine* find this facility indispensable for dealing with the 'combinatorial explosion' that results from file copying. Instead of simply copying a file, one can edit the file in place. Editing operations include the ability to add lines if not present, and to comment out or delete lines of the file in a line-oriented fashion.

What makes file editing controversial is that – although undoubtedly useful, and indeed very widely used – file editing is particularly vulnerable to pre-existing conditions within a file. For example, suppose one is trying to edit a configuration file where duplicate instances are ignored, e.g., /etc/services, and the line for the *ssh* service in the existing file differs slightly from the established norm. Then using the AppendIfNotPresent operation of *Cfengine* to append a line for *ssh* will result in having *two* lines for *ssh* in the file; the prior line will be treated as *different* even though it describes the *same service*. Depending upon how this file is interpreted by the resolver, either the first line or the second line will be used, but not both. In other words, file editing must be utilized with great care to avoid creating latent conditions that can manifest later as behavior problems. One solution to the file-editing quandary is to provide operators that treat files as database relations with local and global consistency constraints [31]. Convergent operators for such 'higher-level languages' are as yet an unsolved problem, which *Cfengine* only approximates.

One effective strategy for dealing with editing pre-conditions is to copy in a 'baseline' version of the file before any edits occur. This assures that edits will be performed on a file with previously known and predictable contents. If this is done, file editing can be as reliable as generating files from databases. Unfortunately, there is a downside to this strategy when used within *Cfengine*; the file being edited goes through two 'gratuitous' rewrites every time the tool is invoked to check the configuration. This can lead to anomalous behavior if a user is trying to use a program that consults the file at the time the file is being updated.

### 13.3. *Change control*

Disciplined use of *Cfengine* requires a different regimen of practice that of *ISConf*, but for the exact same reasons [35]. One cannot ever be sure that *all* machines in a network are in complete compliance with a *Cfengine* configuration, any more than one can be sure that all machines are synchronized with respect to *ISConf*. This lack of confidence that *all* machines have been synchronized leads to somewhat non-intuitive requirements for edits to *Cfengine* configuration files.

The fundamental requirement for distributed uniformity is the *principle of management continuity:* once a file is 'managed' (copied, edited, etc.) by *Cfengine*, it must continue to be managed by *Cfengine* in some fashion, for as long as it remains constrained by policy. It is a normal condition for some hosts to be down, physically absent (e.g., laptops), or just powered off. Thus there is no way that one can be sure that one has 'reverted' a file to an unmanaged (or perhaps one should say 'pre-managed') state. Although not unique to *Cfengine*, this is part of the *Cfengine* ethos of 'living with uncertainty' [12].

### 14. Cloning

Another configuration management strategy arises from the tradition of file copying. Often, in creating a system that should perform identically to another, it is easier to copy *all* system files than to modify just the configuration files. This is called *cloning a system*. Then a manager (or a script, or a tool) can change only the files that have customized contents on each host, e.g., the files representing the identity of the host as a network entity.

Simple forms of cloning include *Norton Ghost*, which simply copies a master version of a system onto a client and allows simple one-time customizations via scripting. After the cloning and customization operations, there is no capability for further control or customization. To make changes, one must change the master device and/or customization script and then reclone each client.

Cloning can be expensive in time and potential downtime. To make a clone of an isolated system, all system information must flow over the network, and this flow generally occurs at a substantively lower rate than when compressed files are copied from disk and uncompressed, depending on network conditions.

Advanced cloning approaches are intimately related to monitoring technologies. Radmind [38] and its predecessors (including Synctree [54]) take an immunological approach to cloning that is based upon the TripWire [49,50] monitoring system. After a reference configuration has been determined, configuration files that are part of this reference are regularly monitored. If their contents deviate from the reference, they are replaced automatically.

Cloning works efficiently when the network to be managed is mostly homogeneous. Each hard class of stations requires at least one clone master. In the presence of high heterogeneity, the number of master systems becomes unmanageable, while the deviations from the master that must be managed become complex and unwieldy. This is similar to the combinatorial explosion that occurs when using file distribution to manage heterogeneous systems.

## 15. Configuration languages

By far the most robust mechanisms for configuration management involve generating the configuration of each host from some intermediate specification. The exact form of this intermediate specification remains, however, a matter of some controversy. Networks are complex entities, and the way we specify their configuration involves many tricks and shortcuts to avoid verbosity and produce a description of configuration that is both readable and maintainable.

### 15.1. *Levels of specification*

Configuration management involves three or more levels of specification or detail. A *policy* is (for the purposes of this chapter) the highest-level specification available for behavior, typically in a natural language such as English. The *configuration* is the lowest-level specification; the sum total of the physical and logical characteristics of each device that (hopefully) produce that behavior. Between these two extremes, there are myriad levels of specification that (for lack of a better term) we will call *intermediate code*: file templates, scripts, declarative specifications, etc. The typical system administrator 'codes' a high-level policy from English into some form of intermediate code, and then invokes an interpreter to realize that code as an actual configuration.

There is some confusion over levels of specification in the literature. For example, it is common for user manuals and papers to fail to distinguish between the high-level policy (that describes intent) and the intermediate specification that describes how to assure that intent. One finds constant reference to intermediate code as if it were the policy itself, and not a derivative work. For example, does *rdist*'s or *Cfengine*'s input file describe a high-level policy? We claim that these do not; they are instead an intermediate description of part of some high-level policy written in natural language.

Confusing actual policy and its representation or realization as intermediate code leads to several difficulties during the routine practice of configuration management. First, the intermediate code describes 'what' to do to assure a configuration, but seldom describes 'why' things are done in specific ways. If we treat the intermediate code as documentation, but do not record the high-level policy upon which it is based, many of the 'why's are lost. For example, intermediate code might describe that "users are not allowed to log into the name server" but does not describe 'why' that policy is a good idea. Worse, the intermediate code does not always clearly distinguish between choices made for policy reasons and those that are arbitrary and not based upon policy at all. For example, one might decide for extremely good reasons how to limit access to one service, and for no reason at all, make the mistake of limiting access to another service in the same way.

It is normal and natural for the best configuration management strategy to change over time as management goals change. Without a conscious step to document the actual goals of management in human-readable form, these transitions can be difficult and expensive, leading to substantive lags between the need for a change in strategy and its implementation, as well as increased cost both from delays in transitioning to a new strategy and from a need to reverse-engineer policy from practice.

### 15.2. *Managing complexity*

Another common theme is how one manages the complexity of a configuration file. This has led to several mechanisms for shortening configuration specifications and improving readability. Some mechanisms for limiting complexity include data inheritance, procedural embedding, and environmental acquisition.

*Data inheritance* (or *instance inheritance*) is a simple shorthand utilized by many configuration management languages to simplify specification of complex configurations. Configuration parameters are often specified for a network of computers as a tree of possibilities. The first few levels of the tree are classes of workstations. For example, we might have a class `workstation`, with a subclass `development`. The idea of data inheritance is simple. Any parameter that is specified for a `workstation` becomes an automatic default for a `development` workstation; overrides for subclasses are specified as needed. In other words, default values are 'inherited' by subclasses from superclasses. This is somewhat the opposite of normal inheritance, in which *constraints* are passed down to subclasses. Data inheritance was introduced in *distr* [26] and is utilized in many configuration management tools, notably Arusha [47].

*Procedural embedding* takes advantage of the similarity between code and data in a dynamically typed language such as python. ARUSHA [47] allows the practicioner to specify code as data; the code will be executed and the results substituted before interpretation. Often a simple procedure can substitute for a long series of specifications, e.g., when configuring a DHCP server.

*Environmental acquisition* is yet a third form of language simplification. In this scheme, environmental parameters are determined via association with peers. For example, suppose one host is a member of a specific subnet and one specifies the service architecture of *one* client on that network. Then the other clients on the same network could infer *their* service architectures 'by association', by looking at the service architectures of their peers. As explained in [55], environmental acquisition can be understood by considering that if we have a 'red car', all of the components of the body are red. Thus, specifying the color of any one body component specifies the color of all components. The downside to environmental acquisition is that mistakes in class membership and conflicts between examples cannot be resolved.

### 15.3. *Common information models*

File copying, generation, or editing are bulky procedures compared to the simple process of setting a configuration parameter. To ease the process of setting parameters on network devices, the Simple Network Management Protocol (SNMP) [57] allows individual parameters to be set by name or number. The Common Information Model (CIM) [42] is an attempt to generalize SNMP-like management to computing systems. The basic configuration of a computing system is encoded as parameters in a management information base (MIB), which can be changed remotely by SNMP. Much of the configuration and behavior of a host can be managed by utilizing this interface; it encompasses most system parameters but fails to address some complexities of software and package management. This approach is implemented in several commercial products for enterprise configuration management.

## 16. Simplifying configuration management

It comes as no surprise that configuration management is difficult and even – in some sense – computationally intractable. One of the major outgrowths of theoretical reasoning about configuration management is an understanding of why configuration management is difficult and steps that one can take to make it a simpler and less costly process [69]. The system administrator survives configuration management by making the problem easier whenever possible. We limit the problem domain by limiting reachable states of the systems being configured. The way in which we do this varies somewhat, but the aim is always simplicity:
   1. By using a particular sequence of operations to construct each host, and not varying the order of the sequence. E.g., script management systems such as *ISConf* enforce this.
   2. By exploiting convergence to express complex changes as simple states.
   3. By exploiting orthogonality to break the configuration problem up into independent parts with uncoupled behavior and smaller state spaces.
   4. By exploiting statelessness and idempotence to prune undesired states from the statespace of an otherwise unconstrained host.
   5. By constructing orthogonal subsystems through use of service separation and virtuality.
   6. By utilizing standards to enforce homogeneity.
Note that most of these are outside the scope of what is traditionally called 'configuration management' and instead are part of what one would call 'infrastructure architecture' [71]. But in these methods, we find some compelling solutions to common configuration management problems.

### 16.1. *Limiting assumable states*

All approaches to configuration attempt to limit the states a system can achieve. Script management enforces a script ordering that limits the resulting state of the system. Convergent management attempts to reduce non-conforming states to conforming ones, hence reducing the size of the statespace. In both cases, the goal is to make undesirable states unreachable or unachievable.

A not-so-obvious way of limiting assumable states is to promptly retire legacy hardware so that it does not remain in service concurrently with newer hardware. This limits states by limiting the overall heterogeneity of the network, which is a help in reducing the cost of management and troubleshooting.

## 16.2. *Exploiting orthogonality with closures*

Another simple way configuration management is made tractable is by viewing the system to be configured as a composition of orthogonal and independent subsystems. If we can arrange things so that subsystem configurations are independent, we can forget about potential conflicts between systems.

A closure [31] is an independent subsystem of an otherwise complex and interdependent system. It is a domain of 'semantic predictability'; its configuration describes precisely what it will do and is free of couplings or side effects. Identifying and utilizing closures is the key to reducing the complexity of managing systems; the configuration of the closure can be treated as independent of that of other subsystems. Systems composed of closures are inherently easier to configure properly than those containing complex interdependencies between components.

A closure is not so much created, as much as it is discovered and documented. Closures already exist in every network. For example, an independent DHCP server represents a closure; its configuration is independent of that of any other network element. Likewise, appliances such as network fileservers form kinds of closures. Building new high-level closures, however, is quite complex [68], and requires a bottom-up architecture of independent subsystems to work well.

## 16.3. *Service isolation*

One way of achieving orthogonality is through service isolation. *Service isolation* refers to the practice of running each network service on a dedicated host that supports no other services. There are several advantages to service isolation:

1. Each service runs in a completely private and customized environment.
2. There is no chance of application dependencies interfering with execution, because there is only one application and set of dependencies.

Thus configuration management is greatly simplified, compared to configuring a host that must provide multiple services. Thus service isolation simplifies configuration by avoiding potential dependency conflicts, but costs more than running a single server.

**16.4.** *Virtualization*

One can reduce the costs of service isolation by use of *virtualization*. *Virtualization* is the process of allowing one physical host to emulate many virtual hosts with independent configurations. There are several forms of virtualization. One may virtualize the function of a whole operating system environment to form an 'information appliance' [67] that functions independently from all other servers and is free of service conflicts. One may instead virtualize some subpart of the execution environment. For example, the PDS system [1] virtualizes the loading of dynamic libraries by applications, *without* virtualizing the operating system itself. The result is that each program runs in a custom operating environment in which it obtains access only to the library versions that it desires, and no version conflicts are possible *in one application's execution environment.*

**16.5.** *Standardization of environment*

A final practice with some effect upon configuration management is that of standardization. The Linux Standard Base (LSB) [53] attempts to standardize the *locations* of common configuration files, as well as the link order and contents of system libraries. Applications expecting this standard are more likely to run on systems conforming to the standard.

The standard consists of three parts:
1. A specification of the contents of specific files and libraries.
2. An *environment validator* that checks files for compliance.
3. A *binary application validator* that checks that applications load proper libraries and make system calls appropriately.

LSB is based upon a logic of 'transitive validation' (Figure 18). If an environment $E1$ passes the environment validator, and an application $A$ passes the binary validator, and $A$ functions correctly in $E1$, and another environment $E2$ passes the environment validator, then $A$ should function correctly in $E2$. This means that if LSB functions as designing, testing an application once suffices to guarantee its function in all validated environments.

Alas, LSB falls short of perfection for a few theoretical reasons. First, there is no way of absolutely being sure that binary code conforms to system call conventions (especially when subroutine arguments are computed rather than literal). This is equivalent to the intractable problem of assuring program correctness in general. Thus there is no way to address this problem.

A second issue with LSB is that the environment validator only validates the 'standard' environment, and does not take into account changes that might be made at runtime, e.g., adding a dynamic library to the library path. If this is done, the environment validation may become invalid without the knowledge of the LSB validator.

However, the important point to take from this is that *standards reduce system complexity, and thus reduce potential conflicts and other issues in configuration management.* Thus standardization is not to be ignored as a major cost-saving factor in configuration management.

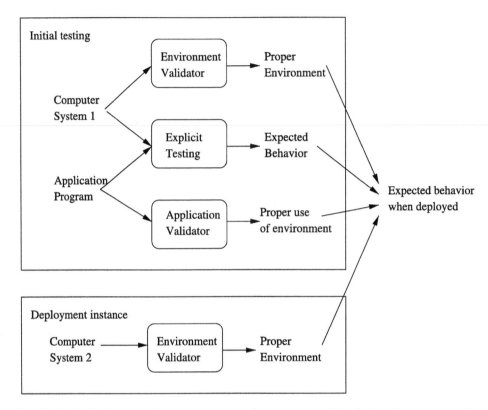

Fig. 18. Standardization and validation lead to claims of correct program behavior in environments in which explicit testing has not occurred.

## 17. Critique

In this chapter, we have outlined many different techniques, practices, and methods for solving the configuration management problem. In these practices, there emerge some central trends and patterns from which one can characterize the field.

First, the goal of current configuration management strategies is not simply to provide appropriate network behavior, but to minimize the cost of doing so. There are several factors that can minimize costs, including:

1. Limiting hardware heterogeneity:
   (a) By promptly retiring legacy systems.
   (b) By purchasing large numbers of identical computers.
2. Limiting gratuitous complexity in configurations:
   (a) By negotiating and implementing highly homogeneous policies for use of computers.
   (b) By opposing policies that contribute to higher cost without added value.
3. Limiting the possible configuration states of each computer:
   (a) By always starting from a blank disk and replaying the same scripts in order.

   (b)  By generating consistent configuration files from databases.
   (c)  By use of idempotent and stateless operations to reduce potential states of each
        target machine.
4. Splitting statespaces into orthogonal subspaces:
   (a)  By identifying behaviorally independent subsystems and closures.
   (b)  By synthesizing independence through service independence and virtualization.
5. Pre-validation of results of deployment:
   (a)  By validating results for each target platform.
   (b)  By combining environment and application standards and validators to exploit
        transitive validation properties of software.

This list of strategies is a fairly good summary of the state of the art. Note that in the list
of strategies, the choice of tool for configuration management is not even mentioned. It is
instead a side-effect of adopting best practices, which can be summed up in one imperative:
*minimize complexity*.

   We have thus come full circle. We started by discussing tools and how they work, as
part of a configuration management strategy. We are now at the level where the tools are
no longer important, and are replaced by the goals that they accomplish. These are goals
of practice, not tool choices, and many of them require human interaction and decision
making to reduce cost and maximize value. The true art of configuration management
is in making the strategic decisions that define mission in a way that reduces cost. For
example, one administrator recently computed a savings of roughly $100,000 by replacing
all desktop computers in a hospital with linux thin clients. This savings had nothing to
do at all with the tools used to manage the network. The complexity of the network itself
was reduced, leading to a reduction in cost *by design* rather than *by practice*, i.e. when one
manages the whole enterprise as a single community of cooperating humans and computers
[13].

## 18. Current challenges

We have seen above that the practice of configuration management arises from two traditions: high-level policy definition and low-level scripting. At the present time, the two
traditions continue as relatively separate practices with little overlap. A majority of the
world uses some form of scripting, including convergent scripting in *Cfengine*, to accomplish configuration management. Very few sites have adopted overarching tools that generate configuration from high-level policies, and in a recent informal poll of configuration
management users, most users of high-level ('generative') tools wrote their own instead of
employing an existing tool.

   In our view, the principal challenge of configuration management is that the cost of management is only very indirectly related to its value [36]. The cost of high-level strategies
seems higher, and their value cannot be quantified. Meanwhile, the average practicioner
has limited ability to justify approaches that cost significant startup time but have unquan-
tifiable benefits. For example, it takes significant effort to re-engineer software packages
so that their installation is declarative rather than imperative. Thus, one main challenge of

configuration management is to quantify, in real money, the cost and benefit of using each strategy, and compare that with the up-front cost of employing it.

A second challenge is that of transition planning. It can be very costly to switch configuration management strategies from one to another. One principal factor in this cost is that of undocumented practices. The goal of configuration management is to avoid software bugs; a bug that is never expressed as behavior is not important.

A third challenge is to determine how configuration management may interoperate well with other management goals. The boundaries between configuration management, package management, user management, resource management, and monitoring are not sharp. Thus, it is often necessary to employ strategies for these with overlapping domains of control, leading to several potential problems. Package management and configuration management commonly conflict and many attempts have been made to combine these activities. Tensions between package and configuration management lead to strategies that are package-centric (viewing the system configuration as similar to another package) [72,73] and configuration-centric (viewing the package management system as part of configuration) [63]. There is no clear best solution to the integration problem at this time.

A topic for future research is that of safety during the configuration management process. When changes are being made, systems being configured often assume states that – for a time – are unsafe, in the sense that there is heightened vulnerability to hacking and security compromises. Current tools do not even possess the ability to analyze safety constraints or to assure that systems never enter unsafe states. Convergent operators show some promise, because each operator can assure safety, leading to safety of the overall result.

Finally, the integration of monitoring and configuration management promises to solve a pressing problem that no current configuration management strategy addresses. Current strategies *verify* configuration but do not *validate* it. Integration of monitoring and configuration management promises to solve this problem in the near future; currently tools such as *BCFG2* do limited monitoring to inform the SA of what machines might need attention.

## 19. Conclusion

The art of system and network configuration has come a long way from its humble roots in scripting of manual system administration, but there is still a long path remaining, to construct management strategies that integrate seamlessly with other management tasks. In the future, it is doubtful that the term 'configuration management' will retain its current meaning. As 'system management' comes to encompass more integrated approaches to managing and maintaining system components, the specific subtask of managing configurations is likely to be absorbed into the greater task of managing human–computer communities [11]. This will have the positive side-effect, however, of integrating such currently disparate activities as validation and verification, as mentioned above. The move toward holistic management has already begun and we look forward to a bold future in which configuration takes second place to community behavior in meeting the computing needs of the future.

# References

[1] B. Alpern, J. Auerbach, V. Bala, T. Frauenhofer, T. Mummert and M. Pigott, *PDS: A virtual execution environment for software deployment*, Proc. Virtual Execution Environments 2005, ACM Press (2005).

[2] P. Anderson, *Towards a high-level machine configuration system*, Proc. LISA-VIII, USENIX Association (1994).

[3] P. Anderson, *System configuration*, SAGE short topics in system administration, USENIX Association (2006).

[4] P. Anderson, P. Goldsack and J. Patterson, *SmartFrog meets LCFG: Autonomous reconfiguration with central policy control*, Proc. LISA-XVII, USENIX Association, San Diego, CA (2003).

[5] E. Bailey, *Maximum RPM*, Red Hat Press (1997).

[6] D. Bianco, T. Priest and D. Cordner, *Cicero: A package installation system for an integrated computing environment*, http://ice-www.larc.nasa.gov/ICE/doc/Cicero/cicero.html.

[7] D.N. Blank-Edelman, *Perl for System Administration*, O'Reilly, Inc. (2000).

[8] F.R. Brooks, *The Mythical Man-Month*, Addison-Wesley, Inc. (1982).

[9] M. Burgess, *A site configuration engine*, Computing Systems **8** (1995).

[10] M. Burgess, *Computer immunology*, Proc. LISA-XII, USENIX Association (1998).

[11] M. Burgess, *Theoretical system administration*, Proc. LISA-XIV, USENIX Association, New Orleans, LA (2000).

[12] M. Burgess, *On the theory of system administration*, Science of Computer Programming **49** (2003), 1.

[13] M. Burgess, *Analytical Network and System Administration: Managing Human–Computer Systems*, Wiley (2004).

[14] M. Burgess, *Configurable immunity for evolving human–computer systems*, Science of Computer Programming **51** (2004), 197–213.

[15] M. Burgess, *An approach to understanding policy based on autonomy and voluntary cooperation*, Lecture Notes on Comput. Sci., Vol. 3775 (2005), 97–108.

[16] M. Burgess and K. Begnum, *Voluntary cooperation in pervasive computing services*, Proc. LISA-XIX, USENIX Association (2005).

[17] M. Burgess and A. Couch, *Modelling next generation configuration management tools*, Proc. LISA-XX, USENIX Association (2006).

[18] M. Burgess and R. Ralston, *Distributed resource administration using Cfengine*, Software: Practice and Experience **27** (1997).

[19] M. Burgess and T. Reitan, *A risk analysis of disk backup or repository maintenance*, Science of Computer Programming **64** (3) (2006), 312–331.

[20] K. Caputo, *CMM Implementation Guide: Choreographing Software Process Improvement*, Addison–Wesley–Longman, Inc. (1998).

[21] Carnegie Mellon Software Engineering Institute, *The Capability Maturity Model: Guidelines for Improving the Software Process*, Addison–Wesley–Longman, Inc. (1995).

[22] W. Colyer and W. Wong, *Depot: A tool for managing software environments*, Proc. LISA-VI, USENIX Association (1992).

[23] L. Cons and P. Poznanski, *Pan: A high-level configuration language*, Proc. LISA-XVI, USENIX Association, Philadelphia, PA (2002).

[24] M. Cooper, *Overhauling Rdist for the '90's*, Proc. LISA-VI, USENIX Association (1992).

[25] A. Couch, *SLINK: Simple, effective filesystem maintenance abstractions for community-based administration*, Proc. LISA-X, USENIX Association (1996).

[26] A. Couch, *Chaos out of order: A simple, scalable file distribution facility for 'intentionally heterogeneous' networks*, Proc. LISA-XI, USENIX Association (1997).

[27] A. Couch, *An expectant chat about script maturity*, Proc. LISA-XIV, USENIX Association (2000).

[28] A. Couch and P. Anderson, *What is this thing called system configuration*, an invited talk at LISA-2004, USENIX Association (2004).

[29] A. Couch and N. Daniels, *The maelstrom: Network service debugging via 'ineffective procedures'*, Proc. LISA-XV, USENIX Association (2001).

[30] A. Couch and M. Gilfix, *It's elementary, dear Watson: Applying logic programming to convergent system management processes*, Proc. LISA-XIII, USENIX Association (1999).

[31] A. Couch, J. Hart, E. Idhaw and D. Kallas, *Seeking closure in an open world: A behavioral agent approach to configuration management*, Proc. LISA-XVII, USENIX Association, San Diego, CA (2003).

[32] A. Couch and G. Owen, *Managing large software repositories with SLINK*, Proc. SANS-95 (1995).

[33] A. Couch and Y. Sun, *Global impact analysis of dynamic library dependencies*, Proc. LISA-XV, USENIX Association, San Diego, CA (2001).

[34] A. Couch and Y. Sun, *On the algebraic structure of convergence*, Proc. DSOM'03, Elsevier, Heidelberg, DE (2003).

[35] A. Couch and Y. Sun, *On observed reproducibility in network configuration management*, Science of Computer Programming, Special Issue on Network and System Administration, Elsevier, Inc. (2004).

[36] A. Couch, N. Wu and H. Susanto, *Toward a cost model for system administration*, Proc. LISA-XVIII, USENIX Association, San Diego, CA (2005).

[37] Cowan et al., *Timing the application of security patches for optimal uptime*, Proc. LISA 2002, USENIX Association (2002).

[38] W. Craig and P.M. McNeal, *Radmind: The integration of filesystem integrity checking with filesystem management*, Proc. LISA-XVII, USENIX Association (2003), 1–6.

[39] I. Crnkovic and M. Larsson, eds, *Building Reliable Component-Based Software Systems*, Artech House (2002).

[40] N. Desai, R. Bradshaw, R. Evard and A. Lusk, *Bcfg: A configuration management tool for heterogeneous environments*, Proceedings of the 5th IEEE International Conference on Cluster Computing (CLUSTER03), IEEE Computer Society (2003), 500–503.

[41] N. Desai et al., *A case study in configuration management tool deployment*, Proc. LISA-XVII, USENIX Association (2005), 39–46.

[42] Distributed Management Task Force, Inc., *Common information model (CIM) standards*, http://www.dmtf.org/standards/cim/.

[43] R. Evard, *An analysis of UNIX machine configuration*, Proc. LISA-XI, USENIX Association (1997).

[44] J. Finke, *An improved approach for generating configuration files from a database*, Proc. LISA-XIV, USENIX Association (2000).

[45] J. Finke, *Generating configuration files: The director's cut*, Proc. LISA-XVII, USENIX Association, San Diego, CA (2003).

[46] J. Hart and J. D'Amelia, *An analysis of RPM validation drift*, Proc. LISA-XVI, USENIX Association, Philadelphia, PA (2002).

[47] M. Holgate and W. Partain, *The Arusha project: A framework for collaborative Unix system administration*, Proc. LISA-XV, USENIX Association, San Diego, CA (2001).

[48] L. Kanies, *Isconf: Theory, practice, and beyond*, Proc. LISA-XVII, USENIX Association, San Diego, CA (2003).

[49] G. Kim and E. Spafford, *Monitoring file system integrity on UNIX platforms*, InfoSecurity News **4** (1993).

[50] G. Kim and E. Spafford, *Experiences with TripWire: Using integrity Checkers for Intrusion Detection*, Proc. System Administration, Networking, and Security-III, USENIX Association (1994).

[51] C. Kubicki, *The system administration maturity model – SAMM*, Proc. LISA-VII, USENIX Association (1993).

[52] A. Leon, *Software Configuration Management Handbook*, 2nd edn, Artech House Publishers (2004).

[53] The Linux Standard Base Project, *The linux standard base*, http://www.linuxbase.org.

[54] J. Lockard and J. Larke, *Synctree for single point installation, upgrades, and OS patches*, Proc. LISA-XII, USENIX Association (1998).

[55] M. Logan, M. Felleisen and D. Blank-Edelman, *Environmental acquisition in network management*, Proc. LISA-XVI, USENIX Association, Philadelphia, PA (2002).

[56] K. Manheimer, B. Warsaw, S. Clark and W. Rowe, *The depot: A framework for sharing software installation across organizational and UNIX platform boundaries*, Proc. LISA-IV, USENIX Association (1990).

[57] D.R. Mauro and K.J. Schmidt, *Essential SNMP*, 2nd edn, O'Reilly and Associates (2005).

[58] T. Oetiker, *Templatetree II: The post-installation setup tool*, Proc. LISA-XV, USENIX Association, San Diego, CA (2001).

[59] A. Oram and S. Talbot, *Managing Projects with Make*, 2nd edn, O'Reilly and Associates (1991).

[60] R. Osterlund, *PIKT: problem informant/killer tool*, Proc. LISA-XIV, USENIX Association (2000).

[61] D. Patterson, *A simple model of the cost of downtime*, Proc. LISA 2002, USENIX Association (2002).

[62]  Reductive Labs, Inc., *Puppet*, http://reductivelabs.com/projects/puppet/.

[63]  D. Ressman and J. Valds, *Use of cfengine for automated, multi-platform software and patch distribution*, Proc. LISA-XIV, New Orleans, LA (2000).

[64]  M.D. Roth, *Preventing wheel reinvention: The psgconf system configuration framework*, Proc. LISA-XVII, USENIX Association, San Diego, CA (2003).

[65]  J.P. Rouillard and R.B. Martin, *Depot-lite: A mechanism for managing software*, Proc. LISA-VIII, USENIX Association (1994).

[66]  F.E. Sandnes, *Scheduling partially ordered events in a randomised framework – empirical results and implications for automatic configuration management*, Proc. LISA-XV, USENIX Association, San Diego, CA (2001).

[67]  C. Sapuntzakis, D. Brumley, R. Chandra, N. Zeldovich, J. Chow, J. Norris, M.S. Lam and M. Rosenblum, *Virtual appliances for deploying and maintaining software*, Proc. LISA-XVII, USENIX Association, San Diego, CA (2003).

[68]  S. Schwartzberg and A. Couch, *Experience in implementing an HTTP service closure*, Proc. LISA-XVIII, USENIX Association, San Diego, CA (2004).

[69]  Y. Sun, *The Complexity of System Configuration Management*, Ph.D. Thesis, Tufts University (2006).

[70]  W.F. Tichy, *RCS – A system for version control*, Software – Practice and Experience **15** (1985), 637–654.

[71]  S. Traugott, *Infrastructures.Org Website*, http://www.infrastructures.org.

[72]  S. Traugott and L. Brown, *Why order matters: Turing equivalence in automated systems administration*, Proc. LISA-XVI, USENIX Association, Philadelphia, PA (2002).

[73]  S. Traugott and J. Huddleston, *Bootstrapping an infrastructure*, Proc LISA-XII, USENIX Association, Boston, MA (1998).

[74]  Y.-M. Wang, C. Verbowski, J. Dunagan, Y. Chen, C. Yuan, H.J. Wang and Z. Zhang, *STRIDER: A black-box, state-based approach to change and configuration management and support*, Proc. LISA-XVII, USENIX Association, San Diego, CA (2003).

[75]  J. Watkins, *Testing IT: An Off-the-Shelf Software Testing Process*, Cambridge University Press (2001).

[76]  P. Wilson, *The Definitive Guide to Windows Installer*, Apress, Inc. (2004).

[77]  W.C. Wong, *Local disk depot – customizing the software environment*, Proc. LISA-VII, USENIX Association (1993).

# 2. The Technology

## 2.1. *Commentary*

Network and System Administration is often perceived as a technology-centric enterprise. One thing is clear: dealing with a vast array of frequently changing technologies is a key part of a system administrator's lot. In this part the authors describe a handful of technologies, from the old to the new.

Unix has played a major role in developing system administration technologies and has bridged the gap between large systems and personal computing successfully, owing to its openness and great flexibility. However, mainframe business solutions developed by IBM in the 1960s are still very much in use today, albeit by only a few, and these offer benefits such as low-level redundancy and reliability at a high price. Stav sketches the relationship between the mainframes and better-known Unix. More well-known operating systems such as Windows and Unix have been omitted as they are dealt with in a vast array of training manuals and textbooks. VMS, AS400 and mainframe OS390, now z/OS are still used in some spheres of industry, particularly in the banking world.

Dealing with device diversity has always been one of the important challenges of system administration. Many devices had only the most rudimentary operating systems. In the beginning it was thought that special protocols (e.g., SNMP) could be designed to apply external intelligence to dumb devices; today it is common to find Unix-like images as embedded operating systems on many devices, thus allowing great flexibility of management.

Diversity is the price one pays for vibrant commercial development and this leads to other technologies to manage the lower level technologies, resulting in a many-layered approach. For some, the answer has been to attempt standardization, through organizations such as POSIX, IETF, and the DMTF etc. For others, the diversity is an evolutionary approach that leads to more rapid benefits and can be sewn together by meta-technologies. As Alvin Toffler wrote in his book Future Shock: *"As technology becomes more sophisticated, the cost of introducing variations declines."* [1].

E-mail is one of the most arcane and Byzantine systems on the Internet today. It is typical of many technologies that were designed in isolation as playful academic exercises during the early days of Unix and the network and which took root in social consciousness. At the other end of the spectrum, we see Web services entering as a form of ad hoc "standardization" today. Using XML as a protocol design tool, and the Hypertext Transfer Protocol as the transport layer, everything that was once old becomes new again in Web services.

## Reference

[1]  A. Toffler, *Future Shock*, Random House (1970).

# – 2.2 –

# Unix and z/OS

Kristoffer Stav

*IBM Corporation*

## 1. Introduction

This chapter describes differences between the Unix and IBM mainframe environments. In spite of rumors of their death, mainframes continue to be used in applications where superior Input/Output (I/O) is essential. Mainframes play a central role in many parts of our society, like i.a. in finance, insurance, airline booking, on-line services and government services. These machines are just not visible to the end users.

During the past ten years, Unix has grown to take on tasks that have traditionally been the domain of consolidated mainframe computers. The cultures of Unix and Windows dominate information technology and mainframe systems programmers have become isolated with concepts that belong to the mainframe environments. This chapter can be thought of as a mapping between different terminologies, concepts and abbreviations, and a high-level overview of the major differences between two operating system environments: Unix and the IBM z/OS.

A mainframe computer is a monolithic, consolidated computer. The name 'mainframe' originates from the fact that all of the computer components are mounted in a single frame rack. Today several different architectures are called 'mainframes'. They are often intensive computers that are expected to perform high availability tasks, with wide band input/output (I/O) and are generally used for batch and transaction processing. They are, however, general purpose computers. Mainframes as fast I/O machines should be distinguished from 'supercomputers', which are generally understood to be fast at high performing CPU processing. This chapter considers a popular mainframe platform and contrasts it with the currently more well-known Unix system.

HANDBOOK OF NETWORK AND SYSTEM ADMINISTRATION
Edited by Jan Bergstra and Mark Burgess

IBM mainframe servers are currently branded as 'System z', where z illustrates 'zero downtime'. The hardware architecture is called z/Architecture. Multiple operating systems can run on the z/Architecture, including z/OS, z/VM, TPF and GNU/Linux. The most used operating system on System z machines is z/OS. z/OS is often called MVS, which z/OS is derived from. Unix in this article is a term used collectively for the family of Unix-like operating systems, one of which is GNU/Linux. It should be kept in mind that z/OS also has a fully fledged Unix environment included.


## 2. Some history

A brief summary of the history is given to set the context of the two alternatives.


### 2.1. *The architectural evolution of System z*

In the spring of 1964, IBM introduced System/360 [5], page 27, [1], page 4, [7], page 1-1. This line of machines was the first general purpose computers in the world. The System/360 had a 24-bit architecture with one central processor and real storage addressing. The 24-bit processor was able to address 16 MBytes of memory. The name '360' referred to all directions of the compass, and symbolized universal applicability. The first major architecture upgrade was in 1970, the System/370. This family of processors was still able to address only 16 MBytes of memory, and this was becoming a limitation for the systems in the late sixties. This led to the development of virtual storage. By 1972 the line of System/370 computers got a new set of registers, the control registers. These registers were primarily added to support virtual storage addressing. This gave the running programs a virtual address space of 16 Mbytes each. This was a breakthrough in the history of computers. Another important feature with System/370 was the support for two central processors.

The mainframe continued to grow. The size of Nucleus (the operating system kernel) and its components increased and maximum 16 MBytes in each address space became a limitation in the late 1970s. In the 1980s, Extended Architecture (XA) with 31-bit addressing became the standard. Now, the central processors were able to address 2 GBytes of storage. With the arrival of 370/XA, IBM had designed a new and more sophisticated I/O architecture, called the channel subsystem architecture [2], page 379. In 1983, the first four-way Symmetrical Multiprocessor (SMP) mainframe was introduced.

The next step on the evolution ladder was the Enterprise Systems Architecture/390 (ESA/390) in 1990. The history repeated itself: the need of more addressable memory was an issue once more. The storage technology was improved by expanding the set of registers, and thus gave the ability to add an extra memory system called Expanded Storage. In 1994 Parallel Sysplex was introduced, and made it possible to cluster the IBM mainframes in a very advanced manner.

With z/Architecture and 64-bit addressing in 2000, the processors are capable of addressing 16 Exabytes. The architectures have always been upward compatible (called backward compatible in the Unix world). The current mainframe central processors are able to execute 24-bit and 31-bit code without the need of recompiling. There are still many organizations that still have old applications running, in which they have made considerable

Table 1

| Decade | Architecture | OS | Addressing |
|--------|--------------|----|-----------:|
| 1964 | S/360 | MVT | 24-bit |
| 1970s | S/370 | MVS/370 | 24-bit |
| 1980s | S/370-XA | MVS/XA | 31-bit |
| 1990s | S/390-ESA | MVS/ESA /OS/390 | 31-bit |
| 2000 | z/Arch. | z/OS | 64-bit |

investments and thus the backward compatibility is essential to protect these investments. The 'traditional' applications are able to coexist with 64-bit code in the z/OS environment.

There has been a major architectural expansion every decade. Table 1 gives a simplified overview [4], page 4.

z/OS is a 64-bit operating system introduced in March 2001, and runs on the System z family of computers. z/OS is based on OS/390, which again is based on the strengths of Multiple Virtual Storage (MVS) from the mid-1960s. z/OS is even today occasionally referred to as MVS by system programmers and in literature. z/OS has a fully fledged Unix environment integrated into the operating system, called Unix System Services (USS), which is POSIX compliant, has HFSes, common Unix utilities and tools, and a Unix shell. The next subsection gives a short historical presentation of Unix.

### 2.2. *Unix*

The development of Unix started in the early 1970s at AT&T Bell Labs on a DEC PDP-7, initially by Ken Thompson and Dennis Ritchie [3]. It was designed to be a portable operating system. The C programming language was developed in order to rewrite most of the system at high level, and this contributed greatly to its portability. Unix was made available cheaply for universities, and throughout the 1970s and 1980s it became popular in academic environments. In the mid-1980s, AT&T started to see a commercial value in Unix. During the 80s, Unix branched out in two variants because the commercial version, System V, did not include the source code any more. The other variant, called Berkeley Software Distribution continued to be developed by researchers at Berkeley. There exists a myriad of Unix dialects today, e.g. HP-UX, AIX, Solaris, *BSD, GNU/Linux and IRIX.

### 3. User interfaces

z/OS generally requires less user intervention than Unix systems, although there is currently an interest in 'autonomic' computing which aims to make Unix and Windows operating systems less human-dependent. The z/OS is highly automated, and a lot of routines exist for handling unexpected situations without operator help. In fact, almost 50 percent of the operating system code belongs to recovery and monitoring routines. The traditional user interface to z/OS was the 3270-terminal; these were dedicated computers attached to the mainframe.

These days the 3270-terminals are replaced by 3270 emulators running on PCs, like IBM Personal Communications (for Windows) or x3270 (for most Unix operating systems, including GNU/Linux). Users can interact with z/OS through the 3270 interface, using Time Share Option (TSO), or Interactive Systems Productivity Facility (ISPF). TSO is a command line environment for executing programs, editing, printing, managing data and monitoring the system. TSO is rarely used any more, and largely replaced by the more user friendly and many driven ISPF.

In Unix, users have traditionally interacted with the systems through a command line interpreter called a shell. This is still the most powerful interface to many system services, though today Graphical User Interfaces (GUI) are growing in popularity for many tasks. A shell is command driven, and the original shell was the Bourne shell 'sh'. The shell is still the primary user interface for Unix administration, but most Unix systems offers a GUI, which makes the interaction with the operating system more user friendly for novice users. This is why many regards Unix systems as more user friendly than z/OS.

## 4. Files and data-sets

The differences in system data organization deserve an explanation. For a person who is used to the Unix-way of organizing data, the way z/OS does it may seem a little strange. In the Unix-world, file data are stored in a sequential byte oriented manner. z/OS organizes data in data-sets, which are record oriented. A record is simply a chunk of bytes, that can either be of a fixed or variable size. A Unix file does not have a common internal structure, from an operating systems point of view. The file is just a sequence of bytes. The internal structure is organized by the application or the user.

Data stored in a data-set can be organized in several ways. The most common are sequential data-sets, where the records are organized in a sequential manner. Partitioned data-set (PDS) and Partition data-set extended (PDSE) contains a set of sequential members. Each of these sequential members is accessed through a directory in the data-set. A PDS is also called a library, and used among other things to store parameters, source programs and job control language (JCL) statements. JCL will be explained later.

Virtual Storage Access Method (VSAM) is yet an access method, used to organize records in a more flexible way than the PDS and sequential data-sets. Examples of VSAM data-sets are Key Sequence data-sets (KSDS), Entry Sequence data-sets (ESDS), Relative Record data-sets (RRDS) and Linear data-sets (LDS). VSAMs are used by applications, not by users. VSAM data-sets cannot be edited manually through an ISPF editor.

The way data-sets are organized also differs from how files are organized in Unix. The data-set structure in z/OS is not hierarchical, like Unix systems are, hence data-sets are not stored in directories. Information of the data-sets is stored in catalogues, such as data-set name and on which volume the data-set is stored. A system has one master catalogue, and multiple user catalogues to organize the data-sets. Data-set names must be unique, since no hierarchical structure (path) separates one data-set from another.

Naming conventions are also different. Names of data-sets must be in upper case, with maximum number of characters 44. A data-set name is divided into qualifiers, separated by periods. Each qualifier can contain maximum eight characters. The first qualifier is called the high-level qualifier or alias, and indicates which catalogue the data-set resides in. A member in a PDS is given in parenthesis. An example of a PDS data-set is SYS1.PARMLIB(IEASYS00) with a member called IEASYS00. PARMLIB is the most important library for z/OS and contains system control parameters and settings. SYS1.PARMLIB can be compared to Unix's etc-directory.

Here follows an example of a source file to illustrate the differences. In Unix, the file is denoted by a file path and a file name, like this:

- /home/bob/project/source.c.

In z/OS the source code can be stored in a data-set like this:

- BOB.PROJECT.SOURCE.C.

This data-set is stored in a catalogue called BOB.

File security is managed in two completely different ways in Unix and z/OS. In Unix systems, security attributes are tightly attached and stored with the file on the file system. The default Unix file security scheme is robust and simple. In z/OS file security is managed by a centralized system component, a security server. This authorized component controls authorization and authentication for all resources in the system, including data-sets. All access to these resources must go through this facility. The security server offers more security control than any Unix system does, and has more fine grained security settings than the Unix approach.

## 5. I/O operations

The way I/O operations are managed differs a lot between z/OS and Unix systems. The z/Architecture is tailor-built for high-throughput and performs I/O operations in a very efficient way [6]. I/O management is offloaded to dedicated processors, called System Assist Processors (SAP) and other specialized hardware and software components, while general processors can concentrate on user related work in parallel. Big level-two caches, fast data buses and many I/O interconnects make sure that a large number of I/O operations can be handled simultaneously in a controlled and efficient manner.

## 6. Task- and resource management

The dispatcher and queue manager in z/OS is fairly sophisticated compared to most Unix systems, and executes different workloads in parallel. Average processor utilization in z/OS during normal operations is typically 80–100 percent, in contrast to maximum 20 percent on a Unix system.

Job control language (JCL) [4], page 128, is a language to control and execute jobs in z/OS. It is used to control execution order of programs, prepare and allocate resources, and define input and output data-sets. This is an area where the mainframe differs a lot from the Unix variants. z/OS 'pre-allocate' resources before programs are executed. This applies for disk usage (data-sets), processor power and memory. In contrast, Unix systems allocate resources dynamically during runtime. This makes z/OS very predictable, and it is possible to set specific goals for different classes of work. The dispatcher will, together with

Workload Manager (WLM), Job Entry Subsystem (JES) and Intelligent Resource Director (IRD) do whatever necessary dynamically to meet these requirements. Setting these kinds of goals is not possible in Unix. Because of the advanced queue and resource management in z/OS, it is capable of running many different types of work in parallel in a controlled and efficient way. In contrast, Unix systems tend to host a single or a few services per OS instance, hence the difference in the resource utilization. While the mainframe can host multiple databases by a single z/OS image, in the Unix world, more servers are needed, generally one server per database or application. In the next section we will see what kind of work z/OS and Unix typically do.

## 7. Platform applicability and typical workloads

Mainframes have traditionally been used mainly for hosting big databases, transaction handling and for batch processing. The current line of System z mainframes has generally become much more function rich and versatile compared to its predecessors. It is capable of hosting almost any kind of services running under z/OS. Many mainframe installations today run traditional CICS and IMS transactions concurrently with modern enterprise applications using J2EE technology and/or service oriented architecture (SOA) concepts. These machines have become more and more universal, and are still being the highest ranking alternative for security, scalability and availability.

z/OS is typically used for:
- Handling large amounts of users simultaneously
- Batch processing
- Online transaction processing
- Hosting (multiple) bigger databases
- Enterprise application servers

Multiple operating systems can run in parallel on the same System z machine, with a virtualization technology built into the hardware, which divides the hardware resources into logical partitions (LPARs). Each LPAR is assigned a dedicated portion of the real storage (memory), and processors and I/O channels can be shared dynamically across the LPARs or be dedicated to the LPARs.

Unix systems are becoming more stable and secure. Unix systems are capable of handling many users in parallel, but it is typically not capable of handling the same amounts as mainframe systems, particularly if the requests require database access.

Unix systems are typically used as:
- Application servers
- Hosting single databases
- Powerful workstations
- Internet and infrastructure services like mail, web servers, DNS and firewalls
- Processing intensive workloads

Please note that the typical usage stated above are a very simplified view of both z/OS and Unix systems. It is important to understand that the platforms are not limited to these characteristics.

Unix systems may outperform the mainframe on compute intensive work, like weather forecasting, solving complex mathematical equations, etc., where raw processor power is the single most important factor. These machines are often referred to as 'supercomputers'. Linux is a popular operating system in such environments because of its lightweight kernel, and flexible and modular design. The fact that GNU/Linux is open source software makes it possible to customize the kernel for almost whatever purpose, thus it is very popular in academic and research environments. Linux is widely used in commercial environments as well. The fact that Linux is cost effective, stable and portable makes it popular as a server operating system, also on the System z servers. Read more about Linux on the mainframe in the following section.

## 8. Linux on the mainframe

In today's market, Linux is the fastest growing server operating system, and one of Linux's big strengths is platform interdependency. It runs on all kinds of processors, from small embedded computers to big mainframe installations. Linux can run on the System z servers natively in an LPAR or under z/VM. z/VM is a mainframe operating system that virtualizes the z/Architecture. It provides a very sophisticated virtualization technology that has evolved for more than 35 years.

Linux makes the System z hardware platform an alternative to small distributed servers, and combines the strengths of Linux with the high availability and scalability characteristics of the z/Architecture hardware. Linux complement z/OS in supporting diverse workload on the same physical box; it gives more choices for the z platform.

Some of the benefits of running Linux on System z are better hardware utilization and infrastructure simplification. A System z server is capable of running hundreds of Linux images simultaneously. z/VM makes it possible to set up virtual LANs and virtual network switches between the guest operating systems, and data transfer across these virtual network interconnects is as fast as moving data from one place in memory to another. This kind of IP connectivity has a major performance gain compared to data transportation over a physical wire, and this is especially beneficial in infrastructures where the applications are hosted on the same engine as the databases, which very often applies.

## 9. The future of the mainframe

The history of the mainframe goes back over 40 years, more than any other computer platform. The current line of System z servers has evolved from the System 360 in the mid sixties. The mainframes dominated the IT industry during the 1970s and the 1980s. In the late 1980s and the 1990s cheap microprocessors-computers and distributed computing became increasingly popular. In fact, some people predicted the mainframe's death, when they saw more and more Unix systems in IT infrastructures throughout the 1990s. But this prediction failed; the mainframe is still a strong platform and the reasons are evident.

IBM invests considerable resources in future development and innovation in the System z, both on hardware-, os- and middleware level. In addition to still improve mainframe characteristics like availability, security, scalability and performance, IBM sees the System z server as an ideal hub for service oriented architecture (SOA) applications and the platform for future software development and deployment. IBM's middleware portfolio for developing and enabling a service oriented infrastructure is big, and will grow and increase in the coming years.

Another strong future asset of the mainframe is server virtualization for infrastructure simplification. Consolidating hundreds of servers on a single mainframe simplifies management, reduces power consumption and floor space, and resource utilization is improved considerably.

In order to keep the z platform strong in the coming decades, recruiting is necessary. Educating young people, becoming IT specialists with mainframe skills and expertise, in addition to foster a mainframe community is needed for the future growth of the platform. IBM has realized this, and is committed to educating students on a worldwide basis in cooperation with universities and colleges around the world.

A current and future characteristic of System z servers is to maintain and enhance the granularity and scaling potential, to make the platform applicable for small businesses as well as big enterprises, both in terms of fitting capacity and pricing.

## 10. Typical platforms characteristics

A very brief listing summarizes the typical Unix and mainframe characteristics.

### 10.1. *Unix*

- POSIX idea: keep it small and simple.
- Distributed approach, portability over efficiency.
- Platform independence.
- Openness.
- Treats devices as files.

### 10.2. *Mainframe*

- High I/O bandwidth for transaction handling, batch processing and managing large amounts of data and users concurrently.
- Availability, redundancy on all levels.
- Security.
- Integrity.
- Centralized approach.
- Very resilient operating system. Approx. 50 percent of operating system code is for failure detection and recovery automation.
- Block representation of data.

## 11.  Key terms

Table 2 compares some terminology between the two alternatives with comments [1], page 122.

Table 2

| Mainframe | Unix | Comments |
|---|---|---|
| DASD (direct access storage device) | Hard disk | z/OS does not see a physical disk or partitions, but a set of virtualized mainframe-standard disks (typically 3390 DASDs) regardless of physical disk device |
| Data set | File | See Section 4 |
| Main storage/real storage/ central memory/storage | Memory/RAM | Non-persistent memory |
| Task | Process | Running instance of a program |
| Subtask | Thread | Threading is widely used in Unix systems, but not that much used in traditional mainframe applications |
| Task switching | Context switching | |
| Started task (often a z/OS subsystem) | Daemon | Long running unit of work |
| SYSIN, SYSOUT, SYSUT1, SYSUT2 | stdin, stdout | Source for input data and destination for output data |
| TASKLIB, STEPLIB, JOBLIB, LPALST, and linklist | In user's PATH | Search for programs |
| Supervisor/maint | Root/administrator/superuser | Priveliged user |
| Nucleus | Kernel | The operating system 'core' |
| Upward compatible | Backward compatible | You will often find 'upward compatible' in the mainframe literature, which is exactly the same as 'backward compatible' in the Unix literature |
| Auxiliary storage | Secondary storage | Disk systems, tape drives, printers, etc. |
| IPL (initial program load) | Boot | Bring up and initialize the system |
| Footprint, CPC (Central Processor Complex), CPU, Engine | 'Box' or 'chassis' | A physical machine |

Table 2
Continued

| Mainframe | Unix | Comments |
|---|---|---|
| EBCDIC | ASCII | Native text encoding standard used |
| Processing unit (PU) | Central Processing Unit | |
| SYS1.PARMLIB (including others) | /etc/ | Default repository for system wide configurations |
| Assembler, PL/S | Assembler, C | Default kernel development language |
| Record oriented (Blocks of data) | Byte oriented (Streams of data) | Traditional data format on file system level [1, p. 130] |
| Control blocks stored in each address space, called the 'common area' and consist of System Queue Area (SQA), Link Pack Area (LPA) and Common Area (CSA) | /proc/ (represented as virtual files) | Control and information repository for the kernel, where important kernel/ nucleus structures are held during runtime |
| Supervisor call (SVC) | System call | User processes invoking operating system services, which often run in privileged mode |
| SDSF | ps, kill | Program management |
| ISPF editor | emacs, ed, vi, sed … | Text editors |

## References

[1]  M. Ebbers et al., *Introduction to the New Mainframe: z/OS Basics*, IBM (2006).
[2]  E. Irv, *The Architecture of Computer Hardware and System Software, An Information Technology Approach*, Wiley (2000).
[3]  D.M. Ritchie and K. Thompson, *The Unix time-sharing system*, Communications of the ACM **17** (7) (1974).
[4]  P. Rogers et al., *Abcs of z/os system programming*, IBM (2003).
[5]  K. Stav, *Clustering of servers, a comparison of the different platform implementations*, University of Oslo (2005).
[6]  D.J. Stigliani et al., *IBM eserver z900 I/O Subsystem*, 40, IBM (2002).
[7]  *z/Architecture Principles of Operation*, (5), IBM (2005).

# – 2.3 –

# Email

## C.P.J. Koymans[1], J. Scheerder[2]

[1]*Informatics Institute, University of Amsterdam, Kruislaan 403, 1098 SJ, Amsterdam, The Netherlands*
*E-mail: ckoymans@science.uva.nl*
[2]*E-mail: js@xs4all.nl*

> *It is surprisingly hard to write an electronic mail system without messing up.*
> *Wietse Venema*

## 1. Introduction

This chapter of the handbook deals with electronic mail, mostly in the familiar form commonly found on the Internet. Email is discussed in terms of its architecture, the protocols that constitute it, related technology, and the surrounding problems of both technological and non-technological nature. Where appropriate other forms of electronic mail are addressed.

## 2. Mail history

As one of the oldest and most valuable applications of computers and internetworking, the now 40 years old electronic mail has a rich history which will be described in the following subsections.

---

HANDBOOK OF NETWORK AND SYSTEM ADMINISTRATION
Edited by Jan Bergstra and Mark Burgess

## 2.1. *Pre-Internet mail*

Contrary to popular belief, email[1] did not arrive on the scene with the Internet. Email pre-dates the Internet, and even the ARPANET. The first email system, a messaging system for multiple users on a single time-sharing mainframe system, MITs Compatible Time-Sharing System, was deployed in 1965. Email message exchange between distinct computers by means of a computer network started in 1971. RFC 196 [47], by Dick Watson, dated July 1971, already described a (probably never implemented) email system, while Ray Tomlinson sent the first actual network email message in late 1971. The vehicle for these early email exchanges was ARPANET where messages piggybacked on the already available File Transfer Protocol (FTP). Email was not yet a first-class-citizen, but it was a very powerful mechanism for communication.

Quoting Ray Tomlinson [43] on the infancy of network email:

> The early uses were not terribly different from the current uses: The exceptions are that there was only plain text in the messages and there was no spam.

## 2.2. *Internet mail's predecessors: UUCP, BITNET and DECnet*

The early 1970s saw the birth and procreation of Unix. As an asynchronous way of loosely interconnecting otherwise unconnected computers, the Unix to Unix Copy Protocol (UUCP) was invented in 1978 at Bell Labs. UUCP provides data exchange between computers on the basis of ad-hoc connectivity, such as direct dial-up connections between computer systems.

Based upon UUCP, store-and-forward protocols for message exchange and message routing were designed and globally implemented. This included not only (or even primarily) email messaging: UUCPs first-class citizen, one could say, was in fact not email, but Usenet News. This is the email and news as we know them today, although both have long since transitioned to data transfer using Internet transmission protocols rather than UUCP.

Similar to UUCP, the 'Because It's Time Network' (BITNET) interconnected IBM mainframes in academia, and similar store-and-forward message exchange services were started on BITNET in 1981.

Unlike UUCP, and much more like today's Internet, the then prevalent computer company Digital Equipment Corporation offered inter-computer network connectivity for its own computers by the name of DECnet, which started in 1975. DECnet included full message exchange and message routing facilities.

## 2.3. *The start of Internet mail*

In the early 1980s, the Internet mail infrastructure was created. Most important here were the efforts made by Postel and colleagues in creating SMTP, the Simple Mail Transfer

---

[1]We follow Knuth in writing email instead of e-mail [26].

Protocol, within the IETF. In 1982, while the ARPANET was still transitioning from the old NCP protocol to TCP/IP, the first widespread implementation of mail over SMTP was implemented by Eric Allman. Sendmail, the followup of delivermail, was shipped with 4.1c BSD Unix, which promoted email to a first-class citizen on the then emerging TCP/IP-based Internet.

At that time, several message exchange networks existed already. In order to cope with all these different networks and addressing systems, an elaborate system for rewriting was introduced into sendmail. This powerful rewrite engine is still present in modern versions of sendmail, although there is a tendency to replace it by table-driven lookups in modern MTAs (see Section 3.3.2) like postfix, qmail and the upcoming sendmail X.

A *gateway* between two messaging systems is a system that can accept and send messages for both messaging systems, and can direct messages between them, manipulating them as needed. A main Internet mail relay in these early days may very well have handled Internet mail, UUCP mail, BITNET mail and DECnet mail, gatewaying between all those. Due to differences in several details, like addressing, this involved a lot of rewriting magic, which sendmail handles particularly well.

From now on, when we mention email (or Internet mail), we mean SMTP-based mail running over TCP/IP.

## 3. Mail architecture

In this section we will discuss the main architectural components for an email infrastructure. These components are addresses, protocols, agents and message formats.

### 3.1. *Mail addresses*

During the development of mail, several addressing schemes have been used. UUCP used full path addresses like the famous 'bang path' `mcvax!moskvax!kremvax!chernenko` used by Piet Beertema in his 1984 Usenet hoax, archived by Google [2]. DECnet addresses took the form *host::user*, where host was supposed to be unique throughout DECnet. Several other addressing schemes existed for local or private networks.

The oldest mail address format was the one used in ARPANET, which was later also used in BITNET, being *user@host*. Again, this presupposes a flat namespace, where every host throughout ARPANET had a unique name. In the beginning that was reasonable, but with the explosive growth of the Internet no longer maintainable. Hence a new, hierarchical scheme was desperately needed.

**3.1.1.** *Mail hierarchy and DNS*    The solution to the problem of the large namespace came with the invention of DNS, the Domain Name System, in 1983. The domain name system offers a distributed database, where administration of parts of the hierarchically organized namespace can be delegated to autonomous servers, which have authority over the delegated subtree. The labels used for the different layers of the hierarchy are separated by dots. Each layer only needs to check that the labels used for its own sublayers are locally unique. Email addresses now take the form *user@host.some.domain*, where an arbitrary number of layers is allowed after the @-sign.

Another extension was added to DNS in 1986: MX records. Traditionally, mail has always used a store-and-forward mechanism to route and deliver mail. There is no need for a direct connection between the sending machine and the machine where the mail should finally be delivered. MX records provide an abstraction layer for email addresses. For instance mail to *user@some.domain* could be instructed towards the next hop by way of an MX record, for example pointing to relay.some.domain. This relaying machine could accept and store the mail in order to forward it on to its, possible final, destination.

It is even possible to forward to non-Internet destinations like machines in the UUCP network. This way users on machines that are not connected to the Internet can have Internet-style mail addresses. This option should not be confused with the use of pseudo domains in addresses like *user@host.UUCP*. The UUCP top level domain does not exist in DNS, and only lives inside rewrite rules of sendmail or similar programs.

It is an established 'best practice' to make use of DNSs possibilities to create a hierarchical layer whenever this is possible within an organisation. System engineers should arguably also try to mimic this same hierarchy when the email infrastructure is set up, even in the case of a centralised administration. This results in more flexibility, scalability and the option to reorganise the infrastructure more easily or treat certain groups in special ways.

### 3.2. *Mail protocols*

Now that addressing has been taken care of, we can start looking at the protocols used for mail transport and pickup. As already mentioned, email became a first-class citizen only after the specification of SMTP in August 1982 in RFC 821 [37]. This protocol is used for mail transport. Later on protocols like POP[2] and IMAP[3] were introduced to facilitate a network based interaction with mail readers.

### 3.2.1. *(Extended) Simple Mail Transfer Protocol*
SMTP was indeed a very simple protocol intended for mail transport. It lacked some of the more advanced features of the X.400 message service. In the original specification only a few commands were defined.

The most important commands are

- The Hello command (HELO) is used to introduce the sending SMTP client to the receiving SMTP server. In the response to this command the server introduces itself to the client.

---

[2]Mostly version 3. We will write POP instead of POP3.
[3]Mostly version 4rev1. We will write IMAP instead of IMAP4rev1.

- The MAIL command (MAIL FROM:) is used to start a mail transaction by specifying the envelope sender in the form of a valid return address, the so-called reverse-path. The special case 'MAIL FROM: <>' is used in error messages and other situations where a return address should not be used.
- One or more RCPT commands (RCPT TO:) follow the MAIL command in order to specify the envelope recipient(s) of the mail message. These addresses are used to further route the message through the email system until it reaches the final destination and can be delivered.
- The DATA command introduces the opportunity to send the actual message. The end of the data is signalled by using a single dot ('.') by itself on a line of text. The mail message itself also has some structure which is outlined in Section 3.4.1.
- The QUIT command is used for graceful termination of the SMTP session.

In a later stage several extensions to the SMTP protocol have been defined, leading to Extended SMTP (ESMTP), or SMTP Service Extensions [24] as it is officially called. One of the extensions is an extended greeting (EHLO), which is used to give the server the opportunity to specify new ESMTP options in its response.

One of the options that can be specified in this way is '8BITMIME', which allows a client to send 8-bit data in the DATA section of the SMTP protocol, which has to be preserved by the server. This is a deviation from the classical restriction that the DATA section may contain only 7-bit US ASCII.

The SMTP protocol, except for the DATA section, specifies what is called the envelope of the mail message. It is only used for routing (see Section 4) and error processing (see Section 4.7) and is not part of the message proper. The DATA section contains the mail message itself, which is built up from mail headers (see Section 3.4.1) and the mail body itself. The body of mail messages is essentially free-format, although there are some do's and don'ts, see Section 8.

The most recent specification of (E)SMTP can be found in RFC2821 [23].

**3.2.2.** *POP and IMAP*   Traditionally mailboxes were accessed by a user using a mail reader that opened the user's mailbox as a local file. Later mechanisms were introduced to be able to access a mailbox stored on a server using a mail reader that accesses it over the network. Several versions of the 'Post Office Protocol (POP)' [4,35,40] have been defined. It is mostly used to transfer complete messages from a server to a client to be read and reacted upon online or offline. Most of the time the message is deleted afterwards on the server. This mechanism has an obvious drawback once people started to access their mail from different clients all over the net. This way there was not a single consistent view of a user's mailboxes. This is where IMAP (Internet Message Access Protocol) comes to the rescue. The most recent specification is 'Internet Message Access Protocol – version 4rev1'[7]. The idea here is that the complete mail store resides and stays on the server and that clients access it in order to present it to the reader. Modifications, such as message status updates, can be applied to the messages stored on the server, and clients can use caching strategies for performance optimisation and for enabling off-line access.

### 3.3. *Mail agents (Table 1)*

Internet mail infrastructure makes use of several software components in order to implement mail handling, which are being referred to as 'mail agents'.[4] Traditionally people make a distinction between mail user agents (MUA), mail transport agents (MTA) and mail delivery agents (MDA). For our discussion it is useful to extend these with mail submission agents (MSA), mail access agents (MAA) and mail retrieval agents (MRA) as in Figure 1.

**3.3.1.** *Mail User Agents*   The task of the mail user agent is to present an interface to the user for reading, composing and replying to email. In order to read mail, a MUA can either access mail on a local or remote store via the filesystem or contact a mail access agent (see Section 3.3.5) in order to transfer mail using the POP or IMAP protocol. After composing or replying to email, a MUA injects or submits the constructed message by way of a message submission agent (see Section 3.3.3) into the mail infrastructure. Either the MUA talks SMTP directly to the MSA or calls a helper program to do the work.

Examples of MUAs are Outlook, Thunderbird and Eudora (GUI-based) or mutt and pine (terminal-based). See also Section 5.2.

Table 1
Mail agents

| Acronym | Meaning |
|---------|---------|
| MUA | Mail User Agent |
| MTA | Mail Transport Agent |
| MDA | Mail Delivery Agent |
| MSA | Mail Submission Agent |
| MAA | Mail Access Agent |
| MRA | Mail Retrieval Agent |

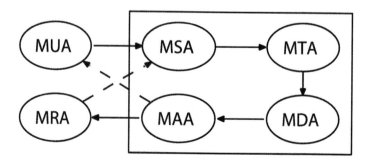

Fig. 1. Mail agents.

---

[4]Sometimes these components are called 'message agents' and the letter 'M' in abbreviations like MUA, MTA and so on might be read as 'message' instead of 'mail'.

**3.3.2.** *Mail Transport Agents*   The heart of the email infrastructure is formed by the mail transport agents,[5] that take care of mail routing (see Section 4) and mail transport across the Internet. MTAs talk SMTP to each other to accomplish this task. MTAs need to understand email addressing formats, might do gatewaying to non-Internet mail systems, must implement a queueing system for mail in transit, handle bounces and other error conditions and take care of forwarding, aliases and special delivery to programs or mailing lists. It is important that MTAs implement some security mechanism and/or access control in order to thwart abuse, see Sections 6 and 8.1. At the ingress MTAs talk to MUAs (or sometimes MSAs) and at the egress they deliver mail to MDAs.

Examples of MTAs are Sendmail, Postfix, qmail and Exim. More on these mail servers can be found in Section 5.1.

**3.3.3.** *Mail Submission Agents*   Not all mail user agents are able to inject email into the system in a completely satisfactory way. For instance, domains used in the SMTP-envelope should be fully qualified domain names. And also the syntax of header fields inside the message from the SMTP-DATA transfer might need some extra attention. This is where the mail submission agent plays its role, as an intermediate step between MUA and MTA. One often sees that this MSA-functionality is built into the mail transfer agent itself.

More on the specific tasks of mail submission agents can be found in RFC 2476 [16].

**3.3.4.** *Mail Delivery Agents*   When an email has reached its final destination, it should be put in the user's incoming mailbox, often called 'INBOX'. Traditionally this was a plain file in mbox format (see Section 3.4), but mailbox storage can also be implemented differently, e.g., as mail directories or mail databases. It is up to the mail delivery agent to perform the task of storing incoming mail in a permanent place for subsequent access by the user. An MDA can operate as simple as the old `/bin/mail`-program that just appends incoming mail to the user's INBOX or as sophisticated as the `procmail`-program that exercises all kinds of operations like user filtering, header manipulation or forwarding on the incoming mail before final delivery.

LMTP, as defined in RFC 2033 [32], is a protocol designed for the purpose of passing messages by a mail delivery system to the final delivery agent. This final delivery agent may run on the local host, but it can also run on a remote host. LMTP can be viewed as a simplification of the SMTP protocol, with the queueing mechanism taken out. Using it provides a little extra functionality, since ESMTP extensions can be catered for. Additionally, LMTP can be used to abstract from actual mail delivery in such a way that mail acceptance and mail delivery become many-to-many relations instead of one-to-one relations, thus offering virtually boundless scalability. It is supported by the main MTAs mentioned before.

**3.3.5.** *Mail Access Agents*   After a mail message has been sent by a user, transported by MTAs and delivered by an MDA, it waits for the intended recipient to access his email. Most of the time the recipient's user agent does not have direct access to the mail store and contacts a mail access agent to interact with the message store on its behalf. The

---

[5]Sometimes 'transport agents' are referred to as 'transfer agents'.

MAA talks with the user agent using an access protocol like POP or IMAP, as discussed in Section 3.2.2.

Examples of MAAs are Courier, Cyrus, UW IMAP, Dovecot and Qpopper.

**3.3.6.** *Mail Retrieval Agents*    Sometimes it is not a user agent that directly (or indirectly via an MAA) checks for new incoming mail, but an automated agent that works on behalf of the user, called a mail retrieval agent. The MRA accesses the mail store without explicit user intervention, using an MAA when needed. After local processing it might reinject the mail into the mail system for further routing and processing. It is often used to access secondary mailboxes users might have on different servers in order to integrate them with the primary mailbox.

As can be seen in Fig. 1, as illustrated with the dashed lines, this might lead to several iterations of the mail cycle, hopefully not creating non-terminating loops. Typical examples of MRAs are fetchmail and getmail.

### 3.4. *Mail message formats*

A companion document to RFC 2821 on SMTP is RFC 2822 [39], titled 'Internet Message Format'.[6] In this specification the format of the actual message itself, which is transferred during the SMTP DATA phase, is defined. The first part of a message consists of a number of mail header lines, followed by an empty line, followed by the message body. Lines consist of US-ASCII (1–127) characters and are terminated by the combination CRLF of a carriage return followed by a line feed, as is prescribed customary in text-oriented Internet protocols. Lines are allowed to contain up to 998 characters, but it is good practice to limit lines to 78 characters in order for the mail to be readible on 80 character wide terminal displays.

**3.4.1.** *Mail headers*    Each header line consists of a field name (made out of printable ASCII characters (33–126)[7], a colon(':'), a space character ('␣'), and the field body, in this order. To keep header lines readable they may be 'folded' over separate physical lines by preceding existing whitespace in the field body with a linebreak, effectively creating continuation lines that start with whitespace. The body of a message is essentially free-format, although the MIME specification introduces new subformats.

Mail headers contain important meta-information for the content of the message. The headers contain information like the date of mail composition (*Date:*), originator fields like author (*From:*) and actual sender (*Sender:*), destination address fields like recipient (*To:*) and carbon copy recipients (*Cc:*), identification fields like a unique message identifier (*Message-ID:*) and referencing information (*In-Reply-To:*, *References:*), informational fields like subject (*Subject:*), trace fields which contain parts of the SMTP-envelope like intermediate MTAs (*Received:*) and return envelope-sender address (*Return-Path:*), but also a whole range of (optional) fields added by user agent or filtering software.

---

[6]This title is a little bit more compact than the title of its predecessor, RFC 822, being 'Standard for the format of ARPA Internet text messages', but it still does not explicitly mention Mail.

[7]With the exception of 58, which represents a colon.

**3.4.2.** *MIME*   In a series of five RFCs [12–15,29] 'Multipurpose Internet Mail Extensions' were defined in order to alleviate the problems with the SMTP specification allowing only 7-bit ASCII inside headers and bodies of emails, while there was growing demand for support of international character sets and multimedia content transfer within email. Internationalisation could have been solved for the body part by implementing 8BITMIME, but this does not work for the header part or for really arbitrary binary content (including NUL, CR and LF).

MIME introduces a special '*Content-Type:*' header to be able to specify the MIME-type and MIME-subtype of following mail content. Because of the applications of MIME technology outside of email these are also called Internet Media (sub)types. Examples of type/subtype are text/plain and image/jpeg.

In order to work within the limits of SMTP, MIME adds a content transfer encoding header (*Content-Transfer-Encoding:*) to the message. Two useful encodings for transferring arbitrary binary data are 'quoted printable' and 'base64'. Base64 is best suited for arbitrary binary data, where each 3 bytes are represented by 4 bytes in the printable ASCII range. Quoted printable is best suited for (international) texts with an occasional high ASCII byte, which is represented by a three character sequence '=XY', with XY the hexadecimal representation of the wanted byte. MIME introduces a special 'encoded-word' syntax for use in header fields, which looks like `=?ISO-8859-1?Q?Keld_J=F8rn_Simonsen?=`, where the byte '=F8' should be displayed within the ISO-8859-1 character set. The 'Q' tells us that quoted printable was used for the encoding.

In MIME-based email it is necessary to specify a character set[8] for type/subtype 'text/plain' via the 'charset'-parameter. The default character set is US-ASCII, but sometimes mail user agents do not specify a character set even though non-US-ASCII characters occur, or even specify a wrong character set. Furthermore web-based MUAs, also known as webmail services, often have trouble determining which character set to use.

**3.4.3.** *Mailbox storage format*   The format of the message itself, as it is transported over SMTP, differs in principle from the way mail messages are stored in permanent or temporary storage. Implementations are free to choose their own formats. Mail can be put in flat files, possibly with local end-of-line conventions and special separator lines like 'From␣', like in the classic example of the 'mbox' format. Mail can also be put in directories with one file per mail message, like in MH folders or the more modern and robust Maildir as used by the qmail mailer. Mail can also be stored in a database system, like the Cyrus software does.

## 3.5. *Terminology*

It is a good idea to wrap up this section on Mail Architecture with the most important notions that have been defined in relation with mail transport and delivery. For extended information and terminology, see RFC 2821 [23].

---

[8]This is a misnomer, because in fact a character set encoding is specified. This is different from the character set itself (or the content transfer encoding for that matter) [48].

A *Mail Object* is a complete mail message consisting of a 'Mail Envelope' and 'Mail Content'. It contains all information necessary to route a message to its final destination and to handle erroneous conditions.

The *Mail Envelope* or 'SMTP Envelope' consists of the content of the SMTP protocol units that transmit the originator address, one or more recipient addresses and possibly protocol extensions.

The *Mail Content* or 'Mail Data', or even 'SMTP Data', is the part of the Mail Object sent within the SMTP DATA protocol unit. Mail Content itself is again made up out of headers and a body. Headers and the body are separated by a single empty line.

The *Mail Header* contains meta-information related to the message, as explained in Section 3.4.1.

The *Mail Body* contains the message proper and is structured according to a user's own insight, except where the body is part of a MIME message.

## 4. Mail routing

One of the important tasks of MTAs is to route a mail message to the correct destination. Usually mail routing requires a few hops, where mail takes several hops towards a central mail relay inside the network of the sender, then takes a giant leap towards the mail relay inside the network of the receiver, where it finally trickles down to the machine that is supposed to handle final mail delivery.

### 4.1. *Mail address resolution*

In order to route a mail message only the domain part (after the @-sign) of the envelope recipient mail address is relevant. The domain part is looked up in DNS to find out whether an MX record is defined for it. The mailhosts returned are tried, in order of their priority, to route the mail to the next hop. If no MX record is found, mail should be delivered to the IP address associated with the domain part by an A record. Although this is the basic idea for mail delivery, MTAs use rulesets or table lookups to overrule this method with local policies.

### 4.2. *Mail aliases and forwarding*

When an email reaches its final destination, as far as the domain part is concerned, local delivery takes place by looking at the mailbox part before the @-sign. Before the mail is handed over to a delivery agent, the MTA finds out whether an alias for the local user or mailbox is defined in an alias database. This may again induce further processing and possibly offsite delivery of the email. Besides the alias mechanism there is also a per user option to forward mail to another address. Some care has to be taken that these mechanisms do not lead to aliasing or forwarding loops. Local loops are easily broken, but loops with more than one MTA involved can be quite nasty. See also Section 4.9.

Another important purpose of the alias mechanism is the ability to deliver email to a program instead of storing it in the mail store. For security reasons it is forbidden to mail directly to programs, but the administrator of a mail domain can enable this possibility indirectly by using an alias, for example to support mailing lists, see Section 4.3.

## 4.3. *Mailing lists*

Mailing lists are used to expand one email address into a lot of email addresses in order to reach a community of users interested in the same topic. It is possible to implement this behaviour by using the alias mechanism just described. In order to have more flexibility, to be able to rewrite or extend the mail message, to be able to archive messages and so on, an implementation often uses the alias file to deliver the email to a program that takes the responsibility of expanding the mailing list address into the list of subscribers to the list while also performing additional administrative tasks. With very large mailing lists, a subscriber is often itself the address of a mailing list exploder. An important property of properly configured mailing lists is that it replaces the original envelope sender address with a predetermined envelope sender address of the mailing list owner, this in contrast with what happens during normal exploding or forwarding. The idea here is that error messages are sent to the mailing list owner in stead of to the mail originator, because the mailing list owner is the one able to act upon these errors.

## 4.4. *Mail relaying*

If a machine accepts incoming SMTP mail and forwards it through an outgoing SMTP connection to another MTA, this machine is called a relay. Within organisations this mechanism is often used to transport mail from or to a central machine, which operates as the main mail relay for a certain domain. Mail routing between domains is mostly taken care of in one hop, but sometimes the MX mechanism supplies a fallback mechanism for mail delivery in case direct delivery is (temporarily) not possible. This is an intended and useful way of relaying.

If a machine accepts arbitrary incoming SMTP connections with envelope recipients to arbitrary other domains, this machine is called an 'open relay'. An open relay can easily be abused by spammers and other abusers to inject large volumes of unwanted mail. More about this can be read in Section 8.1 on mail abuse.

## 4.5. *Virtual domains*

A mail server can be configured to accept email not only for its principal domain, but also for many other domains. These domains are called virtual domains. The mechanism by which mail arrives at the correct server is again the MX mechanism. What happens after that depends on the kind of virtual hosting. One possibility is to map virtual mail addresses to addresses in the real principal domain or even in foreign domains by making use of some virtual domain table lookup. Another possibility is to have real separate mail stores for your virtual domains, but this requires a virtual domain aware POP or IMAP server to make the mail store accessible on a per domain basis.

### 4.6. *Delivery status notification*

RFCs 1891–1894 [30,31,44,45] define an extension to SMTP (see Section 3.2.1) that is meant to be used as a more concise and complete way of generating success and failure messages related to mail transport and delivery. Mail clients can ask per recipient for a combination of notifications of success, failure or delay, while specifying whether the full message or only the message headers should be returned.

### 4.7. *Mail error notification*

An important aspect of mail handling is error notification in case something prohibits normal delivery of email. Because error notification uses the mail infrastructure itself, great care has to be taken that no mailing loops arise, see also Section 4.9. Many error conditions exist, ranging from syntax errors in commands or parameters to problems with delivery like non-existing mailboxes, storage limit excess, or authorisation errors. In most cases errors are processed by the sending mail transport agent, which generates a (new) bounce email addressed towards the envelope sender of the email. Every mail receiving domain is required to have a 'postmaster' mailbox or alias which ultimately should be read by a human responsible for fixing errors and problems with the mail system. The postmaster in charge of the sending mail transport agent gets a copy of this bounce message with the mail body of the original mail suppressed.

In case the specially constructed error message generates a bounce itself, a so-called 'double bounce', this double bounce should be delivered to a local postmaster able to resolve any configuration errors. Triple bounces should never occur.

In sending bounces a special convention is used for the envelope sender address, being <>. This special address can not occur as a normal recipient address, but should always be accepted as a valid sender address.

### 4.8. *Mail address rewriting*

In the course of mail processing it is often the case that addresses, sender and recipient, envelope and header, have to be rewritten. This is evident for envelope recipient addresses in case of alias expansion or forwarding. This is also evident for calculating return addresses for envelope senders. A need for rewriting header addresses occurs when domains want to rewrite addresses into preferred canonical forms. Often these rewrites can be implemented with the help of table lookups. The pre-X versions of sendmail have a very powerful rewriting engine based on regular-expression-like pattern matching on tokenised address strings. The art of crafting one's own rewriting rulesets and rules, as practised by sendmail magicians, has been dying slowly and steadily though, and these days sendmail's configuration files are usually generated by m4-based macros from high level option specifications. Additionally, the need for a general rewriting engine for address formats has largely disappeared because of standardisation on Internet style mail addressing.

**4.8.1.** *Masquerading*    Masquerading is a special form of address rewriting that is applied in order to hide the specific details of the mail infrastructure of an organisation for outgoing and incoming mail. At the boundary of a mail domain any mail addresses referring to internal hosts or domains in outgoing mail are rewritten to refer to a top level domain only. Incoming mail, addressed to users at this unified top level domain, can be routed to the specific internal destinations by using rewriting or table lookup.

**4.8.2.** *Canonical mapping*    Canonical mapping is a special form of rewriting of the local part of a mail address in order to format these local parts according to conventions an organisation wishes to use in the outside world. In most cases it is used to rewrite a login account name to a FirstName.Surname or Initials.Surname convention. Canonical mapping is often combined with masquerading to completely transform internal addresses to addresses exposed to the outside world at the boundary mail relays.

For any canonicalisation of addresses care needs to be taken to avoid potential conflicts. The larger the user base, the more potential for conflict. Furthermore, the canonicalisation scheme is important: the lesser distinguishing characteristics appear in the canonical form, the more likely a clash will be. For example, FirstName.Surname can be expected to generate fewer clashes than Initials.Surname, since two persons sharing a surname might very well have identical initials, even though their first names differ. John and Jane Doe have unique canonical addresses in the former canonicalisation scheme, but not in the latter.

Canonical mapping almost always uses table lookups, because there is usually no regular pattern in the substitution.

**4.9.** *Mail loops*

One of the dangers that mail systems have to be careful about is the occurrence of a mailing loop. Most mailing loops are generated by automailers, be it bouncers, vacation programs or user scripts. There is no foolproof system to avoid mailing loops, but several precautions can be taken to avoid them. Let us first look at some scenarios.

**4.9.1.** *Mail loop scenarios*

SCENARIO 1. The classic example of a mailing loop is created when a user has two email addresses and forwards email from the first to the second address and vice versa. Some hop limit count will finally bounce the message to the original sender. If the user not only forwards the mail but also delivers a local copy, multiple copies of the email, equal to the hop count, will be delivered to the combined addresses.

Things get worse when a user has three accounts and forwards from any account to both others. Because of message duplication we get an exponential growth leading, depending on the hop limit, to a practically unlimited number of bounces.

SCENARIO 2. Much more subtle loops than in the classic scenario can occur when user scripts, like procmail based filters, generate new email messages automatically. Consider a mail from sender *originator@from.domain* to receiver *forwarded@to.domain* that is in fact forwarded to final recipient *recipient@actualto.domain*. Moreover suppose that for some reason mail to the final recipient is not deliverable for some (possibly temporary) reason. Let us look at the different situations when forwarding takes place in the pre-delivery phase or the post-delivery phase.

In the case of pre-delivery forwarding, the incoming mail is forwarded by the MTA for *to.domain*. Since MTA-level forwarding keeps the envelope from address intact, the generated bounce is directed to the *originator@from.domain* address, which leads to no problems.

But in the case of post-delivery forwarding, for instance by a procmail receipt, the forwarded message is in fact reinjected as essentially the same message, but with a new envelope. Specifically, the envelope from address after reinjection is *forwarded@to.domain* instead of *originator@from.domain*. The generated bounce is sent to *forwarded@to.domain*, which forwards this by reinjection to *recipient@actualto.domain*, once again replacing the envelope from address by *forwarded@to.domain*. The bounce message, forwarded this way as a normal message, bounces in its turn – so a second bounce gets directed to *forwarded@to.domain*. Repeat ad infinitum, a surprising mail loop.

**4.9.2.** *Mail loop avoidance*   In order to detect and avoid mailing loops, several mechanisms have been implemented. The most obvious and most important one is a hop count, similar to the TTL field in IP packets. Each time a message passes a relay, a new *Received:* header is added to the message indicating that another hop has been traversed. If this hop count gets bigger than a certain configurable limit (often 20 or something), the mail will be bounced. For this bounce the hop count starts again at 0, so it is important to never promote a bounce message to a normal message when forwarding, because this counteracts the hop limit procedure. The same holds for a double bounce in relation to the corresponding bounce.

A second mechanism, used to protect against mailing loops originating from post-delivery automailers, is the addition of a *Delivered-To:* header by the local delivery mechanism on each final delivery. This header was first introduced in the qmail software, later on adopted by postfix and also available in newer sendmail versions. This header has no official RFC-status, but is quite effective against duplicate final delivery, thereby breaking loops.

Also user level software (scripts in procmail receipts) often include their own loop protection by adding unofficial headers like an *X-Loop:* header. All those mechanisms rely on the fact that no new mail messages are generated except in the case of transition from normal to bounce and from bounce to double bounce messages.

Where new messages are generated, such as in the second loop example in which the mail loop was caused by two-level forwarding by means of message reinjection, those aware of the hazards involved with their forwarding strategy may avoid the loop by applying additional procmail magic to force the reuse of the old envelope from upon reinjection:

```
# Extract original envelope from
:0 h
ENVELOPEFROM= | formail -x Return-Path
# Forward (reinject), reusing original envelope from
:0
| $SENDMAIL -oi -f $ENVELOPEFROM recipient@actualto.domain
```

In general, loop prevention goes hand in hand with the specific cause of potential loops, and may therefore require an in-depth understanding of the specific message handling details for all parties concerned.

## 5. Mail servers and clients

### 5.1. *Mail servers*

In this section a number of mail servers will be mentioned that play the role of MTA in an Internet mail (SMTP) based environment as their main task. We exclude Microsoft Exchange and Lotus Notes from this discussion. They contain SMTP only as gateways to proprietary mail protocol implementations, further leading to non-standard mail access options like MAPI.

**5.1.1.** *Classic Sendmail and Sendmail X*   Classic sendmail was written by Eric Allman and has evolved from his delivermail program whose main purpose was to support switching between different mail transport mechanisms available around 1980. Sendmail is one of the first and still most commonly used mail transport agents. During the year 2005 sendmail has been starting a transition to a new fast, reliable, secure, modular version called Sendmail X, which should replace the slower, monolithic and more difficult to secure previous sendmail versions. This new architecture for sendmail bears much resemblance to Daniel Bernstein's qmail [3] and Wietse Venema's Postfix [46], MTAs that were introduced in 1997 and 1998 respectively.

Sendmail is still very popular, for three reasons. The first reason is that people have a long experience with operating sendmail and have put a lot of time and effort in sendmail configuration and tuning. This substantial investment is often not easily transferred to a new environment. A second reason is the limitless power of sendmail's full rewrite engine, unique to sendmail, and much more flexible than table lookups. Some sites use this power for all kinds of mail processing not easy to mimic in a new setup. The third reason is the availability of the milter library, used for filtering mail in the middle of sendmail MTA processing.

**5.1.2.** *Postfix, qmail and Exim*   Both postfix and qmail are MTAs that aim for a completely different architecture from sendmail's. A central notion in both software packages is that of a queue manager that assigns work coming from a number of queues, for incoming, outgoing, deferred or erroneous messages, to a number of processing agents, for local or remote delivery and for address rewriting and lookup. This architecture accounts for the modularity and scalability of mail processing these MTAs are known for.

Starting in 1995 Philip Hazel's Exim, based on ideas from an earlier program called Smail-3, was developed as a stand-in replacement for sendmail. Its properties are a more user friendly configuration than sendmail, a good security record, extensive facilities for checking incoming mail and extendability.

## 5.2. *Mail clients*

Email clients abound. We only discuss the different classes of email clients, with a few examples. An obvious distinction can be made between clients with a graphical user interface and clients with a terminal-based text interface. A more important distinction is whether the client is native (speaking IMAP, POP and/or SMTP itself) or is in fact accessing a proxy, using for instance a web interface, to do the mail handling.

**5.2.1.** *Native clients*   Examples in this category are Eudora, Mozilla Thunderbird, Evolution and KMail as graphical clients and mutt or pine as text based clients. Microsoft Outlook is also a messaging client, using something similar, but not quite identical, to standard Internet mail.

Graphical clients are often considered easier to use, but have also given rise to a habit of sending HTML-based mail instead of plain text mail, even in situations where plain text is the essential content of the message, unnecessarily complicating the task of reading, understanding and replying to email messages.

Being able to use a terminal based client offers the capability to access your email in situations where only minimal support is available and a graphical client is not present. Email clients like mutt are quite advanced and support access to your mail via IMAP, securely if needed, and with support for content encryption via PGP, GnuPG or S/MIME, see also Section 6.

**5.2.2.** *Proxy clients: Webmail*   In certain situations no access is available to a graphical mail client and also not to terminal emulation and a text mail client. One example of this is most Internet cafes. Already in 1995 a startup called Hotmail, acquired in 1998 by Microsoft, was offering email services through a browser interface as a first step to make the Internet experience browser centric, leading many people into the belief that the Internet and the Web are synonyms.

The almost universal availability of browser access is the main advantage of webmail. Disadvantages are that offline preparation of email is relatively clumsy, many commercial providers of webmail offer rather limited mail storage quota and organisation of mail on a local hard drive is not possible. Webmail, at this point, is no universal replacement for a true email client.

One of the more popular, open and extensive webmail server packages is SquirrelMail [6].

**5.2.3.** *User filtering*    Whatever email client one is using, an important feature is the ability to do user filtering. Server side filtering is certainly an option, made available to users on an opt-in basis, but most flexibility is retained by using client side filtering. Marking messages as belonging to a certain category is a server side feature that helps in client side filtering. Filtering is used to organise mail into folders, to recognize and act upon spam, junk mail or mail containing malware and to auto-reply to certain kinds of messages, like when on vacation.

On Unix systems mail filtering is often implemented as a separate program acting on behalf of the mail recipient. Procmail is a well-known and advanced mail filter and delivery agent. Email clients often have built-in filtering rules for detecting junk email, most of them based on Bayesian filtering, as pioneered by SpamCop [17] and Microsoft Research [28]. More on spam and junk mail in Section 8.1.1.

Sieve [42] is a mail filtering language designed for filtering email messages at final delivery time. It is designed to be implementable on either a mail client or mail server. Dovecot's mail delivery agent, LDA, implements it, as does the Cyrus IMAP server. In the past, the Mulberry MUA allowed users to create user-level Sieve filters in such a way that the resulting filters could be pushed to the server, thus combining direct user manipulation of mail filters with actual delivery-time mail filtering.

## 6. Mail security

Pondering the Tomlinson quote (Section 2) one could say that email in the early days did not need security because it was used in an open and collaborative environment. Times have changed. Malware and other nuisances have substantially turned email's playground more hostile, so it needs to exercise caution too.

When discussing security five aspects are often distinguished: *authentication, authorization, integrity, confidentiality* and *non-repudiation* (accountability).

In this section we will look at mail security from the angles of *transport protection* by way of encrypted and (asymmetrically) authenticated connections and *content protection* by way of encryption and signing of message bodies.

### 6.1. *Transport security*

When transporting email over the Internet via SMTP or accessing an MAA to collect email by IMAP or POP, it would be nice to know that the client is talking to the correct, intended mail server: typically, one wants to disclose authentication information to the proper server only, not to just any system that (maliciously?) tries to assume its identity. On the other hand, a server may wish to know what client it is talking to. Differently put: the client and server may need to certify each others authenticity.

For IMAP and POP it is essential to establish the user's identity prior to granting access to the user's mailbox, that is to authenticate the user prior to authorizing access. For SMTP, client authentication might help against spam and other nuisances. Without SMTP authentication, anyone can inject any message under any identity, so any message can be sent by anyone. Without message authenticity there is no non-repudiation.

Traditionally authentication is performed using a plain text username and password that can easily be intercepted in transit. The risks of *spoofing* and *password sniffing* make traditional authentication (and, therefore, authorization) highly vulnerable for potential abuse.

However, the risks do not stop there. In fact, it is not just the authentication details of a traditional SMTP, POP or IMAP session that can easily be 'sniffed'. The full details, including the potentially confidential message content, are open to prying eyes this way: confidentiality is poor. Additionally, the sessions may even be intercepted and tampered with, putting data integrity at risk.

In the web world mechanisms for resolving these issues have existed since Netscape specified its Secure Socket Layer. Using certificates a server can prove its identity to clients and, if asked for, clients can prove their identity with certificates of their own. An encryption key is negotiated to make the connection confidential. This has the extra benefit of protecting a username/password protocol for authentication from snooping. This web world mechanism can also easily be applied to mail protocols.

**6.1.1.** *SSL, TLS and STARTTLS* SSL (Secure Socket Layer) has been invented by Netscape for use in e-business applications on the Web. SSL version 3.0 has later been adopted by the IETF and renamed, with minor modifications, to TLS (Transport Layer Security) version 1.0 in 1999 [10]. It provides mechanisms for host identity verification as well as data encryption in order to prevent spoofing, sniffing and tampering. This mechanism was used to protect Hyper Text Transfer Protocol (HTTP) transmissions, turning it into https, secure http, with default TCP port 443 instead of the usual port 80.

These same techniques can also be applied to IMAP and POP, leading to secure imap (imaps; default port 993 instead of 143) and secure pop (pop3s; default port 995 instead of 110). Somehow, understandably so considering Section 6.3, the same mechanism for SMTP, secure smtp (smtps; default port 465 instead of 25) is hardly used and was never formally acknowledged. IANA [20] even lists 'URL Rendesvous Directory for SSM' (urd) as using 465/tcp.

An alternative mechanism, STARTTLS [18] (an SMTP service extension), is used to 'upgrade' an existing, unprotected TCP connection to a TLS-based, protected TCP connection. STARTTLS is also available in many IMAP and POP implementations, usually in the form of an advertised capacity, with 'sensitive' features, such as authentication, becoming available only after successfully establishing encryption for the existing connection first. That way, only the standard SMTP port needs to be assigned and published, while still offering the facility of network transport encryption. No dedicated port to provide the encrypted variant for SMTP or other protocols is needed when using STARTTLS. A special port is assigned by IANA (submission 587/tcp) for connections to MSAs where authentication via a secure connection is preferred. Most mail systems still use port 25 for submission and transport without any protection or authentication.

**6.2.** *Authentication frameworks*

For IMAP and POP, after a secure connection is established between client and server, the usual builtin commands specifying user and password can be used safely. SMTP does not have a standard authentication command. In RFC 2554 [34] an 'SMTP Service Extension for Authentication' was defined, using the 'AUTH' command which is based on the 'Simple Authentication and Security Layer' (SASL [33]) mechanism. SASL enables several authentication schemes, like Kerberos, One Time Passwords and GSSAPI-based schemes, and is currently widely used.

**6.3.** *Mail content security*

Another matter is when users feel the urge to protect the content of their mail messages from prying eyes.

When that need arises, one should take notice of the fact that the very store-and-forward nature of the SMTP protocol itself acts contrary. Even with full transport security at the network level for an SMTP transaction between two hosts, the messages will still, typically, be stored in order to be queued for further forwarding. This process will repeat itself for any 'hop' on the way to the final message destination. Any host on this delivery path gets the full message, without protection. Any mail host kan keep copies of the messages it passes on, and it can even change the messages when doing so.

When confidentiality, integrity and/or authenticity of the message itself is important, the MUA therefore needs to apply some extra measures in order to encrypt and/or sign the message content, using MIME to deliver the encrypted message and/or the signature. The two most prominent standards implementing privacy enhanced mail, not to be confused with the first 'secure email standard' called Privacy-Enhanced Mail (PEM [1,21,22,27]), are OpenPGP/MIME and S/MIME. Another standard is MIME Object Security Standard (MOSS [8]). For an overview of the different mechanisms and the pitfalls of the simple sign-&-encrypt habit see [9].

**6.3.1.** *PGP, GnuPG and OpenPGP*    PGP stands for 'Pretty Good Privacy' and was developed by Phil Zimmermann to provide privacy and authentication of documents in general. Many subtly different versions of PGP have been released during the years. This led to the development of the OpenPGP [5] standard and an open source implementation called the Gnu Privacy Guard (GnuPG [25]). Both PGP and GnuPG have been used to encrypt and sign messages inline. Later the MIME standard was used to separate the content of the mail message from the signature or to convey an ASCII-armored encrypted message [11]. MIME types 'multipart/signed' and 'multipart/encrypted' are used for this purpose.

The trust model proposed by OpenPGP is that of a web of trust. There is no centralised hierarchy of authorities, but a loosely coupled system of users signing other user's public keys. Keyservers collect these public keys with attached signatures in order to distribute these on demand to MUAs who need them to send secure email.

**6.3.2.** *S/MIME*    S/MIME stands for Secure/MIME [38], which describes how to add cryptographic signature and encryption services to MIME data. It is based on the 'Cryptographic Message Syntax' (CMS [19]), which is derived from PKCS #7 as specified by RSA Laboratories. S/MIME uses 'application/pkcs7-mime' and 'application/pkcs7-signature' as MIME types for encryption and signing.

The trust model used in S/MIME is based on X.509 certificates and a centralised hierarchy of Certification Authorities.

# 7. Relations to other technologies

On the Internet several other technologies exist that have an intimate relationship with email, or that serve a similar purpose.

## 7.1. *Directory services*

Email cannot exist without proper functioning directory services. A crucial directory service used by email is DNS, without which no Internet application could survive. Specific use of DNS is made by mail transport agents by looking at MX records to find the next hop mail server.

Directory services are needed by mail upon local delivery as well. When accepting a message for a purported local user, the mail system has to check whether that user exists. Traditionally, the mail system consulted a local user database for this purpose. Other directory service mechanisms, such as LDAP, are gaining importance though.

Email software may additionally use directory services (again, typically LDAP) for authentication, routing, aliasing or forwarding.

## 7.2. *Usenet news*

Email is designed as a medium for interpersonal message exchange: one on one, basically. *Usenet news*, or simply *news*, is a similar medium, with the fundamental difference that messages are directed to a collective rather than a person. News *articles* are very similar in format and structure to email messages. In fact, one and the same message format RFC applies to both email and news messages. However, instead of sent to a person's email address, these articles are *posted* to a *newsgroup*. Anyone who is interested can fetch the article from any newsserver that carries the newsgroup in question.

In addition to the extensive similarities between email messages and news articles, email and news share more common ground. Both are based on asynchronous message exchange technologies, using a store-and-forward approach for message transport. With this much similarity, the existence of gateways between email and news, spanning both worlds, should not be too surprising.

**7.3.** *Instant messaging*

A different form of interpersonal communication is gaining ground in the form of instant messaging. Although it serves similar purposes, instant messaging is very different from Internet mail. The differences give rise to a different usage pattern.

Key differences are that instant messaging is inherently synchronous. The text typed is displayed instantly elsewhere. Email is asynchronous: compose a message, send it, and the message is delivered and may at some point get the correspondent's attention.

As a side effect of asynchronicity, email is a persistent medium. Messages are actually stored, and users typically have to discard messages explicitly.

Instant messaging is volatile by nature, in contrast. Conversation takes place in an open window, scrolls by – and is gone, effectively, when out of sight. Logging can be used, of course, to gain persistence as an afterthought.

Another distinction lies in the intended audience. Although email can be used to reach more than a single person by supplying multiple recipient addresses, the email paradigm is one on one message exchange. Many forms of instant messaging are however designed to provide a shared communication channel for any number of participants.

**7.4.** *Email and calendar integration*

Historically, email was often the first facility that became available to users for information exchange, making it an obvious albeit unintended vehicle for arbitrary data exchange. File exchange is one of these unintended applications. Another data exchange function has historically been piggybacked upon mail: calendar synchronisation. There exists mail clients that use the mail transport mechanism to exchange scheduling/reservation notices for the purpose of calendar bookkeeping.

There are a few ways, however, in which mail technology is not particularly well suited to this task. First, mail is essentially asynchronous. A distinct period of time will elapse between sending and receiving a message. It can take a substantial number of messages back and forth before an agreement, for example to make an appointment or to reserve a time slot, can be reached, even between two parties.

The number of message exchanges to reach such an agreement explodes the more parties are involved. Person A sends a proposal for a particular time slot to B and C. B answers with a confirmation, but C answers ten minutes later with a denial and a suggestion for an alternative. Repeat, lather, rinse . . . .

## 8. Real world email

### 8.1. *Mail abuse*

**8.1.1.** *Spam*    In the physical world sending something to a person costs a certain amount of money, and sending something to a thousand persons costs roughly thousand times that amount.

Sending an email message is easy and almost without cost. Email also scales relatively well: sending an email message to a large number of recipients is still easy and cheap.

Observing these basic economics the massive email abuse for direct marketing purposes that can increasingly be witnessed is unsurprising in hindsight. This massive flood of *spam*, bulk amounts of unsollicited messages, often carrying commercial content, has not widely been welcomed with open arms.

Cheap and easy to send this spam may be, but it is a considerable nuisance to many. The frequent routine of deleting spam takes up precious time, and the network capacity and storage space needed to handle the extra resources taken by spam can also add up to non-negligible costs.

To alleviate the users' burden, filtering software has widely been deployed to weed out the spam. Currently, the most effective rule for filtering spam is obtained by applying Bayes' theorem to calculate the probability that a message is spam based on the words that occur inside the message. In order to apply this theorem, a database of word probabilities is learned by classifying mail as either spam or ham (the opposite of spam). This classification can be done by the users themselves on receipt of email or by pre-classification of known spam.

Bayesian filtering of email was pioneered by Sahami and others from Microsoft Research [41]. In the same year, 1998, an implementation of this technique was already available in the SpamCop software [17,36].

**8.1.2.** *Worms*   The widespread adaptation of fully networked, yet insufficiently secure computer systems has inflicted email as well. Being one of the ways in which information is carried between computer systems, email has seen its share of security issues.

Poor security design in popular email clients added fuel to this fire. As a result, hordes of infected systems are used for any purpose, including spam propagation and propagation of messages containing 'worms'. A mail worm is a message that contains an attachment that is an executable program, which is executed when the user tries to view the attachment using an insecure mail program. Typically, the worm plunders the user's address book, spreads itself – often using entries from that address book as sender address – and makes modifications to the running system that enable some or other remotely initiated operation.

**8.1.3.** *Spoofing and sender authenticity*   Based upon the assumption that the spam problem is caused by a lack of accountability, a few mechanisms have been proposed that are meant to provide a more reliable sender identification. It should be noted that these efforts are unrelated to the approaches described in Section 6.

In two similar approaches, Sender Policy Framework (SPF) and Sender ID, it is also assumed that there is a strong relation between the network address of the system from which a message originates and the (domain part of the) sender mail address. This assumption holds water, when a person's service provider (that operates his mail service) and his access provider (that facilitates his network connectivity) coincide. In the real world, someone may very well use one Internet Service Provider, and another Internet Access Provider. So we believe this assumption to be unfounded. Any mail address can occur on messages from any host on any network with perfect validity, in principle.

**8.1.4.** *Filtering*   Spam and worms confront the user with unwanted, potentially even dangerous, messages. To help guard the user, filtering technology can be applied. Popular packages for the detection of spam and worm/virus/trojan/rootkit attachments are SpamAssassin and ClamAV, respectively.

In addition to common user annoyances, corporations may very well impose policy restrictions on email use and apply mail filters to enforce these policies. This way, the use of mail attachments, for example, may be banned altogether or restricted to certain file types, or files up to a certain maximum size.

Filtering can be invasive, and when applied naively prove countereffective. For example, a virus scanner may refuse encrypted mail messages. On the one hand, that may be a good thing. Nothing goes through unless checked. On the other hand, this policy may unintentionally give rise to the dissemination of confidential information. If the only way to send information is to do it insecurely, then people will be forced to do so.

**8.1.5.** *Black- and whitelisting*   Immediate pass- or block-filtering based upon a few simple criteria has become increasingly popular recently. The basic idea here is that some obvious particular message property is looked up in a list. When the particular property is found in the list, that indicates either that the message is undesirable, in which case the list in question is known as a blacklist, or that the message is acceptable, in which case the list is called a whitelist.

Properties typically used in black- and whitelisting are the envelope sender, the envelope recipient or the IP address of the host that's offering the message.

Black- and whitelist lookups are often, but not necessarily, queries of an external resource. Typically, these queries are implemented as DNS lookups, where a non-null response indicates a positive match. This is commonly referred to as real-time black- and whitelisting. Many real-time black- and whitelisting services exist, both non-profit as well as for-profit.

**8.1.6.** *Greylisting and other approaches*   A somewhat recent approach to fight mail abuse builds upon the observation that spammers really need simple and cheap ways to spread massive amounts of messages. To keep it simple and cheap, the observation continues, spammers do not really play by the rules imposed by the SMTP protocol.

One approach based upon this observation can be seen in the `greet_pause` option introduced in Sendmail v. 8.13. Not caring what the other end has to say, spammers typically do not pay attention to its remarks and immediately start sending SMTP commands to inject their message once connected. With the `greet_pause` option set, Sendmail refuses to accept messages offered by a peer before it even got the chance to greet ('HELO', 'EHLO').

Similarly, Postfix' `reject_unauth_pipelining` parameter makes Postfix refuse messages from peers that send SMTP commands ahead of time.

Greylisting leverages the SMTP protocol to a somewhat deeper extent. When stumbling upon a transient (non-fatal) error while offering a message, the SMTP protocol requires that the message gets offered again after a brief pause, but within reasonable time. Even without qualifying the notions of 'brief' and 'reasonable' here any further, it should be obvious that this introduces a little complication for the sending party. Greylisting looks at three parameters, which together are called a 'mail relationship': the IP address of the host attempting a delivery, the envelope sender, and the envelope recipient. If a mail relationship is either new or too recent, service is refused with a transient error message. However, mail relationship is timestamped and remembered. Now when a message – probably the same message – is offered with a mail relationship that is sufficiently old according to its timestamp the message is accepted without further ado.

Greylisting effectively puts a stop to the typical fire-and-forget tactic often used by spammers at the cost of a small, but one-time only delay for valid mail correspondences. Even if spammers actually perform the extra work of offering their messages again at a later time, that means that the cost of spamming has substantially increased. Furthermore, between the first refusal and the second offering real-time blacklists and dynamic signature databases of spam-detection mechanisms may very well have been updated for it. This way, greylisting turns out as a measure that not only fights spam technologically, but it also affects the basic economics of spamming.

## 8.2. *Mail usage*

**8.2.1.** *Conventions*    A few conventions have historically developed. Early mail users, working on fixed size character terminals, found ways to communicate effectively within the confinements of their working environment.

A few habits deserve mentioning. *Author attribution* was conscientiously used to indicate which part of a mail message was attributed to which author. In conjunction with that, a strong discipline of quotation was maintained: when replying to a message, any existing quoted text was marked with a designated quote prefix (>). Existing quoted material became prefixed with >> that way, and so on. This made it possible to unambiguously determine which text parts led back to which authors.

Another notable email habit was to write remarks in context, using proper attribution and quotation of course. Any remark to be made immediately followed the relevant (quoted) text part in the original message, thus interweaving remarks with quoted material from the original message. Proper context also implied that only the relevant parts of the original message, providing sufficient context, were quoted in replies.

This way, messages were minimal in size, sufficiently self-contained as well as unambiguously interpretable.

**8.2.2.** *Etiquette*    To help keep email communicaton effective, some basic rules are often proposed. The established, yet besieged, 'best practices' described in Section 8.2.1 are usually found back in guidelines on email etiquette. A lot more suggestions, that all fall in the category of things that should really be painfully obvious, but somehow turn out otherwise, can be made and float around the Internet.

## 9. Conclusion

Several protocols make up electronic mail as we know it, each acting one or more of the set of roles to be played. Technology related to electronic mail, and a slew of problems surrounding electronic mail of both technological and non-technological nature have been discussed.

## References

[1] D. Balenson, *Privacy enhancement for internet electronic mail: Part iii: Algorithms, modes, and identifiers*, http://www.ietf.org/rfc/rfc1423.txt (1993).

[2] P. Beertema, *USSR on Usenet*, http://groups.google.com/group/eunet.general/msg/cf080ae70583a625.

[3] D.J. Bernstein, *qmail: the Internet's MTA of choice*, http://cr.yp.to/qmail.html.

[4] M. Butler, J. Postel, D. Chase, J. Goldberger and J. Reynolds, *Post Office Protocol: Version 2*, http://www.ietf.org/rfc/rfc0937.txt (1985).

[5] J. Callas, L. Donnerhacke, H. Finney and R. Thayer, *OpenPGP Message Format*, http://www.ietf.org/rfc/rfc2440.txt (1998).

[6] R. Castello, *SquirrelMail – Webmail for Nuts!*, http://www.squirrelmail.org/.

[7] M. Crispin, *Internet Message Access Protocol – Version 4rev1*, http://www.ietf.org/rfc/rfc3501.txt (2003).

[8] S. Crocker, N. Freed, J. Galvin and S. Murphy, *Mime Object Security Services*, http://www.ietf.org/rfc/rfc1848.txt (1995).

[9] D. Davis, *Defective Sign Encrypt in S/MIME, PKCS7, MOSS, PEM, PGP, and XML*, http://citeseer.ist.psu.edu/443200.html.

[10] T. Dierks and C. Allen, *The TLS protocol version 1.0*, http://www.ietf.org/rfc/rfc2246.txt (1999).

[11] M. Elkins, D.D. Torto, R. Levien and T. Roessler, *MIME Security with OpenPGP*, http://www.ietf.org/rfc/rfc3156.txt (2001).

[12] N. Freed and N. Borenstein, *Multipurpose Internet Mail Extensions (MIME) part one: Format of internet message bodies*, http://www.ietf.org/rfc/rfc2045.txt (1996).

[13] N. Freed and N. Borenstein, *Multipurpose Internet Mail Extensions (MIME) part two: Media types*, http://www.ietf.org/rfc/rfc2046.txt (1996).

[14] N. Freed and N. Borenstein, *Multipurpose Internet Mail Extensions (MIME) part five: Conformance criteria and examples*, http://www.ietf.org/rfc/rfc2049.txt (1996).

[15] N. Freed, J. Klensin and J. Postel, *Multipurpose Internet Mail Extensions (MIME) part four: Registration procedures*, http://www.ietf.org/rfc/rfc2048.txt (1996).

[16] R. Gellens and J. Klensin, *Message Submission*, http://www.ietf.org/rfc/rfc2476.txt (1998).

[17] J. Haight, *Spamcop.net – Beware of cheap imitations*, http://www.spamcop.net/.

[18] P. Hoffman, *SMTP Service Extension for Secure SMTP over TLS*, http://www.ietf.org/rfc/rfc2487.txt (1999).

[19] R. Housley, *Cryptographic Message Syntax (CMS)*, http://www.ietf.org/rfc/rfc3852.txt (2004).

[20] IANA, *IANA home page*, http://www.iana.org/.

[21] B. Kaliski, *Privacy Enhancement for Internet Electronic Mail: Part IV: Key Certification and Related Services*, http://www.ietf.org/rfc/rfc1424.txt (1993).

[22] S. Kent, *Privacy Enhancement for Internet Electronic Mail: Part II: Certificate-Based Key Management*, http://www.ietf.org/rfc/rfc1422.txt (1993).

[23] J. Klensin, *Simple Mail Transfer Protocol*, http://www.ietf.org/rfc/rfc2821.txt (2001).

[24] J. Klensin, N. Freed, M. Rose, E. Stefferud and D. Crocker, *SMTP Service Extensions*, http://www.ietf.org/rfc/rfc1869.txt (1995).

[25] W. Koch et al., *The GNU Privacy Guard – gnupg.org*, http://www.gnupg.org/.

[26] D. Knuth, *Email (let's drop the hyphen)*, http://www-cs-faculty.stanford.edu/~knuth/email.html.

[27] J. Linn, *Privacy Enhancement for Internet Electronic Mail: Part I: Message Encryption and Authentication Procedures*, http://www.ietf.org/rfc/rfc1421.txt (1993).

[28] Microsoft, *Microsoft Research Home*, http://research.microsoft.com/.

[29] K. Moore, *MIME (Multipurpose Internet Mail Extensions) part three: Message header extensions for non-ascii text*, http://www.ietf.org/rfc/rfc2047.txt (1996).

[30] K. Moore, *SMTP Service Extension for Delivery Status Notifications*, http://www.ietf.org/rfc/rfc1891.txt (1996).

[31] K. Moore and G. Vaudreuil, *An Extensible Message Format for Delivery Status Notifications*, http://www.ietf.org/rfc/rfc1894.txt (1996).

[32] J. Myers, *Local Mail Transfer Protocol*, http://www.ietf.org/rfc/rfc2033.txt (1996).

[33] J. Myers, *Simple Authentication and Security Layer (SASL)*, http://www.ietf.org/rfc/rfc2222.txt (1997).

[34] J. Myers, *SMTP Service Extension for Authentication*, http://www.ietf.org/rfc/rfc2554.txt (1999).

[35] J. Myers and M. Rose, *Post Office Protocol – Version 3*, http://www.ietf.org/rfc/rfc1939.txt (1996).

[36] P. Pantel and D. Lin, *Spamcop: A Spam Classification & Organization Program*, Learning for Text Categorization: Papers from the 1998 Workshop, AAAI Technical Report WS-98-05, Madison, Wisconsin (1998), http://citeseer.ist.psu.edu/pantel98spamcop.html.

[37] J. Postel, *Simple Mail Transfer Protocol*, http://www.ietf.org/rfc/rfc0821.txt (1982).

[38] B. Ramsdell, *Secure/multipurpose Internet Mail Extensions (S/MIME) Version 3.1 Message Specification*, http://www.ietf.org/rfc/rfc3851.txt (2004).

[39] P. Resnick, *Internet Message Format*, http://www.ietf.org/rfc/rfc2822.txt (2001).

[40] J. Reynolds, *Post Office Protocol*, http://www.ietf.org/rfc/rfc0918.txt (1984).

[41] M. Sahami, S. Dumais, D. Heckerman and E. Horvitz, *A Bayesian Approach to Filtering junk E-Mail*, Learning for Text Categorization: Papers from the 1998 Workshop, AAAI Technical Report WS-98-05, Madison, Wisconsin (1998), http://citeseer.ist.psu.edu/sahami98bayesian.html.

[42] T. Showalter, *Sieve: A Mail Filtering Language*, http://www.ietf.org/rfc/rfc3028.txt (2001).

[43] R. Tomlinson, *The First Network Email*, http://openmap.bbn.com/~tomlinso/ray/firstmailframe.html.

[44] G. Vaudreuil, *The Multipart/Report Content Type for the Reporting of Mail System Administrative Messages*, http://www.ietf.org/rfc/rfc1892.txt (1996).

[45] G. Vaudreuil, *Enhanced Mail System Status Codes*, http://www.ietf.org/rfc/rfc1893.txt (1996).

[46] W. Venema et al., *The Postfix Home Page*, http://www.postfix.org/.

[47] R. Watson, *Mail Box Protocol*, http://www.ietf.org/rfc/rfc0196.txt (1971).

[48] A.F.K. Whistler and M. Davis, *Character Encoding Model*, http://www.unicode.org/reports/tr17/.

# – 2.4 –

# XML-Based Network Management

Mi-Jung Choi, James W. Hong

*Department of Computer Science and Engineering, POSTECH, Pohang, Korea*
*E-mail: {mjchoi,jwkhong}@postech.ac.kr*

## 1. What is XML-based network management?

In this section, we first explain the definition of XML-based network management (XNM) and illustrate merits and demerits of XNM. We also introduce previous research effort on XNM as a related work.

### 1.1. *Definition of XNM*

XML [24] is an extensible markup language, with which one can define own set of tags for describing data. It is also regarded as a meta-language, which describes information about data. XML enables data to carry its meta-information in the form of an XML document. The meta-information is used for searching, filtering and processing efficiently. The main benefits of XML are simplicity, extensibility and openness. The simplicity of XML relies on tree structured text-based encoding of data, so one can easily create, read and modify XML documents with even the most primitive text processing tools. Although simple, XML is extensible enough to express complex data structures. The extensibility of XML allows different applications to represent their own structures of data.

Because XML is an open standard, there is a wide selection of both freely available and commercial tools for handling and processing it. By supplementing XML with several advanced features such as XML Schema [27], XSL [25], DOM [26], XPath [29], XQuery

HANDBOOK OF NETWORK AND SYSTEM ADMINISTRATION
Edited by Jan Bergstra and Mark Burgess

[30], and SOAP [28], XML technology is not only emerging as the language of choice, but also as the universal information-processing platform.

Network management is the process and techniques of remotely and locally monitoring and configuring networks. Network management systems consist of manager (management client) and agent (management server) systems. XNM is a collection of network management methods that applies XML technologies to network management. Therefore, XNM employs the advantages of XML technologies and one can easily develop network management system using XML technologies. In the next subsection, advantages and disadvantages of XNM are examined.

## 1.2. *Merits and demerits of XNM*

The application area of XML is very broad: data presentation on the Internet, multimedia presentation, data saving of specific programs, electronic commerce, academic uses, and etc. In all these areas, XML proves itself to be an excellent solution to solve technical challenges. Using XML in network management presents many advantages [17]:

- The XML Schema defines the structure of management information in a flexible manner.
- Widely deployed protocols such as HTTP can reliably transfer management data.
- DOM APIs easily access and manipulate management data from applications.
- XPath expressions efficiently address the objects within management data documents.
- XSL processes management data easily and generates HTML documents for a variety of user interface views.
- Web Service Description Language (WSDL) [31] and SOAP [28] define web services for powerful high-level management operations.

However, the current XML-based network management has some weaknesses as well. The current efforts to apply XML to network management are accomplished in a limited network management area, such as configuration management, or performed in the development process of a network management system using a few simple XML technologies. Also the previous XML-based network management does not properly provide a general architecture of XML-based network management system (XNMS) from the aspect of manager and agent. Network bandwidth of transferring XML data is large because XML is text-based. The processing of XML is considered heavy, so it is not yet proven that XML is applicable to embedded systems. Many are curious whether the XML-based manager can process information delivered from multiple managed devices. The development experience of an XNMS is insufficient.

Integration efforts of XML-based network management with existing SNMP agents are the first step towards the integrated network management. The specification translation from SNMP SMI to XML DTD or XML Schema is widely in progress. The XML/SNMP gateway approach between an XML-based manager and SNMP agents is also considered to manage the legacy SNMP agents. Therefore, to provide an integrated network management of the XML-based manager and existing SNMP agents, the architecture and implementation of the gateway need to be provided.

## 1.3. *Related work*

In the past several years, some research and development has been done in the area of XML-based network management. Researchers and developers in the network and system management area have begun to investigate the use of XML technologies in order to solve the problems that exist in Internet management standards such as SNMP. Much work has been done in the area of accommodating widely deployed SNMP devices, specifically in the specification and interaction translation for the gateways between XML-based managers and SNMP-based devices [13,15,21,35]. Jens Müller implemented an SNMP/XML gateway as a Java Servlet that allows fetching of such XML documents on the fly through HTTP [20]. Avaya Labs has developed an XML-based management interface for SNMP enabled devices [2]. Y.J. Oh et al. developed several interaction translation methods between an XML/SNMP gateway and an XML-based manager, based on the previously developed specification translation algorithm [15,35].

Some work focused on extending Web-based management to exploit the advantages of XML technologies [8,11–13,19,32]. Web-based Enterprise Management (WBEM) uses XML only for object and operation encoding, and is currently being updated to include emerging standards, such as SOAP [32]. Web-based Integrated Network Management Architecture (WIMA) showed that XML is especially convenient for distributed and integrated management, for dealing with multiple information models and for supporting high-level semantics [12]. H.T. Ju et al. extended the use of embedded Web server for element management to Web-based network management architecture platform by adding XML functionalities [11]. L.R. Menten also showed his experience in application of XML technologies for device management ranging from small resource-limited data networking to high capacity server-hosted control plane monitoring systems [13]. He proposed an XML-based architecture of network management system and described the DOM-based management transactions to cover HTTP/SOAP client, SNMP manager, script client and CLI client. V. Cridlig et al. proposed an integrated security framework for XML-based network management [8]. This framework is based on role-based access control, where requests are issued on behalf of a role, not of an user. M.H. Sqalli et al. presented load balancing approaches to XML-based network management, to distribute the load across multiple parallel Java parallel virtual machine (JPVM) gateways [19].

While other work focused specifically on configuration management of network devices and systems [6,7,10,18,19,34], Juniper networks [18] and Cisco [6] manage the configuration information of their network devices equipped with their own XML-based agents. The Network Configuration (Netconf) [10] Working Group (WG) is chartered to produce a protocol suitable for configuration management of network devices. The Netconf protocol uses XML for data encoding and a simple RPC-based mechanism to facilitate communication between a manager and an agent. YencaP is a Python-based Netconf implementation by V. Cridlig et al. [7]. YencaP provides a Netconf management suite consisted of Netconf agent and manager. S.M. Yoo et al. proposed a mechanism to manage configuration information more effectively using Netconf protocol and Web services technologies [34]. Using Universal Description, Discovery and Integration (UDDI) [14] helps a Netconf manager to quickly and intelligently recognize required parameters of network devices to operate configuration tasks.

**1.4.** *Chapter organization*

XML has been applied to numerous information processing application areas for various objectives and has earned excellent reputation for simplicity, extensibility and openness. Even though potential functionalities and advantages of XML are well known for various Internet applications, XML is not the favorable choice to use in the network management area yet. The lack of a clear guideline on how to apply XML technologies to network management is blamed for insignificant portions of XML usage. This chapter first reviews XML technologies that can be used in developing XML-based management systems, and then discusses on various methods to show the benefits of applying XML related technologies to network management. In order for the reader to easily understand the applicability, we have divided network management into the following areas: information modeling, instrumentation, management protocol, analysis and presentation. This chapter describes how XML technologies can be applied to each area.

That is, we present an architectural framework for the development of XML-based network management systems. The framework provides a detailed architecture for an XML-based manager, agent and gateway. It also provides methods for interacting between a manager and agent. This framework can be used as a guideline for developing XML-based management systems. We have validated this architecture by implementing and using it in managing network devices and systems.

The organization of the chapter is as follows. Section 2 presents an overview of relevant XML technologies. Section 3 discusses methods for applying XML technologies to network management tasks. Section 4 presents our architectural framework for developing XML-based network management systems based on the agent, manager and gateway. Section 5 describes our implementation experiences of XNMS. Finally, we summarize this chapter in Section 6.

## 2. XML technologies

In this section, a set of XML-related technologies that can be applied to developing network management systems is briefly introduced for better understanding of XML. The eXtensible Markup Language (XML) [24] is the World Wide Web Consortium (W3C) standard based on SGML. XML uses a text-based syntax that can be recognized by both computers and human beings and offers data portability and reusability across different platforms and devices. XML is rapidly becoming the strategic instrument for defining corporate data across a number of application domains. The properties of XML markup make XML suitable for representing data, concepts, and contexts in an open, platform-, vendor- and language-neutral manner. The technology map of XML is depicted in Figure 1.

There are two fundamental approaches to define XML document structure: Document Type Definition (DTD) and XML Schemas [27]. Either approach is sufficient for most documents, yet each one offers unique benefits. DTD is in widespread use and shares extensive tool support while XML Schemas provide powerful advanced features such as open content models, namespace integration and rich data typing.

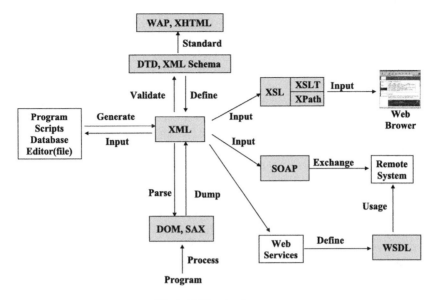

Fig. 1. XML technology map.

The Document Object Model (DOM) [26] specifies the means by which we can access and manipulate XML documents. Without DOM, XML is no more than a storage system for data that can be accessed by various proprietary methods. With DOM, XML begins to approach its promise as a truly universal, cross-platform, application-independent programming language. DOM allows reconstruction of the complete document from the input XML document, and also allows access to any part of the document. The Simple API for XML (SAX) [22] is an event-driven and serial-access mechanism for accessing XML documents. While a DOM parser parses the XML document and creates a DOM tree, keeping the entire structure in memory at the same time, SAX reads the XML document in sequential order and generates an event for a specific element. Therefore, if the application calls for sequential access to XML documents, SAX can be much faster than the other methods with heavy system overhead. However, it does not provide the hierarchical information that a DOM parser provides. A SAX parser generates events such as the start of an element and the end of an element, while accessing the XML document. By capturing these events, applications can process operations, such as gaining the name and attributes of the element.

The Extensible Stylesheet Language (XSL) [25] is a mark-up language to display XML documents on the Web. XML documents describe only the contents and their structure. An XSL style sheet specifies the presentation of a class of XML documents by describing how an instance of the class is transformed into an XML document using a formatting vocabulary. XSL enables XML to separate the contents from the presentation. XSL consists of two parts: a language to transform XML documents and XML vocabularies to specify formatting semantics. The style sheet technology to transform documents is the XSL Transformation (XSLT) [23], which is a subset of XSL technologies that fully supports the transformation of an XML document from one format into another, such as HTML or another custom XML document type. The reason for publishing the XSLT specification separately from XSL is that XML documents can be displayed, providing an easy display format for end users by transforming XML documents without formatting semantics.

The XML Path Language (XPath) [29] is an expression language that is used to address parts of an XML document. The syntax used by XPath is designed for use in URIs and XML attribute values, which requires it to be very concise. The name of XPath is based on the idea of using a path notation to address an XML document. XPath operates under the assumption that a document has been parsed into a tree of nodes. XPath is a language to identify particular sections of an XML document. XPath has a compact, non-XML syntax to be used within URIs and XML attribute values, and operates on the abstract, logical structure of an XML document. Each node in the XML document is indicated by its position, type and content using XPath.

The XML Query Language (XQuery) [30], a query language for XML, is designed to be broadly applicable to all types of XML data sources, such as structured and semi-structured documents, relational databases and object repositories. XQuery uses XPath for path expression. Further, XQuery provides such features as filtering documents, joining across multiple data sources, and grouping the contents. XUpdate [33] is an update language, which provides open and flexible update facilities to insert, update, and delete data in XML documents. The XUpdate language is expressed as a well-formed XML document and uses XPath for selecting elements and conditional processing.

The SOAP [28] is a lightweight protocol for exchanging information in a distributed environment. It is an XML-based protocol that consists of three parts: an envelope that defines a framework for describing the content of a message and how to process it, a set of encoding rules for expressing instances of application-defined data types, and a convention for representing remote procedure calls and responses. SOAP defines the use of XML and HTTP or SMTP to access services, objects, and servers in a platform- and language-independent manner.

The Web Services Description Language (WSDL) [31] is an XML-based language used to define Web Services [9] and describe how to access them. A WSDL document is a collection of one or more service definitions. The document contains a root XML element named definitions, which contains the service definitions. The definition element can also contain an optional targetNamespace attribute, which specifies the URI associated with the service definitions. WSDL defines services as collections of network endpoints which are defined as ports in WSDL. WSDL separates the service ports and their associated messages from the network protocol binding. The combination of a binding and a network address results in a port, and a service is defined as a collection of those ports.

## 3. Applicability of XML technologies to management tasks

In this section, we examine how XML technologies can be applied to network management. Network management involves the following essential tasks: modeling management information, instrumenting management information in managed resources, communicating between manager and agent, analyzing the collected data and presenting the analysis results to users.

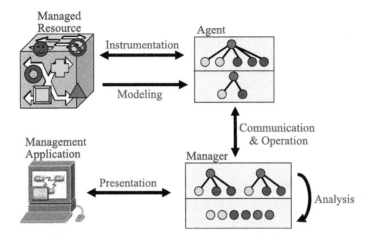

Fig. 2. Basic management tasks in a management system.

As mentioned in Section 1.1, network management systems consist of manager and agent systems. They perform various management tasks to process management information. Agent's tasks involve accessing attributes of managed objects, event reporting and processing the management requests. Under normal conditions, agent's tasks are so simple that the management overhead of the agent can be negligible. On the other hand, manager systems perform complex management tasks to satisfy management objectives while they depend on management applications. The basic management tasks in a management system are depicted in Figure 2.

### 3.1. *Information modeling*

XML is a meta-language, a language to describe other context free languages, that enables applications such as network and system management to define its own tags. XML allows the user to define a customized markup language specific to an application. The expressive power of XML makes it possible to represent a wide range of information models presented in existing management standards and deployed solutions. The XML DTDs and Schemas define the data structure of XML documents. The XML Schemas is devised to overcome the limits of DTDs that cannot define a new data type. It is also flexible and extensible, allowing new tags to be added without changing the existing document structure. The XML Schema supports the presentation of management information by adding new data types to a variety of data types (44 kinds of basic types), and defining the document structure of management information. Therefore, the use of XML Schemas maximizes XMLs key advantages in the management information model over other alternatives.

Compared with SNMP SMI and WBEM CIM, the XML Schema has many advantages in management information modeling. The XML Schema is easy to learn. Various powerful and convenient XML editor tools are available for developers. Another major advantage is that XML can be utilized for various purposes, including validation. By use of XML authoring tools, a developer can generate sample XML data, a database schema and other forms of data structure based on the XML Schema. The disadvantage of XML Schema for management information modeling is that there is no single widely adopted standard model. However, it is convenient to translate from the existing management information models such as SMI, CIM, GDMO to the XML Schemas. Research in this area is widely being conducted [2,20,21,35].

### 3.2. *Instrumentation*

Tasks of management information instrumentation include operations to guarantee consistency between managed objects and managed resources in an agent system. The update operation from the state of managed resources to the variable of managed objects should be executed if the state of managed resources is changed. The mapping operation from the value of managed objects to the state of managed resource should be performed if a manager system modifies the variable of managed objects by management operations. The managed resources might be a hardware which has control or state register as interface mechanism, or execution program which has program interface as interface mechanism. They are implemented without aid from XML technologies. However, managed objects can rely on XML technologies. This means that the managed objects are in the form of XML document.

Management information instrumentation can be viewed as a group of processes for XML document interpretation, creation, and modification. When it comes to mapping, it is possible to access the managed resource after the interpretation of the XML documents, the variables of managed object. Also for the update, the state of managed resource is described in managed object with the modifications to XML documents. There are two standard XML document interpretation technologies: DOM and SAX. They make it easy to write programs that interpret, create and modify XML documents.

Most network devices provide multiple management interface mechanisms, CLI, SNMP, Web, and etc. The management interfaces have been implemented separately, resulting no common interface or data format between managed objects and managed resources. XML can be used as middleware between managed objects and managed resources. By introducing a common interface and data format, the development cost can be reduced, and the consistency issue from multiple accesses to a single managed resource can be resolved.

### 3.3. *Management protocol*

A management protocol must deal with the specifications of management operations and transport protocol for the exchange of management information. It should also define the syntax and semantics for the protocol data unit. With respect to management protocols, XML-based network management follows the model of transferring data over HTTP. Further, it uses XML as management information encoding syntax. This means management data is transferred over the HTTP payload in the form of an XML document. The management information in XML document format is distributed through HTTP over TCP. In contrast to UDP, TCP makes it possible to transmit reliable management data and large application-messages without limits to message size. A large amount of management data can be transferred between the manager and the agent such as an entire table. The XML data through HTTP can be compressed by setting some options in the HTTP header. By compressing data, a significant amount of network overhead can be reduced while keeping almost the same latency at the end-host for both compressed and decompressed data. The network overhead can also be minimized by sending a single message containing multiple sets of management information and by compressing messages with the HTTP header option. The drawback in using this method is the increase in latency from having to establish a TCP connection and perform compression and decompression.

Another important issue regarding the management protocol is in addressing managed objects. When a manager requests management information, it must specify a unique name of the managed object to be retrieved. XML-based network management uses the XPath standard for addressing managed objects. This solution offers the following advantages. First, XPath is a standard Web technology for addressing parts of an XML document. It is already used in conjunction with the XSLT [22] to identify pieces of an XML document targeted for transformations. XPath is also widely supported. Second, the manager can effectively send a query to the managed objects of the agent. XPath expressions are composed of element names, attributes and built-in functions. Given the hierarchical nature of XML, one of the simplest expressions is to follow the tree structure to point at a particular element.

A management protocol defines the management operations. Management operations consist of creation, deletion, retrieval, modification, filtering, scoping, notification, and etc. These high-level management operations can be defined through WSDL and called via SOAP. For example, row creation and deletion in SNMP through Row Status objects can be considerately complicated. However, these tasks can be achieved by convenient higher-level operations in XML-based network management.

### 3.4. *Analysis*

The DOM and SAX APIs can be used to access management data for applications. Most XML parsers implement these standard APIs to access the contents of XML documents. With assist from DOM, one can access any element of the XML document and modify its content and structure from XML input parameters. One can perform such management operations as analyzing statistical traffic data and filtering important events among the notifications by manipulating the XML document using the standard DOM interface. It is easy to extract the necessary management data and analyze them. Therefore, the management information can be analyzed using the DOM API. Items within management data documents can be addressed via XPath expressions.

Because XML data can be easily stored to and accessed from databases (DB) with the support of various tools, DB processing in network management functionality can be easily achieved. Data presented in the DOM tree can be stored to DB and simply transferred to applications. This method reduces the development cost and time, as well as the program overhead. Moreover, there are plenty of tools available for XML, such as XML Parser, DOM, XPath, SAX, etc.

### 3.5. *Presentation*

After analyzing XML data, the next step is to present the analysis results to user. XML, unlike HTML, separates the contents of the document from the display. XML data which is validated by XML DTDs or XML Schemas is transformed to HTML or another XML document through XSL and XSLT. Using the software supporting the transformation from XML to HTML or other display format makes it possible to provide a Web-based management user interface (Web-MUI). Therefore, if management information is expressed in XML format, Web-MUI can be easily generated from the XML document.

## 4. XML-based network management architecture

In this section, we present in-depth details of an XML-based network management architecture that can be used to develop a network or systems management system [5]. The architecture consists of an XML-based agent, an XML-based manager and an XML/SNMP gateway.

### 4.1. *Architecture of XML-based agent*

In this section, we first present the architecture of an XML-based agent and describe its components. The XML-based agent needs to define the management information then retrieve the management information from the managed resources, and transfer to the XML-based manager. In this section, we also explain the management tasks of the XML-based agent from aspects of information modeling, instrumentation and management protocol.

Figure 3 illustrates the architecture of an XML-based agent. The XML-based agent includes an HTTP Server and SOAP Server as basic components. Other components are SOAP RPC operations, which contain DOM Interfaces performing basic management operations. The basic operations use the XML Parser and XML Path Language (XPath) Handler. The XML Parser module allows an agent to parse and access the contents of the XML message using an XPath expression. Also, the agent needs to send notifications to the manager, and the manager needs to receive and handle the notifications. To send notifications to the manager, the XML-based agent needs a SOAP Client module and an HTTP Client module. Consequently, the manager needs a SOAP Server module and an HTTP Server module.

The XML Parser parses an XML document, selects the specified node, and reads management data. In order to send up-to-date information, the XML-based agent gathers information from the Management Backend Interface. The XML Parser updates the selected node value through the Management Backend Interface before replying to the manager.

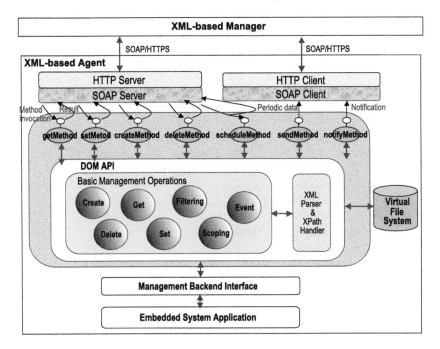

Fig. 3.  Architecture of XML-based agent.

Virtual File System stores management information in an XML format. The DOM API is used to process management information and instrument this information from real device resources through the Management Backend Interface.

The XML-based agent needs to define management information, retrieve the management information from the managed resources, and transfer the management information to the XML-based manager. In this section, we explain the management tasks of the XML-based agent on aspects of information modeling, instrumentation and management protocol.

**4.1.1.** *Information modeling in agent*   We define management information using the XML Schema. An example of management information is SNMP MIB II that can be applied to all network devices. Each node of SNMP MIB converts into an element of the XML Schema: the name of object into XML tag, 'syntax' into the data type definition and 'access' into the attribute for example. Figure 4 shows an example of an XML Schema in system group of MIB II, particularly the 'sysDescr' object of system group.

**4.1.2.** *Instrumentation in agent*   A DOM tree is a virtual repository of management data and provides a manipulation point to managed objects. A DOM parser provides a random access using an element name or an XPath expression and various manipulations of an XML document. The attributes of the DOM interface, such as Node and Text, are mapped to managed objects. When the XML-based agent receives a request message, the SOAP operation module selects specified nodes in the DOM tree using the XPath Handler. For

```
<xsd:element name="system">
  <xsd:complexType>
    <xsd:all>
      <xsd:element ref="sysDescr" minOccurs="0"/>
      <xsd:element ref="sysObjectID" minOccurs="0"/>
      <xsd:element ref="sysUpTime" minOccurs="0"/>
      <xsd:element ref="sysContact" minOccurs="0"/>
      <xsd:element ref="sysName" minOccurs="0"/>
      <xsd:element ref="sysLocation" minOccurs="0"/>
      <xsd:element ref="sysServices" minOccurs="0"/>
    </xsd:all>
  </xsd:complexType>
</xsd:element>
<xsd:element name="sysDescr">
  <xsd:complexType>
    <xsd:simpleContent>
      <xsd:restriction base="DisplayString_0_255">
        <xsd:attribute name="access" type
          ="xsd:string" use="fixed" value="read-only"/>
      </xsd:restriction>
    </xsd:simpleContent>
  </xsd:complexType>
</xsd:element>
```

Fig. 4. XML schema of MIB II-system group.

the selected nodes, the agent retrieves management data from the DOM tree through the DOM interface, and sends the data to the XML-based manager. In order to send up-to-date information, the DOM tree updates the node value for the selected node through the Management Backend Interface before replying to the agent. This type of update of the DOM node value from real resources is called a pull-based update. The DOM tree requests a value to the embedded system application through the Management Backend Interface and receives a response.

For certain nodes, it is not necessary to update the DOM node value before replying because this value is up-to-date. In this case, the Management Backend Interface module is responsible for the update. Periodically, or when events occur, the Management Backend Interface updates the DOM tree node value. This type of update is called a push-based update. For frequently changing data, such as traffic measurement data, the pull-based update is more appropriate than the push-based update, and static or semi-dynamic information can benefit from using a push-based update rather than a pull-based one. In case of pull-based update, the DOM node is updated by replacing the text of the node value with the Processing Instruction (PI) node – a standard node type of DOM.

**4.1.3.** *Management protocol in agent*   The management protocol defines transport protocol, management operation and protocol data unit. First, the XML-based agent transfers

XML data via HTTP. Next, the XML-based agent defines management operations and management data using SOAP. At the basic functionality level, SOAP can be used as a simple messaging protocol and can also be extended to an RPC protocol. The XML-based agent exchanges an XML-encoded message with an XML-based manager. This means that the request message from the manager and the response message from the agent are all formatted as an XML document. SOAP provides a standard method to transfer XML-encoded messages over HTTP between the manager and agent.

Management operations include managed object creation, deletion, retrieval, modification, filtering, scoping, notification, etc. These management operations can be defined using SOAP RPC operations, such as getMethod, setMethod, createMethod, deleteMethod, sendMethod and modifyMethod. Brief explanations of these operations are as follows.

- *getMethod*: used to retrieve information from the management information XML file using the Get operation in DOM Interface.
- *setMethod*: used when the contents of the management information are modified without changing the information structure. This uses the Set operation in the DOM Interface.
- *createMethod*: When a new management information is added, this method is invoked. This calls the Create operation in DOM Interface.
- *deleteMethod*: deletes managed objects and changes the structure of management information.
- *sendMethod*: The agent calls this method to send the periodic monitoring data to the manager.
- *notifyMethod*: The agent sends notifications to the manager via this method.

These operations can also be defined using WSDL. As described in Table 1, WSDL defines the main elements as follows:

- message: An abstract, typed definition of the data being communicated
- operation: An abstract description of an action supported by the service
- portType: An abstract set of operations supported by one or more endpoints
- binding: A concrete protocol and data format specification for a particular port type
- port: A single endpoint defined as a combination of a binding and a network address

The WSDL elements in Table 1(a) describe the getMethod operations published by the XML-based agent as a service named SoapInterfaceService. Table 1(b) describes a request message binding SOAP to call getMethod operation in the XML-based agent. Table 1(c) is the response message to the request (b). The XML-based agent receives the getMethod message, it calls the SoapInterfaceService whose location is 'http://hostname:8080/axis/SoapInterface.jws'. This operation includes following parameters: 'sourceFile', and 'xpath'. The parameter 'sourceFile' is a source file which contains the management information. The parameter 'xpath' is an XPath expression to access the specific part of the management information to retrieve it.

## 4.2. *Architecture of XML-based manager*

The XML-based manager needs to define management information added to the management information of the XML-based agent, retrieve the management information from the

Table 1
WSDL definition of getMethod operation

(a) WSDL elements

| Definitions | Name | Remarks | |
|---|---|---|---|
| Message | getMethod | parameter sourceFile (type:string) parameter xpath (type:string) | |
| Message | getResponse | parameter getResponse (type:string) | |
| PortType | SoapInterface | operation get | input message getMethod output message getResponse |
| Binding | SoapInterface– SoapBinding | operation get | input getMethod output getResponse |
| Service | SoapInterfaceService | port SoapInterface | address location http://hostname:8080/axis/SoapInterface.jws |

(b) SOAP request message: getMethod
```
<?xml version="1.0" encoding="UTF-8"?>
<SOAP-ENV:Envelope
   xmlns:SOAP-ENV="http://schemas.xmlsoap.org/soap/envelope/"
   xmlns:SOAP-ENC="http://schemas.xmlsoap.org/soap/encoding/"
   xmlns:xsi="http://www.w3.org/2001/XMLSchema-instance"
   xmlns:xsd="http://www.w3.org/2001/XMLSchema"
   xmlns:xconf="http://tempuri.org">
<SOAP-ENV:Body id="1" SOAP-ENV:encodingStyle="http://schemas.xmlsoap.org/soap/encoding/">
   <xconf:getMethod>
     <SourceFile>RFC1213-MIB.xml</sourceFile>
     <xpath>//sysDescr</xpath>
   </xconf:getMethod>
</SOAP-ENV:Body>
</SOAP-ENV:Envelope>
```

(c) SOAP response message: getMethod
```
<?xml version="1.0" encoding="UTF-8"?>
<SOAP-ENV:Envelope
   xmlns:SOAP-ENV="http://schemas.xmlsoap.org/soap/envelope/"
   xmlns:SOAP-ENC="http://schemas.xmlsoap.org/soap/encoding/"
   xmlns:xsi="http://www.w3.org/2001/XMLSchema-instance"
   xmlns:xsd="http://www.w3.org/2001/XMLSchema"
   xmlns:xconf="http://tempuri.org">
<SOAP-ENV: Body id="1" SOAP-ENV:encodingStyle="http://schemas.xmlsoap.org/soap/encoding/">
   <xconf:getResponse>
     <getResponse><sysDescr access="read-write">IBM RISC System/6000
       Machine Type: 0x0401 Processor id: 000106335900
     The Base Operating System AIX version: 03.02.0000.0000
       TCPIP Applications version: 03.02.0000.0000</sysDescr>
     </getResponse>
   </xconf:getResponse>
</SOAP-ENV:Body>
</SOAP-ENV:Envelope>
```

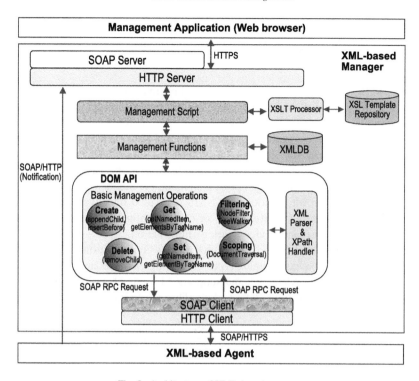

Fig. 5. Architecture of XML-based manager.

agent, then analyze the retrieved information, and present to the user. In this section, we explain the management tasks of the XML-based manager on the aspects of information modeling, management protocol, analysis and presentation.

Figure 5 illustrates the architecture of an XML-based manager. The manager includes HTTP Server and Client, SOAP Server and Client, Management Script, Management Functions Module, DOM Interface Module, XSLT Processor, XMLDB and XSLT Template Repository. The HTTP and SOAP Server/Client modules are necessary to communicate with XML-based agents and to perform management operations. The DOM API is necessary to analyze and process the management information, the XMLDB module is used to store long-term analysis data such as monitoring data. The XSLT processor combines XML document with Extensible Stylesheet Language (XSL) file and generates HTML document for the presentation.

An HTTP Server is used to provide administrators with a Web-MUI and for receiving requests from the management application and passing them to the Management Functions through the Management Script. The SOAP Server and HTTP Server are used to receive asynchronous messages for notifications from the devices via SOAP/HTTP. The SOAP Client and HTTP Client play a role in the interface module of the device and exchanges synchronous management information with the agent. The Management Functions such as object management, state management, event reporting, log control, alarm reporting, and etc. can be divided into several basic management operations such as creation, deletion, get, set, filtering and scoping. The Management Functions module uses the DOM interface to implement the management application functions because management information is represented in XML data.

**4.2.1.** *Information modeling in manager*   XML Schema can be used to define management information. Figure 6 is drawn using an editing tool called XML Spy. It shows the management information of XNMS defined in an XML Schema format. The XML Schema presented in dotted lines in Figure 6 represents optional management information and the '$0 \ldots \infty$' expression under the elements means that the number of elements ranges from zero to infinity. The basic cardinality is one when nothing is specified. A dotted box with no cardinality specification means an optional element, the number of this element is zero or one. The management information of XNMS is divided into two parts: a server part and a device part. The server part consists of server configurations, administrator lists, XML/SNMP gateway lists, and device topology information.

The device part consists of device information, device configurations, logging, and monitoring. The XML Schema has complex types containing elements and attributes and simple types containing none. XML Schema supports 44 kinds of built-in simple types, including string, integer, float, time, date, and etc. DeviceInfoType is defined as a complex

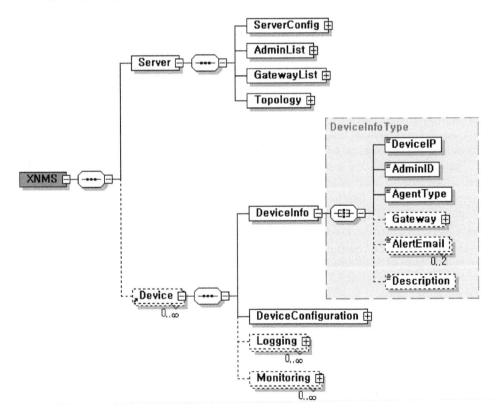

Fig. 6. Management information model of XML-based manager.

type and the type of DeviceInfo is set to DeviceInfoType. Each element, such as DeviceIP, AdminID, or AgentType, has its own data type. We add at least one attribute to the device information as a primary key for search. DeviceInfoType has the attribute string type and DeviceID unique key. If the type of an agent (AgentType) is an XML-based agent, the information from Gateway is not necessary.

The DeviceConfiguration section is defined as XML Schema converted from the MIB definition of each device as described in Section 4.1.1. This is for consistency of the management information between the XML-based manager and the SNMP agent interacting through the XML/SNMP gateway. We translate MIB II to XML Schema for basic management information. If a new device embedding an SNMP agent is added, our MIB2XML translator [35] converts the private MIB definition to XML Schema and the translated XML Schema is added to XML Schema of XML-based manager as a child element of Device-Configuration in Figure 6.

**4.2.2.** *Management protocol in manager*   SOAP is a protocol to exchange XML-based messages over HTTP or SMTP. At the basic functionality level, SOAP can be used as a simple messaging protocol and can also be extended to an RPC protocol. The XML-based manager exchanges an XML-encoded message with communicating peers, such as an XML-based agent and an XML/SNMP gateway. This means that the request message from the manager and the response message from the gateway are all formatted as an XML document. SOAP provides a standard method to transfer XML-encoded messages over HTTP between the manager and the gateway and between the manager and agent.

We define several SOAP RPC messages between an XML-based manager and an XML/SNMP gateway and between an XML-based manager and an XML-based agent using WSDL as shown in Section 5.1.3. WSDL is an XML-based language to define Web Services and describe how to access them. SOAP is becoming the de facto standard protocol for Web Services. Therefore, we apply the WSDL Schema for service description instead of a proprietary XML structure using the XML Schema. The manager finds the definition of a method to invoke using this WSDL document.

**4.2.3.** *Analysis in manager*   The Management Functions such as object management, state management, event reporting, log control, alarm reporting, and etc. can be divided into several basic management operations such as creation, deletion, get, set, filtering, and scoping. The Management Functions use the DOM interface to implement the management application functions because management information is represented in XML data.

The DOM Core defines 4 types, 20 fundamental interfaces, and 6 extended interfaces. In this section, we merely explain some fundamental interfaces to understand the following APIs for basic operations, such as creation, deletion, retrieval, and modification of objects. The DOM level 3 Core Specification includes interfaces such as Document, Node, Element, NamedNodeMap and NodeList. Table 2 shows DOM interfaces corresponding to the management operations and functions.

In addition, the DOM level 2 Traversal and Range Specification defines the framework for filtering and navigating in the filtered tree. It basically defines 4 interfaces: TreeWalker, NodeIterator, NodeFilter and DocumentTraversal interfaces. Filtering conditions are defined in the concrete class implementing NodeFilter interface and DocumentTraversal re-

Table 2
DOM interface for management operations

| Operations | DOM interfaces |
| --- | --- |
| Creation | interface Document: Node<br>• Element createElement(in DOMString tagName) → Creates an element of the type specified. Note that the instance returned implements the *Element* interface so attributes can be specified directly on the returned object.<br><br>interface Node<br>• Node appendChild(in Node newChild) raises(DOMException) → Adds the node *newChild* to the end of the list of children of this node. If the *newChild* is already in the tree, it is first removed.<br>• Node insertBefore(in Node newChild, in Node refChild) raises(DOMException) → Inserts the node *newChild* before the existing child node *refChild*. If *refChild* is null, insert *newChild* at the end of the list of children. |
| Deletion | interface Node<br>• Node removeChild(Node oldChild) raises(DOMException) → Removes the child node indicated by *oldChild* from the list of children, and returns it. |
| Navigation/retrieval | interface Node<br>• readonly attribute Node parentNode<br>• readonly attribute Node firstChild<br>• readonly attribute Node lastChild<br>• readonly attribute Node previousSibling<br>• readonly attribute Node nextSibling<br>• readonly attribute NodeList childNodes<br><br>interface NamedNodeMap<br>• Node getNamedItem(in DOMString name) → Retrieves a node specified by name.<br><br>interface Document<br>• NodeList getElementsByTagName(in DOMString tagname) → Returns a *NodeList* of all the Elements with a given tag name in the order in which they are encountered in a preorder traversal of the *Document* tree. |
| Setting values/modification | interface Node<br>• attribute DOMString nodeValue → The value of this node, depending on its type; When it is defined to be null, setting it has no effect.<br><br>interface Element: Node<br>• void setAttribute(in DOMString name, in DOMString value) raises(DOMException) → Adds a new attribute. If an attribute with that name is already present in the element, its value is changed to be that of the value parameter. This value is a simple string. |

turns traversal object such as TreeWalker or NodeIterator representing filtered-out nodes. This basic operation provides easy-to-use, robust, selective traversal of document's contents. Moreover, these DocumentTraversal API such as createNodeIterator() and create-TreeWalker() and some other APIs of DOM core API provides scoping functionality basically with the root node of the containment tree.

We can analyze management data of XML document format using the DOM Interface. As mentioned in Section 3.4, we use native XML DB to store management information. This reduces the manipulation of the XML document using the DOM Interface for inserting management information to the DB, because the XML document can be directly inserted into the native XML DB. When an analysis is requested, we extract the data from the DB by filtering and scoping using the DOM Interface and calculating the data.

**4.2.4.** *Presentation in manager*   After the management information is analyzed, the analysis result is presented to administrators. We adopted XSLT to accomplish the XML-based manager presentation. XSLT is an XML-based language that can be used to transform one class of XML document to another. An XML document can be transformed so it can be rendered on a variety of formats fitting different display requirements.

We classify types of presentation in the XML-based manager. Static information such as server configuration data, which is not specific to managed devices, can be rendered using pre-defined XSLT. Another type of presentation is to generate a dynamic web page for device configuration, which has various items and styles to present according to devices and their MIBs.

The XML-based manager maintains XML Schemas for the managed devices, and also XSL templates for the XML documents conforming to XML Schema. The XSLT template for each MIB is generated by the XSLT Generator of the XML/SNMP Gateway, and downloaded to the XML-based manager whenever the MIB is translated. Table objects defined in the MIB is presented as HTML table view using the template. The XML-based manager reduces the in-line code to control presentation logic and HTML code for work iterating whenever a device is added or an MIB module is recompiled.

**4.3.** *Architecture of XML/SNMP gateway*

An XML/SNMP gateway provides a method to manage networks equipped with SNMP agents by an XBM manager. The XBM manager transfers XML messages using SOAP/HTTP, and the SNMP agent passes SNMP messages using SNMP. The gateway converts and relays management data between the two management applications. In this section, we describe management tasks of an XML/SNMP Gateway from the specification translation and interaction translation perspectives.

Figure 7 illustrates the architecture of an XML/SNMP gateway, that is, a SOAP-based architecture of XBM manager and the gateway. In this architecture, the SOAP Client in the manager generates an XML-encoded SOAP request, such as get and set. Then the HTTP Client sends the HTTP POST request including the SOAP request in its body to the HTTP Server in the gateway. The SOAP Server parses the HTTP message into a properly formatted RPC call and invokes an appropriate method published by the gateway. This

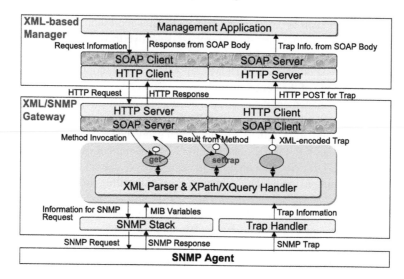

Fig. 7. SOAP-based architecture of manager and gateway.

SOAP Server receives the result of the method and generates a well-formed SOAP response message. The response message backtracks to the SOAP Client in the manager. Finally, the manager application receives the result of the method invocation. Whenever the Trap Handler receives a notification message from the SNMP agent, it invokes the DOM event for the Trap nodes in the DOM tree. For notification delivery, the SOAP Client in the gateway sends an asynchronous event message defined in the XML Trap Schema to the SOAP Server in the manager.

**4.3.1.** *Specification translation*   To interact between the XML-based manager and the SNMP agent, the XML/SNMP gateway first translates management information of SNMP SMI to XML Schema. In this section, we describe an SNMP MIB to XML Schema translation algorithm [35] for specification translation of the gateway.

In the XML/SNMP gateway, the translator converts each node of SNMP MIB into an element of the XML Schema, and the name of the MIB node into the name of the element. Interior clauses of the MIB node such as 'access', 'status', etc. are translated into the attributes of the XML element. Table 3 shows the conversion of document structure between SNMP SMI and XML Schema. The example of a specification translation result in Table 4 shows how a MIB definition is translated into XML Schema. The <syntax> clause of each node is applied to the data type definition of the element in the XML Schema. We define an additional attribute of type "ID", which has an "oid" value of a node. An "oid" attribute of type "ID" enables a random access to a particular node by its "oid" value in the DOM tree.

**4.3.2.** *Interaction translation*   Y.J. Oh et al. proposed three interaction translation methods [15] for an XML/SNMP gateway: XML Parser-based, HTTP-based and SOAP-based. The DOM-based translation provides a method for XML-based manager to directly access

Table 3
Document structure conversion

| SNMP SMI | XML Schema | DOM interface |
|---|---|---|
| MIB Module | XML document | Document |
| MIB Module name | Root Element name | Element::tagName |
| Leaf Node (macro definition) | Element with child text node(s) | Element |
| Node name | Element name | Element::tagName |
| Clauses of MIB node | Attributes of an element | Attr |
| Object Identifier (OID) | Attribute of type 'ID' | Attr |

Table 4
Example of specification translation

| MIB II | Translation result |
|---|---|
| **sysUpTime** | `<xsd:element name="`**sysUpTime**`"><xsd:complexType>` |
| OBJECT-TYPE | `<xsd:simpleContent><xsd:restriction base="TimeTicks">` |
| **SYNTAX** TimeTicks | `<xsd:attribute name="`**oid**`" type="xsd:string" use="fixed" value="1.3.6.1.2.1.1.3"/>` |
| **ACCESS** read-only | `<xsd:attribute name="`**access**`" type="xsd:string" use="fixed" value="read-only"/>` |
| **STATUS** mandatory | `<xsd:attribute name="`**status**`" type="xsd:string" use="fixed" value="mandatory"/>` |
| **DESCRIPTION** | `<xsd:attribute name="`**description**`" type="xsd:string" use="fixed" value="The` |
| "The time..." | `time....."/>` |
| ::= {system 3} | `</xsd:restriction></xsd:simpleContent></xsd:complexType></xsd:element>` |

and to manipulate management information through the standard DOM interfaces. The HTTP-based is a URI extension based translation applying XPath and XQuery, which enables to easily define detailed request messages and thus provides efficiency improvement in XML/HTTP communication between the manager and the gateway. In the SOAP-based translation, the gateway advertises its translation services to the manager using SOAP.

To integrate the XML-based manager with SNMP agents, we select the SOAP-based interaction translation method in this chapter. We define services into three big operations, namely get, set and trap. Each operation has a parameter and a return value, which is also defined as an XML element between the manager and the gateway. In addition to the essential elements, such as get, set and trap, the SNMP GetBulk operation or other complex types of requests can be defined by extending the get or set operation with XPath or XQuery. As described in Section 2, XPath, XQuery and XUpdate provide an efficient means to indicate the managed objects to be retrieved. XPath, XQuery and XUpdate can also be applied in the SOAP request message as a parameter of each method.

## 5. Implementation experience

We have implemented an XNMS based on the proposed architecture in Section 4. The manager uses various XML technologies to provide Web-based user interfaces, to communicate between manager and agent, and to process management information. We implemented the XNMS on a Linux OS using Java language. We used the Apache software

[1] that provides APIs that were implemented with JAVA. XNMS uses the following APIs: Xerces as an XML Parser, Xalan as an XPath handler and an XSLT processor, Xindice as XMLDB and AXIS as SOAP. XMLDB supports XML technologies, such as XPath, XQuery and XUpdate to directly handle XML documents via the DOM parser.

The XML-based agent is applied to the IP sharing device. The IP sharing device equipped with our XML-based agent runs on an embedded Linux based on the linux2.2.13-7 kernel using Motorola's MPC850DE processor with a 16 MB ROM. We used a powerpc-linux-gcc compiler. We implemented MIB-II information and basic configuration information in the IP sharing devices. The agent uses gSOAP [16] implemented with C/C++ languages to exchange SOAP messages. The gSOAP generates a stub, a skeleton and a WSDL definition using the header file which declares RPC operations. We selected the libxml which is more lightweight than most XML parsers in the agent because we considered the low computing resources of the IP sharing device.

The XML/SNMP gateway has been implemented on a Linux server. We used the Apache Tomcat 4.0 for the Web server and Servlet engine. We used Xerces for an XML Parser, Xalan for an XPath/XSLT processor. We also used Innovation's HTTP Client for the HTTP Client and OpenNMS's joeSNMP 0.2.6 for the SNMP Handler and the Trap Handler. These are all Java-based. We used Apache Axis for the SOAP engine, which is the SOAP and WSDL compliant. We deployed the existing translation functions into Web Services using a Java Web Service (JWS) provided by Axis. This deployment method automatically generates a WSDL file and proxy/skeleton codes for the SOAP RPCs. In order to enable the existing Java class as a Web Service, we simply copy the Java file into the Axis Web application, using the extension '.jws' instead of '.java' We deployed the main class namely *SoapInterface* into Web Services as an interface to the SOAP-based XML/SNMP gateway. We installed the Axis SOAP engine both on the manager and the gateway.

The management functionalities of the XNMS can be easily and quickly developed from the support of the standard API and the database. We were able to easily develop analysis management functionalities using standard DOM interfaces. DOM interface results in fast and easy development of the XNMS and saves development time and cost. Building a user interface composed of a number of HTML pages is repetitive and time-consuming work in an application development. We reduced a significant amount of designing Web-MUI using reusable XSLT and generated an XSLT template from an MIB definition. We used freely available codes to process the XML.

## 6. Concluding remarks

In this chapter, we presented how XML technologies can be applied to network management tasks: modeling management information, instrumenting managed resources, management operations, communication between manager and agent, analyzing the data and presenting the analysis result to users.

This chapter also presented an architectural framework for the development of XML-based network management systems. This framework can be used as a guideline for developing management systems. For validating our proposed architecture, we developed an XML-based network management system (XNMS) [5] consisting the XML-based agent, the XML-based manager, and the XML/SNMP gateway. Also, we developed an XML-based configuration management system (X-CONF) for distributed systems [3] and XML-based configuration management system (XCMS) for IP network devices [4]. Our XNMS fully utilizes XML technologies, such as the XML Schema, DOM, XPath, XQuery and XSLT, to perform network management tasks. We were able to reduce the development cost of the management system through the support of standard APIs for processing XML documents.

XML technology is viewed by many as a revolutionary approach for solving the problems that exist in current standards and practices for network and systems management. More work is needed to prove this by developing not only standards but also applying to manage real networks and systems.

# References

[1] Apache Group, *Apache*, http://www.apache.org/.

[2] Avaya Labs, *XML based management interface for SNMP enabled devices*, http://www.research. avayalabs.com/user/mazum/Projects/XML/.

[3] H.M. Choi, M.J. Choi and J.W. Hong, *Design and implementation of XML-based configuration management system for distributed systems*, Proc. of the IEEE/IFIP Network Operations and Management Symposium (NOMS 2004), Seoul, Korea (2004), 831–844.

[4] M.J. Choi, H.M. Choi, H.T. Ju and J.W. Hong, *XML-based configuration management for IP network devices*, IEEE Communications Magazine **42** (2004), 84–91.

[5] M.-J. Choi, J.W. Hong and H.-T. Ju, *XML-based network management for IP networks*, ETRI Journal **25** (2003), 445–463.

[6] Cisco Systems, *Cisco configuration registrar*, http://www.cisco.com/univercd/cc/td/doc/product/rtrmgmt/ie2100/cnfg_reg/index.htm.

[7] V. Cridlig, H. Abdelnur, J. Bourdellon and R. State, *A netconf network management suite: ENSUITE*, Proc. of IPOM 2005, Barcelona, Spain (2005), 152–161.

[8] V. Cridlig, R. State and O. Festor, *An integrated security framework for XML based management*, Proc. Of the IFIP/IEEE International Symposium on Integrated Network Management (IM 2005), Nice, France (2005), 581–600.

[9] F. Curbera, M. Duftler, R. Khalaf, W. Nagy, N. Mukhi and S. Weerawarana, *Unraveling the Web services web: an introduction to SOAP, WSDL, and UDDI*, IEEE Internet Computing **6** (2002), 86–93.

[10] IETF, *Network configuration* (Netconf), http://www.ietf.org/html.charters/Netconf-charter.html.

[11] H.T. Ju, M.J. Choi, S.H. Han, Y.J. Oh, J.H. Yoon, H.J. Lee and J.W. Hong, *An embedded web server architecture for XML-based network management*, Proc. of IEEE/IFIP Network Operations and Management Symposium (NOMS 2002), Florence, Italy (2002), 1–14.

[12] J.P. Martin-Flatin, *Web-based management of IP networks and systems*, Ph.D. Thesis, Swiss Federal Institute of Technology, Lausanne (EPFL) (2000).

[13] L.R. Menten, *Experiences in the application of XML for device management*, IEEE Communications Magazine **42** (2004), 92–100.

[14] OASIS, *Universal description, discovery and integration (UDDI)*, http://www.uddi.org/.

[15] Y.J. Oh, H.T. Ju, M.J. Choi and J.W. Hong, *Interaction translation methods for XML/SNMP gateway*, Proc. of DSOM 2002, Montreal Canada (2002), 54–65.

[16] A. Robert, *gSOAP: Generator tools for coding SOAP/XML web service and client applications in C and C++*, http://www.cs.fsu.edu/~engelen/soap.htm/.

[17] J. Schonwalder, A. Pras and J.P. Martin-Flatin, *On the future of Internet management technologies*, IEEE Communications Magazine **October** (2003), 90–97.

[18] P. Shafer and R. Enns, *JUNOScript: An XML-based network management API*, http://www.ietf.org/internet-drafts/draft-shafer-js-xml-api-00.txt (2002).

[19] M.H. Sqalli and S. Sirajuddin, *Static weighted load-balancing for XML-based network management using JPVM*, Proc. of MMNS 2005, Barcelona, Spain (2005), 228–241.

[20] F. Strauss et al. *A library to access SMI MIB information*, http://www.ibr.cs.tu-bs.de/projects/libsmi/.

[21] F. Strauss and T. Klie, *Towards XML oriented internet management*, Proc. of IFIP/IEEE International Symposium on Integrated Network Management (IM 2003), Colorado Springs, USA (2003), 505–518.

[22] W3C, *Simple API for XML version 2.0*, W3C Recommendation (1999).

[23] W3C, *XSL transformations version 1.0*, W3C Recommendation (1999).

[24] W3C, *Extensible markup language (XML) 1.0*, W3C Recommendation (2000).

[25] W3C, *Extensible stylesheet language (XSL) version 1.0*, W3C Recommendation (2000).

[26] W3C, *Document object model (DOM) level 2 core specification*, W3C Recommendation (2000).

[27] W3C, *XML schema part 0,1,2*, W3C Recommendation (2001).

[28] W3C, *SOAP version 1.2 part 0: primer*, W3C Working Draft (2001).

[29] W3C, *XML path language (XPath) version 2.0*, W3C Working Draft (2002).

[30] W3C, *XQuery 1.0: An XML query language*, W3C Working Draft (2002).

[31] W3C, *Web services description language (WSDL) version 1.2*, W3C Working Draft (2002).

[32] WBEM, *WBEM Initiative*, http://www.dmtf.org/wbem/.

[33] XML:DB, *XUpdate*, Working Draft – 2000-09-14, http://www.xmldb.org/xupdate/xupdate-wd.html.

[34] S.M. Yoo, H.T. Ju and J.W. Hong, *Web services based configuration management for IP network devices*, Proc. of MMNS 2005, Barcelona, Spain (2005), 254–265.

[35] J.H. Yoon, H.T. Ju and J.W. Hong, *Development of SNMP-XML translator and gateway for XML-based integrated network management*, International Journal of Network Management (IJNM) **13** (2003), 259–276.

# – 2.5 –

# Open Technology

J. Scheerder[1], C.P.J. Koymans[2]

[1]*E-mail*: *js@xs4all.nl*

[2]*Informatics Institute, University of Amsterdam, The Netherlands*
*E-mail*: *ckoymans@science.uva.nl*

*Value your freedom or you will lose it, teaches history. 'Don't bother us with politics',*
*respond those who don't want to learn.*
*Richard M. Stallman*

## 1. Introduction

This chapter discusses Open Technology, a concept that encompasses notions of Open Standards, Open Source, and Open Software.

The word 'open' has established itself as an umbrella under which many banners are waved. Much of this stems from the so-called Free Software movement [9,17,19], in which 'free' denotes 'without restrictions' (*libre*, or 'free as in 'free speech'') rather than 'without charge' (*gratis*, or 'free as in 'free beer'') – as extensively advocated by Richard Stallman [18].

The 'open' adjective is also proudly presented in a host of software projects, some of significant prominence in their field: OpenBSD, OpenGL, OpenLDAP, OpenOffice.org, OpenPGP, OpenSSH, OpenSSL and OpenVPN, to name but a few. A number of standards, as defined by standardisation organisations, also boasts the 'open' attribute.

That's a lot of buzz. Let's look at what's under these banners.

---

HANDBOOK OF NETWORK AND SYSTEM ADMINISTRATION
Edited by Jan Bergstra and Mark Burgess

## 2. Open

Openness plays a significant role in many contexts. We discuss a few of these contexts.

### 2.1. *Open standards*

A *standard* is a specification for a technology or methodology. An *open standard* is such a specification that is publically available, in such a way that anyone who wants to can obtain it and use it to create an implementation of the specified technology or methodology. A particular product can, by definition, never be a standard; calling a product a standard is a basic category mistake. While not a standard in itself, it can however of course be an *implementation* of a standard. For software systems, a particular implementation, often called a *reference implementation* in this case, can play a role in the definition of a standard.

On the Internet for example, interoperability between the vast hordes of connected systems, consisting of many different makes and models of systems running a mindbogglingly vast arsenal of different software and software versions, is achieved on the basis of the so-called *Internet Standards*: standards ratified by the Internet Engineering Task Force (IETF) after a process of extensive peer review [2].

An example of an open standard is Postel's 'Transmission Control Protocol' [15] that defines TCP, which is the very mechanism used by high-level protocols such as SMTP (mail transfer, which is of course yet another open standard [12]) and HTTP (datatransfer on the world wide web – [3]).

**2.1.1.** *Standardisation and standards compliance*   *Standardisation* is the process of building functionality based upon standards, using standards as building blocks. Technology is said to comply, or to be *compliant*, to a particular standard if it correctly and faithfully implements that particular standard to both its letter and its spirit. *Standards conformance testing* is the verification of an implementation against a particular standard, or set of standards. Software that adheres to a corpus of standards, which is to say to those parts of that corpus relevant to the software's functionality, is often called *standards compliant*. Typically, the Internet Standards are the corpus of standards intended when standards compliance is considered.

The litmus test of standards compliance of a particular piece of software is substitutability: standards-based systems function as freely exchangeable components. For example, in a standards-compliant mail system the SMTP server can be exchanged for another SMTP server without penalty, as long as the replacement SMTP server works correctly in the sense that it also properly implements the SMTP standards. Standards-based software can freely be substituted for other standards-based software. Standardisation levels the playing ground for software this way, allowing full freedom of choice between standards-compliant software systems, and as a side effect seriously alleviating critical software dependency and legacy issues.

**2.1.2.** *Closed technology, lock-in and legacy problems*   Technology is called *closed* or *proprietary* if it is not open. Closeness can wear many coats and colours; a specification (or an implementation serving as an implicit specification) may not be published at all, but it might also be subjected to legal restrictions. There is no implicit reason why closed, non-standards based, software should be freely interchangeable, and in practise we can observe that it is not. The situation where a software system cannot freely be replaced by another software system is commonly called *lock-in*, often in the forms *vendor lock-in* and *lock-in!product*. Lock-in is usually considered undesirable in itself: not having the option to exchange the software system for another, its vendor can attach an arbitrary price tag and impose arbitrary conditions (force the use of other products, for example) and restrictions (impose arbitrary interoperability barriers, for example). Furthermore, lock-in poses continuity risks. If a vendor changes a software system one is locked-in to in a such way that its functionality changes significantly, or if the vendor drops the product altogether, one is out of options, and left empty-handed in the end.

Without underlying standards, standardisation – at least in its true sense, in which it relates to the application of actual standards – cannot take place. However, in the very real world of current IT practice the word 'standardisation' is often an oxymoron, commonly commandeered to refer to its very opposite [13]: the selection of a particular product, or a particular line of products by a particular vendor, rather than bearing any relation to an actual standard.

## 2.2. *Open source*

**2.2.1.** *What is open source?*   Software is *open source* if its source code is available to the general public without restrictions that limit studying it, changing it, or improving upon it. In legal terms, open software is published under an Open Source *license*. Several such licenses, and licensing strategies, exist.

The Open Source Initiative (OSI)[1] is a non-profit organisation 'dedicated to managing and promoting the Open Source Definition for the good of the community' [14]. It registers, classifies (and certifies) software licenses, and seeks to explicitly define criteria and meta-criteria for Open Source. To do so, OSI publishes a document called 'The Open Source Definition', based upon work by Bruce Perens.

OSIs definition demands, paraphrased here:

- No restrictions on redistribution of the software;
- Source code must be included, or at least be easily obtainable, without charge and without obfuscation;
- Modification and creation of derived work is explicitly permitted;
- Distribution of modified source code may be restricted only if orthogonal modifications ('patch files') are explicitly allowed, and if distribution of software built from modified source code is explicitly allowed. For such modified software, the license may require that modifications are clearly reflected in name or version number;
- The license must not discriminate against persons or groups in any way;
- The license may pose no restrictions on the way the software's use is allowed;
- When the software is redistributed, the license still applies and travels along;

---

[1] This should not be confused with ISOs OSI, Open Systems Interconnection from the International Organisation for Standardisation.

- The license is not specific to a particular product; if it is extracted from a particular distribution and redistributed, the license still applies;
- The license must be technology-neutral, imposing no conditions on the way it is accepted.

In the case that the design of a particular system is unspecified or undisclosed an open implementation, accessible to anyone who wants, may even play the role of an implicit specification. The behaviour of the system can be inspected, and its design may, in principle, be extracted from that particular incarnation.

**2.2.2.** *Examples of open source licenses*   The two most popular licenses, or license schemes, are:

- The BSD License [11] is actually a boilerplate text to serve as a licensing scheme. It is probably the most senior Open Source license, and huge amounts of software distributions, including widely used BSD Unix family members, use it. The BSD license typically poses no restriction whatsoever on use and modification of the distributed source code, although it typically does require proper attribution.
- The GNU General Public License (GPL) is a license that 'is intended to guarantee your freedom to share and change free software' [5]. The Linux kernel, as well as a significant amount of the 'userland' software constituting most Linux (or, GNU/Linux, as some would have us say for that reason) is distributed under version 2 of the GPL, and is likely to stay that way. The GPL 'update' to GPLv3 [6] is intended to replace version 2 for the bulk of the GNU software tree. The GPL has restrictions on reuse and redistribution of distributed software, requiring changes and derived work to be published under the GPL as well. This is why the GPL is sometimes referred to as a 'viral' license. Version 3 poses some more restrictions, mainly in the area of 'digital rights management', excluding the use of GPLv3 code in contexts in which the software is used to limit the users' possibilities to manipulate data freely.
  The GNU Lesser General Public License (LGPL) [7], originally called the GNU Library General Public License, is a GPL-variant designed to offer some leeway for using libraries in proprietary software.

Other popular licences, not further discussed here, are the Apache License [4], used by the Apache Software Foundation[2] projects, including the world's dominant Apache webserver, and the Mozilla Public License [8], which applies to the widely popular software family originating from the Mozilla project,[3] including the Firefox browser and the Thunderbird mail/news client.

### 2.3. *Open software*

Even if the specifications of a particular technology are freely available, and even if the source code of an implementation of a particular technology is freely available as open source software there still can be ways in which it is not open.

---

[2]See http://www.apache.org/.

[3]See http://www.mozilla.org/.

*Portability* is the property of software to be easily transferable between various systems. Consider, for example, a particular software package that relies heavily upon details of the underlying operating system it is built on, or on specific details of the build system. Such a software package may be published as open source software, and implement an open standard, yet may be usable on a single computer only – its developer's workstation.

On the other hand, the more generic in nature the requirements a software package poses are, the more portable the software will be. For example, a software package that merely assumes the presence of a POSIX [10] environment[4] will typically build and run trivially on any convincingly POSIX-like system.

Open software feeds on solid portability. Portable software easily migrates between different environments, with every new environment, and every new application offering the possibility of new insights, further improvement – or simply bringing previously undetected errors in design or implementation (bugs) to light.

Portability is widely seen as a direct beneficial influence on software quality; the argument is a simple chain argument that connects portability to code clarity and portability to the notion of code understandability, and, finally, code quality. This is especially valid in the security arena.

There is more to portability than just portable source code. In order to make changes one needs a reasonable understanding of a software package's architecture. Documentation may be important to disclose that architecture sufficiently. Furthermore, the way the software is actually built is important. Also, the intentions, the project goal, of a software package can be important to those wishing to add to it.

## 2.4. *Open security*

Often, security is seen as protecting secrets. However, it is typically openness rather than secrecy that benefits security most.

**2.4.1.** *Inspection*   The notion of security is based on the distinction between *intended* and *unintended* use of facilities and information. Security is about protecting facilities and information from unintended use, often called *abuse* in this context.

It is, unsurprisingly, very hard to determine the full scope of possible (ab)use offered by a particular technology, and it gets harder the lesser is known about the technology.

The possibility of *actually inspecting* systems is of vital importance when one wishes to know what the system actually does, and if one wishes to analyse the consequences of having that system present. This analysis, however, is needed to assess the risks associated to the system. Furthermore, an insight into the scope of (ab)use of a system provides a context for its deployment, and will help determine operational policies surrounding it.

By definition, a closed system is a 'black box' that discloses its inner workings indirectly at best. In plain words, one cannot tell what a closed system actually does, and one cannot even begin to assess whether or not it will do what it should do.

---

[4]Also commonly known as 'The Single UNIX Specification', of which Version 3 is an IEEE standard (STD 1003.1).

From the viewpoint of security, being able to assess whether a system's workings are not in conflict with existing security policies is essential. Open technology offers full transparency, and allows for any analysis. Closed technology offers much narrower possibilities in that respect.

The expression 'security by obscurity' is often used by security practitioners to refer unfavourably to closed architectures. From the arguments above, it is evident that restricting access to content offers no particular protection against a flawed design.

**2.4.2.** *Exploits.*   It is not uncommon for technology abuse to feed on technological errors. A bug in a program may, contrary to intentions, expose an entire computer system or network, allowing an attacker to leverage this particular malfunction to gain access, escalate privileges, bypass access restrictions, and so on.

For example, a simple 'buffer overflow' in a mail program's message display code may cause arbitrary programs to be executed on a person's computer. A message that is especially crafted to trigger and take advantage of this bug need only be sent to a person using the mail program that suffers this bug. Upon opening (displaying) the message, havoc is wreaked.

Software bugs constitute security problems. The fewer and the less harmful the bugs, the better for security.

**2.4.3.** *Code quality.*   Now if one can actually inspect both a program's operational behaviour as well as its inner workings, its source code, bugs can be fixed by a wider audience of inspectors. One can even actively look for possible bugs, and mitigate them before they cause damage. Open technology is, by definition, open to public scrutiny. This arguably has positive implications for code quality, and as a direct implication for program security. As Eric S. Raymond famously put it in '*The Cathedral and the Bazaar*': "Given enough eyeballs, all bugs are shallow [16]." Conversely, with no eyeballs whatsoever bugs might very well turn out to be surprisingly deep. The implied ramifications for security are obvious.

Additionally, the aforementioned improvement of code quality as a result of requiring portability, and for portability's sake, program clarity, deserves another mention here.

## 3. Open software as a scientific publication

Interesting parallels can be observed between the process of scientific publication, and the process of creating open software: Open software publication.

Eric Raymond's provocative stance evoked an interesting response from 'observer and researcher on the open source process' Nikolai Bezroukov to write up a critique [1]. In direct argument with '*The Cathedral and the Bazaar*', Bezroukov seeks to deflate the 'distortion of reality' he claims Raymond perpetrates, judging *The Cathedral* to convey an overly optimistic and unrealistic view of open source.

After these initial skirmishes, things calm down to an analysis in which Raymond's position is put in perspective. In doing so, Bezroukov outlines a number of practical pitfalls, and does not hesitate to see eye to eye with the grim realities that are *also* present in the realm of open source. In the end, the 'open' bit of the process emerges largely unscathed however, with a final word of warning against unrealistic expectations.

In this rumble one particular interesting digression emerged. Bezroukov invokes an analogy between the open source and scientific communities, stating that "the OSS [*Open Source Software, ed.*] community is more like a regular scientific community than some OSS apologists would like to acknowledge". The analogy, Bezroukov argues, actually even carries so far as to enclose even the typical problems the scientific communities suffer.

Similarities are found in the kind of participants both tend to have: both have professional participants, as well as non-professional ones, with significant developments from both professionals as well as non-professionals. Those participating are, in the end, not in it for the money; people participate in science, as well as in open technology, for intrinsic other reasons. The core processes display a clear similarity. As Bezroukov states, "[...] creation of a program is similar to the creation of applied theory. I would like to classify programming as a special kind, or at least a close relative, of scientific [*activity, ed.*]".

We submit that Bezroukov's analysis in this respect makes sense. Publishing software in the context of open technology essentially means putting it up for public scrutiny. Peers can review it, and Newton's process of 'standing on the shoulder of giants' is in full force.

## 4. Conclusion

Open Technology thrives and feeds on Open Standards and Open Source, and is better characterised as a process and attitude similar to the scientific process than by technological aspects. Openness has a vital influence on security and reduces costs by levelling the playing field for product selection, and helps battle inflexibility and legacy-problems.

## References

[1] N. Bezroukov, *Open source software development as a special type of academic research (critique of vulgar raymondism)*, http://www.firstmonday.org/issues/issue4_10/bezroukov/ (1999).

[2] S. Bradner, *The internet standards process – revision 3*, http://www.ietf.org/rfc/rfc2026.txt (1996).

[3] R.G.J. Fielding, J. Mogul, H. Frystyk, L. Masinter, P. Leach and T. Berners-Lee, *Hypertext transfer protocol – http/1.1*, http://www.ietf.org/rfc/rfc2616.txt (1999).

[4] A.S. Foundation, *Apache license*, http://www.apache.org/licenses/LICENSE-2.0.html (2004).

[5] F.S. Foundation, *GNU general public license (v2)*, http://www.gnu.org/copyleft/gpl.html (1991).

[6] F.S. Foundation, *GNU general public license (v3)*, http://gplv3.fsf.org/ (2006).

[7] F.S. Foundation, *GNU lesser general public license*, http://www.gnu.org/copyleft/lesser.html (2006).

[8] M. Foundation, *Mozilla public license*, http://www.mozilla.org/MPL/MPL-1.1.html.

[9] FSF, *The free software foundation*, http://www.fsf.org/.

[10] O. Group, *JTC1/SC22/WG15 – POSIX*, http://www.open-std.org/jtc1/sc22/WG15/.

[11] O.S. Initiative, *The BSD license*, http://www.opensource.org/licenses/bsd-license.php (1998).

[12] J. Klensin, *Simple mail transfer protocol*, http://www.ietf.org/rfc/rfc2821.txt (2001).

[13] G. Orwell, *Nineteen-Eighty-Four*, Secker & Warburg (1949).

[14] OSI, *Open source initiative*, http://www.opensource.org/.

[15] J. Postel, *Transmission control protocol*, http://www.ietf.org/rfc/rfc0793.txt (1981).

[16] E.S. Raymond, *The Cathedral and the Bazaar*, http://www.catb.org/esr/writings/cathedral-bazaar/cathedral-bazaar/ (2000).

[17]  W.F. Software Movement, *Free software movement*, http://en.wikipedia.org/wiki/Free_software_movement.

[18]  R.M. Stallman, *Why software should be free*, http://www.gnu.org/philosophy/shouldbefree.html (1992).

[19]  M.H. Weaver, *Philosophy of the GNU Project*, http://www.gnu.org/philosophy/philosophy.html (2005).

– 2.6 –

# System Backup: Methodologies, Algorithms and Efficiency Models

Æleen Frisch

*Exponential Consulting, 340 Quinnipiac St., Bldg. 40, Wallingford, CT 06492, USA*
*E-mail: aefrisch@lorentzian.com*

## 1. Introduction

Preventing and remediating data loss has been one of the prime responsibilities of system administrators for almost as long as there have been computers with permanent storage. Indeed, in the earliest days of time-sharing systems, an effective backup plan was crucial given the greater instability of operating system and other software – and attendant system crashes – than are typical now. However, recent advances in storage technologies and widespread changes in computer usage patterns have served to complicate traditional backup procedures and policies, and the area remains one of significant importance and active research.

There are many reasons why system backups remain an ongoing area of discussion and concern rather than being a solved problem. Much of the complexity comes from heterogeneity of computer environments: different hardware types, from different vendors, running different operating systems, purposed in a variety of ways (e.g., from workstations to a multitude of special-purpose servers), containing data in varying formats, stored both locally and remotely, and having a spectrum of purposes and importance. This chapter considers system backups in terms of a set of mutually orthogonal concerns, including media selection, archiving algorithms, backup software architecture, reliability issues, and policy efficiency and optimization.

HANDBOOK OF NETWORK AND SYSTEM ADMINISTRATION
Edited by Jan Bergstra and Mark Burgess

## 1.1. *Causes of data loss*

Completed backups function as insurance against data loss. In designing a backup procedure, it is often helpful to identify the causes of data loss within a specific environment.
Smith's extensive survey of computer data loss information for 2003 [62] yielded some
interesting statistics, summarized in Table 1. The table lists percentages for the various
categories in terms of both the fraction of losses that they represent and the fraction of
all computer systems affected by them. In this context, data loss refers exclusively to data
stored on disk which becomes unreadable or unavailable.

Hardware failure results in largest fraction of losses at about 40%, affecting about 2.5%
of all computer systems annually. Human error is the next largest category at about 30%
(affecting slightly under 2% of systems). The other major precipitators of data loss include
software corruption, viruses and data loss which is a side effect of the theft or accidental
destruction of the computer system itself by natural forces.[1]

Figure 1 plots estimated values for these same categories at approximately 10 year intervals. While this data is more anecdotal than scientific,[2] the chart nevertheless illustrates
some interesting trends. For example, hardware failure is a larger percentage of total losses
than in earlier periods. Note that this does not mean that there are more instances of loss
data due to hardware failure, but merely that the fraction of those associated with this risk
has increased.

Human error has been a significant factor in data loss continuously. However, its contribution to the overall loss rate peaked sometime in the 1990s, which coincides with the rapid
expansion of the computer user base due to widespread deployment of personal computer-
class machines to people with little or no computer experience. The fact that human error
results in a smaller percentage of data loss currently is due to improvements in software
reliability and file system design more than to the modest increase in sophistication of the
average user.

Widespread use and portability of computer systems as well as the existence of the Internet have introduced new potential for data loss along with enhancements to convenience
and productivity. For example, hardware destruction due to fire, flood and the like was
only a very occasional event before computers were removed from protected data centers. Nowadays, little attention is given in general to where computers are placed unless
computer itself is costly.

Identifying the likely sources of data loss within an environment is an important step in
designing an effective backup plan. For example, if human error is the most important risk,
then backup procedures which enable rapid restoration of single files are to be preferred
over ones which allow entire file systems to be rebuilt most efficiently. Thus, such a data
loss analysis is a worthwhile endeavor for any system administrator responsible for system
backups.

---

[1]Data loss from computer damage caused by user errors – e.g., dropping the computer system or spilling
beverages on it – is included in the 'human error' category.

[2]Specific data of the type used by Smith is not available for any but the most recent years. The chart illustrates
my estimation of overall loss rates due to the various causes for the environments with which I was familiar during
the various eras.

Table 1
Causes of data loss in 2003

| Cause | % losses | % systems |
|---|---|---|
| Hardware failure | 40.2 | 2.4 |
| Human error | 29.2 | 1.8 |
| Software corruption | 12.8 | 0.8 |
| Viruses | 6.4 | 0.4 |
| Theft | 8.7 | 0.5 |
| Hardware destruction | 2.7 | 0.2 |
| All losses | 100.0 | 6.0 |

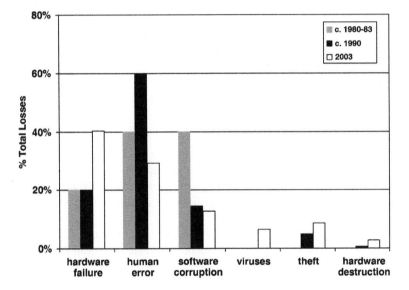

Fig. 1. Electronic data loss categorized by primary cause.

## 2. General backup considerations

There is an extensive literature on backups dating from the mid-1970s. [5,22,27,34,36,41, 45,46,51,63,72] are among the most important general works. Practical discussions also abound, including ones for UNIX and similar operating systems [22,46] and the Windows environment [34].

Certain truisms are well established and universally accepted at this point. They include the following axioms.

AXIOM 1. *System administrators bear the ultimate responsibility for seeing that effective backups are accomplished within their administrative sphere.*

AXIOM 2. *Effective backup procedures require explicit advance planning.*

AXIOM 3. *The most effective backup strategies consider the problem in the context of networks of computers rather than individual systems in isolation.*

The first principle stems from two facts of life. First, backups have always been and remain part of the system administrator's job. Even when the system administrator does not perform the actual backup operations himself, he is still generally responsible for the design and overall effectiveness of the scheme. Second, ordinary users have demonstrated over time that they will not perform backups with any degree of regularity or reliability (see, e.g., [38]). Moreover, personal experience of data loss does little or nothing to change user behavior. As a result, backup remains a centralized administrative function in an age in which the trend is toward diffuse, decentralized control.

The second and third principles require an in depth analysis of the site's[3] data and the potential ways in which it might be lost and need to be restored. The sorts of questions that need to be considered include the following:

### 2.1. *What data needs to be backed up?*

With the exception of temporary scratch files, virtually all computer data needs to be protected against loss by backups. In order to plan effectively, however, it is necessary to move beyond this trivial statement and to classify the different types of data generated and held by the organization since these distinct categories of information can have widely differing backup and restoration requirements.

For example, while operating system files are essential to proper system functioning, they are typically backed up only rarely since they are generally available on the original operating system distribution media and since they seldom change. At the other extreme, files containing users' electronic mail messages require frequent backups since their contents are not replicated elsewhere, are not reproducible (as, for example, results of various numerical calculations might be), and change frequently.

### 2.2. *How often does the data change?*

Backups are a form of maintenance work. It shares the characteristic of such jobs that new work is being generated even as the current operation runs to completion. Obviously, files that are modified frequently need to be backed up frequently as well. Moreover, and more importantly, determining and understanding specific, numerical patterns of modification within the file system is a key part of designing an efficient backup schedule. The most sophisticated offerings among the available backup software automate a significant portion of this task.

---

[3]Or the subset of the site that is administered as a unit for backup purposes.

## 2.3. *How might the data be lost?*

As we have seen, files can be lost or become corrupted due to a variety of causes. If some causes are much more likely than others in a given environment, then the backup plan ought to take this factor into account. For example, if the most likely of loss is hardware failure, then backups containing images of complete disk partitions will provide the most rapid recovery path. On the other hand, if accidental deletion by a user is the most likely cause of loss, then a system optimized for restoring individual files make more sense. In this way, an analysis of the most significant threats to the data within a system or site can again be crucial to developing an appropriate backup strategy.

## 2.4. *Where are the files that need to be backed up and how will they be identified?*

The time required to locate the data that needs to be saved for any given backup operation can significantly affect the overall efficiency. Ideally, the time of the most recent backup is recorded for each item as it is saved, enabling relevant items to be selected automatically for future backup operations.

Historically, this data has been recorded in a variety of locations, most commonly within a field in the file header (metadata), or within a database maintained by the backup software. Current software generally uses the latter approach, and the most sophisticated packages employ a relational database for storing per-item backup data. However, even very simple schemes can be effective in some circumstances. For example, a simple timestamp file can be used to record the most recent backup time on a small, isolated system.

The following sample archiving commands illustrate the preceding strategies. The first command comes from the VAX/VMS operating system, and it backs up all files which have been created or modified since the previous backup. The second command comes from a traditional UNIX system, and it performs a second-level incremental backup, saving all files created or modified since the most recent first-level incremental backup, as recorded in its database (*/etc/dumpdates*). The third command, from a Linux system, locates and saves all files which have modification dates which are more recent than that of the */usr/local/etc/last_backup* file.

```
BACKUP/RECORD DMA0:[*...]/\
   SINCE=BACKUP MTA0:19JUN84.BCK          VMS

dump 2uf /dev/rmt0 /dev/disk2e             Traditional UNIX

tar -c -f /dev/nst0 -newer-mtime/\
   usr/local/etc/last_backup &&touch/\
   usr/local/etc/last_backup               Linux (GNU tar)
```

Each of the commands also records the time of the current backup operation (via the **/RECORD** and **u** options and the **touch** command, respectively).

**2.5.** *What resources are available to perform backup operations?*

Backup procedures cannot be developed in isolation. Rather, they must take into account the resources which are available for performing the backup operations. These include not only the available software and hardware devices which will be used to record the data, but also the additional resources which they require. Thus, if the software performs data compression before recording the files, then there must be sufficient available CPU capacity on the backup server to do so. Similarly, there must be enough network bandwidth available to deliver the data from its originating systems to those where the recording devices reside at a rate which maximizes the latter's efficiency. Finally, human resources are sometimes essential to the success of a backup plan. For example, if backups requiring multiple media must be performed overnight and the recording device handles only a single item, then someone must be available to insert the second and later volumes in order for the operation to succeed.

As these examples make clear, capacity planning is an essential part of developing an effective backup plan. Leber [34] provides the most detailed treatment of this topic, relevant to any backup problem (even though the practical aspects of her work focus on the Windows environment).

Capacity planning requires a quantitative knowledge of the various parameters affecting backup operations. Two most important of these are of the total amount of data which must be saved and the amount of time available for doing so.

The total amount of data, $D$, is simply the sum of the relevant data from each of the computer systems to be backed up, typically measured in gigabytes:

$$D\,(\text{GB}) = \sum_{systems} (1+i)\,data. \tag{1}$$

The initial factor of $1+i$ takes into account that some of the data will be modified during the time period required to back all of it up once; $i$ is the fraction of the data that will be modified over that period. For example, if a system has 10 GB of data to be backed up every week, and 10% of data will change over the same time period, then its total backup data is $(1+0.1)\cdot 10 = 11$ GB.

The available backup window, $W$, is the amount of time when the resources are available for performing backups. Thus, $W$ needs to take into account the availability of the data on the individual systems, the availability of the recording devices, and the availability of sufficient network bandwidth to transfer the data from the former to the latter. In a first effort, the first of these is assumed to dominate, and the other two factors are initially neglected. Data availability is also assumed to be identical on different systems, an assumption which is generally reasonable, although most sites have at least a few deviating systems. $W$ is typically measured in hours.

The required backup rate, $R$, is defined as $D$ divided by $W$, yielding a result in units of gigabytes per hour:

$$R\left(\frac{\text{GB}}{\text{h}}\right) = \frac{D}{W}. \tag{2}$$

For example, if there is 800 GB of data that is available for backup only between the hours of 02:00 and 05:00 weekdays, then a backup rate of about 53 GB/h is needed to successfully do so. Whether this is in fact possible depends on the characteristics of the environment, including:
- The rate at which the required data can be identified and processed by the backup software.
- The achieved transfer rates of the data from their disk storage devices.
- The speed of target backup devices.
- The available bandwidth of the data transport mechanism (typically the network). Note that this factor is relevant only when backup operations target a remote backup server.

Currently, backup efficiency is typically dominated by the latter two factors.[4] The available recording capacity can be computed as the sum of the speed ($r$) and availability window ($w$) for the various backup devices; the following expression includes the conversion factor[5] necessary to express this quantity in gigabytes:

$$C\ (\text{GB}) = \sum_{devices} w\ (\text{h}) \cdot r \left(\frac{\text{MB}}{\text{s}}\right) \cdot \frac{3600}{1000}. \tag{3}$$

Clearly, if $D \leqslant C$, then there is sufficient device capacity to perform the required operation, assuming that the availabilities of the backup devices and the data coincide (or can be made to coincide by the use of an intermediate storage location).

Recording capacity per hour can be similarly calculated by omitting the individual device availability window term in Equation (3). The result can then be directly compared with the required backup rate. In our example, a rate 53 GB/h could be provided in a single drive by most tape technologies other than 4 mm (see Table 2).

The required backup rate can also be compared to available network bandwidth (B):

$$R \cdot \frac{1000}{3600} \cdot 8 \leqslant B_{\text{available}} \left(\frac{\text{Mb}}{\text{s}}\right). \tag{4}$$

Equation (4) converts $R$ two megabits per second, a unit which is directly comparable to typical measures of network bandwidth. In our example, 53 GB/h corresponds to about 118 Mb/s, indicating that a single backup server on even an idle 100BaseT network would not suffice to accomplish the work required. However, multiple servers on different subnets (provided the originating systems for the data were similarly subdivided) or a gigabit Ethernet network would work with the specified parameters (assuming a modest network load during the backup window). Since a 100BaseT network has a peak capacity of 45 GB/h, increasing the length of the backup window $W$, by a factor of 1.5 or 2 depending on typical network usage during the backup window, is another option for handling the load in that environment.

---

[4] Previously, on UNIX systems, the poor performance of backup software, especially when encountering open files, could also have a significant effect on backup performance, but this has not been the case for many years.

[5] In these expressions, $1K = 10^3$ since network speeds always use this factor and device manufacturers usually do.

The simple analysis serves to introduce the main outlines of capacity planning. However, much more sophisticated analyses of these processes are possible, examples of which will be considered later in this article.

Monetary cost is also an important factor in backup planning in some situations. For example, in environments where the cost of backup media must be carefully managed, maximizing each volume's use must be incorporated into the backup plan since doing so can significantly lower costs.

Finally, backup capacity analyses likes the ones outlined here represent idealized circumstances. Achieved device and network performance may fall short of their theoretical peaks for many reasons. For example, it may not be possible to deliver data to the tape drive at a rate sufficient to sustain its top speed. High end backup software typically includes features designed to maximize achieved data transfer rates from originating systems to backup devices for the site as a whole.

### 2.6. *How quickly does lost data need to be restored?*

Whether lost data needs to be recovered in the absolute minimum amount of time also affects backup planning and strategies. At one extreme, backup archives will need to remain online when near instantaneous restoration is required, while nearby storage will be sufficient when some delay prior to restoration is acceptable. Similarly, efficient restoration of individual files requires an equally efficient scheme for locating the desired file – and sometimes the desired version of the file – within the backup archives.

### 2.7. *Where will the data be restored?*

Backup operations are simplest when the data contained within the archive will always be restored to the same system from which it originated, using the same device upon which it was recorded. In fact, these conditions generally do hold true for most backup operations. When they do not, however, then restoring the data can be more of a challenge. There are many sources of incompatibilities between backup devices and target computer systems:

- Hardware differences between, say, tape drives from different manufacturers can cause a tape written on one to be unreadable on the other (and this can occasionally happen even with two different drives from the same manufacturer). The most common source of incompatibilities between drives of the same type from different manufacturers is in their hardware compression schemes.
- Fundamental data formats can differ between the originating system and the one where the data will be restored (i.e., big vs. little endian). However, such incompatibilities are easily resolved via byte swapping utilities and similar features built into current backup software.
- Identical or compatible backup software must be available on the target system when a proprietary format is used for the backup archive.
- When the operating systems used by the source and target systems are quite different, the data format of being restored file may be incompatible with the latter even when

the same application software is present. This sort of incompatibility still crops up, for example, with some database applications.

A related question concerns the specific means by which file restorations are accomplished. Once again, different considerations come into play depending on whether users are provided with the ability to restore their own files or they must rely on the system administrator to do so.

**2.8.** *How long does backup data need to be retained (and why)?*

The useful lifetime of most backup data is rather short, being potentially limited to the period before the same data is archived again. However, backup data is typically retained for much longer than this minimum, for a variety of reasons ranging from the (sensible) desire for multiple redundant copies of important data, to very conservative estimates of how long data might retain some relevancy, to official decisions to retain archive copies of important data for periods of several years or indefinitely.

Considerations like those in the preceding section come into play for backup archives which are retained for lengthy periods of time. In addition, proper storage is intimately related to long-term accessibility for data that will be restored on its originating system (this topic is discussed in detail later in this article).

**2.8.1.** *Legal considerations and implications*   The current trend in many parts of the world is for governments to regulate certain kinds of record-keeping more and more stringently, especially in sensitive corporate environments. In the United States, many recently enacted statutes related to corporate record-keeping have had wide-ranging ramifications for information technology in general and system administration specifically. Since backups ultimately process all of the computer-based data within an organization, such laws apply to their creation, handling and storage.

In the United States, the following are among the most important recent federal acts which affect backup data:

- The Health Insurance Portability and Accountability Act (HIPAA) specifies requirements for the protection of individuals' health-related records, requiring extensive security safeguards, including all aspects of physical and electronic access [25].
- The Gramm–Leach–Bliley Act applies to financial institutions and regulates the collection and disclosure of customer data. It includes requirements for information safeguards which must be implemented [23].
- The Sarbanes–Oxley Act of 2002 (also known as SOX 404 compliance) lays down requirements for financial reporting and record keeping for public companies, including data retention and security [47].

Going beyond regulations regarding data security and storage, a policy of retaining backup data has additional ramifications in the event of any legal action that may arise. Put simply, when an organization archives and stores its data as a matter of policy, it must be "prepared for the consequences resulting from both its existence and its content".[6] For ex-

---

[6]This statement was made by Phil Cox in the context of collecting auditing data [12], but it in fact holds true more generally for all sorts of backed up files.

ample, a Massachusetts court held that a company must surrender its backup archives when requested by its opponent in a civil action:

> While the court certainly recognizes the significant cost associated with restoring and producing responsive communications from these tapes, it agrees [that] this is one of the risks taken on by companies which have made the decision to avail themselves of the technology now available to the business world. ... To permit a corporation ... to reap the business benefits of such technology and simultaneously use that technology as a shield in litigation would lead to incongruous and unfair results [37].

Similar decisions exist in the case law of other states. There are many overviews of computer data and its uses in discovery and litigation written for attorneys. Such discussions highlight the legal issues relating to backups implicitly as they discuss their potential role in legal actions (see, e.g., [44,69]).

### 2.9. *Where will the backup data be stored?*

The ideal storage location for various sorts of backup data is obviously related to some of the preceding questions. For example, data that might need to be restored quickly ought to be stored relatively near to the restoration device. On the other hand, backups designed for long-term archiving is usually best stored off-site, in a secure, climate controlled environment.

### 2.10. *Best practices*

The most effective backup plans take all the preceding questions and issues into account when deciding on the plan's various elements: the backup software, media selection, number and placement of backup servers, the schedule of backup operations, the storage locations for archived data, and facilities and procedures for various types of data restoration, and the like.

A number of best practice principles have emerged from the collective experience of system administrators designing and implementing backup plans over the years, resulting in many papers that purport to describe them. The best such overview is still Zwicky [73], supplemented by [22,36]. Zwicky's discussion is organized around several overarching meta-goals: minimizing the costs of both performing backups and restoring files while maximizing the reliability of archived data and the flexible deployment of available computing resources as a whole.

The following comprise the most useful best practices advice – codified collective common sense – related to backup operations:

*Organizing backup operations:*

- Centralizing backup control and responsibility is usually the most efficient choice.
- Perform backups as frequently as feasible, given the constraints of the environment.

- Schedule backup operations at times of minimum system activity and network load.
- Automate routine backup operations, eliminating the need for human assistance.
- Implement separate backup operations for different purposes: everyday data protection, full system restorations, long term archiving.

*Recording archive data:*
- Purchase the fastest and highest capacity backup devices that you can afford.
- Compress backup data in some fashion (ideally, via hardware compression at the backup device).
- Verify the backup data after it is written. Periodically verify that media written on one device can be read successfully on others.
- Write protect backup media.
- Consider both security and environmental conditions when selecting storage locations for backup media.

*Planning for the long term:*
- Build excess capacity into the backup plan at all points. For example, include more backup devices and longer backup windows than the bare minimum to allow for temporary surges in backup requirements, future expansion, and protection against device failure.
- Plan for technology changes when considering long term data archiving. Doing so will lead you to consider the data format for such data, as well as the archive and media formats. Retain one or more working drive units for all archived media types (even in the face of advancing technology).

*Preparing for restoration operations:*
- Keep backup indexes/catalogues online to aid in locating required data.
- Implement a system which allows users to perform their own ad hoc restores.
- Perform test restoration operations periodically, including full system restorations.

## 3. Operational practicalities

While formulating backup plans requires taken into account a variety of media and device characteristics and system and network metrics, the mechanics of day-to-day backup operations are dominated by the backup software and the media chosen for recording the data.

### 3.1. *Backup software*

All current operating systems provide one or more software programs designed to perform backups. Indeed, this is been the case since at least the early 1970s. For example, the TOPS-10 operating system which ran on PDP 10 hardware from Digital Equipment Corp., considered by many to be the first production timesharing system, included a utility called

BACKUP. This program created a backup archive, known as the save set, as well as an optional separate index file [15].

Here is an example of BACKUP in action:

```
.R BACKUP
/BACKUP MTA1: DSKB:[10,225]*.*
/INDEX DSKC:BAKFIL.LST
/START
```

Once the BACKUP utility is initiated, it accepts a series of commands specifying the characteristics of the desired backup operation. In this case, the contents of a user's home directory will be written to tape, and an index file will be created on a different disk.

The very basic backup facility provided by TOPS-10 evolved into the full featured utility of the same name provided by the VAX/VMS operating system [70]. It was capable of performing full and incremental backups, could selecting files based on a variety of criteria, allowed save sets to span multiple tapes, and provided three distinct backup types: normal save sets, so-called image backups including the additional data needed to reinitialize an entire disk, and stand-alone backups suitable for restoring the system disk (although the latter were created with a separate command available only when the operating system was placed into a special, non-multiuser mode).

UNIX systems have traditionally provided several different utilities suitable for performing backups. The **cpio** and **tar** commands are general file archiving programs which may be used for systematic backup processes (typically driven by shell scripts). The **dump** command is a specialized program designed to perform full and incremental backups of individual disk partitions. It employs a simple database for keeping track of items' most recent backup times, an innovation which is now standard for backup software.

Windows NT-derived operating systems also include a backup utility capable of performing a variety of operations on the local system.

**3.1.1.** *Enterprise backup software*   One characteristic that is common to all of the vendor-provided backup utilities, regardless of manufacturer, is a focus on backing up the local computer system. Over time, such solutions proved insufficient for a variety of reasons, including the desire to share expensive tape devices among several computer systems while still performing backup operations locally and the need to centralize the backup function as the computer systems-to-administrator ratio increased over time. In addition, the backup software itself was not always capable of handling all of the supported file types present on the system (e.g., sparse files, UNIX special files).

The limitations and deficiencies of the simple backup utilities provided with the operating system created an opportunity for commercial backup solutions, and many companies attempted to provide them. Over time, the most widely used commercial backup packages have been Legato NetWorker [20] (now owned by EMC Software), IBM's Tivoli Storage Manager (TSM) [29] and Veritas Netbackup [64] (now owned by Symantec).[7]

Such high-end commercial backup packages included a variety of features which have become standard in such offerings:

---

[7]In the Macintosh world, the market leader has been and remains Retrospect [19].

- A client–server architecture, typically consisting of one or more central controller/scheduler systems communicating with lightweight agent software on each client system to be backed up. Packages include support for heterogeneous client systems.
- Abstracted definitions for the various components of backup operations (including defaults), independent of actual operation scheduling.
- Considerable flexibility in defining backup set contents, from entire disk partitions to arbitrary sets of files and/or filename patterns, supplemented by exclusion lists.
- The ability to backing up all file types (e.g., device files, sparse files) and attributes (e.g., access control lists, extended attributes).
- Intelligently handling of open files encountered during a backup operation.
- The ability to use of multiple backup servers.
- Automated utilization of multi-unit backup devices such as stackers, jukeboxes, libraries and silos.
- Support for multiplexed backup operations in which multiple data streams are interleaved on a single tape device.
- Facilities for automatic archiving of inactive files to alternate on-line or near-line storage devices.
- An integrated database manager (or integration with an external database system), enabling complex querying of the backup catalogues.

**3.1.2.** *Open source backup software*   While commercial backup packages are extremely sophisticated and full featured, they are also quite expensive. This trait, combined with researchers' ideas for backup technology advances, has motivated the development of open source enterprise backup software. The best-known and most widely used of these are Amanda, the "Advanced Maryland Automated Network Disk Archiver" [1,59,60] and Bacula [3] (the latter name is obviously a play on Dracula, chosen because the software "comes by night and sucks the vital essence from your computers").

Amanda employs the operating system-provided backup utilities for performing the actual archiving operations, a choice which both simplified its development and provides automatic archive accessibility and portability, including potential restoration even on systems where Amanda is not installed. In addition to providing enterprise backup features in an open source software package, Amanda also includes a sophisticated, automated incremental level selection mechanism designed to minimize the sizes of backup operations as well as simple yet effective security controls. Amanda's major limitations are its limitation of a single backup server system (which places on upper limits on its scalability) and lack of support for multivolume save sets, although the latter restriction has been removed in the most recent versions.

Bacula's designers took advantage of the insights provided by Amanda's evolution and use. It is full featured package using its own archive format and software. Bacula supports multiple backup servers and multi-volume archives. Its design centers around a highly modularized structure. Its configuration mechanism specifies the separate components required for archiving a given set of data individually, bringing them together into 'jobs', corresponding to instantiated scheduled backup operations. Such structures make the Bacula system easy to learn and minimize the ramp up time required to put it into production.

Specific features of Amanda and Bacula will be considered in detail during discussion of backup algorithms later in this article.

**3.1.3.** *Future software trends*   Current discussions of backups and backup technologies tend to broaden the term's scope and applicability, placing the process within the larger context of general data management. Many discussions having such a focus now employ the term "Information Lifecycle Management" (ILM) [16,24] to designate this topic, most prominently in the popular press and commercial software marketing literature. At some level, such a focus is sensible, given that backups have always served multiple purposes (e.g., data protection and historical archiving), and that evolving storage technologies have resulted in a continuum of online and near-online options rather than the polarized opposites of disk versus tape. Integrating backup processes with legal requirements for data security, privacy and retention is also a challenge that many organizations will need to take on.

It is still too early to tell whether thinking in terms of Information Lifecycle Management will become the norm. Nevertheless, the design of future software versions is already being influenced in this direction. Sometimes, this means designing – and marketing – backup software in the context of suite of related products, each focused on a different phase and/or aspect of the data lifecycle. In other cases, additional features are added to existing software, some of which are tangential to traditional backup operations. Whether such tendencies will be generally beneficial remains to be seen.

**3.2.** *Backup media*

In contemporary usage, backups are redundant copies of data for which the primary source is stored permanently on the computer, typically in hard disk files. However, backups in the sense of physical manifestations of digital information have necessarily existed as long as computers themselves. In this sense, decks of punched cards and paper tape were the first backup media for computer programs, and calculation results were 'backed up' by punched cards generated by the computer and later by line printer output.

Mainframe class computers traditionally used various formats of magnetic tape as their backup media, while the earliest personal computers used floppy disks for both permanent storage and backups. Interestingly, tape is still the principal medium for enterprise backups, although current drives and tapes have only a vague family resemblance to their earlier ancestors. In this arena, other technologies have also become important, including magneto-optical drives, DVD-ROM and hard disks.

Similarly, various media types have taken turns dominating backups within the personal computer sphere over the years, as well as for backups performed by individuals in corporate and educational organizations outside of the site's backup procedures. Floppy disks have given way to a variety of removable hard drive technologies, ranging from the Bernoulli and Syquest drives of the late 1980s through the successive generations of Iomega Zip and Jaz drives throughout the 1990s to contemporary pen and flash drives. In recent years, CD-ROM and DVD-ROM are also commonly used for backups and permanent archiving. In general, media for these kinds of backups are selected by convenience and price more than on the basis of capacity or performance.

The ever-increasing need for backup media with larger capacity and faster speed in the smallest practical form factor has meant that new recording technologies originally developed for audio or video are rapidly adapted to data storage. For example, cassette tapes designed for recording audio were adapted for computer use and were widely deployed in home computers by the late 1970s. Similarly, see Rodriguez [50] for the relatively early example of data adoption of 8 mm video tape. Typically, however, the use of new recording technology for data generally lags its original use by a few years, often because of the need to develop higher-quality media to meet the more demanding reliability and longevity standards for computer data compared to audio and/or video data.

Current backup media and devices can be divided into four general classes:

Magnetic tape  The first magnetic tapes designed for computers consisted of thin metal strips. Modern versions employ magnetically active metal particles suspended within a polymer-based binder supported by a film backing material. Current tapes are enclosed in plastic cartridges containing both the tape and the reel(s) on which it is wound.

Tape technologies are distinguished by the width of the magnetic tape and the internal cartridge configuration. 4 mm and 8 mm tape widths are housed in dual reel cartridges for DAT/DDS and AIT/Super AIT tapes, respectively. Digital Linear Tape (DLT) and Linear Tape-Open (LTO) cartridges use a single reel of half-inch tape. Within a specific tape type, variations differ by tape length and/or bit density, both of which contribute to total capacity.

The main advantages of magnetic tape for backup storage is their large capacities coupled with fast drive speeds. The chief disadvantage of this media type is the fact that it must be accessed sequentially – the tape must be physically moved to the location of a particular file – which can make restoration operations slow.

DVD-ROM  Although recordable CD-ROMs have been used for backup operations for several years, their relatively small capacities did not make them a serious alternative to magnetic tape for enterprise backup operations. However, the advent of reasonably priced DVD burners and double layer recordable DVD media have made them viable choices for backups in some environments. Although there are multiple DVD recording formats, most current devices can handle all of them, and so format incompatibility is not a prohibitive concern.

The advantage of DVDs for backup storage are its long lifetime (when stored properly) and the fact that it is a random access device, speeding restore operations.

Hard disk  Backing up to hard disks is a practice that became increasingly common as hard disk prices dropped dramatically and network accessible storage via Network Attached Storage (NAS) devices and Storage Area Networks (SANs) came into more widespread use. Preston [45] provides one of the earliest discussions of using SANs as the backup archive storage location.

Disk-to-disk backups have a number of advantages over other kinds of backup media. For example, hard disks are by far the fastest backup devices, both for recording the data to be archived and for restoring files from backups. They also allow many backup copies to be accessible in the same way as the original data. However, hard disks are the most expensive media option. They are less convenient to use for offsite archive copies.

Magneto-optical Magneto-optical (MO) media have the advantage that, unlike any of the other backup media types just mentioned, the act of reading the media does not degrade its quality because different technologies are used for recording and reading the data (they are written magnetically but are read optically). Current MO media use a 5¼″ form factor (although some early versions used the 3½″ size). They offer a good tradeoff between media capacity and price and a longer lifetime than magnetic tape.

Table 2 lists the most popular media technologies in current use for backups within corporate and educational enterprises, including media capacities and lifetimes, drive speeds and approximate costs for each technology variant (the latter are prices in the USA in US dollars). The listed minimum lifetime figures assume proper media storage and correspond to 90–95% statistical confidence levels (references for this data are presented in the discussion of this topic within the text).

Figure 2 illustrates recent trends in backup media capacities, drive speeds and associated costs by comparing current values for the top performing items in each category with what was available four years ago in 2002 (the data for the latter comes from [22]). The graphs on the left side in the figure quantify the technology improvements in both media capacity and drive speed over this time period. Costs have also fallen in both cases. The highest capacity storage media now costs less per GB than the lower capacity media available in 2002. The cost of the highest performing drives are also generally less than they were four years ago; the one exception is for 8 mm tape, where drives employing a new standard were released only a month prior to this writing. A comparison of the costs of drives with identical characteristics from the present and 2002 would show even more dramatic cost reductions.

### 3.2.1. *Backup devices in enterprise contexts*
The earlier discussion of capacity planning touched on the fact that overall backup efficiency depends on achieved data archiving and recording rates rather than theoretical device and network bandwidth maximums. For this reason, a variety of strategies are used to maximize data collection and recording efficiency. Some of the most common include the following:

- Use of multiunit physical or logical tape devices. For example, backup software often allows media pools to be defined, consisting of multiple physical tape drives, possibly distributed across multiple computer systems. When a backup operation requires a tape drive, the request goes to the pool, from which it is automatically assigned to an available device in a manner which is transparent to the backup software. In this way, existing tape devices can be potentially utilized for archiving data from any originating client system.

Table 2
Popular backup media

| Media technology | Media characteristics | | | Drive[a] characteristics | |
|---|---|---|---|---|---|
| | Uncompressed capacity (GB) | Price ($) | Minimum lifetime (years) | Speed (MB/s) | Price ($) |
| **4 mm tape** | | | | | |
| DDS-4 (150 m length) | 20 | 10 | 10 | 2.4 | 500 |
| DAT 72 (170 m length) | 36 | 35 | 10 | 3.5 | 600 |
| **8 mm tape** | | | | | |
| AIT-3 | 100 | 65 | 10 | 12 | 1700 |
| AIT-4 | 200 | 65 | 10 | 24 | 2100 |
| AIT-5 | 400 | 80 | 10 | 24 | 3000 |
| SAIT | 500 | 150 | 10 | 30 | 6000 |
| **Digital linear tape** | | | | | |
| DLT-V4 (DLT VS160) | 80 | 45 | 10 | 8 | 1000 |
| SDLT 320 | 160 | 50 | 10 | 36 | 3000 |
| SDLT 600 | 320 | 150 | 10 | 36 | 3500 |
| **Linear tape-open** | | | | | |
| LTO-2 (Ultrium-2) | 200 | 64 | 10 | 40 | 1700 |
| LTO-3 (Ultrium-3) | 400 | 85 | 10 | 80 | 3300 |
| **DVD-ROM** | | | | | |
| DVD-R | 4.7 | 1 | 40 | 21[b] | 100 |
| DVD-R DL (double layer) | 8.5 | 4 | 40 | 10.6[b] | 165 |
| **$5\frac{1}{4}''$ magneto-optical** | | | | | |
| MO | 9.1 | 85 | 15 | 6 | 2000 |
| UDO WORM | 30 | 55 | 15 | 10 | 2800 |
| **Hard disk** | | | | | |
| SATA 150 | 500 | 750 | 5 | 150 | 75 |
| SATA 300 | 250 | 970 | 5 | 300 | 75 |
| Ultra160 SCSI | 180 | 700 | 5 | 160 | 150 |
| Ultra320 SCSI | 180 | 700 | 5 | 320 | 165 |

[a]The drive speed and price are the speed and cost of the controller in the case of hard disks; the media price is the cost of the hard drive itself. All price data is in US dollars and represents the approximate US street price for each item and corresponds to the lowest published price available from major online sources (PC Connection, CDW and Imation) as of this writing. Media prices are for single units; somewhat to significantly lower prices are typically available when media is purchased in bulk.

[b]Corresponding to drive speeds of 16× and 8× for single layer and double layer (single-sided) media, respectively.

A related feature of client backup software is support for tape and other media libraries: standalone hardware units that include one or more recording devices along with integrated storage for a large collection of media. Such libraries are mechanically sophisticated and are capable of retrieving and loading/unloading any given media

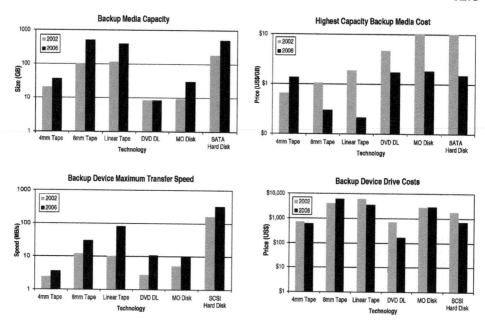

Fig. 2. Recent backup media capacity, drive speed and cost trends.

volume as directed by the software. The individual volumes are often labeled with barcodes, and the unit includes an integrated barcode reader.

- Collecting and transporting the data to be archived to the backup server system where the recording device is located can become a bottleneck in some cases. If the data does not arrive at a sufficient rate, then the recording device will be starved for work. In such cases, a two-stage backup process often ameliorates the situation. In the first phase, backup data is collected and written to a staging area on disk on the backup server. Data is then recorded on backup media when enough has been accumulated to service the device at near peak transfer rates and when system resources permits (for example, other system activity is low, the tape drive is free, and so on).

- Another method of ensuring that a drive has sufficient data to write to tape at its peak speed is to multiplex multiple backup sets onto a single tape as they arrive at the backup server. This process, described in [17], is illustrated in Figure 3. The backup server receives data from four client systems simultaneously, and it writes that data to tape immediately, using two tape drives. In the scenario in the figure, system B's data is being written to the second (lower) tape device, and the maximum recording speed is determined by the rate at which data arise from the client.

The other three client systems are also sending backup data to the server, and these three data streams are interleaved as they are written to the first (upper) tape device. In this way, the data bandwidth from these three clients is aggregated into a single write stream, resulting in higher throughput for the recording operation.

Preston presents a convincing, concrete example of the achievable results from such operations [45], successfully backing up over a terabyte of data in an eight-

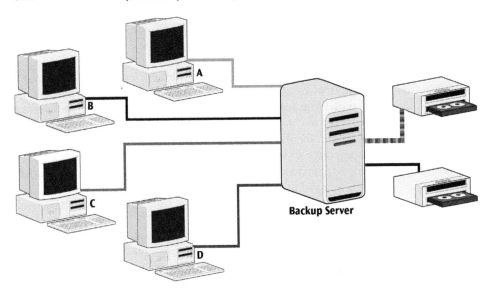

Fig. 3. Normal and multiplexed backup data streams in a client–server environment.

hour backup window (using 1998 technology), using the multiplexing features in the then-current version of the Legato Networker backup software.[8]

While multiplexing is a powerful feature for maximizing data recording efficiency, it complicates the restoration process considerably. For example, it is possible for the data from a single file to be written as multiple, nonsequential chunks on tape. In the worst case, a single file's data can be separated into pieces residing on different tape volumes. Nevertheless, for many organizations, the performance benefits of multiplexing outweigh any restoration and conveniences.

- Recent work by Mu and coworkers [39] attempts to integrate SAN technology with traditional tape-based backup procedures. Noting that readily available inexpensive disks have brought SANs within the budgets of many more sites, and that disk write speeds are much faster than those for tape, these researchers created a Virtual Tape System (VTS), a hardware solution that allows SAN disk resources to be deployed as virtual tapes for backup operations. The VTS is logically positioned between the backup software and the SAN. Its job is to translate the SCSI-based sequential commands traditionally sent to tape devices into block-based commands suitable for use with disk storage. The facility also has the capability of recording these disk-based virtual tapes to actual physical tapes, allowing the SAN to function as a very fast, random-access tape library and as a staging area for actual tape archiving.

**3.2.2.** *Media storage considerations*    Table 2 includes minimum lifetime values for the various media types. These values are estimations based upon measurements made by the US National Institute of Standards and Technology (NIST) [8,40,67,68]. They assume that

---

[8]Preston's environment included a Gigabit Ethernet network dedicated to backups and a tape library containing multiple Quantum DLT7000 drives (having a top speed of about 10 MB/s).

media are stored properly, characteristics which include both the environment in which they are stored and their positioning and other physical properties. Indeed, in the absence of proper care, media degrade very quickly indeed [61].

In general, long media storage lifetimes depend on controlling a variety of relevant factors. Media failure arises from many causes, and its enemies are heat, humidity, light, gravity and ultimately time. With regard to the latter, chemical changes in fundamental components of the media are inevitable, and all media will eventually fail. While such events cannot be prevented forever, they can be forestalled.

The first step in ensuring that backup media have the longest possible lifetimes involves the media selection process. Purchasing data grade media is essential to long-term data stability, despite the significant additional cost over the ordinary versions. In some cases, there are public standards for such media (for example, ISO 18927:2002 [30] for data grade optical media).

Magnetic tape is composed of metal particles suspended in a polymer-based binder layer, positioned on top of a nonmagnetic substrate. The magnetic particles are used to record the data. In addition to keeping these particles in fixed positions, the binder also provides general stability and a smooth surface, enabling the tape to move effectively within the drive. The binder also contains lubricants in reservoir pockets which serve to reduce friction as the tape travels. In some cases, the substrate is backed with a friction reducing coating which also helps to maintain the tape structural coherence.

Tapes can become unreadable due to changes in the metal particles, the binder or the backing. Of these, the binder is the most sensitive, and it typically dominates tape shelf life under ideal storage conditions. The long polymer molecules in the binder can be broken apart into shorter segments by water molecules, resulting in shifting of the metal particles and loss of lubricant. This process, known as hydrolysis, obviously pursuits more rapidly in humid conditions, but it is also sped up by heat. The other typical causes of tape failures are oxidation of the metal particles, which causes them to change magnetic value and/or to lose their ability to be written or even read, and deformation of the tape structure, both of which are primarily caused by excessive heat.

Optical media like DVD-ROMs are composed of layers of polycarbonate material (plastic) sandwiching a dye layer, upon which data is recorded via a laser, and a layer having a reflective metal coating – gold, silver or a gold–silver alloy – which serves to enable the laser to navigate within the media. The various layers are held together with an adhesive. Double layer media contain two data and reflection layers, and both single and double layer media can contain data on one or both sides (although the former is far more common for backup purposes).

Over time, the dye layer in optical media will become opaque, preventing the laser from penetrating it. Both heat and light significantly speed up this inevitable (and irreversible) organic reaction.[9]

Recommended storage conditions are most stringent for magnetic tapes, where the recommendations are a temperature range of $65 \pm 3°F$ ($18 \pm 2°C$) and a humidity range of

---

[9]Tests on CD-ROM media revealed that a phthalocyanine dye layer combined with a gold–silver alloy reflective layer provided the most stability among the manufacturing variations [8]. At the time of the study, the specific dyes used in DVD-ROM media were not known.

$40 \pm 5\%$. Optical media are more forgiving, and the lifetime estimations assumed storage conditions of $77°F$ ($25°C$) and 50% relative humidity [8,40,67].

The storage positions of common backup media can also affect their lifetimes. For example, tape should be stored with the radius of their reel(s) in the vertical plane (i.e., perpendicular to the ground). Tapes which are designed to be retained for periods of years may also require retensioning periodically (perhaps biannually), a process which ensures that the tape is distributed uniformly about the reel. Like tapes, optical media should also be stored upright, a position which minimizes potential surface area distortion.

Finally, luck is also a factor in media lifetimes. Media testing results can be expected to apply to upwards of 95% of the applicable media, but there are obviously no guarantees. The following best practices can reduce the likelihood of data loss items in long-term storage to the bare minimum:

- Use only data quality media.
- Verify media after recording.
- Prepare two copies of all volumes.
- Store media properly and in appropriately controlled environments.
- Attempt to read both copies of stored data on a periodic basis (perhaps annually or semi-annually, depending on the importance and irreproducibility of the data). If one of the copies fails, create a replacement using the other copy.

## 4. Backup strategies and algorithms

Chervenak and coworkers' 1998 taxonomy of backup strategies and options [10] provides an effective method of characterizing differing backup approaches. They identified a variety of distinguishing factors by which backup software can be classified and evaluated:

- Full versus incremental backups, including variations on the latter.
- File vs. device-based data selection methods.
- Operations requiring an active file systems, performable on active file systems, and/or employing disk block-based snapshots.
- Support for multiple concurrent backup operations on distinct computer systems.
- Availability of software-based data compression.
- Achievable restoration speeds based on backup characteristics.
- Completeness of tape management features.
- Inclusion of full system restoration and/or disaster recovery backup operations in addition to everyday data protection.

These items are still generally useful for comparing and characterizing backup software. The most important of them will be discussed in the following sections. In addition, traits arising from more recent developments and research, such as disk block-based data selection and emerging client–server software architecture variations, will be considered.

### 4.1. *Item selection: full vs. incremental backups*

The distinction between full and incremental backups is as old as the concept itself. A full backup operation is one that unconditionally archives every file within its scope, be that a disk partition, directory subtree, or a more or less arbitrary set of files. An incremental backup operation is one that saves only those files which have been created or modified since the time of the most recent backup; modern software also includes the ability to remove deleted files when restoring an incremental save set subsequent to the restoration of a full backup.

Incremental backups are typically distinguished by levels:[10] first-level incrementals include all file system changes with respect to the previous full backup operation. Such backups are also known as differential backups in some contexts. Similarly, second-level incrementals include changes with respect to the most recent first-level incremental backup operation. Current backup software supports many levels of incrementals. Full backup operations are also sometimes referred to as level zero backups, incorporating them within the numbered system of incremental operations.

In practice, the number of incremental levels actually used is generally limited in an effort to find the best trade-off that minimizes both the incremental backup sizes and resource requirements and that time required to perform a complete recovery in the event of a disk failure (an operation which must include every distinct incremental backup level).

A simple but widely used backup scheme employs full backups plus two levels of incrementals (indeed, this scheme was the recommended one within the VAX/VMS documentation [70]). Full backups are performed monthly and are retained for one year. Level 1 incrementals, saving all changes since the preceding full backup operation, are performed weekly and are retained for one month. Level 2 incrementals are performed on every day not used for the preceding backup types, and these backups are saved for seven days.

When multiple incremental levels are used in this way, it is also common to utilize noncontiguous level numbers (Zwicky [73], e.g., recommends this). In general, a level $n$ incremental backup saves all changes made since the most recent backup of the same or lower level. Thus, if backups typically use incremental levels 1 and 4, it is possible to perform an ad hoc level 2 incremental backup if necessary. Such an operation would create an archive encompassing all changes to the file system since the most recent full backup or level 1 incremental backup without disrupting the usual sequence.

Variations on this traditional incremental scheme will be discussed later in this section.

**4.1.1.** *Amanda: intelligent incrementals*   The Amanda open source backup package includes support for a full and incremental backup operations. The software is distinctive in that it provides a very intelligent incremental backup facility in which level selection is automated and self-regulating, based on available backup capacity and observed file system modification rates.

Amanda adjusts incremental levels in order to minimize the size of individual backup requirements as constrained by the values of its fundamental parameters (as configured by the system administrator). There are three key metrics:

- The length of the backup cycle: the time period during which a full backup must be performed for all designated data (Amanda's default is ten days).
- The number of backup operations performed during the backup cycle (typically, one operation is performed per day).

---

[10]Although they do not have to be: for example, continuous incremental backups archive file system changes with respect to the most recent backup operation without requiring any defined levels.

• The average percentage of the data which changes between runs.

The system administrator is responsible for providing reasonable values for the first two parameters, given the amount of data which must be backed up and the constraints arising from available resources. In order to select sensible values, an estimate for the change rate is also necessary.

The amount of data which must be archived in each backup operation – denoted by $S$ – is given by the expression in Equation (5), where $D$ is the total amount of data to be backed up in the course of the backup cycle of $n$ days, and $p$ is the average change rate (expressed in decimal form):

$$
\begin{aligned}
S &= \left(\frac{D}{n}\right)\left(1 + \sum_{j=1}^{n-1} jp - \sum_{k=2}^{n-1}\left((n-1) - k + 1\right)p^k\right) \\
&\approx \frac{D}{n} + \frac{Dp(n-1)}{2} - (n-2)p^2\left(\frac{D}{n}\right).
\end{aligned}
\tag{5}
$$

The first equation computes the size of a daily backup as a three term factor of one day's division of the total data, $D/n$. The first term, the factor 1, is the fraction of the total data which must be backed up every day in order to complete a full backup within the specified backup cycle. The second term corresponds to the maximal incremental changes occurring for the daily full backup fractions from the other days in the backup cycle. The third term is a correction which takes into account the small amount of overlap in the set of modified files on successive days (i.e., for a 10% change rate, the amount of incremental data which must be saved on the third day after its full backup is somewhat less than the maximum of 30% due to files which are modified multiple times).

The final expression in Equation (5) is a reasonable approximation to the initial expression for realistic values of the change rate $p$; it computes the some of the arithmetic series for the first summation, and truncates the second summation after its first term since $p^3 \ll p$.

Figure 4 illustrates the makeup of an Amanda daily backup set with two different change rates for seven-day backup cycles. Note that the data for Day $N$ represents a full backup of the relevant portion of the total data, while the amounts for all earlier days are incrementals with respect to their previous full backup.

The figure corresponding to a 15% change rate clearly illustrates the need for automated incremental level increases since the incremental portions corresponding to the three most distant full backups dominate. Note that the sizes of the two Day $N$ slices should be interpreted as having the same area.

The following expression computes $n$, given a fixed daily backup capacity for $S$, from Equation (5) (neglecting the final overlap term):

$$
n = \frac{x \pm \sqrt{x^2 - 2p}}{p}, \quad \text{where } x = p/2 + S/D.
\tag{6}
$$

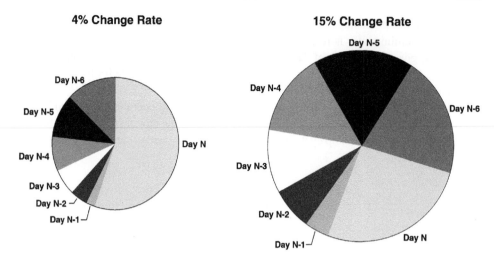

Fig. 4. Amanda daily backup composition for varying change rates.

Alternatively, it may be desirable to select the backup cycle length which will minimize the daily backup size. Doing so uses the following expression for the minimum required per-run backup capacity, derive from the preceding equations:

$$\min(S) = \left(2\sqrt{p/2} - p/2\right)T \tag{7}$$

Minimizing Equation (7) for a given value of $p$ yields the following expression for $n$:

$$n = \sqrt{2/p} \quad \text{when} \quad \frac{\partial S}{\partial n} = 0. \tag{8}$$

Given reasonable values for its required parameters and its measurement of the actual rate of change, Amanda introduces additional incremental levels (i.e., above level 1) when doing so will result in significantly improved resource usage. Its algorithm is constrained by three additional configurable parameters [59,60]:

*bumpsize* The minimum required backup size savings before an additional incremental level will be introduced (the default is a modest 10 megabytes). In recent Amanda versions, *bumpsize* is superseded by *bumppercent*, which expresses the minimum savings as a percentage of the backup size (the default value is zero, which reinstates the effect of *bumpsize*).

*bumpdays* The minimum number of days to remain at each successive incremental level (the default is 2).

*bumpmult* The factor with which to multiply *bumpsize* for each successive incremental level (the default is a factor of 1.5).

Using these features, Amanda becomes quite proficient at making maximal use of available backup resources.

**4.1.2.** *Z-scheme: stream-based incrementals*    Kurmas and Chervenak [33] have designed another incremental backup scheme with a name to minimize the required backup size and necessary resources without significantly increasing restoration times. Clearly, a continuous incremental approach – which these researchers term pure incremental – requires the fewest resources and consumes the smallest amount of media. However, this scheme has the potential for lengthening the required restoration time significantly, especially for frequently changed files. The effect is even more pronounced when restoring entire file systems.

Kurmas and Chervenak proposed an incremental method, which they call Z-scheme, consisting of several concurrent backup operations to distinct backup data streams. This method is centered around a parameter known as the base ($b$). Files requiring backup are assigned to data streams based on the number of days since their most recent modification; a file will be written to stream $i$ if it was last modified $b^i$ days ago. For testing purposes, the value of 8 was used for $b$, and backup operations using four different algorithms were performed for eight months. Table 3 summarizes the results of these tests, indicating the total amount of media used in the whole-file system restoration time for each.

Although the Z-Scheme algorithm is neither the most frugal with media nor the fastest to restore an entire file system, it performs well compared to the best in each category, and it clearly represents the best trade-off between the two factors.

## 4.2. *Client–server software architectures*

As was asserted earlier, users are unlikely to back up their own data effectively or regularly. The Maxtor Corporation is estimated that only 34% of users attempt to perform backups at all, and that only 24% of users perform them properly on a regular basis [38]. Statistics like these argue strongly for the centralization of backup responsibility and authority, and such considerations lead naturally to software based on a client/server architecture. Organizations with data scattered across a myriad of workstations belonging to unsophisticated users benefit greatly from such software, as do ones employing heterogeneous computer environments [17].

The simplest client–server backup software consists of a central controller running on the backup server and lightweight agent software running on each Backup client. The software on the backup server is typically responsible for scheduling and initiating backup

Table 3
The Z-scheme algorithm compared to other backup strategies

| Algorithm | Media consumed (relative to pure incrementals) | File system restoration time (relative to full) |
|---|---|---|
| Daily full backups | 33.3 | 1.0 |
| Pure incrementals | 1.0 | 0.3 |
| Incremental levels | 2.1 | 0.8 |
| Z-scheme | 1.4 | 0.7 |

operations and for overseeing the recording of backup data to the appropriate storage media. The client software response to requests and commands from the backup server and locates and transmits the data to be backed up.

In recent years, many variations to this prototypical structure have emerged.

**4.2.1.** *Bacula: module-based client–server software*  The Bacula open source software package uses a highly modularized architecture, separating many of the controller tasks into distinct processes, which may be optionally run on different server systems. The division of functionality occurs at natural task boundaries.

The following are the main components of the Bacula software package [3]:

Director daemon  This server is responsible for scheduling and overseeing backup operations. It also handles authentication between the various Bacula modules.

File daemon  This process runs on all backup client systems, and is responsible for locating and retrieving data for backup operations, under the direction of a director daemon.

Storage daemon  This server is present on systems where storage devices are located. It is responsible for writing backup data to the desired backup media.

Bacula also uses a database backend for storing its catalog data. The software is capable of interfacing to several open source database packages.

Finally, Bacula provides several administrative interfaces so that configuration and management operations can be performed from any convenient system and are not restricted to the backup server running the director daemon.

Another strength of the package is that backup operations are specified in highly abstracted form. All of the components needed to specify a backup job are configured independently, as part of the configuration of the relevant daemon. Available backup types, client systems, sets of files, operation schedules, storage types, locations and pools, and other parameters (such as relative operational priority) are all defined separately, and backup job definitions bring together the desired components from the relevant modules.

**4.2.2.** *Avoiding duplication of effort*  Given the large numbers of very similar user workstations present in many environments, eliminating duplication within the total collection of backup data can result in considerably reduced resource use [71]. On the other hand, failing to do so can have catastrophic consequences for network bandwidth consumption when centralized backups are implemented. Workman and coworkers estimate that of the two terabytes of data on their 3500 separate systems, at least 61% is common to two or more systems. Their backup approach takes this into account by formulating rule sets for backup file inclusion and exclusion for various system classes (including typical client and server systems).

Carpenter has recently produced a sophisticated variation on this theme [9]. This approach identifies linkages and dependencies between the various files on a computer system. Its ultimate goal is to be able to restore the complete environment required to access any given file. For example, restoration of the full functionality for a word processing document requires that the corresponding application be present on the system, including any libraries needed by the latter. Carpenter's work formulates an algorithm for identifying the clusters of files that are required any specific item to fully function. Actual backup operations take advantage of incremental techniques to minimize the amount of data that must be stored at any given time.

**4.2.3.** *Handling intermittent connectivity*   Computer systems which are only intermittently connected to the organizational network pose a special challenge for centralized backup systems. Pluta and coworkers have designed a backup facility designed to address such systems [43], known as the NCSA Backup Tracking System (BTS).

The BTS system is designed to identify new systems at the time of their initial connection to the network, and to take appropriate action with respect to backing up their data based on their specific computer types. For example, laptop computers are backed up immediately upon detection provided that available resources allow for it. In contrast, a backup operation in is merely scheduled for workstation systems. The system also distinguishes between VIP and non-VIP users, assigning different priorities to backup operations based upon the user's VIP status level. BTS keeps track of the backup operations performed for all systems it knows about using an external database.

BTS uses a continuous incremental backup strategy (with optional versioning capabilities). The actual backup operations are performed by the Tivoli software. The system requires client software on participating systems (which can be installed at detection). Architecturally, the BTS server functionality is partitioned into an enterprise management server which examines potential client systems and manages BTS data, and a backup database server which handles maintains the backup catalogue database.

**4.2.4.** *Peer-to-peer approaches*   At the opposite extreme from Bacula's highly modularized structure are the various peer-to-peer strategies that have become prevalent in the last few years (a development that was inevitable given the popularity of peer-to-peer networking services). Such approaches are designed to take advantage of the excess storage capacity present on systems throughout the network; they also assume relatively low network bandwidth usage, which is a reasonable assumption in many environments. They also promise full data protection at significantly lower costs and administrative effort over centralized solutions as well as increased convenience for users [13]. Such software depends on the existence of a peer-to-peer routing infrastructure.

Such approaches are inevitably cooperative in nature, but their designers remain sanguine about the chances of success even in the face of human nature. For example, Cox and coworkers assert that:

> It is feasible to provide a convenient, low-cost backup service using unreliable, untrusted, and self-interested storage hosts [13].

Pastiche [14] is one of the most promising and fully developed of these types of backup programs. Its designers have spent a lot of effort building in constraints which prevent 'freeloaders' from taking advantage of the backup in the structure without contributing to its overall resources. Solving this problem involves creating incentives for users to share their storage by constructing cyclic exchange relationships from demand graphs based on nodes' desired storage locations.

Internet-based cooperative backup storage schemes have also been proposed, most recently by Lillibridge, Elnikety and coworkers [18,35] (drawing on previous work [2,11]).

These researchers have also provided important discussions of the security considerations inherent in all such backup schemes, which range from selfishness and non-cooperation, to gratuitous, spite-motivated interruptions of backup services, to deliberate attempts to interfere with data integrity via man-in-the-middle attacks. The lack of solutions to these serious issues makes current peer-to-peer backup strategies unsuitable for production environments.

### 4.3. *Disk block-based backup approaches*

UNIX backup software is traditionally distinguished by the method it uses to traverse a file system in order to locate data to be backed up. General-purpose archiving utilities like **tar** and **cpio** follow the directory structure while the **dump** utility examines the file system inodes (the fundamental data structure corresponding to file system objects). Despite their differences, however, both approaches rely on logical structures to identify the relevant data.

An alternative strategy involves identifying the relevant data at the disk block level, without referencing the various file system objects to which may belong. This is a fundamentally more low-level approach.

Hutchinson and coworkers explore the performance differences between backup operations that examine logical file system entities and ones that focus on raw disk blocks, via a network appliance that includes implementations of both sorts of backup schemes [28] (specifically, the Write Anywhere File Layout file system (WAFL) [26]). They compared the implementation of **dump** provided with the device to saving snapshots of physical disk partition. Their results found that the disk block-based operations were 16% faster for backups and 27% faster for restores. The latter approach also made more efficient use of multiple tape drives. On the other hand, the backup data recorded using the disk block-based approach was not portable to any other environment.

Kaczmarski and coworkers propose a related strategy that is a synthesis of backup, disaster recovery, and storage management (including automated backup device migration) [31]. With respect to backups, their goal is to minimize the associated network traffic. Their scheme uses a 'progressive incremental' approach: continuous incrementals combined with a tape reclamation process in which data is migrated between volumes, and new full backup archives are constructed from previous full backup operations and later incrementals. In addition, incremental data processing is not restricted to files as a whole; when less than 40% of the data within a file has changed, a block-based incremental strategies used.

The system bases its decision-making on backup-related data attributes which are specified for individual files, including the desired backup frequency, data retention policy, number of versions to maintain, backup archive destination, and related settings. The Tivoli backup packages used to perform the data backup and manipulation operations.

Seung-Ju and coworkers have implemented what is perhaps the most extreme version of a disk block-based backup solution [57]. Theirs is a hardware only approach in which data is saved directly from disk device to disk device, resulting in an online backup facility with virtually no overhead.

**4.3.1.** *Versioning solutions* The last backup system considered in the preceding section points the way to another potential solution to traditional backup limitations: backups which are built into the file system itself. These so-called versioning file systems maintain multiple copies of modified files, and, if there was sufficient available storage, could potentially hold the entire modification history for file online.

The VAX/VMS operating system provided multiple versions for every file, identified by a terminal version number component within the file name separated from the extension by a semicolon. Experience in this environment confirms that the disk space wastage potential of such systems is vast. Nevertheless, similar constructs have been implemented more recently, most notably the Elephant file system [55,56] and the Plan 9 file system [42,48].

The Venti backup system [49] operates according to related principles as it maintains a disk block-based write once archival repository of modifications to the file system. It uses a very clever addressing scheme in which blocks are addressed via a hash of their contents, ensuring that any given data block will only ever be written once, a process which its developers term 'coalescing duplicate blocks'. Such a system can be viewed as a maximally efficient block-based backup system, using hard disk as its backup media. However, the repository itself can be used as a source for traditional backup operations using conventional software.

## 4.4. *Reliability and security concerns*

Ensuring backup archive reliability has concerned both system administrators and the designers of backup software from the beginning. For example, a very early backup program on the Multics operating system wrote multiple copies of files to the backup tape as a matter of course [63]. The VAX/VMS BACKUP utility had the capability of restarting file write operations in the event of a tape error (a practice which made the resulting backup tapes notoriously difficult for other programs to read successfully).

**4.4.1.** *Open files* Encountering an open file during the backup operation has traditionally been a problem for backup software. In some circumstances file locks made it impossible for the file to be copied to the backup archive, and backup software varied greatly in how robust it was with detecting and handling such situations. A number of strategies were implemented over the years, including simply skipping items with a lock was detected (although the latter process could take up to thirty seconds in the worst cases), and attempting to copy an open file and then archiving the latter.

Snapshots provide the most effective strategy for dealing with open files, and the more general problem of ensuring that the source data for a backup operation is in a consistent state. Snapshots employ the virtual memory subsystem's copy-on-write infrastructure [65] as well as techniques developed for process and system checkpointing [66] in order to capture the exact state of the file system as specific moment in time [21]. The general strategy is to associate an area of disk space with a specific disk partition; the former is used to record the original versions of all data blocks modify after snapshot initialization via copy-on-write operations. Backups of the snapshot device will automatically retrieve data as it was at the moment of creation. Snapshot facilities are provided by a variety of operating system and other storage management environments. For example, the Linux logical volume manager (LVM2) includes a snapshot facility.

**4.4.2.** *Boundary conditions*    Unusual file system objects have been another source of difficulty for backup programs historically (see [58] for an early discussion of these issues). Such items generally fall into three main classes: ordinary object types which simply do not happen to be supported by the backup software (e.g., sparse files), objects which, while not technically illegal, push the boundaries of normal operating system expectations in some way (e.g., filenames containing unexpected characters such as new lines), and objects whose contents are attributes are inconsistent over the course of a backup operation (e.g., items which change type from, say, an ordinary file to a directory between the time that the backup software first encounters them and actually attempts to archive).

Zwicky's 1991 study of such anomalies [74] revealed significant shortcomings in available backup software on UNIX systems, although many of the items tested are seldom if ever encountered on production systems. A recent follow-up study [75] found the situation to be much improved, with remaining problems generally characterizable as 'relatively mild'.

**4.4.3.** *Revocable backups*    Earlier discussion touched on the need for backup software to correctly handle item deletions when restoring archives from incremental backup operations. Boneh and Lipton carried this idea further via their concept of revocable backups [4], goal of which is to allow users to permanently remove files from the system, including all backup copies. Somewhat ironically, the system uses cryptographic techniques to erase data rather than to protect it directly.

Revocable backups accomplish this by encrypting every file on every backup tape with a unique random key. When the user chooses to revoke a file, the corresponding key is expunged in an unrecoverable manner, causing the files on the backup tapes to be blocked. Keys can also be made to expire after some period of time in order to exercise control over multiple versions of same file.

When the keys themselves are backed up, the process involves a master key which encrypts the key file. In this way, the key file is recoverable only by the system administrator, ensuring that revoked files remain inaccessible to ordinary users even given the existence of backup copies of previous versions of the key file.

## 5. Numerical models

Studies performing numerical and quantitative modeling a backup processes are surprisingly rare and so far seem to be limited to 2 research groups. For a discussion of the principles underlying this kind of research, see [6].

### 5.1. *Early work*

The earliest numerical models of backups were performed in the era when floppy disks were still commonly used as backup media in addition to magnetic tape [52–54]. For example, Sandoh, Kawai and coworkers computed optimal times for full and incremental backups by modeling backup operations as a function of the time to complete backup, the risk of losing data, the cost of performing backup operations and hard disk failure rates (with the latter approximated as an exponential distribution). Their approach attempted to minimize data loss via backup operations given system resource constraints, and their study succeeded in demonstrating that optimal times could be derived from their functional expressions [52].

Another, more recent study attempts to determine the optimal time to inform users that a backup operation is required [32], modeling potential data loss in a computing environment dominated by long running computations and in which file system changes result almost exclusively from the results generated by these jobs. Their model predicts the most effective warning interval for informing a user that a backup operation must be performed at the completion of the current calculation. Perhaps their most interesting assumption, possibly based on differing user behavior in different cultures and computing environments, is that the user will heed the warning when it is given.

## 5.2. *Recent work: optimizing backup frequency and timing*

In an important and exemplary study, Burgess and Reitan [7] present a mathematical model for selecting frequency and timing of backup operations based on time-based expressions for overall system activity and data change rates. Assumed risk is defined as sum of time to remediate loss (i.e., recover lost data) and the cost associated with performing backup operations, with the time to repair ultimately proportional to accumulated change.

The following expression mathematically describes these relationships:

$$R(n, t_1, t_2, \ldots, t_n) = \gamma \sum_{i=1}^{n} \eta(\Delta t_i)^2 + C(\Delta t_i), \tag{9}$$

where:
- $R$   Risk: a function of the number ($n$) and times ($t_i$) of backup operations.
- $\eta$   Data change density function (over time).
- $\Delta t_i$   The interval between successive backup operations.
- $C$   The cost associated with a backup operation.
- $\gamma$   Proportionality factor subsuming the relationships between the expected number of arrivals and file loss and other terms.

The second term of Equation (9) can also include a startup overhead value, but neglecting it does not change the results of the subsequent analysis.

Deriving and evaluating this expression involves a number of assumptions, of which the following are the most important:
- Data loss risk is proportional to system activity as well as to the rate of data change. This is a reasonable presupposition since most loss is activity-based rather than due to hardware failure.
- The various $C(\Delta t_i)$'s do not vary much. Thus, the final term reduces to $nC$.

- The time required to complete a backup operation ≪ the interval between backup operations.

Minimizing $R$ yields the following important result: backup plans involving a fixed number of backups ($n$) with some time period should space the operations evenly within the change density:

$$\eta(\Delta t_i) \approx \eta(\Delta t_j), \quad \forall i, j. \tag{10}$$

In simple terms, this means that backups should be positioned so that equal amounts of data modification occurs within each internal and not necessarily evenly in time. This conclusion runs counter to common backup practice.

Alternatively, an optimal value of $n$ – number of backup operations within a set time period – can be computed for given backup costs and total accumulated change ($N$):

$$n = N\sqrt{\frac{\gamma}{C}}. \tag{11}$$

Observe that $\gamma$ reflects the manner in which loss and activity are related. As risk of loss increases, the number of required backup operations must also increase. Conversely, as the cost of a backup operations increase, $n$ necessarily decreases.

Providing a specific function to model the expected arrivals can yield quantitative results for the backup starting point within its time interval of length $P$. For example, the following expression is a reasonable approximation for $q(t)$:

$$q(t) = \left(\frac{3}{2} + \sin\left(5\pi t - \frac{2}{3}\pi\right)\right)\left(1 + \sin(2\pi t)\right)e^{-\pi(t-1/4)}. \tag{12}$$

Using this expression and then minimizing $R$ within the interval yields a starting time of $0.32P$, which corresponds to 13:41 for a 24-hour interval running beginning at 06:00 and having peak activity at 12:00.

This analysis provides many promising directions for future work, including considering risk as having independent relationships to system activity and the rate of change, extending the analysis to resource deprived and other severely stressed systems, and considering the same operations in a distributed environment.

## References

[1] Amanda Team, *Advanced Maryland automatic network disk archiver (Amanda)* (1992–present), www.amanda.org.
[2] R. Anderson, *The eternity service*, Proceedings of Pragocrypt'96 CTU Publishing House, Prague, Czech Republic (1996), 242–252.
[3] Bacula Team, *Bacula* (2002–present), www.bacula.org.
[4] D. Boneh and R. Lipton, *A revocable backup system*, Proceedings of the Sixth USENIX UNIX Security Symposium (San Jose, CA), USENIX Association, Berkeley, CA (1996), 91–96.
[5] M. Burgess, *Principles of Network and System Administration*, Wiley, New York (2000).

[6] M. Burgess, *Analytic Network and System Administration: Managing Human–Computer Systems*, Wiley, New York (2004).

[7] M. Burgess and T. Reitan, *A risk analysis of disk backup or repository maintenance*, Science of Computer Programming **64** (2007), 312–331.

[8] F.R. Byers, *Care and handling of CDs and DVDs: A guide for librarians and archivists*, NIST Special Publication, NIST, U.S. Dept. of Commerce, Gaithersburg, MD (2003), 500–252.

[9] M.A. Carpenter, *File clustering for efficient backup systems*, Undergraduate Honors Thesis #169, Florida State University, Computer Science Dept. (2006), dscholarship.lib.fsu.edu/undergrad/169.

[10] A.L. Chervenak, V. Vellanki and Z. Kurmas, *Protecting file systems: A survey of backup techniques*, Proceedings of the Joint NASA and IEEE Mass Storage Conference, IEEE, New York (1998).

[11] I. Clarke, O. Sandberg, B. Wiley and T.W. Hong, *Freenet: A distributed anonymous information storage and retrieval system*, Designing Privacy Enhancing Technologies: International Workshop on Design Issues in Anonymity and Unobservability, Lecture Notes in Comp. Sci., Vol. 2009, Springer-Verlag, New York (2001), 46.

[12] P. Cox, *Tutorial on Windows 2000 security [T2]*, The 14th Large Installation Systems Administration Conference (LISA 2000), New Orleans, LA (2000).

[13] L.P. Cox, *Collaborative backup for self-interested host*, Ph.D. Thesis, Computer Science and Engineering Dept., University of Michigan (2005).

[14] L.P. Cox, C.D. Murray and B.D. Noble, *Pastiche: Making backup cheap and easy*, ACM SIGOPS Operating Systems Review. (Special issue.) OSDI'02: Proceedings of the 5th Symposium on Operating Systems Design and Implementation, Vol. 36 (2002), 285–298.

[15] *DEC System 10 Operating System Commands*, Digital Equipment Corp., Maynard, MA (1971).

[16] S. Duplessie, N. Marrone and S. Kenniston, *The new buzzwords: Information lifecycle management*, Computerworld Storage Networking World Online (2003), www.computerworld.com/hardwaretopics/storage/story/0,10801,79885,00.html.

[17] P. Eitel, *Backup and storage management in distributed heterogeneous environments*, Thirteenth IEEE Symposium on Mass Storage Systems, IEEE, New York (1994), 124.

[18] S. Elnikety, M. Lillibridge, M. Burrows and W. Zwaenepoel, *Cooperative backup system*, USENIX Conference on File and Storage Technologies (2002).

[19] EMC Insignia, *Restrospect* c. (1985–present), www.emcinsignia.com.

[20] EMC Software, *Legato NetWorker* c. (1987–present), software.emc.com.

[21] R. Fitzgerald and R.F. Rashid, *The integration of virtual memory management and interprocess communication in accent*, ACM Transactions on Computer Systems **4** (2) (1986), 147–177.

[22] Æ. Frisch, *Essential System Administration*, 3rd. edn, O'Reilly, Sebastopol, CA (2002), Chapter 11.

[23] Gramm–Leach–Bliley Financial Services Modernization Act, Pub. L. No. 106-102, 113 Stat. 1338 (1999).

[24] J. Gruener, *What's in store: Debunking informamtion life-cycle management promises*, Computerworld Storage Networking World Online (2004), www.computerworld.com/hardwaretopics/storage/story/0.10801,90254,00.html.

[25] Health Insurance Portability and Accountability Act, Pub. L. No. 104-191, 110 Stat. 1936 (1996), HIPAA.

[26] D. Hitz, J. Lau and M. Malcolm, *File system design for a file server appliance*, Proceedings of the 1994 Winter, USENIX Technical Conference USENIX Association, Berkeley, CA (1994), 235–245.

[27] C.B. Hommell, *System backup in a distributed responsibility environment*, Proceedings of the Large Installation System Administration Workshop, USENIX Association, Berkeley, CA (1987), 8.

[28] N.C. Hutchinson, S. Manley, M. Federwisch, G. Harris, D. Hitz, S. Kleiman and S. O'Malley, *Logical vs. physical file system backup*, Proceedings of the 3rd Symposium on Operating Systems Design and Implementation (New Orleans, LA), USENIX Association, Berkeley, CA (1999), 22–25.

[29] IBM, *Tivoli Storage Manager* (1993–present), www-306.ibm.com/software/tivoli/products/storage-mgr/.

[30] ISO 18927:2002, *Imaging materials – recordable compact disc system – Method for estimating the life expectancy based on the effects of temperature and relative humidity*, *NIST*, Gaithersburg, MD (2004), Followup in progress; see www.itl.nist.gov/div895/loc/overview.html.

[31] M. Kaczmarski, T. Jiang and D.A. Pease, *Beyond backup toward storage management*, IBM Systems Journal **42** (2) (2003), 322.

[32] H. Kawai and H. Sandoh, *An efficient backup warning policy for a hard disk*, Computers and Mathematics with Applications **46** (2003), 1055–1063.

[33] Z. Kurmas and A.L. Chervenak, *Evaluating backup algorithms*, Proceedings of the IEEE Symposium on Mass Storage Systems, IEEE, New York (2000), 235–242.

[34] J. Leber, *Windows NT Backup & Restore*, O'Reilly, Sebastopol, CA (1998).

[35] M. Lillibridge, S. Elnikety, A. Birrell, M. Burrows and M. Isard, *A cooperative internet backup scheme*, Proceedings of the Annual Technical Conference (San Antonio, TX), USENIX Association, Berkeley, CA (2003), 29–42.

[36] T.A. Limoncelli and C. Hogan, *The Practice of System and Network Administration*, Addison–Wesley, Boston (2002), *Backups*: Chapter 21; *Disaster Recovery*: Chapter 8.

[37] Linnen v. A.H. Robins Co., 1999 WL 462015, MA Superior Court (1999).

[38] Maxtor Corporation, *Most computer users walk a thin rope*, Home Data Protection (2005).

[39] F. Mu, J.-w. Shu, B. Li and W. Zheng, *A virtual tape system based on storage area networks*, GCC 2004 Workshops, Lecture Notes in Comput. Sci., Vol. 3252, Springer-Verlag, Berlin, Heidelberg (2004) 2004, 278–285.

[40] NIST, *Data preservation program*, Information Technology Laboratory, National Institute of Standards and Technology (2003), www.itl.nist.gov/div895/isis/datastorage.html.

[41] P.E. Pareseghian, *A simple incremental file backup system*, Proceedings of the Workshop on Large Installation Systems Administration, USENIX Association, Berkeley, CA (1988), 41.

[42] R. Pike, D. Presotto, S. Dorward, B. Flandrena, K. Thompson, H. Trickey and P. Winterbottom, *Plan 9 from Bell Labs*, Computing Systems **8** (3) (1995), 221–254.

[43] G. Pluta, L. Brumbaugh, W. Yurcik and J. Tucek, *Who moved my data? A backup tracking system for dynamic workstation environments*, Proceedings of the 18th Large Installation System Administration Conference (Atlanta, GA), USENIX Association, Berkeley, CA (2004), 177–186.

[44] J.H.A. Pooley and D.M. Shaw, *Finding out what's there: Technical and legal aspects of discovery*, Tex. Intell. Prop. L. J. 4 (1995), 57, www.utexas.edu/law/journals/tiplj/vol4iss1/pooley.htm.

[45] W.C. Preston, *Using gigabit ethernet to backup six terabytes*, Proceedings of the Twelth Systems Administration Conference (LISA XII), USENIX Association, Berkeley, CA (1998), 87.

[46] W.C. Preston, *UNIX Backup and Recovery*, O'Reilly, Sebastopol, CA (1999).

[47] Public Company Accounting Reform and Investor Protection Act of 2002, Pub. L. No. 107-204, 116 Stat. 745 (2002), Sarbanes-Oxley Act of 2002.

[48] S. Quinlan, *A cache worm file system*, Software – Practice and Experience **21** (12) (1991), 1289–1299.

[49] S. Quinlan and S. Dorward, *Venti: A new approach to archival storage*, Proceedings of the FAST'02 Conference on File and Storage Technologies (Monterey, CA), USENIX Association, Berkeley, CA (2002), 89–102.

[50] J.A. Rodriguez, *Rotary heads: New directions for tape backup*, Compcon Spring'88: Thirty-Third IEEE Computer Society International Conference (San Francisco), IEEE, New York (1988), 122–126.

[51] S.M. Romig, *Backup at Ohio State, Take 2*, Proceedings of the Fourth Large Installation Systems Administration Conference, USENIX Association, Berkeley, CA (1990), 137.

[52] H. Sandoh, N. Kaio and H. Kawai, *On backup policies for hard computer disks*, Reliability Engineering and System Safety **37** (1992), 29–32.

[53] H. Sandoh and H. Kawai, *An optimal N-job backup policy maximizing availability for a computer hard disk*, J. Operations Research Society of Japan **34** (1991), 383–390.

[54] H. Sandoh, H. Kawai and T. Ibaraki, *An optimal backup policy for a hard computer disk depending on age under availability criterion*, Computers and Mathematics with Applications **24** (1992), 57–62.

[55] D.J. Santry, M.J. Feeley, N.C. Hutchinson and C.A, Veitch, *Elephant: The file system that never forgets*, Hot Topics in Operating Systems (1999), 2.

[56] D.S. Santry, M.J. Feeley, N.C. Hutchinson, A.C. Veitch, R.W. Carton and J. Ofir, *Deciding when to forget in the Elephant file system*, Proceedings of the 17th ACM Symposium on Operating Systems Principles (Kiawah Island Resort, SC), ACM, New York (1999), 110–123.

[57] J. Seung-Ju, K. Gil-Yong and K. Hae-Jin, *Design and evaluation of high-speed on-line backup scheme for mid-range UNIX computer system*, Proceedings of TENCON'94, IEEE Region 10's Ninth Annual International Conference (Theme: "Frontiers of Computer Technology"), Vol. 2, IEEE, New York (1994), 1102–1106.

[58] S. Shumway, *Issues in on-line backup*, Proceedings of the Fifth Large Installation Systems Administration Conference (LISA V), USENIX Association, Berkeley, CA (1991), 81.

[59] J. da Silva and Ó. Gudmundsson, *The Amanda network backup manager*, Proceedings of the Sixth Systems Administration Conference (LISA'93, Monterey, CA), USENIX Association, Berkeley, CA (1993), 171.

[60] J. da Silva, Ó. Gudmundsson and D. Mossé, *Performance of a Parallel Network Backup Manager*, Proceedings of the Summer 1992 USENIX Technical Conference, USENIX Association, Berkeley, CA (1992), 217–225.

[61] O. Slattery, R. Lu, J. Zheng, F. Byers and X. Tang, *Stability comparison of recordable optical discs – A study of error rates in Harsh conditions*, Journal of Research of the National Institute of Standarts and Technology **109** (5) (2004), 517–524.

[62] D.M. Smith, *The cost of lost data*, Graziadio Business Report: Journal of Contemporary Business Practice **6** (3) (2003), gbr.pepperdine.edu/033/dataloss.html.

[63] J. Stern, *Backup and recovery of on-line information in a computer utility*, SM Thesis, Electrical Engineering and Computer Science Dept., MIT (1974).

[64] Symantec, *[Veritas] Netbackup* (1987–present), www.symantec.com.

[65] F.J. Thayer Fabrega and J.D. Guttman, *Copy on write*, Technical report, The Mitre Corp. (1995).

[66] M.M. Theimer, K.A. Lantz and D.R. Cheriton, Preemptible remote execution facilities for the V-system, Proceedings of the Tenth ACM Symposium on Operating System Principles (Orchas Island, WA), ACM, New York (1985), 2–11.

[67] J.W.C. Van Bogart, *Mag tape life expectancy 10–30 years*, National Media Laboratory, Washington, DC (1995), palimpsest.stanford.edu/bytopic/electronic-records/electronic-storage-media/bogart.html.

[68] J.W.C. Van Bogart, *Magnetic tape storage and handling*, Commission on Preservation and Access, Washington, DC, National Media Laboratory, St. Paul, MN (1995), www.clir.org/pubs/reports/pub54/index.html.

[69] K.J. Withers, *Computer based discovery in federal civil litigation*, Federal Courts Law Review (2000), www.legaltechnologygroup.com/WhitePapers/ComputerBasedDiscoveryInFederalCivilLitigation.pdf.

[70] *VAX/VMS System Management and Operations Guide*, Digital Equipment Corp., Maynard, MA (1982).

[71] K.M. Workman, E. Waud, S. Downs and M. Featherston, *Designing an optimized enterprise Windows nt/95/98 client backup solution with intellegent data management*, Proceedings of the Large Installation System Administration of Windows NT Conference, USENIX Association, Berkeley, CA (1998).

[72] E.D. Zwicky, *Backup at Ohio State*, Proceedings of the Workshop on Large Installation Systems Administration, USENIX Association, Berkeley, CA (1988), 43.

[73] E.D. Zwicky, *Analysing backup systems*, Proceedings of the 9th Annual Sun User Group Conference and Exhibit (SUG91) (1991), 287–300.

[74] E.D. Zwicky, *Torture-testing backup and archive programs: Things you ought to know but probably would rather not*, Proceedings of the USENIX Association, Berkeley, CA (1991).

[75] E.D. Zwicky, *Further torture: more testing of backup and archive programs*, Proceedings of the 17th Annual Large Installation Systems Administration Conference (LISA 2003, San Diego, CA), USENIX Association, Berkeley, CA (2003), 7.

# – 2.7 –

# What Can Web Services Bring to Integrated Management?

Aiko Pras[1], Jean-Philippe Martin-Flatin[2]

[1] *Electrical Engineering, Mathematics and Computer Science, Design and Analysis of Communication Systems Group, University of Twente, P.O. Box 217, 7500 AE Enschede, The Netherlands*
*E-mail: pras@cs.utwente.nl*

[2] *NetExpert, Chemin de la Crétaux 2, 1196 Gland, Switzerland*
*E-mail: jp.martin@ieee.org*

**Abstract**

Since the turn of the millennium, Web services have pervaded the middleware industry, despite their technical limitations, their ongoing standardization, the resulting lack of stable standards, the different meanings of the term *service* to different people, and the fact that marketing forces have blurred the technical reality that hides behind this term. What is so special about Web services and their underlying Service-Oriented Architecture (SOA)?

As a major technology on the software market, Web services deserve to be investigated from an integrated management perspective. What does this middleware technology bring to management applications? Is it yet another way of doing exactly the same thing? Or does it enable management software architects to open up uncharted territories for building a new generation of management solutions? Can it help us manage networks, systems and services with unprecedented flexibility, robustness and/or scalability?

In this chapter, we first present the technical motivation for shifting toward Web services. In Sections 2, 3 and 4, we describe different facets of this middleware technology: its architecture, its protocols, and the main standards that make up the so-called *Web services stack*. In Sections 5 and 6, we show that Web services can be used in integrated management in an evolutionary manner. The example studied here is network monitoring. We propose a fine-grained Web service mapping of SNMP operations and managed objects, and show that the same management tasks can be performed with this new middleware to implement network monitoring. In Section 6, we go back to the rationale behind SOA and compare it with fine-grained Web services. This leads us to propose another perspective, more revolutionary, where services are coarse-grained, wrapping up autonomous programs rather than getting and setting managed objects. We analyze the challenges posed by these coarse-grained Web services to manage-

HANDBOOK OF NETWORK AND SYSTEM ADMINISTRATION
Edited by Jan Bergstra and Mark Burgess

ment applications, and the changes that they require in today's management habits. Finally, we give examples showing how management platforms could gradually migrate toward SOA in the future.

## 1. Motivation

The charm of novelty should not obliterate the fact that it is unwise to change a working solution. Before jumping on the bandwagon of Web services, supposedly the latest and greatest technology in the middleware market, let us review the current state of affairs in integrated management and investigate whether we need a new solution at all.

### 1.1. *Command line interface*

Our first observation is that today's management of networks, systems and services still relies to a great extent on protocols and technologies that were developed several decades ago. For instance, the most frequently used management tool still seems to be the Command Line Interface (CLI), which was introduced in the early days of computing (and later networking) to manage computer systems and network devices. In the Internet world, network operators still often rely on this rather crude interface to configure their equipment. The CLI is not only used interactively (in *attended mode*) when Network Operations Center (NOC) staff log into remote systems to manually monitor and modify their operation, but also in *unattended mode* when management scripts automatically connect to remote machines (e.g., using `expect` [47]) to check their configuration and operation and possibly alter them. In large organizations, these scripts are essential to manage and control the Information Technology (IT) infrastructure in a cost-effective and reliable manner.

Unfortunately, different CLIs use different syntaxes: there is no widely adopted standard. Not only do CLIs vary from vendor to vendor, but also from product line to product line for a given vendor, and sometimes even between versions of a given product. Scripts are therefore highly customized and equipment specific, and considerable investments are needed to maintain them as networks, systems and services evolve and come to support new features.

### 1.2. *SNMP protocol*

In the 1970s and early 1980s, when large networks began to be built all over the world, this interoperability issue became increasingly serious. In the 1980s, the International Organization for Standardization (ISO) and the Internet Engineering Task Force (IETF) proposed to address it by defining standard management protocols.

To date, the most successful of these protocols has been the IETF's Simple Network Management Protocol (SNMP), which is accompanied by a standard describing how to specify managed objects – the Structure of Management Information (SMI) – and a document defining standard managed objects organized in a tree structure – the Management

Information Base (MIB-II). As the latter defines only a small number of managed objects, additional MIB modules soon appeared.

The first versions of these SNMP standards were developed in the late 1980s and became IETF standards in 1990 (they are now known as SNMPv1 and SMIv1 [86]). Soon afterward, they were implemented in many products and tools, and used on a very large scale, much larger than expected. In view of this success, the original plan to eventually migrate to ISO management protocols was abandoned.

Instead, the IETF decided to create a second version of the SNMP standards. This version should have better information modeling capabilities, improved performance and, most importantly, stronger security.

The work on information modeling progressed slowly but smoothly through the IETF standard track; SMIv2, the new version of the Structure of Management Information, became a proposed standard in 1993, a draft standard in 1996, and a standard in 1999 [57]. In addition, a new Packet Data Unit (PDU) was defined to retrieve bulk data (`get-bulk`).

Unfortunately, the original developers of SNMPv2 disagreed on the security model and failed to find a compromise. In 1994, two competing versions of SNMPv2 were standardized (SNMPv2p and SNMPv2u), which utterly confused the network management market. In 1996, a compromise was standardized (SNMPv2c) [86] with only trivial support for security ('community' string, i.e. the clear-text identification scheme implemented in SNMPv1). SNMPv2c brought several improvements to SNMPv1 and was widely adopted, but it did not address the market's main expectation: the support for professional-grade security.

In 1996, a new IETF Working Group was created, with different people, to address security issues in SNMP. This resulted in SNMPv3, which progressed slowly through the standard track and eventually became an IETF standard in 2003.

By the time this secure version of SNMP was released, the market had lost confidence in the IETF's ability to make SNMP evolve in a timely manner to meet its growing demands. To date, two decades after work on SNMP began in 1987, there is still no standard mechanism in SNMP for distributing management across multiple managers in a portable way ... As a result, management platforms have to use proprietary 'extensions' of SNMP to manage large networks.

In last resort, a new IETF group was formed in 2001, called the Evolution of SNMP (EoS) Working Group, to resume standardization work on SNMP. But it failed to build some momentum and the group was eventually dismantled in 2003, without any achievement.

### 1.3. *Waiting for a new solution*

SNMPv1 was undoubtedly a great success, but it failed to evolve [80]. The frequent use of SNMP to manage today's networks should not hide the fact that many organizations also use the CLI and *ad hoc* scripts to manage their network infrastructure. This not only defeats the point of an open and standard management protocol, but also prevents the wide use of policy-based management, because CLI-based scripts alter the configuration settings supposedly enforced by Policy Decisions Points (PDPs, i.e. policy managers in IETF jargon). Today, the market is clearly waiting for a new solution for network management.

SNMP never encountered much success in systems and service management. Since its inception, service management has been dominated by proprietary solutions; it still is. The situation was similar in systems management until the early 2000s. Since then, the Web-Based Enterprise Management (WBEM) management architecture of the Distributed Management Task Force (DMTF) has built some momentum and is increasingly adopted, notably in storage area networks. The DMTF has focused mostly on information modeling so far. The communication model of WBEM can be seen as an early version of SOAP.

For large enterprises and corporations, the main problem to date is the integration of network, systems and service management. These businesses generally run several management platforms in parallel, each dedicated to a management plane or a technology (e.g., one for network management, another for systems management, a third one for service management, a fourth one for the trouble-ticket management system and a fifth one for managing MPLS-based VPNs[1]). In these environments, management platforms are either independent (they do not communicate) or integrated in an *ad hoc* manner by systems integrators (e.g., events from different platforms are translated and forwarded to a single event correlator). For many years, the market has been waiting for a technology that would facilitate the integration of these platforms. Researchers have been working on these issues for years, but turning prototypes into reliable and durable products that can be deployed on a very large scale still poses major issues. The market is ripe now for a new technology that could make this integration easier.

### 1.4. *From specific to standard technologies*

For many years, the IETF encouraged the engineers involved in its working groups to devise a taylor-made solution for each problem. This led to a plethora of protocols. SNMP was one of them: as network bandwidth, Central Processing Unit (CPU) power and memory were precious few resources in the late 1980s, their usage had to be fine-tuned so as not to waste anything.

Since then, the context has changed [53]. First, the market has learned the hard way what it costs to hire, train and retain highly specialized engineers. Second, considerably more network and computing resources are available today: there is no need anymore for saving every possible bit exchanged between distant machines. Third, software production constraints are not the same: in industry, a short time-to-market and reasonable development costs are considerably more important than an efficient use of equipment resources.

For integrated management, the main implications of this evolution are twofold. First, it is now widely accepted that using dedicated protocols for exchanging management information does not make sense anymore: wherever possible, we should reuse existing and general-purpose transfer protocols (pipes) and standard ways of invoking service primitives. Second, instead of using specific technologies for building distributed management applications, we should strive to use standard middleware.

Two platform-independent middleware technologies encountered a large success in the 1990s: the Common Object Request Broker Architecture (CORBA) [68] and Java 2 Enterprise Edition (J2EE) [13]; the latter is also known as Enterprise Java Beans (EJBs). Both

---

[1]MPLS stands for Multi-Protocol Label Switching, VPN for Virtual Private Network.

made there way to some management platforms for distributing management across several machines, but none of them had a major impact on the integrated management market. Why should Web services do any better?

### 1.5. *Web services*

The main difference between Web services and previous types of middleware is that the former support a loose form of integration (*loose coupling*) of the different pieces of a distributed application (*macro-components*[2]), whereas the latter rely on their tight integration (*tight coupling*).

The macro-components of a loosely integrated application can work very differently internally: they just need to be able to share interfaces and communicate via these interfaces. What happens behind each interface is assumed to be a local matter. Conversely, the macro-components of a tightly integrated application are internally homogeneous. They are all CORBA objects, or EJBs, etc.

The concept of loose integration is not new. Not only that, but nothing in CORBA (or J2EE) prevents application designers from supporting loose integration, where CORBA objects (or EJBs) serve as interfaces to entire applications; indeed, examples of wrapper CORBA objects (or EJBs) are described in the literature, and some can be found in real-life. But their presence is marginal in the market because this is not how the object-oriented middleware industry has developed.

What is new with Web services is that loose integration is almost mandatory, because Web services are grossly inefficient for engineering the internals of an application. Reengineering a CORBA application with Web services operating at the object level would be absurd: performance would collapse. What makes sense is to wrap up an existing CORBA application with one or several Web service(s), to be able to invoke it from a non-CORBA application.

What is so great about loose integration?

First, customers love it because it preserves their legacy systems, which allows them to cut on reengineering costs. In the past, people working in distributed computing and enterprise application integration had the choice between proprietary middleware (e.g., .NET or DCOM, the Distributed Object Component Model) or interoperable middleware (e.g., CORBA or J2EE). These different types of middleware all assumed a tight integration of software; everything had to be an object: a CORBA object, an EJB, a DCOM component, etc. Using such middleware thus required either application reengineering or brand new developments. Only the wealthiest could afford this on a large scale: stock markets, telecommunication operators and service providers prior to the Internet bubble burst, etc. With Web services, legacy systems are no longer considered a problem that requires an expensive reengineering solution. On the contrary, they are a given, they are part of the solution. When distributed computing and enterprise application integration make use of Web services, not everything needs to be a Web service.

---

[2]In Section 6.2, 'Loose coupling', we will present examples of macro-components for management applications.

Second, vendors also love loose integration because it allows them to secure lucrative niche markets. Externally, each macro-component of the distributed application looks like a Web service. But internally, it can be partially or entirely proprietary. Interoperability is only assured by interfaces: vendors need not be entirely compatible with other vendors. The business model promoted by this approach to distributed computing and enterprise application integration is much closer to vendors' wishes than CORBA and J2EE.

In addition to loose integration, Web services present a number of generic advantages that pertain to middleware, eXtensible Markup Language (XML) or the use of well-known technologies [1,53]. For instance, they can be used with many programming languages and many development platforms, and they are included in all major operating systems. Even calling Web services from a Microsoft Excel spreadsheet is easy, and the idea of building simple management scripts within spreadsheets can be appealing [43]. Last but not least, as there are many tools and many skilled developers in this area, implementing Web services-based management applications is usually easier and less expensive than developing SNMP-based applications.

## 2. Introduction to Web services

Let us now delve into the technical intricacies of Web services. This technology is defined in a large and growing collection of documents, collectively known as the *Web services stack*, specified by several consortia. In this section, we present the big picture behind Web services, describe the three basic building blocks (SOAP, WSDL and UDDI) and summarize the architecture defined by the World Wide Web Consortium (W3C). In the next two sections, we will present a selection of standards that are also part of the Web services stack and deal with more advanced features: management, security, transactions, etc.

### 2.1. *Communication patterns and roles*

Web services implement the Producer–Consumer design pattern [31]. When a Web service consumer binds to a Web service provider (see Figure 1), the consumer sends a request to the provider and later receives a response from this provider. This communication is generally synchronous: even though SOAP messages are one-way, SOAP is often used for synchronous Remote Procedure Calls (RPCs), so the application running the Web service consumer must block and wait until it receives a response [44]. Asynchronous versions of SOAP also exist [74] but are less frequent.

The first difficulty here is that the consumer usually knows the name of the service (e.g., it can be hard-coded in the application), but ignores how to contact the provider for that service. This indirection makes it possible to decouple the concepts of logical service and physical service. For instance, a given service can be supported by provider P1 today and provider P2 tomorrow, in a transparent manner for the applications. But this flexibility comes at a price: it requires a mechanism to enable the consumer to go from the name of a service to the provider for that service.

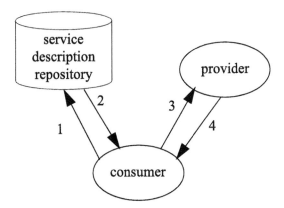

Fig. 1. Finding a service offered by just one provider.

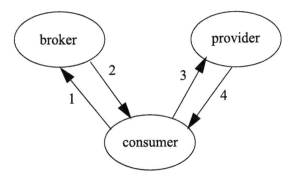

Fig. 2. Finding a service offered by multiple providers.

With Web services, the solution to this problem is illustrated by Figure 1. First, the consumer contacts a service description repository and requests a specific service. Second, the repository returns a handle to the service provider. Third, the consumer can now directly bind to the provider and send a Web service request. Fourth, the provider sends back a Web service response.

The second difficulty is that a given service may be supported not by one provider at a time, as before, but by multiple providers in parallel. For instance, different providers may support a variable Quality of Service (QoS) [94], or charge different prices for the same service.

In this case, the consumer typically uses a broker instead of a repository (see Figure 2). First, the consumer contacts a trusted broker and requests the best offering for a given service, with parameters that specify what 'best' means. Second, the broker builds an ordered list of offers (by contacting repositories, providers or other brokers, or by using locally cached information). Third, the consumer selects an offer and sends a Web service request to that provider. Fourth, the provider sends back a Web service response.

## 2.2. *SOAP protocol*

SOAP [32,33] is a protocol that allows Web applications to exchange structured information in a distributed environment. It can be viewed as a simple mechanism for turning service invocations into XML messages, transferring these messages across the network, and translating them into service invocations at the receiving end [1].

SOAP can be used on top of a variety of transfer protocols: HyperText Transfer Protocol (HTTP) [29], Blocks Extensible Exchange Protocol (BEEP) [69], etc. The SOAP binding used by most applications to date specifies how to carry a SOAP message within an HTTP entity-body; with this binding, SOAP can be viewed as a way to structure XML data in an HTTP pipe between two applications running on distant machines.

A SOAP message is an *envelope* that contains a *header* and a *body*. The header is optional and carries metadata; the body is mandatory and includes the actual application payload. Both the header and the body may be split into *blocks* that can be encoded differently. A SOAP message is used for one-way transmission between a *SOAP sender* and a *SOAP receiver*, possibly via *SOAP intermediaries*. Multiple SOAP messages can be combined by applications to support more complex interaction patterns such as request-response. SOAP also specifies how to make RPCs using XML.

In its original definition, the SOAP acronym stood for Simple Object Access Protocol. In the version standardized by the W3C, SOAP is no longer an acronym.

## 2.3. *Web services description language*

The Web Services Description Language (WSDL) [17,16] is a standard XML language for describing Web services. Each Web service is specified by a *WSDL description* that separates the description of the abstract functionality offered by a service (*type* and *interface* components) from concrete details of the service description (*binding* and *service* components); the latter defines how and where this functionality is offered. Multiple WSDL descriptions can be published in a single WSDL file, which is sometimes called a *WSDL repository*.

The current version of the standard is WSDL 2.0, released in March 2006. The previous version, WSDL 1.1, was standardized in March 2001. Both are widely used today. These two versions of WSDL are quite different. For instance, the *message* component was made obsolete in WSDL 2.0. The *port type* component in WSDL 1.1 evolved into the *interface* component in WSDL 2.0. The *port* component in WSDL 1.1 was renamed *endpoint* in WSDL 2.0. In Section 5.4, we will present the different components of a WSDL description while we study a detailed example.

If we compare Web services with CORBA, the WSDL language corresponds to the CORBA Interface Definition Language (IDL); a WSDL repository is similar to an Interface Repository in CORBA; applications can discover Web services in a WSDL repository and invoke them dynamically, just as CORBA applications can discover an object interface on the fly and invoke it using the Dynamic Invocation Interface. However, the underlying development paradigm is quite different. CORBA requires a developer to create an interface before implementing clients and servers that match this interface, whereas WSDL descriptions may be provided after the initial creation of the service. Moreover, it is not necessary to store all WSDL descriptions in a designated repository (whereas it is mandatory in CORBA): the Web service provider may also choose to serve a WSDL description at the physical location where the service is offered.

### 2.4. *Universal description, discovery and integration*

Universal Description, Discovery and Integration (UDDI) [88,89] is a standard technology for looking up Web services in a registry. A UDDI registry offers "a standard mechanism to classify, catalog and manage Web services, so that they can be discovered and consumed" [21]. Here are typical scenarios for using a UDDI registry [21]:

- Find Web services based on an abstract interface definition.
- Determine the security and transport protocols supported by a given Web service.
- Search for services based on keywords.
- Cache technical information about a Web service, and update this cached information at run-time.

Three versions of UDDI have been specified to date. UDDIv1 is now considered historic. The market is migrating from UDDIv2 [6] to UDDIv3 [21]; both are currently used. The main novelties in UDDIv3 include the support for private UDDI registries [1] and registry interaction and versioning [88].

The XML schema that underlies UDDI registries consists of six elements [21]:

- *businessEntity* describes an organization that provides Web services; this information is similar to the yellow pages of a telephone directory: name, description, contact people, etc.;
- *businessService* describes a collection of related Web services offered by an organization described by a `businessEntity`; this information is similar to the taxonomic entries found in the white pages of a telephone directory;
- *bindingTemplate* provides the technical information necessary to use a given Web service advertised in a `businessService`; it includes either the access point (e.g., a Uniform Resource Locator, URL) of this Web service or an indirection mechanism that leads to the access point; conceptually, this information is similar to the green pages of a telephone directory;
- *tModel* (technical model) describes a reusable concept: a protocol used by several Web services, a specification, a namespace, etc.; references to the `tModel`'s that represent these concepts are placed in a `bindingTemplate`; as a result, `tModel`'s can be reused by multiple `bindingTemplate`'s;
- *publisherAssertion* describes a relationship between a `businessEntity` and another `businessEntity`; this is particularly useful when a group of `businessEntity`'s represent a community whose members would like to publish their relationships in a UDDI registry (e.g., corporations with their subsidiaries, or industry consortia with their members);
- *subscription* enables clients to register their interest in receiving information about changes made in a UDDI registry. These changes can be scoped based on preferences provided by the client.

There are well-known problems with UDDI [1,55]: performance, scalability, taxonomies, etc. In practice, this technology is not widely used for publishing and discovering Web services; when it is used, it is often 'extended' with proprietary enhancements. Web service discovery is still a very active research area, particularly in the field of semantic Web services (see Section 3.7).

### 2.5. *Web services architecture*

The Web Services Architecture (WSA) was an attempt by the W3C to "lay the conceptual foundation for establishing interoperable Web services" [9]. This document defines the concept of Web service, architectural models (the message-oriented model, the service-oriented model, the resource-oriented model and the policy model), relationships that are reminiscent of Unified Modeling Language (UML) relationships, and a hodgepodge of concepts grouped under the name 'stakeholder's perspectives': SOA, discovery, security, Peer to Peer (P2P), reliability, etc. Management is quickly mentioned in Section 3.9 of [9], but the W3C seems to have paid little attention to this issue thus far.

A more interesting attempt to define an architecture for Web services can be found in Section 5.3 of [1].

## 3. Advanced Web services

### 3.1. *Standardization*

Because interoperability is crucial to Web services, their standardization has been of key importance since their inception. So far, two consortia have been particularly active in this field: the W3C and the Organization for the Advancement of Structured Information Standards (OASIS). More recently, other industrial consortia have worked in this field and produced standards or so-called 'best practices', including the Web Services Interoperability Organization (WS-I), the Open Grid Forum (OGF), the Distributed Management Task Force (DMTF) and Parlay.

### 3.2. *Interoperability*

In April 2006, WS-I specified the Basic Profile 1.1 [4], a set of implementation guidelines to help people build interoperable Web services. This document contains "clarifications, refinements, interpretations and amplifications of those specifications which promote interoperability". The main focus is on SOAP messaging and service description. In particular, the guidelines for using SOAP over HTTP are important. A number of Web services standards refer to this Basic Profile.

### 3.3. *Composition*

A *composite Web service* is a group of Web services that collectively offer some functionality. Each Web service taking part in a given composite Web service is known as a *participant*. Each participant can itself be a composite Web service, which can lead to a nested structure of services of arbitrary depth. The process of bundling Web services together to form a composite Web service is known as *service composition*. A Web service that is not composite is called *atomic* (or *basic* [1]).

There are many ways of specifying composite Web services, expressing constraints between participants, and controlling their synchronization and execution. They are usually grouped into two categories: static and dynamic. Static composition is the simplest form of composition: composite Web services are defined in advance, once and for all.

A generally more useful, but also considerably more complex, form of composition is the dynamic composition of Web services. In this case, Web services are composed at run-time: the participants of a composite Web service are chosen dynamically based on a variety of concerns: availability, load-balancing, cost, QoS, etc. The main challenge here is to build composite Web services semi-automatically or in a fully-automated manner. Different techniques have been proposed in the literature [79]; they borrow heavily from other existing and well-established fields (e.g., workflow systems or artificial intelligence planning).

An interesting issue in Web service composition is how to compute the QoS of a composite Web service when we know the QoS offered by each participant Web service [72]. Complex composite services can lead to situations where the error bars become ridiculously large when simple multiplicative techniques are used; reducing these error bars is non-trivial.

Whether a Web service is atomic, statically composed or dynamically composed is transparent to its consumers. Service composition is a key characteristic of Web services: their success in e-business owes a great deal to service composition.

## 3.4. *Orchestration and choreography*

For many years, the terms *orchestration* and *choreography* were used interchangeably in the Web services community. There is still no consensus on their exact meaning. The W3C proposes to distinguish them as follows [34].

**3.4.1.** *Orchestration*   In the realm of Web services, orchestration is about controlling the execution of a composite Web service, viewed as a business process. It specifies how to control, from a central point, the execution and synchronization of all the participant Web services (e.g., by triggering the execution of Web service WS2 when Web service WS1 completes).

Orchestration is usually implemented by an executable process that interacts with all Web services involved. Orchestration languages make it possible to define the order in which the participant Web services should be executed, and to express the relationships between Web services in a workflow manner.

The most famous orchestration language to date is probably WS-BPEL (Web Service Business Process Execution Language) [2], an OASIS draft standard formerly known as BPEL4WS (Business Process Execution Language for Web Services) [22] and already widely used. The main objectives of WS-BPEL are to describe process interfaces for business protocols and define executable process models.

**3.4.2.** *Choreography* Choreography is a declarative way of defining how participants in a collaboration (e.g., a composite Web service) should work together, from a global viewpoint, to achieve a common business goal. The objective here is to define a common observable behavior by indicating in a contract where information exchanges occur, what rules govern the ordering of messages exchanged between participants, what constraints are imposed on these message exchanges, and when the jointly agreed ordering rules are satisfied [45].

The need for a high-level contract, independent of execution and implementations, stems from the fact that enterprises are often reluctant to delegate control of their business processes to their partners when they engage in collaborations. Choreographies enable partners to agree on the rules of collaboration for a composite Web service without specifying how each participant Web service should work. Each participant is then free to implement its own portion of the choreography as determined by the global view [45]. For instance, one participant may use WS-BPEL for orchestrating business processes, while another participant may use J2EE.

The choreography language standardized by the W3C is called the Web Services Choreography Description Language (WS-CDL) [45]. With this language, it is supposedly easy to determine whether each local implementation is compliant with the global view, without knowing the internals of these local implementations.

### 3.5. *Transactions*

Classic transactions are short-lived operations that exhibit the four ACID properties: Atomicity, Consistency, Isolation and Durability. These properties are fulfilled by transactional systems using a coordinator and a two-phase protocol called *two-phase commit*. During the first phase, the coordinator contacts each participant in the transaction and asks them to make local changes in a durable manner, so that they can either be rolled back (i.e., cancelled) or committed (i.e., confirmed) later. During the second phase, we have two options. If a failure occurred on any of the participants during phase one, the coordinator sends an abort to each participant and all changes are rolled back locally by the participants; the transaction then fails. Otherwise, the coordinator sends a commit to each participant and all changes are committed locally by the participants; the transaction completes successfully. This two-phase protocol is blocking: once they have completed phase 1 successfully, the participants block until the coordinator sends them a commit (phase 2).

This *modus operandi* is not applicable to Web services [48,71]. First, Web services are not necessarily blocking (e.g., when they rely on an asynchronous implementation of SOAP). Second, in the realm of Web services, transactions are usually long-lived; using the above-mentioned protocol may cause resources to remain unavailable for other transactions over extended periods of time, thereby impeding concurrency. Third, when composite Web services are used across multiple organizations, security policies often prevent external entities from hard-locking a local database (allowing it would open the door to denial-ofservice attacks, for instance). Hence another mechanism is needed.

Several techniques were devised in the past [48] to release early the resources allocated by a long-lived transaction, and allow other transactions to run in the meantime. In case of subsequent failure, compensation mechanisms make it possible to bring the transactional system to the desired state (e.g., a payment by credit card can be reimbursed). However, these compensations do not guarantee all ACID properties. This is the approach generally adopted for Web services [1]. Another possibility is to subdivide long-lived transactions into independent short-lived transactions that can run independently as classic atomic transactions.

Detailing the support for transactions in a Web service environment would require an entire book. The OASIS consortium is the main standardization actor in this arena. It adopted an approach similar to Papazoglou's [71], where transactions are part of a bigger framework that defines business protocols, coordination protocols, transactions, orchestration of business processes, choreography, etc. Several building blocks have already been standardized: the business transaction protocol (WS-BTP), atomic transactions (WS-Atomic-Transaction), business activities (WS-BusinessActivity), coordination protocols (WS-Coordination), transactions (WS-Transaction), etc. These standards are still evolving and should not be considered stable yet. Readers interested in learning more about transactions in the context of Web services are referred to the Web site of the OASIS Web Services Transaction Technical Committee [66].

### 3.6. *Security*

WS-Security specifies how to secure the SOAP protocol. By supporting end-to-end application-level security, it nicely complements HTTPS, the secure version of HTTP, which can only secure communication on a hop-by-hop basis (with HTTPS, each intermediary can decrypt and re-encrypt the HTTP message) [1]. Whereas the entire payload is encrypted with HTTPS, WS-Security makes it possible to encrypt only one block of the SOAP body (e.g., bank details).

WS-Security was standardized by OASIS in 2004 (version 1.0) and upgraded in 2006 (version 1.1). It consists of a main document, known as SOAP Message Security, and several companion documents.

SOAP Message Security 1.1 [60] specifies SOAP extensions for sending security tokens as part of a SOAP message, guaranteeing message integrity (payload) and ensuring message confidentiality. This specification makes it possible to secure Web services with various security models: Public Key Infrastructure (PKI), Kerberos, Secure Sockets Layer (SSL), etc. *Per se*, it does not provide a complete security solution for Web services; instead, "it can be used in conjunction with other Web service extensions and higher-level application-specific protocols to accommodate a wide variety of security models and security technologies" [60].

In version 1.1, there are six companion documents: Username Token Profile (to identify the requestor by using a username, and optionally authenticate it with a password), X.509 Certificate Token Profile (to support X.509 certificates), the SAML Token Profile (to use Security Assertion Markup Language assertions as security tokens in SOAP headers), the Kerberos Token Profile (to use Kerberos tokens and tickets), the REL Token Profile (to use the ISO Rights Expression Language for licenses) and the SwA Profile (to secure SOAP messages with Attachments).

### 3.7. Semantic Web services

In 2001, Berners-Lee launched the Semantic Web with a powerful statement that has become famous: "The Semantic Web will bring structure to the meaningful content of Web pages" [7]. The idea was to add more metadata to Web pages (using XML) to enable software agents to parse them automatically and 'understand' their hierarchically structured contents (containment hierarchy), as opposed to seeing only flat structures and keywords out of context. The main building blocks defined so far for the Semantic Web are the Resource Description Framework (RDF) [51] and the Web Ontology Language (OWL) [58].

In the realm of Web services, this vision translated into the concept of semantic Web services [59]. Whereas Web services focus mostly on interoperability, semantic Web services complement interoperability with automation, dynamic service discovery and dynamic service matching. Semantic Web services address the variety of representations (e.g., different names for the same concept, or different signatures for semantically equivalent operations) by using automated or semi-automated ontology-matching mechanisms [70]. This makes it possible, for instance, to compose Web services at run-time with increased flexibility. OWL-S (Web Ontology Language for Services) [52] is probably the most famous language to date for specifying semantic Web services.

## 4. Web services for management, management of Web services

In this section, we review the main standards pertaining to (i) the use of Web services for managing networks, systems and services, and (ii) the management of Web services.

### 4.1. OASIS: Web services for distributed management (WSDM)

Unlike the W3C, OASIS has already paid much attention to management issues in the Web services arena. In 2005, the Web Services Distributed Management (WSDM) Technical Committee [64] standardized version 1.0 of a group of specifications collectively called WSDM. In August 2006, version 1.1 was released. To date, these standards are probably the most relevant to the integrated management community as far as Web services are concerned. Let us describe the main ones.

Management Using Web Services (MUWS) [11,12] is a two-part series of specifications that define how to use Web services in distributed systems management. First, MUWS defines a terminology and roles. As prescribed by WSDL and WS-Addressing [10], a Web service is viewed as an aggregate of *endpoints*. Each endpoint binds a Web service interface (described by a WSDL portType element) to an address (URL). Each interface describes the messages that can be exchanged and their format [92]. A *manageable resource* is a managed entity (e.g., a network device or a system). A *manageability endpoint* is bound to a manageable resource and can be accessed by a Web service consumer. It acts as a gateway between Web services and a given manageable resource. It can interact with the manageable resource either directly or via an agent (e.g., an SNMP agent). A *manageability consumer* is a Web service consumer, a Web-enabled managing entity (e.g., a management application running on a remote network management system). After discovering Web service endpoints, it can exchange SOAP messages with these endpoints to request management information (monitoring in polling mode), to subscribe to events (e.g., if a PC needs to report that the temperature of its mother board is too high), or to configure the manageable resource associated with a manageability endpoint [11].

Second, MUWS defines some rules that govern communication between manageability endpoints and consumers (i.e., between managing and managed entities). To discover the manageability endpoint that enables a given manageable resource to be accessed by a Web service, a manageability consumer must first retrieve an Endpoint Reference [10]; next, it can optionally retrieve a WSDL file, a policy, etc. Once the discovery phase is over, the manageability consumer (managing entity) can exchange messages with a manageability endpoint (managed entity) by using information found in the Endpoint Reference [11].

Third, MUWS specifies messaging formats to achieve interoperability among multiple implementations of MUWS (e.g., when management is distributed hierarchically over different domains [83] controlled by different managing entities) [12].

Management of Web Services (MOWS) [92] specifies how to manage Web services. It defines the concept of Manageability Reference (which enables a manageability consumer to discover all the manageability endpoints of a manageable resource), and a list of manageability capabilities (metrics for measuring the use and performance of Web services, operational state of a managed Web service, processing state of a request to a managed Web service, etc.) [92].

## 4.2. *DMTF: WS-Management*

WS-Management [23] was standardized by the DMTF in April 2006 to promote Web services-based interoperability between management applications and managed resources. It is independent of the WBEM management architecture and the Common Information Model (CIM).

First, WS-Management defines a terminology and roles. A *managed resource* is a managed entity (e.g., a PC or a service). A *resource class* is the information model of a managed resource; it defines the representation of management operations and properties. A *resource instance* is an instance of a resource class (e.g., a CIM object).

Second, WS-Management specifies mechanisms (i) to get, put, create, and delete resource instances; (ii) to enumerate the contents of containers and collections (e.g., logs); (iii) to subscribe to events sent by managed resources; and (iv) to execute specific management methods with strongly typed parameters [23].

The approach promoted by WS-Management corresponds to what we call fine-grained Web services in this chapter. In Section 5, we will see Web services (this time in the SNMP world) that similarly perform a get on an Object IDentifier (OID), i.e. a managed object.

The relationship between WSDM and WS-Management is not crystal clear. They do overlap but the scope of WSDM is much wider than that of WS-Management, so they are not really competitors (at least not for now). It seems that WS-Management could eventually provide a migration path for exchanging CIM objects between managing and managed entities, from the DMTFs pre-Web services and pre-SOAP 'CIM operations over HTTP' to a Web services-based communication model for WBEM. In this case, WS-Management would only target the WBEM community, not the Web services community at large. This is however mere speculation and remains to be seen.

### 4.3. *OASIS: WS-Resource*

In 2001, the Global Grid Forum (GGF), now known as the OGF, began working on the modeling of Grid resources. This led to the creation of the Open Grid Services Infrastructure (OGSI) Working Group. In 2002 and 2003, the concepts of Grid services and Web services gradually converged, as the specifics of Grids were separated from features that belonged in mainstream standardization. A number of GGF group members got involved in W3C and OASIS groups, and several ideas that had emerged within the GGF percolated to these groups. GGF focused on its core business (Grid-specific activities) and more generic standardization efforts moved to other standards bodies.

In early 2004, a group of vendors issued a draft specification for modelling and accessing persistent resources using Web services: Web Services Resource Framework (WSRF) version 1.0. It was inspired by OGSI and played an important role in the convergence of Web and Grid services. Shortly afterward, this standardization effort was transferred to OASIS, which created the WSRF Technical Committee [65] to deal with it. This led to the phasing out of OGSI and its replacement by a series of documents collectively known as WSRF. In April 2006, version 1.2 of WSRF was released by OASIS. It includes WS-Resource [30], WS-ResourceProperties, WS-ResourceLifetime, WS-ServiceGroup and WS-BaseFaults (for more details, see Banks [5]).

### 4.4. *Parlay-X*

The multi-vendor consortium called Parlay has defined a number of open service-provisioning Application Programming Interfaces (APIs) to control various aspects of Intelligent Networks. At the same time the Open Service Access (OSA) group of the 3rd Generation Partnership Project (3GPP) was working on similar APIs that allowed applications to control a Universal Mobile Telecommunications System (UMTS) network. These two groups joined their activities and now common Parlay/OSA API standards exist. These APIs require detailed knowledge of the Intelligent Networks, however, which means that only a limited number of experts are able to use these APIs. The Parlay group therefore defined some easier to use APIs, at a higher level of abstraction. These APIs, called Parlay-X, are defined in terms of Web services. Examples of such services include 'connect A to B', 'give status of X (on/off)', 'send SMS[3]', 'give location of mobile handset' and "recharge pre-paid card". The users of a Parlay-X server do not need to know the details of the underlying network; the Parlay-X server translates the Web services calls into Parlay/OSA calls that are far more difficult to understand. It is interesting to note that Web services technologies were applied by the Parlay group long before the IETF or the Internet Research Task Force (IRTF) even started to think about these technologies.

---

[3] Short Message System.

### 4.5. *IETF and IRTF-NMRG*

In the summer of 2002, the Internet Architecture Board (IAB) organized a one-off Network Management Workshop to discuss future technologies for Internet management. One of the conclusions of this workshop was that time had come to investigate alternative network management technologies, in particularly those that take advantage of the XML technology [80].

Web services are a specific form of XML technology. The 11th meeting of the IRTF Network Management Research Group (NMRG) [38], which was organized three months after the IAB workshop, therefore discussed the possible shift toward XML and Web services-based management. Although many questions were raised during that meeting, the most interesting question was that of performance; several attendees expressed their concern that the anticipated high demands of Web services on network and agent resources would hinder, or even prohibit, the application of this technology in the field of network management. At that time, no studies were known that compared the performance of Web services to that of SNMP. Most attendees therefore preferred to take a simpler approach not based on Web services (e.g., JUNOScript for Juniper). The idea behind JUNOScript is to provide management applications access to the agent's management data, using a lightweight RPC mechanism encoded in XML. As opposed to SNMP, JUNOScript uses a connection-oriented transport mechanism (e.g., `ssh` or `telnet`). An advantage of JUNOScript is that related management interactions can be grouped into sessions, which makes locking and recovery relatively simple. In addition, contrary to SNMP, management information is not limited in size and can be exchanged reliably. JUNOScript provided the basis for a new IETF Working Group called NetConf. Details about NetConf can be found in an another chapter of this book.

Within the IRTF-NMRG, several researchers continued to study how to use Web services in network, systems and service management. Some of the results of this work are presented in the next section.

## 5. Fine-grained Web services for integrated management

At first sight, an intuitive approach to introduce Web services in network management would be to map existing management operations (e.g., the SNMP operations `get`, `set` and `inform`) onto Web services. Is this efficient? Is this the way Web services should be used?

### 5.1. *Four approaches*

Mapping the current habits in SNMP-based network management onto Web services leads to four possible approaches, depending on the genericity and the transparency chosen [84]. *Genericity* characterizes whether the Web service is generic (e.g., a Web service that performs a `get` operation for an OID passed as parameter) or specific to an OID (e.g., `getIfInOctets`). *Transparency* is related to the parameters of the Web service, which

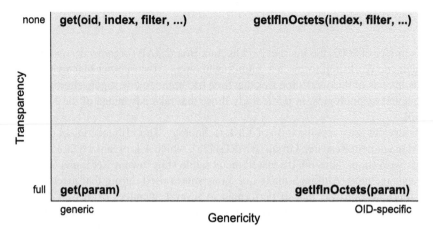

Fig. 3. Fine-grained Web services: transparency vs. genericity.

can be either defined at the WSDL level or kept as an opaque string that is defined as part of a higher-level XML schema; this will be referred to as Web services that have either transparent or non-transparent parameters. The four resulting forms of Web service are depicted in Figure 3.

### 5.2. *Parameter transparency*

Web service operations are defined as part of the abstract interface of WSDL descriptions. An operation may define the exchange of a single message, such as a `trap`, or multiple messages, such as `getRequest` and `getResponse`. For each message, the parameters are defined in so-called `<part>` elements. One approach consists in passing multiple parameters and defining the syntax of each parameter in a separate `<part>` element. An example of this is a `getRequest` message, which defines three parameters: `oid` (object identifier), `index` and `filter`. This example of non-transparent parameters is showed in Figure 4; note that for the sake of simplicity, all parts are of type `string` in this example.

Another approach is to define just a single parameter, and leave the interpretation of this parameter to the (manager-agent) application. In this case, only a single `<part>` element is needed. The syntax of the parameter is of type `string`, which means that the parameter is transparently conveyed by the WSDL layer and no checking is performed at that layer (see Figure 5).

An advantage of having parameter transparency is that a clear separation is made between the management information and the protocol that exchanges this information. In this case, it is possible to change the structure of the management information without altering the definition of management operations. From a standardization perspective, this is a good selling point. However, this is not a mandatory feature as WSDL supports other constructs (e.g., the `import` statement) to divide a specification over several documents, which can then be processed independently along the standardization track.

```
WSDL Description

    ►<message name="getRequest">
       <part name="oid" type="string"/>
       <part name="index" type="string"/>
       <part name="filter" type="string"/>
       ...
     </message>


     <interface name="getInterface">
       <operation name="get">
         <input message="getRequest"/>
         <output message="getResponse"/>
       </operation>
     </interface>

  ....
```

Fig. 4. Operation with non-transparent parameters.

```
WSDL Description

    ►<message name="getRequest">
       <part name="param" type="string"/>
     </message>

     <interface name="getInterface">
       <operation name="get">
         <input message="getRequest"/>
         <output message="getResponse"/>
       </operation>
     </interface>

  ....
```

Fig. 5. Operation with transparent parameters.

In case of transparent parameters, the application can exchange management data in the form of XML documents. In this case, an XML parser is needed to extract the parameters from the document. When the application receives such a document, XPath [20] expressions may be used to select the required information. On the manager side, special functionality is needed to create such XML documents. As specific parsing is needed in the application, transparent parameters should only be used by experienced users who need this flexibility. For a PC user, in his home environment, who wants to include some management information in an Excel spreadsheet, this is too complicated. In such a case, a simple approach is required in which the user does not need to parse XML documents with tools like XPath and XQuery. Instead, parsing and checking should be performed at the WSDL layer, thus by the spreadsheet itself. Non-transparent parameters are therefore better suited for simple management applications, as found in home or small-enterprise environments.

### 5.3. *Genericity*

Another approach is to vary the genericity of operations, as illustrated by this example. Assume a managed system provides host-specific information such as SysLocation and SysUptime, as well as statistical information for the network interfaces such as IfInOctets, IfOutOctets, IfInErrors and IfOutErrors. As the system can have multiple interfaces, the four interface-related OIDs can be provided multiple times. Figure 6 shows this information in the form of a containment tree.

To retrieve this information, management operations need to be defined. If these operations are defined at a low level of genericity, a dedicated operation is required for each variable. The resulting operations then look like getSysUptime or getIfOutOctets (see Figure 6). Note that the operations for retrieving network interface information, need to have a parameter supplied to identify the interface.

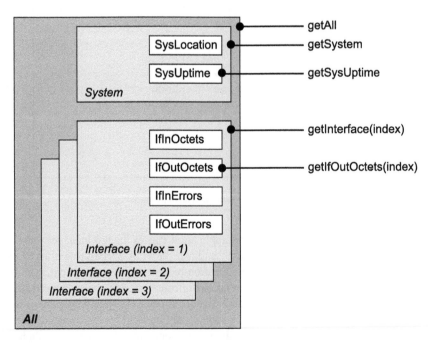

Fig. 6. Containment diagram.

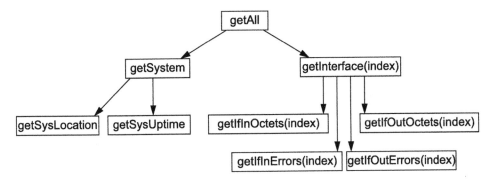

Fig. 7. Containment tree.

Management operations can also be defined at a higher level of genericity. For example, a getSystem operation can be defined to retrieve SysLocation as well as SysUptime, and a getInterface(index) operation to retrieve all management information for a specific interface. If we push this rationale even further, a single getAll operation can retrieve all information contained in the managed system.

An important characteristic of this approach is that the naming of the operations precisely defines the functionality. Therefore it is very easy for the manager to determine which operation to call to retrieve certain information. Note that the number of parameters associated with each operation may be small: no OID is needed because the name of the operation already identifies the object. Only in case multiple instances of the same object class exist (e.g., the Interface object in Figure 6) does the manager have to provide an instance identifier.

Like in other management architectures such as OSI Management (the Open Systems Interconnection management architecture devised by the ISO) [35] or TMN (the Telecommunications Management Network devised by the International Telecommunication Union, ITU) [35], the containment hierarchy can be presented as a tree (see Figure 7). Fine-grained operations can be used to retrieve the leaves of the tree, and coarse-grained operations to retrieve the leaves as well as their higher-level nodes. In SNMP, conversely, only the leaves of the MIB trees are accessible. In general, however, it is not possible to select in a single operation objects from different trees; if the manager wants to retrieve, for example, the OIDs System and IfInErrors, two separate SNMP operations are needed.

To restrict an operation to specific objects, it is possible to use Web services in combination with filtering. Theoretically, filtering allows usage of getAll to retrieve every possible variable, provided that an appropriate filter (and possible index) is given. In combination with filtering, getAll can thus be considered an extreme form of genericity, useful for all kinds of task. In practice, however, filtering can be expensive. When it is performed on the agent side, it may consume too many CPU cycles and memory. When filtering is performed on the manager side, bandwidth may be wasted and the management platform may become overloaded. Although filtering is very flexible and powerful, it should be used with care.

**5.4.** *Example of Web services-based management*

To illustrate the use of Web services for integrated management, let us now study an example of Web services-based network monitoring system. The example is about the SNMP Interface table (ifTable), which is part of the Interfaces Group MIB (IF-MIB) [56]. The ifTable provides information on the operation and use of all network interfaces available in the managed entity. These interfaces can be physical interfaces (e.g., Ethernet and Wireless Local Area Network cards) or virtual interfaces (e.g., tunnels or the loopback interface). For each interface, the table includes a separate row; the number of rows is therefore equal to the number of interfaces. The table consists of 22 columns and includes information such as the interface index (ifIndex), the interface description (ifDescr), the interface type (ifType), etc. Figure 8 shows a summary of this table.

Before we can come up with a WSDL description for retrieving the Interface table, we must choose the genericity of the WSDL operations. Three levels of genericity are possible, ranging from high genericity to low genericity.

At the highest level of genericity, a single operation can be defined to retrieve all objects of the ifTable. We call this Web service operation getIfTable.

Instead of retrieving the entire ifTable in a single call, it is also possible to define Web service operations that retrieve only a single row or a single column. The operation that retrieves a single row of the Interface table is called getIfRow. This operation gets all information related to a single network interface and requires as parameter the identifier of that interface; a possible choice for this parameter is the value of ifIndex. The operation that retrieves a single column is called getIfColumn. To identify which column should be retrieved, a parameter is needed that can take values such as ifIndex, ifDescr, ifType, etc. Alternatively, it is possible to define separate operations per column; in that case, the resulting operations would be getIfIndex, getIfDescr, getIfType, etc.

Finally, at the lowest level of genericity, separate operations are defined to retrieve each individual ifTable object. This approach is somehow comparable to the approach used by the SNMP get operation (although the latter allows multiple objects to be retrieved in a single message). Depending on the required parameter transparency, again two choices are possible. The first possibility is to have a generic getIfCell operation; this operation requires as input parameters an interface identifier to select the row (interface), as well as a parameter to select the column (e.g., ifIndex, ifDescr or ifType). The

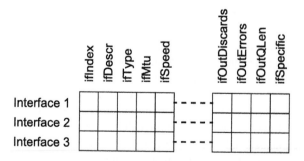

Fig. 8. Interface table.

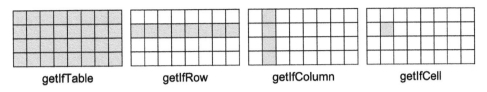

getIfTable         getIfRow         getIfColumn         getIfCell

Fig. 9. Different choices to retrieve table objects.

second possibility is to have special operations for each column; these operations require as input parameter the identifier of the required interface. An example of such an operation is getIfType (interface1).

Figure 9 summarizes the choices that can be made to retrieve table information; they range from high genericity (getIfTable) to low genericity (getIfCell). Note that this figure does not show the choices that are possible with respect to parameter transparency.

Let us now discuss the main parts of the WSDL descriptions of these operations. These descriptions are a mix of WSDL 2.0 and WSDL 1.1, as WSDL 2.0 was not yet finalized when this work was performed. In the gSOAP toolkit that we used [25], the WSDL descriptions consist of the following (container) elements:

- The <types> element (WSDL 2.0), to define new data types.
- The <message> elements (WSDL 1.1), to specify the data that belongs to each message.
- The <interface> element (WSDL 2.0), to combine one or more messages and ease the generation of programming stubs.
- The <binding> element (WSDL 2.0), to associate the interface element with a transport protocol.
- The <service> element (WSDL 2.0), to associate the interface element with a web address (URL).

**5.4.1.** *The <types> element* XML supports two types for defining elements: <simpleType> and <complexType>. A <simpleType> element contains a single value of a predefined form (e.g., an integer or a string). A <complexType> element groups other elements (e.g., an address element can contain street, postal code and city sub-elements).

The WSDL descriptions of our examples use <complexType> elements to group certain objects within the ifTable. Which object are taken together depends on the genericity of the Web service. For example, all columns of the ifTable can be grouped together, yielding a single <complexType> that can be used to retrieve entire ifTable rows. This complex type is called ifEntry in our example (see Figure 10), and can be used by the getIfTableas well as getIfRow operations. Note that for the getIfRow operation, the <sequence> element should be left out, as only one row can be retrieved at a time.

**5.4.2.** *The <message> elements* The <message> elements are used in WSDL 1.1 to describe the information that is exchanged between a Web service producer and its con-

```
<types>
  <complexType name="ifEntry">
    <sequence>
      <element name="ifIndex" type="xsd:unsignedInt"/>
      <element name="ifDescr" type="xsd:string"/>
      <element name="ifType" type="xsd:unsignedInt"/>
      <element name="ifMtu" type="xsd:unsignedInt"/>
      <element name="ifSpeed" type="xsd:unsignedInt"/>
      <element name="ifPhysAddress" type="xsd:string"/>
      <element name="ifAdminStatus" type="xsd:unsignedInt"/>
      <element name="ifOperStatus" type="xsd:unsignedInt"/>
      <element name="ifInOctets" type="xsd:unsignedInt"/>
      <element name="ifInUcastPkts" type="xsd:unsignedInt"/>
      <element name="ifInDiscards" type="xsd:unsignedInt"/>
      <element name="ifInErrors" type="xsd:unsignedInt"/>
      <element name="ifOutOctets" type="xsd:unsignedInt"/>
      <element name="ifOutUcastPkts" type="xsd:unsignedInt"/>
      <element name="ifOutDiscards" type="xsd:unsignedInt"/>
      <element name="ifOutErrors" type="xsd:unsignedInt"/>
      <element name="ifOutQLen" type="xsd:unsignedInt"/>
      <element name="ifSpecific" type="xsd:string"/>
    </sequence>
  </complexType>
  ...
</types>
```

Fig. 10.  The ifEntry type.

```
<message name="getIfTableRequest">
  <part name="community" type="xsd:string"/>
</message>

<message name="getIfTableResponse">
  <part name="sizeTable" type="xsd:int"/>
  <part name="ifEntry" type="utMon:ifEntry"/>
</message>
```

Fig. 11.  Message definitions for getIfTable.

sumers. There are <message> elements for request (input) and response (output) messages. These messages consist of zero or more <part> elements. In request messages, the <part> elements represent the parameters of the Web service. In response messages, these elements describe the response data.

The getIfTable Web service supports a single operation: retrieving the complete table. Figure 11 shows the <message> elements for this operation. The figure shows a request message that contains a 'community' string; this string is used for authentication purposes in SNMPv1 or SNMPv2c. The response message contains an element of type

```
<message name="getIfIndexRequest">
  <part name="index" type="xsd:unsignedInt"/>
  <part name="community" type="xsd:string"/>
</message>

<message name="getIfIndexResponse">
  <part name="ifIndex" type="xsd:unsignedInt"/>
</message>

<message name="getIfDescrRequest">
  <part name="index" type="xsd:unsignedInt"/>
  <part name="community" type="xsd:string"/>
</message>

<message name="getIfDescrResponse">
  <part name="ifDescr" type="xsd:string"/>
</message>
```

Fig. 12. Message definitions for getIfCell.

ifEntry, which was defined in Figure 10, as well as an integer representing the number of table rows.

All other Web services in this section support multiple operations. The getIfRow Web service can be defined to support two operations: retrieve a specified row, and retrieve a list of valid row index numbers (which correspond to the network interfaces in the agent system). The getIfColumn and getIfCell Web services can be defined in terms of operations (one per ifTable column).

Figure 12 shows the messages for two of the operations used by the getIfCell Web service. The request messages have an index element for referencing the correct cell.

**5.4.3.** *The <interface> element*   An <interface> element defines the operations that are supported by the Web service. Each operation consists of one or more messages; the <interface> element therefore groups the previously defined <message> elements into <operation> elements. In addition, the <interface> element may define <description> elements. In general, four types of operation exist: *one way, request-response, solicit response* and *notification*. In our example, they are all of the *request-response* type. Figure 13 shows the <interface> element for the getIfTable Web service.

The other Web services support multiple operations. Figure 14 shows part of the <interface> element for the getIfColumn Web service.

**5.4.4.** *The <binding> element*   A <binding> element specifies which protocol is used to transport the Web service information, and how this information is encoded. Similar to the <interface> element, it also includes the operations that are supported by the Web service, as well as the request and response messages associated with each operation. In

```
<interface name="getIfTableServicePortType">
  <operation name="getIfTable">
    <documentation>function utMon\_\_getIfTable</documentation>
    <input message="tns:getIfTableRequest"/>
    <output message="tns:getIfTableResponse"/>
  </operation>
</interface>
```

Fig. 13. Interface definition for getIfTable.

```
<interface name="getIfColumnServiceInterface">
  <operation name="getIfIndex">
    <documentation>function utMon\_\_getIfIndex</documentation>
    <input message="tns:getIfIndexRequest"/>
    <output message="tns:uIntArray"/>
  </operation>
  <operation name="getIfDescr">
    <documentation>function utMon\_\_getIfDescr</documentation>
    <input message="tns:getIfDescrRequest"/>
    <output message="tns:stringArray"/>
  </operation>
  <operation name="getIfType">
    <documentation>function utMon\_\_getIfType</documentation>
    <input message="tns:getIfTypeRequest"/>
    <output message="tns:uIntArray"/>
  </operation>
  ...
</interface>
```

Fig. 14. Interface definition for getIfColumn.

principle, there are many choices for the transport protocol; in practice, SOAP over HTTP is by far the most popular choice [85]. Figure 15 shows part of the <binding> element for the getIfColumn Web service.

**5.4.5.** *The <service> element*  The <service> element is used to give the Web service a name and specify its location (a URL). Figure 16 shows this element for the example of the getIfRow Web service. The name is getIfRowService, the location is http://yourhost.com/ and the transport protocol is defined by the getIfRowServiceBinding.

**5.5.** *Performance*

One of the main arguments against using Web services in network management is that the performance of Web services is inferior to that of SNMP. We show in this section that this argument is invalid. Although there are scenarios in which SNMP provides better

```
<binding name="getIfColumnServiceBinding"
      type="tns:getIfColumnServiceInterface">
  <SOAP:binding style="rpc"
      transport="http://schemas.xmlsoap.org/soap/http"/>

<operation name="getIfIndex">
  <SOAP:operation soapAction=""/>
  <input>
    <SOAP:body use="encoded" namespace="urn:utMon"
      encodingStyle="http://schemas.xmlsoap.org/soap/encoding/"/>
  </input>
  <output>
    <SOAP:body use="encoded" namespace="urn:utMon"
      encodingStyle="http://schemas.xmlsoap.org/soap/encoding/"/>
  </output>
</operation>

<operation name="getIfDescr">
  <SOAP:operation soapAction=""/>
  <input>
    <SOAP:body use="encoded" namespace="urn:utMon"
      encodingStyle="http://schemas.xmlsoap.org/soap/encoding/"/>
  </input>
  <output>
    <SOAP:body use="encoded" namespace="urn:utMon"
      encodingStyle="http://schemas.xmlsoap.org/soap/encoding/"/>
  </output>
</operation>
...
</binding>
```

Fig. 15. Binding definition for getIfColumn.

```
<service name="getIfRowService">
  <documentation>getIfRow service</documentation>
  <endpoint name="getIfRowService"
      binding="tns:getIfRowServiceBinding">
    <SOAP:address location="http://yourhost.com/"/>
  </endpoint>
</service>
```

Fig. 16. Service element definition for getIfRow.

performance, there are other scenarios in which Web services perform better. In general, the following statements can be made:

- In case a single managed object should be retrieved, SNMP is more efficient than Web services.

- In case many objects should be retrieved, Web services may be more efficient than SNMP.
- The encoding and decoding of SNMP PDUs or WSDL messages is less expensive (in terms of processing time) than retrieving management data from within the kernel. The choice between Basic Encoding Rules (BER) or XML encoding is therefore not the main factor that determines performance.
- Performance primarily depends on the quality of the implementation. Good implementations perform considerably better than poor implementations. This holds for SNMP as well as Web services implementations.

In the literature, several studies can be found on the performance of Web services-based management. Choi and Hong published several papers in which they discuss the design of XML-to-SNMP gateways. As part of their research, the performance of XML and SNMP-based management software was investigated [19,18]. To determine bandwidth, they measured the XML traffic on one side of the gateway, and the corresponding SNMP traffic on the other side. In their test setup, separate SNMP messages were generated for every managed object that had to be retrieved; the possibility to request multiple managed objects via a single SNMP message was not part of their test scenario. The authors also measured the latency and resource (CPU and memory) usage of the gateway. They concluded that, for their test setup, XML performed better than SNMP.

Whereas the previous authors compared the performance of SNMP to XML, Neisse and Granville focused on SNMP and Web services [61]. As the previous authors, they implemented a gateway and measured traffic on both sides of this gateway. Two gateway translation schemes were used: protocol level and managed object level. In the first scheme, a direct mapping exists between every SNMP and Web services message. In the second scheme, a single, high-level Web service operation (e.g., getIfTable) maps onto multiple SNMP messages. These authors also investigated the performance impact of using HTTPS and zlib compression [95]. The Web services toolkit was NuSOAP [63]. For protocol-level translation, they concluded that Web services always require substantial more bandwidth than SNMP. For translation at the managed-object level, they found that Web services perform better than SNMP if many managed objects were retrieved at a time.

Another interesting study on the performance of Web services and SNMP was conducted by Pavlou et al. [76,77]. Their study is much broader than a performance study and also includes CORBA-based approaches. They consider the retrieval of Transmission Control Protocol (TCP) MIB variables and measure bandwidth and round-trip time. Unlike in the previous two studies, no gateways are used, but a dedicated Web services agent was implemented using the Systinet WASP toolkit [91]. The performance of this agent was compared to a Net-SNMP agent [62]. The authors neither investigated alternate SNMP agents, nor the effect of fetching actual management data in the system. They concluded that Web services causes more overhead than SNMP.

In the remainder of this section, we discuss a study performed in 2004 at University of Twente by one of the authors of this chapter; more details can be found in [78]. For this study, measurements were performed on a Pentium 800 MHz PC running Debian Linux. The PC was connected via a 100 Mbit/s Ethernet card. Four Web services prototypes were build: one for retrieving single OIDs of the ifTable, one for retrieving ifTable rows, one for retrieving ifTable columns and one for retrieving the entire ifTable (see Section 5.4). To retrieve management data from the system, we decided to use the same code for the SNMP and Web services prototypes. In this way, differences between the measurements would only be caused by differences in SNMP and Web services handling. For this reason, the Web services prototypes incorporated parts of the Net-SNMP open-source package. All SNMP-specific code was removed and replaced by Web services-specific code. This code was generated using the gSOAP toolkit version 2.3.8 [25,26], which turned out to be quite efficient. It generates a dedicated skeleton and does not use generic XML parsers such as the Document Object Model (DOM) or the Simple API for XML (SAX). For compression, we used `zlib` [95]. The performance of the Web services prototypes was compared to more than twenty SNMP implementations, including agents on end-systems (Net-SNMP, Microsoft Windows XP agent, NuDesign and SNMP Research's CIA agent) as well as agents within routers and switches (Cisco, Cabletron, HP, IBM and Nortel).

**5.5.1.** *Bandwidth usage*   One of the main performance parameters is bandwidth usage. For SNMP, it can be derived analytically as the PDU format of SNMP is standardized. For Web services, there were no standards when we conducted this study, so we could not derive analytical formulae that would give lower and upper bounds for the bandwidth needed to retrieve `ifTable` data. In fact, the required bandwidth depends on the specific WSDL description, which may be different from case to case. This section therefore only discusses the bandwidth requirements of the prototype that fetches the entire `ifTable` in a single interaction; the WSDL structure of that prototype was presented in Section 5.4.

Figure 17 shows the bandwidth requirements of SNMPv1 and Web services as a function of the size of the `ifTable`. Bandwidth usage is determined at the application (SNMP/SOAP) layer; at the IP layer, SNMP requires 56 additional octets, whereas Web services require between 273 and 715 octets, depending on the number of retrieved managed objects and the compression mechanism applied. In case of SNMPv3, an additional 40 to 250 octets are needed.

Figure 17 illustrates that SNMP consumes far less bandwidth than Web services, particularly in cases where only a few managed objects should be retrieved. But, even in cases where a large number of objects should be retrieved, SNMP remains two (`get`) to four (`getBulk`) times better than Web services. The situation changes, however, when compression is used. Figure 17 shows that, for Web services, compression reduces bandwidth consumption by a factor 2 when only a single object is retrieved; if 50 objects are retrieved, by a factor 4; for 150 objects, by a factor 8.

The threshold where compressed Web services begin to outperform SNMP is 70 managed objects. This number is not ridiculously high: it is easily reached when interface data needs to be retrieved for hundreds of users, for example in case of access switches or Digital Subscribe Line Access Multiplexers (DSLAMs). Figure 17 also shows that the bandwidth consumption of compressed SNMP, as defined by the IETF EoS Working Group, is approximately 75% lower than non-compressed SNMPv1. This form of SNMP compression is known under the name ObjectID Prefix Compression (OPC).

**5.5.2.** *CPU time*   Figure 18 shows the amount of CPU time needed for coding (decoding plus the subsequent encoding) an SNMP message (encoded with BER) or Web services

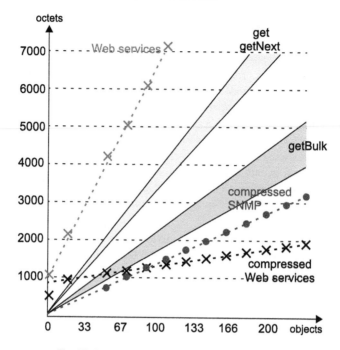

Fig. 17. Bandwidth usage of SNMP and Web services.

(encoded in XML) encoded messages. As we can see, the different curves can be approx-imated by piecewise linear functions. For an SNMP message carrying a single managed object, the BER coding time is roughly 0.06 ms. A similar XML-encoded message re-quires 0.44 ms, which is seven times more. Coding time increases if the number of man-aged objects in the message increases. Coding an SNMP message that contains 216 objects requires 0.8 ms; coding a similar Web services message requires 2.5 ms. We can therefore conclude that XML encoding requires 3 to 7 times more CPU time than BER encoding. Figure 18 also shows the effect of compression: compared to XML coding, Web services compression turns out to be quite expensive (by a factor 3 to 5). In cases where bandwidth is cheap and CPU time expensive, Web services messages should not be compressed.

To determine how much coding contributes to the complete message handling time, the time to retrieve the actual data from the system was also measured. For ifTable, this retrieval requires, among other things, a system call. We found out that retrieving data is relatively expensive; fetching the value of a single object usually takes between 1.2 and 2.0 ms. This time is considerably larger than the time needed for coding.

The results are depicted in Figure 19; to facilitate the comparison, the BER and XML coding times are mentioned as well. The figure shows that data retrieval times for Net-SNMP increase quite fast; five managed objects already require more than 6 ms; for re-trieving 54 managed objects, 65 ms were needed; and retrieving 270 managed objects from the operating system took 500 ms! These staggering times can only be explained by the fact that Net-SNMP performs a new system call every time it finds a new ifTable object within the get or getBulk request; Net-SNMP does not implement caching. The con-

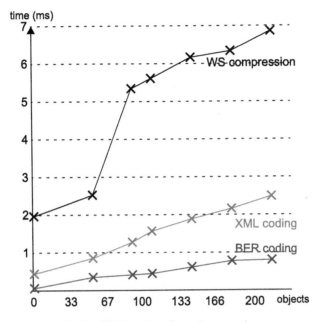

Fig. 18. CPU time for coding and compression.

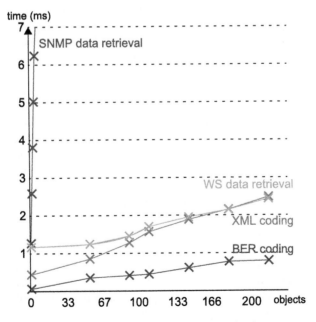

Fig. 19. CPU time for coding and data retrieval.

| | instructions | data | |
|---|---|---|---|
| | | static | dynamic |
| SNMP | 1972 KB | 128 KB | 70 - 160 KB |
| Web services | 580 KB | 470 B | 4 KB |

Fig. 20. Memory requirements.

clusion must therefore be that data retrieval is far more expensive than BER encoding; CPU usage is not determined by the protocol handling, but by the quality of the implementation.

**5.5.3.** *Memory usage*  Memory is needed for holding program instructions ('instruction memory') and data ('data memory'). For data memory, a distinction needs to be made between static and dynamic memory. Static memory is allocated at the start of the program and remains constant throughout the program lifetime. Dynamic memory is allocated after a request is received and released after the response is transmitted. If multiple requests arrive simultaneously, dynamic memory is allocated multiple times.

Figure 20 shows the memory requirements of Net-SNMP and our Web services prototype. We see that Net-SNMP requires roughly three times as much instruction memory than the Web services prototype. Net-SNMP requires 20 to 40 times more data memory, depending on the number of objects contained in the request. These large numbers, much larger than their Web services counterparts, should not be misinterpreted: the functionality of Net-SNMP is much richer than that of our Web services prototype. For instance, Net-SNMP supports three different protocol versions, includes encryption, authentication and access control, and is written in a platform-independent way.

**5.5.4.** *Round-trip time*  In the previous subsections, the performance of the getIfTable Web service was compared to that of Net-SNMP. This subsection compares the performance of this Web services prototype to other SNMP implementations. As we could not add code to existing agents code for measuring BER encoding and data retrieval times, we measured instead the round-trip time, as a function of the number of retrieved managed objects. This time can be measured outside the agent and thus does not require any agent modification.

Figure 21 shows the results for 12 different SNMP agents, as well as the getIfTable Web service prototype. It is interesting to note that the round-trip time for normal Web services increases faster than that of compressed Web services. For most SNMP measurements, we used getBulk with max-repetition values of 1, 22, 66 and 270; for agents that do not support getBulk, a single get request was issued, asking for 1, 22, 66 or 270 objects. Not all agents were able to deal with very large size messages; the size of the response message carrying 270 managed objects is such that, on an Ethernet Local Area Network, the message must be fragmented by the IP protocol. As the hardware for the various agents varied, the round-trip times showed in Figure 21 should be used with great care

|          | 1    | 22    | 66    | 270   |
|---------:|:----:|:-----:|:-----:|:-----:|
| WS       | 1,7  | 2,6   | 10,3  | 36,5  |
| WS-Comp  | 3,3  | 4,3   | 5,6   | 11,8  |
| SNMP-1   | 1,0  | 2,5   |       |       |
| SNMP-2   | 1,0  | 2,7   |       |       |
| SNMP-3   | 1,1  | 3,2   | 6,9   | 14,9  |
| SNMP-4   | 2,0  | 15,2  |       |       |
| SNMP-5   | 3,7  | 18,3  |       |       |
| SNMP-6   | 1,3  | 18,9  | 53,0  |       |
| SNMP-7   | 2,9  | 58,0  | 167,0 | 695,0 |
| SNMP-8   | 3,1  | 58,2  | 168,8 | 700,0 |
| SNMP-9   | 3,5  | 62,0  | 183,8 | 767,0 |
| SNMP-10  | 5,0  | 80,9  |       |       |
| SNMP-11  | 12,0 | 86,9  | 244,3 |       |
| SNMP-12  | 12,4 | 257,9 |       |       |

Fig. 21. Round-trip time (in ms).

and only considered purely indicative. Under slightly different conditions, these times may be quite different. For example, the addition of a second Ethernet card may have a severe impact on round-trip times. Moreover, these times can change if managed objects other than ifTable OIDs are retrieved simultaneously. In addition, the first SNMP interaction often takes considerably more time than subsequent interactions.

Despite these remarks, the overall trend is clear. With most of the SNMP agents that we tested, the round-trip time heavily depends on the number of managed objects that are retrieved. It seems that several SNMP agents would benefit from some form of caching; after more than 15 years of experience in SNMP agent implementation, there is still room for performance improvements. From a latency point of view, the choice between BER and XML encoding does not seem to be of great importance. Our Web services prototype performs reasonably well and, for numerous managed objects, even outperforms several commercial SNMP agents.

## 6. Toward coarse-grained Web services for integrated management

In the previous section, Web services were used in SNMP-based network management as getters and setters of fine-grained MIB OIDs. Similarly, in a WBEM environment, they could be used as getters and setters of CIM object properties, or they could be mapped to fine-grained CIM operations. This is the granularity at which Web services-enabled management platforms generally operate today. For instance, HP's Web Service Management

Framework [14] specifies how to manipulate fine-grained managed objects with Web services. But is this the way Web services are meant to be used as management application middleware?

Web services did not come out of the blue, context free: they were devised with SOA in mind. SOA is an IT architectural style, to use Fielding's terminology [28], and Web services implement it. In this section, we go back to the rationale behind SOA, consider management applications from this perspective, and show that Web services could be used differently in integrated management. We illustrate this claim with examples of coarse-grained Web services for integrated management.

### 6.1. *Service-oriented architecture*

Defining SOA poses two problems. First, there is no single and widely shared definition. No one clearly defined this concept before it began pervading the software industry. This concept grew and spread by hearsay in the software engineering community, as a conceptualization of what Web services were trying to achieve. The definitions proposed by the W3C and OASIS came late, and despite the credit of these organizations in the community, none of their definitions are widely accepted.

Second, since 2004–2005, SOA has turned into a buzzword in the hands of marketing people. Today, there are almost as many definitions of this acronym as vendors in the software industry. There are also countless examples of software tools and packages that have been re-branded 'SOA-enabled' without supporting all the features that engineers generally associate with this acronym.

In an attempt to distinguish between a stable technical meaning and ever-changing marketing blurb, some authors use the term Service-Oriented Computing (SOC) [72,81] to refer, among engineers, to the distributed computing paradigm, leaving SOA to hype. Other authors consider that SOC is a distributed computing paradigm whereas SOA is a software architecture; the subtle difference between the two is not always clear. In this chapter, we consider these two acronyms synonymous and use SOA in its technical sense.

So, what is SOA about? Because no one 'owns' its definition, we reviewed the literature and browsed a number of Web sites to form an opinion.

In WSA [9] (defined in Section 2.5), the W3C defines SOA as "a form of distributed systems architecture" that is "typically characterized" (note the implied absence of consensus) by the following properties:

- *logical view:* the service is an abstract view of actual programs, databases, business processes, etc.;
- *message orientation:* the service is defined in terms of messages exchanged between consumers and providers;
- *description orientation:* the service is described by machine-parsable metadata to facilitate its automated discovery;
- *granularity:* services generally use a small number of operations with relatively large and complex messages;
- *network orientation:* services are often used over a network, but this is not mandatory;

- *platform neutral:* messages are sent in a standard format and delivered through interfaces that are independent of any platform.

In August 2006, OASIS released its Reference Model for SOA (SOA-RM) [50]. This specification was long awaited by the industry. Unfortunately, this version 1.0 is quite verbose and rather unclear, and often proposes overly abstract definitions; we hope that OASIS will solve these teething problems in subsequent versions. Difficult to read as it is, this document still clarifies three important points:

- "SOA is a paradigm for organizing and utilizing distributed capabilities that may be under the control of different ownership domains";
- a service is accessed via an interface;
- a service is opaque: its implementation is hidden from the service consumer.

Channabasavaiah et al. [15] are more practical and define SOA as "an application architecture within which all functions are defined as independent services with well-defined invokable interfaces, which can be called in defined sequences to form business processes". They highlight three aspects:

- all functions are defined as services;
- all services are independent and operate as black boxes;
- consumers just see interfaces; they ignore whether the services are local or remote, what interconnection scheme or protocol is used, etc.

IBMs online introduction to SOA [37] lists the following characteristics:

- SOA is an IT architectural style;
- integrated services can be local or geographically distant;
- services can assemble themselves on demand; they "coalesce to accomplish a specific business task, enabling (an organization) to adapt to changing conditions and requirements, in some cases even without human intervention"; coalescence refers to automated service composition at run-time;
- services are self-contained (i.e. opaque) and have well-defined interfaces to let consumers know how to interact with them;
- with SOA, the macro-components of an application are loosely coupled; this loose coupling enables the combination of services into diverse applications;
- all interactions between service consumers and providers are carried out based on service contracts.

Erl [27] identifies four core properties and four optional properties for SOA:

- services are autonomous (core);
- loose coupling (core);
- abstraction: the underlying logic is invisible to service consumers (core);
- services share a formal contract (core);
- services are reusable;
- services are composable;
- services are stateless;
- services are discoverable.

O'Brien et al. [67] complement this list of properties with the following two:

- *Services have a network-addressable interface:* "Service (consumers) must be able to invoke a service across the network. When a service (consumer) and service provider are on the same machine, it may be possible to access the service through a local interface and not through the network. However, the service must also support remote requests".

- *Services are location transparent:* "Service (consumers) do not have to access a service using its absolute network address. (They) dynamically discover the location of a service looking up a registry. This feature allows services to move from one location to another without affecting the (consumers)".

Singh and Huhns [81] characterize SOA as follows:

- *loose coupling*;
- *implementation neutrality:* what matters is the interface; implementation details are hidden; in particular, services do not depend on programming languages;
- *flexible configurability:* services are bound to one another late in the process;
- *long lifetime:* interacting entities must be able to cope with long-lived associations; services should live "long enough to be discovered, to be relied upon, and to engender trust in their behavior";
- *coarse granularity:* actions and interactions between participants should not be modeled at a detailed level;
- *teams:* participants should be autonomous; instead of a "participant commanding its partners", distributed computing in open systems should be "a matter of business partners working as a team".

For these authors, the main benefits of SOA are fourfold [81]:

- services "provide higher-level abstractions for organizing applications in large-scale, open environments";
- these abstractions are standardized;
- standards "make it possible to develop general-purpose tools to manage the entire system lifecycle, including design, development, debugging, monitoring";
- standards feed other standards.

All these definitions (and many others not mentioned here) draw a fairly consistent picture. SOA is a software paradigm, an IT architectural style for distributing computing with the following properties:

- *Loose coupling:* In SOA, distributed applications make use of loosely coupled services.
- *Coarse-grained:* Services have a coarse granularity, they fulfill some functionality on their own.
- *Autonomous:* Services are stateless and can be used by many applications or other services.
- *Interfaces and contracts:* Services are opaque entities accessed via interfaces on the basis of a high-level contract; implementation details are hidden.
- *Network transparency:* Services can be local or remotely accessed via the network. The location of the service provide may change over time.
- *Multiple owners:* Services can be owned by different organizations.
- *Service discovery:* Services are discovered dynamically, possibly automatically, at runtime; service providers are not hard-coded in applications.
- *Service composition:* Services can be composed transparently, possibly dynamically at run-time.

There is also a large consensus about three things. First, Web services are just one technology implementing SOA. Other technologies may come in the future that could implement SOA differently, perhaps more efficiently. For instance, one could imagine that SOA be implemented by a new variant of CORBA that would (i) define services as high-level CORBA objects wrapping up CORBA and non-CORBA legacy applications, (ii) use the Internet Inter-ORB Protocol (IIOP) instead of SOAP and HTTP to make service consumers and providers communicate, (iii) advertise services in the form of interfaces (expressed in CORBA IDL) in standard external registries (as opposed to CORBA-specific registries), (iv) enable these services to be discovered and composed dynamically at run-time using semantic metadata, and (v) solve the poor interoperability and the performance problems of multi-vendor inter-ORB communication.

Second, nothing in SOA is new. For instance, interfaces that make it possible to access opaque entities were already defined in Open Distributed Processing [39]. What is new is that we now have the technology to make it work: "SOAs are garnering interest, not because they are a novel approach, but because Web services are becoming sufficiently mature for developers to realize and reevaluate their potential. The evolution of Web services is at a point where designers can start to see how to implement true SOAs. They can abstract a service enough to enable its dynamic and automatic selection, and Web services are finally offering a technology that is rich and flexible enough to make SOAs a reality" [49].

Third, a key difference between Web services and distributed object middleware is that Web services are not objects that need other objects for doing something useful: they are autonomous applications fulfilling some functionality on their own. As a result, Web services should not be regarded as yet another type of object-oriented middleware competing with CORBA or J2EE [90]: they are supposed to operate at a higher level of abstraction.

It should be noted that this coarse-grained approach to middleware poses problems to the software industry, because hordes of engineers are used to thinking in terms of tightly coupled objects, and a plethora of tools work at the object level [90]. Even some academics struggle to adjust their mindframe [8].

### 6.2. *Comparison between SOA and current practice*

Now that we better understand the SOA approach to distributed computing, let us compare, one by one, the principles that underpin Web services with the current practice in integrated management. By 'practice', we mean the management platforms that enterprises actually use to manage their networks, systems and services, as opposed to the latest commercial offerings from vendors who may claim to support SOA.

**6.2.1.** *Applications make use of services*     SOA and integrated management both have the concept of *service*; they give it slightly different but reasonably consistent meanings. For both, we have low-level services that abstract certain parts of the IT infrastructure (e.g., a network service for an MPLS-based VPN) and high-level services that are user-centered and achieve a specific business goal (e.g., a business service for online book ordering).

A Web service does not necessarily pertain to Service-Level Management (SLM): it may be a building block for fulfilling network management tasks, or systems management tasks. In this case, the SOA concept of service corresponds roughly to the concept of management task in integrated management.

In integrated management, a high-level service embodies a contract between an end-user (a person) and a service provider (a vendor). In TMN, a network service is captured by the Service Management Layer (SML), while the contract between a user and a vendor belongs to the Business Management Layer [35]. The SML is where network and business services are tied together. In SOA, the service contract binds a service consumer to a service producer.

Both in SOA and integrated management, services can be characterized by a Service-Level Agreement (SLA), which captures the end-users' perception of whether the IT infrastructure works well for achieving a specific business goal. High-level services serve as intermediaries between business views and IT views; they enable end-users to abstract away the infrastructure and consider it as a black box.

**6.2.2.** *Loose coupling*   Integrated management platforms use both tight and loose forms of integration. We can distinguish five cases.

First, communication between managed entities (agents) and managing entities (managers) is loosely coupled. It relies on standard management protocols (e.g., SNMP) that hide implementation details by manipulating abstractions of real-life resources known as managed objects.

Second, the *macro-components* of a monitoring application (event correlator, event filter, event aggregator, data analyzer, data filter, data aggregator, data poller, topology discovery engine, network map manager, etc. [53]) are tightly integrated. They are provided by a single vendor. Communication between these macro-components is proprietary, using all sorts of programming tricks to boost performance (e.g., the maximum number of events that can be processed per time unit).

Third, when management is distributed over multiple domains (e.g., for scalability reasons in large enterprises, or budget reasons in corporations with financially independent subsidiaries), manager-to-manager communication is proprietary and relies on tight coupling in the SNMP world; in the WBEM world, it is usually loosely coupled and based on standard management protocols. Note that exchanging management information with external third parties requires sophisticated firewalls in the case of tight coupling, and simpler firewalls in the case of loose coupling.

Fourth, communication between the monitoring application and the data repository is sometimes loosely coupled, sometimes tightly coupled. The data repository is logically unique but can be physically distributed; it consists typically of a relational database for monitoring data and a Lightweight Directory Access Protocol (LDAP) repository for configuration data; sometimes it may also include an object base, or an XML repository, or a knowledge base, etc. An example of loose coupling[4] is when the monitoring application accesses a relational database via a standard interface such as Open Data Base Connectivity

---

[4]Note that we use 'loose coupling' in a flexible manner here, compared to SOA: by using SQL, the application cannot interact with a generic data repository, but only with a relational database.

(ODBC) using a standard query language such as the Structured Query Language (SQL); the actual data repository used underneath (Oracle database, Sybase database, etc.) is then transparent to the application. An example of tight coupling is when the application uses a proprietary query language (e.g., a so-called 'enhanced' version of SQL, with proprietary extensions to the language). Vendor lock-in always implies tight coupling.

Fifth, in large enterprises or corporations, people not only use monitoring applications, but also data mining applications that perform trend analysis, retrospective pattern analysis, etc. Examples include Enterprise Resource Planning (ERP) to anticipate the saturation of resources, security analysis to search for attack patterns, event analysis to identify the important cases that need to be fed to a case-based reasoning engine, etc. Communication between the data repository and these data mining applications is usually loosely coupled: people use standard interfaces and protocols for exchanging data because these applications are generally provided by third-party vendors. For performance reasons, data mining applications may occasionally be tightly coupled to the data repository.

**6.2.3.** *Services are coarse-grained*   How do the SOA concepts of 'fine-grained services' and 'coarse-grained services' translate into integrated management? In the SOA approach to distributed computing, we saw that applications make use of services, which isolate independent pieces of the application that can be reused elsewhere (e.g., by other applications or other services). Some of these services are conceptually at a high level of abstraction (e.g., "Balance the load of virtual server VS among the physical servers PS1, PS2, PS3 and PS4"), others at a low level (e.g., "Set the administrative status of CPU C to down"). Functionality is thus split between the application and reusable services of varying levels of abstraction.

In integrated management, this corresponds to management tasks in the functional model [53]. Functionality is split into management tasks, some of which are low level, some of which are high level. Some of these functions are generic (e.g., see the standard TMN management functions [41]) and can in principle be separated from the application. The common practice in telecommunications is that management applications are based on CORBA and implement a mixture of OSI Management, TMN and SNMP. Generic functions are accessed via CORBA interfaces as independent low-level services, which can be considered coarse-grained because they do not simply operate at the managed object level. In enterprises, however, the situation is very different: today's management platforms are based on SNMP or WBEM (which do not define standard generic functions), not on OSI Management or TMN; they are typically implemented in C, C++, Java or C#, and based neither on CORBA nor on J2EE. As a result, they do not handle separate services.

In short, in the enterprise world, management tasks are not implemented in the form of independent coarse-grained services. This is a major difference with SOA.

From outside, management applications appear to operate at only two scales: the managed object level (when interacting with managed entities), and the management application level (the monitoring application globally supports some functionality as a big black box). There are no intermediaries that could easily be mapped to Web services.

Internally, management applications actually operate at more scales. When they are coded in an object-oriented programming language, these scales are: managed object, object, software component, thread, process and complete application. With procedural programming languages, these scales are: managed object, function, thread, process and complete application.

**6.2.4.** *Services are autonomous*   We saw in Section 1.2 that large enterprises generally run several management platforms in parallel. These applications are autonomous in the SOA sense, but they are not packaged as services. They usually do not communicate directly (there are exceptions), but they all (at least supposedly) send meaningful events to the same event correlator, which is in charge of integrating management for the entire enterprise. In some cases, this logically unique event correlator can be physically distributed [54].

Network management platforms (e.g., HP Openview and IBM Tivoli) often accept equipment-specific graphical front-ends (so-called 'plug-ins') provided by third-party vendors. These extensions can be seen as autonomous services that are statically discovered and integrated.

**6.2.5.** *Interfaces and contracts*   Integrated management applications work with two types of interface: interfaces to their internal components, and interfaces to managed entities (managed objects). The former do not hide implementation details (on the contrary, these components can expect other components to support programming language-specific features such as Java exceptions, for instance), but the latter do (managed objects are standard and hide the idiosyncrasies of the underlying managed resource).

In SOA, services accessed via interfaces form the basis for a contract. In integrated management, contracts are captured in the form of SLAs.

**6.2.6.** *Network transparency*   In SOA, network transparency means that services can be local or remote. This transparency has no equivalent in management platforms. For obvious confidentiality and robustness reasons, no one wants a monitoring application to discover automatically, and later use, an event correlator provided by an unknown organization on the opposite side of the planet!

**6.2.7.** *Multiple owners*   In integrated management, the monitoring and data mining applications are generally run by the same organization, in which case there is only one owner. In large organizations, these applications can be distributed over a hierarchy of domains; different management applications are then run by different teams, possibly attached to different profit centers; but they all really belong to the same organization, the same owner from an SOA standpoint.

When organizations outsource partially or totally the management of their IT infrastructure and/or the management of their services, there are clearly multiple owners. Relationships between them are controlled by business contracts, and not just Web services contracts.

Extranets and subsidiaries are typical examples where we have different owners. In an extranet, a corporation works closely with a large set of subcontractors and/or retailers. This requires the IT infrastructures of different owners to be tightly coupled. The situation is similar in corporations that have financially independent subsidiaries. Although the latter are different enterprises, different owners, they have strong incentives for cooperating as if they were a single organization, a single owner in SOA parlance.

**6.2.8.** *Service discovery*   In today's integrated management applications, only very simple forms of service discovery are generally implemented, primarily for retrieving configuration settings. In systems management, LDAP repositories are commonly used.

**6.2.9.** *Service composition*   Integrated management platforms generally found in businesses today do not use dynamic service composition at run-time. Management tasks are composed statically in the monitoring application or the data mining applications, and cast in iron in static code.

**6.3.** *Analysis*

The first outcome of our comparison is that the management platforms deployed in real life are far from implementing SOA natively. Instead of simply using Web services to exchange managed objects or events, as suggested by recent standardization activities such as WSDM or WS-Management, we should rethink (i.e., re-architect and re-design) management applications to make them SOA compliant. This is a classic mission for software architects.

The main differences between SOA and today's practice in integrated management are (i) the absence of autonomous coarse-grained management tasks in management platforms, (ii) the tight coupling of the monitoring application (most if not all of its macro-components communicate via proprietary mechanisms), and (iii) the quasi absence of service discovery (except for retrieving configuration settings).

A fourth difference, the lack of dynamic service composition at run-time, is less of a problem because, strictly speaking, service composition is not a requirement in SOA: it is simply desirable, and the odds are that when the first three problems are solved, people will find ways of composing these new services.

Reconciling the first three differences identified above requires modifying not only the way management applications are structured, but also the management architectures that underpin them (notably SNMP and WBEM).

**6.3.1.** *Changes in the management architecture*   In integrated management, it is customary to decompose a *management architecture* (also known as *management framework*) into four models [30]: (i) an organizational model, which defines the different entities involved in a management application and their roles; (ii) an information model, which specifies how resources are modeled in the management domain; (iii) a communication model, which defines how to transfer data (e.g., a communication pipe and service primitives for exchanging data across this pipe); and (iv) a functional model, which defines functions (management tasks) that do real work; these functions can be fine-grained (e.g., a generic class for managing time series) or coarse-grained (e.g., an event correlator). Migrating to SOA requires changes in all but one of these models.

First, the *communication model* needs to migrate to SOAP. This migration would be straightforward for WBEM, whose communication model is close to SOAP. For SNMP, as we saw in Section 5, it would imply the adoption of another standard communication pipe (typically HTTP, possibly another one) and new communication primitives (SOAP RPCs).

Second, SOA challenges the *organizational model* of the management architectures used to date: SNMP, WBEM, TMN, OSI Management, etc. The management roles (defined in the manager–agent paradigm [75]) need to be revisited to make integrated management comply with SOA. First and foremost, the concept of coarse-grained autonomous service needs to be introduced. Migrating to SOA is also an invitation to stop interpreting the manager–agent paradigm as meaning "the manager should do everything and the agent nothing", as already advocated by Management by Delegation [93] in the 1990s but never quite enforced.

As far as the *functional model* is concerned, SOA encourages a new breakdown of management tasks among all actors involved. Instead of bringing all monitoring data and all events to the monitoring application and doing all the processing there, in a large monolithic application, SOA suggests to define intermediary tasks and wrap them as independent and reusable services that can run on different machines. This too is somewhat reminiscent of Management by Delegation [93], the Script MIB [46] in SNMP or the Command Sequencer [40] in OSI Management, except that Web services should be full-fledged applications, not just small scripts with a few lines of code.

**6.3.2.** *Changes in the management applications*   Assuming that the main management architectures used to date (SNMP and WBEM) are updated, what should be changed in management applications to make them implement autonomous coarse-grained services, loosely coupled services and service discovery?

The first issue (autonomous coarse-grained services) is about delegation granularity in a management application, i.e. how this application is subdivided into management tasks and how these tasks are allotted to different entities for execution. A few years ago, one of the authors of this chapter used organization theory to identify four criteria for classifying management architectures: semantic richness of the information model, degree of granularity, degree of automation of management, and degree of specification of a task [53]. Based on that, we defined an 'enhanced' taxonomy of network and systems management paradigms. For the delegation granularity, four scales were proposed in our taxonomy: no delegation, delegation by domain, delegation by microtask and delegation by macrotask. Fine-grained Web services correspond to microtasks; coarse-grained Web services correspond to macrotasks. To be SOA compliant, management applications should therefore support a new form of delegation: macrotasks.

More specifically, we propose that management applications use Web services at three scales, three levels of granularity. First, *simple management tasks* are slightly coarser-grained than Web services operating at the managed-object level. Instead of merely getting or setting an OID, they can execute small programs on a remote managed entity. Their execution is instantaneous. The Web service providers have a low footprint on the machine where they run. Second, *elaborate management tasks* operate at a higher level of granularity and consist of programs that do something significant. Their execution is not instantaneous and may require important CPU and memory resources. Third, *macro-components* (see examples in Section 6.2, 'Loose coupling') communicate via Web services instead of proprietary mechanisms. This last change makes it easy to distribute them across several machines, which complements domain-based distributed hierarchical management.

Once we have autonomous coarse-grained services, the second issue (loosely coupled services) is easy to deal with: management applications just need to use Web services and SOAP for communicating with these services. This guarantees a loose coupling of the management tasks.

To address the third issue (service discovery), we propose to let managing and managed entities discover one another at run-time, using Web services. This enables managed entities to be associated dynamically with a domain, instead of being configured once and for all. This change is also in line with recent standardization activities (e.g., WSDM).

**6.4.** *How to make management platforms SOA compliant*

Let us now go through a series of examples that illustrate how an SOA-enabled management platform could work.

**6.4.1.** *Type 1: Simple management tasks as Web services* Two examples of simple management tasks are the invocation of non-trivial actions and the execution of simple scripts.

EXAMPLE 1 (*Remote execution of a non-trivial action*). Non-trivial actions can be seen as RPCs bundled with a bit of code. The reboot of a remote device is a good example of action that can be invoked by a management application as a Web service delegated to a third party (neither the manager, nor the agent in SNMP parlance). In this Web service, we first try to do a clean ('graceful') reboot by setting an OID or invoking an operation on a CIM object. If it does not work, we then trigger a power-cycle of the machine. This operation can translate into a number of finer-grained operations, depending on the target machine to be rebooted. For instance, to power-cycle a remote PC that is not responding to any query anymore, we can look for an electromagnet that controls its power. If we find one (e.g., in a configuration database or a registry), we then set an OID or invoke a CIM operation on this electromagnet, depending on the management architecture that it supports, to force the power-cycle. A few moments later, the Web service checks whether the PC has resumed its activities; it sends back a response to the management application, with status `succeeded` or `failed`.

Using a Web service for this task allows the management application to abstract away this complexity, which is equipment specific. Binding a given piece of equipment to the right reboot Web service then becomes a matter of configuration or publication in a registry.

EXAMPLE 2 (*Simple script*). An example of simple script is the retrieval of routing tables from a group of routers. If multiple routing protocols are supported by a router, all of its routing tables need to be downloaded. This information can be subsequently exploited by the management application for debugging transient routing instabilities in the network. Alternatively, this Web service can also be used on a regular basis for performing routing-based event masking, for instance in networks where routes change frequently ('route flapping') and network topology-based event masking does not work satisfactorily.

In this example, the Web service providers can be the routers themselves; another possibility is to have a single Web service provider that runs on a third-party machine (e.g., an

SNMP gateway or proxy) and interacts via SNMP with the different routers concerned by this task [73].

**6.4.2.** *Type 2: Elaborate management tasks as Web services*    We now go one level up in abstraction and present four examples of elaborate management tasks: core problem investigation, ad hoc management, policy enforcement check and long-lasting problem investigation.

EXAMPLE 1 (*Core problem investigation*). In a management platform, when the event correlator has completed the root-cause analysis phase, we are left with a small set of core problems that need to be solved as soon as possible. When the solution to the problem is not obvious, we then enter a phase known as problem investigation. Different technologies can be used during this phase: an Event–Condition–Action (ECA) rule-based engine, a case-based reasoning engine, etc. On this occasion, the management application can launch programs[5] to retrieve more management information to further investigate, and hopefully solve, each outstanding problem.

Assuming that we do not have performance constraints that impose to keep all processing local,[6] these programs can be delegated to other machines in the form of Web services, thereby contributing to balance the load of the management application across multiple machines and to improve scalability. They run autonomously, gather data independently, access the data repository on their own, etc.

For instance, when the response time of a Web server is unacceptably long for an undetermined reason, it can be useful to monitor in great detail the traffic between this server and all the hosts interacting with it. This can be achieved by a Web service monitoring, say, for 30 minutes all inbound and outbound traffic for this server and analyzing this traffic in detail afterward, using fast data mining techniques. This task can involve many fine-grained operations; for instance, it can retrieve Remote MONitoring (RMON) MIB [86] OIDs in bulk at short time intervals, or it can put a network interface in promiscuous mode and sniff all traffic on the network. The duration of this debugging activity can be a parameter of the Web service operation. Upon successful completion, the Web service returns the cause of the problem; otherwise, it may be able to indicate whether it found something suspicious.

EXAMPLE 2 (*Ad hoc management*). *Ad hoc* management tasks [53] are not run regularly by the monitoring application. Instead, they are executed on an *ad hoc* basis to investigate problems as they show up. These tasks can be triggered interactively (e.g., by staff in a NOC), or automatically by the management application (see Example 1 with the slow Web server). These tasks are good candidates for Web services: it makes sense to delegate these self-contained and independent programs to third-party machines that do not have the same constraints, in terms of resource availability, as the main management platform that runs the bulk of the management application.[7]

---

[5] These programs can be executable binaries, byte-code for virtual machines, scripts, etc.

[6] For instance, that would be the case if we had to interact with thousands of finite state machines only accessible from the address space of the event correlator.

[7] We do not assume a centralized management platform here. When management is distributed hierarchically over multiple managers, the problem is the same as centralized management: each manager can be a bottleneck for its own sub-domain.

For instance, if the administrator suspects that the enterprise is currently undergoing a serious attack, he/she may launch a security analysis program that performs pattern matching on log files, analyzes network traffic, etc. This program will run until it is interrupted, and report the attacks that it detects (or strongly suspects) to the administrator using SMS messaging, paging, e-mail, etc.

EXAMPLE 3 (*Policy enforcement check*). Another interesting case of elaborate management task is to check that policies used in policy-based management [82] are enforced durably. For instance, let us consider a classic source of problems for network operators: `telnet`, the main enemy of policies. The bottom line of this scenario is invariably the same, with variations in the details.

A technician in a NOC is in charge of debugging a serious problem. Customers are complaining, the helpdesk is overwhelmed, his boss called him several times and is now standing next to him, watching over his shoulder, grumbling, doing nothing but stressing our technician. The latter feels he is alone on earth. All the management tools at his disposal indicate that there is an avalanche of problems, but they all fail to find the cause of these problems. After checking hundreds of configuration settings on many devices, our technician decides to stop services and reconfigure network cards one by one, hoping that this will eventually point to the problem. Using `telnet`, he modifies the configuration of many operational routers, trying all sorts of things... Doing so, he gradually starts to build a picture of what is going on: he had a similar problem several years ago, a misconfigured router was reinjecting routes when it should not. Today's problem is in fact very different, but our technician is now convinced he knows what is going on. Against all odds, after an hour of frantic typing, the situation starts to improve. Everything is still slow, but there seems to be a way out of this serious problem. And when he modifies the 27th router, bingo, the problem is gone and everything comes back to normal! His boss applauds, thanks the Lord for sending him such a great man, goes back to his office and phones his own boss, to make sure he does not get the blame for this major outage...

Is the incident closed? Yes, because things work again. Is the problem solved? No, because our technician still does not know what the problem was. Is the situation back to normal? Not at all: first, the same problem could resurface at any time; second, by changing the configuration of dozens of devices, our technician has placed several timed bombs in the network, without his knowing it. These timed bombs are big potential problems that are just waiting for an occasion to go off. And why so? Because he cannot roll back all the unnecessary changes that he made. Under stress, he did many things that he cannot remember anymore; and because he is dealing with operational networks, he is now very cautious about modifying anything. Yet he did several changes that he may soon regret. He changed configuration settings that are supposed to be entirely controlled by PDPs, and that will disappear in a few hours because they are overwritten everyday, at the same time, during slack hours. Will the problem resume immediately after the next policy update? He also changed many configuration settings that have nothing to do with the actual problem. Will they cause new problems in the future?

All network operators, all corporations, and presumably most large enterprises have been through this one day. This problem is as old as `telnet`. Under stress, NOC staff can modify (and break) just about everything, irrespective of their skills. One solution is to regularly copy the configuration settings (notably those that enforce policies and are generated automatically) of all network devices in a data store, at least once a day, and analyze every day the changes that were made in the past 24 hours. This process can be partially automated; it is a self-contained and coarse-grained activity, independent of the rest of the management application; it does not have to run on the main management platform. In other words, it should lend itself very well to being ported to Web services. In fact, not only some, but all configuration settings that are under the control of a PDP should systematically be checked like this, by Web services or other means.

In this case, the Web service providers should probably be third-party machines, not operational routers, to avoid temporarily saturating them with management tasks.

EXAMPLE 4 (*Long-lasting problem investigation*). If we go back to Example 1, some of the outstanding core problems may not be solved within a predefined time (e.g., 10 minutes). In such cases, it is appropriate to systematically outsource the resolution of these problems to third-party machines, which take over these investigations and run them as background tasks, possibly taking hours or days to come to a conclusion. Transferring these problem investigations to external entities can be achieved in a loosely coupled and portable manner using Web services. The objective here is to conserve precious few resources on the machine running the bulk of the management application, while not giving up on important but hard problems.

**6.4.3.** *Type 3: Macro-components as Web services*   This time, we break up the management application into very large chunks that can be physically distributed over multiple machines. The idea is to implement the network transparency recommended by SOA to improve the scalability of the management application.

EXAMPLE 1 (*Distributed monitoring application*). In today's management platforms, the main building block of the management application is the monitoring application. We propose to break up this large, monolithic, single-vendor monitoring application into large chunks (macro-components) wrapped up with coarse-grained Web services. These macro-components can be provided by several vendors and integrated on a per-organization basis, depending on site-specific needs and customer-specific price conditions.

This integration seldom requires dynamic binding at run-time, because presumably the macro-components that make up management software will not change often. This restructuring of the monitoring application brings more competition among vendors and can drive costs down (feature-rich management applications capable of integrating the management of networks, systems and services are still incredibly expensive).

For instance, one vendor can provide elaborate graphical views of the network with 3D navigation, while another provides an event correlator with a smart rule description language, and a third provides the rest of the monitoring application. Standard Web service-based communication would enable these loosely-coupled services to work together.

EXAMPLE 2 (*Integration of management*). Web services can also be useful to integrate the different management platforms in charge of network management, systems management, service management, the trouble-ticket system, technology-specific platforms, etc.

Doing so, they have the potential to nicely complement today's management solutions, for instance when the management platforms used in an enterprise cannot use the same event correlator due to integration problems (a serious problem that prevents management from being integrated).

**6.4.4.** *Type 4: Web services between managing/managed entities*   The fourth type of coarse-grained Web services is different from the previous three. Instead of focusing on the delegation granularity, it addresses the issue of service discovery. We present three examples: dynamic discovery of the manager, transparent redundancy of the managing entity, and dynamic discovery of the top-level manager.

EXAMPLE 1 (*Dynamic discovery of the manager*).   In this example, instead of configuring the address of the managing entity in each agent of its domain, administrators rely on Web services for managed entities to discover their managing entity dynamically at run-time. This discovery typically uses a standard registry, either configured in the managed entity or discovered at boot time (e.g., with a multicast operation).

EXAMPLE 2 (*Transparent redundancy during discovery*).   When managed entities think they bind directly to their managing entity, they can in fact connect to an intermediary Web service in charge of improving the robustness of management. This intermediary can transparently support two redundant managers with a hot stand-by (i.e., two machines that receive and analyze the same management data in parallel). The primary is active, the secondary is passive. If the primary fails, the secondary can take over transparently at very short notice. Alternatively, this intermediary can balance the load between multiple managers (which are all active). This can be useful in environments where managers experience important bursts of activity.

EXAMPLE 3 (*Dynamic discovery of the top-level manager*).   In distributed hierarchical management (a technique commonly adopted in large enterprises and corporations [53]), the management domain is under the responsibility of a managing entity known as the *top-level manager*. The domain is logically split into sub-domains, usually for scalability reasons, and each sub-domain is under the responsibility of a managing entity called the *sub-level manager*.

In this scenario, instead of configuring the address of the top-level manager in each sublevel manager, administrators rely on Web services for the sub-level managers to discover automatically the top-level manager of their domain. This operation typically uses a registry.

**6.5.** *Migrating to SOA: Is it worth the effort?*

After studying the changes that the migration to SOA requires in management applications, let us ponder over these changes. Are they reasonable? Do the advantages of SOA outweigh the drawbacks of reengineering existing management applications?

**6.5.1.** *We need something new*   In the last decade, we have seen a number of proposals that challenged the organizational models of existing management architectures: Management by Delegation [93], mobile code [3], intelligent agents [42], multi-agent systems [36] and more recently P2P [87] and Web services. So far, none of them have succeeded in durably influencing the architecture and design of the management applications used in real life. Why should Web services succeed where other technologies failed?

As we saw in Section 1.2, the main problem with today's management platforms is that we are prisoners of habits that date back to an era when networks were small and simple to manage, resources were scarce, dependencies were not highly dynamic, event correlation was simple because we did not bother to integrate the management of networks, systems and services, etc.

Today, architecting an integrated management application for a corporation is very difficult, and the tools available on the market often fall short of customers' expectations. Designing these complex applications boils down to designing large and highly dynamic distributed applications. This requires the state of the art in software engineering, and the reengineering of integration management applications. Migrating to SOA can be the occasion for this clean slate approach.

Another important reason for adopting coarse-grained Web services is that they offer a solution for integrating multi-vendor dedicated management platforms (e.g., a platform dedicated to network management with another platform specialized in service management). This integration is a weak point of many of today's management platforms, and a number of corporations or large enterprises currently struggle to achieve this integration.

**6.5.2.** *No guarantee of success*   Businesswise, migrating integrated management platforms to SOA is not a guaranteed win. There is much uncertainty. The market of management applications needs to change its habits, and whether this will happen is difficult to predict. Decision makers in the IT industry are told every year that a new 'revolution' has just occurred, so they tend to be skeptical. Convincing them will not be easy.

Oddly enough, the main non-technical driver for change may be the comfort offered by homogeneity: if everyone in the software industry adopts SOA, then there is a strong incentive for management applications to also migrate to SOA: few people enjoy being the last dinosaur in town...

**6.5.3.** *Paradox*   Migrating management platforms to Web services leads to an apparent paradox. On the one hand, SOA is about preserving investments in legacy systems by wrapping up existing applications with Web services. The goal here is to avoid the high cost of reengineering applications, especially complex ones. Based on this principle, using Web services in management applications should not require profound changes in these applications: we should not have to reengineer them. On the other hand, using Web services as getters and setters of managed objects defeats the point of coarse-grained and autonomous services in SOA. What is the catch?

This paradox is only apparent. It is due to a confusion between applications and services in SOA parlance. Services are meant to be reusable by several applications or other services; applications themselves are not. Management applications are not used as services by other applications: they are at the top of the food chain, so to say. Once SOA-enabled, applications can make use of different Web services without being entirely reengineered. But there is no free lunch in IT: they need to be made SOA-enabled in the first place.

**6.5.4.** *Performance, latency*    The performance of Web service-based management solutions is still very much an open issue. Web services use SOAP messages underneath. This messaging technology, based on the exchange of XML documents, is notoriously inefficient [44]. In corporations and large organizations, integrated management platforms have drastic requirements in terms of performance. Breaking the management application into Web services-enabled macro-components, elaborate management tasks and simple management tasks may turn out to be totally impractical for performance reasons.

The question is particularly acute for communication between macro-components. For instance, wrapping up the event filter, the event aggregator and the event correlator with different Web services could lead to a sharp decrease in performance, because these macro-components need to exchange many events per time unit in large environments, latency must be kept low, and SOAP message exchanges are slow compared to the current tightly coupled practice. A performance analysis of management platforms using coarse-grained Web services is outside the scope of this chapter, but this matter deserves attention in follow-up work.

## 7. Conclusion

In this chapter, we summarized the main standards pertaining to Web services and presented different ways of using this middleware technology in integrated management. Fine-grained Web services are a direct mapping of current habits in management: each Web service operates at the managed object level. This is the approach currently adopted by standards bodies (e.g., in WSDM and WS-Management). We showed that it is efficient when many compressed managed objects are transferred in bulk. Unfortunately, it does not comply with SOA, the distributed computing architectural style that underpins Web services. Coarse-grained Web services, conversely, are SOA compliant and work at a higher level of abstraction. Each Web service corresponds to a high-level management task (e.g., balancing the load of a virtual resource across multiple physical resources) or a macro-component of a management application (e.g., an event correlator). Coarse-grained Web services are autonomous (each of them fulfills some functionality on its own), loosely coupled with the rest of the management application, and coarse-grained. We distinguished several categories of Web service for integrated management and illustrated them by examples.

In the future, will integrated management adopt the vision of Web services promoted by SOA? Answering this question is difficult because the market is only partially driven by technical considerations. From a business standpoint, adopting Web services makes sense because the software industry is massively adopting this technology, so there are many tools available and many experts in the field. From an engineering standpoint, Web services make it possible to organize management applications differently. This is appealing for architecting integrated management applications aimed at large organizations. It also opens new horizons regarding the organizational model of standard management architectures: we now have the technology to do without the "manager does everything, agent does nothing" principle. Web services make it possible to break up management platforms into independent macro-components, packaged as services, which can bring more competition among vendors and drive costs down. For instance, one vendor may provide elaborate views with Graphical User Interfaces (GUIs) while another may provide the event correlator. They also facilitate the integration of network, systems and service management tasks. But the performance of coarse-grained Web services for building management solutions is still an open issue that deserves thorough investigations. The same problem was true of CORBA and J2EE a few years ago, and satisfactory solutions appeared a few years later. Can we count on this for Web services too?

## Acknowledgements

The work presented in this paper was supported in part by the EMANICS Network of Excellence (#26854) of the IST Program funded by the European Commission. Early versions of this material were published in [24,55,78].

## References

[1] G. Alonso, F. Casati, H. Kuno and V. Machiraju, *Web Services: Concepts, Architectures and Applications*, Springer-Verlag (2004).

[2] A. Alves, A. Arkin, S. Askary et al., eds, *Web services business process execution language version 2.0*, Public Review Draft, OASIS (2006), http://docs.oasisopen.org/wsbpel/2.0/wsbpel-specification-draft.html.

[3] M. Baldi and G.P. Picco, *Evaluating the tradeoffs of mobile code design paradigms in network management applications*, Proc. 20th International Conference on Software Engineering (ICSE 1998), R. Kemmerer and K. Futatsugi, eds, Kyoto, Japan (1998), 146–155.

[4] K. Ballinger, D. Ehnebuske, C. Ferris et al., eds, *Basic profile version 1.1*, Final Material, Web Services Interoperability Organization (2006), http://www.ws-i.org/Profiles/BasicProfile-1.1-2006-04-10.html.

[5] T. Banks, *Web Services Resource Framework* (WSRF) – *Primer v1.2*, Committee Draft 02 (2006), http://docs.oasis-open.org/wsrf/wsrf-primer-1.2-primer-cd-02.pdf.

[6] T. Bellwood, *UDDI version 2.04 API specification*, UDDI Committee Specification, OASIS (2002), http://uddi.org/pubs/ProgrammersAPI-V2.04-Published-20020719.pdf.

[7] T. Berners-Lee, J. Hendler and O. Lassila, *The semantic Web*, Scientific American (2001).

[8] K.P. Birman, *Like it or not, Web services are distributed objects*, Communications of the ACM **47** (2004), 60–62.

[9] D. Booth, H. Haas, F. McCabe et al., eds, *Web services architecture*, W3C Working Group Note, World Wide Web Consortium (2004), http://www.w3.org/TR/2004/NOTE-ws-arch-20040211/.

[10] D. Box, E. Christensen, F. Curbera et al., *Web services addressing (WS-addressing)*, W3C Member Submission (2004), http://www.w3.org/Submission/2004/SUBM-ws-addressing-20040810/.

[11] V. Bullard and W. Vambenepe, eds, *Web services distributed management: Management using Web services (MUWS 1.1) part 1*, OASIS Standard (2006), http://docs.oasis-open.org/wsdm/wsdm-muws1-1.1-spec-os-01.htm.

[12] V. Bullard and W. Vambenepe, eds, *Web services distributed management: Management using Web services (MUWS 1.1) part 2*, OASIS Standard (2006), http://docs.oasis-open.org/wsdm/wsdm-muws2-1.1-spec-os-01.htm.

[13] B. Burke and R. Monson-Haefel, *Enterprise JavaBeans 3.0*, 5th edn, O'Reilly (2006).

[14] N. Catania, P. Kumar, B. Murray et al., *Overview – Web services management framework, version 2.0* (2003), http://devresource.hp.com/drc/specifications/ wsmf/WSMF-Overview.jsp.

[15] K. Channabasavaiah, K. Holley and E. Tuggle, *Migrating to a service-oriented architecture, part 1*, IBM (2003), http://www-128.ibm.com/developerworks/webservices/library/ws-migratesoa/.

[16] R. Chinnici, H. Haas and A.A. Lewis, eds, *Web services description language (WSDL) version 2.0 part 2: Adjuncts*, W3C Candidate Recommendation (2006), http://www.w3.org/TR/2006/CR-wsdl20-adjuncts-20060327/.

[17] R. Chinnici, J.J. Moreau, A. Ryman et al., eds, *Web services description language (WSDL) version 2.0 part 1: Core language*, W3C Candidate Recommendation (2006), http://www.w3.org/TR/2006/CR-wsdl20-20060327/.

[18] M. Choi and J.W. Hong, *Performance evaluation of XML-based network management*, Presentation at the *16th IRTF-NMRG Meeting*, Seoul, Korea (2004), http://www.ibr.cs.tubs.de/projects/nmrg/meetings/2004/seoul/choi.pdf.

[19] M. Choi, J.W. Hong and H. Ju, *XML-based network management for IP networks*, ETRI Journal **25** (2003), 445–463.

[20] J. Clark and S. DeRose, *XML path language (XPath) version 1.0*, W3C Recommendation (1999), http://www.w3.org/TR/1999/REC-xpath-19991116/.

[21] L. Clement, A. Hately, C. von Riegen et al., eds, *UDDI version 3.0.2*, UDDI Spec Technical Committee Draft, OASIS (2004), http://www.oasis-open.org/committees/uddi-spec/doc/spec/v3/uddi-v3.0.2-20041019.htm.

[22] F. Curbera, Y. Goland, J. Klein et al., *Business process execution language for Web services (BPEL4WS) version 1.1* (2003), http://www-106.ibm.com/ developerworks/library/ws-bpel/.

[23] DMTF, *Web services for management (WS-management), version 1.0.0a*, DSP0226 (2006), http://www.dmtf.org/standards/published_documents/DSP0226.pdf.

[24] T. Drevers, R. v.d. Meent and A. Pras, *Prototyping Web services based network monitoring*, Proc. of the 10th Open European Summer School and IFIP WG6.3 Workshop (EUNICE 2004), J. Harjo, D. Moltchanov and B. Silverajan, eds, Tampere, Finland (2004), 135–142.

[25] R. v. Engelen, *gSOAP Web services toolkit* (2003), http://www.cs.fsu.edu/~engelen/ soap.html.

[26] R. v. Engelen and K.A. Gallivany, *The gSOAP toolkit for Web services and peer-to-peer computing networks*, Proc. of the 2nd IEEE/ACM International Symposium on Cluster Computing and the Grid (CCGrid 2002), Berlin, Germany (2002), 128–135.

[27] T. Erl, *Service-Oriented Architecture: Concepts and Technology*, Prentice Hall PTR (2005).

[28] R.T. Fielding, *Architectural styles and the design of network-based software architectures*, Ph.D. Thesis, University of California, Irvine, CA, USA (2000), http://www.ics.uci.edu/~fielding/pubs/dissertation/top.htm.

[29] R. Fielding, J. Gettys, J. Mogul et al., eds, *RFC 2616: Hypertext transfer protocol – HTTP/1.1*, IETF (1999).

[30] S. Graham, A. Karmarkar, J. Mischkinsky et al., eds, *Web services resource 1.2 (WS-resource)*, OASIS Standard (2006).

[31] M. Grand, *Patterns in Java, Volume 1: A Catalog of Reusable Design Patterns Illustrated with UML*, Wiley (1998).

[32] M. Gudgin, M. Hadley, J.J. Moreau et al., eds, *SOAP 1.2 part 1: Messaging framework*, W3C Candidate Recommendation, World Wide Web Consortium (2002), http://www.w3.org/TR/soap12-part1/.

[33] M. Gudgin, M. Hadley, J.J. Moreau et al., eds, *SOAP 1.2 part 2: Adjuncts*, W3C Candidate Recommendation, World Wide Web Consortium (2002), http://www.w3.org/TR/soap12-part2/.

[34] H. Haas and C. Barreto, *Foundations and future of Web services*, Tutorial presented at the 3rd IEEE European Conference on Web Services (ECOWS 2005), Växjö, Sweden (2005).

[35] H.G. Hegering, S. Abeck and B. Neumair, *Integrated Management of Networked Systems: Concepts, Architectures, and Their Operational Application*, Morgan Kaufmann (1999).

[36] M.N. Huhns and M.P. Singh, eds, *Readings in Agents*, Morgan Kaufmann (1998).

[37] IBM developerWorks, *New to SOA and Web services*, http://www-128.ibm.com/developerworks/webservices/newto/.

[38] IRTF-NMRG, *Minutes of the 11th IRTF-NMRG meeting*, Osnabrück, Germany (2002), http://www.ibr.cs.tu-bs.de/projects/nmrg/minutes/minutes011.txt.

[39] ISO/IEC 10746-1, *Information technology – Open distributed processing – Reference model: Overview*, International Standard (1998).

[40] ITU-T, *Recommendation X.753, Information technology – Open systems interconnection – Systems management: Command sequencer for systems management*, ITU (1997).

[41] ITU-T, *Recommendation M.3400, Maintenance: Telecommunications management network – TMN management functions*, ITU (2000).

[42] N.R. Jennings and M.J. Wooldridge, eds, *Agent Technology: Foundations, Applications, and Markets*, Springer-Verlag (1998).

[43] P. Kalyanasundaram, A.S. Sethi, C.M. Sherwin et al., *A spreadsheet-based scripting environment for SNMP*, Integrated Network Management V – Proc. of the 5th IFIP/IEEE International Symposium on Integrated Network Management (IM 1997), A. Lazar, R. Saracco and R. Stadler, eds, San Diego, CA, USA (1997), 752–765.

[44] J. Kangasharju, S. Tarkoma and K. Raatikainen, *Comparing SOAP performance for various encodings, protocols, and connections*, Proc. of the 8th International Conference on Personal Wireless Communications (PWC 2003), Venice, Italy, Lecture Notes in Comput. Sci., Vol. 2775, Springer-Verlag (2003), 397–406.

[45] N. Kavantzas, D. Burdett and G. Ritzinger, eds, *Web services choreography description language version 1.0*, W3C Candidate Recommendation, World Wide Web Consortium (2005), http://www.w3.org/TR/2005/CR-ws-cdl-10-20051109/.

[46] D. Levi and J. Schoenwaelder, eds, *RFC 3165: Definitions of managed objects for the delegation of management scripts*, IETF (2001).

[47] D. Libes, *Exploring Expect: A Tcl-based Toolkit for Automating Interactive Programs*, O'Reilly (1994).

[48] M. Little, *Transactions and Web services*, Communications of the ACM **46** (2003), 49–54.

[49] K. Ma, *Web services: What's real and what's not?*, IEEE Professional **7** (2005), 14–21.

[50] C.M. MacKenzie, K. Laskey, F. McCabe et al., eds, *Reference model for service oriented architecture 1.0*, Committee Specification 1, OASIS (2006).

[51] F. Manola and E. Miller, *RDF Primer*, W3C Recommendation (2004), http://www.w3.org/TR/2004/REC-rdf-primer-20040210/.

[52] D. Martin, M. Burstein, D. McDermott et al., *Bringing semantics to Web services with OWL-S*, World Wide Web **10** (2007).

[53] J.-P. Martin-Flatin, *Web-Based Management of IP Networks and Systems*, Wiley (2003).

[54] J.-P. Martin-Flatin, *Distributed event correlation and self-managed systems*, Proc. 1st International Workshop on Self-* Properties in Complex Information Systems (Self-Star 2004), O. Babaoglu et al., eds, Bertinoro, Italy (2004), 61–64.

[55] J.-P. Martin-Flatin, P.-A. Doffoel and M. Jeckle, *Web services for integrated management: A case study*, Web Services – Proc. of the 2nd European Conference on Web Services (ECOWS 2004), Erfurt, Germany, L.J. Zhang and M. Jeckle, eds, Lecture Notes in Comput. Sci., Vol. 3250, Springer-Verlag (2004), 239–253.

[56] K. McCloghrie and F. Kastenholz, eds, *RFC 2863: The interfaces group MIB*, IETF (2000).

[57] K. McCloghrie, D. Perkins and J. Schoenwaelder, eds, *RFC 2578: Structure of management information version 2 (SMIv2)*, IETF (1999).

[58] D.L. McGuinness and F. van Harmelen, *OWL Web ontology language overview*, W3C Recommendation (2004), http://www.w3.org/TR/2004/REC-owl-features-20040210/.

[59] S. McIlraith, T.C. Son and H. Zeng, *Semantic Web services*, IEEE Intelligent Systems **16** (2001), 46–53.

[60] A. Nadalin, C. Kaler, R. Monzillo et al., *Web services security: SOAP message security 1.1 (WS-security)*, OASIS Standard Specification (2006).

[61] R. Neisse, R. Lemos Vianna, L. Zambenedetti Granville et al., *Implementation and bandwidth consumption evaluation of SNMP to Web services gateways*, Proc. of the 9th IEEE/IFIP Network Operations and Management Symposium (NOMS 2004), Seoul, Korea (2004), 715–728.

[62] Net-SNMP home page, http://net-snmp.sourceforge.net/.

[63] NuSOAP – SOAP toolkit for PHP, http://sourceforge.net/projects/nusoap.

[64] OASIS Web Services Distributed Management (WSDM) Technical Committee, http://www.oasis-open.org/committees/tc_home.php?wg_abbrev=wsdm.

[65] OASIS Web Services Resource Framework (WSRF) Technical Committee, http://www.oasis-open.org/committees/tc_home.php?wg_abbrev=wsrf.

[66] OASIS Web Services Transaction (WS-TX) Technical Committee, http://www.oasis-open.org/committees/tc_home.php?wg_abbrev=ws-tx.

[67] L. O'Brien, L. Bass and P. Merson, *Quality Attributes and service-oriented architectures*, Technical Report CMU/SEI-2005-TN-014, Software Engineering Institute, Carnegie Mellon University, PA, USA (2005), http://www.sei.cmu.edu/pub/documents/05.reports/pdf/05tn014.pdf.

[68] R. Orfali, D. Harkey and J. Edwards, *Client/Server Survival Guide*, 3rd edn, Wiley (1999).

[69] E. O'Tuathail and M. Rose, *RFC 4227: Using the simple object access protocol (SOAP) in blocks extensible exchange protocol (BEEP)*, IETF (2006).

[70] C. Pahl, *A conceptual framework for semantic Web services development and deployment*, Web Services – Proc. of the 2nd European Conference on Web Services (ECOWS 2004), Erfurt, Germany, L.J. Zhang and M. Jeckle, eds, Lecture Notes in Comput. Sci., Vol. 3250, Springer-Verlag (2004), 270–284.

[71] M.P. Papazoglou, *Web services and business transactions*, World Wide Web **6** (2003), 49–91.

[72] M.P. Papazoglou and D. Georgakopoulos, eds, *Special issue on "Service-Oriented Computing"*, Communications of the ACM **46** (2003).

[73] M.P. Papazoglou and W.J. van den Heuvel, *Web services management: A survey*, IEEE Internet Computing **9** (2005), 58–64.

[74] S. Parastatidis, S. Woodman, J. Webber et al., *Asynchronous messaging between Web services using SSDL*, IEEE Internet Computing **10** (2006), 26–39.

[75] G. Pavlou, *OSI systems management, Internet SNMP, and ODP/OMG CORBA as technologies for telecommunications network management*, Telecommunications Network Management: Technologies and Implementations, S. Aidarous and T. Plevyak, eds, IEEE Press (1998), 63–110, Chapter 2.

[76] G. Pavlou, P. Flegkas and S. Gouveris, *Performance evaluation of Web services as management technology*, Presentation at the 15th IRTF-NMRG Meeting, Bremen, Germany (2004).

[77] G. Pavlou, P. Flegkas, S. Gouveris et al., *On management technologies and the potential of Web services*, IEEE Communications **42** (2004).

[78] A. Pras, T. Drevers, R. v.d. Meent and D. Quartel, *Comparing the performance of SNMP and Web services-based management*, IEEE e-Transactions on Network and Service Management **1** (2004).

[79] J. Rao and X. Su, *A survey of automated Web service composition methods*, Proc. of the 1st International Workshop on Semantic Web Services and Web Process Composition (SWSWPC 2004), San Diego, CA, USA, J. Cardoso and A. Sneth, eds, Lecture Notes in Comput. Sci., Vol. 3387, Springer-Verlag (2004), 43–54.

[80] J. Schönwälder, A. Pras and J.P. Martin-Flatin, *On the future of internet management technologies*, IEEE Communications Magazine **41** (2003), 90–97.

[81] M. Singh and M.N. Huhns, *Service-Oriented Computing: Semantics, Processes, Agents*, Wiley (2005).

[82] M. Sloman, *Policy driven management for distributed systems*, Journal of Network and Systems Management **2** (1994), 333–360.

[83] M. Sloman and K. Twidle, *Domains: A framework for structuring management policy*, Network and Distributed Systems Management, M. Sloman, ed., Addison-Wesley (1994), 433–453, Chapter 16.

[84] J. v. Sloten, A. Pras and M.J. v. Sinderen, *On the standardisation of Web service management operations*, Proc. of the 10th Open European Summer School and IFIP WG6.3 Workshop (EUNICE 2004), J. Harjo, D. Moltchanov and B. Silverajan, eds, Tampere, Finland (2004), 143–150.

[85] Soapware.org, http://www.soapware.org/.

[86] W. Stallings, *SNMP, SNMPv2, SNMPv3, and RMON 1 and 2*, 3rd edn, Addison–Wesley (1999).

[87] R. Steinmetz and K. Wehrle, eds, *Peer-to-Peer Systems and Applications*, Lecture Notes in Comput. Sci., Vol. 3485, Springer-Verlag (2005).

[88] The Stencil Group, *The evolution of UDDI – UDDI.org white paper* (2002), http://www.uddi.org/pubs/the_evolution_of_uddi_20020719.pdf.

[89] UDDI.org, *UDDI technical white paper* (2000), http://www.uddi.org/pubs/Iru_UDDI_Technical_White_Paper.pdf.

[90] W. Vogels, *Web services are not distributed objects*, Internet Computing **7** (2003), 59–66.

[91] WASP UDDI, http://www.systinet.com/products/wasp_uddi/overview/.

[92] K. Wilson and I. Sedukhin, eds, *Web services distributed management: Management of Web services (WSDM-MOWS) 1.1*, OASIS Standard, 01 (2006), http://docs.oasis-open.org/wsdm/wsdm-mows-1.1-spec-os-01.htm.

[93] Y. Yemini, G. Goldszmidt and S. Yemini, *Network management by delegation*, Proc. of the 2nd IFIP International Symposium on Integrated Management (ISINM 2001), I. Krishnan and W. Zimmer, eds, Washington, DC, USA (1991), 95–107.

[94] C. Zhou, L.T. Chia and B.S. Lee, *Semantics in service discovery and QoS measurement*, IEEE Professional **7** (2005), 29–34.

[95] Zlib, http://www.zlib.org/.

# – 2.8 –

# Internet Management Protocols

Jürgen Schönwälder

*Jacobs University Bremen, Germany*

## 1. Introduction

Operating a large communication network requires tools to assist in the configuration, monitoring and troubleshooting of network elements such as switches, routers or firewalls. In addition, it is necessary to collect event reports to identify and track failures or to provide a log of network activities. Finally, it is necessary to collect measurement data for network planning and billing purposes.

With a handful devices, many of the activities mentioned above can be supported in an ad-hoc approach by using whatever functionality and interface is provided by a given device. However, when the number of network elements and services increases, it becomes essential to implement more advanced management approaches and deploy specific management tools. Since communication networks often consist of a large number of devices from different vendors, the need for standardized management interfaces becomes apparent.

One particularly important management interface for standardization and interoperability is the interface between network elements and their management systems. Network elements are typically closed systems and it is not possible to install management software on the devices them-self. This makes network elements such as routers and switches fundamentally different from general purpose computing systems. Interfaces between management systems have seen less need for standardization since management systems usually run on full featured host systems and it is therefore easier to use general middleware technologies as a means to provide connectivity between management systems.

---

HANDBOOK OF NETWORK AND SYSTEM ADMINISTRATION
Edited by Jan Bergstra and Mark Burgess

This chapter surveys Internet management protocols that are mainly used between network elements and their management systems. Since it is impossible to discuss all protocols in detail, the discussion in this chapter will focus on the requirements that have driven the design of the protocols and the architectural aspects through which the protocols address the requirements. By focusing the discussion on requirements and architectural concepts, it becomes possible to document basic principles that are not restricted to the current set of Internet management protocols, but which will likely apply to future management protocols as well.

## 2. Management protocol requirements

Before discussing Internet management protocols, it is useful to look at the requirements these protocols should address. The approach taken to identify the requirements is to discuss the main functions management protocols have to support.

### 2.1. *Configuration requirements*

Network configuration management has the goal to define and control the behavior of network elements such that a desired network behavior is realized in a reliable and cost-effective manner. To understand the complexity of this task, it is important to note that the behavior of a network element is not only defined by its configuration, but also by the interaction with other elements in the network. It is therefore necessary to distinguish between *configuration state* and *operational state*. The configuration state is installed by management operations while the operational state is typically installed by control protocols or self-configuration mechanisms.

As an example, consider how the packet forwarding behavior of an IP router is determined. A simple router on a stub network likely has static configuration state in the form of a forwarding table which determines how IP packets are forwarded. This forwarding table is usually established once and subsequently changed manually when the network topology changes. A more complex router on a more dynamic network usually takes part in a routing protocol exchange which is used to compute suitable forwarding tables, taking into account the current state of the network. The forwarding behavior of a router participating in a routing protocol therefore is determined by operational state which was obtained at runtime and influenced by neighboring network elements. Note that the way the router participates in the routing protocol exchange is likely determined itself by configuration state, which sets the parameters and constraints under which shortest paths are computed.

REQUIREMENT 1. A configuration management protocol must be able to distinguish between configuration state and operational state.

The example above introduces another important aspect in network configuration management: since devices typically exchange control information to establish operational state, configuration state changes on a single device can affect the operational state in many other devices. As a consequence, configuration state changes can have side effects that may affect a whole network.

With this in mind, it becomes clear that it is not sufficient to look solely at the configuration state of a device in order to identify why a certain network element does not behave as desired. To understand why an element misbehaves, it is necessary to look at the combined configuration state and operational state and it is desirable that devices report configuration information in a way which allows to distinguish between configuration state and operational state. In addition, it is desirable that the device also records sufficient information which allows an operator to figure out why and when operational state was established. For completeness, it should be noted that network elements maintain statistics in addition to configuration and operations state.

Configuration changes in a network may originate from different systems. In addition, it might be necessary to change several systems for a logical change to be complete. This leads to the following three requirements:

REQUIREMENT 2. A configuration management protocol must provide primitives to prevent errors due to concurrent configuration changes.

REQUIREMENT 3. A configuration management protocol must provide primitives to apply configuration changes to a set of network elements in a robust and transaction-oriented way.

REQUIREMENT 4. It is important to distinguish between the distribution of configurations and the activation of a certain configuration. Devices should be able to hold multiple configurations.

Devices usually maintain different configurations. It is common to distinguish between the currently running configuration and the configuration that should be loaded at startup time. Some devices also support an open ended collection of different configurations. This leads to the following two requirements:

REQUIREMENT 5. A configuration management protocol must be able to distinguish between several configurations.

REQUIREMENT 6. A configuration management protocol must be clear about the persistence of configuration changes.

Since different management systems and operators may have access to the configuration of a device, it is important to log information that can be used to trace when and why a certain configuration change has been made.

REQUIREMENT 7. A configuration management protocol must be able to report configuration change events to help tracing back configuration changes.

Not all devices can be expected to have the resources necessary to track all configuration changes throughout their lifetime. It is therefore common practice to keep copies of device configurations on host computers and it is thus desirable to be able to use standard version control systems to track changes.

REQUIREMENT 8. A full configuration dump and a full configuration restore are primitive operations frequently used by operators and must be supported appropriately.

REQUIREMENT 9. A configuration management protocol must represent configuration state and operational state in a form which allows to use existing standard comparison and versioning tools.

Network operators are generally interested in providing smooth services. The number of service interruptions must be minimized and stability should be maintained as much as possible. A smart device should therefore be able to load a new configuration without requiring a full device reset or the loss of all operational state.

REQUIREMENT 10. Configurations must be described such that devices can determine a set of operations to bring the devices from a given configuration state to the desired configuration state, minimizing the impact caused by the configuration change itself on networks and systems.

## 2.2. Monitoring requirements

After configuring the devices making up a network, it is essential to monitor that they are functioning correctly and to collect same basic usage statistics such as link utilizations. The setup of monitoring tools can be greatly simplified if devices reveal their capabilities.

REQUIREMENT 11. Monitoring protocols should support the discovery of the capabilities of a device.

Networks may contain a large number of manageable devices. Over time, monitoring requirements also tend to increase in terms of the number of data items to be monitored per device. As a consequence, monitoring protocols should be highly scalable.

REQUIREMENT 12. Monitoring protocols must scale to a large number of devices as well as a large number of data items to be monitored.

Devices often contain large numbers of similar counters and in many cases not all of them are relevant for monitoring purposes. It is therefore crucial to be able to select which subsets of management data need to be retrieved for monitoring purposes, sometimes up to the level of an individual counter.

REQUIREMENT 13. It must be possible to perform monitoring operations on selected subsets of management data.

REQUIREMENT 14. Monitoring protocols must support a naming scheme which is capable to identify instances up to the granularity of a specific counter.

Finally, it should be noted that devices are build for some primary function (e.g., a router is built primarily to forward IP packets) and monitoring functions should thus have little negative impact on the primary functions of a device.

REQUIREMENT 15. Monitoring protocols should have low impact on the primary functions of a device. In particular, it must be possible that multiple applications monitor different aspects of a single device without undue performance penalties for the device.

### 2.3. *Event notification requirements*

Operationally relevant status changes and in particular failures typically happen asynchronously. Devices therefore need a mechanism to report such events in a timely manner. As will be seen later, there are multiple protocols to transport event notifications. Regardless of the protocol, it is important to be able to identify some key parameters about an event.

REQUIREMENT 16. Event notifications must have sufficient information to identify the event source, the time the event occurred, the part of the system that originated the event notification, and a broad severity classification.

Since event notifications may trigger complex procedures or work-flows, it is necessary to ensure the integrity and authenticity of an event notification. In some cases, it is even required to prove the integrity of events stored in a permanent event log.

REQUIREMENT 17. It is desirable to be able to verify the integrity of event notifications and the authenticity of the event source.

Events may originate from many different devices and processes in a network and it is likely that formats and protocols are not always consistent. Furthermore, event notifications may have to pass middleboxes such as firewalls and network address translators and there might be multiple levels of filtering and different data collectors for different purposes.

REQUIREMENT 18. Event notifications may pass through a chain of relays and collectors which should be (a) transparent for the event notification originator and (b) not change the content of event notifications in any way.

One particular design issue that needs to be addressed when designing an event notification system are effective mechanisms to deal with so-called event storms. Event storms happen for example after a power outage or a cut of a fiber when multiple devices and processes detect an error and issue event notifications. Event storms look very similar to denial of service attacks, except that the event messages actually originate from a valid source.

REQUIREMENT 19. Notification senders should provide effective throttling mechanisms in order to deal with notification storms.

A reliable event notification transport is often desirable. However, it should be noted that choosing a reliable transport layer protocol does not by itself provide a reliable notification delivery system since the transport might simply fail while the network is under stress or attack or the notification receiver might not understand the notification. For a highly reliable event notification system, it is therefore essential to provide a local notification logging facility and a communication protocol providing confirmed delivery of event notifications.

REQUIREMENT 20. Highly reliable event notification systems must provide confirmed event notification protocols and logging facilities to store event notifications at the event source.

Finally, it has proven to be valuable that event notifications can be read by both machines and humans.

REQUIREMENT 21. Event notifications should include machine readable structured data as well as human readable event descriptions.

## 2.4. *Testing and troubleshooting requirements*

Troubleshooting and testing is a common activity, especially when new services and networks are deployed. Management protocols should support these activities.

REQUIREMENT 22. A management protocol must allow the invocation of test functions for trouble shooting purposes.

The ability to invoke test functions from different locations in a given network topology can lead to specific insights about the root cause of observed problems very quickly. In many situations, it is desirable to execute test functions periodically so that alarms can be triggered if tests fail of an expected heartbeat message is not received.

REQUIREMENT 23. Scheduling functions should be provided to execute testing functions periodically or following a calendar schedule.

## 2.5. *Measurement data collection requirements*

It is import to collect accurate measurement data sets describing how networks are being used. Such information is, among other things, useful for usage-based accounting, network planning and traffic engineering, quality of service monitoring, troubleshooting, or attack and intrusion detection.

To collect measurement data, it is necessary to define observations points where meters are attached to a network in order to collect measurement data sets. These data sets are then exported by an exporting process, which transfers the data sets to a collecting process.

There are several specific measurement data collection requirements. The first one deals with scalability:

REQUIREMENT 24. Measurement data collection protocols must scale to hundreds of exporting processes. Furthermore, data exporting processes must be able to export to multiple data collecting processes.

Scalability and efficiency is a key requirement. Since network bandwidth tends to grow faster than processor speeds, it can be expected that only statistical sampling techniques can be deployed on very high-speed optical backbone networks in the future. Related to scalability is the requirement for measurement data transfer protocols to be congestion aware.

REQUIREMENT 25. Measurement data transfer protocols must be reliable and congestion aware.

Congestion during the data transfer may lead to significant data loss and may affect other network services. Data loss can, however, occur at many places, namely at the metering process, the export process, during the data transfer, or at the collecting process. To provide a complete picture, it is therefore necessary to properly track all data losses.

REQUIREMENT 26. It must be possible to obtain an accurate indication about lost measurement data records.

Since metering processes often attach timestamps to data sets, it is necessary to select timestamps with a sufficiently high granularity and to synchronize the clocks on the meters.

REQUIREMENT 27. Metering processes must be able to assign timestamps with a suitable resolution. Furthermore, metering processes must be able to synchronize their clocks with the necessary precision.

Early protocols for exchanging measurement data sets assumed a specific hardwired data model. This approach has proven problematic several times, leading to the following requirement:

REQUIREMENT 28. The data format used by the data transfer protocol must be extensible.

Finally, it should be noted that measurement data sets and protocols that transfer them need proper protection.

REQUIREMENT 29. Measurement data transfer protocols must provide data integrity, authenticity of the data transferred from an exporting process to a collecting process, and confidentiality of the measurement data.

For additional details on measurement data collection requirements, see RFC 3917 [43].

### 2.6. *Security requirements*

Network management information in general is sensitive. This seems obvious in the context of configuration, but may be less obvious in the context of monitoring. But as security attacks are getting more sophisticated, it is important to increase the protection of data that can identify targets for attacks.

REQUIREMENT 30. There is a need for secure data transport, authentication and access control.

Operationally, it seems that the usage of cryptographically strong security mechanisms do not cause too many problems if the required processing power is available. However, the required key management can be time consuming. Hence, it is important that key and credential management functions integrate well.

REQUIREMENT 31. Any required key and credential management functions should integrate well with existing key and credential management infrastructures.

Once you have authenticated principals, it is natural to apply access control to limit the accessible information to the subset needed to perform a specific function.

REQUIREMENT 32. The granularity of access control must match operational needs. Typical requirements are a role-based access control model and the principle of least privilege, where a user can be given only the minimum access necessary to perform a required task.

Unfortunately, it is often difficult to define precisely which information needs to be accessible for a given management application to function correctly. Many applications resort to some probing and discovery algorithms and they use best guess heuristics in cases where some information is not directly accessible.

### 2.7. *Non-functional requirements*

Several non-functional requirements should be taken into account as well. The first one concerns human readability of protocol messages. The experience with SNMP (which uses binary encoded messages) tells us that the usage of specially encoded messages significantly increases the time programmers spend in writing new applications.

REQUIREMENT 33. Human readable protocol messages significantly reduce integration efforts.

The availability of C/C++/Java libraries with well-defined stable APIs does not seem to make it easy enough for developers and operators to create new applications quickly. For network management purposes, good integration into high-level scripting languages is therefore an essential requirement.

REQUIREMENT 34. Management protocols should be easy to access from high-level scripting languages.

On managed devices, it has been observed that the development of management interfaces sometimes happens in a relatively uncoordinated way. This leads often to duplicate implementation and maintenance efforts which increase costs and inconsistencies across management interfaces which decrease the operator satisfaction.

REQUIREMENT 35. Implementation costs on network devices must be minimized and duplicated implementation efforts for different management protocols should be avoided.

Management systems usually contain a significant amount of code that is specific to a certain operator or network type. Furthermore, many network management tasks involve work-flows (e.g., the ordering of repair parts or a temporary allocation of bandwidth from other providers) that need some level of information technology support. Management applications and protocol interfaces should therefore be easy to integrate with standard software tools.

REQUIREMENT 36. Management protocol interfaces should integrate well with standard tools such as revision management tools, text analysis tools, report generation tools, trouble ticketing tools, databases, or in general work-flow support tools.

## 3. Architectural concepts

Before discussing concrete protocols, it is useful to introduce some architectural concepts that are being used later in this chapter to describe to various Internet management protocol frameworks.

### 3.1. *Protocol engines, functions and entities*

The architecture of a management protocol can usually be divided in a *protocol engine* handling protocol messages and *protocol functions* that realize the protocol's functionality. The protocol engine and the protocol functions together form a *protocol entity*, as shown in Figure 1.

*Protocol engine.* The protocol engine is primarily concerned with the handling of the message formats used by a protocol. New protocols usually have a single message format. But during the design of a new protocol, it must already be anticipated that message formats

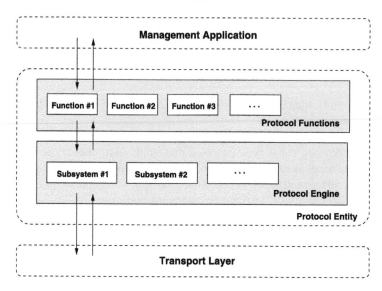

Fig. 1. Protocol engines, functions and entities.

need changes over time if a protocol is successfully deployed and new requirements must be addressed. Some of the older Internet management protocols have already been gone through this evolution while some of the newer protocols are too young to have experienced a need for different message formats. From an architectural point of view, however, it is important to plan for message format changes that should be handled by the protocol engine.

An important service provided by the protocol engine is the identification of the principal invoking a protocol function. This usually requires cryptographic mechanisms and hence the protocol engine must deal with message security processing and provide mechanisms to ensure message integrity, data origin authentication, replay protection and privacy. There are two commonly used approaches to achieve this:

1. The first approach is to design specific security processing features into the protocol engine. This has the advantage that the protocol is self-contained and reduces its dependency on external specifications and infrastructure. The downside is that the specification becomes more complex and more difficult to review and that implementation and testing costs increase.

2. The second approach, which is becoming increasingly popular, is the usage of secure transport mechanism. The protocol engine then simply passes authenticated identities through the protocol engine. This has the advantage that prior work on the specification, testing and implementation of security protocols can be leveraged easily.

A protocol engine can also provide access control services to the protocol functions. While the actual call to check access control rules might originate from protocol functions realized outside of the engine, it is usually desirable to maintain the access control decision logic within the protocol engine to ensure that different protocol functions realize a consistent access control policy.

Finally, a protocol engine has to deal with the mapping of the protocol messages to underlying message transports. It seems that in the network management space, it has become increasingly popular to define different transports mappings for a given management protocol and to mark one of them as mandatory to implement to achieve a baseline of interoperability.

*Protocol functions.*    The protocol functions realize the functionality of a management protocol. A protocol engine usually determines the protocol operation that should be invoked after parsing an incoming message and then searches for a protocol function that has registered to handle the protocol operation. Such a registration can be conceptual and rather static in implementations or it can be something which is dynamic and runtime extensible.

Management protocols differ quite a bit in the number of protocol functions they support. In some management protocols, the number of protocol functions is rather fixed and there is strong consensus that adding new protocol functions should be avoided while other protocols are designed to be extensible at the protocol functions layer.

## 3.2. *Subsystems and models*

The internals of a protocol engine can be decomposed into a set of *subsystems*. A subsystem provides a well-defined functionality which is accessible through a defined subsystem interface. By documenting the subsystems and their interfaces, the data flow between the components of a protocol engine can be understood. Furthermore, defined subsystem interfaces reduce the number of surprises when a protocol design gets extended over time.

Within a subsystem, there might be one or multiple *models* that implement the subsystem's interface. Subsystems designed to support multiple models make an architecture extensible. However, there are also subsystems that by design do not support multiple models since they for example organize the data flow between subsystems and models.

## 3.3. *Naming and addressing*

Management protocols manipulate managed objects, which are abstractions of real resources. The term 'managed object' has to be understood in a broad sense; it does not imply that a managed object is an object-oriented data structure. A managed object can be as simple as a simple typed variable or a part in a complex hierarchically structured document.

A particularly important architectural design decision is the *naming*, that is the identification of managed objects. There are several different approaches to name managed objects and the power, flexibility, or simplicity of the naming system has direct impact on the implementation costs and the runtime costs of a management protocol. As a consequence, the selection of a naming mechanism should be treated as an important architectural design decision.

The names of managed objects are usually implicitly scoped. Within the scope, a name provides a unique identifier for a managed object. One commonly used approach is to scope the names relative to the protocol engine that provides access to the managed object. This means that a globally scoped name within a management domain can be constructed out of the identity of the protocol engine and the name of the managed object. Hence, it is important to define how protocol engines are addressed. Some management protocols introduce their own protocol engine addressing scheme while other management protocols use transport endpoint addresses as a shortcut to identify the protocol engine. Architecturally, the first approach is clearly preferable since it decouples the protocol engine from the transport endpoints. Practically, it turns out that such a decoupling has its price since an additional mapping is needed. The selection of a suitable protocol engine addressing scheme therefore becomes an engineering trade-off between architectural purity and implementation and deployment costs.

## 4. Internet management protocols

This section describes several management protocols that were specifically designed for managing the Internet. The protocols discussed below are all specified and standardized by the Internet Engineering Task Force (IETF).

### 4.1. *Simple Network Management Protocol (SNMP)*

The core Internet architecture was developed and tested in the late 1970s by a small group of researchers [11]. During the early 1980s, several important features such as the domain name system were added to the architecture [12]. At the end of the 1980s, the Internet had grown to a world-wide network connecting all major universities and research institutions. Network management functions were at this time realized on an ad-hoc basis through explicit management.

In 1988, it became apparent that additional technology is needed in order to provide better tools to operate the growing Internet infrastructure [8]. The debate held in 1988 shows a fundamental principle in the IETF approach: The IETF seeks to make pragmatic decisions in order to address short-term technical problems. Rather than spending much time on architectural discussions, concrete proposals which are backed up by interoperable implementations have a good chance to be adopted by the IETF and finally in the market place.

The Simple Network Management Protocol (SNMP) [5], the Structure of Management Information (SMI) [46–48] and an initial Management Information Base (MIB-II) [37] were standardized in 1990 and 1991. One of the fundamental goals of the original SNMP framework was to minimize the number and complexity of management functions realized by the management agent itself [5]. The desire to minimize the complexity of management functions in the network elements reflects the constraints on processing power available in the early 1990s and the hardware/software architecture used in most networking devices at that time. Much has changed in the last 10 years in this aspect. Current network elements, such as routers and switches, implement all the primary and time critical network functions in hardware. Furthermore, the processing speed of the microprocessors embedded in networking devices has increased in magnitudes during the last ten years. But it is also important to understand that the network traffic handled by these devices did grow exponentially. The question whether todays and tomorrows network elements have the computational power to realize much more complex management functions is not easy to answer.

The evolution of the original SNMP framework and the MIB modules during the last two decades did provide security, some efficiency improvements, and enhancements of the data definition language. Today, there are more than 100 MIB modules on the standardization track within the IETF and there is an even larger and growing number of enterprise-specific MIB modules defined unilaterally by various vendors, research groups, consortia, and the like resulting in an unknown and virtually uncountable number of defined objects [7,52].

In December 2002, the specification of SNMP version 3 (SNMPv3) was published as an Internet Standard [1,6,26,30,40,41,58], retiring all earlier versions of the SNMP protocol. The second and current version of the SMI (SMIv2) was already published in April 1999 as an Internet Standard [34–36].

*Deployment scenarios.*    A typical SNMP deployment scenario is shown in Figure 2. It show a single manager which is talking SNMP to three agents running on different SNMP-enabled devices.

For one device, a so-called SNMP proxy is involved. There can be different reasons for deploying proxies. One reason can be firewalls and in particular network address translators [44]. Another reason can be version mismatches between management stations and devices.[1]

Agents internally often have a master/subagent structure [18]. Subagents can be embedded into line-cards on routers or application processes to provide access to specific management information. The access to management is usually implemented by so-called method functions in the instrumentation.

Many management systems use full fledged database management systems to store management information retrieved from network elements. Applications access the data stored in the database to realize management functions. In addition, they may interact directly with devices by using services provided by the SNMP manager interface.

It should be noted that Figure 2 provides a rather traditional view. There are many situations where the notion of an agent and a manager is broken apart. In particular, it is not unusual to logically separate the forwarding and processing of notifications from the initiation and processing of commands to retrieve or modify data stored on SNMP-enabled devices.

*Architecture.*    One of the biggest achievements of the SNMPv3 framework next to the technical improvements of the protocol itself is the architectural model defined in RFC 3411 [26]. It decomposes an SNMP protocol engine into several subsystems and it defines abstract service interfaces between these subsystems. A subsystem can have multiple concrete models where each model implements the subsystem interface defined by the

---

[1]Devices have picked up support for SNMPv3 much faster than some widely used management applications.

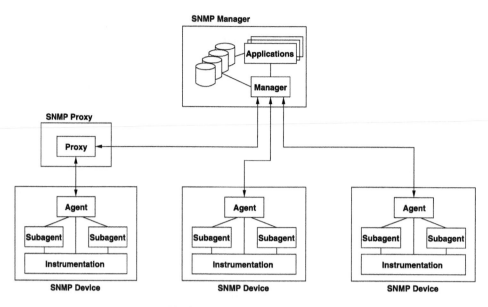

Fig. 2. SNMP deployment scenario.

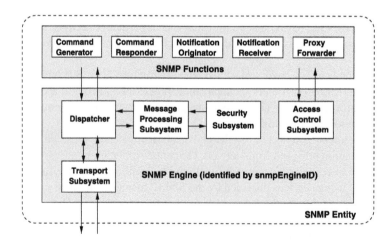

Fig. 3. Subsystem architecture of SNMP.

abstract service interfaces. The architectural model thus provides a framework for future SNMP enhancements and extensions.

Figure 3 shows the decomposition of an SNMP protocol engine into subsystems.[2] SNMP protocol messages enter an SNMP engine via the transport subsystem. Several different

---

[2]Figure 3 does not precisely follow the standard architecture since it includes a transport subsystem which was just recently added to the architectural model [27].

transport models can co-exist in the transport subsystem. The most commonly used transport model is the SNMP over UDP transport model. The transport subsystem passes messages to the dispatcher. The dispatcher, which is a singleton, coordinates the processing in an SNMP engine. It passes incoming messages to the appropriate message processing model in the message processing subsystem. The message processing models take care of the different headers used by the different versions of the SNMP protocol. The message processing model passes messages to a security model in the security subsystem which provides security services such as authentication and privacy of messages. After returning from the security and message processing subsystems, incoming messages are passed by the dispatcher to SNMP functions which implement the requested protocol operations. SNMP functions call the access control subsystem which can provide several different access control models for access control services.

The SNMP framework traditionally follows a variable-oriented approach. Management information is represented by collections of variables. Every variable has a simple primitive type. Variables can appear as simple scalars or they can be organized into so-called conceptual tables. Due to this very granular approach, every variable needs a unique name. The current SNMP architecture uses a four-level naming system. The name of a variable consists of the following components:

1. The identification of the engine of the SNMP entity which hosts the variable. This is called the *context engine identifier*.
2. The identification of the context within an engine which hosts the variable. This is called the *context name*.
3. The identification of the variable type, also commonly called an *object type*. An object type is identified by means of a registered object identifier.
4. The identification of the variable instance, also called an *instance identifier*. An instance identifier is an object identifier suffix which gets appended to the object identifier of the object type.

Like for every naming system, there needs to be a mechanism to lookup names and to explore the name space. The SNMP framework addresses this requirement by restricting protocol operations to a single context. A context is identified by a context engine identifier and a context name. Within the selected context, multiple variables can be read, written, or submitted as part of an event notification.

To explore the name space, SNMP uses a two-level approach:

- Contexts are typically discovered through discovery procedures. SNMP engines can be discovered by probing their transport endpoints. Once a transport endpoint has been found, SNMP messages can be send to the SNMP engine listening on that transport endpoint to discover the engine identifier and existing contexts.
- The variables within a context are discovered by using an iterator which traverses the name space while retrieving variables.

Using an iterator, which retrieves data while traversing the name space, has significant advantages in highly dynamic situations. Management protocols which separate name space traversal operations from data retrieval operations usually are less scalable due to the imposed overhead of managing the name space and synchronization issues in highly dynamic situations.

*Data modeling.*    Data modeling in the SNMP framework is centered around the notion of conceptual tables [34–36]. All columns of a conceptual table have a primitive data type. So-called textual conventions are used to associate special semantics with a primitive type; textual conventions are roughly comparable to typedefs in languages such as C.

Some columns of a conceptual table are marked as index columns. The values of the index columns uniquely identify a row and the index columns, or short the index, thus serve as the primary key for a conceptual table.

The basic concepts resemble relational databases. However, the SNMP framework puts several constraints on the indexing scheme. For example, conceptual tables can only have one index (key) that can be used for fast lookups and the index is subject to several additional constraints imposed by the protocol itself (index size constraints) or by access control mechanisms.

The data modeling framework supports scalars in addition to conceptual tables. For the discussion in this document, it is however sufficient to consider scalars just a degenerate form of a table which has only one row and a fixed index.

In order to assign tables a unique name, an object identifier naming system is used. Object identifier, a primitive ASN.1 data type, essentially denote a path in the global object identifier registration tree. The tree has a virtual root and can be used to uniquely assign names to arbitrary objects [20,54]. In the SNMP framework, tables as well as the table's columns are registered in the tree. Note that the definition of index columns and the registration of tables happen at design time and cannot be changed once a specification has been published. As a consequence, the design of suitable indexing schemes is an important design step in any SNMP data modeling effort.

To fully understand the SNMP framework, it is crucial to properly distinguish between conceptual tables used at the data modeling layer and the registration tree used purely for naming purposes. As the next section explains, the actual protocol operations do neither understand tables nor registration trees and simply operate on an ordered list of variables.

*Protocol operations.*    The SNMP protocol provides a very small number of protocol operations, which is essentially unchanged since 1993 [41]. The protocol operations can be classified into write operations (set), read operations (get, getnext, getbulk), and notify operations (trap, inform). Figure 4 gives an overview which SNMP functions invoke the various operations and which ones are confirmed via a response.

The protocol operations do not operate on conceptual tables nor do they operate on the registration tree. Instead, they solely operate on an ordered collection of variables. The get, set, trap and inform operations all identify variables by their full name (object type identifier plus instance identifier). The getnext and getbulk operations obtain the next variable (getnext) or variables (getbulk) according to lexicographic ordering, which essentially maps the hierarchical name space into a flat ordered collection of variables.

The getnext and getbulk operations combine the traversal of the variable name space with actual data retrieval. This implies that variables can dynamically come and go without any need to register them in a name service.

Something that is often overlooked is the handling of errors and exceptions. SNMP protocol operations can fail with an error code which indicates the reason of the failure and

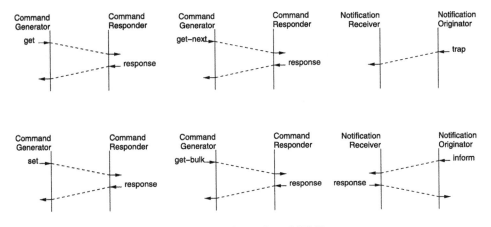

Fig. 4. Protocol operations of SNMP.

which variable caused the failure. In the second version of the SNMP protocol operations, exceptions were introduced to report, on a per variable basis, exceptional conditions without failing the whole protocol operation. Not surprising, most error codes deal with the execution of write operations. However, exceptions are also important for read operations to indicate that specific instances do not exist or that an iterator such as `getnext` or `getbulk` has reached the end of the ordered collection of variables.

*Security.*    A great deal of attention has been given to the security of SNMP as the standardization of a secure version of SNMP proved to be very difficult and time consuming. The SNMPv3 architecture has been designed with a message-based security model in mind where every individual message can be authenticated and also encrypted if so desired. The User-based Security Model (USM) [1] protects against the following threads:

- Modification of information.
- Masquerade.
- Disclosure.
- Message stream modification.

It does not defend against denial of service attacks. The USM authenticates a user, identified by a user name, who is invoking a read or write operation or who is receiving a notification. The user name is mapped into a model independent security name which is passed along inside the SNMP engine together with a security level and the identification of the security model.

The triple of security name, security level, and security model is used by the access control subsystem to decide whether an access to a variable is allowed or denied. The View-based Access Control Model (VACM) [58], the only access control model defined so far in the access control subsystem, provides means to define read/write/notify views on the collection of variables and to restrict access to these views.

One of the original design principles for SNMP was to keep SNMP as independent as possible of other network services so that SNMP can continue to function if the network is not running well. The original security mechanism follow this principle and thus come with an SNMP specific user, key and access control rule management interface. Operators have reported that adding a new user and key management infrastructure has significant costs and that they prefer an approach where security better integrates with existing user, key and access control management infrastructures. The IETF is at the time of this writing working on an additional security model, which utilizes secure session-based transports such as the Secure Shell (SSH) [59] or the Transport Layer Security (TLS) [19] protocol [25].

*Discussion.* The SNMP protocol has been very successful. It is widely deployed in network elements and management systems. However, the main usage of SNMP seems to be monitoring, discovery, and event notification.

SNMP has seen less usage for controlling and configuring devices. By looking at the configuration requirements detailed in Section 2.1, it becomes obvious why. SNMP fails to address most of the configuration related requirements and this is perhaps part of he reason why configuration requirements are so well documented in the RFC series [50,51].

SNMP works well for polling large numbers of devices for a relatively small set of variables. Once the set of variables to be collected increase, SNMP becomes less efficient in terms of required bandwidth but also in terms of overall latency and processing overhead compared to other approaches. The reason is that protocol operations "single step" through the data retrieval, which makes efficient data retrieval and caching on a managed device complicated.

SNMP enjoys significant usage for event notification. However, the binary format of SNMP messages combined with the object identifier naming scheme makes the processing of such event records often difficult in practice and other event notification and logging protocols such as SYSLOG are therefore popular as well.

There were attempts to realize more generic and powerful management services on top of SNMP to delegate management functions [31,32] or to invoke the remote execution of troubleshooting procedures [32]. These approaches have not seen much uptake since the invocation of remote management procedures with read/write/notify primitives is rather unnatural and has relatively high development costs.

### 4.2. *Network configuration protocol (NETCONF)*

In 2002, the Internet Architecture Board (IAB) organized a workshop in order to guide the network management activities in the IETF. The workshop was attended by network operators and protocol developers and resulted into some concrete recommendations [51]. One of the recommendations was to focus resources on the standardization of configuration management mechanisms. Another recommendation was to use the Extensible Markup Language (XML) [2] for data encoding purposes. Some of the discussions related to the future directions are also summarized in [53].

In 2003, the a working group was formed in the IETF to produce a protocol suitable for network configuration. The working group charter mandated that the Extensible Markup Language (XML) [2] be used for data encoding purposes. The protocol produced by this working group is called NETCONF [21] and the subject of this section.

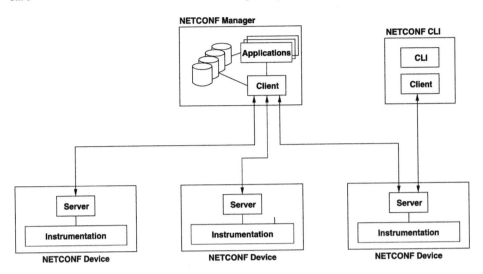

Fig. 5. NETCONF deployment scenario.

*Deployment scenarios.* The driving force behind NETCONF is the need for a programmatic interface to manipulate configuration state. The automation of command line interfaces (CLIs) using programs and scripts has proven to be problematic, especially when it comes to maintenance and versioning issues. Operators have reported that it is time consuming and error prone to maintain programs or scripts that interface with different versions of a command line interface.

Figure 5 shows a NETCONF deployment scenario. It is expected that network-wide configuration or policy systems will use the NETCONF protocol to enforce configuration changes. In a policy framework, the manager involving a NETCONF client might act as a policy decision point while a device involving a NETCONF server might act as a policy enforcement point. Of course, such a setup requires that a policy manager can translate higher-level policies into device configurations; NETCONF only provides the protocol to communicate configuration data.

The right part of Figure 5 shows a CLI which talks NETCONF to a server in order to implement its functionality. It is important to realize that NETCONF must be powerful enough to drive CLIs. Real cost savings can only be achieved if there is a single method to effect configuration changes in a device which can be shared across programmatic and human operator interfaces. This implies that the scope of the NETCONF protocol is actually much broader than just device configuration.

*Architecture.* The NETCONF protocol does not have a very detailed architectural model. The protocol specification [21] describes a simple layered architecture which consists of a transport protocol layer, a remote procedure call (RPC) layer, an operation layer providing well defined RPC calls, and a content layer.

A more structured view of NETCONF, which resembles the SNMP architectural model, is shown in Figure 6. The transport subsystem provides several transport models. The work-

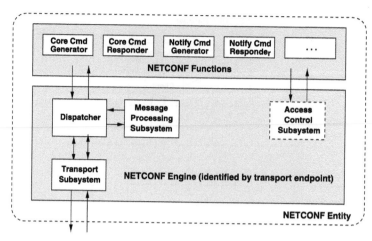

Fig. 6. Subsystem architecture of NETCONF.

ing group so far has defined an SSH transport model [57], a BEEP transport model [28], and a SOAP transport model [24]. The SSH transport is mandatory to implement.

The dispatcher is the central component which organizes the data flow within the NET-CONF engine. It hands messages received from the transport subsystem to the message processing subsystem which currently has only one message processing model. This message processing model handles a capabilities exchange, which happens after a transport has been established, and the framing or RPC messages. The operations being invoked are provided by NETCONF functions. The base NETCONF specification deals mainly with generic operations to retrieve and modify configuration state. An additional document defines operations to subscribe to notification channels and to receive notifications. It is expected that additional operations will be introduced in the future for more specific management purposes.

The base NETCONF specification follows a document-oriented approach. The configuration state of a device is considered to be a structured document that can be retrieved and manipulated. A filtering mechanism has been defined which allows to retrieve a subset of the document representing a configuration.

NETCONF supports multiple configuration datastores. A configuration datastore contains all information needed to get a device from its initial default state into the desired configuration state. The running datastore is always present and describes the currently active configuration. In addition, NETCONF supports the notion of a startup configuration datastore, which is loaded at next re-initialization time, and a candidate datastore, which is a scratch buffer that can be manipulated and later committed to the running datastore.

*Data modeling.* Since NETCONF uses XML to encode network management data and in particular configuration state, it seems obvious to use some XML schema notation to formally specify the format of these XML documents. However, there is no agreement yet which schema language to use. Furthermore, it seems necessary to incorporate additional information in data models that go beyond the capabilities of existing schema languages or requires to resort to hooks in schema languages where additional information can be specified which is ignored by generic tools.

During the development of the NETCONF protocol specifications, which are formally defined using XML schema [22], it was observed that the XML schema notation is difficult to read and verify for humans. Other schema notations such as Relax NG [15] and especially its compact notation [13] seem to be easier to read and write. Still, these schema languages tend to be relatively far away from an implementors view and in the light of integration with SNMP, it might also be possible to adapt a language such as SMIng [55] for NETCONF data modeling purposes.

The NETCONF working group was not chartered to address the data modeling aspects of NETCONF and as such there is no agreement yet how to approach this problem. It is, however, clear that agreement must be reached on how to write data models in order to achieve standardized configuration in a heterogeneous environment.

*Protocol operations.* The NETCONF protocol has a richer set of protocol operations compared to SNMP. It is generally expected that new protocol operations will be added in the future. This essentially means that NETCONF will likely in the future support a command-oriented approach in addition to the document-oriented approach for manipulating configuration state.

Figure 7 shows the protocol operations that have been defined by the NETCONF working group. The first eight operations all operate on configurations stored in configuration datastores. The `lock` and `unlock` operations do coarse grain locking and locks are intended to be short-lived.

The `get` operation is provided to retrieve a device's configuration state and operational state; the `get-config` operation has been provided to only retrieve configuration state. The `close-session` operation initiates a graceful close of a session while the `kill-session` operation forces to terminate a session.

The most powerful and complex operation is the `edit-config` operation. While the `edit-config` operation allows to upload a complete new configuration, it can also be used to modify the current configuration of a datastore by creating, deleting, replacing or merging configuration elements. In essence, `edit-config` works by applying a patch to a datastore in order to generate a new configuration. Since a tree-based representation, is used to express configuration state, it is necessary to describe which branches in the tree should be created, deleted, replaced or merged. NETCONF solves this by adding so called operation attributes to an XML document that is sent as part of the `edit-config` invocation to the NETCONF server.

The `get` and `get-config` operations both support a filter parameter, which can be used to select the parts of an XML document that should be retrieved. The protocol supports a subtree filter mechanisms which selects the branches of an XML tree matching a provided template. As an optional feature, implementations can also support XPATH [14] expressions.

The notification support in NETCONF is based on an event stream abstraction. Clients who want to receive notifications have to subscribe to an event stream. An event stream is an abstraction which allows to handle multiple event sources and includes support for

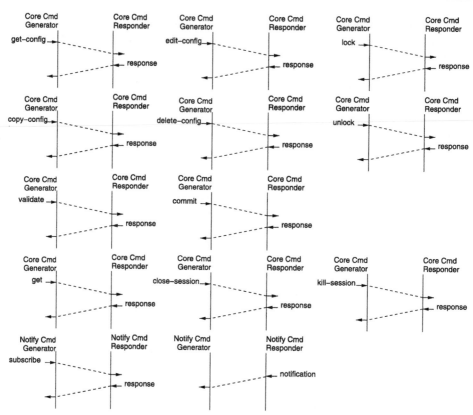

Fig. 7. Protocol operations of NETCONF.

transparently accessing event logs. On systems which maintain an event log, it is possible to subscribe to an event stream at some time in the history and the device will then playback all recorded event notifications at the beginning of the notification stream. A subscription to an event stream establishes a filter which is applied before event notifications are sent to the client. This allows to select only the interesting messages and improves scalability.

*Security.* NETCONF assumes that message security services are provided by the transports. As a consequence, Figure 6 does not contain a security subsystem. While it has been acknowledged by the working group that an access control subsystem and model is needed, standardization work in this area has not yet been started. Some research in this area can be found in [17].

*Discussion.* The design of the NETCONF protocol reflects experiences made with existing proprietary configuration protocols such as Juniper's JunoScript. NETCONF addresses the requirements for configuration management protocols defined in Section 2.1 and is therefore a good choice for this task. The pragmatic approach to layer the NETCONF protocol on top of a secure and reliable transport greatly simplifies the protocol.

The downside of NETCONF, as it is defined today, are the missing pieces, namely a lack of standards for data modeling and the lack of a standardized access control model. Since NETCONF implementations are well underway, it can be expected that these shortcomings will be addressed in the next few years and NETCONF might become a powerful tool for network configuration and trouble shooting.

### 4.3. *System logging protocol (SYSLOG)*

The SYSLOG protocol, originally developed in the 1980s as part of the Berkeley Software Distribution, is used to transmit event notification messages over the Internet. Due to its simplicity and usefulness, the SYSLOG protocol enjoys very wide deployment, even though there was never a specification of the protocol itself. This lead to a number of differences how SYSLOG messages are handled by implementations. In 2001, an informational document was published which describes the de-facto SYSLOG protocol [33] and provides hints how implementations should deal with some of the deployed variations.

The SYSLOG protocol described in [33] has several shortcomings:

- The default transport (UDP) provides only unreliable message delivery.
- The default transport (UDP) does not protect messages. A large number of possible attacks are described in [33], among them message forgery, message modifications, and message replay.
- Some applications require mechanisms to authenticate the source of an event notification and to verify the integrity of the notification.
- SYSLOG messages contain no structured data which makes it difficult for machines to interpret them.

Since the publication of the document describing the de-facto SYSLOG protocol, the IETF has been working on a standards-track SYSLOG protocol addressing the shortcomings mentioned above [23].

*Deployment scenarios.*   The SYSLOG protocol can be deployed in a variety of configurations. Figure 8 shows some of the possible scenarios.

A SYSLOG-enabled device supports at least a SYSLOG sender. The sender can obtain event notifications from various sources. Typical sources are background processes such as routing daemons or processes implementing specific services, such as web server or database server. Another source of event notifications is often the operating system kernel, which usually provides special interfaces to pass kernel notifications to the SYSLOG process forwarding the messages.

Event notifications are collected by so-called SYSLOG collectors, which typically store received notifications in some database and might pass the information on to applications that are interested in processing the event notifications. Stored notifications are often used to investigate why failures happened in order to find root causes. However, stored notifications can also be useful to prove that a certain event did happen or a certain service was provided. It is therefore important that integrity, completeness, and authenticity of messages can be verified when they are fetched from the database.

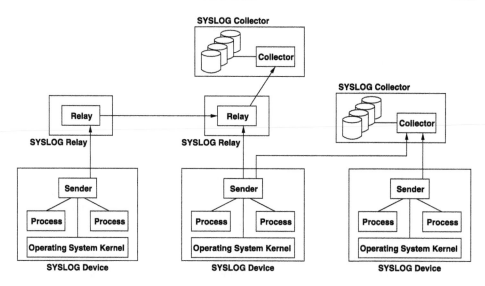

Fig. 8. SYSLOG deployment scenario.

The SYSLOG sender can forward event notifications via the SYSLOG protocol to notification collectors directly as shown in the right part of Figure 8, or it can use relays. Relays
can act as simple forwarders in order to deal with firewalls or network address translators.
They can, however, also be used to aggregate notifications from multiple senders or to filter
notifications before passing them on towards collectors.

*Architecture.*    The SYSLOG protocol is a so called "fire-and-forget" protocol. The originator of an event notification sends the notification to a receiver without getting an ac-
knowledgment in any form whether the message was received, passed on to other systems,
or successfully stored or processed.

SYSLOG receivers typically come in two flavors:
- A *relay* is a receiver that can receive messages and forward them to other receivers.
- A *collector* is a receiver that receives messages without relaying them.

Note that a sender is not aware whether a receiver acts as a relay or a collector. Furthermore, a single receiving entity may act as both a relay and a collector.

Figure 9 presents an architectural view of the SYSLOG protocol. The transport subsystem is responsible for the transmission of messages over the network. The SYSLOG
protocol traditionally uses UDP as a transport. Recent work suggests to run SYSLOG over
TLS [19] in order to protect messages on the wire.

The message processing subsystem currently includes two message processing models.
The original message format is defined by the BSD SYSLOG protocol [33] while the new
message format is the standards-track format developed by the IETF [23].

There are essentially three SYSLOG functions, namely an *originator* generating notifications, a *relay* forwarding notifications, and a *collector* receiving and storing notifications.
As mentioned above, it is possible in the SYSLOG architecture to dispatch a received message to a relay and a collector registered on the same SYSLOG engine.

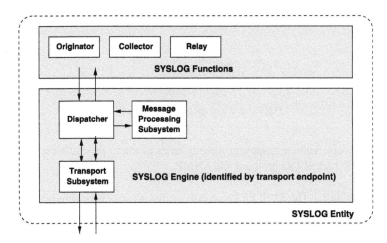

Fig. 9. Subsystem architecture of SYSLOG.

Due to the fire-and-forget nature of the SYSLOG protocol, there is no access control subsystem. An access control subsystem requires that a security subsystem establishes an authenticated identity in order to select appropriate access control rules. With a fire-and-forget protocol, it becomes difficult to establish such an authenticated identity. As a consequence, in SYSLOG there is also no need for a security subsystem.

The original SYSLOG carries only a small number of machine readable elements in notification messages, namely the identification of a facility originating the event, a severity level, a timestamp, an identification of the host originating the event, the name of the process originating the event, and optionally a process identification number. The rest of the message contains an arbitrary text message. The standards-track SYSLOG protocol essentially keeps the original format but adds more machine readable content in the form of structured data elements.

Structured data consists of zero, one, or multiple structured data elements and each structured data element contains a list of parameters in the form of name value pairs. Structured data elements are identified by a name, which is essentially a seven-bit ASCII character string. The name format supports standardized structured data element names as well as enterprise specific structured data element names. The names of the structured data element parameters follow the same rules as structured data element names, but they are scoped by the structured data element name. The values use a simple ASCII encoding with an escape mechanism to handle special characters such as quotes.

*Data modeling.* There is no data modeling framework yet for SYSLOG structured data elements. The specifications so far use plain text to define the format and semantics of structured data elements and their parameters.

Since SYSLOG notifications and SNMP notifications serve similar purposes, it can be expected that mappings will be defined to map SNMP varbind lists to SYSLOG structured data element parameters and also SYSLOG structured data elements into SNMP

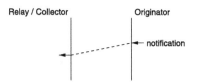

Fig. 10.  Protocol operations of SYSLOG.

notification objects. Similar mappings already exist in many products between the more unstructured BSD SYSLOG protocol and SNMP.

*Protocol operations.*    The SYSLOG protocol operations are fairly simple since the protocol does not expect any bidirectional communication between SYSLOG entities.

Figure 10 shows that a notification is sent to a relay or a collector. A particular feature of SYSLOG is that a SYSLOG entity can act as both a relay and a collector at the same time. This means that the dispatcher of a SYSLOG engine must be able to dispatch a received message to multiple functions, something that does not exist, for example, in the SNMP architecture (a received SNMP notification is either proxied or passed to the notification receiver).

*Security.*    The SYSLOG protocol traditionally provides no security. Not surprisingly, this is a major issue in environments where the management network itself can not be properly secured. Furthermore, there are legislations in some countries that require operators to log certain events for later inspection and it is sometimes required to be able to prove that logged messages are unchanged and originated from the claimed source.

The SYSLOG specifications provide two mechanisms to deal with these security requirements. To provide hop-by-hop message protection, it is possible to run SYSLOG over TLS [38]. To achieve notification originator authentication and integrity checking, it is possible to sign SYSLOG notifications [16]. This works by sending signature blocks as separate SYSLOG messages that sign the hashes of a sequence of recently sent SYSLOG notifications. Note that signature blocks can be logged together with SYSLOG messages and used later to verify the notification's integrity and the identity of the SYSLOG notification originator once the need arises.

*Discussion.*    The SYSLOG protocol has been very successful in achieving wide spread deployment. Many C libraries support a simple API to generate SYSLOG messages and it is not unlikely that the availability of an easy to use and relatively portable API lead to wide-spread acceptance by software developers. From a conceptual point of view, SYSLOG notifications with structured data elements are close to unconfirmed SNMP notifications and it can be expected that mappings will be defined. The concept of SYSLOG relays is close to SNMP proxies, except that SNMP does not directly support a co-located relay and collector (although a fancy configuration of a proxy might achieve a similar effect).

There are some notable differences in the security services supported. SYSLOG provides a mechanism to sign notifications while SNMP provides confirmed notification delivery and access control for outgoing notifications. The SYSLOG signature mechanism might be easy to retrofit into the SNMP protocol, but confirmed notifications are more difficult to add to SYSLOG.

### 4.4. *Flow information export protocol (IPFIX)*

In the 1990s, it became popular to collect traffic statistics about traffic flows in order to track the usage of networks. A flow is a set of IP packets passing an observation point in the network during a certain time interval. All packets belonging to a particular flow have a set of common properties [43].

Flows are a convenient mechanism to aggregate packet streams into something meaningful for the network operator. For example, a network operator might collect flow statistics for specific types of traffic in the network such as all email (SMTP) traffic, all web (HTTP) traffic, or all voice (SIP and RTP) traffic. In addition, the operator might collect flow statistics for the different customers for accounting purposes.

In order to assign packets to flows, it is necessary to derive a set of properties. In general, the set of properties is constructed by applying a function to certain values and characteristics associated with a captured packet [49]:

1. One or more network layer packet header fields, transport layer header fields, or application layer header fields (e.g., IP destination address, destination port, and RTP header fields).
2. One of more characteristics of the packet itself (e.g., number of MPLS labels).
3. One or more fields derived from packet treatment (e.g., output interface).

Some early approaches to collect traffic statistics using the SNMP framework resulted in the Remote Monitoring (RMON) family of MIB modules [56]. While the RMON technology achieved wide deployment for collecting overall link statistics, it did not support the notion of flows very well. The Realtime Traffic Flow Measurement (RTFM) framework [3], centered around a Meter MIB module [4], details the criteria used by a meter to assign packets to flows. The RTFM framework is like the RMON framework based on SNMP technology. Even though the RTFM framework was quite flexible from the very beginning, it only achieved some moderate level of deployment.

In the 1990s, several vendors started efforts to export flow information from their routers using special purpose protocols, most notably Cisco's NetFlow [9], which evolved over time to the current version 9. Cisco's NetFlow has been very successful in achieving significant deployment. This lead to the formation of a working group in the IETF to standardize a flow information export protocol. The evaluation of candidate protocols [29] lead to the selection of Cisco's NetFlow version 9 as a starting point for a standardization effort, which then lead to the IP Flow Information Protocol (IPFIX) [49] framework.

*Deployment scenarios.*    The IPFIX protocol can be used in several different scenarios. Figure 11 shows a deployment scenario where two IPFIX devices (typically IP router) report flow statistics to a single collector. The collector stores the data in a local database. The dashed lines indicate that the collector may include another meter and exporter process to report further aggregated statistics to another collector.

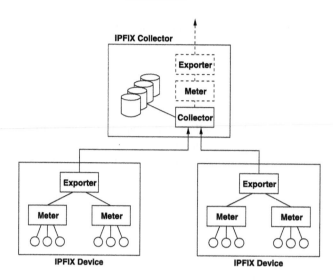

Fig. 11. IPFIX deployment scenario.

IPFIX devices can include several observation points (marked as circles in Figure 11) and meters. It is also possible to have multiple exporting processes on a single device, but this is not indicated in Figure 11. The internal structure typically depends on the configuration of the device. Note that observation points and even complete meters might be running on different line cards on a high-end modular router.

*Architecture.* The IPFIX protocol follows a variable-oriented approach to export flow statistics [49]. Since IPFIX is a unidirectional protocol, the architecture remains fairly simple.

Figure 12 provides an architectural view. Like all other protocols discussed in this chapter, IPFIX has a transport subsystem which supports several different transport models. Since IPFIX is relatively new, there is currently only a single message format and message processing model. A message version number has been designed into the protocol in order to be able to deal with any changes and additions that might be necessary in the future.

There are two IPFIX functions: The exporting process is responsible for generating IPFIX messages and sending them to a collector. The collecting process receives IPFIX messages and processes them. The IPFIX requirements document also mentions concentrators capable to merge multiple IPFIX message streams and proxies, which relay IPFIX message streams [43]. None of these concepts is, however, explicitly discussed in the protocols specifications and this is why they also do not appear in Figure 12.

The data values transmitted in IPFIX messages belong to so called information elements. An information element is primarily identified by its element identifier, which is a numeric identifier. Globally unique element identifiers can be allocated via the Internet Assigned Numbers Authority (IANA). Vendor-specific element identifiers are created by combining the element identifier with an enterprise identifier. The later can be allocated via IANA on a first-come first-serve basis. Information elements can be scoped to a specific meter-

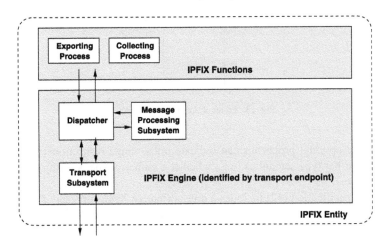

Fig. 12. Subsystem architecture of IPFIX.

ing or exporting process, or they can be scoped to a specific flow measured at a specific observation point, which again is identified by an observation identifier.

IPFIX messages can be exchanged over different transport protocols. The required to implement transport protocol is the Stream Control Transmission Protocol (SCTP) [39]. SCTP provides multiple independent streams in an association. IPFIX explores this by sending meta data and configuration data over SCTP stream zero while multiple additional streams may be allocated to send measurement data records. The optional transports for IPFIX are UDP and TCP.

*Data modeling.* IPFIX uses an ad-hoc informal template mechanism for data modeling [42]. Information elements are defined by defining the following properties:

- The mandatory *name* property is a unique and meaningful name for an information element.
- The mandatory numeric *element identifier* property is used by the IPFIX protocol to identify an information element. The element identifier can be globally unique or it can be scoped by an enterprise identifier.
- The textual *description* property specifies the semantics of an information element.
- The mandatory *data type* property of an information element indicates the type of an information element and its encoding. The basic data types include signed and unsigned integers with different width (8, 16, 32, 64 bits), floating point numbers (IEEE 32-bit and 64-bit formats), a Boolean-type, octet string and unicode strings, several date and time types, and types for IPv4, IPv6, and MAC addresses.
- The mandatory *status* property indicates whether a definition is current, deprecated or obsolete.
- The optional numeric *enterprise identifier* property must be present if the element identifier is not globally unique.

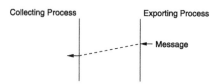

Fig. 13. Protocol operations of IPFIX.

- Additional optional properties can indicate units, range restrictions, references and provide further type semantics (e.g., whether a numeric type is acting as a counter or an identifier).

The IPFIX specification [42] includes a non-normative XML representation of the definitions and a non-normative XML schema for the XML notation.

*Protocol operations.*   The IPFIX protocol is essentially a unidirectional message stream. At the time of this writing, IPFIX only has a single message format which includes a version number, a length field, a sequence number, the observation domain identifier and the export time [10].

The IPFIX message header is followed by one or more sets. Sets can be either data sets, template sets, or options template sets:

- *Template sets* are collections of one or more template records that have been grouped together in an IPFIX message. A template record defines the structure and interpretation of fields in data records. Template sets are identified by a template number which is scoped by the transport session and an observation domain.
- *Data sets* consist of one or more data records of the same type, grouped together in an IPFIX message. Data records refer to previously transported template records (using the template number). The template record defines the structure and the semantics of the values carried in a data record.
- *Options template sets* are collections of one or more option template records that have been grouped together in an IPFIX message. An option template record is a template record including definitions how to scope the applicability of the data record. Option template records provide a mechanism for an exporter to provide additional information to a collector, such as the keys identifying a flow, sampling parameters, or statistics such as the number of packets processed by the meter or the number of packets dropped due to resource constraints.

Template records establish the names and types of the values contained in subsequent data records. Compared to the name-value encoding used by SNMP and SYSLOG, IPFIX has less overhead since names are not repeated in each data record. This can significantly reduce the encoding and bandwidth overhead and makes IPFIX more efficient and scalable.

*Security.*   The data carried by the IPFIX protocol can be used to analyze intrusion attacks or it might be the basis for billing processes. It is therefore necessary to properly maintain the integrity of IPFIX messages and to ensure that data is delivered from an authorized exporting process to an authorized collecting process. Furthermore, it is desirable to prevent disclosure of flow data since flow data can be highly sensitive.

The IPFIX protocol mandates that the Datagram Transport Layer Security (DTLS) protocol [45] is implemented for the SCTP transport. DTLS has been chosen instead of TLS for SCTP since SCTP supports unreliable and partially reliable modes in addition to the default reliable more. For the optional TCP transport, TLS must be implemented to secure IPFIX.

It should be noted that IPFIX does not have any mechanisms for access control. The configuration of an IPFIX exporter therefore implicitly implements an access control policy for flow data.

*Discussion.*    IPFIX is derived from Cisco's NetFlow, a very successful flow information export protocol. Since IPFIX can be seen as an evolutionary improvement of NetFlow, it can be expected that IPFIX also gets widely implemented and deployed. It should, however, be noted that the requirement of the SCTP transport support raises the bar for conforming implementations. Only very recent operating systems provide full SCTP support as part of the operating system kernel.

The data modeling part of the IPFIX framework may be improved. The informal ad-hoc notation makes it difficult to build tools and it would be desirable to achieve some level of integration with the data models used by other management protocols.

## 5. Conclusions

Work on Internet network management protocols started in the late 1980s. The SNMP protocol was created at that time as a temporary solution until full OSI management protocols would take over [8]. In the mid-1990s, it became clear that this will never happen.

The Internet management protocol development and standardization was for a long time almost exclusively focused on the SNMP framework, to a large extend ignoring that many other not well-standardized protocols became widely used to manage networks. In addition, management interfaces originally designed for human interaction, such as command line interfaces, became increasingly used for management automation.

It is only recently that the IETF network management community has started to realize that effective network management requires a toolbox of protocols and that a "one size fits all" approach is unlikely to succeed to address all requirements. This change of mind-set is very important, especially since many of the requirements are also much better understood today than they were some 10 years ago.

The real challenge with a multi-protocol approach is to keep the protocols aligned where necessary. Such alignment concerns mainly security related issues as it is for example operationally very important to harmonize key management procedures and access control lists. With a convergence to secure transports and the movement to support different certificate formats by secure transports, it can be hoped that this problem will be dealt with over time.

A second area where some level of alignment is very important is data modeling. It is crucial that different and potentially overlapping management interfaces operate on the same underlying data structures as this improves consistency and reduces costs. Since data modeling efforts are not as far developed as some of the protocol definitions, it can be expected that work on multi-protocol Internet management data modeling remains an active area for future research and development.


## Acknowledgements

The work on this chapter was supported in part by the EC IST-EMANICS Network of Excellence (#26854).


## References

[1] U. Blumenthal and B. Wijnen, *User-based Security Model (USM) for version 3 of the Simple Network Management Protocol (SNMPv3)*, RFC 3414, Lucent Technologies (2002).

[2] T. Bray, J. Paoli and C.M. Sperberg-McQueen, *Extensible Markup Language (XML) 1.0*, W3C Recommendation, Textuality and Netscape, Microsoft, University of Illinois (1998).

[3] N. Brownlee, *Traffic flow measurement: Architecture*, RFC 2722, The University of Auckland, GTE Laboratories, GTE Internetworking (1999).

[4] N. Brownlee, *Traffic flow measurement: Meter MIB*, RFC 2720, The University of Auckland (1999).

[5] J. Case, M. Fedor, M. Schoffstall and J. Davin, *A simple network management protocol*, RFC 1157, SNMP Research, PSI, MIT (1990).

[6] J. Case, D. Harrington, R. Presuhn and B. Wijnen, *Message Processing and Dispatching for the Simple Network Management Protocol (SNMP)*, RFC 3412, SNMP Research, Enterasys Networks, BMC Software, Lucent Technologies (2002).

[7] J. Case, R. Mundy, D. Partain and B. Stewart, *Introduction and applicability statements for Internet standard management framework*, RFC 3410, SNMP Research, Network Associates Laboratories, Ericsson (2002).

[8] V. Cerf, *IAB recommendations for the development of Internet network management standards*, RFC 1052, Network Information Center, SRI International (1988).

[9] B. Claise, *Cisco Systems NetFlow services export version 9*, RFC 3954, Cisco Systems (2004).

[10] B. Claise, *Specification of the IPFIX protocol for the exchange of IP traffic flow information*, Internet Draft (work in progress) <draft-ietf-ipfix-protocol-24>, Cisco Systems (2006).

[11] D. Clark, *The design philosophy of the DARPA internet protocols*, Proc. SIGCOMM'88 Symposium on Communications Architectures and Protocols (1988).

[12] D. Clark, L. Chapin, V. Cerf, R. Braden and R. Hobby, *Towards the future Internet architecture*, RFC 1287, MIT, BBN, CNRI, ISI, UC Davis (1991).

[13] J. Clark, *RELAX NG compact syntax*, Committee specification, OASIS (2002).

[14] J. Clark and S. DeRose, *XML Path language (XPath) version 1.0*, Recommendation, W3C (1999).

[15] J. Clark and M. Makoto, *RELAX NG specification*, Committee specification, OASIS (2001).

[16] A. Clemm, J. Kelsey and J. Callas, *Signed syslog messages*, Internet Draft (work in progress) <draft-ietf-syslog-sign-21.txt>, PGP Corporation, Cisco Systems (2007).

[17] V. Cridlig, R. State and O. Festor, *An integrated security framework for XML-based management*, Proc. 9th IFIP/IEEE International Symposium on Integrated Network Management, IEEE (2005), 587–600.

[18] M. Daniele, B. Wijnen, M. Ellison and D. Francisco, *Agent Extensibility (AgentX) protocol version 1*, RFC 2741, Digital Equipment Corporation, IBM T.J. Watson Research, Ellison Software Consulting Systems (2000).

[19] T. Dierks and E. Rescorla, *The Transport Layer security (TLS) protocol version 1.1*, RFC 4346, Independent, RTFM (2006).

[20] O. Dubuisson, *ASN.1 – Communication between Heterogeneous Systems*, Morgan Kaufmann (2000).

[21] R. Enns, *NETCONF configuration protocol*, RFC 4741, Juniper Networks (2006).

[22] D.C. Fallside, *XML schema part 0: Primer*, W3C Recommendation, IBM (2001).

[23] R. Gerhards, *The syslog protocol*, Internet Draft (work in progress) <draft-ietf-syslog-protocol-20.txt>, Adiscon GmbH (2007).

[24] T. Goddard, *Using the Network Configuration Protocol (NETCONF) over the Simple Object Access Protocol (SOAP)*, RFC 4743, ICEsoft Technologies Inc. (2006).

[25] D. Harrington, *Transport security model for SNMP*, Internet Draft (work in progress) <draft-ietf-isms-transport-security-model-04.txt>, Huawei Technologies (2007).

[26] D. Harrington, R. Presuhn and B. Wijnen, *An architecture for describing Simple Network Management Protocol (SNMP) management frameworks*, RFC 3411, Enterasys Networks, BMC Software, Lucent Technologies (2002).

[27] D. Harrington and J. Schönwälder, *Transport subsystem for the Simple Network Management Protocol (SNMP)*, Internet Draft (work in progress) <draft-ietf-isms-tmsm-08.txt>, Huawei Technologies, International University Bremen (2007).

[28] E. Lear and K. Crozier, *Using the NETCONF protocol over Blocks Extensible Exchange Protocol (BEEP)*, RFC 4744, Cisco Systems (2007).

[29] S. Leinen, *Evaluation of candidate protocols for IP Flow Information Export (IPFIX)*, RFC 3955, SWITCH (2004).

[30] D. Levi, P. Meyer and B. Stewart, *SNMP applications*, RFC 3413, Nortel Networks, Secure Computing Corporation, Retired (2002).

[31] D. Levi and J. Schönwälder, *Definitions of managed objects for the delegation of management scripts*, RFC 3165, Nortel Networks, TU Braunschweig (2001).

[32] D. Levi and J. Schönwälder, *Definitions of managed objects for scheduling management operations*. RFC 3231, Nortel Networks, TU Braunschweig (2002).

[33] C. Lonvick, *The BSD syslog protocol*, RFC 3164, Cisco Systems (2001).

[34] K. McCloghrie, D. Perkins and J. Schönwälder, *Conformance Statements for SMIv2*, RFC 2580, Cisco Systems, SNMPinfo, TU Braunschweig (1999).

[35] K. McCloghrie, D. Perkins and J. Schönwälder, *Structure of Management Information version 2 (SMIv2)*, RFC 2578, Cisco Systems, SNMPinfo, TU Braunschweig (1999).

[36] K. McCloghrie, D. Perkins and J. Schönwälder, *Textual conventions for SMIv2*, RFC 2579, Cisco Systems, SNMPinfo, TU Braunschweig (1999).

[37] K. McCloghrie and M. Rose, *Management Information Base for network management of TCP/IP-based Internets: MIB-II*, RFC 1213, Hughes LAN Systems, Performance Systems International (1991).

[38] F. Miao and M. Yuzhi, *TLS transport mapping for SYSLOG*, Internet Draft (work in progress) <draft-ietf-syslog-transport-tls-10.txt>, Huawei Technologies (2007).

[39] L. Ong and J. Yoakum, *An introduction to the Stream Control Transmission Protocol (SCTP)*, RFC 3286, Ciena Corporation, Nortel Networks (2002).

[40] R. Presuhn, *Transport mappings for the Simple Network Management Protocol (SNMP)*, RFC 3417, BMC Software (2002).

[41] R. Presuhn, *Version 2 of the protocol operations for the Simple Network Management Protocol (SNMP)*, RFC 3416, BMC Software (2002).

[42] J. Quittek, S. Bryant, B. Claise, P. Aitken and J. Meyer, *Information model for IP flow information export*, Internet Draft (work in progress) <draft-ietf-ipfix-info-15.txt>, NEC, Cisco Systems, PayPal (2007).

[43] J. Quittek, T. Zseby, B. Claise and S. Zander, *Requirements for IP Flow Information Export (IPFIX)*, RFC 3917, NEC Europe, Fraunhofer FOKUS, Cisco Systems, Swinburne University (2004).

[44] D. Raz, J. Schönwälder and B. Sugla, *An SNMP application level gateway for payload address translation*, RFC 2962, Lucent Technologies, TU Braunschweig, ISPSoft Inc. (2000).

[45] E. Rescorla and N. Modadugu, *Datagram transport layer security*, RFC 4347, RTFM, Stanford University (2006).

[46] M. Rose, *A convention for defining traps for use with the SNMP*, RFC 1215, Performance Systems International (1991).

[47] M. Rose and K. McCloghrie, *Structure and identification of management information for TCP/IP-based Internets*, RFC 1155, Performance Systems International, Hughes LAN Systems (1990).

[48] M. Rose and K. McCloghrie, *Concise MIB definitions*, RFC 1212, Performance Systems International, Hughes LAN Systems (1991).

[49] G. Sadasivan, N. Brownlee, B. Claise and J. Quittek, *Architecture for IP flow information export*, Internet Draft (work in progress) <draft-ietf-ipfix-architecture-12>, Cisco Systems, CAIDA, NEC Europe (2006).

[50] L. Sanchez, K. McCloghrie and J. Saperia, *Requirements for configuration management of IP-based networks*, RFC 3139, Megisto, Cisco, JDS Consultant (2001).

[51] J. Schönwälder, *Overview of the 2002 IAB network management workshop*, RFC 3535, International University Bremen (2003).

[52] J. Schönwälder, *Characterization of SNMP MIB modules*, Proc. 9th IFIP/IEEE International Symposium on Integrated Network Management, IEEE (2005), 615–628.

[53] J. Schönwälder, A. Pras and J.P. Martin-Flatin, *On the future of Internet management technologies*, IEEE Communications Magazine **41** (10) (2003), 90–97.

[54] D. Steedman, *Abstract Syntax Notation One (ASN.1): The Tutorial and Reference*, Technology Appraisals (1990).

[55] F. Strauß and J. Schönwälder, *SMIng – next generation structure of management information*, RFC 3780, TU Braunschweig, International University Bremen (2004).

[56] S. Waldbusser, R. Cole, C. Kalbfleisch and D. Romascanu, *Introduction to the Remote Monitoring (RMON) family of MIB modules*, RFC 3577, AT&T, Verio, Avaya (2003).

[57] M. Wasserman and T. Goddard, *Using the NETCONF configuration protocol over Secure Shell (SSH)*, RFC 4742, ThingMagic, IceSoft Technologies (2006).

[58] B. Wijnen, R. Presuhn and K. McCloghrie, *View-based access control Model (VACM) for the Simple Network Management Protocol (SNMP)*, RFC 3415, Lucent Technologies, BMC Software, Cisco Systems (2002).

[59] T. Ylonen and C. Lonvick, *The Secure Shell (SSH) protocol architecture*, RFC 4251, SSH Communications Security Corp, Cisco Systems (2006).

# 3. Networks, Connections and Knowledge

## 3.1. *Commentary*

Networks in nature are an extremely common phenomenon and are amongst the most prolific sources of complexity in science [2,3]. There is every reason to believe that they have been central in yielding many of the major developments in the natural world, and their usefulness in technology is increasingly apparent.

In computer science networks have traditionally been regarded as physical or logical communication channels that connect processes together. Network management nurses both the physical communication infrastructure and the logical patterns of work and data flows. The classical stuff of routing and switching networks, of RIP, OSPF, the OSI model and TCP/IP layers can be found in any number of books today. One of the few networking textbooks containing scholarly references to original work is [1].

Today we understand that networks are everywhere, representing all kinds of relationships. Computer programs are themselves networks of interacting methods and components. Knowledge management itself is about connections that form networks between ideas and ontological primitives. Thus the ability to discuss these relationships, and know how information is located and spreads is a major part of management.

Here, Badonnel et al. discuss probabilistic or *ad hoc* networks and their management. This topic is important, not only for describing the new mobile devices in emergency scenarios or wilderness deployment, but also serves as a model for incorporating general unreliability into network infrastructure. Canright and Engø-Monsen explain how the form and structure of networks can be understood according to principles of social organization, and how this tells us something about information dissemination and the ability to search within the networks.

Knowledge bases and conceptual mappings are desribed using a conceptual model for conceptual models by Strassner, i.e. Ontologies, and Vranken and Koymans describe how such mappings can be used as a universal-translator technology to integrate disparate user applications.

Many aspects of networking are not covered here, including variations on the theme of routing and distributed dissemination algorithms.

# References

[1]  C. Huitema, *Routing in the Internet*, 2nd edn, Prentice Hall (2000).

[2]  M.E.J. Newman, *The structure and function of complex networks*, SIAM Review **45** (2003), 167–256.

[3]  M.E.J. Newman, S.H. Strogatz and D.J. Watts, *Random graphs with arbitrary degree distributions and their applications*, Physical Review E **64** (2001), 026118.

# – 3.2 –

# Management of Ad-Hoc Networks

Remi Badonnel, Radu State, Olivier Festor

*MADYNES Research Team, LORIA – INRIA Lorraine, Campus Scientifique – BP 239,*
*5406 Vandeouvre-lès-Nancy Cedex, France*
*E-mails: {badonnel, state, festor}@loria.fr*

**Abstract**

This chapter presents a state of the art of the management of ad-hoc networks and is structured as follows. We introduce ad-hoc networking and the challenges of ad-hoc networking management in Section 1. We present the main management paradigms and the underlying architectures in Section 2. Several application domains including monitoring, configuration and cooperation control are covered by Section 3. Finally, Section 4 concludes the chapter with a synthesis of research orientations in this domain.

## 1. Introduction

The increasing need of mobility in networks and services leads to new networking paradigms including ad-hoc networking [27,31,39] where the mobile devices can establish direct and multi-hop communications among them.

DEFINITION 1. A mobile ad-hoc network (MANET) is a self-organizing network that can be spontaneously deployed by a set of mobile devices (laptops, phones, personal assistants) without requiring a fixed network infrastructure.

The deployment of an ad-hoc network is performed dynamically by the mobile devices (or nodes) themselves: each mobile node is both a terminal that communicate with the

HANDBOOK OF NETWORK AND SYSTEM ADMINISTRATION
Edited by Jan Bergstra and Mark Burgess

others and a router that can forward packets on behalf of the other nodes. If two network nodes cannot communicate directly because they are not localized in the same one-hop neighborhood, a multi-hop communication is established using intermediate nodes. Ad-hoc networks provide localized and temporal-limited connectivity in dedicated target environments: search and rescue services, intervehicular communications, last mile Internet delivery and geographically challenging environments. The emergence of new applications is favored by the advances in hardware technologies and by the standardization of ad-hoc routing protocols.

Compared to fixed networks, ad-hoc networks permit to reduce the costs of deploying and maintaining the infrastructures but they present other additional constraints such as:

- Dynamic topology: such networks do not require a fixed infrastructure. They are spontaneously deployed from mobile nodes and are then highly dynamic, since nodes might come and go based on user mobility, out-of-reach conditions and energy exhaustion.
- Heterogeneity of devices: a typical ad-hoc network is composed of various devices such as laptops, personal digital assistants, phones and intelligent sensors. These devices do not provide the same hardware and software capabilities but they have to interoperate in order to establish a common network.
- Distributed environment: the network running is based on the cooperation of independent nodes distributed over a given geographical area. The network maintenance tasks are shared among theses nodes in a distributed manner based on synchronization and cooperation mechanisms.
- Resource constraints: bandwidth and energy are scarce resources in ad-hoc networks. The wireless channel offers a limited bandwidth, which must be shared among the network nodes. The mobile devices are strongly dependent on the lifetime of their battery.
- Limited security: ad-hoc networks are more vulnerable than fixed networks because of the nature of the wireless medium and the lack of central coordination. Wireless transmissions can be easily captured by an intruding node. A misbehaving node can perform a denial of service attack by consuming the bandwidth resources and making them unavailable to the other network nodes.

The key concept of ad-hoc networking is the routing mechanism. Ad-hoc networks rely on a routing plane capable to fit with their dynamics and resource constraints. In particular, an ad-hoc routing protocol must be capable to maintain an updated view of the dynamic network topology while minimizing the routing traffic. Numerous ad-hoc routing protocols were proposed in the last ten years. The MANET working group [21] manages the standardization of ad-hoc IP routing protocols at the IETF and classifies them in two major routing approaches: reactive routing and proactive routing.

DEFINITION 2. A reactive routing protocol is a routing protocol which establishes the routes in an on-demand manner only when a route is requested by a node, by initiating a route discovery process.

In the reactive approach, a source node had to flood a request for a route to a destination node in the ad-hoc network. The intermediate nodes are discovered and memorized during the flooding. When the destination node receives the request, it uses this reverse path to contact the source node and to provide the requested route. A typical example of reactive protocol is the Ad-hoc On-demand Distance Vector (AODV) protocol described in [30].

DEFINITION 3. A proactive routing protocol is a routing protocol which periodically maintains the network topology information on each network node, by managing local routing tables.

In the proactive approach, each node maintains a routing table that is regularly updated by exchanging topology information with the other nodes. An example of a proactive protocol is the Optimized Link State routing protocol (OLSR) [15]. Each node performs two main operations: it determines the list of direct-connected neighbor nodes by accomplishing link sensing through periodic emission of beaconing hello messages and exchanges link state information with the other nodes by flooding topology control messages. The OLSR heuristic reduces the control overhead by selecting a subset of nodes called Multi-Point Relays (MPRs) responsible for forwarding broadcast messages during the flooding process.

The two routing approaches do not present the same properties and the choice of the one or the other is done according to the particular application. The reactive routing approach reduces significantly the control overhead since it does not require to flood topology information periodically. However, the proactive routing approach offers a lower latency as the route discovery has not to be done. Hybrid routing protocols provide a trade-off between control overhead and latency.

While the routing plane is progressively converging towards the use of standardized protocols, the management plane raises interesting research challenges towards providing a management framework to ad-hoc networks.

## 1.1. *Management requirements*

A fixed network management model is not directly appropriate for ad-hoc networks. New management paradigms and architectures must be defined to cope with the dynamics and resource constraints of such networks. Managing ad-hoc networks presents us with different research issues which are not encountered in the same way in the common fixed networks:

- Management domain related: since ad-hoc networks are dynamically formed by loosely coupled entities, the notion of administrative domain, responsible for the overall management is difficult to define. Although, from a technical point of view, cooperative and role-based schemes are possible such that the manager/agent roles are assigned dynamically, the management layer has to implement higher level business policies, which are hard to define in blurry shaped administrative domain.
- Relevant management information: what should be monitored and what actions to be taken is still unknown in this case. The challenge is to identify the essential pieces of information relevant in this context and to determine the possible corrective configuration operations in a close-loop management paradigm.

- Management agent reliability and willingness to cooperate: the management agent is constrained by its inherent limits (connectivity, battery, failures) so that the provided monitored data might be biased. Additionally, malicious or non-cooperative nodes might provide false data or no data at all. The key issue is to define management paradigms, where biased or faulty information can be out-weighted or discarded as such.
- Cost of management: resources (essentially network bandwidth and battery power) are already scarce in ad-hoc networks and monitoring and management actions are additional consumers in an already resource-limited context. Defining lightweight schemes, where already available information is used and resource availability and consumption can be minimal, are the main issues.
- Self-organization: the dynamic nature of mobile ad-hoc networks requires that the management plane is able to adapt itself to the heterogeneous capacity of devices and to the environment modifications. The ad-hoc network is deployed in a spontaneous manner, the management architecture should be deployed in the same manner in order to minimize any human intervention.
- Robustness and scalability: the management plane must be easy to maintain and must remain coherent, even when the ad-hoc network size is growing, when the network is merging with an other one or when it is splitting into different ones.

New management paradigms were considered to deal with the research issues posed by ad-hoc networks. They also had to be integrated into the common management approaches of fixed networks.

## 1.2. *Management organizational models*

An ad-hoc network corresponds to a distributed environment where different organizational models can be considered for the management purposes. A management organizational model defines the role (manager or agent) of each component of the management plane and specifies how each component interacts with the others. Typically, we can distinguish four different management organizational models presented on Figure 1:

- Centralized model (Figure 1(a)): a single network manager manages all the nodes in the whole network. The significant advantage of this model is to simplify the management task by considering a single point of control. However, the centralized model can generate a high management traffic and is not robust since the failure of the manager leads to the failure of the whole management system.
- Centralized hierarchical model (Figure 1(b)): the network manager uses local managers as intermediary to perform some management tasks. Each local manager manages a subset of network nodes and is assigned with a given degree of responsibility. The manager is the central point of control since it has the highest degree of responsibility.
- Distributed model (Figure 1(c)): the management is performed by a set of managers communicating according to a peer-to-peer scheme. Each manager has the same degree of responsibility and is fully responsible for managing a subset of nodes in the

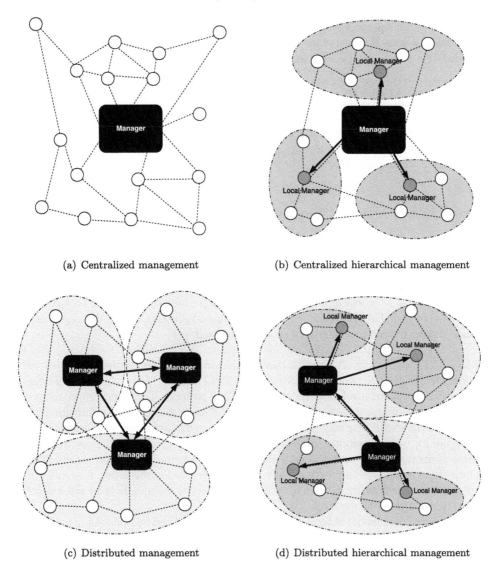

(a) Centralized management

(b) Centralized hierarchical management

(c) Distributed management

(d) Distributed hierarchical management

Fig. 1. Organizational models for ad-hoc network management.

network. The usage of multiple managers reduces the overhead of management messages and improves the robustness of the management system. However, a cooperative mechanism must be defined in order to ensure the coherence of the management operations taken by managers.

- Distributed hierarchical model (Figure 1(d)): this model is a distributed model where each of the managers can delegate part of its management tasks to local managers.

These organizational models are not specific to ad-hoc networking but are recurrent in network management. Because of the nature of ad-hoc networks, the role assignation and the relationships among components are highly dynamic compared to regular fixed networks.

## 2. Management paradigms

This section presents the main management paradigms for mobile ad-hoc networks and describes the underlying architectures able to cope with the constraints of such dynamic and resource scarce networks.

### 2.1. *SNMP-based management*

Current network management systems use a set of available management protocols (CLI [6], SNMP [38], XML-based [18]) to exchange management information between a network manager and the agents. The Simple Network Management Protocol (SNMP) [38] is among the most common used protocol for network monitoring and configuration. Integrating ad-hoc networks into common agent/manager models require the compatibility with this protocol, but also requires in the same time a reorganization of the management plane in order to limit the underlying network overhead.

**2.1.1.** *ANMP approach*    Among the pioneering management approaches for ad-hoc networks, the Ad-Hoc Network Management Protocol (ANMP) [14] architecture defines a centralized hierarchical solution based on the SNMP protocol. The compatibility relies on using the same protocol data unit structure and the same management information base structure as in SNMP. The major difference comes from the architectural model (presented in Figure 2), which combines clusters and hierarchical levels. ANMP comes within the scope of management by delegation. It clusters the management plane and organizes it in a three-level hierarchy composed of: individual managed nodes interacting as agents at the lowest level, clusterheads interacting as local managers at the middle level and a central network manager at the top level.

The hierarchical model reduces the network traffic overhead by collating management data at the intermediate levels, but is limited to three hierarchical levels in order to facilitate the maintenance of the hierarchy in such dynamic networks. Two clustering algorithms are defined at the application layer to organize the management plane:

- The first approach is based on graph-based clustering and forms clusters of one-hop neighbor nodes based on the network topology. In the network topology seen as a graph, each mobile device is represented by a node identified by a unique ID (MAC address, serial number) and a bidirectional link exists between two nodes if and only if they are within the transmission range of each other.

  The algorithm performs in a distributed manner the organization of one-hop clusters where the node with the minimum ID among its neighbors is elected as the clusterhead. Each node implements internal data structures including the list of one-hop neighbors, the list of nodes in the cluster and a ping counter indicating last time a ping was received from the clusterhead.

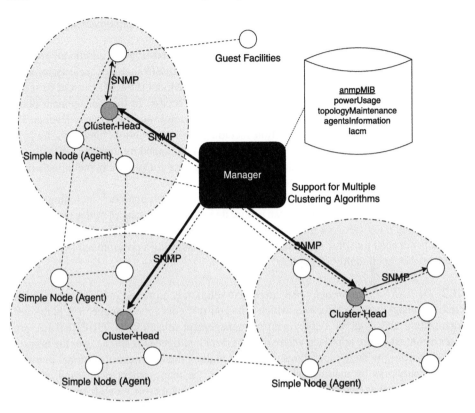

Fig. 2. ANMP architecture.

- The second approach called geographical clustering forms clusters of up to three-hop
  neighbors based on the spatial density. The approach requires that node positions are
  known through a geographical positioning system such as GPS. It is based on two
  algorithms: a centralized periodic processing for cluster definition and a distributed
  maintenance processing to manage node mobility.

  The centralized processing considers the ad-hoc network as a rectangular box and
  calculates horizontal and vertical stripes in order to split the network into rectangular
  clusters of different sizes (up to three times the transmission range) but of homogeneous spatial density. In each of these rectangular clusters, the ad-hoc node located in
  the box center is defined as the clusterhead.

  The clustering processing is executed periodically and a maintenance processing
  manages node mobility between two clustering updates. The maintenance processing
  performs at the cluster level defines an announcement protocol for nodes to leave and
  join a cluster and provides guest facility in each cluster for nearby unmanaged nodes.
ANMP is not limited to the two presented approaches but can be extended to other clustering algorithms. In particular, while ANMP makes a distinction between clustering at
the application layer for management purposes and clustering at the routing layer, it may

be suitable to propose a lightweight management approach by piggybacking an ad-hoc network routing protocol.

A management information base (MIB) is defined by ANMP as an additional group to the standardized MIB-II of SNMP. This group called *anmpMIB* includes security information data to implement the Level-based Access Control Model (LACM) in order to secure management operations and management information access. In the management plane, each node has a clearance level: the higher the level is, the lower the security level is. The group also defines topology information including two information subgroups for each clustering approach, and agent information statistics maintained by managers. Finally, the group describes objects related to device battery and power consumption, including a drain function of the battery used for predictive purposes.

ANMP defines a complete SNMP-compatible management approach including a security model. The clustering schemes can ensure that over 90 percents of nodes are managed while keeping a low network management traffic overhead. ANMP introduces a first identification of a need for some active code facility by considering a programmable framework to extend the functionalities of agents.

**2.1.2. *GUERRILLA approach***   An alternative solution is proposed in [37] by the GUERRILLA management architecture, which relies on two concepts: management plane self-organization and dynamic deployment of management related code. GUERRILLA proposes a context-aware approach where management intelligence is spread over the network nodes according to their capabilities. It aims at making management operational in highly dynamic networks by maintaining connectivity in the management plane (and enabling disconnected management) and by minimizing management costs on node resources. This distributed architecture (presented in Figure 3) is composed of nomadic managers (higher tier) and active probes (lower tier):

- A nomadic manager maintains the connectivity in the management plane with the other managers and updates the management plane configuration in an autonomous way.
- Deployed by a nomadic manager, an active probe is an extension to an SNMP agent. It integrates an execution environment for hosting monitoring scripts and performs the localized management operations.

The clustering algorithm relies on the Cluster-based Topology Control (CLTC) approach which minimizes the consumption of the node resources. Moreover, the nomadic managers implement an auto-configuration mechanism to determine the management actions to be executed in an adaptive manner. In a formal way, a nomadic manager determines the management action $a$ – from a set of possible actions $A$ – that maximizes the objective function $f(a)$ defined by equation (1).

$$f(a) = \frac{U(T(s, a)) - U(s)}{C(a)}. \tag{1}$$

In this equation, $s$ represents the view of the domain state perceived by the manager and $T(s, a)$ is a transition function that provides the new state of the domain predicted by the manager if the action $a$ is executed. $U(s)$ defines a utility function possibly derived

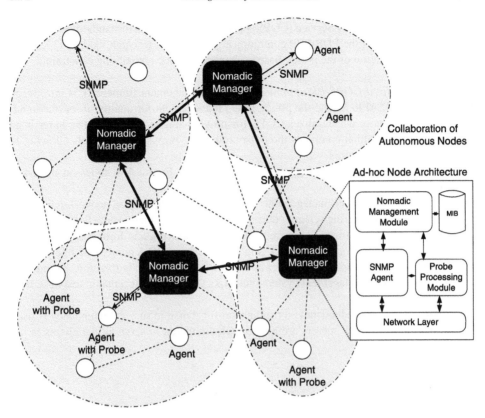

Fig. 3. GUERRILLA architecture.

from policies, while $C(a)$ represents the management cost function. By maximizing the function $f(a)$, the nomadic manager aims at improving the managed environment utility with a minimal cost.

GUERRILLA comes within the scope of management by delegation. It proposes fundamental good assumptions such as self-organization, disconnected management and takes into account heterogeneous devices.

### 2.2. *Policy-based management*

Policy-based management (PBM) consists in defining high-level management objectives based on a set of policies that are enforced in the network. Architectures for policy-based management rely on two main functional components:
- the Policy Decision Point (PDP) component which aims at converting the high-level policies into low-level device-dependent policies and performing their distribution,
- the Policy Enforcement Point (PEP) component which enforce the low-level policies they receive from the PDP component.

The Common Open Policy Service (COPS) protocol was specified to standardize the distribution policy through the PDP-PEP communications. Several research work in [13,26,33] focus on the adaptation of the policy-based scheme to ad-hoc networking constraints.

**2.2.1.** *Decentralized COPS-PR approach*    A first management framework is introduced by Phanse in [33,34] to extend the policy-based management for quality of service in ad-hoc networks. Phanse decentralizes COPS for provisioning (COPS-PR). The latter is an extended protocol which combines the outsourcing and provisioning models allowing to support hybrid architectures.

This distributed hierarchical approach is described on Figure 4. It relies on three key mechanisms:

- Clustering and cluster management: the clustering of the ad-hoc network relies on a simple k-hop clustering scheme: during the initial deployment, some Policy Decision Point nodes called super-PDPs are present in the network and form clusters with the PEP client nodes located in the k-hop neighborhood. The number of intermediate nodes between a super-PDP server and its PEP client nodes is therefore bounded to k-hops and the management system can expect to provide acceptable response time performance.
- Dynamic service redundancy: a mechanism for dynamic service redundancy is used when the super-PDP servers are not enough numbered to serve all the clients in the

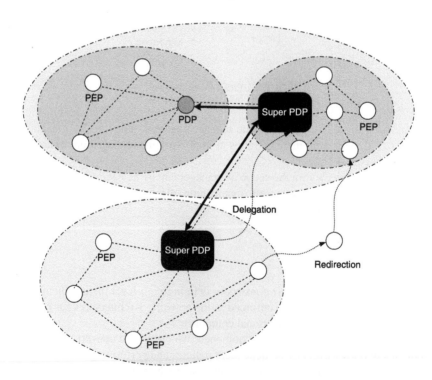

Fig. 4. Policy-based management with decentralized COPS-PR.

network, or when clients are moving from one cluster to another one. It consists of a delegation service and a redirection service. The delegation service answers the lack of super-PDP servers in the network, by allowing a super-PDP server to delegate part of its decisions to another node interacting as a regular PDP server. In this case, the super-PDP server keeps a hierarchical control on the regular PDP servers. The redirection service manages the PEP clients moving among network clusters by allowing a PDP server to dynamically redirect a client to a more appropriate less distant PDP server.

- Service location discovery: a discovery mechanism is proposed to automatically discover new policy servers but it is not yet designed.

The management framework supports multi-domain policy negotiation for PEP client visitors from other domains. An inter-PDP server signaling is introduced in COPS messages to exchange policies among PDP servers from different domains. If the visited PDP server cannot find policies for a visitor PEP client, it queries the PEP client to get its home PDP server address and negotiates the policies with the home PDP server.

Phanse's approach is an innovating solution for policy management in ad-hoc networks, which was prototyped and validated experimentally in a multi-hop environment. However, the considered communication protocol COPS-PR is not widely used in today's network management systems. Some improvements could be possible for the multi-domain policy negotiation mechanism. In particular, the policy could be transfered during the redirection and the PEP clients could come with their own policies.

**2.2.2.** *DRAMA approach*   Burgess evaluates the scalability of policy-based configuration management for ad-hoc networks in [9]. The performance evaluation relies on an analytical framework which takes into account the unreliable nature of the communication medium and the mobility factors in ad-hoc networks.

DRAMA defines a centralized hierarchical policy-based management system for ad-hoc networks in [12,13]. This three-level architecture described on Figure 5 is composed of a set of policy agents having the same basic structure. The Global Policy Agent (GPA) at the top level manages multiple Domain Policy Agents (DPA). A Domain Policy Agent manages in turn multiple Local Policy Agents (LPA). Network management policies are propagated hierarchically from the GPA agent to the DPAs and from the DPAs to the LPAs.

The conceptual architecture is relatively simple, but allows to perform an exhaustive set of experiments: the management system was prototyped and demonstrated in a realistic environment to illustrate several use cases including CPU utilization reporting, server reallocation upon failure, reconfiguration of bandwidth allocation.

**2.2.3.** *Node-to-node approach*   A node-to-node approach is defined in [26] in the framework of corporate ad-hoc networks presenting uniform management and company-wide policies.

DEFINITION 4.  The node-to-node policy distribution defines a fully distributed policy distribution where a network node exchanges and synchronizes the policies with the neighboring nodes (see Figure 6).

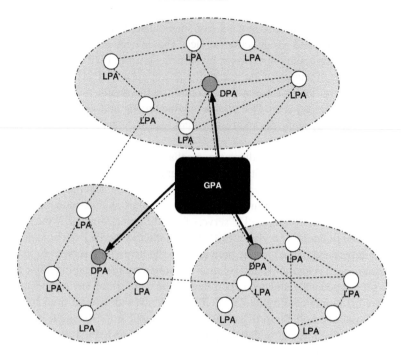

Fig. 5. DRAMA architecture.

This distributed architecture works over the DiffServ quality-of-service model providing the network layer differentiated service: the communications among network nodes can be marked with different priority levels.

In this architecture, each ad-hoc device exports both DiffServ functions: the edge router monitors and marks local application traffic and the core router classifies and schedules routed packets. From the management plane point of view, the ad-hoc devices embed both PDP and PEP functionalities including a local policy repository to store the policy rules applicable to the device.

The key concept of the architecture is based on a two-phase policy distribution using a synchronization mechanism:

- In the first phase of the distribution, the ad-hoc device of the administrator interacts as a Policy Decision Point (PDP) and distributes the policies to the other devices behaving as Policy Enforcement Point (PEP) and storing the policies on their local repository. This phase is not sufficient to configure all the ad-hoc nodes if these were not connected to the network at the distribution time.
- To propagate the policies to all the devices, the second phase consists in a permanent synchronization among two neighboring nodes. An already-configured node can interact as a local PDP to configure a device which has not received the latest policies.

An authority mechanism allows a management hierarchy with delegation. Each manager has an authority level known by all. Network policies are tagged with manager identifiers. The more authoritative policies always replace the less authoritative ones. The approach

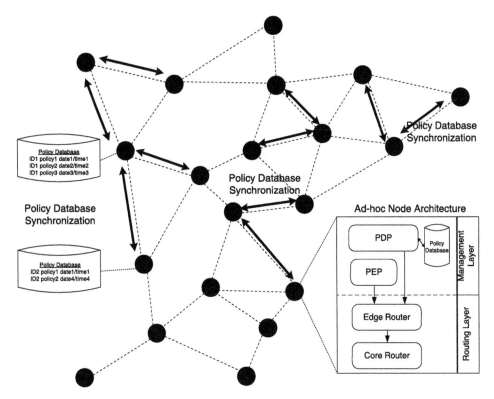

Fig. 6. Node-to-node policy-based management.

describes an interesting device-level management framework for enterprise networks. It assumes that in the future quality-of-service mechanisms will be available at the MAC layer. However, the synchronization protocol and the consistency problems of the policies are not discussed.

### 2.2.4. *Scalability evaluation*

HYPOTHESIS 1. Formally, an ad-hoc network can be represented by a connectivity matrix $A$ where rows/columns stand for an ad-hoc node. A matrix entry $A[i, j]$ indicates whether node $i$ is connected to node $j$. The connectivity among two peers of nodes undergoes time and mobility caused variations in the context of ad-hoc networks. The matrix entry value can be therefore defined by a probability value.

The connectivity matrix serves as an analytical basis for analyzing the network traffic overhead generated by different policy-based management models and for evaluating their scaling behavior.

The organizational models presented in Section 1 are refined for the context of policy-based management and leads to a set of six different management models. Each of these models corresponds to a different degree of decentralization with respect to three main criteria:

- Policy decision: the deployment of policies can be controlled by a central manager or by a set of distributed managers. The final hosts can either strictly respect these policies or can freely modify some aspects of them.
- Policy exchange: the policies can be received from a manager, from an intermediate manager in a hierarchical manner, or from another node in a node-to-node scheme.
- Policy enforcement: the policies can be transmitted by a manager or can be locally enforced by the hosts.

The scalability analysis relies on a key assumption: the probability of a successful configuration depends on the regularity of the maintenance operations, which in turn depends on the reliability of the communications between the policy source and the hosts. In order to estimate the efficiency of the configuration, the average rate of errors generated by an host is compared to the average rate of repair maintenance operations which can be done in function of the communication channel size and reliability. The configuration is successful if and only if all the host errors have been repaired. Therefore, the average rate of possible repair operations must be higher than the average rate of host errors.

The results of this analysis provide a stochastic quantification of the scalability of policy-based management schemes. They confirm the limits of a centralized management due to the communication bottleneck caused by the centralized manager. Distributed management schemes offers a better robustness and scalability but introduce policy convergence issues.

### 2.3. Middleware for management

An ad-hoc network is an heterogeneous set of nodes with diverse capabilities in terms of energy and processing facilities but also in terms of operational protocols and services. A programmable middleware defined by [20] provides a common communication infrastructure capable to align dynamically the software capabilities of network nodes. The middleware defines a programmable scheme where each node is able to share, download and activate loadable plugins offering specific operational functionalities such as an application service or a routing protocol. The key idea is to elect a common set of loadable plugins and to distribute it among the network nodes in order to facilitate their communications.

The middleware for management is organized into clusters, where the cluster head coordinates the election and distribution of loadable plugins (see Figure 7). In a given cluster, each node exposes several loadable plugins and stores locally the executable code in its plugin repository. The cluster head initiates the election process by contacting the cluster nodes and requesting them to advertise the candidate plugins. Based on the set of advertised plugins $P$, the cluster head uses an election algorithm which determines the most appropriate(s) plugin(s) for the given context. The selection of plugins

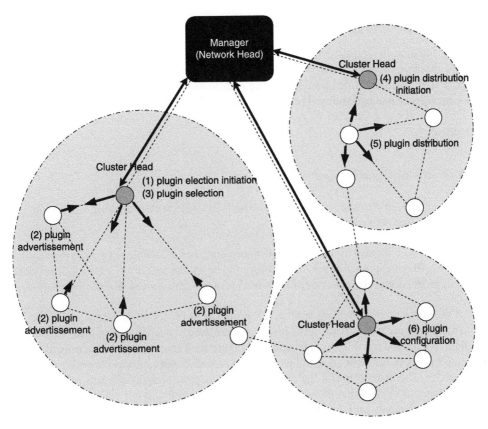

Fig. 7. Middleware for management.

aims at selecting the plugins which minimize the cost function $g(p)$ defined in Equation (2).

$$g(p) = \sum_{i=1}^{n} w_i \times C_i(p),$$

$$\text{where } w_i \in [0, 1], \sum_{i=1}^{n} w_i = 1 \text{ and } C_i \in [0, 10], \forall i \in [1, n].$$

(2)

The characteristics $C_i(p)$ of a given plugin $p$ are rated from 0 to 10 and are assigned with a mathematical weight $w_i$ by the cluster head. This cost function considers the characteristics such as CPU usage, memory consumption and plugin size to converge to a common basis of plugins among the cluster nodes. The cluster head initiates then the plugin flooding by requesting the cluster nodes that own the plugins to distribute them to their one-hop neighbors in the cluster in a peer-to-peer manner. Each cluster mate stores the common plugins into the plugin repository and activates them. The cluster head is capable to configure and to modify on-the-fly the parameters of the current activated plug-ins.

The middleware infrastructure was implemented by combining a lightweight XML-based message-oriented protocol and the Java platform for small and medium devices. The first evaluation performance shows the management middleware is efficient in terms of convergence and scalability. The middleware was limited to a centralized scenario with a single cluster and must be extended to multiple clusters with cluster-to-cluster communications.

### 2.4. *Probabilistic management*

A pure and sound management approach, where all nodes are managed at any time is too strict for ad-hoc networks.

DEFINITION 5. Instead of addressing the management of the whole network, a probabilistic management approach focus on a subset of network devices or on a subset of network parameters to derive the behavior of the whole network with a given probability guarantee.

A probabilistic management approach is proposed in [3] for ad-hoc networks based on probabilistic guarantees. A subset of nodes is considered to be managed in order to provide a light-weight and reliable management (see Figure 8). These nodes are determined based on their network behavior so as to favor subsets of well-connected and network participating nodes. The assumption of the probabilistic management approach was motivated by the simple observation that ad-hoc network management should not comprise the totality of network elements (as it is typically the case in fixed networks).

The requirements on the management plane are relaxed and are limited to a subset of nodes called a spatio-temporal connected component. This component is defined as a subset of the ad-hoc network, such that nodes within such a component have a high probability of being directly connected. The term spatial derives from this neighborhood notion, while the term temporal is related to the temporal behavior of this neighborhood. In a store and forward oriented architecture, such nodes are also capable to inter-communicate at a higher hop-count. The probabilistic management is limited to the first and eventually second largest spatio-temporal connected component. With respect to such a selective management scheme, probabilistic guarantees are determined in function of the percentage of nodes to be managed. The underlying algorithmic describes the extraction of spatio-temporal components from a probability matrix similar to the matrix $A$ defined in Section 2.2.2. The election of manager nodes in each component can be performed with a centrality measure: the most central nodes of a component are elected as managers.

DEFINITION 6. The centrality is a measure capable to quantify the relative importance of a node in a network or a graph structure. Several variants of this measure have been defined in social networking [7].

A spatio-temporal component is a non-oriented connected graph associated to an $m$-dimensional matrix $M$. A typical centrality measure that can be applied on this graph is

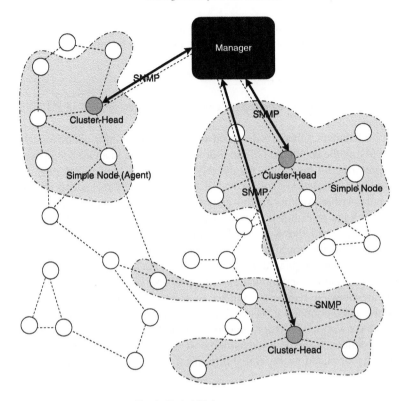

Fig. 8. Probabilistic management.

the degree centrality $c_{\text{degree}}$ defined by Equation (3). A node $i$ is considered as central if it is directly connected to a maximal number of other nodes.

$$c_{\text{degree}}(i) = \sum_{j=1}^{j=m} M(i, j) \quad \text{for a given node } i. \tag{3}$$

The degree centrality is a local measure limited to the one-hop neighborhood information. An alternative centrality measure is the eigenvector centrality defined by Bonacich in [4]. The key idea is to define the centrality measure in a recursive way: the more a node is directly connected to central nodes, the more the node is central. For such a non-oriented connected graph, Bonacich demonstrates that the problem can be reduced to the typical eigenvector problem defined by Equation (4).

$$M \times \vec{x} = \lambda \times \vec{x}. \tag{4}$$

The vectors that solve this equation are called eigenvectors and are paired with corresponding eigenvalues. The principal eigenvector $\vec{v}_{\text{principal}}$ is the solution vector which is paired with the highest eigenvalue. The eigenvector centrality $c_{\text{eigenvector}}$ of node $i$ corresponds to the $i$th element of the principal eigenvector of matrix $M$, as mentioned in Equation (5).

$$c_{\text{eigenvector}}(i) = \vec{v}_{\text{principal}}(i) \quad \text{for a given node } i. \tag{5}$$

A generalized version of the eigenvector centrality has been defined in [5] in the case of oriented connected graphs.

Another probabilistic scheme is designed at the level of the information model in [16]. It aims at improving the acquisition of monitoring information in an ad-hoc network. In this scheme, monitoring information is not directly queried from the network devices but must be processed by a correlation-based probabilistic model. The direct access to raw monitoring data is not possible. An information query must be formulated to the acquisition module which integrates the probabilistic model. This model relies on time-varying multivariate normal distributions and allows to correlate multiple raw monitoring data through statistical techniques in order to improve the information reliability. Disparate and faulty monitoring raw data may generate biased approximated information. The correlation mechanism thus provides information approximations with probabilistic confidences.

The scheme allows to reduce the number of nodes involved in the data monitoring according to the asked confidence. The architecture defines a novel approach to integrate an information base with a correlation probabilistic model and provides a trade-off between information confidence and data acquisition cost. However, the efficiency of the architecture is directly correlated to the availability of efficient probabilistic models for a given monitored metric.

## 3. Application domains

This section details the research work addressing management applications in MANETs including network monitoring, configuration of IP addresses and control of node cooperation.

### 3.1. *Monitoring*

Monitoring ad-hoc networks resides in measuring performance metrics, collecting the measurement data and processing these measurements.

**3.1.1.** *WANMON approach*   A local monitoring approach called WANMon (Wireless Ad-Hoc Network Monitoring Tool) is proposed in [29] for determining the resources usage of an ad-hoc node. The architecture is locally deployed such that an agent is responsible for monitoring the resources of an ad-hoc device. It monitors the resource consumption generated by ad-hoc network-based activities: network usage, power usage, memory usage and CPU usage. In particular, the tool aims at distinguishing both the resources consumed for the node itself and the resources consumed by the routing activity on the behalf of other nodes. WANMon is one of the first monitoring tools dedicated to ad-hoc networking, but presents some limitations. The tool cannot provide real-time information and is limited to a local monitoring activity.

**3.1.2.** *DAMON approach*   DAMON described in [35] defines a Distributed Architecture for MONitoring multi-hop mobile networks. This generic monitoring system is composed of monitoring agents and of data repositories called sinks that store monitored information (see Figure 9). A centralized deployment with a single monitoring sink is supported for small networks, but DAMON provides also in a distributed scheme seamless support for multiple sinks for larger networks. In DAMON, each agent is hosted by an ad-hoc device, monitors the network behavior and sends the monitoring measurements to a data repository. DAMON defines a robust solution adapted to device mobility by supporting sink auto-discovery by agents and resiliency of agents to sink failures. The auto-discovery of repositories relies on a periodic beaconing mechanism initiated by sinks to advertise their presence and broadcast instructions to agents. A network agent receives the beacon messages, maintains a list of available data repositories and associates with the closest sink as its primary repository in order to send the collected information. The distance between the agent and the primary repository is possibly more than one hop. In this case, monitoring data from multiple agents can be aggregated by an intermediate agent to facilitate the data collection. DAMON distinguishes two types of monitoring information: time depen-

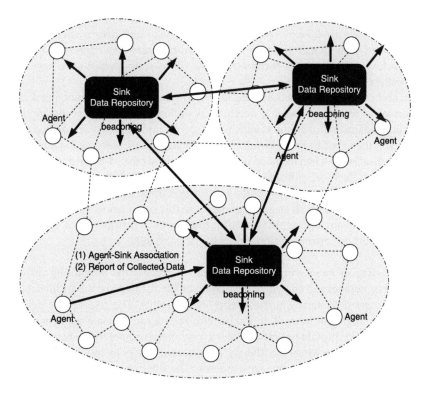

Fig. 9. Distributed monitoring with DAMON.

dent information (list of neighbors, energy left on a device) which require to be frequently transmitted to sinks and time independent information (packet logs, daily traffic statistics).

The resiliency to sink failures relies on a simple switching operation to a different sink when the primary sink undergoes a failure. An agent can detect the failure of the primary sink because of the absence of beaconing messages on a given time period. In this case, the agent updates the list of available data repositories and switches to a new primary sink in order to send the collected information. In order to prevent intermittent switching between two sinks, the agent can modify its primary repository only after the successful reception of a determined number of beaconing messages. The architectural framework of a monitoring agent is composed of a packet classifier module, categorizing in input the packets according to their types and a set of packet handler modules performing specific processing on the received packets such as beacon listening and time dependent information aggregation.

DAMON proposes a distributed scheme for monitoring ad-hoc networks based on multiple data repositories. A complete implementation of DAMON was demonstrated with the on-demand distance vector (AODV) routing protocol. But, the architecture supports a wide range of other protocols and devices. While the solution is robust and maintainable with little effort, the proximity-based association mechanism should be improved to ensure an homogeneous distribution of agents with sinks.

**3.1.3.** *Synchronization of monitoring data* By definition an ad-hoc network is dynamic and temporal such that synchronization mechanisms are required to maintain the coherence and to maximize the life time of monitoring information in such a distributed environment. The problem of time synchronization in sparse ad-hoc networks cannot be solved with regular clock synchronization algorithms because of the restriction of node resources and of the unreliability of network links.

A synchronization method is described in [36] to synchronize information in distributed ad-hoc networks with a low message overhead as well as a low resource consumption. The solution is an adaptation of a time stamps based algorithm using unsynchronized clocks of local nodes. If the monitoring information is not duplicated, the information maintained by a local node disappears when the node is failing or is leaving the ad-hoc network. A dissemination process is therefore defined in [41] to both optimize the information survival within the network and minimize the cost of the information dissemination. A monitoring information should be defined with an expiration time as this information cannot be maintained indefinitely without any update. In particular, the statistical distribution of timescales and the impact on ad-hoc network performances is discussed in [17].

**3.1.4.** *Processing of monitoring data* In order to process the monitoring information from an ad-hoc network, an analytical method based on distributed contrast filtering mechanisms and graph dependency analyzes is proposed in [2]. The goal of this information processing is to highlight network traffic patterns in order to evaluate the network behavior. The contrast filtering can be performed in a centralized manner by the network manager but a distributed scheme where the filtering is processed locally by individual nodes is preferred for scalability and performance purposes. The filtering process aims at making in evidence disparity among nodes. Applied for a given network metric, it consists in a local comparison of the normalized value of the metric among network nodes located in a same neighborhood. A concrete usage relies on applying the contrast filter on the routed packet number per node in order to reveal routing paths and to detect network nodes interacting in the routing task.

A second method is defined in the same paper based on dependency graphs to estimate node influence in an ad-hoc network. This more refined approach is a direct adaptation to ad-hoc networks of the work described in [8] related to the estimation of nodes influence in fixed wired networks. The method is initially inspired from research in social sciences, where social relationships are studied to detect individuals capable to influence important parts of a social network. In an ad-hoc network, the graph analysis identifies the core ad-hoc nodes participating in the network functioning. The analytical approach in [2] comes within the scope of a more global analytical framework introduced in [9] for modeling ad-hoc networks and for quantifying their behavior. However, the experimental work must be thorough with multiple scenarios implying an extensive set of network parameters.

Corrective management operations can be decided from the monitoring information. For instance in [1], a network-wide correction can be defined based on a local monitoring to improve a quality-of-service differentiated service. In this approach, each network node monitors independently the rate of high priority flow. When the rate falls out of a given threshold value, a network node is capable to signal corrective operations to the other nodes such as a selective rejection of lower priority messages in order to achieve quality-of-service guarantees. Many other configuration operations can be envisioned to optimize network performance in a close-loop management framework.

## 3.2. *Configuration*

The configuration of ad-hoc networks [32] must be performed in a self-organized manner to ensure robustness and scalability. A basic expectation in configuring ad-hoc networks lies on the autoconfiguration [25,40] of network devices including the allocation of their IP addresses. While in a fixed wired network the device configuration is often done in a centralized manner using a dynamic host configuration (DHCP) server, such a centralized approach with fixed allocated addressed server cannot be deployed in ad-hoc networks. Alternative address allocation approaches are needed based on a distributed scheme to ensure that two nodes do not have the same address in an ad-hoc network. Allocation approaches are distinguished in [42] into three categories: conflict-free allocation, conflict-detection allocation and best effort allocation.

**3.2.1.** *Conflict-free IP address configuration*   The conflict-free IP address configuration consists in ensuring an IP address of a new node is not already assigned. This is done by using dedicated disjoint pools of addresses. A typical example of conflict-free allocation can be outlined with the IP version 6 stateless autoconfiguration. The latter relies on the statement that dedicated pools of MAC level addresses are assigned to network card producers such as an ethernet card is uniquely identified. This address can be used as a unique identifier to define the IP address in order to avoid an allocation conflict. Indeed, the stateless mechanism allows the mobile node to generate its own unique IP address using a combination of the network prefix and a suffix which is locally computed based on the unique

MAC address. However, this hardware-based addressing scheme does not consider that the physical addresses of a mobile node can be easily modified by reprogramming the network cards.

**3.2.2.** *Conflict-detection IP address configuration*    The conflict-detection IP address configuration consists in selecting a supposed-to-be available IP address and assigning it after having requested a confirmation from the other network nodes that this address is effectively not in use. MANETconf [28] defines a conflict-detection distributed allocation protocol for ad-hoc networks (see Figure 10). No centralized server can ensure the uniqueness of the allocated address. The address allocation relies therefore on a distributed mutual exclusion algorithm performed by the network nodes. Each of the latter maintains the set of the IP addresses already allocated.

The assignment process is performed in an indirect way through an intermediate node called initiator node and is defined by the following process:

- In order to obtain an IP address, a new node chooses in its neighborhood a node already part of the network and requests him to interact as an initiator.
- The initiator node selects an IP address considered to be available for use according to its local view of allocated IP addresses.
- It then performs an address query on the behalf of the new node to the other network nodes.
- At the reception of the address query, a network node sends an affirmative or negative reply if he considers the address is or is not available.

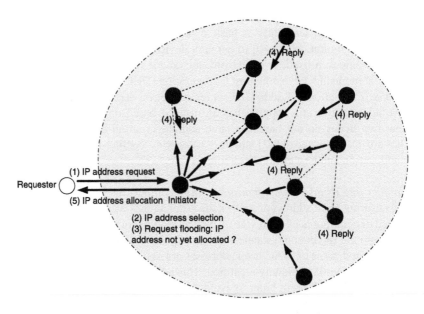

Fig. 10.  IP address configuration with MANETconf.

- If the initiator receives a positive reply from the network nodes, the new node is configured with the chosen address, otherwise a new address must be selected and queried by the initiator node.

MANETconf tolerates network partitioning and network merging. On one hand, the ad-hoc network partitioning is managed by an update mechanism: each time an address query is performed by a network node, the latter is capable to update the set of network nodes in his partition by identifying the nodes which did not send back any reply messages. The set of allocated addresses thus corresponds to the set of nodes part of the current partition. On the other hand, when a network merging occurs, the ad-hoc nodes from different partitions exchange and merge their set of allocated addresses. Conflicting addresses which are allocated to two nodes at the same time are identified and reconfigured by the allocation protocol.

MANETconf offers a robust protocol defining a decentralized dynamic host configuration protocol. It supports network partitioning and merging and also resolves concurrent initiation of address allocations. The solution considers a standalone ad-hoc network but could take advantage of considering an hybrid architecture where a gateway connected to an external network would be available. While MANETconf manages both message losses and node crashes, security issues such as malicious behavior of nodes remain open.

**3.2.3.** *Best-effort IP address configuration*    The best-effort IP address configuration is based on assigning by default a supposed-unused IP address to a node until a conflict is detected. The PROPHET [42] allocation defines such a best-effort scheme using the random properties of a stateful function noted $f(n)$. The stateful function is configured with an initial state value called seed and is capable to randomly generate sequences of different addresses such that the probability of having duplicate addresses is kept low. During the allocation of addresses, the function is propagated towards network nodes with different seed values as follows:

- In order to join the ad-hoc network, a new node contacts a network node and requests a new IP address.
- The node generates the new IP address using the stateful function $f(n)$ configured with its own seed. It generates also a seed value for the new node.
- The latter can in turn use the stateful function configured with this seed value to generate new IP addresses.

This distribution protocol combined with the properties of the stateful function guarantees a low probability of duplicate addresses.

PROPHET offers an allocation scheme adapted to large scale networks by defining a low communication overhead for address distribution. However, the design of the stateful function should be refined. The solution does not fully avoid address duplication and must be complemented by a conflict detection mechanism.

**3.3.** *Cooperation control*

The deployment of an ad-hoc network is based on the cooperation of mobile nodes and on the mutual utilization of their resources. Management mechanisms are therefore required to stimulate the participation of mobile nodes, such as their involvement to share the medium access and to perform packet forwarding on the behalf of the other nodes, but also to prevent the network overload by misbehaving nodes in order to guarantee a fairness cooperation.

**3.3.1.** *Medium access control*   A misbehaving node may not respect the contention resolution mechanisms defined by a wireless medium access control protocol in order to monopolize the bandwidth resources by forcing a not equitable share of the channel. A corrective scheme is defined in [22] over the 802.11 wireless protocol to detect and handle misbehaving nodes at the medium access layer. In this wireless protocol, the contention resolution mechanism requires that a mobile node waits a backoff time interval before starting the transmission of network packets. The backoff interval value is defined by a specific random process ensuring a fair share of the channel. The misbehaving node may force a short backoff interval value such that the channel is systematically assigned to the node before the other ones. The proposed scheme defined by some modifications of the wireless protocol consists in improving the resolution mechanism by implementing some control capabilities to the node which receives the network packets.

- In a regular scenario, the sender node which wants to transmit packets to a receiver node is the node waiting during the shortest backoff interval.
- In the corrective scheme, the receiver node is capable to monitor the packet traffic generated by the sender nodes and to detect among them a misbehaving node. The receiver identifies the deviations from the protocol by monitoring the frequency and duration of the access channel by sender nodes.

If a sender node deviates from the wireless protocol, the receiver node can in a first time penalize the sender by assigning him a larger backoff interval. If the sender node definitively does not respect the protocol, the receiver can refuse all the transmissions initialized by the misbehaving node.

The corrective scheme allows to improve a fair share of the wireless channel and facilitates the detection of misbehaving nodes by modifying the contention resolution mechanisms of the 802.11 wireless protocol. However, the solution focuses on a specific misbehavior detection and does not address for instance the case of misbehaving nodes using multiple physical addresses to acquire more bandwidth resources.

**3.3.2.** *Routing control*   An efficient cooperation model cannot be limited to the medium access layer but must also be defined at the higher network layers. At the routing layer, a control scheme is introduced in [23] in order to detect and limit the impact on the routing process of mobile nodes that agree to forward network packets but fail to perform the task.

This control architecture is composed of two main components both deployed over network nodes: a watchdog capable to monitor failing network nodes and a pathrater capable to mitigate routing misbehavior.

- The watchdog detects misbehaving nodes according to the following mechanism. After a mobile node forwards the packets to an intermediate node, it listens to the intermediate node to check out that this node continues the forwarding process. The watchdog component maintains a buffer of the recently sent packets and compares it to the network packets emitted by the intermediate node. If a network packet is kept

in the buffer more than a predefined timeout, the watchdog considers that the intermediate node fails to perform packed forwarding and sends in this case a notification message to the source node.
- The pathrater component receives the notification messages and maintains a rating of each network node to define their capability to forward network packets. High rating values are equivalent to misbehaving nodes. Both misbehavior rating information and regular link layer reliability information are combined to define a new route in the ad-hoc network.

The routing mechanism is improved by taking into account the behavior of intermediate nodes and by avoiding the definition of new routes through misbehaving nodes. However, the watchdog-pathrater model has some limitations. In particular, the misbehavior detection may be biased because of the physical environment conditions such as ambiguous collisions and limited transmission power.

**3.3.3.** *Economical model*  An economical model is proposed in [10,11] at the service layer in order to both stimulate the cooperation and prevent the network overloading. A virtual currency called nuggets is introduced to enforce service availability and in particular to reward the nodes participating in the packet forwarding service (see Figure 11). Two economical models are defined in this approach: the packet purse model and the packet trade model.

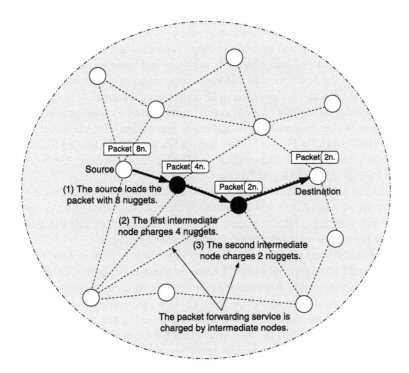

Fig. 11. Cooperation stimulation with a virtual currency.

- In the packet purse model, the packet forwarding service is covered by the source node according to the destination node to reach. The source node loads the network packet with a sufficient number of nuggets. During the network packet route, each intermediate node charges one to several nuggets from the packet for the forwarding service. The deployment of this model is limited by the problem of avoiding an intermediate node to charge all the nuggets loaded to a packet, and also to the difficulty of estimating the number of nuggets required to reach a given destination.
- In the packet trade model, the packet is traded by intermediate nodes in an hop-to-hop manner and does not require to carry nuggets. The packet to be forwarded is processed by an intermediate node in two steps. The packet is first bought from the previous node for a given nugget price and then is sold to the next node for an higher price. In this scheme, the total price of the forwarding process is payed by the destination node and the price of the forwarding process is not required to be estimate by the source node. However, the Packet Trade Model does not prevent a network node to overload the network by flooding.

## 4. Conclusions

Ad-hoc networks are a potential solution for the increasing need of mobility and ubiquitous computing. These self-configuring networks are spontaneously deployed by a set of heterogeneous nodes of various capacities and resources, without requiring any preexisting infrastructure. The dynamic and distributed nature of ad-hoc networks raises new challenges towards monitoring and controlling them. New management paradigms capable to cope with the constraints of such unreliable networks were presented in this chapter. They were completed by the description of different management application domains including network monitoring, auto-configuration of IP addresses and control of node cooperation.

From an organizational point of view, several models can be considered for the purposes of ad-hoc network management. The management approaches were classified according to the organizational model in Figure 12. The approaches aim at reorganizing the management plane with different degree of decentralization. The figure shows the predominance of two antagonist models for managing ad-hoc networks:

- The centralized hierarchical model: the centralized model simplifies the management operations by considering a single point of control but suffers from communication bottleneck with the central manager. The use of intermediate nodes in a hierarchical scheme reduce the management load for the central manager.
- The distributed (flat) model: it provides better scalability performance by using a set of distributed managers. However, it requires a minimum degree of autonomy from the network nodes, such that trust and convergence issues are posed.

From a chronological point of view, Figure 12 shows the publication date of each management approach. These approaches are relatively recent, since the first cited approach is ANMP which was published in 1999. Historically, the concept of ad-hoc networking is not new, since the first infrastructure-less multi-hop network was proposed in the 1970s with the ALOHA project [24]. However, the development of telecommunications has augmented the interest in ad-hoc networking for civil applications such that an impressive

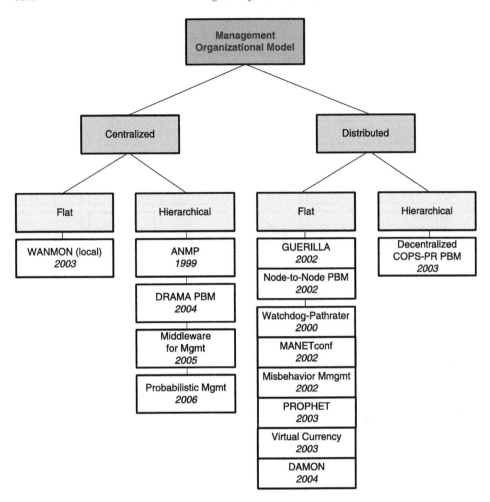

Fig. 12. Classification by management organizational model.

number of routing protocols were proposed in the last ten years. The recent standardization of ad-hoc routing protocols clarifies the context of ad-hoc networking for defining management approaches. It defines an easier way for the piggybacking and the instrumentation of ad-hoc routing protocols for management purposes.

From a functional point of view, the areas covered by each management approach are synthesized in Figure 13. The traditional classification defines five FCAPS functional areas: fault, configuration, accounting, performance and security. Fault management is not a well-covered area. Issues related to observability are challenging: fault management can be hindered by the impossibility to observe a given node in ad-hoc networks. A node that does not reply to legitimate polling in an fixed network can be considered as not functional. In an ad-hoc network, a node might not be reachable because it is moving and is out of reachability, or because it is not functioning properly. Configuration management

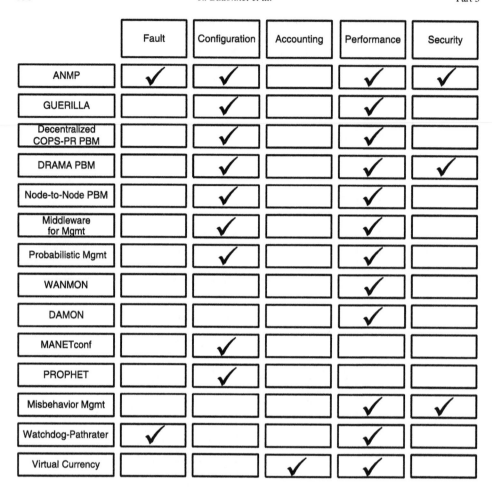

| | Fault | Configuration | Accounting | Performance | Security |
|---|---|---|---|---|---|
| ANMP | ✓ | ✓ | | ✓ | ✓ |
| GUERILLA | | ✓ | | ✓ | |
| Decentralized COPS-PR PBM | | ✓ | | ✓ | |
| DRAMA PBM | | ✓ | | ✓ | ✓ |
| Node-to-Node PBM | | ✓ | | ✓ | |
| Middleware for Mgmt | | ✓ | | ✓ | |
| Probabilistic Mgmt | | ✓ | | ✓ | |
| WANMON | | | | ✓ | |
| DAMON | | | | ✓ | |
| MANETconf | | ✓ | | | |
| PROPHET | | ✓ | | | |
| Misbehavior Mgmt | | | | ✓ | ✓ |
| Watchdog-Pathrater | ✓ | | | ✓ | |
| Virtual Currency | | | ✓ | ✓ | |

Fig. 13. Classification by management functional area.

is a relatively well-covered area. In particular, an analytical model has been defined by [9] to quantify the scalability performance of various configuration scheme in ad-hoc networks. Accounting management is the less covered area: ad-hoc networking does not fit within traditional accounting schemes. An ad-hoc network relies on the common sharing of resources among devices such that incentive model must be preferred to stimulate the cooperation. The performance management and the security management could be further researched when quality-of-service and security mechanisms will be defined in the lower network layers.

Ad-hoc networks are commonly considered as a particular type of networks. However, the constraints of future networks may reverse the tendency: all the optimization approaches defined for ad-hoc networking could be applied to the networks of the future. The growing dynamics of these ones might generate an edge network management crisis. Consequently as it was pointed out by [19], it is of major importance for the future to define a scalable ad-hoc routing protocol that can manage itself automatically in order to tame this crisis.

## References

[1] H. Arora and L. Greenwald, *Toward the use of local monitoring and network-wide correction to achieve QoS guarantees in mobile ad-hoc networks*, Proc. of the IEEE International Conference on Sensor and Ad Hoc Communications and Networks (SECON'04), Santa Clara, CA, USA (2004).

[2] R. Badonnel, R. State and O. Festor, *Management of mobile ad-hoc networks: Evaluating the network behavior*, Proc. of the 9th IFIP/IEEE International Symposium on Integrated Network Management (IM'05), IEEE Communications Society, Nice, France (2005), 17–30.

[3] R. Badonnel, R. State and O. Festor, *Probabilistic management of ad-hoc networks*, Proc. of the 10th IEEE/IFIP Network Operations and Management Symposium (NOMS'06), Vancouver, Canada (2006), to appear.

[4] P. Bonacich, *Factoring and weighing approaches to status scores and clique identification*, Journal of Mathematical Sociology **2** (1972), 113–120.

[5] P. Bonacich and P. Lloyd, *Eigenvector-like measures of centrality for asymmetric relations*, Social Networks **23** (2001), 191–201.

[6] J. Boney, *Cisco IOS in a Nutshell*, O'Reilly (2001).

[7] S.P. Borgatti and M.G. Everett, *A graph-theoric perspective on centrality*, Social Networks (2006).

[8] M. Burgess, *Analytical Network and System Administration. Managing Human–Computer Networks*, Wiley, Chichester (2004).

[9] M. Burgess and G. Canright, *Scalability of peer configuration management in logically ad-hoc networks*, E-Transactions on Network and Service Management (eTNSM) **1** (1) (2004).

[10] L. Buttyan and J.P. Hubaux, *Enforcing service availability in mobile ad-hoc WANs*, Proc. of IEEE/ACM Workshop on Mobile Ad Hoc Networking and Computing (MOBIHOC'00), Boston, MA, USA (2000).

[11] L. Buttyan and J.P. Hubaux, *Stimulating cooperation in self-organizing mobile ad hoc networks*, ACM/Kluwer Mobile Networks and Applications **8** (5) (2003).

[12] R. Chadha and H. Cheng, *Policy-based mobile ad hoc network management for DRAMA*, Proc. of IEEE Military Communications Conference (MILCOM'04), Monterey, CA, USA (2004).

[13] R. Chadha, H. Cheng, Y.-H. Chend and J. Chiang, *Policy-based mobile ad-hoc network management*, Proc. of the 5th IEEE International Workshop on Policies for Distributed Systems and Networks (POLICY'04), New York, USA (2004).

[14] W. Chen, N. Jain and S. Singh, *ANMP: ad-hoc network management protocol*, IEEE Journal on Selected Areas in Communications (JSAC) **17** (8) (1999), 1506–1531.

[15] T. Clausen and P. Jacquet, *Optimized link state routing (OLSR) protocol*, http://www.ietf.org/rfc/rfc3626.txt (2003). IETF Request for Comments 3626.

[16] A. Deshpande, C. Guestrin, S. Madden, J.M. Hellerstein and W. Hong, *Model-driven data acquisition in sensor networks*, Proc. of the 13th International Conference on Very Large Data Bases (VLDB'04) (2004), 588–599.

[17] R. D'Souza, S. Ramanathan and D. Temple Land, *Measuring performance of ad-hoc networks using timescales for information flow*, Proc. of the 22nd IEEE International Conference on Computer Communications (INFOCOM'03), San Francisco, CA, USA (2003).

[18] R. Enns, *NETCONF configuration protocol*, Internet Draft, Internet Engineering Task Force, draft-ietf-netconf-prot-10.txt (2005).

[19] B. Ford, *Unmanaged internet protocol: Taming the edge network management crisis*, Proc. of the 2nd ACM SIGCOMM Workshop on Hot Topics in Networks (HotNets-II), Cambridge, MA, USA (2003).

[20] S. Gouveris, S. Sivavakeesar, G. Pavlou and A. Malatras, *Programmable middleware for the dynamic deployment of services and protocols in ad-hoc networks*, Proc. of the 9th IFIP/IEEE International Symposium on Integrated Network Management (IM'05), IEEE Communications Society, Nice, France (2005), 3–16.

[21] IETF MANET (Mobile ad-hoc networks) Working Group, http://www.ietf.org/html.charters/manet-charter.html.

[22] P. Kyasanur and N. Vaidya, *Detection and handling of MAC layer misbehavior in wireless networks*, Technical report, Coordinated Science Laboratory, University of Illinois at Urbana-Champaign, IL, USA (2002).

[23] S. Marti, T.J. Giuli, K. Lai and M. Baker, *Mitigating routing misbehavior in mobile ad-hoc networks*, Proc. of the IEEE/ACM International Conference on Mobile Computing and Networking (MOBICOM'00) (2000), 255–265.

[24] J.M. McQuillan and D.C. Walden, *The ARPA network design decisions*, Computer Networks **1** (1977), 243–289.

[25] M. Mohsin and R. Prakash, *IP Address assignment in a mobile ad hoc network*, Proc. of IEEE Military Communications Conference (MILCOM'02) **2** (2002), 856–861.

[26] A. Munaretto, S. Mauro, P. Fonseca and N. Agoulmine, *Policy-based management of ad-hoc enterprise networks*, HP Openview University Association 9th Annual Workshop (2002).

[27] C.S.R. Murthy and B.S. Manoj, *Ad-hoc Wireless Networks: Architectures and Protocols*, Prentice Hall, New Jersey, USA (2004).

[28] S. Nesargi and R. Prakash, *MANETconf: Configuration of hosts in a mobile ad-hoc network*, Proc. of the 21st IEEE International Conference on Computer Communications (INFOCOM'02), New York, NY, USA (2002).

[29] D. Ngo and J. Wu, *WANMON: A resource usage monitoring tool for ad-hoc wireless networks*, Proc. of the 28th Annual IEEE Conference on Local Computer Networks (LCN'03), Bonn, Germany, IEEE Computer Society (2003), 738–745.

[30] C. Perkins, E. Belding-Royer and S. Das, *Ad-hoc on-demand distance vector (AODV) Routing*, http://www.ietf.org/rfc/rfc3561.txt (2003), IETF Request for Comments 3561.

[31] C.E. Perkins, *Ad-hoc Networking*, Pearson Education, Addison–Wesley, New Jersey, USA (2000).

[32] C.E. Perkins S., Singh and T. Clausen, *Ad hoc network autoconfiguration: Definition and problem statement*, IETF draft, http://www.ietf.org/internet-drafts/draft-singh-autoconf-adp-00.txt (2004).

[33] K. Phanse, *Policy-based quality of service management in wireless ad-hoc Networks*, Ph.D. Thesis, Faculty of the Virginia Polytechnic Institute and State University (2003).

[34] K. Phanse and L. DaSilva, *Addressing the requirements of QoS management for wireless ad-hoc networks*, International Journal on Computer Communications **26** (12) (2003), 1263–1273.

[35] K. Ramachandran, E. Belding-Royer and K. Almeroth, *DAMON: A distributed architecture for monitoring multi-hop mobile networks*, Proc. of IEEE International Conference on Sensor and Ad Hoc Communications and Networks (SECON'04), Santa Clara, CA, USA (2004).

[36] K. Romer, *Time synchronization in ad-hoc networks*, Proc. of ACM International Symposium on Mobile Ad-Hoc Networking and Computing (MOBIHOC'00), Boston, MA, USA (2000).

[37] C.-C. Shen, C. Jaikaeo, C. Srisathapornphat and Z. Huang, *The GUERRILLA management architecture for ad-hoc networks*, Proc. of IEEE Military Communications Conference (MILCOM'02), Anaheim, CA, USA (2002).

[38] W. Stallings, *SNMP, SNMPv2, SNMPv3, and RMON 1 and 2*, Addison–Wesley Professional (1998).

[39] C.-K. Toh, *Ad-Hoc Mobile Wireless Networks*, Pearson Education, Prentice Hall, NJ, USA (2002).

[40] K. Weniger and M. Zitterbart, *IPv6 autoconfiguration in large scale mobile ad-hoc networks*, Proc. of the European Wireless Conference (European Wireless'02), Florence, Italy (2002).

[41] C. Westphal, *On maximizing the lifetime of distributed information in ad-hoc networks with individual constraints*, Proc. of the 6th ACM International Symposium on Mobile Ad Hoc Networking and Computing (MOBIHOC'05), Urbana-Champaign, IL, USA (2005).

[42] H. Zhou, L. NiMatt and W. Mutka, *Prophet address allocation for large scale MANETs*, Proc. of the 22nd IEEE International Conference on Computer Communications (INFOCOM'03), San Francisco, CA, USA (2003).

# – 3.3 –

# Some Relevant Aspects of Network Analysis and Graph Theory

Geoffrey S. Canright,  Kenth Engø-Monsen

*Telenor R&D, B6d Snarøyveien 30, NO-1331 Fornebu, Norway*
*E-mails: {geoffrey.canright,kenth.engo-monsen}@telenor.com*

## 1. Introduction – basic concepts

Our aim in this chapter is to present some recent developments in a field which is perhaps most appropriately called 'network analysis'. An alternative name for this field is 'graph theory' [30]. However we prefer the term network analysis, as it carries with it a clear connotation of application. The recent developments presented here will be built upon a foundation of basic results and concepts. These latter may be found, in greater detail, in for example [5,42] and [34]. A recent collection of more specific topics, very much in the modern spirit of network analysis, may be found in [10].

Much of our discussion will involve the basic concept of *centrality* in networks. As we will discuss in more detail below, the notion of centrality has important applications in the context of spreading of information on networks. For network and system administrators, information spreading has a clear relevance to the problem of network security. However, we believe there are other applications as well.

In the remainder of this Introduction we present some basic concepts of network analysis. Then, in Section 2, we present an introduction to the concept of centrality on networks. Section 3 then focuses specifically on centrality for networks with symmetric links, and on the close relation between a specific type of centrality (termed *eigenvector centrality*) and information spreading. We carry this discussion to the point of offering specific design recommendations for hindering (or aiding) information spreading on a given network. Then,

HANDBOOK OF NETWORK AND SYSTEM ADMINISTRATION
Edited by Jan Bergstra and Mark Burgess

in Section 4, we address the extension of the notion of centrality to networks with *asymmetric links*. Here the mathematics is considerably more involved. For this reason – and also because progress to date in application in this area has been quite limited – we give more weight to theory, and less weight to application, than in Section 3. We will however discuss briefly a related application (Web link analysis). Also, we will offer some previously unpublished ideas on virus spreading on directed graphs – an application which is directly relevant to system administration. Finally, in Section 5, we mention some recent results – including some unpublished results – on *network visualization*. We strongly believe that good solutions to the problem of visualizing network structure are (i) very useful and (ii) not easily found. Our new results for graph visualization are limited to networks with symmetric links, and are based upon our understanding of network structure in terms of centrality and information spreading.

Now we establish a basic foundation of ideas and terminology for network analysis.

### 1.1. *The network*

**1.1.1.** *Undirected graphs*   Throughout this chapter, we will use the terms 'network' and 'graph' interchangeably. These two terms describe an abstract structure, composed of *nodes* (also called vertices) connected by *links* (also called edges). In fact, in the most simple picture of a network, there is no other information – only the set of nodes, and the specification of the links connecting the nodes (say, by giving a list of all connected pairs). Such a simple network is typically termed an *undirected graph*. It is 'undirected' because (so far) our description has no information on the direction of the links – just presence or absence. We also term as *symmetric* links which do not have a direction. This simple undirected graph may be represented most simply as a list of links (since the list of links give, implicitly, the nodes also – assuming there are no nodes which have no links). Another representation is called the *adjacency matrix A*. This matrix has the same information, stored as follows: if there is a link from node $i$ to node $j$, then, in the adjacency matrix, $A_{ij} = A_{ji} = 1$. All other elements of $A$ are zero. Clearly, for symmetric links, the adjacency matrix is symmetric: $A = A^{\mathrm{T}}$ (where the superscript 'T' means 'transpose').

A simple extension of this simple undirected graph is to allow for the possibility of *weights* (other than 1 or 0) on the links. These weights may represent, for example, capacities of the links, or probability of transmission of a virus. We will call this kind of graph a *weighted undirected graph*. We note here that, for practical purposes (such as numerical calculations), it is often advantageous to store only the nonzero elements of the matrix $A$. Doing so clearly takes us back to the 'list of links' picture we had before – but we can still think of the graph structure (weighted or not) as a matrix – so that operations such as matrix multiplication make perfect sense. We note that matrices of size $N \times N$, but having of order (*constant*) $\times N$ nonzero elements (rather than order $N^2$), are termed *sparse*. Adjacency matrices representing real networks tend to be sparse. The reason is simple: as real networks grow ($N$ being the number of nodes), the average number of links connecting to each node tends to be roughly constant, rather than being proportional to $N$. This makes the total number of links roughly (*constant*) $\times N$, so that the corresponding adjacency matrix is sparse. Sparse matrices are most efficiently represented by not storing the zero elements in memory.

**1.1.2.** *Directed graphs*    Now suppose that the links between the nodes are asymmetric. That is, a link between $i$ and $j$ may be from $i$ to $j$, or from $j$ to $i$. We can schematically represent theses cases as, respectively, $i \rightarrow j$ and $j \rightarrow i$. Such links are termed 'directed' links (also called arcs), and the resulting graph is a *directed graph*. If the links are not weighted (meaning that all weights are 1), then we recover the undirected case by assuming every pair that has any link has in fact links in both directions, thus: $i \leftrightarrow j$.

Directed links are used to represent asymmetric relationships between nodes. For example: Web pages are linked in the 'Web graph' [28] by hyperlinks. These hyperlinks are one-way; so the Web graph is a directed graph. Similarly, software graphs [41] are directed: the nodes are (for example) classes, and the relationships between classes (such as inheritance or invocation) are typically directed. Another example comes from the work of Burgess [12], where the links represent 'commitments' between machines; such commitments are seldom symmetric.

Directed graphs can (like undirected graphs) be represented by an adjacency matrix. We will use the symbol $A$ for this case also. The asymmetry of the links shows up in the asymmetry of the adjacency matrix: for directed graphs, $A \neq A^{\mathrm{T}}$. Finally – just as for the undirected case – one can associate weights (other than 0 or 1) with the links of a directed graph. Thus directed graphs and undirected graphs have many features in common. Nevertheless – as we will see in detail below – the structure of a directed graph differs from that of an undirected graph in a rather fundamental way.

**1.2.** *Path lengths and network diameter*

How then can one describe the structure of a graph? That is, given a complete microscopic description of a graph, what more compact description of this structure can one define that is useful?

Perhaps the simplest possible single characteristic describing a graph is its size. However even this notion is not entirely trivial. A very crude measure of a network's size is the number $N$ of nodes. Clearly, 'big' networks have many nodes. However, one might also say that, for 'big' networks, a typical pair of nodes has a large 'distance' between them. We can sharpen this idea by defining the distance between any two nodes as the length (in 'hops', i.e., number of edges) of the shortest path between them. This definition is quite precise – except for the misleading term 'the' in 'the shortest path' – since there may be many paths which are equally short between the given pair of nodes. Hence a more precise wording for the definition of distance is the length (in hops) of *any* shortest path between the nodes.

The definition of distance may be modified for weighted graphs. Here the 'length' of any path from $i$ to $j$ may (as for undirected graphs) be taken to be simply the sum of weights taken from the links traversed by the path. The distance between $i$ and $j$ is then unchanged (in words): it is the length of any shortest path connecting $i$ and $j$.

**1.2.1.** *Undirected graphs*    We now restrict our discussion to the case of undirected graphs; we will treat directed graphs in a separate subsection, below. We return to the notion of the size of a network, based on distances between nodes. For $N$ nodes, there are

$N(N-1)/2$ pairs of nodes – and each pair has a shortest distance. So we want something like an 'average distance' between nodes. A detailed discussion of several choices for how to take this average may be found in [42]. In this chapter, we will not need a precise notion of average path length (which we will denote by $L$); it is sufficient to assume simply the arithmetic mean of the $N(N-1)/2$ path lengths. (Other definitions give results which are close to this one.) Thus, we now have another notion of the size of a graph – the average distance $L$ between pairs of nodes.

Another common notion of the size of a graph is the graph *diameter D*. The diameter is simply the length of the 'longest shortest path'. That is, of all the $N(N-1)/2$ shortest path lengths, $D$ is the greatest length found in this set. The diameter is clearly larger than the average path length; yet, for large $N$, it typically differs from $L$ only by a constant (i.e., $N$-independent) factor. Hence, if one is interested in the way that path lengths grow with increasing $N$, one may (usually) equally well discuss the diameter $D$ or the average path length $L$. We will normally speak of the latter.

Clearly, for a given 'recipe' for building a network with increasing $N$, we expect the average path length $L$ (which we will sometimes simply call the 'path length') to grow monotonically with $N$. And indeed it does so – almost always. (Here is an exception: for a *complete graph* – i.e., one in which every node links to every other – $L$ is 1, independent of $N$.) However, the monotonic growth can be strong or weak.

Here we must state explicitly an assumption which we make (unless we explicitly state the contrary) throughout this chapter. That is, we assume that any graph under discussion is *connected*. An undirected graph is connected if there is a path from any node to any other. The point is, a connected graph does not 'fall to pieces': it is not in fact more than one graph. Our assumption is important, because for an unconnected graph, the path length between nodes lying in different 'pieces' is either undefined or infinite. Thus, for an unconnected graph, the *average* path length is also either undefined or infinite – in any case, not very useful.

*For a connected, undirected graph there always exists a path from node i to j.*

Our assumption requires then that any 'recipe' for 'growing' a network with increasing $N$ gives a connected graph for every $N$. This requirement is readily satisfied for certain types of highly regular graphs; but it cannot always be guaranteed for graphs with randomness involved in their construction (again see [42] for details). We will not go further into these technicalities – we will simply assume (with theoretical justification) that the recipes we discuss give connected graphs (with high enough probability).

The point of this little digression on connectedness was to allow us to discuss how the average path length grows with $N$, for a given recipe. We consider a very simple case (see the left side of Figure 1), in which all nodes are simply placed on a ring, with links only to nearest neighbors on the ring. It is intuitive that the average path length, for large $N$, is around $N/4$; and adding some next-nearest-neighbor links (right side) will (again, roughly) cut this distance in half [42]. Also, opening the ring into a chain will double this distance. In all of these cases we find the average path length $L$ growing linearly with $N$. This linear growth makes sense, in light of the fact that the nodes lie in a structure which is essentially one-dimensional. (Placing them on $d$-dimensional grids gives a path length growing as $N^{1/d}$.)

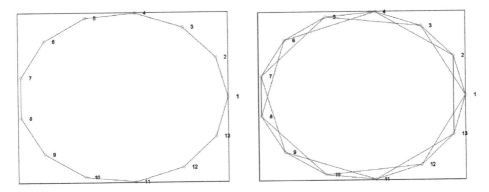

Fig. 1. Two versions of nodes on a ring.

The one-dimensional graph is in a sense the 'worst case', in that it does the poorest job of placing the nodes close to one another (and doing any 'worse' would leave the graph disconnected). The complete graph is an example of the opposite extreme, in which $L$ grows as $N^0$. An important case which is intermediate to these two extremes (but close to $N^0$ growth) is the random graph – in which every possible link is present with a probability $p$. Given certain reasonable caveats (such as $p$ large enough to give connectedness), one finds that the average path length for random graphs grows as $\ln(N)$. This good connectivity is achieved with considerably fewer links than that used in the complete graph (in fact, only about $\ln(N)$ links are needed [42], as opposed to about $N^2/2$ for the complete graph).

Finally, in this regard, we mention the *small-worlds* phenomenon. The basic idea for small worlds came from a sociologist, Stanley Milgram, in the 1960s [31]. Milgram performed letter-passing experiments which supported the idea that, in a United States with hundreds of millions of individuals, the typical path length between any two (where a link represents acquaintance) is about 6 – an astonishingly small number. We see that $L \sim \ln(N)$ offers an explanation for these small distances (in hops – the letters were passed from Kansas and Nebraska to Boston); but there is no reason to believe that the network of acquaintances is set up like a random graph. Watts and Strogatz [42] offered a new class of graphs (i.e., a new 'recipe') which has much less randomness, but still gives (approximately) $L \sim \ln(N)$. We will call these graphs Watts–Strogatz (WS-) graphs. WS-graphs are constructed as follows: begin with a one-dimensional ring, with each node connected to $k/2$ nearest neighbors on each side (giving each node a total of $k$ neighbors). Next, take each link in turn, and with a probability $\beta$, perform a 'random rewiring' which connects *one* end of the link to a randomly chosen node (leaving the other end at its original node). For small $\beta$, the WS recipe introduces only a small amount of randomness – yet it still gives very small path lengths.

This result has sparked a great deal of further research, which we will not attempt to review here. A principal lesson of the WS result may be stated as follows: for networks for which the nodes' connections reflect some reasonable notion of geographic distance – so that it makes sense to speak of the links as being 'short-range' – the addition of a very *small* percentage of *long-range links* can drastically change the average path length. And, stated in this way, we believe the WS result does indeed explain the small-worlds effect. Furthermore, small-worlds effects are *ubiquitous* – one finds them in a huge variety of empirically measured networks, including such things as the Internet, the graph of the World Wide Web, electric power networks, collaboration graphs (of actors and scientists) and (biological) neural networks.

**1.2.2.** *Directed graphs*    Introducing a direction to the links of a network can have profound effects on the properties of that network. Here we give modified definitions of path length and connectedness which are appropriate for directed graphs. First, for directed graphs, we need a more precise definition of path. However the definition is still intuitive: a path from $i$ to $j$ in a directed graph is a series of directed links, always traversed in the direction from 'tail' to 'head' of the arc, beginning with a tail (outlink) at $i$ and ending with a head (inlink) at $j$. Then, with this definition of a path (a 'directed path'), the path length is the same as before (number of hops, or sum of weights).

The notion of connectedness for directed graphs is a bit more complicated. A directed graph is connected (also called *weakly connected*) if, when one ignores the directions of the links (thus making it undirected), it is connected. However, it is clear that there can be node pairs in a connected, directed graph, for which there is *no* (directed) path from one node of the pair to the other. Hence we need a further notion, namely that of being *strongly connected*. A directed graph is strongly connected if there is a directed path from every node to every other node.

We will as usual insist that any directed graph under consideration be connected. However, typically, connected directed graphs are not strongly connected; and we will lose generality if we only study strongly connected directed graphs. Instead, one can find *subgraphs* of a directed graph that are strongly connected. By subgraph, we mean simply a subset of the nodes, plus all links (that are present in the whole graph) connecting these nodes. For example, if one can find, in a given directed graph, any two nodes that point to each other, or for that matter any number of nodes in a ring, then these nodes (plus their mutual links) constitute a strongly connected subgraph of the given graph. We use the word *strongly connected component* (SCC) for 'maximal' strongly connected subgraphs. Here a 'maximal' strongly connected subgraph is simply a strongly connected subgraph which ceases to be strongly connected if any other node from the graph is added to the set defining the subgraph.

We offer some examples for clarification. For a complete graph with $N$ nodes, *any* subgraph is strongly connected. However the complete graph has only one strongly connected component, namely, the entire graph – since, if we choose any smaller subgraph as a candidate SCC, we will find that we can enlarge our candidate without losing the property of being strongly connected.

Now suppose we take two strongly connected graphs $G1$ and $G2$, and connect them with a single link pointing from a node in $G1$ to a node in $G2$. Then the resulting graph is connected; but there are no paths from any node in $G2$ to any node in $G1$. Hence $G1$ and $G2$ are the two SCCs of the composite graph (which we might denote as $G1 \rightarrow G2$). The example of Figure 2 thus illustrates our basic point about directed graphs:

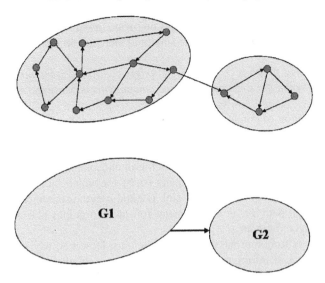

Fig. 2. A directed graph (upper) and its component graph (lower).

*For a directed, connected graph, there is not always a directed path from node i to j – unless the graph is strongly connected.*

A typical directed graph is thus composed of multiple SCCs. Also, it is perhaps clear from the last example that the relationships between the SCCs are asymmetric, i.e., one-way. Thus [14] one can define a 'coarse structure' for a directed graph, in terms of the SCCs. This coarse structure is termed a *component graph* [24]. In the component graph, each SCC is represented as a vertex, and (regardless of the number of links connecting them) each pair of SCCs which is directly linked gets a one-way arc connecting them (and showing the proper direction) in the component graph. Thus, the component graph in the previous example is in fact $G1 \rightarrow G2$; and this remains true, no matter how many links we add which point from (a node in) $G1$ to (a node in) $G2$. (If, in addition, we add just one link from $G2$ to $G1$, then the entire graph becomes strongly connected.)

Some reflection reveals [14] that the component graph must be loop-free. In other words, it is a *directed acyclic graph* or DAG. If we think of any kind of flow or movement (say, of information) which follows the direction of the arcs, then it is clear from the acyclic nature of the component graph that flow is essentially (disregarding 'eddies' *within* the SCCs) one-way. This makes a directed graph profoundly different from an undirected graph. Furthermore, since there are no loops in the component graph, it is also clear that this one-way flow must have one or more end points – and one or more starting points. These 'points' are in fact vertices of the component graph, hence SCCs of the original directed graph. The 'start points' are in other words SCCs which have only outlinks to other SCCs; they are termed *source SCCs*. Similarly, there must be SCCs with only inlinks from other SCCs; they are termed *sink SCCs*. Thus, 'flow' in the component graph is from sources to sinks, via intermediate vertices, but always one-way. In our simple example above with two SCCs, $G1$ is the source and $G2$ is the sink (and there are no intermediate SCCs).

> *Any directed graph is composed of strongly connected components; and*
> *flow between the strongly connected components is one-way (acyclic).*

We see that the notion of connectivity is considerably more involved for directed graphs than for undirected graphs. Now we proceed to the question of path lengths. We immediately see a problem: for a typical directed graph with more than one SCC, there will be pairs of nodes for which the path length is (as for an unconnected graph) either undefined or infinite – because there will be node pairs $(a, b)$ for which there is no directed path from $a$ to $b$. (For example, place $a$ in the sink $G2$ and $b$ in the source $G1$.) Therefore, the definition of average path length and diameter for directed graphs is less straightforward than it is for undirected graphs.

One possibility is to ignore the direction of the arcs. However, we believe that this involves too much loss of information to be very useful – if the directionality of the arcs is meaningful, then it should not be ignored. The World Wide Web (WWW) offers an outstanding practical example of a directed graph. Here, the nodes are Web pages, and the (directed ) links are the hyperlinks from one page to another. A good discussion of the (very coarse) structure of the 'Web graph' is given in [11]; see also Chapter 12 of [5]. We have mentioned above that the WWW qualifies as a small-world graph, in the sense that its average path length grows only logarithmically with the number of nodes (pages). Now we see that it is not an entirely straightforward task to measure this average path length – there are many pairs of pages $a$ and $b$ on the WWW, for which $a$ is not reachable from $b$, or $b$ is not reachable from $a$, or both. As noted above, one solution is to treat the WWW as an undirected graph (see [1] for example). Another is to only look at a strongly connected component (for which, we recall, all nodes are reachable from all others) – perhaps the largest SCC [1]. A third choice is to only average over those pairs for which a directed path exists [2]. This latter strategy actually resembles Milgram's [31]; he did not count those letters which failed to reach their target.

### 1.3. *Node degree distribution*

Next we come to a concept which is heavily used in network analysis, namely the *node degree distribution* or NDD. We consider first the case of an undirected graph.

**1.3.1.** *Undirected graphs*    When links are undirected, then for each node $i$ we can define a *node degree* $k_i$ which is simply the number of links to that node. Now we define $p_k$ to be the fraction of nodes in a graph which have node degree $k$. The set of $p_k$ then defines the node degree distribution. Clearly $0 \leqslant p_k \leqslant 1$; also, $\sum_k p_k = 1$. Thus the $p_k$ have all the properties of a probability distribution, and can in fact be used to describe families of graphs, subject to the constraint that the probability (averaged over the family) for a node to have degree $k$ is $p_k$.

In a *regular graph*, every node has the same degree $K$, so that $p_K = 1$, and all other $p_k = 0$. A complete graph is regular, with $K = N - 1$. Small perturbations on regular graphs give graphs with a peak at $k = K$, and some weight around this peak at $K$. Such graphs are in many ways much like regular graphs; but at the same time, some properties of graphs (such as the average path length) can vary strongly upon only a slight deviation from a regular graph with an ordered structure of short-range links. Here we are thinking of the small-worlds effect [42] discussed above.

This effect is worth discussing in more detail here, as it illustrates a general point – namely, that the node degree distribution is far from adequate as a description of a graph. We illustrate this point with the following example. First, consider $N$ nodes arranged in a ring; but let each node also have a link to both next-nearest neighbors. This is a regular lattice with $K = 4$, so that $p_{K=4} = 1$, and all other $p_k = 0$. As pointed out above, the average path length in this graph is proportional to $N$ – so that it is large for large $N$. Now we 'rewire' this graph according to the WS recipe, with a small $\beta$. The NDD will only change slightly – it will still have $p_{K=4} \approx 1$, with some small weight around $k = 4$. However the path length will change drastically from the regular case: $\ln(N)$ as compared to $N$. Furthermore, we can let $\beta \to 1$. Then we get a random graph – which, for large $N$, will have essentially the same NDD as the WS graph. This random graph will also have a small path length (of order $\ln(N)$); but the structure of the random graph is in many ways very different from that of the WS graph. Finally, we can, if we wish, modify the link structure of the original $K = 4$ regular graph, in such a way that (i) the NDD is the same as for the WS and random graphs, while (ii) all rewired links are short-ranged. This gives us three graphs with the same NDD: the 'noisy' regular graph, the WS graph, and the random graph. The first two have very similar link structure but very different path lengths. The last two (WS and random) have very different link structure, but similar path lengths. These rather interesting variations are not at all revealed by the NDD.

A very important class of graphs are the *scale-free graphs*. These are graphs whose NDD follows (roughly) an inverse power law: $p_k \sim k^{-\alpha}$. They are termed 'scale-free' because the NDD itself is scale free: a power law remains a power law (with the same exponent) under a change of scale of the $k$ axis. This kind of NDD is far from an academic curiosity; in fact scale-free networks are observed in a wide variety of circumstances. Some examples – covering a wide variety of empirical networks, such as the Internet, the WWW, movie actor collaboration graphs, metabolic reaction networks (where the nodes are proteins, and the links indicate interactions), and networks of human sexual contacts – are given in [34,40] and in [4]. Scale-free networks are called 'heavy-tailed', because they decay more slowly than exponentially at large $k$ – so slowly, in fact, that they can have an unbounded variance. In practice, the NDD of real networks must deviate from this ideal form at large $k$ – often showing a power-law decay over a wide range of $k$, followed by an exponential decay at larger $k$. Figure 3 shows an empirical example, taken from a snapshot of the Gnutella peer-to-peer network [25]. We see a smooth and approximately linear region of the NDD in the log–log plot.

Scale-free graphs are also small-world graphs [4,34] – in the sense given above, i.e., that their average path length grows no faster than $\ln(N)$. However, the reverse is not true: not all small-worlds graphs are scale free. An obvious counterexample is the class of Watts–Strogatz graphs, which are small-world graphs, but which has an NDD very different from power-law. A good discussion of the relationships between these two aspects of networks is given in [4].

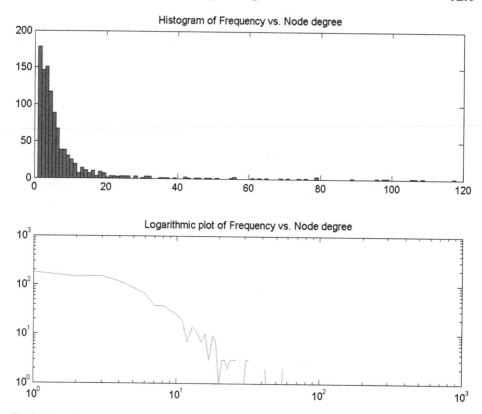

Fig. 3. The node degree distribution of a Gnutella network. The upper panel shows the histogram on linear scale, whereas the lower panel shows the same plot in logarithmic scale.

In summary, we want to emphasize an important *limitation* of the NDD as a measure of graph structure:

> *Graphs that are structurally very different can*
> *have the same node degree distributions.*

Finally, we note that Newman [34] gives a good discussion of the 'typical structure' of the 'most probable graph' for a given NDD. That is, one (in theory) generates all possible graphs with a given NDD, and picks a typical one from the resulting ensemble. In this (limited, theoretical) sense, one can say that the NDD 'implies' a structure for the graph.

**1.3.2.** *Directed graphs*    For directed graphs, each node has two types of links: inlinks and outlinks. Hence one can define, for each node $i$, two types of node degree: the indegree $k_i^{in}$ and the outdegree $k_i^{out}$. Similarly, each type of node degree has its own node degree distribution over the entire directed graph – denoted, respectively, $p_k^{in}$ and $p_k^{out}$. In previous discussion, we have mentioned the WWW as an example of a scale-free (and small-world) graph. Typically, studies [2,11] which report a scale-free structure for the WWW look separately at each of $p_k^{in}$ and $p_k^{out}$, reporting that *each* of them shows a power-law NDD (although with different exponents).

Newman [34] has suggested the study of the joint NDD $p_{jk}$, giving the fraction of nodes having simultaneously indegree $j$ and outdegree $k$. This two-dimensional distribution contains more information than the pair of one-parameter distributions $p_k^{in}$ and $p_k^{out}$. However, we are not aware of any work following up on this suggestion.

### 1.4. *Clustering coefficients*

Next we discuss an aspect of network structure that addresses the question: to what extent does the neighborhood of a node actually 'look like' a neighborhood? To clarify this, consider for instance an acquaintanceship network. Suppose node $j$ has five one-hop neighbors (i.e., acquaintances). Do we expect (at least some pairs of) these five people to be acquainted with one another? The answer is surely yes: acquaintanceship tends to be *clustered*. Friends are not chosen at random; and the likelihood that two of my friends are acquainted with one another is much higher than that for two randomly chosen people.

The *clustering coefficient* of a graph is an attempt to quantify this effect. However, it turns out that there exist multiple forms for this quantity. Two of them are discussed in [34]; we mention them here, following his notation. First we define the term 'triple': this is simply any connected subgraph of three nodes. A little reflection then tells us that any triple has either two links or three. That is, I and my two friends form a triple – which has three links if they also know each other, two if they do not.

Now we come to clustering coefficient definitions. The first of this we call $C^{(1)}$. In words, $C^{(1)}$ measures the fraction of triples that have three links (we call these three-link triples 'triangles'). More precisely,

$$C^{(1)} = 3 \times \left( \frac{\text{number of triangles}}{\text{number of triples}} \right).$$

Here the factor of three is needed because each triangle in fact represents three triples, one based at each node in the triangle. With this definition, $0 \leqslant C^{(1)} \leqslant 1$.

The second definition comes from Watts and Strogatz [42]. First one defines a *local clustering coefficient* around each node $i$:

$$C_i = \frac{\text{number of triangles connected to node } i}{\text{number of triples centered on node } i};$$

then one takes the average of this local clustering coefficient over all nodes:

$$C^{(2)} = \frac{1}{N} \sum_i C_i.$$

Each of these definitions seems to be addressing the same idea; but, as pointed out by Newman, $C^{(1)}$ may be regarded as the ratio of two averages (multiplying top and bottom by

$1/N$), while $C^{(2)}$ is the average of the local ratio. Hence the two definitions give different quantitative results. Furthermore, sometimes the difference is large – and it can occur in either direction (i.e., $C^{(1)} < C^{(2)}$ or $C^{(2)} < C^{(1)}$). See Table II of [34] for a collection of empirical results for these two definitions.

In spite of these complications, we will sometimes refer to 'the' clustering coefficient – thus ignoring these differences. The point is that, for some types of graphs (recipes), either of these definitions tends to vanish as $N$ grows; while for other types, neither of them vanishes, even for very large $N$. Intuitively, we expect 'real' networks to have some degree of neighborhood structure, and hence to fall into the second category. In contrast, random graphs fall into the first category – they have no correlations among the links, and hence no clustering. For this reason, even though random graphs have very small path lengths, they are not considered as realistic models for empirical networks.

Seeking a definition which is more in harmony with what we know and believe about empirical networks, Duncan Watts [42] has proposed a definition of a small-world graph which includes *two* criteria: (i) the average path length $L$ should be like that of a random graph; but (ii) at the same time, the clustering coefficient ($C^{(2)}$ in this case) should be much larger than that for a random graph. That is, by this (widely accepted) definition, a small-worlds network should have a significant degree of neighborhood structure, and *also* enough 'long-range' hops that the network diameter is very small. The WS 'rewired' graphs fit this definition, as do scale-free networks [4] (although $C^{(1)}$ actually vanishes for very large $N$, for certain types of scale-free networks [34]). It seems in fact that an enormous variety of measured networks are small-world graphs by this definition.

*A small-world graph fulfills two criteria:*

(1) *The average path length is similar to that of a random graph.*

(2) *The clustering coefficient is much larger than that of a random graph.*

*Small-world graphs – according to this definition – are ubiquitous among the empirical networks.*

## 2. Centrality on undirected graphs

In the previous section, we reviewed a number of quantitative measures of the structure of whole graphs. In this section, we address a property of individual nodes rather than of an entire graph. This property is *centrality*. The notion of centrality will play a large role in most of the rest of this chapter. Even though it is a single-node property, we will see that it can give us a new view of the structure of the whole network – a view which is quite different from the conventional whole-graph statistics discussed in Section 1, and yet (we believe) highly useful. In short, we will build from the one-node concept of centrality to a coarse-grained view of network structure – one which furthermore has important implications for information spreading, and hence for network security issues.

First we develop the concept of node centrality. We consider undirected graphs, and take an obvious motivating example: a star graph. This graph has one node at the 'center', and $N - 1$ 'leaf' nodes. Each leaf node is connected only to the center of the star. (Note in passing that the clustering is zero here.) Most of us would then agree that, however we define 'centrality', the center node has more 'centrality' than any leaf node, and furthermore

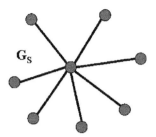

Fig. 4. A star graph.

that all leaf nodes have the same centrality. It also seems reasonable that the centrality of the center node should grow with a growing number of leaves. Our intuition, applied to this simple graph, tells us that the center node must in some sense be considerably more 'important' than the others.

The aim is then to build from this intuition, so as to come to quantitative definitions of node centrality, which can be used when intuition fails – as it rapidly does, for most realistic networks. Unfortunately (or fortunately), there exist many distinct definitions of centrality – even if we restrict ourselves to undirected graphs. We will discuss some of them here. A good (albeit very short) summary of many of the commonly used definitions of centrality, as seen from a social-science perspective, may be found in [8]. We will mention three here, since it is these three that we will make use of in this chapter.

## 2.1. *Degree centrality*

Our first definition is very simply stated: the *degree centrality* of a node $i$ is simply its degree $k_i$. This definition fits our intuition about the star graph. Also, it is very easily calculated.

Is there anything missing then from this definition? Well, it clearly ignores *every* aspect of the network's structure beyond the one-hop neighborhood view from each node. We can easily imagine cases in which this might be a disadvantage, which more than outweighs the advantages coming from the extreme simplicity of degree centrality. For example, we take our star graph ($G_s$), with (say) 50 leaves, and then connect one of the nodes (it matters little which one, so long as it is just one) to a much larger, and much better connected, graph ($G_{big}$), one with (say) many thousands of nodes. Let us furthermore connect the one node of $G_s$ to a node of $G_{big}$ which has low centrality by any reasonable definition (including degree centrality) – i.e., to a 'peripheral' node of $G_{big}$. Finally, we suppose that there is no node in $G_{big}$ with more than (say) 20 neighbors. This example is of course chosen to make a simple point: by most reasonable criteria, the center node $X$ of $G_s$ should be viewed as being rather peripheral – not central – with respect to the composite network $G_s + G_{big}$. Yet the simple measure of degree centrality will rate $X$ as being far more central (by a factor at least 2.5) than any other node in $G_s + G_{big}$!

We note that we are not saying here (or, for that matter, anywhere in this chapter) that any one measure of centrality is always better or worse than any other. In fact, we believe

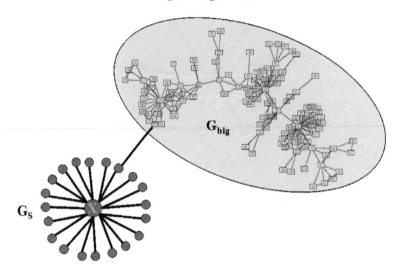

Fig. 5. Locality of degree centrality.

the contrary: that each measure has some domain of valid application. For example, suppose that we are responsible for a computer network, and we are concerned with a failure mechanism which can affect a machine's nearest neighbors, but, for technical reasons (perhaps time scale), cannot 'cascade' further before we can readily intervene. We then want to know which machines can cause the most damage under this scenario; and the answer is clearly given by the degree centrality.

On the other hand, for failures which can 'cascade' more rapidly than the administrator can intervene, it should be clear that one needs to look beyond one hop in order to evaluate which machines in a network represent the most risk to the system as a whole.

Our $G_s + G_{big}$ example is designed to illustrate this point. We can picture this network in terms of a human social network: $X$ is (perhaps) a drug dealer, with many contacts; but the entire dealer + users network lies peripherally to the rest of the society. Thus – unless we are specifically studying phenomena related to drug use – we are likely to need a definition of node centrality which does not pick out node $X$ as being most central.

## 2.2. Betweenness centrality

Our next definition, as we will see, looks well beyond one hop in defining the centrality of a node. In fact, the *betweenness centrality* of a node depends on the entire network topology (by which we mean simply the complete list of links). Also, the name of the definition gives us a clear indication of what is being captured here: a node has high betweenness centrality – also termed simply 'betweenness' – if it lies (in some sense) 'between' lots of other pairs of nodes. Thus, if I have high betweenness, than many node pairs wishing to communicate will be likely to want to do so through me.

Now we render this idea quantitative. We look at each pair of nodes $(i, j)$, and define $g_{ij}$ as the number of shortest paths between $i$ and $j$. The betweenness is then considered for

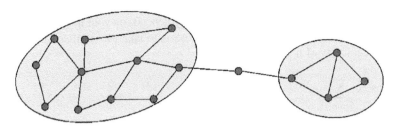

Fig. 6. An illustration of betweenness centrality. The node in between the two shaded regions will display high betweenness centrality, since this node is on every shortest path between the two shaded regions.

some *other* node $k$; so we define $g_{ikj}$ as the number of shortest paths connecting $i$ and $j$ which go through node $k$. The contribution to node $k$'s betweenness coming from the pair $(i, j)$ is then $g_{ikj}/g_{ij}$. Then, the betweenness of node $k$ is simply the sum of this quantity over all pairs of nodes:

$$b_k = \sum_{\substack{\text{pairs } i,j \\ i \neq k; \, j \neq k}} g_{ikj}/g_{ij}.$$

This definition seems complicated; but note that, in the case that there is a unique shortest path between every pair of nodes, this definition for $b_k$ simply reduces to counting the number of shortest paths that node $k$ lies on (but is not an endpoint of). Thus the extra complications arise from the fact that there are often multiple shortest paths for a given pair of nodes.

Girvan and Newman [19] have defined a strictly analogous measure for the betweenness of a *link*. In fact, the same formula as that given above applies, if we simply let the index $k$ represent a link rather than a node (thus rendering the condition ($i \neq k$, $j \neq k$) superfluous).

Betweenness centrality thus measures the importance of a node in the *transmission* (of something) over the network – for example, of information, traffic, viruses, etc. A possible weakness is the choice to count only shortest paths; in many cases, all paths between two nodes have some role in transmission – with the shortest paths having most weight (significance), but not the only weight.

## 2.3. *Eigenvector centrality*

Now we come to a kind of centrality which is in a meaningful sense the 'dual' of betweenness. That is, instead of quantifying to what extent a node sits *between* something, we seek to quantify the extent to which it sits *in the middle of* something. This is closer to the colloquial English meaning of the word 'central' – 'lying near the center'; and, as with betweenness, this definition – *eigenvector centrality* – will also (as we will see) depend on the entire network topology.

We begin as always with a verbal description. We want to capture (quantitatively) the following idea from common folk wisdom: "It's not how many you know, but *who* you know that matters". In other words: your centrality depends on your neighbors' centralities. The simplest way of making these words precise is to take the sum; so, letting $e_i$ be the eigenvector centrality of node $i$, we write

$$e_i \propto \sum_{j=nn(i)} e_j.$$

Here '$j = nn(i)$' means only include nodes $j$ which are nearest neighbors of $i$ in the sum. We do not use an equality here, as it turns out that the resulting equation is in general insoluble. This is because we have made a circular definition: my centrality depends on my neighbors'; but theirs depends on that of *their* neighbors – including mine. Hence we must seek a self-consistent solution; and this requires some factor in front of the summation.

A natural guess is to set my centrality equal to the *average* of my neighbors' centralities:

$$e_i = \left(\frac{1}{k_i}\right) \sum_{j=nn(i)} e_j.$$

This equation is in fact soluble. Unfortunately, the answer is not very interesting: the solution [15] is that all nodes have the same centrality, regardless of the network topology! (This is easily seen: just set $e_i = 1$ for all $i$. Also, we note here that, because of the linearity of this equation for the unknowns $e_i$, we can equally well choose any other value for the common solution.)

What has happened here is that our goal of making centrality depend on neighbors' has been achieved; but the effect is *too* strong. Hence what is desired is something in between degree centrality (which gives all weight to degree, and none to context) and the above result (which is so 'democratic', i.e. gives so much weight to context, that it erases all differences among the nodes). We note that this equation 'punishes' nodes with high degree more than it does nodes with low degree (via the factor $1/k_i$). Let us then instead 'punish' each node equally:

$$e_i = \left(\frac{1}{\lambda}\right) \sum_{j=nn(i)} e_j.$$

Now our reasoning has given us this parameter $\lambda$: what is it? Can it be chosen freely?

In fact the answer is simple. We rewrite the above as:

$$A\mathbf{e} = \lambda\mathbf{e},$$

where $A$ is the adjacency matrix, and $\mathbf{e}$ is the vector of centralities. Here we have used the fact that

$$\sum_{j=nn(i)} e_j = (A\mathbf{e})_i.$$

Thus we have come to an eigenvector equation involving the adjacency matrix $A$. From this we know that there are $N$ possible eigenvalues $\lambda$ to choose from – each with its corresponding eigenvector. We can however quickly narrow down this choice: the adjacency matrix is non-negative and real, which (along with Perron–Frobenius theory [32]) tells us that one, and only one, eigenvector has all positive elements – namely, the eigenvector corresponding to the *principal* (largest) *eigenvalue*. We want a node's centrality to be a positive quantity. Hence we choose the *principal eigenvector* **e** as the solution to the above equation; and its $i$th component $e_i$ is then the *eigenvector centrality* of node $i$.

> *Eigenvector centrality (EVC) is defined as the principal eigenvector of the graph's adjacency matrix. The principal eigenvector contains n scores; one for each node.*

Thus defined, the eigenvector centrality (EVC) [7] of a node has the properties we want: a node's EVC $e_i$ is sensitive to the EVC values of its neighbors – and their values in turn to those of *their* neighbors, etc. However the EVC is not so sensitive that all nodes receive the same value. Instead, we will see that EVC provides highly useful insight into the structure of an undirected network.

Furthermore, the fact that the EVC is obtained from the principal eigenvector allows us to picture the calculation of the EVC via a very simple process. That is: suppose we give all nodes some starting guess for their EVC. This guess can be almost anything; but it is reasonable to give all nodes a positive starting guess – for instance, $e_i = 1$ for all nodes $i$. Now we start a ticking clock. At each tick of the clock, each node sends its current value to all of its neighbors – and then replaces that value with the sum of the values it has just received (in the same tick of the clock) from its neighbors. We let this process continue for many ticks. What happens is that, after enough clock ticks, two things become true: (i) the ratios of the values $e_i/e_j$, for all pairs $i$ and $j$, stops changing, while at the same time (ii) each value $e_i$ increases by the same factor $\lambda$ at each tick.

The point here is that the process occurring at each tick is exactly equivalent to multiplying the vector $e^{(\text{guess})}$ of current guesses by the adjacency matrix $A$. Thus, after enough ticks, the distribution of values converges to some eigenvector of $A$, and this eigenvector simply increases thereafter by the factor $\lambda$ (its eigenvalue) at each tick. Furthermore, we know that – except in pathological cases – the eigenvector that the process converges to is the principal eigenvector, with $\lambda$ the principal eigenvalue. Thus this process in fact represents a way of calculating the EVC (termed the 'power method' [20]), which can in many cases also be a practical approach.

This process also reflects the *idea* behind the definition of EVC: I get my importance values from my neighbors; they get theirs from their neighbors; and we just repeat this sending until we all agree – not on the absolute values, but on the ratios. The absolute values grow with each iteration, because the principal eigenvalue $\lambda$ is normally greater than 1. Thus a practical implementation of the power method must work with a running guess of $\lambda$ (as well as of **e**), and correct by this guess periodically.

Because of the continual growth of the weights, we call this iterative process 'non-weight conserving'. That is, a 'weight conserving' process, operating on a vector of 'weights' (numerical values) distributed over the nodes of the network, would leave the sum of the weights unchanged.

*A weight conserving process will conserve the sum of the weights.*

Let us clarify this with an example of a weight conserving process. It is easy to design one: we take our EVC power method process, and make a single modification: each node $i$ must divide its current weight $w_i$ (we do not write $e_i$ since we will not converge to the EVC) by its degree $k_i$, before sending it out. This ensures that each node sends out precisely the weight it has, and no more – and this is sufficient to make the process weight conserving.

For such a weight conserving process, the dominant eigenvalue is 1. This means that this process can converge to a vector, which then neither grows nor shrinks with repeated iterations – the dominant eigenvector for the matrix representing this process is a 'fixed point' of the iterative process. The process is in fact equivalent to multiplying the current weight vector **w** by a matrix $A^{\text{norm}}$, which is a column normalized version of the adjacency matrix $A$. Here by 'column normalization' we simply mean dividing each column by the sum of its entries. (If the (undirected) graph is connected, then there will be no columns with all zeros.)

We mention this weight conserving process, using a 'normalized' adjacency matrix, for two reasons. First, the idea of column normalization will be important in our discussion of centrality for directed graphs, in the next section. Secondly, this process converges to an interesting eigenvector – which we will now show, since the proof is simple.

The dominant eigenvector for $A^{\text{norm}}$ is in fact a vector $\pi$, where the entry $\pi_i$ for node $i$ is in fact the node degree $k_i$ (or proportional to it). Here is the proof:

$$\left(A^{\text{norm}}\pi\right)_i = \sum_{j=nn(i)} \frac{1}{k_j}\pi_j = \sum_{j=nn(i)} \frac{1}{k_j}k_j = \sum_{j=nn(i)} (1) = k_i = \pi_i = 1 \cdot \pi_i.$$

(For the – very general – conditions ensuring that $\lambda = 1$ is the dominant eigenvalue, see [33].) Thus we find that our weight conserving process using $A^{\text{norm}}$ is in fact a very tedious way of computing the degree centrality of each node! This result is nevertheless interesting [33]. For such weight conserving processes, if we simply adjust the start vector such that its (positive) weights sum to one, then this sum will remain unchanged during the process, so that the weights themselves can be viewed as probabilities. More specifically: suppose we release a 'packet' on a network, and then allow that packet to undergo a random walk – such that each new hop is chosen with equal probability from all the links connected to the current node. If we then accumulate good statistics (i.e., many packets, and/or after enough time), we can speak of the probability for each node to be visited. This probability converges to the vector $\pi$, rescaled so that its entries sum to 1. That is, the probability $p_i$ for visiting node $i$ converges to $k_i/2M$, where $M = (1/2)\sum_i k_i$ is the total number of links in the network.

*In short: the limiting probability for a random walker to visit a*
*node is directly proportional to the node degree.*

Stated differently: nodes of high degree centrality 'attract more weight' under iterated, weight conserving process. The counterpart to this statement, for EVC, is then: for non-weight conserving processes, nodes with high EVC 'attract more weight'. We will flesh out this rather vague statement in the next section, where we will consider non-weight conserving processes such as the spreading of information or viruses.

## 3. EVC, topography and spreading on undirected graphs

In this section we will summarize our own work on information spreading over undirected networks. This summary is based a new kind of structural analysis (reported in [15]), plus an application of this analysis to the basic phenomenon of spreading on networks. The latter work is reported in [17] and in [16]. We have built on the understanding that we have gained by proposing [16] two simple sets of design guidelines – one set for the case in which it is desirable to *help* the spreading, and one set for the case that spreading is to be *hindered*. The latter set has direct application for enhancing the security of computer networks. These ideas have been tested in simulations, reported in [16] and summarized here.

In the previous section, we have emphasized that the EVC of a node tells us something about the importance of that node in non-weight conserving processes. In this section we consider, in a general way, such non-weight conserving processes – that is, processes in which some quantity ('information', 'virus', 'infection') may be spread from one node to *any or all* of its neighbors – without regard to weight conservation. We will most often use, in discussing such processes, terminology appropriate to disease and infection; however, our discussion is general enough that it also applies to the spread of desirable quantities such as innovations or useful information. The common thread is simply that the quantity which spreads can be *duplicated* or *reproduced*: it is this aspect which (we argue) makes EVC a good measure of a node's importance.

### 3.1. *A topographic picture of a network*

From the discussion of Section 2, it should be clear that the notion of node centrality is related to the (so far only nebulously defined) notion of being '*well connected*'. Furthermore, we have seen that degree centrality is a purely local measure of well-connectedness, while EVC (and betweenness) gives measures which are not purely local. In particular, the eigenvector centrality of node $i$ which tells us something about the *neighborhood* of node $i$. We reason that this kind of centrality can be useful for defining *well-connected clusters* in the network, and, based on that, for understanding spreading on the same network.

We mention another property of the EVC which is important for our reasoning: we say that the EVC is 'smooth' as one moves over the graph. (More mathematical arguments for this 'smoothness' are given in [15].) When we say that the EVC is 'smooth', we mean that a hypothetical 'explorer' moving over the graph by following links, and observing the EVC values at each node, will not experience sudden, random, or chaotic changes in EVC from one node to the next. In other words, the EVC at a given node will be significantly

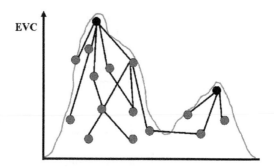

Fig. 7. The EVC-height metaphor applied to a network.

correlated with the EVC of that node's neighbors. This is intuitively reasonable, given the definition of the EVC.

The smoothness of the EVC allows one to think in terms of the 'topography' of the graph. That is, if a node has high EVC, its neighborhood (from smoothness) will also have a somewhat high EVC – so that one can imagine EVC as a smoothly varying 'height', with mountains, valleys, mountaintops, etc. We caution the reader that all standard notions of topography assume that the rippling 'surface' which the topography describes is continuous (and typically two-dimensional, such as the Earth's surface). A graph, on the other hand, is not continuous; nor does it (in general) have a clean correspondence with a discrete versions of a $d$-dimensional space for *any d*. Hence one must use topographic ideas with care. Nevertheless we will appeal often to topographic ideas as aids to the intuition. Our definitions will be inspired by this intuition, but still mathematically precise, and appropriate to the realities of a discrete network.

First we define a 'mountaintop'. This is a point that is higher than all its neighboring points – a definition which can be applied unchanged to the case of a discrete network. That is, if a node's EVC is higher than that of any of its neighbors (so that it is a local maximum of the EVC), we call that node a Center. Next, we know that there must be a mountain for each mountaintop. We will call these mountains *regions*; and they are important entities in our analysis. That is, each node which is not a Center must either belong to some Center's mountain (region), or lie on a 'border' between regions. In fact, our preferred definition of region membership has essentially no nodes on borders between regions. Thus our definition of regions promises to give us just what we wanted: a way to break up the network into well connected clusters (the regions).

Here is our preferred definition for region membership: all those nodes for which a steepest-ascent path terminates at the same local maximum of the EVC belong to the same region. That is, a given node can find which region it belongs to by finding its highest neighbor, and asking that highest neighbor to find *its* highest neighbor, and so on, until the steepest-ascent path terminates at a local maximum of the EVC (i.e., at a Center). All nodes on that path belong to the region of that Center. Also, every node will belong to only one Center, barring the unlikely event that a node has two or more highest neighbors having exactly the same EVC, but belonging to differing regions.

Finally we discuss the idea of 'valleys' between regions. Roughly speaking, a valley is defined topographically by belonging to neither mountainside that it runs between. Hence,

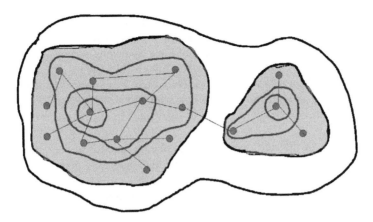

Fig. 8. Our topographic view of a network, with two 'mountains' (regions).

with our definition of region membership, essentially no nodes lie in the valleys. Nevertheless it is useful to think about the 'space' between mountains – it is after all this 'space' that connects the regions, and thus plays an important role in spreading. This 'valley space' is however typically composed only of inter-region links. We call these inter-region links *bridging links*. (And any node which lies precisely on the border may be termed a bridging node.)

Figure 8 offers a pictorial view of these ideas. We show a simple graph with 16 nodes. We draw topographic contours of equal height (EVC). The two Centers, and the mountains (regions) associated with each, are clearly visible in the figure. The figure suggests strongly that the two regions, as defined by our analysis, are better connected internally than they are to one another. Furthermore, from the figure, it is intuitively plausible that spreading (e.g., of a virus) will occur more readily *within* a region than *between* regions. Hence, Figure 8 expresses pictorially the two aims we seek to achieve by using EVC: (i) find the well-connected clusters, and (ii) understand spreading. In the next section we will elaborate on (ii), giving our qualitative arguments for the significance of EVC and regions (as we have defined them) for understanding spreading. We emphasize here that the term 'region' has, everywhere in this paper, a precise mathematical meaning; hence, to discuss subgraphs which lie together in some looser sense, we use terms such as 'neighborhood' or 'area'.

### 3.2. *Topography and the spreading of viruses – qualitative*

Now we apply these ideas to the problem of 'epidemic spreading'. Modeling epidemic spreading on a network involves (among other things) choosing a small set of states that each node can have (susceptible, infected, cured, immune, dead, etc.). Here we study only the simplest form (termed the 'SI model'): each node is either Infected, or Susceptible.

SI MODEL DEFINITION. No node is ever 'cured' of its infection, once infected; and no nodes are immune to the infection.

We believe that there is much to be learned from this simple picture.

In order to understand spreading from a network perspective, we would like somehow to evaluate the nodes in a network in terms of their 'spreading power'. That is, we know that some nodes play an important role in spreading, while others play a less important role. One need only imagine the extreme case of a star: the center of the star is absolutely crucial for spreading of infection over the star; while the leaf nodes are entirely unimportant, having only the one aspect (common to every node in any network) that they can be infected.

Clearly, the case of the star topology has an obvious answer to the question of which nodes have an important role in spreading (have high spreading power). The question is then, how can one generate equally meaningful answers for general and complex topologies, for which the answer is not at all obvious? In this section we will propose and develop a qualitative answer to this question.

Our basic assumption (A) is simple, and may be expressed in a single sentence:

ASSUMPTION (A). Eigenvector centrality (EVC) is a good measure of spreading power.

This idea is reasonable. That is, the spreading power of a node should depend both on its own degree, and on the spreading power of its neighbors – and this is the same verbal criterion we give for the eigenvector centrality. We have tested Assumption (A), via both simulations and theory [17,16]. In this section we will give qualitative arguments which support this assumption; we will then go on to explore the implications. We will see that we can develop a fairly detailed picture of how epidemic spreading occurs over a network, based on (A) and our structural analysis – in short, based on the ideas embodied in Figure 8.

First we note that, because of smoothness, a node with high EVC is embedded in a neighborhood with high EVC. Similarly, nodes of low EVC lie in neighbourhoods of low EVC. Thus, the smoothness of EVC allows us to speak of the EVC of a *neighborhood*. Furthermore, if we take our basic Assumption (A) to be true, then we can speak of the spreading power of a neighborhood. That is, there are no isolated nodes with high spreading power. Instead, there are neighborhoods with high spreading power.

We then suppose that an infection has reached a node with modest spreading power. Suppose further that this node is not a local maximum of EVC; instead, it will have a neighbor or neighbors of even higher spreading power. The same comment applies to these neighbors, until one reaches the local maximum of EVC/spreading power.

Now, given that there are neighborhoods, we can discuss spreading in terms of neighborhoods rather than in terms of single nodes. It follows from the meaning of spreading power that a neighborhood characterized by high spreading power will have more rapid spreading than one characterized by low spreading power. Furthermore, we note that these different types of neighborhoods (high and low) are smoothly joined by areas of intermediate spreading power (and speed).

It follows from all this that, if an infection starts in a neighborhood of low spreading power, it will tend to spread to a neighborhood of higher spreading power. That is: spreading is faster *towards* neighborhoods of high spreading power, because spreading is faster *in* such neighborhoods. Then, upon reaching the neighborhood of the nearest local maximum of spreading power, the infection rate will also reach a maximum (with respect to time).

Fig. 9. Schematic picture of epidemic spreading in one region. The two regions of Figure 8 are now viewed from the 'side', as if they were really mountains. (Left) Infection spreads from an initially infected node (black) toward more central ('higher') regions. (Right) Spreading of the infection reaches a high rate when the most central neighborhood (red) is reached; thereafter the rest of the region is infected efficiently.

Finally, as the high neighborhood saturates, the infection moves back 'downhill', spreading out in all 'directions' from the nearly saturated high neighborhood, and saturating low neighborhoods.

We note that this discussion fits naturally with our topographic picture of network topology. Putting the previous paragraph in this language, then, we get the following: infection of a hillside will tend to move uphill, while the infection rate grows with height. The top of the mountain, once reached, is rapidly infected; and the infected top then efficiently infects all of the remaining adjoining hillsides. Finally, and at a lower rate, the foot of the mountain is saturated.

Figure 9 expresses these ideas pictorially. The figure shows our two-region example of Figure 7, but viewed from the 'side' – as if each node truly has a height. The initial infection occurs at the black node in the left region. It then spreads primarily uphill, with the rate of spreading increasing with increasing 'height' (=EVC, which tells us, by our assumption, the spreading power). The spreading of the infection reaches a maximum rate when the most central nodes in the region are reached; it then 'takes off', and infects the rest of the region.

We see that this qualitative picture addresses nicely the various stages of the classic *S* curve of innovation diffusion [38]. The early, flat part of the *S* is the early infection of a low area; during this period, the infection moves uphill, but slowly. The *S* curve begins to take off as the infection reaches the higher part of the mountain. Then there is a period of rapid growth while the top of the mountain is saturated, along with the neighboring hillsides. Finally, the infection rate slows down again, as the remaining uninfected low-lying areas become infected.

We again summarize these ideas with a figure. Figure 10 shows a typical *S* curve for infection, in the case (SI, as we study here) that immunity is not possible. Above this *S* curve, we plot the expected centrality of the newly infected nodes over time. According to our arguments above, relatively few nodes are infected before the most central node is reached – even as the centrality of the infection front is steadily rising. The takeoff of the infection then roughly coincides with the infection of the most central neighborhood. Hence, the part of Figure 10 to the left of the dashed line corresponds to the left half of Figure 9; similarly, the right-hand parts of the two figures correspond.

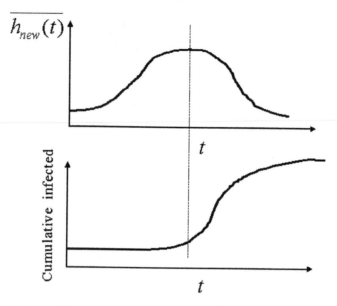

Fig. 10. Schematic picture of the progress of an infection. (Lower) The cumulative number of infected nodes over time. The infection 'takes off' when it reaches highly central nodes. (Upper) Our prediction of the average height (centrality) of newly infected nodes over time. The vertical dashed line indicates roughly where in time the most central node's neighborhood is reached.

One might object that this picture is too simple, in the following sense. Our picture gives an $S$ curve *for a single mountain.* Yet we know that a network is often composed of several regions (mountains). The question is then, why should such multi-region networks exhibit a single $S$ curve?

Our answer here is that such networks need not necessarily exhibit a single $S$ curve. That is, our arguments predict that each region – defined around a local maximum of the EVC – will have a single $S$ curve. Then – assuming that each node belongs to a single region, as occurs with our preferred rule for region membership – the cumulative infection curve for the whole network is simply the sum of the infection curves for each region. These latter single-region curves will be $S$ curves. Thus, depending on the relative timing of these various single-region curves, the network as a whole may, or may not, exhibit a single $S$ curve. For example, if the initial infection is from a peripheral node which is close to only one region, then that region may take off well before neighboring regions. On the other hand, if the initial infection is in a valley which adjoins several mountains, then they may all exhibit takeoff roughly simultaneously – with the result being a sum of roughly synchronized $S$ curves, hence a single $S$ curve.

Let us now summarize and enumerate the predictions we take from this qualitative picture.

(a) *Each region has an S curve.*

(b) *The number of takeoff/plateau occurrences in the cumulative curve for the whole network may be more than one; but it will not be more than the number of regions in the network.*

(c) *For each region – assuming (which will be typical) that the initial infection is not a very central node – growth will at first be slow.*

(d) *For each region (same assumption) initial growth will be towards higher EVC.*

(e) *For each region, when the infection reaches the neighborhood of high centrality, growth "takes off".*

(f) *An observable consequence of (e) is then that, for each region, the most central node will be infected at, or after, the takeoff – but not before.*

(g) *For each region, the final stage of growth (saturation) will be characterized by low centrality.*


### 3.3. *Simulations of epidemic spreading*

We have run simple simulations [17,16] to test this qualitative picture. As noted above, we have implemented an SI model. That is, each node is Susceptible or Infected. Once infected, it remains so, and retains the ability to infect its neighbors indefinitely. We have used a variety of sample networks, extracted from data obtained in previous studies. Here we give a sampling of the results from [17,16], in order to make this chapter more self-contained. Results reported here will be taken from simulations performed on: (i) seven distinct snapshots [25] of the Gnutella peer-to-peer file-sharing network, taken in late 2001; and (ii) a snapshot of the collaboration graph for the Santa Fe Institute [19]. (In this collaboration graph, the nodes are researchers, and they are linked if they have published one or more papers together.) We will denote these graphs as (i) $G1$–$G7$; and (ii) SFI. We would describe the Gnutella graphs as hybrid social/computer networks, since they represent an overlay network on the Internet which mediates contacts between people; while a collaboration graph is purely a social network. Most computer networks are in fact hybrid social/computer networks [13]; Gnutella is simply a self-organized version of this.

Our procedure is as follows. We initially infect a single node. Each link $ij$ in the graph is assumed to be symmetric (consistent with our use of EVC), and to have a constant probability $p$, per unit time, of transmitting the infection from node $i$ to node $j$ (or from $j$ to $i$), whenever the former is infected and the latter is not. All simulations were run to the point of 100% saturation. Thus, the ultimate $x$ and $y$ coordinates of each cumulative infection curve give, respectively, the time needed to 100% infection, and the number of nodes in the graph.

We note that our analysis is essentially the same if we make the more realistic assumption that the probability $p_{ij}$ (per unit time) of transmitting the infection over the link $ij$ is not the same for all links. One needs simply to adjust the weights in the adjacency matrix $A$ accordingly; the resulting EVC values will take these differing link weights into account. Thus virtually all of our discussion will focus on the case $p_{ij} = p$, the same for all links.

**3.3.1.** *Gnutella graphs* The Gnutella graphs, like many self-organized graphs, have a power-law node degree distribution, and are thus well connected. Consistent with this, our analysis returned either one or two regions for each of the seven snapshots. We discuss these two cases (one or two regions) in turn. Each snapshot is taken from [25], and has a number of nodes on the order of 900–1000.

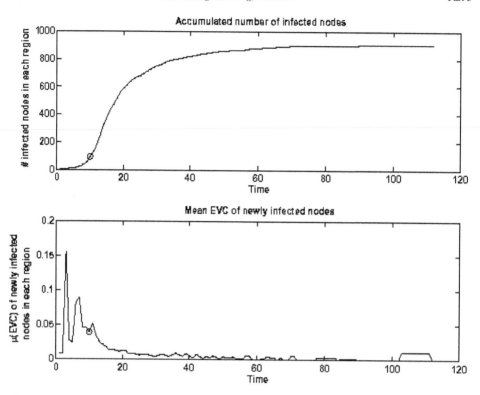

Fig. 11. (Top) Cumulative infection for the Gnutella graph $G3$. The circle marks the time at which the most central node is infected. (Bottom) Average EVC ('height') of newly infected nodes at each time step.

**3.3.1.1.** *Single region*    The snapshot termed '$G3$' was found by our analysis to consist of a single region. Figure 11 shows a typical result for the graph $G3$, with link infection probability $p = 0.05$. The upper part of the figure shows the cumulative $S$ curve of infection, with the most central node becoming infected near the 'knee' of the $S$ curve. The lower part of Figure 11 shows a quantity termed $\mu$ – that is, the average EVC value for newly infected nodes at each time. Our qualitative arguments (Figure 10) say that this quantity should first grow, and subsequently fall off, and that the main peak of $\mu$ should coincide with the takeoff in the $S$ curve. We see, from comparing the two parts of Figure 11, that the most central nodes are infected roughly between time 2 and time 20 – coinciding with the period of maximum growth in the $S$ curve. Thus, this figure supports all of our predictions (a)–(g) above, with the minor exception that there are some fluctuations superimposed on the growth and subsequent fall of $\mu$ over time.

**3.3.1.2.** *Two regions*    Figure 12 shows typical behavior for the graph $G1$, which consists of two regions. We see that the two regions go through takeoff roughly simultaneously. This is not surprising, as the two regions have many bridging links between them. The result is that the sum of the regional infection curves is a single $S$ curve. We also see that each region behaves essentially the same as did the single-region graph $G3$. For instance,

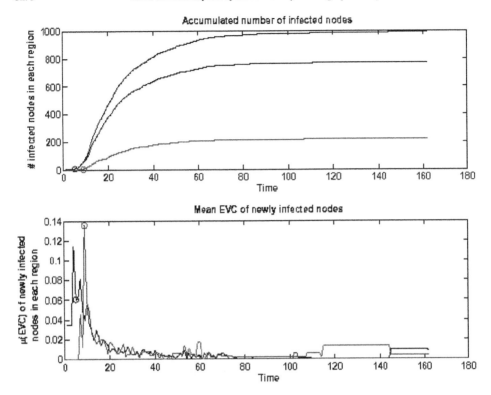

Fig. 12. (Top) Cumulative infection curves, and (bottom) $\mu$ curves, for a two-region graph $G1$, with $p = 0.04$. The upper plot shows results for each region (black and blue), plus their sum (also black). The lower plot gives $\mu$ curves for each region. Note that here, and in all subsequent multiregion figures, we mark the time of infection of the highest-EVC node for each region with a small colored circle.

the new centrality ($\mu$) curves for each region first rise, and then fall, with their main peak (before time 20) coinciding with the period of most rapid growth for the respective regions (and for the whole graph). These results thus add further support to our predictions (a)–(g).

**3.3.2.** *Santa Fe Institute*    The SFI graph gives three regions under our analysis. The spreading behavior varied considerably on this graph, depending both on the initially infected node, and on the stochastic outcomes for repeated trials with the same starting node.

First we present a typical case for the SFI graph. Figure 13 shows the behavior for infection probability $p = 0.04$, and a randomly chosen start node. The interesting feature of this simulation is that the cumulative curve shows very clearly two takeoffs, and two plateaux. That is, it resembles strongly the sum of two $S$ curves. And yet it is easy to see how this comes about, from our region decomposition (upper half of Figure 13): the blue and red curves take off roughly simultaneously, while the largest (black) region takes off only after the other regions are saturated.

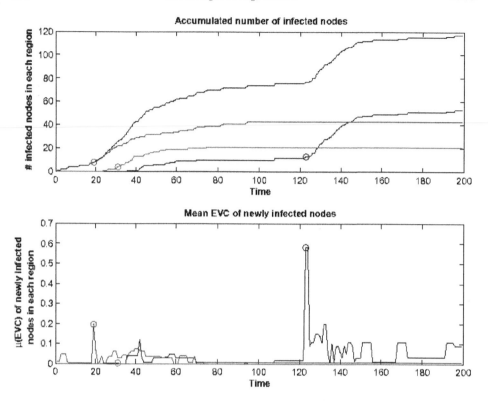

Fig. 13. Spreading on the SFI graph, $p = 0.04$, with a randomly chosen start node. Each of the three regions (as found by our analysis) are plotted with a distinct color. Also, in the upper plot, we show the total cumulative number of infected nodes in black; this is the sum of the other three curves in the upper plot. Note that the $\mu$ curve for the smallest (red) region is not visible on the scale of the lower plot.

The $\mu$ curves show roughly the expected behavior – the qualification being that they are rather noisy. Nevertheless the main peak of each $\mu$ curve corresponds to the main rise of the corresponding region's $S$ curve.

We present one more example from the SFI graph. Figure 14 shows a simulation with a different start node from Figures 13, and with $p = 5\%$. The message from Figure 14 is clear: the cumulative infection curve shows clearly three distinct $S$ curves – takeoff followed by plateau – one after the other. It is also clear from our regional infection curves that each region is responsible for one of the $S$ curves in the total infection curve: the smallest (red) region takes off first, followed by the blue region, and finally the largest (black) region. In each case, the time of infection for the most central node of the region undergoing takeoff lies very close to the knee of the takeoff. And, in each case, the peak of the $\mu$ curve coincides roughly with the knee of the takeoff.

We note that the behavior seen in Figure 14 is neither very rare nor very common. The most common behavior, from examination of over 50 simulations with this graph, is most like that seen in Figure 13; but we have seen behavior which is most like a single $S$ curve (rare), as well as behavior like that seen in Figures 13 and 14.

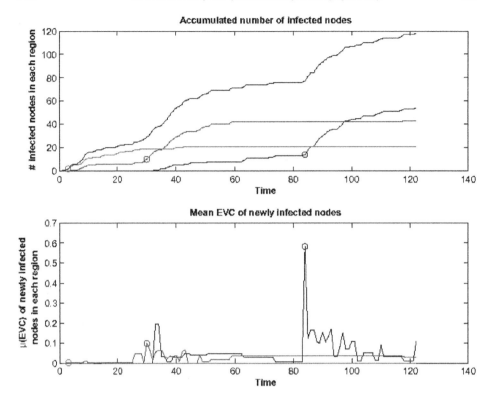

Fig. 14. Spreading on the SFI graph, $p = 0.05$. In this case, the existence of three regions in the graph is clearly seen in the total cumulative infection curve.

Thus the behavior of spreading on the SFI graph gives the strongest confirmation yet of our prediction b: that, for a graph with $r$ regions, one may see up to $r$, but not more than $r$, distinct $S$ curves in the total cumulative infection curve. All of our observations, on the various graphs, are consistent with our predictions; but it is only in the SFI graph that we clearly see all of the (multiple) regions found by our analysis to be present in the graph.

One very coarse measure of the well-connectedness of a network is the number of regions in it, with a high degree of connectivity corresponding to a small number of regions, and with a large number of regions implying poor connectivity. By this very coarse criterion, the SFI graph is as well connected as another three-region graph studied by us (the 'HIO graph', reported in [17] and in [18]). Based on our spreading simulations on this HIO graph (reported in [17]), we would say that in fact the SFI graph is less well connected than the HIO graph. We base this statement on the observation that, for the HIO graph, we never found cases where the different regions took off at widely different times – the three regions are better connected to one another than is the case for the SFI graph. (In detail: the three regions define three *pairs* of regions; and these three pairs are bridged by 57, 50 and 7 bridge links.)

Examination of the SFI graph itself (Figure 15) renders this conclusion rather obvious: the three regions are in fact connected in a linear chain. (The three pairs of regions have

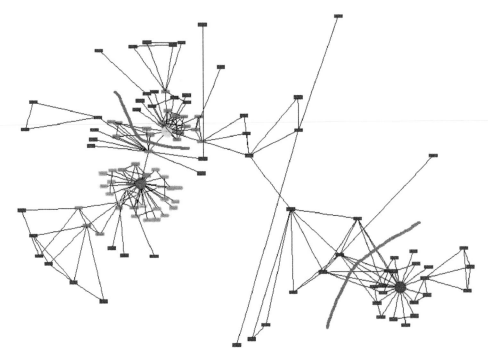

Fig. 15. A visualization of the collaboration graph for the Santa Fe Institute. Colors of nodes code for eigenvector centrality, with warm colors implying higher EVC. Also, the most central node in each region is denoted with a round circle. Thick red lines show boundaries between regions. The leftmost region (with a magenta center) is the largest, and has black curves in Figures 13 and 14; the rightmost region is the smallest (red curves in Figures 13 and 14); and the middle region has blue curves in Figures 13 and 14. Hence, in the text, we refer to these three regions, respectively, as the 'black region', the 'blue region' and the 'red region'.

6, 10 and 0 bridging links.) Thus, it is not surprising that we can, in some simulation runs, clearly see the takeoff of each region, well separated in time from that of the other regions. The HIO graph, in contrast, has many links between each pair of regions. We note also in this context that the Gnutella graphs are very well connected; those that resolve to two regions always have numerous links between the two.

Finally, we note that the SFI graph, while clearly conforming to our assertion b (we see up to three regions), does not wholly agree with assertion a (that each region will have a clear $S$ curve). For instance, in Figure 13, the black region has a small 'premature' takeoff around time 40, giving a visible plateau before the 'main' takeoff after time 120. Similarly, both the blue and the black regions show such 'premature' takeoffs in Figure 14. These observations are not exceptional: the blue region in particular is prone to such behavior, showing two stages of growth in over half of the simulations we have run on the SFI graph. This weakens the support for assertion a; we would then say that a single $S$ curve for a single region is a rule that is followed in most cases, but not all. However these two-stage takeoffs do not contradict the other predictions on our list. For instance, there is always a corresponding 'premature hump' in the $\mu$ curve for the region in question.

It is clear from Figure 15 where the 'premature takeoffs' arise. The black region includes a subcluster which is easily identified visually in Figure 15 (close to the next region). This subcluster is connected to the remainder of the black region by a single link. These same statements hold true for the blue region – it has a subcluster, joined to it by a single link, but close to the rightmost region. Hence, in Figure 14, for example, the infection starts on the 'far side' of the red region. Then it spreads to the adjacent subcluster of the blue region, but does not reach the most central nodes of that region until after about 20 more units of time (about $1/p$). Soon after the center of the blue region is infected, the subcluster of the black region begins to be infected; but again there is a delay (about twice $1/p$) before reaching the main center of the black region.

### 3.4. *Theory*

In [17] we have developed a mathematical theory for the qualitative ideas expressed here. We have focused on two aspects there, which we will simply summarize here.

**3.4.1.** *Definition of spreading power*   The first problem is to try to quantify and make precise our Assumption (A). Since (A) relates two quantities – spreading power and EVC – and the latter is precisely defined, the task is then to define the former, and then to seek a relation between the two.

Such a relation is intuitively reasonable. A node which is connected to many well-connected nodes should have higher spreading power, and higher EVC, than a node which is connected to equally many, but poorly connected, nodes. We have offered a precise definition of spreading power in [17]. Our reasoning has two steps: first we define an 'infection coefficient' $C(i, j)$ between any pair of nodes $i$ and $j$. This is simply a weighted sum of all non-self-retracing paths between $i$ and $j$, with lower weight given to longer paths. Thus many short paths between two nodes gives them a high infection coefficient. Our definition is symmetric, so that $C(i, j) = C(j, i)$.

Next we define the spreading power of node $i$ to be simply the sum over all other nodes $j$ of its infection coefficient $C(i, j)$ with respect to $j$. As long as the graph is connected, every node will have a nonzero $C(i, j)$ with every other, thus contributing to the sum. Hence each node has the same number of terms in the sum; but the nodes with many large infection coefficients will of course get a higher spreading power.

We then show in [17] that one can make a strong connection between this definition of spreading power and the EVC, if one can ignore the restriction to non-self-retracing paths in the definition. We restrict the sum to non-self-retracing paths because self-retracing paths do not contribute to infection in the SI case. This restriction makes the obtaining of analytical results harder.

**3.4.2.** *Mathematical theory of SI spreading*   We have given in [17] exact equations for the propagation of an infection, for arbitrary starting node, in the SI case. These equations are stochastic – expressed in terms of probabilities – due to the probabilistic model for spreading over links. They are not generally solvable, even in the deterministic case when $p = 1$. The problem in the latter case is again the need to exclude non-self-retracing paths.

However we have performed an expansion in powers of $p$ for the time evolution of the infection probability vector. This expansion shows that the dominant terms are those obtained by naively applying the adjacency matrix (i.e., ignoring self-retracing paths because they are longer, hence higher order in $p$). The connection to EVC is then made: naively applying the adjacency matrix gives weights (infection probabilities) which approach a distribution proportional to the EVC. Hence we get some confirmation for our claim that, in the initial stages of an infection, the front moves towards higher EVC.

An interesting observation that remains to be explained is the insensitivity of the time evolution (see [17]) to the probability $p$. We note that short paths receive the most weight in two limits: $p \to 0$ and $p \to 1$. The observed insensitivity to $p$ suggests that such short paths dominate for all $p$ between these limits; but we know of no proof for this suggestion.

### 3.5. *Some ideas for design and improvement of networks*

Now we apply the ideas discussed above to the problem of design of networks [13]. Our ideas have some obvious implications for design. We begin with a discussion of how to *help* spreading, by modifying the topology of a given network.

**3.5.1.** *Measures to improve spreading*    We frame our ideas in terms of our topographic picture. Suppose that we wish to design, or modify the design of, a network, so as to improve its efficiency with respect to spreading. It is reasonable, based on our picture, to assume that a *single region* is the optimal topology for efficient spreading. Hence we offer here two ideas which might push a given (multi-region) network topology in this direction:

*To improve spreading:*

1. *We can add more bridge links between the regions.*
2. *As an extreme case of 1, we can connect the Centers of the regions.*

Idea 2 is a 'greedy' version of idea 1. In fact, the greediest version of idea 2 is to connect *all* Centers to *all*, thus forming a complete subgraph among the Centers. However, such greedy approaches may in practice be difficult or impossible. There remains then the general idea 1 of building more bridges between the regions. Here we see however no reason for not taking the greediest practical version of this idea. That is: build the bridges between nodes of high centrality on both sides – preferably, as high as possible. Our picture strongly suggests that this is the best strategy for modifying topology so as to help spreading.

We note that the greediest strategy is almost guaranteed to give a single-region topology as a result. Our reasoning is simple. First, the existing Centers cannot all be Centers after they are all connected one to another – because two adjacent nodes cannot both be local maxima of the EVC (or of anything else). Therefore, either new Centers turn up among the remaining nodes as a result of the topology modification, or only one Center survives the modification. In the latter case we have one region. The former case, we argue, is unlikely: we note that the EVC of the *existing* Centers is (plausibly) strengthened (raised) by the modification *more* than the EVC of other nodes. That is, we believe that connecting existing centers in a complete subgraph will 'lift them up' with respect to the other nodes, as well as bringing them closer together. If this 'lifting' idea is correct, then we end up with a single Center and a single region.

We have tested the 'greediest' strategy, using the SFI graph as our starting point. We connected the three Centers of this graph (Figure 15) pairwise. The result was a single region, with the highest Center of the unmodified graph at its Center. We then ran 1000 SI spreading simulations on this modified graph, with $p = 10\%$, and collected the results, in the form of time to saturation for each run.

In our first test, we used a random start node (first infected node) in each of the 1000 runs. For the unmodified SFI graph, we found a mean saturation time of 83.8 time units. For the modified, single-region graph, the corresponding result was 64.0 time units – a reduction by 24%. Note that this large reduction was achieved by adding only three links to a graph which originally had about 200 links among 118 nodes.

Figure 16 gives a view of the distribution of saturation times for these two sets of runs. We show both the binned data and a fit to a gamma probability distribution function. Here we see that the variance as well as the mean of the distribution is reduced by the topology improvement.

Our next set of tests used a start node located close to the highest Center (the only Center in the modified graph). These tests thus involve another form of 'help', namely, choosing a strategically located start node. We wish then to see if the new topology gives a

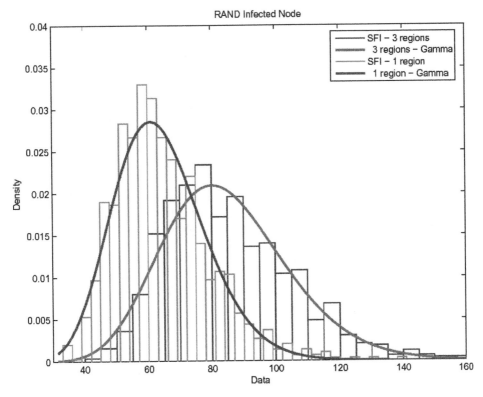

Fig. 16. Distribution of saturation times for 1000 runs on the SFI graph, with a random start node. Left (blue): with the addition of three links connecting the three Centers. Right (red): original graph, with three regions.

Table 1

Mean time to saturation for the SFI graph, with and without two kinds of assistance: connecting the Centers of the three regions, and choosing a well placed start node

|  | Random start node | High-EVC start node |
|---|---|---|
| Original graph | 83.8 | 68.9 |
| Connect Centers | 64.0 | 56.0 |

significant difference even in the face of this strategic start. Our results, for mean saturation time over 1000 runs, were: 68.9 time units for the unmodified SFI graph, and 56.0 units for the modified graph. Thus, connecting the three Centers of the SFI graph reduced the mean saturation time by almost 20%, even though a highly advantageous node was used to start the infection. Also, as in Figure 16, the variance was reduced by the topology improvement. We summarize these results in Table 1. From Table 1 we get some idea of the relative efficacy of the two kinds of help. That is, our base case is the unmodified graph with random start node (83.8). Choosing a strategic start node gave about an 18% reduction in mean saturation time. Improving the topology, without controlling the start node, gave almost 24% reduction. Thus we see that, for these limited tests, both kinds of help give roughly the same improvement in spreading rate, with a slight advantage coming from the greedy topology change. Finally, note that using *both* kinds of help in combination gave a reduction of mean saturation time of about 1/3. We note that both of these kinds of help are based on our topology analysis, and cannot be implemented without it.

**3.5.2.** *Measures to prevent spreading*    Now we address the problem of designing, or re-designing, a network topology so as to hinder spreading. Here life is more complicated than in the helping case. The reason for this is that we build networks in order to support and facilitate communication. Hence we cannot simply seek the extreme, 'perfect' solution – because the ideal solution for hindering spreading is one region per node, i.e., discon-nect all nodes from all others! Instead we must consider *incremental* changes to a given network. We consider two types of 'inoculation' strategies – inoculating nodes (which is equivalent to removing them, as far as spreading of the virus is concerned), or inoculating links (which is also equivalent to removing them). Again we offer a list of ideas:

    *To hinder spreading:*
1. *We can inoculate the Centers – along with, perhaps, a small neighborhood around them.*
2. *We could instead find a ring of nodes surrounding each Center (at a radius of perhaps two or three hops) and inoculate the ring.*
3. *We can inoculate bridge links.*
4. *We can inoculate nodes at the ends of bridge links.*

We note that ideas 1 and 2 are applicable even in the case that only a single region is present. Ideas 3 and 4 may be used when multiple regions are found. Note that inoculating a bridge link (idea 3) is not the same as inoculating the two nodes which the link joins (idea 4): inoculating a node effectively removes that node and *all* links connected to it, while inoculating a link removes only that link.

Also, with link inoculation, one has the same considerations as with link addition – namely, the *height* of the link matters. We define the 'link EVC' to be the arithmetic mean of the EVC values of the nodes on the ends of the link. Ideas 3 and 4 are then almost certainly most effective if the bridging links chosen for inoculation have a relatively high link EVC.

In order to get some idea of the merits of these four ideas, we have tested two of them: ideas 1 and 3. (Ideas 1 and 2 are similar in spirit, as are ideas 3 and 4.) First we give results for idea 3. We again start with the SFI graph, and measure effects on saturation time, averaged over 1000 runs, with a random start node for each run. We have tested two strategies for bridge link removal: (i) remove the $k$ bridge links between each region pair that have the *lowest* link EVC; and (ii) remove the $k$ bridge links between each region pair that have the *highest* link EVC. As may be seen by inspection of Figure 15, the highest and middle regions have 6 bridge links joining them, while the middle and lowest regions have ten (and the highest and lowest regions have no bridging links between them). Hence we have tested idea 3, in these two versions, for $k = 1$ and for $k = 3$ (which is possible without disconnecting the graph). Our confidence that strategy (ii) – remove the $k$ *highest* bridge links – will have much more effect is borne out by our results. For example, we have tested the strategy of removing the $k = 3$ highest bridge links. (We find that the effect of removing the lowest three bridge links is negligible.) The results for removing the highest bridge links, in contrast, show a significant retardation of saturation time as a result of removing the top three bridge links from between each pair of regions.

We have also, for comparison purposes, removed the same number of links, but at random. For example, for the case $k = 1$, we remove one bridge link between each connected pair of regions, hence two total; so, for comparison, we do the same, but choosing the two links randomly (with a new choice of removed links for each of the 1000 runs).

Table 2 summarizes the results of our bridge-link removal experiments, both for $k = 1$ and for $k = 3$. In each case we see that removing the highest bridges has a significantly larger retarding effect than removing the lowest – and that the latter effect is both extremely small, and not significantly different from removing random links. We have examined growth curves for some of the bridge-link removal experiments. These curves reveal a surprising result: the number of regions actually *decreases* (for $k = 3$) when we deliberately remove bridge links. Specifically, removing the lowest three links between region pairs changes three regions to two; and removing the highest three links changes three regions to one.

Figure 17 shows fairly representative infection growth curves for the $k = 3$ case, contrasting random link removal (top) with removal of bridge links with highest EVC (bot-

Table 2
Mean saturation times for two sets of bridge-removal experiments on the SFI graph

|                | $k = 1$ | $k = 3$ |
|----------------|---------|---------|
| Reference      | 82.9    | 83.3    |
| Remove random  | 84.3    | 87.1    |
| Remove lowest  | 84.4    | 85.8    |
| Remove highest | 87.7    | 96.5    |

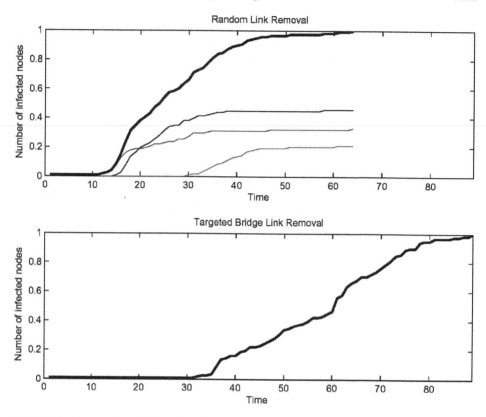

Fig. 17. Infection curves for the SFI graph. (Top) Six randomly chosen links have been removed. (Bottom) The three highest links between the black and blue regions, and also those between the blue and red regions, have been removed. The result is a graph with a single region.

tom). The upper part of Figure 17 offers few surprises. In the bottom part, we see a single region, which nevertheless reaches saturation more slowly than the three-region graph of the upper plot. It is also clear, from the rather irregular form of the growth curve in the bottom part of Figure 17, that the single region is not internally well connected.

Thus, our attempt to isolate the three regions even more thoroughly, by removing the 'best' bridge links between them, has seemingly backfired! Some reflection offers an explanation. It is clear from examining Figure 15 that removing the three highest links between the highest and middle regions will strongly affect the centrality (EVC) of the nodes near those links. In particular, the Center (round yellow node in Figure 15) of the middle region (giving blue growth curves in Figures 13 and 14) will have its EVC significantly lowered by removing the strongest links from its region to the highest region. If its EVC is significantly weakened, it (plausibly) ceases to be a local maximum of the EVC, so that its region is 'swallowed' by the larger one. A similar effect can make the lowest Center also cease to be a Center. This picture is supported by the fact that the Center of the modified, one-region graph (bottom curve in Figure 17) is the same Center as that of the highest region in the original graph (round, magenta node in Figure 15).

Next we test the idea of inoculating the Centers. We did not use the SFI graph for these tests, because the graph breaks into large pieces if one inoculates (removes) either of the two high-EVC Centers (see Figure 15). Hence we used the HIO graph from [18]. This graph also has three regions, and hence three Centers which can be inoculated. We find that, as with the SFI graph, removing these Centers also isolates some nodes – but (unlike with the SFI graph) only a few. Hence we simply adjusted our spreading simulation so as to ignore these few isolated nodes (13 in number, in the case of inoculating all three Centers).

As noted earlier, we regard the HIO graph as being better connected than the SFI graph (even though both resolve to three regions). Hence we might expect the HIO graph to be more resilient in the face of attacks on its connectedness – in the form of inoculations. In fact, we only found a slight increase in average saturation time, even when we inoculated all three Centers. For this test on the HIO graph, we found the mean saturation time to increase from 71.5 to 74.8 – an increase of about 5%. We know of no valid way to compare this result quantitatively with the corresponding case for the SFI graph – because, in the latter case, so many nodes are disconnected by removing the Centers that the mean saturation time must be regarded as infinite (or meaningless).

It is also of interest to compare our two approaches to hindering spreading (inoculate Centers, or inoculate bridges) on the same graph. Our results so far show a big effect from inoculating bridges for the SFI graph, and a (relatively) small effect for inoculating Centers for the HIO graph. Can we conclude from this that inoculating bridges is more effective than inoculating Centers? Or is the difference simply due to the fact that the SFI graph is more poorly connected, and hence more easily inoculated? To answer these questions, we have run a test of idea 3 on the HIO graph. Specifically, we ran the $k = 3$ case, removing 9 links total. The results are given in Table 3. Here we see that the $k = 3$ strategy, which worked well on the SFI graph when targeted to the highest-EVC links, has almost no effect (around 1%) for the HIO graph. This is true even when we remove the highest links, and remove 9 of them (as opposed to 6 for the SFI case). Hence we find yet another concrete interpretation of the notion of well-connectedness: the HIO graph is better connected than the SFI graph, in the sense that is it much less strongly affected by either bridge link removal or Center inoculation (which causes the SFI graph to break down).

We also see that removing the three Centers from the HIO graph has a much larger effect (5% compared to 1%) than removing the three highest bridge links between each pair of regions. (Note however that removing a Center involves removing many more than three links.) This result is reminiscent of a well-known result for scale-free graphs [3]. Scale-free graphs have a power-law degree distribution, and are known, by various criteria, to be very well connected. In [3] it was shown however that removing a small percentage of nodes – but those with the highest node degree – had a large effect on the connectivity of the graph

Table 3
Mean saturation times for two sets of bridge-removal experiments on the HIO graph

|                  | $k = 3$ |
| --- | --- |
| Remove random   | 71.2 |
| Remove lowest   | 72.1 |
| Remove highest  | 71.9 |

(as measured by average path length); while removing the same fraction of nodes, but randomly chosen, had essentially no effect. That is, in such well-connected networks, the best strategy is to vaccinate the most connected nodes – a conclusion that is also supported, at this general level of wording, by our results here. In our center-inoculation experiments, we remove again a very small fraction of well-connected nodes, and observe a large effect. However our node removal criterion is neither the node degree, nor even simply the EVC: instead we focus on the EVC of the node *relative to the region it lies in*. Also, our measure of effect is not path length, but rather the saturation time – a direct measure of spreading efficiency. Hence our approach is significantly different from that in [3] (and related work such as [21]); but our general conclusion is similar.

## 4. Directed graphs

Now we address questions similar to those of the previous section – but with the single difference that we assume the graph is directed. We recall the discussion in Section 1, where we described in some detail the 'generic' structure of a directed graph. Our aim in this section is then to build on that picture, while seeking generalizations of the notions of centrality which have been applied to undirected graphs (and discussed in Section 2). Also, we will seek to find *applications* for our definitions of centrality.

We briefly recapitulate our picture of a generic directed graph. Paths are assumed to be directed, i.e., following the arrows. In contrast, if we explicitly specify an 'undirected path', we mean one in which the direction of the arrows is ignored. We will work only with weakly connected directed graphs here – that is, directed graphs for which there is an undirected path from any node $i$ to any other node $j$. Such graphs are typically composed of several strongly connected components (SCCs); in any such component $C$, there is a directed path from any node $i \in C$ to any other node $j \in C$. The SCCs are then connected by exclusively one-way connections; and the 'component graph', in which each SCC is simply represented as a node, is a directed acyclic graph or DAG. If we imagine 'flows' in such graphs (following the arrows), we see then that such flows are, at the level of the component graph (i.e., in terms of SCCs), simply *one-way*: flows start at *source* SCCs (which have no inlinks), proceed via intermediate SCCs (having both in- and outlinks), and end at *sink* SCCs (having no outlinks). In this 'flow' picture, each SCC is then itself a kind of 'eddy' in the flow, since flows can circulate indefinitely in an SCC (assuming of course that the SCC has at least two nodes).

Now we develop notions of centrality for directed graphs.

### 4.1. *Centrality for directed graphs*

The notion of degree centrality is readily extended to directed graphs; we simply extend the single idea to two, since each node $i$ in a directed graph has both an indegree and an outdegree ($k_i^{in}$ and $k_i^{out}$). Otherwise the notion is very much like that for undirected graphs; in particular, it is purely local.

The notion of betweenness, as defined in Section 2 for undirected graphs, may be useful also for directed graphs. In fact, the generalization is immediate. Betweenness is defined in terms of counting shortest paths between nodes; hence one only need add the condition that the shortest paths are valid directed paths. To our knowledge however there has been very little work done with betweenness on directed graphs.

What about eigenvector centrality (EVC)? We have made heavy use of this centrality measure in for undirected graphs; what is its analog for directed graphs?

The answer is not simple: in fact, there is no one measure which uniquely corresponds to EVC if one ignores the direction of the links.

*For directed graphs there are* many *centrality measures,* all *of which reduce to the EVC if one ignores the direction of the links.*

**4.1.1.** *'Backward' centrality*  The most obvious generalization of the EVC is simply to take the dominant eigenvector $\mathbf{d}$ of the adjacency matrix $A$, and then let $d_i$ be the 'directed eigenvector centrality' for node $i$. (We reserve the symbol $\mathbf{e}$ for the principal eigenvector of a symmetric $A$ – that is, for the EVC vector.) This definition appears in fact to be entirely reasonable; let us see what it involves.

When the links are directed, we prefer to call the adjacency matrix the 'backwards operator' $\widehat{B}$. The reason for this is that right multiplication of a weight vector $\mathbf{w}$ with the matrix $A$ – plus the assumption that the element $A_{ij}$ in the $i$th row and $j$th column of $A$ is 1 if $i$ points to $j$ (i.e., $A_{ij} = A_{i \to j}$) – gives the result that all weights are sent *against* the arrows, i.e., backwards. (This is easily seen for a two-node directed graph, with one link: $1 \to 2$.)

While these conventions are not unique, they are convenient, and we will use them here. The important point is that, for directed graphs – regardless of choice of convention – there exists both a Forward operator $\widehat{F}$ (sending weights forward) and a Backward operator $\widehat{B}$; and they are not the same.

*Given our conventions,* $\widehat{F} = A^{\mathrm{T}}$ *and* $\widehat{B} = A$.

Thus, we find that the most obvious generalization of the EVC to directed graphs (along with our choice of conventions) tells us to take each node's centrality from the corresponding element of the dominant eigenvector of a Backward operator. Now we ask, what does this eigenvector look like? We will give a brief and intuitive answer here – saving a more detailed discussion for the more intuitively easy Forward operator (and referring the reader to [14] for further details). The dominant eigenvector $\mathbf{b}$ for the Backward operator $\widehat{B}$ has the property that no element of $\mathbf{b}$ is negative. This is good – we wish to avoid dealing with the idea of negative centrality. However, in general, *many* of the elements $b_i$ are zero – thus giving us *zero centrality* from many nodes in a directed graph. This is an undesirable property: if many or even most of the nodes of a graph are given zero centrality, then the centrality measure loses its utility.

We can see intuitively why these zero weights occur. We recall that the dominant eigenvector may be seen as the limiting weight vector, obtained after many iterations of multiplication by the corresponding operator (in this case, $\widehat{B}$). The Backward operator simply sends weights against the arrows. If we consider an SCC with inlinks but no outlinks – that is, a sink SCC (call it $S$) – sending weight against the arrows amounts to sending weight *from* the SCC $S$, but never *to* it. Thus it is plausible that, after some iterations, $S$ ends up with zero weight. Furthermore, there will likely be SCCs which depend on 'input' from the SCC $S$ for their weight. These SCCs are thus also in danger of ending up with zero weight after further iterations of $\widehat{B}$ – as are any other SCCs which are 'downstream' of *these* (etc.). In short: it seems that repeated iterations of $\widehat{B}$ will tend to send most or all of the weight to the source SCCs – which are at the 'end of the line' for backwards flow. SCCs at the 'upstream' side of the flow will tend to 'dry up' and so receive zero weight. Thus we get:

> *As a first, rough, guess: the dominant eigenvector* **b** *for the backward operator* $\widehat{B}$
> *has positive weight at sources, and the rest of the nodes get zero weight.*

This intuitive picture is mostly correct [14]. We will give a more correct and detailed picture below, in our discussion of the Forward operator $\widehat{F}$.

**4.1.2.** *'Forward' centrality*   Our picture for the Forward operator $\widehat{F}$ is very much like that for $\widehat{B}$. However we will offer more detail here as the reader's (and the authors') intuition works more readily when the flow is with the arrows. Thus, our intuition says that repeated operation with $\widehat{F}$ will tend to send all weight to the downstream (sink) SCCs, leaving upstream (source, and intermediate) SCCs with little or no weight. Now let us make this picture more precise.

The crucial factor that is missing from our picture is the 'eddying' property of SCCs. In fact, weight propagation within an SCC does not just circulate – it also, in general, involves *growth* of the circulating weights. This follows from the non-normalized, hence non-weight conserving, nature of both $\widehat{F}$ and $\widehat{B}$. Furthermore, the rate of growth which is seen after many iterations, for any given SCC, may be stated quite precisely: it is the largest eigenvalue of that SCC, *when it is considered as an isolated directed graph*. These eigenvalues are almost always greater than one. (Some exceptions: (i) a one-node SCC, whose only eigenvalue is 0; (ii) a directed ring of nodes, without shortcutting links, which has a dominant eigenvalue of 1.)

Now however we seem to have a contradiction: we say that weights vanish in upstream SCCs; yet we also say that weights grow with a factor $\lambda > 1$ for these same SCCs. And in fact the key to understanding the asymptotic weight distributions for both $\widehat{F}$ and $\widehat{B}$ is to accept and reconcile *both* of these statements: the weights grow in most SCCs, and yet they can also vanish in most SCCs.

The reason that both statements can be true is simply that weights can grow in an *absolute* sense, but vanish in a *relative* sense. More precisely: consider two SCCs 1 and 2, with SCC1 growing at rate $\lambda_1 > 1$ and SCC2 at rate $\lambda_2 > \lambda_1 > 1$. After $n$ iterations, SCC1 has grown, relative to SCC2, by a factor $(\lambda_1/\lambda_2)^n$. This factor becomes vanishingly small for large $n$. Hence SCC1, in this picture, has weights which do grow under iteration – but they grow too slowly, and so vanish on the scale of the whole graph.

Our picture is now almost complete. Let us think through an example, still using Forward propagation. Suppose SCC $\Sigma$ has the dominant eigenvalue $\Lambda$, and that this eigenvalue is larger than that of any other SCC. Thus $\Lambda$ is the dominant eigenvalue for the entire graph (because Perron–Frobenius theory [32] tells us that the set of eigenvalues of the whole graph is simply the union of the sets of eigenvalues of the SCCs). Clearly, the SCC $\Sigma$ will have positive weight in the dominant eigenvector of $\widehat{F}$. Furthermore, because the SCC $\Sigma$ is after all strongly connected, *every* node in $\Sigma$ will have positive ($> 0$) weight.

Now we come to the final aspect of the picture developed more mathematically in [14]. We consider an SCC $X$ which is directly downstream of $\Sigma$. The key (final) point is then that $\Sigma$ is, with each iteration, injecting weight into $X$, and furthermore that this injected weight grows with a factor $\Lambda$ at each iteration. Thus, $\Sigma$ is trying to 'force' $X$ to grow at a rate $\Lambda$ – which is larger than $X$'s 'natural' growth rate, i.e., its dominant eigenvalue $\lambda_X < \Lambda$. Thus we have two possibilities: (i) there is no consistent solution to the difference between these two rates; or (ii) there is a consistent solution, which allows the 'driven' SCC $X$ to grow at the rate $\Lambda$ – consistent with the rate of growth of the injected weights.

Some simple mathematical checks [14] show that (ii) is always the case. Thus we find that $X$ also has nonzero weight in the dominant eigenvector $\mathbf{f}$ of $\widehat{F}$. Similar arguments give nonzero weight in every SCC which is 'driven' by $\Sigma$ – either directly (as in the case of $X$), or indirectly (for example, an SCC into which $X$ injects weight). All of these nonzero weights are positive (as required by Perron–Frobenius [32]). At the same time, any SCC which is *not* downstream of $\Sigma$ will have exactly zero weight in $\mathbf{f}$.

These statements are precise because the notion of being downstream is precise – which follows from the acyclic nature of the component graph, showing the relationships between the various SCCs. For any two given SCCs 1 and 2 of a directed graph, there are in fact [14] three distinct possible values for this relationship between SCC1 and SCC2: we can have (i) 1 is downstream of 2; (ii) 1 is upstream of 2; or (iii) 1 is neither downstream nor upstream of 2. There is no fourth possibility ('both upstream and downstream'), because if the injection is two-way then the two SCCs become one SCC. Furthermore, both possibilities (ii) and (iii) are implied by the statement 'SCC1 is not downstream of SCC2'.

In short:

> *SCC $\Sigma$, along with all SCCs which are downstream of $\Sigma$, has positive weight – for all nodes – in the eigenvector $\mathbf{f}$. All other nodes for all other SCCs (not downstream of $\Sigma$) have zero weight in $\mathbf{f}$.*

These statements are precise; and they replace the less precise statements made earlier.

The same statements hold for the eigenvector $\mathbf{b}$ of the operator $\widehat{B}$ – with the one change that 'downstream' must be reinterpreted, to mean 'with the flow direction of $\widehat{B}'$.

We summarize our picture with a figure. Figure 18 shows a generic directed graph, in terms of its component graph (a DAG of the SCCs). The size (diameter) of each SCC, as it is portrayed in the figure, indicates the size of the largest eigenvalue for that SCC. Thus the SCC corresponding to the 'dominant' SCC $\Sigma$ (in the discussion above – that

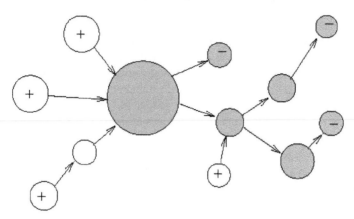

Fig. 18. A schematic view of the distribution of positive (shaded) and zero (unshaded) weights in a generic directed graph. Each SCC is a single bubble here; the size of the bubble indicates the size of the principal eigenvalue (the 'gain') for each SCC. The weights have been generated by a Forward operator. Source SCCs are indicated with $+$, and sink SCCs with $-$.

with largest eigenvalue $\Lambda$) is the largest bubble in the figure. All other shaded SCCs satisfy two conditions: (i) they are downstream from $\Sigma$; and (ii) their largest eigenvalue is less than $\Lambda$. (Of course, *all* other SCCs satisfy (ii).) These SCCs will have positive weight, for all nodes, in the eigenvector **f**. The unshaded SCCs are 'not downstream' with respect to $\Sigma$; they have zero weight in **f**. More precisely: the unshaded SCC farthest to the right (a source) is neither upstream nor downstream of $\Sigma$; all of the other unshaded SCCs are upstream of $\Sigma$.

The picture for the **b** eigenvector is readily visualized from Figure 18. The dominant SCC $\Sigma$ remains shaded. All of the other shaded SCCs in the figure have zero weight in **b**. Also, the 'neither upstream nor downstream' SCC has zero weight in **b**. All of the remaining unshaded SCCs in the figure will have positive weight in **b**, since they are 'downstream' from $\Sigma$ under the action of $\widehat{B}$.

**4.1.3.** *PageRank and 'sink remedies'*    Our two operators $\widehat{F}$ and $\widehat{B}$, giving their respective dominant eigenvectors **f** and **b**, do not exhaust the possible generalizations of EVC to directed graphs. We will discuss one more in this subsection, and two more in the next.

Suppose that we *normalize* the forward operator $\widehat{F}$. That is, we divide each (nonzero) column of $\widehat{F}$ by the column sum of elements, so that operation with $\widehat{F}$ is weight conserving. The resulting operator $\widehat{F}^{(norm)}$ has a dominant eigenvalue equal to 1. Let us call the corresponding dominant eigenvector **p**. What can we say about **p**? Can it also give us a useful measure of centrality?

Our experience with EVC leads us to suspect that normalization of the matrix does not give useful, nonlocal information in the dominant eigenvector. (Recall that, for the undirected case, normalizing the matrix gave simply the node degree in the dominant eigenvector.) This suspicion is however misleading: the vector **p** *does* in fact include nonlocal and useful information.

The simplest way to confirm this claim is to refer to the search engine Google [23]. Google uses, among other things, a measure of the importance of a Web page, in order to rank this page when it turns up in a list of 'hits' from a search. This measure of importance (called the page's PageRank) is obtained from a vector $\mathbf{p}'$ which is close to our vector $\mathbf{p}$. The experience of Google has convincingly shown that the PageRank of a page gives useful information.

The vector $\mathbf{p}'$ is 'close to', but not identical to, the vector $\mathbf{p}$ as defined above. This is because $\mathbf{p}$ has too many zero entries – for reasons similar to those given for the vectors $\mathbf{b}$ and $\mathbf{f}$ above. In contrast, $\mathbf{p}'$ is positive everywhere (i.e., for all Web pages) – thus giving a positive PageRank importance score for every page.

Let us fill out this sketchy discussion, and give the reasons for our simple statements. First we note that, if we first disconnect each SCC from all others, and then normalize its adjacency matrix, then its dominant eigenvalue will of course be 1. However, if we do *not* disconnect the SCCs, and instead normalize the full matrix – giving $\widehat{F}^{(\text{norm})}$ as described above – then the dominant eigenvalue for 'most' SCCs will in fact be less than 1. More precisely: every SCC that is not a sink will have its dominant eigenvalue less than 1. This is because such SCCs, viewed in isolation from the entire graph, are not weight conserving, but rather weight *losing*. That is: the sum over all nodes in non-sink SCC $X$ of the weights sent out upon operation with $\widehat{F}^{(\text{norm})}$ is indeed equal to the sum of the weights these same nodes started with; but the former sum includes those weights sent from $X$ to *other* SCCs. Hence the sum over all nodes in non-sink SCC $X$ of the weights sent out *to nodes in X* upon operation with $\widehat{F}^{(\text{norm})}$ is *less than* the sum of the weights these same nodes started with. So, with each iteration, the SCC $X$ loses weight to downstream nodes. This weight may or may not be 'made up' by inputs from upstream SCCs; but the key fact is that the dominant eigenvalue of *all non-sink SCCs* is less than 1. By similar reasoning, the dominant eigenvalue of all sink SCCs (assuming they have more than one node) is exactly one.

Thus there is, in general, no single SCC $\Sigma$ which has 'the' largest dominant eigenvalue for a directed graph with normalized propagation. Instead, all sink SCCs have 'the' largest dominant eigenvalue – which is 1. Thus [14], by the same reasoning as that given above for the non-normalized cases, *all non-sink SCCs have zero weight in the vector* $\mathbf{p}$.

This problem is solved in the PageRank approach [37] by making the entire graph strongly connected – so that it is a single SCC. That is: the set of Web pages, plus the set of (directed) hyperlinks connecting these pages, form together a *Web graph*, which is far from being a single SCC. The PageRank algorithm then adds (with weight adjusted by a tuning parameter $\varepsilon$) a link from every page to every other – that is, a complete graph with weight $\varepsilon$ is added to the matrix $(1 - \varepsilon) \cdot \widehat{F}^{(\text{norm})}$ – and the vector $\mathbf{p}'$ is the dominant eigenvector of the resulting (weight conserving) matrix. Since this resulting matrix represents a strongly connected graph, the vector $\mathbf{p}'$ is positive everywhere.

$$\text{PageRank operator} = (1 - \varepsilon) \cdot \widehat{F}^{(\text{norm})} + \varepsilon \cdot \widehat{R}.$$

The problem that so many nodes get zero weight in a dominant eigenvector for a typical directed graph (whether or not its matrix is normalized) is often colloquially termed the 'sink problem' – even though, for forward operators, it is the non-sinks that have the biggest problem. A more precise expression would be the 'problem of not being strongly connected'. However this expression is very cumbersome, and unnecessarily so: *every* graph which is not strongly connected has at least one sink SCC (and at least one source SCC) – because its component graph (a directed *acyclic* graph) must have beginning nodes and ending nodes. Hence we will also refer to the 'sink problem'. Similarly, we will term solutions to this problem 'sink remedies'. Brin and Page [37] referred to the adding of a complete graph as representing, in some sense, a 'Web surfer' who (with probability $\varepsilon$) decides not to follow any offered hyperlink from the present page of attention, but rather to hop to a randomly chosen page. Hence we also call this complete graph the 'random surfer operator' $\widehat{R}$; and it is a sink remedy.

Summarizing: both the vectors **p** and **p**$'$ contain nontrivial (nonlocal) information about the nodes' positions in the directed graph. Hence they both qualify as 'EVC-like' measures of centrality for a directed graph. However, only **p**$'$ has the desirable feature that it is positive everywhere; and in fact the vector **p** is typically zero for most of the nodes in the graph – and hence not very useful.

By the same token, we can say that neither **f** nor **b** gives a very useful measure of node centrality. However, it is clear from the discussion of this subsection that one can add a (non-normalized) random surfer operator to either $\widehat{F}$ or $\widehat{B}$, and thus obtain a modified dominant eigenvector **f**$'$ or **b**$'$ which is nonzero everywhere [14]. In general, none of the three vectors **p**$'$, **f**$'$ or **b**$'$ is equal to any of the others.

### 4.1.4. *Authorities and hubs*

Kleinberg has offered a very elegant solution to the sink problem in [27]. This solution has the further advantage that it gives *two new and distinct* centrality measures for each node. That is, the two scores formulated by Kleinberg – 'hub' and 'authority' scores – are new (not equal to any of the measures discussed above) and distinct from one another. We will take these scores as usual from a pair of dominant eigenvectors – the 'hub vector' **h** and the 'authority vector' **a**.

We summarize Kleinberg's reasoning as follows. We want to evaluate the 'goodness' of a node in a directed graph; but we will in fact think in terms of two distinct types of goodness. We start with the EVC philosophy: important nodes have important neighbors. Then we use language appropriate to a directed graph – good nodes are pointed to by good nodes – but we make a distinction (as is possible and appropriate for directed graphs) between pointing and being pointed to. Doing so gives (still using simply words): good-pointed-to nodes are pointed to by good-pointing nodes. More elegantly: good authorities are pointed to by good hubs.

This definition (or pair of definitions) is circular, just as EVC is. We make the circularity obvious by inverting the final statement: good hubs point to good authorities. Thus the hub and authority scores for the nodes are defined in terms of one another. However, just as with the EVC, the solution is not hard; in this case it is a pair of eigenvectors, rather than a single one.

Now we express the words in mathematics. We start with a guess for the vector **a**$^0$ of authority scores for the nodes, and one for the vector **h**$^0$ of hub scores. The new estimate for hub scores is obtained from the old authority scores – the words 'good hubs point to good authorities' tell us this. In mathematics, these words say

$$h \propto \widehat{B}a = Aa,$$

where we leave out the (unknown) constant of proportionality. That is, we take the current guess of authority scores, and propagate them Backwards to get a new guess of hub scores. Similarly,

$$a \propto \widehat{F}h = A^{\mathrm{T}}h$$

(where again we ignore the constant). Putting these two equations together, in two different ways, gives two more:

$$A^{\mathrm{T}}Aa = \lambda_A a$$

and

$$AA^{\mathrm{T}}h = \lambda_H h.$$

That is, the authority scores are taken from the dominant eigenvector **a** of the Authority operator $\widehat{A} = A^{\mathrm{T}}A$, and the hub scores are taken from the dominant eigenvector **h** of the Hub operator $\widehat{H} = AA^{\mathrm{T}}$.

We note that both $\widehat{A}$ and $\widehat{H}$ are symmetric. We can think of each of these matrices as representing a kind of weighted adjacency matrix for a corresponding 'effective (authority or hub) graph'. For example, the element $(\widehat{A})_{ij}$ tells the number of ways one can come from node $i$ to node $j$ by *two* hops – whereof the first hop is against an arrow, and the second is with an arrow. Thus $\widehat{A}$, acting on a set of authority scores, allows nodes to send one another authority scores – analogously to EVC – but via multiple paths of two hops (on the original graph). An analogous statement holds of course for $\widehat{H}$. Note that both effective graphs have nonzero diagonal elements. For example, for every node $i$ having at least one inlink, you can get from $i$ to $i$ by hopping first against an inlink and then back, 'with' the same inlink (making $(\widehat{A})_{ii}$ nonzero); and, similarly, nodes with at least one outlink get nonzero diagonal elements in $\widehat{H}$.

Because each of these two effective graphs is symmetric, there is no 'sink problem' with computing hub and authority scores by Kleinberg's method. Hence there is no need for any sink remedy. We note finally, however, that these effective graphs may be disconnected, even when the original directed graph is weakly connected [14].

### 4.2. *Web link analysis – an illustrative example*

Of the above-discussed approaches to computing centrality of nodes in directed graphs, the PageRank method and Kleinberg's method were developed with a specific application in mind – namely, for quantifying the 'importance' of Web pages, by examining their placement in the topology of the Web graph. This is an interesting approach, which is typically termed 'link analysis' (see for example [9]). The advantage of link analysis is that it is quite readily performed by a machine. That is, it is fairly easy to program a machine to evaluate the importance of a document, based on how that document lies in a Web of links – a sort of 'document community', which places each document in a *context*. In contrast, it is (we believe) much more challenging to ask a machine to examine a single document, out of context, and to evaluate the importance of that document based on its *content*. This latter task is much more readily done by humans; but humans are, as we all know, slow, limited in capacity, and prone to giving idiosyncratic responses. These ideas – plus the phenomenal success of Google – lead us to believe that link analysis, in one form or other, has a significant future role in the problem of evaluating the importance of documents.

The $\widehat{F}$ and $\widehat{B}$ operators have (as far as we are aware) received little or no attention in the literature on Web page ranking. Our own recent work [14] reports some limited tests on these operators, as compared with the Kleinberg and PageRank operators. These tests indicate that these simple operators merit further attention. We note that the $\widehat{F}$ operator is closer in spirit to an Authority ('pointed-to') operator than to a Hub operator, while $\widehat{B}$ is more like a Hub ('pointing') operator. Similarly, we regard PageRank scores as being in the spirit of authority scores.

We noted earlier that many distinct operators give results (eigenvectors) which are distinct for directed graphs, but which all become the same when the arrows are removed (thus making the graph undirected). All of the non-normalized operators we have discussed here fit that statement:

> *For a directed graph, the principal eigenvector for each of the*
> *following un-normalized operators will be reduced to EVC by*
> *symmetrization of the directed graph:*
> $$\widehat{F}, \widehat{B}, \widehat{F}\widehat{B} = \widehat{A} \text{ and } \widehat{B}\widehat{F} = \widehat{H}.$$

The PageRank operator is, in contrast, normalized. Thus, if we make the graph undirected (and remove the random-surfer part $\widehat{R}$), the PageRank operator will give results which are – as shown in Section 2.3 – proportional to the node degree. Thus it is quite interesting to note that the PageRank operator – which gives the very local degree centrality for an undirected graph – gives useful, nonlocal results for a directed graph.

### 4.3. *A novel 'sink remedy'*

Regardless of the application (Web pages, or otherwise – see below), it is clear from the above that many measures of centrality that have been defined for directed graphs need a sink remedy. The random surfer operator is one such remedy.

In [14] we point out two other sink remedies – one we will call 'obvious', and one which we regard as being less obvious. The obvious remedy is simple: given a directed graph, (i) find the set of SCCs, (ii) find all inter-SCC links, and (iii) add (perhaps with some weight $\varepsilon < 1$) the reverse of all inter-SCC links. This renders the entire graph strongly connected, and hence solves the sink problem – but it does so with the addition of far fewer links than those used in the random surfer operator. Thus one might regard this approach as a less extreme disturbance of the original graph.

Here we wish however to devote more attention to the 'less obvious' sink remedy. This approach (called 'pumping', for reasons which will become clear) in fact solves the sink problem without making the graph strongly connected – in fact, without altering the component graph at all. Thus the pumping approach is rather unique.

Nevertheless the main idea can be stated simply, based on the understanding developed above. For concreteness we assume that we want a sink remedy for the Forward operator. The idea is then as follows: (i) find the dominant eigenvalues for every SCC in the graph; (ii) find all source SCCs in the graph; (iii) adjust the link weights in every source SCC, such that (a) every source SCC has the same dominant eigenvalue $\mathit{\Xi}$, and (b) $\mathit{\Xi}$ is larger than any other dominant eigenvalue for any SCC in the graph.

In short:

> We 'pump' all source SCCs up to a common growth rate, such that they can
> jointly 'drive' the rest of the graph – all of which is downstream of them.

Here we have made the idea plausible; in [14] we show that, even for the case when many SCCs share the same, largest, dominant eigenvalue, it is possible to obtain an eigenvector for the pumped graph which is positive definite everywhere, and which has the eigenvalue $\mathit{\Xi}$.

### 4.4. *Application to virus spreading on computer networks*

We have already devoted considerable discussion to the problem of understanding how information or virus spreading takes place over an undirected graph with symmetric links. It may be that, given incomplete information, the assumption that the links which support the spreading are symmetric is a reasonable assumption. However, in this section we want to explore the problem of spreading over a directed graph. All our experience from the above sections tells us that directed graphs may be expected to be significantly different, sometimes in nonintuitive ways, from undirected graphs. This makes the problem interesting in itself.

However there are also *practical* reasons to study spreading on directed graphs. We consider a very practical problem for a system administrator, namely, the spreading of data viruses over a network of computers. Clearly, the physical network, built up from routers and physical links, is symmetric – allowing two-way information flow. However – as pointed out by [26] in 1991 – the network which supports spreading is a logical network which is built on top of the physical network; and this logical network is directed. More specifically, we picture the typical virus as spreading itself by reading the email address list of its current host, replicating itself, and sending copies to all recipients on this address list. The address list is in principle a one-way link: a directed link from current host to recipient. Of course, it is often that case that my email recipients have my address on their email address list – but 'often' is not 'always', and can in fact be rather far from always in real situations. Thus we find that:

*The logical network supporting virus propagation is in principle directed.*

Newman, Forrest, and Balthrop (NFB) [35] performed a direct measurement of an 'email address graph' at a large university in the US. They found that around 23% of the directed links in this graph were actually two-way. More precisely: of all directly connected node pairs in the graph, around 23% had links in both directions. This of course leaves the large majority of directly connected pairs with only one-way links. As Newman and Forrest point out (and as one expects), this percentage of two-way links is much higher than that percentage one would get from randomly laying in directed links. However their measured email graph is clearly very much a directed graph.

The structure of this email graph is strongly reminiscent of the famous 'bow-tie' structure reported by Broder et al. [11] and others [29] for the Web graph – which is not even approximately an undirected graph. That is: the email graph was found to have a large, strongly connected component LC, with about 34% of the roughly 10,000 nodes in the connected graph. Another subset of nodes (about 3%) may be called 'In': this is a set of nodes which allow propagation into the LC, but which cannot be reached from the LC. Hence the In set includes a number of source SCCs. Furthermore, there is an Out set (about 58% of the connected graph) which can be reached from LC, but from which there is no directed path into the LC. This triad In $\rightarrow$ LC $\rightarrow$ Out makes up the bulk of the structure in a typical directed graph, and is the origin of the term 'bow-tie' structure. (These components do not sum to 100%, as there are other nodes – lying in the 'Tendrils' – which are connected to In and/or Out, but with no directed path to or from the LC.)

Clearly, when the email graph has this structure, the LC and Out sets are by far the most likely to be infected by any new computer virus coming into the network from outside. Furthermore, the Out set is likely to be infected from the LC. Hence protective measures should reasonably be focused on the LC. The question is then: which nodes (or links) of the LC are the most important for spreading of a virus?

NFB offered a slightly qualified idea of degree centrality as an answer to this question. That is, they identified the most important nodes in the LC as being those nodes with highest outdegree. (In fact, they lumped together the In component and the LC; but the former was only about 1/10 the latter in their email graph. Thus, nodes with highest outdegree in this 'giant in-component' were treated as most important.) They then showed that 'inoculating' a small number of the most important nodes (according to this measure) gave a marked decrease in the long-time infection fraction for the graph – thus confirming that these nodes are indeed important to spreading.

Our own work on epidemic spreading has been confined to undirected graphs. However, here we would like to sketch an extension of those ideas to the case of directed graphs. It should come as no surprise to the reader that we will seek a measure of a node's importance for spreading which looks beyond the one-hop view implicit in node degree centrality. Our long-range view for undirected graphs gave the eigenvector centrality or EVC. Now we will seek a useful analogy – giving a measure of a node's importance *for spreading* – for a directed graph.

We reason as follows: a node plays an important role in spreading if (i) it is likely to become infected, and (ii) it can subsequently, with high probability, infect many other nodes. The one-hop measure for each of these is, respectively, (i) in-degree, and (ii) out-degree. However, we want (in the spirit of EVC) long-range measures for both of these concepts.

A long-range measure of in-degree should look at those nodes pointing to a node $n$ and ask, how many nodes point to *these* nodes? – and then iterate this process, always looking farther back. Furthermore, double-counting should be avoided when loops arise (as they most certainly will in the LC). The operator which does precisely this is the Forward operator $\widehat{F}$. That is: when we look *back* from node $n$ to find (in a many-hop sense) how many nodes point to $n$, we simply ask those nodes to send their weights *Forward*, to $n$. The in-neighbors of $n$ (i.e., those pointing to $n$) will get weight from their in-neighbors, which get weight from *their* in-neighbors, etc. – always propagating Forward. Furthermore, allowing convergence to the dominant (stable) eigenvector of this process makes (we argue) correct allowance for loops: growth of weight at nodes which get much of their weight from loops feeding back upon them will be damped with respect to growth at nodes which have a truly tree-like, converging structure upstream. Thus, for all these reasons, we will take the 'in-importance' or 'infectibility' of node $n$ to be its component in the dominant eigenvector $\mathbf{f}$ of $\widehat{F}$.

Similarly, we look forward, and so propagate weights Backward (again at long range, and again making allowance for loops) in order to find the 'infecting range' of this same node $n$. Thus we get measure (ii) for node $n$ by taking the $n$th component of the dominant eigenvector $\mathbf{b}$ of the Backward operator $\widehat{B}$. Thus our two measures of importance for node $n$ are: (i) infectibility $= f_n$ (a 'pointed-to' score); (ii) infecting range $= b_n$ (a 'pointing' score). In summary:

*We argue that, for a directed graph, a node's spreading power depends on:*
   (i) *Infectibility – a node's ability to be infected; the 'pointed-to' score* $\mathbf{f}$.
   (ii) *Infecting range – a node's ability to infect other nodes; the 'pointing' score* $\mathbf{b}$.

Two questions immediately arise. First, we know that both the vector $\mathbf{b}$ and the vector $\mathbf{f}$ will have zero entries for many of the nodes, unless we add some kind of sink remedy to the simple operators $\widehat{B}$ and $\widehat{F}$. The first question is then, do we desire a sink remedy for this purpose also – i.e., for assessing the importance of a node for spreading?

The second question is also simply stated: given our two measures of a node's importance for spreading, what is the best way to combine these two measures? (Our question here implicitly allows for the answer that *not* combining the two may be the 'best' way of assessing their importance.) Let us now address these questions in order.

First we consider the question of sink remedies. Consider for instance a node $i$ in the In set – hence, lying upstream of LC – and which furthermore is in an SCC with a small dominant eigenvalue (relative to that of LC). This node (recall our discussion in an earlier section on directed-graph centralities) will have zero weight in $\mathbf{f}$. Thus – even though this node may have many other nodes pointing to it – our simple, intuitively reasonable rule will give node $i$ *zero* infectibility. Does this make sense? Alternatively stated: should we attempt to 'repair' this result, by adding a sink remedy to our approach?

Let us try to sharpen our intuition about this question. We believe it is important here to state explicitly, and to focus on, an obvious fact: any infection of a given email graph comes originally from 'outside'. The graph itself of course can and does *spread* the infection; but it does not *initiate* it.

What do we learn from this obvious point? Let us quantify it: we suppose that infections (viruses) 'trickle' into our graph $G$ from the outside world $W$, at some rate $q$. More specifically, $q$ may be stated to be the probability, per unit time, that any given node is infected from outside (i.e., from $W$). Thus we can compare this quantity $q$ with the quantity $p$ – which (as in our earlier discussion for undirected graphs) is the probability per unit time of virus transmission over an outlink, from an infected node to an uninfected node. (We again treat only the simple case of the same $p$ for all links and the same $q$ for all nodes; the extension to varying $p$ and $q$ is not qualitatively different.) This comparison is direct: both $p$ and $q$ have the same engineering dimensions, so that the expected behavior may be discussed in terms of the dimensionless ratio $q/p$.

Suppose (the very easiest case) that $q/p$ is very large. Then network effects are negligible – each node is more likely to be infected from $W$ than from any neighbor; the distinctions among In, LC, and Out are unimportant; and the cumulative infection percentage grows simply linearly with time (with slope $q$). This case is of course not very interesting. However it makes explicit what was implicit before. That is, both our own studies (described above) on undirected graphs, and the NFB approach [35], assume the opposite limit – that $q/p$ is very small. These studies focus entirely on network effects on spreading; and it is just in this limit ($q/p \to 0$) that network effects are entirely dominant. One cannot of course have $q$ exactly zero; so instead, in simulations, one infects (typically) a single node, and then allows the infection to spread to completion, without any further infection from the World. This is, clearly, the limit $q/p \to 0$.

We believe that, in this limit, the importance of the In set is in fact small. All nodes are of course equally likely to be the first node infected from $W$. However, repeated experiments, each with a randomly chosen first infected node, will reveal the importance of LC in this $q/p \to 0$ limit – precisely because of the large 'gain' (eigenvalue) of LC. This gain will ensure that infection will spread much more rapidly in LC than outside it – thus making (we argue) the nodes in LC much more important (as measured by $\mathbf{f}$ or $\mathbf{b}$) than those in the other graph components, in any average over many experiments.

Now we consider explicitly the infectibility – one aspect of a node's importance – as measured by $\mathbf{f}$. Our arguments say that we can accept, in the limit $q/p \to 0$, zero *effective* infectibility for nodes in In (assuming always that the largest eigenvalue is found in LC). We have also seen that the nodes in In have exactly the same importance (of either type) as those in LC, in the limit $q/p \to \infty$. We then ask, what do we do for general $q/p$?

We believe that part of the answer lies in what has already been said: our directed graph $G$ has in fact been modified by adding one more node $W$ (see Figure 19). This World node $W$ has an outlink to every other node in $G$, and no inlinks. Thus $W$ is a source, and also a one-node SCC. If we do nothing further with $W$, it will have no effect on the eigenvector $\mathbf{f}$: its only eigenvalue is zero, and it is upstream of the dominant component LC.

However we want $W$ to have some effect on $\mathbf{f}$ – at least, for sufficiently large $q/p$, we know it must have a large effect. We suggest here that this can be accomplished by giving $W$ a self-loop – thus '*pumping*' the SCC $W$ – with a weight $w$ on the self-loop, so that the

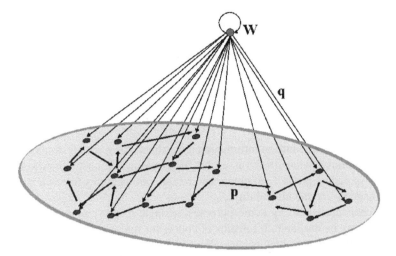

Fig. 19.  A directed graph modified with the World node $W$, and where $q$ denotes the infection rate from the outside world $W$ into the graph, and $p$ denotes the infection rate among the nodes within the graph. Virus propagation
within the graph transmits over outlinks of infected nodes to uninfected nodes.

gain of $W$ is $w$. It seems clear that $w$ must be of the form $c(q/p)$, where $c$ is an unknown
constant. This makes $W$ dominant for large $q/p$, and negligible for very small $q/p$. Since
we know (as yet) nothing about an appropriate value for the coefficient $c$, we will often
ignore it, by implicitly treating it as equal to 1 (while recognizing that future work may
indicate a better value for $c$).

With this simple topological addition to $G$, we get what we believe are the correct effects: LC is dominant for small $q/p$, and $W$ is dominant for large $q/p$. Furthermore, we
have implicitly answered our original question: we see no need for a sink remedy in computing $\mathbf{f}$, which will then be used in evaluating the 'infectibility' of each node. The addition
of the mega-source node $W$, with a pumped gain $c(q/p)$, seems sufficient for this purpose.
Of course, we have only sketched our reasoning here; it remains to be tested by studies like
those we have reported for undirected graphs.

What about the 'infecting range vector' $\mathbf{b}$? Is the same topology appropriate here?

We think not. Our reasoning is simple. The World node $W$ will always get a very large
weight in $\mathbf{b}$ – since every node sends its weight to $W$ under action of $\widehat{B}$. This effect is
neither useful nor desirable: we have no information about the true network structure of
the World which is represented by $W$; we have no interest in finding the infecting range
of $W$; and finally, the large weight of $W$ in $\mathbf{b}$ simply obscures the effects we do wish
to measure – namely, the infecting range of the various nodes in $G$, as compared to one
another on a common scale. For all of these reason, we suggest to simply remove $W$ in the
determination of the eigenvector $\mathbf{b}$.

Neither do we propose a sink remedy in the determination of $\mathbf{b}$. Without a sink remedy,
the infecting range of all SCCs in the Out set will be zero (assuming, again, that the gain
of the LC is the largest gain of all SCCs). This may seem an extreme simplification; but
for now, in this discussion of untested ideas, we let it stand.

After some effort, then, we have an answer to our first question:

*Do not add any sink remedy to the calculation of the eigenvectors* **b** *and* **f**. *Instead, add the World node W in the case of the calculation of infectibility (via* **f**)*; and use the original, unaltered graph topology in determining the infecting range from* **b**.

Now we come to our second question: how to combine these two measures of importance?

It is clear that the two kinds of information are distinct. Also, each is of interest in itself. Therefore, it seems obvious that it is useful for a system administrator who has access to the virus spreading graph $G$ to obtain and retain both measures for each node $n$ – the infecting range $b_n$ and the infectibility $f_n$.

It also seems clear that, for some purposes, a single measure of a node's 'spreading importance' may be desirable. It remains of course for future work to examine this idea. However we will offer one comment here. We note that a naive 'tree' picture of a graph – as seen from a given node $n$ – suggests that probabilities multiply as they move through $n$. For instance, five inlinks to $n$ seem to give five ways for (after two hops) infecting *each* of $n$'s outneighbors, via $n$. Thus, $n$'s importance seems to demand a weight proportional to the *product* $k_n^{in} \times k_n^{out}$. However we believe that this line of thought fails when one looks many hops (in principle, an infinite number of hops) forward and back from $n$, as one does in calculating **f** and **b**. Each of these operations will visit the same set of nodes (all of them), and traverse every loop in the graph $G$. Both of these effects are entirely incompatible with a naïve tree-like approach to the graph. Hence we do not propose combining our two importances by multiplication (that is, taking $f_n \times b_n$).

Instead, we suggest that, when a single combined measure of 'spreading importance' for each node is desired, one simply take the sum $\sigma_n = f_n + b_n$. The quantity $\sigma_n$ has the following properties: it is of course monotonically increasing in both $f_n$ and $b_n$; it is (almost certainly) largest in LC; but furthermore it is nonzero at every node that is at least in In or Out. This last point follows from the fact that every such node is downstream of LC under one of the operations $\widehat{F}$ or $\widehat{B}$. More explicitly: In will have weight in **b**, and Out will have weight in **f**; hence every node in the graph (In → LC → Out) will have nonzero $\sigma_n$.

We offer a remark in this context on the idea of *smoothness*. We have argued above (and in more detail in [15]) that the EVC is smooth as one moves over the network. A careful examination of that argument shows that it may be generalized to our vectors **f** and **b**. It is plausible that these quantities are 'less smooth' than the EVC **e**, in a sense that we express with the following argument. Suppose we take a directed graph $G$ and simply erase all arrowheads – making it the corresponding undirected graph $G^{(u)}$. This is equivalent to adding a large number of new directed links – one for every node pair which has only a one-way link in $G$. This, we argue, (i) makes **f** = **b** (= **e**), but further (ii) makes them both smoother than they were before the addition of the extra links. That is, we argue that graphs with a sparse link structure give – unless they have special symmetries – less smooth versions of **f** and **b** than do graphs with dense link structure. Furthermore, the change from $G$ to $G^{(u)}$ not only increases the density of links – this change also cannot decrease the symmetry of the graph. Thus, based on these arguments, we expect **e** to be in general smoother than **f** or **b** (for the same graph).

We also suggest that $\sigma_n$ is likely smoother than either $f_n$ or $b_n$. That is, summing the two should tend to cancel out the fluctuations in them, to some degree. Thus we (speculatively) add this property of $\sigma_n$ to those mentioned above – making it much like the EVC, in that it is smooth, nonzero almost everywhere (that is, on the bulk of the connected part of a 'typical' directed graph), and gives a useful measure of the importance of a node to the spreading of an infection. Thus:

> *For a directed graph, we propose $\sigma_n = f_n + b_n$ as a promising single measure of a node's spreading power, which may be taken as an analog of the EVC $e_n$ for an undirected graph.*

We wish to make one final observation in this section. It is built on all that we have said so far; but it concerns the spreading process itself, rather than the simple generation of a 'score' which evaluates a node's importance in spreading.

Recall our prescription for determining $\mathbf{f}$, and thereby a node's infectibility score. We argued that one should include the World node $W$, giving it a self-gain $c(q/p)$. Now we point out one fact which follows from our earlier discussion of the $\mathbf{f}$ vector's properties: the node $W$ will have *no effect on any node's score* until its gain $c(q/p)$ is greater than the gain $\Lambda$ of the SCC with the largest gain (which we take to be LC). That is, all upstream SCCs from LC will have zero weight in $\mathbf{f}$ until we pump $W$ (which is upstream of all other nodes) so much that its gain exceeds $\Lambda$. Only above the threshold value $c(q/p) = \Lambda$ will the presence of $W$ begin to have an effect – giving nonzero weight in fact in every node, including all those which had zero weight (those in In) for $c(q/p)$ less than the threshold value $\Lambda$.

In other words: we predicted earlier that setting the gain of $W$ to be $c(q/p)$ would move the importance scores from the appropriate small-$q$ values to the appropriate large-$q$ values (uniform) with increasing $q/p$. Now we point out that this evolution is not totally smooth: the importance scores, as a function of increasing $q$ and at fixed $p$, will in fact stay unchanged over a finite range of $q$. The infectibility scores of nodes like those in In will only begin to grow (above zero) when $c(q/p)$ first exceeds the threshold value $\Lambda$.

So far we have simply stated correct conclusions from our analysis. Now we make a small leap. That is: we suppose that our mapping of node importance scores over the graph $G$ is well related to the actual infection patterns that would be observed in repeated infection experiments on $G$, at a given $p$ and $q$. This simple supposition then implies the following:

> *The pattern of infection will itself remain unchanged*
> *over a finite range of q (for small q), and will only begin*
> *to change (toward uniformity) when q exceeds a threshold value.*

This is an interesting, nontrivial, and testable prediction which we obtain from our arguments and reasoning. We see good reason to test this hypothesis, since our reasoning is simple and plausible. Hence we look forward to future work toward the goal of testing for threshold effects, as a function of $q/p$, in the spreading of viruses on directed graphs. This work should be a part of a larger effort, aimed at understanding virus spreading on real email graphs, and on other directed network structures of interest to system administrators.

## 5. Visualization

We believe that the problem – or, better stated, *challenge* – of visualizing a given graph is of great importance and interest. Any graph may be described in terms of a huge variety of different numerical measures – and we have discussed a number of them in this chapter. However, we believe that, as long as *human beings* are involved in understanding the structure of networks – i.e., for some time to come! – the possibility of forming a visual image of a graph or network will always be of great value. Numerical measures give unique and extremely useful information; but properly developed graph visualization can contain, and convey to a viewer, a great deal of useful (even if not always explicit) information and understanding of the graph's structure.

Hence we address the problem of graph visualization in this, final, section. It is clear that there is no single answer to this problem – that is, there is no best visualization for a given graph. Different visualizations convey different kinds of information. For a more extensive overview of different approaches to visualization, we refer the reader to [6], to UCINet [22] and to Pajek [36].

In this section we will confine our discussion to visualizations which express ideas from our own interests and analysis: using EVC to find regions in a given undirected graph, and the resulting connection to spreading. This work has so far been confined entirely to undirected graphs; hence we will not discuss visualization for directed graphs.

Our 'regions' approach to analyzing the structure of an undirected graph has a certain inflexibility, which has both positive and negative aspects. The positive side is that the graph itself determines how many regions are given – the answer is inherent in the structure of the graph. Hence one can have a certain confidence in this number, which cannot be simply dialed in by the analyst. This inflexibility can however be inconvenient as well: it can readily occur that a graph with millions of nodes gives only one or two regions according to our topographic analysis; and then the problem of visualizing the result is not any easier after the analysis than it was before the analysis (i.e., simply based on the raw topology).

Thus, a desirable further development of our regions analysis would be techniques for *refinement* – such that the analyst could dial in the number of levels of resolution (regions, subregions, sub-subregions, etc.) desired – while letting the graph determine the number of regions or subregions at any given level of refinement. The technique of Girvan and Newman [19] does allow for such refinement, and so gives a hierarchical picture of a given network which is quite useful for visualization. A hierarchical picture allows one to choose a level of resolution, to zoom in or out – options which are extremely useful for visualizing graphs with very many nodes.

We note that our topographic approach gives, for each node, a 'height'. Hence, the resulting topographic picture allows for a three-dimensional approach to visualization, if we find a suitable approach to assigning $xy$-coordinates to each node. We have here explored (in a limited fashion) two possible approaches to this '$xy$-problem': (a) assign $xy$-coordinates

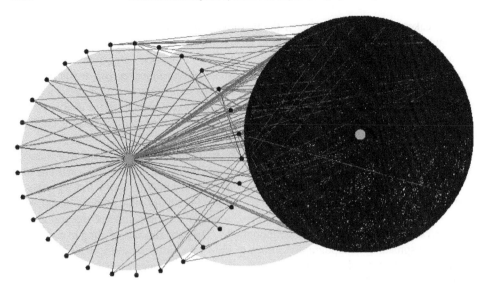

Fig. 20. A simple visualization of the Gnutella graph $G6$, showing the size of the two regions and the extent of their interconnection.

based on our knowledge of the topography – which mountain a node $n$ belongs to, how high it lies on that mountain, etc.; or (b) use a refinement method which may be iterated to conclusion, such that the graph may be represented as a tree. A tree graph is fairly trivially laid out in two dimensions; hence we view the $xy$-problem as essentially solved when we can refine the structure to a tree.

Figure 20 shows an early visualization, based on our regions analysis, of a Gnutella graph with two regions [39]. This visualization is neither (a) nor (b): no height information is used (a), and there is no refinement (b). What we have done is to show the two regions, represented as two rings, with nodes for each region placed on the corresponding ring, the respective Centers placed at the geometric center of each ring, and with all links displayed. In this 'Archipelago' visualization, the size of each region (in terms of number of nodes) is clear, as is the extent of interconnection of the two regions. We note that the Gnutella graphs (from 2001) that we have studied have shown a scale free node degree distribution; hence we are not surprised to find either one region, or two which are very strongly interconnected.

An example of a more topographic visualization, using approach (a), has been presented in an earlier section, in Figure 15. We recall that this graph is the collaboration graph of the Santa Fe Institute (SFI), taken from [19]; our visualization has been reported in [17] and in [16]. We note that, for effecting this visualization, all links have been retained. However, fictitious link strengths have been chosen, based entirely on topographic information. Our goal here was to persuade the nodes to lie in a way dictated by our topographic picture. Thus, each link strength has a factor which is monotonic in the centrality of the two nodes it connects – so that two connected, very central nodes get pulled tightly together, while two very peripheral nodes are only weakly drawn together. This tends (as we see from Figure 15) to allow the peripheral nodes to drift in fact out to the $xy$-periphery. Another factor weakens a link even further if the two nodes are in two distinct regions – thus encouraging nodes from the same region to cluster together in the $xy$ plane. Once the link strengths are set by these factors, we use a standard force-balance algorithm to determine the $xy$-coordinates of all the nodes. A force-balance algorithm (see [6]) treats the links as springs, with the spring constant determined by the link strength; it then seeks to find a static mechanical equilibrium for the resulting system by seeking a local (or ideally global) minimum of the energy of the mechanical system. Such minimization problems are however in general very hard for a large number of nodes.

Next we present some results using approach (b): refinement of the graph to a tree. We have explored, to a limited extent, one approach to refining our regions analysis. The idea is again topographically motivated. A topographic surface may be characterized either in terms of its contours of equal height (as typically seen in topographic contour maps), or in terms of lines which are perpendicular to these contours – namely, lines of steepest ascent. The latter define the 'ridge tops' of the topography. They are also – according to our results connecting spreading to topography – lines of most rapid information spreading. Finally, the steepest-ascent graph presents two very practical advantages over contours of equal height: it is very easily calculated (in fact it is a byproduct of our regions analysis); and it is a tree.

We elaborate on these points briefly. In order to find which region a node $n$ belongs to, we find its steepest-ascent EVC path. This path begins at the node $n$, moves to $n$'s highest EVC neighbor $n_1$, then to $n_1$'s highest EVC neighbor $n_2$, and so on, until it reaches a node which is higher (in EVC) than all its neighbors – the Center $C$ of the region, or top of the mountain. It is clear from this that $n_1$'s steepest-ascent path is a part of $n$'s steepest-ascent path; and similarly, if $n$ has lower neighbors, then $n$'s steepest-ascent EVC path *may* (or may not) be part of the steepest-ascent EVC path for one or more of them. The point here is that steepest-ascent paths are 'reused', over and over again, because the choice (looking up) is virtually always unique. Hence the collection of all steepest-ascent paths forms, for each region, a single, well-defined tree. The collection of all steepest-ascent EVC paths, for the entire graph, we call the steepest-ascent graph or SAG.

> *The EVC-based steepest-ascent graph (SAG) consists of all the*
> *steepest-ascent EVC-paths from the entire, original network.*

Since the SAG gives a tree for each region, we can view (for each region) each branch of the corresponding tree as a subregion. Topographically speaking, these branches of the tree are the dominant 'ridges' of the mountain on which they lie. Hence this subregion approach chooses to define the mountain in terms of its highest point, and the ridge structure leading up to this highest point. Further branchings of the subregion (as one moves downhill) give rise to further refinements: sub-subregions, etc. This branching picture, for all levels of refinement, is consistent with the ridge metaphor given here.

Now we show the visual representation of these ideas – i.e., of the SAG. Figure 21 shows the one-region Gnutella graph $G3$, as laid out in 2D using the force-balance approach found in the Netdraw [22] tool (with all links given equal strength). We see from this visualization that some nodes are rather peripheral, and that many others are not. It is very difficult to extract any other information from this visualization.

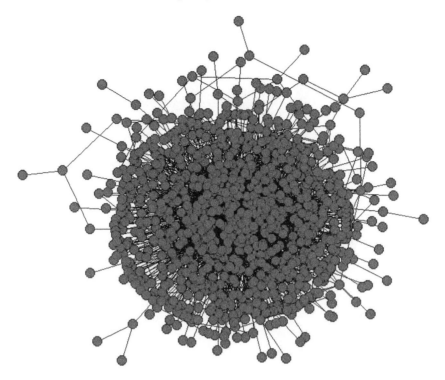

Fig. 21. A standard force-balance visualization of the Gnutella graph $G3$.

Figure 22 shows the same graph as in Figure 21, but now represented as its SAG. Our analysis gives only a single region for this graph; however we see clearly two main subregions in the visualization – plus a great deal of other substructure. Comparing Figures 21 and 22 shows one of the significant advantages of the SAG approach (or any other fully hierarchical approach): it allows one to display only a strategically chosen fraction of all the links in the graph, while at the same time allowing most or even all of the nodes to be clearly seen.

We note that the nodes in Figure 22 were laid out in 2D using the same method as in Figure 21 – namely, the force-balance method of Netdraw [22]. The finding of mechanical equilibrium is enormously easier for the tree structure of the SAG (Figure 22) than it is for the full graph (Figure 21). This is another (practical) advantage of omitting a large fraction of the links from the visualization. Of course, the determination of the heights of all the nodes, and the resulting SAG, incorporates information from the *full* graph topology – but the visualization is greatly facilitated by choosing to omit very many links after this information is obtained.

Figure 23 shows another Gnutella graph, again visualized using the standard Netdraw force-balance approach. This graph appears to be less 'dense' than the graph of Figure 21; but otherwise there is little insight to be gained from this visualization.

Figure 24 shows the same graph – using the same layout ($xy$-coordinates) as those of Figure 23, but coloring the nodes according to our regions analysis (which gave two re-

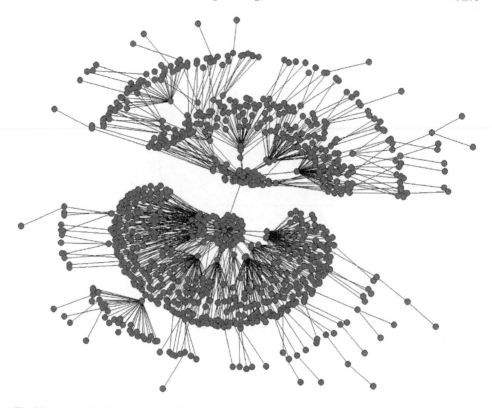

Fig. 22. A hierarchical visualization of the same graph ($G3$) as in Figure 21. Our analysis gives one region; the tree structure is that obtained from the steepest-ascent graph (SAG).

gions). It is encouraging that the blue nodes tend to lie together in the picture, as do the red nodes. Also, it seems clear that there are more blue nodes than red nodes. Thus Figure 24 displays *some* useful information which has been obtained by our analysis.

Figure 25 then gives the full hierarchical refinement for each region of this graph. We omit all inter-region links; hence the two regions are displayed essentially independently. We think that the progress in the conveying of structural information, as one moves from Figure 23 to Figure 24, and then to Figure 25, is striking. We expect that the method shown in Figure 25 could be even further improved by exploiting, and displaying, information about relationships between regions: some nodes, lying at the periphery of each region, are connected to bridging links, and so are far less peripheral than other, 'truly peripheral' nodes. We expect to be able to include this kind of information in future improvements of the SAG approach.

Nevertheless we feel that the picture of Figure 25 already contains a great deal of useful information for a system administrator who is concerned with the spreading of viruses. We see the nodes which have the most spreading power lying at the center of each region, with the least important nodes lying at the periphery, and nodes of intermediate importance placed appropriately. We also see lines of (likely) fastest virus spreading; these are the

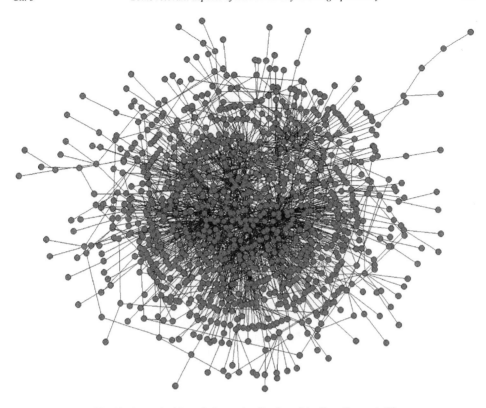

Fig. 23. A standard force-balance visualization of the Gnutella graph $G2$.

displayed links, i.e., the links of the SAG. Addition of bridging links would thus give a fairly complete and detailed picture of the vulnerability of this network to virus spreading. Hence we expect this kind of visualization to be useful in cases where the network can be mapped (more or less completely), and where information security is a concern.

Finally, we present another visualization approach, distinct from the SAG approach but also involving refinement. We can call this approach 'subregion visualization'. That is, we suppose that the information of most interest involves the regions and subregions – their number, size and interrelationships. Thus we drop all information from further refinements (sub-subregions, etc.). We then define the size of a subregion to be the number of nodes it contains. We can also define the strength of inter-subregion links (which, we note, do not lie on the SAG) as follows: it is the sum of the 'average centralities' of each inter-subregion link, where the average centrality of a link is simply the average centrality of the two nodes connected by the link. Thus the inter-subregion effective link strength contains information both about the number of such links, and on the heights of the nodes they connect.

With these definitions we have a graph with $s$ nodes, where $s$ is the number of sub-regions. This graph (like those for SAG visualization) has been broken into $r$ connected components (where $r$ is the number of regions), by again ignoring inter-region links. The result – for the same graph as that portrayed in Figures 23 through 25 – is shown in Fig-

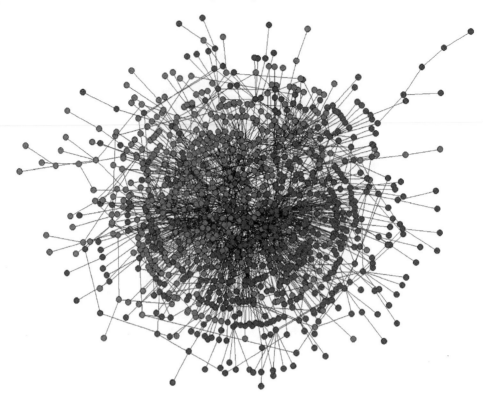

Fig. 24. The same as Figure 23; but we have colored nodes belonging to different regions (as determined by our EVC analysis) with different colors.

ure 26. (Here the numbers for each subregion give the node number at that subregion's 'head' – not the number of nodes in the subregion.) It is easy in this figure to pick out the largest subregions and their most central nodes, as well as the strongest effective links between these subregions. Both kinds of information may be useful for protecting against the spreading of viruses. That is, those features which first capture the eye's attention in Figure 26 are just those regions which should likely receive the most attention in devising strategies for preventing virus spreading.

We note again that the subregion visualization method, as shown in Figure 26, would be enhanced by the addition of information on bridging links. This is a very straightforward task for this approach (and is not much more difficult for the SAG approach). Hence we plan to extend the methods reported here in this way in the immediate future. We emphasize that the work reported in this subsection is very much a work in progress.

## 6. Summary

In this chapter, we have studied an abstract structure called a *graph* (or equivalently, a *network*). A graph is a simple object: a set of nodes, connected by links – which are directed

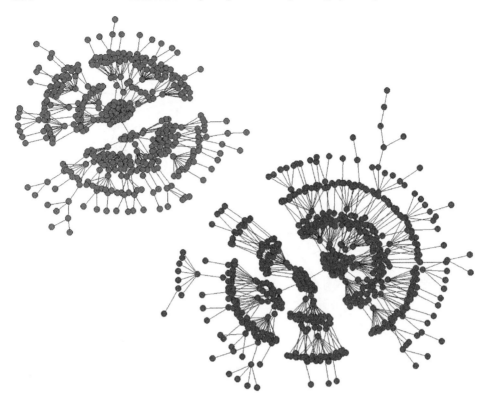

Fig. 25. The same graph as in Figures 23 and 24; but now we have represented each region as its steepest-ascent-graph tree.

or symmetric. In spite of this simplicity, graphs are extremely useful in describing a huge variety of phenomena involving connections. In particular, they are useful for describing and managing networks of computers, including logical or overlay networks which are built on top of the physical network. The latter may often be justifiably called social networks – since the overlay (logical) links are laid down by people. An excellent example of this is the 'email address network' discussed in Section 4.4. It is this logical graph which determines how (for example) a computer virus may be spread over the physical network; hence it is worth study for that reason (among others).

We have emphasized that there are fundamental and nontrivial differences between a directed graph (such as the email graph) and an undirected graph. Thus our introductory discussion of graph properties has repeatedly consisted of two sections, one for each type of graph.

We have also offered our own ideas regarding a highly nontrivial question, namely: how can one describe, in a useful way, the '*structure*' of a graph (beyond simply stating all microscopic information about nodes and links)? Here, because of the word 'useful' in the question posed, the answer is not unique: it depends on the reasons one may have for studying the graph. Our own approach is strongly skewed towards understanding *spread-*

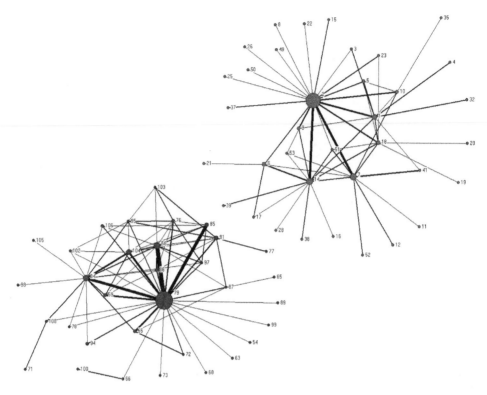

Fig. 26. A 'subregion visualization' of the graph of Figures 23–25. Here, for each region, we have taken one level of refinement (subregion) and represented it as a node of the graph. Size of nodes and thickness of links indicate, respectively, number of nodes in the subregion, and strength of inter-subregion links (see text for details).

*ing* (e.g., of viruses) on the graph. For this purpose, we have argued that it is very useful to (i) find the *eigenvector centrality* (EVC) of the nodes (when the graph is undirected), and then (ii) obtain from the EVC a very coarse-grained *topographical view* of the network, in terms of a few 'mountains' (regions), with bridging links crossing the valleys. We have presented a body of evidence, taken from simulations run on empirical networks, that this view is useful for understanding spreading. Also, we have shown that *design improvements* based on this topographic view can strongly affect the spreading – either by hindering it (as is desirable for viruses), or by helping it (as may be desirable for useful information).

We have sketched, in some detail, an extension of these ideas to directed graphs. This discussion (Section 4.4) has not been published previously. So far, this work is just at the idea stage; but we believe that these ideas are promising, and look forward to testing them in the near future.

Finally – returning to our theme of the structure of a graph – we have included a section (Section 5) on the problem of *visualization* of a graph. Here too we have some unpublished results – in particular, the use of the steepest-ascent graph (SAG), taken from the topographic picture, for visualizing the structure of an undirected graph. We believe that a glance at Figures 21 and 22 (comparing the two), or at Figures 23 and 25, will illustrate the promise of these ideas. The SAG visualization is of course also skewed towards our interest in spreading phenomena – thus, it should be especially useful in helping to understand, monitor, and control spreading over the network in question.

Finally, we note that the new ideas of Section 4.4 lend themselves readily to a topographic picture for directed graphs. Hence we hope, in future work, to develop visualization techniques like those presented in Section 5, for directed graphs.

## Acknowledgement

This work was partially supported by the Future & Emerging Technologies unit of the European Commission through Project BISON (IST-2001-38923), and partially supported by the European Commission within the 6th Framework Programme under contract 001907 (DELIS).

## References

[1] L.A. Adamic, *The small world Web*, Proc. 3rd European Conf. Research and Advanced Technology for Digital Libraries, ECDL (1999).

[2] R. Albert, H. Jeong and A.-L. Barabási, *Diameter of the World-Wide Web*, Nature **401** (1999), 130–131.

[3] R. Albert, H. Jeong and A.-L. Barabási, *Error and attack tolerance of complex networks*, Nature **406** (2000), 378–382.

[4] L.A.N. Amaral, A. Scala, M. Barthélémy and H.E. Stanley, *Classes of small-world networks*, Proc. Natl. Acad. Sci. USA. **97** (21) (2000), 11149–11152.

[5] A.-L. Barabási, *Linked: The New Science of Networks*, Perseus Publishing, Cambridge, MA, USA (2002).

[6] G. Di Battista, P. Eades, R. Tamassia and I.G. Tollis, *Graph Drawing: Algorithms for the Visualization of Graphs*, Prentice Hall PTR, Upper Saddle River, NJ, USA (1998).

[7] P. Bonacich, *Power and centrality: A family of measures*, American Journal of Sociology **92** (1987), 1170–1182.

[8] S. Borgatti, http://www.analytictech.com/networks/centrali.htm.

[9] A. Borodin, J.S. Rosenthal, G.O. Roberts and P. Tsaparas, *Link analysis ranking: Algorithms, theory and experiments*, ACM Transactions on Internet Technology **5** (2005), 231–297.

[10] S. Bornholdt and H.G. Schuster, eds, *Handbook of Graphs and Networks: From the Genome to the Internet*, Wiley–VCH, Weinheim (2003).

[11] A. Broder, R. Kumar, F. Maghoul, P. Raghavan, R. Stata, A. Tomkins and J. Wiener, *Graph structure in the web*, Proceedings of the 9th International World Wide Web Conference (2000), 247–256.

[12] M. Burgess, *An approach to understanding policy based on autonomy and voluntary cooperation*, Lecture Notes on Comput. Sci., Vol. 3775 (2005), 97–108.

[13] M. Burgess, G. Canright and K. Engø-Monsen, *A graph-theoretical model of computer security: From file sharing to social engineering*, Int. J. Inf. Secur. **3** (2004), 70–85.

[14] M. Burgess, G. Canright and K. Engø-Monsen, *Importance functions for directed graphs*, Submitted for publication.

[15] G. Canright and K. Engø-Monsen, *Roles in networks*, Science of Computer Programming **53** (2004), 195.

[16] G.S. Canright and K. Engø-Monsen, *Epidemic spreading over networks: A view from neighbourhoods*, Telektronikk **101** (2005), 65–85.

[17] G. Canright and K. Engø-Monsen, *Spreading on networks: A topographic view*, Complexus **3** (2006), 131–146.

[18]  G. Canright, K. Engø-Monsen, Å. Weltzien and F. Pourbayat, *Diffusion in social networks and disruptive innovations*, IADIS e-Commerce 2004 Proceedings, Lisbon (2004).

[19]  M. Girvan and M. Newman, *Community structure in social and biological networks*, Proc. Natl. Acad. Sci. USA **99** (2002), 8271–8276.

[20]  G.H. Golub and C.H. Van Loan, *Matrix Computations*, 2nd edn, The Johns Hopkins University Press (1989).

[21]  P. Holme, B.J. Kim, C.N. Yoon and S.K. Han, *Attack vulnerability of complex networks*, Phys. Rev. E **65** (2002), 056109.

[22]  http://www.analytictech.com/ucinet.htm.

[23]  http://www.google.com/.

[24]  http://www.mcb.mcgill.ca/~bryant/MS240/Notes/5_5_Digraphs.pdf.

[25]  M.A. Jovanovic, F.S. Annexstein and K.A. Berman, *Scalability issues in peer-to-peer networks: A case study of Gnutella*, Technical report, University of Cincinnati (2001).

[26]  J.O. Kephart and S.R. White, *Directed-graph epidemiological models of computer viruses*, Proceedings of the 1991 IEEE Computer Society Symposium on Research in Security and Privacy, Oakland, CA (1991), 343–359.

[27]  J.M. Kleinberg, *Authoritative sources in a hyperlinked environment*, Journal of the ACM **46** (1999), 604.

[28]  R. Kumar, P. Raghavan, S. Rajagopalan, D. Sivakumar, A.S. Tomkins and E. Upfal, *The Web as a graph*, Proc. 19th ACM SIGACT-SIGMOD-AIGART Symp. Principles of Database Systems, PODS (2000).

[29]  S. Leonardi, D. Donato, L. Laura and S. Millozzi, *Large scale properties of the web graph*, European Journal of Physics B **38** (2004), 239–243.

[30]  J.A. McHugh, *Algorithmic Graph Theory*, Prentice Hall, Upper Saddle River, NJ (1990).

[31]  S. Milgram, *The small world problem*, Psychology Today (1967), 60–67.

[32]  H. Minc, *Nonnegative Matrices*, Wiley Interscience, New York (1987).

[33]  R. Motwani and P. Raghavan, *Randomized Algorithms*, Cambridge University Press, Cambridge, UK (1995).

[34]  M.E.J. Newman, *The structure and function of complex networks*, SIAM Review **45** (2003), 167–256.

[35]  M.E.J. Newman, S. Forrest and J. Balthrop, *Email networks and the spread of computer viruses*, Phys. Rev. E **66** (2002), 035101.

[36]  W. de Nooy, A. Mrvar and V. Batagelj, *Exploratory Social Network Analysis with Pajek*, Cambridge University Press (2005). See also http://vlado.fmf.uni-lj.si/pub/networks/pajek/.

[37]  L. Page, S. Brin, R. Motwani and T. Winograd, *The pagerank citation ranking: Bringing order to the web.* Technical report, Stanford University, Stanford, CA (1998).

[38]  E.M. Rogers, *Diffusion of Innovations*, 3rd edn, Free Press, New York (1983).

[39]  T. Stang, F. Pourbayat, M. Burgess, G. Canright, K. Engø and Å. Weltzien, *Archipelago: A network security analysis tool*, Proceedings of the 17th Conference on Systems Administration (LISA 2003), San Diego, CA, USA (2003).

[40]  S.H. Strogatz, *Exploring complex networks*, Nature **410** (2001), 268–276.

[41]  S. Valverde and R. Solé, *Hierarchical small-worlds in software architecture*, Working Paper SFI/03-07-044 Santa Fe Institute (2003).

[42]  D. Watts, *Small Worlds: The Dynamics of Networks between Order and Randomness*, Princeton University Press, Princeton, NJ (1999).

# – 3.4 –

# Knowledge Engineering Using Ontologies

John Strassner

*Waterford Institute of Technology, Waterford, Ireland and Autonomic Computing, Motorola Labs,*
*1301 East Algonquin Road, Mail Stop IL02-2240, Schaumburg, Illinois 60010, USA*
*E-mail: john.strassner@motorola.com*

**Abstract**

While ontologies are widely used in various disciplines, such as knowledge engineering and artificial intelligence, their use in network and systems administration and management is very limited. This chapter will explain how the use of ontologies will make network and system administration more efficient and powerful, so that it can better handle the needs of next generation networks and systems.

## 1. Introduction and background

Network and system administration is unique, in that it enables multiple technologies, applications and products to cooperate to solve a given task. Unfortunately, there is no single standard for representing information that can be shared among heterogeneous applications and information sources. This prevents the sharing and reuse of information, which leads to more complex systems as well as potential problems in realizing the significance and relevance of data.

Existing Business Support Systems (BSSs) and Operational Support Systems (OSSs) are designed in a *stovepipe* fashion [35]. This enables the OSS or BSS integrator to use systems and components that are best suited to their specific management approach. How-

HANDBOOK OF NETWORK AND SYSTEM ADMINISTRATION
Edited by Jan Bergstra and Mark Burgess

ever, this approach hampers interoperability, because each management component uses its own view of the managed environment, thereby creating information islands. An example of an industry OSS is shown in Figure 1. This particular OSS exemplifies the underlying *complexity* of integrating the many different subsystems in a modern OSS.

It is easily seen that this can *not* be solved by new protocols or middleware. The use of different GUIs simply complicates matters, since a GUI inherently shows *its* own application-centric view of the portion of the network that it is managing, and *not* the entire system. The use of different protocols does not help, because different protocols have different limitations in the data that they can query for, the commands that they can transmit, and the data that they can represent.

This is a block diagram of a *real* OSS (Operational Support System) that currently exists. It epitomizes 'stovepipe' design, which is the design of an overall system that uses components that were *not* built to seamlessly integrate with each other. Referring to Figure 1, two important problems are exhibited:

- connections are *brittle*: changing any one component affects a set of others that are directly connected, which in turns affects a new set of components, *ad nauseum*,
- the learning curve for understanding this architecture, let alone maintaining and upgrading pieces of it, is enormous,
- the lack of common information prohibits different components from sharing and reusing common data.

These are all different forms of a single problem – *complexity*. Legacy approaches have tried to solve this problem by building a set of application-specific data models that represent the characteristics of devices and management applications. This is shown in Figure 2.

In the above OSS, the inventory solution is shown as being designed in a vacuum. While this enables it to represent data in the best way for it to operate, this is *not* the best way for the system as a whole to use this data. This is because many different applications all share a view of the same customer. Unfortunately, if they are designed in disparate ways without regard to each other, then the same data is likely to be represented in different formats. In Figure 2, we see that the Inventory Management and the Configuration Management applications name the customer in different ways. Furthermore, while both use the concept of an ID, one calls it 'CUSTID' and the other calls it simply 'ID'. In addition, the former represents its ID as an integer, while the latter uses a string. These and other problems make the finding, let alone sharing, of information more difficult. Now imagine how this problem is magnified – first, because different inventory, configuration, and other OSS and BSS applications of the same type will be used, and second, because of the large number of distinct types of applications that are required.

This simple example illustrates the tendency of each application to build its own view of the data its own way. This creates stovepipes, and makes it very hard (if not impossible) for data to be shared. Note that the above was a very simplistic example. In general, a *mapping* (which can be quite complicated) is required to associate the same or related data from different applications, since they have varying attribute names and values having different attributes in different formats to each other. This results in a set of architectural and integration issues listed in Figure 2.

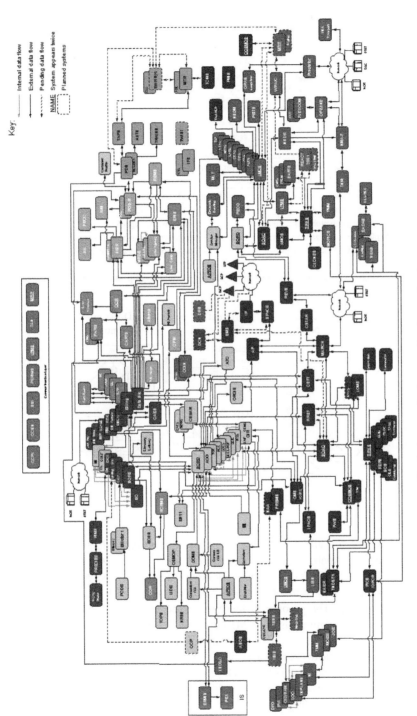

Fig. 1. An example of an operational support system.

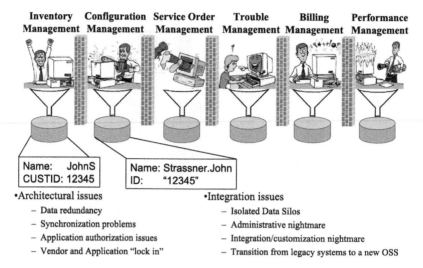

Fig. 2.  Infrastructural shortcomings resulting from stovepipe architectures.

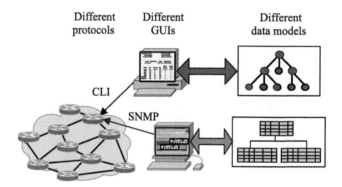

Fig. 3.  The lack of a unified way to represent data semantics.

Unfortunately, even if all OSS and BSS application vendors could agree on a common way to represent information, there is another source of complexity – the programming model of the device. In general, each network device has its own programming model or models. For example, it is common for some routers to support CLI for provisioning and SNMP for monitoring. The problem is that there is no underlying common information model that relates CLI commands to SNMP commands. Without this common model, it is impossible to build a closed loop system, because the administrator does not know what SNMP commands to issue to check if a set of CLI commands acted as expected. The second problem, which causes the first, is that each programming model is built using an object model that prescribes its own view of the world upon the applications and systems that use it. Hence, if the underlying object models used are different, how can they effectively share and reuse data? These issues are shown in Figure 3 [33,35].

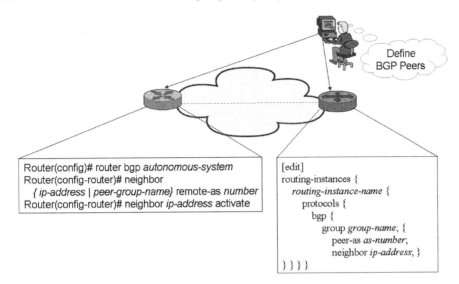

Fig. 4. Different programming models of two routers.

Figure 4 shows an example of the complexity caused by different programming models of network devices – a system administrator defining two different routers from two different manufacturers as BGP peers. BGP is a standard, so this should be easy. But this is not the case.

Looking at how this is implemented, we see some important differences:
- the syntax of the language used by each vendor is different,
- the semantics of the language used by each vendor is different,
- the router on the left uses MODES, which are not present in the router on the right, which implies their programming models are different,
- the router on the right supports virtual routing (via its routing-instances command) which the router on the left does not support.

There are other differences as well in their functionality that are not shown. In effect, by using these two different routers, the operator has built two different stovepipes in its network. Now, administrators are forced to learn the vendor-specific differences of each router. This is how the industry builds OSSs and BSSs today.

One should not underestimate these problems of different device OS versions. Script-based configuration solutions, such as PERL or EXPECT, cannot easily cope with syntax changes. Command signature changes, such as additional or reordered parameters, exacerbate the problem and commonly occur as network device OSs are versioned.

Therefore, if this cannot be automated (due to its complexity), how can it be solved? Certainly not manually, since besides being the source of errors, humans may not even be present, which means that problems can perpetuate and become worse in the system. To better understand this problem, consider the following analogy: the language spoken by a device corresponds to a language spoken by a person, and different dialects of a language correspond to different capabilities in different versions of the OS. Most vendors change syntax and semantics of their languages with major and sometimes minor releases of their device operating systems, thereby producing another language dialect. How can a single administrator, or even a small number of administrators, learn such a large number of different languages for managing the different types of network devices in a network? More importantly, if there is an end-to-end path through the network consisting of multiple devices, how can this path be configured and managed without a single 'über-language'? Remember that most systems use vendor-specific command line interfaces (CLIs) for configuration, but rely on SNMP or other standards for monitoring (mostly because the Network and Element Management systems use SNMP). This is because most vendors do not support SNMP configuration, and even if they do, that functionality is limited. CLIs are almost always more powerful because they are typically used in the development of the product. In contrast, network management vendors typically use SNMP because they can (in theory) avoid the problems of parsing different CLIs. This is only partly true, since SNMP supports vendor-specific management information. Hence, the problem of translating multiple vendor-specific languages to and from SNMP exists.

Recently, interest in using a single information model [33,36,37,40] has increased. The idea is to use this single information model to define the management information definitions and representations that *all* OSS and BSS components will use. This is shown in Figure 5.

The top layer represents a single, enterprise-wide information model, from which multiple standards-based data models are derived (e.g., one for relational databases, and one for directories). Since most vendors add their own extensions, the lowest layer provides a second model mapping to build a high-performance vendor-specific implementation (e.g., translate SQL92 standard commands to a vendor proprietary version of SQL). Note that this is especially important when an object containing multiple attributes is split up, so that some of the object's attributes are in one data store and the rest of the object's attributes are in a different data store. Without this set of hierarchical relationships, data consistency and coherency is lost.

This works with legacy applications by forming a single mediation layer. In other words, each existing vendor-specific application, with its existing stovepipe data model, is mapped to the above data model, enabling applications to share and reuse data. In the BGP peering example shown in Figure 4, a single information model representing common functionality (peering using BGP) is defined; software can then be built that enables this single common information model to be translated to vendor-specific implementations that sup-

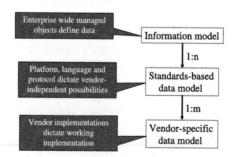

Fig. 5. The relationship between information models and data models.

port the different languages of each router manufacturer. Hence, we reduce the complexity of representing information from $n^2$ (where each vendor's language is mapped to every other language) to $n$ mappings (to a common data model).

However, most existing approaches are still being driven by the capabilities of the device itself. The current state-of-the-art using this approach is represented in Figure 6. This reflects the relatively common practice to use multiple management systems from the same vendor to manage different devices manufactured by that vendor. A good example of this is quality of service (QoS). Many manufacturers build QoS policy servers or QoS tools that work with a subset of the commands of a given version of a device operating system for a subset of the products made by that manufacturer. Other manufacturers build generic applications that are able to implement a subset of available commands for a specific subset of different devices from different manufacturers. The problem with both of these is shown in Figure 6.

Once again, a stovepipe design mentality is embedded in the approach, and that stovepipe approach makes sharing and reusing management information difficult, if not impossible. Looking at the first two translation layers (from IPsec and MPLS VPNs to different device models), we are missing the common abstraction of a VPN. Similarly, the second, third and fourth translation layers are missing the common abstraction of MPLS. Using heterogeneous devices exacerbates this problem, since different languages and programming models are introduced.

In addition, there is no connection between management software (which is used to control network devices and services) and management applications, such as billing, that provide revenue. Most importantly, however, there is no standard way to tie the definition of business services to the definition of network configuration. Thus, there is no means to ensure that the network delivers the services that the business needs at any given time. In other words, current networks are *unaware* of business needs in the first place.

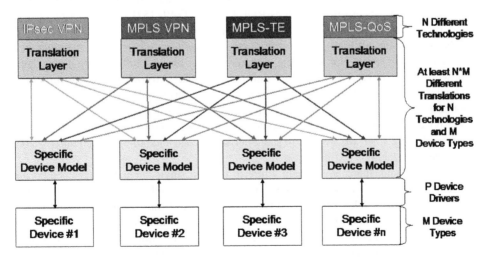

Fig. 6. Different management applications for different functions.

Even if the problem in Figure 6 above is fixed by the use of a common information model, unfortunately we find that information and data models in and of themselves are insufficient to solve the problem of data harmonization described above. Two reasons which ontologies are uniquely positioned to solve are that information and data models, in and of themselves:

- lack the ability to link different representations of the same data together (the "Strassner.John" vs. "JohnS" problem shown in Figure 2),
- are unable to establish semantic relationships between information (not just describing whether they are synonyms and antonyms, but other relationships that compare different entities).

For example, there is no ability in UML [25] to represent:

- time (specifically, the use of *tense* in a sentence),
- relationships, such as 'is similar to', that are needed to relate different vocabularies (such as commands from different vendors) to each other,
- formal semantics of sets,
- introduction of synonyms, antonyms, homonyms, etc.,
- ability to link many representations of an entity to each other (e.g., precise and imprecise),
- groups whose components are *not* individually identifiable, only quantifiable,
- richer sentence semantics, through representations of assertions, queries, and sentences about sentences as *objects*.

Furthermore, while UML is itself a language, it does not have the ability to model the languages used in devices. This is one reason why network management and administration is still so difficult. One obvious approach is to use a single 'über-language' to represent the syntax and semantics of the different device languages being used. This is a very complex task, since in general each device has its own vocabulary, terminology and grammar. Even standards, such as SNMP, fail because of the large variance in constructing MIB-based information, not to mention the presence of public vs. private MIBs *at the root of the MIB tree*. In addition, newer efforts such as RDF [53] and OWL [52] still have a number of shortcomings, such as lack of tense and indirect objects.

The above points all focus on the *inability to share and reuse knowledge due to the lack of a common language*. This has been pointed out as far back as 1991, where it was stated that with respect to representing and acquiring knowledge, "Knowledge-acquisition tools and current development methodologies will not make this problem go away because the root of the problem is that knowledge is inherently complex and the task of capturing it is correspondingly complex... Building qualitatively bigger knowledge-based systems will be possible only when we are able to share our knowledge and build on each other's labor and experience." [24].

That being said, there is a certain allure to using a single 'über-language' based approach. That allure is the promise to be able to translate into the different languages used by device vendors – a kind of 'device Esperanto'. However, just as there will never be a single programming language that is used by all, how could there be a single management language? I argue that such a device Esperanto will never be built – not because it is difficult (which indeed is true) but because *there is no motivating business reason for the device vendors to do so*. Hence, this chapter describes an emerging new approach to solving this problem, where information and data models are augmented with ontologies to provide the ability to *learn* and *reason* about data.


## 2. Ontologies to the rescue

Ontologies have their root in philosophy. In philosophy, the study of ontology is a branch of metaphysics that deals with nature of being. In computer science and knowledge engineering, ontologies define theories of what exists. We are interested in ontologies for two reasons:

- Ontologies offer a formal mechanism for defining a better understanding of facts.
- Shared ontologies ensure that different components and applications communicate about different aspects of the same entity in a standard way.

Ontologies are used to model declarative knowledge. By this, we mean knowledge in which the relationships between the data are declared, or stated, and then one or more automatic mechanisms are used to answer queries about the data.

There are numerous definitions of ontologies. In order to arrive at a definition for this Handbook, we will first examine the most popular of these. Then, we will pull together the strengths of these and mold them to the needs of network and system administration.


### 2.1. *Why ontologies?*

Knowledge exists in many different forms. Humans, and the systems and applications that they use, will not 'give up' knowledge that they understand or are used to manipulating. Remember our OSS, where the operator has chosen to use multiple systems and components that were not built to interoperate. This lack of interoperability has ensured that different applications and devices develop their own private knowledge bases, because there is no easy way to share their knowledge with other applications. This is a serious problem, that requires a much more innovative solution than just defining some new tags and representing management data in XML, because that will not be able to represent the *semantics* of these heterogeneous knowledge bases. In addition, such an approach would not be able to delineate the set of assumptions made regarding how the application was using that knowledge. The cost of this duplication of effort has been high and will become prohibitive as we build larger and larger systems. Hence, we must find ways of preserving existing knowledge bases while sharing, reusing, and extending them in a standard, interoperable manner.

Information and data models will not and must not go away – they serve a vital need in being able to represent both facts as well as behavior. Representing knowledge in ontologies enables a formal, extensible set of mechanisms to be used to define mappings between the terms of knowledge bases and their intended meanings. In this way, we can avoid trying to define a single 'über-language' that has no underlying business reason, and instead achieve knowledge interoperability by using a set of ontologies to precisely and unambiguously identify syntactic and semantic areas of interoperability between each vendor-specific language and programming model used. The former refers to reusability in parsing data, while the latter refers to mappings between terms, which requires content analysis.

## 2.2. *Existing ontology definitions*

In [24], Neches et al. gave the following definition:

> An ontology defines the basic terms and relations comprising the vocabulary of a topic area as well as the rules for combining terms and relations to define extensions to the vocabulary.

This definition describes how to build and extend an ontology. The extension of an ontology is crucial, as we cannot anticipate new devices or their languages. Hence, we need a mechanism that is inherently extensible. In addition, it defines knowledge as the terms of the ontology plus additional knowledge that can be inferred from using the ontology.

In [10], Gruber provided the following (in)famous definition of an ontology:

> An ontology is an explicit specification of a conceptualization.

The problem with this definition is that people new to the ontology community are often confused by its brevity. [14] provides a much more readable expansion of what is meant by this definition; the following are important excerpts from this that expand on the above definition:

> A body of formally represented knowledge is based on a conceptualization: the objects, concepts, and other entities that are assumed to exist in some area of interest and the relationships that hold among them (Genesereth & Nilsson, 1987). A conceptualization is an abstract, simplified view of the world that we wish to represent for some purpose. Every knowledge base, knowledge-based system, or knowledge-level agent is committed to some conceptualization, explicitly or implicitly... An **ontology** is an explicit specification of a conceptualization. The term is borrowed from philosophy, where an Ontology is a systematic account of Existence... When the knowledge of a domain is represented in a declarative formalism, the set of objects..., and the describable relationships among them, are reflected in the representational vocabulary with which a knowledge-based program represents knowledge... We use common ontologies to describe ontological commitments for a set of agents so that they can communicate about a domain of discourse without necessarily operating on a globally shared theory. We say that an agent **commits** to an ontology if its observable actions are consistent with the definitions in the ontology.

This more complete definition expands the original definition of [24] by adding first, the notion of a *shared* vocabulary and second, the notion of an ontological commitment. This latter is an agreement by each entity to use the shared vocabulary in a consistent manner.

In [11], the notion of an ontological commitment is further defined as follows:

> An ontological commitment is a function that links terms of the ontology vocabulary with a conceptualization.

In the ontological community, ontological commitments tend to be minimized. This is because in order to make a statement in which the existence of one thing is implied by asserting the existence of a second thing, an ontological commitment is necessary (because the existing of the first thing may not have been realized or expected).

In [42], Studer et al., refine Gruber's definition as follows:

> An ontology is a formal, explicit specification of a shared conceptualization. Conceptualization refers to an abstract model of some phenomenon in the world by having identified the relevant concepts of that phenomenon. Explicit means that the type of concepts used, and the constraints on their use are explicitly defined. Formal refers to the fact that the ontology should be machine readable. Shared reflects the notion that an ontology captures consensual knowledge...

This definition tightens up the use of 'formal' and 'explicit', and emphasizes 'consensual'. It defines a new measure for ontologies: being machine readable. However, it loses the concept of ontological commitments. In [42], Studer realizes this, and tries to differentiate between consensual taxonomies and ontologies. In general, the ontology community categorizes them according to the *restriction on domain semantics* that are present, as follows:

- Unrestricted: defined by a natural language.
- Semi-restricted: defined by a restricted and/or structured form of a natural language.
- Restricted: defined using a formal language.
- Rigorously restricted: defined using a formal language with complete semantics and definitions, wherein it can be proved that: (1) all valid conclusions can be deduced using only the permitted deduction rules, and (2) invalid conclusions cannot be deduced.

### 2.3. *Our ontology definition*

In order to build our definition of ontology, let us revisit the principal motivation for using ontologies in network and system administration.

Since it is not practical to have a single language to represent all of the different device languages, we instead want to have software that can identify one or more meanings that correspond to each of the commands in each of the devices. This enables us to develop a more in-depth understanding of each command by referring to its set of meanings, and use these meanings to map to other commands that have similar meanings.

Ontologies will be used to supply these meanings. Hence, we need:

- A set of ontologies, one for each programming model of each device used in the system. Since a device can support multiple programming models and/or multiple languages, we thus need an ontology for each distinct combination of language and programming model.
- Each ontology must define its set of objects, concepts and other entities, as well as the relationships that hold among them, using the *same set of declarative formalisms*.
- A set of logical statements that define the terms, rules for relating terms, and the relationships between the terms, of the shared vocabulary of a given ontology.
- A similar set of logical statements that enable multiple ontologies to be related to each other.
- The same formal method for extending the shared vocabulary must be used by all users of each of the ontologies.
- A *life cycle* for ontologies must be rigorously defined. A life cycle is defined as the period of time between when an entity is created and when that entity is either retired from service or destroyed.

In summary, the underlying principal of my definition is: "if something is known, it should be able to be represented in a machine-interpretable manner". Therefore, the working definitions of a generic ontology as well as ontologies for network and system administration are defined as follows:

> An ontology is a formal, explicit specification of a shared, machine-readable vocabulary and meanings, in the form of various entities and relationships between them, to describe knowledge about the contents of one or more related subject domains throughout the life cycle of its existence. These entities and relationships are used to represent knowledge in the set of related subject domains. Formal refers to the fact that the ontology should be representable in a formal grammar. Explicit means that the entities and relationships used, and the constraints on their use, are precisely and unambiguously defined in a declarative language suitable for knowledge representation. Shared means that all users of an ontology will represent a concept using the same or equivalent set of entities and relationships. Subject domain refers to the content of the universe of discourse being represented by the ontology.

Given the above definition, we can now define ontologies for network and system administration as:

> An ontology for network and system administration is a particular type of ontology whose subject domain is constrained to the administration of networks and systems. Administration is defined as the set of management functions required to create, set up, monitor, adjust, tear down, and keep the network or system operational. One or more ontologies must be defined for each device in the network or system that has a different programming model.

We also need to define how ontologies can be combined or related to each other:

> An ontology commitment represents a selection of the best mapping between the terms in an ontology and their meanings. Hence, ontologies can be combined and/or related to each other by defining a set of mappings that define precisely and unambiguously how one node in one ontology is related to another node in another ontology.

Ontologies are thus one form of *knowledge representation*. They relate to cognitive science, where explanations for perceptions, conclusions, and so forth are proposed. Hence, how knowledge is represented is fundamental to the understanding of cognitive issues, such as what a concept means, what implications it has, and how it is represented affects any decisions that may arise from that representation.

Agreeing on a consensual definition for how to represent knowledge is not as easy as it appears. While people often think of representation as a visible presentation of a concept or idea, the formal study of representation [22,29] is much more demanding. In [29], Pamer defines five features of a representational system:

- What the represented world is (what we perceive).
- What the representing world is (the theoretical structures used to formalize our perception).
- What aspects of the represented world are being modeled.
- What aspects of the representing world are doing the modeling.
- What are the correspondences between the two worlds.

For example, suppose my represented world consists of an ice cube, a glass of water at room temperature, and a pot of boiling water. If I choose to represent these three objects using a common representational notion of their temperature, then I will necessarily lose other aspects of the environment in which they exist, such as their size, shape, and so forth.

The point of this example is that in almost all known representational systems, the representing concepts are not able to capture all of the information about the entities that they are representing. The process of representing entities in a domain will typically produce a loss of information. Furthermore, the choice of a representational system will make some information very easy to find at the expense of making other information very difficult to find.

These are acceptable tradeoffs. The principle of *abstraction* relies of mechanisms to *reduce or eliminate* unwanted details so that one can focus on only a few concepts at any given time. In this book, we are thus interested in ontology as a means to focus on how to better understand knowledge as applied to network and system administration.

## 3. An overview of current ontologies

There are many different types of ontologies, built from many different tools, to solve many different problems. This section will provide an overview to help the reader decide what type of ontologies best suit their need.

### 3.1. *What type of ontology language do you need?*

The functionality of an ontology is directly dependent on how knowledge is represented, and what types of reasoning capabilities are required. This in turn depends on which constructs are available for representing knowledge, and the language(s) used to implement the ontology.

**3.1.1.** *Knowledge representation approaches that we will not consider*   In order to represent knowledge, we need a representation language. While one can certainly use natural languages (such as English or Norwegian), there are three main problems with doing so. The first is the degree of *ambiguity* that is present in such languages. For example, a 'bank' in English can refer to a financial institution, the action of moving laterally, an arrangement of similar objects in a row or column, a long ridge (as in a 'long bank of earth'), a slope in a turn of a road, trust ('bank on education'), and others. The second is that a natural language is *too* expressive (i.e., too flexible) to perform efficient reasoning. For the purposes of this chapter, reasoning means "to be able to deduce what must be true, given what is known". This requires an unambiguous starting point that describes exactly what knowledge you have. In other words, reasoning requires precision of meaning.

There are other types of representation languages that we will not consider. Database technologies, as well as UML, are not appropriate for the types of ontologies that we are considering in this chapter. This is because they both lack formal mechanisms for expressing semantics. The reader may object, as the reader may be familiar with the ontology definition metamodel [27], which is an effort to build a MOF2 Metamodel, UML2 profile, and any additional information needed to support the development of ontologies using UML modeling tools, the implementation of ontologies in OWL, and other factors. The problems, however, with this include:

- the lack of *explicit* semantics – there is no definition of lexical semantics, and there are no definitions of sentence elements and their ordering,
- the inability to define relationships, such as 'is similar to', that are needed to relate different vocabularies (such as commands from different vendors) to each other,
- tense – past, present and future reasoning cannot be done. This is important, because network and system administration requires traceability of configuration changes made. Hence, it is important to be able to reason about changes in past and present configurations, and what needs to be done in the future,
- modals – translation of English into network administration cannot be done directly, meaning that even restricted natural language interfaces cannot fully express management tasks,
- episodic sequences – the ability to formalize the set of actions performed in a configuration. Interaction diagrams fall short of this, because of their lack of semantics and their inability to specify irregular patterns.

The second problem occurs when different terms need to be equated in a relative manner. Network and system administration usually works with multiple representations of the same information; two examples of this are standards-based models, such as the DMTF CIM and the TMF SID, and vendor programming commands, such as SNMP and a command-line interface. In general, this is a graph-matching problem, where one concept in one model (which can be represented as one or more model elements in the model) must be equated to another concept in the other model. Regardless of whether the models share a common vocabulary (which most network management devices and applications do not), then a mapping between these two sub-graphs (most likely with constraints to take account of operational and behavioral dependencies) to define the semantic similarity between them needs to be constructed. Furthermore, we cannot capture the *semantics* associated with different representations of management information (e.g., "the use of a particular feature requires hardware having a particular set of physical capabilities along with a specific operating system version").

Another type of representation mechanism that we will not explore is the use of Frames *by themselves*. A frame represents some object, class, or concept. Frames have named properties called slots. The values of these slots can be primitive objects (e.g., numbers or strings), other frames, and other characteristics that help define the meaning of the enclosing frame. The components of a slot are facets, such as the value facet (which enables the slot to hold a value) and the daemon facet (which enables the slot to compute a value). However, unlike traditional objects in object-oriented languages, frames:

- have flexibility, in that new slots can be added at any time,
- daemons can be associated with slots to dynamically compute their value (note that this is different compared to most languages, like Java, in which methods are *not* linked to a *particular* variable),
- frames are inherently reflexive, and represent their own structure in terms of frames,
- frames can have *reasoning mechanisms* built into their slot interface.

The problem with frames *by themselves* is that reasoning depends on the *mode of the representation*, so there is less *reuse* of knowledge. This makes it much more difficult to represent the knowledge from all possible points-of-view that you might need. Also, the knowledge representation must be designed around the indexing, and the implicit knowledge is lost if the code is separated from the knowledge base. When using logic-based languages (either description logics or first-order logic (with or without frames)), the knowledge representation is independent of mode, the indexing is independent of the knowledge representation, and the implicit knowledge is independent of the knowledge base.

This is illustrated using the simple example shown in Figure 7 below. The simple frame-and-slot representation on the left defines three frames – one each for the three objects Armani, Fendi and Dog. Armani has the animal type 'dog' defined, as well as his mother ('Fendi') defined. Fendi is also of type 'dog', but her mother is named 'Versace'. Finally, a 'Dog' is a type of mammal, whose size is 'small'. This set of frames enables us to query for Armani, and discover that his mother is Fendi. However, if we query for Fendi, we will discover that her mother is Versace, but we will *not* discover that she is the mother of Armani, because that information is not provided. In order to get this information, we need to add a separate index going from the mother attribute of each frame to the objects of which they are mothers.

Compare this to the logic-based approach. This is read as follows: Armani is an instance of Dog, whose mother is Fendi. The advantage of this representation is that *any* of the attribute values of this object can be queried on. Hence, we could query on Armani and determine who his mother is; more importantly, we could query *mother* and get all of the animals bearing this relationship. The ability of logic-based languages to provide this type of comprehensive indexing automatically enables very efficient reuse of knowledge.

The frame only approach separates knowledge from the knowledge base. For example, if you wanted to define a generic formula for computing the approximate weight of a dog using a set of other attributes, there is no place to store that slot in the three frames on the

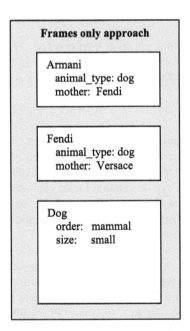
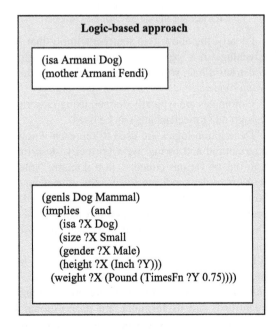

Fig. 7.  Difference between frames and logic-based knowledge representation.

left. You instead would need to create code that attaches to the weight slot with instructions. Unfortunately, this means that if you later separate the knowledge base from the code base, you lose that information.

In the logic-based approach, we can *augment* the description of objects with rules. This rule reads as follows: "If an entity is a small, male dog whose height is Y inches, then its weight will be approximately 0.75·Y pounds". Since this rule is part of the definition, the knowledge stays in the knowledge base even when you separate it from the code.

**3.1.2.** *Knowledge representation approaches that we will consider*  Since network and system administration requires specificity of semantics, our goal is to build ontologies that are either restricted or rigorously restricted.

The three most popular means for representing knowledge are predicate logic approaches [9], AI techniques (mainly frames and first-order logic [21]) and description logics [1]. In each case, the proper choice of ontology depends on the degree of knowledge representation and the power of reasoning provided.

All three of these approaches use logic-based languages, which provide basic constructs for representing and facilitating reasoning mechanisms. As we have seen, logic-based languages offer a simplified, more efficient approach compared to natural language, by first identifying common concepts in a domain, followed by defining logical relationships to formulate rules about those common concepts. For example, "if $x$ caused $y$, then $x$ temporally precedes $y$". This avoids having to individually interpret multiple similar, but different, sentences stating this fact. Logic-based languages also provide a formal calculus of meaning. For example, consider the sentences:

"It is not a fact that all men are taller than all women."
"It is a fact that all men are taller than two feet."

Clearly, the above two sentences imply that some women are taller than two feet. The advantage of a logic-based language is that this can be expressed *based on the logical constants alone*, without knowing what the other, non-logical words (i.e., *men*, *taller* and *women*) mean.

Ontologies are typically defined using Description Logics [48] or First-Order Logic [9], though other mechanisms can be used.

Description logics are used to represent knowledge, and are used to represent terms in a structured and formal way. Formally, description logics refers to explanations and definitions of various concepts in a domain. Note that description logics can be translated into first-order predicate logic – the advantage of this is that description logics can often be used to structure and codify the logic-based semantics of first-order logic. A DL supports hierarchical classification through the use of a well-defined notion of subsumption. Description logics are also heavily used in the semantic network.

**3.1.3.** *Using predicate logic approaches to represent ontologies*  Predicate logic approaches are usually based on first-order logic. This is a type of logic that extends propositional logic, whose formulae are propositional variables. In contrast, predicate logic is quantified, and its formulae are propositional functions, not just variables. Practically speaking, this means that predicate logic can define propositions such as "for all $x, \ldots$" or "for some $x, \ldots$".

The two most important examples of this approach are KIF [16] and CycL [3]. KIF is a language designed for use in the interchange of knowledge among disparate computer systems. Hence, it is very difficult to construct an ontology using KIF. CycL [20] was developed within the CyC project for specifying a large common sense ontology that should provide AI to computers. Both KIF and CycL are similar – they each extend first-order logic with second-order concepts (including extensions to handle equality, default reasoning and some second-order features), and they both use a LISP-like syntax for their calculi. However, CycL is more powerful (i.e., it has richer modeling primitives) because its purpose was to *model* ontologies, whereas the purpose of KIF was to exchange knowledge between ontologies.

This type of approach is suitable for an experienced developer knowledgeable in logic programming, but not for the general user.

**3.1.4.** *Using frames and first-order logic to represent ontologies*　In contrast to the use of predicates, the frame based approach uses classes (i.e., frames), some of which have properties called slots (attributes). Attributes are only defined to classes they are defined for. Note that the same attribute may be redefined for different classes.

Gruber was the first to model ontologies using frames and first-order logic [10]. In this approach, there are five key elements: concepts (represented as either metaclasses or classes, usually organized in a taxonomy), instances, relations (which represent associations between different concepts), attributes (relation between a concept and a property of that concept), functions (a special type of relation in which the last element of the relation is unique), and axioms (facts that are always assumed true, whether or not they can be formally defined by other components). Two examples are Ontolingua [26] and FLogic [17]. Ontolingua is based on KIF, and also includes the Frame Ontology [45] (a knowledge-based ontology for modeling ontologies using a frame-based approach). JTP [15] is a theorem prover that has been implemented for Ontolingua ontologies. FLogic (short for Frame Logic) is a language that integrates features from object-oriented programming, first-order logic and frame-based languages. There are a number of inference engines that can be used with FLogic ontologies, such as Ontobroker [7] and FLORID [8].

This type of approach is still for developers that are knowledgeable in logic programming, but the support tools and products built around these types of languages make them more productive.

**3.1.5.** *Using description logics to represent ontologies*　Description logics (DL) model an application domain in terms of concepts (classes), roles (relations) and individuals (objects). The domain is a set of individuals; a concept is a description of a group of individuals that share common characteristics; roles model relationships between, or attributes of, individuals; individuals can be asserted to be instances of particular concepts and pairs of individuals can be asserted to be instances of particular roles.

The two important uses of DLs are inheritance and automatic classification. However, they are intrinsically static (i.e., it is very different to represent processes and workflows) and there is no explicit notion of variables and rules.

A DL consists of two parts, the TBox and the ABox. The TBox defines a terminology, and is built through declarations that describe the properties a concept or role has. The

ABox contains assertions that define individuals in the domain. The power of description logic is derived from the automatic determination of subsumption between compositional descriptions (i.e., given two concepts $x$ and $y$, if $x$ subsumes $y$, then every instance of $y$ is an instance of $x$). *Subsumption* and *classification* are used for the construction, maintenance and use of an ontology. Note that this provides well-defined semantics: if terms are placed in a child–parent relationship in a taxonomy of concepts, they are in this relationship because of subsumption. Retrieval of a concept is easy, since in description logic, the definition language and the query language are the same thing. To retrieve the individuals satisfying a concept, or to find the subsuming or subsumed concept descriptions of a concept, one describes the concept in the same way as one would define it.

One of the better known examples of a DL language is LOOM [47]. It was built as an environment for the construction of expert systems. LOOM is descended from the KL-ONE family of DL languages; consequently, it has two 'sub-languages', a description sub-language (for describing domain models through objects and relations in the TBox) and an assertional language (for asserting facts about individuals in the ABox). Significantly, LOOM supports rules and predicate calculus, as well as procedural programming, making it very powerful. The LOOM classifier supports classification (through subsumption) as well as deductive reasoning (through backward chaining).

There are good arguments for using either frames and first-order logic or for using DL. The basic problem with a pure DL language is that it is, in general, *too expressive* to enable complete reasoning. However, most DLs can be reduced to first-order logic, which enables the use of first-order logic theorem provers.

**3.1.6.** *Markup languages to represent ontologies*    There are many different markup languages. For the sake of brevity, only three will be discussed: RDF(S), DAML + OIL and OWL. All three of these languages are based on XML. More importantly, all three are used in the Semantic Web [32].

RDF [19] was originally developed by the W3C as a semantic network based language to describe Web resources. A *semantic network* is defined as a knowledge representation formalism which describes objects and their relationships in terms of a network (in the form of a directed graph) consisting of labeled arcs and nodes. The World Wide Web was originally built for human consumption, which means that its data is usually not machine-understandable. RDF proposed a new approach – the use of metadata to describe the data contained on the Web. RDF is a foundation for processing metadata – through the use and management of metadata, it provides increased interoperability between applications that exchange machine-understandable information. Metadata literally means 'data about data'. In ontologies, this usually includes who or what created the data, and when, how, and by whom it was formatted, reformatted, received, created, accessed, modified and/or deleted. Metadata is captured using three types of objects: resources, properties and statements. One of the goals of RDF is to make it possible to specify semantics for data based on XML in a standardized, interoperable manner. RDF can be characterized as a simple frame system. A reasoning mechanism could be built on top of this frame system, but is not provided by RDF.

RDF(S) is shorthand for the combination of RDF and the RDF Schema [31]. RDF provides a way to express simple statements about resources, using named properties and values. However, RDF user communities also need the ability to define the *vocabularies* they intend to use in those statements. This vocabulary is defined by a new specification, called RDF Schema. RDF Schema does not provide a vocabulary of application-specific classes and properties *per se*, since it could never fill the needs of all applications; rather, it provides the facilities needed to *describe* such classes and properties, and to indicate which classes and properties are expected to be used together. The strategy of RDF(S) is "…to acknowledge that there are many techniques through which the meaning of classes and properties can be described. Richer vocabulary or 'ontology' languages such as DAML + OIL, W3Cs [OWL] language, inference rule languages and other formalisms (for example temporal logics) will each contribute to our ability to capture meaningful generalizations about data in the Web". Basically, RDF(S) combines semantic networks with some, but not all, of the richness of frame-based knowledge representation systems. In particular, it does not contain the formal semantics for creating RDF(S).

DAML + OIL [4] was an extension of OIL, and was created by a joint US–EU committee working on the DARPA project DAML. IN a nutshell, DAML + OIL adds decision logics based primitives to RDF(S). Compared with RDF(S), DAML + OIL semantics are axiomatized using KIF, and hence inference engines can be used or even built for DAML + OIL. This also enables the JTP theorem prover [15] to be used. Finally, external classifiers based on decision logics such as FaCT [6] and RACER [30] (which is also a reasoner) can be used with DAML + OIL as well.

OWL [28] is a W3C standard, and was created for the semantic web. It is derived from and supercedes DAML + OIL. It covers most (but not all) of the features of DAML + OIL, and renames most of the DAML + OIL primitives to make them more approachable. OWL has a layered structure – OWL Full (OWL DL syntax plus RDF), OWL DL (first-order logic only, roughly equivalent to DAML + OIL) and OWL Lite (a subset of OWL DL). OWL contains three types of objects: concepts, individuals and properties. Since OWL is similar to DAM + OIL, inference engines used for DAML + OIL can also be used with OWL (e.g., FaCT and RACER). An inference engine for use with OWL is EULER [5].

### 3.2. *What type of ontology do you need?*

There are a large number of different types of ontologies. The following is taken from synthesizing the works of [23] and [18] with my own thinking.

In [23], Mizoguchi et al. proposed four basic types of ontologies:
- Content ontologies:
  - Domain Ontologies.
  - Task Ontologies.
  - General Ontologies.
- Tell and Ask Ontologies.
- Indexing Ontologies.
- Meta-Ontology.

Content ontologies are designed to reuse knowledge. They consist of domain ontologies, which focus on a particular set of related objects, activities or fields; task ontologies, which focus on describing different activities using sentence elements; and general ontologies, which include other concepts not covered. Tell and ask ontologies focus on sharing knowledge, while indexing ontologies are specifically designed for querying. Mizoguchi defines meta-ontologies as ontologies designed for representation knowledge.

In [18], ontologies were classified according to the richness of the ontology structure and the knowledge expressed by the ontology. This is shown in Figure 8.

In this categorization:

- *Controlled vocabularies.* This is a finite list of terms, such as those present in a catalog. There is no guarantee that each term is unique and unambiguous.
- *Glossary.* This is a list of terms and meanings, in which the meanings are specified in a natural language. This provides more information than controlled vocabularies, since humans can read the natural language (of course, machines cannot necessarily read these). There is no guarantee that each term is unique and unambiguous, and hence these are not adequate for computer agents.
- *Thesauri.* These provide some additional semantics in their relations between terms, such as synonym and antonym relationships. Typically, thesauri do not provide an explicit hierarchy (although with narrower and broader term specifications, one could deduce a hierarchy).
- *Informal is-a.* This is an informal notion of generalization and specialization, typically found in web applications. They are typified by broad, shallow hierarchies. The term 'informal' is used to differentiate this type of hierarchy from hierarchies formed by strict subclassing, which is usually the minimal point for an ontology to be officially recognized.
- *Formal is-a.* This is a formal notion of generalization and specialization. In these systems, if A is a superclass of B, then if an object is an instance of B, that object must be an instance of A. Strict subclass hierarchies are necessary for exploitation of inheritance.
- *Formal instance.* This differentiates between classification schemes that only include class names and schemes that provide formal identity of all content.
- *Frames.* This adds the notions of formal property definitions, which are subject to inheritance and other semantic relationships.
- *Value restrictions.* This enables restrictions to be placed on what values a property can take on (e.g., datatype as well as range).
- *General logical constraints.* This enhances the above category by enabling the value of a property to be based on a mathematical equation using values from other prop-

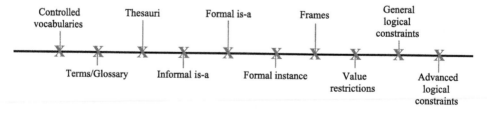

Fig. 8. Categorization of ontologies by Lassila and McGuinness.

erties. This is typified by allowing the specification of arbitrary logical statements to determine the value of a property.

- *Advanced logical constraints.* This enhances the above category by enabling the value of a property to be based on the specification of first order logic constraints between terms of the ontology.

In addition to two classification schemes that we have reviewed, there are some other notable ontologies that are best classified as belonging to multiple groups of these schemes. Five important ones are:

- Dublin core, which is a metadata ontology that provides a vocabulary for describing the content of on-line information sources [44]. Metadata is not well described in either of the above two classification schemes.
- Linguistic ontologies, such as WordNet [51], which are on-line lexical reference systems that enable dictionary terms to be searched conceptually, instead of alphabetically, since they include words, their meaning and the relations between them (such as synonyms, hypernyms, antonyms, and so forth). More importantly, it organizes lexical information in terms of word meanings, rather than word forms, and hence resembles a thesaurus. So, this type of ontology falls into multiple domains, and has some new features.
- Generic ontologies of Mizoguchi are not so easily categorized. The Cyc ontology contains a very large number of generic terms, which makes them valid over multiple domains.
- The IEEE Standard Upper Ontology [46] describe very general concepts which are used to provide high-level concepts that can link all root terms of all existing ontologies. This is because an upper ontology is limited to concepts that are meta, generic, abstract and philosophical, and therefore are general enough to address (at a high level) a broad range of domain areas.
- Representational ontologies span multiple domains. Such ontologies provide representational entities without stating what should be represented. For example, the Frame Ontology [45] defines concepts such as frames, slots, and slot constraints, which enables other ontologies to be built using frame-based conventions.

In summary, the type of ontology that you need is a direct function of the requirements of the management information that you need to represent. There is no one answer that is always right – your requirements will vary based on the needs of your application.

### 3.3. *Tools for building ontologies*

Network and system administration applications require two basic types of ontology tools. They are:

- Ontology development tools, for creating, editing and managing ontologies that can be queried using one or more inference engines.
- Ontology merging tools, for either combining multiple ontologies, or identifying nodes in one ontology that are semantically similar to nodes in other ontologies.

**3.3.1.** *Ontology development tools*    Ontology development tools can be split into two groups – those tools that are independent of an ontology language, and those that are tightly dependent on one.

Protégé-2000 [12] is the *de facto* standard for all tools. Protégé-2000 is a Java-based stand-alone application that has an extensible architecture, which uses the concept of plug-ins [13] to add and integrate new functionality. Three types of plug-ins exist, organized by the functionality that they offer:

- Slot widgets, which provide custom data creation, editing, deleting and management capabilities including formatting.
- Tab plug-ins, which provide significant added functionality available under a new tab in the ontology editor.
- Backends, which provide import, export, storage and retrieval facilities.

The core of the Protégé-2000 application is based around an ontology editor, which uses first-order logic and frames. The knowledge model can be extended by means of metaclasses. It offers custom support for designing how instances are created, and has basic support for ontology merging and querying. Finally, it has basic inference support using PAL (a subset of KIF), but also allows other inference engines to be used via its tab plug-in mechanism.

Ontolingua [49] was designed to build ontologies in a collaborative fashion. The idea was to achieve consensus on common shared ontologies by geographically distributed groups of users. This enables the group to use the expertise of each of its members to not just construct an ontology, but more importantly, to be able to examine, evaluate and edit it as part of the development process.

The architecture of the Ontolingua Server provides enables new ontologies to be created, existing ontologies to be modified, and optionally for ontologies to be accessible as part of a library of ontologies. This is supported through a set of ontology-related web applications. Ontologies are stored as Lisp text files, but users interface with Ontolingua using web forms. Users interact with the Ontolingua Server remotely (using a web browser) via HTTP and HTML collaboratively through features such as notifications and comparisons. Remote application can manage ontologies stored at the server over the web using a network API that extends the Generic Frame Protocol with a network interface. Finally, a user can translate an ontology into a format used by a specific application, which enables ontologies to be imported and exported. Ontolingua uses a knowledge representation language of the same name, which is based on a combination of frames and first-order logic. The definition of classes, axioms and other building blocks is done using KIF. It does include an inference engine, but can use other inference engines by exporting the ontology to a different language for which a reasoner exists.

**3.3.2.** *Ontology merging tools*    The purpose of an ontology merge tool is to enable the user to identify similarities and differences between different ontologies, and to optionally create a new ontology that includes terms from them.

The PROMPT [50] plug-in is integrated in Protégé-2000, and is a very complete tool for comparing and merging ontologies in a semi-automated process. It consists of four integrated tools. The first provides a report (both using an explorer-like tree view or a table view showing the frames side-by-side) of the differences found in the two ontologies (e.g., for your current ontology, new frames are underlined, deleted frames are crossed out, and moved and changed frames are shown in different fonts and colors). When a class is selected, the class as well as a table comparing the current frame to its old version is shown.

PROMPT supports merging ontologies by making suggestions, determining conflicts, and proposing conflict–resolution strategies; the initial suggestions are based on the linguistic similarity of the frame names. After the user selects an operation to perform, PROMPT determines the conflicts in the merged ontology that the operation has caused, and then proposes resolutions. Once the basic merging is complete, PROMPT then examines the overall structure of the ontology and proposes other operations that the user should perform.

Ontologies can be reused through its project metaphor, which enables projects as well as ontologies to be included in a project. This enables small ontologies specific to a particular set of tasks to be developed, and later reused into larger projects. It also enables PROMPT to support extracting a part of an ontology, by enabling one or more frames to be selected and copied into a project. PROMPT supplies a list of conflicts that arise when ontologies are merged.

Chimaera [43] supports merging ontologies together and diagnosing individual or multiple ontologies. It supports users in such tasks as loading knowledge bases in differing formats, reorganizing taxonomies, resolving name conflicts, browsing ontologies, and editing terms. It is built on top of the Ontolingua Server, and can handle many different inputs through the Ontolingua translation mechanisms. Its approach is slightly different than that of PROMPT – it loads the ontologies to be merged into a single ontology, and then supports the user in merging them. Like PROMPT, some of these support steps are semi-automatic (e.g., providing a list of concepts that are equivalent in both ontologies, as well as suggesting concepts that can be merged). However, it does not guide the user through the set of operations that can be performed in the merging process like PROMPT does.

## 4. Ontology-based knowledge fusion

This section will describe a novel work in progress that represents a different use of ontologies from that described in section – one specifically tailored to solve the current interoperability problem that exists in network administration.

Given the above definition of ontology for network and system administration, we can now start outlining a solution to the data fusion problem, and show how ontologies can play a significant role in solving this problem.

Remember that network device languages are *programming languages,* and not fully-fledged conversational languages. In other words, their grammar and syntax are much simpler than spoken languages. Thus, if a device Esperanto cannot be built, then the next best thing is to be able to provide a framework which can represent the syntax, semantics, and terminology of device languages. If this can be done, then a *language translation* can be built between the management system and each device language being used, as well as between different device languages.

The problem statement can be formulated as follows: how do we ensure that a function executes across multiple devices, where that function is mapped to one or more commands

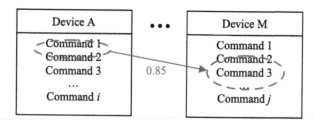

Fig. 9. Mapping a command from one device to a set of commands in another device.

in each device. Furthermore, we assume that in general, the command structure of each device is different.

Ontologies can help from a linguistics point-of-view to solve this problem. We use them to define a *shared* vocabulary, along with rules for defining the structure of the common language. This structure focuses on the structure and formation of words in the vocabulary (morphology), sentence structure (syntax), and meaning and meaning change (semantics).

Figure 9 shows an example illustrating this problem. We have a set of M total devices, each having a different number of commands (*i* commands for Device A, and *j* commands for Device M). So, our first objective is to build relationships between one command in one device to a set of commands in another device.

In this example, command 1 of Device A (denoted as A.1) is semantically related to the set of commands 2 and 3 of Device B (denoted as {B.2, B.3} with 85% equivalence (denoted as the '0.85' annotation under the arrow). Note that this is a *uni-directional* mapping. Once a set of these relationships are defined, we can build another mapping that abstracts this to a set of common functions. For example, enabling BGP peering, as shown in Figure 4, proceeds as first identifying the high-level function (BGP) to be performed, and then mapping this function to the appropriate set of commands in each device. This is shown in Figure 10.

In Figure 10, we see that the same high-level function (BGP peering) is mapped to two different sets of commands. This reflects the fundamental fact that different devices have different programming structures, and perform the same high-level task using different commands. Significantly, this means that sometimes, the semantics of performing these tasks are different, and must be taken into account. Hence, we need an ontology merging tool (or set of tools) to find these semantic differences.

The next task is to determine if it is *safe* to execute these commands. Often, there are many different sets of device commands that could be used to implement a given high-level function. In general, each set of device commands will have a set of side-effects. Hence, the system must be able to evaluate the impact of these side effects, so as to choose the best command set to implement. One way to do this is to feed the output of the ontology difference process to a policy based management system [34]. In this approach, the output of the ontology examination process shown in Figure 10 is sent for further analysis by a policy server. The function of the policy server is to define how to implement the device configuration in an optimal fashion.

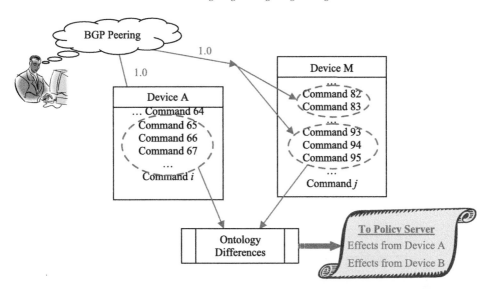

Fig. 10. Mapping a high-level function to different command sets in different devices.

## 5. Using knowledge fusion in autonomics

There are many applications that can use knowledge fusion. A very popular set of applications, having to do with the semantic web, can benefit from knowledge fusion, as this will provide a greater ability to reason about data using machines. The Semantic Web represents data as a mesh of information that can be processed in a machine-efficient manner. The problem with the majority of data on the Web is that it is difficult to use in its current form, because there is no global system for representing and understanding the syntax and semantics of data, let alone reusing data in different contexts. Please see the chapter 'Application Integration Using Semantic Web Services' in this handbook.

Another application is network and system administration. This section will describe a novel architecture for autonomic networking, called the FOCALE architecture. [39] FOCALE was developed specifically for managing network components and systems. The control loop of FOCALE is shown in Figure 11.

Figure 11 shows an abstract version of the FOCALE control loop. The Foundation portion consists of a set of State Machines, which are used to orchestrate the behavior of the managed element or system that is being controlled. The loop starts by observing information from the system and the environment – this is done to determine the current state of the managed element. This is then compared to the desired state (as defined by the foundation state machine(s). If the current state is equal to the desired state, monitoring continues; otherwise, the autonomic system determines a set of actions to take to reconfigure the system to try and move the state of the managed element to its desired state. While this loop is running, the system learns from the decisions made and increases the knowledge base of the system. Concurrently, the system has the ability to reason when necessary. The FOCALE architecture will be described in more detail below.

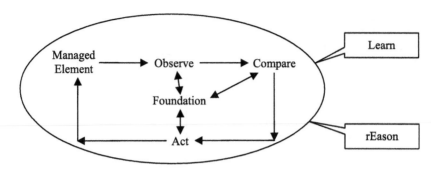

Fig. 11. The FOCALE control loop.

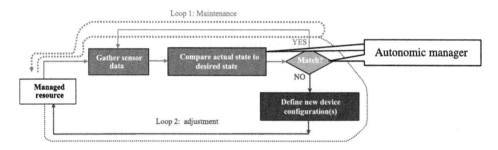

Fig. 12. Separation of control loops.

Figure 12 shows that in reality, there are two control loops. The top (green) control loop is used when no anomalies are found (i.e., when either the current state is equal to the actual state, or when the state of the managed element is moving towards its intended goal). The bottom (blue) control loop is used when one or more reconfiguration actions must be performed. Note that most alternative architectures propose a single loop; FOCALE uses multiple control loops to provide better and more flexible management. Put another way, why should a control loop designed for something going right be useful for something that went wrong?

Figure 13 shows a mediation layer that enables a legacy Managed Element with no inherent autonomic capabilities to communicate with autonomic communication systems. A model-based translation layer (MBTL), working with an autonomic manager, enables the Managed Element to communicate with other autonomic communication systems. The purpose of the MBTL is to enable the multiple heterogeneous languages and programming models that exist in networks (e.g., CLIs and SNMP private information) to be mapped to a single common lingua franca. Conceptually, the MBTL is a programmable engine, with different blades dedicated to understanding one or more languages. The MBTL translates vendor-specific data to a common XML based language that the autonomic manager uses to analyze the data, from which the actual state of the Managed Element is determined. This is then compared to the desired state of the Managed Element. If the two are different, the autonomic manager directs the issuance of any reconfiguration actions required; these actions are converted to vendor-specific commands by the MBTL to effect the reconfigu-

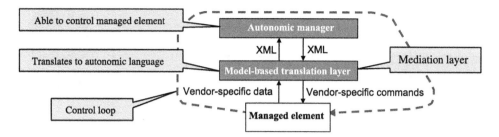

Fig. 13. Generic autonomic control.

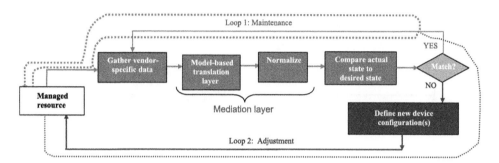

Fig. 14. The basic FOCALE control loop.

ration.

Figure 14 shows the basic FOCALE control loop, where the MBTL and its normalization layer are explicitly shown. Not explicitly shown is the normalization in the reconfiguration step, due to the complexity of drawing this step.

Figure 15 shows the complete FOCALE architecture. A Managed Resource (which could be as simple as a device interface, or as complex as an entire network, or anything in between) is under autonomic control. This control orchestrates the behavior of the managed resource using information models, ontology tools, and policy management. Information models are used to map vendor-specific data from the managed resource to a vendor-neutral format that facilitates their comparison. This is accomplished using ontologies, which define a *common lexicon* for the system. In general, the objective is to see if the actual state of the managed resources matches its desired state. If it does, the top maintenance loop is taken; if it does not, the bottom reconfiguration loop is taken. Significantly, ontologies play another role – they enable data from different sources to be compared using its common vocabulary and attached semantic meaning. For example, SNMP and CLI data can be used to ascertain the actual state of the managed resource, even though these are different languages having different descriptions of data. Policies are used to control the state transitions of a managed entity, and processes are used to implement the goal of the policy.

In Figure 14, the autonomic manager is limited to performing the state comparison and computation of the reconfiguration action(s), if required, *inside* the control loop. While this meets the needs of static systems, it fails to meet the needs of systems that change dy-

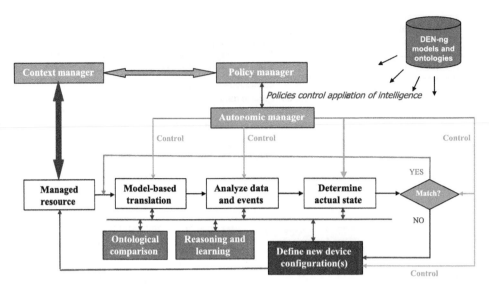

Fig. 15. The complete FOCALE architecture.

namically, since the functions inside the control loop cannot change. Of course, networks change dynamically, which is why autonomic network management and the FOCALE control loop is fundamentally different than other management architectures. This is why, in Figure 15, the management function is placed *outside* of the control loop. This enables independent control over the constituent elements of the control loop.

In the FOCALE architecture, the autonomic manager is responsible for controlling and adjusting (as necessary) the functionality of each of the components of the control loop to meet the current syntactic and semantic needs of the operation being performed. For example, the autonomic manager may determine that, due to the type of data that needs to be processed, a multi-level threshold match is more appropriate than looking for a particular attribute value, and hence changes the functionality of the match function appropriately. It can even break or suspend the current control loop in order to obtain new information. Knowledge fusion plays a significant role in supplying the autonomic manager the underlying meaning of data that it acquires, as well as enabling it to examine how the data is related to other events and information.

The Policy Manager is responsible for translating business requirements written in a restricted natural language (e.g., English) into a form that can be used to configure network resources [38]. For example, the business rule "John gets Gold Service" will be translated into a set of configuration commands that will be applied to all affected devices. This translation is done via the Policy Continuum [34,41], which forms a continuum of policies in which each policy captures the requirements of a particular constituency. For example, business people can define an SLA in business terms; this is then transformed into architectural requirements and finally network configuration commands. Note that each policy of the SLA does not 'disappear' in this approach – rather, they function as an integrated whole (or continuum) to enable the translation of business requirements to network configuration commands. Again, knowledge fusion plays a significant role – it supplies the lexicon used to effect the translation between different levels of the continuum.

Moving the autonomic manager from inside the loop to outside the loop (so that it can control all loop components) is a significant improvement over the state-of-the-art. For example, assume that computational resources are being expended on monitoring an interface, which suddenly goes down. Once the interface is down, it will stay down until it is fixed. FOCALE recognizes this, stops monitoring this interface, and instead starts monitoring other entities that can help establish the root cause of why the interface went down. This re-purposing of computational resources is directed by the AM, which controls each architectural component.

The Context Manager uses the DEN-ng Context Model to connect dynamically occurring events (both internal and external to the managed resource being controlled) to the autonomic system. In this approach, *context* determines the working set of *policies* that can be invoked; this working set of policies defines the set of *roles* and *profiles* that can be assumed by the set of ManagedEntities involved in defining context [2].

The bus shown in Figure 15 enables the benefits of knowledge fusion to be applied to all elements of the control loop. The Ontological Comparison element embodies the semantic similarity comparisons illustrated in Figure 9 and Figure 10. These data are then used by the reasoning and learning algorithms as follows. The reasoning algorithm is used primarily for hypothesis generation and theory maintenance. For example, when an SNMP alarm is received, its relation to customers and service level agreements is not known. The reasoning component will construct one or more hypotheses that assert a relationship between the alarm and customers; the autonomic manager will then try to verify each hypothesis by gathering additional data (guided by the models and ontologies). The learning component is used for two different purposes. First, it will continuously monitor the operation of the autonomic system, and adjust its knowledge base accordingly (e.g., the success and failure of the control asserted by the autonomic manager, and the relationships inferred between data under different circumstances). Second, it can 'short-circuit' the control loop by recognizing a past occurrence and instructing the control loop to stop its processing and instead perform the indicated operations.

In summary, just as object-oriented class hierarchies form a taxonomy of objects that can be programmed to provide a function, information and data models form a taxonomy of objects that define facts, and ontologies form a taxonomy of objects that define meaning of these facts. This combination enables facts gathered by the system (as represented by information and data models) to be understood (using ontologies) and acted upon (using code).

## 6. Summary

Ontologies represent a powerful set of tools to attach semantic meaning to data. This semantic meaning has several purposes:

- It forms the basis for a *shared vocabulary*.
- It enables data to be *understood*, so that different data can be related to each other, even if those data are in different formats using different languages.

- It enables the efficient *reuse of knowledge*.

There are a large number of ontologies and ontology tools that are available. The functionality of an ontology is directly dependent on how knowledge is represented, and what types of reasoning capabilities are required. This in turn depends on which constructs are available for representing knowledge, and the language(s) used to implement the ontology. Guidance on this choice was given in this chapter, both through the definition of ontology as it relates to network and system administration, as well as through the description of a set of representative ontologies and their development tools and environments. In general, it is recommended that a system using either the combination of frames and first-order logic, or one using description logics, is recommended. This is because these two approaches contain semantic mechanisms to develop a common vocabulary and set of meanings and relationships that can be customized to a particular set of application needs.

Finally, the importance of knowledge fusion was shown by describing its use in a novel autonomic architecture, called FOCALE. FOCALE uses knowledge fusion for understanding the meaning of data that it acquires, as well as for semantic inference. Ontologies are also used to provide a common lexicon.

## References

[1] F. Baader, ed., *The Description Logic Handbook* (2003), ISBN 0521781760.
[2] R. Carroll, J. Strassner, G. Cox and S. Van der Meer, *Policy and profile: Enabling self-knowledge for autonomic systems*, DSOM (2006).
[3] CycL, http://www.cyc.com/cycdoc/ref/cycl-syntax.html.
[4] DAML + OILs, http://www.daml.org/language/.
[5] Euler, http://www.agfa.com/w3c/euler/.
[6] FaCTs, http://www.cs.man.ac.uk/~horrocks/FaCT/.
[7] D. Fensel, J. Angele, S. Decker, M. Erdmann, H. Schnurr, S. Staab, R. Struder and A. Witt, *Ontobroker: Semantic-based access to information sources* (1999).
[8] J. Frohn, R. Himmeröder, P. Kandzia and C. Schlepphorst, *How to write F-logic programs in FLORID* (2000).
[9] J. Gallier, *Constructive logics, Part I: A tutorial on proof systems and typed λ-calculi*, Theoretical Computer Science **110** (1993), 249–339.
[10] T.R. Gruber, *A translation approach to portable ontology specification*, Knowledge Acquisition **5** (2) (1993), 199–220.
[11] N. Guarino, *Formal ontology in information systems*, First International Conference on Formal Ontology in Information Systems, Trento, Italy, 6–8 June, N. Guarino, ed., IOS Press (1998), 4.
[12] http://protege.stanford.edu/.
[13] http://protege.stanford.edu/plugins.html.
[14] http://www-ksl.stanford.edu/kst/what-is-an-ontology.html.
[15] JTP, http://www.ksl.stanford.edu/software/JTP.
[16] KIF, http://logic.stanford.edu/kif/dpans.html.
[17] M. Kifer and G. Lausen, *FLogic: a higher-order language for reasoning about objects, inheritance and scheme* (1997).
[18] O. Lassila and D. McGuinness, *The role of frame-based representation on the semantic web*, Technical Report KSL-01-02, Knowledge Systems Laboratory, Stanford University (2001).
[19] O. Lassila and R. Swick, *Resource description framework (RDF) model and syntax specification*, http://www.w3.org/TR/1999/REC-rdf-syntax-19990222/.
[20] D. Lenat and R. Guha, *Building Large Knowledge-Based Systems: Representation and Inference in the Cyc Project*, Addison–Wesley (1990), ISBN 0201517523.

[21] J. Lloyd, *Foundations of Logic Programming*, Springer-Verlag (1998).

[22] A.B. Markman, *Knowledge Representation*, Lawrence Erlbaum Associates (1998), ISBN 0805824413.

[23] R. Mizoguchi, J. Vanwelkenhuysen and M. Ikeda, *Task ontology for reuse of problem solving knowledge*, KBKS 95.

[24] R. Neches, R. Fikes, T. Finin, T. Gruber, R. Patil, T. Senator and W. Swartout, *Enabling technology for knowledge sharing*, AI Magazine **12** (3) (1991), 36–56.

[25] OMG, *Unified Modeling Language Specification*, version 1.5 (2003).

[26] Ontolingua, http://www.ksl.stanford.edu/software/ontolingua/.

[27] Ontology Definition Metamodel, public RFP is www.omg.org/docs/ad/03-03-40.txt; responses are available from the OMGs member site only.

[28] OWL, http://www.w3.org/2004/OWL/.

[29] S. Palmer, *Fundamental aspects of cognitive representation*, Cognition and Categorization, E. Rosch and B.B. Lloyd, eds, Lawrence Erlbaum Associates, Hillsdale, NJ (1978).

[30] RACERs, http://www.racer-systems.com/.

[31] RDF Schema, http://www.w3.org/TR/rdf-schema/.

[32] Semantic web, http://www.w3.org/2001/sw/.

[33] J. Strassner, *A new paradigm for network management – business driven device management*, SSGRRs summer session (2002).

[34] J. Strassner, *Policy-Based Network Management*, Morgan Kaufman Publishers (2003), ISBN 1-55860-859-1.

[35] J. Strassner, *Autonomic networking – theory and practice*, IEEE Tutorial (2004).

[36] J. Strassner, *A model-driven architecture for telecommunications systems using DEN-ng*, ICETE conference (2004).

[37] J. Strassner, *Knowledge management issues for autonomic systems*, TAKMA conference (2005).

[38] J. Strassner, *Seamless mobility – A compelling blend of ubiquitous computing and autonomic computing*, Dagstuhl Workshop on Autonomic Networking (2006).

[39] J. Strassner, N. Agoulmine and E. Lehtihet, *FOCALE – a novel autonomic computing architecture*, LAACS (2006).

[40] J. Strassner and B. Menich, *Philosophy and methodology for knowledge discovery in autonomic computing systems*, PMKD conference (2005).

[41] J. Strassner, D. Raymer, E. Lehtihet and S. Van der Meer, *End-to-end model-driven policy based network management*, Policy Conference (2006).

[42] R. Studer, V. Benjamins and D. Fensel, *Knowledge engineering: Principles and methods*, IEEE Transactions on Data and Knowledge Engineering **25** (1–2) (1998), 161–197.

[43] The CHIMAERA, http://www.ksl.stanford.edu/software/chimaera/.

[44] The Dublin Core, http://dublincore.org/.

[45] The Frame Ontology, http://www-ksl-svc.stanford.edu:5915/FRAME-EDITOR/UID-1&sid=ANONYMOUS&user-id=ALIEN.

[46] The IEEE Standard Upper Ontology, http://suo.ieee.org.

[47] The LOOM tutorial, http://www.isi.edu/isd/LOOM/documentation/tutorial2.1.html.

[48] The official DL homepage is http://dl.kr.org.

[49] The Ontolingua, http://www.ksl.stanford.edu/software/ontolingua/.

[50] The PROMPT, http://protege.stanford.edu/plugins/prompt/prompt.html.

[51] The WordNet, http://wordnet.princeton.edu/.

[52] W3C, http://www.w3.org/2004/OWL/.

[53] W3C, *RDF: Resource Description Framework*, www.w3c.org/rdf.

# – 3.5 –

# Application Integration using Semantic Web Services

## Jos Vrancken[1], Karst Koymans[2]

[1]*Delft University of Technology, Faculty of Technology, Policy and Management, The Netherlands*
*E-mail: josv@tbm.tudelft.nl*

[2]*University of Amsterdam, Informatics Institute, The Netherlands*
*E-mail: ckoymans@science.uva.nl*

## 1. Introduction

*Application integration* is needed when organizations make use of existing, or *legacy* applications in building new applications. Application integration means making legacy applications accessible in a uniform way and making the information they produce compatible. In order to be able to combine information from different applications and to use it within new applications, the information will have to be unified, syntactically and semantically. *Syntactically* means that data formats have to be unified, with ideally a single format for each type of information. *Semantically* means that the interpretation, or meaning, of information from different applications must be unified.

The way *interaction* between applications (both new and legacy) takes place, also has to be integrated: if one legacy application behaves as a server within the client–server paradigm, and another one does asynchronous messaging, this will have to be unified to support interaction with both applications in a uniform way.

Finally, the necessary organization to manage and maintain applications will have to be unified, otherwise the results of the integration effort will not last.

HANDBOOK OF NETWORK AND SYSTEM ADMINISTRATION
Edited by Jan Bergstra and Mark Burgess

This chapter is about a specific technology for the purpose of information integration, called semantic Web services. Plain Web services are a well-known and well-established technology, used mainly for the syntactic integration of information and, to some degree, for the unification of interaction. Semantic Web services, as the name suggests, are a further development of Web services, namely the result of applying semantic Web technology to Web services.

Semantic Web technology deals with making more of the meaning of information processable by machines. Its seminal example was the vision of transforming the Web into a semantic Web [5], where for instance searching can be done in a semantically enhanced way, rather than exclusively syntactically. Although this technology is not yet completely mature, especially with respect to tool support [10], it is already a good idea to learn about it. When dealing with legacy applications, information integration is unavoidable. It is a good idea to do it in a standardized way. Semantic Web technology is an emerging standard with good prospects for gaining wider adoption. There are not that many alternatives to it, and certainly none with an equal degree of standardization. Semantic Web technology also builds on the well-established XML family of information representation standards [7].

It goes without saying that the level of meaning humans can attach to information, will never be reached by computers. That is not the goal of semantic Web technology. What it does, is to express just a bit of the meaning of information, which is sufficient to constitute a clear advantage over the usual, purely syntactic way information is handled by machines, which makes it likely that semantic Web technology and its application to Web services will further mature in the coming years.

Why would this technology be relevant to system administrators? That is because system administrators are involved in the strategic management of an organization's information assets and application portfolio, in which the integration of legacy applications is a key activity. Semantic Web technology and semantic Web services are particularly applicable to the problem of integrating legacy applications, which implies their relevance to system administrators.

The next section will describe a number of technologies needed in Section 3, in which a step-by-step approach will be given for the integration of legacy applications.

## 2. Framework and basic technologies

This section describes an architectural framework for the use of IT in organizations, and several basic technologies needed in application integration, i.e. the bus interface, interaction patterns, Web services, semantic Web technology and the enterprise service bus. First, we will treat a model property that occurs in all of these subjects: layeredness.

### 2.1. *Layeredness*

Layeredness is a property of models that applies when part of a model is essentially dependent on some other part even though the two parts deal with essentially different issues.

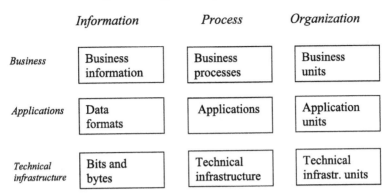

Fig. 1. Framework for the use of IT in organizations.

The framework for the use of IT, shown in Figure 1, is vertically layered. Looking, for instance, at the middle column, the business processes in the top layer vitally depend on adequate applications in the middle layer and the applications cannot live without an adequate technical infrastructure in the bottom layer. Yet the three items, business processes, applications and technical infrastructure components, are very different objects. Likewise, in the leftmost column, meaningful business information cannot exist without data formats and formatted information cannot exist without a representation in bytes and bits, assuming one would not like to avoid IT altogether. The dependence is inevitable but dependence is not an attractive property: it can easily cause change propagation. Using layeredness, this dependence can be greatly reduced.

Layeredness means that the items in different layers deal with essentially different issues. This is best recognized by the typical terminology used in each layer. A second characteristic of layers is a crisscross relationship between adjacent layers. In a pair of adjacent layers, items in the upper layer make use of different items in the layer below, and the items in the layer below offer services to, or form part of, several items in the layer above. So there is no one-to-one relationship between items in different layers. The reduced dependence and reduced change propagation between layers is a consequence of the different terminology used in different layers: changes in one layer are most often changes formulated in the typical terminology of that layer and therefore have no meaning and no consequences in other layers. The crisscross nature of the relationship between the items in two adjacent layers, also expresses this reduced dependence: for instance, applications in the middle layer in Figure 1 may run on different platforms in the infrastructure layer without any functional difference, while the facilities in the infrastructure layer are applicable in various applications in the middle layer.

## 2.2. *A framework for the use of IT*

An overview of the primary aspects of the use of IT in an organization is depicted in Figure 1. It is a well-known architectural framework for IT [17], serving many different purposes, including application integration.

Table 1
Open Office document representation

| Open Office XML document format |
| --- |
| XML |
| Unicode |
| Bits and bytes |
| Physical representation of bits |

Vertically the framework is a layered model. In the columns, three different aspects are addressed: *information, process* and *organization*.

The business layer is the uppermost layer and is about the primary activities of an organization. The second layer, usually called the applications layer, covers software components that serve business processes and have a business related meaning. Below the applications layer we find the technical infrastructure layer. This layer is about software and hardware facilities that are not business related, but generic. It goes without saying that the sharp distinction in Figure 1 does not correspond to an equally sharp distinction in reality. Yet the properties typical for a layered model, i.e. which components are dependent on which other components and which notions are present in the description of a component's functionality, are important for making the whole manageable.

Layeredness can also be recognized within the information assets of an organization (the leftmost column in Figure 1). In the information column, the relation between layers is different from, yet similar to, the relation in the process column. In the latter, the top side of a layer corresponds with functionality and the bottom side with implementation. In the information column, the top side corresponds with *meaning* and the lower side with *representation*: information has a *meaning* and it is *represented* in a more primitive form of information. Often, the representation of complex information is done in a number of layers, as illustrated of the Open Office document format in Table 1.

### 2.3. *The bus interface*

Key in the approach to application integration described here is a pattern which is applicable in many cases and which we will call the *bus interface*. It deals with the transition from many bilateral interfaces between, in this case, applications to far fewer interfaces to a bus. The bus then becomes the grand interface for the applications, that is why we call it the *bus interface*. Many organizations moved from stand-alone applications to connecting applications using bilateral interfaces (Figure 2), forced by short term business needs and available technology, but soon resulting in an unmanageable tangle. Obviously, the number of interfaces to maintain is then $O(n^2)$, where $n$ is the number of applications. It is possible to reduce this to $O(n)$ interfaces by using the bus interface (Figure 3). This has strong advantages in terms of reduced maintenance effort, improved modularity and new opportunities.

The idea is that applications are no longer connected directly, but are connected to the bus, thereby becoming accessible from any other application in a uniform way. Two of the

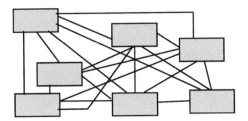

Fig. 2. Bilateral interfaces.

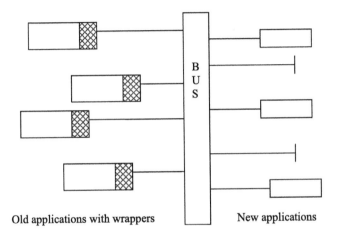

Old applications with wrappers       New applications

Fig. 3. The bus interface.

great benefits of the bus interface, in addition to the reduced maintenance effort, are the increased modularity of and independence between applications. The interface between two applications can no longer be specific for that pair of applications: via the bus, it has to become unified.

What parts does the bus interface have? Experience shows that the framework of Figure 1 can also be applied to the bus interface, which reveals its essential parts. The bus also has an information, a process and an organizational aspect.

The information aspect has a semantic or business related layer, an application layer and an infrastructure layer. The business layer for information describes unified semantics for information in the business processes. The application layer gives unified data formats for each type of information (e.g. vector graphics, raster graphics, recorded sound, formatted text or geographic maps). The infrastructure layer in the case of information describes the byte level representation of information and other infrastructural properties of information, such as volumes and timing requirements.

The process aspect deals with the, often quite complex, interactions between applications. In general they will have to be reduced to well-known interaction patterns, leaving more complex interactions to be handled by the requesting application.

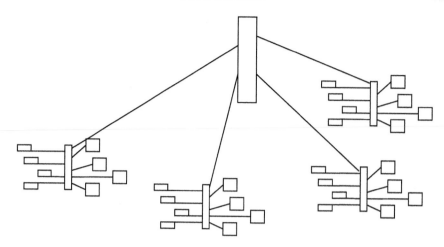

Fig. 4. Recursive application of the bus interface.

The wrappers (shaded in Figure 3) do the necessary conversions in order to have unified information formats, semantics and interaction patterns within the bus.

The bus interface has several attractive properties. First it can be implemented in an evolutionary manner. From a tangle of bilaterally connected applications, the interfaces can be removed one by one and replaced by connections to the bus. There is no need to do all the interfaces at once. The old and new interface can be simultaneously operational for a while if necessary. Thus, the bus and the unification of interfaces can be introduced in a gradual, step-by-step fashion.

A second attractive property is that the bus can be implemented distributively, thus preventing the bus from becoming a performance bottleneck. The bus may exist only virtually. In effect, most of the bilateral interface lines in Figure 3 may very well remain, but the information that passes through these lines and the interaction that takes place will be far more uniform in the new situation.

Finally, the bus interface pattern can be applied recursively (Figure 4). This means that if several organizations want to cooperate and have applied such a bus interface internally, all the applications of all the organizations involved can then be made interoperable by just connecting the individual organizational busses (that we might call *legacy busses*) to a supra-organizational bus.

A more detailed description of how to construct the bus interface can be found in Section 3.

### 2.4. *Interaction patterns*

We use the term *interaction* to denote the interaction between two applications, not the interaction between an application and a human user.

Besides the information formats that an application sends and expects to receive, the sequence of actions it takes and expects is vital for correct communication between applications. A very simple and very common pattern is synchronous request-response: a client sends a request to a server, who is waiting for requests. The server determines a response and sends it to the client, who is waiting for the response and meanwhile doing nothing else. This is the pattern on which the client–server paradigm is built. There are many other interaction patterns, for instance asynchronous messaging: both parties can take the initiative to send a message and then to go on with other tasks. This pattern is common in peer-to-peer systems. An example of a more complex pattern is the pattern needed for a transaction, which is a series of actions that have to take place, either all of them or none of them.

Interaction patterns describe the process aspect of communication between applications: i.e. the series of actions without regard to the actual content of the information exchanged in the actions. The bus interface will also have to deal with interaction patterns. Here too, some unification is desirable, and again the wrappers play an important role in the necessary conversions.

The importance of interaction patterns is illustrated by [13], which makes a strong case for asynchronous messaging as the preferred basic interaction pattern in application integration. The reason is that asynchronous messaging is one of the most basic interaction patterns, and therefore most suitable to express other patterns. For instance, the synchronous request–response of the client–server paradigm can easily be implemented using asynchronous messaging: the client sends an asynchronous message and then just waits for the answer. The other way around, expressing asynchronous messaging starting from synchronous messaging, is must probably not possible at all. If a separate process is created to do the waiting and allow the sender to carry on with other tasks, the sender would have to communicate asynchronously with this process. A polling mechanism and intermediate answers do not solve the problem either, because the sender still cannot avoid waiting when the receiver happens to be busy.

## 2.5. *Web services and SOA*

Web services [11] constitute a well-established standard way of implementing services, in the sense given above: services are components with a layered, standardized interface, that act only in response to a request, and with a tendency to be stateless and coarse-grained. The various interactions and standards that are typical for Web services are illustrated in Figure 5.

Server and client are two software components. They interact, according to the SOAP protocol, with the initiative lying with the client. The information they exchange is represented in XML. In essence a SOAP message deals with a procedure call and SOAP offers a format to represent procedure name, parameter values and return value, thus defining a unified format for calling procedures on various platforms, written in various languages. The language may be object-oriented, so the procedure called may be a method of an object. Originally this was the primary use of SOAP, which stood for Simple Object Access Protocol. Now that the requirement of object-orientedness has been dropped, *SOAP* is no longer an acronym.

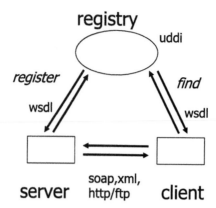

Fig. 5. Web services parts, functions and standards.

Low-level connectivity is implemented via HTTP or FTP, with HTTP having the advantage of being less sensitive to firewall barriers, owing to its simple request/response protocol.

Once created, a service offered by a server can be registered in a Web services registry, built according to the UDDI standard (Universal Description, Discovery and Integration). This is a standard format used to represent information about a service: a description of the organization involved and of the service's functionality, both in human readable form, and of its calling protocol in machine readable form. Potential clients can find the details of the service in this registry. It should be noted that registration and finding are both actions carried out by humans: the application programmers involved; this does not yet take place automatically, but these actions are far less frequent than the use of a service.

In several ways Web services are an improvement over CORBA and other kinds of distributed objects middleware such as DCOM. The HTTP protocol is far less hampered by firewalls and Web services are by nature more coarse-grained and less dependent on state, while the object interaction facilitated by distributed objects middleware is by nature fine-grained and dependent on object state. Thus Web services are currently far more popular than its alternatives.

A SOA (Service Oriented Architecture [11]) is an architecture for an organization's software environment that strives to represent components as services, as far as possible. It does not necessarily mean that the services must be implemented as Web services, although Web services are the founding example for SOA and are still the most suitable implementation for it. The service concept leads to highly generic interfaces, avoiding bilateral dependencies between pairs of applications: a service is a general facility for any client application. In addition, the interfaces are layered for their information and interaction aspect. All this together leads to the high degree of modularization and flexibility for which SOA is renowned.

**2.6.** *Semantic Web technology*

Computers are blind to the one aspect of information that is most important to humans: its meaning. Semantic Web technology [2,14] is about making more of the meaning of information accessible to machines. The name was derived from a seminal paper on the future of the World Wide Web [5]. The idea was to enhance Web pages with semantics, to be used by various applications reading the pages. For instance, search engines, that are now purely syntactic and statistical in nature, would then be able to distinguish between the different meanings of words, to replace search terms by synonyms, etc.

The idea of making semantics accessible to machines has old roots in artificial intelligence and database design [4]. Although revived by its applicability to the Web, it is certainly not only applicable there. For instance, it is highly relevant to application integration.

Semantic Web technology features a number of formalisms or languages that can be used to express meaning. They vary in expressiveness as a result of the trade-off between expressiveness and computational feasibility. More expressive formalisms, such as OWL, can offer inference rules and allow reasoning with data, but this is far more time consuming than just following references as done in the less expressive formalisms. In addition, the different formalisms form a layered model, starting with Unicode and XML. The more expressive formalism in the higher layers are built on top of the less expressive ones.

OWL, the Web Ontology Language, is a formalism designed to create *ontologies*, sets of notions with semantic relationships between them. OWL is an extension of RDF-Schema (RDFS), which itself is an extension of RDF, the Resource Description Framework. RDF can be used to expresses a limited form of meaning using triples: <object, property, object>. The triples allow the RDF user to build graphs of related notions (illustrated in [5]), but the main action a computer can do with such an RDF graph is just following references. OWL can express more complex relationships between notions. Within OWL, there are three variants:

- OWL Lite, the least expressive form. Inferencing, i.e. making logical deductions, is computationally efficient in OWL Lite;
- OWL DL, where DL stands for *description logic*. This variant is more expressive than OWL Lite and inferencing is still decidable but no efficient algorithm for inferencing within OWL DL is known;
- OWL Full, with the best expressiveness. Inferencing is no longer decidable.

The making of OWL and the many design considerations involved, is thoroughly described in [14]. An extensive overview of ontologies, almost an ontology of ontologies, can be found in [16].

It may seem attractive to build an ontology for a business area as broad as possible. An attempt has even been made to make a single ontology for all information exchange on the Internet. This was found to be infeasible. In building an ontology, it is again important to take an incremental approach, to start small and to extend it gradually. Effective ontologies can only be made for well-delineated communities. It is possible to combine two separately developed ontologies into a single one, either by building a new one, replacing all the notions in the two given ontologies, or by constructing an interface ontology, with the necessary mappings between the two given ontologies.

Table 2
The semantic formalisms stack

| OWL |
| --- |

| RDF-Schema |
| RDF |
| XML |
| Unicode |

It is important to have the right expectations for semantic Web technology. On the one hand, the expressiveness of the formalisms in Table 2 comes nowhere near the depth, detailedness and situation-dependence of meaning that humans can attach to information. That is not what the technology pretends to accomplish. On the other hand, it can help us to take a substantial step forward beyond the current, purely syntactic means of representing information, in which even a simple replacement of one term by a synonym or a translation into a different language cannot be done automatically.

The fact that OWL is expressed in RDF, which in turn is expressed in verbose XML, means that OWL code is not very readable. In practice, it is usually not written or read directly by humans but only via tools that offer a graphical user interface to OWL. An extensive, highly accessible tutorial and example of OWL, using one of the most common tools for OWL, can be found at [8].

To illustrate the expressiveness of OWL, without the burden of the XML syntax of OWL, we give an example in plain English. It is an adaptation of an example found in [14].

EXAMPLE OF OWL EXPRESSIVENESS. In RDFS, we can define classes *Company*, *Person*, *Employee* and *Unemployed* and we can express that *Employee* is a *subclass* of *Person*. We can express that *IBM* and *Microsoft* are *instances* of class *Company*. We can declare that *WorksAt* is a *property* relating the classes *Person* and *Company*. We can state that *HasAge* is a *property* relating class *Person* to integers. We can state that *Peter* is an instance of *Employee*, that he *WorksAT IBM* and that he *HasAge* 37.

With OWL, we can additionally express that *IBM* and *Microsoft* are different members of class *Company* and that they do not share members of class *Person*. We can express that *Employs* is the inverse of *WorksAt* and that the class *Unemployed* consists of exactly those members of *Person* that have no right hand side value for the property *WorksAt*. We can declare a class *MultiplyEmployed* that consists of those members of class *Person* that have at least two different right hand side values for property *WorksAt*. We can express that *Age* is a functional property, which means that in *p HasAge n,* there is exactly one value of *n* for each *p*.

### 2.7. *Semantic Web services*

Above we saw that the description of a web service in a registry is still only human readable. The information in the WSDL format, about the proper way to call a service, is still purely syntactic. This is where semantic Web technology can enhance Web services, by adding machine processable meaning to services and to the call parameters and return value. In this way, services can be discovered and called automatically and composition of services can be made automatically [6]. Negotiations and trust protocols can be executed without human intervention.

There are several initiatives in the area of adding semantics to Web services using semantic Web technology. The OWL-S consortium is developing an extension of OWL, specifically for defining semantically enhanced Web services [9,10,1]. The Web services modeling ontology group at the Digital Enterprise Research Institute is developing an ontology to model Web services [12], while the Semantic Web Services Architecture committee has developed a set of architectural models to serve as a framework for semantic Web services technologies [6].

## 2.8. *Enterprise service bus*

In the interaction between servers and clients over a network, a number of activities and functions are not service-specific but can be offered as an infrastructural facility, i.e. as an enhancement of the network. Such a facility in the case of Web services is called an Enterprise Service Bus (ESB). It deals with functions such as routing messages, hiding network and platform heterogeneity, security, data conversion, failure recovery, load spreading and transactional integrity. ESB is an improvement over the traditional broker architectures with proprietary protocols and a hub-and-spoke connection structure with its inherent lack of scalability. ESB is open standards based and specific for Web services. Its strongly distributed implementation gives it high scalability. Several commercial products offer ESB functionality. At points where open standards are still lacking or are not widely adopted, these commercial products resort to proprietary solutions. This should be taken into consideration when selecting this kind of product.

## 3. Integrating applications

The integration process can now be described in a number of steps, related to the structure of the bus interface, given in the framework of Figure 1. As already pointed out, the steps do not need to be taken consecutively and do not need to be completed before the next step can be taken. On the contrary, it is highly recommended to apply a cyclical process, where each step is applied to only a small subset of the relevant items, be it applications or information assets or key notions. The fact that it is possible to apply a cyclical, incremental process, is an essential advantage of this approach.

Note: to avoid the complexity of incremental processes in the description of the steps below, the descriptions refer to the complete step.

Step 1: *Make a catalogue of all information assets of the organization at hand.*

This inventory gives an overview of existing, planned or desired information resources, described in a way independent of the systems involved in producing, storing and distributing them. Such a catalogue can serve many different purposes. It fits with an activity called by various names such as *information management* or *information strategy* [3]. For the purpose of application integration, the catalogue helps in two aspects: one, a single data format (preferably XML based) can be selected or defined for each type of information in the catalogue, and two, an ontology of key notions can be derived. The data formats serve the purpose of unifying the different data formats for the same type of information that usually occur in legacy applications. The ontology serves the purpose of unifying the semantics of the information assets.

Step 2: *Make Web services of all the legacy applications.*

To this end, for each legacy application, a wrapper is built that provides the application with a Web services interface. This wrapper implements the unified data formats and unified semantics defined in the first step. In addition to the data formats, some unification also has to take place with regard to the interaction patterns. Several things can be done here. One may restrict oneself to simple interaction patterns that can be directly expressed in SOAP (such as synchronous request-response or asynchronous messaging). More complex interaction patterns can be reduced to simpler steps, leaving the responsibility for the proper interaction with the new client application. Or, the wrappers may implement complex interactions, preferably doing some unification of the interaction patterns that occur in the different legacy applications.

Step 3: *Build a registry of all the services in the organization.*

The registry will contain the usual descriptions, according to the UDDI standard, of the services of step 2. Interfaces to the services are described using WSDL. As soon as tool support allows the enhancement of registry information with machine processable semantics, this can be done. Otherwise, the semantic components in this approach can be used to help humans to find their way in this registry of services.

Step 4 (optional): *Implement an Enterprise Service Bus.*

This step is not necessary but can be helpful if one of the ESB products on the market does fit the organization's specific characteristics. Once installed, new and legacy applications can be made accessible via the ESB.

Step 5: *Set up the necessary maintenance and support organization.*

This organizational unit is responsible for maintaining all the products mentioned above, i.e. information catalogue, data formats, ontology, application wrappers, services registry and ESB. In addition, it should support the users of these products, especially application developers.

These five steps effectively build the bus interface and constitute an approach to legacy management which is both thorough and lasting, yet can be introduced very gradually.

More on tool support can be found in [15], the conclusion of this case study from 2004 being that tool support for the information catalogue is already quite adequate, but that semantic Web services are still hampered by the immaturity of the tools.

## 4. Summary

In many organizations that have existed for a number of years, one can observe that new applications often do not replace existing, *legacy* applications, but are built on top of them. This requires that the legacy applications have to be integrated with the new applications, which means integrating them with regard to information, process and organization. Information integration means the syntactic and semantic unification of the data in the different applications. Process integration means unification of the way different applications interact. Organizational integration means unification of the way the applications and the infrastructural facilities to support them, are managed and maintained.

A common way to do the information and process integration is to transform applications into Web services and to make them accessible via a bus interface. The different data formats and data semantics found in the legacy applications are unified into one data format per type of information, with unified semantics, which becomes the data format and semantics offered via the bus. The same holds for the different interaction patterns found in legacy applications: for each type of interaction, one interaction pattern can be applied for all applications accessed via the bus.

Currently, a Web service definition is still largely syntactic in nature and all semantic parts in its description are only human readable. Semantic Web technology can be applied not only to define the unified semantics mentioned above, but also to semantically enhance the definition of Web services.

Although semantic Web technology and its application to Web services are emerging standards, it is already a good idea to apply them. The semantic unification in application integration is a necessary step. Then an emerging standard, especially one that has few if any alternatives, is to be recommended over any proprietary solutions.

## References

[1] H.P. Alesso and C.F. Smith, *Developing Semantic Web Services*, A K Peters Ltd., Wellesley, MA, USA (2005), ISBN 1568812124.

[2] G. Antoniou and F. van Harmelen, *A Semantic Web Primer*, The MIT Press, Cambridge, MA, USA (2004), ISBN 0262012103.

[3] L.M. Applegate, R.D. Austin and F.W. McFarlan, *Corporate Information Strategy and Management: Text and Cases*, McGraw-Hill, New York, NY, USA (2002), ISBN 0072456728.

[4] J. ter Bekke, *Semantic Data Modeling*, Prentice Hall, Upper Saddle River, NJ, USA (1992), ISBN 0138060509.

[5] T. Berners Lee, J. Hendler and O. Lassila, *The semantic web*, Scientific American (2001), 34–43.

[6] M. Burstein, C. Bussler, M. Zaremba and T. Finin, *A semantic Web services architecture*, IEEE Internet Computing, September–October (2005), 72–81.

[7] M. Choi and J.W. Hong, *XML-based Network Management*, Handbook of Systems and Network Administration, Elsevier, Amsterdam (2007), 173–196, Chapter 2.4 in this Handbook.

[8] CO-ODE, *The Cooperative Ontologies Programme*, http://www.co-ode.org/.

[9] DAML, http://www.daml.org/services/owl-s/.

[10] D. Elenius, G. Denker, D. Martin, F. Gilham, J. Khouri, S. Sadaati and R. Senanayake, *The OWL-S editor – A development tool for semantic Web services*, Lecture Notes in Comput Sci., Vol. 3532 Springer, Berlin/Heidelberg (2005), 78–92.

[11] Th. Erl, *Service-Oriented Architecture: A Field Guide to Integrating XML and Web Services*, Prentice Hall, Upper Saddle River, NJ, USA (2004), ISBN 0131428985.

[12] ESSI WSMO Working Group, http://www.wsmo.org/.

[13] M. Fowler, *Patterns of Enterprise Application Architecture*, Addison–Wesley, Boston, MA, USA (2002), ISBN 0321127420.

[14] I. Horrocks, P.F. Patel-Schneider and F. van Harmelen, *From SHIQ and RDF to OWL: The making of a Web ontology language*, Journal of Web Semantics 1 (2003), 7–26.

[15] O.C. Krüse and J.L.M. Vrancken, *Enhancing information management in the water sector*, Proceedings SCI2005, Orlando, USA (2005), 194–197.

[16] J. Strassner, *Knowledge Engineering Using Ontologies*, Handbook of Systems and Network Administration, Elsevier, Amsterdam (2007), 425–455, Chapter 3.4 in this Handbook.

[17] D. Tapscott and A. Caston, *Paradigm Shift: The New Promise of Information Technology*, McGraw-Hill Education, New York, NY, USA (1992).

# 4. Policy and Constraint

## 4.1. *Commentary*

A policy is a set of decisions, usually about the goals and operational constraints of a system, which is documented in some form, and which specify its properties, the programme of its regular behaviour, and its responses to foreseen eventualities.

Policy can be expressed at many levels, starting at the high level with natural language, through program code at an intermediate level, to various forms of logic, and most recently in the form of networks of 'promises'. The subject of policy based management is huge but is skillfully reviewed by Bandara et al.

Security is an area where policy has long been understood to be of central importance. Burgess' somewhat parodical definition of a secure system, for instance, is intrinsically related to policy: *"A secure system is one in which every possible threat has been analyzed and where all the risks have been assessed and accepted as a matter of policy."* [1]. The sentence taunts us to achieve such a state of divinity. Bishop explains in his chapter some of the attempts to lock down system security properties using policies, under the assumption that a policy can be complete and that it can describe a legal behaviour for every possible degree of freedom. He describes the security models which have made historically important contributions to the discussion of security design.

Unlike in the natural world, computer software has no fixed set of laws that computer systems have to obey. Policies are not quite laws, because they are usually incomplete, but they are similar. To a large extent system architects and users decide the boundaries of how a system will behave, but there is also basic uncertainty in behaviour through unknown degrees of freedom. This is especially true of computers that are networked to a larger community of systems where different policies apply.

### Reference

[1]  M. Burgess, *Analytical Network and System Administration – Managing Human–Computer Systems*, Wiley, Chichester (2004).

# – 4.2 –

# Security Management and Policies

Matt Bishop

*Department of Computer Science, University of California at Davis,*
*One Shields Ave., Davis, CA 95616-8562, USA*
*E-mail: bishop@cs.ucdavis.edu*

## 1. Introduction

Managing security of a system depends on what the term 'security' means for that particular system. For example, a military system typically treats confidentiality of information as the most critical component of security. Availability is of less importance, and indeed may be restricted to only a few people. A web-based firm, that sells merchandise through its web sites, considers information about its products to be completely non-confidential, but information about the purchasers confidential. Unlike the military system, the web-based firm places an extremely high value on availability. If its web connection is terminated by a denial of service attack, the company loses money. If a portion of the population that is connected to the Internet fails to get through to the company, it loses money. Availability is a critical element of its security.

Security policies define 'security'. Each institution, each site has its own notion of what 'security' means, and the policy defines what is allowed and what is not. Intuitively, it seems that discussion about policy must focus on specific cases, rather than on generic policies. However, security policies often share common elements. For example, elements of three *security services*: underlie all security policies. These services are confidentiality, integrity and availability. We can examine each of these in turn, and make general statements about policies that constrain the sharing of information with respect to these services.

---

HANDBOOK OF NETWORK AND SYSTEM ADMINISTRATION
Edited by Jan Bergstra and Mark Burgess

Ideally, the policy is *complete*, so every action, and every state of the system, is either allowed (secure) or not allowed (non-secure). Unfortunately, unless the policies are specified mathematically, the imprecisions and inherent ambiguities in the natural languages creates confusion. For example, the following seems to be a reasonable policy component: "system administrators shall not view email without the signed approval of the president of the company". But consider the effects of such a policy. When is something 'email'? The SMTP protocol defines how email is transmitted through a network. Does the ban on email viewing also ban any form of network monitoring? If a system administrator needs to view a user's file, and that file happens to contain a saved email message, has the system administrator violated policy? Both these actions are reasonable under the right circumstances. Companies typically monitor networks for evidence of intrusion, for example, or a system administrator may need to recover the contents of a user's file for that user. This inherent ambiguity and imprecision in the term 'email' exemplifies the problems of working with policies expressed in English.

### 1.1. *Policies and policy models*

Rather than discuss a particular policy, then, we want to examine a class of policies. To do this, we look at a *policy model*, which is an abstraction of (a set of) security policies, and focuses on the specific elements of the security services that those policies are to supply. In many cases, we express the policy models mathematically, so we can make precise statements about the effect of policies that conform to these abstract policy models. We can also reason about the effects of specific requirements and environmental properties on policies based upon the model.

But, in security, the details matter. So while we learn much from analyzing policy models, the application of the model to particular situations produces specific policies. We can then compare the actual policies to the model and see where we must focus our analysis, specifically on the parts of the actual policy that add details to the model. As an example, a model may ban writing to an object. When the policy is constructed, the policy may allow the creation of a directory. Whether that violates the model depends upon whether creating a directory is an instantiation of 'writing' or some other operation. We shall elaborate upon this example in the next section.

### 1.2. *Overview of chapter*

Computer security textbooks present policy models by classifying them as models for confidentiality, integrity, and availability. But specifics of these services often affect the construction of policies just as much as the services themselves. For example, one class of policies deals with different levels of confidentiality and integrity; these are the *multilevel security and integrity* policies such as the Bell–LaPadula model and the Biba model. A second class of policies focuses on restricting privilege and power through the application of the *principle of separation of privilege*, or (as it is usually framed), *separation of duty*, which requires that multiple conditions be satisfied before something can be done.

The Clark–Wilson model provides perhaps the most explicit focus on this. A third class examines ownership of information as opposed to files. Although this question was first studied in the context of organizations, it describes problems in digital rights management. Originator-controlled access control models deal with these questions, and we will examine why they are difficult to implement. Finally, user rights may depend upon the user, or upon the function of the job that the user performs. The latter situation eases the problem of managing rights, because permissions are tied to need, which allows the system to implement the privilege of least principle more closely. Role-based access control enables this type of control.

This chapter discusses these policy models. Because each is an ideal, we will present some of the issues that arise in the application of these models, and how the policies involved make allowance for practical issues. We then discuss how applicable these models are in practice. We conclude with some thoughts on the role of policy models.

## 2. Policy models

Policy models present abstract systems that define a set of entities, a set of rights, and a set of rules that describe how the entities interact with one another given a particular assignment of rights. Before proceeding further, we explain several key terms and ideas that are common to all models.

**Entities.** Models divide entities into two basic types. A *subject* is an active entity, and represents users and processes (for example). An *object* is a passive entity, and represents files and devices (for example). Types need not be fixed. A model may provide rules under which a subject becomes an object, and *vice versa*. Usually, though, types are fixed and rules allow subjects to be acted upon as though they were objects. All the models discussed below are of the latter type.

**Rights.** A *right* is a relation that describes how subjects can access entities. For example, if a subject wants to view a file, the model may require that the subject have *read* rights over the object. The model itself defines the specific rights of interest, and the actual system instantiates those rights. This interpretation may raise subtle issues. For example, on a Linux system, a user being able to read a file means that the use can view the contents of the file. But one process being able to read another process can mean either that the first process can view the second process' memory space (as a debugger can print the values of variables in a subordinate process), or that the first process can receive messages from the second process (via signaling or interprocess communication facilities). With respect to the model, the interpretation of the right is irrelevant; what matters is how the abstract right alters the arrangement of rights in the system, and how they affect the flow of information throughout the system.

**Control.** Some policies allow a user to control access to objects; these policies are called *discretionary access control policies* (or, more simply, the policies have *discretionary* controls). Other policies control access to objects based upon attributes of the object or the subject requesting he access. These policies are *mandatory access control policies* and are said to have *mandatory* controls. An example of the former is the access controls on a Linux or Windows system, because the owner of the file can change its permissions. An example of the latter is the ability of the Linux superuser to access files, because regardless of the settings of the permissions of a file, the superuser can access the file.

Mandatory and discretionary controls are usually combined. When a subject requests access to an object, the rules first apply the mandatory controls. If the mandatory controls deny access, then the subject cannot access the object. If the mandatory controls allow access, then the rules apply the discretionary controls, which determine whether access should be granted. This is in accord with the principle of fail–safe defaults [22], which says that access is to be denied unless *explicitly* granted.

A third type of control, *originator-controlled access control policies*, gives control of access rights to the originator (usually the creator) of an entity, or of the information in that entity. This policy model is different from mandatory and discretionary controls, although it embodies aspects of both. Because of its complexity, we discuss it separately below.

We present five policy models, each of which emphasizes some unique aspect of those properties commonly considered part of security. This organization is non-traditional because we do not classify these models along the axes of confidentiality, availability and integrity. As these elements are typically present in all security policies, to some degree, we chose to emphasize other aspects of the policies.

## 2.1. *Multi-level security*

Multi-level security models constrain access on two components: a *rule* and an *access permission*. Each entity is assigned a *security level*. Mandatory controls constrain interaction between entities at different levels according to some set of rules. If the mandatory controls do not prevent access, then discretionary access controls determine whether the request for access should be granted.

For example, suppose Anne is at security level A, and a book is at security level B. If the model says that subjects at level A cannot read objects at level B, then Anne cannot read the book even if the owner of the book wants her to. The mandatory controls inhibit her access. However, if the mandatory controls allow Anne to access the book, then the system checks that the owner of the book has given Anne permission to access the book. If so, she is granted access; if not, she is denied access.

The most familiar model of multi-level security is the *Bell–LaPadula model* [3]. First, we present a simple version of the model to establish the concepts; then, we present the standard version of the model to show its complexity, and to indicate many practical problems that arise when implementing and administering a system that follows the model.

**2.1.1.** *The simplified Bell–LaPadula model*   Most people are familiar with the notion of *classified material*. This material is available only to people who are authorized to see it. We speak of *classifications* for objects, and *clearances* for subjects; so, in our previous example, Anne *had clearance at* security level A (or *was cleared for* security level A) and the book *had a classification* of security level B (or *was classified at* security level B). The **label** associated with a subject states the clearance of the subject, and the label associated with an object states the classification of the object.

In practice, material may take on any of several levels of classification. For our purposes, we define four, including a level of no classification:

1. *Unclassified* (U) refers to publicly available information. Anyone can read this information, and there are no restrictions on its distribution.
2. *Confidential* (C) refers to the lowest security level of classified information.
3. *Secret* (S) refers to the next higher security level of classified information.
4. *Top Secret* (TS) refers to the highest security level of classified information.

Note that the levels are ordered, with *top secret* being the highest and *unclassified* the lowest. This property is critical to what follows, because the goal of the Bell–LaPadula model is to prevent information from *flowing down*. For example, information classified *top secret* should never flow to a subject cleared for at most *secret*.

Consider a person cleared for security level *secret*. She can read any documents at that level. She can also read documents classified at lower levels, because she is cleared at a higher level than those documents. Similarly, she cannot read documents at the *top secret* level, because she is not cleared to that level, and *top secret* is 'higher' than *secret*, the maximum security level she is cleared for. This leads to the first rule:

SIMPLE SECURITY PROPERTY. A subject cannot read any object with a security level higher than that of the subject. This is sometimes called the 'no reads up' rule.

However, information can flow through mechanisms other than reading. Theresa has top secret clearance. Suppose she writes a paper containing top secret information, but she classifies the paper as confidential. Then Cathy, who has confidential clearance only, can read the paper and learn the top secret information. This "writing down" enables information to flow from one security level to a lower security level, violating the goal of the model. This leads to the second rule:

SIMPLE SECURITY PROPERTY. A subject cannot write to any object with a security level lower than that of the subject. This is sometimes called the 'no writes down' rule.

The simple security property and *-property are rules enforced by the system, without users being able to intervene. They are mandatory. The third rule allows the owner of objects to control access, should the mandatory controls determine that access does not violate those rules:

DISCRETIONARY SECURITY PROPERTY. A subject cannot access any object unless the discretionary access control mechanism allows the type of access requested.

To return to our book example, suppose Anne is cleared for top secret, and the book is classified confidential. Cathy, who owns the book, is also cleared for confidential. Anne wants to read the book. The simple security property does not disallow this. So, the system checks whether Cathy has given Anne permission to read to book. If so, Anne can read the book. If not, Anne's request is denied. Note that the mandatory mechanism can *block* access, but not *require* access.

These three properties provide a basis for asserting that a system implementing the model perfectly is secure. One can show:

BASIC SECURITY THEOREM. *A system with a secure initial state, and in which all actions obey the simple security property, the \*-property, and the discretionary security property, always remains in secure states.*

How practical this result is will be explored after we examine the full Bell–LaPadula model.

**2.1.2.** *The Bell–LaPadula model*   The problem with the simplified version of the Bell–LaPadula model presented above is its assumption that all *secret* entities are equal. In fact, they are not. Some entities contain information about a country's nuclear program. Others contain information about a country's agricultural problems. A third group may contain documents describing financial transactions. Merely because one is cleared to see the information about a country's nuclear program typically does not mean that person should be able to see documents describing unrelated financial transactions.

This leads to the idea of *categories*. A category is a modifier to the security level indicating a further restriction of access. This restriction is expressed by placing the entity into one or more categories, and collecting those categories into a set. For example, a document about attacks against nuclear reactors might be in the category NUC; a document about the amount of money spent to analyze a proposed anti-poverty program might be in the category FIN; and a document about the cost of thwarting attacks against nuclear reactors might be in both categories, Assuming these documents are all *secret* (S), they would have the labels

(S, {NUC}), (S, {FIN})

and

(S, {NUC, FIN})

respectively.

This leads to an extension of the term 'security level'. From here on, a *security level* refers to pairs such as

(S, {NUC})

where the first element is one of unclassified, confidential, secret, and top secret, and the second element is a category set. Some authors call this a *security compartment* and reserve security level for the first element of the pair; others reserve the term 'security compartment' for the categories. We use the terms of the original paper [3].

In order to be able to read a document, Anne needs to be cleared into the appropriate security level. In the above example, clearing her into secret is not enough. She must also be cleared into the appropriate categories. So, just as a document's classification now consists of a level and a category set, so too does a subject's clearance consist of a level and a category set.

Suppose the book that Anne wishes to read is classified

(C, {NUC, FIN}).

Anne is cleared for

(S, {NUC}).

Although S is higher than C, Anne is not cleared to read anything in category FIN. If she were able to read the book, she would see something in that category; so her request for access is denied.

However, if Anne were cleared for

(S, {NUC, FIN, AGR})

then she would be able to read documents in the categories NUC and FIN, provided those documents were at secret or lower. As the book is in those categories, and at confidential, which is lower than secret, Anne would be able to read the book.

In order to allow us to express this relationship simply, we introduce a notion of *dominating*, written *dom*.

***dom.*** A security level $(L, C)$ *dominates* a security level $(L', C')$ when $L' \leqslant L$ and $C' \subseteq C$. This is written $(L, C)$ *dom* $(L', C')$.

Continuing our example while using this notion, Anne can read the book if, and only if, her security clearance dominates the book's classification. If Anne's clearance is

(S, {NUC})

then while $C \leqslant S$, {NUC, FIN} $\subseteq$ {NUC}, so Anne's clearance does not dominate the book's classification. She cannot read it. On the other hand, if Anne's clearance is

(S, {NUC, FIN, AGR})

then $C \leqslant S$ and {NUC, FIN} $\subseteq$ {NUC, FIN, AGR}, so Anne's clearance *does* dominate the book's classification and thus she can read it.

This leads to a revised version of the simple security property:

SIMPLE SECURITY PROPERTY. A subject cannot read an object unless the subject's security level dominates the object's security level. As before, this is sometimes called the 'no reads up' rule.

The wording of this property is important. It does *not* say that if the subject's security level dominates the object's security level, then the subject can read the object. That is a *necessary* condition, but it is not sufficient. The discretionary access control mechanism must sill give the subject access to the entity.

Similar reasoning leads to an augmented version of the *-property:

SIMPLE SECURITY PROPERTY. A subject cannot write to an object unless the object's security level dominates the subject's security level. As before, this is sometimes called the 'no writes down' rule.

The discretionary security property remains unchanged, as it deals solely with entities and not with security levels. The Basic Security Theorem also holds with these versions of the properties.

Some observations will prove useful later on. First, *dom* is reflexive as $(L, C)$ *dom* $(L, C)$. This means that if a subject and an object are at the same security level, the simple security property and the *-property permit the subject to read and write the object. (Of course, the discretionary controls must also allow these for the subject to be able to read and write the object!) *dom* is also antisymmetric, because if $(L, C)$ *dom* $(L', C')$ and $(L', C')$ *dom* $(L, C)$, then $(L, C) = (L', C')$. This means that if one security level dominates another, and the second also dominates the first, then those levels are the same. Finally, *dom* is transitive, so if $(L, C)$ *dom* $(L', C')$ and $(L', C')$ *dom* $(L'', C'')$, then $(L, C)$ *dom* $(L'', C'')$. This means that if Anne's clearance dominates Paul's clearance, and Paul can read a document, so can Anne.

Second, two security levels may, or may not, be related by the *dom* relationship. For example,

$$(S, \{NUC, FIN, AGR\}) \ dom \ (S, \{NUC, FIN\})$$

as we saw above. But neither

$$(S, \{NUC\}) \ dom \ (S, \{FIN\})$$

nor

$$(S, \{FIN\}) \ dom \ (S, \{NUC\}).$$

This means that a person cleared into either class cannot read or write a document in the other class. For this reason, the relation *dom* is called a *partial ordering*, because while one can compare some security levels using it, other security levels cannot be so compared.

However, two security levels prove to be useful. The first is the security level that all other security levels dominate. This is called *System Low*. In the examples we have used thus far, the security level that *System Low* corresponds to is

$$(U, \emptyset).$$

The second is the security level that dominates all other security levels. This is called *System High*. Again, in the examples so far, the security level corresponding to this would be

$$(TS, \{NUC, FIN, AGR\}).$$

Two immediate consequences of the existence of these levels are that any subject cleared to *System High* can read any object, and any subject cleared to *System Low* can write to any object.

A third consequence of these two security levels is a method to guide the assignment of classifications and clearances. Suppose we need to assign a clearance to someone to read both financial documents and documents on attacks against nuclear power plants. The former are classified

(C, {FIN}),

the latter

(S, {NUC}).

The obvious clearance to assign is System High, but that is excessive; the assignee could read anything, not just documents in those security levels. But clearly an appropriate clearance exists. To find it, simply pick the higher of S and C (which is S), and take the union of the two sets involved ({ FIN } ∪ {NUC} = {FIN, NUC}). Then assign the clearance

(S, {FIN, NUC}).

This, in essence, computes the *least upper bound* of the classifications, which is the lowest clearance that dominates both classifications.

Similarly, suppose we want Anne to be able to write information to documents at the security levels

(C, {FIN, NUC})

and

(S, {NUC}).

As a subject can write to objects whose security levels dominate those of the subject, clearly a System Low clearance would work. But again, that allows the assignee to write to anything, not just documents in the two security levels. To find a class that restricts the subject's writing as much as possible, simply take the lesser of C and S (which is C), and take the intersection of the category sets ({FIN, NUC} ∩ { NUC } = NUC). Then assign the clearance

(C, {NUC}).

The result is called the *greatest lower bound* of the classifications, which is the highest clearance that both classifications dominate.

In formulas:

**greatest lower bound:** $glb((L, C), (L', C')) = (min(L, L', ), C \cap C')$,

**least upper bound:** $lub((L, C), (L', C')) = (max(L, L',') , C \cup C')$.

That these two security levels exist for every pair of security levels, combined with the characteristics of *dom*, make the structure of security levels and *dom* a lattice. Hence this type of model is sometimes called a *lattice model* [12].

Two practical problems arise from the rigidity of a subject's clearance. The first is the inability of a superior subject to communicate with inferior subjects. The second is an inability for *any* information to flow down, even if that information is sanitized. We deal with these separately.

Suppose a manager is cleared into the security level

(TS, {NUC, FIN}).

Her subordinates, alas, are only cleared into the level

(S, {NUC}).

As her security level dominates theirs, she can read anything they send her (or listen to them), but – according to the model presented thus far – cannot write anything to them (or talk to them)! Clearly, this situation is absurd. So the model provides flexibility through the use of a subject's *maximum security level* and *current security level*. A subject can raise its own security level to its assigned maximum security level; the security level that the subject is working at is the current security level.

Suppose our manager, who is working at her maximum security level of

(TS, {NUC, FIN})

security level, must talk to the program librarian, whose clearance is at most

(S, {NUC}).

The manager drops her current security level to

(S, {NUC}),

and proceeds with the conversation. This is valid under the model, because the manager may drop her security level below her maximum security level. Afterwords, she can raise her level back to her maximum.

Note that the subject changing its current security level is trusted to abide by the simple security property and the *-property. That is, information which its maximum security level gives it access to is not to be shared with entities whose security levels are dominated by that maximum security level.

The second problem is called the *declassification problem*. It arises when some information needs to be declassified, or have its security level lowered. This problem arises because classified information sometimes becomes known through other channels, or becomes safe to reveal. The best-known example is perhaps the British breaking the Enigma cryptosystem in World War II, which remained classified until 1972, or the American breaking the

Purple cryptosystem in the same war, which became known shortly after the war. The usual way to declassify information is to have a trusted entity examine the document, remove all references or portions still considered sensitive, and then change the security level of the document accordingly. The effectiveness of this *sanitization procedure* depends entirely on the entity doing the declassification. That problem depends on the context and information involved and is beyond the scope of this chapter. What is not beyond our scope is whether declassification is allowed at all.

The *principle of tranquility* constrains entities changing security levels. It comes in two forms. The *principle of strong tranquility* states that entities do not change their security levels during the lifetime of the system. So, declassification is not allowed (or is an exception to the rules, the more usual case). The *principle of weak tranquility* states that entities may change their security levels, but only in ways that do not violate the simple security property and the *-property. So, for example, if a document is classified

$$(S, \{NUC, FIN\})$$

and its security level is changed to

$$(S, \{NUC\}),$$

the system does not exhibit strong tranquility (because an entity's level changed). If making the change did not allow any subjects that could not read the document before the change to read the document after the change, and did not allow any subjects that could not write to the document before the change to read the document after the change, then the system is exhibiting weak tranquility; otherwise, it does not exhibit tranquility.

We next examine an integrity-based version of multilevel security, and then discuss specific system administration issues.

**2.1.3.** *The Biba model*    Although the Bell–LaPadula model touches upon integrity with the *-property, its primary purpose is confidentiality: to inhibit the flow of information about a system. The Biba model [4] uses a multilevel structure to prevent illicit changes to entities. Hence, it is an *integrity model*.

Like the Bell–LaPadula model, Biba's *strict integrity policy model* (almost universally called 'Biba's model') is a multilevel model. *Integrity levels* are analogous to the Bell–LaPadula notion of security levels. However, they are based on a notion of trust, not secrecy.

In Biba's model, if level A dominates level B, the entities cleared into (classified in) A are more trustworthy that those cleared into (classified in) B. To understand what this means, consider three newspapers, the Truth Times (which always prints accurate information), the Credulous Chronicle (which sometimes prints the accurate information and sometimes prints inaccurate information) and the Rotten Rag (which never prints accurate information). If reporters from the Credulous Chronicle rely on the Rotten Rag for a story, the more trustworthy entity (Credulous Chronicle) will be 'contaminated' by information from a less trustworthy entity (Rotten Rag). Thus, an entity reading information from another entity with less integrity risks contaminating the first entity, and should be forbidden. However, if the Credulous Chronicle gleans information from the Truth Times, the information the Chronicle gets is certain to be accurate and so may be used without fear of compromising the Chronicle's integrity. This argues that an entity reading information from another entity at a higher integrity level cannot reduce its own integrity. So, the analogue to the Bell–LaPadula simple security property is the simple integrity property:

SIMPLE INTEGRITY PROPERTY. A subject cannot read an object unless the object's integrity level dominates the subject's integrity level. This is sometimes called the 'no reads down' rule.

A similar rule applies for writing. If the Credulous Chronicle sends information tot he Truth Times, the Times reporters will not know whether the information meets their standards of accuracy. True, the data meets the Credulous Chronicle's standards; but that standard is considerably lower than the Times' standard. Hence the Times reporters will not use it, for fear of compromising their reputation (and trustworthiness). But if the Rotten Rag reporters received that same letter, and used that information, either the information would be accurate (which would improve the Rotten Rag's reputation by being the first bit of accurate information that journal ever printed) or the information would be inaccurate (which would not hurt the Rag's reputation, because it would be the same quality as the rest of the Rag's stories). In either case, the Rag's trustworthiness is no worse if it used the information.

Summarizing, if the more trustworthy entity writes information to a less trusted destination, the second entity does not risk lowering its trustworthiness. But if a less trusted entity writes information to a more trusted destination, that destination risks lowering its trustworthiness level. This suggests an equivalent to the Bell–LaPadula *-property:

*-INTEGRITY PROPERTY. A subject cannot write an object unless the subject's integrity level dominates the objects's integrity level. This is sometimes called the 'no writes up' rule.

If the Bell–LaPadula model and the Biba model are implemented on the same system, the relationship between the security levels and integrity levels affects how information flows about the system in unexpected ways. Suppose the security and integrity levels are the same; that is, they use the same labels, and have the same (partial) ordering. Consider a subject at security level

(TS, {FIN}).

What objects can it read and write?

Reading first. Under the Bell–LaPadula model, the simple security property requires that all objects which the subject can read be placed at security levels that the subject's security level dominates. Under the Biba model, the simple integrity property requires that all objects which the subject can read be placed at integrity levels that dominate the subject's integrity level. As the security and integrity levels are the same, the objects must be at the subject's level (by antisymmetry). Similarly, for a subject to be able to write to an object, both must be at the same level.

At this point, we need to consider some practical issues arising from the implementation of these multilevel security models.

**2.1.4.** *System administration issues*    Several issues of system administration arise when one tries to implement the Bell–LaPadula, or indeed any, multi-level security or integrity model. We focus here on a security model; the issues for integrity models are similar.

According to the Bell–LaPadula model, writing up is allowed when it does not violate either the simple security property or the *-property. However, in practise, allowing writing to a higher security level can weaken the integrity of the system. For example, a system may keep authentication data at a security level that dominates those of all users. If users could write up, they might be able to alter that data, and thereby be able to masquerade as more privileged users. Indeed, if the initial security level of a user were also stored at that dominating security level, a user could alter her own initial security level, log out, and then log back in at System High! For this reason, systems typically bound the higher security levels that users can write to.

EXAMPLE.  The Data General B2 UNIX system (DG/UX) [11] divides its security levels into three regions. The highest is the 'Administrative Region' and contains user authentication and authorization information, log files, security label definitions, and system databases. User processes cannot access any data in this region. Administrative servers execute with security levels in this region, but sanitize data before passing it to user processes. The next region, the 'User Region', contains user processes and files. All user processes execute with security levels in this region. The lowest region is the 'Virus Prevention Region' and consists of site executables, trusted data, and other executables. The name of the region comes from the inability of user processes to infect the executables in the virus prevention region, because an infected user executable would have to 'write down' to a system executable in that region – and that violates the *-property.

But not completely. If users can *only* write to their own security level, they become unable to communicate or work with other users with slightly different clearances. Thus, most systems weaken the rules controlling writing down.

EXAMPLE.  The DG/UX system has two methods of implementing the mandatory controls for reading and writing. The first is the conventional one in which each process and each object has a single label, and the mandatory rules are applied by comparing the labels of the subject and object involved. The second method involves *tuples*, and allows a wider range of reading and writing accesses. A tuple consists of up to three *ranges*. A range consists of an *upper bound* and a *lower bound*. When a subject tries to write an object, the subject's label is compared to the object's range. If the subject's label lies between the upper and lower bound of the object's range, write access is granted; otherwise, not. The object may have up to three associated ranges, one for each of the three regions described earlier.

One additional complication arises when a multi-level system uses a model originating with a non-secure system. Consider the UNIX model, in which processes share a common area in which to write temporary directories. For example, the UNIX C compiler creates temporary files in the directory '/tmp'. When the UNIX model is moved into the multi-level security arena, such a shared directory is untenable because it allows processes to communicate through the shared directory. This technique is called a 'covert channel' and is an example of the confinement problem, which is the problem of preventing information that is considered confidential from leaking [18].

EXAMPLE. Heidi is running a process at System High, and wants to communicate with Lara, whose clearance is System Low. Heidi creates a temporary file named '1' in the shared directory to send a 1 bit, and '0' in the shared directory to send a 0 bit. Every minute, Lara checks for such a file. If she sees it, she records the bit, and then writes to a file named '2' in the temporary directory. When Heidi sees that the '2' file has been updated, she deletes the file she had created in the temporary directory, and creates a new one with the appropriate bit value. In this way, Heidi and Lara can communicate (write) without sharing a file. This is a 'covert storage channel'.

To prevent this illicit communication, systems use a *multi-level directory*. Conceptually, the directory name and security label are combined. Thus, if a process with clearance at System High accesses the shared directory named 'x', the actual directory accessed is determined by the combination of the name 'x' and the label associated with the level System High.

EXAMPLE. The DG/UX system maps the combination of a shared directory name and security label into a 'hidden directory'. To continue the previous example, when Heidi creates the temporary file '1' in the shared directory '/tmp', it is actually created in the directory '/tmp/system_high'.[1] When Lara looks for the file '/tmp/1', the system translates that reference into '/tmp/system_low/1'. Thus, Lara will never see Heidi's file. Note that the subdirectories have the appropriate labels, so even if Lara can refer to the subdirectory 'system_high', her process cannot access it.

The last system administration issue considered here is the propagation of labels to objects. When a process creates an object, the object's label is that of the process. However, when an existing unlabeled file, or file system, is added to the system, it must be given a label. (This occurs when, for example, a multi-level system uses NFS to mount a shared file system. As NFS does not support security labels, all files on that file system are unlabeled.)

A complication is the use of aliases, sometimes called links. A *direct alias* or *hard link* is an alternate name for a file. An *indirect alias* or *symbolic link* is a special file the contents of which name another file; the operating system interprets a reference to the indirect alias as a reference to the file it names. Two questions arise. First, must the symbolic link have the same label as the file to which it refers? The second is how directory classifications play into accessing the file – for example, if a link to a file classified System Low is created in a System High directory. In that case, a process with System Low clearance may be able to access the file along one path, but not along another.

---

[1] This subdirectory of '/tmp' is associated with the security label corresponding to System High. The directory names 'system_high' and 'system_low' are for illustration only.

EXAMPLE. The DG/UX system solves these problems by allowing security labels to be
inherited. Associated with every file on the system is either an *explicit label*, one explicitly
assigned by the system, or an *implicit label*, one not explicitly assigned but implied by the
location and nature of the file. Five rules control the assignment of labels.

1. The root directory of *every* file system has an *explicit* label. Specifically, if the root
   directory does not have a label at the time of mounting, it is explicitly assigned the
   label of the mount point.
2. An object without a security label is treated as though it had the security label of its
   parent. This is really an *implicit* label, and is treated just like an explicit label.
3. If the label of an object's parent changes, and the object has an implicit label, the
   object's label is made explicit *before* the change. This means that the child retains the
   label of the parent before the change.
4. When a direct alias to an object is created, if the object has an implicit label, that
   label is converted to an explicit label. As the DG/UX system uses UNIX semantics,
   the direct alias is simply an alternate name for the file, and the label is associated with
   the object and not the name.
5. An indirect alias is treated as a file for labeling purposes; that is, it has its own (explicit or implicit) label. When the system resolves the indirect alias, the label of the
   result is the label of the *target* of the symbolic link. Note that this means all components of the name in the symbolic link must be accessible to the process accessing
   the target through the symbolic link.

These rules ensure the consistency of labels within the DG/UX file system when dealing
with files imported from structures that do not support labels.

## 2.2. *Integrity and separation of duty*

Multi-level security models focus on the classification of entities. This captures the properties of some environments well, especially military and government environments. However, it fails miserably to describe many other environments.

Consider bank accounts. The security level of the account is less important than the
accuracy (integrity) of the account. So, we might apply Biba's model. Unfortunately, Biba's
model speaks to the trustworthiness of the account, and the information in it. It does not
speak directly to what that trustworthiness means, and in this case characterizing what
conditions must be met for the account (and, by implication, the banking institution) to be
considered trustworthy. Our next model deals with this.

A bank account $X$ contains a balance $b_X$ of money. The customer makes deposits $d_X$
and withdrawals $w_X$, and after these the account contains the balance $e_X$. These quantities
should satisfy the formula

$$e_X = b_X + d_X - w_X.$$

This requirement is called an *integrity constraint*. If it ever fails to hold, the integrity of the
account is compromised.

Consider how a transfer between accounts $X$ and $Y$ takes place. First, money is withdrawn from account $X$. At this point, account $X$ is balanced, but account $Y$ has not yet received the money. Hence the two accounts are unbalanced. Then money is deposited in account $Y$, balancing the accounts. If the transfer is interrupted after the money is withdrawn from $X$ but before it is deposited in $Y$, the customer has lost money. Indeed, if integrity constraints had to hold at all times, the transfer itself would violate them. Hence, we view actions in terms of *transactions*, uninterruptable sequences of events.[2] In this case, the transaction would be the sequence of withdrawing the money from account $X$ and depositing it into account $Y$.

The *Clark–Wilson model* [8] is built around the notions of transactions and integrity. The model views the system as consisting of two types of entities. *Constrained data items* (*CDIs*) are bound by the integrity constraints. When all CDIs satisfy the integrity constraints, the system is said to be in a *valid state*. *Unconstrained data items* (*UDIs*) are not. In our bank example, the bank accounts would be examples of CDIs, because they must satisfy the integrity constraints at all times (that is, they must balance). However, the free toasters that the bank gives to people who open new accounts are UDIs, because the integrity constraints do not apply to them.

The model defines two procedures. The *integrity verification procedures* (*IVPs*) check that the data satisfies the integrity constraints. IVPs change nothing; they merely report. The *transformation procedures* (*TPs*) alter CDIs. If the system is in a valid state before a TP is applied, then the resulting state will also be a valid state.

The Clark–Wilson model defines five certification rules and four enforcement rules. The certification rules describe properties of the CDIs, IVPs and TPs. The enforcement rules describe the minimum mechanisms needed to enforce the certification rules. The enforcement rules do not prescribe particular mechanisms; rather, they say what the chosen mechanisms must do. Because the two sets of rules are intertwined, we present them together.

The enforcement rules are:

C1. An IVP, when run, must ensure that all CDIs are in a valid state.

C2. A TP transforms some set of CDIs in a valid state into another valid state.

Rule C1 ensures that when the CDIs are checked by an IVP, the IVP validates that the current state of the system is valid. Rule C2 implies that a TP works on some set of CDIs. This defines a relation *certified* on the set of TPs and CDIs. The elements of this relation are $(t_i, \{c_{i1}, \ldots, c_{in}\})$, where the transformation procedure $t_i$ can operate on the CDIs $c_{i1}, \ldots, c_{in}$. This relation constrains the use of $t_i$. Going back to our bank example, the transformation procedure that moves money from one account to another would be certified on the bank account CDIs, but not on other CDIs such as address information. Enforcement rule E1 rule requires that the system enforce this relation:

E1. The system must maintain the certified relation and enforce it by ensuring each TP manipulates only those CDIs for which it is certified.

However, some users will not be allowed to run certain TPs on certain CDIs. Again, our bank requires that only officers of the bank transfer money to and from overseas accounts. Such a transfer is a TP that is certified to run on the bank account CDIs. But an ordinary

---

[2]In practice, the requirement is simply that the transaction either complete or, if aborted, restore the system to the state existing before the transaction began.

teller is not allowed to run that TP. The second enforcement rule constrains the use of TPs in this fashion.

E2.  The system must associate with each user, and each TP that the user is allowed to run, the set of CDIs to which that user may apply that TP.

If the user attempts to run a TP with which she is not associated, the system must refuse. Similarly, if the user attempts to run a TP with which she is associated, but on a CDI not in the set to which that user may apply that TP, the TP cannot run. This rule defines a second relation, called *allowed*, over the sets of users, TPs and CDIs.

At this point the principle of *separation of duty* comes into play. Separation of duty is based upon Ben Frankin's maxim that "three can keep a secret, if two of them are dead". If violating the integrity of the system takes only one person, then that person can keep the violation secret. But if that violation takes two people, the probability of their keeping it secret is less.[3] The principle therefore requires at least two people to approve of an action. Of course, not *every* action needs two people; when the bank accepts an over-the-counter deposit, only the teller need enter it. But certain critical actions, for example issuing a bank check for more than $100,000, would require two signatures.

C3.  The allowed relation must meet any requirements imposed by the principle of separation of duty.

The Clark–Wilson model does not say what the principle requires. That is left to the interpretation of the model. But the Clark–Wilson model requires that the allowed relation respect the appropriate interpretation of the principle. In our banking example, a single user (the teller) could transfer some amount of money from one account to the other. But the TP to issue a $100,000 bank check would require two (or more) users to execute the appropriate TP.

Rules E2 and C3 imply that the system can identify the users executing TPs. The 'users' in this context are processes that correspond to humans, so the system must bind the external identity of these people to the processes. Rule E3 captures this binding process, authentication.

E3.  The system must authenticate each user trying to execute a TP.

The model does not state *how* the user must be authenticated. That is left up to the interpretation of the model, and the policy would use whatever means is appropriate to provide the degree of confidence required for the user to be allowed to execute the TP. In our bank example, a bank messenger could be authenticated by a password, but a bank vice president would need some stronger authentication method, for example a password and a biometric check.

Part of integrity is the ability to verify that the system works as planned. In the context of this model, the ability to reconstruct the actions of the system, and previous states, allows an auditor to check for problems. This leads to certification rule C4:

C4. Each TP must append enough information to reconstruct the operation to an append-only CDI.

The 'append-only CDI' is a log. The log cannot be erased or cleared. It can only be added to. This enables the auditor to reconstruct the complete history of the system.

Some TPs require data provided from an external source. For example, in order to deposit money, the 'deposit money' TP must accept a figure entered by the bank teller. This

---

[3]While not mathematically provable, this is a result of human nature.

item is a UDI, because it is not part of the system when typed. For example, if the TP does not check the input, a teller could enter arbitrary data, causing the TP to act in an undefined manner. This cannot be allowed, because it compromises the integrity of the system. Yet, obviously, the system must accept external input. Similarly, TPs may need other sources of data that are not CDIs. The final certification rule speaks to these.

    C5. A TP that accepts a UDI as input may perform only valid transformations or no transformation for all possible values of the UDI. The transformation either converts the UDI to a CDI or rejects it.

In the above example, the TP would read the teller's entry and validate it using appropriate criteria (such as it being a number). If the entry failed, the TP would discard it. Otherwise, it would convert the input to a CDI and process it.

The final rule applies the principle of separation of privilege to the model itself. Suppose a bank teller could not only execute TPs to balance accounts, but was also a programmer who could write programs and certify them to be TPs. Then a malevolent teller could write a program that implemented the apocryphal salami attack, certify it as correct, install it, and execute it herself. This violates the principle. On the other hand, if the certifier and executor must be two distinct individuals, either they must be conspirators (in which case Franklin's maxim applies), or the teller must hope that the certifier misses the attacking code she has put in the TP. Hence:

    E4. Only the certifier of a TP may change the list of users and CDIs associated with that TP. No certifier of a TP, or of any user or CDI associated with that TP, may ever execute that TP on those entities.

This protects systems implementing this model from the above attack. Note the temporal element of this rule: 'ever'. This rule limits the roles that certifiers can play in executing TPs.

### 2.2.1. *Example application: Managing privileged programs*

A typical computer system has programs allowing users to temporarily change privilege. In UNIX and Linux systems, the *setuid* programs serve this function.[4] This example applies the Clark–Wilson model to the management of these programs.

To instantiate the model, we first must determine the entities. Our policy will constrain the programs that grant privilege (called 'priv-granters'), so those are CDIs. Users will be the processes on the system. The TPs are those programs which manipulate the priv-granters; call these programs 'priv-controllers'. Examples of such priv-controllers are programs to change ordinary programs to priv-granters, or to delete priv-granters, or to change the set of privileges that a priv-granter gives, or to whom it gives them. Any configuration files that control what the priv-controllers are also CDIs. Note that the programs used to edit the configuration files are priv-controllers, and hence TPs. The IVPs that check the CDIs simply examine the priv-granters to ensure their permissions are properly set and they have not been altered unexpectedly (call these 'priv-checkers'), and check the configuration files. The set of priv-controller, priv-granter, and priv-checker programs are called the 'priv' programs.

---

[4]Other programs, such as *setgid* programs and servers, may also serve this function. We focus on *setuid* programs for convenience; a similar analysis applies to *setgid* programs and servers.

This structure implies that *only* the priv-controllers may make changes to any of the priv-controllers, priv-granters, and priv-checkers, because otherwise a non-TP would be able to change a CDI – something that violates the assumptions underlying the model.

This particular site divides privileges into two classes. The first class is called 'administrative privileges', and in order to modify a priv-granter either to grant these privileges or that grants these privileges, two system administrators must agree to the change. All priv-controllers are considered to deal with administrative privileges. The second class is called 'user privileges' and one system administrator may create priv-granters which change these or grant a user extra privileges (but not, of course, administrative privileges). The division arose because someone had once accidentally given a user administrative privileges, which caused problems.

As an aside, we note that the priv-controllers themselves are privileged programs, and enable authorized users to manipulate the priv-granters. Without them, the authorized users could not do so. Therefore, they are also priv-granters, and hence CDIs. This dual nature is a reflection of the Clark–Wilson model, in which the TPs are themselves instantiations of CDIs.

We instantiate the rules in the same order as above.

C1. The priv-checkers check the ownership, permissions, and contents of the priv-granters to ensure they are valid.

In order to determine whether the current values are valid, the system must maintain a protected database of those values. The database itself can only be changed by priv-controllers, when they manipulate a priv-granter. Hence this database itself is a CDI.

C2. Each priv-controller acts on some subset of the set of priv-granters. Changes to the priv-granters must keep them in a valid state.

The set of priv-granters that each priv-controller can affect depends upon the function of that priv-controller. For example, a priv-controller that changes the rights of a priv-granter to a device may not work on a priv-granter that does not access a device. But a priv-controller that deletes priv-granters might work on all such programs. This emphasizes that our 'example' is not tied to a specific UNIX or Linux system; indeed, it applies to all of them and must be instantiated further to apply to any particular one. The complete instantiation depends upon the details of the site involved. One site may not allow priv-granters for devices to be deleted, but may instead require them to be archived, for example.

E1. The system must maintain a database that control which priv-controllers can operate on which priv-granters. Each priv-controller must use this database to determine whether it can operate on the requested priv-granter.

This database embodies the *certified* relation for this example. A (possibly different) database must control which users can execute each priv-controller on which priv-granters:

E2. The system must maintain a database that associates a user with each priv-controller and each set of priv-granters. Each entry specifies the set of priv-granters on which a user may execute a particular priv-controller. No user may execute any priv-controller on any priv-granter that has no corresponding entry in the database.

This embodies the *allowed* relation. Notice that the sets of priv-granters associated with each priv-controller in this database is a (possibly improper) subset of the set of priv-granters associated with each priv-controller in the previous database.

Recall that two system administrators must agree on any changes to administrator privileges, but one can implement changes to other types of privileges. This implements the principle of separation of privilege:

C3.  The relations in the *allowed* database must require the use of priv-controllers that need two administrators to approve[5] before changing priv-granters that deal with administrative procedures, whereas it may allow the use of priv-controllers that take a single administrator's approval for all changes to other priv-granters.

Of course, the individuals who execute priv-controllers must be identified, so the priv-controllers know which entries in the *allowed* database to check. Enforcement rule 3 ensures this happens.

E3.  Each user who attempts to run a priv-controller or priv-granter must have authenticated himself or herself.

In order to reverse any changes made in error or illicitly, the system must log all changes to the priv-controllers and priv-granters. This also allows a reconstruction of what happened.

C4.  Each priv-controller must log any changes to priv-controllers and priv-granters. The information recorded must be sufficient to allow the altered program to be restored to its previous state.

An obvious implication is that when a priv-controller deletes a program, it cannot simply remove it from the system. Instead, it must copy it to backup media, or otherwise ensure that the deleted program can be restored later. In practice, this probably means that the program being deleted must be stored somewhere on the system until the next backup. In this case, the backups cannot be discarded.

The priv-controllers take as input the priv programs and configuration files. If a priv-controller requires other input, that priv-controller must check the input carefully, and convert it to a configuration file before using it. This is an application of the fifth certification rule:

C5.  Any priv-controller that takes as input user data (either directly or from a file) must check the input for consistency and sanity. If the input is acceptable, the priv-controller must convert it to a configuration file before using it; otherwise, the input is to be discarded.

Finally, the system must apply the principle of separation of privilege to the system administrators:

E4.  System administrators may either certify a priv-controller for use on a set of priv-programs, or may execute it on programs in that set, but not both. Furthermore, once an administrator is assigned a function (certify or execute for a given priv-program on a specific set), that administrator may never change to the other function.

From this application of the Clark–Wilson model, the requirements that the site must impose on its users are:

1. Only priv-controllers may make changes to priv-controllers, priv-granters and priv-checkers. [Model.]
2. There must be a 'consistency checker' to check that the priv-granters and priv-controllers are the only privileged programs, and that they are properly set (ownership and other file attributes set as required). [Rule C1.]

---

[5]For example, by requiring each to enter a distinct password.

3. There must be a database that controls which users can run which priv-controllers and priv-granters. Further, the database must restrict the objects on which these programs can run, when and as appropriate. [Rules C2, C3, E1, E2.]
4. The system must provide an authentication mechanism suitable to identify users to the granularity needed by the database described in requirement 3. [Rule E3.]
5. The system must have an append-only object; for example, a WORM drive or equivalent. The log file will be written on this object. [Rule C4.]
6. The site must be able to restore any changed or deleted priv program. [Rule C5.]
7. The system must keep a database of the certifiers and executors of each priv-controller and the set of priv programs that a user may change. Once an administrator is a certifier for such a pair, that administrator may never become an executor for that pair. Similarly, once an administrator is an executor for such a pair, that administrator may never become a certifier for that pair. [Rule E4.]

**2.2.2.** *System administration issues*   When implementing the Clark–Wilson model, applying the enforcement rules are straightforward, with only one complication. Enforcement rule 4, which deals with the certifiers of transformation procedures and of associated entities, presents a temporal difficulty. It states that the certifier may *never* have execute permissions over the entities or procedures certified. The reason is separation of duty; however, this effectively removes the certifier from using a procedure, or operating on a set of entities. The implementers of the model must ensure that the environment allows this limit, and that the sites involved can accurately track who certified what.

The application of the seemingly simple certification rules is anything but simple. Certification rule 4, requiring logging of information, is simple. To implement certification rule 1, the managers must define what a 'valid state' is. Certification rule 3 requires them to define how the principle of 'separation of duty' applies to the data, procedures and people. In many situations, this is difficult, especially when the number of people is limited, or geographically separated.[6] These three rules, however, can be handled by procedural controls and proper management.

Certification rules 2 and 5 speak to the certification of transformation procedures. These requirements present both practical and theoretical problems because they speak to the integrity not only of the transformation procedures but also of the people performing the certification, and of the tools and procedures used to provide the information that the certifiers use.

Consider rule 2, which requires that the transformation procedure transform CDIs from one valid state to another. This states that the certifier must ensure that under *all* circumstances, the transformation procedure will *never* place into an invalid state a CDI that begins in a valid state. For procedures implemented using software, this first requires that the software be validated. Validating software is an open problem, widely discussed in the literature. Frequent reports of compromised systems, vulnerabilities such as buffer overflows and denial of service, the effectiveness of malicious logic such as computer worms and viruses, and the lack of the widespread use of high assurance software attests to the

---

[6]In theory, the latter problem is obviated by telecommuting and remote access. In practice, without well-defined policies and procedures, communication problems often occur, and those create other problems.

lack of robust software. Unfortunately, even were the software robust and mathematically proven to transform CDIs properly, the software itself relies on underlying layers such as the operating system, libraries, and hardware – and errors could occur in those layers. Ultimately, 'proof' in the mathematical sense requires considering all these layers, and relies upon assumptions that must be validated in other ways.

Certification rule 5 also speaks to this problem. Any transformation procedure depends upon input, and if the input is from an untrusted source, the transformation procedure's operation is no longer trustworthy. So, the transformation procedure must validate the input to ensure it is valid. An example would be checking that the input does not exceed the length of the input buffer. In order to ensure that the transformation procedure does this, though, it must be completely analyzed to find all points at which it relies on external data, which is also a form of 'input'. Then the data must be analyzed to determine whether it is certified (a CDI) or not (a UDI). If the latter, the data must also be validated. Again, this requires considerable analysis and time.

Certification rules 2 and 5 also apply to procedures, because not all controls are technological ones. If procedure requires that a user have a smart card to access a system, the process to give smart cards is a transformation procedure, one CDI being the names of those authorized to get the smart cards. The implementation of this procedure requires that no unauthorized names be added to the list of those authorized to get the smart cards, and that the smart cards are only handed out to those on the list. These procedures are often easier to vet than the software, but may be more susceptible to corruption by carelessness or error as well as maliciousness than is software.

### 2.3. *Creator retaining control*

Sometimes controlling access to the object is not so important as controlling access to the information contained in the object. Suppose a company is being audited. The auditors need access to confidential, proprietary information. The company can give the auditors the right to read the files containing that information. That does not prevent the auditors from copying the information to their own files (because they can read the information and hence copy it), and then giving others access to those copies of the information. The company wants to retain control of any copies of its confidential, proprietary information.

This type of policy is known as an *originator-controlled access control policy*, or ORCON policy[7] [14]. With it, the data itself is protected, so even if the auditors made their own copies, the company (the *originator* of the data) controls access to it. The owner of the file cannot grant access without the consent of the originator.

A key issue is the definition of 'originator'. The originator may be an individual, for example the president of the company. But it may also be the organization itself. As an example, the president may devise a new policy controlling access to the confidential information. She then distributes it to her aides, and informs them they may not distribute the document any further. At this point, the president is the originator, because she created the document and controls all access to the information in it.

---

[7]In older literature, it is also called an ORGCON (for 'organization controlled') policy.

The president then resigns for personal reasons. At this point, the company lawyers examine the document, and inform the aides that the auditors must see the document. The auditors are from a different firm, so this will require the document being released to the auditors' firm under a strict promise of confidentiality, in which the auditors agree not to divulge the information to anyone not in their firm. At this point, the (ex-)president is no longer the originator for purposes of the model. The firm is. The firm, not the former president, controls access to the contents of the document. If the auditors need to divulge the contents to anyone not belonging to their firm (for example, to a lawyer specializing in tax law), they cannot do so without the consent of the company.

ORCON policies are critical in many fields. They are most prominent under a different name: digital rights management (DRM). When a multimedia company sells a DVD of a movie, the purchaser does not have the right to reproduce or broadcast the contents of the DVD. The purchaser has the right to watch the movie. With the advent of computers and programs to play and copy DVDs, the multimedia companies wish to protect their product, the movie. The movie is the information in the ORCON model, and the multimedia company (the copyright holder) is the originator. The ORCON policy model, if implemented fully, would allow the originator to control who had access, and the type of access, to their information (the movie).

Formally, an ORCON policy model consists of subjects and objects, ad do other models. A subject $X$ may mark an object as ORCON. The subject $X$ may then disclose the object to another subject $Y$ under the following conditions:

1. The subject $Y$ may not disclose the contents of that object to any other subject without the consent of subject $X$.
2. Any copies of the object have the same restriction placed upon them.

The astute reader will note that this model has elements of both mandatory and discretionary controls. The mandatory component is that the owner of a file cannot grant access to it. The discretionary component is that the originator can. In theory, a mandatory access control can capture an ORCON control. But in practice, this does not work because one needs far too many security levels. There is also a philosophical problem, in that security levels of mandatory access controls are typically centrally defined and administered. ORCON controls, by their very nature, are defined and administered by individual organizations.

Thus, any implementation of ORCON models must satisfy three properties:

1. The originator of data may change the rights of any object containing that data.
2. The owner of any object containing ORCON data may not change the rights on that object.
3. When an object containing ORCON data is copied, the rights of the object are bound to the copy (that is, copying does not change any rights).

Implementing an ORCON policy is difficult for one reason. Current systems tie rights to objects and not information. But ORCON models control information, not objects. Hence ORCON requires new mechanisms to be added to provide the requisite control.

In the next two sections, we shall examine two different controls. One has been simulated, but never directly implemented. The other was, with disastrous effects.

**2.3.1.** *Example: Propagated access control lists*    Propagated access control lists (PACLs) are a modification of traditional access control lists (ACLs) designed to implement an ORCON policy [24]. The key idea is that the PACL associated with an object constrains the privileges of a subject that accesses that object. That way, if a subject reads an object, the subject cannot write the contents of an object to anyone disallowed by the PACL of the object.

Let $s$ be a subject, $o$ an object, $\mathrm{PACL}_s(s)$ the PACL of subject $s$ and $\mathrm{PACL}_o(o)$ the PACL of object $o$. Then the following rules control the propagation of rights:

1. If $s$ creates $o$, then $\mathrm{PACL}_o(o) = \mathrm{PACL}_s(s)$. If $o$ has multiple creators $s_1, \ldots, s_n$, then

$$\mathrm{PACL}_o(o) = \bigcap_{i=1}^{n} \mathrm{PACL}_s(s_i)$$

2. Let $s'$ be a subject distinct from $s$. If $s'$ reads $o$, $\mathrm{PACL}_s(s')$ becomes $\mathrm{PACL}_s(s') \cap \mathrm{PACL}_o(o)$.

As an example, suppose Tom creates the file *data*. Then $\mathrm{PACL}_o(data) = \mathrm{PACL}_s(\text{Tom})$. He allows Dick read access to *data*. After Dick reads that file, Dick's new PACL becomes $\mathrm{PACL}_s(\text{Dick}) \cap \mathrm{PACL}_o(data)$, or $\mathrm{PACL}_s(\text{Dick}) \cap \mathrm{PACL}_s(\text{Tom})$, where $\mathrm{PACL}_s(\text{Dick})$ is Dick's PACL before the access. Now, if Dick tries to make a copy of *data*, any file he creates will have a PACL of $\mathrm{PACL}_s(\text{Dick}) \cap \mathrm{PACL}_s(\text{Tom})$, which constrains Dick to propagate the contents of data only to those approved by Tom, the originator of *data*.

Additionally, one can augment PACLs with discretionary access controls. If this is done, the PACLs are evaluated first. If the PACL allows access, *then* the discretionary rules are applied.

PACLs have not been implemented in computer systems because, in a general purpose computer, users would quickly be unable to share anything. If, however, the computer is intended to implement an ORCON policy with one user as the creator, then these controls would be acceptable. In modern systems, though, the restrictiveness PACLs impose would be unacceptable.

**2.3.2.** *Example: The rootkit DRM tool*    Digital rights management is an important area because of the proliferation of multimedia. The purchaser of a CD of music cannot broadcast the music, or play it at a public performance, without permission. However, the purchaser can make a private copy of it for her own use, for example 'ripping' it to a computer. Some copyright holders want to impose additional conditions. For example, Sony BMI wanted to restrict the number of times their CDs could be copied to prevent owners from making copies and giving them to friends [6].

The approach taken was to require the purchasers to install a proprietary music player on their Windows system. Among other things, the installation process modified the Windows kernel in order to prevent illicit copying. The mechanisms used were concealed, presumably to prevent the owner from deleting them.[8] These techniques, while relevant to system administration, are irrelevant to this discussion. What is relevant is that the modifications to the kernel were an attempt to implement an ORCON policy.

---

[8]The techniques involved were those used by attackers when installing rootkits [6].

What followed showed the difficulty of implementing an ORCON policy on a standard Windows system. An unmodified system allowed copying of the music (information) without restriction. This was the default policy – a standard discretionary access control policy. But the modified system combined an ORCON policy with one in which the owner could control access. The two policies conflicted. The conflict was resolved by the installation mechanism, which placed controls in the kernel that overrode the discretionary access control.

Apple Computer's iTunes store provides similar restrictions. The songs may be copied to up to five authorized computers, and may be burned to CDs for private use [2]. Apple enforces these through the iTunes program that plays the music downloaded from the store. Interestingly, one can load CDs purchased from elsewhere into iTunes, and iTunes does not enforce the restrictions on that music. Again, this is an ORCON policy, because the originator (here, Apple Computer) controls how the information (music) is accessed.

**2.3.3.** *System administration issues*   The difficulty of implementing ORCON controls through technological means only underlies the difficulty of digital rights management. Procedural controls are necessary, and these generally are beyond the scope of the duties of system administrators because they entail management of personnel and human (and business) relations. For example, among other restrictions, the Sony BMI license required deletion of all copies of the music if the owner filed for bankruptcy!

The key point is that system administrators should check any (purported) implementation of ORCON, and understand exactly how it works. Modern implementations of access control lists typically are not adequate to support ORCON.

**2.4.** *Job function*

Basing privileges on identity can lead to problems when privileges are associated with a job. Suppose Matt is a professor who administers his own computer system. As a professor, he needs access to student files in order to grade them. But he does not need access to the network management tool, because a professor does not administer the network. When he works as a system administrator, he does not need access to student files because a system administrator does not grade homework. But he does need access to the network management tool, because a system administrator may need to observe network traffic and tweak various parameters. This illustrates that access privileges often depend on what a person does rather than who he is.

This suggests adding a level of indirection to the assignment of privileges. First, the possible jobs that users on the system will perform are analyzed, and assigned privileges (ideally, in accordance with the principle of least privilege [22]). Second, the users are assigned to differing jobs. These jobs, called *roles*, dictate what rights each user has. Note that if a user changes job functions, her rights also change.

EXAMPLE. Debbie works as the bookkeeper of a large store. Her job requires her to have complete access to the financial records on the store computer. Her system uses role-based access control, so her ability to access the financial records is associated with the *book-keeper* role, to which Debbie is assigned. After several years, she is promoted to Head Customer Server, managing the Customer Evasion Department. She is shifted from the *bookkeeper* role to the *evader* role. Her access privileges change automatically, because she is no longer in the *bookkeeper* role. She no longer has any of the *bookkeeper* rights, but acquires all those that the *evader* role has. All the system administrator must do is ensure her role is changed from *bookkeeper* to *evader*.

Naomi works for Debbie's competitor. Like Debbie, she is a bookkeeper for her store, and requires access to the financial records on the store computer. Unlike Debbie's system, Naomi's system does not use role-based access control; her access is based upon her identity. Like Debbie, after several years she is promoted to Head Customer Annoyer, managing the Consumer Dissatisfaction Department. But unlike Debbie's situation, the system administrator must examine Naomi's privileges individually to determine which ones must be changed to support Naomi's new job. Should Naomi's duties change again, another re-examination will be necessary.

Role-based access control is essentially a formalization of the notion of *groups*. A group is simply a collection of users to which common access control privileges can be assigned. Many operating systems allow users to belong to one or more groups. Each object also has a group owner, and access control permissions are assigned to the group owner. But groups are *arbitrary* collections of users, whereas roles depend solely on job function. Further, because roles are tied to job function and an individual can only do one job at a time, a user can generally be in only one role at a time. The same is not true for membership in groups.

The fact that roles are tied to job function allows one to develop a rigorous formalism showing limits on how roles can be assigned in order to meet specific constraints. As an example, we focus on separation of duty, a concept used in Clark–Wilson above (see Section 2.2). First, we define some terms.

Let $S$ be the set of subjects, and $s \in S$ be a subject. The set of roles that $s$ is authorized to assume is *authr*$(s)$. Let $R$ be the set of roles, and let $r_1, r_2 \in R$. The predicate *meauth*$(r_1)$ is the set of roles that are mutually excluded by $r_1$; this means that a subject that can assume a role in *meauth*$(r_1)$ can never assume the role $r_1$, and vice versa. This gives us a way to express the principle of separation of duty, which says that if $r_1$ and $r_2$ are mutually exclusive, then if $s$ can assume role $r_1$, it cannot assume role $r_2$:

$$(\forall r_1, r_2 \in R)\big[r_2 \in meauth(r_1) \rightarrow \big[(\forall s \in S)\big[r_1 \in authr(s) \rightarrow r_2 \notin authr(s)\big]\big]\big].$$

Role-based access control can be integrated with any mechanism that uses subjects, such as the Bell–LaPadula model; the subjects are roles rather than individuals.

EXAMPLE. The DG/UX system described earlier has roles as well as users. Users need not assume any role, in which case their accesses are based solely upon their clearance. When a user assumes a role, her privileges are augmented with those of the role. Example roles are *lpadmin*, for administering the printers and *netadmin*, for administering the networks. Users do not log out; instead, they issue a special command to enter into the new role.

**2.4.1.** *System administration issues*  When a system supports roles, system administrators must determine who is assigned to each role. Often this is dictated by the jobs for which each user was hired. In some cases, the system administrator must determine what technical functions a particular job requires, because people are usually hired based on imprecise job descriptions. If the system administrator can define roles, then she does so on the basis of needed functionality. The needed but undefined roles usually become apparent during the lifetime of the system.

Many systems, such as Linux and UNIX-based systems, do not implement roles directly. Instead, they have role accounts – that is, accounts that are essentially roles. These accounts do not augment the rights of existing users, as do the DG/UX roles. These accounts instead have their own sets of rights. To assume a role, the user logs into that account as though it were an ordinary user. To change roles, the user logs out of the current role account and into the new role account. Alternate implementations such as administrative accounts [5] have been used to implement role accounts as well.

The difficulty in using role accounts is that they are typically shared – and that is considered poor administrative practice because of the inability to tie an action to an individual. A common approach in the UNIX/Linux world is to create numerous login names, each with a unique password, and all having the same user identification number (UID), upon which the system bases access control decisions. The problem with this approach is that any logging is based on the UID, but entered into the log by login name, so in most cases logging will occur under the wrong name.[9]

Some systems have developed the notion of a *login identifier*. This identifier is associated with an account tied to an individual user. If the user later changes to a role account, the login identifier remains attached to the process giving access to the role account. The login identifier determines the name placed in log files. Assuming that no user can log directly into the role accounts, and can only switch to them after logging into an ordinary account, this method works well.

One can also use groups to implement role-based access control, as implied above. This can be simple if the privileges involved are not system-level privileges. For example, if a role is 'member of project X', make the files related to project X be owned by an account to which no user has access, and put all users into the group that can access those files. The system may impose limits that create differences between the group and roles; for example, some systems do not allow users to drop group privileges, or to belong to multiple groups. System administrators should be aware of these.

At the system level, careful planning is required, and in some cases the planning and re-organization required is prohibitive. Consider the *root* account in UNIX or Linux systems, or the *Administrator* account in Windows systems. These accounts subsume a number of functionalities. For example, the *root* account manages user accounts, edits system files, installs sources and executables, reboots and starts daemons, and handles backups. But many of these functions can be handled by judicious assignment of group privileges. For example, if the semantics of the system support being able to change the ownership of a file from the owner to another, then a user with appropriate group permissions can create a new user account. Managing functions for which *root* is required takes judicious use of

---

[9]Specifically, the *first* name with that UID in the password database will normally be used, unless the password database itself is being accessed. In that case, *any* of the names could be used.

*setuid* programs with execute permissions for the appropriate group. For example, allowing a group of non-*root* users to start daemons requires a privileged tool that can start the daemons and that is executable by members of the group, and no others.

The system administrator considering implementing privileged roles using groups must determine the minimum set of groups needed (in accordance with the principle of least privilege [22]), the membership of those groups, their functionality, and how to implement that functionality. Typically, the implementation is the most difficult part, because the model must be fit into a system that does not cleanly delineate what each job function requires. The administrator must unravel the interactions required to separate the jobs to be undertaken by different roles.

### 2.5. *Summary*

The five models presented in this section show policies that meet very different requirements. The Bell–LaPadula model, the archetypal multilevel secure model, covers the flow of information among entities, and restricts access of subjects based on attributes associated with the subject and object. The Biba model restricts access similarly, but its goal is to prevent objects from being altered by less trustworthy subjects, and subjects from taking action based on less trustworthy objects – a simple, multilevel, integrity model. Both are oriented towards government or military use, because multilevel mechanisms are commonly used in those environments, and much less common in other environments such as commercial companies and academia.

The Clark–Wilson model was developed for a commercial environment, in which integrity is paramount. Two aspects of this integrity are worth noting. The first is that information is altered only in authorized ways. In this, the fundamental concept of integrity is similar to that of the Biba model, although the specific integrity constraints of the Clark–Wilson model are more general than those of the Biba model. The second is certification, in which a party attests that the programs and people meet specific conditions (such as separation of duty). There is no analogue to the program attestation in the Biba model; rather, it is assumed. The closest analogue to the attestation about people in the Bell–LaPadula model is the notion of 'trusted users' to whom the *-property does not apply – a very different concept. The Clark–Wilson model mimics practices in many commercial communities, such as the financial community, very closely.

ORCON policy models describe a class of policies that differ from the previous two. While Bell–LaPadula and Biba apply controls to the container (file) holding information, ORCON policies apply controls to the information itself. Further, the creator of the information applies those constraints to all future copies of the information. Contrast that with the Bell–LaPadula and Biba models, in which the owner of the container (file) controls access to the information, and can change that access without the permission or knowledge of the creator. While important in many situations, such as digital rights management, implementing this type of control requires both technical and procedural means.

Unlike the other four policy models, role-based access control focuses on the entities' functions rather than the information to be protected. It is a recognition that people play different roles in their jobs and lives; a systems administrator is also a user, a professor may be a parent, and so forth. The constraints that policies impose are often tied to an entity's *current* function rather than to the entity's identity. In that case, the indirection that roles provide enables the models to take into account the changing nature of what an entity does.

## 3. Applications of the models

The use of policy models requires that they be *instantiated* to actual policies. As the above examples show, in many cases a policy derived directly from the policy model is too constrictive for direct application, or requires the use of both technical and procedural controls.

Begin with the use of the Bell–LaPadula model. Two aspects arise: first, the application of the *-property; and second, the interpretation of the model when applied to a system.

The first aspect is straightforward. If a subject at a lower level could alter information at a higher level, the *confidentiality* of that information is not compromised but the *integrity* is. This emphasizes that the model protects integrity only incidentally. Because this is unacceptable in most realistic environments, the instantiation of the model in most environments forbids various forms of writing up as well. The use of label ranges in the DG/UX system, discussed above, is an example of a restriction that limits writing up. When a system allows writing up outside such a restriction, the constraint often restricts the writing to appending. This is the modification that the DG/UX system uses to allow logs to be in the administrative label range.

As another example, the UNIVAC 1108 implemented the simple security property, but not the *-property, of the Bell–LaPadula model. Fred Cohen used this system to demonstrate the effectiveness of computer viruses [9,10]. As the controls did not inhibit writing, the viruses spread rapidly. The conclusion that mandatory controls do not inhibit the spread of computer viruses is unwarranted, but the conclusion that mandatory controls that do not inhibit writing will also not inhibit the spread of computer viruses is justified [15].

The second aspect, interpretation of the model, points out the difficulty of applying models in general to specific systems or situations. Return to the example of the temporary directory discussed above. The semantics of the UNIX system define 'creating a directory' as different than 'writing'; the two involve different system calls (*mkdir* creates a directory; *write* writes to a file), and one cannot issue a 'write' system call to alter a directory.[10] Thus, a naïve application of the Bell–LaPadula policy model would not consider file system structures as a channel for communication, and miss this channel.

A proper application of the policy model requires interpreting 'write' as 'sending information', and 'read' as 'receiving information'. This means that *all* channels of communication must be identified and placed into the framework of the model. In practice, this is difficult because it requires the analysis of all shared resources, and the ability to draw inferences from system behavior.

EXAMPLE. Paul Kocher demonstrated an interesting attack involving the derivation of cryptographic private keys from timing [16]. Many public key cryptosystems, such as the RSA cryptosystem [21], are based on the notion of *modular exponentiation* in which

---

[10]This is primarily a reliability issue, because creating a file requires that the directory be updated.

the cryptographic key is used as an exponent. Let $a$ be the message, $n$ the modulus, and $z = z_{k-1}z_{k-2}\cdots z_1 z_0$ the private key, which is to be kept secret. The following algorithm implements one technique used to perform this operation:

```
1   x := 1;  atmp := a;
2   for i := 0 to k − 1 do begin
3       if z_i = 1 then
4           x := (x ∗ atmp) mod n;
5       atmp := (atmp ∗ atmp) mod n;
6   end;
7   result := x;
```

If the $z_i$ bit is 0, then the multiplication in line 4 does not occur. As multiplication operations are typically slow (especially when dealing with large numbers), the difference in time between a $z$ with all bits set to 1 and one with many 0 bits is noticeable. By taking careful measurements, Kocher was able to deduce information about the number of 0 and 1 bits in the private key. While this did not reveal the key, it did greatly reduce the search space, enough so that a brute-force search of the remaining key space was feasible.

Kocher and his colleagues later developed a similar attack exploiting the use of power [17].

The non-deducibility and non-interference models generalize the Bell–LaPadula model. The idea of both is to prevent a user from deducing information and activity at a higher level by examining outputs. A *non-interference secure* system meets this requirement, but also requires that the user not be able to deduce anything about the higher level activity [19]. A system is said to be *non-deducible secure* if the outputs the user sees depend only on the inputs that the user could see [13]. An example of the difference occurs when a user is able to deduce that higher level commands are being issued but not what those commands, or inputs, are; the system is non-deducible secure (because nothing about the higher level inputs can be deduced), but not non-interference secure, because the user can deduce that higher level commands were issued. These models are used in the analysis of system designs when high assurance is required. They also have implications for the theory of composability of policy models [19,20].

The Biba model deals solely with integrity, and confidentiality only as necessary to provide integrity. For that reason, applying the Biba model is essentially the same as applying the Bell–LaPadula model with the obvious modifications. More interesting is *when* to apply that model.

In general, government organizations are well suited to the application of multilevel models because they can manage multiple categories and security classifications. The delineation of data is not so clear in industry. For example, should proprietary data have a higher or lower classification than customer information? Their incomparability suggests that both should be in categories, leaving the question of level open. Further, separation of duty requirements are notoriously difficult to capture in the Biba model, because the goal of the Biba model is to compartmentalize subjects and objects. The effect of the principle of separation of duty is to require multiple subjects to co-operate to achieve a particular end. This leads to a key observation.

The Biba model deals with entities (users and data). Many commercial firms also focus on *process*: how is the data manipulated, and by whom? How can the company ensure the data is manipulated correctly, or at least in authorized manners by authorized people? Biba elides these questions into the assumptions underlying the model, reserving the model to determining what operations are allowed, which entities can perform the operations, and upon which entities they may be performed. Like Bell–LaPadula, Biba also has notions of trusted entities, but provides no guidance to verify that the trust is properly placed, or that the actions of those trusted entities do not inappropriately violate the rules. Unfortunately, these differences between the model and the needs of commercial firms means that Biba is little used in practice.

The Clark–Wilson model was developed to address these gaps. It specifically requires the certification of entities (CDIs), of procedures (TPs), and distinguishes between them and untrusted entities (UDIs). The model gives specific rules for certification, and additional enforcement rules describing how the system and environment must constrain the application of the procedures in order to prevent compromise of integrity. The model was developed for commercial practice, and captures how, in principle, many commercial firms work.

The issue of its application, as noted above, lies in the effectiveness of the certification. If the certification is done thoroughly and results in a high level of assurance with respect to the functionality of the transformation procedures and the integrity validation procedures, then the integrity suggested by the model will be realized. But, if the certification is done poorly, the integrity suggested by the system may not be realized. One important issue is that of the insider, the trusted or certified user. If that user is corrupt, the model will not prevent her from damaging the integrity of the TPs or CDIs – but the insider problem applies to all models, because once an entity is trusted, it is an exception to the rules imposed by the model, and can violate them. The insider problem has yet to be understood fully, let alone solved.

Unlike the previous three models, which control access based upon the file containing the information, the ORCON model controls access based on the creator of the information itself, and so must bind the rights to the data in such a way that only the creator, or her designated agent, can alter those rights. The problem is that any program accessing ORCON data must be certified to abide by the ORCON rules unless the underlying system enforces them. Otherwise, the program could simply make a bit-for-bit copy of the information, *excluding* any control information. Both hardware and cryptographic mechanisms have been suggested to enforce ORCON (usually in the context of digital rights management). Currently, though, procedural controls (including legal controls) implement ORCON on general-purpose systems.

Of the five models discussed in this chapter, role-based access control is the most intuitive. It ties privileges to job needs, rather than identity. In most cases, that allows system administrators to minimize the number of rights that each user has. In cases where rights must be tied to identity, the system administrator will understand why, and be able to justify this connection.

Role-based access control may be used along with any of the other models described above. In essence, the 'subject' or 'originator' of the other models is simply the user's identity paired with the user's job function; for example, user 'Matt' in the role of 'faculty member', or user 'Anne' in the role of 'system developer'. As Matt and Anne change roles, their rights change also. If the underlying system does not support roles, this type of access control can often be emulated using groups (with some limits that usually depend on the semantics of the system). Hence it is perhaps the most widely used of these models.

## 4. Conclusion

The goal of this chapter has been to describe some common and influential policy models. Other policy models speak to different types of policies. For example, the Chinese Wall model deals with conflicts of interest, and models rules for some commercial firms [7]. The Clinical Information Systems Security model describes rules governing control of medical information [1]. Traducement deals with accountability, in particular in the recording of real estate and other legal documents [23]. These models grew out of specific policies and situations.

Policy models are important for several reasons. They allow us to abstract properties of specific systems and policies, so we can focus on the goals of the policies and the ideas underlying how those goals are to be achieved. They enable us to prove properties of types of policies, and perhaps more importantly, what the policies cannot provide. They allow us to focus on the application of the model to new situations by examining the assumptions of the model, and determining how well the situation satisfies those assumptions. Finally, models allow us to examine our policies, to see where they are lacking and how to augment them to provide the protection we desire.

## References

[1] R. Anderson, *A security policy model for clinical information systems*, Proceedings of the 1996 IEEE Symposium on Security and Privacy (1996), 34–48.

[2] Apple Corp., *iTunes music store terms of service*, http://www.apple.com/legal/itunes/us/service.html (2005).

[3] D. Bell and L. LaPadula, *Secure computer system: Unified exposition and multics interpretation*, Technical Report MTR-2997, Rev. 1, The MITRE Corporation, Bedford, MA (1975).

[4] K. Biba, *Integrity considerations for secure computer systems*, Technical Report MTR-3153, The MITRE Corporation, Bedford, MA (1977).

[5] M. Bishop, *Collaboration using roles*, Software – Practice and Experience **20** (5) (1990), 485–498.

[6] M. Bishop and D. Frincke, *Who owns your computer?*, IEEE Security and Privacy **4** (2) (2006), 61–63.

[7] D. Brewer and M. Nash, *The Chinese wall security policy*, Proceedings of the 1989 IEEE Symposium on Security and Privacy (1989), 206–214.

[8] D. Clark and D. Wilson, *A comparison of commercial and military security policies*, Proceedings of the 1987 IEEE Symposium on Security and Privacy (1987), 184–194.

[9] F. Cohen, *Computer viruses: Theory and experiments*, Proceedings of the 7th DOD/NBS Computer Security Conference (1984), 240–263.

[10] F. Cohen, *Computer viruses: Theory and experiments*, Computers and Security **6** (1) (1987), 22–35.

[11] Data General Corporation, *Managing Security on the DG/UX® System*, Manual 093-701138-04, Data General Corp., Westboro, MA (1996).

[12] D. Denning, *A lattice model of secure information flow*, Communications of the ACM **19** (5) (1976), 236–243.

[13] J.A. Gougen and J. Meseguer, *Security policies and security models*, Proceedings of the 1982 Symposium on Privacy and Security (1982), 11–20.

[14] R. Graubert, *On the need for a third form of access control*, Proceedings of the 12th National Computer Security Conference (1989), 296–304.

[15] H. Israel, *Computer viruses: Myth or reality?*, Proceedings of the 10th National Computer Security Conference (1987), 226–230.

[16] P. Kocher, *Timing attacks on implementations of Diffie–Hellman, RSA, DSS, and other Systems*, Advances in Cryptology – Proceedings of CRYPTO'96 (1996), 104–113.

[17] P. Kocher, J. Jaffe and B. Jun, *Differential power analysis*, Advances in Cryptology – Proceedings of CRYPTO'99 (1999), 388–397.

[18] B. Lampson, *A note on the confinement problem*, Communications of the ACM **16** (10) (1973), 613–615.

[19] D. McCullough, *Specifications for multi-level security and a hook-up theorem*, Proceedings of the 1987 IEEE Symposium on Security and Privacy (1987), 161–166.

[20] D. McCullough, *Non-interference and the composability of security properties*, Proceedings of the 1988 IEEE Symposium on Security and Privacy (1988), 177–186.

[21] R. Rivest, A. Shamir and L. Adleman, *A method for obtaining digital signatures and public-key cryptosystems*, Communications of the ACM **21** (2) (1978), 120–126.

[22] J. Saltzer and M. Schroeder, *The protection of information in computer systems*, Proceedings of the IEEE **63** (9) (1975), 1278–1308.

[23] T. Walcott and M. Bishop, *Traducement: A model for record security*, ACM Transactions on Information Systems Security **7** (4) (2004), 576–590.

[24] D. Wichers, D. Cook, R. Olsson, J. Corssley, P. Kerchen, K. Levitt and R. Lo, *PACLs: An access control list approach to anti-vival security*, Proceedings of 13th National Computer Security Conference (1990), 340–349.

# – 4.3 –

# Policy-Based Management

Arosha Bandara, Nicodemos Damianou, Emil Lupu,
Morris Sloman, Naranker Dulay

*Department of Computing, Imperial College London,*
*180 Queensgate, London SW7 2AZ, UK*
*E-mails: {a.k.bandara,n.damianou,e.c.lupu,m.sloman,n.dulay}@imperial.ac.uk*

**Abstract**

Policies are rules governing the choices in behaviour of a system. They are often used as a means of implementing flexible and adaptive systems for management of Internet services, distributed systems and security systems. There is also a need for a common specification of security policy for large-scale, multi-organisational systems where access control is implemented in a variety of heterogeneous components. In this chapter we survey both security and management policy specification approaches. We also cover the issues relating to detecting and resolving conflicts which can arise in the policies and approaches for refining high-level goals and service level agreements into implementable policies. The chapter briefly outlines some of the research issues that have to be solved for large-scale adoption of policy-based systems.

## 1. Overview

As technology pervades every aspect of the modern enterprise, from communications and information management to supply chain management and even manufacturing, the applications, services and devices used in these business activities are becoming increasingly complex. In such an environment, it is important that systems administrators are able to effectively manage the underlying technology such that the security, performance and availability requirements of the enterprise are met.

The complexity of management is caused by number of factors. Firstly, the systems to be managed will span a number of layers – from networks, servers and storage devices at

HANDBOOK OF NETWORK AND SYSTEM ADMINISTRATION
Edited by Jan Bergstra and Mark Burgess

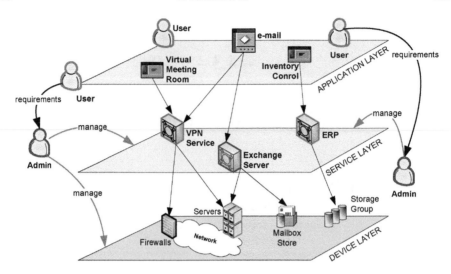

Fig. 1. Managing enterprise systems.

the lowest level; to services and applications at the higher levels (Figure 1). To manage such systems the administrator is forced to deal with configuration, deployment and monitoring activities in a highly heterogeneous environment. Secondly, because of the range of technologies involved and the physical distribution required in global enterprises, managing these systems will be the responsibility of multiple human administrators. Making sure that the management activities of one administrator do not override or interfere with the activities of other administrators is non-trivial. In order to survive, any enterprise has to change as the world changes around it. Therefore, in order to cope with changes in the business needs of users, administrators must be able to easily change the configuration and behaviour of the systems they are managing.

Policy-based approaches to systems management are gaining widespread interest because they allow the separation of the rules that govern the behavioural choices of a system from the functionality provided by that system. This means that it is possible to adapt the behaviour of a system without the need to recode any of the underlying functionality; and changes can be applied without the need to stop and restart the system. Such features provide system administrators with the capability to manage systems in a very flexible manner.

Whilst there has been research into policies and policy-based management for over a decade, there is still some ambiguity in the definition of a policy. Some researchers view policies as goals to be translated into a set of management operations; whereas others view them as event triggered rules that specify the operations to be performed if the system is in a given state (event–condition–action rules). Additionally, the term policy is used in security management to refer to rules that specify permissions and prohibitions in a managed system. Whilst, all of these definitions are valid in the context in which they were made, they are not general enough to be applied to systems management in the large.

Therefore, in this chapter, we define policies as *"rules governing the choices in behaviour of a system"* [84]. As such, policies can take a number of different forms:

- Management policies define the conditions on which manager agents perform actions on the system to change the way it performs, or the strategy for allocation of resources. Typical policies include what resources to allocate to particular users, for example, to allocate 40% of available bandwidth for sales managers to have a video conference between 15:00 and 17:00 every Friday; when to perform back up of the critical data bases; or the conditions on which to install software. Typically management policies are implemented as obligations in the form of event-triggered condition–action (ECA) rules.

- Behavioural policies define the rules by which agents within the system interact. For example, the conditions under which they will exchange particular state information; or what filtering/transformations should be applied to email forwarded over a wireless link to a mobile user. These are also often implemented as obligations.

- Security authorisation policies are concerned with defining the conditions under which users or agents can (or cannot) access particular resources. Firewall rules are an example of authorisation policy in the form of filtering rules for forwarding packets based on conditions such as source or destination IP addresses, port addresses or type of traffic. Security management policies are concerned with how the system should adapt to unusual events such as the actions to perform when a login failure is detected or what resources should be shut down when a security attack is detected. The heterogeneity of security mechanisms used to implement access control makes security management an important and difficult task.

- Routing policies are specific to routers in a network and are used to define the conditions for choosing particular paths within the network for forwarding packets. For example traffic from military installations should not be sent over insecure links, or traffic requiring low-delay should not be sent over satellite links. Some routers can also perform similar traffic filtering to that performed by firewalls.

In this survey we will concentrate on Management/Behavioural policies which are similar and security policies, particularly those related to authorisation, as this is such an important issue in large-scale networked systems. Security management is covered by management policies. Routing policies are highly specialised and have been well described in a recent survey [89] so will not be described here.

Policies are persistent so a one-off command to perform an action is not a policy. Scripts and mobile agents, based on interpreted languages such as Java, can also be used to support adaptability as well as to introduce new functionality into distributed network components. Policies define choices in behaviour in terms of the conditions under which predefined operations or actions can be invoked rather than changing the functionality of the actual operations themselves. In today's Internet-based environments security concerns tend to increase when mobile code mechanisms are introduced to enable such adaptation, and so many researchers favour a more constrained form of rule-based policy adaptation.

In order to make policy-based management usable in real-world applications it is important to be able to analyse policies to ensure consistency and to ensure that defined properties are preserved in the configuration of the managed system, e.g. in network quality of service management, to check that traffic marked in the same way is not allocated to different queues. Although the functionality that policies implement can be realised through general scripting languages, policy languages adopt a more succinct, declarative, form in order to facilitate analysis. Many policy languages have been proposed in the literature, but policy analysis techniques still remain poorly explored. Unless such techniques are developed and implemented in enterprise operational systems support tools, the additional expense required to deploy policy-based management will have poor short term return on investment and will remain difficult to justify.

In addition to analysing policies for consistency, administrators will need help in translating the business needs of the enterprise into policies that can be enforced by the underlying systems. Often these requirements can be expressed as high-level policies (e.g., all project teams are allowed to use the virtual meeting room application) which then must be transformed into lower-level policies, expressed in terms of management operations supported by the system. In some cases the users' requirements can be satisfied by specifying policies that perform management operations at the application layer; but other times it may be necessary to write policies that interact with the device layer of the system. This process is called policy refinement and is an area of policy-based systems research that has recently gained increased attention.

## 1.1. *Chapter outline*

This chapter provides a survey of the work on policy specification for network and security management. We begin in Section 2 by looking at a general architecture for policy-based management systems. This is followed by a description of the various models proposed for policy specification in Section 3. In Section 4 we cover some approaches to security policy specification, including logic-based languages, role-based access control and various access control and trust specification techniques. In Section 5 we focus on approaches to management and behavioural policy specification, including event-triggered policies and configuration management policies. In Section 6 we present an overview of frameworks that have been developed to support policy-based management techniques and in the following section we describe the Ponder policy specification language, which combines many of the above concepts and can be used for both security and management policies. Finally in Section 8 we present a discussion of research issues relating to policy-based management followed by a summary of the chapter in Section 9.

Online references for much of the work on policy and many of the papers described here can be found from http://www-dse.doc.ic.ac.uk/Research/policies.

## 2. General features and architecture

The research literature presents many proposals for implementing policy-based management systems, each with different approaches to the specification, deployment and enforcement of policies. Often these differences arise because the system is tailored to a specific type of policy, such as security authorisation policies or network routing policies. In most cases the differences manifest themselves in the form of different language constructs for specifying policy rules.

However, there are a number of components within a typical policy rule that are common to security authorisation and management/behavioural obligation policies. Additionally, most approaches to policy-based management provide some capability for organising large sets of policies. In this section we present details of these common components of policy-based management systems, together with a general architecture for specifying, deploying and enforcing policies.

## 2.1. Policy components

*Subject* defines the scope of the policy in terms of users or agents which are authorised to perform an action. For an obligation policy the subject is the agent which interprets the policy and performs the action specified. In some policy specifications the subject may not be explicitly defined in that it can be inferred from context in which the policy is specified or it relies on an administrator to assign the policy to an agent.

*Target* defines the scope of a policy in terms of the resources or objects on which the actions defined in the policy are permitted (for an authorisation) or should be performed (for obligations). Large-scale systems may contain millions of users and resources. It is not practical to specify policies relating to individual entities, so many systems specify subjects and/or targets in terms of groups of entities. Routing and firewall policies generally perform pattern matching on source and destination addresses in order to decide how to treat a packet. Whether this corresponds to a subject or target depends on whether it is an incoming or outgoing packet and whether it is part of a request or response message.

*Actions* specify what must be performed for obligations and what is permitted for authorisations. Typically, actions may be one or more operations specified by an object or service interface specification. However in some systems an obligation action is a script or sequencing rules for a number of operations. Some authorisation policies combine an action with a specific target object and call it a permission e.g. a read operation on the staff-list file is defined as a permission which can be assigned to specific subjects within the policy.

*Conditions* define the specific predicates which must be evaluated before the action can be performed. This could related to a time period, e.g., between 15:00 and 17:00 on Fridays or relate to attributes of the subjects or targets themselves. For example the policy only applies when the subject is at a particular location such as in the office. All types of policies potentially can have conditions specified as part of the rule.

*Triggers* define events which initiate the actions in an obligation policy. An event could be generated by a component failure e.g. DoC access router failed or at a particular time e.g. at 23:59 every weekday. In some systems, the triggering event is implicit so is not explicitly defined in the policy e.g. the arrival of a packet at a router or a page request at a web server. Authorisation policies are evaluated whenever an action is performed and so are not triggered by an explicit event.

## 2.2. *Policy organisation*

Large organisations will have numerous policies and so it necessary to provide concepts for grouping and structuring policies. Many policy specification notations have some form of syntactic grouping of policies which relate to a particular application or task e.g. the policies related to registering a new user in the system, the set of policies for backing up personal workstations. There may be some common conditions which can be used for multiple policies in the group.

It is also useful to group the policies pertaining to the rights and duties of a role or position within an organisation such as a network operator, nurse in a ward or mobile computing 'visitor' in a hotel. This concept is particularly developed in role-based access control in which roles group access permissions. However the idea of a role with both authorisations (rights) and obligations (duties) is important in management applications as well and has been widely used in defining organisational structures and hierarchies.

Some systems provide support for specifying nested organisational structures which consist of instances of related roles which could represent a ward in a hospital, department in a university, branch of a bank or members of a committee. The structure may include policies pertaining to the relationships between roles or how the structure interacts with its environment.

## 2.3. *Policy-based management system architecture*

A typical policy-based management system consists of a policy management tool, a policy repository, a management information base (MIB), management agents, access controllers, an event monitoring system and the managed resources themselves. The relationship between these various components is shown in Figure 2.

The policy management tool allows the administrator to create, edit, analyse and deploy policies through the policy repository. Additionally, the management tool provides an interface to the management information base (MIB), which stores the specifications of the managed resources in the system. In a typical system, the MIB will be populated using a discovery service that can detect the managed devices, services and applications in the enterprise environment. The policy management tool can also be used to deploy managed resources into the system. In recent years, a significant effort has been devoted to standardising the information models used for specifying policies and managed resources. Outcomes of this work include the Common Information Model (CIM) specification produced by the IETF/DMTF [43].

Policy implementation requires a Policy Execution Point (PEP) which actually enforces policies e.g. by deciding whether to forward a message, permit an invocation or set the priority in a packet to a particular level. This action is the result of interpreting policies which is done by a Policy Decision Point (PDP). Some legacy components may not have the capability of interpreting the policies so the PDP and PEP may be implemented as separate entities with the PEP querying the PDP for policy decisions. However in more recent systems, a combined PFP/PEP would be implemented as a management agent. A typical

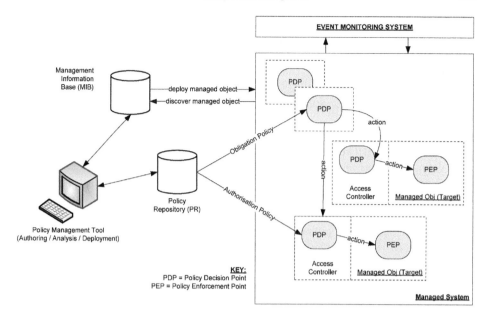

Fig. 2. Typical policy management implementation framework.

system would have many distributed management agents enforcing the obligation and authorisation policies. Obligation policies are interpreted by the subject management agents whereas authorisation policies are usually enforced by access controllers at the target, as discussed below. Organisational constructs, such as the roles or domains described in the previous section can be used as place holders in policy specifications, thus allowing late binding of policies to management agents.

In order to enforce authorisation security policies, the policy-based management system uses access controllers which may consist of separate or combined PDP/PEP. These are components that intercept any action invocations on a managed object and use policy rules to decide if the action should be permitted or not. In order to make an access control decision, with each invocation, the controller needs to be provided with the identity of the requestor (subject) as well as the action itself.

The event monitoring system is responsible for disseminating events generated by managed objects to policy decision points that have been loaded with obligation policies that will be triggered by the events. In typical implementations, the event system is implemented using a publish/subscribe architecture, with a network of distributed event routers.

## 3. Policy models

When deploying policy-based management in large-scale systems it is important to have models for relating the policy rules to the managed resources and users of the system. Such models can often simplify the policy specification notations used and offer opportunities for automated analysis of policy specifications for inconsistencies. In this section we describe some of the policy models proposed by the research community. We will go on to describe policy specification approaches that use some of these models in later sections.

**3.1.** *Role-Based Access Control (RBAC) specification*

Roles permit the grouping of permissions related to a position in an organisation such as finance director, network operator, ward-nurse or physician. This allows permissions to be defined in terms of positions rather than individual people, so policies do not have to be changed when people are reassigned to different positions within the organisation. Another motivation for RBAC has been to reuse permissions across roles through a form of inheritance whereby one role (often a superior in the organisation) can inherit the rights of another role and thus avoid the need to repeat the specification of permissions.

Sandhu et al. [81] have specified four conceptual models in an effort to standardise RBAC. We discuss these models in order to provide an overview of the features supported by RBAC implementations. $RBAC_0$ contains *users*, *roles*, *permissions* and *sessions*. Permissions are attached to roles and users can be assigned to roles to assume those permissions. A user can establish a session to activate a subset of the roles to which the user is assigned. $RBAC_1$ includes $RBAC_0$ and introduces *role hierarchies* [80]. Hierarchies are a means of structuring roles to reflect an organisation's lines of authority and responsibility, and specify inheritance between roles. Role *inheritance* enables reuse of permissions by allowing a senior role to inherit permissions from a junior role. For example the finance director of a company inherits the permissions of the accounts manager, as the latter is the junior role. Although the propagation of permissions along role hierarchies further simplifies administration by considerably reducing the number of permissions in the system, it is not always desirable. Organisational hierarchies do not usually correspond to permission-inheritance hierarchies. The person in the senior role may not have the specific skills needed for the more junior role. For example, a managing director would not usually be able to perform the functions of a systems administrator much lower down in the organisational hierarchy. Such situations lead to exceptions and complicate the specification of role hierarchies [68]. Another problem with RBAC is that it does not cater for multiple instances of a role with similar policies but relating to different target objects. For example there may be two different instances of a network operator role, responsible for the North and South regions of the network respectively.

$RBAC_2$ includes $RBAC_0$ and introduces *constraints* to restrict the assignment of users or permissions to roles, or the activation of roles in sessions. Constraints are used to specify application-dependent conditions, and satisfy well-defined control principles such as the principles of least-permission and separation of duties. Finally, $RBAC_3$ combines both $RBAC_1$ and $RBAC_2$, and provides both role hierarchies and constraints. In recent work Sandhu et al. [78] propose an updated set of RBAC models in an effort to formalise RBAC. The models are called: flat RBAC, hierarchical RBAC, constrained RBAC and symmetrical RBAC, and correspond to the $RBAC_0$–$RBAC_3$ models. Although the updated models define the basic features that must be implemented by an RBAC system more precisely, their description remains informal. A number of variations of RBAC models have been developed, and several proposals have been presented to extend the model with the notion of relationships between the roles [15], as well as with the idea of a team, to allow for team-based access control where a set of related roles belonging to a team are activated simultaneously [95].

## 3.2. *IETF/DMTF Policy Core Information Model (PCIM)*

The area of network policy specification has recently seen a lot of attention both from the research and the commercial communities. *Network policy* comprises rules that define the relationship between clients using network resources and the network elements that provide those resources. The main interest in network policies is to manage and control the quality of service (QoS) experienced by networked applications and users, by configuring network elements using policy rules. The most notable work in this area is the Internet Engineering Task Force (IETF) policy model, which considers policies as rules that specify actions to be performed in response to defined conditions:

```
if <condition(s)> then <action(s)>
```

The condition-part of the rule can be a simple or compound expression specified in either conjunctive or disjunctive normal form. The action-part of the rule can be a set of actions that must be executed when the conditions are true. Although this type of policy rule prescribes similar semantics to an obligation of the form event–condition–action, there is no explicit event specification to trigger the execution of the actions. Instead it is assumed that an implicit event such as a particular traffic flow, or a user request will trigger the policy rule. The IETF approach does not have an explicit specification of authorisation policy, but simple admission control policies can be specified by using an action to either allow or deny a message or request to be forwarded if the condition of the policy rule is satisfied. The following are simple examples of the types of rules administrators may want to specify. The first rule assures the bandwidth between two servers that share a database, directory and other information. The second rule gives high priority to multicast traffic for the corporate management sub-network on Monday nights from 6:00 PM to 11:00 PM, for important (sports) broadcasts:

```
if ((sourceIPAdress = 192.168.12.17 AND
    destinationIPAdress = 192.168.24.8) OR
    (sourceIPAdress = 192.168.24.8 AND
    destinationIPAdress = 192.168.12.17)) then
    set Rate := 400Kbps

if ((sourceIPSubnet = 224.0.0.0/240.0.0.0) AND
    (timeOfDay = 1800-2300) AND
    (dayofweek = Monday)) then
    set Priority := 5
```

The IETF do not define a specific language to express network policies but rather a generic object-oriented information model for representing policy information following the rule-based approach described above, and early attempts at defining a language [91] have been abandoned. The *policy core information model* (PCIM) [69] extends the common information model (CIM) [43] defined by the Distributed Management Task Force (DMTF) with classes to represent policy information. The CIM defines generic objects such as managed system elements, logical and physical elements, systems, service, users, etc., and provides abstractions and representations of the entities involved in a managed environment including their properties, operation and relationships. The information model defines how to represent managed objects and policies in a system but does not define how to actually specify policies. Apart from the PCIM, the IETF are defining an information model to represent policies that administer, manage and control access to network QoS resources for integrated and differentiated services [85]. The philosophy of the IETF is that business policies expressed in high-level languages, combined with the network topology and the QoS methodology to be followed will be refined to the policy information model, which can then be mapped to a number of different network device configurations. Strassner shows how this approach can be applied to policy-based network management in the context of work to develop the Next Generation Operations Support Systems (NGOSS) [90]. Vendors following the IETF approach are using graphical tools to specify policy in a tabular format and automate the translation to PCIM. We provide an overview of commercial tools in Section 6.

Figure 3 shows the classes defined in the PCIM and their main associations. Policy rules can be grouped into nested policy groups to define policies that are related in any application specific way, although no mechanism exists for parameterising rules or policy groups. Note that both the actions and conditions can be stored separately in a policy repository and reused in many policy rules. A special type of condition is the time-period over which the policy is valid. The *PolicyTimePeriodCondition* class covers a very complex specification of time constraints.

Policy rules can be associated with a priority value to resolve conflicts between rules. However, this approach does not easily scale to large networks with a large number of rules specified by a number of different administrators. In addition policy rules can be tagged with one or more roles.

A role represents a functional characteristic or capability of a resource to which policies are applied, such as backbone interface, frame relay interface, BGP-capable router, web-server, firewall, etc. Roles are essentially used as a mechanism for associating policies with the network elements to which the policies apply.

An advantage of the information modelling approach followed by the IETF is that the model can be easily mapped to structured specifications such as XML, which can then be used for policy analysis as well as distribution of policies across networks. The mapping of CIM to XML is already undertaken within the DMTF [44]. The IETF define a mapping of the PCIM to a form that can be implemented in an LDAP directory [92].

Other approaches to network policy specification try to extend the IETF rule-based approach to specify traffic control using a concrete language. An example is the *path-based policy language* (PPL) from the Naval postgraduate school described in [89]. The language is designed to support both the differentiated as well as the integrated services model and

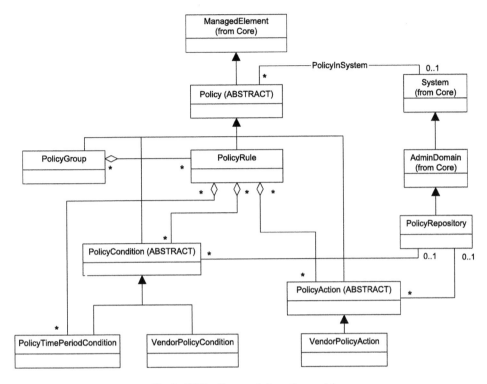

Fig. 3. IETF policy core information model.

is based on the idea of providing better control over the traffic in a network by constraining the path (i.e. the links) the traffic must take. The rules of the language have the following format:

```
policyID <userID> @{paths} {target} {conditions}
  [{action_item}]
action_item = [{condition}:] {actions}
```

*Action_items* in a PPL rule correspond to the *if-condition-then-action* rule of the IETF approach. The informal semantics of the rule is: "*policyID created by <userID> dictates that target class of traffic may use paths only if {conditions} is true after action_items are performed*". The following are examples of PPL rules from [89]:

```
Policy1 <net_manager> @ {<1,2,5>}
  {class = {faculty}} {*} {priority := 1}
Policy2 <Betty> @ {<1,*,5>}
  {traffic_class = {accounting}}
  {day != Friday : priority := 5}
```

*Policy1* states that the path starting at node 1, traversing to node 2, and ending at node 5 will provide high priority for *faculty* users. *Policy2* uses the wild-card character to specify a partial path. It states that, on all paths from node 1 to node 5, accounting class traffic will be lowered to priority 5 unless it is a Friday. In this policy the *action_items* field is used with temporal information to influence the priority of a class of traffic.

Note that the use of the *userID* is not needed in the specification of the rules, and unnecessarily complicates the grammar. The ID of the creator of a policy, as well as information such as the time of the creation, or other priority labels attached to a rule are better specified as meta-information that could be used for policy analysis. PPL does not provide any way of composing policies in groups, and there is no use of roles.

### 3.3. *Open Distributed Programming Reference Model (ODP-RM)*

The group working on the International Standards Organisation (ISO) Open Distributed Programming Reference Model (ODP-RM) are defining an *enterprise language* as part of the RM-ODP Enterprise Viewpoint [52], which incorporates concepts such as policies and roles within a community. A *community* in RM-ODP terminology is defined as a *configuration* of objects formed to meet an *objective*. The objective is expressed as a contract, which specifies how the objective can be met, and a configuration is a collection of objects with defined relationships between them. The community is defined in terms of the following elements:

- the enterprise objects comprising the community,
- the roles fulfilled by each of those objects and the relationships between them,
- the policies governing the interactions between enterprise objects fulfilling roles,
- the policies governing the creation, usage and deletion of resources,
- the policies governing the configuration of enterprise objects and assignment of roles to enterprise objects,
- the policies relating to the environment contract governing the system.

Policies constrain the behaviour of enterprise objects that fulfil actor roles in communities and are designed to meet the objective of the community. Policy specifications define what behaviour is allowed or not allowed and often contain prescriptions of what to do when a rule is violated. Policies in the ODP enterprise language thus cover the concepts of obligation, permission and prohibition.

The ODP enterprise language is really a set of abstract concepts rather than a language that can be used to specify enterprise policies and roles. Recently, there have been a number of attempts to define precise languages that implement the abstract concepts of the enterprise language. These approaches concentrate on using UML to graphically depict the static structure of the enterprise viewpoint language as exemplified by [88] (see Figure 4), as well as languages to express policies based on those UML models. Steen et al. [87,88] propose a language to support the enterprise viewpoint where policy statements are specified using the grammar shown below. Each statement applies to a role, the subject of the policy, and represents either a permission, an obligation or a prohibition for that role. The grammar of the language is concise, however it does not allow composition of policies or constraints for groups of policies. Constraints cannot be specified to restrict the

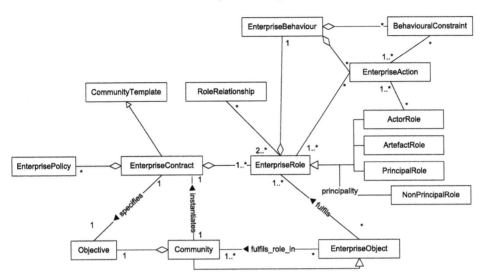

Fig. 4. A UML meta-model of the enterprise viewpoint language.

activation/deactivation of roles or the assignment of users and permissions in roles. Note that the Object Constraint Language (OCL) [71] is used to express the logical conditions in the before-, if- and where- clauses defined in the grammar.

```
[R?] A <role> is (permitted | obliged | forbidden)
     to (do <action> [before <condition>] |
        satisfy <condition>)
     [, if <condition>][, where <condition>]
     [, otherwise see <number>].
```

The authors specify the semantics of the policy language by translating it to Object-Z, an object-oriented extension of the specification language Z. The following examples from [88] demonstrate the use of the proposed language.

```
[R1] A Borrower is permitted to do Borrow(item:Item),
        if(fines < 5*pound).
[R2] A UGBorrower
        is forbidden to do Borrow(item:Item),
        where item.isKindOf(Periodical).
[R3] A Borrower is obliged to do Return(item:Item)
        before (today > dueDate),
        if (loans→exists(loan | loan.item = item)),
        where (dueDate = loans→select(loan |
                loan.item = item).dueDate),
        otherwise see R4.
```

Policy statements *R1* and *R2* specify a permission and a prohibition respectively. *R1* permits a member of the *Borrower* role to borrow an item if the fines of that borrower are less than 5 pounds. *R2* forbids an undergraduate student belonging to the *UGBorrower* role to borrow periodicals. *R3* is an obligation specifying that a borrower must return an item by the *dueDate* of that item. *R3* is conditional upon the item to be returned actually being on loan to the borrower, as specified by the *if*-clause. The *where*-clause constrains the logic variable *dueDate* to be equal to the *dueDate* of the loan in question, and the *before*-clause contains a condition upon which the obligation should have been fulfilled. Obligations do not contain explicit specifications of the events upon which the actions must be executed, which make their implementation difficult. Note that the *otherwise*-clause is an exception mechanism that indicates what will happen when the obligation is violated. The actions to be executed on a violation are specified in another policy (*R4*).

## 4. Security and trust specification policies

Access control is concerned with permitting only authorised users (subjects) to access services and resources (targets). It limits the activity of legitimate users who have been successfully authenticated. Authorisation or access control policy defines the high-level rules specifying the conditions under which subjects are permitted to access targets [77]. However, in many systems there is no real policy specification, only the implementation in terms of low-level mechanisms such as access control lists. The study of access control has identified a number of useful access control models, which provide a formal representation of security policies and allow the proof of properties about an access control system. Access control policies have been traditionally divided into discretionary and mandatory policies.

**Discretionary access control** (DAC) policies restrict access to objects based on the identity of the subjects and/or groups to which they belong, and are discretionary in the sense that a subject can pass it's access permissions on to another subject. The notion of delegation of access rights is thus an important part of any system supporting DAC. Basic definitions of DAC policies use the access matrix model as a framework for reasoning about the permitted accesses. In the access matrix model the state of the system is defined by a triple $(S, O, A)$, where $S$ is the set of subjects, $O$ is the set of objects and $A$ is the access matrix where rows correspond to subjects, columns correspond to objects and entry $A[s, o]$ reports the privilege of $s$ on $o$. Discretionary policies do not enforce any control on the flow of information, once this information is acquired by a process, making it possible for processes to leak information to users not allowed to read it.

**Mandatory access control** (MAC) policies enforce access control on the basis of fixed regulations mandated by a central authority, as typified by the Bell–LaPadula, lattice-based model [17]. Lattice-based models were defined to deal with the issue of data confidentiality, and concentrate on restricting information flow in computer systems. This is achieved by assigning a security classification to each subject (an active entity that can execute actions) and each object (a passive entity storing information) in the system. Subjects and objects form a lattice based on their classification, which is used to enforce some fixed mandatory policies regarding the actions that subjects can execute on objects. The conceptual framework of the Bell–LaPadula model forms the basis of other derived models one of which is the Biba model. The Biba model uses similar controls as those used in the Bell–LaPadula model for providing *integrity* of data [19].

**Non-discretionary access control** (NDAC) identifies the situations in which authority is vested in some users, but there are explicit controls on delegation and propagation of authority [1]. Any global and persistent access control policy relying on access control decision information not directly controlled by the security administrator is non-discretionary. Administrative policies [79] determine who is authorised to modify the allowed access rights and exist only within discretionary policies. Whereas, in mandatory policies, the access control is determined entirely on the basis of the security classification of subjects and objects. Administrative policies can be divided into: (i) *Centralised* where a single authoriser (or group) is allowed to grant and revoke authorisations to the users. (ii) *Hierarchical* where a central authoriser is responsible for assigning administrative responsibilities to other administrators. The administrators can then grant and revoke access authorisations to the users of the system according to the organisation chart. (iii) *Cooperative* where special authorisations on given resources cannot be granted by a single authoriser but needs cooperation of several authorisers. (iv) *Ownership* where a user is considered the owner of the objects he/she creates. The owner can grant and revoke access rights for other users to that object, and (v) *Decentralised* where the owner or administrator of an object can also grant other users the privilege of administering authorisations on the object.

Over the years other sophisticated security models have been proposed to formalise security policies required for commercial applications. The *Clark–Wilson model* [36] is a well-known one for commercial data processing practises. Its main goal is to ensure the integrity of an organisation's accounting system and to improve its robustness against insider fraud. The Clark–Wilson model recommends the enforcement of two main principles, namely the *principle of well-formed transactions* where data manipulation can occur only in constrained ways that preserve and ensure the integrity of data, and the *principle of separation of duty*. The latter reduces the possibility of fraud or damaging errors by partitioning the tasks and associated privileges so cooperation of multiple users is required to complete sensitive tasks. Authorised users are assigned privileges that do not lead to execution of conflicting tasks. This principle has since been adopted as an important constraint in security systems.

A security policy model that specifies clear and concise access rules for clinical information systems can be found in [7]. This model is based on access control lists and the authors claim it can express Bell–LaPadula and other lattice-based models. Finally the *Chinese-wall policy* [24] was developed as a formal model of a security policy applicable to financial information systems, to prevent information flows that cause conflict of interest for individual consultants. The basis of the model is that people are only allowed to access information that is not held to conflict with any other information that they already possess. The model attempts to balance commercial discretion with mandatory controls, and is based on a hierarchical organisation of data. It thus falls in the category of lattice-based access control models.

### 4.1. *Logic-based languages*

Logic-based languages have proved attractive for the specification of access-control policy, as they have a well-understood formalism, which is amenable to analysis. However they can be difficult to use and are not always directly translatable into efficient implementation. This section starts by presenting some of these specification languages classified by the type of logic used in their definition. We go on to discuss languages that were specifically developed to support the Role-Based Access Control (RBAC) model.

**4.1.1.** *First-order logic*    There are several examples that illustrate the application of first order logic to the specification of security policy. These include the logical notation introduced in [32]; the Role Definition Language (RDL) presented in [50] and RSL99 [4]. Because all these approaches are based on the Role Based Access Control (RBAC) model, we will defer a more detailed discussion to Section 2.2 where they can be considered together with other RBAC specification techniques.

In addition to these RBAC examples, there are some examples of the application of Z to defining security policies for a system. Z is a formal specification language that combines features of first-order predicate logic with set theory [86]. In [23], the use of Z to specify and validate the security model for the NATO Air Command and Control System is described. The aim of this work was to develop a model for both discretionary and mandatory access controls based on the Bell–LaPadula model mentioned previously.

One of the main problems encountered when using first-order logic for policy specification arises when negation is used together with recursive rules. This leads to logic programs that cannot be decided and cannot be evaluated in a flounder-free manner [41]. Although it is possible to avoid the use of negation or recursion, this is not practical since it diminishes significantly the expressive power of the logical language. In the next section, we discuss an alternative solution, stratified logic, which helps overcome the problems associated with using recursion and negation together.

**4.1.2.** *Stratified logic*    When using first-order logic-based languages to describe policies, many approaches considered here are described in terms of stratified logic, which permits a constrained use of recursion and negation while disallowing those combinations which lead to undecidable programs. A stratified program is one where it is possible to order the clauses such that for any clause containing a negated literal in its body, there is a clause later in the program that defines the negated literal. Another way of describing stratified theories makes use of directed dependency graphs. These are graphs that comprise a node for each predicate symbol appearing in the program and a directed edge from the node representing any predicate that appears in the body of a clause to the node representing the predicate defined in the head of the clause. The edges are labelled positively or negatively, where a negative symbol indicates that the predicate at the tail end of the edge appear in negated form in a clause of the program. Using this technique, a program is stratified if the dependency graph contains no cycles having a negative edge.

The concepts of stratified theories and stratification were originally developed in the context of databases [31] and were later adopted into the area of logic programming as described in [8]. Programs that use stratified logic can use negation to extend their expressive power and can be evaluated in a flounder free manner. Indeed there are numerous studies that identify stratified logic as a class of first order logic that supports logic programs that are decidable [41,53]. Moreover, such programs are decidable in polynomial time [55]. A more detailed analysis of the computational complexity and expressive power of stratified logic can be found in [41].

The *authorisation specification language* (ASL) [55] is an example of a stratified first-order logic language for specifying access control policies. Authorisation rules identify the actions authorised for specific users, groups or roles, but cannot be composed into roles to provide for reusability, i.e. there is no explicit mechanism for assigning authorisations to roles; instead this is specified as part of the condition of authorisation rules. Although the language provides support for role-based access control, it does not scale well to large systems because there is no way of grouping rules into structures for reusability. The following is an example of an authorisation rule in ASL, which states that all subjects belonging to group *Employees* but not to *Soft-Developers* are authorised to read *file1*.

```
cando(file1, s, +read) ← in(s, Employees) &
                          ¬in(s, Soft-Developers)
```

The *cando* predicate can also be used to specify negative authorisations; the sign in front of the action in the *cando* predicate indicates the modality of the authorisation, e.g., `-read` would indicate read not permitted. However, there is no explicit specification of delegation and no way of specifying authorisation rules for groups of target objects that are not related by type. A *dercando* predicate is defined in the language to specify derived authorisations based on the existence or absence of *cando* rules (i.e. other authorisations in the system). In addition, two predicates *do* and *done*, can be used to specify history-dependent authorisations based on actions previously executed by a subject. The language includes a form of meta-policies called *integrity rules* to specify application-dependent conditions that limit the range of acceptable access control policies. In a recent paper [54] the language has been extended with predicates used to evaluate hierarchical or other relationships between the elements of a system such as the membership of users in groups, inclusion relationships between objects or supervision relationship between users.

Barker [11] adopts a similar approach to express a range of access control policies using stratified clause-form logic, with emphasis on RBAC policies. According to the author, this form of logic is appropriate for the specification of access control policies mostly due to its simple high-level declarative nature. The function-free logic adopted by Barker, defines a normal clause as an expression of the following form: $H \leftarrow L1, L2, \ldots, Lm \; (m \geqslant 0)$. The head of the clause, $H$, is an atom and $L1, L2, \ldots, Lm$ is a conjunction of literals that constitutes the body of the clause. If the conjunction of literals $L1, L2, \ldots, Lm$ is true (proved) then $H$ is true (proved). A literal is an atomic formula or its negation and a normal theory is defined as a finite set of normal clauses. As described previously, a stratified theory extends a normal theory by eliminating some forms of 'recursion-via-negation', which makes the computation of the theory more efficient. The negation of literals is used to specify negative permissions.

In [13] the authors show how policies specified in stratified logic can be automatically translated into a subset of SQL to protect a relational database from unauthorised read and update requests. The following example from [12] demonstrates their approach:

```
permitted(U,P,O) ← ura(U,R1),activate(U,R1),
                   senior-to(R1,R2),rpa(R2,P,O)
```

The above clause specifies that user $U$ has the permission $P$ on object $O$ if $U$ is assigned to a role $R1$, $U$ is active in $R1$, and $R1$ inherits the $P$ permission on $O$ from $R2$. This expression assumes that the following predicates have been defined: *activate*$(U, R)$ to denote that $U$ is active in $R$, *ura*$(U, R)$ to assign user $U$ to role $R$, *rpa*$(R, P, O)$ to assign permission $P$ on object $O$ to role $R$, and *senior-to(R1, R2)* to denote that role $R1$ is senior to $R2$.

**4.1.3.** *Deontic logic*   Deontic logic was developed starting in the 1950s by Von Wright [104], Castaneda [30] and [5] by extending modal logic with operators for permission, obligation and prohibition. Known as Standard Deontic Logic (SDL), traditionally it has been used in analysing the structure of normative law and normative reasoning in law. Because SDL provides a means of analysing and identifying ambiguities in sets of legal rules, there are many examples of the application of SDL to represent legislative documents [56,83]. An excellent overview of the applications of SDL can be found in [105].

Before looking at the details of how SDL has been applied to policy specification, it would be useful to summarise the basic axioms of the language. These are as follows:

**[SDL0]** Tautologies of propositional calculus

**[SDL1]** $O(p \rightarrow q) \rightarrow (Op \rightarrow Oq)$

If there is an obligation that $p$ implies $q$, then an obligation to do $p$ implies an obligation to do $q$.

**[SDL2]** $Op \rightarrow Pp$

If there is an obligation to do $p$ then $p$ is permitted.

**[SDL3]** $Pp \leftrightarrow \neg O \neg p$

Iff $p$ is permitted then there is no obligation to not do $p$. In other words, iff $p$ is permitted then there is no refrain policy with respect to $p$.

**[SDL4]** $Fp \leftrightarrow \neg Pp$

Iff $p$ is forbidden then there is no permission to do $p$.

**[SDL5]** $p, (p \rightarrow q)/q$(Modus Ponens)

If we can show that $p$ holds and that $q$ is implied by $p$, then it is possible to infer that $q$ must hold.

**[SDL6]** $p/Op$ (O-necessitation)

If we can show that $p$ holds, then it is possible to infer that the obligation to do $p$ also holds.

At first glance it would appear that SDL is ideally suited to specifying policy because it provides operators for all the common policy constructs like permission, prohibition (cf. positive and negative authorisation) and obligation. However, closer examination of some of the axioms reveals inconsistencies between the definition of policy rules and the behaviour of SDL. For example, SDL2 indicates that an obligation can imply a permission and

SDL3 indicates that a permission implies no obligation to not do an action. However, these implications between permissions and obligations do not exist in many systems in which the obligation and authorisation policies are specified independently and implemented in different ways.

Some of the earliest work using deontic logic for security policy representation can be found in [47]. The focus of this work was to develop a means of specifying confidentiality policies together with conditional norms.

In [33], SDL is used to represent security policies with the aim of detecting conflicts in the policy specifications. This approach is based on translating the SDL representation into first-order predicate logic before performing the necessary conflict detection analysis. In an extension to this work, [38] also describes how delegation can be represented using deontic logic notation.

Ortalo describes a language to express security policies in information systems based deontic logic [72]. In his approach he accepts the axiom $Pp = \neg O \neg p$ ("permitted $p$ is equivalent to not $p$ being not obliged") as a suitable definition of permission. As mentioned previously, this axiom is not appropriate for the modelling of obligation and authorisation policies because the two need to be separated.

An inherent problem with the deontic logic approach is the existence of a number of paradoxes. For example, Ross' paradox can be stated as $Op \rightarrow O(p \wedge q)$, i.e. if the system is obliged to perform the action *send message*, then it is obliged to perform the action *send message* or the action *delete message*. Although there is some work that offers resolutions to these paradoxes [74], the existence of paradoxes can make it confusing to discuss policy specifications using deontic logic notations.

## 4.2. *Role-based security languages*

**4.2.1.** *Role definition language* The Role Definition Language RDL [50] is based on Horn clauses and was developed as part of the Cambridge University Oasis architecture for secure interworking services. RDL is based on sets of rules that indicate the conditions under which a client may obtain a name or role, where a role is synonymous to a named group. The conditions for entry to a role are described in terms of credentials that establish a client's suitability to enter the role, together with constraints on the parameters of those credentials. The work on RDL also falls into the category of *certificate-based access control*, which is adopted by trust-management systems described separately in Section 2.3.4. The following is an example of an authorisation rule in RDL, which establishes the right for clients assigned to the *SeniorHaematologist* role to invoke the *append* method if they also possess a certificate called *LoggedOn(km, s)* issued by the *Login service*, where *s* is a trusted server, i.e. if they have been logged on as *km* on a trusted machine.

```
append(haematology-field, y, x)
  ← SeniorHaematologist(x)
    ∧ Login.LoggedOn(km, s):s in TrustedServers
```

Roles in RDL are also considered as credentials, and can be used to assign clients to other roles as in the following example where a user $x$, who belongs to both the *Haematologist* and the *SeniorDoctor* roles, is also a *SeniorHaematologist*:

```
SeniorHaematologist(x)
    ← Haematologist(x) ∧ SeniorDoctor(x)
```

RDL has an ill-defined notion of delegation, whereby roles can be delegated instead of individual access rights in order to enable the assignment of users to certain roles. The notion of *election* is introduced to enable a client to delegate a role that they do not themselves possess, to other clients. We believe that assignment of users to roles should be controlled with user-assignment constraints instead. The following example from [50] specifies that the *chief examiner* may elect any logged on user who belongs to the group *Staff* to be an *examiner*, for the examination subject *e*.

```
Examiner(p,e)
    ← Login.LoggedOn(p,s) ◁ ChiefExaminer : p in Staff
```

**4.2.2.** *Specifying role constraints*   Chen et al. [32] introduce a language based on set theory for specifying *RBAC state-related constraints*, which can be translated to a first-order predicate-logic language. They define an RBAC system state as the collection of all the attribute sets describing roles, users, privileges, sessions as well as assignments of users to roles, permissions to roles and roles to sessions. They use this model to specify constraints for RBAC in two ways: (i) by treating them as invariants that should hold at all times, and (ii) by treating them as preconditions for functions such as assigning a role to a user. They define a set of global functions to model all operations performed in an RBAC system, and specify constraints which include: conflicting roles for some users, conflicting roles for sessions of some users, and prerequisite roles for some roles with respect to other users. The following example from [32] can be used as an invariant or as a precondition to a user–role assignment, to indicate that assigned roles must not be conflicting with each other:

```
Role set: R = {r₁ , r₂ ,..., rₙ }
User set: U = {u1 , u2 ,..., um }
Check-condition:
   oneelement(R) ∈ role-set(oneelement(U))
       → allother(R) ∩ role-set(oneelement(U)) = ∅
```

The two non-deterministic functions, `oneelement` and `allother`, are introduced in the language to replace explicit quantifiers. `Oneelement` selects one element from the given set, and `allother` returns a set by taking out one element from its input.

*RSL99* [4] is another role specification language which extends the ideas introduced in [32] and can be used for specifying separation of duty properties in role-based systems. The language covers both static and dynamic separation of duty constraints, and its grammar is simple, although the expressions are rather complicated and inelegant.

Since time is not defined as part of the state of an RBAC system as defined in [32], the above languages cannot specify temporal constraints. The specification of temporal constraints on role activations is addressed in the work by Bertino et al. [18], which defines a temporal model called *TRBAC*. They propose an expression language that can be used to specify two types of temporal constraints: (i) periodic activation and deactivation of roles using periodic expressions, and (ii) specification of temporal dependencies among role activations and deactivations using role triggers.

```
(PE1)  ([1/1/2000, ∞], Night-time,
          VH:activate doctor-on-night-duty)
(PE2)  ([1/1/2000, ∞], Day-time,
          VH:deactivate doctor-on-night-duty)
(RT1)  (activate doctor-on-night-duty
          → H:activate nurse-on-night-duty)
(RT2)  (deactivate doctor-on-night-duty
          → H:deactivate nurse-on-night-duty)
```

A role can be activated/deactivated by means of role triggers specified as rules to automatically detect activations/deactivations of roles. Role triggers can also be time-based or external requests to allow administrators to explicitly activate/deactivate a role, and are specified in the form of prioritised event expressions. The example above, from [18], shows two periodic expressions (*PE1* and *PE2*) and two role triggers (*RT1* and *RT2*).

The periodic expressions state that the role *doctor-on-night-duty* must be active during the night. The role triggers state that the role *nurse-on-night-duty* must be active whenever the role *doctor-on-night-duty* is active. In the example, the symbols *VH* and *H* stand for very high and high respectively and denote priorities for the execution of the rules.

### 4.3. *Other security specification approaches*

In this section we cover other approaches to specifying security policy which includes an event-based language, the use of XML, a graphical approach and finally trust policy specification.

**4.3.1.** *SPL*   The *security policy language* (SPL) [75] is an event-driven policy language that supports access-control, history-based and obligation-based policies. SPL is implemented by an event monitor that, for each event, decides whether to allow, disallow or ignore the event. Events in SPL are synonymous with action calls on target objects, and can be queried to determine the subject who initiated the event, the target on which the event is called, and attribute values of the subject, target and the event itself. SPL supports two types of sets to group the objects on which policies apply: groups and categories. Groups are sets defined by explicit insertion and removal of their elements, and categories are sets defined by classification of entities according to their properties. The building blocks of policies in SPL are constraint rules which can be composed using a specific tri-value algebra with three logic operators: *and*, *or* and *not*. A simple constraint rule is comprised of two logical binary expressions, one to establish the domain of applicability and another to decide on the acceptability of the event. The following extract from [75] shows examples of simple rules and their composition. Note that conflicts between positive and negative authorisation policies are avoided by using the tri-value algebra to prioritise policies when they are combined as demonstrated by the last composite rule of the example. The keyword *ce* in the examples is used to refer to the current event.

```
// Every event on an object owned by
the author of the event is allowed
OwnerRule: ce.target.owner = ce.author :: true;

// Payment order approvals cannot be done by
the owner of payment order
DutySep: ce.target.type = "paymentOrder" &
         ce.action.name = "approve"
      :: ce.author != ce.target.owner;

// Implicit deny rule.
deny: true :: false;

// Simple rule conjunction, with default deny value
OwnerRule AND DutySep OR deny;

// DutySep has a higher priority then OwnerRule
DutySep OR (DutySep AND OwnerRule);
```

SPL defines two abstract sets called *PastEvents* and *FutureEvents* to specify history-based policies and a restricted form of obligation policy. The type of obligation supported by SPL is a conditional form of obligation, which is triggered by a pre-condition event:

```
Principal_O must do Action_O if Principal_T
   has done Action_T
```

Since the above is not enforceable, they transform it into a policy with a dependency on a future event as shown below, which can be supported in a way similar to that of history-based policies:

```
Principal_T cannot do Action_T if Principal_O
   will not do Action_O
```

SPL obligations are thus additional constraints on the access control system, which can be enforced by security monitors [76], and not obligations for managers or agents to execute specific actions on the occurrence of system events, independent of the access control system.

The notion of a policy is used in SPL to group set definitions and rules together to specify security policies that can be parameterised; policies are defined as classes that allow parameterised instantiation. Instantiation of a policy in SPL also means activation of the policy instance, so no control over the policy life cycle is provided. Further re-use of specifications is supported through inheritance between policies. A policy can inherit the specifications of another policy and override certain rules or sets. Policy constructs can also be used to model roles, in which case specific sets in the policy identify the users allowed to play the role. Rules or other nested policies inside a role specify the access rights associated with the role. SPL provides the ability to hierarchically compose policies by instantiating them inside other policies, thus enabling the specification of libraries of common security policies that can be used as building blocks for more complex policies. The authors claim that this hierarchical composition also helps restrict the scope of conflicts between policies, however this is not clear, as there may be conflicts across policy hierarchies. Note that SPL does not cater for specification of delegation of access rights between subjects, and there is no explicit support for specifying roles. The following example from [75] defines an *InvoiceManagement* policy, which allows members of the *clerks* team to access objects of type *invoice*. The actual policy that permits the access is specified as *ACL* separately and instantiated within *InvoiceManagement* using the keyword *new*:

```
policy InvoiceManagement {
    // Clerks would usually be a role but for
    simplicity here it is a group team clerks ;
    // Invoices are all object of type invoice
    collection invoices =
          AllObjects@{ .doctype = "invoice" };
    // In this simple policy clerks can perform every
    action on invoices
    DoInvoices: new ACL(clerks, invoices, AllActions);
    ?usingACL: DoInvoices; }
```

**4.3.2.** *XACML*   XACML [70] is an XML specification for expressing policies for information access over the Internet, initially defined by the Organisation for the Advancement of Structured Information Standards (OASIS) technical committee (unrelated to the Oasis work at Cambridge described previously) and now being adopted as a W3C standard. The language permits access control rules to be defined for a variety of applications from protecting database entries and XML document contents to physical assets. Similar to existing policy languages, XACML is used to specify a *subject–target–action–condition* oriented policy. The notion of subject comprises identity, group, and role and the granularity of target objects is as fine as single elements within a document. The language supports both roles and groups, which are defined as collections of attributes relevant to a principal. XACML includes conditional authorisation policies, as well as policies with external post-conditions to specify actions that must be executed prior to permitting an access. e.g. *"A physician may read any record and write any medical element for which he or she is the designated primary care physician, provided an email notice is sent to the patient or the parent/guardian, in case the patient is under 16"*. An example of a policy specified in XACML is shown below:

```
<?xml version="1.0" encoding="UTF-8"?>
<Policy PolicyId="SamplePolicy"
        RuleCombiningAlgId
          ="urn:oasis:names:tc:xacml:1.0:rule-combining-
                              algorithm:permit-overrides">
<!-- This Policy only applies to requests on
 the SampleServer -->
 <Target>
  <Subjects> <AnySubject/> </Subjects>
  <Resources>
   <ResourceMatch MatchId
      ="urn:oasis:names:tc:xacml:1.0:function:string-equal">
    <AttributeValue DataType
       ="... XMLSchema\#string">
        SampleServer</AttributeValue>
    <ResourceAttributeDesignator DataType
       =http://www.w3.org/2001/XMLSchema\#string
                AttributeId="urn:oasis:names:tc:xacml:1.0:
                              resource:resource-id"/>
   </ResourceMatch>
  </Resources>
  <Actions> <AnyAction/> </Actions>
 </Target>
 <!-- Rule to see if we should allow
 the Subject to login -->
 <Rule RuleId="LoginRule" Effect="Permit">
 <!-- Only use this Rule if the action is login -->
  <Target>
   <Subjects> <AnySubject/> </Subjects>
   <Resources> <AnyResource/> </Resources>
   <Actions>
   <ActionMatch MatchId
      ="urn:oasis:names:tc:xacml:1.0:function:string-equal">
    <AttributeValue DataType
       ="... XMLSchema\#string">login</AttributeValue>
    <ActionAttributeDesignator DataType
       ="http://www.w3.org/2001/XMLSchema\#string"
                      AttributeId="ServerAction"/>
   </ActionMatch>
   </Actions>
  </Target>
  [CONTD.]
 <!-- Only allow logins from 9am to 5pm -->
 <Condition FunctionId
    ="urn:oasis:names:tc:xacml:1.0:function:and">
```

```
  <Apply FunctionId
     ="... function:time-greater-than-or-equal">
   <Apply FunctionId="... time-one-and-only">
   <EnvironmentAttributeSelector DataType
      ="... XMLSchema\#time"
        AttributeId="urn:oasis:names:tc:xacml:1.0:
                       environment:current-time"/>
   </Apply>
   <AttributeValue DataType
      ="... XMLSchema\#time">09:00:00</AttributeValue>
   </Apply>
   <Apply FunctionId="... time-less-than-or-equal">
    <Apply FunctionId="... time-one-and-only">
     <EnvironmentAttributeSelector DataType
        ="... XMLSchema\#time AttributeId
        ="... current-time"/>
    </Apply>
    <AttributeValue DataType
       ="... XMLSchema#time">17:00:00</AttributeValue>
   </Apply>
  </Condition>
 </Rule>
</Policy>
```

The above policy only applies to requests for the server called 'SampleServer'. The Policy has a Rule with a Target that requires an action of 'login' and a Condition that applies only if the Subject is trying to log in between 9 AM and 5 PM. Although XACML supports a fine granularity of access control specification, the policy is rather verbose and not really aimed at human interpretation. XACML may be used in conjunction with SAML (security assertion and markup language) assertions and messages, and can thus also be applied to certificate-based authorisations. We discuss certificate-based authorisations in the following section. The work on XACML includes an architecture for enforcing policies which extends the IETF policy architecture described in Section 3.

**4.3.3.** *SAML*   The Security Assertion Mark-up Language (SAML) is the result of an initiative by Oasis to develop a notation that would support the specification and exchange of security information. One of the major design goals of SAML is to support single sign-on features that allow users to access multiple resources through a single authentication operation. An SAML document consists of a set of assertions, provided by a trusted authentication service, that specify information regarding authentication and attributes that apply to a particular subject. For example, a PEP could send an SAML document containing credential assertions obtained from a request to a PDP. The PDP would evaluate policies and send back an authorisation decision assertion, which will be used by the PEP to permit or deny access to the requested resource. An example of an SAML document fragment is presented below:

```
<saml:Assertion MajorVersion="1" MinorVersion="0"
     AssertionID="186CB370-5C81-4716-8F65-
     F0B4FC4B4A0B" Issuer="www.example.com" IssueInstant
                         ="2001-05-31T13:20:00-05:00">

<saml:Conditions NotBefore="2001-05-31T13:20:00-05:00"
     NotAfter="2001-05-31T13:25:00-05:00"/>

  <saml:AuthenticationStatement
       AuthenticationMethod="password"
    AuthenticationInstant="2001-05-31T13:21:00-05:00">
       <saml:Subject>
         <saml:NameIdentifier>
              <SecurityDomain>"www.example.com"
              </SecurityDomain>
              <Name>"cn=Alice,co=example,ou=sales"</Name>
         </saml:NameIdentifier>
       </saml:Subject>
  </saml:AuthenticationStatement>

</saml:Assertion>
```

The SAML framework supports policy specification, deployment and enforcement. However, it is designed to support access control decision-making at the implementation level of a system and does not provide any features that aid policy analysis or refinement. Beyond the example described here, SAML has general applicability. Fundamentally it encodes assertions and predefines 3 types of assertions: attribute assertions, authentication assertions and authorisation assertions. Authorisation decisions can encode whether a principal has been authorised to access a resource, authentication assertions encode whether the principal has been authenticated (cf. single sign-on).

**4.3.4.** *Trust specification*    Applications such as e-commerce and other Internet-enabled services require connectivity between entities that do not know each other. In such situations, the traditional assumptions for establishing and enforcing access control do not hold; subjects of requests can be remote, previously unknown users, making the separation between authentication and access control difficult. A possible solution to this problem is the use of digital certificates or credentials representing statements certified by trusted entities, which can be used to establish properties of their holder (e.g., identity, accreditation). Access control makes the decision of whether or not a party can execute an action based on properties that the party may have, and can prove by presenting one or more certificates. Such an approach is often called certificate-based authorisation and is adopted for the specification of trust. Trust management frameworks combine authentication with authorisation [49] and are used for applications such as web-based labelling, signed email, active networks and e-commerce.

In [20,21], two trust management applications are presented: the *PolicyMaker* and its successor *KeyNote*. Both of these applications are used to answer signed queries of the form *"does a set of requested actions r, supported by credential set C, comply with policy P?"*, where the credentials can be public key certificates with anonymous identity. Both policies and credentials are predicates specified as simple C-like and regular expressions. In this context a policy is a trust assertion that is made by the local system and is unconditionally trusted by the system. Although trust management systems provide an interesting framework for reasoning about trust between unknown parties, assigning authorisations to keys may result in authorisations that are difficult to manage [77]. In addition, providing a common solution to both authentication and access control makes the system more complex.

The *trust policy language* (TPL) by IBM [51] provides a clearer separation between the authentication of subjects based on certificates and the assignment of permissions to those subjects which have been successfully authenticated. With TPL, the credentials result in a client being assigned to a role which has a specific set of permissions, where a role is defined as a group of entities that can represent specific organisational units (e.g., employees, managers, auditors). The assignment of access rights to roles is outside the scope of TPL; the philosophy of the work on TPL is to extend role-based access control mechanisms by mapping unknown users to well defined roles. Although the certificate is intended to be format-independent, the current implementation of the system uses X.509v3 certificates, and defines the language in XML, which makes the syntax rather verbose. Note that unlike KeyNote, TPL permits negative certificates interpreted as suggestions not to trust a user or not to assign a user to a given role. The following example is taken from [49] to demonstrate the use of TPL.

```
<POLICY>
 <GROUP NAME="self"> </GROUP>
 <GROUP NAME="partners">
    <RULE>
    <INCLUSION ID="partner" TYPE="partner" FROM "self"/>
    </RULE>
 </GROUP>
 <GROUP NAME="departments">
    <RULE>
    <INCLUSION ID="partner" TYPE="partner" FROM="partners"/>
    </RULE>
 </GROUP>
 <GROUP NAME="customers">
    <RULE>
    <INCLUSION ID="customer" TYPE="employee" FROM=
                  "departments"/>
    <FUNCTION>
        <GT><FIELD ID="customer" NAME="rank"/></FIELD>
        <CONST>3</CONST></GT>
    </FUNCTION>
    </RULE>
 </GROUP>
</POLICY>
```

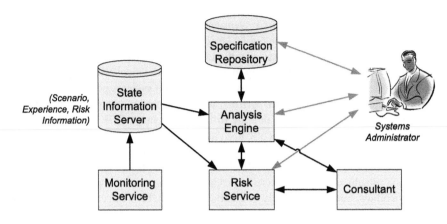

Fig. 5. SULTAN trust analysis framework.

In summary, the policy states that a customer of a retailer company is an employee of a department of a partner company. The first group defined is the originating *retailer*. Then, it is stated that entities having partner certificates, signed by the original retailer, are placed in the group *partners*. The group *department* is defined as any user having a partner certificate signed by the partners group. Finally, the *customer* group consists of anyone that has an employee certificate signed by a member of the departments group who has a rank greater than 3. The trust specification approaches described thus far have been based on the trust implicitly put in the issuers of credentials. More recent work by Grandison aims at calculating explicit trust based on knowledge and recommendations from others. To this end, the *SULTAN trust management framework* (see Figure 5) provides a language and analysis module for managing trust specifications [48]. The language supports conditional trust assertions of the form {trust(Tr, Te, As, L) ← Constraints}, which states that the Tr trusts/distrusts Te to perform the actions, As, at a trust/distrust level L if the optional constraints are satisfied. The specification language also allows trust relationships to be transferred through recommendation rules.

The SULTAN Trust Management Suite consists of four primary components: (1) The Specification Editor – which integrates an editor for trust specifications, a compiler for the SULTAN specifications, and auxiliary tools for storage, retrieval and translation of the specifications, (2) The Analysis Tool, which integrates a Query Command Builder, basic editor and front-end to Sicstus Prolog, (3) The Risk Service, which allows for the calculation of risk and the retrieval of risk information, and (4) The Monitoring Service, which updates information relating to experience or reputation of the entities involved in the trust relationships. The Specification Editor is used by the system administrator to encode the initial set of trust requirements and to analyse the specifications entered. This information is then stored in the system. Over time, applications present SULTAN with new information regarding the state of the system, experiences or even risk information. Whenever a decision is to be made, the SULTAN system may be consulted to provide information that will enable one to make a better decision. We will not be discussing SULTAN trust consultation in this paper. The architecture that supports this model is shown in Figure 5.

## 5. Management and control policies

### 5.1. *Event–condition–action rules*

There has been considerable recent interest in using policies for specifying dynamically adaptable management strategies that can be easily modified to change the management approach without recoding the management system. Most of the policy-based management approaches use condition–action rules, either with or without event triggering, although some also use interpreted scripting languages. We describe these various approaches in this section.

**5.1.1.** *Policy description language (PDL)*   The *policy description language* (PDL) is an event-based language from Bell Labs [62] that uses the *event–condition–action* rule paradigm of active databases to define a policy as a function that maps a series of events into a set of actions. The language can be described as a real-time specialised production rule system to define policies. The syntax of PDL is simple and policies are described by a collection of two types of expressions: *policy rules* and *policy defined event propositions*. Policy rules are expressions of the form:

```
event causes action if condition
```

Which reads: If the event occurs and the condition is true, then the action is executed. Policy defined event propositions are expressions of the form:

```
event triggers policy-defined-event if condition
```

Which reads: If the event occurs under the condition, the policy-defined-event is triggered.

Events can be primitive or complex, and there are two types of primitive events: policy defined events, which are only generated by policy defined event propositions, and system events, which are generated by the environment. Primitive event classes can define attributes, and instances of the classes take actual values for those attributes that can be referenced by other events, actions or conditions within the same rule. Primitive events can be composed to form complex events that enable policies to be enforced under any of the following situations:

- If two events $e1$ and $e2$ occur simultaneously.
- If an event $e$ does not occur.
- If an event $e2$ immediately follows an event $e1$.
- If an event $e2$ occurs after an event $e1$.

The following example from [59] makes use of some of the different features of the language to define a policy for a service provider network which rejects call requests when there is an excessive number of network signalling timeouts over the calls made (i.e. over-

load state) until the time-out rate goes down to a reasonable number. The policy has three
policy defined event propositions and one policy rule proposition.

> *Events*: normal_mode: policy defined event,
>     restricted_mode : policy defined event
>     call_made: system event, time_out: system event,
>     power_on: system event
>
> *Actions*: restrict_calls, accept_all_calls
>
> *Policy description*:
> // when the system starts the primitive event
>     normal_mode is triggered. i.e. the
> // system starts in normal mode
>     power_on **triggers** normal_mode
>
> // when in normal_mode, a sequence of call_made or
>     time_out events will trigger
> // restrict_mode if the overload threshold is
>     exceeded. t is the overload ratio
> // of signalling timeouts over the calls made.
>     The ∧ sign denotes a sequence of
> // zero or more events.
>     normal_mode, ∧(call_made | time_out)
>       **triggers** restricted_mode
>       **if** Count(time_out) > t*Count(call_made)
>
>     restricted_mode **causes** restrict_calls
> // when in overlaod mode, a sequence of call_made or
>     time_out events will
> // trigger normal_mode if the normal threshold is
>     exceeded. t' is considered to //
> be a reasonable timeout rate
>     restricted_mode, ∧(call_made | time_out)
>       **triggers** normal_mode
>       **if** Count(time_out) < t'*Count(call_made)
>
> // Assumes only one callMade or timeOut event per
>     epoch normal_mode **causes** accept_all_calls

Despite its expressiveness, PDL does not support access control policies, nor does it support the composition of policy rules into roles, or other grouping structures. The language
has clearly defined semantics and together with an implementation of PDL policy enforcement. Work on conflict resolution for policies written in PDL is described in [34], and
extensions to the language to specify workflows for network management can be found in
[59]. The language has been used to program Lucent switching products [103] and proves
to be powerful in a variety of network operations and management scenarios.

**5.1.2.** *Event–trigger–rules*   Work done at the University of Florida presents a specification scheme similar to the ECA rule based approaches already discussed here, with some key differences [93]. The development of this approach, called Event–Trigger–Rule (ETR) paradigm, is motivated by the need for rule-based processing capabilities in the distributed environment of electronic commerce enterprises [93].

The ETR paradigm is a generalisation of the ECA approach where the event specification and conditions and actions of the rule are specified as separate entities. Specifying a trigger then associates the event and rule together into a policy. This is in contrast to the ECA rule specification approach, where the event specification, associated conditions and actions be combined into a single rule.

In the ETR approach, events can be classified into 3 types – method associated events, explicit events and timer events. Method events are associated with a particular method invocation and can be raised either before, after or on-commit of the method. This distinction is referred to as the coupling mode of the event. Each of these coupling modes raises synchronous events that will cause a rule to be evaluated before execution of the program continues. Additional coupling modes are instead-of (raises a synchronous event that allows the rule to replace the method invocation) and decoupled (raises an asynchronous event). Explicit events are those raised by the application during execution and timer events are those associated with a particular time of interest. The following illustrates a method associated event specification.

```
            IN     InventoryManager
         EVENT     update_quantity_event(String item,
                                         int quantity)
          TYPE     METHOD
 COUPLING_MODE     BEFORE
     OPERATION     UpdateQuantity(String item, int quantity)
```

A rule specifies some operations that should be performed if certain conditions apply. The conditional part of an ETR rule is defined as a guarded expression, where the guard is used to control evaluation of the conditional expression. This allows the entire rule to be skipped if any part of the guard expression evaluated to false, thus avoiding potential exception conditions (e.g., if required variables are not initialised). Additionally, the rule specifies an action block (cf. a 'then' block) and an alternative action block (cf. an 'else' block). The complete syntax of a rule specification is presented below.

```
          RULE     rule_name(parameter list)
[      RETURNS     return_type]
[  DESCRIPTION     description_text]
[         TYPE     DYNAMIC/STATIC]
[        STATE     ACTIVE/SUSPENDED]
[      RULEVAR     rule variable declarations]
[    CONDITION     guarded expression]
[       ACTION     operation block]
[    ALTACTION     operation block]
[    EXCEPTION     exception and handling block]
```

When specifying a rule, it is possible to define local variables using the RULEVAR clause and also handle errors using the EXCEPTION clause. The STATE clause specifies if the rule will be active or suspended after its definition. A suspended rule will not be triggered until it is made active. The specification syntax also provides optimisation hints to the runtime environment using the TYPE clause. A dynamic rule can be changed at runtime whereas a static rule is less likely to be changed. This information is used when generating the runtime representation of the rule to provide optimal performance.

The final component of the ETR approach is the trigger. Triggers are used to specify which event(s) causes the processing of a particular rule, as illustrated in the following:

```
       TRIGGER      trigger_name(parameter list)
  TRIGGEREVENT      set of event connected by OR
[ EVENTHISTORY      event expression]
    RULESTRUC      set of rules
[      RETURNS      return_type: rule_in_RULESTRUC]
```

The TRIGGEREVENT clause is used to specify the set of events, combined using an OR connective, that will cause the rule(s) specified in RULESTRUC to be evaluated. The event specification can be augmented using the EVENTHISTORY clause to define other event expressions that need to have occurred prior to the one defined in the TRIGGEREVENT clause. When specifying the rules to be triggered in the RULESTRUC clause, it is possible to combine several rules using one of 4 constructs: sequential (rules are triggered one after the other), parallel (rules are triggered concurrently), AND-synchronised (all members of a set of rules must complete evaluation before another, specified, rule is triggered), and OR-synchronised (any two members of a set of rules must complete evaluation before another, specified, rule is triggered).

The literature that discusses the ETR approach presents several applications of this technique, including the development of a knowledge management network [61] and of a dynamic business process management service described in [93].

Based on its similarity to the ECA rule approaches like PDL, it is easy to see how the ETR approach could be used to specify obligation policies in a distributed system. However, because of the manner in which events are defined and the ability to associate them to method invocations, it is also possible to specify authorisation policies, albeit less succinctly, using this notation. Additionally, by separating the event specifications from the rules, the ETR approach allows the user to reuse the events in multiple triggers and thus associate them with different rules as necessary. Despite the ability to specify different types of policy, and reuse parts of the specification in multiple rules, this approach does not support other useful features like policy extension (defining policies that inherit features from some parent policy) or policy groupings (organising policies that relate to the same activity together).

### 5.2. *Agent interaction policies*

The Foundation for Intelligent Physical Agents (FIPA) has defined a set of use cases and abstract architectural elements that can be used to guide the specification of policy mechanisms in concrete Agent Platform architectures [46]. The objective is to define constraints on resources that agents can use, other agents with which they can interact, what messages can be exchanged, what quality of service with respect to encryption, non-repudiation etc. should be used when interacting. Their use cases cover a wide range of authorisation, obligation, delegation scenarios, identify the need for policy composition and they are also looking at including concepts of conversation policies within the FIPA work on interaction protocols.

Conversations are sequences of messages involving two or more agents intended to bring about a particular set of (perhaps jointly held) goals, and conversation policies are declarative specifications that govern specific instances of communications between agents using an agent communication language (ACL). The conversation policies could be represented as sets of fine-grained constraints on ACL usage, which define the computational process models that are implemented in agents. The main reason for conversational policies are that they provide a level of analysis that abstracts from the actual propositional content, ACL, and implementation of individual conversations. Conversational policies also help ensure reliable communication between agents whilst simplifying any inference required in determining which communicative act or other action should be made in response to a message [46].

Currently the FIPA work is at a requirements analysis and concept definition stage and they have not defined any notations for policy specification. Other approaches to policy specifications that originate from the agent systems community include KaOS and Rei which are described in Section 6. A comparative evaluation of these approaches can be found in [97].

### 5.3. *Configuration management*

Policy-based management is also applied to *configuration management* and uses monitoring software to enable automation of network and system administration through the event–condition–action paradigm; policy-based configuration languages associate the occurrence of specified events or conditions, with responses to be carried out by an agent. *Cfengine* is a language-based administration system targeted primarily at Unix, and to a lesser extent Windows, operating systems connected via a TCP/IP network [25]. Cfengine grew out of the need to replace complex shell scripts used for the automation of administration tasks on Unix systems and allows the creation of single, central configuration file which describe how every host on the network should be configured. It uses the idea of classes to group hosts and dissect a distributed environment into overlapping sets. Host-classes are essentially labels which document the attributes of different systems. The following classes are meaningful in the context of a particular host: (i) the identity of the machine, including hostname, address, network, (ii) the operating system and architecture of the host (iii) an abstract user-defined group to which the host belongs (iv) the result of any proposition about the system, including the time or date. Policies are specified for classes of hosts and define a sequence of actions regarding the configuration of a host. The following example demonstrates the use of the language for configuration management [26]:

```
files:
    (linux|solaris).Hr12.OnTheHour.!exception\_host::
        /etc/passwd mode=0644 action=fixall inform=true
```

The first line simply defines the name *files* for the action. The second line identifies the target of the policy, i.e. all the hosts falling within the classification, the condition for execution of the policy, which is a time interval, and a trigger which specifies that the action must be executed *on the hour*. The command-line specifies that the cfengine agent, which is always the subject of the policy, must search for all password files with an invalid mode, fix them, and inform the administrator. The class membership expression specifies all hosts which are of type *linux* or *solaris*, during the time interval from 12:00 AM to 12:59 AM, apart from a host labelled with the class *exception_host*. Policies are stored in a central repository, accessible to every host, and an active cfengine agent on each host executes the policies which apply only to that host.

Cfengine is a powerful and concise declarative scripting language, suitable for system administrators to automates common administrative tasks on Unix systems. However, it cannot be used to specify authorisation policies and lacks support for object oriented concepts such as inheritance and parameterised instantiation. Its creators admit the need for extensions to enable enterprise-level policy specification [28].

Others, focus on the specification of policies using the full power of a general purpose scripting or interpreted language (e.g., TCL or Java) which can be loaded into network components or agents to implement policies. Such approaches are often leveraging the mechanisms in the area of active networks [94] to enable the control of resources at a very low level. For example Bos et al. [22] use C-programs to specify application policies for resource management in netlets, which are small virtual networks within a larger virtual network. In general for all of these approaches, the security concerns are increased, and malicious or improperly tested code can potentially damage the network. In addition, it is difficult to determine whether two computer programs specifying two different policies are contradictory or conflict with each other in any way. A comparison of different approaches to implementing policies as scripts can be found in [65].

### 5.4. *Law-governed interaction (LGI)*

*Law-governed interaction* [66] is another approach to defining policies for distributed heterogeneous systems. The motivation for this work is to provide a framework for controlling the flow of messages between multiple agents and thus control their interactions. In order to define the message delivery policy rules, referred to as *laws* in this approach, are specified using a simple Prolog notation. These rules are then interpreted by trusted agents, called controllers that act as proxies for every agent in the system.

In the LGI approach, the enforcement of the law is triggered by the occurrence of events that are specified as part of each law. Known as controlled events, these could occur at any point in time and the system described in the literature does not assume any a priori knowledge of how events may be generated by the systems. A law is defined as a prescription for the behaviour of the system when it is in a given state, known as the control-state, and a controlled-event occurs. This prescription for the behaviour of the system, referred to as the ruling of the law, is denoted by the sequence of primitive operations that must be performed when the law is enforced.

A law is specified with a set of rules that define the final ruling upon the occurrence of particular events in a given state. The LGI approach specifies a do(...) operation that is used to add primitive operations to the final ruling of the law. The following primitive operations and controlled-events are specified in the literature.

     sent(x, m, y): this event occurs when agent *x*, sends a message *m*, to agent *y*.

     forward(x, m, y): this operation is performed when the law being enforced by agent *x* rules that message *m* should be sent to agent *y*.

     arrived(x, m, y): this event occurs when message *m*, sent by agent *x*, arrives at agent *y*.

     deliver(x, m, y): this operation is performed when the law being enforced by agent *y* rules that the message *m*, sent by agent *x*, should be delivered to agent *y*.

Of course this set can be extended to suit the particular domain in which LGI is being applied. For example, a simplified token ring protocol law specification presented in [66]. As shown below, the set of primitive operations is extended to include recant(token), which causes the agent to clear the token property from its control state; and set(token), which adds the token to the control state of the agent. Additionally, a predicate next(D) is used which is defined to hold when D is the next agent in the ring. Note that the @ symbol is used to indicate that the predicate is to be evaluated with respect to the control state of the agent.

```
R1: sent(S,M,r) :- token@,do(forward(S,M,r)).
{To send a message to r a node must have the term
token in its cState}

R2: arrived(S,M,r) :- do(deliver(t4)).
{Any message that arrives at r is delivered.}

R3: sent(S,yourTurn,D) :- token@, next(D)@,
                          do(recant(token)),
                          do(forward(S,yourTurn,D)).
{The owner of the token can give it up by sending the
yourTurn message to the next object on the ring}

R4: arrived(S,yourTurn,D)
       :- do(set(token)),
          do(deliver(memo(yourTurn))).
{When the yourTurn message arrives at an object, the
token is added to it, and an appropriate memo is
delivered.}
```

In this specification, *R1* states that if the forward(...) operation is added to the ruling and the token is present in the control state of the agent, the sent(...) event is generated. *R2* is used to specify the policy that any message that arrives at a recipient agent is delivered. *R3* specifies that in order for the `yourTurn` message to be `sent(...)`, the destination D must be the next node in the ring, the current agent must have the token and the law must prescribe that the agent `recant(...)` the token and `forward(...)` the `yourTurn` message. Finally, *R4* specifies that the `arrived(...)` event is generated once the token is `set(...)` into the control state of the recipient agent and the `deliver(yourTurn)` operation has been added to the ruling of the law.

In this approach, permissions and prohibition are specified as a set of rules that are similar to positive and negative authorisations. Additionally, the approach supports a common global set of constraints, similar to obligation policies, which are implemented by means of filters in every node that check that all interactions are consistent with a global law [67]. There are other examples of the use of LGI systems to specify business rules in electronic commerce applications [99] and to support the provision of security policies in heterogeneous systems [98].

## 6. Policy-based management frameworks

In addition to the logic-based policy specification approaches described in the previous sections, a number of higher-level policy languages and frameworks have been developed to support a variety of management applications. These range from systems that provide policy-based services for distributed agent environments to those that use policies for network management. A notable feature that distinguishes these approaches from the formal ones described above is that, in most cases, they combine a policy specification language with a framework for specifying the properties of the managed system and a means of deploying and enforcing policies on the managed objects. In this section we present an overview of a number of these high-level languages and frameworks, and in each case discuss how they support policy analysis and refinement.

### 6.1. *Policy middleware for autonomic computing (PMAC)*

The policy middleware for autonomic computing (PMAC) framework has been developed as part of IBM's autonomic computing initiative [2]. The framework consists of a policy editing tool, a federator, an autonomic manager and a managed resource (Figure 6). Policies specified using the editor are published to the federator which acts as a publish/subscribe hub for distributing policies to the relevant autonomic managers. The principal component in this framework is the autonomic manager, which is responsible for providing a local policy-based control loop to manage the resource assigned to it. Managed resources provide two interfaces, referred to as *sensors* and *effectors*, which respectively represent the attributes that can be read from the resource and the management operations that can be performed to change the state of the resource. Since an autonomic manager can be treated as a managed component, it also provides an interface to its sensors and effectors for use by other autonomic managers.

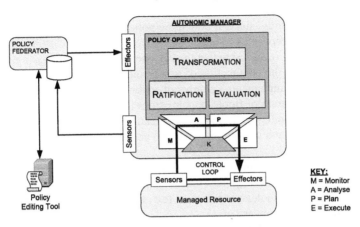

Fig. 6. Policy Management for Autonomic Computing (PMAC) framework.

Policies in the PMAC framework are specified as Event–Condition–Action (ECA) rules and their structure is defined using an XML schema. Additionally, the framework includes a general constraint specification language, also specified using XML, which can be used in the ECA policy rules to define the condition clause. The use of XML schema based definitions for the policy and constraint language allows the PMAC framework to extend the set of types supported by the language by simply including off-the-shelf XML definitions which come with pre-defined syntax and semantics. The problem with using an XML representation for policy is that the specifications tend to be quite verbose and not easily interpreted by the user. This problem is addressed in the PMAC framework by providing a *Simple Policy Language (SPL)* which supports the specification of ECA rules. At present, there is no documentation regarding the syntax of SPL, but it is expected to be similar to the obligation policy syntax of the Ponder framework.

PMAC provides support for policy analysis and conflict resolution using a process called *Policy Ratification* [3]. Essentially this process provides the administrator with information about the impact of a new policy on the existing set of policies. The ratification process consists for four domain-independent, generic operations: dominance checking, conflict checking, coverage checking and consistent priority assignment.

*Dominance checking* determines if there is a policy in the system that is subsumed (or dominated) by another. For example a policy that specifies that *'password length > 5'* is dominated by one that states *'password length > 8'* since the latter policy makes the former redundant. Generally it can be stated that a policy P1 is dominated by policy P2, if the constraint of policy P1 logically implies the constraint of P2.

One of the main challenges in policy analysis is in dealing with the constraints that control the applicability of a policy and identifying the situations in which two policies have constraints which are satisfied at the same time. The *Conflict checking* operation in PMAC is used to identify policies whose constraint statements are simultaneously true, and therefore may cause a problem if the actions defined in those policies are incompatible with each other. For example a policy that defines that "Joe has access to the database between 9 AM and 5 PM" would be in conflict with a policy that states "Joe should not be granted access to the database on weekends" since, if both these policies are enforced, Joe is both granted and refused permission between 9 AM and 5 PM at weekends. PMAC addresses the conflict detection requirement by recognising that the key operation in detecting a potential conflict is to determine whether the conjunction of two Boolean expressions can be satisfied. To this end, the framework implements a number of analytical algorithms for checking the satisfiability of different classes of Boolean expressions [3]. Whilst this approach is effective at detecting potential conflicts in a policy specification, since the analysis does not use a model of the system behaviour, it cannot take into account changes in system state caused by enforcing policies.

*Coverage checking* allows an administrator to ensure that for a given range of input parameters there is at least one policy that is applicable. This is achieved by checking whether a disjunction of Boolean expressions (e.g. the constraints of the policies) implies another (e.g. the input parameter range).

Finally, the PMAC framework assumes that problems identified through the above ratification operations will be resolved by the administrator marking policies as inactive or assigning some relative priority to them. In this latter case there is the possibility that introducing a new policy requires that existing priority values are reassigned so that the overall ordering of the policy rules remains consistent. The fourth ratification operation, *consistent priority assignments* automates this priority assignment process using a modified version of an algorithm used to insert items into an ordered list.

Policy refinement is supported in PMAC through the definition of transformation rules. These take the form "if certain conditions on a policy are satisfied (e.g., if the Role is WindowsSecurity) then update a certain portion of the policy with a new entity (e.g., replace the Condition and the Action by a new Boolean expression)". Since these transformation rules are implemented as policy rules that operate on the policies themselves, they can be executed using the same enforcement system as used by any other PMAC policy. One drawback of this approach is that the transformation rules have to be defined manually by an expert in the application domain and there is no means of verifying that the transformed policy satisfies the original goal of the user.

To summarise, the PMAC framework provides a language for specifying policy and a policy deployment and enforcement architecture. The system has been designed to be extensible through the use of XML schemas and is capable of policy analysis and refinement operations. However, the analysis process does not take into account the behaviour of the managed system and therefore cannot detect inconsistencies that are caused because the enforcement of one policy causes a state change such that some other policies then conflict. Finally, the policy refinement process is defined in terms of explicit transformation rules which must be specified by a domain expert. Therefore it is not possible to verify the correctness of the policies generated by applying the transformation rule with respect to the original policy.

## 6.2. KaOS

KaOS is a collection of component-based agent services developed to support a variety of mobile agent platforms [100]. Since its initial presentation KaOS has been adapted to general purpose grid computing and Web Services environments as well. The principle components of the KaOS framework are the domain service and policy service.

The KaOS policy service is responsible for specification, management, conflict resolution and enforcement of policies. In the initial versions of KaOS, policy specification takes an ontology based approach and policies are represented using the DAML[1] notation [40] but the language has since evolved to use OWL.[2] The ontology definition makes a distinction between authorisation policies and obligation policies and each of these policy types are specialised further using a modality operator. A *positive authorisation* policy defines actions that are permitted when given constraints hold true and a *negative authorisation* policy defines actions that are prohibited under given constraints. Similarly, *positive obligations* define a requirement that a given operation be performed whereas *negative obligations* waive the requirement to perform the action (cf. Ponder refrain policies). The basic policy ontology can be extended to support application specific concepts by defining management operations and their associated parameters.

The framework further comprises a KaOS Policy Administration Tool, that provides support for policy specification, manages ontologies, deploys policies and detects conflicts between policies in a specification. This addresses the problem associated with the verbosity of policies specified using the DAML + OIL notation, which can be hard for users to understand. This is illustrated by the example shown below, taken from [97] which specifies that members of a domain A are prohibited from performing encrypted communication with anyone outside this domain. In this example, the action is specified in lines 1–15 and the prohibition in lines 16–21.

```
1  <daml:Class rdf:ID="ExampleAction">
2     <rdfs:subClassOf rdf:resource
         ="#EncryptedCommunicationAction" />
3     <rdfs:subClassOf>
4        <daml:Restriction>
5           <daml:onProperty rdf:resource="#performedBy" />
6           <daml:toClass rdf:resource="#MembersOfDomainA" />
7        </daml:Restriction>
8     </rdfs:subClassOf>
9     <rdfs:subClassOf>
10       <daml:Restriction>
         <daml:onProperty rdf:resource="#hasDestination" />
11          <daml:toClass rdf:resource="#notMembersOfDomainA" />
         </daml:Restriction>
12    </rdfs:subClassOf>
13 </daml:Class>
14 <policy:NegAuthorizationPolicy rdf:ID="Example">
15    <policy:controls rdf:resource="#ExampleAction" />
16    <policy:hasSiteOfEnforcement rdf:resource=
               "#ActorSite" />
17    <policy:hasPriority>10</policy:hasPriority>
18.   <policy:hasUpdateTimeStamp>4237445645589
         </policy:hasUpdateTimeStamp>
19 </policy:NegAuthorizationPolicy>
```

---

[1] DARPA Agent Markup Language.
[2] Ontology Web Language.

KaOS provides support for detecting and resolving modality conflicts which can arise if two policies with opposite modalities specify the same actions and are enforced simultaneously. To achieve this, KaOS uses extensions to its integrated Java Theorem Prover (JTP) to implement an algorithm that is based on using the subsumption mechanisms between classes. Detected conflicts are then resolved through priority assignment and the automated creation of *harmonisation policies* [100].

Because the KaOS framework has a built-in theorem prover, it is also possible to perform other types of query (e.g. "which entities have read/write access to the examination marks database"). However, as in the case of PMAC, the policy analysis procedure does not account for the behaviour of the managed system and therefore is not able to detect conflicts due to changes in the system state.

The deployment architecture makes use of KaOS domain services to determine the entities to which a policy applies. Additionally, the deployment model has a *directory service* for storing policies, while *guards* interpret policies before passing them on to *enforcers* which translate the policy actions into platform-specific operations.

By using the idea of platform-specific enforcement components in the runtime architecture, KaOS policies can be used in a variety of applications. Indeed, KaOS has been applied in a range of application scenarios, ranging from web services management to grid computing. It supports authorisation and management/behavioural policies and provides tools for specifying these policies, analysing them and deploying them into the managed system. Its ontology-based approach allows application specific information about the managed system and operations to be added easily. However, whilst the analysis procedure does allow detection of modality conflicts and checking other properties, it does not account for policy interactions where enforcing one policy causes conflicts to arise between others. Finally, KaOS does not provide any support for policy refinement.

### 6.3. *Rei*

Rei was developed by Kagal et al., to support policy-based management in pervasive computing applications [57]. It provides three language constructs: policy objects, meta-policies and speech acts. Policy objects are defined as rights, prohibitions, obligations or dispensations which conceptually correspond to the positive and negative authorisations and positive and negative obligation policies described in KaOS and Ponder.

Policies are specified using an ontology-based approach to represent each policy type and therefore policy rules are specified in the verbose OWL representation [73], which is similar to the notation used by KaOS. However, Rei also provides a Prolog-based notation that is more compact and easier to understand. The basic ontology includes the description of actions in terms of an action identifier, the target objects on which the action can be performed, the set of pre-conditions required to perform the action and the post-condition of the action.

The listing below shows an example authorisation policy that specifies that any final year student can print on the colour printer. The policy is defined using the action *print-ColourPage* (Line 1) which has the pre-conditions that the printer has paper and has toner; and the post-condition that there is one less page in the paper tray. This is used in a has(Object, PolicyDef) predicate to specify that Students who are members of the finalYearGroup have the right to perform the action (Line 2)

```
1 action(printOnePageColour, [printerColour],
         (containsCartridge(printerColour),
          availablePaper(printerColour, X), X>1),
          availablePaper(printerColour, X-1))

2 has(Student, right(printOnePageColour,
                     isMember(Student,
                              finalYearGroup)).
```

The Rei system supports policy enforcement by providing a policy decision engine that deduces the rights and obligations of objects in the managed system in response to requests that specify the current state of the system. The use of Prolog to specify policy rules means that Rei can use a Prolog reasoning engine to analyse the policy specification and detect modality conflicts. Since the policy engine only supports deductive reasoning, only runtime conflicts can be detected. By using the ontology to define information about application specific actions, theoretically it should be possible for the analysis process to detect application specific conflicts. For example, an obligation to lock a door would conflict with another obligation that required the door be opened. However, in the literature there are no examples of such conflict analysis being performed on Rei policies.

As mentioned, the Rei language also includes the concepts of speech acts [82] and meta-policies. Speech acts are used to allow objects in the system to dynamically transfer rights and obligations to other entities. This is similar to the idea of delegation discussed in the literature. Meta-policies are rules about other policies and are used to resolve conflicts that are detected by the Rei policy engine at runtime.

Whilst Rei has been successfully used in a number of pervasive computing applications to provide policy-based access control functionality, it does have a number of limitations. Firstly, since the conflict detection method depends on deductive reasoning, it only works when a complete definition of the system state is provided, i.e. at runtime. Whilst this is useful, it would be better if administrators were able to detect potential conflicts in their policies at specification time so that they can minimise the risk of deploying a policy that causes a failure in the system. Other limitations are the lack of support for policy refinement and the absence of any tools to help administrators specify policy rules.

### 6.4. *Commercial frameworks*

The majority of policy-based management frameworks, and associated tools, are research prototypes and there are few examples of policy-based management being applied in a commercial environment. Verma [101] describes a QoS tool used to specify Service Level Agreements (SLAs) and to manipulate SLA related information in a tabular format. The tool transforms high-level policy information into device configurations, and stores them in an LDAP directory. Another tool, presented in [29], focuses solely on template-based refinement of policies from high-level goals.

Existing work within the RBAC community is limited to specifying access control configurations in terms of roles. A centralised tool, presented in [96], translates access control configuration from the RBAC framework to the target's native security mechanism, which is then transported to the target. Another web-based tool, presented in [14], allows administrators to specify roles, role hierarchies and constraints to implement RBAC for networked servers using Web protocols in order to manage access to an organisation's Web information.

In policy-based networking most of the tool support comes from industry and is based on the IETF policy framework. The majority of the commercial tools are specific to quality of service management, but many also include access control configuration. The list of vendor products is very big and space considerations prevent us from discussing each tool in detail. However, because these tools have a significant influence on the adoption of policy-based management solutions, we feel it is necessary to consider some of the major commercial policy-based network management products in greater depth.

It should be noted that the information presented here is based on public-domain documentation, available at the time of writing, from the vendors' website and industry surveys. For the latest information, surveys of commercial tools, together with product comparisons are available on the web (see http://www-dse.doc.ic.ac.uk/Research/policies for more information).

**6.4.1.** *HP openview PolicyXpert* HP's PolicyXpert tool is a multi-platform policy-based management solution, designed for integration into the company's Openview network management suite. In its current release, version 2.1, PolicyXpert supports traffic management actions ranging from priority marking to DiffServ code points. Like many of the other tools considered here, policies are defined using the *if <condition> then <action>* paradigm where conditions can be based on packet information, time of day or higher-level protocol information like HTTP URL or VLAN ID. The tool supports many of the prevalent standards, including COPS, DiffServ and RSVP.

**6.4.2.** *Cisco CiscoAssure* Cisco's policy-based management offering, CiscoAssure Policy Manager, is also aimed at QoS service management. Although policies are specified using the condition/action approach defined by the IETF-CIM standard, the tool policies themselves are stored in a flat-file database [35]. The user interface allows administrators to easily specify multiple conditions for triggering policies. Like the other tools considered here, conditions can be specified using a combination of IP addresses (source and destination), application ports, and the protocol being used (IP, TCP or UDP). Policy actions are applied to routers by using the Cisco Command Line Interface (CLI) language. Multi-vendor interoperability is provided with an implementation of COPS. In addition to supporting QoS related management operations, this tool allows the administrator to define access control policies for the devices being managed.

**6.4.3.** *Allot Communications NetPolicy* NetPolicy aims to provide policy-based management capabilities for a range of Allot Communication's network hardware in addition to

Cisco routers. However, results of tests performed on an early version of this tool concluded that the Cisco support was incomplete [6]. Once again, policies are specified using the condition/action notation, and the conditions can be defined in terms of the packet information parameters mentioned previously. The policy repository is implemented using LDAP and policy information is passed to target devices using either COPS or CLI. Additionally, NetPolicy supports management operations on simple access control lists.

## 7. Ponder policy framework

The Ponder language for specifying Management and Security policies [39] evolved out of work on policy management at Imperial College over a period of about 10 years. Ponder is a declarative, object-oriented language that can be used to specify both security and management policies. Ponder authorisation policies can be implemented using various access control mechanisms for firewalls, operating systems, databases and JAVA [37]. It supports obligation policies that are event triggered condition–action rules for policy-based management of networks and distributed systems. Ponder can also be used for security management activities such as registration of users or logging and auditing events for dealing with access to critical resources or security violations. Key concepts of the language include domains to group the object to which policies apply, roles to group policies relating to a position in an organisation [63], relationships to define interactions between roles and management structures to define a configuration of roles and relationships pertaining to an organisational unit such as a department.

### 7.1. *Domains*

Domains provide a means of grouping objects to which policies apply and can be used to partition the objects in a large system according to geographical boundaries, object type, responsibility and authority or for the convenience of human managers. Membership of a domain is explicit and not defined in terms of a predicate on object attributes. A domain does not encapsulate the objects it contains but merely holds references to objects. A domain is thus very similar in concept to a file system directory but may hold references to any type of object, including a person. A domain, which is a member of another domain, is called a *sub-domain* of the parent domain. A sub-domain is not a subset of the parent domain, in that an object included in a sub-domain is not a *direct* member of the parent domain, but is an *indirect* member, cf., a file in a sub-directory is not a direct member of a parent directory. An object or sub-domain may be a member of multiple parent domains, i.e. domains can overlap.

An advantage of specifying policy scope in terms of domains is that objects can be added and removed from the domains to which policies apply without having to change the policies. Domains have been implemented as directories in an extended LDAP Service.

## 7.2. *Ponder primitive policies*

*Authorisation policies* define what activities a member of the subject domain can perform on the set of objects in the target domain. These are essentially access control policies, to protect resources and services from unauthorised access. A positive authorisation policy defines the actions that subjects are permitted to perform on target objects. A negative authorisation policy specifies the actions that subjects are forbidden to perform on target objects.

The language provides reuse by supporting the definition of policy types to which any policy element can be passed as a formal parameter. Multiple instances can then be created and tailored for the specific environment by passing actual parameters as shown below.

```
type auth+ PolicyOpsT (subject s,
                          target <PolicyT> t) {
          action load(), remove(), enable(),
                                    disable() ; }

inst auth+ switchPolicyOps
  =PolicyOpsT(/NetworkAdmins, Nregion/switches);
inst auth+ routersPolicyOps
  =PolicyOpsT(/QoSAdmins, /Nregion/routers);
```

The two policy instances created from a PolicyOpsT type allow members of /NetworkAdmins and /QoSAdmins (subjects) to load, remove, enable or disable objects of type PolicyT within the /Nregion/switches and /Nregion/routers domains (targets) respectively.

Policies can also be declared directly without using a type as shown in the negative authorisation policy below, which indicates the use of a time-based constraint to limit the applicability of the policy

```
inst auth- /negativeAuth/testRouters {
  subject  /testEngineers/trainee ;
  action   performance_test() ;
  target <routerT> /routers ;
when time.between ("0900", "1700")
  }
```

Trainee test engineers are forbidden to perform performance tests on routers between the hours of 0900 and 1700. The policy is stored within the /negativeAuth domain.

Ponder also supports a number of other basic policies for specifying security policy: *Information filtering* policies can be used to transform input or output parameters in an interaction. For example, a location service might only permit access to detailed location information, such as a person is in a specific room, to users within the department. External users can only determine whether a person is at work or not. *Delegation* policies permit subjects to grant privileges, which they possess (due to an existing authorisation policy), to grantees to perform an action on their behalf, e.g., passing read rights to a printer spooler in order to print a file. *Refrain* policies define the actions that subjects must refrain from performing (must not perform) on target objects even though they may actually be permitted to perform the action. Refrain policies act as restraints on the actions that subjects perform and are implemented by subjects. See [39] for more details and examples of these policies.

Obligation policies are event-triggered condition–action rules, similar to Lucent's PDL, and define the activities subjects (human or automated manager components) must perform on objects in the target domain.

```
inst oblig loginFailure {
    on              3*loginfail(userid) ;
    subject         s = /NRegion/SecAdmin ;
    target <userT>  t = /NRegion/users  {userid} ;
    do              t.disable() -> s.log(userid) ;
}
```

This policy is triggered by 3 consecutive loginfail events with the same userid. The NRegion security administrator (SecAdmin) disables the user with userid in the /NRegion/users domain and then logs the failed userid by means of a local operation performed in the SecAdmin object. The '->' operator is used to separate a sequence of actions in an obligation policy. Names are assigned to both the subject and the target. They can then be reused within the policy. In this example we use them to prefix the actions in order to indicate whether the action is on the interface of the target or local to the subject.

Events can be simple, i.e. an internal timer event, or an external event notified by monitoring service components, e.g., a temperature exceeding a threshold or a component failing. Composite events can be specified using event composition operators.

### 7.3. *Ponder composite policies*

Ponder composite policies facilitate policy management in large, complex enterprises. They provide the ability to group policies and structure them to reflect organisational structure, preserve the natural way system administrators operate or simply provide reusability of common definitions. This simplifies the task of policy administrators.

*Roles* provide a semantic grouping of policies with a common subject, generally pertaining to a position within an organisation. Ponder roles include most of the functionality of RBAC roles described in Section 3, but include obligations. Specifying organisational policies for human managers in terms of manager positions rather than persons permits the assignment of a new person to the manager position without re-specifying the policies referring to the duties and authorisations of that position. A role can also specify the policies that apply to an automated component acting as a subject in the system, e.g., a security manager agent. Organisational positions can be represented as domains and we consider a role to be the set of authorisation, obligation, refrain and delegation policies with the *subject domain* of the role as their subject. A role is just a group of policies in which all the policies have the same subject, which is defined implicitly, as shown below.

```
type role ServiceEngineer (CallsDB callsDb) {
  inst oblig serviceComplaint {
      on    customerComplaint(mobileNo) ;
      do    t.checkSubscriberInfo(mobileNo, userid) ->
                  t.checkPhoneCallList(mobileNo) ->
                  investigate_complaint(userId);
      target t = callsDb ; // calls register }

  inst oblig deactivateAccount { . . . }
  inst auth+ serviceActionsAuth { . . . }
  // other policies
}
```

The role type ServiceEngineer models a service engineer role in a mobile telecommunications service. A service engineer is responsible for responding to customer complaints and service requests. The role type is parameterised with the calls database, a database of subscribers in the system and their calls. The obligation policy serviceComplaint is triggered by a customerComplaint event with the mobile number of the customer given as an event attribute. On this event, the subject of the role must execute a sequence of actions on the calls-database in order check the information of the subscriber whose mobile-number was passed in through the complaint event, check the phone list and then investigate the complaint. Note that the obligation policy does not specify a subject as all policies within the role have the same implicit subject.

Managers acting in organisational positions (roles) interact with each other. A *relationship* groups the policies defining the rights and duties of roles towards each other. It can also include policies related to resources that are shared by the roles within the relationship. It thus provides an abstraction for defining policies that are not the roles themselves but are part of the interaction between the roles. The syntax of a relationship is very similar to that of a role but a relationship can include definitions of the roles participating in the relationship. However roles cannot have nested role definitions. Participating roles can also be defined as parameters within a relationship type definition as shown below.

Many large organisations are structured into units such as branch offices, departments, and hospital wards, which have a similar configuration of roles and policies. Ponder supports the notion of *management structures* to define a configuration in terms of instances of roles, relationships and nested management structures relating to organisational units. For example a management structure *type* would be used to define a branch in a bank or a department in a university and then *instantiated* for particular branches or departments. A management structure is thus a composite policy containing the definition of roles, relationships and other nested management structures as well as instances of these composite policies, has some similarity to an Enterprise Community mentioned in Section 3.

```
type rel ReportingT (ProjectManagerT pm,
                     SecretaryT secr) {
   inst oblig reportWeekly {
        on              timer.day ("monday") ;
        subject         secr ;
        target          pm ;
        do              mailReport() ;
   }
   // . . . other policies
   }
```

The ReportingT relationship type is specified between a ProjectManager role type and a Secretary role type. The obligation policy reportWeekly specifies that the subject of the SecretaryT role must mail a report to the subject of the ProjectManagerT role every Monday. The use of roles in place of subjects and targets implicitly refers to the subject of the corresponding role.

Figure 7 shows a simple management structure for a software development company consisting of a project manager, software developers and a project contact secretary. Ref. [39] gives the definition of the structure.

```
type mstruct BranchT (...) {
   inst  role  projectManager = ProjectManagerT(...);
         role  projectContact = SecretaryT(...);
         role  softDeveloper = SoftDeveloperT(...);
   inst  rel supervise
               = SupervisionT (projectManager,
                                    softDeveloper);
         rel   report = ReportingT (projectContact,
                                         projectManager);
   }
   inst    mstruct branchA = BranchT(...);
           mstruct branchB = BranchT(...);
```

This declares instances of the 3 roles shown in Figure 7. Two relationships govern the interactions between these roles. A supervise relationship between the softDeveloper and the projectManager, and a reporting relationship between the ProjectContact and the projectManager. Two instances of the BranchT type are created for branches within the organisation that exhibit the same role-relationship requirements.

Ponder allows specialisation of policy types, through inheritance. When a type extends another, it inherits all of its elements, adds new elements and overrides elements with the same name. This is particularly useful for specialisation of composite policies. For example

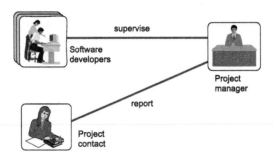

Fig. 7. Simple management structure.

it would be possible to define a new type of mobile systems project manager, from a project manager role with additional policies.

In Ponder a person can be assigned to multiple roles but rights from one role cannot be used to perform actions relating to another role. A person can also have policies that pertain to him/her as an individual and have nothing to do with any roles. In RBAC inheritance is based on policy instances and all policies are defined in terms of roles. This means RBAC requires a much more complicated role structure to separate the policies that are inherited from those that are private.

A compiler has been implemented for the Ponder language. Various back ends have also been implemented to generate firewall rules, Windows access control templates, JAVA security policies [37] and JAVA obligation policy rules for interpretation by a policy agent. We also have a system to automatically disseminate policies to the relevant agents that will interpret them i.e. to subjects for obligation and refrain policies and access control agents for authorisation and filter policies.

## 8. Policy analysis and refinement

Despite significant efforts in developing different policy specification techniques, there remain a number of issues to be addressed. In particular, since one of the objectives of using policy-based systems is to fulfil organisational goals, the ability to refine such goals into concrete policy specifications would be useful. As we have discussed in this chapter, it is desirable to maintain the properties of correctness, consistency and minimality when performing any refinement transformation. Of course, this is only possible if the chosen policy specification technique provides support for checking whether these properties hold. To this end, we have developed a mapping of policy specifications into a more formal, logic representation which is based in on the Event Calculus [60]. In order to exploit the properties of decidability and lower computational complexity, it is intended that the formal representation would be based on first-order stratified logic. This notation is then used in conjunction with the goal elaboration approach, developed by [42], to develop a usable policy refinement technique.

As part of solving the policy refinement problem, it will be necessary to address some of the outstanding issues related to policy analysis and conflict detection. Unless we have a means of checking for conflicts in a policy specification, it will be impossible to maintain the required consistency property in a given refinement. In Ponder modality conflicts arise between positive and negative policies that apply to the same subjects, targets and actions [64]. These can be detected by syntactic analysis of the policies as the conflict can be determined by detecting overlap of subjects, targets and actions. However, the analysis detects only potential conflicts rather than actual conflicts since constraints may limit the applicability of the policy to disjoint sets of circumstances, e.g., different times of day. While modality conflicts can be detected syntactically, other conflicts can only be determined by understanding the actions being performed by the policies. For example, there will be a conflict between two policies that result in the same packet being placed on 2 different queues. Similarly, separation of duty conflicts arise from authorisation policies, which permit the same person to approve payments and sign cheques. Generally, these conflicts are application specific and to detect them it is necessary to specify the conditions that result in conflict. The approach is therefore to specify constraints on the set of policies (i.e. meta-policies) using a suitable notation and then analyse the policy set against these constraints to determine if there are any conflicts [64]. Whether conflicts occur or not may depend on run-time parameters specified in constraints such as time or the current state of the components to which the policy apply. It is thus rather difficult to determine all possible conflicting conditions in advance and so it is still necessary to detect conflicts at run-time. Furthermore, when conflicting policies are detected it is not obvious how to resolve the conflicts automatically. Explicit priority may work in some cases. In some situations, negative authorisation policies should override positive ones, but in other situations the positive authorisation is an exception to a more general negative authorisation. In some situations more specific policies that apply to a department may override general policies applying to the whole organisation. We have been experimenting with meta-policies that define application specific precedence relationships between conflicting policies.

Although some progress has been made in dealing with policy conflicts [3,64,102], significant challenges remain to be addressed. In particular, how can one detect conflicts when arbitrary conditions restrict the applicability of the policies? Sometimes, it is possible to compare restrictions placed by the constraints. For example, it is possible to detect if two time intervals overlap or if the policies apply when subjects are in different states, e.g., active or standby. Other challenges concern the different levels of abstraction at which policy is specified. Conflicts between organisational goals will inevitably lead to conflicts between the policies derived from these goals. Some policies will trigger complex management procedures, which require the execution of actions that may be specified as part of different policies. This renders the task of ensuring the consistency of a policy specification much more complex.

Recent efforts on formalising the specification of policies and the behaviour of managed systems have gone some way to addressing the challenges of policy analysis and refinement [9]. This work proposes a formalism that is based on the standard Event Calculus [60] that models both authorisation and management policy specifications together with the behaviour of the managed system. Event Calculus was chosen as both the policies and the management behaviour we are modelling are event driven. Additionally, since an Event Calculus specification of a system can be generated from a state transition model, users can specify the management behaviour using a familiar high-level notation. In a similar fashion, it is possible to translate policies specified in a high-level policy specification language, like Ponder [39], into an Event Calculus representation that describes the semantics of the policy language. This eliminates the need for the user to become conversant with the details of logic programming and the Event Calculus notation.

In order to analyse the policy specification, the Event Calculus representation supports the specification of rules for detecting a range of consistency properties. This includes modality and application specific conflicts such as conflicts of duty together with policy validation. Additionally, the formal representation supports a number of types of review query, allowing the administrator to obtain different views of the information in the policy specification.

This analysis technique uses a combination of deductive and abductive reasoning. Deductive techniques are primarily used for policy validation and to perform review queries whereas abduction is used to detect conflicts between policies. By using abductive reasoning techniques, it is possible to analyse the policy specifications to identify existing conflicts and provide explanations on how they might arise. Because the abduction process is applied to a specification that models both the systems behaviour and the policy specification it is possible to detect conflicts when the applicability of the policies is constrained on the runtime state of the system. Furthermore, by using abduction, the analysis can be performed even with partial specifications of the system state. Having detected conflicts, it is also necessary to have mechanisms for resolving them. Statically defined resolution techniques such as negative policy overrides, and priority assignments have been suggested in the literature [3,64]. Such approaches suffer from problems of managing consistency in priority assignments and lack of flexibility, making them difficult to use in practise. In order to address this problem, recent research has investigated the application of preference logics for expressing conflict resolution policies in the context of firewall rules [58]. Whilst showing some promise, this approach requires further study before it can be adopted in policy-based management systems.

The Event Calculus formalisation of policy-managed systems can be combined with the KaOS goal elaboration technique [42], to provide a framework for policy refinement. However, the low-level goals derived using this technique cannot be directly used in policies without first identifying the management operations that will achieve them. This identification process is supported by the introducing the concept of a *strategy*. A strategy is the mechanism, by which a given system can achieve a particular goal, i.e., a strategy is the relationship between the system description and the goal. By having a formal specification of the latter two types of information abductive reasoning is used to infer the strategy. The KaOS goal elaboration technique is used because it provides the concept of domain-specific and domain-independent refinement patterns, logically proven goal refinement templates that can be easily reused. Such patterns capture the refinement of goals that are commonly encountered in policy-based management, thus simplifying the refinement process for the user. Additionally the refined policies derived using this process can be encoded in policy refinement patterns that can be later reused when the administrator wishes to satisfy a similar goal. This approach has been applied to refining policies in the context of network QoS management [10].

## 9. Research issues

Policies can be used to support adaptability within many different administrative domains. For example a distributed video conferencing application between cooperating organisations in multiple countries may require application specific policies to be implemented within multiple network service providers. There is thus a need to define representations and techniques from transferring policy between the organisations. However the problem is more complex than just one of policy representation. The policies may require knowledge of network topology but an application will not have this information and even one service network provider will not have this information about another service provider. A similar problem occurs when an application or service needs to adapt to changes in an underlying service which it is using. There is thus a need to negotiate a common policy at a fairly high level of abstraction, i.e. common goals which have predefined mappings to local policies, as it would be unrealistic to expect goals to be automatically mapped into implementable policies. Very little work has been done on policy negotiation but it is possible that some of the concepts from agent plan negotiation may be applicable.

A policy-based management system may itself need to adapt to changes in the system being managed due to failures, new situations arising or new application requirements arising. An example would be a security management system which adapts the current policies when it detects that an attack is in progress. This may require policies which dynamically modify the current policies within the system e.g. by activating or deactivating installed policies or selecting new policies to install in the system, thus the target objects on which policies act are other policies and policy interpreters. The next stage in this type of policy feedback loop would be to have a form of adaptive policy which 'learns' the best strategy to use, but this is very far off at present.

The whole area of ubiquitous and mobile computing introduces many issues of policy related to security and privacy, particularly when these systems are able to track a user's activity and location. Techniques are needed for users to be able to specify the conditions under which their identity may be known and what credentials should be revealed for authentication. However in many circumstances users may not wish their current location and activity to be monitored by a ubiquitous computing environment, although these requirements may change in emergency situations. Ubiquitous computing environments provide many new applications for obligation policies. For example how to deal with messages or information being sent to you based on current context – whether it should be forwarded to someone else to deal with, transformed from text to voice or voice to text, what information should be filtered etc. and intelligent environments could detect activity and perform actions on your behalf to control the temperature, switch on appliances or generate warning messages about dangers.

The main problem with the current tools and techniques for specifying policy, whether logic based or otherwise, is that they require considerable computing expertise. None of them are really designed for non-computing literate people to use in order to define the policies they require with respect to security or how to interact with their ubiquitous environment. Even if policy programmers predefine groups of policies for specific roles or situations, there will be a need to adapt these generic templates to specific requirements or to be able to select from a range of those available. It is only recently that research efforts have been directed towards addressing the issues related to these human-centric views of policy specification [16]. However, there is still considerable work to be done in this area.

## 10. Summary

In this chapter we have presented an overview of the available specification approaches for security, management and enterprise collaboration policies. Policy-based approaches to systems management are gaining widespread interest because they allow the separation of the rules that govern the behavioural choices of a system from the functionality provided by that system. This means that it is possible to adapt the behaviour of a system without the need to recode any of the underlying functionality; and changes can be applied without the need to stop and restart the system. Such features allow administrators to manage systems in a very flexible manner and also provide a potential mechanism for realising the goal of autonomic management of systems.

A common theme across the policy specification notations presented here is that they are focused in a single functional area – routing, access control or management not a combination of them. Broadly speaking, many of the security policy specification languages are based on logic-based approaches whereas management and enterprise collaboration policy specification notations (with some exceptions) adopt a more informal approach. Whilst it is possible to implement access control policies in the form of ECA rules for the reference monitor, this is not a convenient way of represented authorisations and few systems do it in practise. Ponder is one of the few languages that allows both security and management policies to be represented in a declarative form. Additionally, it supports policy templates that can be configured with application specific parameters as needed and meta-policies that can be used to modify the behaviour of a policy at run-time. Ponder also allows policies to be organised according to roles, management structures and groups. Recent work on the Ponder framework has focussed on redesigning the system to allow policies to be deployed and executed on resource constrained computing platforms, such as the mobile devices typically found pervasive computing environments [45].

At present, authorisations and event–condition–action rules are the predominant paradigms used in policy-based management, although the latter come in slightly different flavours. Implementation platforms have slowly matured in recent years and increasingly work focuses on formal analysis and refinement of policies. As policy-based techniques are increasingly used to provide adaptation in pervasive systems, ad-hoc networks and collaborations across administrative domains, recent work such as the Promise Theory [27] also focuses on the collaborative aspects of adaptive systems.

Although an essential technique for providing adaptation and management, policies are only one side of the story. To manage large-scale systems and increase the degree of automation they must work in conjunction with other techniques such as advanced metrics, utility functions and planning-based approaches. There has been some very initial work on these integrated techniques.

## Acknowledgements

The authors' research described in this chapter is continuing through participation in the International Technology Alliance sponsored by the U.S. Army Research Laboratory and the U.K. Ministry of Defence.

## References

[1] M.D. Abrams, *Renewed understanding of access control policies*, Proceedings of 16th National Computer Security Conf., Baltimore, MD, USA (1993), 20–23.

[2] D. Agrawal, S. Calo, J. Giles, K.-W. Lee and D. Verma, *Policy management for networked systems and applications*, Proceedings of 9th IFIP/IEEE International Symposium on Integrated Network Management, IEEE, Nice, France (2005).

[3] D. Agrawal, J. Giles, K.-W. Lee and J. Lobo, *Policy ratification*, Proceedings of 6th IEEE International Workshop on Policies for Distributed Systems and Networks, IEEE, Stockholm, Sweden (2005).

[4] G.-J. Ahn and R. Sandhu, *The RSL99 language for role-based separation of duty constraints*, Proceedings of Fourth ACM Workshop on Role-Based Access Control, ACM Press, Fairfax, VA, USA (1999).

[5] C.E. Alchourron, *Normative Systems*, New York, Wien (1971), xvii+208.

[6] Allot Communications, *Policy based networking architecture*, http://www.allot.com/media/ExternalLink/policymgmt.pdf (2001).

[7] R.J. Anderson, *A security policy model for clinical information systems*, Proceedings of IEEE Symposium on Security and Privacy, Oakland, CA, USA (1996).

[8] K.R. Apt, H.A. Blair and A. Walker, *Towards a Theory of Declarative Knowledge*, Foundations of Deductive Databases, J. Minker, ed., Morgan Kaufmann, San Mateo, CA (1988), 89–148.

[9] A.K. Bandara, *A formal approach to analysis and refinement of policies*, Ph.D. Thesis, Imperial College London, London (2005).

[10] A.K. Bandara, E.C. Lupu, A. Russo, N. Dulay, M.S. Sloman, P. Flegkas, G. Pavlou and M. Charalambides, *Policy refinement for IP differentiated services quality of service management*, IEEE e-Transactions in Networks and Systems Management **3** (2) (2006).

[11] S. Barker, *Security policy specification in logic*, Proceedings of Int. Conf. on Artificial Intelligence (ICAI00), Las Vegas, NV, USA (2000).

[12] S. Barker, *Access control policies as logic programs*, Technical Report, Imperial College of Science, Technology and Medicine, London (2001).

[13] S. Barker and A. Rosenthal, *Flexible security policies in SQL*, Proceedings of Fifteenth Annual IFIP WG 11.3 Working Conf. on Database and Application Security, Niagara on the Lake, Ontario, Canada (2001).

[14] J. Barkley, R. Kuhn, L. Rosenthal, M. Skall and A. Cincotta, *Role-based access control for the Web*, Proceedings of CALS Expo Int. & 21st Century Commerce 1998: Global Business Solutions for the New Millennium, Long Beach, CA, USA (1998).

[15] J.F. Barkley, K. Beznosov and J. Uppal, *Supporting relationships in access control using role based access control*, Proceedings of Fourth ACM Workshop on Role-Based Access Control, Fairfax, VA, USA (1999).

[16] R. Barret, *People and policies: Transforming the human–computer partnership*, Proceedings of 5th IEEE International Workshop on Policies for Distributed Systems and Networks, IBM TJ Watson Research Centre, IEEE Press, New York, USA (2004).

[17] D.E. Bell and L. LaPadula, *Secure computer systems: Mathematical foundations and model*, Technical Report M74-244, MITRE Corporation, Bedford, MA (1973).

[18] E. Bertino, P. Bonatti and E. Ferrari, *TRBAC: A temporal role-based access control model*, Proceedings of 5th ACM Workshop of Role-Based Access Control, Berlin, Germany (2000).

[19] K.J. Biba, *Integrity constraints for secure computer systems*, Technical Report ESD-TR76 -372, USAF Electronic Systems Division, Bedford, MA (1977).

[20] M. Blaze, J. Feigenbaum, J. Ioannidis and A.D. Keromytis, *The role of trust management in distributed systems security*, Secure Internet Programming: Security Issues for Mobile and Distributed Objects, Springer-Verlag, New York, NY, USA (1999), 185–210.

[21] M. Blaze, J. Feigenbaum and A. Keromytis, *Keynote: Trust management for publc-key infrastructures*, Proceedings of Security Protocls Int. Workshop, Springer-Verlag, Cambridge, England (1998).

[22] H. Bos, *Application-specific policies: Beyond the domain boundaries*, Proceedings of Sixth IFIP/IEEE Int. Symposium on Intergrated Network Management (IM'99), Boston, MA, USA (1999).

[23] A. Boswell (1995), *Specification and validation of a security policy model*, IEEE Transacations on Software Engineering **21** (2) (1995).

[24] D.F.C. Brewer and M.J. Nash, *The Chinese wall security policy*, Proceedings of IEEE Symposium on Research in Security and Privacy, IEEE, Oakland, California, USA (1989).

[25] M. Burgess, *A site configuration engine*, USENIX Computing Systems **8** (3) (1995).

[26] M. Burgess, *Recent Developments in CfEngine*, Proceedings of Unix NL Conf., The Hague (2001).

[27] M. Burgess, *An approach to understanding policy based on autonomy and voluntary cooperation*, Proceedings of 16th IFIP/IEEE Distributed Systems: Operations and Management (DSOM 2005), Springer-Verlag, Barcelona, Spain (2005).

[28] M. Burgess and F.E. Sandnes, *Predictable configuration management in a randomized scheduling framework*, Proceedings of IEEE/IFIP Workshop on Distributed Systems Operations and Management (DSOM'2001), Nancy, France (2001).

[29] M. Casassa Mont, A. Baldwin and C. Goh, *POWER prototype: Towards integrated policy-based management*, Technical Report, HP Laboratories Bristol, Bristol, UK (1999).

[30] H. Castaneda, *The Paradoxes of Deontic Logic*, New Studies in Deontic Logic: Norms, Actions and the Foundations of Ethics, Reidel Publishing Company, Hingham, MA, USA (1981), 37–85.

[31] A. Chandra and D. Harel, *Horn clause queries and generalizations*, Journal of Logic Programming **2** (1) (1985), 1–5.

[32] F. Chen and R.S. Sandhu, *Constraints for role-based access control*, Proceedings of First ACM/NIST Role Based Access Control Workshop, ACM Press, Gaithersburg, MD, USA (1995).

[33] L. Cholvy and F. Cuppens, *Analyzing consistency of security policies*, Proceedings of IEEE Symposium on Security and Privacy (S&P97), IEEE Press, Oakland, CA (1997).

[34] J. Chomicki, J. Lobo and S. Naqvi, *A logic programming approach to conflict resolution in policy management*, Proceedings of 7th Int. Conf. on Principles of Knowledge Representation and Reasoning (KR2000), Morgan Kaufmann, Breckenridge, CO, USA (2000).

[35] Cisco, *CiscoAssure policy networking: End-to-end quality of service*, Cisco Systems (1998).

[36] D.D. Clark and D.R. Wilson, *A comparison of commercial and military computer security policies*, Proceedings of IEEE Symposium on Security and Privacy (1987).

[37] A. Corradi, R. Montanari, E. Lupu, M. Sloman and C. Stefanelli, *A flexible access control service for JAVA mobile code*, Proceedings of Annual Computer Security Applications Conf. (ACSAC 2000), IEEE Press, New Orleans, LA, USA (2000).

[38] F. Cuppens and C. Saurel, *Specifying a security policy: A case study*, Proceedings of Ninth IEEE Computer Security Foundations Workshop, IEEE Press, Co. Kerry, Ireland (1996).

[39] N. Damianou, N. Dulay, E.C. Lupu and M.S. Sloman, *The ponder policy specification language*, Proceedings of 2nd IEEE International Workshop on Policies for Distributed Systems and Networks, Springer-Verlag, Bristol, UK (2001).

[40] DAML, *The DARPA agent markup language*, http://www.daml.org (2000).

[41] E. Dantsin, T. Eiter, G. Gottlob and A. Voronkov, *Complexity and expressive power of logic programming*, Proceedings of 12th Annual IEEE Conf. on Computational Complexity (CCC'97), IEEE Press, Ulm, Germany (1997).

[42] R. Darimont and A. van Lamsweerde, *Formal refinement patterns for goal-driven requirements elaboration*, 4th ACM Symposium on the Foundations of Software Engineering (FSE4) (1996), 179–190.

[43] DMTF, *Common information model (CIM) specification, version 2.2* (1999).

[44] DMTF, *Specification for the representation of CIM in XML, version 2.0* (1999).

[45] N. Dulay, S. Heeps, E.C. Lupu, R. Mathur, O. Sharma, M.S. Sloman and J. Sventek, *AMUSE: Autonomic management of ubiquitous e-health systems*, Proceedings of UK e-Science All Hands Meeting 2005 (2005).

[46] FIPA, *Domains and policies specification*, http://www.fipa.org/specs/fipa00089/ (2001).

[47] J. Glasgow, G. Macewen and P. Panangaden, *A logic for reasoning about security*, ACM Transactions on Computer Systems (TOCS) **10** (3) (1992), 226–264.

[48] T. Grandison, *Trust management for Internet applications*, Ph.D. Thesis, Imperial College London, London (2003).

[49] T. Grandison and M. Sloman, *A survey of trust in internet applications*, IEEE Communications Surveys and Tutorials **3** (4) (2000).

[50] R.J. Hayton, J.M. Bacon and K. Moody, *Access control in an open distributed environment*, Proceedings of IEEE Symposium on Security and Privacy, Oakland, CA, USA (1998).

[51] A. Herzberg, Y. Mass, J. Michaeli, D. Naor and Y. Ravid, *Access control meets public key infrastructure, or: Assigning roles to strangers*, Proceedings of IEEE Symposium on Security and Privacy, Oakland, CA, USA (2000).

[52] ISO/IEC, *Information technology – open distributed processing reference model – enterprise viewpoint* (1999).

[53] G. Jager and R.F. Stark, *The defining power of stratified and hierarchical logic programs*, Journal of Logic Programming **15** (1&2) (1993), 55–77.

[54] S. Jajodia, P. Samarati, M.L. Sapino and V.S. Subrahmanian, *Flexible support for multiple access control policies*, ACM Transactions on Database Systems **26** (2) (2000), 214–260.

[55] S. Jajodia, P. Samarati and V.S. Subrahmanian, *A logical language for expressing authorisations*, Proceedings of IEEE Symposium on Security and Privacy, IEEE, Oakland, USA (1997).

[56] A.J.I. Jones and M.J. Sergot, *A formal characterisation of institutionalised power*, Logic Journal of the IGPL **4** (3) (1995), 429–445.

[57] L. Kagal, *Rei: A policy language for the me-centric project*, Technical Report HPL-2002-270, HP Laboratories, Palo Alto, CA, USA (2002).

[58] A. Kakas, A.K. Bandara, A. Russo, E.C. Lupu, M.S. Sloman and N. Dulay, *Reasoning techniques for analysis and refinement of policies for service management*, Technical Report DoC # 2005/7, Imperial College London, London (2005).

[59] M. Kohli and J. Lobo, *Policy based management of telecommunication networks*, Proceedings of Policy Workshop 1999, HP Labs, Bristol, UK (1999).

[60] R.A. Kowalski and M.J. Sergot, *A logic-based calculus of events*, New Generation Computing **4** (1986), 67–95.

[61] M. Lee, *Event and rule services for achieving a Web-based knowledge network*, Ph.D. Thesis, University of Florida (2000).

[62] J. Lobo, R. Bhatia and S. Naqvi, *A policy description language*, Proceedings of 16th National Conf. on Artificial Intelligence, Orlando, FL, USA (1999).

[63] E.C. Lupu and M.S. Sloman, *Towards a role based framework for distributed systems management*, Journal of Network and Systems Management **5** (1) (1997), 5–30.

[64] E.C. Lupu and M.S. Sloman, *Conflicts in policy-based distributed systems management*, IEEE Transactions on Software Engineering – Special Issue on Inconsistency Management **25** (6) (1999), 852–869.

[65] M. Martinez, M. Brunner, J. Quittek, F. Strauß, J. Schönwälder, S. Mertens and T. Klie, *Using the script MIB for policy-based configuration management*, Proceedings of IEEE/IFIP Network Operations and Management Symposium (NOMS 2002), IEEE, Florence, Italy (2002).

[66] N.H. Minsky, *The imposition of protocols over open distributed systems*, IEEE Transactions on Software Engineering **17** (2) (1991), 183–195.

[67] N.H. Minsky and P. Pal, *Law-governed regularities in object systems – part 2: A concrete implementation*, Theory and Practice of Object Systems (TAPOS) **3** (2), Wiley (1997), 87–101.

[68] J. Moffett, *Control principles and role hierarchies*, Proceedings of Third ACM/NIST Role Based Access Control Workshop, ACM Press, Fairfax, VA, USA (1998).

[69] B. Moore, E. Ellesson, J. Strassner and A. Westerinen, *Policy core information model – version 1 specification*, Network Working Group – RFC3060, http://www.ietf.org/rfc/rfc3060.txt (2001).

[70] OASIS, *XACML language proposal, version 0.8* (2001).

[71] OMG, *Object constraint language specification, version 1.3* (1999).

[72] R. Ortalo, *A flexible method for information system security policy specification*, Proceedings of 5th European Symposium on Research in Computer Security (ESORICS 98), Louvain-la-Neuve, Belgium, Springer-Verlag (1998).

[73] OWL, *The ontology Web language*, http://www.w3.org/TR/owl-features/ (2004).

[74] H. Prakken and M.J. Sergot, *Dyadic deontic logic and contrary-to-duty obligations*, Defeasible Deontic Logic: Essays in Nonmonotonic Normative Reasoning, D. Nute, ed., Synthese Library, Vol. 263, Kluwer Academic Publishers, Boston (1997), 223–262.

[75] C. Ribeiro, A. Zuquete and P. Ferreira, *SPL: An access control language for security policies with complex constraints*, Proceedings of Network and Distributed System Security Symposium (NDSS'01), San Diego, CA (2001).

[76] C. Ribeiro, A. Zuquete and P. Ferreira, *Enforcing obligation with security monitors*, Proceedings of Third Int. Conf. on Information and Communications Security (ICICS 2001), Xian, China (2001).

[77] P. Samarati and S. Vimercati, *Access control: Policies, models, and mechanisms*, Foundations of Security Analysis and Design (Tutorial Lectures), R. Focardi and R. Gorrieri, eds, Springer-Verlag (2000), 137–196.

[78] R. Sandhu, D. Ferraiolo and R. Kuhn, *The NIST model for role-based access control: Towards a unified standard*, Proceedings of 5th ACM Workshop on Role-Based Access Control, Berlin, Germany (2000).

[79] R. Sandhu and P. Samarati, *Access control: Principles and practice*, IEEE Communications Magazine **32** (9) (1994), 40–48.

[80] R.S. Sandhu, *Role activation hierarchies*, Proceedings of Third ACM/NIST Role Based Access Control Workshop, ACM Press, Fairfax, VA, USA (1998).

[81] R.S. Sandhu, E.J. Coyne, H.L. Feinstein and C.E. Youman, *Role-based access control models*, IEEE Computer **29** (2) (1996), 38–47.

[82] J.R. Searle, *Speech acts: An essay in the philosophy of language*, Cambridge, Cambridge University Press (1969).

[83] M.J. Sergot, F. Sadri, R.A. Kowalski, F. Kriwaczek, P. Hammond and H.T. Cory, *The British Nationality Act as a logic program*, Communications of the ACM **29** (1986), 370–386.

[84] M.S. Sloman, *Policy driven management for distributed systems*, Journal of Network and Systems Management **2** (4) (1994), 333–360.

[85] Y. Snir, Y. Ramberg, J. Strassner, R. Cohen and B. Moore, *Policy QoS Information Model* (2001).

[86] J.M. Spivey, *An introduction to Z and formal specifications*, IEE/BCS Software Engineering Journal **4** (1) (1989), 40–50.

[87] M.W.A. Steen and J. Derrick, *Formalising ODP enterprise policies*, Proceedings of 3rd Int. Enterprise Distributed Object Computing Conf. (EDOC'99), IEEE Publishing, University of Mannheim, Germany (1999).

[88] M.W.A. Steen and J. Derrick, *ODP Enterprise viewpoint specification*, Computer Standards and Interfaces **22** (2000), 65–189.

[89] G.N. Stone, B. Lundy and G.G. Xie, *Network Policy languages: A survery and a new approach*, IEEE Network (2001), 10–20.

[90] J. Strassner, *Policy-Based Network Management*, Morgan Kaufmann, San Francisco (2004).

[91] J. Strassner and E. Ellesson, *Terminology for describing network policy and services (version 00)* (1998).

[92] J. Strassner, E. Ellesson, B. Moore and R. Moats, *Policy Core LDAP Schema* (2002).

[93] S.Y.W. Su, H. Lam, M. Lee, S. Bai and Z.-J.M. Shen, *An information infrastructure and e-services for supporting Internet-based scalable e-business enterprises*, Proceedings of 5th IEEE Annual Enterprise Distributed Object Conf. (EDOC2001), IEEE Computer Society, Seattle, WA (2001).

[94] D.L. Tennenhouse, J.M. Smith, W.D. Sincoskie, D.J. Wetherall and G.J. Minden, *A survey of active network research*, IEEE Communications Magazine **35** (1) (1997), 80–86.

[95] R.K. Thomas, *Team-based access control (TMAC): A primitive for applying role-based access controls in collaborative environments*, Proceedings of Second ACM/NIST Role Based Access Control Workshop, ACM Press, Fairfax, VA, USA (1997).

[96] D. Thomsen, D. O'Brien and J. Bogle, *Role based access control framework for network enterprises*, Proceedings of 14th Annual Computer Security Applications Conf. (1998).

[97] G. Tonti, J. Bradshaw, R. Jeffers, R. Montanari, N.S. Suri and A. Uszok, *Semantic Web languages for policy representation and reasoning: A comparison of KAoS, Rei, and Ponder*, Proceedings of 2nd International Semantic Web Conference, Springer-Verlag, FL, USA (2003).

[98] V. Ungureanu and N.H. Minsky, *Unified support for heterogeneous security policies in distributed systems*, Proceedings of 7th USENIX Security Symposium, San Antonio, TX, USA (1998).

[99] V. Ungureanu and N.H. Minsky, *Establishing business rules for inter-enterprise electronic commerce*, Proceedings of 14th Int. Symposium on Distributed Computing (DISC 2000), Springer-Verlag, Toledo, Spain (2000).

[100] A. Uszok, J. Bradshaw, R. Jeffers, N. Suri, P. Hayes, M. Breedy, L. Bunch, M. Johnson, S. Kulkarni and J. Lott, *KAoS policy and domain services: Toward a description-logic approach to policy representatin, deconfliction and enforcement*, Proceedings of 4th IEEE Workshop on Policies for Networks and Distributed Systems (Policy 2003), IEEE, Lake Como, Italy (2003).

[101] D. Verma, M. Beigi and R. Jennings, *Policy based SLA management in enterprise networks*, Proceedings of Policy Workshop 2001, HP Labs, Springer-Verlag, Bristol, UK (2001).

[102] D.C. Verma, *Policy-Based Networking: Architecture and Algorithms*, New Riders Publishing (2001).

[103] A. Virmani, J. Lobo and M. Kohli, *Netmon: Network management for the SARAS softswitch*, Proceedings of 2000 IEEE/IFIP Network Operations and Management Seminar (NOMS 2000), Hawaii (2000).

[104] G.H. von Wright, *Deontic logic*, Mind **60** (1951), 1–15.

[105] R.J. Wieringa and J.-J.C. Meyer, *Applications of deontic logic in computer science: A concise overview*, Proceedings of Practical Reasoning and Rationality (PRR 98), Wiley, Brighton, UK (1998).

# 5. Computational Theories of System Administration

## 5.1. *Commentary*

The notion of computational complexity is one of the classic topics in computer science. It attempts to determine the algorithmic resources needed to search through a set of conceivable solutions to a problem. Applying this perspective to system administration is a relatively new idea. The chapters presented here contain the results of recent research in this area, and much remains to be done.

In Burgess and Kristiansen's chapter, we find a general introduction to complexity theory to help bridge the cultural gap between system administration researchers and classical algorithmic computer science. The authors sketch out the problem of complexity and what it means to the field. They also derive results for one interpretation of the complexity of system administration by searching for complete (already-composed) operations with predictable final outcomes satisfying certain properties. They find that these algorithms can be solved efficiently in memory but not in time. In the worst case a verified solution is most likely harder than NP!

Sun and Couch formulate a very similar problem using a different formalization while obtaining similar results. They consider stepwise operations that have repeatable outcomes, and then consider the composition of these to solve system requirements. They find that the problem of finding a suitable composition of operations towards a reliable outcome is NP hard.

Bergstra and Bethke close by connecting predictability of specification to programming by studying program semantics. The issue of semantics haunts the fringes of system administration in several ways. Reliable scripting is an important idea for system predictability and hence security, but there is also a deeper sense in which programming semantics enter: system managers really want to specify system behaviors not program code or configurations, but they can only do this through programming or configuration interfaces, hence these must be predictable. Programs and configurations are only stepping stones; the relationship between intention and result is often tenuous or unclear. Bergstra and Bethke discuss the problems surrounding programming semantics and hence address the fundamental mapping between design and behavior.

# – 5.2 –

# On the Complexity of Change and Configuration Management

Mark Burgess[1], Lars Kristiansen[1,2]

[1] *University College Oslo, Faculty of Engineering, Room 707, Cort Adelers Gate 30, NO-0254 Oslo, Norway*
*E-mail: mark.burgess@iu.hio.no*

[2] *Department of Mathematics, University of Oslo, PO Box 1053, Blindern, NO-0316 Oslo, Norway*
*E-mail: larsk@math.uio.no*

## 1. Introduction

How hard is it to build and tune a computer system so that one can predict its operational behaviour from an original specification?

In computer programming, the boundaries of this problem have been considered at length (see the chapter by Bergstra and Bethke in this volume, for instance). In system administration, the framework for the problem is both larger and potentially much less reliable. System administration is not only a set of procedures executed on a computer, it also involves the interaction of humans in a more more pervasive way. It raises questions of a pilosophical nature: are we really in control of computer behaviour, or are there limitations to what we can achieve? In computer programming, one does not normally take into account the random arrival of input, or the interactions between independent tasks which share common resources. These are matters that are of central importance in system administration.

Suppose we were indeed able to rule or govern the behaviour of a computer system, by choosing the values of all the right registers, how hard would it be to find out what values or patterns of data were required to map the specification of those values to a desired

HANDBOOK OF NETWORK AND SYSTEM ADMINISTRATION
Edited by Jan Bergstra and Mark Burgess

behaviour? How much effort would we need to put in to solve an arbitrarily posed problem of this kind? What are our chances of being able to optimize it, or simply do it at all?

These are questions we discuss in this chapter. To begin formulating an answer we must cross several bridges, from empirical system administration and more formal computational disciplines such as languages and algorithmic complexity. We discuss what it means to talk about the difficulty of arranging a particular computer configuration and why we believe the choice of configuration has something to do with its ultimate behaviour.

There are many measures one might use to discuss the complexity of a solution to a problem: the number of lines of code required to solve the problem, the number of hours of planning needed, the CPU cycles expended during the task, the memory resources used to run the algorithms, etc. All of these are potential measures of difficulty or complexity, but to answer the question in a comprehensible way, we have to pose the question in a way that can be analysed.

None of the measures above have any meaning to system administration until we can identify a set of acceptable tests and determine the answer to a question: does the solution pass the test? Yes or no? So, how shall we proceed? Our story is a complicated one; it requires both a broad background and lengthy argumentation. To make it comprehensible to the widest possible audience, we have chosen to use a pedagogical style with a moderate amount of introductory background. We hope that this makes our discussion as self-contained as possible. Perhaps half of the material will be known to about half our audience, but we doubt that these halves will overlap.

We base our discussion on classical computational complexity theory [15,20] and the modern theory of configuration management in terms of operations [7,12,13]. Many readers will have heard of phrases like "NP complete" used to describe hard problems, and understand that therein lie dæmons and dragons concerning the difficulty of a task. But what does this really mean? Similarly, readers will have heard about the problem of configuration management and computer maintenance. Why is this any different from computer programming? We begin from the context of computer management.

## 1.1. Configuration and systems

The problem of configuring (and sometimes maintaining) a computer system from an original specification has been discussed in the literature in a variety of settings [5–7,9,11–13]. Much software has been written over the years purporting to solve the problem in some manner [3,17,21,27,29].

We understand intuitively that a *configuration* is an assemblage of components and parts that makes up a complete system. The term appears both in system adminisration and in software engineering with analogous meanings. According to the IEEE software engineering glossary [18], "Configuration Management is the process of identifying and defining the items in the system, controlling the change of these items throughout their life-cycle, recording and reporting the status of items and change requests, and verifying the completeness and correctness of items".

Although the goal of 'configuration management' in software engineering is somewhat different than in system administration, this definition still fits quite well the general idea for system administration too, but let us be more specific.

Suppose we start with a blank computer and install an operating system from source. We are now able to switch on the computer and give it instructions, but this is rarely enough to make it into a useable device within some organization or home. In addition, we must add software, set a number of defaults and preferences such as how to print and how to access information from a network, and then we must consider what users may log in, share data, what permissions resources must have, which services we should run from the host or make use of from outside.

In biology, this is analogous to the procedure in which generic stem cells are differentiated into dedicated bodily cells in a growing foetus. In other words, what will the purpose of the computer be? And how should its running state be managed so that it will fulfill that purpose for the time we require and the specific (or unspecific) list of users who require it? The task of configuration management is to turn our expectations of behaviour into a number of operations on the working computer such that it will fulfill our needs.

This all sounds eminently reasonable, but can it really be done? Of course, arranging the pieces of a puzzle is a straightforward task of writing data to a disk or memory device, so what is the problem? The problem, it turns out, is not so much implementing these patterns themselves, but in finding them and in deciding what they mean.

## 1.2. *How to get there from here*

Suppose we know what policy, configuration, state, or pattern of data, we want to implement on a computer, i.e. we have solved a problem and found a specification that we want to implement. We must know translate this desired state into a set of *configuration operations* (or operators). There are then two ways of proceeding with the operations we need for configuration implementation. They are roughly analogous to procedural programming and logic programming [10].

- We specify the starting point, e.g. a blank computer, and we specify a sequential (imperative) programme of steps to transform it into the desired state.
- We specify the final state and leave it up to the (declarative) program to compute any suitable algorithm for getting there.

In neither case is it guaranteed that we shall be able to reach the configuration state we had in mind at the start, but let us set aside that problem for now.

Consider an example. If we fix a bit string, like a file mode, using a numerical value, there is little ambiguity in the procedure. It seems like a straightforward problem to write some configuration to a computing device (e.g. `chmod 755 filename`). It is straightforward, easy, uncomplex. There is little difference between the two alternatives for such a trivial change.

But now consider the expression of a high level behaviour, e.g. `InstallPackage(webserver)`. This is no longer straightforward, because it describes a configuration change only at a medium level, not all the way down to the bits. This is like saying: "Make me prettier!". It is not a uniquely defined or reproducible goal. We therefore need to define this at a low level – and now it does matter which of these two approaches we choose. Someone might tell you that it is their goal to make you prettier, but you cannot guarantee their behaviour from this assertion. You might trust them more, if they told you about what end result they were going to guarantee.

An imperative beautician might specify: first I will put blue on your eyes, then I will put red on your lips, and finally I will take a centimetre off your hair. The declarative beautician might say: I will make sure that your eyelids are blue and your lips are red and that your hair is 10 cm long. In the first case, the operation could go drastically wrong if the configuration object was wearing sun-glasses and only has 2 cm long hair to begin with.

The problem is that, if we base our actions on assumed *pre-conditions*, rather than on desired *post-conditions*, procedures are fragile and one needs to be certain about that initial configuration. See the chapter by Sun and Couch in this volume for a detailed discussion of the imperative approach.

### 1.3. *Static or dynamic configuration?*

There is another more important complication to this simple viewpoint: the view described above is too static. A computer system is not static like a picture or painting, it is more like a board-game with players who are changing the state of a board at all times. The configuration problem described above is only analogous to setting out the pieces at time $t = 0$. Once we let the computer run, interact with users and change its operational state, the pieces move and the configuration changes [5]. Do we want them to move? Who are we playing against? Is it the system versus the users, or is it a maintenance agent versus gremlins? We need to understand why the system is changing and how much of it we want to keep constant.

In fact, we do not necessarily care about the exact pattern of every byte and register on the computer – only certain patterns will affect the behaviour that we are interested in, others will have no effect on it. For instance, if we decide to run a Web service that responds withing a maximum of 5 milliseconds, we do not need to specify the contents of every text file and picture file on the computer in order to achieve that goal. There will be a few key files that affect the functionality; all other files are irrelevant (symmtrical with respect to the goal of making a Web service). Indeed, our specification of the behavioural configuration does not usually extend to the *data content* of user files and images – that is determined by a different set of policies, specifications or goals.

There is thus a *separation of concerns*, between different parties and their interests in the state of the system which complicates the matter in system administration. In a computer program, there is one overriding programmer that decides authoritatively over a domain and all of its subdomains. In system administration, the separation is less clear.

Separation of concerns is a common problem in computer science, and it is generally handled by encapsulating independent concerns in independent data structures using some kind of high level coding of the computer's state. Object Orientation and Aspect Orientation are examples of this in computer programming. At the level of system resources and their administration, this is partially achieved using data structures abstractions like *files*, *directories* and *processes*. However, there is no operating system in use today that achieves a clean separation of operational and configuration concerns, or even complete separation between subsystems. This makes system administration more difficult and partially explains the currrent revival of interest in virtual machine technologies which allow individual services to be isolated.

### 1.4. *Goals, policies and behaviours*

An important development which changed thinking about configuration management was the notion of policy based management [3,25,26] (see Chapter 4.3 by Bandara et al. in this volume).[1]

We understand policy as some supposed catalogue of decisions, whose outcomes have been pre-determined, like a code-book of behaviours. When an expected situation arises, there is already an answer to the problem encoded in policy and we merely have to look up the appropriate behaviour. This idea is quite general and has been applied in many arenas, both for runtime behaviour and configuration choices. A policy can be thought of as a somewhat advanced notion of *preferences*, in which the semantics of certain preferences have been defined in advance.

In configuration management, policy allowed a move away from explaining configuration in the imperative terms of an algorithmic recipe of *configuration operations*, towards the specification of declarative guidelines for the end goal itself. This conceptual leap was important because it refocused the attention on final behaviour rather than intermediate state. Policy about the final state is easily associated with post-condition based configuration. The semantics of post-condition policy are easier to fix than an attempt to construct a scenario by imperative recipe.

Of course, by focusing on goal-oriented, post-condition policies one does not remove the need to find an algorithm for implementing the end result. However, this approach furnishes us with a neat framework for discussing the *complexity* of solving a configuration problem. We can make a much more direct attack on the problem of configuration management's algorithmic complexity by studying ways of achieving a fixed final result, satisfying certain properties.

## 2. Configuration operations as an automated form of system administration

### 2.1. *The fundamental hypothesis*

What does configuration management entail? It is more than simply implementing changes. Picking some arbitrary configuration for a system is easy (one could roll a die or set all values in the system to zero, etc.), but picking a configuration that satisfies a highly particular set of constraints could be a very non-trivial request – and policy, after all, is all about setting constraints on what is allowed or desired.

The often unstated assumption about configuration management is the following (see Figure 1):

---

[1] In fact this was two separate developments at approximately the same time which later came together into a common conceptual framework. The term Policy Based Management is due to Sloman et al.

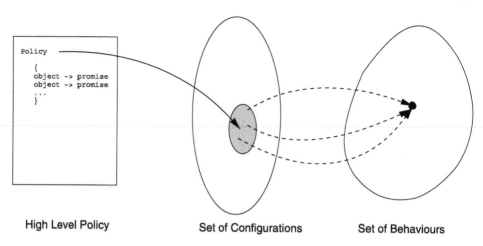

**High Level Policy**                     **Set of Configurations**                     **Set of Behaviours**

Fig. 1.  The main hypothesis in configuration management is that there is an association between high level policy, low level configuration and eventual behaviours.

HYPOTHESIS 1 (*Policy guided behaviour*).  A high level policy can be used to determine a number of equivalent low level configurations on a computer system. There will then be a direct association between a "correctly configured computer" and a "correctly behaving computer", where "correct" means "policy or specification compliant".

It should be stated straight away that this hypothesis remains to be proven. Most system administrators have a gut feeling that it is approximately true, but there is no sense in which we have a formal proof of this. The reason is that computer systems are not simply deterministic automata. They receive unpredictable input from users through their input/output channels, and the software that executes on the systems is not usually considered to be a part of the configuration itself, thus the behaviour of the system is not fully determined.

In practice what one normally means by this hypothesis is that, for a constant configuration, the system will behave *predictably* on average. However, it is important to note an important objection: it is far from clear that an arbitrary policy will be stable – or by implication, that an arbitrary configuration will be *sustainable* over time, without a sufficiently invasive maintenance process.

An approximate understanding of the hypothesis has been developed in ref. [5], using a post-condition view of configuration. There it was shown that, by limiting the set of definable policies to those that can be faithfully represented and maintained over time, one can associate probable computer behaviour with an initial policy.

$$\text{Policy} \quad \mapsto \quad \frac{\text{Configuration}}{\text{Equivalences}} \quad \mapsto \quad \text{Behaviour.} \tag{1}$$

We read this as follows. Any system can be understood as a balance between the freedoms to change and the constraints which limit that change [6]. A policy implies a number of constraints which partially determine a configuration of the system up to certain equivalences (which are implied by the freedoms that policy does not constrain) and it is further assumed that those equivalence lead to equivalent behaviour.

This fulfils the intention of the hypothesis, but it does so by restricting the notion of policy in order to lead to make it true. If we want sustainable behaviour, it seems that it can be achieved for a subset of possible policies for which this property holds, but we do not really know how to find that subset.

There will always be fluctuations in a configuration, due to environmental effects, and these might be relevant or irrelevant to the observed behaviour. Thus the best one can hope to sustain (by correcting policy deviant errors continuously) is a time-average behaviour. In effect, the model likens configuration management to Shannon's problem of error correction in a noisy channel [2,23].

An alternative approach to studying the problem has been suggested in ref. [13], based on a pre-conditional model of configuration changes. In that case, there are no such restrictions on policy other than reachability from an initially known state. In that case, it is quite unclear whether there is a predictable behaviour for every possible composition of configuration operations. The task is then to weed out those reproducible behaviours that one would like to see.

The hypothesis has characteristics akin to the general problem of programming semantics (see chapter by Bergstra and Bethke), in which the policy is a program, the configuration is an intermediate code that is then executed on a (virtual) machine. The main difference here is that the execution machine is not necessarily either constant or reliable over the time interval for which policy is expected to be maintained. This is partially due to the presence of interactions with the external environment [5]. Hence we must verify each of the outcomes by explicit measurement to be certain of the persistence of the resulting behavioural promises.

### 2.2. *Configuration defined*

We discuss now the essence of configuration management in terms of a specific model, using the idea of patterns and configuration operators (or functions) [6,7]. The description of discrete patterns is well known in computer science as the theory of formal grammars [19]. Any pattern must be formed from a number of distinct symbols (the set of which is called the *alphabet* $\Sigma$ of the language), and the specific rules to which a pattern conforms determine its *language L*. Patterns divide the set of languages into four levels of grammatical complexity known as the Chomsky hierarchy [19], which we shall not discuss further in this essay. We then define a general configuration as follows:

DEFINITION 1 (*Configuration*). A configuration is a discrete pattern that is exhibited by a computer system. It is a string belonging to some language $L$, with alphabet $\Sigma$.

In this essay, we shall ultimately confine ourselves to the simplest binary alphabet consisting of the symbols $\Sigma = \{0, 1\}$, from which we can code any other. However, we must acknowledge that there are many levels of coding that are represented in a computer system.

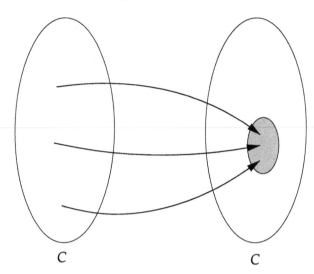

Fig. 2. A change in configuration is generally a mapping from some general configurations into a subset of policy-compliant configurations.

A configuration can, in other words, be thought of as a complicated state: it is an instance of a pattern $q$ found in a specified region of system's resources. We think of any such region as being a system 'variable', such as the contents of a register, a file or an abstract container such as a Process Control Block inside a process dispatcher.

These definitions are clear enough, but they are too abstract to say much about configuration management, as we understand it. What we are really interested in is the ability to describe *changes* of configuration, i.e. the reorganisation of resources into new patterns. To make the discussion fit for analysis, we need a formal representation for that too.

A change from one configuration state $q$ to another state $q'$ takes place as a result of an *operation*, and in the current paper we will view as one iteration of a function or mapping

$$f : \mathcal{C} \to \mathcal{C}, \tag{2}$$

where $\mathcal{C}$ is the set of all possible configurations (see Figure 2). A *policy* is then, according to (1), a specification of some range of configurations that we want to allow [5], and the set of policy compliant configuration $\mathcal{P}$ will be a (normally very small) subset of $\mathcal{C}$, i.e. $\mathcal{P} \subseteq \mathcal{C}$.

If we want to maintain the eqivalence between actual state and the idealised state implied by policy, we need to repair or maintain it (perhaps repeatedly). We can think of a repair strategy as being an operation (or function) $f_r$ which, when repeated a sufficiently number of times, will take the system from an arbitrary configuration to a policy compliant configuration, that is, for all sufficiently large $k \in \mathbb{N}$ we have[2]

$$f_r^k(x) \in \mathcal{P} \quad \text{for any } x \in \mathcal{C}. \tag{3}$$

---

[2] $f^k(x)$ denotes the result of repeatedly applying the function $f$, i.e. $f^0(x) = x$ and $f^{n+1}(x) = f^n(f(x))$.

Before discussing the details of such (configuration) operations, we must spend a section clearing up the role played by configuration coding, or transformations of alphabet, in the complexity of configuration management.

However, before leaving this section, we must briefly note another assumption that we are making about configuration. We are not being completely clear about the distinction between *configuration* and *state*. The concept of state is fairly well understood in computer science, particularly in relation to the theory of languages and automata. A working computer contains state in every part of its memory, but it would be wrong to think of a real computer as a closed Turing machine that deterministically rewrites the contents of all its possible states according to a fixed plan.

In configuration management, we usually only try to determine a part of the computer's total state – that part which contains the installed software and its customisations, settings and preferences. Such state is recorded in software and in configuration files and databases as options on their behaviours. We do not typically imagine trying to control the internals of different programs' states, or model the operational history of processes that learn from their past experiences, e.g. customer transactions or the contents of databases that we do not believe have any *direct* affect on the behaviour of the system. Such operational state does however enter into the observation of a system, as a reflection of its behaviour.

In a sense, we glibly assume that the state of the system can be divided into two parts: a part that affects behaviour, and a part that is changing, but which has no direct influence on behaviour.

HYPOTHESIS 2 (*Separation of state*). A computer's total state can be divided into two parts: configuration state which determines behaviour, and operational state which is merely the result of behaviour.

As with the first hypothesis, it is not clear that this must always be true. What about the contents of software itself? Software performs operations on the state of the system of a kind that we are not even modelling, despite being not so fundamentally different from configuration operations. Software is, in a real sense, a part of configuration state as it affects behaviour, and yet we do not control its contents or behaviour in detail.

The hypothesis above is, in fact, directly analogous to the division one makes in describing the physics of dynamical systems [6]. It is a separation of state into 'slow' and 'fast' changing variables. What we are really postulating is the existence of two separable *time-scales* – a slow time-scale over which humans can make high level decisions, and a fast time-scale at which the computer performs its own manipulations. We hope that these are separable, else the model breaks down.

The interface between the fast and the slow is software. In practice, we only decide to install or not install certain packages of software, and there is nothing to stop a program from using the contents of fast operational state for affecting and modifying behaviour, which we then try to regulate by making slow changes of configuration. Clearly, we are assuming an idealization in order to make limited progress.

We shall suppress further discussion of this point below and focus only on the supposed slowly-changing *configuration* or *configuration state*. Part of the real problem with configuration management, and the difficulty in fully believing in our first hypothesis derives from the uncertainties described here.

### 2.3. *The importance of coding*

System administration is performed at a high level of abstraction, but the aim of configuration management is to eventually map these high level concerns into precise operations on the resources of the computer [4–6]. By a *resource* we mean either an abstract data object or a hardware capability, represented through its various control registers.

The basic configuration states of a computer system and its resources are therefore coded within its memory. At a fundamental level, the memory takes the form of a finite array of *bits*, of length $n$ whose configuration value at any given moment is a single instance of the set of all such strings $\mathbb{B}^n$, to which we return presently. This is the *configuration space* of the system. We denote this lowest level representation of the resources as $R^1$, in anticipation of higher level codings to follow.

Written onto the configuration space of bits is a pattern of values, zero or one, which changes over time; thus, a configuration may be defined as follows: A level 1 configuration $q_1(x, t)$, on $R^1$, is a pattern of Boolean values associated with each element of the configuration space, at a fixed snapshot in time $t$:

$$q_1 : R^1 \to \{0, 1\}. \tag{4}$$

Readers can follow this formalization in more detail in ref. [5]. A configuration at some point changes in time as a *chain*, or *process* [16].

Although we understand that computer systems are bit-configurations, operated on by low level rules, the viewpoint ignores an important issue, which is the layered coding of information used to represent user-level data. Most data and processes require quite complicated representations that require one to deal with high level objects such as files (containing sub-languages) and directories. It is at these levels that both the usage and the administration of the system take place, so it is this level to which one must address a model of administration.

We accomplish the transition from low level to high level by successive levels of grouping smaller objects into larger ones, such as the hierarchy of objects expressed in Table 1. For simplicity, one can assume that all the objects have the same size. This is not the case in practice, but the assumption serves to avoid unnecessary additional specification and does not alter the argument, only the details.

These items represent effective properties of the system, i.e. new alphabets of non-overlapping objects. At level $\ell$, the system may be considered $\ell$-dimensional, since there are up to $\ell$ independent degrees of freedom. (A degree of freedom is a capacity for independent change.)

Suppose we make $R^\ell$ the set of $D_\ell$ objects of level $\ell$, and let $Q_\ell$ be the set of $d_\ell$ bits within each element of $R^\ell$ (see Figure 3). A high level coding of the configuration space $R^1$, is then understood simple as a mapping $L$, which coarse-grains a set of $D_\ell$

Table 1

A separation of scales in a computer system. At a certain level, human users begin to interact with the system, and thus form part of it

| Level | Example objects |
| --- | --- |
| 6 | Groups, departments, LANs |
| 5 | Users, virtual machines, agents |
| 4 | Compound objects, files and attributes |
| 3 | int, float, char etc. |
| 2 | bytes, words |
| 1 | bits (RAM, ROM, disk) |

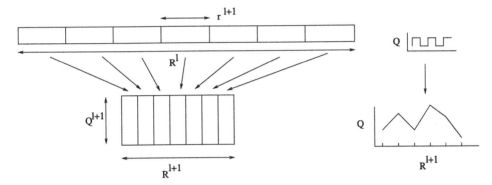

Fig. 3. A partitioning of the configuration space into higher level objects, is a mapping from one to two dimensions. The vertical scale is the range of the coding level; as the level increases so does the range of values taken by objects at that level. For the purpose of discussing rates of change, these values can be represented as numerical values.

lower level objects into $D_{\ell+1}$ higher level objects, of size $d_{\ell+1}$ [5]:

$$L_\ell : R^\ell \to R^{\ell+1}. \tag{5}$$

Now, these higher level codings are constructed in much the same way that languages are constructed from grammar rules [22]. For example, most structures can be captured by the notion of a context free grammar $G = (N, \Sigma, P, S)$, where $N$ is an alphabet of non-terminal symbols, $\Sigma$ is an alphabet of terminal symbols, $S$ is a starting symbol from $N$ and $P$ is a set of production schemas. Each $p \in P$ has the form $n \to \omega$, where $n \in N$ and $\omega \in (N \cup \Sigma)^*$. Context free grammars can be extended to allow a general regular expression over the total alphabet in place of $\omega$. We shall use this idea later in Section 4.1.

A high level configuration is thus no longer viewed in terms of elemental bits, but in terms of high level objects such as numbers, symbols or file objects. The potential range of values represented by such objects increases exponentially in proportion to their size. In practice, the values used by different software systems will be considerably less than this.

This mapping takes bit patterns into a higher alphabet of effective symbols. Thus the set of all possible configurations $\mathcal{C}$ referred to earlier is an ambiguous quantity in terms of symbols, since it depends on the level at which we choose to view it. The only unambiguously general way of discussing it is to view it in terms of its atomic bits, i.e. to replace $\mathcal{C}$ in the discussion with $\mathbb{B}$.

It is natural to ask the question: how much of the complexity of configuration management arises from the coding itself? If the coding is the major part of the complexity then the problem is artificially hard. We shall discuss this further below, but one can immediately be sceptical of this idea. If transforming the coding of the system itself were a hard problem, in the sense of algorithmic complexity then it would never have been implemented on computers, as it would take too long. But these computers exist and work more or less well. Thus is seems unlikely that the coding could have an important bearing on the matter. We return to this in Section 4.1.

### 2.4. *Some examples*

Notation can be an important cultural bridge. In this essay, we work with two different notations so that readers coming from different backgrounds will recognise the text. In the definitions below, we describe an operator expression often as a tuple which gives the output as a number of expressions that show how the input is combined to yield the output. We also write this as a vector $\vec{\alpha}$. The variables $x_n$ also form a vector $\vec{x}$.

An alternative approach which can be used is to express the operator as a matrix which then multiplies a vector tuple of inputs and whose product yields the output vector. In refs [5–7] the notation $\widehat{O}_f$ is used for operators. Here we shall keep to a representational notation, such that, if we think of a function $f$ then this has an operator representation $[\![f]\!]$; thus an operator expression from these references of the form

$$\vec{\alpha} = \hat{e}\vec{x}. \tag{6}$$

would be written

$$\langle \alpha_1, \ldots \rangle = [\![e]\!]\left(\langle x_1, \ldots, x_n \rangle\right) \tag{7}$$

here. A potential limitation of the matrix form used below and in refs. [5–7] is that we must have $\dim(\vec{\alpha}) = \dim(\vec{x})$, but this is sufficient to discuss appropriate fragments of the conifguration.

**2.4.1.** *Web server configuration*   An example of a configuration $Q \in \mathcal{C}$, which has no natural expression at the level of bits is the configuration of a web service, using a piece of software, e.g. an HTTP web sever.

A web server is a piece of software whose behaviour is partly progammed, which partly depends on the operating system (or virtual machine) on which it runs, and which admits a number of alternative behaviours that are configurable in it 'configuration file', which is an ASCII file called something like `httpd.conf`. This file contains directives such as the following examples:

```
Port 80
HostnameLookups on
LoadModule access_module mod_access.so
LoadModule auth_module mod_auth.so
```

These lines tell the software which communications port to use, whether or not it should try to look up clients in a host-name database and which additional behaviour-modifying modules to load and use.

A policy might constrain this configuration by asking two promises from the file:

- A web server promise to use port 80.
- A web server promise that `HostnameLookups` will be off.

In the latter case, the excerpt above would have to be edited in order to change value of the second line. For this we need some specialized operator that brings the system back into line with the promise compliant state $p$

$$[\![\text{HostnameLookups(off)}]\!]\,\vec{q} = \vec{p} \tag{8}$$

by editing the file. This operator should then be able to search through the file's internal representation for the line concerned, read it, check its value and, if necessary, rewrite it so that it complies with the policy constraint. The assumption then is that this so-configured text will cause the web server to behave as advertised, i.e. stop trying to resolve the names of hosts.

This example is just one of many similar cases in use today. The current trend in modern computing systems is to use context free languages, such as XML, to represent configuration information in a more structure form. The principal of is the same, but the language belongs to a different place in the Chomsky hierarchy, that of context free languages. This leads to additional computation.

Before leaving this example, notice a matter of some importance about causality. The configuration decisions cached and catalogued in this file apply only to the web server process that reads and uses it, but the resulting behaviour of the web server can affect many other parallel processes in the system – and not just those that use its services. The use of port 80 might block its use by another program. The choice to look up host names in a directory could place a burden on load on the directory service. In other words, it is far from clear that the main hypothesis (that correct policy leads to correct configuration leads to correct behaviour) is true if one follows all of the causal threads of this case.

**2.4.2.** *Example: a trivial fragment in* $\mathbb{B}^3$     Consider now the particular example of Unix file permissions. Under Unix, file access rights have a rather simple representation in terms of bit strings. A user's file permissions are stored in three bits 'r', 'w' and 'x', standing for read, write and execute access rights respectively. (Similar bits are used for groups and for other users, but it is sufficient to consider the user or owner of a file here.) Here are some examples:

```
rwx
100 Read only
110 Read-write
```

```
010 Write only
001 Execute only
```

If one of these bits is set to '1', permission to perform the appropriate action is granted, otherwise it is denied. The operator which changes these permissions is contained within the Unix command 'chmod': the command '+r', for instance, would set the 'r' bit, whereas the command '−r' would unset it. We need a notation to describe this. Here we shall use the notation $[\![X]\!]$ to denote a representation of the operator $X$:

$$[\![+\mathrm{r}]\!] \quad \Longleftrightarrow \quad \text{'chmod}+\mathrm{r'}. \tag{9}$$

These operations can be represented as *operators*, but how shall we represent them?

Consider first the bit strings that represent the configuration of a file object. Here are two different ways of representing these bit strings: a *vector* representation

$$\begin{pmatrix} 1 \\ r \\ w \\ x \end{pmatrix} = \begin{pmatrix} 1 \\ 1 \\ 1 \\ 0 \end{pmatrix} \tag{10}$$

and a representation as a tuple of *Boolean expressions*

$$\langle r, w, x \rangle = \langle 1, 1, 0 \rangle. \tag{11}$$

In the vector case, there is an extra '1', whose role becomes apparent during the matrix multiplication below. What do the operators look like to, say, open a file object for reading?

Regardless of representation, such an operator must set the 'read bit', while leaving the others unchanged. For the first representation, we can construct this by matrix multiplication. We write this in all its notations:

$$\vec{p} = [\![+\mathrm{r}]\!](\vec{q}) = [\![\text{chmod}+\mathrm{r}]\!](\vec{q}) = \begin{pmatrix} 1 & 0 & 0 & 0 \\ 1 & 0 & 0 & 0 \\ 0 & 0 & 1 & 0 \\ 0 & 0 & 0 & 1 \end{pmatrix} \begin{pmatrix} 1 \\ r \\ w \\ x \end{pmatrix} = \begin{pmatrix} 1 \\ 1 \\ w \\ x \end{pmatrix}. \tag{12}$$

We start with generic values $r, w, x \in \{1, 0\}$ and end up with policy conformant values. The 'r' bit is set to 1, regardless of its previous value, and all other bits are kept the same.

**2.4.3.** *File permissions*   For the second representation, we can use a logical OR operation:

$$\begin{aligned} \langle p_r, p_w, p_x \rangle &= \text{chmod}+\mathrm{r}\big(\langle r, w, x \rangle\big) \\ &= [\![+\mathrm{r}]\!]\langle r, w, x \rangle \\ &= \langle r, w, x \rangle \,\text{OR}\, \langle 1, 0, 0 \rangle \\ &= \langle 1, w, x \rangle. \end{aligned} \tag{13}$$

Representation of data lies at the heart of modern computing. Digital computers use bits at the bottom-most layer and programmers build layers of data types on top of these in order to organise and normalise information. Thus, our problem reduces to the manipulation of strings of bits by operators. We shall refer to this as alphabetic configuration management, since each bit string of length $n$ represents, in essence, the assignment of a symbol in a $2^n$ long alphabet $\Sigma$.

## 2.5. *Implementability and maintainability*

At we mentioned in Section 1.2, we can attempt to implement policy in two ways: by either starting from a known state and specifying the specific sequence of changes to get to the result, or we can specify the final state and look for a set of operators which are not hindered from getting to that state. The former method is fragile to single obstacles in a chain of pre-conditions from the initial state to the final state; one must deal with the problem of operator composition.

We shall not consider this approach, as it is discussed in the chapter by Sun and Couch. Rather we shall define configuration in terms of goals, or post-conditions. The idea is then that one should quickly arrive at the policy-compliant state and remain there. If the operator is applied to an already compliant system, it must not lead to any change. Thus the key to describing configuration management in these terms is to look for special properties of the operators.

There is actually a deeper point to this. We claim that maintainence is something that must be repeated because we cannot entirely predict all of the changes that will occur in a configuration fragment. But if we are repeating operations with unclear knowledge of the initial state, how do we know that those operations are not actually harming the system?

Suppose during the maintenance of the Sisteen Chapel, a painter is asked to add a fresh coat of paint to Michelangelo's elaborate ceiling decorations. If the task is done only once, it would last a few years but the ceiling would soon need painting again. However, if the artist painted continually, layer upon layer, the paint would become so thick as to fill the room. This operation might even interfere with other operations, such as cleaning the floor. This is not the desired result.

A model of interest is the so-called 'immunity model' [3,7] of maintenance. The approach used in the immunity model is to break policy into a number of tasks or issues that are *primitive* in the following sense:

- A number of operational *types* is defined, that identify high level system properties. These are coded with configuration symbols that do not overlap.
- An operation of a given type now leads to a single change of state in the bit string the represents that type.

Operations hence form a collection agents for independent changes, with distinct types, which can be applied in an order independent fashion (since the states themselves do not overlap the operators must commute).

Thus, regardless of the logistics of implementation, this post-conditional approach to configuration requires a spanning set of *operators*, which act on alphabetic patterns of configuration data at the appropriate level of coding. To capture the idea that we should end up with a predictable result, there are two properties of interest that we can give to these operators:

- *Idempotence*: the repetition of an operation leads to no change after the initial application. i.e. $f(f(x)) = f(x)$, or for any state $\vec{q}$, $[\![O]\!]^2\vec{q} = [\![O]\!]\vec{q}$. This contains the essence of non-repetition, but it is a property without an anchor – it does not lead to a predictable post-condition, unless we also know the initial state. So this property has the special status of a bridge between the two approaches to confuguration management.
- *Convergence at a fixed point*: the first application of an operation leads to the specified policy result, regardless of the initial state. Thereafter, no further changes should take place. i.e. for any initial state $\vec{q}$ and a policy-compliant state $\vec{p}$, $[\![O]\!]^2\vec{q} = [\![O]\!]\vec{q} = \vec{p}$. This is the approach preferred by cfengine [3,8].

Other properties might also be of interest to us. For instance, *injectivity* of an operator tells us whether it has a unique inverse (is it a one-to-many map). This is a property that some system administrators value, since it means that operations can be 'rolled back' to their former state, after a change.

The approach we pursue below is to search for operators that are themselves equipped with the properties above.

### 2.6. *The algorithmic complexity of configuration management*

Thus we arrive at the key question we wish to address in this chapter: what is the computational *complexity* of configuration management and what do we mean by this?

As we have seen, one can easily take individual examples of configuration *operations*, as we did in the previous section, and say that configuration management is easy. The algorithms required to make changes are clearly simple things. If they were not, computers would not exist as we know them. Implementation is simply a matter of writing data into data structures, at some appropriate level of coding, e.g. changing the permissions of a file, editing its contents, starting and stopping programs etc. Implementing a constant, predecided policy is therefore straightforward. But how do we determine whether a policy exists that leads to the behaviour we require?

If we formulate this question in terms of pre-conditioned sequences, then we must deal with a lot of combinatoric complexity. See the chapter by Sun and Couch for a discussion of this problem. If, on the other hand, we use an approach based on operators that are assumed to be orthogonal and fixed-point convergent, the question is then: can we identify operators that verifiably have the desired policy behaviour in a reasonable length of time?

If we have to determine the implementation method in advance, as part of our estimation of complexity, then the complexity of the task grows considerably from the application of a single transformation to a search in a potentially huge space of possible operations. The result we shall find is the result of a somewhat dumb and blind, brute-force search through this space, by a systematic and somewhat dogged agent. Clearly, one could narrow the search and reduce the complexity if one could inject some intelligence in the form of additional constraints. But that is not what we want to do. We want to know how difficult the whole problem is, from the beginning.

The question we are asking then is not how long does the configuration management software take to run: rather it is, do we know how we write the configuration management software in the first place? The difficulty with system administration lies in determining the correct configuration, satisfying the constraints of policy in an efficient manner.

In the remainder of this chapter, we wish to show that even simple alphabetic configuration problems are, in general at least NP hard.

## 3. A gentle introduction to complexity theory

Let us take a step backward in order to move forward and study some very simple problems to understand the meaning of complexity.

### 3.1. *What is a 'problem'?*

In complexity theory a *problem* is a question to which the answer is YES or NO. The question has an *input*, and *solving* the problem amounts to providing the correct answer to the question when a particular input is given. Let us study some examples.

> (Problem **SEARCH**) *Input*: a natural number $n$ and a list of numbers $\ell_1, \ldots, \ell_k$. *Question*: Does the number $n$ occur in the list $\ell_1, \ldots, \ell_k$?

The input should be expressed in some fixed alphabet according to some sensible convention. For the problem $A$ we may e.g. use the alphabet

$$0, 1, 2, 3, 4, 5, 6, 7, 8, 9, !, /$$

and the convention that the input should be given on the form $n!\ell_1/\ldots/\ell_k$ where the natural numbers $n, \ell_1, \ldots, \ell_k$ are given by decimal digits without leading zeros. Thus, e.g. $27!17/0/23961/27/61$ and $31!71/9/961$ are admissible; for the former input, the answer to the question is YES, for the latter, the answer is NO.

The input to a problem should always be a string of symbols over some fixed alphabet, and the representation of the input should always follow some fixed convention. However, as long as we make natural and reasonable choices, it does not matter which particular alphabet, or which particular convention, we choose. These choices have no essential mathematical implications. Thus, when a problem is stated in complexity theory, the exact alphabet and convention are often suppressed. The reader is free to chose his[3] favourite representation in his favourite alphabet. (He has to be reasonable, though. A typical unreasonable representation of the input of the problem **SEARCH**, will be that the $m$th number in the list should be given with $2^m$ leading zeros.)

---

[3] We assume that women are too complex to be represented here.

Intuitively, the problem **SEARCH** is very easy to solve. We just have to traverse the list $\ell_1, \ldots, \ell_k$; if the number we are looking for, that is $n$, is there, the answer is YES; if the number is not there, the answer is NO. Let us look at an apparently harder problem.

> (Problem **SPLIT**) *Input*: a set of natural numbers $X$. *Question*: Is it possible to split $X$ into two sets $Y$ and $Z$ such that the sum of the numbers in $Y$ equals the sum of the numbers in $Z$?

If the input the problem is the set $\{1, 2, 3, 4, 5, 7\}$, the answer is the questions is YES because $1 + 2 + 3 + 5 = 4 + 7$. Now, what is the answer when the input is

$$\{3, 6, 9, 10, 20, 30, 173, 174, 175, 1027, 1058, 1132, 1719, 2480\}?$$

No doubt, intuitively problem **SPLIT** seems harder to solve than problem **SEARCH**.

### 3.2. *The class* P

DEFINITION 2 (*The class* P). Let $f$ be any function from $\mathbb{N}$ to $\mathbb{N}$ and let $|x|$ denote the number of alphabet symbols required to represent the input $x$. An algorithm taking $x$ as input *works in time* $f(|x|)$ if $f(|x|)$ bounds the number of basic steps executed by the algorithm. We will say that an algorithm *works in polynomial time* if there exists a polynomial $p$ such that the algorithm works in time $p(|x|)$. A problem belongs to the class P if (and only if) the problem can be solved by an algorithm working in polynomial time. We use standard terminology when say that the problems in P are the problems *solvable*, or *decidable*, in *polynomial time*.

The class of problems solvable in polynomial time, called P, is well known from the literature, and the class will be introduced at an early stage in any basic course in complexity theory. The problems in the class are seen as the tractable problems, that is, if a problem belongs to P, a computer can solve the problem in reasonable time; if a problem does not belong to P, the problem will be intractable in the sense that it require more time to solve the problem than any practical application can afford.

At this stage, the practically-minded reader should ask how reasonable it is to identify the tractable problems and the polynomial time decidable problems, and the mathematically-minded reader should find the definition of P somewhat suspicious. (Hopefully, the two groups of readers are not disjoint.)

Unfortunately for the practically-minded reader, it is beyond the scope of this paper to discuss his question any further, but such a discussion can be found in numerous text books and survey papers, e.g. see [20] and [15]. These discussions will assure the reader that the formally defined class P capture the informal class of tractable problems surprisingly well; if we can prove that a natural problem belongs to P, we can be pretty sure that we can implement an efficient algorithm solving the problem; if we can prove that a problem does not belong to P, no real-life computer can deal with the problem no matter how powerful it might be. Of course, no one asserts that algorithm executing, let us say, $|x|^{(2^{100})}$ basic steps terminates in reasonable time, still, such an algorithm runs in polynomial time. However, it turns out that most *natural* problem that can be solved in polynomial time, actually can be solved in time $p(|x|)$ where $p$ is a polynomial of low degree, typically 2 or 3.

The mathematically-minded reader should worry since the definition of the class P is a bit sloppy. The definition does not say what a *basic step in the execution of an algorithm* is. Neither does the definitions say what an *algorithm* is. Well, in order to help the reader to grasp the essence of polynomial time computability, we have chosen to rely on slightly informal notions that, strictly speaking, should not appear in a rigorous definition. We do so because any computer scientist or programmer knows what an algorithm is. It is an efficient and unambiguous receipt for transforming input into output. It is something you can implement on your desktop computer if the computer is equipped with sufficiently memory. The reader probably has a very strong sense of what algorithm is even if he cannot come up with the exact formal definition, like e.g. most educated people know what a real number is even if they are not familiar the technical definitions occurring in the mathematical literature.

In complexity theory, algorithms are usually defined by by Turing machines. A Turing machine is a formally defined mathematical entity that can be viewed as a mechanical device manipulating symbols on an infinite tape. The informal notion of an algorithm are captured by Turing machines in the following sense: (i) *any procedure we intuitively conceive as an algorithm, can be implemented on a Turing machine* and (ii) *any computation carried out by a Turing machine is an algorithm in the intuitive sense*. It is straightforward to define the basic steps of a Turing machine since you can point to something, declare that *this is a basic step* and this is it, e.g. to write a symbol on the tape counts as one step, to move the machine's scanning head one position counts as one step, etc. It is not that easy to identify the basic steps when we are dealing with informal algorithms, but in every algorithm there will be some operations that are natural candidates, e.g. if the algorithm is given as a Java program, the basic instructions you find in the Java manual are the natural candidates; if the algorithm is a clerical routine meant to be carried out by pen and paper, to write down a symbol is a natural candidate. P is a robust class of problems, and as long as our definition of a step in an algorithm is reasonable, the class will not be sensitive to the details of the definition. For sure, it might very well be the case that a particular problem is solvable in time, let us say, $(x+1)^2 + 17$ if one step amounts to carry out one instruction in a high level programming language, but not in time $(x+1)^2 + 17$ if one step amounts to carry out one instruction on a rudimentary device like a Turing machine, still, the problem would be solvable in *polynomial time* by any reasonable rudimentary device you can imagine, maybe in time $(x+2)^3$, maybe in time $x^{23} + 1000$.

Thus, we have chosen to skip the detour via Turing machines, and thus some dull technical definition, and rely on the intuitive notion of an algorithm in our definition of the class P. (We will define several complexity classes that are closely related to P, see Figure 4.)

### 3.3. *Proof techniques*

Before trying to write a program solving a particular problem, a programmer should ask: is the problem is in P? If the answer is no, the problem is essentially intractable for any computer no matter how powerful, and the programmer's task is in some sense insurmountable.

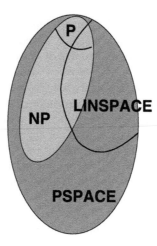

Fig. 4. The complexity classes in the space of algorithms.

(It is beyond the scope of this paper what a programmer should do in such a situation, but he will have to resort to various advanced techniques and methods for finding approximate solutions.) Now, how can a programmer establish if the problem he encounters is in P or not?

To prove that a problem is in P, you simply have to find an algorithm solving the problem in polynomial time. It might be hard to find the algorithm, it might be hard to prove that the algorithm indeed solves the problem, and it might be hard to prove that the algorithm indeed works in polynomial time, but it is easy to understand the overall structure of such a proof. The definition of P says that the problem is in P if and only if there exists an algorithm solving the problem in polynomial time. If you can provide such an algorithm, you have proved that the problem is in P.

In the remainder of this section we will discuss how we can prove that a problem does not belong to P. We will demonstrate the standard proof technique.

Imagine you cannot come up with a polynomial time algorithm solving a particular problem $X$ in polynomial time. Can you conclude that $X$ does not belong to P? Of course not. Even if you are an experienced and clever programmer, you cannot conclude that such an algorithm does not exist just because you cannot come up with one. How should you argue, then? There are infinitely many algorithms, and you cannot go through each and one of them and prove that this does not work, and this does not work, and *this* does not work, and so on; on your dying day there will still be infinitely many algorithms left to consider. (Besides, imagine what a boring life.) Now, what you need is a problem that you already know does not long to P. Let us denote this problem by $H$ for HARD, and you may e.g. find $H$ in a textbook where you also find a theorem stating that

> no algorithm can solve $H$ in polynomial time.                              (†)

Then you can try to prove that

> if there is an algorithm solving $X$ in polynomial time,
>
> then there is an algorithm solving $H$ in polynomial time.        (‡)

If you succeed, you can conclude beyond any doubt and with mathematical certainty that there is no algorithm solving $X$ in polynomial time, that is, $X$ does not belong to P. Your argument has the structure

$$\frac{\text{If } p \text{ then } q. \qquad \text{Not } q.}{\text{Not } p.} \qquad \begin{array}{l} \text{(premises)} \\[4pt] \text{(conclusion).} \end{array}$$

Numerous mathematical proofs embody this structure. The obviously sound inference is of course frequently used in daily life reasoning too, and it has even got its own name as the medieval philosophers baptised the inference *modus tollens*.

So far, so good. If you can prove (‡), then you are done, but how should you proceed to prove (‡). This is where the pivotal notion of polynomial time reducibility enters the stage.

DEFINITION 3 (*Polynomial time reducibility*). A function $f$ *polynomial time reduces* a problem $A$ to a problem $B$ if (and only if)
1. the answer to question $A$ with input $x$ is the same as the answer to question $B$ with input $f(x)$,
2. the function $f$ in computable in polynomial time.

If there exists a function $f$ which polynomial-time reduces the problem $A$ to the problem $B$, we will say that $A$ is *polynomial timed reducible* to $B$.

If you can find a function $f$ such that $f$ polynomial time reduces $H$ to $X$, then (‡) follows. Given such a reduction function, you can solve problem $H$ on input $x$ by first computing the value $v = f(x)$ and thereafter solve the problem $X$ on input $v$. (The answer to problem $X$ on input $v$ will be the same as the answer to problem $H$ on input $x$.) The number of basic steps needed to compute $f(x)$ is bounded by a polynomial in the length of the input, let us say $p(|x|)$. The length of the output, that is $v$, will also be bounded by a polynomial $p'(|x|)$ since the length of the output of any computation taking place in polynomial time will be bounded by a polynomial in the input. If the number of basic steps required to solve $X$ also is bounded by a polynomial in the length of the input, let us say $q(|v|)$, then the whole computation will take place in polynomial time. We have

$$\overbrace{\underbrace{p(|x|)}_{\substack{\text{compute} \\ v=f(x)}} + \underbrace{q(|v|)}_{\substack{\text{solve } X \\ \text{on input } v}}}^{\text{solve } H \text{ on input } x} \leqslant p(|x|) + q\big(p'(|x|)\big)$$

and the expression $p(|x|) + q(p'(|x|))$ is a polynomial in the length of the input. Thus, if there is a algorithm solving $X$ in polynomial time, there will also be one solving $H$ in polynomial time, i.e. (‡) holds, and when (‡) holds, it follows that $X$ does not belong to P.

We conclude this section by a brief summary: To prove that a problem $X$ belongs to P, you should simply provide an algorithm solving $X$ in polynomial time. To prove that a problem $X$ does not belong to P, you should (i) find a problem $H$ that does not belong to P and (ii) prove that $H$ is polynomial time reducible to $X$.

### 3.4. *A hard problem*

Let us formulate one of the main insights of the previous section as a theorem.

THEOREM 1. *Let A and B be problems such that* (i) *A does not belong to* P, *and* (ii) *A is polynomial time reducible to B. Then, B does not belong to* P.

In Theorem 1 we find the contours of an essential property of any sensible measure: *If A is far away, and B is at least as far away as A, then B is also far away.* We are measuring complexity. If a problem $A$ is polynomial time reducible to a problem $B$, then it will be as least as hard to solve $B$ as it will be to solve $A$ (modulo polynomial time). Thus, if $A$ is far away from the tractable, so is also $B$.

We will apply Theorem 1 to argue that several natural problems on operators are intractable. Hence, the $B$ in the theorem will be instantiated by different problems on operators whereas the $A$ might e.g. be instantiated by a celebrated hard problem on Boolean algebra, namely the problem **SAT**.

> (Problem **SAT**) *Input*: a Boolean expression $\alpha$. *Question*: Is $\alpha$ satisfiable?

Presumably most of the readers will to some extent be familiar with Boolean algebra, still, let us re-examine the subject briefly to assure that everyone understands the problem **SAT**. First we define the Boolean expressions.

DEFINITION 4 (*Boolean expressions*). We have an infinite supply of *Boolean variables* $x_0, x_1, x_2, \ldots$.
- A Boolean variable is a Boolean expression.
- The *constants* 0 and 1 are Boolean expressions.
- $(\alpha \cdot \beta)$ is a Boolean expression if $\alpha$ and $\beta$ are Boolean expressions.
- $(\alpha + \beta)$ is a Boolean expression if $\alpha$ and $\beta$ are Boolean expressions.
- $\overline{\alpha}$ is a Boolean expression if $\alpha$ is a Boolean expression.

Further, we will use standard terminology and say that $(\alpha \cdot \beta)$ is *the product* of $\alpha$ and $\beta$; that $(\alpha + \beta)$ is *the sum* of $\alpha$ and $\beta$; that $\overline{\alpha}$ is *the complement* of $\alpha$.

As we say that $(\alpha \cdot \beta)$ and $(\alpha + \beta)$ are respectively a *product* and a *sum*, it would also be natural to say $\cdot$ and $+$ are respectively the *multiplication* operator and the addition *operator*. This is standard terminology, but the reader should be aware that the two operators also frequently are referred to as the AND-operator and the OR-operator in the literature. The reader might even be familiar with propositional logic where we have two truth values

**false** and **true** and three connectives AND, OR and NOT. Boolean algebra is nothing but propositional logic, and

- 0 and 1 correspond to respectively **false** and **true**,
- $\cdot$ and $+$ correspond to respectively AND and OR,
- $\overline{\alpha}$ corresponds to NOT $\alpha$.

Definition 4 gives the form of an Boolean expression. According to the definition e.g.

$$\left(x_3 + \overline{(x_{17} \cdot x_2)}\right) \quad \text{and} \quad \left(\overline{\left(x_9 + \overline{(1 + x_2)}\right)} \cdot (x_4 \cdot \overline{x}_4)\right) \tag{14}$$

are Boolean expression. Now, a Boolean variable either takes the value 0 or the value 1, and given an assignment of 0-1-values to the variable, we can compute the value of a Boolean expression. To do so, we proceed as we do when we compute the value of an arithmetical expression, but we stick to the following rules:

$$
\begin{array}{lll}
0 + 0 = 0 & 0 \times 0 = 0 & \\
0 + 1 = 1 & 0 \times 1 = 0 & \overline{0} = 1 \\
1 + 0 = 1 & 1 \times 0 = 0 & \overline{1} = 0. \\
1 + 1 = 1 & 1 \times 1 = 1 &
\end{array}
$$

Hence, since the sum and the product work as in ordinary arithmetic except that $1 + 1$ equals 1, the value of a Boolean expression will always turn out to be 0 or 1, e.g. given the assignment

$$x_3 := 0, \qquad x_{17} := 1, \qquad x_2 := 1$$

we have $(x_3 + \overline{(x_{17} \cdot x_2)}) = 0$; whereas given the assignment

$$x_3 := 1, \qquad x_{17} := 0, \qquad x_2 := 1 \tag{15}$$

we have $(x_3 + \overline{(x_{17} \cdot x_2)}) = 1$.

DEFINITION 5 (*Satisfiability*). A Boolean expression $\alpha$ is *satisfiable* if there exists an assignment to the variables in $\alpha$ such that $\alpha = 1$.

Hence, the expression $(x_3 + \overline{(x_{17} \cdot x_2)})$ is satisfiable since the expression equals 1 under the assignment (15). The rightmost expression in (14) is not satisfiable.

This should settle what the problem **SAT** is all about, but why should the problem be hard, on the surface it might look like an innocent little problem? To get some intuition, the reader should observe that the number of possible assignments is exponential in the number of variables in the expression. If there are $n$ variables in an expression, there will be $2^n$ different assignments, and to solve the problem **SAT** we might be forced to consider each and one them. We will simply not be able to do so when $n$ is large, if $n = 100$, we will not be done in a million years even if we checked a million assignments per second.

But why should we be forced to checked each and one of the $2^n$ assignments? Well, there are algorithms solving **SAT** that in certain respects are far more efficient than the straightforward stupid brute force search for an satisfying assignment, however, to make a long story short, none has ever been able to come up with an algorithm solving the problem in polynomial time, and if you can come up with one, there is a million dollar award waiting for you. The problem is known to be NP-complete, that is, if **SAT** is solvable in polynomial time, every problem in the complexity class called NP will be solvable in polynomial time, and that is considered as highly unlikely among all sane complexity-theorists, and probably among a few insane ones too. Though we lack a mathematical proof, there is absolutely no evidence there exists an algorithm solving **SAT** in polynomial time.

We will treat the class NP more thoroughly in Section 5, still, let us mention right now that the standard definition says that NP is the class of problems solvable by a *non-deterministic Turing machine* working in polynomial time. For reasons explained above, a slightly sloppy, but rather informative alternative definition would be the class of problems solvable by a *non-deterministic algorithm* working in polynomial time (but a deterministic algorithm would have to use exponential time). The hardest problems in NP are the NP-complete ones. In addition to **SAT**, hundreds of problems from science, mathematics and daily life are known to be NP-complete, among them our example **SPLIT** discussed in Section 3.1. If you either can prove that P = NP or that P ≠ NP, the million dollar award mentioned above is yours. Those of you in urgent need of money, should take our advice and not waste your time on the former option. Start working right away on the latter.

## 4. Analysis of some configuration problems

### 4.1. *The complexity of the system coding*

We asked the question earlier: what is the effect of the high level coding on the complexity of configuration management? This is an important question because we want to argue that a model of bit strings is adequate to discuss the complexity of the whole problem. We can now make a plausible answer to this question.

The complexity of coding has two facets: (i) the complexity of the synactic parsing and translation, (ii) the semantic translation – how do we find identify the behaviour associated with the strings in the high level coding?

For (i), the following result is fundamental in parsing theory [24,30].

THEOREM 2. *Given a context free grammar G and a string s, deciding whether the string s is an element of the language $\mathcal{L}_G$ generated by G can be done in $O(|G| \times |s|^3)$ deterministic time and $O(|G| \times s^2)$ space.*

This complexity does not increase even for extended context free grammars. In other words, within polynomial time, we can translate the syntax of any reasonable high level coding, so this will not affect the answer as to whether configuration management is hard or not.

For (ii), the second of these points is not a question we can answer with any certainty, but we believe that whatever the answer is, any contribution to behaviour as a result of the coding belongs in the definition of the problem itself and is therefore a valid part of the answer we seek. With this assumption, we can proceed to view configuration management as a problem of determining operators represented in $\mathbb{B}^n$.

There is a large literature on coding, and its algorithmic complexity. See refs. [22,28] for a sample. The usual problem addressed in coding is that posed by Shannon for information transmitted over a noisy channel, namely given a transmitted code word and a noisy or uncertain channel, what is the complexity of determining the original transmitted word, e.g. by maximum probability? This problem is known to be NP complete. The configuration problem is essentially related to this scenario [2] in two stages: (i) given a result in the configuration alphabet observed at the end of the policy-configuration channel, what policy does this correspond to? and (ii) given an observed behaviour at the end of the configuration-behaviour channel, what is the maximum likelihood configuration that led to this?

## 4.2. *Operators formalised*

Our goal now is to undertake a complexity-theoretic analysis of some problems occurring naturally in the theory of system administration and configuration management, using the model of operators over bit strings. A complexity-theoretic analysis requires that we formalise the operators such that we know exactly what we are talking about and can state the problems with sufficiently of mathematical precision. In this section will carry out such a formalisation. How well our formalisation captures the operators as they appear in the theory of system administration, will not be discussed any further, but based on the discussion above, we believe our formalisation to be reasonable.

Recall that $\mathbb{B}^n$ is the set of all bit strings of length $n$, e.g.

$$\mathbb{B}^4 = \{0000, 0001, 0010, 0011, 0100, 0101, 0110,$$

$$0111, 1000, 1001, 1010, 1011, 1100, 1101, 1110, 1111\}.$$

If $b \in \mathbb{B}^n$, then $b_i$ denotes the $i$th bit in $b$. We will count the bits in a string from the left to the right, thus if $b = 0111$, then $b_1 = 0$, $b_2 = 1$, $b_3 = 1$ and $b_4 = 1$.

An operator is nothing but a total function with domain $\mathbb{B}^n$ and range $\mathbb{B}^n$. So, an operator transforms a string of bits into another string of bits of the same length. Let us study a few examples. The operator INV inverts the bits of a sting, thus we have e.g. INV$(0111) = 1000$ and INV$(1001) = 0110$. The operator RST sets all the bit in a string to 0, and we have e.g. RST$(0111) = 0000$ and RST$(1001) = 0000$. The operator SUC increases the binary number given by a string by 1 modulo 16. We have SUC$(0000) = 0001$, SUC$(0001) = 0010$, SUC$(0010) = 0011, \ldots,$ SUC$(1111) = 0000$.

Even if there are only finitely many operators over the domain $\mathbb{B}^n$, there are still a lot of them. The exact number of operator over the domain $\mathbb{B}^n$ is $2^{n2^n}$. There are more that $10^{19}$ operators over the domain $\mathbb{B}^4$, and there are more operators over the domain $\mathbb{B}^8$ than there are elementary particles in the universe. We need a language strong enough to express all the operators over the domain $\mathbb{B}^n$ in a convenient and uniform way. That is our motivation for introducing the operator expressions.

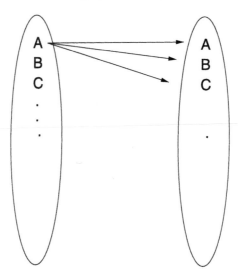

Fig. 5. The collection of all possible mappings of $\sigma$ configurations is $\sigma^\sigma$ large.

Why are there $2^{n2^n}$ operators over the domain $\mathbb{B}^n$? Let $A$ and $B$ be a finite sets containing respectively $m$ and $n$ elements. The number of functions with domain $A$ and range $B$ will be $n^m$. The set $\mathbb{B}^n$ contains $2^n$ elements, and an operator over the domain $\mathbb{B}^n$ is a function with domain $\mathbb{B}^n$ and range $\mathbb{B}^n$. Thus, by applying the number-theoretic equality $(a^b)^c = a^{bc}$, we see that the number of operators over the domain $\mathbb{B}^n$ will be

$$\left(2^n\right)^{\left(2^n\right)} = 2^{n2^n}.$$

Another way of saying this is that $2^n$ is the size $\sigma$ of an effective alphabet of configurations $\Sigma$, that can be represented in $n$ bits, so $2^{n2^n}$ is $\sigma^\sigma$ the number of "commands" or algorithms one can form to transform one string in this alphabet to another (see Figure 5).

DEFINITION 6 (*Operator expressions*). An *operator expression* $e$ is a sequence $\langle \alpha_1, \alpha_2, \ldots, \alpha_n \rangle$ of Boolean expressions. If there are $n$ Boolean expression in the sequence, the only variables allowed in the expressions are $x_1, x_2, \ldots, x_n$.

An operator expression $e \equiv \langle \alpha_1, \alpha_2, \ldots, \alpha_n \rangle$ defines an operator over the domain $\mathbb{B}^n$. As explained, we use $[\![e]\!]$ to denote the operator defined by the operator expression $e$. The value $[\![e]\!](b)$, that is, the value of $[\![e]\!](b)$ in the particular argument $b \in \mathbb{B}^n$ is given by the value of the sequence $\langle \alpha_1, \ldots, \alpha_n \rangle$ under the assignment $x_1 := b_1, \ldots, x_n := b_n$. The first bit of $[\![e]\!](b)$ is given by the value of $\alpha_1$ under the assignment, the second bit by the value of $\alpha_2$, and so on. We better look at a few examples.

For instance, let $e \equiv \langle \overline{x}_1, \overline{x}_2, \overline{x}_3, \overline{x}_4 \rangle$, then $[\![e]\!]$ will be the operator INV inverting a bit string of length 4. Further, we can compute the value $[\![e]\!](0011)$ by computing each of the Boolean expressions in the sequence $\langle \overline{0}, \overline{0}, \overline{1}, \overline{1} \rangle$.

A more sophisticated example will be if

$$e \equiv \langle ((\overline{x}_1 \cdot ((x_2 \cdot x_3) \cdot x_4)) + (x_1 \cdot \overline{((x_2 \cdot x_3) \cdot x_4)})),$$
$$(\overline{x}_2 \cdot (x_3 \cdot x_4)) + (x_2 \cdot \overline{(x_3 \cdot x_4)}), (\overline{x}_3 \cdot x_4) + (x_3 \cdot \overline{x}_4), \overline{x}_4 \rangle.$$

Now, let us compute the value $[\![e]\!](0011)$. First we make the assignment

$$x_1 := 0, \qquad x_2 := 0, \qquad x_3 := 1, \qquad x_4 := 1$$

and then we evaluate the Boolean expressions in the sequence one by one. We have

$$((\overline{x}_1 \cdot ((x_2 \cdot x_3) \cdot x_4)) + (x_1 \cdot \overline{((x_2 \cdot x_3) \cdot x_4)}))$$
$$= ((\overline{0} \cdot ((0 \cdot 1) \cdot 1)) + (0 \cdot \overline{((0 \cdot 1) \cdot 1)})) = 0$$

and thus the first bit in the value is 0; we have

$$(\overline{x}_2 \cdot (x_3 \cdot x_4)) + (x_2 \cdot \overline{(x_3 \cdot x_4)}) = (\overline{0} \cdot (1 \cdot 1)) + (0 \cdot \overline{(1 \cdot 1)}) = 1$$

and thus the second bit in the value is 1; we have

$$(\overline{x}_3 \cdot x_4) + (x_3 \cdot \overline{x}_4) = (\overline{1} \cdot 1) + (1 \cdot \overline{1}) = 0$$

and the third bit turns out to be 0; the forth bit also turns out to be 0 since $\overline{x}_4 = \overline{1} = 0$. Hence, $[\![e]\!](0011) = 0100$. It might be an instructive exercise for the reader to verify that $[\![e]\!]$ indeed is the operator SUC discussed above, and that $[\![e]\!](0000) = 0001$, $[\![e]\!](0001) = 0010$, and so on. In such a situation we might use the notation $[\![e]\!] = $ SUC and say that the expression $e$ *defines* the operator SUC. Observe that many expression will define the same operator, e.g. the operator RST can are defined by the expression $\langle 0, 0, 0, 0 \rangle$, but also by the expression $\langle (x_1 \cdot \overline{x}_1), (x_1 \cdot \overline{x}_1), (x_1 \cdot \overline{x}_1), (x_1 \cdot \overline{x}_1) \rangle$ and the expression $\langle \overline{1}, (x_1 \cdot \overline{x}_1), (x_4 \cdot \overline{x}_4), 0 \rangle$, indeed, there will be infinitely many expressions defining any operator.

### 4.3. *Some problems on operators:* **IDM, INJ** *and* **CON**

Having formalised the notion of an operator, we are now ready to state some problem (see Figure 6).

> (Problem **IDM**) *Input*: an operator expression $e$. *Question*: Is $[\![e]\!]$ idempotent, i.e. do we have $[\![e]\!]([\![e]\!](x)) = [\![e]\!](x)$ for all $x \in \mathbb{B}^n$?
>
> (Problem **INJ**) *Input*: an operator expression $e$. *Question*: Is $[\![e]\!]$ an injection, i.e. do we have $[\![e]\!](x) \neq [\![e]\!](y)$ if $x \neq y$?

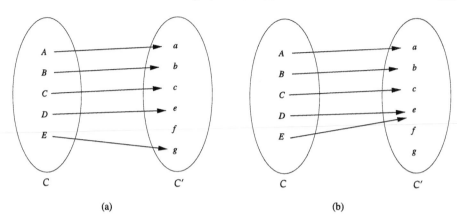

Fig. 6. Operator mapping properties. (a) An injective map has a unique inverse; (b) A non-injective map does not have a unique inverse. In both cases a operator transformation could eliminate parts of the configuration space from the domain of possibility, i.e. both represent potential constraints.

> (Problem **CON**) *Input*: an operator expression $e$. *Question*: Is $[\![e]\!]$ convergent, i.e. does there, for every $x \in \mathbb{B}^n$, exist $k \in \mathbb{N}$ such that $[\![e]\!]^k(x) = [\![e]\!]^{k+1}(x)$?

> (Problem **CONx**) *Input*: a pair $\langle e, b \rangle$ where $e$ is an operator expression and $b \in \mathbb{B}^n$. *Question*: Is $[\![e]\!]$ convergent in $b$, i.e. does there exist a $k \in \mathbb{N}$ such that $[\![e]\!]^k(b) = [\![e]\!]^{k+1}(b)$?

The problem **CON** corresponds roughly to cfengine's notion of convergence at a fixed point, since it claims that (after a certain number of changes have taken place by repeated application of the operator) the result will converge to a specific location for any input.[4] Hence we are particularly interested in this case for configuration management [7]. What we mean by the complexity in this context is determining whether the policy-operator expression $e$ is convergent (i.e. approaches a fixed and predictable result $b$) given that it starts in *any given starting state*.

Put another way: we declare the answer $b$ to be policy. This tells us how we would like the system to look after the repeated operation of $[\![e]\!]$. Then we take the trial method $[\![e]\!]$, defined by the operator expression $e$, and we ask: is this policy implementable for every possible start configuration $x$? If not, then we have not truly solved the problem with $[\![e]\!]$.

We will focus on the four problems stated above, but we can easily imagine other interesting problems on operators, e.g.

> (Problem *EQUAL*) *Input*: a pair $\langle e, e' \rangle$ where $e$ and $e'$ are operator expressions. *Question*: Does $e$ define the same operator as $e'$, i.e. do we have $[\![e]\!](x) = [\![e']\!](x)$ for all $x \in \mathbb{B}^n$?

---

[4] It does not specify what the final location is, so the outcome is not as predictable as we would like. Ideally we want $[\![e]\!]^k(x) = [\![e]\!]^{k+1}(x) = $ policy for all $x$, but from a complexity theoretic viewpoint it is the convergence that is hard to verify, not the point at which the operator converges.

(Problem *INVERSE*) Input: a pair $\langle e, e' \rangle$ where $e$ and $e'$ are operator expressions. *Question*: Is $[\![e']\!]$ the inverse operator of $[\![e]\!]$, i.e. do we have $[\![e]\!]([\![e']\!](x)) = x$ for all $x \in \mathbb{B}^n$?

(Problem *SAME LIMIT*) Input: a pair $\langle e, e' \rangle$ where $e$ and $e'$ are operator expressions. *Question*: Do $[\![e]\!]$ and $[\![e']\!]$ converge to the same limit, i.e. will there for any $x \in \mathbb{B}^n$ exist $k, \ell \in \mathbb{N}$ such that

$$[\![e]\!]^k(x) = [\![e]\!]^{k+1}(x) = [\![e']\!]^\ell(x) = [\![e']\!]^{\ell+1}(x)?$$

(Problem *COMMUTE*) Input: a pair $\langle e, e' \rangle$ where $e$ and $e'$ are operator expressions. *Question*: Do $[\![e]\!]$ and $[\![e']\!]$ commute, i.e. do we have $[\![e]\!]([\![e']\!](x)) = [\![e']\!]([\![e]\!](x))$ for all $x \in \mathbb{B}^n$?

### 4.4. *The complexity of* **IDM** *and* **INJ**

In this section we will show that any algorithm solving the problem **IDM**, or the problem **INJ**, in polynomial time, yields an algorithm solving **SAT** in polynomial time, and if **SAT** does not belong to P, it follows from our results that neither **IDM** nor **INJ** belong to P. Unfortunately, we cannot prove that the problems do not belong to P because we cannot prove that **SAT** does not belong to P, but that is, as we discussed above, highly unlikely, and anyway, even if the world should be so strange that there is an algorithm solving **SAT** in polynomial time, as long as we do not know that algorithm, we are better off living our lives as if **SAT** does not belong to P.

If $A$ denotes a problem, then $coA$ denotes the co-problem. The problem $coA$ is the problem which on any input has the opposite answer of the problem $A$, that is, the answer to $coA$ on input $x$ is YES if, and only if, the answer to $A$ on input $x$ is NO. Hence, $co$**IDM** is the problem

      ($co$**IDM**) *Input*: an operator expression $e$.
      *Question*: Do we have $[\![e]\!]([\![e]\!](b)) \neq [\![e]\!](b)$ for some $b \in \mathbb{B}^n$?

and $co$**INJ** is the problem

      ($co$**INJ**) *Input*: an operator expression $e$.
      *Question*: Do we have $[\![e]\!](b) = [\![e]\!](b')$ and $b \neq b'$ for some $b, b' \in \mathbb{B}^n$?

We will give a very detailed and meticulous proof of the next theorem.

THEOREM 3. *SAT is polynomial time reducible to co**IDM**.*

PROOF. Definition 3 says that **SAT** is polynomial time reducible to $co$**IDM** if there exists a function $f$ such that:
1. the answer to the question **SAT** on input $x$ is the same as the answer to the question $co$**IDM** on input $f(x)$,

  2. the function $f$ in computable in polynomial time.

When we can provided such a function $f$, we have proved the theorem. The input to **SAT** is a Boolean expression whereas the input to $co\textbf{IDM}$ is an operator expression. Thus, the function $f$ should transform an arbitrary Boolean expression $\alpha$ to an operator expression $e = f(\alpha)$ such that

  (1) if $\alpha$ is satisfiable, then $[\![e]\!]([\![e]\!](b)) \neq [\![e]\!](b)$ for some $b \in \mathbb{B}^n$,

  (2) if $\alpha$ is not satisfiable, then $[\![e]\!]([\![e]\!](b)) = b$ for all $b \in \mathbb{B}^n$.

When we have found such a function $f$, and argued that $f$ is computable in polynomial time, we are done.

  We can, without loss of generality, assume that if a Boolean expression contains $n$ different variables, then every variable in the expression occurs in the list $x_1, \ldots, x_n$.[5] Let INV be the operator inverting the bits of string, and let the operator $\mathrm{OP}_\alpha : \mathbb{B}^n \to \mathbb{B}^n$ be such that

$$\mathrm{OP}_\alpha(b) = \begin{cases} \mathrm{INV}(b) & \text{if } \alpha = 1 \text{ under either} \\ & \text{the assignment } x_1 := b_1, \ldots, x_n := b_n \\ & \text{or the assignment } x_1 := \overline{b}_1, \ldots, x_n := \overline{b}_n, \\ b & \text{otherwise.} \end{cases}$$

The Boolean expression $\alpha$ is a parameter to the operator, and for each $\alpha$ we have an operator $\mathrm{OP}_\alpha$. Let us see how $\mathrm{OP}_\alpha$ behaves in (i) the case when $\alpha$ is satisfiable and (ii) the case when $\alpha$ is not satisfiable. In the case (i) when $\alpha$ is satisfiable there will be a least on bit string $b$ such that $\mathrm{OP}_\alpha(b) = \mathrm{INV}(b)$. Observe that if $\mathrm{OP}_\alpha(b) = \mathrm{INV}(b)$, then we also have $\mathrm{OP}_\alpha(\mathrm{INV}(b)) = \mathrm{INV}(\mathrm{INV}(b))$. Further, for any bit sting $c$ we have $\mathrm{INV}(\mathrm{INV}(c)) = c$. Thus, if $\alpha$ is satisfiable, there will be at least one bit sting $b$ such that

$$\mathrm{OP}_\alpha\big(\mathrm{OP}_\alpha(b)\big) = \mathrm{INV}\big(\mathrm{INV}(b)\big) = b \neq \mathrm{INV}(b) = \mathrm{OP}_\alpha(b)$$

that is, we have $\mathrm{OP}_\alpha(\mathrm{OP}_\alpha(b)) \neq \mathrm{OP}_\alpha(b)$ for some $b \in \mathbb{B}^n$. In the case (ii) when $\alpha$ is not satisfiable we have $\mathrm{OP}_\alpha(b) = b$ for all $b \in \mathbb{B}^n$, and hence also $\mathrm{OP}_\alpha(\mathrm{OP}_\alpha(b)) = \mathrm{OP}_\alpha(b)$ for all $b \in \mathbb{B}^n$.

  This shows that if $e$ is an operator expression such that $[\![e]\!] = \mathrm{OP}_\alpha$, then (1) and (2) will be satisfied. We will proceed as follows. First we will provide a function $f$ transforming a Boolean expression $\alpha$ into an operator expression $e$ such that $[\![e]\!] = \mathrm{OP}_\alpha$. Thereafter we will provide an algorithm computing $f$ and argue that the algorithm runs in polynomial time. That will prove the theorem.

*The definition of the reduction function $f$:* Let $\hat{\alpha}$ denote the Boolean expression we get when we replace each variable in $\alpha$ by its complement, that is, we replace each occurrence $x_i$ by $\overline{x}_i$. Further, let

$$\beta_i \equiv \left( (\overline{x}_i \cdot (\alpha + \hat{\alpha})) + (x_i \cdot \overline{(\alpha + \hat{\alpha})}) \right)$$

for $i = 1, \ldots, n$, and finally let $f(\alpha) = \langle \beta_1, \ldots, \beta_n \rangle$.

---

[5] This is a harmless assumption making life a little bit easier. Let us say that there a three different variables in an expression and that these three variables are $x_9, x_{341}, x_5$. By renaming variables in an obvious way we can find an equivalent expression where no variables except $x_1, x_2$ and $x_3$ occur.

The observation that

$$\beta_i = \begin{cases} \overline{x}_i & \text{if } \alpha = 1 \text{ under either} \\ & \text{the assignment } x_1 := b_1, \ldots, x_n := b_n \\ & \text{or the assignment } x_1 := \overline{b}_1, \ldots, x_n := \overline{b}_n, \\ x_i & \text{otherwise,} \end{cases}$$

makes it easy to see that we indeed have $[\![e]\!] = \text{OP}_\alpha$ whenever $e = f(\alpha)$. Thus, we can devote the remainder of the proof to argue that $f$ is computable in polynomial time.

There is straightforward algorithm for computing $f$. The input to the algorithm is of course a Boolean expression $\alpha$. First the algorithm counts the number of different variables in $\alpha$. Let us say there are $n$ different variables and that the variables occurring in $\alpha$ is $x_1, \ldots, x_n$.[6] After the counting, the algorithm prints (to the output) the marker symbol $\langle$, thereafter the algorithm prints

$$\left( \left( \overline{x}_1 \cdot (\alpha + \hat{\alpha}) \right) + \left( x_1 \cdot \overline{(\alpha + \hat{\alpha})} \right) \right),$$

thereafter the algorithm prints

$$\left( \left( \overline{x}_2 \cdot (\alpha + \hat{\alpha}) \right) + \left( x_2 \cdot \overline{(\alpha + \hat{\alpha})} \right) \right),$$

and so on, finally the algorithm prints

$$\left( \left( \overline{x}_n \cdot (\alpha + \hat{\alpha}) \right) + \left( x_n \cdot \overline{(\alpha + \hat{\alpha})} \right) \right) \rangle$$

and terminates. We will argue that this straightforward algorithm runs in polynomial time, that is, that the number of basic steps the algorithm carries out is bounded by a polynomial in $|\alpha|$ where $|\alpha|$ is the number of symbols required to represent $\alpha$.

First, we argue the number of symbols the algorithm prints, is bounded by a polynomial in $|\alpha|$. In addition to printing the two marker symbols $\langle, \rangle$ and $n - 1$ copies of the comma, the algorithm prints $n$ copies of $\beta_i$, that is, $n$ copies of

$$\left( \left( \overline{x}_i \cdot (\alpha + \hat{\alpha}) \right) + \left( x_i \cdot \overline{(\alpha + \hat{\alpha})} \right) \right).$$

For any reasonable representation of $\beta_i$, there will be the case that the number of symbols required to represent $x_i$ is bounded by $|\alpha|$; the number of symbols required to represent $\overline{x}_i$ is bounded by $2|\alpha|$; the number of symbols required to represent $\hat{\alpha}$ is bounded by $2|\alpha|$. Thus, there exist fixed numbers $k$ and $\ell$ such that the number of symbols required to represent $\beta_i$ is bounded by $k|\alpha| + \ell$, and the number of symbols the algorithm prints will be bounded by $2 + (n - 1) + n(k|\alpha| + \ell)$. Now, $n$ is the number of (different) variables in $\alpha$, and there cannot be more variables in $\alpha$ than there are symbols in $\alpha$, hence, we also have $n \leqslant |\alpha|$, and the number of symbols printed will be bounded by

$$2 + (n - 1) + n\left(k|\alpha| + \ell\right) \leqslant 2 + |\alpha| + |\alpha|\left(k|\alpha| + \ell\right).$$

---

[6]If there are other variables occurring in $\alpha$, the algorithm can simply rename the variables such that no other variables than $x_1, \ldots, x_n$ occur in $\alpha$.

Let $p(x) = 2 + x + x(kx + \ell)$ and $p(|\alpha|)$ is a bound on the number of symbols printed by the algorithm.

We want to argue that the algorithm runs in polynomial time, and there is some counting going on before the algorithm starts printing, but there will definitely be a polynomial $q$ such that the number of steps required to complete this counting is bounded by $q(|\alpha|)$. Now, how many step will be required to write $p(|\alpha|)$ symbols? Well, no matter how we count the steps, and no matter how we implement the algorithm, as long as we avoid doing something outright strange or abnormal, there will exist fixed numbers $a$ and $b$ such that the overall number of steps executed is bounded by $q(|\alpha|) + (p(|\alpha|) + a)^b$. Thus, the algorithm runs in polynomial time. This completes the proof the theorem.                                    □

The proof of the next theorem is analogous to the proof of the previous one. We will give a more laconic proof this time, and it might be helpful for those of readers who are less proficient in complexity theory, to study the proof of Theorem 3 thoroughly before proceeding to the proof of Theorem 4.

THEOREM 4. *SAT is polynomial time reducible to coINJ.*

PROOF. The input to **SAT** is a Boolean expression whereas the input to *co***INJ** is an operator expression. We will provide polynomial time computable reduction function $f$ transforming a Boolean expression $\alpha$ into an operator expression $e = f(\alpha)$ such that
- if $\alpha$ is satisfiable, then there exist $b, b' \in \mathbb{B}^n$ such that $[\![e]\!](b) = [\![e]\!](b')$ and $b \neq b'$,
- if $\alpha$ is not satisfiable, then $[\![e]\!](b) \neq [\![e]\!](b')$ whenever $b \neq b'$.

We can without loss of generality assume that the variables occurring in $\alpha$ are $x_1, \ldots, x_n$. Let $0^n$ be the bit string of length $n$ where every bit is 0, and let the operator OP$: \mathbb{B}^n \to \mathbb{B}^n$ be such that

$$\mathrm{OP}_\alpha(b) = \begin{cases} 0^n & \text{if } \alpha = 1 \text{ under either} \\ & \text{the assignment } x_1 := b_1, \ldots, x_n := b_n \\ & \text{or the assignment } x_1 := \overline{b}_1, \ldots, x_n := \overline{b}_n, \\ b & \text{otherwise.} \end{cases}$$

In the case when $\alpha$ is satisfiable we have $\mathrm{OP}_\alpha(b) = 0^n$ and $\mathrm{OP}_\alpha(\mathrm{INV}(b)) = 0^n$ for some $b \in \mathbb{B}^n$, and hence, since $b \neq \mathrm{INV}(b)$, there exist $b, b' \in \mathbb{B}^n$ such that $\mathrm{OP}_\alpha(b) = \mathrm{OP}_\alpha(b')$ and $b \neq b'$. In the case when $\alpha$ is no satisfiable we have $\mathrm{OP}_\alpha(b) = b$ for any $b \in \mathbb{B}^n$, and hence, $\mathrm{OP}_\alpha(b) \neq \mathrm{OP}_\alpha(b')$ whenever $b \neq b'$. This shows that it is sufficient to define $f$ such that $[\![f(\alpha)]\!] = \mathrm{OP}_\alpha$.

*The definition of the reduction function* $f$: Let $\hat{\alpha}$ denote the Boolean expression $\alpha$ where each occurrence of $x_i$ is replaced by $\overline{x}_i$. Let $\beta_i \equiv (x_i \cdot \overline{(\alpha + \hat{\alpha})})$ and finally let $f(\alpha) = \langle \beta_1, \ldots, \beta_n \rangle$. This completes the definition of $f$.

We have $[\![f(\alpha)]\!] = \mathrm{OP}_\alpha$ since

$$\beta_i = \begin{cases} 0 & \text{if } \alpha = 1 \text{ under either} \\ & \text{the assignment } x_1 := b_1, \ldots, x_n := b_n \\ & \text{or the assignment } x_1 := \overline{b}_1, \ldots, x_n := \overline{b}_n, \\ x_i & \text{otherwise.} \end{cases}$$

The proof that $f$ is computable in polynomial time is similar to the corresponding part of the proof of the preceding theorem.                    □

COROLLARY 1.  (i) *If **SAT** does not belong to* P, *then **IDM** does not belong to* P. (ii) *If **SAT** does not belong to* P, *then **IDM** does not belong to* P.

PROOF.  We prove (i). First we observe that

**IDM** belongs to P if and only if its *co***IDM** belongs to P.                    (∗)

This is completely trivial. If you have an algorithm solving a problem in polynomial time, then it is very easy to construct an algorithm solving the co-problem in polynomial time. The same algorithm answering NO in place of YES, and YES in place of NO, will do the job.

Theorem 3 states that **SAT** is polynomial time reducible to *co***IDM**. By Theorem 1, it follows that if **SAT** does not belong to P, then neither will *co***IDM** do so. Now, (i) follows by (∗).

The proof of (ii) is of course similar to the proof of (i).                    □

So far our discussion shows that the problems **IDM** and **INJ** are most likely to not be in P. In the next sections will undertake a further complexity theoretic analysis and see if we can say something more about the inherent computational complexity of these problems.

## 5. More complexity theory

Complexity can be measured in terms of both time and space resources.

### 5.1. *The space classes* LINSPACE *and* PSPACE

The definitions of the complexity classes LINSPACE and PSPACE are analogous to the definition of the class P, but in contrast to imposing a bound on the number of steps executed by an algorithm, we impose a bound on the storage resources required to execute an algorithm.

DEFINITION 7 (*The class* LINSPACE *and the class* PSPACE).  Let $f$ be any function from $\mathbb{N}$ to $\mathbb{N}$ and let $|x|$ denote the number of alphabet symbols required to represent the input $x$. An algorithm taking $x$ as input *runs*, or *works*, *in space* $f(|x|)$ if $f(|x|)$ bounds the number of storage units required to execute the algorithm.

An algorithm *works in polynomial space* if there exists a polynomial $p$ such that the algorithm works in space $p(|x|)$. A problem belongs to the class PSPACE if (and only if) the problem can be solved by an algorithm working in polynomial space. Further, an algorithm *works in linear space* if there exists a fixed number $k$ such that the algorithm works in space $k|x|$. A problem belongs to the class LINSPACE if (and only if) the problem can be solved by an algorithm working in polynomial space.

We use standard terminology when we say that the problems in LINSPACE, respectively PSPACE, are the problems *solvable*, or *decidable*, in *linear*, respectively *polynomial, space.*

When we defined P, we relied on our intuitive notion of an *algorithm* and, in order to impose resource bounds on the algorithms, we resorted to another intuitive, but closely related, notion: *the number of basic steps in an execution.* In the definitions of LINSPACE and PSPACE, we are again relying on our intuitive notion of an algorithm, but now the resource bounds are stated in terms of another informal notion: *the number of storage units required by an execution.* After we defined P, we had a long discussion concluding that P is robust class, and no matter how we define *algorithm* and *basic step*, as long as we are reasonable, the class remains the same. The class P is simply not sensitive to the details in the formal definitions of *algorithm* and *basic step*, nor is the class very sensitive to the conventions for representing inputs. This robustness of P can hardly be overestimated and shows that class P is a natural entity well suited for mathematical studies. Fortunately, both LINSPACE and PSPACE are in the same boat. Neither of these classes turns out to be sensitive to various details in the definitions, in particular, they are not very sensitive to how we define an *algorithm* and the related notion of *the number of storage units required by an execution.*

Hence, we might view an algorithm as a program written in an object oriented programming language and the number of storage units required as the number of objects created during the execution of the program. However, be reasonable: if we take such a view, we naturally cannot permit the program to store data other places than in the objects, e.g. on files.

Personally we tend to think of an algorithm as a clerical routine carried out by hand. We imagine a clerk writing and erasing symbols in an infinite spreadsheet where each cell have room for exactly one symbol. The symbols should be take from a finite alphabet, but all sorts of symbols can be there (digits, Latin letters, marker symbols, Greek letters, etc.). The number of storage units required is of course the number of cells the clerk have to touch.

In a standard textbook one finds complexity-theoretic spaces classes defined by Turing machines. A Turing machine has an infinite tape that is divided into cells, and the number of storage units required for a computation will be defined as the number of cells the Turing machine's scanning head will visit during a run. Such a formal definition gives associations to the clerk manipulating symbols in a spreadsheet: the clerk is replaced by a mechanical device, and the two-dimensional spreadsheet is replaced by a one-dimensional tape.

## 5.2. *Example: SAT is in* LINSPACE

Let us gain some familiarity with the newly introduced space classes by verifying that the problem **SAT** belongs to LINSPACE. We will discuss how the clerk mentioned above, can solve the problem in linear space. The input is a Boolean expression, e.g.

$$\left(x_3 \cdot \overline{(x_{17} + x_2)}\right). \tag{16}$$

Now, the clerk needs to write down the input in his spreadsheet, and there is only room for one symbol in each cell, thus, he cannot use subscript or overline groups of symbols. We have to come up with a suitable convention for representing the input, and we can e.g. represent the Boolean expression in (16) by

$$( \ \{3\} \ . \ c( \ ( \ \{17\} \ + \ \{2\} \ ) \ ) \ ) . \tag{17}$$

This is a straightforward and obvious convention where we represent the complement $\overline{\alpha}$ of a Boolean expression $\alpha$ by $c(\alpha)$ and a variable by its subscript enclosed in braces. Any reasonable convention will do, and we will stick to this one.

The clerk will write down the input as a sequence of symbols. Immediately after the input, he will write a marker symbol !, and immediately after the marker, for each variable occurring in the input, he will write a copy of the symbol 0. When he is done, we have the configuration

$$( \ \{3\} \ . \ c( \ ( \ \{17\} \ + \ \{2\} \ ) \ ) \ ) \ ! \ 0 \ 0 \ 0 . \tag{18}$$

One can see how the clerk might proceed to solve the problem in linear space. The symbols at the right-hand side of the marker can be interpreted as three Boolean values, and the clerk will assign these values to the three Boolean variables in the input expression. The assignment can e.g. be carried out by writing the values in suitable cells below the expression:

```
( {3} . c( ( {17} + {2} ) ) ) ! 0 0 0
  0              0       0
```

Now the clerk can proceed in a fairly obvious way and compute the value of the whole expression by computing values of subexpressions. Under the given assignment, the value of ( {17} + {2} ) is 0, and the value will be written in a suitable cell, e.g. in the cell right below the cell where the subexpression starts:

```
( {3} . c( ( {17} + {2} ) ) ) ! 0 0 0
  0        0 0       0
```

Next, the clerk computes the value of c( ( {17} + {2} ) ) by reading off the value of ( {17} + {2} ), the result is e.g. placed right below the c:

```
( {3} . c( ( {17} + {2} ) ) ) ! 0 0 0
  0     1  0 0       0
```

And thus the clerk proceeds until he ends up with the value of the entire expression. If this value is 1, the clerk's job is done, and he can answer YES (the Boolean expression is satisfiable); if the value is 0, he will erase the bottom line, and start all over again from a configuration like (18), but of course, with a new assignment. The clerk will check the assignment

$$( \ \{3\} \ . \ c( \ ( \ \{17\} \ + \ \{2\} \ ) \ ) \ ) \ ) \ ! \ 0 \ 0 \ 1$$

Next he checks the assignment

$$( \ \{3\} \ . \ c( \ ( \ \{17\} \ + \ \{2\} \ ) \ ) \ ) \ ) \ ! \ 0 \ 1 \ 0$$

and so on. He is instructed to view the bit string at the right-hand side of the marker as a binary number, and he gets the subsequent assignments by increasing this number by 1. If he encounters an assignment satisfying the expression, he will quit and say YES (the expression is satisfiable), if not, he will go on until he has checked the assignment 1 1 1, and then he will quit and say NO (the expression is not satisfiable).

It should be fairly obvious from our discussion that the number of cells the clerk will touch, is bounded by a multiple in the number of cells required to represent the input. The bit sequence to the right of the marker will be shorter than the input to the left, and the bits written beneath the input will not require as much space as the input itself. Hence, if $|x|$ is the length of input, the clerk will not touch more than $3|x| + 1$ cells? Well, actually he might touch a few more. We should be careful. This clerk is a strange being. On the one hand, he is extremely unknowledgeable and forgetful, on the other hand, he is a devil at following unambiguous rules and carry them out the letter.[7] We cannot just tell him to count the number of variables in the input, or just tell him to increase a binary number by 1. We have to give him instructions of how to do so, and carrying out these instruction the clerk will need some auxiliary space. Exactly how much space he need, will of course depend on the instructions; maybe he needs to write down a few marker symbols, maybe he needs to write down some numbers, anyway, there is definitely possible to provide instructions such that the number of auxiliary cells required is bounded by e.g. $2|x| + 1000$. Further, if we just claim *the existence* of fixed numbers $k$ and $\ell$ such that number of cells required for the entire computation is bounded by $k|x| + \ell$, then we should be on the safe side. But then, the number of cells required is bounded $(k + \ell)|x|$ where $k + \ell$ is a fixed number.[8] Hence, **SAT** is in LINSPACE.

### 5.3. *Non-determinism and the class* NP

All the algorithms we have encountered so far have been *deterministic algorithms*. We have just called them algorithms since the standard and natural notion of an algorithm is the notion of a deterministic algorithm.

---

[7]Someone might call him a true bureaucrat, but beware, in contrast to a bureaucrat, the clerk is not inclined to act inconsistently.

[8]We assume that the input always has some length, i.e. that $|x| > 0$.

What is a *non-deterministic algorithm*? The very idea of a non-deterministic algorithm seems suspicious and might even sound like a contradiction in terms. Algorithms are supposed to be exhaustive and unambiguous recipes for carrying out tasks mechanically, whereas non-determinism alludes to something arbitrary and non-mechanical.[9]

Now, whereas a deterministic algorithm is a recipe that always tell you what to do next, a non-deterministic algorithm is a recipe that leaves you with several options. Extend a standard programming language, like C or a Java, by the construction

> **if** ? **then** program 1 **else** program 2

and you have a language capable of expressing any non-deterministic algorithm. When this if-then-else construction is executed, one, and only one, of the two program will be executed, but we do not know which one. The clerk can carry out a non-deterministic algorithm in his spreadsheet the way as he carries out a deterministic one, but occasionally he will find his instructions ambiguous in the sense that he could either do $A$ or $B$, and to decide what to do, he flips a coin.

An algorithm solves a problem if (and only if) the algorithm answer the question correctly, that is, if the answer to the problem on input $x$ is YES, respectively NO, the algorithm should terminate on input $x$ and say YES, respectively NO. Now, a non-deterministic algorithm has several possible executions on one and the same input, and one of the executions on input $x$ might result in a YES whereas another results in a NO. Thus, it is not obvious that it makes any sense saying that a non-deterministic algorithms solves a problems, neither is it obvious that it makes sense to say that a non-deterministic algorithm works in polynomial time, and the reader should study the next definition carefully.

DEFINITION 8 (*The class* NP). A *non-deterministic algorithm solves a problem* if (and only if)
  (1) if the answer to the question on input $x$ is NO, then any execution of the algorithm on input $x$ will give output NO,
  (2) if the answer to the question on input $x$ is YES, then there exists an execution of the algorithm on input $x$ giving output YES.

Let $f$ be any function from $\mathbb{N}$ to $\mathbb{N}$ and let $|x|$ denote the number of alphabet symbols required to represent the input $x$. A non-deterministic algorithm taking $x$ as input *runs*, or *works, in time* $f(|x|)$ if $f(|x|)$ bounds the number of basic steps in any execution of the algorithm on input $x$.

We will say that a non-deterministic algorithm *works in polynomial time* if there exists a polynomial $p$ such that the non-deterministic algorithm works in time $p(|x|)$.

A problem belongs to the class NP if (and only if) the problem can be solved by an non-deterministic algorithm working in polynomial time. We use standard terminology when say that the problems in NP are the problems *solvable*, or *decidable*, in *non-deterministic polynomial time*.

---

[9]Certain people seem inevitably attracted to a word like 'non-determinism', like they seem attracted to words like 'non-reductionism', 'non-monotone' and 'non-linear'. Perhaps because such words give associations to something elevated and sophisticated as opposed to their counterparts 'monotone', 'linear' and 'determinism'.

It should be obvious to any proficient computer programmer that any problem solved by a non-deterministic algorithm also can be solved by a deterministic one. On a particular input $x$ there will be finitely many executions of the non-deterministic algorithm, and by applying a standard technique called back-tracking, a deterministic algorithm can systematically check out each and one of the executions. If at least one of the executions results in a YES, the answer to the problem of input $x$ is also YES, and if every execution results in a NO, the answer is NO.

Hence, if non-deterministic algorithm solves a problem, there will also be a deterministic algorithm solving the problem. So what is all the fuss about? If non-determinism adds no computational power, why do we bother to introduce non-determinism at all? This is why: If a non-deterministic algorithm solves a *problem in polynomial time*, there will not necessarily be a deterministic algorithm solving the problem *problem in polynomial time*.

We can always simulate a non-deterministic algorithm by a deterministic one, but to simulate a non-deterministic algorithm running in polynomial time, will in general require exponential time. It is very likely that the class of problems solvable in deterministic polynomial time, that is P, is strictly included in the class of problems solvable in non-deterministic polynomial time, that is NP, but as mentioned above, this is one of the most famous open problems of mathematics.

### 5.4. *Example: SPLIT and SAT are in* NP

In Section 3.1 we discussed the problem **SPLIT**, and we rhetorically ask for the solution of the problem on the input

$$\{3, 6, 9, 10, 20, 30, 173, 174, 175, 1027, 1058, 1132, 1719, 2480\}. \tag{19}$$

Is it possible to split the numbers in the set into two piles such that the sum of the numbers in one pile equals the sum of the numbers in the other? We asked this question to make the reader realise that **SPLIT** is a hard problem. The reader that tried to answer the question, probably got bored after a while because he had to check out various ways to split the set, and there are awfully may ways to split the set, $2^{14}$ ways to be exact, that is more than 15 thousand possibilities. Maybe the reader did not find it necessary to check all the $2^{14}$ splittings one by one, but he probably realised that he would be forced to check quite a few of them, and then he eventually gave up. No one has ever found a fast deterministic algorithm for solving the problem **SPLIT**, and the problem seems intractable. However, it is easy to come up with a non-deterministic procedure solving the problem very quickly:

> **Input:** a set $X$ of natural numbers.
> **foreach** $a \in X$ **do**
>     **if** ? **then** put $a$ in $Y$ **else** put $a$ in $Z$
>     $y :=$ the sum of the numbers in $Y$
>     $z :=$ the sum of the numbers in $Z$
>     **if** $y = z$ **then** SAY YES **else** SAY NO

Given the input in (19), the algorithm will run through the numbers

$$3, 6, 9, 10, 20, 30, 173, 174, 175, 1027, 1058, 1132, 1719, 2480$$

and non-deterministically put some of them in $Y$ and some of them in $Z$. It is possible that $30, 174, 1027, 1058, 1719$ end up in $Y$, and that $3, 6, 9, 10, 20, 173, 175, 1132, 2480$ end up in $Z$. Then the output will be YES since

$$30 + 174 + 1027 + 1058 + 1719$$
$$= 4008$$
$$= 3 + 6 + 9 + 10 + 20 + 173 + 175 + 1132 + 2480.$$

It is also e.g. possible that 3 and 9 ends up in $Y$ and the rest of the numbers in $Z$. Then the output will be NO.

When we recall the definition of what it means for a non-deterministic algorithm to solve a problem, it is obvious that the given algorithm solves **SPLIT**. The algorithm is very fast and it is easy to argue that any execution of the algorithm runs in polynomial time in the length of the input. Hence the problem **SPLIT** is NP.

The problem **SAT** is also in NP. In Section 5.2 we argued that **SAT** is in LINSPACE by showing how a clerk with a spreadsheet could solve the problem in linear space. The clerk was instructed to systematically run through all possible assignments and check if one of them satisfied the Boolean expression given as input. Now, if we in contrast to instructing the clerk to check all assignments, instruct him to non-deterministically generate a single assignment and check that particular assignment, then we have a non-deterministic algorithm solving **SAT**. (If the assignments he is led to by flipping a coin, satisfies the input expression, he should say YES, otherwise, he should say NO.) But then, just checking a sole assignment, the clerk will be done very quick, and it is easy to see that the number of basic steps he has to carry out is bounded by a polynomial in the length of the input. Hence, the problem **SAT** is in NP.

### 5.5. *The class co-NP*

DEFINITION 9. Let COMP be any complexity class (e.g. P, NP, LINSPACE). We define the complexity class *co*-COMP by

> $coA$ is in *co*-COMP if and only if $A$ is in COMP.

If we have a deterministic algorithm solving a problem $A$, we can trivially modify the algorithm to solve the problem $coA$, just modify the algorithm to give output NO in place of YES and output YES in place of NO. The modified algorithm requires exactly the same amount of time and space resources as the initial algorithm, and thus the problem $coA$ will be in exactly the same deterministic complexity classes as the problem $A$. Thus, e.g.

P = co-P and LINSPACE = CO-LINSPACE since P and LINSPACE are deterministic complexity classes, that is, they are complexity classes defined by the resource requirements of deterministic algorithms.

Recall the definition of what it means for a non-deterministic algorithm to solve a problem: An non-deterministic algorithm solves a problem if and only if

(1) if the answer to the question on input $x$ is NO, then any execution of the algorithm on input $x$ will give output NO,

(2) if the answer to the question on input $x$ is YES, then there exists an execution of the algorithm on input $x$ giving output YES.

Notice the asymmetry: The answer to the problem is YES when *one* execution gives output YES (even if many other executions give output NO); the answer to the problem is NO when *all* executions give output NO (and not a single execution gives the output YES). This asymmetry implies that we cannot modify a non-deterministic algorithm solving a problem $A$ to solve the problem co-$A$ by swapping its outputs YES and NO, which again implies that the complexity class NP probably is different from the complexity class co-NP.

To determine the relationship between NP and co-NP is one of the famous open problems of complexity theory. It is likely that the two classes are different, but the research community will perhaps not be as unanimous in this respect as with respect to the problem $P \stackrel{?}{=} NP$. Since P = co-P, it cannot be the case that we both have NP $\neq$ co-NP and P = NP. Thus, to prove that NP is different from co-NP, is a way of proving that P is different from NP. (However, if NP should equal co-NP, it might still be the case that P is different from NP.)

## 5.6. *More on the relationship between the complexity classes*

It is proven (see Figure 7) that

$$P \subseteq NP \subseteq PSPACE. \tag{20}$$

It is an open problem whether P $\neq$ PSPACE, and hence, it is also an open problem if P $\neq$ NP and if NP $\neq$ PSPACE. The problems are considered open as no one has come up with rigorous mathematical proofs of these inequalities, however, it is considered highly likely that all the inequalities hold, and we have absolutely no experience indicating otherwise. Thus, the likely situation is that all the inclusions in (20) are strict, that we indeed have

$$P \subset NP \subset PSPACE. \tag{21}$$

Now, if we believe that (21) holds, we are also bound to believe that

$$P \subset co\text{-}NP \subset PSPACE. \tag{22}$$

Why are we bound to believe that (22) holds? Well, PSPACE is a deterministic complexity class, and hence, if a problem $A$ belongs to PSPACE, the co-problem $coA$ does also

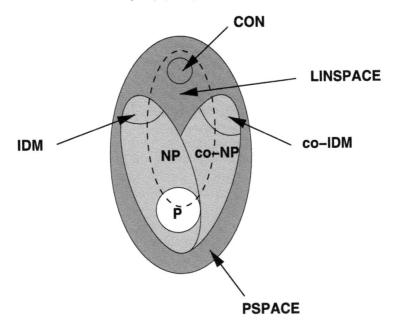

Fig. 7. Locations of problems in the space classes.

belong to PSPACE. Hence, we are bound to believe that CO-NP is included in PSPACE. Further, we cannot believe that CO-NP = PSPACE since this entails that NP = PSPACE (and we do not believe that). Hence we have to believe that CO-NP is strictly included in PSPACE. A symmetric argument shows that we also are bound to believe that P is strictly included in CO-NP.

The relationship between LINSPACE and the other complexity classes is slightly complicated. It is proved that LINSPACE ⊂ PSPACE. Further, it is proved that LINSPACE is different from P, that is LINSPACE ≠ P. Thus, it follows that we have one of the following situations:

1. LINSPACE ⊂ P (LINSPACE is strictly included in P),
2. P ⊂ LINSPACE (P is strictly included in LINSPACE),
3. P ⊄ LINSPACE and LINSPACE ⊄ P (neither of the classes is contained in the other).

Any of three situations is compatible with what we can prove, namely that LINSPACE cannot equal P, however, situation 1 implies that P = PSPACE and we do not believe that, so we have to rule out 1. (The argument 1 implies that P = PSPACE is non-trivial, and we omit the details.) Situation 2 does not seem to be incompatible with anything else we believe in, but the situation is still considered unlikely, and the general opinion among complexity theorists is that situation 3 is realised.

It is also proved that LINSPACE ≠ NP, and the likely situation is that neither of classes is included in the other, that is, NP ⊄ LINSPACE and LINSPACE ⊄ NP. The same can be said about the relationship between LINSPACE and CO-NP.

### 5.7. *The importance of the complexity classes*

The complexity class P delineates a border between the tractable and the intractable. If a problem belongs to P, real-life computing machinery can predictably solve the problem, if a problem does not belong to P, such machinery cannot deal with the problem. (See Section 3.2.) This makes P an important complexity class. To realise why NP is an important complexity class, we need the next definition.

DEFINITION 10.  A problem $A$ is NP-*hard* if (and only if) if every problem in NP is polynomial time reducible to $A$. Further, a problem $A$ is NP-*complete* if (and only if)
  1.  $A$ is NP-hard.
  2.  $A$ belongs to the class NP.

The NP-complete problems are the hardest problems in the class NP, and by studying the definition the reader will realise that if a single NP-complete problem can be solved in polynomial time, then every problem in NP can be solved in polynomial time. Now, since it is very likely that P $\neq$ NP, it is also very likely that no polynomial-time algorithm can solve an NP-complete problem, and it is not only very like, but for sure absolutely true, that no one in the history of computing has come upon a polynomial-time algorithm solving an NP-complete problem. Thus, NP-complete problems are intractable problems, that is, problems computing machinery cannot deal with. Besides, numerous natural problems from engineering, mathematics, science and daily life have turned out to be NP-complete, and thus NP becomes a very important complexity class. By proving that a problem is NP-complete, and a lot of natural problems will be NP-complete, we have proved that the problem is intractable. This is for sure a useful thing to know about a problem.

The notion of PSPACE-completeness is defined analogous to the notion of NP-completeness.

DEFINITION 11.  A problem $A$ is PSPACE-hard if (and only if) if every problem in PSPACE is polynomial time reducible to $A$. Further, a problem $A$ is PSPACE-complete if (and only if)
  1.  $A$ is PSPACE-hard.
  2.  $A$ belongs to the class PSPACE.

The PSPACE-complete problems are the hardest problems in PSPACE, and they should be even harder to solve than NP-complete problems. As far as we know NP $\neq$ PSPACE, and then even a *non-deterministic* algorithm will require *exponential* time to solve a PSPACE-complete problem.

Computers can to a certain extent deal with NP-complete problems. They cannot deal with such problems inn their full generality, but techniques and methods have been developed to find approximate and partial solutions. If a programmer realises he is facing an NP-complete, the situation is for sure not ideal, but the situation is not hopeless as he might be able to make his software respond satisfactorily on normal inputs. (There is a vast theory on how programmers can approach intractable problems.) If a programmer realises he is facing an PSPACE-complete, he is worse off, and it would probably be far more difficult to make the software behave satisfactorily.[10]

There are not nearly as many PSPACE-complete natural problems as there are NP-complete natural problems, but a significant number of interesting natural problems have turned out to be PSPACE-complete, and PSPACE is an important complexity class.

The class co-NP gains its importance from being the co-class of NP. What should we say about LINSPACE? In what sense is LINSPACE an important complexity class? Well, the class is surrounded by other important complexity classes, and that makes the class interesting, particularly since the relationship between LINSPACE and these classes is not settled. Besides, it is of course nice to know that it is possible to solve a problem in very limited space, still, from an applied point of view, LINSPACE might not be as important as P, NP and PSPACE. (Note that LINSPACE is not closed under polynomial time reductions, that is, a problem lying outside LINSPACE might be polynomial time reducible a problem in LINSPACE.)

In the literature the reader might find numerous complexity classes that we have not found time to discuss in this paper, but the ones we have discussed, are all very important and well-known complexity classes. The most prominent class that we have ignored is called LOGSPACE. (It is proved that LOGSPACE $\subseteq$ P, and it is an open problem if LOGSPACE $\neq$ P.) Other fairly well-known classes are known as *NC*, NLOGSPACE, NLINSPACE, ETIME, NETIME and $P^{\#}$.

## 6. More on operators and configuration management

### 6.1. *More on the complexity of IDM and INJ*

Some ideas needed to prove Theorem 5 and Theorem 6, are explained in Section 5.2, and the reader might benefit from studying the details of the section thoroughly before he embarks on the proofs of the theorems.

THEOREM 5. *The problem INJ is in* LINSPACE.

PROOF. In Section 5.2 we showed how a clerk calculating in a spreadsheet could solve the problem **SAT** in linear space. The input to the problem was a Boolean expression $\alpha$, and the clerk generated systematically all possible assignments to the variables of $\alpha$, and for each assignment, the clerk evaluated $\alpha$.

Now, the input to **INJ** is an operator expression $e$, and an operator expression is nothing but a sequence $\langle \alpha_1, \ldots, \alpha_n \rangle$ where each $\alpha_i$ is a Boolean formula over the variables $x_1, \ldots, x_n$. We can argue as in Section 5.2 that a clerk working in linear space, can assign values to $x_1, \ldots, x_n$, and for each assignment, he evaluate the expressions $\alpha_1, \ldots, \alpha_n$. The evaluation of $\alpha_1, \ldots, \alpha_n$ under a particular assignment yields a string of $n$ bits, and this

---

[10] An important mission of theoretical computer science is to tell engineers, programmers and applied computer scientists what they *cannot do*. This might be more important than telling them what they *can do* since they might know a lot about that already.

string is of course nothing but the result of applying the operator $[\![e]\!]$ to a particular argument in $\mathbb{B}^n$. Thus, it should be clear that a clerk can carry out the following routine in linear space:

> **Input:** an operator expression $e$.
> $n :=$ the number of variables in $e$
> Let $b^0, b^1, \ldots, b^{2^n-1}$ be the obvious enumeration of the elements in $\mathbb{B}^n$
> **for** $i := 0 \ldots 2^n - 2$ **do**
>   **begin**
>   $b := [\![e]\!](b^i)$
>   **for** $j := i + 1 \ldots 2^n - 1$ **do**
>     **begin**
>     $c := [\![e]\!](b^j)$
>     **if** $b = c$ **then** SAY NO AND QUIT **else** PROCEED
>     **end**
>   **end**
> SAY YES

The routine is simply a brute force check. For each bit string $b^i$, the clerk check if $[\![e]\!](b^i) = [\![e]\!](b^j)$ for some bit string $b^j$ where $j > i$. (Obviously, there is no need check when $j < i$, but it would not have harmed if we did so.) If the clerk finds two bit strings witnessing that the operator given by the input is not an injection, he says NO and quits the routine. If such witnesses never turn up, the clerk says YES after the checking is done.

This proves that **INJ** is in LINSPACE. $\qquad\qquad\square$

THEOREM 6. *The problem* **IDM** *is in* LINSPACE.

PROOF. This proof is similar to the proof of Theorem 5, but slightly simpler. Again, we give a routine meant to be carried out by a clerk working in a spreadsheet:

> **Input:** an operator expression $e$.
> $n :=$ the number of variables in $e$
> Let $b^0, b^1, \ldots, b^{2^n-1}$ be the obvious enumeration of the elements in $\mathbb{B}^n$
> **for** $i := 0 \ldots 2^n - 1$ **do**
>   **begin**
>   $c := [\![e]\!](b^i)$
>   $d := [\![e]\!](c)$
>   **if** $c = d$ **then** PROCEED **else** SAY NO AND QUIT
>   **end**
> SAY YES

The clerk is instructed to check for absolutely every bit string $b \in \mathbb{B}^n$ and see if the string yields a counterexample to the equation $[\![e]\!](b) = [\![e]\!]([\![e]\!](b))$. If a counterexample turns up, the clerk will interrupt the routine and give output NO, and if no counterexample turns up, he will give output YES. Following the lines of the reasoning in Section 5.2, the reader can verify that it is possible for the clerk to carry out this routine in linear space. Thus, the problem **IDM** is in LINSPACE.                                                                                    □

THEOREM 7. *The problem coINJ is* NP-*complete.*

PROOF. We are going to prove that *co***INJ** is NP-complete. How should we proceed? Well, we have to consult our definitions to find out what it means that the problem *co***INJ** is NP-complete. Definition 10 states that

> *co***INJ** is NP-*complete* if and only if (1) *co***INJ** is NP-hard and (2) *co***INJ** belongs to the class NP.

Thus, we have to prove two things, i.e. we have to prove that (1) *co***INJ** is NP-hard, and we have to prove that (2) *co***INJ** belongs to the class NP.

First we prove (1). According to the definition of NP-hardness (Definition 10) we have to prove that every problem in NP is polynomial time reducible to *co***INJ**. In the proof we will apply the following claim.

CLAIM. *The problem **SAT** is* NP-*hard.*

A proof of Claim can be found in any standard textbook on basic complexity theory.

Pick an arbitrary problem $A$ in the class NP. Since we know that **SAT** is NP-hard, we know that $A$ is polynomial time reducible to **SAT**, and hence, according to Definition 3 is there exists $g$ such that:
- The answer to question $A$ with input $x$ is the same as the answer to question **SAT** with input $g(x)$.
- The function $g$ is computable in polynomial time.

Further, Theorem 3 states that the problem **SAT** is polynomial time reducible to *co***INJ**, and hence, again according to Definition 3, there exists $h$ such that:
- The answer to question **SAT** with input $x$ is the same as the answer to question *co***INJ** with input $h(x)$.
- The function $h$ is computable in polynomial time.

Now, let $f(x) = h(g(x))$. The reader can easily check that:
- The answer to question $A$ with input $x$ is the same as the answer to question *co***INJ** with input $f(x)$.
- The function $f$ in computable in polynomial time.

Hence, again by Definition 3, the problem $A$ is polynomial time reducible to *co***INJ**. Now, $A$ was arbitrarily chosen from the class NP, and we have not assumed that the problem possess any particular properties beyond belonging to NP. Hence, any problem in NP will be polynomial time reducible to *co***INJ** which means that the problem *co***INJ** is NP-hard.

We will now turn to the proof of (2), that is, we will prove that *co***INJ** belongs to NP, and we will do so by providing an non-deterministic algorithm solving *co***INJ** in polynomial time. If such an algorithm exists, the definition of NP says that $co$**INJ** is in NP. Let us remind ourself of how the problem $co$**INJ** is stated.

($co$**INJ**) *Input*: an operator expression $e$.
*Question*: Do we have $[\![e]\!](b) = [\![e]\!](b')$ and $b \neq b'$ for some $b, b' \in \mathbb{B}^n$?

It is obvious that whenever an operator expression $e$ and two bit string $b, b' \in \mathbb{B}^n$ are available, then a deterministic algorithm can check quite fast if it will be the case that $[\![e]\!](b) = [\![e]\!](b')$ and $b \neq b'$, and it is beyond any doubt that such an algorithm will work in polynomial time in the length of the input expression $e$. Now, $n$, that is the length of bit strings $b$ and $b'$, will definitely be shorter than the length of the input, and thus, a non-deterministic algorithm working in polynomial (even linear) time in the length of input, might non-deterministically generate two candidates for $b$ and $b'$. Hence, it is easy to see that there exists an non-deterministic polynomial time algorithm solving $co$**INJ**: simply generate non-deterministically the two bit strings $b$ and $b'$, and thereafter check if it is the case that $[\![e]\!](b) = [\![e]\!](b')$ and $b \neq b'$. If it is the case, the answer to the problem is YES, else the answer is NO.                                                                                                                 □

THEOREM 8. *The problem co***IDM** *is* NP-*complete.*

PROOF. The proof of this theorem is analogous to the proof of Theorem 7. We omit the details.                                                                                                                 □

COROLLARY 2. (i) *The problem* **INJ** *is in* co-NP, *furthermore, any problem in* co-NP *is polynomial time reducible to* **INJ**, *i.e.,* **INJ** *is* co-NP-*complete.* (ii) *The problem* **IDM** *is in* co-NP, *furthermore, any problem in* co-NP *is polynomial time reducible to* **INJ**, *i.e.,* **IDM** *is* co-NP-*complete.*

PROOF. (i) follows straightforwardly from Theorem 7, and (ii) follows straightforwardly from Theorem 8.                                                                                                                 □

## 6.2. *The complexity of* **CON** *and* **CONx**

Next we turn to the so-called 'convergent' operations, in the classical sense of the phrase. Ideally we want an operator that not only converges, but which converges always to the same value in the fixed-point sense of cfengine [7], i.e. $[\![e]\!]^k(x) = [\![e]\!]^{k+1}(x) =$ "policy" for all $x$. However, this problem is at least as hard as the general convergent case, since we must first check for convergence and then additionally check whether the value is correct, so we do not need to mention it again.

**6.2.1.** *A brief discussion*    Note the difference between the problem **CON** and the problem **CONx**. If you are asked to solve the problem **CONx**, you will be given a pair $\langle e, b \rangle$ where $b \in \mathbb{B}^n$ and $e$ is an operator expression of the form $\langle \alpha_1, \ldots, \alpha_n \rangle$, and then, it is your task to decide if the operator $[\![e]\!]$ converges in the particular argument $b$, that is, you should decide if there exists a $k$ such that $[\![e]\!]^k(b) = [\![e]\!]^{k+1}(b)$. If you are asked to solve the problem **CON**, you will be given nothing but an operator expression of $e$ of the form $\langle \alpha_1, \ldots, \alpha_n \rangle$, and it is your task to decide if the operator $[\![e]\!]$ converges for every $x \in \mathbb{B}^n$. This sounds like a formidable task as you have to check if the operator converge for each of the $2^n$ elements in $\mathbb{B}^n$. Thus, at a first glance the problem **CON** might seem harder that the problem **CONx**, and it certainly does not seem any easier.

In fact even **CONx** is a hard problem, indeed, the problem turns out to be PSPACE-hard. So far we have done our best to avoid Turing machines, but now we are forced to introduce them as we will prove that **CONx** is PSPACE-hard by constructing operators simulating Turing machines.

**6.2.2.** *Turing machines*    A Turing machine might be seen as a mechanical device shuffling symbols around an infinite tape. The tape is divided into cells, or squares; the tape starts somewhere but does not end anywhere. Thus, we can talk about the first cell of the tape, the second cell of the tape, and in general the $n$th cell of the tape. Each cell has room for exactly one symbol taken from a finite *alphabet*. The alphabet varies form machine to machine, but it is always finite, and it always contains a dedicated symbol called the *blank* and denoted **B**. A cell containing **B** will be called *a blank cell*. At any particular moment in time the machine will be in one, and only one, of a finite number of possible states, and a device called the machine's *head*, will be scanning one, and only one, of the cells on the tape. The symbol in the scanned cell is called the *scanned symbol*. The combination of the current state and the scanned symbol decides what the Turing machine will do next, and the possible actions are indeed very limited:

  (i)   write a symbol in the scanned cell and enter a new state,
  (ii)  move the head one position to the left and enter a new state,
  (iii) move the head one position to the right and enter a new state.
The exact action to be executed is determined by a finite set of instructions of the form

$$a, q, A, r$$

where $a$ is an alphabet symbol, $q$ and $r$ are states, and $A$ is either an alphabet symbol, or a left arrow ($\leftarrow$), or a right arrow ($\rightarrow$). If the machine is in the state $q$ and is scanning the symbol $a$, then the machine will carry out the action $A$ and enter state $r$. If $A$ is an alphabet symbol, the action will simply be to write that symbol in the scanned square; if $A$ is an arrow, the action will simply be to move the scanning head one position in the direction of the arrow. If the Turing machine is *deterministic*, there will never be more than one instruction in the set that is applicable in a particular situation, if the Turing machine is *non-deterministic*, several instructions might be applicable, and the machine will execute one of them. It might also be the case that none of the instructions are applicable, then the machine will simply 'hang'.

When a Turing machine starts, the leftmost cell on the tape, that is, the first cell, will be blank, and the input will be found, symbol by symbol, in the consecutive cells. The input will not contain any blanks, and the head will be scanning the first symbol of the input.

The remaining cells of the infinite tape are all blank. E.g., if the input *abbaa* is given to a machine, the tape will be in the configuration

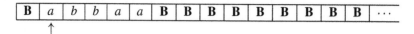

when the execution starts. The arrow marks the scanned cell.

Turing machines can perform tasks like computing functions and deciding problems by manipulating the symbols on the tape. The instructions tell a machine how to write and delete symbols and how to move the head back and forth. To delete a symbol from a particular cell, does of course amount to write the blank symbol into that cell. When a new symbol is written into a cell, the old symbol vanishes and is lost forever. Let us study a Turing machine copying a string of *a*'s. The string might be of arbitrary length and is given to the machine as input, so the Turing machine's task is to bring the tape from the configuration

$$\mathbf{B}\underbrace{a\ldots a}_{n}\mathbf{BBBBBBBBBBBBB}\ldots$$

to the configuration

$$\mathbf{B}\underbrace{a\ldots a}_{n}\mathbf{B}\underbrace{a\ldots a}_{n}\mathbf{BBBBBBBBB}\ldots$$

for any $n \in \mathbb{N}$. To define a particular Turing machine, we need to give (i) the alphabet, (ii) the states, (iii) the instructions and (iv) the start state, that is, the state the Turing machine shall be in when the execution starts. The machine copying a string of *a*'s, is given in Figure 8.

We might visualise the execution of a Turing machine (at a particular input) as a sequence of configurations $c_0, c_1, c_2, \ldots$. A single configuration is a pair consisting of

- a finite sequence of alphabet symbols where exactly one symbol is underlined,
- a state.

(i) Alphabet: $\{a, X, \mathbf{B}\}$
(ii) States: $\{q_0, q_1, q_2, q_3, q_4, q_5\}$
(iii) Instructions:

| | | | | | | | | | |
|---|---|---|---|---|---|---|---|---|---|
| $\mathbf{B}$ | $q_1$ | $\mathbf{B}$ | $q_0$ | (inst. 1) | $\mathbf{B}$ | $q_3$ | $a$ | $q_4$ | (inst. 7) |
| $a$ | $q_1$ | $X$ | $q_2$ | (inst. 2) | $a$ | $q_4$ | $\leftarrow$ | $q_4$ | (inst. 8) |
| $X$ | $q_2$ | $\rightarrow$ | $q_2$ | (inst. 3) | $\mathbf{B}$ | $q_4$ | $\leftarrow$ | $q_4$ | (inst. 9) |
| $a$ | $q_2$ | $\rightarrow$ | $q_2$ | (inst. 4) | $X$ | $q_4$ | $a$ | $q_5$ | (inst. 10) |
| $\mathbf{B}$ | $q_2$ | $\rightarrow$ | $q_3$ | (inst. 5) | $a$ | $q_5$ | $\rightarrow$ | $q_1$ | (inst. 11) |
| $a$ | $q_3$ | $\rightarrow$ | $q_3$ | (inst. 6) | | | | | |

(iv) Start state: $q_1$

Fig. 8. A Turing machine for copying a string of *a*'s.

The state of the configuration gives the current state of the Turing machine, and the sequence of alphabet symbols gives the current content of the tape, symbol by symbol, from the left to the right. If there are $n$ symbols in the sequence, the sequence gives the content of the $n$ leftmost cells of the tape. The remaining cells on the infinite tape will all be blank. The underlined symbol is of course the scanned symbol.

The execution of our copy machine on the input $aa$ results in the sequence of configurations in Figure 9.

We will not spend any time considering the various ways a Turing machine might yield output. We are only interested in deterministic terminating Turing machines solving problems, and the answer to a problem is just YES or NO, and hence, all we need to provide is a convention for Turing machine to say YES/NO. In this respect we follow the standard literature and equip the Turing machine with two dedicated states: the *accept* state $q_A$ and the *reject* state $q_R$. Furthermore, a Turing machine solving a problem $A$ should starting with the input to $A$ on its tape end up in

- the accept state $q_A$ if the answer to $A$ on input $x$ is YES,
- the reject state $q_R$ if the answer to $A$ on input $x$ is NO.

After the machine has entered the accept state or the reject state, no instruction should be applicable, that is, the machine should halt.

| Config.: | Tape: | State: | Comments: |
|----------|-------|--------|-----------|
| $c_0$ | B$\underline{a}a$B | $q_1$ | initial config. |
| $c_1$ | B$\underline{X}a$B | $q_2$ | by (inst. 2) |
| $c_2$ | B$X\underline{a}$B | $q_2$ | by (inst. 3) |
| $c_3$ | B$Xa\underline{B}$B | $q_2$ | by (inst. 4) |
| $c_4$ | B$XaB\underline{B}$B | $q_3$ | by (inst. 5) |
| $c_5$ | B$XaB\underline{a}$B | $q_4$ | by (inst. 7) |
| $c_6$ | B$Xa\underline{B}a$B | $q_4$ | by (inst. 8) |
| $c_7$ | B$X\underline{a}Ba$B | $q_4$ | by (inst. 9) |
| $c_8$ | B$\underline{X}aBa$B | $q_4$ | by (inst. 8) |
| $c_9$ | B$\underline{a}aBa$B | $q_5$ | by (inst. 10) |
| $c_{10}$ | B$a\underline{a}Ba$B | $q_1$ | by (inst. 11) |
| $c_{11}$ | B$a\underline{X}Ba$B | $q_2$ | by (inst. 2) |
| $c_{12}$ | B$aX\underline{B}a$B | $q_2$ | by (inst. 3) |
| $c_{13}$ | B$aXB\underline{a}$B | $q_3$ | by (inst. 5) |
| $c_{14}$ | B$aXBa\underline{B}$B | $q_3$ | by (inst. 6) |
| $c_{15}$ | B$aXBa\underline{a}$B | $q_4$ | by (inst. 7) |
| $c_{16}$ | B$aXB\underline{a}a$B | $q_4$ | by (inst. 8) |
| $c_{17}$ | B$aX\underline{B}aa$B | $q_4$ | by (inst. 8) |
| $c_{18}$ | B$a\underline{X}Baa$B | $q_4$ | by (inst. 9) |
| $c_{19}$ | B$a\underline{a}Baa$B | $q_5$ | by (inst. 10) |
| $c_{20}$ | B$aa\underline{B}aa$B | $q_1$ | by (inst. 11) |
| $c_{21}$ | B$aa\underline{B}aa$B | $q_0$ | by (inst. 1) |

Fig. 9. A sequence of configurations generated by the machine copying a string of $a$'s. The input is the string $aa$.

**6.2.3.** *Operators simulating Turing machines*   We can view a deterministic Turing machine $M$ as an operator $T_M$, that is, $T_M$ is a unary function transforming one configuration into another one, and when $M$ makes a transition from the configuration $c$ to the configuration $c'$, then $T_M$ makes the same transition, that is, $T_M(c) = c'$. If none of the instructions of $M$'s is applicable to a given configuration $c$, we define $T_M(c) = c$. Thus, if $M$ is a machine generating the sequence of configurations given in Figure 9, we have[11]

$$T_M(\mathbf{B}\underline{aa}\mathbf{B}q_1) = \mathbf{B}\underline{X}a\mathbf{B}q_2$$
$$T_M(\mathbf{B}\underline{X}a\mathbf{B}q_2) = T_M\big(T_M(\mathbf{B}\underline{aa}\mathbf{B}q_1)\big) = \mathbf{B}X\underline{a}\mathbf{B}q_2$$
$$T_M(\mathbf{B}X\underline{a}\mathbf{B}q_2) = T_M\big(T_M(\mathbf{B}\underline{X}a\mathbf{B}q_2)\big) = T_M\big(T_M\big(T_M(\mathbf{B}\underline{aa}\mathbf{B}q_1)\big)\big) = \mathbf{B}Xa\underline{\mathbf{B}}\mathbf{B}q_2$$
$$\vdots$$

etc.

We will call such an operator $T_M$ the *transition function* of $M$. Note that if no instruction of $M$'s is applicable for the configuration $c$, we have $T_M(c) = c$ by definition.

Now, suppose the Turing machine $M$ is solving the problem $A$, that is, when $M$ is executed on input $x$, then $M$ will halt in

- the accept state $q_A$ if the answer to $A$ on the input $x$ is YES,
- the reject $q_R$ if the answer to $A$ on the input $x$ is NO.

Let us modify $M$ slightly such that in contrast to rejecting by entering the state $q_R$, $M$ rejects by moving the head back and forth between two cells forever and ever. Thus, when $M$ is given the input $x$, the execution of $M$ will either end up in the state $q_A$ (accepting $x$) or end up alternating between to configurations $c, c', c, c', c, c', \ldots$ (rejecting $x$). Let $T_M$ be the transition function of this modified version of $M$, and let $c$ be the initial configuration for $M$ at the input $x$. Then, if the answer to $A$ on the input $x$ is YES, there exists $k$ such that $T_M^{k+1}(c) = T_M^k(c)$, and if the answer to $A$ on the input $x$ is NO, no such $k$ exists, in other words, $T_M$ converges in the argument $c$ if and only if the answer to $A$ on the input $x$ is YES.

Now, a configuration can of course be represented as a sequence of bits, and if a Turing machine working in polynomial space generates the configurations $c_0, c_1, c_2, \ldots$, the number of bits required to represent each configuration $c_i$ will be bounded by a polynomial in the length of the input, hence, if $n \geqslant p(|x|)$ where $x$ is the input and $p$ is some fixed polynomial, bit strings taken from $\mathbb{B}^n$ will be large enough to encode any configuration in the sequence $c_0, c_1, c_2, \ldots$. This explains why the next theorem holds.

THEOREM 9. *Let $M$ be a deterministic Turing machine solving a problem in polynomial space. There exists a fixed polynomial $p$ such that for any input $x$ to $M$ there exist a bit string $b \in \mathbb{B}^n$ and an operator $T : \mathbb{B}^n \to \mathbb{B}^n$ where $n = p(|x|)$ such that the following assertions are equivalent:*

- *$T$ converges in the argument $b$,*
- *the execution of $M$ on input $x$ halts in the accept state $q_A$.*

---

[11]We write the state of the configuration immediately after the sequence of alphabet symbols giving the configuration of the tape.

**6.2.4.** **CONx** *is* PSPACE-*hard*    To prove that **CONx** is PSPACE-hard, we have to prove that every problem in PSPACE is polynomial time reducible to **CONx**, that is, we have to prove that for every problem $A$ in PSPACE there exists a polynomial time computable function $f$ such that the answer to the problem $A$ on the input $x$ is the same as the answer to the problem **CONx** on the input $f(x)$. (See Definition 11 and Definition 3.)

If a problem $A$ belongs to PSPACE there does of course exist a deterministic Turing machine $M$ solving $A$ in polynomial space. Now, have a second look at Theorem 9. Read the theorem carefully and do particularly note that

$$\dots \text{for any input } x \text{ to } M \text{ there exist } b \in \mathbb{B}^n \text{ and } T : \mathbb{B}^n \to \mathbb{B}^n \dots$$

The operator $T$ is more or less the transition function for the Turing machine $M$, and the bit string $b$ codes the initial configuration of the execution of $M$ on input $x$. When $x$ and $M$ are given, it will be possible to construct the bit string $b$ and an operator expression $e$ for the operator $T$, that is, $e$ such that $[\![e]\!] = T$. Moreover, the construction can be carried out by an algorithm running in polynomial time in the length of $x$. (The Turing machine $M$ is fixed and should not be viewed as input to the algorithm.) Any trained complexity theorist can easily design such an algorithm, but since the details are involved and cumbersome, we will save the reader from the entire story. The average reader should be satisfied to realise the consequences of having such an algorithm: There exists a polynomial time computable function $f$ such that when $f(x) = \langle e, b \rangle$, then the answer to the problem $A$ on input $x$ is the same as the answer to problem **CONx** on input $\langle e, b \rangle$. Hence, $A$ is polynomial time reducible to **CONx**. But $A$ is an arbitrary problem in PSPACE. Hence, any problem in PSPACE is polynomial time reducible to **CONx**, and thus, the next theorem holds.

THEOREM 10. **CONx** *is* PSPACE-*hard*.

Let us do a brief summary sketching the proof of the theorem once more: Let $A$ be an arbitrary problem in PSPACE. (When we have proved that $A$ is polynomial time reducible to **CONx**, we have proved that **CONx** is PSPACE-hard, see Definition 11.) Let $M$ be a deterministic Turing machine solving the problem $A$ in polynomial space. (Since $A$ is in PSPACE, we know that such a Turing machine exists.) We have a polynomial time computable function $f$ such that when $f(x) = \langle e, b \rangle$, the execution of $M$ on input $x$ halts in the accept state $q_A$ if, and only if, the operator $[\![e]\!]$ converges in the argument $b$. Thus, when $f(x) = \langle e, b \rangle$, the following assertions are equivalent:

- the answer to $A$ on input $x$ is YES,
- the execution of $M$ on input $x$ halts in the state $q_A$,
- the operator $[\![e]\!]$ converges in the argument $b$,
- the answer to the problem **CONx** on input $\langle e, b \rangle$ is YES.

Thus, the answer to the problem $A$ on input $x$ is the same as the answer to the problem **CONx** on input $f(x)$, and since $f$ is a polynomial time computable function, the problem $A$ is polynomial time reducible to **CONx**.

**6.2.5.** *Our last words on the complexity of* **CONx** *and* **CON**    It follows Theorem 10 that if **CONx** can be solved in polynomial time, then any problem in PSPACE can be solved in polynomial time. Thus, it makes very good sense to say that **CONx** is as least as hard any other problem in PSPACE, but exactly how hard is the problem? Is the problem much harder than the other problems in PSPACE, or can the problem itself be solved in polynomial space? Well, it turns out that problem indeed can be solved in linear space and hence belongs to LINSPACE! This might sounds like a contradiction since we know that LINSPACE is strictly included in PSPACE, indeed, it would have been a contradiction if LINSPACE were closed under polynomial time reduction, but LINSPACE is not closed under polynomial time reduction.

THEOREM 11.  ***CONx** and **CON** belongs to* LINSPACE, *and hence, also to* PSPACE.

It is a suitable exercise for the reader to verify that Theorem 11 holds: First, argue along the lines of the example in Section 5.2 that it is possible to solve **CONx** in linear space, then, argue that if **CONx** can be solved linear space, **CON** can also be solved in linear space.

Definition 11 says that a problem is PSPACE-complete if it in addition to be PSPACE-hard, belongs to PSPACE. Thus, the next corollary is a straightforward consequence of Theorems 10 and 11.

COROLLARY 3.  ***CONx** is* PSPACE-*complete*.

What about **CON**? The informal discussion in the beginning of the section concluded that the problem seems harder than **CONx**, and thus, since we know that the problem is in PSPACE, one should expect the problem to be PSPACE-complete. Well, we have not been able to prove that **CON** is PSPACE-complete, and maybe it is not. Hence, we have an open problem seeming quite interesting from a pure complexity-theoretic point of view.

OPEN PROBLEM 1.  Determine the complexity of **CON**. Is the problem PSPACE-complete? Does the problem belong to NP or *co*-NP? Does the problem belong to P?

## 7. Conclusions

### 7.1. *A summary*

Let us ask ourself if we have become any wiser as a result of our analysis, and if so, in what respect? Well, what did we want to know? If we were asked by layman, we could say that we wanted to know the answer the following question: (i) *How hard can it be to configure a computing system?* In some sense or another, this is the question that motivates our investigations; but there are many sensible ways to approach such an informal question.

We might, for instance, discuss configuration management in light of our past experiences; study the practice of system administrators; study the user-friendliness of software tools for configuring systems [1] and so on. We have chosen, however, to approach the question (i) mathematically, and in order to do so, we had to turn (i) into a mathematical question. The first step in the processes was to introduce the notion of an operator, and we said that to configure a system according to a specific policy amounts to recognising certain properties of operators.

Thus, we transformed (i) into the following question: (ii) *How hard is it to decide if an operator is idempotent, injective, fixed-point convergent etc.?* Now, (ii) is far more suited to analysis mathematically. In order to undertake a complexity-theoretic analysis, we had to turn (ii) into an even more specific question. We had to formalise the operators by introducing the operators expressions, etc.

Following this chain of reasoning to the end, we were able to relate the problem of policy-directed operators to other problems of known complexity and therefore determine a series of answers. The most important of these answers (since this is the one that answers the main question) is that the complexity of **CONx** is PSPACE-complete (note that this is harder than the worst case estimate of a similar but differently formulated problem according to the combinatoric formulation of Sun and Couch, but the additional difficulty goes hand in hand with additional benefits).

What does the result mean? Are we sure that our formulation is the right one, or an interesting one? Ask a stupid question, get a stupid answer. Garbage in, garbage out, etc. There are plenty of expressions to express this well-known adage. We believe that the following is the significance of the result:

- In the absence of any special restrictions, and irrespective of the tools used, the problem of planning, solving and deploying a solution with a deterministic behaviour is at least NP hard, even at time $t = 0$. The presence of an environment might already start to mess with the system after this. This means that system administrators and system designers are forced to employ heuristic methods to solve this matter. Heuristic methods are a plausible route for solution, but the price for using them is an intrinsic uncertainty in the result. This uncertainty is probably no worse than the uncertainties introduced by the uncertain environment after $t = 0$.
- By limiting the form of allowed policies, their corresponding configurations and hence expected behaviours to a set with specific properties, one can find solutions with inherently less complexity that are verifiable. We do not currently understand the limitations imposed by this.

So, are we forced to approximate or limit the complexity of configurations in order to have a reasonable chance of understanding and verifying them? There is a certain practical value to approximation, since (in contrast with the determination of an algorithm with lasting or reusable value) a general policy based solution in a changing environment will require change and adaptation regularly – thus one cannot afford to use algorithms that take too long.

It is actually hard to put oneself in the situation we are proposing. We are almost never in the position of not knowing the answer in the way we suggest here. We have many heuristic rules for fixing problems that are quickly implemented, but it is important to realise that we have no way to verify their correctness, except empirically.

Thus perhaps the real question of interest, for the future, is how is the complexity of a problem reduced by constraining it? Might this lead to provable solutions to configuration problems, or are we stuck with empiricism?

**7.2.** *Approximation or bust*

If we want to solve the general system administration problem in terms of operators then we know, either from the arguments here or from the parallel model developed by Sun and Couch, that the problem is at least NP hard. But what if we do not want to solve the general problem? What if we are willing to approximate it, or live with certain restrictions?

It seems intuitive that, if we restrict the problem by imposing additional constraints on the operators, then it might be possible to reduce the inherent complexity. Sun and Couch argue that this is true in restricting to the notion of observable states, for instance; Burgess argued that the fundamental hypothesis can be true by identifying policy as the set of achievable behaviours [5]. Indeed, suppose we take the matrix representation use in the examples earlier: then the search space of these square objects filled with only ones or zeroes is quadratic, and the complexity of solving problems with such operators would fall within the tractable P.

The price paid for such a restriction must be that not all configurations or policies are then implementable. The number of implementable policies would have to be reduced as a result. Just how much they would be reduced in not clear. The answer lies in the grey areas of the fundamental hypothesis, for which we have no detailed answers, but this remains to be determined by future research, at least in idealised systems. Real systems are probably too informal to admit a simple answer.

So the question 'What is an acceptable policy?' takes on a new significance after these studies. It could be that the answer to the question is that it is a policy that we can find in a reasonable length of time, rather than one that satisfies every detail of our wishes.

System administrators often speak of 'best practises' rather than 'optimisations'. This is a disturbing idea for a scientist who immediately wants to know: what is the criterion for best? Perhaps one answer is the closest policy that is achievable within polynomial time by some method? For now, this does not tell us anything new however: even NP hard problems can be verified in polynomial time if one makes the right lucky guess (e.g. if a human administrator or Monte Carlo algorithm performs a short random search and comes up lucky). There is clearly work to be done to clarify these matters.

**7.3.** *Last words*

There are many definitions of complexity in science. We have chosen to look at a single one that is related to the difficulty of problem solving. This is a useful measure because it tells us something about how hard it might be to find a 'guaranteed' solution to a problem. Alternatives exist: for example, the Kolmogorov complexity, which is, in turn, related to Shannon entropy [14], is a measure of how much variability there is in a pattern of data. It is the length of the shortest computer program that could generate the pattern; however it does not tell us how hard it might be to find that program, so it does not help us with the management of problems.

We have chosen to specialise arguments by talking about bit strings. These bit strings are quite easy to imagine in terms of configuration management problems, but the conclusions are, of course, general, as we illustrated in Figure 5. One can always build higher structures out of bit strings in polynomial time.

We have used the algorithmic definition of complexity to talk about the complexity of management problems in Network and System Administration. We have found that, in general, solving problems predictably and in a guaranteed fashion is hard (NP or even PSPACE hard). However, if we can live with certain restrictions (if we can formulate our systems within certain reasonable constraints and assumptions) then the problems can be solved in polynomial time. Of course, one is never guaranteed to be able to find such a representation for the task one wishes to accomplish. We are probably doomed to live with the need for approximation.

A pragmatic strategy for management would therefore be to pursue policies in terms of those operations that are easily computed. The question then remains what one might be missing in such a framework. Alas, the margin is too small to answer that question here.

## References

[1] A. Brown, A. Keller and J. Hellerstein, *A model of configuration complexity and its application to a change management system*, Proceedings of the IXth IFIP/IEEE International Symposium on Integrated Network Management IM'2005, IEEE Press, New York (2005), 631–644.

[2] M. Burgess, *System administration as communication over a noisy channel*, Proceedings of the 3nd International System Administration and Networking Conference (SANE2002) (2002), 36.

[3] M. Burgess, *A site configuration engine*, Computing Systems **8**, MIT Press, Cambridge, MA (1995), 309.

[4] M. Burgess, *Theoretical system administration*, Proceedings of the Fourteenth Systems Administration Conference (LISA XIV), USENIX Association, Berkeley, CA (2000), 1.

[5] M. Burgess, *On the theory of system administration*, Science of Computer Programming **49** (2003), 1.

[6] M. Burgess, *Analytical Network and System Administration – Managing Human–Computer Systems*, Wiley, Chichester (2004).

[7] M. Burgess, *Configurable immunity model of evolving configuration management*, Science of Computer Programming **51** (2004), 197.

[8] M. Burgess, *Configurable immunity for evolving human–computer systems*, Science of Computer Programming **51** (2004), 197.

[9] A. Couch and N. Daniels, *The maelstrom: Network service debugging via "ineffective procedures"*, Proceedings of the Fifteenth Systems Administration Conference (LISA XV), USENIX Association, Berkeley, CA (2001), 63.

[10] A. Couch and M. Gilfix, *It's elementary, dear Watson: Applying logic programming to convergent system management processes*, Proceedings of the Thirteenth Systems Administration Conference (LISA XIII), USENIX Association, Berkeley, CA (1999), 123.

[11] A. Couch, J. Hart, E.G. Idhaw and D. Kallas, *Seeking closure in an open world: A behavioural agent approach to configuration management*, Proceedings of the Seventeenth Systems Administration Conference (LISA XVII), USENIX Association, Berkeley, CA (2003), 129.

[12] A. Couch and Y. Sun, *On the algebraic structure of convergence*, Proc. 14th IFIP/IEEE International Workshop on Distributed Systems: Operations and Management, Lecture Notes in Comput. Sci., Vol. 2867, Heidelberg, Germany (2003), 28–40.

[13] A. Couch and Y. Sun, *On observed reproducibility in network configuration management*, Science of Computer Programming **53** (2004), 215–253.

[14] T.M. Cover and J.A. Thomas, *Elements of Information Theory*, Wiley, New York (1991).

[15] M.R. Garey and D.S. Jonhson, *Computers and Intractability: A Guide to the Theory of NP-Completeness*, W.H. Freemand and Company (1979).

[16] G.R. Grimmett and D.R. Stirzaker, *Probability and Random Processes*, 3rd edn., Oxford Scientific Publications, Oxford (2001).

[17]  B. Hagemark and K. Zadeck, *Site: A language and system for configuring many computers as one computer site*, Proceedings of the Workshop on Large Installation Systems Administration III, USENIX Association, Berkeley, CA (1989), 1.

[18]  IEEE, *Glossary of software engineering terminology, standard 729-1983*, Technical report, IEEE (1983).

[19]  H. Lewis and C. Papadimitriou, *Elements of the Theory of Computation*, 2nd edn., Prentice Hall, New York (1997).

[20]  P. Oddifreddi, *Classical Recursion Theory, Vol. II*, Studies in Logic and the Foundations of Mathematics, Vol. 143, North-Holland, Amsterdam (1999).

[21]  R. Osterlund, *Pikt: Problem informant/killer tool*, Proceedings of the Fourteenth Systems Administration Conference (LISA XIV), USENIX Association, Berkeley, CA (2000), 147.

[22]  R. Rizzi, *Complexity of context-free grammars with exceptions and the inadequacy of grammars as models for XML and SGML* (2002).

[23]  C.E. Shannon and W. Weaver, *The Mathematical Theory of Communication*, University of Illinois Press, Urbana (1949).

[24]  S. Sippu and E. Soisalon-Soininen, *Parsing Theory. Vol. 1: Languages and Parsing*, Springer-Verlag, New York (1988).

[25]  M. Sloman, *Policy driven management for distributed systems*, Journal of Network and Systems Management **2** (1994), 333.

[26]  M.S. Sloman and J. Moffet, *Policy hierarchies for distributed systems management*, Journal of Network and System Management **11** (9) (1993), 1404.

[27]  S. Traugott and J. Huddleston, *Bootstrapping an infrastructure*, Proceedings of the Twelth Systems Administration Conference (LISA XII), USENIX Association, Berkeley, CA (1998), 181.

[28]  A. Vardy, *Algorithmic complexity in coding theory and the minimum distance problem*, Proceedings of the Twenty-Ninth Annual ACM Symposium on Theory of Computing (1997), 92–109.

[29]  Webmin project, http://www.webmin.com.

[30]  D. Wood, *Theory of Computation*, Harper & Row, New York (1987).

# Complexity of System Configuration Management

Yizhan Sun[1], Alva Couch[2]

[1] *Computer Science, Tufts University, 161 College Avenue,*
*Medford, MA 02155, USA*
*E-mail: yizhan.sun@gmail.com*

[2] *Computer Science, Tufts University, 161 College Avenue,*
*Medford, MA 02155, USA*
*E-mail: couch@cs.tufts.edu*

## 1. Introduction

System administration has often been considered to be a 'practice' with no theoretical underpinnings. In this chapter, we discuss a theory of system administration, centered upon configuration management. We formalize and explore its complexity, and demonstrate that configuration management is intractable in the general case. We also explore how system administrators keep configuration tasks tractable in practice. This is a first step toward a theory of system administration and a common language for discussing the theoretical underpinnings of the practice.

Configuration management has been a subject of active debate in the system administration community for several years. Debates have centered upon the languages and tools used to accomplish the task. Complexity analysis shows that part of the difficulty in configuration management is actually independent of the language used to specify configuration or the tools used to accomplish it. Rather, configuration management can best be simplified by modifying the problem and reducing the complexity of the overall system to be managed.

For example, the way one expresses one's desires and policies often has a profound effect upon complexity and cost. This will become clear in the coming sections.

HANDBOOK OF NETWORK AND SYSTEM ADMINISTRATION
Edited by Jan Bergstra and Mark Burgess

## 2. A model of configuration management

The simplest description of configuration management as a process is to model it as a state machine [12,22]. Configurations or observed behaviors comprise the states of a system, while configuration processes accomplish state transitions between behavioral states. This model gives us a framework within which to discuss questions of complexity formally and precisely.

### 2.1. *Closed- vs. open-world models of systems*

First, we must emphasize some assumptions necessary for the discussion. In this chapter, we assume that the computer system under study is a closed-world system.[1] A *closed-world system* is a system that has no interactions with anything outside of the system. In practice, there is no such thing as a closed system, but systems are *effectively closed* if the behavior that can be affected by outside forces is not something that concerns us. For example, one may change the color of a web server, but this change does not affect whether this web server can achieve its functional expectations to provide web service, so the web server can be considered to be in a closed-world system even if the color is determined by outside forces, e.g., an interior decorator. A system is also effectively closed if there are dependencies upon constant properties of the outside world that cannot be violated. For example, in normal circumstances, we can treat a web server and a DNS server as a closed-world system without considering whether electric power will fail, since it is nearly always available.

By contrast, in an *open-world system*, unknown entities from outside the defined boundaries of the system affect the behavior of the system in such a way that whether functional expectations can be satisfied is not solely determined by the states of the system. The behavior of the system is thus unpredictable. In some cases, this can be addressed by increasing system boundaries. E.g., one can draw a boundary around *both* DNS and web services in order to view the pair as a closed system; the configuration of DNS affects the behavior of web service too much for the two entities to be considered as independent.

### 2.2. *Observed behavior*

Noted that the difference between 'closed' and 'open' systems depends somewhat upon what we choose to observe or not to observe about the system. We next must define what we mean by 'observed behavior'. This requires identifying the questions we might ask in determining whether a particular behavior is present.

---

[1]There is some controversy over this definition. Burgess claims that there is no such thing in reality as a closed-world system, so that our definition is too idealized to be useful [7]. But at the same time, practitioners often assume for the purposes of management and troubleshooting that particular subsystems are closed. For the purposes of our argument, closing the system simplifies the problem in a way that does *not* make it tractable; hence the problem of managing open-world systems is at least *more* difficult than the problems discussed here.

One way to characterize behavior is to test it in various ways and record the results. Suppose that we are given a finite set of tests $T$ that we can apply to a configuration.[2] Tests are statements that can be either 'TRUE' or 'FALSE' by some testing method. For example tests might include:

- The system has a hard drive with more than 5 GB of free space.
- The system has at least 128 MB of RAM.
- The system has a network card.
- Port 69 is listed in /etc/services.
- TCP port 80 answers web service requests.
- UDP port 69 rejects tftp requests.

Some of these facts concern hardware configuration, while others concern behavior, both internal and external. A numeric measurement is represented by a *set* of facts, one per possible value. For example, if there can be between one and five server processes on a machine, this would be represented by the five tests as to whether that number was 1, 2, 3, 4, or 5.[3] In practice, all the parameters can take finitely many values. Thus $T$ is a finite set. Also,

AXIOM 1. The observed behavior of a system is a function of its actual configuration.

Here we do not consider dependencies upon other systems. Also, tests are chosen to represent static properties of configuration, not resource bounds, and presume the presence of adequate resources to perform the test.

The *observed behavior* of a system is a subset $\psi$ of $T$, where $t \in \psi$ exactly when the system passes the test $t$. We represent user requirements as a subset $R \subseteq T$. For each test $t$ that should succeed, we add it to $R$; for each test $t \in T$ that should return false, we put its negation into $R$. Any test not mentioned in $R$ is one for which we are not concerned about the outcome. A system *meets its requirements* if $\psi \supseteq R$ (Figure 1).

### 2.3. *Actual state and observed state*

One key step in our argument is to distinguish between the actual state of a system and that part of its state that we can see. This is a complexity-reducing step; the actual states of a system are many in number, but the parts of its behavior that are practical to observe are limited to a smaller number of characteristics. The *actual state* of a system is defined as its configuration information recorded on the machine. This is distinguished from its *observed state*, which is the set of behaviors in which we are interested. For example, there are many, many ways to configure a machine to provide web services, but usually, all we are interested in is the result, not the internal details. The observed behavior of a

---

[2]The set of 'all' possible tests is not finite, but in practice, the set of tests that we focus upon must be finite.

[3]Of course, this means that parameters that can take on many values can only be characterized through a large number of potential tests. We can, however, express ranges of values with a single test. So the statement that the number of web server processes is between 5 and 15 is one test, that represents a range of values. The important issue here is that the tests we might make represent functional states, the states are finite, and the desirable states are often quite limited in number.

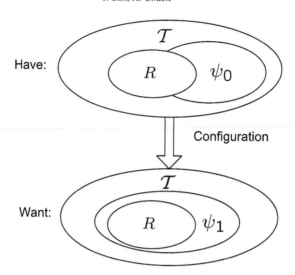

Fig. 1. Observed behavior covers all of user requirements after configuration.

web server includes its behavior in response to external requests, but does not include the ways in which those requests are satisfied internally. In this way, a multitudinous amount of detail is reduced to a manageable size [10].

From Axiom 1, we eliminate possible behavioral effects caused by inadequate resources. The actual state of a system is its configuration, denoted as $s \in S_{all}$ where $S_{all}$ is the set of all possible configurations of a system. $|S_{all}|$ can be arbitrarily large.[4] Axiom 1 also implies that the behavior of a system is always in synchronization with its actual state. We do not consider the cases where the configuration files are modified but server daemons do not reread these files or act accordingly.

The *observed state* of a system is defined as the observed behavior of a system, i.e., $\psi \subseteq T$. We define $\Psi = power(T)$ to be the set of all potential observed states of a system, which is the set of all subsets of $T$. Note that $|\Psi| = |power(T)| = 2^{|T|}$.

We can represent the act of observing state via a test function $\sigma$ that maps from actual state to observed state, i.e., $\sigma : S_{all} \to \Psi$.

## 2.4. *Configuration operations*

Configuration operations act on the actual state of a system, not the observed state.

DEFINITION 1. A *configuration operation* $p$ takes as input a state $s$ and produces a modified state $s' = p(s)$ where $s, s' \in S_{all}$.

Configuration operations might include things like:

---

[4]However, we assume the size of $S$ is finite in practice; even though we are dealing with large scale systems, the total number of possible states is still finite.

- replace `/etc/inetd.conf` with the one at `foo:/bar/repo/inetd.conf`.
- delete all udp protocol lines in `/etc/services`.
- cd into `/usr/src/ssh` and run `make install`.

An operation can be automated or manual, accomplished by a computer program or by a human administrator. Operations can be combined through function composition:

DEFINITION 2. For two operations $p$ and $q$, the operation $q \circ p$ is defined as applying $p$, then $q$, to $s$: $(q \circ p)(s) = q(p(s))$.

In this chapter, sequences of operations are read from right to left, not left to right, to conform to conventions of algebra.

AXIOM 2. For any system, there is a baseline operation $b$ that – when applied to any configuration of the host – transforms it into a baseline configuration with a predictable and repeatable actual state, a 'baseline state' $B$.

There are no other restrictions on $b$, for example $b$ could be 'reformat the hard disk'. Note that since observed behavior is a function of actual state, the baseline state also corresponds to a repeatable observed state as well.

DEFINITION 3. Given a set of $m$ configuration operations $P = \{p_1, p_2, \ldots, p_m\}$, let $P^*$ represent the set of all possible results of composing finite sequences of operations, i.e. $P^* = \{\tilde{p} \mid \tilde{p} = p_{\alpha(1)} \circ p_{\alpha(2)} \circ \cdots \circ p_{\alpha(k-1)} \circ p_{\alpha(k)}$ where $k$ and $\alpha$ are integers and $k \geqslant 1$, $\alpha : \{1, 2, \ldots, k\} \to \{1, 2, \ldots, m\}$, $p_{\alpha(i)} \in P\} \cup \{\varepsilon\}$ where $\tilde{p}$ is a sequence of operations and $\varepsilon$ represents the empty operation ('do nothing').

$P^*$ is the set of all possible results of composing finite sequences of operations from $P$. Sometimes, we will loosely refer to the sequence as the composition of the elements of the sequence.

A sequence of operations causes state transitions in actual/observed state space. Refer to Figure 2.

DEFINITION 4. The set of *reachable states* of a system with respect to a baseline state $B$ and operations $P$ is $S_{\text{reachable}} = \{s \mid s = \tilde{p}(B), \tilde{p} \in P^*\}$.

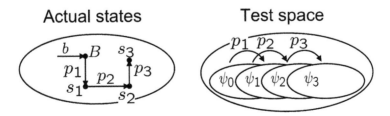

Fig. 2. State transactions caused by a sequence of operations.

The set of reachable states $S_{\text{reachable}}$ is a subset of all possible actual states $S_{\text{all}}$. Since states in $S_{\text{all}}$ that are not reachable are not of interest, we will notate $S_{\text{reachable}}$ as $S$ from now on. There might be other ways to change an actual state other than configuration operations. In this chapter, we do not distinguish these two concepts and assume that only configuration operations can change the actual state of a system. This assumption is reasonable because usually the set of actual states is not arbitrary, but is instead the result of applying configuration operations to some known initial system state.

The direct consequence of Axiom 2 is that an actual state $s$ constructed by starting at the baseline configuration and applying a series of operations from $P$ is completely determined by the sequence of operations applied since the last baseline operator $b$, i.e., all operations prior to the last baseline can be ignored. This defines a set of equivalence classes on $P^*$, where for $\tilde{p}$ and $\tilde{q} \in P^*$, $\tilde{p} \equiv \tilde{q}$ iff $\tilde{p}$ and $\tilde{q}$ agree on operations since the last baseline. This kind of algebraic equivalence can be utilized to predict the properties of the operations [11].

Thus there is a one-to-one correspondence between sequences of operations since the last baseline operation $b$ and configurations $s \in S$. Thus for the rest of the chapter, we assume that $s$ is a sequence of operations preceded by a baseline operation.

In practice, many sequences of operations can have exactly the same effect; thus there are at least two (perhaps poorly understood) equivalence relations on $P^*$: observed and actual. Two machines are *observably equivalent* if their results agree for all tests in a test suite $\mathcal{T}$. They are equivalent in actuality if their configurations are identical. In practice, the latter is impossible; two machines that are identical in configuration cannot even share the same network; they must have differing internet addresses, and thus their configurations must be different in some crucial ways. In this chapter, we define equivalence in terms of observed state to avoid this problem.

### 2.5. *Two configuration management automata*

Configuration operations, by definition, change the state of the system being configured. There are two ways in which we can think about this state change, based upon either observed or actual state.

Two configuration automatons can be defined: the first is based upon actual states of the system and the second is based upon observed states of the system.

$\mathcal{M}_1$: $(S, P, \mathcal{W}_1)$,

$\mathcal{M}_2$: $(\Psi, P, \mathcal{W}_2)$,

where $S$ is the set of all possible actual states (configurations) of the system, $P$ is the set of configuration operations applicable to the system, and $\mathcal{W}_1$ and $\mathcal{W}_2$ are transition rules and defined as follows.

First we consider the state machine based upon actual states. $\mathcal{W}_1$ is the set of all triples $(p, s, s')$ where $s \in S$ is the actual state of the system before an input configuration operation $p \in P$ and $s' \in S$ is the resulting actual state of the system.

The state machine for observed states is similar. $\mathcal{W}_2$ is the set of all triples $(p, \psi, \psi')$. Let $A(\psi)$ represent the set of reachable actual states that correspond to the observed state $\psi$.

**Actual States**          **Test Space**

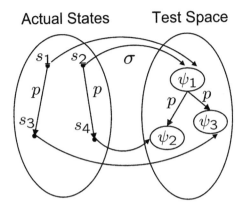

Fig. 3. Non-determinism of configuration operations.

If $s \in A(\psi)$ and $s' \in A(\psi')$ and $s' = p(s)$, then $(p, \psi, \psi') \in \mathcal{W}_2$. Usually, $A(\psi) \subset S$; if $A(\psi) = S$, then all actual states have the same observed state, which is not particularly interesting.

Note that $\mathcal{M}_1$ can be arbitrarily large since $|S|$ is arbitrarily large; $\mathcal{M}_2$ is bounded by $2^{|T|}$.

$\mathcal{M}_1$ is deterministic by Axioms 1 and 2 since the actual state of a system can be represented by its configuration from Axiom 1 and its configuration can be represented by a sequence of operations applied to its baseline state from Axiom 2. $\mathcal{M}_2$ is possibly non-deterministic. Before each operation, the observed state $\psi$ corresponds to a subset of possible actual states $A(\psi) \subset S$. Without further constraints, a typical operation $p$ *seems* non-deterministic, because the actual states that can result from applying $p$ are the set $p(A(\psi)) = \{p(s) \mid s \in A(\psi)\}$. Thus it is possible that $p$, when applied to a system in observed state $\psi$, can produce one of several observed states $\sigma(A(\psi))$ as a result. Without further information, one cannot limit $p(A(\psi))$ to be the particular configuration $s' \in p(A(\psi))$ that is actually in effect after applying $p$ (Figure 3). This uncertainty leads to apparent non-determinism when applying configuration operations.

A common problem that plagues the practice of configuration management is that two systems that initially have the same behavior do not continue to have the same behavior after the same operation is applied to both. There are many causes of this, including latent preconditions [17]. This leads to our study of reproducibility of configuration operations.

## 3. Reproducibility

Testing is expensive and sometimes impractical. The goal of any configuration management strategy is to achieve *reproducibility of effect*: repeating the same configuration operation on different hosts in a large network produces the same behavior on each host. Configuration operations should ideally cause deterministic state transitions from one behavioral state to another.

Reproducibility is difficult to achieve due to the difference between the actual state of a host and its observed state that humans and operations can practically observe. In making a configuration change, it is not practical to examine the whole state of the hard disk beforehand. Much of the actual state of a host is not observed. Latent preconditions arise in parts of host state that humans or operations are not currently considering when making changes.

The majority of the discussion of reproducibility theory is described in [12]; we reiterate that discussion here for completeness. For one host in isolation and for some configuration processes, reproducibility of observed effect for a configuration process is a statically verifiable property of the process. However, when dealing with large populations of hosts, reproducibility is difficult to achieve and can only be verified by explicit testing. Using configuration processes verified to be locally reproducible, we can identify latent preconditions that affect behavior among a population of hosts. Constructing configuration management tools with statically verifiable observed behaviors thus reduces the lifecycle cost of configuration management.

### 3.1. *Local reproducibility*

First we develop the concept of reproducibility for a single host $h$. Even though each configuration operation itself is deterministic, a change in actual state from $s$ to $s' = p(s)$ may not effect a change in the observed state $\psi$; the observed state is an unspecified function of the actual one. This results in situations where the same configuration operation, applied to two configurations in the same observed state but differing actual states, leads to two different observed states as a result. This can occur if prior configuration operations left the two configurations in differing actual states that are observed as identical, but for which further operations expose differences. This apparent non-determinism, in applying a configuration change, is the cause of much downtime in practice; a change can invoke some latent condition that prevents the system from behaving as desired.

Local reproducibility is formally defined as follows:

DEFINITION 5. Suppose we have a system $h$, a set of candidate operations $P$ appropriate to $h$, a set of actual states $P^*(b)$ for $h$, a set of tests $T$ that one can perform on $h$, and a test function $\sigma : P^*(b) \to power(T)$, where $power(T)$ denotes the set of all subsets of $T$. Then the formal system $(h, b, T, P)$ exhibits *observed local reproducibility* (or simply *local reproducibility*) if the following condition holds:

For every pair of actual states $s, s' \in P^*(b)$ with $\sigma(s) = \sigma(s')$ and for every $p \in P$, we have $\sigma(p(s)) = \sigma(p(s'))$.

Local reproducibility means for a single host, two actual states that have the same behavior will continue to have the same behavior if the same operation is applied to both. Another possible definition of local reproducibility is that the configuration automaton based upon observed behavior for a single host is deterministic.

PROPOSITION 1. *The formal system* $(h, b, \mathcal{T}, P)$ *exhibits observed local reproducibility exactly when the state machine* $(\Psi, P, \mathcal{W}_2)$ *with states* $\Psi = \sigma(P^*(b))$, *operations* $P$, *and transition rules*

$$\mathcal{W}_2 = \left\{ \left( p, \sigma(s), \sigma\left(p(s)\right) \right) \mid s \in P^*(b), p \in P \right\} \tag{1}$$

*is deterministic* [12].

This idea is a simple one. If there is observed local reproducibility, this means that operations have predictable local effects; this is exactly the same thing as saying that the state machine determined by the operations has the *same* predictability.

Now we define the local reproducibility of a particular operation $p \in P$.

DEFINITION 6. With respect to the formal system $(h, b, \mathcal{T}, P)$, an operation $p \in P$ exhibits *observed local reproducibility* (or simply *local reproducibility*) if for every $s, s' \in P^*(b)$ with $\sigma(s) = \sigma(s')$, $\sigma(p(s)) = \sigma(p(s'))$. In this case, we say that $p$ is a *locally reproducible operation*.

PROPOSITION 2. *The system* $(h, b, \mathcal{T}, P)$ *exhibits observed local reproducibility exactly when every operation* $p \in P$ *exhibits observed local reproducibility with respect to the system* $(h, b, \mathcal{T}, P)$ [12].

In other words, a set of operations exhibits observed local reproducibility with respect to a baseline $b$ if each operation has a reproducible observed effect on each reachable configuration $s \in P^*(b)$. This is the definition of reproducibility that best models the operation of Cfengine [4–6] and related convergent agents.

Note that local reproducibility of $P$ trivially implies local reproducibility of $P^*$. Although $P^*$ is infinite, $P^*(b)$ is a subset of a finite (though large) set of configurations, as the number of possible configurations is finite.

**3.1.1.** *Properties of locally reproducible operations* Several relatively straightforward propositions demonstrate the properties of locally reproducible operations in more detail. In the following propositions, to ease notation, we will presume the existence of a host $h$, a baseline $b$, a set of possible operations $P$, and a set of tests $\mathcal{T}$. We will presume that $S = P^*(b)$ is the set of reachable configurations. All claims of local reproducibility of an operation refer to Definition 6 and are made in the context of the formal system $(h, b, \mathcal{T}, P)$.

PROPOSITION 3. *The set of operations* $P$ *is locally reproducible if and only if for each operation* $p \in P$ *and each actual state* $s \in S = P^*(b)$, $\sigma(p(s))$ *is a function* $\tau_p$ *of* $\sigma(s)$. *Then we can express the observed state of a configuration after* $p$ *as* $\sigma(p(s)) = \tau_p(\sigma(s))$ [12].

Because locally reproducible operations $p$ correspond with state functions $\tau_p$, they also exhibit the typical properties of functions, notably, that a composition of functions is also a function:

PROPOSITION 4. *A composition of operations that each exhibits observed local reproducibility also exhibits observed local reproducibility* [12].

As composing operations on a configuration is the same as applying them in order, this means that an arbitrary sequence of locally reproducible operations is locally reproducible as well. However, the above does not yet tell us how to implement local reproducibility for the operations $p$ that we might compose.

In particular, some counter-intuitive results arise straightforwardly from the model.

**3.1.2.** *Constructing locally reproducible operations*   It is easy to construct a locally reproducible configuration operation. Each locally reproducible configuration operation $p$ corresponds to a function from initial states $\psi$ to final states $\psi'$, in the context of a set of reachable states $S$. This operation must thus depend upon the value of $\psi$ and avoid conditioning its effects on the values of other variant properties of the host or network. It can, however, depend upon host properties $\gamma$ that do not vary as a result of any configuration operation $p$. The following few results, from [12], serve to illustrate how one can achieve local reproducibility in practice.

PROPOSITION 5. *Suppose an operation $p$ consists only of a sequence of assignments $X :=$ $F(\psi, \gamma)$, where $X$ is a configuration parameter, $\psi$ and $\gamma$ are held constant throughout, and $F$ is a function of $\psi$, $\gamma$ and constants. Then $p$ is locally reproducible* [12].

Conditional statements pose no difficulties (although they contain more than one possible execution path) because repeatability is guaranteed by the fact that both $\psi$ and $\gamma$ are constants.

PROPOSITION 6. *A sequence of conditional assignments of the form*

$$if \left( F(\psi, \gamma) \right) \quad then \ X := G(\psi, \gamma), \tag{2}$$

*where $F$ and $G$ are functions only of $\psi$, $\gamma$ and constants, is locally reproducible* [12].

With the above results in mind, we are ready to relate reproducibility to the structure of configuration operations as programs.

PROPOSITION 7. *Let $p$ be a configuration operation and $s$ a configuration. Let $\psi$ be the observed state of $s$ and let $\gamma$ represent the set of host invariants that do not change during configuration. Suppose that when $p$ is applied to $s$, it:*
   1. *Sets parameters to values that are functions of $\psi$, $\gamma$ and constants.*
   2. *Takes program branches (conditionals) depending only upon the values of functions of $\psi$, $\gamma$ and constants.*
*Where $\psi$ and $\gamma$ are held constant throughout $p$. Then if the operation $p$ terminates successfully on every configuration $s$ in its domain, $p$ exhibits observed local reproducibility* [12].

Note, in particular, that we do not claim that every operation of this form actually ends; we simply claim that it achieves local reproducibility *if it happens to end*. In other words, this does not contradict the theory of undecidability, and it is possible to construct a configuration operation that, by *not* halting, does not achieve local reproducibility. It is rather important, however, that the operation utilize only *static* values of $\psi$ during execution of $p$, and not *dynamic* re-measurements of tests during $p$. One can get away with limited dynamic measurements of $\psi$, provided one is careful not to use two differing measurements of $\psi$ simultaneously:

PROPOSITION 8. *Suppose that $p$ as described in Proposition 7 is also allowed to re-measure the whole observed state $\psi$ at any time during its execution, as well as setting parameters and branching based upon functions of the observed state $\psi$ and host invariants $\gamma$. Only one measurement of $\psi$ is available at a time and any setting must be a function of the most recent measurement. If $p$ terminates, the result is locally reproducible, but not all processes are guaranteed to terminate* [12].

The idea of the proposition is simple. If we allow re-measurement of state, we can break down the operation into a series of components, each of which is locally reproducible, and thus the composition is locally reproducible as well.

### 3.2. *Population reproducibility*

We next turn our attention to assuring reproducibility of operations over a population of hosts. Such reproducibility is extremely important as a cost-saving measure. If we must validate the behavior of each host of a population separately, it becomes very expensive to build large networks. Ideally, we should be able to test one particular host in order to understand the behavior of a population.

DEFINITION 7. If, for any subset of a population of hosts whose configurations are in the same observed state, an operation $p$ results in identical observed final states over the subset, then $p$ exhibits *observed population reproducibility* (or simply *population reproducibility*).

Local reproducibility does not imply population reproducibility.

PROPOSITION 9. *An operation that is locally reproducible on every host to which it applies can fail to exhibit population reproducibility* [12].

As a counterexample, consider an operation applied to hosts with differing operating systems. Suppose that $p$ is the operation of copying a constant xinetd.conf into /etc/. The population reproducibility of this operation has no dependence upon its implementation; it is a property of the operating system. If the operating system does not support the xinetd abstraction, the operation does nothing to behavior. Consider also an operation that exposes a bug in one OS that is not present in another. There are many other such latent variables that cause the same operation on different hosts to yield different outcomes.

While local reproducibility is relatively easy to achieve, population reproducibility remains a pressing problem, and there is no currently available tool that addresses it sufficiently. There is one overarching observation that motivates our approach to population reproducibility:

PROPOSITION 10. *If all configuration processes are verified as locally reproducible, and one observes population differences in behavior in applying these processes, then the variation must be due to latent preconditions in the population rather than artifacts in the processes* [12].

These processes have effects that are observably locally reproducible. If their effects differ for two hosts in a population, then since they are locally reproducible in isolation, they will always differ in the same way regardless of when the operations are applied. Hence some other factor is causing the difference, and the only other variants are host identity and preconditions.

Thus a locally reproducible process that differs in effect over hosts in a population can be used to test for heterogeneity of latent variables within the population.

One way of assuring population reproducibility is to utilize locally reproducible actions that fail to be population reproducible to expose each latent variable in the population. With latent variables exposed, we can form equivalence classes of hosts with the same latent structure, and condition further configuration changes by that structure.

We show a simple algorithm for constructing 'synthetic' population reproducibility to inspire more prudent thoughts on population reproducibility.

Given a set of hosts $\mathcal{H}$, a set of locally reproducible operations $P$ on hosts in $\mathcal{H}$, a set of tests $\mathcal{T}$, and an initially empty set of additional tests $\mathcal{T}'$:

1. Perform an operation sequence $\tilde{p}$ on the population of hosts to which it should apply.
2. Perform all tests in $\mathcal{T} \cup \mathcal{T}'$ on each host $h$ to compute an observed state $\psi_h$.
   (a) If results are the same for all hosts, then $p$ exhibits population reproducibility over the hosts to which it applies.
   (b) If results are different, then form equivalence classes of hosts $\widetilde{\mathcal{H}}$, where for $h, h' \in \tilde{h} \in \widetilde{\mathcal{H}}$, $\psi_h = \psi_{h'}$. For each equivalence class $\tilde{h}$, and a configuration $c$, add the test '$h(c) \in \tilde{h}$' to $\mathcal{T}'$.

The above construction is impractical. There are $|P^*|$ iterations at step 1, and each iteration requires $O(2^{|\mathcal{T}|})$ operations to carry it out. The point is that it is possible to deal with population non-reproducibility incrementally, one deviation at a time, adding tests to $\mathcal{T}'$ one by one. In this way, our state of knowledge grows with the operations we apply and observe.

## 4. Limits on configuration operations

There has recently been much attention to the limits imposed upon configuration operations and how they affect the usability of a set of operations [11]. Some authors maintain that operations should be constructed to be repeatable without consequence once a desirable state has been achieved [4–6,8]. Other authors maintain that operations must, to provide a consistent outcome, be based upon imperative order [23,24] or upon generating the whole configuration as a monolithic entity [1,2]. In this chapter, we precisely define limits upon operations with the intent of discussing how those limits affect composability of operations.

## 4.1. *Limits on configuration operations*

We now turn our attention to the properties that configuration operations can have upon behavior. We can think of a sequence of operations as synonymous with the actual state that results when it is applied to a baseline state. E.g., we think of the sequence of operations $\tilde{p}$ as corresponding with the state $\tilde{p}(B)$, where $B$ is the baseline state. Thus we can write $\sigma(\tilde{p})$ to represent the observed state of the result of the sequence of operations, when applied to the baseline state.

DEFINITION 8. A set of operations $P$ is *observably idempotent* if for any operation $p \in P$, repeating the operation twice in sequence has the same effect as doing it once, i.e., for any $p \in P$ and any $\tilde{q} \in P^*$, $\sigma(p \circ p \circ \tilde{q}) = \sigma(p \circ \tilde{q})$.

The use of $\tilde{q}$ in this definition asserts that the idempotence occurs when applied to any *actual state* of the system from $B$. We could say, equivalently, that $\sigma(p \circ p) = \sigma(p)$ when starting at any state in $S$.

In continuous spaces, an operation is *convergent* if repeating the operation many times can cause the system to move to some target state, i.e. $\lim_{n \to \infty} p^n = $ target state.

By contrast, in system administration idempotent operations take on the discrete character of configuration space.

DEFINITION 9. A set of operations $P$ is *observably convergent* if for any operation $p \in P$ and any $\tilde{q}$ in $P^*$, there is an integer $n > 0$, such that $p$ is observably idempotent when applied to $p^n \circ \tilde{q}$, i.e., $\sigma(p \circ p^n \circ \tilde{q}) = \sigma(p \circ p \circ p^n \circ \tilde{q})$.

For example, consider the task of creating a user in a directory service.
Classical procedural operations would be:
1. Create user record in LDAP.
2. Wait for LDAP directory service to sync up and serve the directory record.
3. Create user's home directory and associate data.
Time must pass between steps 1 and 3.

An equivalent convergent operation to accomplish this task might be an operation $c$ with the following pseudo-code:

**if** (user does not exist)
    put user in LDAP
**else** if (user's home directory does not exist)
    create an appropriate home directory.

While $c$ by itself does not accomplish the task, repeating $c$ several times (while time passes) guarantees that the task will be completed. After this time, repeating $c$ does not change the system behavior. The value of $n$ in the definition of convergence thus depends upon how fast the LDAP update occurs. The key issue is that rather than thinking of system configuration as one large change, a convergent operation makes one small change at a time. Through repetition of the operation, its preconditions are fulfilled by previous runs or other operations, or simply by the passage of time. Thus idempotence is a special case of convergence where $n = 1$.

In dynamic situations, when we do not know the nature of the best solution, we can allow convergent operations to 'discover' the quickest solution to a problem. One kind of convergent operator for which this is true employs 'simulated annealing' [18]. This approach presumes that the 'best' solution lives in a solution space that is not convex [3]; a particular locally best configuration is not necessarily globally best. This corresponds to a potential 'surface' in which the best solution is one of many peaks of the objective function. The simulated annealing approach is to choose at each step whether to move to or away from a peak, in order to allow one chance to switch to a better peak over several trials. For example, in web server response optimization, the behavior of the operator depends upon an external parameter $\rho$, the 'temperature' of the system. Simulated annealing occurs when at each application of the operator, one makes a decision how to proceed based upon the probability $\rho$: with probability $\rho$ move files to a server that seems slower (as observed before the move); with probability $1 - \rho$, move files to a server that seems faster. The annealing process involves letting $\rho \to 0$ as time passes; this is also called a 'cooling schedule'. The result of simulated annealing is often a near-optimal response time [18].

In the initial configuration phase, behavior of the host is completely determined by configuration operations; while in the maintenance phase, especially during resource optimization, the influence of users becomes nontrivial. The effectiveness of convergence then increases as in the above example.

Convergent operations are not necessarily consistent; they might even oppose one another to achieve equilibrium in the overall observed state of the system. Possible interrelationships between convergent operations include:

1. Orthogonal – working on different parts of the system.
2. Collaborative – helping each other.
3. Conflicting – opposing one another.

Convergence requires careful design. Poorly designed operations that do not exhibit convergence can create latent conditions that worsen as the operations are repeatedly applied. For example, it is possible to make a mistake in what one thinks is a convergent edit of a file, that fills the file with garbage over time. When we expect them to work independently, they might interfere with one another; when we want them to work collaboratively, they might conflict with and undo the work of each other.

DEFINITION 10. A set of operations $P$ is *observably stateless* if repeating an operation (where the repetition is not necessarily adjacent) will accomplish the same result, i.e., for any $p \in P$ and any $\tilde{q}, \tilde{r} \in P^*$, $\sigma(p \circ \tilde{q} \circ p \circ \tilde{r}) = \sigma(p \circ \tilde{q} \circ \tilde{r})$.

Unconditional commands that set parameter values are stateless. For example, *set* $M = 0$. But if carefully crafted, stateless operations can also contain conditionals [22].

Statelessness is a property of a set of operations used together, and not a property of any operation considered in isolation. Note that stateless sets of operations can contain conflicting operations; there is no need for them to agree on an outcome. A stateless set of operations has a fixed point (in the sense of the chapter by Mark Burgess and Lars Kristiansen) (and is called 'fixed-point convergent') if, as well as having the property in Definition 10, it also contains no conflicting operations.

For example, for a parameter $x$, $x = 1$ and $x = 2$ are stateless, because only the last one applies. But the set of operations '$x = 1$', '$x = 2$' is not fixed-point convergent in the sense of the chapter by Mark Burgess and Lars Kristiansen.

DEFINITION 11. A set of operations $P$ is *observably commutative* if for any two operations $p, q \in P$, $\sigma(p \circ q \circ \tilde{r}) = \sigma(q \circ p \circ \tilde{r})$, for any $p, q \in P$ and $\tilde{r} \in P^*$.

Note that commutative operations *can* make conflicting changes to the system. For example, $p$: $M = M + 1$ and $q$: $M = M - 1$.

DEFINITION 12. A set of operations $P$ is *observably consistent* or *homogeneous* if operations never undo the changes made by others, i.e., for any $p \in P$ and $\tilde{q} \in P^*$, $\sigma(p \circ \tilde{q}) \supseteq \sigma(\tilde{q})$.

DEFINITION 13. An operation is *atomic* if it has two result states. Either
  1. it has all required preconditions and it asserts known postconditions or
  2. it lacks some precondition and does nothing at all.

### 4.2. *Relationship between limits*

Each of the concepts of idempotence, convergence, and statelessness refer to a condition true of an operation or set of operations. We can understand these concepts by outlining how each one limits an operation or operations. The 'strength' of a condition refers to the amount of limitation imposed; a stronger condition means that less operations meet the conditions.

PROPOSITION 11. *Idempotence is a stronger condition upon operations than convergence*[5] [22].

Idempotence is a special case of convergence by restricting $n = 1$.

PROPOSITION 12. *Statelessness is a stronger condition upon operations than idempotence* [22].

If $\tilde{q} = \varepsilon$ in statelessness, then statelessness becomes idempotence.
Figure 4 depicts the relationships among convergence, idempotence and statelessness.

---

[5]Editorial comment: note the different terminology of these authors. Their notion of convergence refers to that problem CONx of Burgess and Kristiansen, not to 'fixed-point' (or cfengine-like) convergence CON.

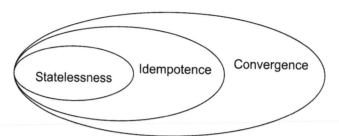

Fig. 4.  Relation among convergence, idempotence and statelessness.

PROPOSITION 13.  *If a set of operations is both idempotent and commutative, it has to be stateless and consistent* [22].

The enemy of tractability of composability seems to be lack of knowledge and control over combinatorial behaviors of configuration operations, except through explicit testing. While we usually know exactly what an operation does when applied to a baseline system, we seldom know precisely what will happen when two operations are composed, i.e., when the second operation is applied to a non-baseline system. The limits on operations we discussed previously all add some forms of control over combinatorial behaviors. Convergence, idempotence, sequence idempotence, statelessness add a structure of equivalence relations between sequences of configuration operations. Commutativity limits potential results of a set of scripts so that different permutations of the same set are equivalent. Consistency rules out conflicting behaviors. The limit of atomicity makes operations tighter, more robust and more secure. And they are used with other limits to enhance their functionality. However, in the next chapter we show that only these limits are not enough to make composability tractable.

## 5.  Complexity of configuration composition

Composability of configuration operations refers to the ability to find a finite sequence of configuration operations that reliably transforms a given initial state of a system to an observed state that satisfies some given user requirements. Composability is a desirable feature in configuration management. A composable approach can greatly reduce costs of configuration and training of new administration staff and encourage collaborations among system administrators. Composability is far from being achieved for current tools, which avoid the issue by either limiting the ways in which operations are composed, or using operations that by nature can be composed.

In [22], composability with or without various limits is proved mathematically to be NP-hard. Given a list of requirements and a configuration operation repository with well defined behavior, the process of finding the optimal sequence of operations to meet the requirement is an NP-hard problem and the problem remains NP-hard regardless of whether operations are:

- idempotent/stateless/convergent,

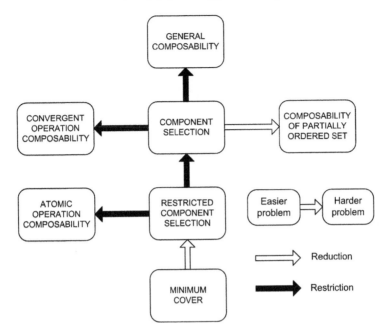

Fig. 5. Summary of proofs.

- consistent/commutative/atomic,
- orderable via a known partial order on the operations (utilizing known dependencies between operations).

Figure 5 shows the summary of proofs.

Here, we only show the main thread of these proofs. The composability problem with no limits imposed upon configuration operations is called the General Composability (GC) problem. We start with a simplified version of GC called Component Selection (CS). We first prove that CS is NP-hard, hence GC is NP-hard (Figure 6).

### 5.1. *Component selection*

DEFINITION 14. Component Selection (CS) is defined as follows:

*Instance*: Set $P = \{p_1, p_2, \ldots, p_m\}$ of $m$ components, set $T = \{t_1, t_2, \ldots, t_n\}$ of $n$ tests, subset $R \subseteq T$ of desired outcomes, let $power(P)$ and $power(T)$ denote the sets of all subsets of $P$ and $T$, respectively; let a test process function be denoted by $\sigma : power(P) \to power(T)$ (that describes the outcome $\sigma(Q)$ for each $Q \subseteq P$ and takes time $O(|T|)$ to compute) and choose a positive integer $K \leqslant |P|$.

*Question*: Does $P$ contain a subset $Q \subseteq P$ with $|Q| \leqslant K$, such that $R \subseteq \sigma(Q)$?

THEOREM 1. *CS is NP-hard.*

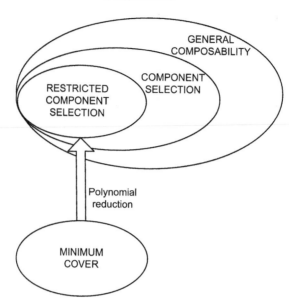

Fig. 6. Composability is NP-hard.

PROOF. We must show that CS is at least as difficult as a known NP-complete problem. Restrict CS to Restricted Component Selection (RCS) by allowing only instances having $R = T$ and $\sigma(Q) = \bigcup_{p_i \in Q} \sigma(\{p_i\})$. We show in the following proof that RCS is NP-complete, thus CS is NP-hard. ☐

Note that making $R = T$ is trivial since tests are artifacts of human choices. It simply restricts all the tests in $T$ relevant to the user's requirement. However restriction $\sigma(Q) = \bigcup_{p_i \in Q} \sigma(\{p_i\})$ is significant. It imposes structure on the test process function $\sigma$. It eliminates the case where the set of user requirements satisfied by a combination of components cannot be satisfied by any of the components alone. It requires the operations must have the property of *compositional predictability*, i.e., the observed behavior of a composition of operations can be predicted without testing by considering the observed behaviors of individual operations in the composition. This is a very strong requirement. It is a stronger condition than consistency.

DEFINITION 15. The definition of Restricted Component Selection (RCS) is as follows:
*Instance*: Set $P = \{p_1, p_2, \ldots, p_m\}$ of $m$ components, let $R = \{t_1, t_2, \ldots, t_n\}$ be a set of $n$ desired outcomes, let $power(R)$ and $power(P)$ be the sets of all subsets of $R$ and $P$, respectively; let a test process function be denoted as $\sigma : power(P) \to power(R)$ whose computation takes time of $O(|R|)$ and $\sigma(Q) = \bigcup_{p_i \in Q} \sigma(\{p_i\})$, and choose a positive integer $K \leqslant |P|$.
*Question*: Does $P$ contain a subset $Q \subseteq P$ with $|Q| \leqslant K$, such that $R = \sigma(Q)$?

THEOREM 2. *RCS is NP-complete.*

PROOF. The proof of NP-completeness uses the Minimum Cover (MC) problem, known to be NP-complete. MC is defined as follows:

DEFINITION 16. *Minimum Cover*:

*Instance*: Collection $C$ of $m$ subsets of a finite set $S$, where $|S| = n$, positive integer $K \leqslant |C|$.

*Question*: Does $C$ contain a cover of $S$ of size of $K$ or less, i.e., a subset $C' \subseteq C$ with $|C'| \leqslant K$ such that $\bigcup_{c_i \in C'} c_i = S$?

First it must be shown that RCS is in NP. Given a subset $Q$ of $P$, determining if $R = \sigma(Q)$ can be done by searching in $\sigma(Q)$ for each element of $R$. Since $\sigma(Q)$ has at most $n$ elements and $R$ has $n$ elements, a simple algorithm requires $O(n^2)$ time, which is polynomial in the length of the instance, thus RCS is in NP.

The transformation function $f$ from any instance $\omega$ of MC to an instance $f(\omega)$ of RCS is defined as follows:

1. Let every $c_i \in C$ correspond to an operator $p_i \in P$.
2. Let $S$ be $R$.
3. For every $c_i = \{s_{i,1}, s_{i,2}, \ldots\} \in C$, let $\sigma(c_i) = \sigma(\{p_i\}) = \{s_{i,1}, s_{i,2}, \ldots\} = \{t_{i,1}, t_{i,2}, \ldots\}$.
4. Let $K$ in MC ($K_{\text{MC}}$) be $K$ in RCS ($K_{\text{RCS}}$).

Step 1 requires time $O(m)$; step 2 requires $O(n)$; step 3 requires time $O(mn)$; step 4 requires $O(1)$. So $f$ is a polynomial function of input.

Now we show:

$$\omega \in \text{MC} \Longleftrightarrow f(\omega) \in \text{RCS}$$

$\longleftarrow$:

Assume $f(\omega) \in \text{RCS}$, then there exists subset $Q \subseteq P$, with $|Q| \leqslant K_{\text{RCS}}$, such that $R = \sigma(Q)$. Let $C' = \{c_i \mid c_i = \sigma(\{p_i\}), p_i \in Q\}$. $\bigcup_{c_i \in C'} c_i = \bigcup_{p_i \in Q} \sigma(\{p_i\}) = \bigcup_{p_i \in Q} \{t_{i,1}, t_{i,2}, \ldots\} = \bigcup_{p_i \in Q} \{s_{i,1}, s_{i,2}, \ldots\} = \sigma(Q)$ by $f$, because $S = R$. Then $S = \sigma(Q) = \bigcup_{c_i \in C'} c_i$, because $K_{\text{MC}} = K_{\text{RCS}}$ and $|C'| = |Q| \leqslant K_{\text{MC}}$. Therefore $\omega \in \text{MC}$.

$\longrightarrow$:

Assume $\omega \in \text{MC}$. Then there exists a subset $C' \subseteq C$ with $|C'| \leqslant K_{\text{MC}}$, such that $S = \bigcup_{c_i \in C'} c_i$. Let $Q = \{p_i \in P \mid c_i \in C'\}$. Then $\sigma(Q) = \bigcup_{p_i \in Q} \sigma(\{p_i\}) = \bigcup_{p_i \in Q} \{s_{i,1}, s_{i,2}, \ldots\} = \bigcup_{p_i \in Q} \{t_{i,1}, t_{i,2}, \ldots\} = \bigcup_{c_i \in C'} c_i$, because $R = S$ by $f$ then $R = S = \bigcup_{c_i \in C'} c_i = \sigma(Q)$, because $K_{\text{RCS}} = K_{\text{MC}}$ and $|Q| = |C'| \leqslant K_{\text{RCS}}$. Therefore $f(\omega) \in \text{RCS}$. $\qquad\square$

## 5.2. *General composability*

Unfortunately, our problem is somewhat *more* difficult than CS, because in our problem, order of operations does matter. We are most interested in what test outcomes result from applying operations from a set of *operations* $P$. Operations differ from components in the above theorem and definition, because the order of application of operations does matter, and repeating an operation is possible with different results than performing it once. We assume that to apply an operation $p_i$ to the system takes only one step and we can test the system in linear time. We also assume that the system begins in a predictable and reproducible baseline state $B$ to which the sequence of operations is applied.

DEFINITION 17. General Composability (GC) is defined as follows:

*Instance*: Set $P = \{p_1, p_2, \ldots, p_m\}$ of $m$ configuration operations, set $T = \{t_1, t_2, \ldots, t_n\}$ of $n$ tests, let $R$ be a set of user requirements $R \subseteq T$, let $\sigma$ be a test process function $\sigma : P^* \to power(T)$, where for $\tilde{p} \in P^*$, $\sigma(\tilde{p})$ represents the tests in $T$ that succeed after applying $\tilde{p}$ to a given repeatable baseline state. Suppose the computation time of $\sigma$ is a linear function of $|T|$, and choose a positive integer $K \leqslant |\mathcal{F}|$, where $\mathcal{F}$ is a polynomial function of $|P|$.

*Question*: Does $P^*$ contain a sequence $\tilde{q}$ of length less than or equal to $K$, such that $T \subseteq \sigma(\tilde{q})$?

For $\tilde{q}$ a sequence of operations, $\sigma(\tilde{q})$ is the result of testing the application of the sequence $\tilde{q}$ to a given baseline state. This should at least satisfy the requirements given in $R$, but may satisfy more requirements.

The reason that we impose integer $K$ to be less than or equal to a polynomial function of the length of the set of operations is that in practice no one implements a sequence of operations with arbitrary length. In practice, there is a finite upper bound on the number of operations and people seldom repeat an operation more than a few times.

PROPOSITION 14. *If operations in $P$ are commutative and idempotent, then CS is reducible to GC.*

PROOF. The difference between the two problems is that in GC, each option for operations is a *sequence*, whereas in CS each option is a *set*. If all operations in a sequence are commutative and idempotent, then there is a 1–1 correspondence between equivalent subsets of sequences and sets, and the reduction simply utilizes this correspondence.     □

THEOREM 3. *GC is NP-hard.*

PROOF. We must show that GC is at least as difficult as another NP-hard problem by restricting it to CS by allowing only instances for which the operations are commutative and idempotent.     □

We know from the proof from Section 4.2 that commutative and idempotent operations must be stateless and consistent, so the above proof shows that even with limits of idempotence, statelessness, consistency and commutativity, composability is still an NP-hard problem.

**5.3.** *Discussion*

Practitioners of system administration argue about limits upon configuration operations. Some maintain that operations should be constructed to be repeatable without consequence once a desirable state has been achieved. Others maintain that operations must be based upon imperative order or upon generating the whole configuration as a monolithic entity. Our work proves that these arguments are not based upon mathematical fact. The problem remains difficult no matter how one limits operations.

We will discuss how current tools and practices avoid composability problems in Section 6.

## 6. Configuration management made tractable

In this section, we discuss some general guidelines used by system administrators to keep configuration management tractable and examine current strategies of configuration management as examples.

Configuration management is intractable in the general case because the complete understanding of the system and its behavior is intractable. People manage complexity in many ways. There is no perfect solution that makes configuration management tractable without making some sacrifice such as accuracy, flexibility, or convenience. The intrinsic complexity cannot be destroyed, but can only be hidden. In the following paragraphs, we summarize these mechanisms used by system administrators to reduce the complexity of configuration management. Note that these mechanisms are often used in combination.

### 6.1. *Experience and documentation*

System administrators use their experience of past solutions and documentation to avoid dependency analysis and composing operations. After validating that it is proper to use past experience or documentation in the current environment, they simply repeat those actions or follow instructions of the documentation without doing dependency analysis and composition.

For example, suppose a system administrator needs to install a network card. The installation guide instructs one to install the driver first, then plug the card in the machine. The system administrator can just follow these instructions without bothering to analyze the interactions and relationships between software and hardware of the network card. The instructions are *sufficient* but there is no claim that they are *necessary*. It is not important to know that if the instructions are not followed, the card will not work; this possibility is not even considered. Often, this leads to cases where procedures are overly constrained for no particularly good reason, simply to avoid a deeper analysis of dependencies.

The drawback of this approach is that experience and documentation are not always 100% accurate and these might not be proper for the current environment system administrators are working with. In a very dynamic environment, it is possible to misconfigure a system merely by following instructions from documentation or experience that do not apply to the system's current state.

**6.2.** *Reduction of state space*

Reduction of state space is a matter of limiting one's choices during configuration. With a set of *n* operations, theoretically, one can compose an infinite number of possible sequences. Thus there exist a large number of possible actual states of the system. However, there are only a few distinct behaviors that we might want the resulting system to exhibit. Limiting the states of the system to match the number of distinct desirable behaviors simplifies the configuration process. If we make a limit that only a few sequences can be chosen, composition of operation sequences becomes tractable simply by choosing from a limited number of sequences. Composability entails just the selection of the one operation available, which is obviously tractable. This is a strong argument for 'over-constraining' the documentation so that the options available do not overwhelm the administrator.

Following instructions from documentation is a case of reduction of state space; only one sequence of operations is available to choose. Also acting accordingly to some shared practices, agreed upon by a group of system administrators for a site, can also efficiently reduce the state space of the system.

By restricting heterogeneity of the network (making machines alike), one can increase the predictability of the system. Solutions on one machine are likely to work on another. Instructions in documentation become more feasible to apply for a large network. Homogeneity can be achieved by always following the same order of procedures or cloning machines.

Standardization is an effective method to proactively and strictly define critical dependencies. The Linux Standard Base (LSB) project [13] seeks to provide a dynamic linking environment within Linux in which vendor-provided software is guaranteed to execute properly. LSB achieves compatibility among systems and applications by enforcing restrictions on critical dependencies include the locations of system files, the name and content of dynamic libraries.

The drawback of reduction of state space is its inflexibility during configuration. Strategically, variety in configuration can pay off in terms of productivity. Moreover, a varied system is less vulnerable to a single type of failure. System administrators must find an appropriate balance between homogeneity and variety.

**6.3.** *Abstraction*

*Abstraction* means to represent complex dependency relationships by simpler mechanisms to hide complexities. System administrators have a simple model of the system. They operate at a higher level of abstraction than the interaction details of different subsystems.

In many software systems, interactions and relationships among packages within the system are abstracted as strings of fields under 'requires' and 'provides'. This dependency information does not come from system administrators' analysis of packages but from the white box knowledge of developers of the system. The true dependency relationship is hidden by this abstraction. For example, RedHat Package Manager (RPM) files and their equivalents use this mechanism to the ease the task of adding software to a Linux system. In the package header, each package is declared to 'provide' zero or more services. These are just strings with no real semantic meaning. A package that needs a service then 'requires' it. This dependency information can be considered as 'documentation' embedded in the system. Just as other documentation, this abstraction can be in error [16].

Another abstraction involves considering dependencies between procedures instead of dependencies between system components. The dependencies between entities of the system is represented by the order in which procedures are performed. For example, in order to install a wireless card, one must install its software before plugging in the card. The complex interactions between the hardware and software of the card is hidden by the order of these two installation procedures.

The drawback of abstraction is that the representation may not correctly reflect the real dependencies that the abstraction represents, and can fail to deal with problems that occur at lower levels than the abstraction. For example, if package names listed in the 'required' field of a package do not include all packages that this package depends on, then a problem occurs. By only looking at the abstracted level of these string names, one cannot solve the problem. Some analysis of implementation details of the package is needed to solve the problem.

### 6.4. *Re-baselining*

Re-baselining the system to a repeatable baseline state can be used to eliminate latent preconditions or prior configuration mistakes that would otherwise lead to future unpredictability and downtime. Re-baselining is a last resort when the complexity of utilizing regular configuration operations has grown to an unmanageable level. The drawback of this approach is that it might be time consuming to rebuild a system from scratch. One must consider the cost of downtime when re-baselining the system.

### 6.5. *Orthogonality*

Two sets of configuration parameters are *orthogonal* if they control the behavior of differing subsystems independently, so that choices for one set do not determine choices for the other. Using orthogonality separates configuration parameter space into independent subsets. Component Selection is in P if the requirements that can be satisfied by each operation are disjoint subsets. The drawback is that it is not always feasible due to interactions and dependencies between subsystems.

### 6.6. *Closure*

The techniques used in closure are closely related to documentation and reduction of state space. A 'closure' is a 'domain of semantic predictability', a structure in which configuration commands or parameter settings have a documented, predictable, and persistent effect upon the external behavior of software and hardware managed by the closure [9]. Using closure systems is like delegating configuration tasks to someone else; either the closure system or the developer of the closure system is responsible to perform necessary operations in order to accomplish system goals.

RedHat Enterprise Linux [21] can be considered as a closure. It is an operating system that accommodates a wide number of third party applications. It is self-patching and self-updated by its vendor.

However, suppose that one has to add a foreign package to the system managed by the closure. Then he must know which parts of RedHat Linux Enterprise will be violated through dependency analysis, so that they can be managed differently after the insertion. For example, it may be necessary to turn off automatic software updating in order to install certain foreign packages.

The drawback of a closure system is its design complexity [10]. A closure cannot exist by itself but can only exist within a system of consistent closures. The design of closures must conquer or smartly avoid the complexities of dependency analysis and composition of operations.

In summary, system administrators utilize the above strategies (except closures) in combination to make configuration tractable. Closure is at an early stage of development. The drawback of each strategy is that it limits the system administrator to a certain way of configuring the systems, for their whole lifecycle, in such a way that the limits can become unmanageable.

## 7. Some theoretical ways to reduce complexity

In this section, we explore how intractable problems are managed in general, based upon computation theory, and how these techniques can be applied to configuration management.

Even though many problems are intractable, humans have lived with them for centuries. Trains need to be scheduled, salesmen must plan their trips for marketing, and thieves have to decide in very limited time to choose which items to pick up, even though *Train Scheduling*, *Traveling Salesman* and *Integer Knapsack* are all NP-complete problems. The methods by which intractable problems are solved in real life can give some insight into how our composability problem might be solved.

There are three approaches of getting around intractability:

- simplification: choosing easy instances to solve,
- approximation: trading optimality for computability,
- memorization and dynamic programming.

In the following sections we will discuss some of these strategies that apply to configuration management.

### 7.1. *Choosing easy instances to solve*

NP-completeness refers to *worst case complexity*. Even if a problem is NP-complete, many subsets of instances can be solved in polynomial time. We suggest three different ways to reduce complexity in configuration management: reduction of state space, using simple operations and forming hierarchies and relationships among operations.

**7.1.1.** *Reduction of state space*   Composability can be made relatively easy by use of a reduced-size state space, as we have seen in the current configuration management strategies. For details, see Section 6.2.

**7.1.2.** *Using simple operations*   Minimum Cover is solvable in polynomial time by matching techniques if all of the candidate sets $c$ in the cover set $C$ have $|c| \leqslant 2$ [14]. This implies that relatively simple operations that only address one or two user requirements can be composed efficiently.

That composability is tractable in this case can be understood by considering the components of graphical user interfaces. In these systems, typical components are widgets with limited function. The dependencies between widgets are obvious. The semantics of widgets are constrained for easy reuse when the widget is used in a different context. For example, the semantics for a button is to trigger an event when it is clicked or released. There are no underlying assumptions or dependencies that are not obvious from the context. The semantics of widgets and the limited domains in which the widgets are used are simple enough that solutions for syntactic composability are also solutions for semantic composability.

Configuration management is also a somewhat 'constrained' domain. Configuration of a site typically includes:

1. Editing the bootstrap scripts.
2. Configuring internal services, typically DNS, LDAP, NFS, Web, etc.
3. Installing and configuring software packages.

We can consider building smaller operations that are designed to work together, engineering them to a common framework, and then scaling up.

**7.1.3.** *Forming hierarchies and relationships*   Minimum Cover is in $P$ if each element of $C$ except the largest element is laminarily 'contained' in some other element, i.e., $c_i \subseteq c_j$, if $|c_i| \leqslant |c_j|$. In our composability problem, nesting means operations are sorted with the increasing order of the number of objectives that they can achieve; each operation can accomplish at least the objectives of the previous one. Composability of operations is trivial by always selecting the least upper bound of the objectives. Also, *minimum cover* can be reduced to *Integer Knapsack* via surrogate relaxation [20]. The *Integer Knapsack* problem is still NP-complete, however, it is polynomially solvable if the values of the items are in increasing order, and each value is greater than the sum of the values before it. In this case we can solve the problem quickly by a simple greedy algorithm. Besides, the complexity of *Minimum Cover* can be reduced using divide-and-conquer method if subsets are organized into disjoint sequences of subsets.

These simplifications of NP-completeness suggest that we can reduce complexity by forming hierarchies and relationships within the set of operations. In the following, we propose a system managed by closures [9].

In the theory of closures, configuration parameters are grouped into two distinct sets: exterior parameters and interior parameters. Exterior parameters of a closure completely determine the behavior of the closure. Interior parameters are those parameters that are invisible to an observer of behaviors. For example, the port number of a web server is exterior; the name of its root directory may not be. The former is necessary to pass any behavioral test; the latter may change without affecting external behavior at all. One closure contains another through parameter dominance, i.e., the exterior parameters of the dominated closure must be present in the parameter space of the dominant closure. Thus the dominant closure controls the behavior of dominated closure. The parameter space of a closure is either disjoint from that of all other closures or it is dominated by some other closure. The system is composed by several disjoint sets of closures; each set is a chain of nesting closures.

A system can be divided into relatively large independent (orthogonal) subsystems, and one can construct the configuration with a few large closures. These closures 'contain' smaller closures through parameter dominance. Smaller closures 'contain' even smaller closures and at the bottom are the simple operations we suggested in the previous subsection.

## 7.2. *Approximation*

In practice, if a tractable algorithm cannot be found to solve a problem exactly, a *near-optimal* solution is used instead; this is called *approximation*.

In many cases, configuration management does not need to be optimal. Any sequence of operations that will accomplish the result will do. One criterion of increasing importance is cost. It is not worthwhile to apply much effort in searching for the optimal solution when a near-optimal solution is 'good enough' with low cost, especially when we are composing mostly simple and orthogonal operations. The reason for this is that one can embody a near-optimal sequence – as long as the length of the sequence is polynomial to the number of available operations – into a script that can be executed efficiently. In other words, scripting precludes optimality. Moreover, to avoid latent preconditions, a *repeatable* composition is often more important than *optimal* composition. However, even we can accept non-optimal solutions, a general polynomial algorithm for searching for a sequence of operations that satisfies requirements at all can be difficult to construct.

In addition, most scripts are complex entities, the cost of managing them and maintaining them is sometimes significant; keeping the scripts to a minimum might save for the long run. Moreover, composition of operations is often repeated in similar situations, it makes sense to spend more effort to find the optimal solution once and amortize the cost over all repetitions.

## 7.3. *Dynamic programming*

*Integer Knapsack* can be solved in pseudo-polynomial time with dynamic programming. In dynamic programming, we maintain memory of subproblem solutions, trading space for time. The two key ingredients that make dynamic programming applicable are optimal substructure (a global optimal solution contains within it optimal solution to subproblems) and overlapping subproblems (subproblems are revisited by the algorithm over and over again).

System configuration can often be composed from configurations of relatively independent subsystems. For example, configuration of a typical departmental site includes a series of configurations of server subsystems, i.e., web service, file system service, domain service, mail service, etc. The optimal solution for configuration of the whole system is often a union of optimal solutions for configurations of subsystems. This implies optimal substructure. One might argue that when resource competition between different subsystems occurs, optimal substructure does not exist, since the interests of subsystems are not consistent. This argument mainly affects performance of dynamic programming. In our composability problem, performance or security requirement can be coded within the requirements. We search for the shortest sequence of configuration operations that achieve the requirements. The shortest sequence of the whole is composed by the shortest sequences of subsystems.

Subproblems such as deploying a specific service are repeated constantly in configuring a large network. For example, file editing is used almost everywhere in configuration since services are typically carried out by various daemons which read configuration files for instructions. Another example, disk partitioning of a group of hosts is repeated on every individual host. Thus we have the ingredient of overlapping of subproblems.

The idea of dynamic programming is to memoize the previous computed solutions of subproblems and save them for future use. If we generalize this idea, complicated configuration management can by simplified by forming a set of 'best practices', i.e., a set of solutions that have been used and proven to be effective. If a community of administrators agrees to utilize the same initial baseline state for each kind of host and the same set of operations for configuring hosts at multiple sites, system administrators can then amortize the cost of forming the practices by aggregating effective practices in a global knowledge base [15,19]. The salient feature of best practices is that they keep latent states to a minimum and thus reduce the amount that a system administrator has to remember in maintaining a large network. Without some form of consistent practice, routine problem-solving causes a state explosion in which randomly selected modifications are applied to hosts for each kind of problem. For example, we could decide to solve the problem of large logfiles by backing them up to several *different* locations, depending upon host. Then, finding a specific logfile takes much longer as one must search over all possibilities, rather than codifying and utilizing one consistent place to store them. However one should always use 'best practices' cautiously by asking questions such as "in what sense is a practice best?" "When and for whom?"

## 8. Conclusion

We have developed a theoretical framework to study and examine the complexity of configuration management. Two kinds of automatons were constructed. One is based upon the actual configuration and the other is based upon observed behavior. By reducing the problem to considering what can be observed, we limit complexity and approach the behavior of human administrators performing tasks; they are as well limited by what they can observe. This limitation allows us to quantify just how difficult the problem can become, and to discuss ways in which it can be simplified.

The nondeterminism of operations on observed states is one of the most challenging problems of configuration management. In the discussion of reproducibility of configuration operations, we have shown that for one host in isolation and for some configuration processes, reproducibility of observed effect for a configuration process is a statically verifiable property of the process. However, reproducibility of populations of hosts can only be verified by explicit testing. Using configuration processes verified to be locally reproducible, we can identify latent preconditions that affect behavior among a population of hosts.

Much attention has been paid to the limits imposed upon configuration operations and how they affects usability of a set of operations. Based upon our theoretical framework, we formally defined many limits of configuration operations including: idempotence, statelessness, convergence, commutativity, consistency and atomicity. We showed in our composability theory that only these limits are not enough to make composability tractable.

Using commutativity and idempotence we reduced a known NP-complete problem to composability of configuration operations. Thus the general composability problem without any limit is NP-hard. We also studied how other limits affect the complexity of composition. The conclusion was that composition remains NP-hard regardless of whether the operations have those limits.

By the contextualization of the configuration process and review of current configuration strategies, we summarized how configuration management is made tractable in practice. These mechanisms include the use of experience and documentation, reduction of state space, abstraction, re-baselining, the use of orthogonality, and the use of closures. For each mechanism, we discussed its drawbacks. We also made observations on how the current configuration strategies simplify configuration management. Many ways of reducing complexity of a problem have been explored in computation theory. We suggested ways to apply various techniques of reducing complexity to configuration management including choosing easy instances, approximation, and using dynamic programming. At the end of this discussion, one quandary remains. If configuration management is so mathematically difficult, then how is it possible at all? This is – fortunately – an easy question to answer. We assume in our work that the behavior of a system is unknown and unpredictable; the model we use is that of a 'black box'. In realistic situations, the box is neither closed nor black. The system administrator uses knowledge of past experience, as well as detailed understanding of the internals of systems, to achieve a result. What our work shows is not that configuration management is impossible, but that a 'gray box' understanding is necessary in order to make it tractable.

## References

[1] P. Anderson, *Towards a high level machine configuration system*, Proceedings of the Eighth Large Installation System Administration Conference (LISA VIII), USENIX Association, Berkeley, CA (1994), 19.
[2] P. Anderson, P. Goldsack and J. Patterson, *Smartfrog meets lcfg: Autonomous reconfiguration with central policy control*, Proceedings of the Seventeenth Large Installation System Administration Conference (LISA XVII), USENIX Association, San Diego, CA (2003).
[3] S. Boyd and L. Vandenberghe, *Convex Optimization*, Cambridge University Press (2004).
[4] M. Burgess, *A site configuration engine*, Computing Systems **8** (1995), 309.

[5] M. Burgess, *Computer immunology*, Proceedings of the Twelfth Large Installation System Administration Conference (LISA XII), USENIX Association, Berkeley, CA (1998), 283.

[6] M. Burgess, *Theoretical system administration*, Proceedings of the Fourteenth Large Installation System Administration Conference (LISA XIV), USENIX Association, Berkeley, CA (2000), 1.

[7] M. Burgess and A. Couch, *Autonomic computing approximated by fixed-point promises*, Proceedings of the Twentieth Large Installation System Administration Conference (LISA 06), USENIX Association, Washington, DC (2006), 101.

[8] M. Burgess and R. Ralston, *Distributed resource administration using cfengine*, Software Practice and Experience **27** (1997), 1083.

[9] A. Couch, J. Hart, E.G. Idhaw and D. Kallas, *Seeking closure in an open world: a behavioural agent approach to configuration management*, Proceedings of the Seventeenth Large Installation System Administration Conference (LISA XVII), USENIX Association, Berkeley, CA (2003), 129.

[10] A. Couch and S. Schwartzberg, *Experience in implementing an http service closure*, Proceedings of the Eighteenth Large Installation System Administration Conference (LISA XVIII), USENIX Association, Berkeley, CA (2004), 213.

[11] A. Couch and Y. Sun, *On the algebraic structure of convergence*, LNCS, Proc. 14th IFIP/IEEE International Workshop on Distributed Systems: Operations and Management, Heidelberg, Germany (2003), 28–40.

[12] A. Couch and Y. Sun, *On observed reproducibility in network configuration management*, Science of Computer Programming **53** (2004), 215–253.

[13] Free Standards Group, *Lsb – linux standards base*, http://www.linuxbase.org.

[14] M. Garey and D. Johnson, *Computers and Intractability, A Guide to the Theory of NP-Completeness*, Freeman, New York, NY (1979).

[15] G. Halprin, ed., *Sa-bok (the systems administration body of knowledge)*, http://www.sysadmin.com.au/sa-bok.html.

[16] J. Hart and J. D'Amelia, *An analysis of rpm validation drift*, Proceedings of the Sixteenth Large Installation System Administration Conference (LISA 02), USENIX Association, Berkeley, CA (2002), 155–166.

[17] L. Kanies, *Practical and theoretical experience with isconf and cfengine*, Proceedings of the Seventeenth Large Installation System Administration Conference (LISA XVII), USENIX Association, San Diego, CA (2003).

[18] S. Kirkpatrick, C. Gelatt Jr. and M.P. Vecchi, *Optimization by simulated annealing*, Science (1983), 168.

[19] R. Kolstad, ed., *The sysadmin book of knowledge gateway*, http://ace.delos.com/taxongate.

[20] L. Lorena and F. Lopes, *A surrogate heuristic for set covering problem*, European Journal of Operational Research **79** (1994), 138–150.

[21] Redhat, *Redhat enterprise linux*, http://www.redhat.com/enus/USA/rhel.

[22] Y. Sun, *Complexity of system configuration management*, Ph.D. Thesis (2006).

[23] S. Traugott, *Why order matters: Turing equivalence in automated system administration*, Proceedings of the Sixteenth Large Installation System Administration Conference (LISA XVI), USENIX Association, Berkeley (2002), 99.

[24] S. Traugott and J. Huddleston, *Bootstrapping an infrastructure*, Proceedings of the Twelfth Large Installation System Administration Conference (LISA XII), USENIX Association, Berkeley, CA (1998), 181.

# – 5.4 –

# Predictable and Reliable Program Code: Virtual Machine-based Projection Semantics

J.A. Bergstra[1,2], I. Bethke[3]

[1] *University of Amsterdam, Faculty of Science, Programming Research Group,*
*Kruislaan 403, 1098 SJ Amsterdam, The Netherlands*
*E-mail: janb@science.uva.nl*

[2] *Utrecht University, Department of Philosophy, Applied Logic Group, Utrecht, The Netherlands*
*E-mail: janb@phil.uu.nl*

[3] *University of Amsterdam, Faculty of Science, Programming Research Group,*
*Kruislaan 403, 1098 SJ Amsterdam, The Netherlands*
*E-mail: inge@science.uva.nl*

## 1. Introduction

Network and systems administration seems to be a subject not based on theory, or at least not primarily based on theory. Mark Burgess [14–17] provides aspects of theory and so do Alva Couch and Yizhan Sun in [20], but the gap between these papers and conventional foundational theory of computing is quite large. Theoretical computer science has its empty areas as well, but we hold that many critical issues in network and systems administration cannot be reliably addressed without the application of semantic theory, be it tailor made theories.

The line between system administration and other aspects of computing is never clear. System administrators are often amongst the first users to test new technologies, and are involved in building new applications in a production environment. When things go wrong, the system administrator is often one of the first people to be involved in problem finding.

Many problems in the security and reliability of mission-critical software can result from misleading program semantics. The possibility for mistakes in coding to weaken a

HANDBOOK OF NETWORK AND SYSTEM ADMINISTRATION
Edited by Jan Bergstra and Mark Burgess

system or result in unexpected behaviour was one of the motivations for the introduction of the Ada language. Here we present a simple theory developing a reliable syntax whose execution semantics are unambiguous. Of all the languages used by system administrators and operators in deploying services (e.g. PHP, Perl, TCL, Scheme etc.) few can be said to have such a clear connection between syntax and behaviour. The system world would do well to foster this kind of predictability in future technologies.

Since the introduction of time sharing operating systems, command languages or scripts have been used to manage the running of programs. Job Control Language was introduced to manage program execution for batch control. In the 1980s, the DOS environment on the IBM PC mimicked the idea of scripts with its simple 'batch files', used to start applications from the command line, but these were never as sophisticated as those coming from timesharing systems.

The Unix Bourne shell, and later C shell, were a turning point for system scripting. They allowed – and continue to allow – a user to combine independent programs using 'pipes' or communication channels which behave like a Virtual Private Network interface between applications. Later, the inefficiencies and irregularities of syntax led to the development of operating system independent languages such as Awk, then Perl and Tcl which have only partially replaced the shell.

Today there is a need for a new generation of languages for system management, in a variety of contexts. One example is configuration management. Configuration management is currently an active area – see the chapter by Alva Couch in this book. A prototype language that is widely used for here is Cfengine [13], which takes a declarative view that has been likened to Prolog [19]. Cfengine uses an irregular syntax but aims to give clear semantics based on fixed point behaviour [16].

As languages become high level, the details of their semantics become important in order to understand their behaviour. Today, scripting languages like shell, Perl, PHP and Tcl are very low level and leave the higher-level semantics entirely to imperative, declarative, functional and object oriented programmers. Python and Ruby are bridging this gap.

In this chapter we wish to consider this subject – the semantics of scripting languages – as part of the foundations of system and network administration and provide a theoretical account to elementary higher-level object oriented scripting languages using the field of semantic frameworks. Our aim is to highlight the role of layers of language abstraction and their effect on semantics of language statements. We show how these languages can be built up from low-level primitives through virtualisation layers to high-level constructs, so that we might build new languages that are 'correct by construction'. This is a large field of research, so we have modest aims for our chapter, but we hope that this will inspire more work of an applied nature to feed the current debates and controversies over semantics.

We start with providing a survey of semantic theories of programming and propose the viewpoint that all of these theories have the intrinsic problem of asking their readers to think in terms of mathematical domains that may fail to be self evident for the practical minds working in network and systems administration. Then we provide an alternative semantic theory that we have developed under the name 'projection semantics'. Projection semantics is a way to state that compilers determine the meaning of programs. A projection is a theoretical account of a possibly slow but conceptually convincing compiler, or rather of the transformation that the compiler is supposed to produce. The classical assumption that such a projection should be validated against a denotational or mathematical semantic theory is dropped! Semantic intuitions are supposed to reside at the level of the projections, i.e. conceptual abstractions of what compilers do or might do.

A second aspect of the approach that we advocate is that it is execution architecture based. With this we mean that in giving conceptual models for programs and their meaning it is essential that an operating context, or execution architecture is taken as the point of departure. Here we will use the execution architecture of [11] which proposes that a program executes in a context where it interacts with both local (auxiliary) and external services. The external services represent the world to which the running program ultimately directs its activity – i.e. whether its operation is supposed to achieve intended effects – while the services local to the execution architecture represent transformations on auxiliary and volatile data which are thrown away as soon as the execution of the program halts and which cannot be inspected by any other system component than the running program. Because this viewpoint is reasonably close to a possible description of a virtual machine our projection semantics is called *virtual machine based*. Perhaps a more informative name is *execution architecture based* projection semantics. Programming with program notations that permit being modelled by these techniques might be called *execution architecture based programming*. Below we will reconstruct a small fragment of object oriented scripting as execution architecture based programming, by providing a projection from a high-level – though very simple – syntax to a low-level one for which an execution architecture is known in detail.

Many matters can be clarified on the basis of projection semantics for an execution architecture based reconstruction of programs and program notations. A further development in the direction of concurrent programming languages, for instance the large variety of thread scheduling mechanisms also called *interleaving strategies*, will capture major user intuitions while being almost irresponsibly naive in comparison to existing dominant concurrency theories.

The instruction sequence language ISLA that is used below finds its semantics in so-called *thread algebras*. Thread algebras are conceptually simplified – and for that same reason technically more complex and less stable – process algebras fully dedicated to the semantic issue to be solved. Therefore our projection semantics may be classified as a special case of process algebra based semantics. Process algebra based semantics of program notations has a history of more than 20 years: see e.g. Milner [37], Vaandrager [48], Cleaveland et al. [18], Andrews [3], Bergstra, Middelburg and Usenko [9] and Mauw and Reniers [34].

## 2. Semantics of programming languages

The study of programs has a profound history in the setting of formal languages and grammars. Originating from theoretical studies of automata and parsing, the path from program texts to program execution has been subject to many investigations and has to some extent been found useful in various practical applications such as compiler-writing systems, type-checkers, code-generators and documentation generators. However, in contrast to the popularity of formal syntax, formal semantics – the field concerned with the rigorous mathematical study of the meaning of programming languages and their models of computation – has been less exploited in practical applications concerning the design and implementation of programming languages. Currently, only few systems generating any kind of implementation from semantic descriptions are available [29]. In particular, semantic results are rarely incorporated in practical systems that help language designers to implement and test a language under development, or to assist programmers in answering their questions about the meaning of some language feature not properly documented in a language's reference manual. Among the reasons for this state of affairs, one stands out. The use of even a properly-implemented and well-maintained system is limited to the input available for it. For significant practical uses of semantic descriptions, the input should be a complete, fully *formal* semantic description of the observable behaviour of a program in contrast to the kind of descriptions usually found in textbooks and manuals, where crucial details are often omitted and informal conventions are introduced in the interest of conciseness. Yet despite the theoretical efforts in establishing the foundations of various frameworks, good pragmatic features needed for the efficient use of semantic descriptions are often lacking. Below we give a recap of the semantic frameworks of current interest. For an extensive survey along with a list of uses that have been made of them, and speculations on the hindrances for greater use, the reader is referred to [40].

Before we give a brief summary of the semantic frameworks of current interest, let us distinguish between *static* and *dynamic* semantics. Static semantics – which appears to be used more often in practical applications than dynamic semantics – concerns the checking for well-formedness of a program without actually executing it. It thus corresponds to its compile-time behaviour and is independent of any input that is provided to the program at run-time. Well-formedness is usually decidable, so static semantics may be treated as a kind of (context-sensitive) syntax specified by attribute grammars. For compiled programs, their well-formedness has already been checked, and their dynamic semantics corresponds to their run-time behaviour. For languages implemented by interpreters, programs usually run without a foregoing overall well-formedness check. Any required checks happen at run-time and can thus be considered part of the dynamic semantics. In the sequel we shall focus entirely on dynamic semantics and do not worry about what semantics is given to programs containing ill-typed expressions, assuming that such programs are filtered out by a preceding static semantics.

The dynamic semantics of a language is given by a mathematical model which represents the possible computations described by the language. The three main classes of approach can be classified as *operational*, *denotational*, or *axiomatic*.

Starting from the early 1960s [35], operational semantics models the computations of programs as a – perhaps infinite – sequence of computational steps between states. Although states in computations generally depend on the syntactical appearance of the executed program, syntactically-distinct programs can have the same sets of states. In order to obtain a reasonable notion of semantic equivalence in operational semantics, some equivalence relation that ignores the syntactic dependence of states has to be introduced; popular choices are bisimulation equivalences [49]. Various ways of modelling computations have been developed yielding different frameworks for operational semantics of programming languages. The most prominent ones are:

(1) *Structural operational semantics* (SOS) – the pioneering work on operational semantics – was proposed by Plotkin in [1]. Here computations are modelled as sequences of – possibly labelled – transitions between states – a mixture of syntax and computed values – involving syntax, computed values, and auxiliary entities. SOS has been widely used to describe process calculi used for specification, but also for the design and description of the programming languages Facile and Ada.

(2) *Natural semantics* was developed by Kahn in the mid-1980s [32]. Terminating computations are modelled as evaluation relations between syntax and computed values, possibly involving auxiliary entities; nonterminating computations are generally ignored. This semantics was used during the design and official definition of Standard ML.

(3) *Reduction semantics* is a framework proposed by Felleisen and Friedman [24] towards the end of the 1980s. This type of operational semantics models computations as sequences of term rewriting steps where the intermediate terms consist again of computed values and auxiliary entities. Reduction semantics was used during the design of Concurrent ML.

(4) *Abstract state machines (ASM)*, previously called 'evolving algebras'[1] [28], was developed by Gurevich in the late 1980s. In contrast to the preceding approaches, states are mathematical structures where data come as abstract objects which are equipped with basic operations – partial functions – and relations. The notion of computation is given by a transition system which determines which functions in a given state have to be updated. ASM was adopted by ISO for the use in the Prolog standard. More recently in the new millenium, ASM has been used to provide specifications of Java and the JVM [46], and of the JCVM [43].

Denotational semantics models each part of a program as its denotation – typically a higher-order function between complete partial orders [27] – representing its observable behaviour. The semantics of loops and recursion usually involves the explicit use of fixed-point operators or a specification as equations whose least solution is to be found. Apart from the original Scott–Strachey style of denotational semantics, we consider here also *monadic semantics* and *predicate transformers*.

(5) *Scott–Strachey semantics* is the original style of denotational semantics developed at the end of the 1960s [42]. Here domains of denotations and auxiliary entities are defined by domain equations, i.e. equations involving complete partial orders with continuous functions. Many standard techniques for representing programming concepts as pure mathematical functions have been established in this framework. For instance, sequencing may be represented by composition of functions, or by the use of continuations. The denotational description of nondeterminism, concurrency, and interleaving has led to the development of so-called power domains [45]. Because of Strachey's fundamental contributions, researchers were able to give semantic definitions to programming languages like Ada, Algol60, Algol68, PL/1, Lisp, Pascal, Scheme and Snobol [4,12,36,39,41,47].

(6) *Monadic semantics* is based on category-theoretic concepts and was developed by Moggi [38] at the end of the 1980s. Denotations in monadic semantics are elements

---

[1]The idea of evolving algebras has probably independently been developed and applied during the design of the wide-spectrum language COLD. For more details see [22,23,31].

of so-called monads, i.e. type constructors that capture various notions of sequential computation. Monad transformers construct monads incrementally and add new aspects of a computation to a given monad. Monadic semantics appears to be currently lacking an example of its use in connection with a practical programming language.

(7) *Predicate transformer semantics* was proposed by Dijkstra [21] in the mid-1970s. Here the semantics of a programming language is defined by assigning to each command in the language a corresponding predicate transformer returning the weakest pre-condition which ensures termination of the command under a certain post-condition. Predicate transformers are required to have properties corresponding to the continuity of functions on Scott-domains. Unlike other semantic formalisms, predicate transformer semantics was not designed as an investigation into the foundations of computations. Rather, it was intended – and still is – to provide programmers with a methodology to develop their programs as 'correct by construction' in a 'calculational style'.

Axiomatic semantics is an approach based on mathematical logic. Its purpose is to provide a set of logical axioms and rules in order to reason about the correctness of computer programs with the rigour of mathematical logic. As usual with axiomatic specifications, there may be insufficiently many properties been given – in which case there may be more than one model – or the set of properties may be inconsistent – in which case there are no models at all. As with predicate transformers, predicates are used as formulae. We consider *Hoare logic*.

(8) *Hoare logic* (also known as *Floyd-Hoare logic*) was developed by Hoare [30] in the late 1960s. Hoare's logic has axioms and inference rules for all the constructs of a simple imperative programming language. In addition to the rules for the simple language in Hoare's original paper, rules for other language constructs – such as concurrency, procedures, jumps, and pointers – have been developed since then by Hoare and many other researchers. Hoare logic was used during the design of Pascal, and the resulting description as the basis for program verification.

A large number of semantic frameworks have been provided during the past three decades. We have classified them mainly as operational, denotational, and axiomatic in a deliberately brief and superficial style, focusing on the most important differences between the frameworks, and ignoring many details. Compared to the amount of effort that has been devoted to the development of these semantic frameworks and their uses listed above, the list of significant practical uses may be considered as relatively small. Hindrances to greater use are user-unfriendliness and the fact that most frameworks do not scale up smoothly from tidy illustrative to full-scale practical languages.

## 3. Projection semantics

A major difficulty with the approaches to programming language semantics that stem from theory is that the audience is asked to think in terms of certain more or less sophisticated semantic domains, typically function spaces or transition systems of some particular kind. This is always felt as an obstacle and textbooks on programming never seem to pay any attention to such mathematical background.

*Projection semantics* (PS) was developed in the setting of *program algebra* in the years 1998 and 1999 [6,7] and is different in that it explains the semantics of a program in a 'higher' language in terms of a translation, here called *projection*, in a known program notation. A program is supposed to have the same meaning as its projection, by definition. The semantics of the projection, i.e., the resulting program after projection, is therefore given in terms of programming language semantics as well, the only progress made being that it is in a notation which one claims to be understood.

As an example we consider the basic program notation based on a parameter set of basic instructions $\{a, b, c, \ldots\}$ representing a command issued by a program during execution to its context. That context may involve classical ingredients such as a printing device but it may also involve the calculation of a value contained in a variable which is often taken simply as something inside a program. The execution of a basic instruction takes place during program execution. It is done by the machine architecture that executes the program and it will have two effects: a possible state change in the execution context and a Boolean value which is produced as a result of processing the instruction – viewed as a request to its environment – and which is subsequently returned to the program under execution where it may be used to decide which is the next instruction to be carried out. In addition to the parameter set, the basic program notation contains test instructions – each basic instruction can be turned into positive or negative test instruction by prefixing it with − or + – and absolute jump instructions ##$k$ with $k \in \mathbb{N}$.

A program is then a finite stream of instructions. Execution starts at the first instruction and ends when the last instruction is executed or if a jump is made to a non-existing instruction. In particular, ##0 represents program termination because the instruction count is taken to start with 1. The working of test instructions is as follows: if the $i$th instruction is +$a$ its execution begins with the execution of $a$. After completion of that action a reply value is at hand. If the reply was `true` the run of the program proceeds with instruction number $i + 1$ – if that exists otherwise execution terminates. Alternatively, if the reply was `false` the execution proceeds with instruction number $i + 2$ – again assuming its existence, leading to termination in the other case as well. Thus a negative reply enacts that an instruction is skipped. Execution of a negative test instruction −$a$ proceeds in a similar way yielding a skip if $a$ returns `true`. Finally, the execution of a jump instruction ##$k$ makes the execution continue with the $k$th instruction if that instruction exists. Otherwise the program execution terminates. If ##$k$ itself is the $k$th instruction the execution is caught in a never ending loop.

An example of a basic program is $a$; +$b$; ##0; $c$; $c$; ##2 which may be read as: first do $a$ and then $b$, and then repeat $c$; $c$; $b$ as long as the last reply for $b$ was negative, and terminate after the first positive reply to an execution of $b$ is observed.

The behavioural semantics of basic programs is defined in [5,7,8]. This semantics is based on the basic *thread algebra* TA. TA has a constant $S$ for termination, a constant $D$ for inaction or divergent behaviour, and a composition mechanism named *post-conditional composition*: the expression $P \trianglelefteq a \trianglerighteq Q$ represents the execution of action $a$, and if `true` was returned, behaviour continues as $P$, and otherwise as $Q$. A shorthand for $P \trianglelefteq a \trianglerighteq P$ is $a \circ P$, and this resembles the usual action prefix in process algebra. Furthermore TA is equipped with a family of approximation operators which can be used to handle infinite behaviour. Now any basic program $X$ gives rise to a (possibly infinite) TA process. For example, the process $P$ corresponding to the above mentioned basic program $a; +b; \#\#0; c; c; \#\#2$ can be defined by the two equations $P = a \circ Q$, $Q = S \trianglelefteq b \trianglerighteq (c \circ c \circ Q)$.

Based on this extremely simple program notation more advanced program features can be developed within projection semantics such as conditional statements, while loops, recursion, object classes, method calls etc. In taking care of a cumulative and bottom up introduction of such complex features while providing appropriate projections into the lower levels of language development, an *intermediate* language can be developed that keeps all definitions rigorous, ensures a clear meaning of higher program constructs and serves as an intermediate step for programs written in a higher-order language.

## 4. Virtual machines and intermediate languages

*Virtual machines* (VMs; see Figure 1) are a popular target for language implementors. They reside as application programs on top of an operating system and export abstract machines that can run programs written in a particular *intermediate language* (IL). As an intermediate step, programs written in a higher-order language (HL) are first translated into a more suitable form before object or machine code is generated for a particular target machine. Masking differences in the hardware and software layers below the virtual machine, any program written in the high-level language and compiled for these virtual machines will run on them. In an ideal situation the frontend of a virtual machine will be entirely independent of the target hardware, while the backend will be sensibly independent of the particular language in which source programs are written. In this way the task of writing compilers for $n$ languages on $m$ machines is factored into $n + m$ part-compilers rather than $n \times m$ compilers.

One of the first virtual machines was the *p-Code machine* invented in the late 1970s as an intermediate form for the ETH Pascal compilers [2] but becoming pervasive as the machine code for the UCSD Pascal system. What had been noted was that a program encoded for an abstract machine may be used in two ways: a compiler backend may compile the code down to the machine language of the actual target machine, or an interpreter may be written

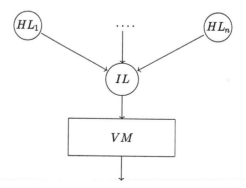

Fig. 1. Outline of a virtual machine design.

which emulates the abstract machine on the target. This interpretative approach surrenders a significant factor of speed, but has the advantage that programs are much more dense in the abstract machine encoding. As a consequence of this technology high-level languages became available for the first time on microcomputers. As an additional benefit, the task of porting a language system to a new machine reduced to the relatively simple task of creating a new interpreter on the new machine.

In the late 1990s Sun Microsystems released their Java [25] language system. This system is also based on an abstract machine – the *Java Virtual Machine* (JVM) [33]. Nowadays, JVMs are available for almost all computing platforms. In mid-2000 Microsoft revealed a new technology based on a wider use of the world wide web for service delivery. This technology became known as the *.NET* system and was designed with the objective of supporting multiple languages. At the heart of this initiative stands the *Common Language Infrastructure* (CLI) virtual machine executing the *Common Intermediate Language* (CIL).

A long-running question in the design of virtual machines has been whether register or stack architectures can be implemented more efficiently with an interpreter. Many designers favour stack architectures since the location of the operands is implicit in the stack location, prominent examples being the stack-based JVM and CLI. In contrast, the operands of register machine instructions must be specified explicitly. A more recent example of this sort of virtual machine is Parrot intended to run dynamic languages. In the sequel we shall only consider stack-based VMs. For a translation from stack-based to register-based code and efficiency comparisons see e.g. [26,44].

## 5. The base language

The base language is the PS answer to the conceptual question: 'what is a program?'. The tricky aspect is that at the same time as the base language is claimed to provide this answer PS predicts that this answer may be in need of refactoring at some future stage. PS runs on the unproven assumption that whatever formal definition of a program is given, there will always be a setting in which this definition needs to be compromised or amended in order to get a comprehensible theory. But at the same time a PS answer to any problem needs full precision definitions. There is no room for an open ended concept of program like 'a text meant to tell a machine what to do', or – in the absence of precise definitions of 'unit' and of 'deployment' – 'a definition of a software component as a unit of deployment'. Likewise, a machine language may not be defined as a program notation meant for machine execution – in the absence of a clear model of the concept of machine execution – and a scripting language is not definable as a language for writing scripts – unless, again the concept of a script is rigorously defined beforehand.

As a start for syntax development *instruction stream language with absolute jumps* (ISLA) (cf. Section 3) may be taken which features among the sequence of the program notations that have been designed on the basis of program algebra under the name PGLD [7]. The basis of the definition of ISLA is formed by the notion of a *basic action*. Basic actions may be used as instructions in ISLA. A basic instruction *a* represents a command issued by the program during execution to its context. That context may involve classical ingredients such as a printing device but it may also involve the operation of a value contained in a variable which is often taken simply as something inside a program. The execution of a basic action $a$ takes place during program execution. It is done by the machine architecture that executes the program and it will have two effects: a possible state change in the execution context and a Boolean value which is produced as a result of processing the action – viewed as a request to its environment – and which is subsequently returned to the program under execution where it may be used to decide which is the next instruction to be carried out.

For each basic action $a$ the following are ISLA instructions: $a$, $+a$ and $-a$. These three versions of the basic actions are called: *void basic action* ($a$), *positive test instruction* ($+a$) and *negative test instruction* ($-a$). A void basic action represents the command to execute the basic action while ignoring the Boolean return value that it produces. Upon completion of the execution of the basic action by the execution environment of the program processing will proceed with the next instruction. The test actions represent the two ways that the Boolean reply can be used to influence the subsequent execution of a program. The only other instructions in ISLA are *absolute* jumps ##$k$ with $k$ a natural number.

An ISLA program is a finite stream of ISLA instructions. Execution by default starts at the first (left most) instruction. It ends if the last instruction is executed or if a jump is made to a nonexisting instruction. In particular, ##0 – which will be abbreviated by ! – represents program termination because the instruction count is taken to start with 1. The working of test instructions is as follows: if the $i$th instruction is $+a$ its execution begins with the execution of $a$. After completion of that action a reply value is at hand. If the reply was `true` the run of the program proceeds with instruction number $i + 1$ if that exists, otherwise execution terminates. Alternatively, if the reply was `false` the execution proceeds with instruction number $i + 2$ again assuming its existence, leading to termination in the other case as well. Thus a negative reply enacts that an instruction is skipped. In the case of a negative test instruction $-a$ execution proceeds with instruction number $i + 2$ at a positive reply and with instruction number $i + 1$ at reply `false`. The execution of a jump instruction ##$k$ makes the execution continue with the $k$th instruction if that exists. Otherwise the program execution terminates. If ##$k$ is itself the $k$th instruction the execution is caught in a never ending loop.

EXAMPLE 1. Typical ISLA programs are:
- $a$; $+b$; !; $c$; ##2 which may be read as first do $a$ and then $b$ and then repeat $c$; $b$ as long as the last reply for $b$ was negative, and terminate after the first positive reply to an execution for $b$ is observed.
- $-a$; ##6; $b$; $c$; !; $e$; $f$ which may be read as first do $a$, and if $a$ returns `true` then do $b$ and $c$, else do $e$ and $f$, and then terminate.

The definition of ISLA in PS has the following intention. At the initial stage of PS development it is reasonable to view ISLA as a definition of the concept of a program. Moreover, any text $P$ is a program provided there is at hand a projection $\phi$ which maps the text to an ISLA instruction sequence $\phi(P)$ which, by definition, represents the meaning of $P$ as a program. The position regarding ISLA is therefore not the untenable assertion that every program is identical to an ISLA instruction sequence but the much more flexible assertion that for some entity $P$ to qualify as a program it must be known how it represents an ISLA instruction sequence. By means of the application of the projection that ISLA program is found and the meaning of the entity $P$ as a program is determined.

An alternative to ISLA is the program notation ISLR which admits basic actions, positive and negative test actions and the termination instruction !. The jumps are different, however. Only forward jumps (#$k$) and backward jumps (\#$k$) (with $k$ a natural number) are admitted. Thus ISLR is ISL with relative jumps. Termination also occurs if a jump is performed outside the range of instructions.[2]

EXAMPLE 2. A typical ISLR program is $+a; \#3; b; !; c; \backslash\#4$ which may be read as first do $a$ and, if the reply to $a$ was negative, continue with $b$ and terminate; otherwise enter an infinite repetition of $c$.

ISLA and ISLR are intertranslatable by the mappings $\psi : \text{ISLA} \to \text{ISLR}$ and $\phi : \text{ISLR} \to$ ISLA which transform jumps depending on the place of their occurrence and leave the other instructions unmodified. The mappings are defined by

- $\psi(u_1; \ldots; u_n) = \psi_1(u_1); \ldots; \psi_n(u_n)$ where

$$\psi_i(\#\#k) = \begin{cases} \#k - i, & \text{if } k \geqslant i, \\ \backslash\#i - k, & \text{otherwise.} \end{cases}$$

- $\phi(u_1; \ldots; u_n) = \phi_1(u_1); \ldots; \phi_n(u_n)$ where $\phi_i(\#k) = \#\#k + i$ and

$$\phi_i(\backslash\#k) = \begin{cases} \#\#i - k, & \text{if } k < i, \\ !, & \text{otherwise.} \end{cases}$$

EXAMPLE 3. The translations of the programs in Examples 1 and 2 are

$$\psi(a; +b; !; c; \#\#2) = \psi_1(a); \psi_2(+b); \psi_3(!); \psi_4(c); \psi_5(\#\#2)$$
$$= a; +b; !; c; \backslash\#3,$$
$$\psi(-a; \#\#6; b; c; !; e; f) = \psi_1(-a); \psi_2(\#\#6); \psi_3(b); \psi_4(c);$$
$$\psi_5(!); \psi_6(e); \psi_7(f)$$
$$= -a; \#4; b; c; !; e; f,$$
$$\phi(+a; \#3; b; !; c; \backslash\#4) = \phi_1(+a); \phi_2(\#3); \phi_3(b); \phi_4(!); \phi_5(c); \phi_6(\backslash\#4)$$
$$= +a; \#\#5; b; !; c; \#\#2.$$

A virtue of ISLR is that it supports *relocatable* programming. Programs may be concatenated without disturbing the meaning of jumps. Placing a program behind another program may be viewed as a relocation which places each instruction at an incremented position. This property can be best exploited if programs are used that always use ! for termination.

---

[2] The program notation ISLR was originally defined in the setting of program algebra in [7] under the name PGLC.

In some cases it is useful to use only a forward jump to the first missing instruction for termination. For programs of that form concatenation corresponds to sequential composition.

Another alternative to ISLA as a base language would be ISLAR, the instruction stream notation with absolute *and* relative jumps. Throughout this paper, however, we shall take ISLA as base language, and projections will be (chained) mappings to ISLA programs. We shall write A $\subseteq$ B for a program notation B extending A and use the operator $\phi$ for projections. In most cases a projection transforms some program notation back to a simpler one. Sometimes the choice of the right order of such steps is important. A formalism for specifying a strategy for chaining projections will not be included below, however. It is assumed that a reader will be easily able to determine the right order of the various projections. A full precision story may need an annotation of all projection functions with domains and codomains. But as it turns out these matters are often clear from the context and merely needed for formalisation rather than for rigorous explanation. Therefore that kind of bookkeeping has been omitted, with the implicit understanding that it can always be introduced if additional clarification is necessary.

## 6. Extensions of ISLA with labels and goto's, conditional constructs and while loops

**6.1.** *Projective syntax for labels and goto's*

Labels have been introduced in the program algebra setting in [7] in the program notation PGLDg. Here, however, labels will have the form [s], with s a non-empty alphanumerical string, i.e., a sequence over the alphabet $\{a, \ldots, z, A, \ldots, Z, 0, \ldots, 9\}$; ##[s] and ##[s][t] are the corresponding *single* or *chained goto* instructions. The intended meaning is that a label is executed as a skip. A single goto stands for a jump to the leftmost instruction containing the corresponding label if that exists and a termination instruction otherwise, and a chained goto of the form ##[s][t] stands for a jump to the leftmost instruction containing the label [s] followed by a jump to the leftmost instruction thereafter containing the label [t] assuming hat they both exist and termination otherwise. The label occurrence that serves as the destination of a single or chained goto is called the *target* occurrence of the goto instruction.

In order to provide a projection for A:GL (program notation A with goto's and labels) the introduction of annotated goto's is useful. ##[s]m (##[s][t]m) represents the goto instruction ##[s] (##[s][t]) in a program where the target label is at position m. ##[s]0 (##[s][t]0) represents the case that no target label exists. The projection from A:GLa (A with annotated goto's and labels) to A and from A:GL to A:GLa are obvious.

PROJECTION 4. Let ISLA $\subseteq$ A.
  (1) $\phi : \text{A:GLa} \rightarrow \text{A}$ is defined by $\phi(u_1; \ldots; u_n) = \phi_1(u_1); \ldots; \phi_n(u_n)$ where
      (a) $\phi_i([s]) = \#\#i + 1$,
      (b) $\phi_i(\#\#[s]m) = \phi_i(\#\#[s][t]m) = \#\#m$.
      The other instructions remain unmodified.
  (2) $\phi : \text{A:GL} \rightarrow \text{A:GLa}$ is defined by $\phi(u_1; \ldots; u_n) = \phi(u_1); \ldots; \phi(u_n)$ where
      $\phi(\#\#[s]) = \#\#[s]m$ and $\phi(\#\#[s][t]) = \#\#[s][t]m$ with m the instruction number of
      the target label if it exists and 0 otherwise. The other instructions remain unmodified.

## 6.2. *Second-level instructions*

When transforming a program it may be necessary to insert one or more instructions. Unfortunately that renders the counting of jumps useless. Of course jump counters may be updated simultaneously thus compensating for the introduction of additional instructions. Doing so has proved to lead to unreadable descriptions of projection functions, however, and the device of two-level instruction counting will be proposed as a more readable alternative.

Instructions will be split in *first-level* instructions and *second-level* instructions. The idea is that expanding projections – projections that replace instructions by non-unit instructions – are given in such way that each instruction sequence replacing a single instruction begins with a first-level instruction while all subsequent instructions are taken at the second level.

Second-level instructions are instructions prefixed with a $\sim$. First-level instructions do not have a prefix; they are made second level by prefixing them with a $\sim$. First- and second-level absolute jumps ##$k$ and $\sim$ ##$k$ will represent a jump to the $k$th first-level instruction if that exist (and termination otherwise) simply ignoring the second-level instructions. The extension of an ISLA based program notation A with second-level instructions is denoted with A:SL. For any A $\supseteq$ ISLA, the projection $\phi$ : A:SL $\rightarrow$ A removes second-level instruction markers and updates the jumps.

PROJECTION 5. Let ISLA $\subseteq$ A.
$\phi$ : A:SL $\rightarrow$ A is defined by $\phi(u_1; \ldots; u_n) = \phi(u_1); \ldots; \phi(u_n)$ where

$$\phi(\text{\#\#}k) = \begin{cases} !, & \text{if } k > \max, \\ \text{\#\#}k + l, & \text{otherwise.} \end{cases}$$

Here max is the number of first-level instructions occurring in $u_1; \ldots; u_n$ and $l$ is the number of second-level instructions preceding the $k$th first-level instruction. The other first-level instructions remain unmodified. Moreover, $\phi(\sim \text{\#\#}k) = \phi(\text{\#\#}k)$ and $\phi(\sim u) = u$ for all other first-level instructions $u$.

EXAMPLE 6. $\phi(\sim a; \sim +b; \sim \text{\#\#}2; \text{\#\#}1) = a; +b; !; \text{\#\#}4$.

The only use of A:SL is as a target notation for projections from an ISLA-based program notation. There will be a number of examples of this in the sequel.

NOTATION 7. For notational convenience we introduce the following abbreviations. We write
- $a$##$jmp$ for $a; \sim$ ##$jmp$, $+a$##$jmp$ for $+a; \sim$ ##$jmp$, and $-a$##$jmp$ for $-a; \sim$ ##$jmp$, and likewise
- $\sim a$##$jmp$ for $\sim a; \sim$ ##$jmp$, $\sim +a$##$jmp$ for $\sim +a; \sim$ ##$jmp$ and $\sim -a$##$k$ for $\sim -a; \sim$ ##$jmp$

with $jmp \in \mathbb{N} \cup \{[s], [s][t] \mid s, t \in \{a, \ldots, z, A, \ldots, Z, 0, \ldots, 9\}^+\}$.

**6.3.** *Projective syntax for the conditional construct*

Conditional constructs can be added to ISLA by using four new forms of instruction: for each basic action $a$, $+a\{$ and $-a\{$ are *conditional header* instructions, further $\}\{$ is the *separator* instruction and $\}$ is the *end of construct* instruction. The idea is that in $+a\{; X; \}\{; Y; \}$ after performing $a$, at a positive reply $X$ is performed and at a negative reply $Y$ is performed. The program notation combining an ISLA based program notation A and these conditional instructions is denoted with A:C. A projection for the conditional construct instructions is found using second-level instructions. Due to the use of second-level instructions the projection can be given by replacing instructions without the need to update jump counters elsewhere in the program. The projection to second-level instructions will be such that e.g.

$$\phi(b; +a\{; c; -d; \#\#0; \}\{; +e; \#\#4; \}; f)$$

$$= b; -a\#\#7; c; -d; \#\#0; \#\#10; +e; \#\#4; \#\#10; f.$$

The general pattern of this projection is as follows: the conditional header instructions are mapped onto a sequence of two instructions, one performing the test and the second instruction containing the second-level jump to the position following the projected separator instruction. The separator instruction is mapped to a jump to the position following the projected end of construct instruction, and the end of construct instruction is mapped to a skip, i.e., to a jump to the position thereafter. For this projection to work out some parsing of the program is needed in order to find out which separator instruction matches with which conditional construct header instruction and which end of construct instruction.

The annotated brace instructions needed for the projection of the conditional construct instructions are as follows: $+a\{k, -a\{k, \}\{k, k\}$ with $k$ a natural number. This number indicates the position of instructions that contain the corresponding opening, separating or closing braces, or 0 if such an instruction cannot be found in the program. A:Ca is the extension of a program notation A with annotated versions of the special conditional statements. An annotated version of the program given above is

$$b; +a\{6; c; -d; \#\#0; \}\{9; \{9; +e; \#\#4; 6\}; f.$$

The projection from A:Ca to A:SL is now again obvious.

PROJECTION 8. Let ISLA $\subseteq$ A.
$\phi : \text{A:Ca} \to \text{A:SL}$ is defined by $\phi(u_1; \ldots; u_n) = \phi_1(u_1); \ldots; \phi_n(u_n)$ where

$$\phi_i(+a\{k) = -a\#\#k + 1,$$

$$\phi_i(-a\{k) = +a\#\#k + 1,$$

$$\phi_i(\}\{k) = \#\#k + 1,$$

$$\phi_i(k\}) = \#\#i + 1,$$

for $k > 0$ and,

$$\phi_i(+a\{0) = -a\#\#0,$$

$$\phi_i(-a\{0) = +a\#\#0,$$

$$\phi_i(\}\{0) = \#\#0,$$

$$\phi_i(0\}) = \#\#0.$$

The other instructions remain unmodified.

The projection of A:C to A:Ca involves a global inspection of the entire A:C program. There is room for confusion, for instance, in the case

$$P = a; \}; +b\{; c; \}\{; d; \}\{; e; \{.$$

In spite of the fact that the program $P$ is not a plausible outcome of programming in ISLA:C, an annotated version of it can be established as

$$a; 0\}; +b\{5; c; \}\{0; d; \}\{0; e; \{0,$$

which contains all semantic information needed for a projection.

PROJECTION 9.  Let ISLA $\subseteq$ A.
$\phi$ : A:C $\rightarrow$ A:Ca is defined by $\phi(u_1; \ldots; u_n) = \phi(u_1); \ldots; \phi(u_n)$ where
(1) $\phi(+a\{) = +a\{k$ and $\phi(-a\{) = -a\{k$
    with $k$ the instruction number of the corresponding separator instruction if it exists and 0 otherwise,
(2) $\phi(\}\{) = \}\{k$
    with $k$ and the instruction number of the corresponding end of construct instruction if it exists and 0 otherwise,
(3) $\phi(\}) = k\}$
    with $k$ the instruction number of the corresponding separator instruction if it exist and 0 otherwise.
The other instructions remain unmodified.

### 6.4. *Projective syntax for while loops*

The following three instructions support the incorporation of while loops in programs projectible to ISLA: $+a\{*, -a\{*$ (*while loop header* instructions) and $*\}$ (*while loop end* instruction). This gives the program notation A:W. As in the case of conditional constructs, while constructs will be projected using annotated versions.

A:Wa allows the annotated while loop instructions $+a\{*k, -a\{*k$ and $k*\}$. The projection of annotated while loop instructions is again obvious.

PROJECTION 10. Let $ISLA \subseteq A$.
$\phi : A{:}Wa \to A{:}SL$ is defined by $\phi(u_1; \dots; u_n) = \phi_1(u_1); \dots; \phi_n(u_n)$ where

$$\phi_i(+a\{*k) = -a\#\#k + 1,$$

$$\phi_i(-a\{*k) = +a\#\#k + 1,$$

$$\phi_i(k*\}) = \#\#k,$$

for $k > 0$ and,

$$\phi_i(+a\{*0) = -a\#\#0,$$

$$\phi_i(-a\{*0) = +a\#\#0,$$

$$\phi_i(0*\}) = \#\#0.$$

The other instructions remain unmodified.

The introduction of annotated while braces by means of an annotating projection follows the same lines as in the case of conditional statement instructions.

PROJECTION 11. Let $ISLA \subseteq A$.
$\phi : A{:}W \to A{:}Wa$ is defined by $\phi(u_1; \dots; u_n) = \phi(u_1); \dots; \phi(u_n)$ where
 (1) $\phi(+a\{*) = +a\{*k$ and $\phi(-a\{*) = -a\{*k$
      with $k$ the instruction number of the corresponding while loop end instruction if it exists and 0 otherwise, and
 (2) $\phi(*\}) = k*\}$
      with $k$ the instruction number of the corresponding while loop header instruction if it exist and 0 otherwise.
The other instructions remain unmodified.

Notice that a projection of the extension A:CW of an ISLA-based program notation A obtained by a simultaneous introduction of conditional and while constructs can be given by a concatenation of the projections defined above by starting either with conditional or while instructions. Thus

$$A{:}CW = (A{:}C){:}W \to (A{:}C){:}Wa \to (A{:}C){:}SL \to A{:}C \to \cdots \to A.$$

We end this section with an example containing conditional and while construct instructions.

EXAMPLE 12. Let $P = +a\{; b; +c\{*; d; *\}; \}\{; e; f; \}$. $P$ can be read as do $a$ and continue with $e$ and $f$ and terminate if $a$ returns `false`; otherwise do $b$ and repeat $c; d$ as long as the last reply for $c$ was positive, and terminate after the first negative reply to an execution for $c$. We may view $P$ as a program in (ISLA:C):W. The annotated version

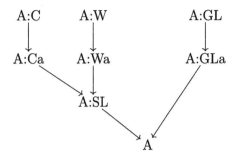

Fig. 2.  Basic extensions of an ISLA based program notation A.

of $P$ given by the above projections is $+a\{; b; +c\{*5; d; 3*\}; \}\{; e; f; \}$ and projection to (ISLA:C):SL yields

$$+a\{; b; -c\#\#6; d; \#\#3; \}\{; e; f; \}.$$

Projecting to ISLA:C we obtain

$$+a\{; b; -c; \#\#7; d; \#\#3; \}\{; e; f; \}.$$

We now proceed with the projection of the conditional constructs. The annotated version is

$$+a\{7; b; -c; \#\#7; d; \#\#3; \}\{10; e; f; 7\}$$

and projection to ISLA:SL yields

$$-a\#\#8; b; -c; \#\#7; d; \#\#3; \#\#11; e; f; \#\#11.$$

Finally, projecting to ISLA we obtain

$$-a; \#\#9; b; -c; \#\#8; d; \#\#4; !; e; f; !.$$

The projections discussed in this section are summarised in Figure 2.

## 7. Extension with constructs involving recursion

In some cases a program needs to make use of a data structure for its computation. This data structure is active during the computation only and helps the control mechanisms of the program. Such a data structure is called a *service* or *state machine* [10]. A formalisation of this matter involves a refinement of the syntax for basic actions. Any service action used in a program is now supposed to consist of two parts, respectively the *focus* and the *method*, glued together with a period that is not permitted to occur in either one of them. For the service only the method matters because the focus is used to identify to which service (or other system component) an atomic instruction is directed. We denote the extension of a program in notation A by the instructions of service SER by A/SER. A service action $a$ used in the extension will be written as $ser.a$. An example is given below.

NNV is a very simple data structure holding a single natural number – initially 0 – which can be set and returned. The only basic actions performed by NNV are

(1) $set(n)$. The set action always succeeds setting the value to $n$.

(2) $eq(n)$.[3] This action fails if the value is unequal to $n$; otherwise it returns *true*.

With a natural number value a *case statement*

$$nnv.case(m)\{; X_0; \}\{; X_1; \}\{; \cdots; \}\{; X_{m-1}; \}\{; X_m; \}$$

becomes available providing a multi-way branching based on a natural number. The idea here is that if the value $i$ of NNV lies in the range 0 to $m$, then $X_i$ is performed and execution continues after the last closing brace, otherwise this construct is skipped. The program notation combining an ISLA based program notation A with NNV based case statements is denoted with $A:CS_{nnv}$. Its projection can be given by the use of annotations and second-level instructions.

The annotated brace instructions needed for the projection are $nnv.case(m)\{k$ and $i\}\{k$ with natural numbers $k, i$ where $k$ indicates the instruction number of the corresponding separator or end of construct instruction and $i$ the value to be tested. $k$ is set to 0 in the case of ill formed program syntax. $A:CS_{nnv}a$ denotes the extension with annotated case statements.

PROJECTION 13. Let ISLA $\subseteq$ A.

$\phi: A:CS_{nnv}a \to A/NNV:SL$ is defined by $\phi(u_1; \ldots; u_n) = \phi_1(u_1); \ldots; \phi_n(u_n)$ where

$$\phi_i(nnv.case(m)\{k) = -nnv.eq(0)\#\#k + 1,$$

$$\phi_i(i\}\{k) = -nnv.eq(i)\#\#k + 1,$$

$$\phi_i(\}) = \#\#i + 1,$$

for $k > 0$ and,

$$\phi_i(nnv.case(m)\{0) = \#\#0,$$

$$\phi_i(i\}\{0) = \#\#0.$$

The other instructions remain unmodified.

The projection of $A:CS_{nnv}$ to $A:CS_{nnv}a$ involves again a complete inspection of the A:CS program.

PROJECTION 14. Let ISLA $\subseteq$ A.

$\phi: A:CS_{nnv} \to A:CS_{nnv}a$ is defined by $\phi(u_1; \ldots; u_n) = \phi(u_1); \ldots; \phi(u_n)$ where

---

[3] Since basic actions are supposed to return Boolean values, we have replaced the return operation by an equality indicator function.

(1) $\phi(nnv.case(m)\{) = nnv.case(m)\{k$

with $k$ the instruction number of the corresponding separator or end of construct instruction if it exists and 0 otherwise,

(2) $\phi(\}\{) = i\}\{k$

with $i$ the test value and $k$ and the instruction number of the corresponding separator or end of construct instruction if it exists and 0 otherwise.

The other instructions remain unmodified.

A more advanced data structure is the *stack*. The key application of a stack is to obtain projections for program constructs involving recursion.

The basic actions performed by a stack S are

(1) *pop*. This instruction fails if the stack is empty. Otherwise it succeeds and removes the top from the stack while giving a positive reply.

(2) *push(n)*. The push action always succeeds placing the natural number $n$ on top of the stack and producing a positive reply.

(3) *topeq(n)*. This action returns `true` if the top of the stack equals $n$; otherwise it returns `false`.

Similar to the case statement for natural number values one can now introduce case statements

$$s.case(m)\{; X_0; \}\{; X_1; \}\{; \cdots; \}\{; X_{m-1}; \}\{; X_m; \}$$

based on the topmost element of the stack. Annotation and projection are defined in the same fashion. The program notation combining an ISLA based program notation A with stack based case statements is denoted with $A:CS_s$; its annotated version is denoted $A:CS_s a$.

PROJECTION 15. Let ISLA $\subseteq$ A.

$\phi : A:CS_s a \rightarrow A/S:SL$ is defined by $\phi(u_1; \ldots; u_n) = \phi_1(u_1); \ldots; \phi_n(u_n)$ where

$$\phi_i(s.case(m)\{k) = -s.topeq(0)\#\#k + 1,$$

$$\phi_i(i)\{k) = -s.topeq(i)\#\#k + 1,$$

$$\phi_i(\}) = \#\#i + 1,$$

for $k > 0$ and,

$$\phi_i(s.case(m)\{0) = \#\#0,$$

$$\phi_i(i)\{0) = \#\#0.$$

The other instructions remain unmodified.

PROJECTION 16. Let ISLA $\subseteq$ A.

$\phi : A:CS_s \rightarrow A:CS_s a$ is defined by $\phi(u_1; \ldots; u_n) = \phi(u_1); \ldots; \phi(u_n)$ where

(1)  $\phi(s.case(m)\{) = s.case(m)\{k$

with $k$ the instruction number of the corresponding separator or end of construct instruction if it exists and 0 otherwise,

(2)  $\phi(\}\{) = i\}\{k$

with $i$ the test value and $k$ and the instruction number of the corresponding separator or end of construct instruction if it exists and 0 otherwise.

The other instructions remain unmodified.

In preparation of the incorporation of recursion, *dynamic jumps* ##*s.pop* – where the counter depends on the top of the current stack – will be introduced. The program notation that combines an extension A of ISLA with dynamic jumps is denoted with A:DJ. The projection to $A/S:CS_s:SL$ takes into account that in a program $u_1; \ldots; \#\#s.pop; \ldots; u_n$ at most $n$ different values of the jump counter are needed.

PROJECTION 17.  Let ISLA $\subseteq$ A.
$\phi : A{:}DJ \to A/S{:}CS_s{:}SL$ is defined by $\phi(u_1; \ldots; u_n) = \phi(u_1); \ldots; \phi(u_n)$ where

$$\phi(\#\#s.pop) = s.case(n)\{; \sim s.pop\#\#0; \sim\}\{; \cdots; \sim\}\{; \sim s.pop\#\#n; \sim\}.$$

All other instructions remain unmodified.

Having dynamic jumps available, we can now introduce recursion. The instruction pair *R*##*k*, ##*R* represents recursion in the following way: if a *returning jump instruction R*##*k* occurring at position $i$ is executed a jump to instruction $k$ is made, moreover the instruction counter $i + 1$ is placed on the stack. Whenever a *return instruction* ##*R* is executed a jump is performed to the top of the stack which is simultaneously popped.

A:R is the program notation A augmented with returning jump instructions of the form *R*##*k* and return instructions ##*R*. A projection to $A/S{:}SL{:}DJ$ can be given in the following way.

PROJECTION 18.  Let ISLA $\subseteq$ A.
$\phi : A{:}R \to A/S{:}SL{:}DJ$ is defined by $\phi(u_1; \ldots; u_n) = \phi_1(u_1); \ldots; \phi_n(u_n)$ where

$$\phi_i(R\#\#k) = s.push(i + 1)\#\#k,$$

$$\phi_i(\#\#R) = \#\#s.pop.$$

All other instructions remain unmodified.

The part of the computation taking place between a returning jump and its corresponding return instruction may be called a subcomputation. Recursion using returning jumps and return instructions can be made more expressive by means of parameters, i.e., values that serve as inputs to a subcomputation. They are put in place before executing a returning jump, and the program part executed as a subcomputation 'knows' where to find these parameters. Various strategies can be imagined for arranging the transfer of parameters during returning jump and return instructions. Here we have chosen for an automatic parameter transfer to and from a stack to local spots.

A *stack with value pool* is a data structure that maintains a stack and a value pool. The stack stores values of primitive types and instruction counters which do not count as values. The value pool consists of finitely many pairs with the first element being the *spot* relative to the value pool and the second element its content. There are two distinguished spots: *this* and *that*. *this* contains the instances for object oriented instance method calls and *that* the results of subcomputations. In addition to the usual stack actions a stack with value pool can perform the following actions which transfer values between stack and pool:

(4) *store(s)*. The precondition of this basic action is that the stack is non-empty; otherwise it fails. If the stack is non-empty its top is removed and made to be the content of spot *s*. The previous content of *s*, if any, is lost in the process.

(5) *load(s)*. This basic action takes the content of the spot *s*, which must have been introduced at an earlier stage, and pushes that on the stack. A spot is defined if it has been assigned a value by means of a *store* instruction at some stage. If the spot is still undefined the action fails. This is a copy instruction because the value is still the content of *s* as well after successful execution.

We now assume that the $m$ parameters of a recursion – and in the case of an object oriented instance method call also an instance parameter that plays the role of the target instance – are initially stored consecutively on a stack. From here they are placed in the value pool at spots $arg1, \ldots, argm$ – and possibly *this* – but only after having copied the contents of these spots to a second auxiliary stack in order to allow recovery of the original contents. Just after the returning jump and just before the return instruction the top of the auxiliary stack contains the parameters in decreasing order. The projection of the return instruction will then involve the recovery of the several spots before the returning jump from the data that were placed on the auxiliary stack.

In the sequel we shall write $Rm\#\#k$ and $IRm\#\#k$ ($m, k \in \mathbb{N}$) for returning jump instructions with $m$ parameters and a possible instance. Given a program notation A the extension by parametrised returning jumps and return instructions is denoted A:Rp. The projection uses two stacks sharing a single value pool called $svp$ and $svp_{aux}$, respectively. In order to recover the copied spots it is necessary to replace returning jumps by instruction sequences containing 2 first-level instructions. As a consequence jump counters have to be updated.

PROJECTION 19. Let ISLA $\subseteq$ A.
$\phi: \text{A:RP} \to \text{A/SVP:SL:DJ}$ is defined by $\phi(u_1; \ldots; u_n) = \phi_1(u_1); \ldots; \phi_n(u_n)$ where

$$\phi_i(Rm\#\#k) = svp_{aux}.load(arg1);$$

$$\sim svp_{aux}.load(arg2); \ldots; \sim svp_{aux}.load(argm);$$

$$\sim svp.store(arg1); \ldots; \sim svp.store(argm);$$

$$\sim svp.push(i+1)\#\#k;$$

$$svp_{aux}.store(argm);$$

$$\sim svp_{aux}.store(argm-1); \ldots; \sim svp_{aux}.store(arg1),$$

$$\phi_i(IRm\#\#k) = svp_{aux}.load(arg1);$$

$$\sim svp_{aux}.load(arg2); \dots; \sim svp_{aux}.load(argm);$$

$$\sim svp_{aux}.load(this);$$

$$\sim svp.store(this); \sim svp.store(arg1); \dots; \sim svp.store(argm);$$

$$\sim svp.push(i + 1)\#\#k;$$

$$svp_{aux}.store(this);$$

$$\sim svp_{aux}.store(argm); \dots; \sim svp_{aux}.store(arg1)$$

$$\phi_i(\#\#R) = \#\#svp.pop,$$

$$\phi_i(\#\#l) = \#\#l + r,$$

where $r$ is the number of (instance) returning jumps preceding $u_l$. All other instructions remain unmodified.

The two stacks used in the projection above van be combined into a single stack with value pool by considering a stack $svp$ that performs in addition to the usual stack and transfer actions also the action

(6) *down(n)*. This action places the top just below the $n$th position from above after removing the top. Thus after *down(n)* the top has migrated to the $(n + 1)$st position from above. If that is impossible the action fails, a negative reply is given and no change is made to the stack.

Allowing for *down* actions, one can replace in Projection 19 instructions of the form $svp_{aux}.load(x)$ by $svp.load(x); down(m)$ if no instance parameter is involved and by $svp.load(x); down(m + 1)$ otherwise; instructions of the form $svp_{aux}.store(x)$ have to be stripped off their subscript.

It is important to notice that for the projection of a specific program only a finite number of *down(n)* instructions is used. This implies that the stack manipulations do not reach the expressive power of a full stackwalk, but rather may be simulated (per program) from a stack and a finite memory service, or more easily obtained using a bounded value buffer.

The projections of recursion instructions are summarised in Figure 3. The final projection to A/S can be obtained by chaining the appropriate projections discussed in the preceding sections.

## 8. Intermediate-level programs

The design of instructions can be viewed as proceeding in three phases. Until this point only so-called low-level instructions have been designed. Low-level instructions have either fixed projections or their projection trivially depends on the program context. Program context dependence occurs for instance with a closing brace instruction for which the projection may depend on the index of its corresponding opening brace instruction. The second design phase introduces an intermediate language ILN for high-level object oriented programming constructs which makes use of an *operation interface*.

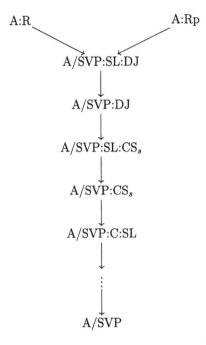

Fig. 3.  Projections of recursion and parametrised recursion with automatic parameter transfer using a stack with value pool.

An operation interface OI is a data structure that maintains in addition to a stack and a value pool a heap which memorises instance data and the class of objects. The stack stores values of primitive types, instruction counters and references to objects.

Assuming also the existence of a family of operators, the basic actions performed by OI are in addition to the stack and value pool actions:

(7) *compute*$(f, k)$ (with $f$ the name of an operator with arity $k$). It is assumed that $k$ values $v_1, \ldots, v_k$ are stored consecutively on the stack. If there are too few values on top of the stack the action fails and the state is unchanged. If there are sufficiently many values the reply is positive and the value $f(v_1, \ldots, v_k)$ is placed on the stack after removing all $k$ arguments from it.

(8) *create*. This basic action allocates a new reference and puts it on top of the stack. If a new reference is not available this action fails.

(9) *getfield*$(F)$. The precondition of this action is that the top of the stack is a reference denoting an instance with field $F$; otherwise it fails. If the precondition is met, the content of field $F$ of the instance denoted by the top is pushed from the heap on the stack after having removed the top.

(10) *setfield*$(F)$. The precondition of this action is that the top of the stack is a reference followed by a value or a reference; otherwise it fails. If the precondition is met, the reference and the value are popped from the stack and field $F$ of the instance denoted by the popped reference is set to the popped value.

(11) *instanceof(C)*. The precondition of this action is that the top of the stack is a reference; otherwise it fails. If the precondition is met, this action returns `true` if the class of the instance denoted by the reference is $C$, otherwise it returns `false`. Afterwards the reference is taken from the stack.

(12) *setclass(C)*. Again, the precondition is that the top of the stack is a reference; otherwise the action fails. If the precondition is met, the class of the instance denoted by the reference is set to $C$. Afterwards the reference is popped.

In order to make this data structure less abstract, we provide a more detailed and concrete description.

Let us denote with *BinSeq* the finite sequences of 0's and 1's. Three subsets of *BinSeq* are then at the basis of the concrete operation interface:

- *Loc* is a finite or infinite set of so-called *locations* – in practical cases it must be finite, in some formal modelling cases an infinite set of locations may be more amenable.
- *Ic* is the finite or infinite collection of instruction counters. Here the most plausible choice is that *Ic* contains binary forms of all natural numbers.
- *Pval* is the collection of primitive values. This is again a finite set in most practical cases, just like *Loc*.

On the stack one finds

- references: taking the form $rw$ with $w \in Loc$, e.g. $r00110$; the set of $rw$'s with $w \in Loc$ is denoted *Ref* below,
- instruction counters: taking the form $iw$ with $w \in Ic$, and
- primitive values: taking the form $w$ with $w \in Pval$.

In any state of computation the spots have either no content – the initial state for all spots – or, if a spot has a content, it is either a reference $rw$ or a primitive value.

Several collections of names are used: *ClassNames*, *FieldNames* and *SpotNames*. Any convention will do and these sets may overlap without generating confusion. A state of the concrete operation interface consists now of the following sets:

- *Obj* is a subset of *Loc*. *Obj* contains those locations that contain an object. Initially a computation will start with *Obj* empty.
- *Classification* is a subset of $Obj \times ClassNames$. We may regard *Classification* in fact as a partial function from *Obj* to *ClassNames*.
- *Fields* is a subset of $Obj \times FieldNames \times (Ref \cup Pval)$. Again we assume that *Fields* is a partial function from $Obj \times FieldNames$ to $Ref \cup Pval$.
- *Spots* is a subset of $SpotNames \times (Ref \cup Pval)$ and hence a partial function from *SpotNames* to $Ref \cup Pval$.

It is important to make a distinction between vital objects and garbage objects. Vital objects are those that can be reached from spots containing a reference and references positioned anywhere on the stack via zero or more selections of fields containing a reference. Non-vital objects are called garbage objects. After each method execution garbage objects are removed and so are their outgoing fields and classifications.

Now the operations of the heap can be interpreted as follows:

- *create*. If $Obj = Loc$ then return `false`; otherwise find $w \in Loc - Obj$, place $rw$ on the top of the stack, put $w$ in *Obj*, and then return `true`.

- *getfield(f)*. If the top of the stack is not a reference or if the top of the stack contains $rw$, say, but *Fields(rw, f)* is not defined return `false`; otherwise remove the top of the stack, place *Fields(rw,f)* on the stack and return `true`.
- *setfield(f)*. If the top of the stack is a reference $rw$, say, and if it lies on top of a value $v$ in *Ref* ∪ *Pval*, pop the two uppermost items from the stack, define *Fields(rw, f)* = $v$ and return `true`; otherwise return `false`.
- *instanceof(C)*. If the top of the stack is not a reference return `false`; if the top of the stack contains $rw$, say, pop the stack, return `true` if *Classification(rw)* = $C$, and return `false` otherwise.
- *setclass(C)*. If the top of the stack is not a reference return `false`; if the top of the stack contains $rw$, say, define *Classification(rw)* = $C$, pop the stack and return `true`.

In this concrete operation interface a reference takes the form $rw$ and spots may also be considered references. In other models references may work differently so this description of a concrete operation interface cannot be taken as an analysis of what constitutes a reference in general.

In preparation of the projection of ILN we introduce a *dynamic chained goto* instruction ##[*instanceof*][M] where the first goto depends on the class of the instance *this*. The projection from A:DCG (A with dynamic chained goto's) to A/OI:C:GL:SL can be given in the following way.

PROJECTION 20. Let ISLA $\subseteq$ A.
$\phi$ : A:DCG $\rightarrow$ A/OI:GL:SL is defined by $\phi(u_1; \ldots; u_n) = \phi(u_1); \ldots; \phi(u_n)$ where

$$\phi(\#\#[instanceof][M]) = oi.load(this); \sim +instanceof(C_1)\#\#[C_1][M];$$
$$\sim oi.load(this); \sim +instanceof(C_2)\#\#[C_2][M];$$
$$\vdots$$
$$\sim oi.load(this); \sim +instanceof(C_k)\#\#[C_k][M]$$

where $C1, \ldots, Ck$ are the classes introduced by *setclass* instructions in $u_1; \ldots; u_n$. All other instructions remain unmodified.

We now proceed with the second design phase. A typical ILN program consists of a sequence of labelled class parts. E.g., an ILN program with 4 classes looks as follows:

$$[C1]; CB1; \ldots; [C4]; CB4.$$

The subprograms $CB1, \ldots, CB4$ are so-called class bodies, which consist of zero or more labelled method body parts with end markers

$$[M1]; MB1; end; \ldots; [Mk]; MBk; end.$$

In addition to the usual labels, ILN has the following instruction set.

**new(C)** This instruction pushes a new reference to an instance of class $C$ onto the stack.

**stack push**  The instruction $E \Rightarrow$ represents pushing an entity with name $E$ onto the stack. $E$ may be a single non-empty alphanumerical string or of the form $E.F$, i.e., a string consisting of two alphanumerical parts glued together with a period.

**top from stack**  The instruction $\Rightarrow E$ takes the top from the stack and places that entity on the place denoted with $E$. Again, $E$ may be a single non-empty alphanumerical string or of the form $E.F$.

**method end marker**  The method end marker *end* serves as the end of a method.

**function call**  A function call $fc(f, n)$ represents the call of the function $f$ from a given operator family with arity $n$. The algorithmic content of an operator call is not given by the program that contains it.

**class method call**  A class method call instruction has the form $mc(C, M, n)$. Here $M$ is a method, $C$ a class and $n$ is the number of arguments. We tacitly assume that the names occurring in the stack push and pop instructions of $M$ are amongst $arg1, \ldots, argn$.

**instance method call**  An instance method call instruction has the form $mc(M, n)$. Here $M$ is a method and $n$ is the number of arguments. We tacitly assume that the names occurring in the stack push and pop instructions of $M$ are amongst *this*, $arg1, \ldots, argn$.

**if header**  An if-then-else construction has the form $if(E); X; else; Y; endif$ separated by *else* and completed by an *endif* instruction. If the value of $E$ is *true* then the instruction stream $X$ is performed; otherwise $Y$ is executed.

**while header**  A while construction has the form $while(E); X; endwhile$ and performs the instruction stream $X$ as long as the value of $E$ is *true*.

**separator and end of construct**  *else* and the end markers *endif* and *endwhile* serve as separator and end markers for the conditional construct and the while loop.

ILN instructions can be projected to ISLA/OI:DCG:GL:Rp:CW.

PROJECTION 21.  $\phi : \text{ILN} \to \text{ISLA/OI:DCG:GL:Rp:CW}$ is defined by $\phi(u_1; \ldots; u_n) = \phi_1(u_1); \ldots; \phi_n(u_n)$ where

$$\phi_i(new(C)) = oi.create;\ oi.setclass(C)$$

$$\phi_i(E \Rightarrow) = oi.load(E)$$

$$\phi_i(\Rightarrow E) = oi.store(E)$$

$$\phi_i(E.F \Rightarrow) = oi.load(E);\ oi.getfield(F)$$

$$\phi_i(\Rightarrow E.F) = oi.load(E);\ oi.setfield(F)$$

$$\phi_i(end) = \#\#R$$

$$\phi_i(fc(f, m)) = oi.compute(f, m)$$

$$\phi_i(mc(C, M, m)) = Rm\#\#3k_1 + 2k_2 + k_3 + i + 2;\ \#\#3k_1 + 2k_2 + k_3 + i + 3;$$
$$\#\#[C][M]$$

$$\phi_i(mc(M, m)) = IRm\#\#3k_1 + 2k_2 + k_3 + i + 2;\ \#\#3k_1 + 2k_2 + k_3 + i + 3;$$
$$\#\#[instanceof][M]$$

$\phi_i(if(E)) = oi.load(E); oi.compute(== true, 1); +oi.topeq(1)\{; oi.pop$

$\phi_i(while(E)) = oi.load(E); oi.compute(== true, 1); +oi.topeq(1)\{*; oi.pop$

$\phi_i(else) = \}\{; oi.pop$

$\phi_i(endif) = \}$

$\phi_i(endwhile(E)) = oi.load(E); oi.compute(== true, 1); *\}; oi.pop.$

Here $k_1$ is the number of if and while headers and endwhile instructions, $k_2$ the number of method calls and $k_3$ the number of creation, field pushing and popping, and else instructions occurring in $u_1; \ldots; u_{i-1}$. Moreover, $== true$ is the name of the operator that returns 1 if the argument has value *true* and 0 otherwise. In a setting with relative jumps one can replace the complex absolute jumps by relative jumps of length 2. Labels remain unmodified.

A simple ILN program $P$ and its projection are given below. $P$ consists of a single class part labelled $[C]$ with method parts $[CC]$, $[get]$ and $[M]$. $[CC]$ is a class constructor initialising the $F$ field of any object of this class with a random Boolean value, $[get]$ is an instance method returning the value of the $F$ field of the calling object, and $[M]$ is a static method creating two objects of this class – $X$ and $Y$ – by calling the constructor for $X$ but setting the $F$ field of $Y$ to the complement of the $F$ field of $X$ by using the constant operators *true* and *false*.

> $[C];$
>
> > $[CC]; fc(random, 0); \Rightarrow this.F; end;$
> >
> > $[get]; arg1.F \Rightarrow; \Rightarrow that; end;$
> >
> > $[M];$ new $C; \Rightarrow X; X \Rightarrow; mc(CC, 0);$
> >
> > > new $C; \Rightarrow Y;$
> > >
> > > $X \Rightarrow; mc(C, get, 1);$
> > >
> > > $if(that); fc(false, 0); \Rightarrow Y.F; else; fc(true, 0); \Rightarrow Y.F; endif.$

Projection into ISLA/OI:DCG:GL:Rp:CW yields

> $[C];$
>
> > $[CC]; oi.compute(random, 0); oi.load(this); oi.setfield(F); \#\#R;$
> >
> > $[get]; oi.load(arg1); oi.getfield(F); oi.store(that); \#\#R;$
> >
> > $[M]; oi.create; oi.setclass(C); oi.store(X); oi.load(X); IR0\#\#19; \#\#20;$
> >
> > > $\#\#[instanceof][CC];$
> > >
> > > $oi.create; oi.setclass(C); oi.store(Y);$

$oi.load(X)$; $R1\#\#26$; $\#\#27$; $\#\#[C][get]$;

$oi.load(that)$; $oi.compute(== true, 1)$;

$+s.topeq(1)\{;\ oi.pop;\ fc(false, 0);\ \Rightarrow Y.F;$

$\}\{;\ oi.pop; fc(true, 0);\ \Rightarrow Y.F;\ \}$

## 9. A high-level program notation HLN

In the third phase we deviate from the projective syntax paradigm and introduce instructions at a level close to the high-level program notations used by human programmers and
computer software users.

In this section a toy high-level program notation HLN is developed. Its meaning is given
by a compiler projection into ILN. Like ILN programs HLN programs are supposed to have
a certain fixed structure. In this structure a program is a series of class parts as depicted
in Figure 4. Class parts consist of a number of segments. These segments can be of two
kinds: instance field parts and method declarations. The compiler projection will transform
the list of instance field parts in a class to a class constructor for that class with label $[CC]$.
Method declarations – which have the form $M(E1, \dots, En)\{X\}$ and are named different
from $CC$ – will be projected to class method parts.

HLN has the following expressions.

$$exp := E \mid E.F \mid f(exp1, \dots, expn) \mid I.M(exp1, \dots, expn) \mid$$
$$C.M(exp1, \dots, expn)$$

with $f$ an $n$-ary operator from a given operator family, $I$ a class instance calling an $n$-ary

---

$class\ C\{$

    $instance\ fields\{$

      $F1 = exp1;$

      $\vdots$

      $Fn = expn;$

      $\}$

      $M1(E1, \dots, En)\{X\}$

      $\vdots$

$\}$

---

Fig. 4. The structure of class parts in HLN.

instance method and $C.M(exp1, \ldots, expn)$ an $n$-ary class method call. Assignments and statements are of the form

$$asg := E = exp \mid E.F = exp \mid E = \text{new } C \mid E.F = \text{new } C$$

$$stm := asg \mid return\ exp \mid I.M(exp1, \ldots, expn) \mid C.M(exp1, \ldots, expn) \mid$$

$$if\ (exp)\ blk\ else\ blk \mid while\ (exp)\ blk$$

$$blk := \{stm1; \ldots; stmn\}$$

The projections to ILN can be given by

$exp$:

$\phi(E) = E \Rightarrow$

$\phi(E.F) = E.F \Rightarrow$

$\phi(f(exp1, \ldots, expn)) = \phi(exp1); \ldots; \phi(expn); fc(f, n)$

$\phi(I.M(exp1, \ldots, expn)) = \phi(exp1); \ldots; \phi(expn); I \Rightarrow; mc(M, n); that \Rightarrow$

$\phi(C.M(exp1, \ldots, expn)) = \phi(exp1); \ldots; \phi(expn); mc(C, M, n); that \Rightarrow$

$asg$:

$\phi(E = exp) = \phi(exp); \Rightarrow E$

$\phi(E.F = exp) = \phi(exp); \Rightarrow E.F$

$\phi(E = \text{new } C) = \text{new } C; mc(CC, 0); \Rightarrow E$

$\phi(E.F = \text{new } C) = \text{new } C; mc(CC, 0); \Rightarrow E.F$

$stm$:

$\phi(return\ exp) = \phi(exp); \Rightarrow that$

$\phi(I.M(exp1, \ldots, expn)) = \phi(exp1); \ldots; \phi(expn); I \Rightarrow; mc(M, n)$

$\phi(C.M(exp1, \ldots, expn)) = \phi(exp1); \ldots; \phi(expn); mc(C, M, n)$

$\phi(if\ (exp)\ blk\ else\ blk) = if\ (\phi(exp)); \phi(blk); else; \phi(blk); endif$

$\phi(while\ (exp)\ blk) = while\ (\phi(exp)); \phi(blk); endwhile(\phi(exp))$

$blk$:

$\phi(\{stm1; \ldots; stmn\}) = \phi(stm1); \ldots; \phi(stmn)$.

The meaning of class parts can finally be given by

$$\phi(class\ C\{X\}) = [C]; \phi(\{X\})$$

$$\phi(instancefields\{F1 = exp1; \ldots; Fn = expn\}) = [CC];$$

$$\phi(\mathit{this}.F1 = \mathit{exp}1);$$

$$\cdots$$

$$\phi(\mathit{this}.Fn = \mathit{expn});$$

$$end$$

$$\phi(M(E1,\ldots,En)\{X\}) = [M];$$

$$E1 \Rightarrow; \ldots; En \Rightarrow;$$

$$\phi(E1 = arg1);$$

$$\cdots$$

$$\phi(En = argn);$$

$$\phi(\{X\});$$

$$\Rightarrow En; \ldots; \Rightarrow E1;$$

$$end$$

Note that the projection of HLN to ILN is dynamically correct. That is, the elements on the stack meet the requirements of the current instructions: whenever an instruction counter is expected, the top of the stack will contain one, and whenever $n$ values are expected, the $n$ uppermost elements will be values.

We end this section with a typical HLN program and its projection into ILN. Let $P$ be the following program

```
class A{
    instancefields{F = 0; }
    get(){return this.F; }
    set(E){this.F = E; }
    inc(E){this.set(this.get() + E); }
}
class B{
    get(E){return E.F; }
    test(){
        X = new A;
        Y = new A;
        X.set(1);
        X.inc(1);
```

$$Y.F = B.get(X);$$

$$\}.$$

Assuming that the constants 0, 1 and addition $+$ are predefined operators, $P$ can be projected into ILN by

$[A]$;

$\quad$ $[CC]$; $fc(0, 0)$; $\Rightarrow$ *this.F*; *end*;

$\quad$ $[get]$; *this.F* $\Rightarrow$; $\Rightarrow$ *that*; *end*;

$\quad$ $[set]$; $E \Rightarrow$; $arg1 \Rightarrow$; $\Rightarrow E$; $E \Rightarrow$; $\Rightarrow$ *this.F*; $\Rightarrow E$; *end*;

$\quad$ $[inc]$; $E \Rightarrow$; $arg1 \Rightarrow$; $\Rightarrow E$; *this* $\Rightarrow$; $mc(get, 0)$; *that* $\Rightarrow$; $E \Rightarrow$; $fc(+, 2)$;

$\qquad$ *this* $\Rightarrow$; $mc(set, 1)$; *end*;

$[B]$;

$\quad$ $[get]$; $E \Rightarrow$; $arg1 \Rightarrow$; $\Rightarrow E$; $E.F \Rightarrow$; $\Rightarrow$ *that*; $\Rightarrow E$; *end*;

$\quad$ $[test]$;

$\qquad$ *new A*; $mc(CC, 0)$; $\Rightarrow X$;

$\qquad$ *new A*; $mc(CC, 0)$; $\Rightarrow Y$;

$\qquad$ $fc(1, 0)$; $X \Rightarrow$; $mc(set, 1)$;

$\qquad$ $fc(1, 0)$; $X \Rightarrow$; $mc(inc, 1)$;

$\qquad$ $X \Rightarrow$; $mc(B, get, 1)$; *that* $\Rightarrow$; $\Rightarrow Y.F$;

$\qquad$ *end*.

## 10.  Concluding remarks

At this point a simple program notation for object oriented programming has been provided admitting a projection semantics. Many more features exist and one might say that the project of syntax design has only been touched upon. Still the claim is made that the above considerations provide a basis for the design of reliable syntax for much more involved program notations. Projection semantics provides a scientifically well-founded and rigorous yet simple approach to the semantics of programming languages. Future work will have to clarify that this framework is industrially viable for the high-level design and analysis of complex systems, and for natural refinements of models to executable and reliable code.

## Acknowledgements

This paper benefited greatly from the insightful comments made by Mark Burgess and Kees Middelburg.

## References

[1] L. Aceto, W.J. Fokkink and C. Verhoef, *Structural operational semantics*, Handbook of Process Algebra, North-Holland (2001), 197–292.

[2] U. Ammann, *Code generation for a Pascal compiler*, Software – Practice and Experience **7** (1977), 391–423.

[3] J.H. Andrews, *Process-algebraic foundations of aspect-oriented programming*, 3rd International Conference on Metalevel Architectures and Separation of Crosscutting Concerns (REFLECTION 2001), Lecture Notes in Comput. Sci., Vol. 2192, Springer-Verlag (2001), 187–209.

[4] H. Bekić, D. Bjørner, W. Henhapl, C.B. Jones and P. Lucas, *A formal definition of a PL/1 subset*, IBM Laboratory Vienna, TR25.139 (1974).

[5] J.A. Bergstra and I. Bethke, *Polarized process algebra with reactive composition*, Theoretical Computer Science **343** (3) (2005), 285–304.

[6] J.A. Bergstra and M.E. Loots, *Program algebra for component code*, Formal Aspects of Computing **12** (1) (2000), 1–17.

[7] J.A. Bergstra and M.E. Loots, *Program algebra for sequential code*, Journal of Logic and Algebraic Programming **51** (2) (2002), 125–156.

[8] J.A. Bergstra and C.A. Middelburg, *A thread algebra with multi-levelr strategic interleaving*, New Computational Paradigms: First Conference on Computability in Europe, CiE 2005, Amsterdam, The Netherlands, June 8–12, 2005, Lecture Notes in Comput Sci., Vol. 3526, Springer-Verlag (2005), 35–48.

[9] J.A. Bergstra, C.A. Middelburg and Y.S. Usenko, *Discrete time process algebra and the semantics of SDL*, Handbook of Process Algebra, North-Holland (2001), 1209–1268.

[10] J.A. Bergstra and A. Ponse, *Combining programs and services*, Journal of Logic and Algebraic Programming **51** (2) (2002), 175–192.

[11] J.A. Bergstra and A. Ponse, *Execution architectures for program algebra*, Journal of Applied Logic **5** (1) (2007), 170–192.

[12] D. Bjørner and O. Oest, eds, *Towards a Formal Description of Ada*, Lecture Notes in Comput. Sci., Vol. 98, Springer-Verlag (1980).

[13] M. Burgess, *A site configuration engine*, Computing Systems **8** (2) (1995), 309–337.

[14] M. Burgess, *Principles of Network and System Administration*, Wiley (2000).

[15] M. Burgess, *On the theory of system administration*, Science of Computer Programming **49** (1–3) (2003), 1–46.

[16] M. Burgess, *Configurable immunity for evolving human – computer systems*, Science of Computer Programming **51** (3) (2004), 197–213.

[17] M. Burgess, *Analytical Network and System Administration – Managing Human–Computer Systems*, Wiley (2004).

[18] R. Cleaveland, X. Du and S.A. Smolka, *GCCS: A graphical coordination language for system specification*, Proceedings of the Fourth International Conference on Coordination Models and Languages (COORDINATION 2000), Lecture Notes in Comput. Sci., Vol. 1906, Springer-Verlag (2000), 284–298.

[19] A. Couch and M. Gilfix, *It's elementary, dear Watson: Applying logic programming to convergent system management processes*, Proceedings of the 13th USENIX Conference on System administration, Seattle, Washington, USENIX Association, Berkely, CA, USA (1999), 123–138.

[20] A. Couch and Y. Sun, *On observable reproducibility in network configuration management*, Science of Computer Programming **53** (2) (2004), 215–253.

[21] E.W. Dijkstra, *Guarded commands, non-determinacy and formal derivations of programs*, Commun. ACM **18** (1975), 453–457.

[22] L.M.G. Feijs and H.B.M. Jonkers, *Formal Specification and Design*, Cambridge Tracts in Theoretical Computer Science **35**, Cambridge University Press (1992).

[23] L.M.G. Feijs, H.B.M. Jonkers and C.A. Middelburg, *Notations for Software Design*, FACIT Series Springer-Verlag (1994).

[24] M. Felleisen and D.P. Friedman, *Control operators, the SECD machine, and the λ-calculus*, Formal Description of Programming Concepts III, Proc. IFIP TC2 Working Conference, Gl. Avernaes, 1986, North-Holland (1987), 193–217.

[25] J. Gosling, B. Joy and G. Steele, *The Java Language Specification*, Addison–Wesley, Reading, MA (1997).

[26] D. Gregg, A. Beatty, K. Casey, B. Davis and A. Nisbet, *The case for virtual register machines*, Science of Computer Programming **57** (3) (2005), 319–338.

[27] C.A. Gunter and D.S. Scott, *Semantic domains*, Handbook of Theoretical Computer Science, Vol. B: Formal Models and Semantics, Elsevier Science Publishers, Amsterdam and MIT Press (1990), 633–674.

[28] Y. Gurevich, *Evolving algebras 1993: Lipari guide*, Specification and Validation Methods, Oxford University Press (1995), 9–36.

[29] J. Heering and P. Klint, *Semantics of programming languages: A tool-oriented approach*, ACM SIGPLAN Notices **35** (3) (2000), 39–48.

[30] C.A.R. Hoare, *An axiomatic basis for computer programming*, Commun. ACM **12** (1969), 576–580.

[31] H.B.M. Jonkers, *An Introduction to COLD-K*, Algebraic Methods: Theory, Tools and Applications, Lecture Notes in Comput. Sci., Vol. 394, Springer-Verlag (1989), 139–205.

[32] G. Kahn, *Natural semantics*, STACS'87, Proc. Symp. on Theoretical Aspects of Computer Science, Lecture Notes in Comput. Sci., Vol. 247, Springer-Verlag (1987), 22–39.

[33] T. Lindholm and F. Yellin, *The Java Virtual Machine Specification*, Addison–Wesley, Reading, MA (1997).

[34] S. Mauw and M.A. Reniers, *A process algebra for interworkings*, Handbook of Process Algebra, North-Holland (2001), 1269–1328.

[35] J. McCarthy, *Towards a mathematical science of computation*, Information Processing 1962, Proceedings of the IFIP Congress 62, North-Holland (1962), 21–28.

[36] R. Milne and C. Strachey, *A Theory of Programming Language Semantics*, Chapman and Hall (1976).

[37] R. Milner, *A Calculus of Communicating Systems*, Lecture Notes in Comput. Sci., Vol. 92, Springer-Verlag (1980).

[38] E. Moggi, *Notions of computation and monads*, Information and Control **93** (1991), 55–92.

[39] P.D. Mosses, *Mathematical semantics of Algol60*, Technical Report Technical Monograph PRG-12, Programming Research Group, University of Oxford (1974).

[40] P.D. Mosses, *The varieties of programming language semantics and their uses*, Perspectives of System Informatics, 4th International Andrei Ershov Memorial Conference, PSI 2001, Novosibirsk (2001), 165–190.

[41] S. Muchnick and U. Pleban, *A semantic comparison of LISP en SCHEME*, Proceedings of the 1980 ACM Symposium on LISP and Functional Programming (1980), 56–64.

[42] D.S. Scott and C. Strachey, *Toward a mathematical semantics for computer languages*, Proceedings of the Symposium on Computers and Automata, Microwave Research Institute Symposia Series 21, Polytechnic Institute of Brooklyn Press (1971), 19–46.

[43] I.A. Severoni, *Operational semantics of the Java card virtual machine*, Journal of Logic and Algebraic Programming **58** (2004), 3–25.

[44] Y. Shi, D. Gregg, A. Beatty and M.A. Ertl, *Virtual machine showdown: Stack versus registers*, VEE '05: Proceedings of the 1st ACM/USENIX International Conference on Virtual Execution Environments, Chicago, IL, USA, ACM Press, USA (2005), 153–163.

[45] M.B. Smyth, *Power domains*, Journal of Computer and System Sciences **16** (1) (1978), 23–36.

[46] R. Stärk, J. Schmid and E. Börger, *Java and the Java Virtual Machine, Definition, Verification, Validation*, Springer-Verlag (2001).

[47] R.D. Tennent, *A denotational description of the pr ogramming language Pascal*, Technical Report 77-47, Dept. of Computing and Information Sciences, Queen's University, Kingston, Ontario, Canada (1977).

[48] F.W. Vaandrager, *Process algebra semantics of POOL*, Applications of Process Algebra, Cambridge Tracts in Theoretical Computer Science, Vol. 17, Cambridge University Press (1990), 173–236.

[49] R.J. van Glabbeek, *The linear time – branching time spectrum I. The semantics of concrete, sequential processes*, Handbook of Process Algebra, North-Holland (2001), 3–100.

# 6. System Science

## 6.1. *Commentary*

Science if usually understood to be a search for two kinds of knowledge: theoretical knowledge which can be known precisely but has only an approximate relationship to reality, and empirical knowledge which definitely concerns reality but can only be measured with uncertainty. The admission of approximation and uncertainty is therefore fundamental in science. This is a relatively new idea in computing, whose traditions derive from mathematics and logic.

In this part of the handbook we look at the application of modelling techniques more familiar from physical science to both system design and business modelling. Confronted with 'modelling' most computer scientists would say 'Unified Modelling Language (UML)?' But UML is at best a simple ontological model for framing knowledge, it cannot cope with the complexities of agreements, valuations, measurements and observations etc. These things are the subject of this part.

Reitan brings the established field of system reliability to bear on computer science. This probabilistic approach to system modelling is more familiar in electrical engineering and the nuclear power industry. Here we see it applied directly to mission criticality in services.

# – 6.2 –

# System Administration and the Scientific Method

Mark Burgess

*University College Oslo, Faculty of Engineering, Room 707, Cort Adelers Gate 30, NO-0254 Oslo, Norway*
*E-mail: mark.burgess@iu.hio.no*

## 1. Introduction

Networks and computer systems are objects to be observed and understood by a process of scientific inquiry. Even in isolation, computers are too complicated to predict from purely deterministic principles of logic and reason; then, once we place them in a non-deterministic environment of users and networks, the task becomes impossible. The result is that network or system manager is in the position of the natural scientist observing unknown phenomena and learning about behavior that he or she strives to explain.

Computer Science looks traditionally to logic and mathematics rather than to natural philosophy to describe computers, examining the consistency of assumptions rather than the reality of their implementation. Clearly such studies are valuable, but they do not represent the *practice* of computer systems. Scientists have long since abandoned such fundamental approaches to studying the real world. Trying to describe the behavior of the human body in terms of atoms and their basic interactions, for instance, is simply not a tractable problem.

Science has learned to eat humble pie and make do with *suitably idealized approximation*. To defend itself from unreasonable criticism, it has then developed a theory for gauging the uncertainty of the conclusions it reaches. In the engineering world, this scientific approach is not only useful but essential. We can distinguish two complementary approaches to understanding the world. Put very simply:

- Empiricism: to observe, model and explain.

HANDBOOK OF NETWORK AND SYSTEM ADMINISTRATION
Edited by Jan Bergstra and Mark Burgess

- Mathematics: to assume, formulate and deduce.

Science, or natural philosophy, is the discourse between *observation* and *rational approximation*. It is about finding a suitably idealized description of the world by which we might summarize its essence and use the information to good effect. It begins with empiricism, or observation. Mathematics and logic, on the other hand, are precise constructs based on rewriting assumptions according to unambiguous rules. What goes in, comes out in a new and enlightening form.

There are two reasons for modelling: the first is to make sure that one's assumptions actually lead to the outcomes one imagines, through a plausible chain of cause and effect, and also to within an acceptable margin of error. The second is to make predictions about new scenarios that we have never seen, based on past observation and modelling. The world, of course, never obeys simple laws and models exactly: all models are 'wrong', in a sense, but that need not matter as long as we know how wrong they are likely to be.

System administration is about the planning, deployment and maintenance of human–computer systems. The human element of the system is important here, since it leads to significant uncertainty. Humans make approximation essential.

## 2. Modelling

We have good reason to model systems: we would like to improve their designs, reliabilities, efficiencies and integrities of purpose. Computers are, above all, technological innovations, i.e. they are tools that are designed to perform a function. We must see them in this role.

This makes them subtly different from natural phenomena, which simply exist for better or for worse. Thus, when applying the methods of science to describe technological innovations, we want to understand and account for how the prejudice of demanding 'success in fulfilling its purpose' alters our objectivity and judgments about a device.

For example, we commonly talk about:

- Optimization: the tuning of a device parameter in such a way that it maximizes some derived quantity, such as service level or the number of business transactions per second.
- Policy: the making of essentially ad hoc decisions about what the technology is supposed to do, how it should behave, its goals and the acceptable parameters for its functioning. We cannot make value judgments such as good or bad, right or wrong, without a statement of the rules of the world in which we place our technology. Such things cannot be derived from any deeper knowledge, they must simply be decided and defined.

Unlike the natural world, which has no agenda, the technological world has many, possibly conflicting goals and we must take this into account in modelling. In some respects, policy defines the laws of our technological universe.

The two approaches are distinct in their relationship to modelling, as noted by the philosopher of science David Hume:

- Empiricism: what we observe is real, but we cannot be certain that we have observed is the 'truth'. Another observation might disprove any conclusion we might draw from observation. Our observation might even be prone to error, or noise.

- Mathematics and logic: we can prove statements to be true or false, but only within a construct that is, at best, only based on reality. The result is not a truth about the real world, however. It is only a truth about our model.

We can, of course, try to validate the results of a model-construct by experiment, but then we are in the realm of empiricism again, where observation is uncertain. The best we can do is to obtain a bulk of evidence in support of a point of view. A good scientist does not believe that a theory is correct, only successful in describing the uncertain observations he or she makes.

One of the most pragmatic approaches to science has been Karl Popper's notion of falsification: namely, that it is sufficient to find one counter-example to a theory to invalidate it, but one can never fully validate a model. Thus models remain models, with intrinsic uncertainty, and never ascend to the position of truth.

Many philosophers have pondered these issues [11]. These distinctions are important to the research scientist as well as to the engineer, as one must routinely protect oneself against abuses of science by marketeers. Phrases such as 'scientifically proven' are clearly nonsense, and yet we read them on a daily basis.

## 3. Measurements and scales

Whether our goal is to reason about the world, or to measure it, we need scales of measurement. In terms of computer science engineering dimensions can be thought of essentially as a *type theory* of measurement. All measurements are simply pure numbers, but they are derived from observations made using different devices, over different time scales. For example, both bytes per second and bits per second are data rates, but are they the same? Clearly not.

How old are you? I am five. Five what? Years? Months? Kilogrammes? Metres? Just as it makes no sense to assign a text string to a number, so we must be able to assign like quantities in measurement. If we wish to relate height in metres to age, then we need a theory and a formula for that relationship, which makes sense.

We call the units of measurement *engineering dimensions*, and the square brackets are normally used to denote the dimensions of an object:

$$[\text{data rate}] = \frac{\text{bits}}{\text{seconds}}. \tag{1}$$

Dimensions are useful because they help us to identify the correctness and form of relationships. Suppose we suspect that the weight $W$ (kilogrammes) of fibre optic cable is related to its length $L$ (metres). Dimensions tell us that they cannot be equal without a multiplying constant of proportionality:

$$W = \alpha L,$$

$$[W] = [\alpha][L], \tag{2}$$

$$\text{kilogrammes} = \frac{\text{kilogrammes}}{\text{metres}} \times \text{metres}.$$

Thus $\alpha$ must have units kilogrammes per metre. Similarly, if $a = b + c$, then $a$, $b$ and $c$ must all have the same dimensions. These are elementary observations, but important.

## 4. Experiments and data collection

To gather information about a system we must observe it and classify our observations into useful categories. Some observations are quantitative, while others are qualitative: Qualitative distinctions present a challenge to deal with: is the horse red or brown? Is a computer PC or a workstation? Where exactly is the borderline between the two? Clearly one must choose classifications with some care to avoid unnecessary ambiguity.

In many cases, one deals with the performance of systems that have measurable data or work rates, using *time series*. A time series is s sequence of measured values at different times, see Figure 1.

A measurement is represented as a cross on this time series, because it is a single event in space and time. The diagram shows a dotted line interpolating the measurements. When and why is it appropriate to assume such a line?

- If we believe the value $q$ is an evolving value that we have only managed to sample at limited certain intervals due to limitations on our resources (music sampling is an example of this).
- If the time scale for changes in $q$ is, say milliseconds, and we measure every hour, then we can clearly see a trend, even if the actual process taking place is not continuous. At the time scale of hours, the individual events seem to form a continuous curve. Relative time scales are highly important.

In these cases, we can assume the existence of a possibly smooth function $q(t)$. If we believe that we are not justified in assuming that $q$ evolves through a number of intermediate values between each observation, then we should not draw a curve through the points, unless we wish to approximate a *trend*. A trend is always a smooth characterization of change, but it is understood to be an indicator rather than a observation about the system.

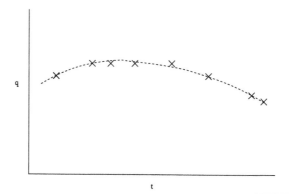

Fig. 1. A series of values $q$ marked by crosses are measured at different times $t$. The dotted line interpolates a proposed smooth curve passing through the points. When is it appropriate to assume such a curve?

## 5. Errors and uncertainties

Scientists speak often about the 'theory of errors', since a major source of doubt about results, historically, has been due to human inconsistency. However, even if we could eliminate inconsistency and mistake from the process of measurement, e.g. by using reliable machines to assist us, there are still reasons for doubting the correctness of measurements. A more accurate name is therefore the 'theory of uncertainties'.

Many uncertainties occur because the act of measurement takes place in an environment over which we have no control. The environment can influence the system we are measuring of the measuring instrument itself.

For example, if we use a computer to measure a precise value at a precise time, the unfortuitous arrival of a disk interrupt could delay the registration of a measurement by some milliseconds and lead to an error, of the same order, in the time. Since one can never be certain that this will not happen, there is an intrinsic uncertainty in the value of the time, due to activities beyond one's control.

Similarly, the value being measured could be affected by the measurement process. Suppose one is measuring the amount of memory in use: by starting our measurement program, we alter the amount of memory used by the system, and this change might not be exactly predictable, due to differences in paging algorithms.

We identify two types of 'error' or 'uncertainty':

- Random error: inconsistently fluctuating values that might add or subtract to a measured value, making its value too large or too small at random.
- Systematic error: a constant shift in the value of a measurement due to a consistent pattern of error. For example, forgetting to calibrate the zero mark on a measuring device could make all values measured wrong by the same amount.

With regard to time series, we can identify two kinds of uncertainty:

- Scatter: uncertainty in value (vertical position in Figure 1).
- Jitter: uncertainty in time (horizontal position in Figure 1).

One way to estimate random error is to take multiple measurements. If the value we measure varies on repeated measurement, we should collect sufficient data to be able to determine whether there is any pattern to the variation (see Figures 2 and 3). Data points that are measured repeatedly are generally plotted as a mean value with an 'error bar' that denotes the estimated uncertainty. The makes it easy to gauge the uncertainty by inspection.

There is a number of possibilities that might arise:

- We always obtain same measured value: the variation or uncertainty in the value is not measurable within the framework of the experiment. Note that this does not necessarily mean there is no uncertainty, because the measuring apparatus itself has a limited resolution. A single value is populated in the histogram $N(q)$.
- We obtain a distribution of values with a mean or expectation value. The expectation value is thus indicative of a 'true' value. A preferred value can be seen in the histogram $N(q)$.
- There is no discernable pattern to the values, the value is random without preference. The histogram $N(q)$ is flat, or has no significant peaks.

Several myths abound about the statistics of measurements. It is often assumed that the distribution of measured values forms a Gaussian distribution or approximating his-

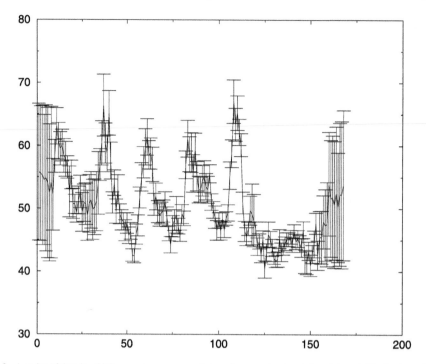

Fig. 2. A series of data in which several measurements have been made at each time interval. The horizontal axis measures granular time, and the vertical axis measure the value $q(t)$. The mean value is plotted as a solid line, with an error bar at each measurement showing $\pm\sigma$.

togram [26,54]. Without qualification, this is an erroneous assumption to make. However, for the case of random errors in independent measurements, the Gaussian model is reasonable and derivable from a mathematical model.

Any distribution of values forms a histogram (or limiting histogram forming a smooth curve) whose axes are $q$, the actual value measured, and $N(q)$, the number of times the value $q$ occurred in all measurements (see Figure 4). Since we associate meaning with patterns of behavior, a peak in the histogram signifies a repeated pattern of behavior in a system.

A single peak indicates the possibility of a 'true' value in the measurements. However, we must still explain the reason for the scatter.

- Is it purely due to random environmental interference?
- Does the pattern signify an actual, genuine change in the value measured from the system over time, combined with environmental uncertainty?
- Does the distribution represent different sub-populations of events within the total mass, possibly implying multiple explanations for the results?

Without further information, we cannot tell the reason for scatter. A rational approach to understanding scatter is to formulate a *cause tree* documenting all of the possible causes of variation, and assigning probable cause [11].

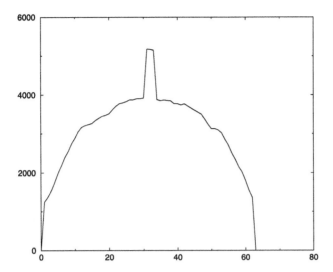

Fig. 3. The histogram of the deviation of the actual data points in Figure 2. The horizontal axis shows position about the mean (in the centre), and the vertical axis shows frequency.

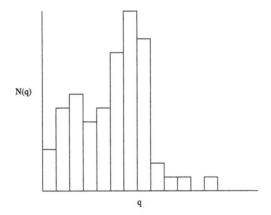

Fig. 4. A histogram of values of $q$ shows the frequency with which certain values have been measured. A pattern here signifies important information about whether there is anything deeper going on in the system, or whether the values are truly without meaning.

Another myth is that the use of a standard deviation assumes a Gaussian distribution. This is false. The standard deviation is merely the second moment, or characteristic, of the distribution (first moment is the expectation value). It is a characteristic scale for the scatter, whose interpretation must be determined in each case. It does not help us to determine whether the distribution is symmetrical about the mean, for instance, but one can use it in the spirit that one uses RMS voltages to represent a useful measure of varying voltages from alternating current power sources.

It has a special significance for Gaussian distributions, owing to the fact that the vast majority of the values in the Gaussian form are bounded by two standard deviations on either side of the mean. However, we can still use the mean and standard deviation of values in $(q, N(q))$ to characterize the scatter. In particular, one can use it as an estimate of the uncertainty in the result. to quote a measured value, one therefore quotes the mean, plus/minus a standard deviation.

$$\text{Measured } q \equiv \langle q \rangle \pm \Delta q \equiv \langle q \rangle \pm \sigma_q, \tag{3}$$

where $\langle q \rangle = E(q) = \sum_{i=1}^{N} q_i$ and $\sigma \equiv \sqrt{\sum_{i=1}^{N}(q_i - \langle q \rangle)^2/N}$, for sample size $N = \sum_q N(q)$.[1] We note the limitations of this, when $N(q)$ is highly asymmetrical about the mean.

Some authors use the standard error of the mean $\sigma/\sqrt{n}$, in $n$ repeated trials, as the uncertainty. This makes the assumption that there is a single 'correct' value to be determined from the measurements, and avoids overestimating the uncertainty in that case. This is a special case, however, and one cannot make such a reduction in general.

Taking account of errors and uncertainties in algebraic expressions is a challenge, but there is a simple approach, which again works best on an approximately Gaussian random uncertainty model. Each individual source of error is quoted as in Equation (3). If several measured quantities are combined into a function of several independent variables, then each is measured independently, and the uncertainties are combined by a Pythagorean, orthogonal sum. For example, suppose we have a function of several quantities $f(q_1, q_2, \ldots, q_n)$, and the functional form of $f$ is known, then we can construct:

$$\Delta f = \sqrt{\left(\frac{\partial f}{\partial q_1}\Delta q_1\right)^2 + \left(\frac{\partial f}{\partial q_2}\Delta q_2\right)^2 + \cdots + \left(\frac{\partial f}{\partial q_n}\Delta q_n\right)^2}. \tag{4}$$

## 6. Hypotheses

Understanding data requires, in essence, constructive guesswork. Science is an art as much as it is a discipline: one must have the imagination to think of likely causes in order to be able to test whether the guess is correct or not. Our main aim is to find the probable cause of an observed phenomenon or source of uncertainty. If we can identify the cause, we have a chance of being able to control it, improve upon performance, optimize the result etc.

One way of mapping out probable causes is to use a causality diagram, or cause tree [11, 33] (see Figure 5). This is essentially the same a 'fault tree' [1,56] is a way of identifying the possible channels of cause and effect in a system. Each branch can be assigned a certain probability which may be combined with rules logical rules for combination of probabilities into a final estimate of the probability of the outcome (top box). The most interesting aspect of this kind of exercise is in identifying the most likely pathways in the tree, since these are the dominant causes of the effect. In the case of uncertainty estimation, this gives us a clue as to how to reduce the total uncertainty in the answer.

---

[1]Note that it is common practice to use $N - 1$ in the denominator of the standard deviation for a small sample to prevent underestimation by numerical accuracy.

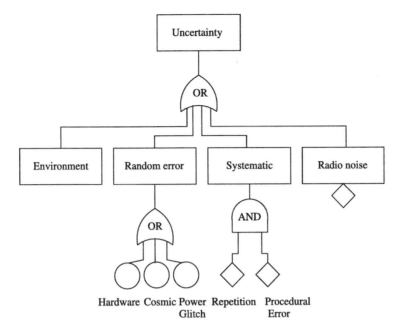

Fig. 5. Ordering the hierarchy of possible causes as a cause or fault tree.

## 7. Discrete and continuous modelling

The distinction between qualitative and quantitative descriptions of phenomena mirrors a similar dichotomy of representation in modelling systems.

Certain problems lend themselves to *discrete* variable modelling, while others require a *continuum description*. Continuum descriptions of phenomena are always approximations, because there is always a limit to the amount of actual detail that can be perceived, e.g. smooth shapes in engineering must eventually make way for the atomic lumpiness of matter; similarly, smooth time-variation in computers must eventually give way to the resolution of the system clock.

- Continuous: infinitely detailed, using calculus of variations to describe change. In a mathematical sense, continuous means smooth or *differentiable*, e.g. probabilities form a continuum of values in the interval $[0, 1]$.
- Discrete: a countable set of values can be measured. Derivatives do not usually have a straightforward meaning in this case. A discrete variable takes values in a finite set, e.g. $q \in \{Windows, Linux, MacOS\}$.

Stochastic or event-driven systems are cases where we have discrete arrivals, but the arrivals can occur within a continuum of time. Moreover, the average number of arrivals per second is a pseudo-continuous variable, since an average is a rational number.

How we model and represent change in a system depends on which of these representations we choose.

- Continuous: characterizations can often be formulated as differential equations, with smooth derivatives that develop according to well known rules based on the rates of change.

$$\frac{dq}{dt} = \lim_{dt \to 0} \frac{q(t + dt) - q(t)}{dt}.$$ (5)

- Discrete: Changes jump between the discrete set of states $q_i$ (see Figure 6). A transition matrix may be used to characterize the allowed transitions as an automaton. One may speak of the conditional probability of making a transition from $q_1$ to $q_2$, in a stochastic system,

$$P(q_i|q_j) = \begin{pmatrix} P(q_1|q_1) & P(q_2|q_1) & \cdots & P(q_n|q_1) \\ P(q_2|q_1) & P(q_2|q_2) & \cdots & P(q_n|q_2) \\ \vdots & \cdots & \ddots & \\ P(q_1|q_n) & \cdots & & P(q_n|q_n) \end{pmatrix}.$$ (6)

If this probability is 1 for each possible transition, then the system is said to be deterministic.

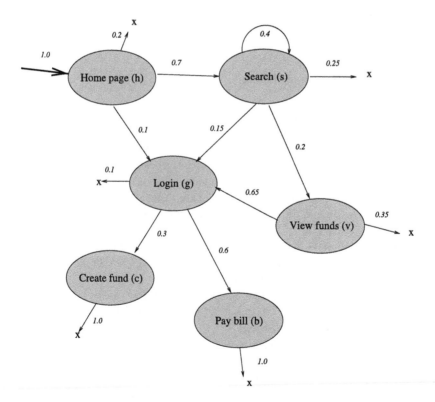

Fig. 6. A transition diagram as a graph. The matrix of probabilities can be assembled from the diagram.

## 8. Modelling responses

In computer systems we are often interested in the response of systems to input. This can be handled both in a continuum and discrete kind of model.

- Continuum: we consider the arrival of a signal current $J(t)$, as a function of time, and measure the response of the system to its arrival. The continuum analogue of the transition matrix is the *response function*, which is formally Green's function of the system. The state $q(t)$ of the system at time $t_f$ is then

$$q(t_f) = \int_{t_i}^{t_f} R(t, t') J(t') \, dt', \tag{7}$$

  given that the signal began arriving at time $t_i$.
- Discrete: here we consider the arrival of discrete symbols belonging to an alphabet $\Sigma = \{A, B, \ldots\}$. These might represent user commands or protocol transactions, etc. The system state $q$ is now determined by a transition matrix, and by the particular symbol that arrives. The result is a set of *labelled transitions*:

$$q_i \xrightarrow{A} q_f. \tag{8}$$

  This is to be interpreted as: if the system is in state $q_i$ and receives symbol $A$, from the alphabet, then a transition is made to state $q_f$.

We would like to have ways of identifying patterns of input in order to associate meanings to them.

In a continuum approximation, we can simplify the details of individual events into a continuum of behavior and attempt to *fit* a smooth function. For example, waves impinging on a receiver might properly consist of the arrival of individual photons, but we normally represent the average behavior as a smooth function such as

$$q(x, t) = \sin(kx - \omega t). \tag{9}$$

This kind of approximation is very useful, because we can write down the entire behavior in a simple closed form that can be manipulated according to known rules of algebra. As a mathematical model, we do not normally model the uncertainty of observations, but treat this closed form as a theoretical 'truth'.

In the case of a discrete alphabetic stream of symbols, the *theory of languages* provides a classification of patterns. The Chomsky hierarchy classifies the levels of sophistication in a pattern, called a grammar, by associating each level of pattern with an automaton that can decipher it. More often than not, a complete specification of a grammar is complicated and the most basic pattern structure, a *regular expression* is used to classify patterns of text.

The Chomsky hierarchy is important, for example, in configuration management, where one would like to edit the contents of a file or database. A file editor that understands structures like recursively bracketed blocks of symbols is clearly able to provide a much more sophisticated understanding than a file editor which is only able to match simple strings of characters. Regular expressions (and regular languages) are helpful in summarizing the essence of structured interactions, like protocols. An extremely simple representation of a TCP stream might therefore be written,

$$\text{"}[SYN][ACK]*.*[FIN]\text{"},\tag{10}$$

where the '.' (dot) character stands for any symbol in the alphabet.

## 9. Configuration management and maintenance

As an example of these modelling forms, let us consider the highly important example of computer configuration and maintenance. Placing a system in a certain state and trying to keep it there is an important part of management policy. One must take into account the stochastic environment, which tends to push the system into neighboring states from which one must recover.

To model this, we use our two approaches:
- Discrete: each change is an operation that transforms a state $q$ into a new state $q'$, by some transition matrix.
- Continuous: look at the long-term trends $q(t)$ and find the rates at which changes happen. What does that mean about the need for management?

Changes are introduced
- By us (intentional change).
- By accident (random error).

Later we shall relate changes to information theory in order to discuss reliability.

As always, we use continuous models to discuss rates of change, and discrete models to discuss the microscopic details of individual changes.

### 9.1. Discrete configuration

Configuration management is the discrete form of regulation theory. Policy can be thought of as a specification of the state in which we wish the system to reside. Since changes and errors can be introduced by users and administrators at any time, one must ensure that this is the case by performing regular maintenance.

Configuration management includes both hardware and software configuration in
- The set up of the system.
- The maintenance of the system.

Suppose we represent file or other system objects by a vector. We can think of this as a vector of bits [12]:

$$\vec{q} = \begin{pmatrix} q_1 \\ q_2 \\ q_3 \end{pmatrix},\tag{11}$$

where e.g. $q_1$ is a file, $q_2$ is a process, $q_3$ is an access permission etc. One can make any abstraction that is convenient.

To make a change $q \to q'$, we need a transition *matrix* or *operator*. This is a convenient representation of the problem. The operation of repairing a broken state is written:

$$\vec{q}_{\text{fixed}} = \widehat{O}_{\text{repair}} \vec{q}_{\text{broken}}, \tag{12}$$

i.e. in explicit form one can write larger matrix generalizations of

$$\begin{pmatrix} 1 \\ q'_1 \\ q'_2 \end{pmatrix} = \begin{pmatrix} 1 & 0 & 0 \\ 0 & ? & ? \\ 0 & ? & ? \end{pmatrix} \begin{pmatrix} 1 \\ q_1 \\ q_2 \end{pmatrix} \tag{13}$$

going up to $q_n$, for arbitrary $n$. Each change or repair is an operation mediated by operator $\widehat{O}$. We include a 1 as the top-most element of the vector to enable us to set the Boolean states to 1 by copying this value.

Suppose we start with a system 'out of the box', in state $q_0$. We decide what we want and then set this up by performing a string of operations on it, in order to obtain a policy state.

$$\vec{q}_{\text{policy}} = \widehat{O}_n \cdots \widehat{O}_3 \widehat{O}_2 \widehat{O}_1 \vec{q}_0, \tag{14}$$

e.g. $\widehat{O}_1$ is edit /etc/passwd, $\widehat{O}_2$ is 'install software', etc. This notation allows us to discuss the algebra of changes.

One must now be careful. Do the operators commute? I.e. is order important in the application of the changes? If two operations commute

$$[\widehat{O}_1, \widehat{O}_2] \equiv \widehat{O}_1 \widehat{O}_2 - \widehat{O}_2 \widehat{O}_1 = 0 \tag{15}$$

then order does not matter. If the result is non-zero, we might get the wrong answer unless the order is guaranteed.

If the order of a collection of operators matters, then we must know what state we start out from, in order to know what state we will end up in. But if the order does not matter, we have the possibility of arriving at the same end result no matter how we apply the changes.

For example, the configuration management software cfengine works on a philosophy of trying to provide operators that commute [7,12], so that one can repair systems in real time, with the minimum of interference.

The software IsConf takes the approach that, when an error is detected, one should reset the system back to $q_0$ and reproduce the exact order of operations [55]. These two approaches are not as different as they sound.

### 9.2. *Discrete convergence*

What additional properties, or constraints, should we impose on the maintenance operations $\widehat{O}_i$ in order to make maintenance predictable and reproducible? Stability is clearly a powerful concept, related to maintenance. The need to maintain stability in the face of random change means that one must place additional restrictions on the operators.

The stochastic nature of environment can be modelled as a series of random operators arriving and altering the desired state to a new state. Maintenance should therefore be considered as 'undoing' all of these operations relative to the current state. Configuration maintenance can then be automated by simplifying continuously applying operations $\widehat{C}$ repeatedly, in any order.

Undoing operations caused by environmental change might make one think of looking for the inverse operators that reverse the ordering of changes. However, finding inverse operators for the changes is a difficult problem, and we need not find an inverse in the normal sense.

An alternative is to create an operator that takes us from any state to our policy state, and then no further i.e. if we apply operator to a non-policy state, we get the policy state. If we apply it to the policy state, nothing happens.

$$\widehat{C}\vec{q}_{\text{broken}} = \vec{q}_{\text{policy}},$$
$$\widehat{C}\vec{q}_{\text{policy}} = \vec{q}_{\text{policy}}. \tag{16}$$

This property is called convergence [8], and has the effect of making the policy state *absorbing*, with respect to the operation. It is a generalization of the concept of idempotence for operators [22,30] which includes a 'memory' of the preferred state.

An example of a convergent edit operation would be e.g. $\widehat{C}$ is "add mark to passwd file if user mark does not already exist". What is new is that the operators need to 'observe' the state of the system before acting, so that they are aware of their location in the set of possible states.

### 9.3. *Continuous maintenance*

Continuous models are approximations over long time-scales or large numbers of operations.

They allow one to make general statements about the rate of damage versus the rate of repair. How quickly do we have to act to maintain trends in behavior, i.e. policy?

Formally, one obtains a continuum model from a discrete model using a *local averaging procedure* to find a continuum picture. This only works for numerical variables, so it is natural to consider the stochastic point processes which lead to change and their associated counting processes. Alternatively, one can consider an enumeration of the states whose which can be plotted as a time series (as in Figure 2). In other words, we could the numbers of primitive operators or changes in order to gain insight into the trends.

The continuum model is thus found by dividing time up into regions and using the average value from each region as the smooth trend. Interpolating a smooth curve through these gives one the approximate picture seen when zooming out to long times.

A change in each region of the system can be written relative to the local mean value $\langle q \rangle_t$:

$$q(t) = \langle q(t) \rangle + \delta q(t), \tag{17}$$

i.e. signal = trend + fluctuation.

**9.4.** *The maintenance theorem*

Based on a fluctuation model of the stochastic state of a system, one can consider the rate at which repairs must be implemented in order to maintain stability over a certain time-scale. The theorem says that we should associate policy with $\langle q \rangle_t$, i.e. the average state. We should allow some 'slack' into the policy because we cannot define the precise state realistically. So we define policy as the acceptable trends in system development, which means we allow a slack of the order $\delta q$.

Now, to ensure that the system does not deviate by more than $\delta q$, we need to repair the change within a time $\Delta t$, where:

$$\delta q \simeq \frac{dq}{dt} \Delta t \ll q \tag{18}$$

i.e. when the system is dominated by a smooth average development. For example, security policy might be adapting slowly over weeks (small $dq/dt$), but a rapid change $\delta q$ should be repaired quickly to maintain the integrity of that policy.

The continuous modelling a system state is important in all cases in which large numbers of transitions are involved. This includes anomaly detection and network monitoring.

## 10. The concept of a system

The concept of a system is common to a wide range of activities and scenarios, from machines to human enterprises. In general terms, a system can be thought of organized effort to fulfill a role. As a tool, information technology is integrated into other systems in almost every sphere of application, from hospitals to businesses.

In terms of modelling, we need some way of formalizing the activities in a system, measuring and parameterizing the results. Once again, we are led to consider the two approaches to modelling: the discrete and the continuous. The discrete approach excels at detailed interactions, while the continuum approach deals best with trends and rates of change over longer time-scales.

What constitutes a system? A system is a balance between *freedoms* and *constraints*. A freedom is a potential for change within the system. Freedom to change is usually represented by a parameter which can be varied over a range of values. A constraint is a limitation on the possible changes which can occur on variables or parameters in a system. This often takes the form of a rule, or parameter inequality.

Symmetries are descriptions of what can be changed in a system without affecting the system's function. Determining what is *not* important to the functioning of a system is a way of identifying degrees of freedom that could be manipulated for strategic advantage.

If we wish to describe the behavior of a system from an analytical viewpoint, we need to be able to write down a number of variables which capture its behavior. Ideally, this characterization would be numerical since quantitative descriptions are more reliable than qualitative ones, though this might not always be feasible.

It is reasonable to distinguish *static* or *dynamic* systems. A system that is static undergoes no measurable change over the timescales of interest, whereas a dynamical system is undergoing noticeable change for at least part of the time. This is not to say that a static system is uninteresting. System architectures are static, but transactional processes and resource usage are dynamical.

For instance, a library or database archive is a static organization of data (a data structure). Data models represent aspects of a systematic organization that are not subject to frequent change. A number of such models has been developed for network and system management: for example, DEN-ng [50,51], eTOM [53].

Principles such as database normalization [24] are used to rationalize and optimize discrete structures.

An *algorithm*, or protocol, is a discrete encapsulation of a *process* in terms of operational steps. Flow models and networks can also be used to apply continuum approximations to modelling schemes for processes include Petri nets [25,40] and UML [20,52] to name only a couple [6].

Inventory resource models [29] and schedules are also continuum approximations to discrete processes, which consider methods of optimization.

A final categorization of dynamical systems fall is to consider their descriptive completeness, and hence the uncertainty involved in their modelling. One speaks of *open systems* (partial systems) and *closed systems* (complete, independent systems).

An open system is essentially a *sub-system* of a larger system. An open system received input from its environment and generates output, i.e. it communicates with its environment. A system is said to be open because input changes the state of the system's internal variables and output changes the state of the environment. Since one cannot normally predict the exact behavior of what goes on outside of a black box (it might itself depend on many complicated variables), any study of an open system tends to be incomplete. An open system can be internally *deterministic*, in terms of its operation, meaning that it follows strict rules and algorithms, but its behavior is not necessarily determined, since the environment is an unknown.

A closed system is an idealization: a system which is complete, in the sense of being isolated from its environment. A closed system receives no input and normally produces no output. Computer systems can only be approximately closed for short periods of time. The essential point is that a closed system is neither affected by, nor affects its environment. In thermodynamics, a closed system always tends to a steady state. Over short periods, under controlled conditions, this might be a useful concept in analyzing computer sub-systems, but only as an idealization. In order to speak of a closed system, we have to know the behavior of all the variables which characterize the system. A closed system is said to be completely *determined*.

The principal difference between these is how the principal of causality can be applied. Since most systems are best approximated by open systems, one must turn to stochastic models in which causality is not always directly traceable. Statistical methods and uncertainties must then be used.

## 11. Diagrammatic modelling

Diagrammatic methods of modelling are manifold [11], from entity relation diagrams, through flow diagrams and graphs. Diagrams are themselves systems with the freedom

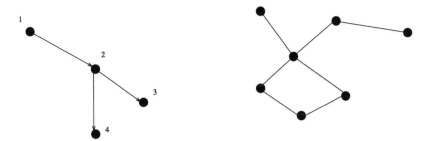

Fig. 7. Directed and undirected graphs.

of space, location, color, shape, direction, and constraints on size, location, topology, etc. Graphs are of special interest in the modelling of networks of components.

The use of graphs is a powerful way of analyzing relationships between discrete objects (see the chapter by Canright and Engø-Monsen in this volume). One may represent both directed and un-directed relationships, e.g.

- Directed $A \rightarrow B$: dominates, depends on, promises to, etc.
- $A \leftrightarrow B$: is associated with, is correlated with, is adjacent to.

A graph is formally a pair $\Gamma = \langle N, L \rangle$, where $N$ is a set of nodes and $L$ is a set of directed links. The degree of a node $k$ is the number of links it has. For directed graphs one separates in-degree and out-degree, according to the direction of the arrows. Graphs are conveniently represented by their adjacency matrices $A$.

For example, the first graph in Figure 7 is represented by an adjacency matrix of the form:

$$A = \begin{pmatrix} 0 & 1 & 0 & 0 \\ 0 & 0 & 1 & 1 \\ 0 & 0 & 0 & 0 \\ 0 & 0 & 0 & 0 \end{pmatrix}. \tag{19}$$

The set of modes labels the rows and columns, and the links are represented by 1s in the matrix.

### 11.1. *Connectivity*

The concept of connectivity is useful in networks of communicating hosts [11]. The connectivity in a graph may be defined as a fraction of the total number of possible directed connections. Let $\vec{h}$ be a vector of $n$ host-nodes in a network. Let $\vec{h}$ be a vector of node availabilities, i.e. the $i$th component $h_i$, $i = 1, \ldots, n$, is 1 if the host-node $i$ is fully available, and zero if it is completely unavailable. Then, if $A$ is the adjacency matrix, the scalar connectivity $\chi$ may be defined as

$$\chi = \frac{1}{n(n-1)} \vec{h}^{\mathrm{T}} A \vec{h}. \tag{20}$$

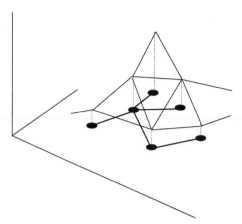

Fig. 8. A vector represents a 'height' function based on the discrete points of a graph. This is easily visualized as a rough surface.

$\chi$ has a maximum value of 1 when all nodes are connected bi-directionally and are fully available. It has a value of zero if the nodes are either unavailable or disconnected. This is a useful scalar measure of a network that is easily computed.

### 11.2. *Continuum functions on graphs*

A vector of values $\vec{h}$ can be thought of as defining a function, of graph nodes, i.e. a surface (see Figure 8).

Such a vector can be used to characterize local properties of the graph in a neighborhood of a node. One measure of particular interest in the eigenvector centrality. The relative centrality of a node $i$ is given by the $i$th component of the principal eigenvector belonging to the adjacency matrix, i.e. let $\rho$ be the largest positive eigenvector of the adjacency matrix $A$, then

$$A\vec{I} = \rho\vec{I}, \tag{21}$$

where $I_i$ is the centrality of the $i$th node.

The value of $I_i$ represents a fixed-point importance ranking for the graph, which is based on topology alone. It is therefore a complementary *vector* measure to the scalar connectivity which measures how well connected a node is. The most well-connected nodes are those which are bottlenecks for communication, as which are the most vulnerable nodes to attack by an agent wishing to destroy the network [16].

### 11.3. *Probabilistic networks*

We do not always have full information about the topology of a graph. In that case, one can collect statistics about the degree distributions and make a statistical estimate of the probability for being able to travel to any node in the graph from any other [11,43]. Consider a graph of $n$ nodes with degree distribution $p(k)$, where $\sum_k p(k) = 1$. The condition for 'percolation' through the network is:

$$\langle k \rangle^2 + \sum_k k(k-2)\, p(k) > \log(n). \tag{22}$$

This can be understood as follows. If a graph contains a giant component, it is of order $n$ and the size of the next largest component is typically $O(\log n)$; thus, according to the theory of random graphs the margin for error in estimating a giant component is of order $\pm \log n$. In the criterion above, the criterion for a cluster that is much greater than unity is that the right-hand side is greater than zero. To this we now add the magnitude of the uncertainty in order to reduce the likelihood of an incorrect conclusion.

The utility of this relation lies in being able to gauge the security or penetrability of a network based entirely on a knowledge of average numbers of connections. Although it is not a precise way of modelling, it gives a simple indicator for large and difficult networks.

### 11.4. *Fault trees*

Fault trees, such as that in Figure 5 are a useful and quantitative way of analyzing networks. They enable one to combine denumerable causes of failure quantitatively, within the framework of probability theory. One uses logic gate notation to indicate whether multiple branches of the tree are needed simultaneously (AND) or only independently (OR); these individual probability estimates can then be combined as follows. Let $p$ be the probability of a failure in a component.

$$\begin{aligned} p_{\text{AND}} &= \prod_{i=1}^{N} p_i, \\ p_{\text{OR}} &= 1 - \prod_{i=1}^{N} (1 - p_i). \end{aligned} \tag{23}$$

Fault tree analysis is beyond the scope of this review, however see the excellent book by Rausand and Høyland [33], and the chapter in this volume by Reitan for more details.

## 12. Discrete streams: information

The theory of languages provides a classification of the discrete models for classifying strings of discrete events (symbols) [38]. This theory, while interesting, has few direct engineering applications. A continuum approximation to this theory was developed by Shannon in the 1940s however, for specific application in the stochastic world of telecommunications [49].

This is somewhat analogous to the statistical theory of graphs in the previous section. Rather than considering the detailed microscopics of transmissions, which would pose an intractable problem in most cases, one can attempt to make the best of things by formulating a theory based on statistics of symbol distributions.

The theory of communication, now known as *information theory*, is a discussion about symbols in a stochastic framework: it considers how symbols can be misunderstood, or lost in transit, it considers errors of communication and their correction, the limits of symbolic compression, the smallest amount of information required to exactly reproduce a state of configuration, and so on.

There are several applications of information theory in the field of network and system administration: clearly the transmission of data by network is an obvious one; however, even at a high level, the most systems can be viewed as a propagation of state from an input to an output, and therefore information theory should also be able to say something about the integrity of systems in general.

The essence of information theory lies in the discrete time-series in Figure 9. A message or system process can be regarded as a stream of discrete symbols (e.g. operators which are causing change, or data packets of different kinds etc.) which are arriving at some kind of detector. The origin of the symbols could be from the system itself, from an external program or even a remote network source.

There are several aims in this model:

- To characterize the average properties of the data stream. This allows one to perform what is essentially inventory control on the symbols.
- To understand what happens when we expose the symbols to an environment which can introduce errors into the data, by randomly altering symbols.

One is thus interested in the extent to which the symbols which were transmitted by the sender (e.g. the configuration state that was asserted by a manager) actually arrives at the system intact (actually remains valid as time passes) as one transmits the data over a channel (as the system is propagated into the future in the presence of noise).

The analogy in the preceding paragraph is the essence of average state and its maintenance, i.e. it is the essence of a system's *policy state*. It links the idea of a stochastic arrival process with that of a stream of symbols of different *types*.

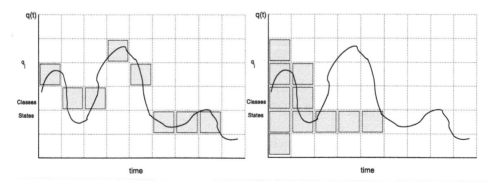

Fig. 9. The statistical properties of a time series of discrete symbols can be collected and used to define a theory of information.

Consider the arrival of different symbols $q$ at different times. The arrivals $q$ fall into a number of classes or types, labelled by $i = 1, \ldots, C$. If one collects data in the histogram in Figure 9 for long enough, one can turn the frequency counts for each symbol type $i$ into a probability $p_i$ for the occurrence of that symbol type over time, such that $\sum_{i=1}^{C} p_i = 1$.

Based on this, it is possible to compute a scalar measure of the information content of a string of symbols, called the Shannon entropy:

$$H = -\sum_{i=1}^{C} p_i \log_2 p_i, \tag{24}$$

where $i$ runs over the alphabet of symbol types. It is shown in ref. [11] that, in the limit of long messages, $H$ is equal to the average number of bits per symbol that is needed to uniquely reproduce the message, i.e. to distinguish it from another messages based on the same alphabet.

The entropy does not remember the actual message, it provides only a statistical characterization of the amount of bulk information per symbol of the original message.

The entropy has a minimum of zero value when a message consists of a continuous stream of the same symbol, and a maximum value of $\log_2 C$ when all of the symbols in the alphabet are used with equal frequency.

Interestingly, the entropy provides a limit on the extent to which a message can be compressed without altering its content.

Entropy can also be used in an analytical technique of optimization. By maximizing the entropy of a distribution of symbols, one finds the 'most random' or flattest or most widely distributed configuration of possible types, subject to additional constraints. For example, the normal distribution can be understood as the flattest distribution of points with a definite mean value and fixed root-mean square uncertainty (variance). Maximize the Lagrangian, with multipliers $\alpha$ and $\beta$ enforcing the normalization of probability and the constant variance:

$$L = -\sum_{i=1}^{C} p_i \ln p_i - \alpha \left( \sum_{i=1}^{C} p_i - 1 \right) - \beta \left( \sum_{i=1}^{C} \frac{p_i (q_i - q_i)^2}{\sigma^2 - 1} \right), \tag{25}$$

$$\frac{\partial L}{\partial p_i} = \frac{\partial L}{\partial \alpha} = \frac{\partial L}{\partial \beta} = 0, \tag{26}$$

$$p_i \propto \exp\left( -\frac{\beta (q_i - \bar{q})^2}{\sigma^2} \right). \tag{27}$$

Consider the transmission of a message from a sender to a receiver. We can call the message *policy*. Can we measure the integrity of the message as it is propagated through the system, over a noisy channel [9]?

Suppose one knows the original message, what is probability of it still being the same after some time along its journey?

We can compare the information (i.e. the entropy) at the start and end of the journey. The conditional entropy $H(O|I)$ is the average information at the output, given that the output

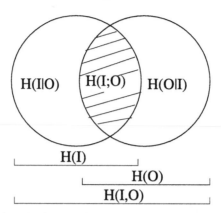

Fig. 10. The mutual information related to the other entropies. Here the quantities are shown schematically as measures on an abstract space of all possible messages.

is causally connected with the input, i.e. the two are related by more than just coincidence. Similarly $H(I|O)$ is the converse. The overlap between these two $H(I; O)$ (see Figure 10) is the amount if information that is common to both the input and the output, by more than mere coincidence [23]. Mutual information is therefore a measure of the integrity of one's system.

A well known and important formula that results from the informational entropy is the Shannon formula for transmission of a *bandwidth limited* continuous channel, with Gaussian noise. In this case, the representation in Figure 9 can be understood as follows. The total height of the graph represents a bandwidth $B$ in Hertz, which could be infinitely divided (and therefore carry infinite information) if it were not for the fact that noise $N$ (Watts) limits the resolution of the signal $S$ (Watts). The Gaussian spread of noise makes the resolution finite (think of these as the squares or graduations on the diagram): the greater the noise, the larger the squares – i.e. the fewer that will fit into the bandwidth. Thus the question is how many squares can one fit into the bandwidth: this tells us the capacity $C$ of the channel in bits per second. The answer [11,23] is the classic formula

$$C = B \log_2\left(1 + \frac{S}{N}\right). \tag{28}$$

The connection between system administration and information theory is not just fortuitous, it is important for several reasons. It relates the matter of system management to a well-known body of knowledge about communication, and a model that is established. It provides a quantitative way of addressing the issue of system integrity using error correction. It allows us to see policy as the transmission of *intent* over time and in the presence of an unpredictable environment, as a system talking to itself over a noisy channel.

## 13. Continuous streams: queues

Service provision is usually modelled as a relationship between a client and a server (or service provider). A stream of requests arrives randomly at the input of a system, at a rate of $\lambda$ transactions per second, and they are added to a queue, whereupon they are serviced at a rate $\mu$ transactions per second by a processor.

Generalizations of this model include the possibility of multiple (parallel) queues and multiple servers. In each case one considers a single incoming stream of transaction requests; one then studies how to cope with this stream of work.

As transaction demand is a random process, queues are classified according to the type of arrival process for requests, the type of completion process and the number of servers. In the simplest form, Kendall notation of the form $A/S/n$ is used to classify different queueing models.

- $A$: the arrival process distribution, e.g. Poisson arrival times, deterministic arrivals, or general spectrum arrivals.
- $S$: the service completion distribution for job processing, e.g. Poisson renewals, etc.
- $n$: the number of servers processing the incoming queue.

### 13.1. *The Poisson assumption*

The distribution of inter-arrival times for both $A$ and $S$ is normally considered to be a Poisson distribution in discrete time. One writes this $M/M/n$. This assumption is made in order to simplify the analysis. $M$ stands for 'memoryless' because the Poisson distribution, taken over non-overlapping time intervals is a Markov process whose behavior, at any discrete time $t$, is quite independent of what has happened in the past, i.e. it has no memory of the past. The provides a huge simplification of the analysis.

Another reason for the Poisson assumption is that, in the limit of large numbers of independent arrivals, one would expect the limiting distribution to have a Poisson form. The argument, on very general grounds is based on the fact that a Poisson distribution is the limiting form of any Bernoulli process in which an event either occurs or does not occur at time $t$. Suppose the probability of obtaining a result is fixed and is equal to $p$ on each independent observation, then the probability of obtaining $q$ positive arrivals in $n$ observations is

$$P(q) = {}^nC_q p^q (1 - p)^{n-q}, \tag{29}$$

where ${}^nC_r = n!/(n - r)!r!$ are the binomial coefficients. This is a binomial distribution whose mean value is $\bar{q} = np$. Now, suppose the probability of observing an arrival is scarce i.e. $p \to 0$, but we consider the limit of long times or many observations $n \to \infty$, taking the limit in such a way as to make $np \to \lambda$, where $\lambda$ is a constant. Then, noting the limits,

$$
{}^nC_r \leqslant \frac{n^r}{r!},
$$

$$
{}^nC_r \geqslant \frac{n^r}{r^r}, \tag{30}
$$

$$
\lim_{\alpha \to \infty} (1 - x)^\alpha \to e^{-\alpha} \quad \text{where } x < 1.
$$

One observes that

$$P(q = k) \leqslant \frac{n^k}{k!} p^k (1 - p)^{n-k} \rightarrow \frac{(np)^k}{k!} (1 - p)^{n-k} \rightarrow \frac{\lambda^k}{k!} e^{-\lambda}, \tag{31}$$

which is the the Poisson distribution, for which one verifies that $\sum_{k=0}^{\infty} P(q = k) = 1$.

This widely held belief is somewhat controversial, however, as measurements of network traffic have shown evidence of considerable 'burstiness', or long-tailed behavior [37,46,58, 59]. Other work indicates that these contradictory measurements would in fact settle into a Poisson distribution if only enough measurements were taken [18,32]. However, it is estimated that something of the order of $10^{10}$ transactions might be needed to see this limiting form emerge.

Regardless of this controversy, the Poisson model survives in queueing theory for its overriding simplicity.

### 13.2. *M/M/s queues*

The basic ideas about queueing can be surmised from the simplest model of a single server, memoryless queue: $M/M/1$. Then, let $n$ be the number of unprocessed requests in the queue at time $t$, and suppose that requests for transactions arrived at a rate of $\lambda$ per second, and can be processed at a rate of $\mu$ per second.

We can treat this simple case as a continuum flow approximation, using expectation values.

Consider any time $t$: the system makes a transition from a state of queue length $n - 1$ to $n$ at a continuous rate $\lambda$. Similarly, when it is in state $n$, it makes transitions to a state of length $n$ at a rate $\mu$. The expectation of $\lambda$ arrivals per second, when the queue length is in a state $n - 1$ (for any $n$) must therefore lead to the expectation of having $\mu$ completions when the state is $n$, in order to balance the queue, so we write, on balance $\lambda p_{n-1} = \mu p_n$ or

$$p_n = \rho p_{n-1}, \tag{32}$$

where $\rho = \lambda/\mu < 1$ is called the traffic intensity [11,34]. This is a recurrence relation that can be solved for the entire distribution $p_n$, for all $n$. One finds that

$$p_n = (1 - \rho)\rho^n, \tag{33}$$

and hence, the expected length of the queue is

$$\langle n \rangle = \sum_{n=0}^{\infty} p_n n = \frac{\rho}{1 - \rho}. \tag{34}$$

Clearly as the traffic intensity $\rho$ approaches unity, or $\lambda \rightarrow \mu$, the queue length becomes infinite, as the server loses the ability to cope.

This situation improves somewhat for $s$ servers ($M/M/s$), where one finds a much more complicated expression. In simplified form one has

$$\langle n \rangle = s\rho + P(n \geqslant s)\frac{\rho}{1-\rho}, \tag{35}$$

where the probability that the queue length exceeds the number of servers $P(n \geqslant s)$ is of order $\rho^2$ for small load $\rho$, which naturally leads to smaller queues.

It is possible to show that a single queue with $s$ servers is at least as efficient as $s$ separate queues with their own server. This satisfies the intuition that a single queue can be kept moving by any spare capacity in any of its $s$ servers, whereas an empty queue that is separated from the rest will simply be wasted capacity, while the others struggle with the load.

### 13.3. *Flow laws in the continuum approximation*

Dimensional analysis provides us with some simple continuum relationships for the 'work-flows'. Consider the following definitions:

$$\text{Arrival rate } \lambda = \frac{\text{No. of arrivals}}{\text{Time}} = \frac{A}{T}, \tag{36}$$

$$\text{Throughput } \mu = \frac{\text{No. of completions}}{\text{Time}} = \frac{C}{T}, \tag{37}$$

$$\text{Utilization } U = \frac{\text{Busy time}}{\text{Total time}} = \frac{B}{T}, \tag{38}$$

$$\text{Mean service time } S = \frac{\text{Busy time}}{\text{No. of completions}} = \frac{B}{C}. \tag{39}$$

The utilization $U$ tells us the mean level at which resources are being scheduled in the system. The 'utilization law' notes simply that:

$$U = \frac{B}{T} = \frac{C}{T} \times \frac{B}{C} \tag{40}$$

or

$$U = \mu S. \tag{41}$$

So utilization is proportional to the rate at which jobs are completed and the mean time to complete a job. Thus it can be interpreted as the probability that there is at least one job in the system.

Suppose a web server receives hits at a mean rate of 1.25 hits per second, and the server takes an average of 2 milliseconds to reply. The law tells us that the utilization of the server is

$$U = 1.25 \times 0.002 = 0.0025 = 0.25\%. \tag{42}$$

This indicates to us that the system could probably work at four hundred times this rate before saturation occurs, since $400 \times 0.25 = 100\%$. This conclusion is highly simplistic, but that is the price for having a simple formula!

Another dimensional truism is known as Little's law of queue size. It says that the mean number of jobs in a queue $\langle n \rangle$, is equal to the product of the mean arrival rate $\lambda$ (jobs per second) and the mean response time $R$ (seconds) incurred by the queue:

$$\langle n \rangle = \lambda R. \tag{43}$$

Note that this equation has the generic form $V = IR$, similar to Ohm's law in electricity. This an analogy is a direct consequence of a simple balanced flow model.

In the $M/M/1$ queue, it is useful to characterize the expected response time of the service centre. In other words, what is the likely time a client will have to wait in order to have a task completed? From Little's law, we know that the average number of tasks in a queue is the product of the average response time and the average arrival rate, so

$$R = \frac{Q_n}{\lambda} = \frac{\langle n \rangle}{\lambda} = \frac{1}{\mu(1 - \rho)} = \frac{1}{(\mu - \lambda)}. \tag{44}$$

Notice that this is finite as long as $\lambda \ll \mu$, but as $\lambda \to \mu$, the response time becomes unbounded.

Load balancing over queues is a topic for more advanced queueing theory. Weighted fair queueing and algorithms like round-robin etc., can be used to spread the load of an incoming queue amongst multiple servers, to best exploit their availability. Readers are referred to refs. [11,21,34,57] for more discussion on this.

## 14. Workflow and scalability

What architecture should one employ in network and system management? An obvious answer is to centralize control in one convenient place. However, this often results in a bottleneck, or a constriction of the workflow. The entire world cannot ring the central office in Ireland with their help-desk queries. It would cost too much, and it would place too much burden in one place.

Centralization might seem like a rational way to manage systems, but it results in *single points of failure*, i.e. places where the system is vulnerable to destruction by random or malicious failures.

The opposite of centralization is a completely peer to peer distribution of nodes, with communication channels in some meshed network with nearest neighbors. Each node can relay messages that need to traverse the network beyond its immediate horizons. This results in a cheap solution, but one in which each node has control over its own faculties. A different principal is needed to manage systems which collaborate voluntarily [13,17].

There are six architectures between a star topology and peer-to-peer autonomy. These are described in refs [14,15]. Three of these models are shown in Figure 11.

The logical and physical layers can both be described in these terms, with different consequences. Centralization has the effect of generating bottlenecks for both logical and

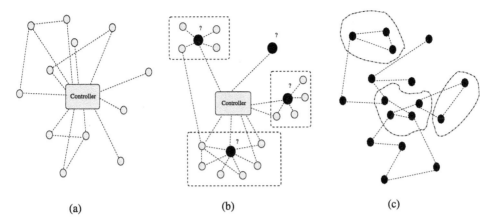

(a)                                    (b)                                        (c)

Fig. 11. Three forms of management between a number of networked host nodes. The lines show a logical network of communication between the nodes. Dark nodes can determine their own behavior, i.e. they have political freedom, while light nodes are controlled. The models display a progression from centralized control to complete de-centralization of policy (see refs [14,15]).

physical layers of a system. Distribution is the only strategy for scaling, but its cost is in a lack of perceived order. Psychologically, many administrators feel uncomfortable with the idea of releasing the leash of control that one perceives in a centralized system. The actual uncertainty associated with any of the workflow models is, however, dependent on too many factors to be able to say that one method or the other is better.

### 14.1. *Continuum approximation*

To examine the scalability, we consider a simple rate approximation based on average flows. We assume a simple linear relationship between the probability of successful mainte-nance and the rates of communication with the policy- and enforcement-sources. This need not be an accurate description of reality in order to lead to the correct scaling laws [15]. Let us suppose that a change of configuration $\Delta Q$ is proportional to an average rate of information flow $I$, over a time $\Delta t$; that is

$$\Delta Q = I \Delta t. \tag{45}$$

This equation says that $I$ represents the time-averaged flow over the interval of time for which is acts. As we are interested in the limiting behavior for long times, this is sufficient for the job.

In the following we will use $p$ to denote the average (over time, and over all nodes) probability that configuration management information flow (repair current) is not avail-able to a node. This unavailability may come from either link or node unreliability. We can lump all the unreliability into the links (see above) and so write

$$p = \left(1 - \langle A_{ij} \rangle \right), \tag{46}$$

where $\langle A_{ij} \rangle$ denotes both time and node-pair average. Each node then can only receive repair current during the fraction $(1 - p)$ of the total elapsed time.

The policy maintenance can be transmitted and/or enforced at a maximum rate given by the channel capacity of the source. Thus, in case (a) the currents must simply obey Kirchoff's law:

$$I_{\text{controller}} = I_1 + I_2 + \cdots + I_N. \tag{47}$$

The controller current cannot exceed its maximum capacity, which we denote by $C_S$. We assume that the controller puts out a 'repair current' at its full capacity and that all nodes are average nodes. This gives that

$$I_{\text{repair}} = \frac{C_S}{N}. \tag{48}$$

The total current is limited only by the bottleneck of queued messages at the controller, thus the throughput per node is only $1/N$ of the total capacity.

As we move towards a peer model (c), the repair current becomes increasingly internalized in the nodes, and the need for concentrating the error currents from all the nodes at a single location is obviated. The models therefore tend towards a peer to peer architecture in which the number of nodes $N$ becomes irrelevant and the the error correction rate stabilizes to a constant value.

## 14.2. *Reliability*

A common definition of reliability is the *mean time before failure* (MTBF). This is the average amount of time we expect to elapse between serious failures of the system. Another way of expressing this is to use the *average uptime*, or the amount of time for which the system is responsive (waiting no more than a fixed length of time for a response).

$$\rho = \frac{\text{Mean uptime}}{\text{Total elapsed time}}. \tag{49}$$

This is essentially a measure of the time-utilization of a device. Some like to define this in terms of the Mean Time Before Failure (MTBF) and the Mean Time To Repair (MTTR), i.e.

$$\rho = \frac{\text{MTBF}}{\text{MTBF} + \text{MTTR}}. \tag{50}$$

This is clearly a number between 0 and 1. Many network device vendors quote these values with the number of 9's it yields, e.g. 0.99999.

Ad hoc networks are networks whose adjacency matrices are subject to a strong, apparently random time variation. If we look at the average adjacency matrices, over time, then we can represent the probability of connectivity in the network as an adjacency matrix of probabilities. For our purposes, the important property of the ad hoc net is the intermittency of the links, due to the components' mobility.

An ad hoc network is thus represented by a symmetric matrix of probabilities for adjacency. Thus the time average of the adjacency matrix (for, e.g., four components) may be written as

$$\langle A \rangle = \begin{pmatrix} 0 & p_{12} & p_{13} & p_{14} \\ p_{21} & 0 & p_{23} & p_{24} \\ p_{31} & p_{32} & 0 & p_{34} \\ p_{41} & p_{42} & p_{43} & 0 \end{pmatrix}. \tag{51}$$

One can move the probabilities (uncertainties) for availability from the host vectors to the connectivity matrix and vice versa; for example:

$$\begin{pmatrix} p_1 \\ p_2 \\ p_3 \end{pmatrix}^{\mathrm{T}} \begin{pmatrix} 0 & 1 & 1 \\ 1 & 0 & 1 \\ 1 & 1 & 0 \end{pmatrix} \begin{pmatrix} p_1 \\ p_2 \\ p_3 \end{pmatrix} = \begin{pmatrix} 1 \\ 1 \\ 1 \end{pmatrix}^{\mathrm{T}} \begin{pmatrix} 0 & p_1 p_2 & p_1 p_3 \\ p_2 p_1 & 0 & p_2 p_3 \\ p_3 p_1 & p_3 p_2 & 0 \end{pmatrix} \begin{pmatrix} 1 \\ 1 \\ 1 \end{pmatrix}. \tag{52}$$

This equivalence tells us that a given host in the network can always regard the failures in hosts as being failures within the communications channel, or network, between them (though not vice versa). Thus all collaborating networks of clients and servers (peers) are ad hoc, probabilistic networks.

### 14.3. *Redundancy and folk theorems*

The only strategy certain against component failure in system is redundancy, or parallelism. Systems are a combination of serial dependencies and parallel flows. Several theorems exist showing the probabilities of failure in the presence of serialism and parallelism [11, 33]. The most important one, perhaps, can be stated without proof as follows:

> *Low level component redundancy is never a worse strategy*
>
> *for reliability than high-level system redundancy.*

In other words, a single computer with redundant components is almost always better, and never worse than several redundant computers.

This can be understood as follows. Consider the example, taken from the excellent ref. [33]. If one starts with a basic system, whose architecture is based on necessity, parallelism can be added to secure redundancy in effective and ineffective ways. The folk theorem tells us that low-level redundancy is never worse than high-level redundancy; this is seen intuitively in Figure 12.

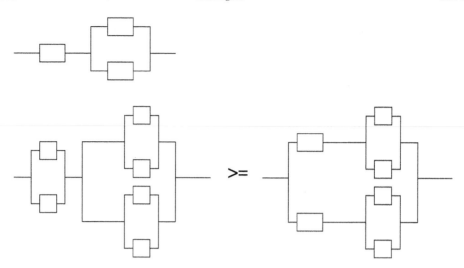

Fig. 12. The upper architecture is parallelized in two ways below. In the left hand case redundancy is applied at the low-level component layer. Disabling any single component results in the minimum disruption to the architecture. In the right hand solution, disabling the serial component from either the upper or lower arm results in the entire arm not working.

### 14.4. *Human error*

How does one measure and quantify the likelihood of human error? One method is clearly to use the empirical approach of fault trees [1,56]. Another interesting approach has been employed in ref. [31], using ideas from reliability theory applied to the task network [33].

In this work, the authors presume that human error is related to the complexity of the tasks they have to perform. Task complexity boils down to a number of things:

- Execution complexity – manual errors in implementation.
- Memory complexity – how many things does a person have to remember?
- Parallelism – how does time-sharing/multitasking affect a person?

One can use estimations of complexity to argue that the probability of error is proportional to some function of these. The effect of errors in networks is a complicated issue around which a considerable body of theory exists [11,33].

Task graphs can be viewed in two ways: in terms of open components that propagate input to output, or as larger closed systems of components with loops and cyclic connections. In the latter case, traditional network theory is less tractable. Newer methods such as network centrality can be useful tools for identifying vulnerable spots. Using these methods on a task or dependency graph, one can find vulnerable spots and cut-sets quite easily.

### 15. Policy modelling and promise theory

Promise theory is a representation of policy, from the viewpoint of autonomous agents – i.e. agents which make their own decisions, guided by a form of policy. Its strength lies, amongst other things, in making normally hidden assumptions explicit, so as to avoid conflicts of policy. Several approaches have been used in the past to resolve such conflicts [4,5, 28,36,39,44,47]. Other formalisms that might be considered for describing policy include process algebras [6] such as the $\pi$-calculus [45] or Petri nets [25] or process algebras with probabilistic extensions [35]. However, these formalisms are more like control theory – they are about understanding events in fully specified systems, with synchronization and resource constraints. They do not capture the behavior of a system given an imperfect knowledge or specification, such as that one usually faces in a management scenario. There is thus additional uncertainty to be considered.

In ref. [10], policy is identified as a specification of the average configuration of the system over persistent times. An important aspect of this definition is that it allows for error tolerances, necessitated by randomly occurring events that corrupt policy in the system management loop. There is a probabilistic or stochastic element to system behavior and hence policy can only be an average property in general. We do not require a full definition of host policy here from ref. [10]. It suffices to define the following.

Let us define the policy of an individual computing device (agent or component) as a representative specification of the desired average configuration of a host by appropriate constraints [10].

Promise theory takes a *service viewpoint* of policy and uses a graphical language to compose system properties and analyze them. The promise theory manifesto is to begin with detailed, *typed* promises, from which the algebra of cooperation can be understood, and then gradually to eliminate the types through transformations which elucidate large scale structure and reliability.

A key feature of promises is that they make a clean separation between what is being constrained and to whom a constraint applies. This is another area in which confusions can arise in policy theory. The form of a promise is:

$$n_1 \xrightarrow{\pi} n_2, \tag{53}$$

where $n_1$ is the node making a promise, $n_2$ is the recipient of the promise and $\pi$ is the typed promise body, which describes the interpretation and limitation of the promise. A promise body is a constraint on behavior, asserted from one agent to another. We shall associate agents with components in order to discuss control theory. Promises fall into approximately four notable categories (see Table 1).

Basic service promises form any number of types, e.g. 'provide web service in less than 5 milliseconds' or 'tell you anything about X'. Some other examples include:

Table 1

| Promise type | Notation | Interpretation |
| --- | --- | --- |
| Basic | $n_1 \xrightarrow{\pi} n_2$ | Provide service/flow |
| Cooperative | $n_1 \xrightarrow{C(\pi)} n_2$ | Imitate/Follow |
| Use | $n_1 \xrightarrow{U(\pi)} n_2$ | Use/Accept from |
| Conditional | $n_1 \xrightarrow{\pi_1/\pi_2} n_2$ | Promise 'queue': $\pi_1$ if $\pi_2$ |

- $X \xrightarrow{q \leqslant q_0} Y$: agent $X$ promises to never exceed the limit $q \leqslant q_0$.
- $X \xrightarrow{q = q_0} X$: agent $X$ promises to satisfy $q = q_0$.
- $X \xrightarrow{\ell \subseteq L} Y$: $X$ promises to keep $\ell$ to a sub-language of the language $L$.
- $X \xrightarrow{S} Y$: agent $X$ offers service $S$ to $Y$.
- $X \xrightarrow{R} Y$: agent $X$ promises to relay $R$ to $Y$.
- $X \xrightarrow{\neg R} Y$: agent $X$ promises to never relay $R$ to $Y$.
- $X \xrightarrow{S,t} Y$: agent $X$ promises to respond with service $S$ to $Y$ within $t$ seconds.

They are somewhat analogous to service level agreements.

Cooperative promises are used in order to build concensus amongst individual agents with respect to a basic promise type. If $\pi$ is a promise type, then $C(\pi)$ is a corresponding promise type between agents to collaborate or imitate one another's behavior with respect to $\pi$.

A promise is said to be *broken* if an agent makes two different promises of the same type at the same time (this is different from a promise that has expired or been changed).

Consider how promise theory compares to control theory with a simple example (see Figure 13). Consider how one would model a basic feedback loop consisting of a number of components.

A basic control theory network consists of a number of components (called controllers) connected together in a functional view: some observable value is injected into a network and a corresponding value is extracted elsewhere. The ingress of the observable is at the behest of an external agent; the egress occurs deterministically in the manner of a flow.

Control theory uses a scalar value with a compatible engineering dimension throughout (like an electric current or voltage). Promise theory, on the other hand, involves promises of several different types, based possibly on possibly structured information models. A single promise can be a declaration of intent to keep a general tuple of data within a given range of values. Promises are thus *typed* and their attributes can belong to discrete languages [38].

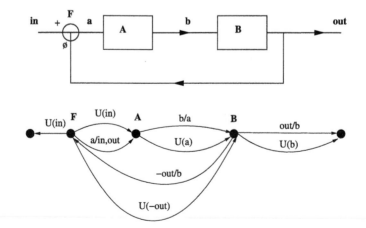

Fig. 13. Corresponding diagrams of control theory (above) and promise theory (below).

In the upper diagram, one represents the following deterministic relations, in real time parameterization:

$$A: \text{out}(t) = \int B(\tau)b(t-\tau)\,d\tau,$$

$$B: b(t) = \int A(\tau)a(t-\tau)\,d\tau, \qquad (54)$$

$$F: a(t) = \text{in}(t) - \text{out}(t).$$

In the promise diagram, the basic service promises are:
- $F \xrightarrow{\pi_a} A$ where $\pi_a : a(t) = \text{in}(t) - \text{out}(t)$ if out is promised.
- $A \xrightarrow{\pi_b} A$ where $\pi_b : b(t) = \int A(\tau)a(t-\tau)\,d\tau$ if $a$ is promised.
- $B \xrightarrow{\pi_{\text{out}}} A$ where $\pi_{\text{out}} : \text{out}(t) = \int B(\tau)b(t-\tau)\,d\tau$, if $b$ is promised.

There are several things to note about this rough translation. First of all, a promise looks very much like a service and (at this stage), a promise asserted is assumed to be a promise held. However, the autonomy of nodes makes this diagram much more complicated than the control diagram, since there is no a priori obligation for the components to work together. Rather we must ensure this by explicit means.

The promises above are conditional promises: i.e. they can only be made if the agents or components making them are in receipt of a dependent promise. The so-called 'use' promises are required because we have turned the control scenario around from being mandatory compliance of components to voluntary cooperation. No autonomous component is actually obliged to play a role in this network, and each must assure its compliance to dependent interests by pledging to use the services promised to them in order to fulfill their own promise.

## 16. Games – rational decision theory

The ability to model systems is an important step to understanding them, and therefore to being able to control them. Models that explain the causality of changing metrics of a system are called type I models [10]. Another kind of modelling that goes one step further is a type II model: this introduces a user-level, semantic interpretation of the system based on a set of values – i.e. it introduces a concept of policy, or user satisfaction. A type II model is about evaluating strategies for effectively achieving the goals of the system, i.e. planning and maintaining policy. Type II models consider the *utility* of decisions towards the goals of the system [42].

The theory of type II models is the theory of decision making: how to make rational decisions, i.e. how to justify conclusions by formalizing questions and answers. This is one of the ways in which we can optimize goals described by policy. Decisions which are motivated by goals are made *relative* to those goals so we must specify policy to eliminate arbitrariness in decision.

**16.1.** *Classification of alternatives*

Decision-making requires us to categorize or classify the available options. We can think of each category or option as a different symbol in an alphabet $\Sigma$, much as we did in Section 9. Decisions can involve conflicting interests, between different individuals. How do these compete in with one another, in the eyes of a rational observer? I.e. what decision would a purely impartial observer make?

The simplest kinds of decisions are discrete yes/no Boolean decisions, which can be combined with operators AND/NOT/OR, as in Section 11.4 (Figure 5). We know these from computer programming. They are a simplistic discrete view to decision representation, which is appropriate for a deterministic machine with a black and white view of the world.

Humans place different values on different kinds of issues, however. Continuous, non-deterministic decisions must be decided in terms of the value conferred to the decision maker, as measured in some form of currency. This is usually called a 'payoff' of 'utility'. Just as money, in markets, is only worth what people think it is worth – it has no absolute value – so the value of policy and decision is subjective and context dependent.

Placing a range of numerical values on decisions allows one to evaluate and describe when one situation is 'better than' or 'worse than' another. This is a simple matter in a utility viewpoint. To build up a similar structure in terms of logical assertions requires an extensive apparatus [19].

One can decide, for instance, that there is more than one 'right' answer to a question. It is then necessary to make a value judgment about the 'best' of optimal answer. We can then weight the importance of the possibilities using the expected payoff a result yields back to us: with this approach, one can end up with a single optimal answer or a mixture of several answers that result in the biggest payoff or utility.

**16.2.** *Mathematical games*

The theory of games, developed and generally attributed to von Neumann [27,41,42] is about competition between different goals in a common space. Opposing goals are thought of as different 'players' in a game. which compete with one another to attack and defend one anothers' positions. Two player games are generally sufficient for the kinds of models that can be dealt with in system administration. As with all models, there are discrete and continuous representations of game.

The discrete or extensive form of games involves making a decision tree of every possible action for every player in response to every possible move in a problem. The game tree is not unlike a fault or threat tree, described earlier (Figure 5). This level of detail is generally intractable from the viewpoint of modelling. As each decision confers some payoff to each player, ideally one would be able to add up the payoffs along each path, through the game, to find the best compromises at each branching point, given that both players are trying to do the same. The advantage of this approach is that one can retain the order of decisions. However, order does not always matter.

Such a two-player game is a tug-of-war between two players. Both players would like to win, and try their hardest, but they are prevented from having their own way by the other. The result is a compromise, or equilibrium.

The discrete modelling of the extensive form is difficult, so one generally approaches games from an averaged continuum approximation known as the strategic form of the game. This is also called the 'normal' form of the game. To reach this approximation, one lumps together actions that form single pathways into *classes* of play-behavior for the players. It is somewhat analogous to high-level management, rather than micro-management. The order of individual decisions is lost, as both players are now considered to act at the same time, in the manner of a duel.

Each player, in a strategic game, has a matrix $\Pi$ of payoffs for playing a particular strategy, given the strategies played by the opponent (see Figure 14).

- Player 1 has alternative strategies: $A, B, C$.
- Player 2 has alternative strategies: $\alpha, \beta, \gamma$.

If player 1 plays $A$ and player 2 plays strategy $\alpha$, then player 1 will receive payoff 1 and player 2 will receive payoff 3, and so on. Players can also use "mixed strategies" or linear combinations of these basic possibilities:

$$S_1 = c_1 A + c_2 B + c_3 C + \cdots. \tag{55}$$

A particular admixture of strategies be viewed as being mixed during the course one game, or over several separate 'plays' of the entire game.

An example game matrix $\Pi_{(1,2)}$ is shown in Table 2. The contents of the body of the matrix are the payoffs: these need to be chosen 'rationally'. This choice or evaluation of utility is the crux of rational decision making. It is not the fact that one arranges them in a matrix, whether they are discrete or continuous values. The procedure for analyzing the payoff matrix could not be considered rational if the basis for the analysis were not itself rational: this evaluation of payoffs can often be subjective, so one has to be careful.

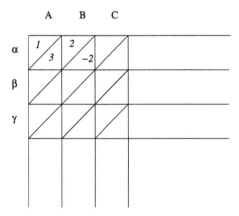

Fig. 14. The twin payoff matrices for a strategic form game, written in a single table.

Table 2

|  | Security hole | Bug in function | Missing function |
| --- | --- | --- | --- |
| Upgrade version now | (10, 5) | (10, 0) | (5, −5) |
| Test then upgrade | (5, 5) | (3, 9) | (0, 8) |
| Keep parallel versions | (−10, 5) | (−1, 10) | (0, 10) |

**16.3.** *Making the optimal decision: solution methods*

A solution of a game is a determination of optimal strategies for each player and a computation of the payoff-value received by each player when they behave rationally by trying to maximize their own payoff.

As both players are trying to maximize their payoff at the same time, the result will be a compromise. Saddle-points and other equilibria are of particular interest here.

Games are broadly classified into two kinds:

- If $\Pi_1 + \Pi_2 = 0$ a game is said to be a *zero sum game*, and the game is solved by von Neumann's 'minimax method'. This case is special because there is a 'best answer' when the game has a saddle-point, and otherwise there is a mixture or combination of answers that gives best result.
- If $\Pi_1 \neq -\Pi_2$ have to use more general equilibrium ideas – e.g. Nash equilibrium.

Zero-sum games are solved by virtue of the saddle-point, minimax theorem: take the payoff matrix for either one of the players (e.g. $\Pi_1$) and compare:

$$\max_{1}\min_{2} \Pi_1 \equiv v_1,$$

$$\min_{2}\max_{2} \Pi_1 \equiv v_2.$$

(56)

If $v_1 = v_2$ then there is a pure strategy solution, i.e. a 'best' answer. Else there is a mixture of strategies for both players that yields an equilibrium. For example, let

$$\Pi_1 = \begin{pmatrix} 1 & -3 & -2 \\ 2 & 5 & 4 \\ 2 & 3 & 2 \end{pmatrix}.$$

(57)

Then

$$\max_{\updownarrow}\min_{\leftrightarrow} \Pi_1 = \max \begin{pmatrix} -3 \\ 2 \\ 2 \end{pmatrix} = 2.$$

(58)

Subsequently,

$$\min_{\leftrightarrow}\max_{\updownarrow} \Pi_1 = \min(2, 5, 4) = 2.$$

(59)

Thus $v_1 = v_2 = 2$ and the relevant strategies are: $(\updownarrow, \leftrightarrow) = (2, 1)$ or $(3, 1)$.

Dominated strategies are a particularly easy way of reducing the number of rational options by elimination. This method works both for zero-sum and non-zero-sum games. Sometimes the payoff matrix can be simplified for both players if one of the players has a weak strategy, i.e. one that there is no reason to play because it performs worse than all others. e.g.

$$\begin{pmatrix} 50 & -10 & -20 \\ 20 & 40 & -44 \\ 0 & -10 & -40 \end{pmatrix} \rightarrow \begin{pmatrix} 50 & -10 \\ 20 & 40 \\ 0 & -10 \end{pmatrix}. \tag{60}$$

The Nash equilibrium is the most popular solution concept for non-zero sum games. It is equivalent to minimax for zero-sum games. Computer software such as GAMBIT can be used to locate equilibria.

The Nash solution concept is based on the idea of bargaining – or economic agreement. If players can win by working together in the long-term, we call games cooperative, e.g. imagine two devices sharing wireless bandwidth either in a friendly way or competing aggressively to use the maximum share.

$$\Pi = \begin{pmatrix} & \text{Compete} & \text{Share} \\ \hline \text{Compete} & ? & ? \\ \text{Share} & ? & ? \end{pmatrix}. \tag{61}$$

Does it pay them to share nicely or to compete aggressively? E.g. should you work with a competitor or compete and risk losing the fight? To get the whole picture, we need to model payoffs. This might be a complex mixture of policy-motivated expressions, joined into a common currency. This idea leads one to so-called *dilemma games*, of which the 'prisoner's dilemma' is the most famous [2,3,48].

Consider a choice which involves making a promise to a competitor, such as delivering on a Service Level Agreement, or some other policy decision. The agreement has to be mutual, i.e. both players have something of value that the other wants, and thus they are in a bargaining position. Player 1 offers his promise to 2; player 2 offers her promise to 1. Border Gateway Protocol (BGP) peering relationships are of this type, for example.

Either party offering a promise can either 'cooperate' or 'defect' from the obligation implied by the agreement. What do we do if we have this? Let 'C' mean cooperate and 'D' mean 'defect' from cooperation. The payoff matrix is given in Figure 15.

- $R$ is the reward for mutual cooperation.
- $T$ is the value of a temptation to 'defect'.
- $S$ is the sucker's payoff, for getting exploited.
- $P$ is the punishment for both losing the gamble of trust.

We assume $T > R > P > S$ and we see both players gamble on each others' kindness. The result is that:

- If both cooperate, both get maximum reward.
- If one cooperates and the other defects, the defector wins by exploiting the other.
- If both try to exploit each other, they both lose with a minimum payoff.

If we treat each game as a single interaction and play the game many times, so each player can only see *previous* rounds, then the strategy of 'tit for tat' is found to be most effective

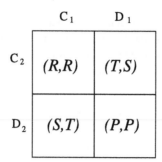

Fig. 15.  The basic form of a dilemma game: a non-zero sum bargaining problem.

at stabilizing a players' returns. The dilemma scenario is used to argue for the significance of strategy in conflict/cooperation scenarios. There is a large literature on this topic.

The dilemma game is the basis for discussions of economics in the system. It is a representation of the maxim 'you scratch my back and I'll scratch yours', i.e. trade of services for mutual gain, such as peering in BGP. The game framework allows one to negotiate the equilibrium contract between the parties, motivated on rational grounds. This kind of analysis has been used in auto-discovery protocols, using auctions for electing network representatives amongst equal peers.

## 17.  Conclusions

A mixed strategy of rationality and causality enables system administrators to formalize problems of planning and management, through a variety of methods. These methods make contact with diverse fields of research. Theory is an increasingly difficult pill to sell and swallow in this age of increasing mathematical illiteracy, and yet we need it more than ever. This lightning overview brings together only a few of the ideas required in network and system administration.

## Acknowledgement

This work was supported in part by the EC IST-EMANICS Network of Excellence (#26854).

## References

[1] R. Apthorpe, *A probabilistic approach to estimating computer system reliability*, Proceedings of the Fifteenth Systems Administration Conference (LISA XV), USENIX Association, Berkeley, CA (2001), 31.
[2] R. Axelrod, *The Evolution of Co-operation*, Penguin Books (1990) (1984).
[3] R. Axelrod, *The Complexity of Cooperation: Agent-based Models of Competition and Collaboration*, Princeton Studies in Complexity, Princeton (1997).

[4] A.K. Bandara, E.C. Lupu, J. Moffett and A. Russo, *Using event calculus to formalise policy specification and analysis*, Proceedings of the 4th IEEE Workshop on Policies for Distributed Systems and Networks (2003).

[5] A.K. Bandara, E.C. Lupu, J. Moffett and A. Russo, *A goal-based approach to policy refinement*, Proceedings of the 5th IEEE Workshop on Policies for Distributed Systems and Networks (2004).

[6] J. Bergstra, A. Ponse and S.M. Smolka, *The Handbook of Process Algebra*, Elsevier, Amsterdam (2001).

[7] M. Burgess, *A site configuration engine*, Computing systems, MIT Press, Cambridge, MA (1995), 309.

[8] M. Burgess, *Computer immunology*, Proceedings of the Twelth Systems Administration Conference (LISA XII), USENIX Association, Berkeley, CA (1998), 283.

[9] M. Burgess, *System administration as communication over a noisy channel*, Proceedings of the 3rd International System Administration and Networking Conference (SANE2002) (2002), 36.

[10] M. Burgess, *On the theory of system administration*, Science of Computer Programming **49** (2003), 1.

[11] M. Burgess, *Analytical Network and System Administration – Managing Human–Computer Systems*, Wiley, Chichester (2004).

[12] M. Burgess, *Cfengine's immunity model of evolving configuration management*, Science of Computer Programming **51** (2004), 197.

[13] M. Burgess, *An approach to policy based in autonomy and voluntary cooperation*, Lecture Notes in Comput. Sci., Vol. 3775, Springer-Verlag, Berlin (2005), 97–108.

[14] M. Burgess and G. Canright, *Scalability of peer configuration management in partially reliable and ad hoc networks*, Proceedings of the VIII IFIP/IEEE IM Conference on Network Management (2003), 293.

[15] M. Burgess and G. Canright, *Scaling behaviour of peer configuration in logically ad hoc networks*, IEEE eTransactions on Network and Service Management **1** (2004), 1.

[16] M. Burgess, G. Canright and K. Engø, *A graph theoretical model of computer security: From file access to social engineering*, International Journal of Information Security **3** (2004), 70–85.

[17] M. Burgess and S. Fagernes, *Pervasive computing management ii: Voluntary cooperation*, IEEE eTransactions on Network and Service Management (submitted).

[18] J. Cao, W.S. Cleveland, D. Lin and D.X. Sun, *Non-linear Estimation and Classification*, Springer-Verlag (2002), Chapter: Internet Traffic Tends Toward Poisson and Independent as the Load Increases.

[19] B.F. Chellas, *Modal Logic: An Introduction*, Cambridge University Press, Cambridge (1980).

[20] T. Clark, A. Evans and S. Kent, *A meta-model facility for a family of UML constraint languages*, Object Modeling with the OCL: The Rationale Behind the Object Constraint Language, T. Clark and J. Warmer, eds, Springer-Verlag, Berlin (2002), 4–20.

[21] R.B. Cooper, *Stochastic Models*, Handbooks in Operations Research and Management Science, Vol. 2, Elsevier, Amsterdam (1990), Chapter: Queueing Theory.

[22] A. Couch and Y. Sun, *On observed reproducibility in network configuration management*, Science of Computer Programming **53** (2004), 215–253.

[23] T.M. Cover and J.A. Thomas, *Elements of Information Theory*, Wiley, New York (1991).

[24] C.J. Date, *Introduction to Database Systems*, 7th edn, Addison–Wesley, Reading, MA (1999).

[25] R. David and H. Alla, *Petri nets for modelling of dynamic systems – A survey*, Automatica **30** (1994), 175–202.

[26] R.H. Dieck, *Measurement and Uncertainty (Methods and Applications)*, 3rd. edn., Instrument, Systems and Automation Society, Triangle Park, NC (2002).

[27] M. Dresher, *The Mathematics of Games of Strategy*, Dover, New York (1961).

[28] J. Glasgow, G. MacEwan and P. Panagaden, *A logic for reasoning about security*, ACM Transactions on Computer Systems **10** (1992), 226–264.

[29] S.C. Graves, A.H.G. Rinnooy Kan and P.H. Zipkin, eds, *Logistics of Production and Inventory*, Handbooks in Operations Research and Management Science, Vol. 4, Elsevier, Amsterdam (1993).

[30] G.R. Grimmett and D.R. Stirzaker, *Probability and Random Processes*, 3rd edn., Oxford Scientific Publications, Oxford (2001).

[31] J. Hellerstein, K. Katircioglu and M. Surendra, *A framework for applying inventory control to capacity management for utility computing*, Proceedings of the IX IFIP/IEEE International Symposium on Integrated Network Management IM'2005, IEEE Press, New York (2005), 237–250.

[32] K.I. Hopcroft, E. Jakeman and J.O. Matthews, *Discrete scale-free distributions and associated limit theorems*, Journal of Mathematical Physics **A37** (2004), L635–L642.

[33]  A. Høyland and M. Rausand, *System Reliability Theory: Models and Statistical Methods*, Wiley, New York (1994).

[34]  R. Jain, *The Art of Computer Systems Performance Analysis*, Wiley, New York (1991).

[35]  B. Jonsson, W. Yi and K.G. Larsen, *Probabilistic extensions of process algebras*, The Handbook of Process Algebra, Elsevier, Amsterdam (2001), 685.

[36]  A.L. Lafuente and U. Montanari, *Quantitative mu-calculus and ctl defined over constraint semirings*, Electronic Notes on Theoretical Computing Systems QAPL (2005), 1–30.

[37]  W.E. Leland, M. Taqqu, W. Willinger and D. Wilson, *On the self-similar nature of ethernet traffic*, IEEE/ACM Transactions on Networking (1994), 1–15.

[38]  H. Lewis and C. Papadimitriou, *Elements of the Theory of Computation*, 2nd edn, Prentice Hall, New York (1997).

[39]  E. Lupu and M. Sloman, *Conflict analysis for management policies*, Proceedings of the V International Symposium on Integrated Network Management IM'97, Chapman and Hall (1997), 1–14.

[40]  J.F. Meyer, A. Movaghar and W.H. Sanders, *Stochastic activity networks: Structure, behavior and application*, Proceedings of the International Conference on Timed Petri Nets (1985), 106.

[41]  R.B. Myerson, *Game Theory: Analysis of Conflict*, Harvard University Press, Cambridge, MA (1991).

[42]  J.V. Neumann and O. Morgenstern, *Theory of Games and Economic Behaviour*, Princeton University Press, Princeton (1944).

[43]  M.E.J. Newman, S.H. Strogatz and D.J. Watts, *Random graphs with arbitrary degree distributions and their applications*, Physical Review E **64** (2001), 026118.

[44]  R. Ortalo, *A flexible method for information system security policy specifications*, Lecture Notes in Comput. Sci., Vol. 1485, Springer-Verlag, Berlin (1998), 67–85.

[45]  J. Parrow, *An introduction to the $\pi$-calculus*, The Handbook of Process Algebra, Elsevier, Amsterdam (2001), 479.

[46]  V. Paxson and S. Floyd, *Wide area traffic: The failure of Poisson modelling*, IEEE/ACM Transactions on Networking **3** (3) (1995), 226.

[47]  H. Prakken and M. Sergot, *Dyadic deontic logic and contrary-to-duty obligations*, Defeasible Deontic Logic: Essays in Nonmonotonic Normative Reasoning, Synthese Library, Vol. 263, Kluwer Academic Publisher (1997).

[48]  A. Rapoport, *Two-Person Game Theory*, Dover, New York (1966).

[49]  C.E. Shannon and W. Weaver, *The Mathematical Theory of Communication*, University of Illinois Press, Urbana (1949).

[50]  J. Strassner, *Police Based Management, Solutions for the Next Generation*, Morgan Kaufmann/Elsevier (2004).

[51]  J. Strassner, *A model driven architecture for telecommunications systems using den-ng*, Proceedings of the 2nd International Conference on E-Business and Telecommunication Networks (ICETE) (2004), 118–128.

[52]  J.A. Sykes and P. Gupta, *UML modeling support for early reuse decisions in component-based development*, Unified Modeling Language: Systems Analysis, Design and Development Issues, K. Siau and T. Halpin, eds, Idea Publishing Group (2001), 75–88.

[53]  TM forum, http://www.tmforum.org (2005).

[54]  J. Topping, *Errors of Observation and Their Treatment*, Chapman and Hall (1972).

[55]  S. Traugott and J. Huddleston, *Bootstrapping an infrastructure*, Proceedings of the Twelth Systems Administration Conference (LISA XII), USENIX Association, Berkeley, CA (1998), 181.

[56]  U.S. Nuclear Regulatory Commission NRC, *Fault Tree Handbook*, NUREG-0492, Springfield (1981).

[57]  J. Walrand, *Stochastic Models*, Handbooks in Operations Research and Management Science, Vol. 2, Elsevier (1990), Chapter: Queueing Networks.

[58]  W. Willinger and V. Paxson, *Where mathematics meets the internet*, Notices of the Am. Math. Soc. **45** (8) (1998), 961.

[59]  W. Willinger, V. Paxson and M.S. Taqqu, *Self-similarity and heavy tails: Structural modelling of network traffic*, A Practical Guide to Heavy Tails: Statistical Techniques and Applications (1996), 27–53.

# – 6.3 –

# System Administration and Micro-Economic Modelling

University College Oslo, Faculty of Engineering, Room 707, Cort Adelers Gate 30, NO-0254 Oslo, Norway
E-mail: mark.burgess@iu.hio.no

## 1. Introduction

The integration of computer technology into a business process is an obvious and desirable goal for an organization or company that relies on information services. In this chapter we ask: is it possible to understand business modelling and analytical system administration in the same light? In other words, how do we bring a rational scientific method to bear on the problem of building a business around an Information Technology (IT) infrastructure, or building an IT infrastructure for an existing business? Perhaps surprisingly, this overlaps only marginally with the stories told in economics.

The human component of both businesses and human–computer systems requires us to consider both rational decision making (optimization based on objective criteria) and intangible choices (arbitrary choices made as policy for reasons that are not part of a scientific framework, e.g. ethics or image) and choices to which no quantifiable value can be assigned. Some choices are made as a matter of policy for reasons that are difficult to view rationally. For example:

- Standards of 'best practice'. Why are they best?
- Modelling a business process as an information system.
- Logistics of production and inventory.
- Identification of relevant scales of operation.
- Capacity planning.

HANDBOOK OF NETWORK AND SYSTEM ADMINISTRATION
Edited by Jan Bergstra and Mark Burgess
© 2007 Elsevier B.V. All rights reserved

- Time management.
- Strategic thinking: activity patterns like type I models, and decision planning like type II models (see the previous chapter in the volume).
- Gauging and minimizing risks and uncertainties.
- Finding policies as stable equilibria.

One strategy for integrating business thinking into network and system management is to view activities in terms of services, rather than as low-level protocol interactions. This has become a popular paradigm since the ITIL/BS15000 [12] and now ISO20000 language of service provision and Service Level Agreements etc. See also the work in the Telemanagement Forum with eTOM and DEN-ng [62,63].

Although the individual details of internal procedures are important, there is a more over-arching need to understand the logistics of combining such procedures in the environment of an economic motor. Businesses are not just part of a slavish service chain, or factory production line: businesses are creative enterprises that also use information technology to innovate and generate new ideas, like a creative palette. It is not necessarily clear that a service model for aligning the goals of the business with IT infrastructure can express this creative need. We wish to turn some of these loose ideas into more rigorous ideas so that one can apply the techniques of scientific modelling and rational decision making to them.

Not all models that are related to aspects of a business are 'business models' in the fuller sense of the word. There are good reasons to separate off specialized aspects of the whole and consider them separately, just as one might in software modelling. A business model claims to supply the raison d'être of a business, along with its capacity for long term profitability. Like a computer system, a business is a box surrounded by an environment with which it is interacting [39].

Modelling something as complex as a business in full is probably not a practical proposition, using today's technology. Perhaps in the future one might model businesses as one today models the weather, but we are some way off having this technology today. A key difference with computer management is that businesses tend to seek growth rather than stability. We shall also looks at inventory models, efficiency models and so on, which pertain to smaller and more specialized parts of the whole.

The plan for this article is as follows: We begin by discussing the goals of businesses, in the sense of a system with a definable purpose or function. As with traditional scientific thinking, we do not hope to capture every nuance, rather we expect to capture a suitably idealized approximation of a business and its operations, in the hope of learning about its main features and mechanism in a general way. In the usual scientific way, we must therefore decide or discover the appropriate scales for observation and measurement. Once a simple model has been identified, it can be embellished in order to approach a high level of realism.

Much of business modelling involves human issues and procedures. We review some of the standards for increasing predictability in human (manual) operations, which are dealt with in other chapters. The essential autonomy of businesses is key to understanding how they interact, so we review some of the work on contracts and agreements.

Businesses are also stochastic systems, i.e. they are too complex to be modelled in every aspect, and they are embedded in an environment whose behavior we cannot fully predict.

We must therefore model the risk, or probability, of achieving the goals we set for the system, as well as an estimation of uncertainties. Key points are:

- Modelling business as a human–computer system.
- Identifying the principle for optimization.
- Managing internals, such as inventory and configuration.
- Planning volume services and controlling uncertainties.

## 2. Heuristic goals of business

Whether explicitly defined or not, every business has a *business model* that defines the services, costs and strategies that allows it to survive in a competitive environment. These strategies include both service roles and innovation.

Since businesses are pragmatic rather than idealistic enterprises, it is commonly accepted that there must exist a *business reason* for any change – i.e. what *value* does an activity offer a business? Ideas, standards and innovations must be perceived as being valuable to future profits in order to be adopted. There are many examples of good ideas that fell by the wayside because businesses were unimpressed. Value, however, is not a one-dimensional quantity: it could mean money, reputation, potential for growth or any other quantifier which translates into the belief of future profit. Profit itself is a somewhat unpredictable matter: as the fluctuations of markets in global stock exchanges attest, money does not have a fixed value: at any given moment, it is worth exactly what the concensus of the markets *believes* it is worth.

Service delivery is a key part of a business model. IT services are increasingly a part of this, but the principles for IT service provision are similar to those for the provision of any service. Today, one is seeing a paradigm shift in describing systems, towards the *Service Oriented Architecture* (SOA). This is a way of composing systems through interacting services, or service agreements. It is an architecture that is supported by the best practices scheme ITIL [12], and has found wide acceptance already. We shall focus particularly on the service oriented viewpoint for this reason.

GOAL 1 (Service design). *A business must define the scope and quality of its services, ensure that they are deliverable and manageable, at the right cost.*

How we define the 'right cost' for a service is a matter of some subtlety, since a business that is functioning well can operate with several kinds of currency. Hard monetary value is only one aspect of a business: reputation and goodwill are other kinds of service, which some providers of web services make use of. For example, a web site which performs a community service for free could attract attention and be regarded as good advertising for a sponsor. A business model has to account for such hidden forms of currency by relating the intangible value to a tangible one (see for instance ref. [17]).

It is normal to speak of businesses today as being service-based. The service model is a good model of interacting autonomous enterprises [18]. This does not mean that supply or manufacturing industries are excluded, only that one should imagine supply businesses as providing a service to their customers. Thus the service paradigm is a convenient umbrella for business concepts.

GOAL 2 (Capacity management). *What rates of service delivery are required and how can they be achieved with a planned infrastructure?*

Capacity management must deal with the realities of uncertain supply and demand. Should one plan excess capacity, or try to adapt capacity requirements to suit the current market?

From a service perspective, capacity management can be made more predictable by making an agreement or 'treaty' between the autonomous parties. The aim of an agreement is to declare the intention to behave in a predictable way, while at the same time defining what will happen if one's promised behavior is not achieved.

- A Service Level Objective (SLO) is the level of ambition that a provider aspires to provide to a customer.
- A Service Level Agreement (SLA) is a formalized agreement, i.e. a contract, between a provider and a client. It is generally a collection of Service Level Objectives with additional embellishments such as contingency plans for failure to provide the service.

GOAL 3 (Business relationships). *What is the structure of the organization with respect to suppliers and clients?*

From a functional point of view, any organization is somewhat like a computer program, with subroutines and closed objects that communicate through some form of protocol. Outsourcing is a form of subroutine must also be taken into account in a business plan. Outsourcing means that one buys a required sub-service from a third party, rather than providing the service in-house. This could lead to an additional level of management complexity, and even a higher level of uncertainty in the worst case. Such relationships are often critical dependencies for a business, so their failure to deliver could be a complete show-stopper. We must therefore model failures.

GOAL 4 (Risk and security management). *What are the risks of the business failure, either as a result of IT or other disasters, and what is the cost of securing against them?*

Security and reliability go hand in hand. Security is an intangible that is based on reliability with respect to policy. To achieve 'security' one must first have a security policy (defining what the individual means by security) and then apply counter-measures that eliminate the uncertainties surrounding these issues. One can only say one has a secure system when every uncertainty is either eliminated or defined as being 'acceptable risk'.

Systems are generally of two types: deterministic and non-deterministic. Another way of expressing these alternatives is to speak of reactive 'event processing' or scheduled 'batch processing'. The main difference between these two methods is where the uncertainty lies. In batch processing one trades uncertainty in resource requirements for uncertainty in response time; i.e. we do not guarantee to respond 'immediately' (e.g. within 1 millisecond), but we know that we can do it by the end of the day, working in shifts. There is a thus variety of models for describing how to achieve objectives according to the logistics and quality of different approaches.

GOAL 5 (Logistics of production). *What physical and resource constraints and are there to providing the service?*

Quality control is a subject that applies identically to system administration and business administration. It requires a proper understanding of the operation of a system:

- A risk model for the system is required, so that one identifies the most likely threats to the success of the business or system (e.g. what happens if there is illness amongst key personnel, or a major power failure or natural distaster such as a flood?). Identifying these threats is important for the next point.
- Completeness of procedures: a response procedure should exist and be implementable for each event that can occur within the system. Responses to events should be mandatory and be executed within a minimum time-frame, determined by the scales identified above.
- Monitoring of performance: a non-intrusive method of monitoring performance should be available to all levels of the system or business, so that a natural feedback, with a minimum of overhead can be provided. System or Business Performance Indicators should address both internal procedures and external factors, which influence the markets for a business or the environment in which the activity of the system takes place.
- Securing security and efficiency in performance trough redundancy, load balancing, resource management etc.

Finally there are issues that are not so easily described in scientific terms, and hence we call them 'irrational' or unverifiable.

GOAL 6 (Arbitrary business policy). *Decisions which cannot be made using a rational method of decision e.g. ethical or moral prerogatives that are not easily quantified or verified.*

## 3. De facto standards for human-centric procedures

Industry standards exist for setting basic quality assurances for IT business [11,12]. The standards are often presented as existing in order to protect business relationships such as outsourcing, in which a company isolates and gives autonomy to a part of its operation. Standards cover issues such as:

- Management's role.
- Documentation requirements.
- Competence acquisition (training and hiring).
- Configuration management.
- Location and scope.
- Monitoring and review.
- Service Level Agreements with clients.
- Complaints procedures: Fault recovery, Incident management, Problem resolution.
- Budgeting and inventory.

- Business relationships: dependencies (outsourcing).
- Legal constraints.

The IT Infrastructure Library (ITIL) was developed by the Central Computer and Telecommunications Agency of the British government and has been adopted by many companies. It attempts to provide a 'cohesive set of best practices' for businesses and organizations. A British Standard which summarizes the issues in ITIL is BS15000 [12] (see the contribution in this volume by Scheffel). The IT Service Management Forum (ITSMF) is now the custodian of the standard [46].

The enhanced Telecom Operations Map (eTOM) is claimed to be the most widely used and accepted standard for business process in the telecom industry. The eTOM is analogous to and overlaps with ITIL and describes business processes required by a service provider as well as how they interact. eTOM provides more of a top-down hierarchical view of an enterprise whose aim is to provide service to external customers. There is considerable overlap between eTOM and ITIL, as is testified by the eTOM/ITIL interpreters guide [45, 64], but the eTOM vision is motivated very much by the traditional telecommunications hierarchy. It attempts to describe an organization as a database. The reader is referred to the chapter by Brenner et al., in this volume for more details on these matters.

Management frameworks have been discussed by several corporate research milieux [4, 42]. The authors emphasize the difference between IT Management (service level management) and IT Governance (strategic planning and integration of IT in the business process). Information models for business interactions have also been discussed [58,62].

## 4. System, environment and scales

In the previous chapter in this volume [20], systems are described in terms of their degrees of freedom and their constraints. One should look for the freedoms, constraints and natural scales of measurement that are involved in the operation of a system.

### 4.1. *Freedoms and constraints*

There is a plethora of ways one might model freedoms and constraints in the context of a business model. Needs analyzes can tell us about the kinds of variables at out disposal. The activities of customers and internal parts of the business are another way of identifying freedoms. Essentially the freedoms are the resources that are available to the enterprise, either directly or indirectly (e.g. by purchase or outsourcing).

Constraints on an enterprise include debts to be paid, sustainability or minimum profitability, physical and market limitations, equipment and processing etc. 'Use cases', in the sense of software engineering, can be used to represent individual issues within the whole business. A quantitative model must show how freedoms and constraints should be balanced.

Most managers would like to live in a world that was clockwork and deterministic, as this would allow the power to control, however this is far from the case. The identification of random processes is important because this signals the extent to which the larger environment impinges on business activities. Random processes are essentially processes which cannot be fully understood without a complete model of the 'rest of the world'. Since such a model is impractical, one speaks of randomness, probabilities and expectations.

Event modelling is crucial in models and lead us to consider queueing systems, where service requests arrive as a stream of discrete 'orders' or units. Arrival and renewal processes are then needed to understand and optimize the strategies for dealing with demand [28,68].

## 4.2. *Scales*

An identification of approximate numbers is the first step towards determining whether a business can sustain itself. The basic scales or numbers should be the average values, since most processes in a real world environment are stochastic in nature. Moreover, whenever one deals with averages, some attention should be given to the distribution of values about the mean.

For example: what is the approximate demand for a service over a fiscal year? What is the rate of arrival of payments at different times of year? What are the estimated costs of running the business, e.g. the price per kilowatt hour of electricity, the rent of your data centre, wages for staff? What is the typical size and monetary value of a sale? How long does it take to complete and fulfill an order? How much profit can one make per order transaction? Given that short-term losses can be offset by long-term profits, over what time-scale could one afford to make a loss? Give that there might be inhomogeneous demand and supply, over what time scale are the prices of services and resources approximately constant?

Comparing time scales is particularly useful way of seeing whether one has adequate resources to carry out a task. An organization that cannot deliver on its goals will not survive long.

If one thinks in terms of queues and arrival processes, we can pose the questions about service processing in a suitably analytical way: what is the time scale of trends and fluctuations in the arrival of service requests [9]? How large are fluctuations about the mean values and what is the scatter and jitter in them?

What total number of transactions can one expect in an interval of time e.g. a year? (A common model of sparse events is the Poisson arrival model, but this is known to work only for very large numbers of sparse arrivals, within the measurement interval, in general.)

Finally, from queue theory one ought to be able to estimate, what is the average response time of a business to a customer, or a server to a client?

In these questions, we already see that the same issues are directly applicable in both businesses and computer systems. This should not be a surprise, since the operation of a business becomes essentially machine-like once it has been defined.

## 4.3. *Risk modelling*

Fault tree analysis [1,44,61,66] is a way of modelling the deviations from goals. It is a kind of anomaly detection and associated response. Anomalies are categorized into a tree of likely occurrences, and the overall risk can be evaluated as a probability. After this, the result must be translated into loss of revenues etc., by embedding the probable failure in the wider context of the income/cost model. What are the potential risks to loss of income or reputation? What is the contingency plan for recovery? The security management code of practice addresses this issue [11].

## 5. Business models

There is little written about the modelling of businesses, possibly because there is no con-cencus of agreement about what constitutes a business model. In refs. [27,38,39], the authors suggest the following as a basis for a business plan.
1. A value proposition.
2. Customer and market segment documentation.
3. Cost structure and potential for profit.
4. Interval value chains.
5. Positioning with a value network.
6. A strategy for positioning and competition.

In a value proposition one describes exactly what portfolio of products or services a business will deliver. A product description is required and a statement of the problem solved, or need quenched. The market segment addresses which part of a potential market-place is to be covered by the proposal, e.g. old or young customers, large or small businesses etc. This kind of profiling is important for estimating the potential for profit, as well as the possible risk of failure.

The economics of planning a competitive business are potentially complex. In particular, one must be clear about the role of the product in relation to existing competition.

- Goods and services $S_1$ and $S_2$ are said to be *substitutes* if a price increase for $S_1$ leads to increased demand for $S_2$, i.e. if one replaceable service becomes more competitive.
- Goods and services $S_1$ and $S_2$ are said to be *complements* if a price increase in $S_1$ leads to a decrease in demand for related services $S_2$. In other words. an increase in price reduces the demand for a service and its supporting services.

The estimates of potential failure in a business are thus dependent on the network of products and dependent services surrounding the product. If one is entirely in the business of performing outsourced work, then a decrease in demand for the primary service would mean a corresponding loss for one's own company.

The estimate of 'added value' is also a complex matter, subject to fickle re-evaluations by customers and markets. Exactly how much a customer is willing to pay for a service depends on psychological as well as economic and marketing skills. Some products belong together, e.g. operating systems and computers, and the promise of one implies a need for the other. In general whole networks of relationships are involved in the web of commerce.

The potential for profit is a matter for educated guesswork. There is always a risk of loss. Customers might not pay their bills on time (or at all, if they go bankrupt). Businesses require a margin for error, and an estimation of the time-scales over which they can survive losses.

Finally, a competitive strategy will put all of these elements together and position a service or product in a market-place with *survivability* as a key concern.

No company can afford to stand still. Businesses need to turn some of their profits into development and other investments in a future. This is a part of long-term planning (again a matter of separating time-scales). Sufficient margins have to be allowed for this.

## 6. Service relationships

The service oriented view of businesses and IT systems is, at the time of writing, very much in vogue [10,48]. The service oriented approach is both technically challenging and intellectually interesting, since it is a model that naturally supports *voluntary cooperation* [22,23]. The Information Technology Infrastructure Library is a set of heuristics or 'best practices' for standardizing service-oriented operations.

A service-oriented architecture is interesting because it emphasizes the *autonomy* of the providers; hence it is capable of modelling issues like inter-company dependency (outsourcing) and free corporate collaboration in a market economy. Modelling such scenarios has been problematic until recently, however *Promise theory* offers a natural framework for describing the Service-Oriented Architecture, amongst other things [18]. Promise theory is a graphical approach to constraint planning and solving that lends itself to a service viewpoint. We shall consider only the briefest of introductions here.

### 6.1. *Promise theory*

Suppose $\Gamma = \langle N, L \rangle$ is a graph, where $N$ is a collection of nodes or vertices and $L$ is a collection of links or edges between them. The nodes represent autonomous agents, which are the recipients and providers of certain *promises*. A promise is a labelled edge in the graph from one node to another. Clearly a service from one node to another is one kind of promise that can be made. One writes

$$n_1 \xrightarrow{\pi} n_2, \tag{1}$$

meaning that node $n_1$ promises $\pi$ to node $n_2$. The notation can simply be drawn as a labelled arrow in the graph (see Figure 1). The meaning of the promise body $\pi$ is to be specified. It belongs to a collection of possible types of promise. A service promise could be of type 'World Wide Web' or 'Simple Network Management Protocol' for instance, and include all configurable aspects of these services as part of the promise body. In addition to a *type*, the promise body therefore contains a constraint, which indicates exactly what is being promised, out of the realm of possibilities for the given type. A constraint is somewhat like a Service Level Agreement then: e.g. a promise of World Wide Web Service could involve a response time. The promise body could then include a desired distribution of response times from $n_1$ to $n_2$, such as '90% of requests will be serviced in less than 5 milliseconds'.

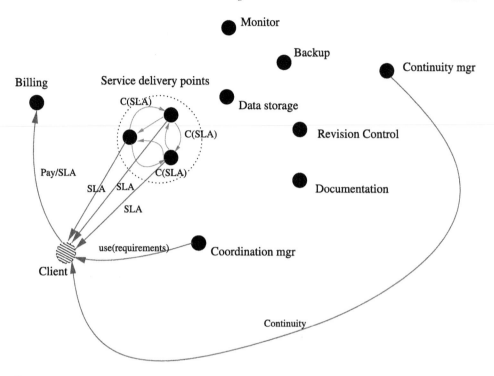

Fig. 1. An example promise graph with labelled edges representing typed promises in an ITIL like environment.

Promise theory can be applied at many levels. Even at the simplest level, it provides a graphical documentation about the services provided between different entities within a collaborative network. At a deeper level, one can derive properties such as *roles*. For example, if many agents make the same kind of promise then they naturally play the same kind of functional role in the graph [22]. Promise theory hence admits techniques for identifying roles and functional entities within the graph. It also ties services and other promises to the notion of a two-person bargaining game [52].

### 6.2. *Agreements, contracts and relationships*

A contract is a tool for both establishing cooperative goals between independent parties or agents, and for imposing penalties when goals are not met. They are a method of codifying voluntary cooperation between peers in a social network [2] to form cooperative relationships and communities. Service Level Agreements are examples of contracts for service provision between IT peers.

Determination and optimization of contracts mixes both human and technological issues [30,56,59]. In particular, the terms of an agreement can be quite sensitive to the human and technological context in which they are made. A great variety of automated technologies can be involved in automated contract exchange, motivating the need for a *lingua franca* for agreement. This is often achieved with *ontologies* (see the chapter by Strassner in this volume).

Deontic logic, the logic of obligations and counter-obligations, has been used as a framework for making the human language of contracts precise, often in a legal context [31]. In ref. [58] contracts are used to define penalty costs for failing to live up to their obligations. This is a common theme in economic models of promise breaking [26] (see below). A continuum study of this kind of contractual penalty feedback between a pair of peers has been examined in ref. [5]. There the authors noted the potential instabilities in relationships between agents occurs when they were allowed to impose punishments and penalties on one another.

Community policy is often formed through agreements. It can be thought of as a kind of 'law making' either by mutual cooperation of by external law-giver. The Law Governed Interaction project [50,51] considers, for example, how policy seeds a community through the conformance of its participant with a global standard. The external law giver can, on the other hand, be seen as a special case of concensus through voluntary cooperation. In promise theory [18,22,23], which has qualitative similarities to 'commitments' in multi-agent theory [36,69], one assumes that every agent decides for itself how it should behave. Stable communities can then occur spontaneously if the conditions for voluntary cooperation are sufficiently conducive. Here one considers the stability and consistency of a series of agreements in a network, to see whether the sum of individual promises can lead to a consistent union of ideas.

Consider two promises, back to back as in Figure 2. This diagram represents a peering agreement (e.g. such as one might expect in a Border Gateway Protocol (BGP) configuration between two service providers). Alternatively, it could represent promise of service for the promise of payment. Promise types should generally not overlap, so as to avoid the possibility of contradictory or broken promises.

Typed promises form disjointed graphs, but bargaining agreements clearly motivate the unification of promises using a kind of *common currency* or valuation of the promises. One therefore arrives at a natural unification between a service viewpoint and an underlying economics which can motivate the *reliability* or *duration* of promises over time [23]. In promise theory, a *commitment* is a promise that requires an irreversible investment of resources, i.e. it has a non-returnable cost associated with it.

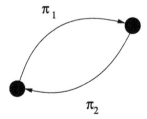

Fig. 2. A trade of promises behaves like a two-person bargaining game.

### 6.3. Contract roles

A business that provides a service makes promises to various parties. These promises are usually made in the form of legally enforceable contracts, but the success or failure of such contracts has little to do with the fact that they are legally binding. Both the provider and the recipient of a service have an interest in maximizing their gain with regard to a business relationship and this is the primus motor for successful contracts.

What is interesting in such a situation is that it is not always a symmetrical exchange. One of the parties is usually in possession of an informational advantage over the other, and who makes their move first can completely change the optimal outcome for the parties. There is a risk involved in making a judgment.

### 6.4. Contract selection and incentives

In economic game theory, the *Principal Agent Model* is a basic model for highlighting the issues surrounding voluntary cooperation between agents making promises to one another [13,55]. It is a simple (indeed simplistic) model of a contract negotiation. The basic idea is that a *principal* (agent) wants to outsource a task to one or more (service) agents, and that the service agents fall into a variety of different types that indicate their relative costs and abilities.

Some agents, for instance, might be better qualified than others and so demand higher wages, while others might be able to do the same job more cheaply. An obvious case is the trend of outsourcing to Asia in which agent types represent the living standards can costs of their respective countries. One has expensive Western agents, and cheap Eastern agents, both of whom are capable of doing the same job. This leads to a decision dilemma for the principle. If the principle offers too little for the task and no cheap agents are available, the work will not be done. If the principal offers a higher rate, then even the cheap agents will pretend to be expensive agents in order to make a higher profit.

Thus, if the principal makes its promise of payment before the agent agrees to perform work, then the principal has a disadvantage. This, of course, is a normal state of affairs. So what can the principal do to try to coerce agents to provide their service for an optimal price?

Information is key in voluntary relationships because there are decisions to be made by each agent (as promise theory predicts). An Asian company could easily pose as a Western company in order to receive a better price for their services. Why would they not do this, given the chance? Economic game theory has addressed this question, in what has now become contract theory.

Consider a principal $P$ and two agents $A_1$ and $A_2$ of different types $\theta_1$ and $\theta_2$, where $\theta_2$ is a type of agent with a high price and high overheads, and $\theta_1$ is a low price, low overhead supplier type.

There are two possible scenarios, depending on who starts the negotiation (see Figure 3). In scenario (a), the principal first promises separate deals for the two agents without knowing their types, then the agents can accept or refuse. In scenario (b), the agents first reveal

Fig. 3. The principal agent model's bilateral promises.

their true types and then the principal offers them a deal accordingly. The dotted lines denote conditional promise responses.

In (a) the principal is at a disadvantage, since the agents have not revealed any information about themselves. In (b), the agents are at a disadvantage since the principal can offer them the minimum rate at which it knows they will work (this is called the first best solution for the principal). In most cases, a client (principal) is in the position of not knowing an agent's overheads, so (a) is the likely scenario. We can analyze (a) further by making a general promise scenario.

The potential promises made by the principal are about wages (money) $m_1, m_2$ to the agents, whereas the potential promises made by the agents are about 'effort' $s$ or the agreed Service Level for the principal. In a contract situation, both agents will make promises conditional on what the other delivers:

$$P \xrightarrow{m_i/s_i} A_i,$$

$$A_i \xrightarrow{s_i/m_i} P,$$

(2)

where $i = 1, 2$. Let now the value of a promise to an agent is denoted by $v_A$. Then the economic value of the these promises may be written:

$$v_P\left(P \xrightarrow{m_i/s_i} A_i\right) \mapsto -m_i \quad \text{if } s_i,$$

$$v_A\left(P \xrightarrow{m_i/s_i} A_i\right) \mapsto +m_i \quad \text{if } s_i,$$

$$v_P\left(A \xrightarrow{s_i/m_i} P\right) \mapsto +s_i \quad \text{if } m_i,$$

$$v_A\left(A \xrightarrow{s_i/m_i} P\right) \mapsto -C(s_i).$$

(3)

Here $C(s_i)$ represents the cost of exerting the effort or service level $s_i$ to the agent, i.e. this promise is in fact a commitment once made. The cost is normally represented by some convex polynomial function of $s_i$ such that greater effort leads to a more than proportionally high cost (this is the opposite of economies of scale), i.e. it hurts to work hard, so the agent want to be paid much more for harder work than for light work. For definiteness, we define the simplest quadratic form for this exposition:

$$C_i(s_i) = \frac{1}{2}\theta_i s_i^2.$$

(4)

Notice that the cost is proportional to the type or 'overhead' parameter, which might be real or fake.

The valuations of each agent's promises (given and received) now provide us with an optimization problem, which each autonomous agent valuates in its own terms. This is a so-called *adverse-selection* game whose outcome essentially decides whether the proposed promises will be made [23].

The total payoff for the principal is the sum of its promise valuations. It amounts to the difference between the effort received from the two agents and the wages paid to them:

$$\pi_{\text{principal}} = \sum_{\text{promises}} v_P = \sum_{i=1}^{2}(s_i - m_i).$$

(5)

The vector of values $\vec{m}$ with components $m_i$ represents a *contract specification* in relation to the promises above [13,55]. The payoff to the body of contracting agents is

$$\pi_{\text{agent}-i} = \sum_{\text{promises}} v_A = m_i - C(s_i),$$

(6)

i.e. the wages received minuses the cost of expending the service level effort.

The problem for each autonomous agent, including the principal, is to maximize its own payoff in this exchange. Agents are normally thought of as being risk-averse. That is, they choose to avoid options that could reduce their payoff. Incentives can help the principal convince the agent of a low risk (when the agent does not mind bearing all of the risk of a transaction it is said that the problem is solved by 'selling the firm', i.e. the principal no longer needs to be involved). The solutions of this simple model offer a basic insight into incentive management.

**6.4.1.** *'First best' solution*   In the first best case, the agents announce their types and make clear that they will only accept a contract that will not give them a negative payoff:

$$m_i - C_i(s_i) \geqslant 0.$$

(7)

This is known as the *participation constraint*. However, since the principal knows the agent's overheads it can simply choose $m_i$ to render this an equality, thus bleeding all of the agents' profit with a minimum wage. This kind of price discrimination is normally forbidden by law – a principal is not allowed to offer a different price for the same deal to different agents. It is however a possible solution to the problem.

Meanwhile the principal maximizes, subject to the agent's constraint for participating. The Lagrangian for this is

$$L = \sum_i (s_i - m_i) + \alpha_i \left( m_i - \frac{1}{2}\theta_i s_i^2 \right), \tag{8}$$

where $\alpha_i$ are Lagrange multipliers. Solving:

$$\frac{\partial L}{\partial m_i} = -1 + \alpha_i = 0,$$

$$\left. \frac{\partial L}{\partial s_i} \right|_{e^*} = 1 - \alpha_i \theta_i s_i = 0 \tag{9}$$

i.e. $s_i^* = 1/\theta_i$. This gives the level of effort that the agent is bullied into providing, for minimum wage $m_i^* = 1/(2\theta_i)$.

This solution assumes that the agents have told the truth about their overheads, or that their true costs have somehow been determined by the principal. It is easy to see however that if the principal simply guesses a contract without knowing the real agent types, it can be tricked. A cheap agent can claim an expensive contract, giving it wages minus its cheaper costs at the lower effort level, giving it a non-zero profit:

$$\pi_1 = m_2^* - \frac{1}{2}\theta_1 s_2^*$$

$$= \frac{1}{2\theta_2} \left( 1 - \frac{\theta_1}{\theta_2} \right) > 0. \tag{10}$$

Ironically the contract $m_i$ awards the higher wages to the cheaper agent, who also gets the bulk of the work if work is to be divided amongst several agents of the same type. This solution places the principal in the position of advantage over the contractors. Of course, the solution assumes that agents of the appropriate types exist and are available for carrying out this work. A principle might have to cover a task by employing several agents of different types in practice. In that case, they would be optimally assigned according to the proportions above.

**6.4.2.** *Second best solution – adverse selection*   In the second best solution, the agents do not reveal their abilities or costs to the principal up front. It is up to the principal to prepare a contract to a general mass of agents who might be of different types. We can assume that the principal has done some market research so it knows the different types $\theta_i$ and it knows roughly the probabilities $p_i$ which they occur. We can now show that it is possible for the principal to come up with a contract that dissuades the cheaper agents from pretending to have high overheads so as to trick the principal into offering the higher wage. The strategy is to use the right incentives. Thus we must add an additional *incentive constraint* with the principal, which says that an agent will obtain the maximum payoff by not pretending to be of a different type, i.e. the principal requires that

$$m_1 - C_1(s_1) \geqslant m_2 - C_1(s_2),$$
$$m_2 - C_2(s_2) \geqslant m_1 - C_2(s_1). \tag{11}$$

This constraint can be satisfied. Now, in this scenario, there might be many agents bidding for work, and the principal knows only the probable distribution of these $p_i$, thus it can now compute its *expected payoff*:

$$\langle \pi_{\text{principal}} \rangle = \sum_i p_i(s_i - m_i), \tag{12}$$

subject to the constraints above and the assumed participation constraint from before $m_i \geqslant C_i(s_i)$.

Let us solve this for two types $\theta_1$ and $\theta_2$. To determine the contract offer $(m_1, m_2)$, the principle has sufficient information from its market research to maximize Equation (12) subject to the following constraints:

$$m_1 \geqslant \frac{1}{2}\theta_1 s_1^2, \tag{13}$$

$$m_1 - \frac{1}{2}\theta_1 s_1^2 \geqslant m_2 - \frac{1}{2}\theta_1 s_2^2, \tag{14}$$

$$m_2 \geqslant \frac{1}{2}\theta_2 s_2^2, \tag{15}$$

$$m_2 - \frac{1}{2}\theta_2 s_2^2 \geqslant m_1 - \frac{1}{2}\theta_2 s_1^1, \tag{16}$$

$$\theta_2 > \theta_1, \tag{17}$$

$$p_1 + p_2 = 1. \tag{18}$$

If we add (14) and (16), then it follows that $s_2 \geqslant s_1$, i.e. the expensive agents must work at least as hard as the cheaper agents in an optimal solution. Moreover, it is easy to see that not all of the constraints can be independent. Substituting Equation (13) into (16), we get

$$m_2 \geqslant \frac{1}{2}\theta_2(s_2^2 - s_1^2) + \frac{1}{2}\theta_1 s_1^2. \tag{19}$$

Thus we may maximize the Lagrangian:

$$L = p_1(s_1 - m_1) + p_2(s_2 - m_2) + \alpha\left(m_1 - \frac{1}{2}\theta_1(s_1^2 - s_2^2) - m_2\right)$$
$$+ \beta\left(m_2 - \frac{1}{2}\theta_2 s_2^2\right). \tag{20}$$

The new equations are therefore:

$$\frac{\partial L}{\partial m_1} = -p_1 + \alpha = 0, \tag{21}$$

$$\frac{\partial L}{\partial m_2} = -p_2 - \alpha + \beta = 0, \tag{22}$$

$$\frac{\partial L}{\partial s_1} = p_1 - \alpha\theta_1 s_1 = 0, \tag{23}$$

$$\frac{\partial L}{\partial s_2} = p_2 - \alpha\theta_1 s_2 - \beta\theta_2 s_2 = 0. \tag{24}$$

Substituting Equation (21) into (23) gives the optimum effort

$$s_1 = s_1^* = \frac{1}{\theta_1} \tag{25}$$

and substituting Equation (22) into (24) gives

$$s_2 = s_2^* = \frac{1}{\theta_2 + (p_1/p_2)(\theta_2 - \theta_1)} < \frac{1}{\theta_2}. \tag{26}$$

To find the contract rates $m_i$, we note that maximization of the Lagrangian implies minimization of $m_i$, so the constraint inequalities become 'shrink wrapped' to equalities in Equation (19) and (13), giving:

$$m_1 = \frac{1}{2\theta_1} + \frac{1}{2}(\theta_2 - \theta_1)\left(\theta_2 + \frac{p_1}{p_2}(\theta_2 - \theta_1)\right)^{-2}, \tag{27}$$

$$m_2 = \frac{1}{2}\theta_2\left(\theta_2 + \frac{p_1}{p_2}(\theta_2 - \theta_1)\right)^{-2}. \tag{28}$$

We note that $m_1 > m_2$, thus we pay the cheaper agents a surplus to prevent them from choosing the $\theta_2$ deal. This amount is just enough to make a rational agent choose the deal designed for it.

**6.4.3.** *Comments*    Although this principal agent model is simplistic, it makes the interesting point that optimizations are possible, even when incomplete information prevents a business from achieving a theoretical maximum. The model assumes however, from the underlying promise constraints, that wages will only be paid if the agent does the work it promise. The model is flawed in a number of ways. It assumes that work can be given in a continuum and that the required level of work can be assigned into bundles $s_i$ that are suitable for the available agents. There are many reasons why this simple exchange is too simple. The cost function $C_i(s_i)$ is taken to be convex as an incentive, but in fact many operations will have an opposite experience: economies of scale will allow the agents to have concave costs.

## 7. Business as a network service

The most important concept in a sustainable business process is that the value created by the process should be greater than or equal to the cost of operation.

### 7.1. *A functional view of services*

In a simplistic sense, a business model is something that explains the relationship between input and output. A business is a machine for generating wealth and produce (output), given a certain amount of investment and raw materials (input). The question is how it can be made self-sustaining (see Figure 4).

How can such a system be made? Let us strip the matter down to the most simplistic case. The following is not meant to be a realistic business model, but a realistic business model must contain the elements of this simplistic example.[1]

A self-sustaining process must have resources to begin with, and must produce more than it consumes. This is a peculiar idea, which seems to fly in the face of scientific reason and which readers may wish to ponder at length: how is it possible to sustain economic growth, and what is really being produced? Something from nothing? The answer, of course, comes down to understanding what value is, and why money is not a fixed scale of measurement, but rather a subjective convenience.

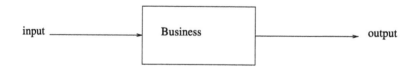

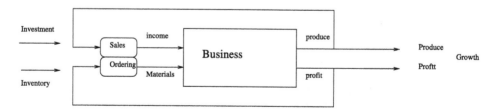

Fig. 4. A functional viewpoint of a business as a self-sustaining process.

---

[1] The strategy of science is to build on toy models. Here we present a model which a business practitioner would probably laugh at. But a scientist would scorn a real business model as being encumbered with irrelevant detail. He success of modelling is in finding the right balance between realism and clarity. Generally one starts with something that is easy to grasp.

Let us set aside these issues and view the monetary value just as an arbitrary scale by which to express input and output. Then our simplest representation of a company is:

$$\text{output} = f(\text{input}), \tag{29}$$

which for the business translates in simplistic terms into:

$$\text{profit} = BusinessValue(\text{income}). \tag{30}$$

In terms of supply and demand, one inputs demand and outputs supply. This basic viewpoint of the business as a black box function with input and output is simplistic but essentially correct. There are various interpretations of this (see Table 1).

A more detailed modelling can be achieved by opening up and filling in the black-boxes, just as in software engineering.

$$(\text{profit}, \text{produce}) = BusinessResult(\text{income}, \text{materials}). \tag{31}$$

Then we take part of the profit to buy materials and sell the produce; thus functionally, we have a prototypical equation can be written in functional terms:

$$(\text{profit}, \text{produce}) \geqslant BusinessResult\big(Income(\text{produce}), Materials(\text{profit})\big). \tag{32}$$

We hope that this two-dimensional mapping of produce and profit does not merely spiral to a fixed point of zero [7].

We do not normally control the input or output fully: it should be thought of as a random process [40]. Control factors such as advertising might influence expected demand at some predictable cost, but only within certain limits. Similarly production performance can be controlled by factors familiar in control theory [41], but only subject to basic limitations on certainty.

How shall we measure income? If the price of business services is constant then one could use either number of business transactions or the actual money paid into the business by external customers. But sometimes business give discounts for quantity etc., so there is a non-linear relationship between transactions and income.

Another issue is how should be coarse grain income? It has not traditionally been considered sensible to consider every individual micro-payment as a separate event. Rather one collects income over an interval such as a day, week or even month and considers the gross amount for that interval as the income. Similarly with payments, this coarse graining allows a certain smoothing of the dynamical processes involved, allowing temporary debts

Table 1

| Input | Output |
| --- | --- |
| Income | Payments |
| Materials | Production output |

to accumulate and be covered within a planning horizon. We can simplify this by writing with fewer assumptions.

$$BusinessResult(\text{transactions}) \geqslant Costs(\text{transactions}). \tag{33}$$

By including costs in the function on the RHS we can also allow for the fact that costs might not be linear in respect of production. Transactions is now a more or less pure variable that measures the demand.

What do we learn from such a simplistic model? Two things emerge from this view. The relationship between surplus and requirement is highlighted. There is an implicit queue involved, i.e. a buffering of workflow both with regard to income and inventory. Not all models will necessarily have two forms of currency, i.e. money and inventory, but in many cases there will be analogues of these separate entities. Thus we need an understanding of how these two currencies are interrelated.

A simple economic answer to what best practice means for business-oriented system administration has been given by Couch et al. [29]. These authors propose to minimize a cost function of the form:

1. Minimize (value − cost).
2. Subject to the condition that cost < budget.
3. Subject to the condition that value > SLA agreed value.

This can be associated with a business aligned strategy, if the concept of value includes all appropriate measures (including reputation etc.).

## 7.2. Deterministic and stochastic demand

Demand is a stochastic process – an arrival process. Indeed, this simple black-box model is a queue. We might imagine that the cost of the business process includes certain economies of scale and therefore falls off to some kind of asymptote with increasing demand. Similarly the result from a small number of sales might be small and rise to a maximum asymptote with increasing demand.

Using linear programming, one can see this on a simple planar diagram (see Figure 5). Any business enterprise needs to incorporate this basic framework. Time scales are important in this interpretation however (see Figure 6). It might be feasible to operate at a loss for short intervals of time, provided we make up for this loss later on.

The arrival of business opportunities is an essentially inhomogeneous random process which one should consider smoothing into trends to eliminate such fluctuations. By smoothing, or local averaging one can look at the trends and use these for seasonal planning [3].

## 7.3. Diagrammatic modelling

The understanding of a system is greatly facilitated by diagrams and illustrations. Humans are particularly good at comprehending scenarios where several senses are involved. Visualization is one of the most important tools for comprehension.

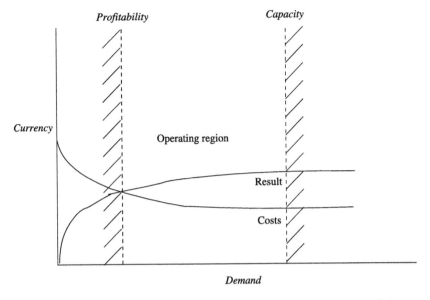

Fig. 5. The operating region of a business lies in between profitability and capacity limits.

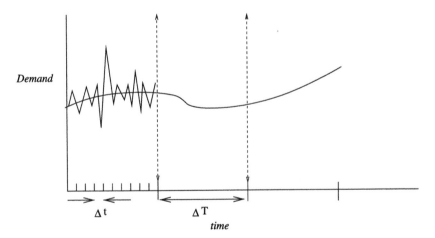

Fig. 6. A time series view of the demand. The unpredictability of demand makes all its derived quantities random processes.

Illustration generally begins by assembling the entities and relationships between them. There are many approaches to this, and one is unlikely to find a single diagram sufficient to describe what is happening in a business system. A more realistic expectation is to use a variety of diagrammatic techniques to model specific issues.

Two extreme design strategies for building systems have emerged, and have developed with increasing refinements and compromises.

- *Top-down*: The name 'top down' is motivated by the traditional way of drawing hierarchical structure, with high level (low detail) at the top, and increasing low-level detail at the bottom. A top-down analysis is *goal driven*: one proceeds by describing the goals of the system, and by systematically breaking these up into components, by a process of *functional decomposition* or *normalization*.
- *Bottom-up*: In a 'bottom up' design, one begins by building a 'library' of the components that are probably required in order to build the system. This approach is thus *driven by specialization*. One then tries to assemble the components into larger structures, like building blocks, and fit them to the solution of the problem. This approach is useful for building a solution from 'off the shelf', existing tools.

The difference between these strategies is often one of pragmatism. In general a mixture of these two extreme viewpoints is warranted. A top down design is usually only possible if one is starting with a blank slate. It might not be possible to implement one's wishes from a top-down viewpoint with an existing set of constraints and components.

A 'bottom up' design is a design based on existing constraints, namely the components or resources that are to hand. An advantage of building from the bottom up is that one solves each problem only once. In a top-down strategy one could conceivably encounter the same problem in different branches of the structure, and attempt to solve these instances independently. This could lead to inconsistent behavior. The process of 'normalization' [17] of a system is about eliminating such inconsistencies.

- (Computer system – 'top down'.) In a the design of a new computer system, one examines the problem to be solved for its users (banking system, accounts), then one finds software packages which solve these problems, then one chooses a platform on which to run the software (Windows, Macintosh, Unix), and finally one buys the and deploys the hardware that will run those systems.
- (Enterprise – 'top down'.) In the organization of a maintenance crew, one looks at the problems which exist and breaks these down into a number of independent tasks (plumbing, electrical, ventilation). Each of these tasks is broken down into independent tasks (diagnosis, repair) and finally individuals are assigned to these tasks.
- (Computer system – 'bottom up'.) First one buys reliable hardware, often with operating system already installed, and installs it for all the users; then one looks for a software package that will run on that system and installs that. Finally, the users are taught to use the system.
- (Enterprise – 'bottom up'.) The enterprise bosses look at everyone they have working for them and catalogues their skills. These are the basic components of the organization. They are then grouped into teams which can cooperate to solve problems like diagnosis and repair. Finally these teams are assigned tasks from the list of plumbing, electrical work and ventilation.

In system administration, especially configuration management, these two strategies are both widely used in different ways. One may either implement primitive tools (bottom up) that are designed to automatically satisfy the constraints of a system, or one can use trial and error to find a top down approach that satisfies the constraints.

An entity-relation diagram, with associated data model, is a way of understanding information *types* and *relationships*, such as the approach used in the DEN-ng model [62].

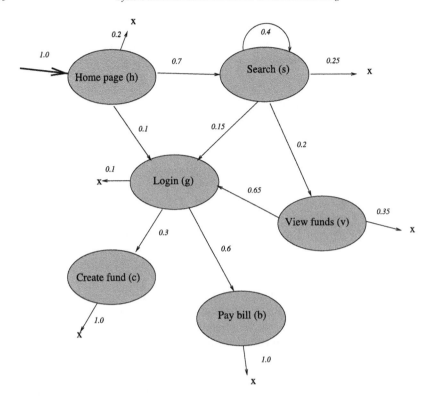

Fig. 7. State transition diagram for a web-based service. The transition probabilities weight the connections.

However, such a diagram will not tell us anything about the efficiency of flows of work between the entities: for that, one needs something like a queueing network [68].

Transition diagrams provide a way of mapping out the activities in a system, in order to apply graphical methods of optimization [49]. Strictly speaking, they involve the assumption of a Markov process. More complex scenarios can be modelled with other kinds of learning network [53,54].

This tells us how often the system spends its time the different states. By relating these states to the resources that are in use when the states are active, one can identify the most used resources during the interactive process. A centrality analysis of this graph can yield important information about resource allocation. For a deeper discussion of graph analyzes, see the contribution by Canright and Engø-Monsen in this volume [25].

### 7.4. *Continuum workflow models: scalability*

Flow models take over where modelling of individual transactions becomes intractable. Continuous flows of information, goods or services are a convenient fiction for applying probabilistic methods. One replaces actual measurements with expectation values and events with probabilities.

Flow models use network or graph theory to analyze outcomes and consequences of flows. The six basic models of networking scalability are explained in [21]. See also the brief treatment in the foregoing chapter.

### 7.5. *The hidden costs of ownership*

A common way of expressing actual costs associated with components, in an enterprise, is the concept of *total cost of ownership*. Formally, this is the sum of all costs associated with the equipment, including the initial purchase price and licensing, over the lifetime of the equipment.

Let $C$ be the number of customers, $\lambda$ be the arrival rate of transactions per customer, and $P$ be the price of a transaction. Then we can say that, on average. the total income $I$, over a period is proportional to the product of these multiplied by the time interval $\Delta t$:

$$\text{Income } I \propto \langle C \rangle \langle \lambda \rangle \langle P \rangle \Delta t. \tag{34}$$

As usual, we use the notation $\langle X \rangle$ to mean the mean or expectation value of $X$. Thus business income from an online service will depend on the processing capacity [59]. Without proper modelling, it is easy to argue incorrectly that the correct strategy for equipment purchasing is to buy the fastest processor for the cheapest price. This is too simplistic, as we now show with an example.

Computer logic chips work by dumping large amounts of charge to ground through resistive materials, and because there is physical motion of mechanical components such as disks and fans, almost all of the power consumption of a computer ends up as heat eventually. Power consumption of modern computers is a major expense. Moreover, the cost of cooling the equipment is also proportional to the total power output of the environment, and thus to the transaction rate of data processing.

Recently several large companies have boasted the use of off-the-shelf PC or Macintosh hardware for high performance computing servers. This sounds like an idealistic and attractive idea, but the economics of using cheap equipment in data centres are not be as clear cut as one would believe. Measuring power consumption in a data-centre environment is a non-trivial task: power companies charge by delivered current, but that always over-estimates the actual power consumed, owing to *power factor* phase relationships in the alternating supply. Money can be saved even by investing in power factor correcting devices such as Uninterruptible Power Supplies, or more expensive hardware.

Inexperienced organizations think of the purchase cost rather than the cost of ownership, in production. Consider a simple example, based on numerical estimates from the Oslo region, here in Norway.[2] We shall measure currency in 'mu' or 'money units'.

Let us suppose that we have a company considering populating a server farm with either cheap, off-the-shelf computer servers, or with high quality blade-racks. See Table 2.

One hundred servers, with associated switching equipment costs of the order of 60,000 mu for off-the-shelf hardware and 101,000 mu for the quality hardware.

---

[2]The costs are approximate and based on Norwegian kroner divided by a factor of ten for more ready comparison with dollars and Euros.

Table 2

| Cost | Shelf hardware | Quality hardware |
|---|---|---|
| Purchase server | 500 mu | 1000 mu |
| Purchase switch | 10000 mu | 1000 mu |
| Power usage per server | 0.1 kW | 0.05 kW |
| Rent per 100 units per year | 2400 | 600 |
| Failures | 15% | 1% |

Electricity costs about 0.035 mu per kilowatt hour. Our cheap servers produce about 0.1 kilowatts of heat per unit, while the quality servers produce half this amount per unit. The cost of cooling is proportional to the heat produced. Thus it is proportional to the power consumption. Let us suppose then that cooling adds fifty percent to the total power consumption. Thus, in a data centre with a hundred servers, we are paying electricity for a year, to the tune of

$$E_{shelf} = 100 \times 0.035 \times 1.5 \times 0.1 \times (24 \times 365) = 4599 \text{ mu},$$

$$E_{quality} = 100 \times 0.035.5 \times 0.05 \times (24 \times 365) = 2300 \text{ mu}.$$

(35)

Thus the 41,000 money units saved on purchase cost leads to 2300 money units per year of additional power costs. It would take 18 years to justify the additional cost of these servers, by this reckoning. However, one must also add to this the cost of storage, cooling, management and reliability (replacements and repairs).

Cheap off-the-shelf PCs are large and bulky and need storage and cabling, and additional switching equipment. In additional rented infrastructure, one can expect at least:

$$R_{shelf} = 2400,$$

$$R_{quality} = 600.$$

(36)

Failure costs, equivalent to 15 of the cheap PCs per year occur, but only 1 of the more expensive machines fails, so

$$F_{shelf} = 15 \times 500 = 7500,$$

$$F_{quality} = 1 \times 1000.$$

(37)

The total 'saving' by buying off-the-shelf hardware ($\Delta$ = quality minus shelf) is thus

$$\Delta S = \Delta R + \Delta F + \Delta E$$

$$= -1800 - 6500 - 2300$$

$$= -10600 \text{ mu/year}.$$

(38)

In $41000/10600 = 3.9$ years, one has recovered the cost of the quality servers. This could be the actual lifetime of the cheap off-the-shelf PCs, thus, it is not clear that one actually saves any money by buying cheap commodity hardware.

We have not included the costs of administrating and lost revenue due to down-time etc. There are additional costs involved here which could tend to favor more expensive hardware. Soon, the rate of development in chip speeds etc will slow down and environmental levies will be placed on the disposal or recycling of computing equipment and this cost will also be paid by the consumer. Thus the present style of throwing away equipment after a few years will eventually have to change. Investment in quality could be the strategy of the future.

## 8. Demand and capacity: 'provisioning'

The Americanization 'provisioning' has become common parlance for the activity of providing the necessary utilities to cope with demand, i.e. the provision of service capacity.

All we can say with certainty, about a system, is that the rate of production in the business cannot exceed some maximum capacity $C$. If a business is merely a fixed machine whose costs are fixed, then the capacity of the system can be fixed. The problem then is to maximize capacity to cope the maximum expected demand (see Figure 8). This is often done by arranging for a certain over-capacity. However, if costs are proportional to some monotonic function of capacity then one could envisage an enterprise saving money by adjusting their available capacity (re-deploying equipment, hiring and firing etc.). See Figure 9.

### 8.1. *Service provision models*

Network services are the new paradigm for almost every new enterprise and service today. Even production industries can be regarded as services producing goods within a network of suppliers and consumers. The service model is simply a way of thinking. The advance which accompanied this point of view was the idea that delivery of a service is a contractual agreement between autonomous parties, and that customers are allowed to expect a certain Quality of Service (QoS) in order to maintain relations [6].

Quality levels are negotiated using Service Level Agreements (SLA). In the literature, these concepts have multiplied into a plethora of Three Letter Abbreviations, including Service Level Objectives (SLO), Service Level Indicators (SLI) etc. A service level agreement is a contract, and thus one imagines that failure to live up to the expectations of the agreement will lead to reprisals, such as unpaid bills or lawsuits.

One is therefore interesting in dimensioning service capacity to be able to fulfill the contracts that have been signed. There are various approaches to this.

- Over-provisioning: by making service capacity much better than demand, one allows for any margins of error by always having extra capacity. This implies a certain level of waste.

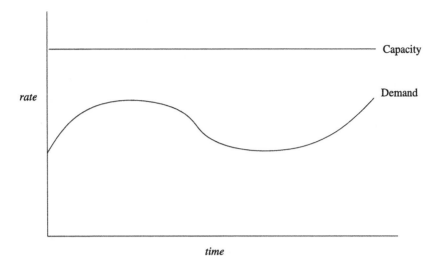

Fig. 8. An over capacity – does this gap between capacity and demand cost anything for the company?

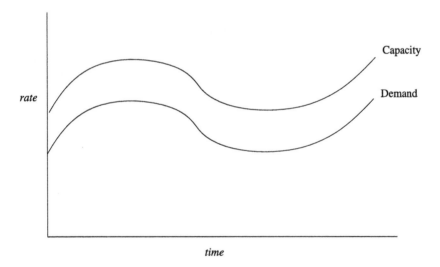

Fig. 9. Capacity adjusted to follow demand – in principe more efficient.

- Optimization: calculating the optimal resource needs using a models and data measurements.
- Inventory management: use of inventory control methods to provide a more dynamic resource management based on periodic maintenance of stocks (this is somewhat analogous to periodic system maintenance regimes).

The service paradigm allows us to couple models to well-known analyzes like queueing theory, which deals with everything from packet networks to vehicle traffic.

In terms of business, or enterprise organization we would like to have some way of calculating reasonable boundaries and levels for service, based on what we know of our capabilities. An interesting approach to this has been suggested by Sauvé et al. [59]. There one considers a Service Level Agreement to consist of two essential items per service type:

- A minimum service level (availability).
- An average response time $\langle t \rangle$.

In addition two input values are required: a longest response before users 'defect', or give up on delivery and go elsewhere, and a probability distribution for the response time in relation to the threshold, so that the probability of being over threshold can be computed. Using this basic information, one can create a cost model for service provision, including over-capacity. By minimizing a cost function, one can determine the optimum number of servers required to support a given Service Level Agreement, or the maximum values in an agreement that can currently be supported by current hardware.

### 8.2. *Quality of Service*

The concept of Quality of Service (QoS) has moved centre stage in network management as service providers seek to offer measurable guarantees to their customers in the form of Service Level Agreements (SLA). The Quality of Service builds on terms like QoD (Quality of Devices), QoE (Quality of Experience), QoB (Quality of Business), and any number of variations to discuss the issue of service provision. Each of these is trying to capture the essence of a usable measure, or 'value', that can be sold to customers. It has been suggested that service quality must be a function of 'Quality of Devices', since service is a function of devices:

$$\text{QoS} = f(\text{QoD}). \tag{39}$$

This makes clear sense: it follows from the laws of causality. The question remains, however, what kind of function should this be, and would a knowledge of the function help us to understand the limits of predictability of service levels?

Quality of Service advertises a qualitative characterization, where one would prefer a quantitative one. One must therefore pin down the meaning of the term.

One measure of service quality is the average response time. From queueing theory, this is given by:

$$T = \frac{1}{\mu - \lambda}, \tag{40}$$

where $\lambda$ is the mean arrival rate and $\mu$ is the mean service rate. If we accept that the arrival and service rates are in fact distributions, with uncertainty characterized by $\lambda \pm \Delta\lambda$ and $\mu \pm \Delta\mu$, then the Pythagoras rule for independent errors tells us that the uncertainty in average response time is correspondingly:

$$\Delta T = \sqrt{\left(\frac{\partial T}{\partial \lambda}\right)^2 (\Delta\lambda)^2 + \left(\frac{\partial T}{\partial \mu}\right)^2 (\Delta\mu)^2}, \tag{41}$$

i.e.

$$\Delta T^2 = \frac{\Delta \lambda^2 + \Delta \mu^2}{(\mu - \lambda)^2}.$$  (42)

Another key measurable is the average service rate $R$, measured in *bits per second*. We interpret Quality of Service to mean the level of predictability in the *quantity of service*; i.e. we identify:

$$\text{Quantity of Service} \pm 1/\text{Quality of Service} \to R \pm \Delta R.$$  (43)

Thus, if quality of service is high, the uncertainty in service level is small and guarantees will be reliable; and if it is low, the uncertainty is high and guarantees are likely to be unreliable. This viewpoint ties Quality of Services to more traditional quantities, such as reliability and certainty.

Ref. [56] also attempts a causal approach. This paper captures the spirit of a causal approach by suggesting the link between service quality to its dependent variables, but it does not take the idea to its logical conclusion by actually defining the functional relationships or how they relate to the service quality. In ref. [8], a heuristic overview with similar ideas was presented, and in ref. [56] an acceptance of stochastic ideas was acknowledged in modelling using network simulations.

The method of Fault Tree Analysis [1,66] can be used to construct an overview of the causal processes. The theory of errors and uncertainties [35,65] may be used to estimate the probabilities.

Fault tree analysis can be used to gauge both the likelihood of error and the uncertainty in service provision (see Figure 10). Probabilities can be used to represent either the likelihood of a service failure (i.e. the likelihood that an SLA is not fulfilled), or the probability of deviation from expected (average) service levels. In the latter case, there this gives us a way of deriving the limits of certainty in service provision.

For instance, suppose we have – at the very lowest level of the system – that the data rate for the transmission medium is given by Shannon's Gaussian noise formula:

$$C(B, S, N) = B \log_2 \left(1 + \frac{S}{N}\right),$$  (44)

where $C$ is the channel capacity in bits per second, $B$ is the bandwidth in Hertz, $S$ is the signal power and $N$ is the noise power. We now wish to discuss the effect of temperature on the capacity of a metallic conductor (copper wire), for which it is known that thermal noise has the form

$$N_{\text{metal}}(T) = N_0(T + 273),$$  (45)

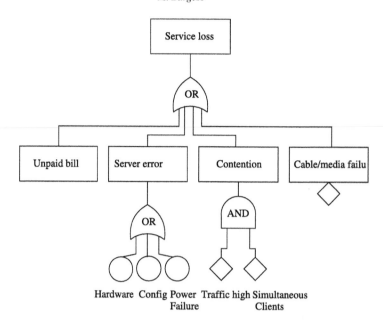

Fig. 10. Fault trees can capture the essence of the hierarchical relationship between quality of components and final service. Here is a typical example of how fault trees are used to gauge reliability. Logical AND and OR gates reflect the interdependency of lower level components in a system.

where $N_0$ is a constant and $T$ is the temperature in Celsius.[3] The complete formula is given by straightforward substitution:

$$C(B, S, T) = B \log_2\left(1 + \frac{S}{N_0(T + 273)}\right). \tag{46}$$

With the full expression capturing all of the dependencies in the dependency tree, through substitutions, the theory of errors and uncertainties can now be applied to the list of independent parameters.

### 8.3. *Non-linearity: reactionary models of service dependency*

Service provision is, in some contexts, thought of as the delivery of fixed-rate, policy-controlled transactions between willing parties. A contract of service, called a Service Level Agreement (SLA) documents the promise that the service provider makes to the client. The SLA is a form of policy for the server to uphold [60,67]; what distinguishes it from other policy rules is that it is not deterministically enforceable, since the parties have no direct control over one another.

---

[3]Note that the uncertainty due to thermal noise is not negligible in high capacity copper wires: it can represent tens of Megabytes that might be sold to a customer.

Any service provider's policy is subject to environmental uncertainties and, for that reason, the promise can never be more than an expectation or hopeful prediction of average service levels [16]. To verify that a promise has been kept, the client must monitor the service level. Service Level Agreements are often complex. They combine several measured values into composite metrics that are then compared to a template [30,32,37,56,57]. This has important implications for the stability of policy relationships.

Feedback regulation of policy is an idea that has received more attention recently, due to interest in autonomous system operation; it has been discussed in a variety of circumstances [14,15,33,34] for different service types ranging from web services to configuration management. Feedback regulation occurs regardless of whether it is carried out autonomously by machines or by humans, whenever there is a dynamical interplay between monitoring and SLA.

From the theory of dynamical systems, it is known that the combination of several dynamical variables using AND (multiplication) and OR (addition) in feedback loops can lead to 'chaotic' or non-linear behavior. Such combinatoric ideas have been discussed to provide more expressive policies for service level agreements [30,32,37,56,57]; for example, in ref. [56], the authors discuss 'aggregated Quality of Service parameters' of the form:

$$S \text{ depends\_on } (SS_1 \textbf{ AND } SS_2) \tag{47}$$

as well as more general rules like:

$$\text{QoS}_A(S) = f\big(\text{QoS}_B(SS_1) \textbf{ AND } \text{QoS}_B(SS_2)\big). \tag{48}$$

They are basing service quality promises on a function of measured values. However, there are problems with combining dynamical variables non-linearly on finite precision systems. It is known that such systems can spiral out of control, or simply ebb to zero for certain ranges of parameters (for a review, see [43]). It is therefore important to understand how such behavior emerges in dynamically regulated policies and what can be done to secure stable operation.

In ref. [5], the authors attempt to answer the basic question: whether or not a reactive policy, based on mutual observation of several data points, can lead to stable behavior or whether escalation of reprisals by the parties can spiral out of control. The studies made on continuum approximation models indicate that feedback regulation of SLA-like policies is a highly complex matter that must be parameterized with caution. Even simple non-linear policies have no stable fixed point that can be associated reliable service levels, other than the trivial case of no service. If two parties enter into a policy agreement of this type, they will either undergo a long fibrillation of changing behavior, or they will spiral down to offering zero service levels. Thus the automation of policy should be wary of these issues.

## 9. Inventory models

Most traditional businesses involve the actual sale of goods which need to be stored in some kind of warehouse. The name *inventory* is used for stocks of revenue generating goods. For example, an Internet bookseller, Like Amazon, needs a warehouse to keep its stock, pending retail sale to customers. There is a cost involved in storing goods, both in terms of expected returns and in terms of warehouse rent. The longer inventory is sitting in a warehouse, the less profitable it can be.

However, even a computer centre needs certain supplies to function. Computers, monitors, mice, keyboards, CDROMs, hard-disks, network cables etc., are all examples of potential inventory that a system administrator should consider keeping as inventory.

There are three considered motives for storing inventory, rather than ordering on demand (or so-called Just In Time) models [47].

1. Transaction motive: by ordering large or predictable quantities of the merchandise, one can minimize transaction costs, or administrative overhead $k$.
2. Precautionary motive: Keeping inventory is a way of minimizing the risks of uncertainty. It provides a buffer against late deliveries or sudden demand.
3. Speculative motive: If the price of stock is expected to rise in the near future, preordering and stockpiling can be a way to save money.

Modern computer systems do something similar in their 'caches', where the cost in not immediately monetary but a cost in time. Recently some authors have attempted to map inventory models onto service based computing models [42]. Although these issues are motivations for inventory models, the models themselves only account for the first of these reasons.

Inventory models are essentially primitive form of queueing theory, where the focus is shifted from responding as quickly as possible to spreading the load evenly. They were invented to deal with actual warehouses (see the reviews in volume of ref. [47] for lucid introductions) and the logistics of planning, but they can be applied to Internet commerce and also elementary transaction performance models for networking. The over-simplifications involved in the most basic models are usually tolerated precisely because of their simplicity.

## 9.1. *Basic continuum model of averages*

The basic model of inventory is known as the Economic Order Quantity (EOQ) model [3,47] and assumes a completely deterministic process of predictable supply and demand. While this is clearly not true, by any stretch of the imagination, it is plausibly true over long times for average quantities. It can therefore be thought of as a 'mean field approximation'.

We make the following assumptions:

1. The inventory of merchandise is all of a single type and stored at a single location.
2. We expect inventory to cycle continuously with a provision of service that never stops (a deterministic queue). So we ignore start up costs, and we are interested in the costs per inventory cycle-period, i.e. the time period $T$ of the inventory cycle is a natural timescale for the model.
3. The model is deterministic, i.e. the rate of consumption (or the *expected demand*, in a sales interpretation) for merchandise is known with certainty, and is $\lambda$ inventory units per unit time. It is like a leaking bucket.

4. One cycle of the model is one single (average) order of (average) size $q$, taking (average) time $t$.
5. Shortages of inventory do not occur, nor does over-production, i.e. demand is always matched to supply, on average.
6. Delivery time is instantaneous, i.e. immeasurably short compared to an average timescale $T$ for ordering.
7. We do not take into account the provision for discounts for old stock, i.e. merchandise is as valuable today as it was when it entered the inventory.

As always, the assumptions underpinning a model are the key to knowing how and when we should take its results seriously.

Let us assume that each order of goods is made with average quantity size $q$, which is clearly constant over the time interval for which the average is defined. The cost of each order, of size $q$, has a fixed average transaction price $k$ (money units) for and an overhead that comes from the cost rate of *holding* the merchandise in the warehouse $h$ (money units per unit time) over time.

Finally, let the actual inventory, as a function of time be written $I(t)$ (see Figure 11). From our constraining assumptions, the inventory function can only have the simple form:

$$I(t) = q - \lambda t, \quad 0 \leqslant t \leqslant T. \tag{49}$$

The cost per unit time of keeping merchandise is

$$C(T) = \frac{1}{T}\left(k + h \int_0^T I(t)\,dt\right) \tag{50}$$

and it measured in dimensions of money units per period. From Figure 11, we note that $T = q/\lambda$, just from the geometry of the triangles. This integral can be calculated trivially,

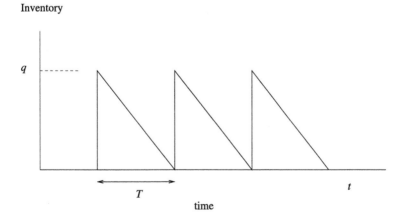

Fig. 11. The basic inventory model with continuous cyclic operation shows how the stock is produced 'suddenly' and used at a constant rate $\lambda$ until it is exhausted.

or one can use the half-base-times-height rule for the area of a triangle:

$$\int_0^T I(t)\, dt = \left( qT - \frac{1}{2}\lambda T^2 \right) \Big/ T = \frac{1}{2}q^2\lambda = \frac{1}{2}qT. \tag{51}$$

Thus, we have a cost per cycle/order function that may be expressed either as a function of $q$ or of $T$:

$$C(T) = \left( k/T + \frac{1}{2}h\lambda T \right), \tag{52}$$

$$C(q) = \left( k\lambda/q + \frac{1}{2}hq \right). \tag{53}$$

In either variable, we see that the cost per order is a non-linear function that is the super-position of a hyperbola and a straight line (see Figure 12).

The cost function has a minimum, which may be found by differentiation by $q$ or $T$, at the values $q^*$ and $T^*$ respectively:

$$q^* = \lambda T^* = \sqrt{2k\lambda/h} \tag{54}$$

at which point

$$C(q^*) = C(T^*) = \sqrt{2k\lambda h}. \tag{55}$$

Thus, if we have optimized the average performance of this systems, the average cost per order grows with fixed order price with the holding cost.

The result displays a symmetry between order cost and holding price and demand:

$$\langle k, \lambda, h \rangle. \tag{56}$$

How do we explain this symmetry causally, and what does mean? We must first explain the interpretation of the variables, and understand their forms.

- $T^*$ is the cheapest time interval for customers to place their orders. If we make $T$ less than this, Equation (50) tells us that the transaction price will dominate the cost, i.e. the customers will pay the basic processing fees too often. If we make the time larger than this, the holding time dominates the cost, as the storage cost of the inventory increases and is passed on the customer.
- $q^*$ is the cheapest order size. It is related to the time by the constant demand rate, so they are not independent variables. If $q$ exceeds the optimum, i.e. customers greedily order too much at a time then the order period increases, and vice versa.

The optimum cost increases as either the unit cost, storage cost or demands increase. This is because the larger costs force the consumer to slow down the rate ordering (increase $T^*$).

The steady-state, average assumptions of the model allow us to eliminate time as a variable at optimal operation.

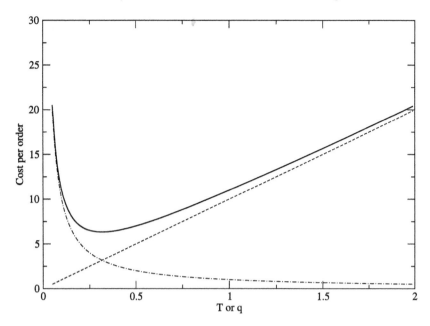

Fig. 12. The solid line cost per order (or per cycle) function, is the superposition of a linear growth and hyperbolic cost per unit time. In either parameterization, has the form of a lifted hyperbola. There is a clear minimum at an abscissa point $T^*$ or $q^*$, which represents the minimum cost value.

How do we know that we can tune the system to perform optimally? On average we can imagine that it is reasonable to compensate for non-optimal during a period by suitable monitoring.

Suppose we do not manage to compensate, a deviation is characterized by the beautiful dimensionless relation:

$$\frac{C(q)}{C(q^*)} = \frac{1}{2}\left(\frac{q}{q^*} + \frac{q^*}{q}\right). \tag{57}$$

Suppose we achieve $q = \frac{1}{2}q^*$, then this ratio is 1.25. In other words, missing the target by a hundred percent of the achieved goal leads to only a twenty-five percent increase in costs. This is quite insensitive.

### 9.2. *Applying the model*

The inventory model is not a complete business model, in the sense of dealing with income and expenditure. It is a sub-model that deals with optimizing the costs of holding stocks and supplies. The burden of applying these formulae, as always, is in finding reliable and realistic data for the inputs.

The optimal rate of ordering is not about how much the inventory costs to purchase. It is about the administration costs. This, if $P$ is the purchase price per unit for inventory, the total cost of the order is simply $qP$, but the cost of ordering and keeping the inventory $C(q)$ is an overhead which is to be minimized.

- $k$ is the administrative cost of placing an order, including the actual time taken converted using the man-hour rate.
- $h$ is the cost of keeping inventory which includes storage, insurance, any interest paid over the period on money borrowed etc.

1. Suppose that the yearly consumption of PCs is about 20 per year. An order of PCs, which includes decision-making, discussion and processing accounts for 5 man-hours of time at 50 money units per hour, hence $k = 250$ money units.

   The cost of storage for these PCs, including cleaning and insurance per unit, per year is $h = 100$ money units per unit per year.

   The optimal order size is now,

$$q^* = \sqrt{\frac{2 \times 250 \times 20}{100}} = 10 \text{ units} \tag{58}$$

and

$$T^* = \frac{1}{2} \text{ years.} \tag{59}$$

   Thus, two purchases of ten units, over a year is the optimal solution.

2. For a second example, consider the consumption of disk space on a mail server. As users accumulate e-mail, or as an ISP accumulates customers with fixed quotas. (In fact dealing with fixed quotas is much easier to analyze because we can more easily estimate the consumption, without having to take into account how much users tidy up old mail etc.)

   We shall take disks to be the unit of inventory. Suppose the expected number of new customers (consumers) per year is 100 per year. If we assume that a typical hard-disk can accommodate 100 users, then this amounts to a usage of $\lambda = 1$ disk unit per year.

   The processing cost for an order of disks is a half man-hour at 50 money units per hour, i.e. $k = 25$.

   The holding cost for disk space (assuming that they are installed on arrival) includes power consumption, replacement disks on error etc. Suppose this amounts to a small amount of power consumption, plus a ten percent chance of having to buy a new disk at 500 money units. Let us take $h = 55$ money units per disk per year as the holding price. In fact, if we allow for the rapidly increasing capacity of disks, we might make $h$ larger to account for the loss of disk space from ordering too early.

   The optimal order size is now:

$$q^* = \sqrt{\frac{2 \times 25 \times 1}{55}} = 0.95 \text{ units.} \tag{60}$$

This tells us that there is no benefit to mass ordering disks before they are needed. It is okay to order them one at a time, since they are relatively expensive to own, and relatively cheap to order.

### 9.3. *Extended model*

In reality, the production of merchandise takes a certain amount of time. to produce or assemble the inventory. We can make a trivial modification of the model to include a finite and linear production schedule, with the simplifying assumption that the rate of production $\psi$ is always great than the rate of consumption $\lambda$.

We shall once again assume an order size of $q$ and a cycle time of $T$. By analyzing the triangular period in Figure 13, we can use the triangle formula to find the area, as before in the integral of Equation (50). The vertical height of the triangle (representing the peak amount of inventory in store, given the stock is being used as it is being produced) is

$$I_{max} = (\psi - \lambda)q/\psi. \tag{61}$$

The length of the base is still $T$. Moreover, by examining the triangle, we may confirm that $T = q/\lambda$ still holds.

The cost per order function is thus now:

$$C(q) = \left(k + \frac{1}{2}\frac{h(\psi - \lambda)Tq}{\psi}\right)\Big/ T$$

$$= \frac{k\lambda}{q} + \frac{1}{2}h(\psi - \lambda)\frac{q}{\psi}. \tag{62}$$

Inventory

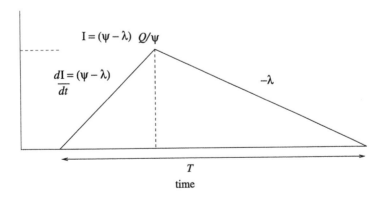

Fig. 13. The extended inventory model with continuous cyclic operation shows how the stock is produced at a finite rate $\psi$ and used at a constant rate $\lambda$ until it is exhausted.

Minimizing this cost function, by differentiation gives an optimal order size:

$$q^* = \sqrt{\frac{2k\lambda\psi}{h(\psi - \lambda)}},$$ (63)

and we recall that $\psi > \lambda$. As $\psi \to \infty$, this model become the basic model.

### 9.4. *Effect of uncertainty in the rates*

Even in an averaged regime, one must expect a certain amount of uncertainty in the rates of production and consumption. Suppose we assume that $\psi$ and $\lambda$ are independent variables subject to small, symmetrically distributed uncertainties:

$$\lambda \to \lambda \pm \Delta\lambda,$$
$$\psi \to \psi \pm \Delta\psi.$$ (64)

Then, we may calculate the compound uncertainty in $q^*$ by the Pythagorean combination of the first-order Taylor expansions, in the usual way.

$$\Delta q^* = \sqrt{\left(\frac{\partial q^*}{\partial \lambda}\right)^2 (\Delta\lambda)^2 + \left(\frac{\partial q^*}{\partial \psi}\right)^2 (\Delta\psi)^2}$$
$$= \frac{1}{2}\frac{q^*}{|\psi - \lambda|}\sqrt{\left(\frac{\Delta\lambda}{\lambda}\right)^2 \psi^2 + \left(\frac{\Delta\psi}{\psi}\right)^2 \lambda^2}.$$ (65)

There is a similarity between the idea of inventory control and the idea of system maintenance. The maintenance theorem of ref. [16] says that a system is maintainable if

$$\frac{1}{\langle q \rangle}\frac{\mathrm{d}\langle q \rangle}{\mathrm{d}t} \ll \frac{1}{T}$$ (66)

where $\langle q \rangle$ is the average level of inventory.

However, we note that the inventory models are steady state models, so the maintenance theorem is satisfied automatically, since $\langle q \rangle$ is constant, by assumption. If, however, we allow for stochastic uncertainties in the model, then Equation (66) becomes essentially

$$\frac{1}{q}\frac{\Delta q}{T} \ll \frac{1}{T}$$ (67)

which translates into $\delta\lambda/\lambda \ll 1$ and $\delta\psi/\psi \ll 1$.

## 10. What is the cost of poor system administration?

There are clearly many areas in which a system administrator's decisions affect the potential for a business to operate. In purchasing decisions and operating costs, there are large amounts of money to be saved. In a globalized environment, in which one has data centres with load balancing, companies could even consider re-routing requests to locations where load is low or power is cheap. Resource utilization management is an important part of cost management.

A competent engineer is one of the most important accessories to any machine or system. It is usually the member of personnel who is most forgotten until needed. Delays in an enterprise's operations could mean lost revenue, lost business or even lost reputation (an ability to attract new income). The repercussions can be serious.

In the present world of virtual issues like CPU speeds and numbers of megabytes, one sometimes forgets the importance of physical issues. What is the effect of heat? How should one avoid earth loops in a server room? Don't hang the leaky air-conditioner over your expensive mainframe, and so on.

### 10.1. *Importance of empiricism*

In the world wars, pilots were said to fly by the seat of their pants, meaning that they were intimately connected to their machinery – that they had learned the bumps and foibles of the machine. An understanding of the empirical nature of a system is just as important today, even if there are no physical vibrations.

How does a system respond under load? How much power does it use? How does the weather affect the performance of computing equipment, or the staff? While these might seem like frivolous questions, they have a very serious side.

In a data centre, the questions above become design issues. A system might thrash and slow down under heavy load. A peak of power consumption might cause fuses to blow or cooling equipment to overload. The weather can affect the air in the centre: if the humidity is too high, condensation can form on cool surfaces; if it is too low (very cold weather), static electricity can cause errors and even power spikes that can damage equipment.

Perhaps we might think that running a server room a few degrees hotter can save some money in cooling power? Or that extra humidity can increase the specific heat capacity of the air and increase the efficiency of the eat transfer? These things can only be understood through rational inquiry and measurement. For example, the specific heat capacity of dry air is 1005 Joules per kilogramme per degree Kelvin, and the heat capacity of air with one hundred percent humidity is 1030 J/kg/K. The difference is only one percent, which is probably far less than normal power fluctuations, and is therefore irrelevant. On the other hand, humidity should be kept between forty to fifty percent to balance the risks of condensation and static electricity.

The scientific method: rational inquiry is a powerful medicine against such problems.

## 10.2. *Return on investment and total cost of ownership*

Return on investment is defined as the income per fiscal year, divided by the cost of the investment. This is often proportional to the throughput of the enterprise. Ineffective use of resources (poor scheduling or load balancing) can limit this. Poor training of staff and inadequate management can also affect it. Automation of basic tasks can increase the up-time of the system and avoid security problems.

## 10.3. *Risk minimization*

Understanding the risks of an organization and being able to plan for them is best achieved with analytical tools (see Figure 14). The identification of points of failure in the system network, the analysis of dependencies and probability that service targets will be met: these are examples of issues that are impossible to plan for without a mixture of empiricism and analysis.

Risk can be used a means of optimization. For example, in ref. [24], the authors use a minimum risk principle to find the best time to make a backup copy of data in a busy data centre. Many administrators will do this when the system is most quiescent; however, an analysis of risk shows that is not the way to minimize risk of loss due to random error. Rather, the answer is about halfway between peak activity and minimum activity (after peak time), since more changes that can be lost occur during peak times (see Figure 15). Whether an optimal solution is desirable is a matter for policy.

## 10.4. *Incident response and fire-fighting*

Incident response is like an inventory model or queueing process. Incidents or events form an arrival, point process that accumulate into a queue or 'inventory' of tasks to be cleared.

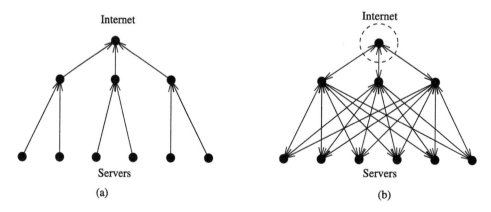

Fig. 14. Load balancing without points of failure is a tricky matter if one is looking for redundancy. A load balancer is a tree – which is a structure with many intrinsic points of failure and bottlenecks. How can these be completely eliminated to make a high availability data centre?

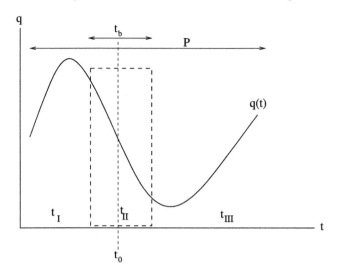

Fig. 15. The 'windshield wiper model' for disk backup scans across files in a time $t_b$ about $t_0$. The wavy line shows the average rate of point change arrivals during a 24 hour period. There are three regions of currently unknown sizes: region I before backup, region II during backup and region III after backup. The file change process $q(t)$ brings $q$ arrivals at time $t$ and provides a probabilistic weight to the expected risk: if more files are arriving, the risk of loss will be higher.

By analogy with the language used in configuration management policy, a business 'incident' could be defined as a deviation from policy, or it could be a complaint or a failure of some part of the business to deliver on a promise. We can model this kind of process with various levels of sophistication.

The illustration in Figure 11 can also be interpreted, somewhat simplistically, as a deterministic expectation of maintenance costs. At the leading edge of the saw-tooth one can imagine a quota of jobs to be complete arises (suddenly). Gradually, these jobs are completed and the inventory is cleared. After a periodic interval, one reschedules the next batch.

Although simplistic, this could be a realistic approach to time management – a way of clearing processes from a help-desk message queue, or some other e-mail queue. One simply chooses $\lambda = 1/T$ to fix the frequency of turn-around. There is an initial logistical cost of scheduling time to answer job messages (e.g. learning and research time to study the problem), and there is a cost associated with having them unanswered, or un-repaired (broken promises, maintenance contract penalties).

The optimum time $T^*$ tells us how to prioritize this process; the optimal inventory order tells us how many jobs $q^*$ to schedule at a time. This is related to the queue length in queueing models.

Another example of this kind of scheduling is periodically scheduled maintenance tasks, like garbage collection and configuration maintenance. where the scheduling order cost is the amount of resources needed to locate and fix potential problems. The holding cost is harder to estimate, but involves lost productivity or due to faults.

Incident management in relation to ITIL and business objectives is advocated in ref. [4]. The authors consider a policy-based prioritization, based on the penalty costs of failing to meet obligations in Service Level Agreements, and using time-discounting as an importance ranking, based on the urgency of the incident. It is a form of anomaly detection. These authors take essentially the analogous view to anomaly definition as is put forward for system administration in refs. [16,19], namely that policy is an expression of the limits of a probability distribution. Prioritization can be defined essentially by the number of standard deviations off 'normal'.

## 11. Conclusions

The administration of systems and business process are clearly related, at an intimate level. The goals of the two are closely aligned. The service-oriented viewpoint of system management allows for an even clearer integration of these issues. Business modelling, especially in an IT framework, can rightfully be regarded as the logical extension of system performance analysis, by a variety of metrics.

One must be cautious in applying any set of analytical tools to a situation of great complexity. The aim of modelling is to offer suitably idealized approximations to the truth, using a variety of possibly ad hoc measures and metrics. What science tells us is that it is not the metrics or scales of measurements themselves that are the bringers of truth, but rather how they are interpreted. Measures and metrics must be *calibrated* to a particular environment, and interpreted *in situ*, with a certain amount of imagination to extract an essence of truth. A successful calibration of a well-conceived model will allow a strict comparison of values and is therefore a reliable thinking aid.

The system administrator plays the role of a service provider and of a maintenance agent. In a world of increasing reliance on computing systems, the engineers who design, tune and maintain the system are the keys to its success. This short review cannot do justice to the wealth of details, but hopefully it captures some of its essence.

## Acknowledgements

The author is grateful to Claudio Bertolini, Alva Couch and David Hausheer for helpful remarks. This work was supported in part by the EC IST-EMANICS Network of Excellence (#26854).

## References

[1] R. Apthorpe, *A probabilistic approach to estimating computer system reliability*, Proceedings of the Fifteenth Systems Administration Conference (LISA XV), USENIX Association, Berkeley, CA (2001), 31.
[2] R. Axelrod, *The Evolution of Co-operation*, Penguin Books (1990) (1984).
[3] S. Axsäter, *Logistics of Production and Inventory*, Handbooks in Operations Research and Management Science, Elsevier (1993).

[4] C. Bartolini and M. Sallé, *Business driven prioritization of service incidents*, Proc. 15th IFIP/IEEE Distributed Systems: Operations and Management (DSOM 2004), Springer-Verlag (2004).

[5] K. Begnum, M. Burgess, T.M. Jonassen and S. Fagernes, *Summary of the stability of service level agreements*, Proceedings of International Policy Workshop (2005).

[6] K. Begnum, M. Burgess, T.M. Jonassen and S. Fagernes, *On the stability of adaptive service level agreements*, eTransactions on Network and System Management **2** (1) (2006), 13–21.

[7] K. Begnum, M. Burgess, T.M. Jonassen and S. Fagernes, *On the stability of service level agreements*, IEEE eTransactions on Network and System Management **2** (2006), to appear.

[8] A. Bouch and M.A. Sasse, *It ain't what you charge, it's the way that you do it: A user perspective of network qos and pricing*, Proceedings of the VI IFIP/IEEE IM Conference on Network Management (1999), 639.

[9] J.P. Bouchard and M. Potters, *The Theory of Financial Risks*, Cambridge University Press, Cambridge (2000).

[10] M. Brenner, *Classifying ITIL processes, a taxonomy under tool support aspects*, Proceedings of the 1st IEEE/IFIP International Workshop on Business-Driven IT Management (2006), 19–28.

[11] British Standard/International Standard Organization, *BS/ISO 17799 Information technology – Code of practice for information security management* (2000).

[12] British Standards Institute, *BS 15000 IT Service Management* (2002).

[13] E. Brousseau and J.-M. Glachant, eds, *The Economics of Contracts Theory and Applications*, Cambridge University Press (2002).

[14] M. Burgess, *Computer immunology*, Proceedings of the Twelfth Systems Administration Conference (LISA XII), USENIX Association, Berkeley, CA (1998), 283.

[15] M. Burgess, *Two-dimensional time-series for anomaly detection and regulation in adaptive systems*, IFIP/IEEE 13th International Workshop on Distributed Systems: Operations and Management (DSOM 2002), Lecture Notes in Comput. Sci., Vol. 2506 (2002), 169.

[16] M. Burgess, O*n the theory of system administration*, Science of Computer Programming **49** (2003), 1.

[17] M. Burgess, *Analytical Network and System Administration – Managing Human–Computer Systems*, Wiley, Chichester (2004).

[18] M. Burgess, *An approach to understanding policy based on autonomy and voluntary cooperation*, IFIP/IEEE 16th International Workshop on Distributed Systems Operations and Management (DSOM), Lecture Notes in Comput. Sci., Vol. 3775 (2005), 97–108.

[19] M. Burgess, *Probabilistic anomaly detection in distributed computer networks*, Science of Computer Programming **60** (1) (2006), 1–26.

[20] M. Burgess, *Handbook of Network and System Administration*, Elsevier (2007), Chapter: System Administration and the Scientific Method.

[21] M. Burgess and G. Canright, *Scaling behaviour of peer configuration in logically ad hoc networks*, IEEE eTransactions on Network and Service Management **1** (2004), 1.

[22] M. Burgess and S. Fagernes, *Pervasive computing management: A model of network policy with local autonomy*, IEEE Transactions on Software Engineering (submitted).

[23] M. Burgess and S. Fagernes, *Voluntary economic cooperation in policy based management*, IEEE Transactions on Software Engineering (submitted).

[24] M. Burgess and T. Reitan, *A risk analysis of disk backup or repository maintenance*, Science of Computer Programming **64** (2007), 312–337.

[25] G. Canright and K. Engø-Monsen, *Handbook of Network and System Administration*, Elsevier (2007), Chapter: Some Aspects of Network Analysis and Graph Theory.

[26] J.D. Carrillo and M. Dewatripont, *Promises, promises*, Technical Report 172782000000000058, UCLA Department of Economics, Levines's Bibliography.

[27] H. Chesbrough and R.S. Rosenbloom, *The role of the business model in capturing value from innovation: Evidence from xerox corporation's technology spinoff companies*, Industrial and Corporate Change **11** (1) (2002).

[28] R.B. Cooper, *Stochastic models*, Handbooks in Operations Research and Management Science, Vol. 2, Elsevier (1990), Chapter: Queueing Theory.

[29] A.L. Couch, N. Wu and H. Susanto, *Towards a cost model for system administration*, Proceedings of the Nineteenth Systems Administration Conference (LISA XIX), USENIX Association, Berkeley, CA (2005), 125–141.

[30] D. Daly, G. Kar and W. Sanders, *Modeling of service-level agreements for composed services*, IFIP/IEEE 13th International Workshop on Distributed Systems: Operations and Management (DSOM 2002) (2002), 4.

[31] A. Daskalopulu and M. Sergot, *The representation of legal contracts*, AI & Society **11** (1997), 6–17.

[32] M. Debusmann and A. Keller, *Sla-driven management of distributed systems using the common information model*, Proceedings of the VIII IFIP/IEEE IM Conference on Network Management (2003), 563.

[33] Y. Diao, J.L. Hellerstein and S. Parekh, *Optimizing quality of service using fuzzy control*, IFIP/IEEE 13th International Workshop on Distributed Systems: Operations and Management (DSOM 2002) (2002), 42.

[34] Y. Diao et al., *Generic on-line discovery of quantitative models for service level management*, Proceedings of the VIII IFIP/IEEE IM Conference on Network Management (2003), 158.

[35] R.H. Dieck, *Measurement and Uncertainty (Methods and Applications)*, 3rd edn., Instrument, Systems and Automation Society (2002).

[36] J. Ferber, *Multi-agent Systems: Introduction to Distributed Artificial Intelligence*, Addison–Wesley (1999).

[37] P. Flegkas et al., *Design and implementation of a policy-based resource management architecture*, Proceedings of the VIII IFIP/IEEE IM Conference on Network Management (2003), 215.

[38] K. Gaarder, *Business models – what are they and how do we design them?* Technical Report R21/2003, Telenor Research and Development (2003).

[39] J. Gordijn, H. Akkermans and H. van Vliet, *Business modelling is not process modelling*, ER Workshops (2000), 40–51.

[40] G.R. Grimmett and D.R. Stirzaker, *Probability and Random Processes*, 3rd edn., Oxford Scientific Publications, Oxford (2001).

[41] J.L. Hellerstein, Y. Diao, S. Parekh and D.M. Tilbury, *Feedback Control of Computing Systems*, IEEE Press/Wiley Interscience (2004).

[42] J. Hellerstein, K. Katircioglu and M. Surendra, *A framework for applying inventory control to capacity management for utility computing*, Proceedings of the IXth IFIP/IEEE International Symposium on Integrated Network Management IM'2005, IEEE (2005), 237–250.

[43] A.V. Holden, ed., *Chaos*, Manchester University Press (1986).

[44] A. Høyland and M. Rausand, *System Reliability Theory: Models and Statistical Methods*, Wiley, New York (1994).

[45] J. Huang, http://www.bptrends.com/publicationfiles/01-05 etom and itil - huang.pdf (2005).

[46] Information Technology Service Management Forum, http://www.itsmf.com.

[47] H.L. Lee and S. Nahmias, *Logistics of Production and Inventory*, Handbooks in Operations Research and Management Science, Vol. 4, Elsevier (1993), Chapter: Single Product, Single Location Models.

[48] C. Mayerl, F. Tröscher and S. Abeck, *Process-oriented integration of applications for a service-oriented it management*, Proceedings of the 1st IEEE/IFIP International Workshop on Business-Driven IT Management (2006), 29–36.

[49] D.A. Menascé and V.A.F. Almeida, *Scaling for E-Business: Technologies, Models, Performance, and Capacity Planning*, Prenctice Hall (2000).

[50] N. Minsky, *Law governed interaction (LGI): A distributed coordination and control mechanism (an introduction, and a reference manual)*, Technical report, Rutgers University (2005).

[51] N. Minsky and P.-P. Pal, *Law-governed interaction: A coordination and control mechanism for heterogeneous distributed systems*, ACM Transactions on Software Engineering and Methodology **9** (2000), 273–305.

[52] J.F. Nash, *Essays on Game Theory*, Edward Elgar, Cheltenham (1996).

[53] J. Pearl, *Probabilistic Reasoning in Intelligent Systems: Networks of Plausible Inference*, Morgan Kaufmann, San Francisco (1988).

[54] J. Pearl, *Causality*, Cambridge University Press, Cambridge (2000).

[55] J. Rasmussen, *Skills, rules, and knowledge; signals, signs and symbols, and other distinctions in humans performance models*, IEEE Transactions on Systems, Man and Cybernetics **13** (1983), 257.

[56] G.B. Rodosek, *Quality aspects in it service management*, IFIP/IEEE 13th International Workshop on Distributed Systems: Operations and Management (DSOM 2002) (2002), 82.

[57] A. Sahai et al., *Automated SLA monitoring for web services*, IFIP/IEEE 13th International Workshop on Distributed Systems: Operations and Management (DSOM 2002) (2002), 28.

[58] M. Sallé and C. Bartolini, *Management by contact*, Proceedings of 8th Network Operations and Management Symposium NOMS 2004, IEEE/IFIP (2004).

[59] J. Sauvé et al., *SLA design from a business perspective*, IFIP/IEEE 16th international workshop on distributed systems operations and management (DSOM), Lecture Notes in Comput. Sci., Vol. 3775 (2005), 73–83.

[60] M. Sloman, *Policy driven management for distributed systems*, Journal of Network and Systems Management **2** (1994), 333.

[61] M. Steinder and A. Sethi, *A survey of fault localization techniques in computer networks*, Science of Computer Programming **53** (2003), 165.

[62] J. Strassner, *Police Based Management, Solutions for the Next Generation*, Morgan Kaufmann/Elsevier (2004).

[63] Telemanagement Forum, http://www.tmforum.org.

[64] Telemanagement Forum, *An interpreter's guide for etom and itil practitioners* (2004).

[65] J. Topping, *Errors of Observation and their Treatment*, Chapman and Hall (1972).

[66] U.S. Nuclear Regulatory Commission NRC, *Fault Tree Handbook*, NUREG-0492, Springfield (1981).

[67] D. Verma et al., *Policy based SLA management in enterprise networks*, Policy Workshop 2001, Springer-Verlag (2001).

[68] J. Walrand, *Stochastic Models*, Handbooks in Operations Research and Management Science, Vol. 2, Elsevier (1990), Chapter: Queueing Networks.

[69] M. Wooldridge, *An Introduction to MultiAgent Systems*, Wiley, Chichester (2002).

# – 6.4 –

# System Reliability

Trond Reitan

*Department of Mathematics, University of Oslo, P.O. Box 1053 Blindern. N-0316 Oslo, Norway*
*E-mail: trond@math.uio.no*

## 1. Introduction

The object of this chapter is to give a brief glimpse into the field of system reliability analysis. System reliability analysis concerns itself with finding the probability for system failure, which is often called the failure rate, or the probability of the system functioning, called the reliability. This has to be calculated using existing knowledge about the architecture of the system and the probability of component failure (or for the component working).

This subject can be of interest to computer system administrators, computer system programmers and decision makers. Three reasons for that will be mentioned here.

Firstly, knowing the probability of a system failure means that sufficient resources for handling system failure can be allocated. Furthermore, knowing which parts of the system that are most critical, means knowing which components to watch with the most care.

Secondly, if one is able to calculate the cost and system reliability for several systems, one can choose a system that minimizes risk, defined as the cost of the system plus the cost of system failure times the probability of system failure;

$$R_{\text{system}} = C_{\text{system}} + C_{\text{failure}} Pr(system\ failure), \tag{1}$$

where $Pr(event)$ stands for the probability of a given event, $C$ stands for cost and $R$ stands for risk. In this case, it is natural to use cost in terms of money used as a measure for the expected loss and thus the risk.

HANDBOOK OF NETWORK AND SYSTEM ADMINISTRATION
Edited by Jan Bergstra and Mark Burgess

Thirdly, experience of system reliability analysis can make one better able to design reliable computer systems.

This text will not provide a full set of tools for doing such analysis, but will hopefully point those interested in the right direction. It will mainly be concerned with finding the system reliability, defined as the probability that the system will function in a given fixed time period, given the probability that each component will function in that period and that no repairs are done in that period. If the latter assumption is to be approximately correct, one needs to look at sufficiently small time periods. Typically the probability of the system functioning will be high in such cases. Often, one can also assume the probability of each component functioning is high.

We will start by going through basic probability theory, in order to have the right tools for further study. Next, the simplest type of systems will be analyzed. The results from that treatment will be used for studying so-called fault trees, before a more general set of system reliability tools will be introduced. Next we will study systems that can be represented by graphs. Dynamic system, that is systems evolving in time, will then be studied. Then tools for studying the importance of each component will be introduced. These tools can hopefully be used by computer decision makers in order to find where systems can be improved. At the end, some advanced topics will be discussed, namely system simulation, component dependency, repairable systems and multistate systems.

For many purposes, analysis using repairable multistate systems, is the realistic way to model how real-world systems actually behave. However, such refined models are difficult to handle analytically and are thus not readily available to most system administrators. This text will thus concentrate on systems and components that either works or fails, and where the components are not repaired before the system fails. Furthermore, it will in most of the text be assumed that the probability of system failure in the time period is so low that it is not of interest to check if the system fails more than one time in the time period.

Most of what will be described in this text, is covered in the seminal work of Barlow and Proschan, see [1]. The text here is mainly inspired by a text written by Natvig [8], though. Some extra depth is added in [4], while [11] handles the case of computer system reliability in great practical detail.

## 2. Probability and risk

### 2.1. *Basic probability*

The main object of basic system reliability analysis is to compute the probability of the system functioning (or failing) over a given time period. This will be computed by specifying how the system is composed in terms of components and the reliability of the components. Thus, in order to do this kind of analysis, one must be able to handle probability calculations.

Probabilities are defined for combinations of the possible outcomes. Such combination of possible outcomes are called events. In this case, typical events will for instance be the system working, component 1 working or component 2 failing while component 1 is working. There are some basic rules for the probability of events.

Probability rules:
1. $0 \leqslant Pr(event) \leqslant 1$. Thus the probability of a given event lies between 0% and 100%.
2. $Pr(some\ event) = 1$. Thus the probability for some kind of event to happen is $1 = 100\%$. For instance, $Pr(system\ works\ or\ fails) = 1$, since all kind of events will be covered by the system either working or failing.
3. $Pr(A\ or\ B) = Pr(A) + Pr(B)$ if event $A$ and event $B$ cannot happen at the same time. Thus $Pr(system\ works\ or\ system\ fails) = Pr(system\ works) + Pr(system\ fails)$. Note that together with rule number 2, this implies that $Pr(system\ fails) = 1 - Pr(system\ works)$. The same is true for components, of course. An example of mutual exclusion is the state of a component. In the easiest case, it either works or does not work. An example of a set of events that are not mutually exclusive is the state of two different components. If there is no strict dependency between the two components, than one component might fail while the other works and vice versa. A special case is if a component is crucial to a system. If so, the combination of the system working and the component failing is mutually exclusive. Note that the event of the component working and the system failing is not mutually exclusive, as long as the system also relies on other components.

In illustrating the rules, it has been assumed that either a component (or system) works or it does not work. Thus there is no room for a component being in a middle state, where it works but works poorly. Such are the problems associated with multistate systems, which we will not study before the end of this chapter. Before that, we will assume so-called binary states, where a component or system either works or fails.

Since the components and the system is described by a binary set of states (working or failing), one only needs one probability for each component and one probability for the system. The probability for failing is one minus the probability for working and vice versa. However, specifying what the component probabilities should be, may not be so easy. There are two distinct schools as to what probability really is in the real world.

The frequentistic approach says that probability of an event is the rate at which the event happens, when the number of events goes toward infinity. If you have a large number of independent trials where you test whether a component fails or works (thus a binomial distribution of success), you can set the probability of component failure as

$$Pr(component\ working) = x/n, \tag{2}$$

where there are $x$ trials where the component has worked and $n$ total trials.

Often you do not have a large number of times when a specific component has failed, and cannot wait until you get such an extensive data set. Then it might be useful to take the other major view of what probability is, namely the subjectivist (Bayesian) view. Here, probability defines how much you believe in a statement, such as 'the component will fail in the next time period'. If you have no data at all concerning a component, the probability will simply be whatever your best judgment says it is most likely to be. When some data is available, like for instance that the component has not failed in the 25 preceding time periods, that subjective probability will be tempered by the data. How this is done mathematically, can be somewhat complicated even for binary events. However, unless one have very specific subjective views about a component, one can use the following formula, based on a Bayesian treatment of the binomial distribution:

$$Pr(component\ working\ given\ data)$$
$$= \frac{sPr(component\ working) + x}{n + s}, \tag{3}$$

where $Pr(component\ working)$ is the subjective probability before the data was processed and $s$ is the number of observations that the subjective probability can be thought to be based on. Quite often, it might be realistic to set $s = 1$ in order for the prior to be influential only when the number of measurements is very low. Note that when $n$ and $x$ becomes large, Equation (3) approaches Equation (2).

Here is one example. Suppose that you are interested in the probability that a specific program does the job it is intended to do. While you feel certain that the program works tolerably well when correctly installed and configured, you are not sure whether you have managed to install and configure it correctly. Your knowledge about the program suggests that $Pr(success) = 75\%$ is realistic, but you are not very certain about this estimate, thus $s = 1$. Assume that the program has not failed the 19 times when you have tried it. Then the updated Bayesian probability for the program doing what it is intended to do is $Pr_{updated}(working) = (0.75 + 19)/20 = 98.75\%$. Note that the frequentistic probability of success would be exactly 100% in this case, thus indicating no chance of failure. There are frequentistic ways to rectify this result, but these methods are a bit more difficult both to calculate and to justify. Thus, this result can serve as a reminder why Bayesian probabilities may be of use in system reliability analysis.

### 2.2. Combining events

Probability rule number 3 gives us a way of combining different events, but only if they are disjoint, which means that they are mutually exclusive. In order to combine events and find the probability for the combination, one needs the concept of a joint probability. A joint probability is the probability that two or more events are true at the same time; $Pr(A, B) = Pr(A\ and\ B)$: For instance, A might be the event of someone sending an email to you, while B might be the event of the email system not working. $Pr(A, B)$ would then give the probability that an email is sent to you while the email system is down.

Specifying joint probabilities can often be quite difficult. If there are two components, one needs to specify the probability for component 1 and 2 working, component 1 working and 2 failing and component 1 failing and 2 working. The case of both failing is covered, since the sum of all possible events is equal to one and the events covered here are disjoint. For three components, there are 7 possible combinations to consider, and for $n$ components there are $2^n - 1$ possible events, a hopeless task for a large system.

Fortunately there is a concept called independence that can help considerably. Two events, such as two components working, are said to be independent, if $Pr(A\ and\ B) = Pr(A)Pr(B)$. It might be easier to think of independence in terms of conditional probabilities. The probability for event $A$ conditioned on event $B$ is defined as $Pr(A|B) = P(A, B)/P(B)$. Independence means that $Pr(A|B) = P(A, B)/P(B) = Pr(A)$. Thus if

$A$ and $B$ are independent events, and we know that the event $B$ has happened, this does not change our probability for event $A$.

Here is one example of a calculation using independent events. Assume that the probability of receiving email a given day is 50% and the probability of the email system failing is 2%. Then, if receiving email is independent of email system failure, the probability of you loosing an email that given day is 1%.

Usually it is assumed that all components in a system are independent. If there is cause for thinking otherwise, it might be that the system is not fully specified. For instance, if experience shows that two email servers goes down at the same time more often than what independence requires, then it might be that the servers have a common component that causes them to fail simultaneously, such as the power supply. It could then be easier to add the power supply to your system description, rather than having to deal with dependent components. Thus, an insufficient system description might cause an artificial component dependency, that can be removed by a better system description.

The concept of independent events such as failures can be expanded to more than two events, thus:

$$Pr(E_1 \text{ and } E_2 \text{ and } \cdots \text{ and } E_n) = Pr(E_1)Pr(E_2)\cdots Pr(E_n) \equiv \prod_{i=1}^{n} Pr(E_i), \quad (4)$$

where each $E_i$ is an event independent of the other events, $\{E_j\}$ for $j \neq i$. This is makes it possible to handle a compound event such as component 1 and 2 working and component 3 and 4 failing.

If one is interested in events such as component 1 or 2 failing, then one also needs an algebra for combining such events. Probability rule number 2 could look like that, but this rule is used for disjoint events, not independent events. However, handling that is easy in the independent, binary case, as $Pr(event) = 1 - Pr(!event)$, where '!' means 'not'. Thus

$$Pr(E_1 \text{ or } E_2 \text{ or } \cdots \text{ or } E_n) = 1 - Pr(!E_1 \text{ and } !E_2 \text{ and } !E_n)$$

$$= 1 - \prod_{i=1}^{n} Pr(!E_i) = 1 - \prod_{i=1}^{n} \left(1 - Pr(E_i)\right)$$

$$\equiv \coprod_{i=1}^{n} Pr(E_i), \quad (5)$$

where the meaning of the symbol $\coprod$ is defined by $\coprod_{i=1}^{n} x_i = 1 - \prod_{i=1}^{n}(1 - x_i)$ and $x_1 \coprod x_2 = 1 - (1 - x_1)(1 - x_2)$. Thus if an email system works if one of two servers works, and these servers functions independently, and $Pr(server\ 1\ works) = Pr(server\ 2\ works) = 90\%$, then $Pr(email\ system\ works) = 1 - 0.1 \times 0.1 = 99\%$. If rule number 3 had been used uncritically in stead, this would have yielded $Pr(email\ system\ works) = 180\%$, a result that is obviously wrong.

## 2.3. *Expectation and risk*

Risk was mentioned in conjunction with Equation (1). There, risk was given an explicit form which fits with binary system analysis. The term is a bit more general, though. Risk is generally defined as expected loss. Expectation is also a well-defined mathematical entity in probability theory. It can be described as the long-run average of some mathematical variable, the same way that frequentistic probability is the long-run rate of events. For a variable $X$ which can have several outcomes $\{x_i\}$, $M$ in number, with probabilities for each outcome being $\{p_i\}$ we have:

$$E(X) = \sum_{i=1}^{M} x_i p_i, \tag{6}$$

where E stands for expectation. Expectation is linear, that is if $a$ and $b$ are constants while $X$ can vary, then $E(a + bX) = a + bE(X)$. When there are only two outcomes, called success and failure, and $X = 1$ indicates success while $X = 0$ indicates failure, then $E(X) = p(success)$ and $E(1 - X) = p(failure)$. The loss in a typical system setting is the cost of the system plus the cost of system failure if the system fails. Thus $R = E(loss) = E(C_{system} + C_{failure}(1 - X)) = C_{system} + C_{failure} p(failure)$ and this describes the origin of Equation (1).

## 3. Parallel and serial systems, duality

We shall now study the simplest of systems, namely those called serial and parallel systems. First off we will look at serial systems.

## 3.1. *Serial systems*

In a serial system, all the components needs to work, in order for the system to work. A couple of examples will illustrate this. In most electronic hardware all of the electronic components must be functioning properly, if the hardware system itself should work. In a computer program that reads a file, processes the information in several routines and writes a new file, the reading routine, writing routine and all of the processing routines must all work properly in order for the program to work properly. In a 3-level architecture, the system only works if the software in each level works and the interface between each of the levels works.

In such a system, where none of the components can do the job of any of the other components, the probability of success, $h$, is

$$h \equiv Pr(system\ works)$$

$$= Pr(C_1\ works\ and\ C_2\ works\ and \cdots and\ C_n\ works), \tag{7}$$

where $h$ is the system reliability and $C_i$ is component $i$. Assuming independence and setting $Pr(C_i) = p_i$, one gets:

$$h = \prod_{i=1}^{n} p_i. \tag{8}$$

This equation is easy to calculate, but constitutes a worst case scenario when it comes to system reliability. All systems where there are components that can take over the job of other components will be more reliable than a serial system. Never the less, for some systems a serial design might be much more practical to implement than any other design.

### 3.2. *Parallel systems*

In the sense of duality, which we will explore more later on, a parallel system is the exact opposite of a serial one. In a parallel system, if at least one component functions, this will ensure that the whole system is working. Thus

$$h = Pr(C_1 \text{ working or} \cdots \text{or } C_n \text{ working}) = \coprod_{i=1}^{n} p_i, \tag{9}$$

where independence is assumed for the last part of the equation. One example of this has been handled already, but can be expanded. Assume that an email system works if one of the 10 email servers works. Assume further that $p_i = 90\%$ for each of the servers and that the servers are all independent (this might be an unrealistic assumption in most cases). Then $h = 1 - 0.1^{10} = 99.99999999\%$! A load-balancing system will ideally be a parallel system.

A component is said to be in series with the system if the system always fails if the component fails. A component is said to be in series with another component if the state of one component only is relevant for the system if the other component works.

Likewise, a component is said to be in parallel with the system, if it is so that if the component works then the system works, even if all the other components have failed. A component is said to be in parallel with another component if the state of one component only is relevant for the system if the other one has failed.

### 3.3. *Duality*

As mentioned earlier, the probability of success is one minus the probability of failure. That has already been used for finding the probabilities for *and/or* combination of events. The expression makes use of the fact that the opposite of say component 1 *or* 2 *or* $\cdots$ *or* $n$ failing is component 1 *and* 2 *and* $\cdots$ *and* $n$ working. Thus the probability for a serial system failing is the same expression one would get in a parallel system for the probability that the system works. Because of this symmetry, serial and parallel systems are called dual systems. Duality means that one can switch between expressing the conditions and probabilities for failure and for success.

This can be useful to keep in mind, as it sometimes can be less cumbersome to study the conditions for a system working while it often is the conditions for system failure that is of most interest.

From now on, we will denote $p_i$ as probability of a component working, while $q_i = 1 - p_i$ is the probability that it fails. $h$ denotes the probability that the system works, while $\psi = 1 - h$ denotes the probability of system failure.

## 4. Fault trees

### 4.1. Standard fault tree analysis

In fault tree analysis, one is interested in expressing the conditions for a system failure. A graphical expression of this is made by combinations of *and* gates and *or* gates, such that the gate fails if the logical combination of events indicates failure.

An example of this is shown in Figure 1. This example is built on the following premises: There are two email servers. Incoming email is sent to both servers if they work, and to the one working if the other one fails. Thus, with a working network and power supply, the conditions for the system failing is that server 1 and 2 both fails. This explains the leftmost *and* gate. The power supply fails if the primary power fails and if either the secondary power fails or the switching mechanism fails. This explains the *or* gate between secondary

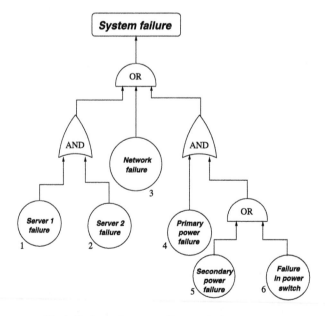

Fig. 1.  Fault tree for two email servers in the same room.

power failure and switching mechanism failure, as well as the *and* gate above it. The system as such fails if either both the servers fails, the network fails or the power supply fails. This explains the uppermost *or* gate.

Having made a fault tree, one can use the combination methods in Equations (4) and (5), starting at the basic event, which are labeled from 1 to 6 in the figure. One then work upward, gate by gate. Thus if $q_1 = q_2 = 10\%$ then the probability in the leftmost *and* gate is $0.1^2 = 1\%$. This is the probability that both servers fail. Working from component 5 and 6, with $q_5 = 1\%$ and $q_6 = 10\%$ one finds that the rightmost *or* gate has a probability of $1 - 0.99 \times 0.9 = 10.9\%$. If $q_4 = 1\%$ then the *and* gate above that gets the probability $0.109 \times 0.01 = 0.109\%$. Finally, if the probability of network failure is $q_3 = 2\%$ then the uppermost *or* gate and thus the probability of system failure becomes $\psi = 1 - 0.99 \times 0.98 \times 0.99891 \approx 3.1\%$.

The largest contributor to this probability is in this case the probability for network failure, so if all things are equal then improving network reliability might be the best investment.

One does not need to specify the probabilities to do calculation. One can make an analytic expression of the probability of system failure from the probabilities of component failure:

$$\psi = q_1 q_2 \coprod q_3 \coprod q_4 \left( q_5 \coprod q_6 \right)$$
$$= 1 - (1 - q_1 q_2)(1 - q_3)\left(1 - q_4\left(1 - (1 - q_5)(1 - q_6)\right)\right). \qquad (10)$$

There can be arbitrary many *and* and *or* gates in a fault tree. As long as there are no repeated component (more on that later), the method described here can be used.

### 4.2. *Dual picture*

In the case of fault trees, the dual picture can easily be derived. As mentioned before, the opposite of event $A$ and $B$ is the event of no $A$ or no $B$. Also, the opposite of event $A$ or $B$ is no $A$ and no $B$. Thus if one has a fault tree, can make a graphical representation showing what is necessary for the system to work, a 'success tree'. This is achieved by switching from components failing to components working and switching *and* gates with *or* gates and vice versa. Thus from the previous example, one gains a tree like that in Figure 2.

If some subset of a system is easier to describe by finding conditions for it working than failing, then one can make a tree describing these conditions, as in Figure 2. Duality can then be used for finding the fault tree of that subsystem, which can then be inserted into the rest of the fault tree for the system.

### 4.3. *Fault tree failures*

As was mentioned earlier, when component failure events are repeated in a fault tree, one runs into difficulties. The reason is that the method for combining events in a fault tree

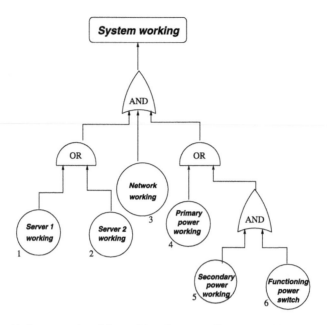

Fig. 2. Graphical representation of the conditions for two email servers working in the same room.

rests on the assumption of independent events. This is obviously not the case when one combines two events that both contain the same component failure event.

As an example, we will study a so-called 2-of-3 system. Such a system is composed of three components. Two of these components are needed in order for the system to work. This means that if two or more components fail, then the system fails. Thus the system and the dual system is the same.

As an example, consider an email system consisting of three servers, where network and power failure is disregarded. The load of each of the servers is so high, that if only one server is left to handle that load, then in a short while the machine will crash. Thus at least two servers need to be functioning at any time.

One can then make a fault tree as that in Figure 3.

Using the same method as before, one gets that the probability in the first *and* is $q_1 q_2$, in the second $q_1 q_3$ and in the third $q_2 q_3$. Using the fault tree method naively, one gets that the probability in the *or* gate and thus the probability for system failure ought to be:

$$\psi_{\text{naive fault tree}} = q_1 q_2 \coprod q_1 q_3 \coprod q_2 q_3$$
$$= 1 - (1 - q_1 q_2)(1 - q_1 q_3)(1 - q_2 q_3). \tag{11}$$

This is not the correct result.

We can find the correct probability by introducing four disjoint events, namely:

(1) Server 1 and 2 fails while server 3 works.
(2) Server 1 and 3 fails while server 2 works.
(3) Server 2 and 3 fails while server 1 works.

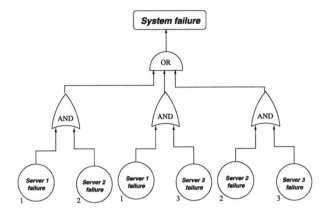

Fig. 3. Fault tree for a 2-of-3 system.

(4) All servers fail.

Then the probability of system failure will be the sum of the probabilities for each of these disjoint events, thus

$$\psi = q_1 q_2 (1 - q_3) + q_1 q_3 (1 - q_2) + q_2 q_3 (1 - q_1) + q_1 q_2 q_3. \tag{12}$$

Equation (12) is not the same as (11). This can be checked by setting $q_1 = q_2 = q_3 = 30\%$. This yields $\psi_{\text{naive fault tree}} = 24.6\%$, while the correct result is $\psi = 21.6\%$. Also note that since a 2-of-3 system is it's own dual system, one gets the same result for the probability of the system working, just with success probabilities in stead of failure probabilities.

The expression in Equation (11) can be used to obtain the correct expression. As will be described in the next section, system failure and success probabilities should be linear in all component probabilities. Writing out the factors in Equation (11) and putting $q_i$ in stead of $q_i^2$ at all places, gives Equation (12). However, there is much room for making mistakes when dealing with fault trees with repeated components, so the fault tree approach is thus not recommended for such systems. However, fault trees without repeating components constitutes a very limited set of systems. The system function approach and the graph system approach, which will be described in the next sections, are methods that can describe a wider variety of systems.

## 5. General definitions

### 5.1. *The structure function*

As was described previously, a fault tree is a good tool for those cases where a system can be described using *and* and *or* gates of events, with no repeated components. There are several examples of systems where such an assumption will not hold, however. For such cases, more general definitions and expressions are needed.

We will first and foremost consider the structure function description in this section. A structure function is a function that describes under what circumstances a system works and fails. If a system works, the function takes the value 1, while if the system fails it takes the value 0. The input is the state of the components, which can be seen as a vector of zeros and ones, where one denotes that a component works, and zero denotes that it has failed.

Since each component can be in two states, and there are $n$ components, that means that the structure function, $\phi(\underline{x})$ is described by a table of $2^n$ entries, where $\underline{x} =$ (*state of component* $1, \ldots,$ *state of component* $n) = (x_1, \ldots, x_n)$. This is a very cumbersome way of describing a function, but it is useful as a theoretical tool.

Here is one example of a structure function, applied to the 2-of-3 system in the previous section:

$$\phi(0, 0, 0) = 0, \quad \phi(0, 0, 1) = 0, \quad \phi(0, 1, 0) = 0, \quad \phi(1, 0, 0) = 0,$$
$$\phi(0, 1, 1) = 1, \quad \phi(1, 0, 1) = 1, \quad \phi(1, 1, 0) = 1, \quad \phi(1, 1, 1) = 1.$$

Not every kind of structure function makes sense, though. For instance, if there is a component $i$ such that $\phi(x_1, \ldots, x_{i-1}, 1, x_{i+1}, \ldots, x_n) = \phi(x_1, \ldots, x_{i-1}, 0, x_{i+1}, \ldots, x_n)$ for any configuration of the remaining $n - 1$ components, than component $i$ is said to be irrelevant. This will be expressed in a more compact manner as $\phi(\underline{x}_{-i}, 1_i) = \phi(\underline{x}_{-i}, 0_i)$. Usually one would not want to study systems with irrelevant components. Also, one would not study any system where the system can go from failing to working by having one component going from working to failing. Thus one wants that $\phi(\underline{x}_{-i}, 1_i) \geqslant \phi(\underline{x}_{-i}, 0_i)$ for all configurations of the remaining $n - 1$ components. If this holds for all the components in the system and the system works if all the component works and fails if all the components fail, then the system is called monotone. A monotone system without irrelevant components is called coherent. Coherence will be assumed for all the systems described here.

For two components, there are only two coherent systems, namely a serial system and a parallel system. For three components, there are five different coherent systems, namely one that is purely serial, one that is purely parallel, one with a single component in series with a parallel subsystem, one with a single component in parallel with a serial subsystem and the last is the 2-of-3 system described earlier. When the number of components increases, experience shows that more and more coherent systems will not be expressible as fault trees without repeating components. All fault trees without repeated components are coherent, as long as the basic events are single component failures and there are no system components left out of the fault tree. The reason is that each node in the fault tree consists of *and* and *or* gates, which combines the events below each gate in a coherent manner.

For any coherent system of $n$ components, one has that

$$\prod_{i=1}^{n} x_i \leqslant \phi(\underline{x}) \leqslant \coprod_{i=1}^{n} x_i. \tag{13}$$

The reason is that if $\prod_{i=1}^{n} x_i = 1$ then all components work, and since the system is monotone, that means that $\phi(\underline{x}) = 1$. For all other configurations, $\prod_{i=1}^{n} x_i = 0$ which is the lower limit for the structure function. Meanwhile $\coprod_{i=1}^{n} x_i = 0$ only if all components fails, and then $\phi(\underline{x}) = 0$ also.

One thing that Equation (13) says is that a purely parallel system will perform better than a purely serial system, as long as the probabilities are the same. In order to show that we need to find the link between the structure function and probabilities, assuming independent components. Before that we need the following results:

$$\phi(\underline{x}) = x_i \phi(\underline{x}_{-i}, 1_i) + (1 - x_i)\phi(\underline{x}_{-i}, 0_i). \tag{14}$$

This equation expresses the fact that we can pick one particular component and express the structure function in terms of whether it works or not. If it works, we get $x_i = 1$ times the structure function for the remaining components with component $i$ set to working. If it fails, we get $1 - x_i = 1$ times the structure function for the remaining components. The equation combines these two cases. Using Equation (14) on all components, yields:

$$\phi(\underline{x}) = \sum_{\underline{y} \in \{0,1\}^n} \prod_{i=1}^{n} x_i^{y_i} (1 - x_i)^{1-y_i} \phi(\underline{y}). \tag{15}$$

This might seem like a useless expression. However, if you take the component state to be stochastic (random), then one can see from Equation (15) that the structure function is a linear combination of the component states, which is good news if one wants to find the expectation of the structure function.

As stated earlier, an expectation of a binary result is simply the probability of the result $X = 1$. Using that the expectation is a linear operator, we get

$$h(\underline{p}) = E\big(\phi(\underline{x})\big) = \phi\big(E(\underline{x})\big) = \phi(\underline{p}). \tag{16}$$

In the case of serial verses parallel systems, this means that $\phi_{\text{serial}}(\underline{p}) \leqslant h(\underline{p}) \leqslant \phi_{\text{parallel}}(\underline{p})$. Thus we have a lower and upper limit for how well a system can perform, given the component reliabilities.

Other useful rules can also be obtained from this general set of assumptions. Suppose that we have a system that we wish to improve. Suppose that we can double the number of components and either set each of the new components in parallel with the original ones or make a new system in parallel with the original one. Which of these two solutions is the best?

Let $\underline{x}$ be the state of the original components and $\underline{y}$ be the state of the inserted components. Let $\phi(\underline{x} \coprod \underline{y})$ be a component-by-component co-product of each state. It can be shown, using monotony, that

$$\phi\Big(\underline{x} \coprod \underline{y}\Big) \geqslant \phi(\underline{x}) \coprod \phi(\underline{y}). \tag{17}$$

This expresses that for any component configuration, a system where the new, inserted components are parallel to the original ones will always perform better than the original system in parallel with the inserted components put in the same configuration. Stated more intuitively, a system with parallel components is better than a system with parallel subsystems. Using Equation (16), Equation (17) gives that

$$h\left(\underline{p}\coprod\underline{p}'\right) \geqslant h(\underline{p})\coprod h(\underline{p}'),\tag{18}$$

where $\underline{p}'$ is the reliabilities of the new components. Thus the system reliability will also be greater in the former setting than in the latter. An example of this can be seen in Figure 5 in the next section. There the basic system is a series of two components. Figure 5(a) represents parallel components while Figure 5(b) represents parallel subsystems. The first system will be better than the second.

In the same way, it can be shown that:

$$\phi(\underline{x}\,\underline{y}) \leqslant \phi(\underline{x})\phi(\underline{y}),$$
$$h(\underline{p}\,\underline{p}') \leqslant h(\underline{p})h(\underline{p}').\tag{19}$$

Thus a serial structure on the component level is always worse or equal to a serial structure on the system level.

These results can give us some simple design guidelines.

### 5.2. Cut sets and path sets

A cut set is a set of components such that if all components in the cut set fail, then the system fails for any state of the other components. A minimal cut set is a cut set such that, if one removes any component from the set, the result will not be a cut set. Thus if the components that are not in the minimal cut set works and a single component in the minimal cut set works, then the system works.

The opposite of a cut set is a path set. If all the components in a path set works, then the system works. A minimal path set is a path set where one can not remove a component from the set and still have a path set.

Going back to the example in Figure 1, one can see that for instance $(1, 2)$, $(2, 3)$, $(3)$, $(1, 4, 6)$, $(4, 5)$ and $(4, 6)$ are all cut sets. However, only $(1, 2)$, $(3)$, $(4, 5)$ and $(4, 6)$ are minimal cut sets.

Similarly using Figure 2, one sees that $(1, 3, 4)$, $(2, 3, 4)$, $(2, 3, 5, 6)$, $(1, 3, 5, 6)$ and $(1, 2, 3, 4, 5)$ are path sets. Only the first four sets will be minimal path sets, though.

Since a system can work if one of the components in a minimal cut sets works and this is true for all minimal cut sets, one can look at a system as a serial system of cut sets, where each cut set is a parallel system. Thus if there are $k$ minimal cut sets and each cut set is labeled $K_j$ then:

$$\psi(\underline{x}) = \prod_{j=1}^{k} \coprod_{i \in K_j} x_i.\tag{20}$$

Thus if all minimal cut sets are known, the structure function is also known. Note that the function found here can be linearized in all components, since $x_i^2 = x_i$ for $x_i = 0$ and $x_i = 1$. Thus when such replacements have been done, one can write the resulting structure function as $\psi(\underline{x}) = \sum_A \delta(A) \prod_{i \in A} x_i$, where $A$ are sets of components and $\delta(A)$ is the coefficient found from Equation (20). This can then be used for finding the system reliability for independent components, $h = \sum_A \delta(A) \prod_{i \in A} p_i$. There are more sophisticated methods for finding reliability from minimal cut sets, but this will be outside the scope of this text.

Similarly as for minimal cut sets, when there are $m$ minimal path sets, $P_j$, then the structure function can be expressed as:

$$\psi(\underline{x}) = \coprod_{j=1}^{m} \prod_{i \in P_j} x_i. \tag{21}$$

Duality means that a cut set for a system will be a path set for the dual system and a minimal cut set for a system will be a minimal path set for the dual system. The same holds the other way around.

Finding minimal cut sets or path sets can be seen as a way to expand upon the fault tree analysis. Having the cut sets means that one has an *or* gate on top with one *and* gate for each minimal cut set and with the components of the minimal cut set below the corresponding *and* gate. However, note that when a component is found in more than one minimal cut set, then the fault trees will have repeated components.

For independent components a knowledge about the structure function and the component reliabilities is enough to find the system reliability. Thus finding the minimal cut sets having the component reliabilities is also enough to find the system reliability.

## 6. Graph systems

### 6.1. *Introduction to graph systems*

Graph systems are systems that can be expressed as graphs, that is figures with vertices and edges. An example is shown in Figure 4. The edges represent the different components in the system. For the system in Figure 4, which is called the bridge system, there are five different labeled components. In this case the edges does not have a direction, and the graph is thus called undirected. Often the components are further emphasized on the edges by a box or circle with the component name or number.

The vertices only represent joining points for the edges, except for two (or more) connection vertices, which are used for determining whether the system works or not. The system is said to be working if there is a path going either way from one of the connector vertices to the other, going where the path contains only working components. If there is no path through working components that joins the two vertices, then the system does not work. In the example in Figure 4 the connector vertices are labeled C1 and C2.

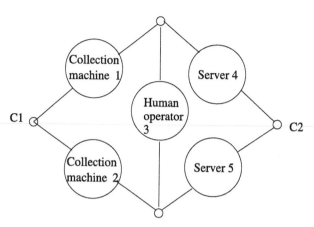

Fig. 4. Example of a system represented by a graph, the bridge system.

## 6.2. *The bridge system*

As an example that leads to Figure 4, consider a database collecting data from different measurement stations out in the periphery. We shall study a system that handles incoming data from these stations. Assume that the database reliability is so high that this component can be ignored. Data is sent via modem to two small collection machines, labeled 1 or 2. Data from collection machine number 1 is then default sent to server 4, which contains a program that inserts the data into the database. Data from machine number 2 is usually sent to server 5.

If a human operator has found that the components 1 and 5 are not working, then he can manually make collection machine number 2 send the incoming files to server 4 in stead. Similarly, the operator can make collection machine 1 send to server 5 if the components 2 and 4 are not working. However, a human component is seldom completely reliable, so we will include the operator into the system, labeled component number 3. Note that if a server gets files from both collection machines, then the system will not work properly. We will assume that the operator will not configure the system so that this will happen.

We can now do a minimal cut set analysis of the situation. Clearly the system will fail if component 1 and 2 fails. If only component 1 fails or only component 2 fails, then there are configurations where the system will work, so $(1, 2)$ is a minimal cut set. Thus the system depends on either machine number 1 or 2 working as specified. It is easy to see that $(4, 5)$ is also a minimal cut set. The set $(1, 3, 5)$ will also be a minimal cut set. To see this assume that all those components are not working properly. Since collection machine 1 does not work, no data will come from that machine. Thus only data from collection machine 2 will have a chance of reaching through to the database. But the default way to the database is through server number 5 which is not working. Since the human operator seems to be inactive (for instance sleeping at the job), the files will not be redirected to server 4. Thus no data will reach through to the database. Similarly $(2, 3, 4)$ is a minimal cut set.

The minimal paths sets are $(1,4)$, $(2,5)$, $(1,3,5)$ and $(2,3,4)$. Note that by interchanging component 2 and 4, one gets the minimal cut sets. This means that the dual system of the bridge system will also be a bridge system, with component 2 and 4 interchanged.

One could, theoretically speaking, put these results into Equation (20) or (21), write the result out as a large sum of different contributions and replace $x_i^2$ with $x_i$ where ever one found such terms. One could then use Equation (16) to find the system reliability.

This is however a very complicated way of doing things, as there is a much easier way of studying this. Note that if component number 3 (the human operator) works, then the system is simply a serial system of machine 1 parallel to machine 2 and server 4 parallel to server 5, see Figure 5(a). If component number 3 does not work, then the system is a parallel system containing a series of component 1 and 4 and a series of component 2 and 5, see Figure 5(b).

The system in Figure 5(a) has a reliability $(p_1 \coprod p_2)(p_4 \coprod p_5)$ since it contains two subsystems in series. The first subsystem is component 1 and 2 in parallel, and has reliability $p_1 \coprod p_2$ while the second subsystem is component 4 and 5 in parallel and has reliability $p_4 \coprod p_5$. Similar argument yields that the system in Figure 5(a) has a reliability $p_1 p_4 \coprod p_2 p_5$. Since component 3 either works or fails, and these are disjoint events, the system reliability can thus be written as

$$ h = p_3 \left( p_1 \coprod p_2 \right)\left( p_4 \coprod p_5 \right) + (1 - p_3)\left( p_1 p_4 \coprod p_2 p_5 \right). \tag{22} $$

As an example, assume that the servers are quite reliable, $p_4 = p_5 = 95\%$. The collection machines are somewhat less reliable, so that $p_1 = p_2 = 80\%$ while the human operator is notoriously unreliable so that $p_3 = 50\%$. Then $h = 0.5 \times (1 - 0.2^2)(1 - 0.05^2) + 0.5 \times (1 - 0.24^2) = 95\%$. Thus the system will fail with a probability of $\psi = 5\%$. If the human component was completely reliable then $\psi = 4.24\%$ while if said component was completely unreliable then $\psi = 5.76\%$. Clearly, component 3 matters in such a case, though maybe not so much that extra resources are justified.

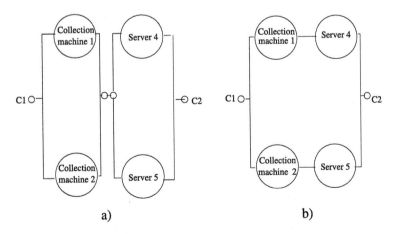

Fig. 5. The bridge system if component 3 either works (a) or fails (b).

## 6.3. *An example of risk analysis*

As an example of decision making in this scenario, assume that the choice stands between three choices. The first choice is to not employ any human operator, thus no extra cost and $p_3 = 0\%$. Giving the task to someone with a lot of other responsibilities on his or her hands who can only watch the system during day-time has in this scenario a probability of failure per day of $p_3 = 50\%$ and the extra cost per day is \$100. The third option is to employ several trustworthy people to watch this task day and night. The assumption here (which is somewhat extreme), is that this choice makes the human component totally reliable, thus $p_3 = 100\%$. The extra cost per day will in this case be \$1000. Assume that a system failure costs \$20000 per day. Using Equation (1), the risk in these three cases are $R_1 = 5.76\% \times \$20000 = \$1152$, $R_2 = 5\% \times \$20000 + \$100 = \$1100$, $R_3 = 4.24\% \times \$20000 + \$1000 = \$1848$. Clearly, in this case, going for the unreliable human operator with a lot else on his or her hands is in fact the best solution. Obviously, this result does not hold if the various costs are changed.

## 6.4. *The general approach*

The bridge system example illustrates a more general way of solving undirected graph systems. There are also other classes of systems, besides graph systems, that can be solved by the method shown here. The primary example of this is the $k$-of-$n$ system. This is a system which works if at least $k$ of the $n$ components works and fails this is not the case. A special case is the 2-of-3 system which we have already studied.

The first element in this method is called s-p-reduction and the first element in that method is called serial reduction. If any components are found serial to each other, then the system reliability of those two components $i$ and $j$ will only rely on the product of them, $p_i p_j$. We can see this by observing that we can replace the two components with a single component which has a probability of success equal to a two component serial system, thus $p_{\text{new component}} = p_i p_j$. This will be done in all places where it is possible. Such a method, called serial reduction, was used on component 1 and 4 and component 2 and 5 in Figure 5(b).

If any components are found in parallel with each other, then they can be replaced by a new component having the reliability $p_i \bigsqcup p_j$. This method, called parallel reduction, was used in Figure 5(a) for component 1 and 2 and for component 4 and 5. After that, a serial reduction was done one these composite components, which resulted in the reliability for Figure 5(a). Similarly a parallel reduction of the serial composite components in Figure 5(b) yielded the reliability of that system.

Repeated usage of serial and parallel reduction as far as it goes, is called s-p-reduction. In the case of the systems in Figure 5, this reduction was done until only one composite component remained, which gave us the reliability of each of those systems. When an s-p-reduction on a system yields one remaining component, then that system is called completely s-p-reducible.

The other operation done was finding a component that made s-p-reduction impossible and handling that component by looking at the event of the component working or not working. This is what Equation (14) does in essence. If this is combined with Equation (16), then one gets that;

$$h = p_i \phi(\underline{p}_{-i}, 1_i) + (1 - p_i) \phi(\underline{p}_{-i}, 0_i). \tag{23}$$

This splits the problem into two lesser problems, namely finding the reliability if the component works and finding the reliability if it does not work. This is often called a pivot decomposition.

The method for solving systems of undirected graphs (with independent components) then simply becomes the following algorithm:

While there remains more than one composite component: Step 1 – Do s-r-reduction as far as it goes. Step 2 – If there is more than one component remaining, do a pivot decomposition such that both systems are coherent (no irrelevant components). If there is only one component left, the system reliability is set to the reliability of that single component.

In the bridge system case, no s-r-reduction was possible in the start. After the pivot decomposition however, both cases were completely s-r-reducible. This means that an s-r-decomposition resulted in a single component. When that was achieved, we stopped the computation.

## 6.5. *Fault trees as undirected graphs*

Note that a fault tree with no repeating components is completely s-r-reducible from the start. This is because *and* and *or* gates represents serial and parallel combinations of single and composite components. Since the whole system can be represented by such gates, s-r-reduction will result in a single composite component, represented by the gate at the top of the tree.

Any system represented by a fault tree with no repeating components can also be represented by an undirected graph. If the dual system is found, as described in Figure 2 then *or* gates can be represented by putting the components in parallel graphically with each other and *and* gates can be represented by putting the components serial to each other graphically. See Figure 6 for the result. Often such a representation can give a more comprehensible view of the system than a fault tree.

Since if a completely s-p-reducible system only contains serial and parallel subsystems and can thus be described by *and* and *or* statements, a completely s-p-reducible undirected graph system can also be represented as a fault tree with no repeated components. This is not so for a system which requires pivot decomposition before complete reduction is achieved. For instance the bridge system cannot be represented as a fault tree with no repeated components.

## 6.6. *Directed graphs and more complex graph based system descriptions*

It should be noted that there are also systems represented by directed graphs, that is graphs where the edges have directions. A system represented in this fashion works if one can

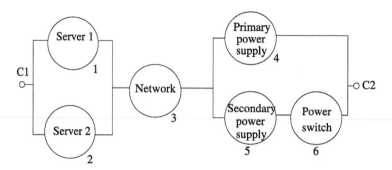

Fig. 6. The example from Figures 1 and 2 as an undirected graph.

go from one specific vertex to one or several other specific vertices by always going in the correct direction and only going through working components. One can think of such systems as signal sending systems where the object is to get a signal from one end of the system to the other and a component will block the signal if it does not work. There are special methods for handling such systems too [5], but these methods are a bit too complicated to be described here.

Even more complex system descriptions could be though of by letting each vertex be a $k$-of-$n$ system, such that signals are transmitted if $k$ of the $n$ in incoming edges are sending a signal. A more complex system description method makes one able to describe a larger class of coherent systems, but there will be less available computational methods and the remaining methods will likely be quite inefficient. Note that a complete listing of system states is the extreme case when it comes to inefficiency, as $2^n$ different states needs to be handled by a reliability analysis. In comparison, the pivot plus s-p-reduction done in the bridge system example yielded one pivot decomposition plus three s-r-reduction steps for each of the resulting systems, thus the number of operations was $7 < 2^5 = 32$.

## 7. Dynamic system analysis

### 7.1. *Why dynamic system analysis*

So far, we have only studied probabilities of component and system failure in a fixed time period. This is useful when one primarily has to think about the risk and cost per, for instance, day or week. We do not consider what happens in later periods, and fixate only on what happens the next period. While this kind of analysis might answer quite a few questions about a system, the question of the time to the next system failure or the probability that a system works at a given time is not answered.

Obviously, since system failure is stochastic, then so is the time $S$ to the next system failure. The same holds for the time to the next component failure for each component, $T_i$.

Note that under the assumption that components are repaired, it is very hard to answer whether the system works or not at a given time. A system that is repaired will work and fail a number of times in a large enough time period. Also, the availability of the system depends on the reparation policy. These factors make it quite difficult to handle systems under repair, so studying such systems will be postponed until the end of this chapter.


### 7.2. *Cumulative distributions*

So far we have studied stochastic variables that only have a binary set of possible outcomes. In order to look at stochastic variables whose outcome is continuous, one needs a way to describe the probability of the outcomes in such cases. Since a continuous stochastic variable often can have any outcome on the real line or on a open interval on the real line, the probability of one single outcome looses much of it's meaning. If there is a (non-countable) infinite number of outcomes then most or even all of the outcomes must have a vanishing probability, in order for rule number two in Section 2 to hold.

In stead, one can study the probability of getting an outcome in a small interval, or often preferably, study the probability of getting an outcome below a given threshold. The latter probability  is called the cumulative function, $F(x)$, and for a stochastic variable, $X$, and an outcome on the real line, $x$, it is defined as

$$F(x) = Pr(X \leqslant x). \tag{24}$$

The probability of an outcome in a small interval $\langle x, x + \Delta x \rangle$ will be $Pr(x \leqslant X \leqslant x + \Delta x) = F(x + \Delta x) - F(x) \approx F'(x)\Delta x = f(x)\Delta x$. $f(x)$ is often called the distribution function of the variable and can be used for finding the probability inside an interval $Pr(x_1 \leqslant X \leqslant x_2) = \int_{x_1}^{x_2} f(x)\,dx = F(x_2) - F(x_1)$. One can also find the expectation of a stochastic variable with a known distribution; $E(X) = \int_{-\infty}^{\infty} xf(x)\,dx$.

The most common distribution in system analysis is the exponential distribution. The reason can be seen from the previous sections, where fixed size time periods and their probabilities were studied. In order for such analysis to be truly useful, we would not only do such analysis for one single time period, but rather use it repeatedly for the coming time periods, without changing the probabilities and thus without taking into account what happened in the previous time periods. It can be shown that the exponential distribution has these qualities [9]. Also, an exponentially distributed time between events is what characterizes a Poisson process, where the number of events in a time period is independent of what happens in other time periods.

The exponential distribution is specified as $F(t) = 1 - e^{-t/\tau}$, where the parameter $\tau$ is often called the characteristic time. The reason for that name is that the expectation of a variable, $T$, distributed in such a way, is $E(T) = \tau$. Thus over many trials, the average time between events will tend toward $\tau$.

A more mathematical glimpse at the exponential distribution can be gained from the so-called rate function, defined as $R(t) = f(t)/(1 - F(t))$. This rate times a small time interval $\Delta t$ determines the probability of an outcome $t \leqslant T \leqslant t + \Delta t$ given that $T \gg t$. Thus for a component, it defines the probability that a component which has worked up to time $t$ breaks down in the next $\Delta t$ time units. For the exponential distribution the rate will be constant, $R(t) = 1/\tau$.

As mentioned earlier, we will not consider repairable systems here, so we will only use this and other distributions to look at the time to the first breakdown of each component and of the system.

One other distribution worth describing is the Weibull distribution, $F(x) = 1 - e^{-(t/\tau)^b}$ where $b = 1$ yields the exponential distribution, $b > 1$ will give a an increasing rate, while $b < 1$ will give a decreasing rate. Thus for components where the rate is expected to be greater than usual, compared to the exponential distribution, for small time intervals and less than usual for large time intervals, $b < 1$ will be the right choice. For the opposite situation, a Weibull distribution with $b > 1$ is more reasonable.

### 7.3. The basics of dynamic systems

In a dynamic system, the probability of a component working at a time $t$, when the time to breakdown is a stochastic variable $T_i$ distributed as $F_i(t)$ will be:

$$p_i(t) = Pr(T_i > t) = 1 - Pr(T_i \leqslant t) = 1 - F_i(t). \tag{25}$$

If $G(t)$ is the distribution of the time to system breakdown $S$, and $\underline{X}(t)$ are the component states at time $t$, then the probability of the system working is

$$h(t) = 1 - G(t) = Pr\big(\phi(\underline{X}(t)) = 1\big) = \phi\big(\underline{p}(t)\big), \tag{26}$$

where the last transition is achieved by noting that probability is the same as expectation for binary events, as was used in Equation (16). Independence was also assumed here.

This equation gives us a tool for finding the distribution of the time to breakdown for a system, even when the different components have different distributions. Note that it requires the structure function, though, so using this directly for large systems will often take a lot of resources.

### 7.4. Some easy examples

As a set of examples, consider an exponential distribution with $\tau_i = 1$ for all three components, thus $p_i(t) = e^{-t}$. We will study three different coherent systems, namely the serial system, the parallel system and the 2-of-3 system.

For a serial system, the analysis is easy to do, as the structure function is just the product of the component reliabilities. Thus $h_{\text{serial}}(t) = 1 - G(t) = e^{-3t}$, so that $G(t) = 1 - e^{-t/\tau_0}$, where $\tau_0 = 1/3$. This means that the time to breakdown will also be exponentially distributed with a characteristic time which is one third of the characteristic time of the component. This seems reasonable, as the system breaks down when one of the three components breaks down.

As a practical example, we can take a look at a 3 tier web site system, consisting of a web server, an application (using PHP) and a database. Assume that all three tiers have a expected time to failure of $\tau_i = 6$ weeks. Then the expected time to system failure will be $\tau_0 = 2$ weeks.

Since the exponential distribution is quite easy to handle in this case, we can also take a look at a asymmetric situation, where $\tau_{\text{webserver}} = 10$ weeks (semi-reliable web server), $\tau_{\text{application}} = 2$ weeks (unreliable software) and $\tau_{\text{database}} = 52$ weeks (reliable database). Then $1/\tau_0 = 1/\tau_{\text{webserver}} + \tau_{\text{application}} + 1/\tau_{\text{database}}$, thus expected time to system failure is $\tau_0 \approx 1.61$ *weeks* $\approx 11.3$ *days*. Notice that the expected time to system failure is just below the expected time to failure of the application software. This illustrates that the most unreliable component is the most influential component in a serial system.

We will now return to the simpler assumption of $\tau_i = 1$.

For a parallel system, we do not have such a nice attribute. We have that $h_{\text{parallel}}(t) = 1 - (1 - p_1(t))(1 - p_2(t))(1 - p_3(t)) = (1 - e^{-t})^3$, which means that although the times to the different breakdowns of the components are exponentially distributed, the life-time of the system will not be exponentially distributed. This attribute is shared by the 2-of-3 system which will have the reliability $h_{\text{2-of-3}}(t) = 3 \times e^{-2t}(1 - e^{-t}) + e^{-3t}$. The three reliabilities are plotted in Figure 7.

As can readily be seen, $h_{\text{parallel}}(t) \geqslant h_{\text{2-of-3}}(t) \geqslant h_{\text{serial}}(t)$ for all $t$.

We have seen that the life-time distribution of the parallel system is not exponential, even though this is the case for the components. In general for $\tau_i = 1$, an $n$ component parallel system will have $G(t) = 1 - h(t) = (1 - e^{-t})^n$, thus the distribution will be $g(t) = n(1 - e^{-t})^{n-1}e^{-t}$. This distribution is shown for different parallel systems in Figure 8.

As can be seen from the figure, increasing the number of components by a factor of ten only seems to push the distribution a fixed set of time further to the right. It can be shown that the mode of this distribution is found at $t_{\text{mode}} = \log(n)$, which tells us the same. The reason is that $t_{\text{mode}}(10n) = \log(10n) = \log(10) + \log(n) = t_{\text{mode}}(n) + \log(10)$. We will not try to find the expectation here, but it does not seem unreasonable to assume that the expectation has the same attribute.

If the base cost of the system is proportional to the number of components, then this cost will quickly overcome the expected cost of system failure when we add parallel components. The reason is that we do not repair the components and the probability of a component working becomes extremely small when the time increases over a certain level. Thus even when we have a thousand parallel components, each with characteristic time $\tau_i = 1$, we cannot hope for the system to last much longer than $S = 10$.

If we studied a serial system with exponentially distributed components, where $\tau_1 \ll \tau_j$ for all $j > 1$, then $h(t) = e^{-t/\tau_1 - t/\tau_2 - \cdots - t/\tau_n} \approx e^{-t/\tau_1}$. Thus only the least reliable component matters for the system reliability. A more matter of fact way of stating this is that a chain is not stronger than it's weakest link.

A similar parallel system where $\tau_1 \gg \tau_j$ for all $j > 1$ would have a reliability $h(t) = 1 - (1 - e^{-t/\tau_1})(1 - e^{-t/\tau_2}) \cdots (1 - e^{-t/\tau_n}) \approx e^{-t/\tau_1}$ for $t \gg \tau_j$ and $j > 1$. Thus in a parallel system, after a while only the most reliable component will matter.

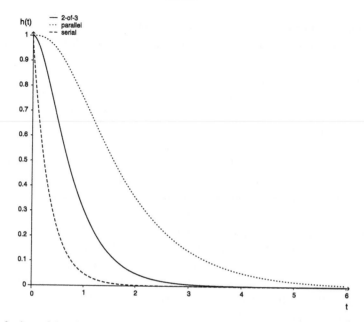

Fig. 7. $h(t)$ for three of the coherent systems having three components, namely serial, 2-of-3 and parallel, for $T_i$ exponentially distributed with $\tau_i = 1$.

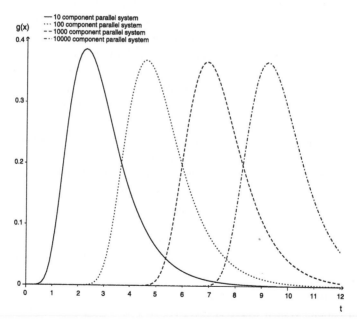

Fig. 8. Four distributions of the time to breakdown of parallel systems where all the times to breakdown of the components are exponentially distributed with $\tau_i = 1$.

## 8. Component importance

### 8.1. *About component importance*

So far we have concentrated on finding the statistical attributes of a single given system. We can then hypothetically find the optimal system, using a continuous risk assessment similar to the fixed time period risk in Equation (1), used on all possible systems. Howerver, this approach will not be practical for large systems. In such cases, we would rather have to do a less ambitious analysis, where we try to find a component in the system where we can concentrate our improvement efforts. In order to find such a component, we need the concept of component importance. There are several definitions of component importance, and we will study a few of them.

### 8.2. *The Birnbaum measure*

The most basic importance definition is called the Birnbaum measure [3] and is defined as:

$$I_{\mathrm{B}}^{(i)}(t) = \frac{\partial h(\underline{p}_i(t))}{\partial p_i(t)}. \tag{27}$$

It can be shown that $I_{\mathrm{B}}^{(i)}(t) = Pr(component\ i\ is\ critical)$, where a component being critical means that the state of the other components is such that the system will work if the component works and fail if the component fails. Thus $\phi(1_i, \underline{X}_{-i}(t)) = 1$ and $\phi(0_i, \underline{X}_{-i}(t)) = 0$.

In a serial system, a component will be critical if all other components are working, thus $I_{\mathrm{B}}^{(i)}(t) = \prod_{j \neq i} p_j(t)$. If we study the least reliable component, so that $p_i(t) < p_j(t)$ for all $i \neq j$, then $I_{\mathrm{B}}^{(i)}(t) > I_{\mathrm{B}}^{(j)}(t)$ for all $i \neq j$. Stated differently, a chain is not stronger than it's weakest link, as was remarked earlier in conjunction with serial systems. This is a more general result, though, as only component reliability at a given time is considered.

In a parallel system, a component will only be critical if all the other components have failed, so that $I_{\mathrm{B}}^{(i)}(t) = \prod_{i \neq j}(1 - p_i(t))$. Thus if $p_i(t) > p_j(t)$ for all $i \neq j$ then $I_{\mathrm{B}}^{(i)}(t) > I_{\mathrm{B}}^{(j)}(t)$ for all $i \neq j$. Stated in words, in a parallel system the most important component is the most reliable component.

This definition of importance does has an attribute that often is not wanted, namely that the importance depends on the time. Thus, we do not gain an insight into what the most important component is for all times. Potentially, it could be that all components scores higher than the other at some time interval. This makes this measure hard to use for practical applications.

A further attribute of this measure is that the importance at a given time is not readily comparable to the importance at a different time. This can be seen from the serial system. If there is one single component that always is the least reliable, then that component will have the most importance. However, the importance measure will decrease as $I_{\mathrm{B}}^{(i)}(t) = \prod_{j \neq i} p_j(t) \to 0$ when the time increases. Thus $I_{\mathrm{B}}^{(i)}(t)$ does not give us an absolute measure, and needs to be compared to $I_{\mathrm{B}}^{(j)}{}'(t)$ for the other components for all times $t$.

### 8.3. *The Barlow–Proschan measure*

In order to rectify the shortcomings of the Birnbaum measure, Barlow and Proschan [2] suggested the following importance measure:

$$I_{\mathrm{B-P}}^{(i)} = Pr(\textit{the ith component causes system failure}). \tag{28}$$

It can be further shown that

$$I_{\mathrm{B-P}}^{(i)} = \int_0^\infty I_{\mathrm{B}}^{(i)}(t) f_i(t)\, dt, \tag{29}$$

where $f_i(t)$ is the distribution function of the time to component breakdown, $T_i$. Thus the Barlow–Proschan measure is a weighted time average of the Birnbaum measure, where the weight is the distribution of the component. Since it is a probability and some component failure has to bring down the system in time, we have that $\sum_{i=1}^n I_{\mathrm{B-P}}^{(i)} = 1$. Thus this measure will be time independent and absolute, in that if we decrease the importance of one component, we know that the importance of the other components will increase.

For a serial system, exponentially distributed breakdown times and Equation (29)

$$I_{\mathrm{B-P}}^{(i)} = \int_0^\infty \left(\prod_{j\neq i} e^{-t/\tau_j}\right) \frac{1}{\tau_i} e^{-t/\tau_i}\, dt = \frac{1}{\tau_i} \int_0^\infty e^{-t\sum_{j=1}^n 1/\tau_j}\, dt = \frac{\tau_0}{\tau_i}, \tag{30}$$

where $\tau_0 = 1/(1/\tau_1 + \cdots + 1/\tau_n)$ is the characteristic time of the serial system. Thus the most important component $i$ will be the one that is least reliable, that is $\tau_i < \tau_j$ for $i \neq j$.

Such a calculation can even be done for the Weibull distribution for a given Weibull form parameter $b$ for all the components. This yields

$$I_{\mathrm{B-P}}^{(i)} = \int_0^\infty \frac{1}{\tau_i^b} e^{-t^b(1/\tau_1^b + \cdots + 1/\tau_n^b)}\, dt = \left(\frac{\tau_0}{\tau_i}\right)^b, \tag{31}$$

where $\tau_0 = (1/\tau_1^b + \cdots + 1/\tau_n^b)^{-1/b}$.

For a parallel system, we get an expression that is not so easy to work with. For three components we get $I_{\mathrm{B-P}}^{(1)} = 1 - 1/(1 + \tau_1/\tau_2) - 1/(1 + \tau_1/\tau_3) + 1/(1 + \tau_1/\tau_2 + \tau_1/\tau_3)$, using Equation (29). If $\tau_1 \gg \tau_2, \tau_3$ this yields $I_{\mathrm{B-P}}^{(1)} \approx 1$. Thus the most important component is the one that is most reliable in this parallel system. (This is really a more general result.)

For the exponential distribution, one can also use the parametrization $F_i(t) = 1 - e^{-\lambda_i t}$, so that $\lambda_i = 1/\tau_i$, which can be more convenient for purposes of calculations. Using this, we can try to look at the example from Figure 6. First we need the derivatives of the structure function, in order to find $I_B^{(i)}(t)$. This will only be done for component 1, in order to illustrate the procedure.

$$\phi(\underline{x}) = \left( x_1 \coprod x_2 \right) x_3 \left( x_4 \coprod (x_5 x_6) \right),$$

$$\frac{\partial \phi}{\partial x_1} = (1 - x_2) x_3 \left( x_4 \coprod (x_5 x_6) \right).$$

Putting $p_i(t) = e^{-\lambda_i t}$ into the expressions for the structure function derivative, the Barlow–Proschan importance measure of component 1 becomes;

$$I_{B-P}^{(1)} = \int_0^\infty \left( 1 - e^{-\lambda_2 t} \right) e^{-\lambda_3 t} \left( e^{-\lambda_4 t} + e^{-(\lambda_5 + \lambda_6)t} - e^{-(\lambda_4 + \lambda_5 + \lambda_6)t} \right) \lambda_1 e^{-\lambda_1 t} \, dt$$

$$= \lambda_1 \left[ \frac{1}{\lambda_1 + \lambda_3 + \lambda_4} - \frac{1}{\lambda_1 + \lambda_2 + \lambda_3 + \lambda_4} + \frac{1}{\lambda_1 + \lambda_3 + \lambda_5 + \lambda_6} \right.$$

$$- \frac{1}{\lambda_1 + \lambda_2 + \lambda_3 + \lambda_5 + \lambda_6} - \frac{1}{\lambda_1 + \lambda_3 + \lambda_4 + \lambda_5 + \lambda_6}$$

$$\left. + \frac{1}{\lambda_1 + \lambda_2 + \lambda_3 + \lambda_4 + \lambda_5 + \lambda_6} \right]. \tag{32}$$

If all the components in this system is equally reliable, thus $\lambda_i = 1/\tau_i = constant$, then we get $I_{B-P}^{(1)} = 10.0\%$. Similar calculations for the other components yields $I_{B-P}^{(2)} = 10.0\%$, $I_{B-P}^{(3)} = 48.3\%$, $I_{B-P}^{(4)} = 18.3\%$, $I_{B-P}^{(5)} = I_{B-P}^{(6)} = 6.7\%$. Thus the network is the most important component. If improving the network does not cost significantly more than improving other components, then efforts to improve it's reliability ought to be the highest priority. Note that this result only holds for $\lambda_i = constant$. If the network was very reliable, $\lambda_3 \gg \lambda_j$ for $j \neq 3$, then the importance of the network would be vanishing and other components, such as the primary power supply, would be more important. Since the primary power supply is likely not controlled by the email service provider, he or she would probably rather look at the reliability of the two email servers (component 1 and 2).

### 8.4. *The Natvig measure*

The Barlow–Proschan importance measure can be criticized for it's main feature, which is that only components that bring down the system are deemed important. There could be component that contribute greatly to reducing the life-span of the system by breaking down before the system collapses. Such components would not be deemed important by the Barlow–Proschan importance measure.

The Natvig measure [6] tries to rectify this by introducing what is called a minimal repair. A minimal repair of a component is such that when the component breaks down, it is replaced by a component that is in the exact same state as the component that broke down. Thus the further evolution of the component is the same as if the component did not break down after all.

The Natvig importance measure is then defined as

$$I_N^{(i)} = \frac{E(Z_i)}{\sum_{j=1}^n E(Z_j)}, \tag{33}$$

where $E(\cdot)$ stands for expectation, $Z_i$ is the reduction in the system life span when you go from allowing one minimal repair of component $i$ to not allowing it. Because of this definition, the Natvig measure also has the attributes that the sum of the measures over all components is equal to 1. Thus one has an absolute measure of component importance.

It can be shown [7] that, just as for the Barlow–Proschan measure, this measure can also be expressed as an integral over the Birnbaum measure:

$$E(Z_i) = \int_0^\infty I_B^{(i)}(t) p_i(t) \big(-\log p_i(t)\big) \, dt. \tag{34}$$

Where as the Barlow–Proschan measure scaled the Birnbaum measure with the component distribution, the Natvig measure scales the Birnbaum measure with an entropy term.

For a serial system and using the exponential distribution, this yields:

$$E(Z_i) = \int_0^\infty e^{-t \sum_{j \neq i} \lambda_j} e^{-t\lambda_i} t \lambda_i \, dt = \lambda_i \Big/ \sum_{j=1}^n \lambda_j \int_0^\infty \Big(\sum_{j=1}^n \lambda_j\Big) e^{-t \sum_{j=1}^n \lambda_j} t \, dt$$

$$= \frac{\lambda_i}{\lambda_0} E_{\lambda=\lambda_0}(t) = \lambda_i = \frac{\tau_0^2}{\tau_i}, \tag{35}$$

$$I_N^{(i)} = \frac{\tau_0}{\tau_i},$$

where $\tau_0 = 1/(1/\tau_1 + \cdots + 1/\tau_n)$. This is the same result as for the Barlow–Proschan measure.

For a three component parallel system, after much calculation, one gets:

$$I_N^{(1)} = \frac{\tau_1 - \tau_{1,2}^2/\tau_1 - \tau_{1,3}^2/\tau_1 + \tau_0^2/\tau_1}{\tau_1 + \tau_2 + \tau_3 - \tau_{1,2} - \tau_{1,3} - \tau_{2,3} + \tau_0}, \tag{36}$$

where $\tau_{i,j} = 1/(1/\tau_i + 1/\tau_j)$ and $\tau_0$ is defined as before. This is not the same result as for the Barlow–Proschan measure, though for $\tau_1 \gg \tau_2 \gg \tau_3$ one gets $I_N^{(1)} = 1$ like in the Barlow–Proschan case.

The difference between the Natvig measure and the Barlow–Proschan measure can be somewhat hard to see using formulas, even for the three component parallel system. Let us in stead look at a two component parallel structure, such that

$$I_{B-P}^{(1)} = \frac{\tau_0}{\tau_1},$$

$$I_N^{(1)} = \frac{\tau_1 + \tau_0^2/\tau_1}{\tau_1 + \tau_2 - \tau_0}. \tag{37}$$

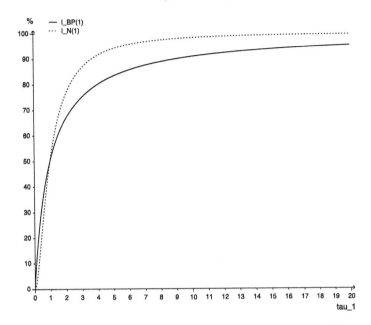

Fig. 9. The Barlow–Proschan and Natvig measure for a two component parallel system, with exponentially distributed components, varying $\tau_1$ and constant $\tau_2 = 1$.

In Figure 9, these two measures have been drawn for a constant $\tau_2 = 1$ and for varying $\tau_1$. As can be seen from the plot, the importance of component 1 for $\tau_1 = 0$ is zero for both measures, and for $\tau_1 \to \infty$ the importance goes toward 100% in both cases. However, the Natvig measure is below the Barlow–Proschan measure for $\tau_1 < 1$ and above for $\tau_1 > 1$. The conclusion would in this case be the same, though, namely to concentrate on making the best component better.

There are some general results concerning the Natvig measure. One such result basically states that if a component is less reliable than the other components and is in series with the system, then that component is most important, using the Natvig measure. If, on the other hand, there are no component in series with the system and there is a component in parallel with the system which is more reliable than the other components, than that component is the most important one.

## 9. Advanced topics

### 9.1. *Dependent components*

As mentioned earlier, dependent components can be harder to study than independent ones. There are several reasons for that. One is that formulating the dependence between components might be hard. In the case of $n$ independent components, there are $n$ different probabilities to specify in order to get a specific system reliability. For dependent components there are $n(n+1)/2$ different probabilities that have to be specified. For large systems, specifying so many probabilities will be a daunting task and often one will not have any real data that makes one able to formulate such dependent probabilities. This does however not mean that analytical tools are not available for this topic, as can be seen in [1] and [4].

The second factor that makes dependent components harder to study, is that independence makes it much easier to compute system reliability. Using independence, we found that we could express the system reliability as the structure function used on the independent probabilities in Equation (16). Without this assumption, one is left with:

$$h = \sum_{\underline{x}\in[0,1]^n} \phi(\underline{x})Pr(\underline{X}=\underline{x}), \tag{38}$$

where $\underline{X}$ is the stochastic state of the components and $\underline{x}$ is a specific state.

This great difficulty in handling dependent components is unfortunate, because there are many realistic real-world cases where components are dependent. For instance the case of a parallel server system, if one server goes down it increases the load of the other servers and will realistically increase the probability that other servers crash, too.

There is also the problem that because of little data, component probabilities often needs to be determined using subjectivist (Bayesian) methods. If there is several components where the same subjectivist thinking applies, then one introduces a further dependence in the analysis, even if the components themselves are independent. This will also be a problem when one can use frequentistic estimation techniques, as the estimates themselves will introduce dependency in further calculations.

While finding the exact system reliability can be hard in the case of dependence, there are methods for finding limits for how low and high the reliability can be. These methods does not handle dependent components in general, but looks at a sub-set called associate components. Associate probabilities of events are such that they generalize the concept of positive dependence to more than two events, so that all increasing functions of the events (components) are positively correlated. Covariance is defined as $cov(X,Y) = E(XY) - E(X)E(Y)$ for two stochastic variable $X$ and $Y$. When the covariance is positive, the two variables are said to be positively correlated. For binary events, this means that $Pr(X \text{ and } Y) > p_1 p_2$.

In most cases of component dependence, such as the parallel email servers or the Bayesian probabilities, the assumption of associate components seems reasonable. The system reliability is still hard to determine exactly, but one gets the following limits for it:

$$\prod_{j=1}^{k} Pr(X_i = 1 \text{ for at least one } i \in K_j)$$

$$\leq h \leq \prod_{j=1}^{p} Pr(X_i = 1 \text{ for all } i \in P_j), \tag{39}$$

where there are $k$ minimal cut sets $K_j$, $p$ minimal path sets $P_j$ and the reliability of each component $i$ independent of the other is $p_i$.

This will be quite hard to use for large systems, but we can study it for the two component parallel system. Here $k = 1$, $K_j = \{1, 2\}$, $p = 2$ and $P_1 = \{1\}$, $P_2 = \{2\}$. Not assuming independence, one gets that $P(\text{component 1 or 2 works}) = P(\text{component 1 works}) + P(\text{component 2 works}) - P(\text{component 1 and 2 works}) \equiv p_1 + p_2 - p_{1,2}$ since we double-count the event that both components work when we just add the probability of each event. Now the probability that both components work has to be less than each of the single component events, so $p_{1,2} \leqslant \min(p_1, p_2)$. Thus Equation (39) yields $p_1 + p_2 - p_{1,2} \leqslant p_1 + p_2 - \min(p_1, p_2) \leqslant h \leqslant p_1 + p_2 - p_1 p_2$. Thus the best case scenario is that the components works independently. For $p_1 = p_2 = p$ this again yields $p \leqslant h \leqslant 2p - p^2$.

Equation (39) can be used for proving that for independent components, one have the following limits:

$$\max\left(\prod_{j=1}^{k} \coprod_{i \in K_j} p_i, \max_{1 \leqslant j \leqslant p} \coprod_{i \in P_j} p_i\right) \leqslant h \leqslant \min\left(\coprod_{j=1}^{p} \prod_{i \in P_j} p_i, \min_{1 \leqslant j \leqslant k} \coprod_{i \in K_j} p_i\right). \quad (40)$$

### 9.2. *Simulation studies*

Study of advanced systems might not be practical or possible to do analytically. For such cases, one needs to do simulations for realistic assumptions on the system in stead. Monte Carlo methods can then be used for determining the system reliability. One can draw the component states $\underline{X}$ from the distribution that one thinks the components follows. This is done $M$ times. Then one determines whether the system works for each drawing from the component state, $\underline{X}^{(i)}$ and calculates the rate of systems working:

$$\hat{h} \approx \frac{1}{M} \sum_{i=1}^{M} I\left(\text{system works in configuration } \underline{X}^{(i)}\right) = \frac{1}{M} \sum_{i=1}^{M} \phi\left(\underline{X}^{(i)}\right), \quad (41)$$

where $I(\text{statement})$ is the indicator function, which is 1 when the statement is true and 0 if it is false. As the reliability of each component is often quite great, a lot of the drawn component states, $\underline{X}^{(i)}$, will contain mostly working components and thus very often the system will be working, too. There are methods that do such simulations more effectively, where one draws such that the number of components working does not directly imply that the system either works or does not work. However, such more effective methods needs extra calculations and are somewhat hard to implement. Such methods will thus not be described here.

Dynamic systems can also be described using simulation, by sampling the time to breakdown of each component in stead.

### 9.3. *Repairable systems*

In the dynamic system analysis, we assumed that components are not repaired once they fail. This is quite often not a realistic assumption. One of the tasks of an computer administrator is to find which component that caused a system collapse and repair that component.

The administrator is also responsible for finding errors that in the future can cause the system to fail in the future, and repair the computer components associated with such an error.

Thus, realistic dynamic analysis might quite often entail looking at repairable systems, see [1], [4] and [10]. This does not make the non-repairable case totally irrelevant, as time to the first system failure can be of interest in any case and quite a lot of errors will not be found before a system collapse.

When studying repairable systems, the analysis of time to breakdown is replaced by analysis of availability at a given time, $A(t) = P(system/component\ working\ at\ time\ t)$. When we look at $t \to \infty$ and $A(t)$ converges to a given limit, we call that limit the limiting availability, $A$.

For a component with exponentially distributed up-time, $T_i \sim \exp(\tau_i)$, and down-time, $U_i \sim \exp(\eta_i)$, the availability will be become

$$A(t) = \frac{\tau_i}{\tau_i + \eta_i} + \frac{\eta_i}{\tau_i + \eta_i}e^{-t/\tau_i - t/\eta_i}. \tag{42}$$

The limiting availability will thus be $\tau_i/(\tau_i + \eta_i)$.

Since expectation and probability is the same for a binary system, we have the tools for finding the availability of independent components:

$$A(t) = E\big(\phi\big(\underline{X}(t)\big)\big) = \phi\big(\underline{A}_i(t)\big) \tag{43}$$

since, for independent components, $E(X_i X_j) = E(X_i)E(X_j)$.

As an example, a two component parallel system with exponentially distributed up-times and down-times, will have the limiting availability $A = A_1 \coprod A_2 = \tau_1/(\tau_1 + \eta_1) + \tau_2/(\tau_2 + \eta_2) - \tau_1\tau_2/((\tau_1 + \eta_1)(\tau_2 + \eta_2))$.

The availability of three three-component systems are shown in Figure 10. This can be compared to Figure 7.

Figure 7 was shown in order to compare very different types of systems. Normally, one do not have the choice between serial and parallel design. However, when one has a system and wonders whether one or two similar systems ought to run in parallel, making a larger system, it can be instructive to look at the different availabilities of such systems. The case of one component, two parallel components and three parallel components is shown in Figure 11.

There are quite some difficulties in studying repairable systems, as analytical expressions for system availability can be hard or impossible to find. The availability will not only depend of the time to breakdown of each component, but also the time to repair following breakdown. This time can again depend on whether the component is being monitored all the time or only in special cases (such as the failure of another component that is constantly watched). It also depends on how one models the time of repair once it has been determined that the component has failed.

There are also some difficulties in determining good measures of component importance in such cases.

Such factors makes it more tempting to study such problems using simulations than using analytical tools.

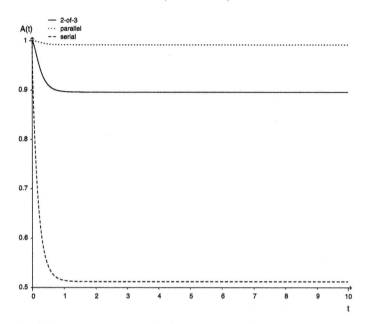

Fig. 10. Availability, $A(t)$, for three coherent systems of three repairable components, for the up-time, $T_i$, and down-time, $U_i$, exponentially distributed with $\tau_i = 1$ and $\eta_i = 1/4$.

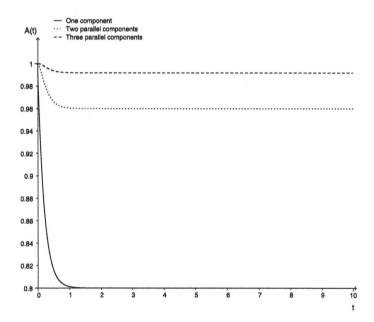

Fig. 11. Availability, $A(t)$, for one component, two parallel components and three parallel components for the up-time, $T_i$, and down-time, $U_i$, exponentially distributed with $\tau_i = 1$ and $\eta_i = 1/4$.

**9.4.** *Multistate systems*

So far we have only studied components that either work or fail, and the same is true for systems. However, some times the system might be in a state where it does some of the tasks that it is supposed to do, but not all. This can also be the case of components, when such components in reality are complex systems in their own right. Thus one would like the system and each component to have more than two possible states.

Usually, one can compare the different states, so that state *A* is worse than state *B* and so forth. One can do some analysis on such systems with a few extra assumptions. However, the calculations can quickly become unmanageable. Thus again, simulation studies might be more appropriate than analytical tools.

**9.5.** *Summary and conclusions*

The statistical treatment of system reliability theory has been studied and illustrated using examples from computer science. The examples have hopefully shown that reliability theory can be used in order to analyze existing computer systems and may act as a guide for finding ways to improve the reliability such systems.

The perhaps most useful guide was described in Equation (17), and states that low-level parallelism is better than high-level parallelism. Likewise, it should be useful to know that serial components is worse than serial systems. Also, a component in series with the rest of the systems ought to be avoided, if this is possible.

The field has so far been in constant development, and there might very well be significant contributions in the future that might act as even better guides for constructing and maintaining computer systems. The existing tools are well equipped for finding the reliability of existing systems, and there are good methods for finding the components which influence the system reliability the most. A few theorems about parallel and serial systems can act as a guide for constructing new systems. However, there may still be undiscovered methods that are even better suited for guiding or determining the process of finding the best computer system for a given task. It is however probable that such developments will spring from the system reliability theory as it stands now.

**Acknowledgements**

I would like to thank Mark Burgess for his support, his help making examples for this chapter and for his constructive criticism. Nils F. Haavardson should also be thanked for proof-reading the chapter. I would also like to thank Bent Natvig for useful discussions concerning system reliability.

**References**

[1]  R.E. Barlow and F. Proschan, *Statistical Theory of Reliability and Life Testing*, Probability Models, Holt, Rinehart and Winston, New York (1975).

[2]  R.E. Barlow and F. Proschan, *Importance of system components and fault tree events*, Stoch. Proc. Appl. **3** (1975), 153–173.

[3]  Z.W. Birnbaum, *Inference for the exponential life distribution*, Multivariate Analysis-II, P.R. Krishnaiah, ed., Academic Press, New York (1969), 581–592.

[4]  A. Høyland and M. Rausand, *System Reliability Theory: Models and Statistical Methods*, Wiley, New York (1994).

[5]  A.B. Huseby and R. Bendheim, *A unified theory of domination and signed domination with application to exact reliability computations*, Statistical Research Report No 3, Institute of Mathematics, University of Oslo (1984).

[6]  B. Natvig, *A suggestion of a new measure of importance of system components*, Stoch. Proc. Appl. **9** (1979), 319–330.

[7]  B. Natvig, *New light on measures of importance of system components*, Scand. J. Statist. **12** (1985), 43–54.

[8]  B. Natvig, *Pålitelighetsanalyse med teknologiske anvendelser*, 3. utgave, Institute of Mathematics, University of Oslo (1998).

[9]  J.A. Rice, *Mathematical Statistics and Data Analysis*, 2nd edn, Wadsworth Publishing Company, Belmont, CA (1995).

[10]  S.E. Rigdon and A.P. Basu, *Statistical Methods for the Reliability of Repairable Systems*, Willey, New York (2000).

[11]  N. Viswanadham, V.V.S. Sarma and M.G. Singh, *Reliability of Computer and Control Systems*, Elsevier, Amsterdam (1987).

# 7. Business and Services

## 7.1. *Commentary*

*"Long live system administration, system administration is dead!"*

At a recent meeting of the Network Management Research Group we asked the group why no one ever speaks any more about the computers at the end of the network. Researchers talk of routers and switches and end-to-end guarantees but never of hosts, users, data and storage. Why not? One of the vendors had a simple answer: "That's just services".

The dominance of the European and Asian telecommunications companies in the network management research arena has formed the terminology of the field in their own image. Service delivery is the new paradigm for computer systems, thanks largely to the success of the Internet. What Unix and Windows administrators think of as Network or System Administration is no longer considered to be any different to managing the software that runs on web servers that are assumed to be the work-horses of the Internet. This change of viewpoint is motivated by the commercialization of the Internet. The driving force has been the Internet Service Providers and companies using the World Wide Web as their palette and medium.

Commercialization has also led to an identity crisis, from the increasing pressure to improve the dependability of services. Mark Weiser, former chief of technology at Xerox PARC once said, "The most profound technologies are those that disappear. They weave themselves into the fabric of every day life until they are indistinguishable from it" [2]. This vision has not yet happened in the data center. Network and System Administration is still, in many ways, a fly by the seat of the pants enterprise, but efforts are underway to better this situation.

The economics of self-configuration and self-maintenance are clear. The benefits of self-healing systems have been apparent for years [1] and were recently made an initiative by companies like IBM (so-called self-management or autonomics). These ideas seem to be replacing a focus on Operational Support Systems (OSS) and 'standardized' bureaucracies like eTOM for running their businesses. The five areas Fault, Configuration, Accounting, Performance and Security (FCAPS) are still considered important, but now businesses want these things for free. There are many coming challenges in the commercial arena of services. Mobility and pervasiveness will have a dramatic effect on how we define business and management: what will 4G networks mean for us?

In this part, our contributors cover aspects of the economics of network service provision from the basic models for charging for services to the management of those services. The decentralized nature of the Internet has led many company principals to seek help from the cheaper but massively growing and highly skilled agent workforces in the Asian subcontinent. This has led to the practice of 'outsourcing' and 'offshoring' with both benefits and dangers to companies. The authors here assess the factors for successful delegation in an economic network. Internet business has shown us more convincingly than ever that communication and commerce are integral parts of the same phenomenon.

## References

[1]  M. Burgess, *Computer immunology*, Proceedings of the Twelfth Systems Administration Conference (LISA XII), USENIX Assoc., Berkeley, CA (1998), 283.
[2]  M. Weiser, *The computer for the 21st century*, Scientific American (1991).

# – 7.2 –

# State-of-the-Art in Economic Management of Internet Services

Burkhard Stiller[1,2], David Hausheer[1]

[1] *University of Zürich, Department of Informatics IFI, Switzerland*
*E-mails: {stiller,hausheer}@ifi.unizh.ch*

[2] *ETH Zürich, Computer Engineering and Networks Laboratory, TIK, Switzerland*

**Abstract**

Managing the heterogeneous and partly broadband-capable infrastructure of tomorrow's Internet requires suitable and scalable technology as well as mechanisms in support of viable economic means of service deployment and provisioning. Such services will advance beyond the pure Internet Protocol (IP) access and need to cover value-added services as well. Currently, an economic unfairness exists in the current Internet and in a converged IP infrastructure also.

This state-of-the-art in traditional network management and related basic economics focuses mainly on the use case of network service provisioning. Newly developed, integrated concepts of network management and economics, and ideas of emerging, urgent and required research questions, technology requirements, and future economic management schemes have been identified and discussed briefly.

Driven by these considerations, the need for developing advanced network management techniques becomes obvious, of course, only based on well-defined roles and models, covering promises and interactions, that will enable multiple parties to access this infrastructure in a technically suitable, efficient and economically fair manner. This state-of-the-art work, which goes partly even one step beyond the current state of network management – in terms of models and mechanisms – has been termed EMIS (Economic Management of Internet Services). This approach shall lead to a common and administrable information infrastructure, which in turn can be operated economically and efficiently under predefined assumptions.

HANDBOOK OF NETWORK AND SYSTEM ADMINISTRATION
Edited by Jan Bergstra and Mark Burgess

## 1. Introduction and motivation

Managing the infrastructure of tomorrow's Internet – based on the Internet Protocol (IP), being broadband-capable, and including wireless access links – requires suitable *technology and mechanisms* as well as valid and viable *economic means*, to lead not only to a generic and open information infrastructure, but also provide a technology-independent access to this infrastructure, and enable multiple parties' access to this infrastructure in an economically fair manner.

In particular, the *economic dimension of network management* has to be an integral part of an overall networking infrastructure, including existing technologies on Layer 1 (Physical Layer), such as SDH (Synchronous Digital Hierarchy) or xDSL (Type x Digital Subscriber Line), Layer 2 (Data Link Layer), such as POTS (Plain Old Telephony System), FR (Frame Relay), ISDN (Integrated Services Digital Network), WLAN (Wireless Local Area Network), Layer 3 – typically IP only – and the wide range of applications on top of IP.

While different technical mechanisms and economic measures depend on various factors and influence providers' charging models, their mechanisms for network and services accounting and monitoring, as well as their management optimization dimensions and technical operations procedures, have to be integrated in a common fashion. Only this approach enables a competitive and open service provisioning solution across domains in the Internet.

Due to many approaches on the technical aspects of network management (cf. [27] for an excellent text book), this overview here addresses especially the *economics of the Internet*, comprising any economic measures and means of the networking technology and data transfer. The Internet in its current form has determined the basis for many successful business models and companies over the last years. These companies typically belong to one or more of the following categories: equipment vendors, software vendors, advertisers and retailers [40]. While in general, those successful companies did not belong to the carrier and IP service provider category, three problems may be considered to be the major root of the Internet profit dilemma [40], thus representing a main motivation for the work in economic management of Internet services:

- Fixed price models for the Internet access, originally technology-wise being bound by dial-up connections, limit the Internet usage and burden the investments required to upgrade the capacity needed to foster a larger usage. The effects of Internet pricing and communications in general may be obtained in a larger and complete overview in [42].
- Although technical means exist to offer Quality-of-Service (QoS)-based IP services, commercial-grade interfaces between Internet Service Providers (ISP) with guarantees on performance and their respective inter-ISP settlements do not exist. The relation between technical mechanisms and Service Level Agreements (SLA), under a certain assumption of a business model, need to be improved and made possible in a multi-domain scenario.
- The service differentiation between QoS-based and best effort services shall lead to different service types to be paid for differently. But since a capacity increase on a link will provide benefits for guaranteed and best effort services at the same time, appropriate solutions are essential.

Additionally, the well-received notion of *convergence* – basically determining the integration of IP and telecommunications on a single platform – or even the adoption of an all-IP infrastructure [1] as the universal communications basis for all services emphasizes this dilemma. Furthermore, due to these dramatic technology changes, revenues for legacy services, e.g., telephony services, decrease dramatically, and the Average Revenue Per User (ARPU) numbers drop. In turn, extraordinary effects on the market of service providers, covering ISPs and the telecommunications industry in support of data transport, are envisioned clearly.

These strong concerns pinpoint sharply the need of well-defined technical and economic management techniques for networks and services that will lead toward a generic information infrastructure provided by many service providers, and that will enable multiple parties access to this infrastructure. The research community has addressed major factors of technical and economic management as separated activities, such as achieving technical efficiency of accounting modules, definitions of charging models independent of technological prerequisites, maintaining viable Service Level Management (SLM) aspects, and determining management-independent broadband services models. However, a complete inter-operation model of those essential components still remains to be designed.

Therefore, the approach termed *EMIS* (*Economic Management of Internet Services*) outlines major pieces of the current state-of-the-art and sketches the potential future approach and corresponding architecture. The development of a customer service management approach and corresponding interfaces as the foundation of the integration of a production environment is essential. In addition, the mapping of different QoS architectures and parameters between organizational boundaries and platforms will lead to an approach to map QoS parameters to the infrastructure, across organizational boundaries, and onto charges to be paid for by the customer. The envisioned future of broadband networks has the potential to become an enabler for attractive new kinds of services and applications in business as well as in private sectors. Therefore, the need for appealing new service offerings and their management solutions is emerging.

Those solutions have to be provided to customers with a guaranteed QoS at an agreed upon or negotiated cost and have to be captured in a legally binding and technically supervised contract, typically combined into an SLA, which includes the more specific Service Level Specification (SLS). While the service model allows for respective tasks to be operated in an integrated manner, the corresponding role model determines the value network, market segments, and service promises. This is complemented by the charging model, which defines the cost structure and profit potential for services and presumably competitive strategies for their offerings. All of those are combined to form the elements of the business model, which itself can be instantiated in a deployment model, stating explicitly the application domain of interest and its specific parameters, values and functions to be applied and originating from the role, service and charging model.

The recently started European Network of Excellence EMANICS (Management of the Internet and Complex Services) [16] addresses particularly this combination of network and economic management.

**1.1.** *Chapter overview*

Section 2 outlines the basics for economic management with respect to the business model (Section 2.1), the role model (Section 2.2), the service model (Section 2.3), the charging model (Section 2.4) and a sample deployment model (Section 2.5).

This is complemented by a specific focus on key mechanisms (Section 3) to support service provisioning, on the technical side namely SLA definitions (Section 3.1.1), SLA negotiations (Section 3.1.2), policy-based management (Section 3.1.3), accounting (Section 3.1.4) and on the economic side cost parameters and economic measures (Section 3.2.1), mobile micro-payment and accounting (Section 3.2.2) and service provisioning templates (Section 3.2.3).

Finally, a brief summary is presented and preliminary conclusions are drawn in Section 4, mainly showing the key steps of integration to be performed.

## 2. Existing models

The basis for this state-of-the-art work is laid by the definition of core models, which determine the key for contractual agreements, technologies, and mechanisms to be applied. Thus, Figure 1 introduces and summarizes the five major models relevant for economic management and outlines their course-grained interdependencies in terms of a three-level hierarchy: This includes the *Business Model* which is determined by means of value propositions defined by the *Service Model* and characterized by a single or multiple interacting actors. Accordingly, the *Charging Model* defines how those are able to achieve economic profits, based on player-specific incentives and behaviors as determined by the *Role Model*. Finally, the *Deployment Model* defines a formalized path to an instantiation of a business model based on a set of concrete and case-specific assumptions.

**2.1.** *Business model*

A vast number of business model definitions is found in economic literature. In the comprehensive study of [43] various existing approaches are outlined, mainly by comparing

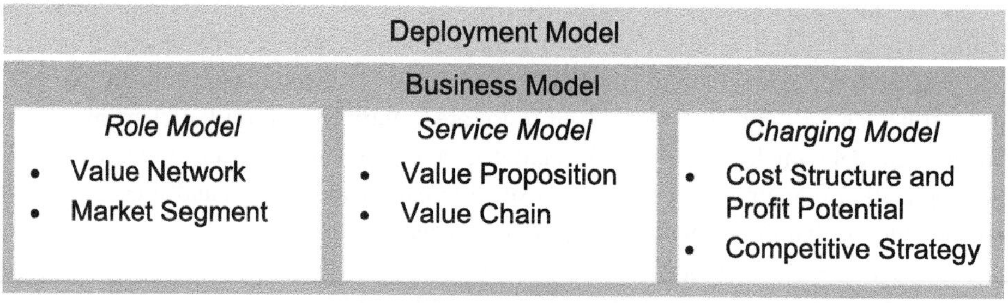

| Deployment Model | | |
|---|---|---|
| Business Model | | |
| *Role Model* | *Service Model* | *Charging Model* |
| • Value Network<br>• Market Segment | • Value Proposition<br>• Value Chain | • Cost Structure and Profit Potential<br>• Competitive Strategy |

Fig. 1. Business modeling overview.

and classifying them according to various dimensions. This investigation was driven originally by a focus on business models for electronic commerce. Even though this narrow perspective was widened again in order to cover approaches not being Information and Communication Technology (ICT)-centered only, this study represents a highly valuable resource in terms of an embracing overview of business models applicable to electronic businesses, thus, being especially applicable to network management areas and technologies of electronic service provisioning. The key instantiation of such an electronic business is the product of transporting data across physical infrastructures, such as offered by ISPs or telecommunication providers. In addition, the data transport itself, typically across multiple provider domains, is provided by the use of the Internet Protocol (IP).

From the wide range of existing approaches to business modeling, [7] determines the most relevant approach applicable to the IP-based data transport and it reveals the following major components typically to be present in business models. Due to the fact that the product offered here, the data transport service, fulfills major criteria of an electronic product, those six components can be distinguished clearly:

- *Value Proposition:* The main objective for any business model is to determine how and by what means value creation is enabled in an organization. Value creation is particularly of interest in fully competitive environments, since it represents the steps in the overall value chain that can be exploited commercially. With respect to the specific context of EMIS, value is assumed to be created mainly by offering basic or complex, composed electronic services to the customer base. Thus, services are perceived as a form of electronic products, providing a certain amount of utility to the service user for which, in turn, the service customer will be charged.
- *Market Segment:* Services are designed in order satisfy customer needs. Depending on the specific market environments, both, service providers and service users may show different incentive settings which determine behavior patterns and, thus, limit the range of envisaged market segments. Accordingly, marketing activities in a Business-to-Business (B2B) context vary from the Business-to-Consumer (B2C) area. In particular, these differences bring about contractual agreements that are adapted to a given market segment. For EMIS, SLAs and SLSs are of very high interest in this context.
- *Value Chain:* In order to create value, usually several steps in an organization are needed. These steps are modeled by means of processes and workflows. Besides activities, also complementary assets have to be considered. As those procedures are assumed to create commercially exploitable value, their actual implementation encapsulates business-critical knowledge. This determines on the one hand and to a certain extent the competitive strategy an actor takes, on the other hand this knowledge will be protected from inspection by organization-external entities.
- *Expected Cost Structure and Profit Potential:* The basic economic principle comprises the rule to pursue activity $A$ if, and only if, related costs do not exceed economic benefits of $A$. Costs include also opportunity costs, meaning costs of not pursuing alternatives to $A$. Therefore, a business model needs to provide assumptions on what economic profits the proposed values might generate, and what costs their provision might cause. The second part clearly demands for cost ac-

counting mechanisms to be in place. Traditional cost accounting systems, however, are designed for production-oriented processes. In the context of EMIS, with its focus on Internet service provisioning, Activity-based Costing (ABC) [23,32] appears to be a better-suited method to determine accurate cost elements on a per-service basis. This is mainly due to the assumption that services are produced once, while they are offered in the same or similar way many times. In this case, service production costs become less important with an increasing number of service requests/deliveries.

- *Value Network:* Similar to considerations with respect to value chain steps, business models usually cover information on how an organization is embedded into a dedicated value network. Value networks feature a wider angle across administrative borders. This includes the respective perspectives of suppliers, customers, competitors, complementors and authorities. Interactions between actors and overall market conditions, thus, gain center stage. While the value chain determines what steps a given organization creates value with, the value network reflects what other players value is created with, against, or for.

- *Competitive Strategy:* The presence of economic profits sooner or later attracts competitors. In fully competitive markets, sustainable profits are provided only, if an organization is able to gain a competitive advantage over its competitors. Therefore, business models rely also on a competitive analysis. According to Porter's five forces model [47], this consists in investigating the bargaining power of suppliers and customers, the threat of new market entrants or substitute services, and market-internal rivalry. Based on the competitive analysis' results, a suitable competitive strategy is chosen. Two basic alternatives exist: cost leadership or differentiation strategy. While the first focuses on cost efficiency ("offer the same services as your competitors, but offer them at a lower price"), the latter envisages to offer greater value to customers than competitors are able to ("offer better services than your competitors, and possibly charge more").

From those business model elements presented, the EMIS business modeling approach has been developed as shown in Figure 1. Thus, these business models are determined by the three models of a role model, a service model and a charging model. Accordingly, business models are put into operation by means of deployment models. Assumptions on costs and revenue potentials are based on expectations gained from market forecasting. The actual incidence of expected numerical values may vary from profits possible and costs incurred. This aspect is reflected explicitly within the scope of the deployment model by means of considerations on risk assessment. The same holds true for the role model as well, where participants defined – typically service providers and users – exchange promises, which determine a commitment that costs a participant work to fulfill them.

In general, those models may be applied for any type of ICT systems. Those examples presented in the following, however, are addressed toward the IP data transport or – generally termed – provisioning of Internet services. This forms, as indicated above, one important instance of an electronic service, thus, determining the approach with a high potential for the integration of economic and technical mechanisms for the management of networks and their services.

## 2.2. *Role model*

Besides the overall business model introduced above, the role model proposes an approach for service promises, rather than network services. It is based specifically on the value network, which determines the information on how an organization is embedded into a wider angle across multiple and administrative borders. Additionally, the market segment under consideration determines those roles to be considered as well as the incentives under which those roles may act or operate.

Therefore, a refinement of a role as an entity within a system – probably a network, a domain, or an IT infrastructure in general – is based on its intended function, such as a service offering with a certain QoS guarantee, not on whether it fulfills its commitments. The roles presented within EMIS – service providers, service customers and service users – are determined as principal actors needed for a generic, multi-domain, multi-service IP QoS delivery and their interactions. These roles are largely in-line with the roles specified in the context of the EU MESCAL project (Management of End-to-end Quality-of-Service Across the Internet at Large) [28,37].

In general, a role should not be understood as being identified with a particular body or entity amongst a number of collaborating entities. Each entity in a system could play several roles, pertaining to different aspects of the system. Thus, roles are identified as being related to specific promises made by those parties. Additionally, key example roles have been identified as being important in the process of provisioning Internet services.

**2.2.1.** *Concept of a service promise*   Services are about committing to behavior that is valuable to a client. The approach can use promise theory [2,3] to define this in terms of interactions between a number of agents. In this case, agents are initially peers that have no pre-decided roles. It is useful to begin with general definitions that can be specialized to the view of network services later.

A *Promise* is defined as an autonomous specification of future behavior by an agent. A respective *Commitment* is defined as a promise that costs an agent work or effort to fulfill.

A concrete promise from one agent to another can be expressed as follows:

$$A_1 \xrightarrow{\pi} A_2,$$

where $\pi$ is called the promise body, e.g.,
- $\pi = P$ – is a promise of $A_1$ to another agent $A_2$ to do $P$.
- $\pi = U(P)$ – is a promise to use/accept or receive the offer of $P$ promised by another agent.

A service is not necessarily something that is provided (given). It can be the opposite – something taken, e.g., a backup service takes data from a client. Thus, the direction of the arrow does not reflect a flow of data. A promise body is a tuple, in general, which contains a type, e.g., a promise of web service or a promise to make a backup, and a constraint, e.g., a web service in less than 5 ms or backup every hour on the hour.

The agents can already be classified into some prototype roles, and this view can be extended below.

A *Service Provider*, as outlined below in Section 2.2.3.1 is any agent $S$ that makes a commitment to one or more external agents. In promise theory, a service is associated with a typed promise made by a service provider. A *Client* is any agent who enters an agreement that includes a promise to use/accept a service. Finally, an *Agreement* is a sustainable bundle of bilateral (two-way) promises between two agents. Both agents make promises to one another. These might include promises of payment.

Note, that a collection of promises is not necessarily consistent or sustainable. An agreement has to be worked out so that it is of mutual benefit to both parties, from each of their perspectives. Thus, a service agreement is an agreement between a service provider and a client, and an SLA (cf. Section 3.1.1) is defined as a subset of a service agreement, which concerns the scope of the commitment made by the service provider and the scope of the promise to use the service by the client.

**2.2.2.** *Concept of a role in promise theory*   Promises provide a notion of a relationship between autonomous parties in a system. Roles are associated with types of promises, i.e. categories of promise bodies $\pi$. In this respect, a collection of agents can be associated with a role, given that the members of the collection make (or receive) the same type of promise to (from) some third party. In other words, we are guided by the typed structure of the graph alone.

For example, web servers or clients, are roles. Roles are effective labels for groups of nodes. In this respect, definition of a group is essential. The notation for a promise from one set of nodes to another set of nodes can be defined as follows:

$$\{n_i\} \xrightarrow{\pi} \{n_j\}.$$

Let $A$ be a set of autonomous agents. The family of sender/receiver subsets $S, R \subseteq A$ of a promise graph, which take part in promises with bodies $\Pi$, define $\Pi$-*Roles*. Any such subset $S$ or $R$ in the graph is said to play a role of (sender/receiver) type $\Pi$ in the promise graph.

The set of all agent nodes $W \subseteq A$ that promises to provide web service to any agent:

$$W \xrightarrow{\text{Web}}$$

plays a role, which is called web server. This is called a role by association, since the members of $W$ are not connected in any other way than they make the same promise. Any combination of promise bodies can be concatenated to define compound roles or role hierarchies.

**2.2.3.** *Key example roles in EMIS*   Based on these considerations a set of typical roles – mainly found in the networking domain of the Internet and its services – are introduced. The value network view focuses here on those perspectives of suppliers, customers, competitors, complementors and authorities, where all of those are considered in two classes of service providers as well as users and customers. The market segment under consideration addresses B2B contexts and B2C ones.

**2.2.3.1.** *Service providers*   In general, service providers offer services to service customers (cf. Section 2.2.3.2). Based on the specific type of service provided, service providers can be further divided into three categories:

- *IP Connectivity Providers (ICP):* IP connectivity providers offer only plain IP connectivity services, i.e. a functionality that provides reachability between hosts in an IP-based network, typically uniquely specified by the IP address space. In order to enable this functionality, ICPs must own and administer an IP network infrastructure. As a result, service customers connected to the same ICP are able to communicate with each other directly, since they belong to the same administrative IP network infrastructure. In order to enable the communication between service customers that are connected to different ICPs, individual ICPs need to establish provider SLAs with each other for the reachability expansion.

- *Application-level Service Providers (ASP):* Compared to the plain IP connectivity services offered by ICPs, ASPs offer higher-level services, such as IP telephony and content streaming services. In contrast to ICPs, ASPs do not need to own and administer necessarily their IP network infrastructure; they need to administer the necessary infrastructure required by the provisioning of services offered, e.g., VoIP gateways, IP video-servers, or content distribution servers. As such, for fulfilling the connectivity aspects of their services, ASPs may rely on the connectivity services offered by ICPs.

- *Physical Connectivity Providers (PCP):* PCPs offer physical connectivity services (up to the link layer) between protocol-compatible equipment in determined geographic locations. It should be noted that connectivity services may also be offered in higher layers (for the network layer, layer 3, e.g., IP), however these services are mainly between specific points as apposed to the IP connectivity services offered by ICPs, which may be between any hosts residing within the dedicated IP address space. PCPs are distinguished further into two main categories according to their target market: Facilities providers and access providers.

  - *Facilities Providers* offer link layer connectivity to other higher-level providers such as ICPs for these to interconnect their own infrastructure and for interconnecting with their peers. Facilities providers may be further differentiated according to the technology they rely upon (e.g. optical fiber, satellite, antennas), deployment means (terrestrial, submarine, aerial) and their size in terms of geographical span and customer base.

  - *Access Providers* provide the connection of the end-customers premises equipment to the ICPs or ASPs equipment. They own and administer appropriate infrastructure, e.g., cables, concentrators. They may be differentiated according to the technology they employ, e.g., POTS (Plain Old Telephony System), FR (Frame Relay), ISDN (Integrated Services Digital Network), xDSL (Type x Digital Subscriber Line), WLAN (Wireless Local Area Network), or Ethernet, as well as their deployment means and their size in terms of covered geographical area and customer base.

These key example roles address the provider's side of the role model. It has to be noted that all those providers may become a customer or user, as soon as a multi-domain case is considered. Thus a type of recursive behavior can be observed.

**2.2.3.2.** *Customers and users*    A *Service Customer* (subscriber) defines a role, which has the legal ability to subscribe to the services offered by ICPs or ASPs. They interact typically with ICPs and ASPs following a customer–provider paradigm, with the purpose to subscribe to services to meet their communication needs and requirements. Nowadays, services are offered on the basis of respective agreements (SLAs), setting the terms and conditions on behalf of ICPs and ASPs, and service customers for providing and requesting/accessing these services, respectively. Service customers can be either end-customers or even entire ICPs and ASPs, who act themselves as customers of other ICP and ASPs in order to, e.g., extend the geographical scope of services they provide to their own customers.

A *Service User* is an entity (human being or a process from a general perspective), which has been named by a service customer and appropriately identified by an ICP and ASP for actually requesting/accessing and using services subscribed to by the service customer.

The use of the services should be in-line with the terms and conditions agreed in the SLA between the service customer and the ICP/ASP. In essence, service users are end-users of services, and they can only exist in association with a service customer. A service user may be associated with more than one service customers using services according to the agreed upon SLAs of respective service customers. For instance, an employee may be acting as a service user of services that its company, the service customer, has subscribed to, as well as a service user of its own subscription as a residential service customer.

## 2.3. *Service model*

The role model presented above – as the first part of the EMIS business model – leads to the generic service model, in which those defined roles are considered as entities within the context of service provisioning. Additionally, the value proposition made – typically the definition of the Internet services offered – is complemented by the value chain, which determines the steps to be undertaken to create the value based on a specific service offering. Therefore, the service model is outlined with its generic characteristics of service provisioning, independent of a concrete scenario or a specific service functionality, and applicable to any kind of service provider defined (ICPs, ASPs, or PCPs – cf. Section 2.2.3.1).

The core view on Internet services comprises two components: the service life cycle, which displays the dynamic behavior of a service, and the service model, which describes the composition of entities and interactions inside a service.

**2.3.1.** *Service life cycle*    The service life cycle is divided into 7 phases, which form a refinement of [27]. In the offering phase (1) the service provider makes an offer for service provisioning to the customer. In the negotiation phase (2) a contract – typically an SLA between provider and customer – is settled. This contract contains a detailed description of the service functionality, the service level, adoptable rates, and contract penalties. During the implementation phase (3) the service functionality is implemented by the provider. Implementation and testing (4) determines the (physical) establishment of the service within a productive environment. After the customer has accepted the contract, the service is put into operation (5). Afterwards the service can switch into the change phase (6), if updates are required, and back again into operation. Upon termination of the service, it runs through the decomposition phase (7).

**2.3.2.** *Common service model details*    The common generic service model, developed in [19] and refined in [20], is applicable to EMIS, since the terminology in the scope of service orientation is defined by a structured analysis of entities, roles, and interactions within the service life cycle. Figure 2 shows the service model, whose elements are explained below, covering major definitions as well.

A *Service* describes a functionality, which a service provider offers to a customer or customer domain at a specified interface, assuring a certain service quality. The functionality offered implies usage as well as management functionality.

Components in the common service model cover a wider range. Thus, the following list of specifications includes key terms applied within the service model. Even though a single term of the service model determines a well-specified definition, only the generic term is instantiated as a definition.

The *Functionality* of a service is understood as a set of interactions taking place between the provider domain and the customer domain. Usage functionality and management functionality are distinguished:

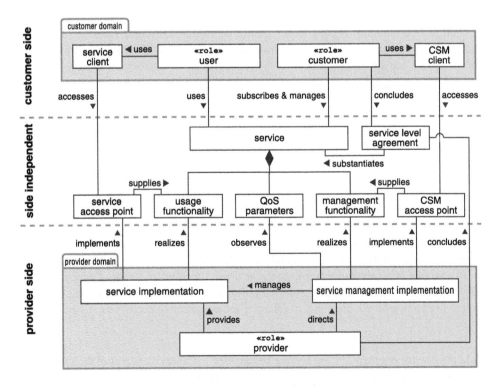

Fig. 2.  Service model – service view.

- *Usage functionality:* The usage functionality contains all interactions between the provider and the customer domain, which represent the essential purpose of the service.
- *Management functionality:* Every interaction taking place between the provider and the customer domain, which is required to operate the service, but cannot be assigned to the essential purpose of the service, belong to the management functionality of this service. This comprises also all charging activities as they are addressed in more detail in the charging model for services (cf. Section 2.4).

The quality for any functionality offered, thus, both kinds of service functionality, are specified by *QoS Parameters*. Usually, when the contract is concluded, bounds for a tolerable QoS are specified by a determination of concrete values.

A *Service Interface* includes all entities, which are required by the user in order to provide the usage functionality. Two kinds of interfaces appear in the service model:

- *Service Access Point (SAP):* The SAP provides the usage functionality of the service to the user as contracted.
- *Customer Service Management (CSM) interface:* The CSM interface provides the management functionality of the service to the customer. A concrete implementation of the CSM interface may be implemented by an application programming interface as well as by e-mail or telephone, by which management information between customer and service provider are being exchanged.

The *Service Level Agreement (SLA)* contains a description of the service functionality as well as definitions of related QoS parameters. Additionally, interfaces for accessing the service functionality by the customer domain are specified. The SLA is settled within the scope of a legal contract. A detailed discussion of SLA specifications follows in Section 3.1.1.

The service model uses *Domains* in order to define and delineate scopes of responsibility:

- *Provider domain:* The provider domain comprises all entities vital for providing the service functionality as specified in the SLA. The service provider is responsible for service provisioning and operates a service implementation and a service management component.
- *Customer domain:* The customer domain includes the user as well as the customer role according to the role model outlined in Section 2.2.

Finally, for deploying the service functionality, a *Client* is or multiple clients are required to access the SAP as well as the CSM interface. The customer domain is in charge of the proper operation of the service client and the CSM client.

The *Service Implementation* implements the usage functionality of the service provided. The *Service Management Implementation* covers all tasks and resources, which are required to ensure a proper planning, installation and operation of the service in line with the SLA.

**2.3.2.1.** *Service aggregation and service hierarchies*    If a service is part of the implementation of the functionality of another service, it is considered to be – from the perspective of the constitutive service – a sub-service. Figure 3 shows the implementation view on the

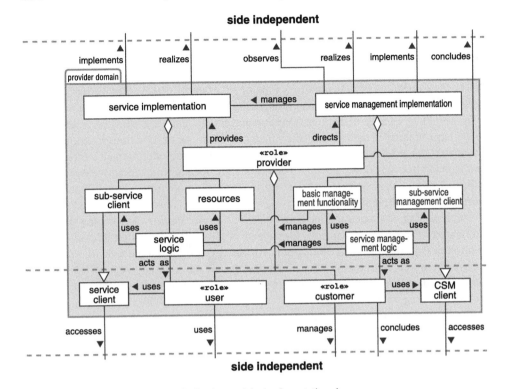

Fig. 3. Service model – implementation view.

provider domain of this service model and points out, how service chains are reflected in this model.

**2.3.2.2.** *Resources*    The generic service model introduced above, abstracts away from the resources within the underlying ICT environment. To complete this view by a resource-orientated perspective, EMIS' viewpoints concerning resources cover devices and applications.

A *Resource* is a physical or logical item, on which the usage or management functionality implementation of a service is at least partially based upon.

- *Hardware resource (Device):* A hardware resource is a physical item, which is part of a service implementation or a service management implementation.
- *Software resource (Application):* A software resource is the installation of a software distribution within a productive ICT environment, which is part of a service implementation or a service management implementation.

**2.4.** *Charging model*

*Charging* determines the activity of bringing into account any type of service usage that has be paid for by a customer to a service provider. Service types cover the full range from pure Internet Protocol (IP) access to Value-added Services (VAS). Thus, a number of prerequisites for an appropriate charging approach are required to be determined and defined, which are addressed below [54].

**2.4.1.** *Prerequisites and functionality*  First of all, *Metering* determines the particular usage of resources within end-systems (hosts) or intermediate systems (routers) on a technical level, including Quality-of-Service (QoS), management, and networking parameters. Due to the large volume of data generated by metering components in various locations within a network and at its access points as well as within VAS provider domains, *Mediation* is intended to filter, aggregate, and correlate raw technical data, which in most cases has been collected by metering. Mediation transforms these data into a form which can be used for storing and further processing.

The next step on processing mediated data will be *Accounting*. Accounting defines all summarized information (typically formalized in terms of Accounting Records, AR) in relation to a customer's service utilization. ARs are expressed in some form of quantifiable metered resource consumption, e.g., for applications, calls, or any type of connections. Basically, these three functions – metering, mediation and accounting – finalize the pure technology-driven set of parameters.

The second set of parameters are defined on the economic level of the problem and include the definition of prices for resources, especially, unit resources, which can be metered. Thus, the activity of *Pricing* – called pricing schemes as well – covers the specification and setting of prices for goods, specifically units of networking resources and units of services in a market situation, which may differ according to country-specific deregulation aspects, competitive situations, or other influencing aspects. This process may combine some technical considerations, e.g., resource consumption, and economic ones, e.g., applying tariffing theory or marketing methods. Prices may be, a.o., calculated on a cost/profit base or on the current market situation. To specialize the detailed algorithms for those prices, *Tariffs* define those ones used to determine a charge – thus, the mapping of technical parameters into economic units, typically, monetary units – for a service usage. They are applied in the charge calculation activity for a given customer to the service he utilizes. For calculating charges, the tariff may contain, e.g., discount strategies, rebate schemes, or marketing information.

Therefore, *Charge Calculation* covers the complete calculation of a price for a given AR and its consolidation into a Charging Record (CR), while mapping technical values into monetary units. Therefore, charge calculation applies a given tariff to the data accounted for and stored in the accounting database. To be able to identify the responsible customer for a set of charging records, the *Identification* allows for the mapping of data within the accounting database onto customer IDs.

Finally, the *Billing* defines the collection of those CRs, summarizing the charging content, and delivering a Billing Record (BR) to be included into an invoice/bill – including an optional list of detailed charges accumulated over a billing period, typically a month, a week, or even a day – to a user. The *Payment* of this bill may be undertaken in a pre-paid or post-paid manner, each of which does pose certain requirements onto the underlying management system as well as network infrastructure. While pre-paid schemes may use pre-paid service cards or credit cards, post-paid schemes may be supported by electronic funds transfers or credit card payments.

Finally, the term *Charging* is used within this work as an overall term, depicting all tasks required to calculate the finalized content of a single or multiple BRs. Sometimes in the literature, the term billing is utilized instead, however, it includes the full handling of invoices and customer relation management tasks, which are of clearly second priority for EMIS, since the technological and economic aspects are addressed in an integrated manner.

**2.4.2.** *Comparison of different charging systems and approaches*    The terminology used in different standardization and working domains – all of which are addressing communications as such, though – varies largely. Therefore, the following paragraph summarizes the key areas besides the IETFs (Internet Engineering Task Force), mainly including the wireless communications domain. As discussed in [33], today's 3G systems in Europe are based on UMTS (Universal Mobile Telecommunication System) Release 99, while its counterpart consists of CDMA2000 (Code Division Multiple Access) 1x Release 0 for circuit-switched services, together with CDMA2000 EV-DO (Evolution-Data Optimized) Release 0 for packet-switched services. Most of the terminology used in these specifications for different 3G (3rd Generation) releases does not match one-to-one with the IP world. Thus, those correlations are shown in Table 1. The right column is based on 3GPP's (3rd Generation Partnership Project) vocabulary definitions [17], which are also used in 3GPP2's online charging specifications.

**2.4.3.** *EMIS charging model*    Thus, the model of charging support in any networking domain is defined as shown in Figure 4, partially extended from [53].

This model consists of those 8 types of components, which address on the bottom the three types of resources to be charged for: computational resources – those originating from any type of traditional or advanced grids as well as from single computers – networking resources – at this stage limited to IP-based parameters such as bandwidth, error rates, or delay – and services, which do exist as resources or value-added services going beyond the pure IP access.

At the second level, all of those terms defined above do interact as shown. While the mediation takes metered data and aggregates them as defined, the accounting component covers the first storage capability of that type of data flow: the accounting data base stores Accounting Records (ARs) – either locally or in remote locations, mainly depending on the scenario. While prices and tariffs are stored in respective databases, the IDs of customers, users, and their grouping and stored in additional customer/user databases.

Additionally, the charge calculation stores, if required, Charging Records (CRs) for every single transaction performed, when ARs are taken out of the accounting database. These CRs are handed to the billing component, which generates bills or invoices, which in turn are stored in Billing Records (BRs) to be handed over to the respective payment implementation of the payment module. Payment itself is outside the scope of EMIS, but it may be operated in the traditional cash-based payment, via an electronic funds transfer, or a credit card payment. Those types of payments are considered as off-line payments and may be complemented by on-line payment methods, which integrate – at least to a cer-

Table 1
Comparison of charging terminology – based on [33] and extended

| All IP-based networks/EMIS | 3G mobile networks |
| --- | --- |
| Metering | Collecting charging information |
| Accounting | Charging |
| Accounting Record (AR) | Charging Data Record (CDR)[a] |
| Charging options | Billing arrangements, payment methods |
| Pre-paid/post-paid charging | Pre-paid/post-paid billing |
| Charge calculation | Charging mechanism |
| Charging Record (CR) | Transferred Account Procedure (TAP)[b] |
| Billing and parts of charging | Rating (parts of rating) |
| Billing Record (BR) | Transferred Account Procedure (TAP)[c] |
| Inter-/multi-domain charging/billing | Accounting |

[a]The details and content of such CDRs depend on the service offered, particularly being dependent on the packet-switched service or the circuit-switched service. Even different names may be used, e.g., M-CDR for a CDR of a mobile service.
[b]A new set of records was developed in terms of TAP2 and TAP2+, which support the process of operators to bill for new services and to provide the additional information required [24].
[c]Partly the use of TAPs includes the input data for billing systems, however, TAPs cover the combination of All IP-based terms Charging Record and Billing Record, mainly addressing wireless services and roaming support.

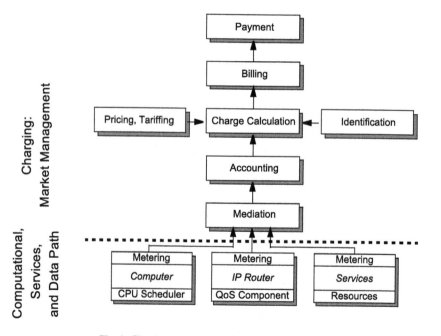

Fig. 4. Charging model – extended view based on [53],

tain extent – the accounting, charge calculation, and billing function – to ensure that any services being used at any point in time will be paid for right away, even during their use.

**2.4.4.** *Brief overview of Internet pricing schemes*   The numerous Internet pricing schemes proposed over the last years (for a closer overview see, e.g., [55]) cover a range from highly dynamic and market-oriented to simple and inflexible approaches. The major milestones proposed so far look as follows [48]:

- *Auctions* provide a very accurate and highly flexible feedback mechanism about the market situation. Unfortunately, implementing auctions on packets would yield a size of expenses which today is considered to be unbearable.
- *Volume-based* approaches reduce the accounting task to the volume as an important parameter in terms of resource usage, but still require a rather accurate monitoring of the amount of data travelling through the network.
- *Classification* is another well-established concept for reducing complexity, e.g., by defining service profiles for individual users, and tags packet demand to be in- or outside this profile. Similar ideas have led to classifying offered services themselves in terms of QoS, thus forming one of the core points of the Differentiated Services framework. Here, as a rule of thumb, higher QoS yields higher prices.
- *Edge pricing* deals with another aspect of complexity reduction, now in terms of locally concentrating the distributed nature of pricing decisions by shifting them to the edge of the ISP. This concept is charming for its simplicity, decoupling complex price negotiations between customers and various ISPs into a series of bilateral ones, as well as for its transparence towards the user.
- *Flat rate* approaches are finally the only ones that are universally accepted to be technically feasible in today's Internet. They consist of a periodical (typically monthly) charge which includes unlimited usage of the Internet service.
- *Cumulus Pricing Scheme (CPS)* proposes a variable rate scheme, which varies over long time-scales only. To bring market forces into play it provides a feedback mechanism, where this feedback is not an immediate one, but requires the accumulation of discrete 'flags' according to user behavior.

While those milestones look as six fixed categories, they do not form a commonly agreed upon classification so far. The major reason for this is based on the fact that many practical pricing schemes for Internet services are constructed out of many different components and those pricing categories are combined.

Furthermore, the problem occurring is growing worse, since marketing terms and company-specific names are utilized to defined 'new' schemes, which in fact show a slightly changed or even different set of parameters being applied, but the core principle hidden behind these names remains similar. Thus, this overview briefly outlines an overview and does not determine a full-fledged and justified classification.

**2.4.4.1.** *General charging requirements*   A closer look into the different charging approaches reveals that any proposal for tariffing Internet services has to cope with several general requirements [48], which show three main Requirement Types (RT):

- *Requirement Type 1: Customer.* Over the last years, there has been an extensive discussion on preferences customers show towards dynamic tariff schemes. E.g., within the INDEX [15], CATI [54] and M3I [38] projects it has turned out that especially transparency and predictability of charges are of major importance to the Internet customer.

- *Requirement Type 2: ISP – Economic.* Economically, the ISP is interested in running the network efficiently, e.g., by maximizing network utilization or total revenue. Thus, pricing schemes represent an important interface to the customer, as prices may be used to indicate well-behavior or misbehavior of the customer or to signal the congestion state of the network, i.e. allowing the ISP to communicate the relationship between current customer behavior and overall system status.
- *Requirement Type 3: ISP – Technical.* Each Internet pricing scheme depends heavily on the existence of tools for technical accounting. There is a vast field of possibilities as to which detail data about the network status are to be obtained since technical conditions may vary enormously.

Thus, a charging approach to be applicable, acceptable, and economically viable needs to optimize those three requirements types for the given situation.

**2.4.5.** *Time scales*   In consequence, the following four different time scales have been identified as being highly relevant for Internet charging schemes [48]. Those time scales are becoming more and more important since the efficiency to be achieved – either for the technology deployed and the charging scheme established – needs to be optimized continuously for any provider and any arbitrary set of services to comply (a) with the business model's goals and (b) with the competitive strategy chosen.

- *Atomic (communication-relevant):* This involves sending packets, round-trip times, and managing feedback between sender(s) and receiver(s).
- *Short-term (application-relevant):* This time-scale is concerned with the usual duration of applications like file-transfer, video-conferencing, or IP phone calls. The tasks of accounting and metering are closely related to these activities.
- *Medium-term (billing-oriented):* The time-scale for performing billing actions depends strongly on the usual human lifestyle habits of humans, e.g., monthly payments of rents, phone charges, or newspaper bills.
- *Long-term (contract-specific):* The largest time-scale in this context is the duration of contracts between customers and ISPs, which usually varies from several months to years. Note that contracts between ISPs may be shorter.

Figure 5 sketches the relationship between these time-scales, the respective management tasks, and communication contents as well as various pricing schemes. Note that the technical support for a particular pricing approach needs to be technically efficient to ensure that a potential economic advantage can be kept in a productive and operational system.

**2.5.** *Deployment model*

Deployment models determine the means by which business models are instantiated. Thus, different application domains or application areas will show different deployment models. To outline briefly the key steps to be undertaken to 'instantiate' a business model in the context of EMIS and for an operational case, the following example of a telecommunications company has been taken.

Telecommunications company $T$, defines the business model being of interest to them by determining the value proposition, addressing the market segment, identifying the value

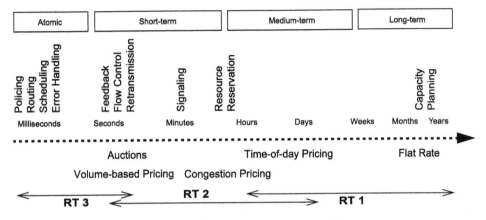

Fig. 5. Charging time scales – based on [53].

chain, summarizing the cost structure and the potential profit, outlining the value network, and investigations on the competitive strategy. Those very minimal steps to be undertaken are exemplified in a small example as follows:

- *Sample value proposition:* The objective for $T$ is to create value by offering a set of voice services, Voice-Over-IP (VoIP) services, and a number of Value-added Services (VAS). This will be based on a competitive assumption, since other telecommunication companies are present in the market. Thus, those services are perceived as the relevant form of electronic products, which provide a certain amount of utility to the service usage. For this usage, in turn, the service customer will be charged.

  Thus, the roles of the provider $T$ and the customers $C$ – originating from the role model – if any mentioned explicitly, are determined. In addition, the services have to be 'instantiated', explicitly naming the relationship between $C$ and those services $S$ – originating from the service model.

- *Sample market segment:* To satisfy customer needs in a particular country and addressing its specific market environments, both, service providers and service users do have different incentive settings, which determine behavior patterns. Accordingly, the marketing activity in this sample addresses the B2C area. Therefore, the SLA is based on an ISP-to-user contract and it contains SLS data for traditional voice services, the VoIP services, and all VAS with a certain level of guarantee.

  Thus, the different customers $C$ may be subdivided into different users $U$ and their groups, covering their different types of service usage behavior, which forms a refinement of the role model in accordance to the specific market segment.

- *Sample value chain:* In a very simplistic case the value chain consists of provisioning activities of $T$, the monitoring activities of $T$ to ensure that quality metrics are met and kept, and to provide for a charging system, in which all three types of services can be accounted for and billed at the end of an accounting period. The pricing for the voice service follows a time-based and region-based approach. The VoIP traffic is priced based on the amount of data leaving the customer (source of origin), which can simply be accounted for by a traffic meter. Finally, the set of VAS is priced on an event basis, in which case certain actions, such as click on pre-determined buttons start a certain application and the respective accounting operation (counting). On one hand, the competitive strategy (as exemplified below) an actor takes, effects the value to be achieved, on the other hand, the this knowledge forms the key business advantage of $T$, as it is hidden from inspection by organization-external entities.

Thus, the charging model has been 'instantiated' for the set of services $C$, probably even based on a per-customer contract, if required for personalization reasons. In the latter case a single service $S$ instantiation may show a different charging scheme to be applied.

- *Sample cost structure and profit potential:* In the example shown, an investigation may have shown that the offer of VoIP services will increase the customer basis of $T$, which will lead to an increased usage of those VAS offered. Thus, the economic benefit of $T$ is given by (a) a larger customer base and (b) a higher utilization of services. A detailed set of assumptions is required to ensure that those conclusions can be justified. Furthermore, the set of cost accounting systems has to be determined, to make sure that electronic service provisioning characteristics are met. Thus, means to obtain accurate cost elements on a per-service basis are essential.

- *Sample value network:* Similar to those considerations with respect to the value chain steps, the deployment models covers the information on how $T$ is embedded into the overall value network of the telecommunications industry. Since the multi-domain view does form the major influencing factor here, technology, regulatory and customer demands have to be considered. Of course, such a broad range of factors can not be simply outlined for the example taken, however, besides telecommunication customers, other telecommunication competitors and complementors as well as governmental authorities determine under which conditions the set of services are beneficial for $T$.

- *Sample competitive strategy:* As soon as economic profits are gained by $T$, competitors are attracted. To ensure that this advantage remains in place, the competitive analysis will form the basic instrument to determine new sectors, new segments, new technology, new services, under which the current position can be strengthened or extended.

This sample sketch of a deployment model outlines that the key relations between the technology and its characteristics, as well as the economics of services provided can be combined.

## 3.  Existing key mechanisms

Mechanisms in support of multiple network management tasks in a distributed system consist out of two different types:
- Technical mechanisms and
- Economic mechanisms.

While the former ones are quite well known in traditional management schemes, such as Service Level Agreements (SLA) and policy-based management approaches, the latter ones determine the new integration into management models. This is driven by various

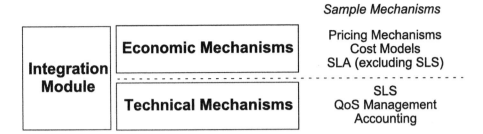

Fig. 6. Layering approaches for mechanisms.

needs, where economic measures, payments, and business model-driven influences enhance the traditional view of network management tasks into the economic management of networks. Both of those layers are seen from an integrative perspective as indicated in Figure 6, which is supported by a vertical integration module.

Integration in that sense determines that the set of parameters to be used, e.g., for a QoS management mechanism (residing in the lower layer) needs to be reflected in the SLS (residing in the upper layer), which in turn is part of the SLA. Additionally, this SLA influences the cost structure to be utilized for a given provider and technology in use, which finally determines in one way or another the pricing mechanism chosen, of course, addressing deployment model's assumptions as indicated in earlier sections.

This two-layered structure is based on a logical agreement at this stage, but it may not remain static in the future, since a conceptual discussion reveals that, on one hand, there seem to be good reasons for including an intermediate layer, in which the integrative view of technical and economic parameters may need to become visible. On the other hand, such an intermediate layer may also become an interface layer only, at which a predefined Application Programming Interface (API) may reside. Thus, the following overview addresses those two categories and summarizes the major and highly important mechanisms known today. Furthermore, open issues and emerging problems to be solved are outlined.

### 3.1. *Technical mechanisms*

To be able to address the technology basics for defining a service quality and delivery action of a service, SLAs form the basics for specifying technical, price-wise, and legal data. SLAs specify services and their characteristics, which shall be followed in the service provisioning process to ensure a customer/user-to-provider relation based in common grounds. In addition, this initial specification has to be followed by a negotiation scheme, under which those two roles will be able to agree on parameters, parameter sets, values and characteristics to be applied explicitly in a given service usage.

This process may be guided by policies, determining rules, which do address context-specific details to be applied for a certain case. Finally, the accounting in the technical sense addresses the collection – and possibly aggregation – of data, which originate from measurement points and define service-specific definitions of resources, actions and events, which have taken place in the service usage – and possibly the set-up and tear-down – phase.

**3.1.1.** *Definition of an SLA specification and its provisioning*   There are several definitions and terms of a what an SLA is and should be, respectively. In the following, some of those definitions are reviewed. If an agreement is considered to be a bilateral promise between a service provider and a customer (and vice versa), an agreement constrains the behavior of both parties, otherwise it is merely a promise.

A *Service Level Agreement (SLA)* is defined as a bilateral bundle of promises between a customer and a service provider, in relation to the provision of a service. Another definition is that an SLA specifies service levels (quality, security) that a provider needs to offer to a customer at a particular service access point (SAP) at a certain cost [13].

Sometimes, also the term Operation Level Agreement (OLA) appears. It refers to an agreement about the quality provided internally within a provider. An OLA includes no legal aspects but just technical parameters and thresholds (i.e., SLS). In case of provider chains, the term Underpinning Contract (UC) is utilized to identify an agreement about service levels between a provider and its sub-providers.

However, there is no common definition of what an SLA entails. There seems to be a consensus that an SLA must cover performance characteristics that are measurable, hence, verifiable. Other legalities might be discussed. These might include reimbursement for loss, breach of copyright, or privacy due to incompetence. Additionally, the price of a service unit is an important part of the agreement. Furthermore, SLAs require a clear and unambiguous language, since they are legal documents. Modal logic has played a role in developing such language syntax.

The clear vision is to formalize an SLA in order to automate the service provisioning as far as possible. The vision of a service provider and also strategic competitive advantage is flow-through provisioning, which means that a customer reviews a set of services offered to him by a provider, selects an appropriate service, chooses one of the available service quality options, then simply places an accordingly formed service order, and waits for its almost instantaneous fulfillment. The service provider, who receives such an order, would have automated provisioning systems that activate, monitor, and manage the ordered services with minimal additional manual input. According to [51], the following issues are warranted:

- Introduction.
- Scope of service.
- Performance, tracking and reporting.
- Problem management.
- Compensation.
- Customer duties and responsibilities.
- Warranties and remedies.
- Security.
- Intellectual property rights and confidential information.
- Legal compliance and resolution of disputes.
- Termination of agreement.
- Signatures.
- Schedules.

Hence, SLAs include a legal and a technical part. For an applicable SLA, the technical part consists of parameters that have to be defined explicitly. However, the semantics of a QoS parameter can be quite ambiguous. Issues appear, such as how to determine the availability, e.g., at the SAP, if the availability of the SAP should be tested from the provider's and customer's side, respectively, and what mechanisms should be used for testing, such as ICMP (Internet Control Message Protocol) or SNMP (Simple Network Management Protocol). This leads to the conclusion that parameters should be described as precise as possible by identifying also basic values and metrics, i.e. how the parameters are calculated. Thus, the following list of sample parameters may serve as an insight to those:

- Availability (e.g., times of day, calender days, or hours).
- Mean response time.
- Mean data rate.
- Time scale over which means will be computed and enforced.
- Probability of achieving mean data rate.

- Probability of achieving mean data response time.
- Accuracy limits.
- Penalties if service is unavailable.
- Penalties if a user exceeds their limit.
- A schedule for meetings.

The representation of an SLA can take many forms. If there is a standard syntax for SLAs, then the potential representation as XML (eXtended Mark-up Language) is simply a mapping. The approach shown in [35] – the Web Service Level Agreement (WSLA) language – proposes to "express the operations of monitoring and managing the service". These tasks "may include third parties (such as Management Service Providers) that contribute to the measurement of metrics, supervision of guarantees, or even the management of deviations of service guarantees". Additionally, the work of [58] describes which "role an SLA plays in the Internet Data Center (IDC) and how it helps assure that one's reputation stays intact. It also includes sample agreements that can be used as templates". In general, there are several approaches one might take:

- *Verbal:* Work on SLAs must somehow end up with this, since the content has to be discussed.
- *Logical:* E.g., some form of temporal logic about what must happen.
- *Promise theory approach:* Will become highly useful for more complicated agreement graphs.

**3.1.2.** *SLA negotiation mechanisms*    A lot of research has been done on negotiations from various perspectives. Researchers of agent technology conducted a strong stream of work to support negotiation activities [6]. The increasing importance of Grid computing and virtual organizations has increased the importance of SLA negotiation mechanisms as well. For example, in the EU project GEMSS (Grid-enabled Medical Simulation Services) [21] the negotiation of SLAs between competing service providers is solved by agent technology. The macro QoS negotiation between a client and multiple service providers is based on the FIPA (Foundation for Intelligent Physical Agents) reverse English auction protocol [18].

The goal of the OASIS ebXML Collaboration Protocol Profile and Agreement Technical Committee [41] is to automate the negotiation process between two negotiation parties for bargaining different technical issues in the context of the Collaboration Protocol Profile

(CPP). As a result, an agreement between two negotiation parties is expressed in the format of a Collaboration Protocol Agreement (CPA).

Traditionally, there are two types of negotiations, which can be distinguished. On one hand, distributive negotiations (also known as zero-sum and competitive negotiations) are classified as win–lose negotiations. For example, one party reaches its goals and the other party must fail to realize it. On the other hand, integrative negotiations (also known as collaborative and cooperative negotiations) are classified as win–win negotiations. The major reason to adopt an integrative negotiation is that integrative negotiation always reduces the likelihood that negotiations will fail, by making it possible to locate options that satisfy parties' ultimate expectations.

The negotiation process consists of several phases:

- Pre-negotiation for understanding the negotiation problem,
- Conduct of negotiation in terms of offers construction and counter-offer, and
- Post-settlement computation of possible offers that dominate the most recent compromise.

Certainly, the negotiation between a customer and a provider can span several iterations, and refers to the negotiation of service levels at the Service Access Point (SAP) [22].

The negotiation differs for individual or mass services. In case of individual services, a customer asks for services and service levels that a provider tries to implement, also if he needs to change his infrastructure or change contracts with his suppliers. Certainly, such negotiations are cost-driven. In case of mass services, a customer selects a service or a service package from the available set of services with predefined service levels. The negotiation process is in such a case very limited, almost non-existent.

SLA negotiation mechanisms address mainly three aspects:

- the negotiation objects,
- the negotiation protocols, and
- the decision making models.

*Negotiation objects* refer to issues over which an agreement must be reached, the identification of attributes which are negotiable or non-negotiable, respectively, and the type of attributes, whether they refer to quality and service level, or price and penalties.

*Negotiation protocols* identify and specify who (especially, which participants) can do what activity (e.g., an event or actions) at what time. There exist various approaches to identify an adequate negotiation protocol. An approach is to use the Contract Net protocol [52], which is a high-level protocol for achieving efficient cooperation in agent-based systems. Another example is COPS-SLS [39], an extension of COPS, that allows for the configuration of domain policies regarding service levels and the automatic negotiation of service levels within the domain policies. WS-Negotiation [29] is an additional approach applicable to Web Service (WS) environments. WS-Negotiation is an independent XML language that can be applied to different types of agreement templates. Negotiation on business information is a much more complex subject, and WS-Negotiation is targeting on the business parameters and legal matters.

Finally, *decision making models* specify the negotiation strategy and the reasoning model in general. They depend on the negotiation objects and protocols.

**3.1.3.** *Definition of common policy-based management basis*    Besides SLAs and their negotiation, a flexible approach on decisions based on current status information or conditions is required to achieve an adaptive management approach. Therefore, the generally admitted concepts in policy-based Information Technology (IT) management [56] are applied.

**3.1.3.1.** *A universal policy architecture*    Basic components of policy-based management cover according to [20,36]:
- *Policy repository:* A policy repository defines the maintenance component for a whole set of policies. It represents a knowledge-base for the IT management, in which policies are stored persistently.
- *Policy Decision Point (PDP):* The PDP is the logical place/location, where the decision whether a policy shall be executed, takes place. Therefore, the condition part of the policy must be evaluated. One central PDP is part of most policy architectures.
- *Policy Enforcement Point (PEP):* The PEP is the logical place/location, where policies are being executed. The PEP has to put the action part of a policy into practice.

As depicted in Figure 7, the policy repository, PDP and PEP in combination build a simple, universal policy architecture. For a communication between PDP and PEP, the Common Open Policy Service (COPS) protocol [14] and its secure transfer COPS over TLS [57] via Transport Level Security have been recommended by the Internet Engineering Task Force (IETF).

PEPs may be distributed over the whole managed IT environment. For example, every network node (e.g., routers or switches) can hold its own PEP. The (central) PDP additionally has to select the PEP, which has to execute the evaluated policy.

Usually, policies are event-triggered and can be executed in parallel, which means that one event may trigger more than one policy at the same time. PDPs are assumed to be without memory. The evaluation of a policy-condition relies exclusively on the Boolean expressions in the condition part of the policy. Policies which have been executed before have no influence on the evaluation process.

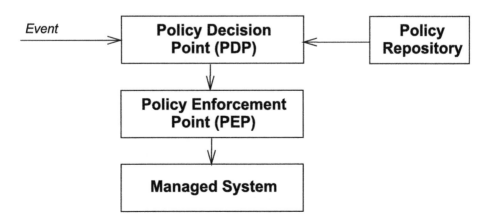

Fig. 7. A common policy architecture.

**3.1.3.2.** *Policy hierarchy* An important feature of policies is the possibility to express management instructions in different levels of abstraction and detail. Hence, policies are applicable for all kinds of management disciplines, i.e. element, network, system, service, customer, or business management. Policies of a lower level are derived by refinement of higher-level policies. In this process the entities and related actions are rendered more precisely or narrowed down.

- *Corporate policies:* Corporate policies (or high-level policies) can be assigned to business and customer management. They are derived directly from business objectives and have a strategic alignment. The level of detail is low and technical aspects are of minor prominence. Often these policies are described informally as text.
- *Functional policies:* Functional policies determine a refinement of corporate policies and are assigned to distributed system and service management. They operate on management platforms and deploy services provided, such as CORBA services [10] or OSI systems management functions [30]. Policies on this level have a formal syntax and are executable.
- *Low-level policies:* Low-level policies build the bottom layer as they are assigned to network and element management. It is essential that this kind of policies operates directly on managed objects or sets of managed objects.

**3.1.3.3.** *Policy languages* The common structure of a policy is derived from the most important publications in the area of policy-based management. Popular policy languages for IT management, like Ponder [11] and ProPoliS [12], follow exactly this structure.

- *Subject* defines who will execute the policy (often specified by a role) and in so far declares the responsibility.
- *Target* specifies objectives, which are affected by the policy.
- *Event* specifies the event, which has to occur in order to start the policy evaluation.
- *Action* defines the action, which will be performed on the specified target.
- *Condition* defines a constraint for the execution of the policy. The condition part can be regarded as a precondition. If no condition part exists, the policy is always executed when the specified event occurs.

**3.1.4.** *Accounting* Accounting, charging and billing have been identified as three essential processes involved in the commercial exploitation of products and services, in particular the Internet and for telecommunication services. Within the Internet Engineering Task Force (IETF) the Authentication, Authorization and Accounting (AAA) working group focuses on developing requirements for authentication, authorization and accounting as applied to network access, and proposed a generic architecture [34].

In general, the provisioning of commercially exploitable services requires that users are correctly authenticated and that they possess sufficient authorization privileges in order to access the service. Once that access is granted, the consumption of the resources and the usage of services has to be adequately accounted for. Generally, *Accounting* can be seen as the collection and aggregation of data about resource consumption and service usage. This includes the fully decentralized control of data gathering (via metering), transport and storage of accounting data. Besides the usage-to-user mapping, the main goal of the underlying accounting system is the storage of accounting data and data management, which is typically done centrally, but may also be performed in a fully decentralized manner [25, 26].

Furthermore, *Accounting Policies* determine which data is stored, how often and in which way data is collected and made available to other functions. The accounting information, usually stored in so-called accounting records, can for instance be used for capacity and trend analysis, load balancing, cost allocation, billing, or auditing. Hence, from the viewpoint of commercial service provisioning, accounting is perceived as the central component in a row of supporting processes.

Accounting relies on identification of users (authentication) and the determination of respective user profiles (authorization). Stored accounting records then feed into downstream processes, embracing well verification of SLA compliance (auditing), calculation of non-monetary values (charging), and finally cleared monetary values (billing), summarizing resource consumption and service usage in condensed form, thus presented in a readable and understandable way for service customers.

For systems being in place today, the generic AAA architecture [34] describes the set of general network components required and their functionality on an abstract level. Concerning this architecture, a layered approach and structure for AAA protocols has been proposed. While initially the RADIUS protocol (Remote Authentication Dial-In User Service) [50] has determined the widely deployed AAA protocol – mainly in support of authentication and providing also accounting functionality [49] – the protocol termed Diameter [4] defines the next generation AAA protocol. It is based on RADIUS, but Diameter is much more flexible, and it provides reliable data transfer, fail-over mechanisms, better error handling, and security mechanisms. The Diameter protocol includes the generic base protocol and various Diameter applications. Diameter allows for the specification of Attribute–Value Pairs (AVPs), which define, a.o., for accounting purposes the parameters to be accounted for.

With respect to accounting records in existing systems and standards, current important accounting record formats can be summarized as follows. They belong to:

- the RADIUS approach [50], defined by a fixed set of AVPs,
- the Diameter approach [4,49], defined by a flexible set of AVPs,
- IPDRs (Internet Protocol Detail Record), defining a unified XML format [31],
- Cisco's NetFlow data [8], or
- IETFs IPFIX proposal [9] for the representation of IP flows.

Finally, other protocols in support of a principle accounting may be considered, which include COPS (Common Open Policy Service), RSVP (Resource Reservation Protocol), or SNMP (Simple Network Management Protocol), however, they are not relevant in practice for a large accounting approach due to a slightly different initial focus and a performance-sensitive negative externality.

As this overview on existing systems shows, accounting has been considered a necessity for commercial telecommunication providers or IT outsourcers in order to charge their resources and services respectively. However, for a number of particular application domains accounting has been rarely implemented, such as for the domain of Grid computing. The virtualization of organizations, resources and services within a Grid environment additionally leads to requirements and questions beyond the common accounting architectures, thus forming a larger task on management of distributed resources beyond networking services.

Generally, the focus of existing Grid accounting systems and tools is on the accounting of physical Grid resources, e.g., computing elements, storage resources, software licenses, and scientific devices. An adequate accounting of virtual, potentially complex services, such as information services or computation services which may be offered within a multi-provider scenario has not been sufficiently addressed so far.

In order to perform accounting for Dynamic Virtual Organizations (DVO), comprising a set of virtual resources and virtual services, a generic service model for DVOs is needed. An important aspect of the service model on the one hand is to reflect the underlying virtualization concept and on the other hand to contain different views of the provisioned virtual resources and services in regard to the level of aggregation. Thus, a comprehensive service model offers the flexibility to reflect the view of a single service provider, a service aggregator, as well as the view of the entire Virtual Organization (VO) by using the same basic principles. Another important aspect which has to be addressed is the specification of adequate accountable units building a basic set of service primitives any service provisioned within a VO is composed of.

## 3.2. *Economic mechanisms*

To be able to address the economic mechanisms for defining service quality and delivery action, economic measures and metrics, as well as cost parameters form the appropriate basis. The definition of those is followed by a detailed view on mobile micro-payment schemes and their respective accounting systems in support of extremely low-value transactions required for various on-line services.

Finally, the analysis of service provisioning templates – under the assumption of SLAs – and their definition addresses (1) a customer- and provider-based SLA approach, where IP network and service providers offer physical connectivity, and (b) a pure service provider-based approach under which providers agree upon an interconnection agreement.

**3.2.1.** *Definition of economic measures and cost parameters*   Any type of service offered in a networking environment – ranging from pure IP access services to Value-added Services (VAS) – is required to be specified and described. While the technical specification may be based on an SLA, as defined above in Section 3.1.1, the economic specification does include relevant tariffing and pricing information as well, as outlined in Section 2.4.4. The description additionally may contain further information, with which the use of the service, the customer's profiles, the market segment and the target number of users are outlined. This type of information can be considered as business-critical, since it defines provider-internal models and assumptions under which the publicly announced service specifications have been calculated in support of a typically economically viable business model.

Thus, a clear understanding of relevant economic measures and cost parameters is essential to determine the key economic drivers for a finalization of any pricing information on any type of service offered.

*Cost parameters* define the different types of costs involved in any type of service provisioning. Accordingly, *economic measures* define the range of potentially quantified metrics applicable, typically technical values derived from, e.g., QoS parameters or metering parameters. Finally, those economic measures are mapped onto service charges, which are defined in terms of monetary units per service unit utilized.

Thus, the economic measures and cost parameters of interest for an IT system are required to map such technical – in a normal case, well defined, quantifiable and standardized in many instances – specifications, to the provider-internal business model assumptions. For those economic measures and cost parameters it will be possible to list most of them in an initial investigation. However, the modeling correlations between those and the service offered, the provider's networking domain, or the legal requirements valid for such an offering cannot be performed in a general manner. Only the detailed provider-specific set of assumptions – including the services, the service mix, the market situation, the competition regions, the demand curves over time, and the set of user classifications – will enable the calculation of a detailed and a meaningful result, typically including benefit/loss information, sensitivity analysis on demand variations, and technical restrictions to be considered.

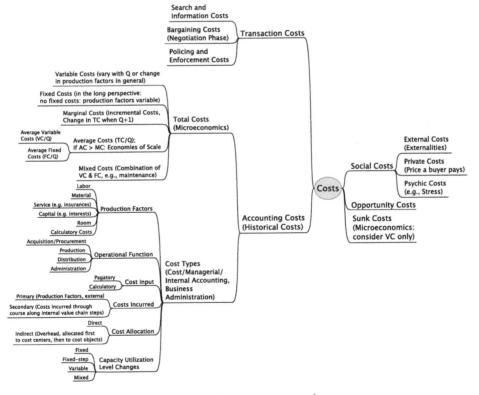

Fig. 8. Overview of cost parameters.[1]

---

[1] AC denotes Average Costs, FC denotes Fixed Costs, MC denotes Marginal Costs, TC denotes Total Costs, VC denotes Variable Costs, Q denotes in microeconomics the output achieved.

To provide an overview on economic measures and cost parameters, Figure 8 outlines the *theoretical view of costs*, which exist in any IT system of relevance. Basically, costs can be subdivided into transaction costs and accounting costs, which are complemented by social costs, opportunity costs and sunk costs. While *transaction costs* address those ones which are related to a specific operation, such as search, information retrieval, bargaining, negotiation, policing and enforcement, *accounting costs* cover microeconomic total costs and their different types, as well as a number of cost types driven by financial accounting approaches (cost, management and internal accounting) and business administration costs.

Besides this theoretical evaluation of costs, the *technology-driven view of costs* can be determined. Thus, to provide for an initial list of cost parameters in the technical domain of network management the following ones are based on the OSI Network Management Model FCAPS (Fault, Configuration, Accounting, Performance, Security) [59]:

1. *Fault Management Costs:* These costs occur for identifying problems and network failures, or costs for problem resolution.
2. *Configuration Management Costs:* These costs occur for hardware and software inventories, or costs for configuration changes.
3. *Accounting Management Costs:* These costs include the collection and storage costs of network usage data.
4. *Performance Management Costs:* These costs are required for network-capacity planning, i.e. network availability, throughput, delay, and eliminating network bottlenecks.
5. *Security Management Costs:* These costs exist due to the definition and execution of security policies, e.g., access control policies, logging, network protection, and risk identification. They may rise dramatically if a network is under attack, and may even include liability costs.

In a wider view these 5 categories can be refined in the context of economic management tasks with the two following cost parameters:

6. *Audit Management Costs:* These costs are required for a compliance check of an SLA or the verification of a contractual fulfillment of a pre-negotiated contract, typically an SLA in the networking domain. These costs could be considered, if required, as special fault management costs for external errors.
7. *Contract Management Costs:* These costs cover contract negotiations and interactions with customers, providers, and peers, which may also include call centers as external interfaces to a network management system. They could be considered, if required, as special configuration management costs for setting up a bilateral or multilateral customer/user-to-provider relation.

Finally, in a much longer term time-scale the following cost parameter may be relevant:

8. *Research Costs:* These costs exist for the design and development of network services in a determined networking technology environment, for a given market segment, and a class of customers.

Categories 1–5 show key characteristics of overhead costs. Network management cost elements usually are not directly allocatable to a specific cost object. In this context, cost objects consist in electronic services that are delivered to service consumers. Thus, in other words, network management costs cannot be attributed directly to a given service delivery, since this specific service delivery cannot be taken as the single or at least the main reason that network management tasks occur in the first place. However, a certain share in overall network management costs might consist of direct costs. This is the case if a network management activity is triggered by a given service delivery. The costs related to that specific network management activity are considered as direct costs. From the range presented in Figure 8, other cost type classifications might apply to network management costs in parallel, such as fixed or secondary costs. Category 8, research costs, is characterized in the same way as network management costs mainly by overhead/indirect costs.

Categories 6 and 7 are included in the transaction cost theory (cf. Figure 8). Contract management (category 7), however, goes beyond pure negotiations. It involves also costs for investigation, searching and – after a contract was concluded – costs incurred when contract determinations need to be ensured/enforced. Thus, audit-related and negotiation-related costs can be seen as a subcategory of transaction costs. Again, this cost type classification should not be seen as the only possible classification.

In order to provide an initial sample list of economic measures and cost parameters, Figure 9 addresses for two examples a non-closing subset of parameters, measures and charges relevant for the pure IP access service for private or home use – the most basic service to be thought of – and a VoIP service for corporate customers – a special VAS of large interest.

**Example IP Access Service (Private/Home Use)**

| Cost Parameters | Economic Measures | Service Charges |
|---|---|---|
| • Variable costs for upstream data traffic<br>• Fixed indirect costs (production factors, transaction costs) | • Transferred Data Volume<br>• Time/Duration<br>• Best-effort QoS guarantees | • 0.10 € / MB<br>• 0.50 € / h<br>• 9.99 € / month |

Service Provider ⟶ Service Consumer

**Example VoIP Service (Business Customers)**

| Cost Parameters | Economic Measures | Service Charges |
|---|---|---|
| • Fixed-step costs (telephony minute bundles; number blocks)<br>• Variable direct costs (IP access service)<br>• Fixed indirect costs (production factors, transaction costs, opportunity costs due to QoS guarantees) | • Time/Duration (internal calls)<br>• Time/Duration (external calls to foreign networks)<br>• QoS guarantees (Availability, Jitter, Delay, Throughput) | • 0.0 € / min (internal calls)<br>• 0.03 € / min (external calls)<br>• 5.00 € / month and line |

Fig. 9. Cost parameters, economic measures and service charges for two sample services.

The set of open problems to be tackled in future research needs to address this type of mapping function between SLA and cost parameters under certain assumptions. It needs to show in which cases the most generic approach of a sensitivity analysis can be developed in terms of a toolbox or a guideline in support of a provider, and – at least hypothetically – for business customers, who would like to evaluate a number of counter offers for their dedicated demands. Furthermore, a potential list of mandatory parameters of interest for a service or service class, as well as optional ones, may be determined based on the investigations of parameter dependencies and correlations.

### 3.2.2. *Mobile micro-payment and accounting systems*

Micro-payments define the set of technology, which combines the action of payment – the transfer of value or a monetary claim from a payer to a payee – with the amount of such values, which are very small by definition. Typically, micro-payment systems belong to the class of electronic payment systems, in which a type of e-money is transferred, initiated, processed and acknowledged fully electronically. Examples of micro-payment systems include Wallie, Minitix and Way2Pay (cf. [44] for further details and systems). This type of e-money [5] – or electronic money – is defined by a complete electronic medium, in which this money is issued by banks and exchanged in a fully electronic manner. Finally, the mobility aspect of those payment systems is added in case of

- the mobile device determining the payment instrument,
- the mobile device determining the payment terminal, or
- the mobile network determining the direct payment channel.

Thus, currently the three systems 'Paiement CB sur mobile', Paybox [46], and Moxmo are examples of mobile payment systems, not necessarily micro-payment systems, though. A nice overview on all aspects of payment systems, micro aspects, and technologies can be obtained from [44].

The mobility aspect of accounting systems (for the set of existing accounting technologies refer to Section 3.1.4), however, remains at this stage a preliminary one. This is based on the fact that a number of decisions have to be taken in advance. Initially, the location of the accounting database is essential for either the reliable access to it as well as for a performance-sensitive access. Furthermore, the size of such an accounting database is highly relevant for the type of device and its storage capabilities. Finally, the security concern of the operator utilizing a mobile accounting system has to be met, of course, meeting customers' demands at the same time.

There is a fundamental difference between the security needed for 'traditional' accounting systems and mobile micro-payment systems. Within traditional accounting systems, using for example protocols like RADIUS [49], Diameter [4] or NetFlow [8], information is exchanged between trusted parties. However, within micro-payment systems information is exchanged between parties that have no direct trust relationship. As a consequence, security aspects like data-integrity and non-repudiation are essential for such systems. Trust not only requires technical measures, but needs to be enforced also by legislation.

For the design of an efficient and technically scalable mobile micro-payment system a set of system and security requirements, application specifics and performance considerations have to be taken into consideration. In addition, such systems should support anonymity, be easy to use, and allow payments to be made across country and payment operator borders. Standardization and interoperability are, therefore, key and important requirements [45].

**3.2.3.** *Service provisioning templates for IP QoS delivery*   For business relationships between principal actors for an IP QoS delivery two different possible business models are distinguished. The first one is largely in-line with the business model specified in the context of the MESCAL project [28,37]. The second business model is based on observations of pure service providers recently emerged in the Internet, such as Skype.

Each of those two examples are described in terms of service provisioning templates, expressing the fact that the business model defines the product to be offered – here the service to be offered – and the concrete interdependencies of roles distinguished – here service users and customers as well as the three different provider roles – explain the interaction template those roles will typically show for the type of service under consideration.

**3.2.3.1.** *Service provisioning Template 1 – MESCAL-compliant*   Most of today's telecommunication companies providing Internet access services are ICPs and ASPs at the same time, meaning that they not only own the IP network infrastructure, but also provide various services on a higher level. Figure 10 shows the template that assumes the integration of ICPs and ASPs. This means that individual end-customers are not only connected to the ICP, but also subscribe to services they are providing. The main focus is on the relationships between end-customers and ICP/ASPs and between ICP/ASPs themselves.

It is assumed, that interactions between ICPs and access providers for connecting end-customers to the IP network infrastructure owned by the ICPs either do not exist (end-customers are connected through means/facilities provided by the ICPs themselves), or if they do, they are completely orthogonal to (they do not affect) the provisioning of services; as such they are outside the scope of this investigation.

In the scenario shown above, the primary interest is in the investigation of SLAs underlying the interactions between stakeholder domains of concern from technical perspectives (e.g., specifications, protocols) and the required functionality in the ICP/ASP domains for supporting these SLAs. Regarding the latter, the emphasis is on SLA fulfillment aspects through traffic engineering and service admission control functions, more specifically, technical aspects of SLAs, referred to as SLSs. SLSs can be described as a documented result of a negotiation that specifies availability, performance, pricing, and other aspects of a – in this case – transport service a provider offers to a customer. The terms customer SLS

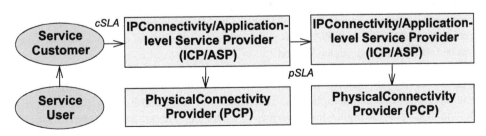

Fig. 10. Service provisioning Template 1 – MESCAL-compliant.

(cSLS) and provider SLS (pSLS) denote, respectively, the SLSs between end-customers and ICP/ASPs and among peering ICP/ASPs themselves.

*Customer SLS (cSLS)*    A cSLS is the technical specification of the SLA between the end-customer and the ICP/ASP. It specifies how the end-customer traffic should be treated within the ICP/ASP's network. The intra-domain cSLS can be taken as an example, where traffic source and destination can be covered by one single provider. A typical cSLS includes the following elements:

- *SLS Id* identifies uniquely the SLS specific to the end-customer.
- *Scope* exists in three types of an SLS scope for end-customers, who would inject traffic as far as a single ICP/ASP network is concerned:
  - the pipe model (1, 1) has one ingress and one egress
  - the hose model (1, $N$) has one ingress and multiple egresses
  - the one-to-any model (1, $*$) has one ingress and a wildcard with egresses.
- *Service schedule* specifies when the cSLS is active, i.e. operating times such as per-day or per-week.
- *Flow descriptor* is essentially the packet stream identifier that includes source address, port number and DSCP (Differentiated Services Code Point).
- *QoS parameter* includes, e.g., bandwidth, delay, jitter, packet loss ratio.
- *Traffic descriptor* describes traffic patterns; a typical example is a token bucket algorithm for identifying in- and out-of-profile packets.
- *Excess treatment* indicates how to treat out-of-profile packets, e.g., dropping, shaping, or remarking.
- *Service availability* denotes the percentage of time the service is available, e.g., 99.9%.

*Provider SLS (pSLS)*    As mentioned earlier, individual ICP/ASPs need to establish provider level SLSs with each other if they would like to expand IP reachability or higher level services to other providers. With the presence of pSLSs between multiple ICP/ASPs, customers are able to subscribe and use services that are provided by remote providers in the Internet.

In essence, pSLSs extend – typically to the end of QoS traffic exchange – respective agreements for transiting or exchanging traffic that exist today between providers in the best-effort Internet. As such, they should be in-line with the specific context of traffic exchange, which is implied by particular business relationships holding between providers. Compared to cSLSs, a significant difference is that the technical service agreement (pSLS) is based on aggregated traffic rather than individual flows. In other words, the task of negotiating pSLSs between ICP/ASPs is to determine how individual cSLSs from customers can be fulfilled in an aggregate fashion.

**3.2.3.2.** *Service provisioning Template 2 – pure service provider-compliant*    In the second service provisioning template, the functionality of ASPs and ICPs are distinct. ASPs, on one hand, provide higher-level services to customers without owning any IP layer infrastructure. ICPs, on the other hand, offer only IP connectivity services to ASPs for supporting higher-level applications, and do not interact directly with the customers. Taking

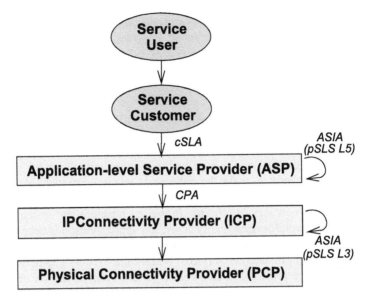

Fig. 11. Service provisioning template 2 – pure service provider-compliant.

Skype as an example, Skype is able to provide IP telephony with only VoIP gateways at the edge of Internet network providers, but it does not necessarily have a physical IP network.

Today, services provided by Skype are based on the best effort Internet without guaranteeing any QoS. In order to offer IP telephony services with QoS guarantees, ASPs need an IP level QoS support (e.g., dedicated bandwidth) from ICPs. ASPs interact vertically with ICPs on the basis of Connectivity Provisioning Agreements (CPA), as shown in Figure 11. Beyond the forwarding and QoS treatment of the service control and data traffic entering the ICP's network from the ASP's sites, the ICP offers to the ASP means to control the connectivity provisioning, such as receiving reports and alarms, invoking admission control at the IP network resources level, or enforcing the required configuration on the ICP's network on service activation.

Similar to the first business model, for expanding the scope and augmenting the portfolio of services offered, ASPs may interact horizontally with each other through setting up business agreements termed ASP Interconnection Agreements (ASIA), which are also called pSLS layer 5 (Session Layer). The underlying ICP also needs layer 3 pSLS between each other in order to expand their IP reachability services. This type of pSLSs is termed IP Interconnection Agreements (IIA). Detailed elements to be included in ASIAs and IIAs are still under study.

## 4. Summary and conclusions

The paradigm shift to Information Technology (IT) service management currently seen is a consequence of various trends such as the liberalization of the telecommunication market, the convergence between computer networks and telecommunication systems, and the need for an end-to-end management. New technical, organizational and economic challenges become apparent and new technologies appear to be full of promise. While the domains of traditional network management and IT management have been considered separately so far, the interrelation and inter-operation of services from both domains does require a joint approach.

Managing new services on many different types of networks with many different players are among those challenges. Alongside is the need for a provisioning of a seamless interoperability and interdependence between services and networks. In this area service providers compete for customers with high quality services, with the ability to deploy services rapidly and efficiently, to make the ordering of services extremely simple, to keep inconvenience for the customer at a minimum, and to provide for the ability to rapidly introduce and roll out new service offerings in various locations and systems. Instead of resources, such as network devices, end systems, and applications, it is necessary to think in terms of services and service quality. The new approach developed and termed EMIS (Economic Management of Internet Services) bridges the existing gap between network and IT management by providing a comprehensive overview on the set of models and mechanisms that are common to both worlds. As such, it sketches the future functionality of the Internet economic management, thereby often going one step beyond the current state-of-the-art.

The concept of a service is clearly a recent advance in the understanding of networking technology as well as information technology. But what is such a service in detail, what is a Service Level Agreement (SLA), and what are respective management mechanisms for those? In order to analyze these questions in a much greater level of detail and gain a common basis of understanding, the key models and terms of EMIS have been identified and determined. While the service model outlines the type of services envisioned and specified, the role model determines players together with their specific behavior in the multi-user and multi-provider domains of service usage and offering. The charging model identifies key activities to be undertaken for being able to pay for a certain service usage a fair and announced price. Finally, business models, as shaped by those three models mentioned, find instantiations for operations in deployment models.

Based on this common ground, the next research and collaboration issue was dedicated to the investigation of key mechanisms, including technical as well as economic management mechanisms. Those technical mechanisms were devoted to the key research issues such as formalizing SLAs, SLA negotiation mechanisms, policy-based management and accounting. Discussing about technical aspects of SLAs and accounting has lead to the research issue of respective economic mechanisms. Of special importance was the definition of economic measures and cost parameters, mobile micro-payment schemes, accounting and the analysis of service provisioning templates in order to move from 'bits to business values'.

These models and mechanisms provide the necessary common basis for further investigations, technical and economic discussions, and developments on top of EMIS. Yet, two further integration steps remain to be addressed in future work: On one hand, the obvious relationship between the charging model and the accounting mechanisms needs to be tackled in the most general case of a multi-service, multi-provider, multi-domain scenario.

The service provisioning templates provided so far determine clearly a first step in this direction. On the other hand, the two sets of mechanisms collected on the technical and economic side are required to be combined to achieve an economic management approach for any type of IP-based data transport services. Due to these detailed lists of parameters – originating from the QoS and RADIUS, Diameter, and general accounting world as well as from the cost and economics measures domain – the key interfaces between them have been initially identified. Of course, this step determines the first one required, however, this basis clarified views and abstracted away from low-level details, which will not be required. Thus, the integration module – as indicated in Figure 6 – will become a key functional aspect of future economic management solutions.

## 5. Abbreviations

| | |
|---|---|
| 3G | 3rd Generation |
| 3GPP | 3rd Generation Partnership Project |
| AAA | Authentication, Authorization and Accounting |
| ABC | Activity-based Costing |
| AC | Average Cost |
| API | Application Programming Interface |
| AR | Accounting Record |
| ASIA | ASP Interconnection Agreements |
| ASP | Application-level Service Provider |
| AVP | Attribute Value Pairs |
| B2B | Business-to-Business |
| B2C | Business-to-Consumer |
| BR | Billing Record |
| CDMA | Code Division Multiple Access |
| COPS | Common Open Policy Service Protocol |
| CORBA | Common Object Request Broker Architecture |
| CPA | Connectivity Provisioning Agreements |
| CPA | Collaboration Protocol Agreement |
| CPP | Collaboration Protocol Profile |
| CPS | Cumulus Pricing Scheme |
| CPU | Central Processing Unit |
| cSLA | Customer Service Level Agreement |
| cSLS | Customer Service Level Specification |
| CSM | Customer Service Management |
| CR | Charging Record |
| DNS | Domain Name System |
| DSCP | Differentiated Services Code Point |
| DVO | Dynamic Virtual Organization |
| EMANICS | European Network of Excellence for the Management of Internet Technologies and Complex Services |

| | |
|---|---|
| EU | European Union |
| EV-DO | Evolution-Data Optimized |
| FC | Fixed Cost |
| FCAPS | Fault, Configuration, Accounting, Performance, Security |
| FIPA | Foundation for Intelligent Physical Agents |
| FR | Frame Relay |
| GEMSS | Grid-enabled Medical Simulation Services |
| h | Hour |
| ICMP | Internet Control Message Protocol |
| ICP | Internet Connectivity Provider |
| ICT | Information and Communication Technology |
| IDC | Internet Data Center |
| IETF | Internet Engineering Task Force |
| IIA | IP Interconnection Agreements |
| IP | Internet Protocol |
| IPDR | Internet Protocol Detail Record |
| ISDN | Integrated Services Digital Network |
| ISP | Internet Service Provider |
| IT | Information Technology |
| ITIL | IT Infrastructure Library |
| M | Mega |
| MB | Mega Byte |
| MC | Marginal Cost |
| MESCAL | Management of End-to-end Quality of Service Across the Internet at Large |
| min | Minute |
| OASIS | Organization for the Advancement of Structured Information Standards |
| OLA | Operation Level Agreements |
| OSI | Open System for Interconnect |
| PCP | Physical Connectivity Provider |
| PDP | Policy Decision Point |
| PEP | Policy Enforcement Point |
| POTS | Plain Old Telephony System |
| pSLA | Provider Service Level Agreement |
| pSLS | Provider Service Level Specification |
| Q | Output (in microeconomics) |
| QoS | Quality-of-Service |
| RADIUS | Remote Authentication Dial-In User Service |
| RSVP | Resource Reservation Protocol |
| RT | Requirement Type |
| SAP | Service Access Point |
| SDH | Synchronous Digital Hierarchy |
| SLA | Service Level Agreement |
| SLM | Service Level Management |

| SLS | Service Level Specification |
|-----|------------------------------|
| SNMP | Simple Network Management Protocol |
| TAP | Transferred Account Procedure |
| TC | Total Cost |
| UMTS | Universal Mobile Telecommunication System |
| UC | Underpinning Contract |
| VAS | Value-added Service |
| VC | Variable Cost |
| VO | Virtual Organization |
| VoIP | Voice-over-IP |
| WLAN | Wireless Local Area Network |
| WSLA | Web Service Level Agreement |
| WS | Web Service |
| xDSL | Type x Digital Subscriber Line |
| XML | eXtended Mark-up Language |

## Acknowledgements

The editors would like to express their gratitude to EMANICS project partners and researchers for providing text and material, particularly at the Ludwigs-Maximilian-Universität München namely T. Schaaf (Sections 2.3 and 3.1.3); at the Oslo University College namely M. Burgess (Sections 2.2 and partly 3.1.1); at the University of Federal Armed Forces Munich namely G. Dreo, M. Göhner and F. Eyermann (Sections partly 3.1.1, 3.1.2, partly 3.1.4, and partly 4); at the University of Surrey namely S. Georgoulas and N. Wang (Sections 2.2 and 3.2.3); at the University of Twente namely A. Pras (partly Section 3.2.2); and at the University of Zürich namely B. Stiller (Sections 1, 2.4, partly 2.5, partly 3.1.4, partly 3.2.2, and partly 4), D. Hausheer (Sections 1 and partly 4), M. Waldburger (Sections 2.1, partly 2.5, partly 3.2.1), P. Kurtansky (partly Section 2.4) and G. Schaffrath.

## References

[1] 3GPP, *Architecture for an all IP network*, TR 23.922 (1999).

[2] M. Burgess, *An approach to understanding policy based on autonomy and voluntary cooperation*, IFIP/IEEE 16th International Workshop on Distributed Systems Operations and Management (DSOM'06), Barcelona, Spain, Lecture Notes in Comput. Sci., Vol. 3775, Springer-Verlag, Heidelberg (2006), 97–108.

[3] M. Burgess and S. Fagernes, *Pervasive computing management: A model of network policy with local autonomy*, IEEE Transactions on Software Engineering (2007), in press.

[4] P. Calhoun, J. Loughney, E. Guttman, G. Zorn and J. Arkko, *Diameter base protocol*, Internet Engineering Task Force (IETF) RFC 3588 (2003), 1–147.

[5] L.J. Camp, M. Sirbu and J.D. Tygar, *Token and notational money in electronic commerce*, 1st USENIX Workshop on Electronic Commerce, New York, USA (1995), 1–12.

[6] A. Chavez, P. Maes and Kasbah, *An agent marketplace for buying and selling goods*, 1st International Conference on the Practical Application of Intelligent Agents and Multi-Agent Technology, London, UK (1996), 75–90.

[7] H. Chesbrough and R.S. Rosenbloom, *The role of the business model in capturing value from information: Evidence from Xerox Corporation's technology Spin-off companies*, Industrial and Corporate Change **11** (2002), 529–555.

[8] Cisco Systems, *Cisco IOS Netflow: Introduction*, http://www.cisco.com/en/US/products/ps6601/products_ios_protocol_group_home.html (2006).

[9] B. Claise, ed., *IPFIX Protocol Specification*, IPFIX Working Group, Internet Engineering Task Force (IETF), draft-ietf-ipfix-protocol-21.txt (2006).

[10] CORBA Services Complete Book, *OMG Specification formal/98-12-09*, Object Management Group (1998).

[11] N. Damianou, N. Dulay, E. Lupu and M. Sloman, *Ponder: A language for specifying security and management policies for distributed systems, the language specification version 2.3*, IC Research Report DoC 2000/1, Imperial College of Science, Technology and Medicine, University of London, Department of Computing (2000).

[12] V. Danciu and B. Kempter, *From processes to policies – Concepts for large scale policy generation*, Managing Next Generation Convergence Networks and Services: Proceedings of the 2004 IEEE/IFIP Network Operations and Management Symposium (NOMS'04), Seoul, Korea (2004), 17–30.

[13] G. Dreo Rodosek, *A framework for IT service management*, Habilitation Thesis, Ludwig Maximilian Universität, Munich, Germany (2002).

[14] D. Durham, J. Boyle, R. Cohen, S. Herzog, R. Rajan and A. Sastry, *The common open policy service (COPS) protocol*, Internet Engineering Task Force (IETF) RFC 2748 (2000), 1–38.

[15] R. Edell and P.P. Varaiya, *Providing Internet access: What we learn from the INDEX* (1999), 18–25.

[16] *EMANICS – Management of the Internet and complex services*, Network of Excellence (NoE), FP6-2004-IST-026854-NoE, http://www.emanics.org/ (2006).

[17] ETSI TR 121 905 V6.10.0 (2005-09), *Digital cellular telecommunications system (Phase 2+), Vocabulary for 3GPP Specifications*, 3GPP TR 21.905 Ver. 6.10.0 Release 6 (2005).

[18] FIPA, *The foundation of intelligent physical agents*, IEEE FIPA Standards Committee, http://www.fipa.org/ (2006).

[19] M. Garschhammer, R. Hauck, H.-G. Hegering, B. Kempter, M. Langer, M. Nerb, I. Radisic, H. Roelle and H. Schmidt, *Towards generic service management concepts – A service model based approach*, 7th International IFIP/IEEE Symposium on Integrated Management (IM 2001), Seattle, WA, USA (2001), 719–732.

[20] M. Garschhammer, R. Hauck, B. Kempter, I. Radisic, H. Roelle and H. Schmidt, *The MNM service model – refined views on generic service management*, Journal of Communications and Networks (2001), 297–306.

[21] GEMSS Project, *Grid-enabled medical simulation services*, 5th Framework EU Project, IST Program, No. IST-2001-37153, http://www.gemss.de (2005).

[22] J. Gerke, *A generic peer-to-peer architecture for Internet services*, Dissertation, ETH Zurich, Switzerland (2006).

[23] J. Gerlach, B. Neumann, E. Moldauer, M. Argo and D. Frisby, *Determining the cost of IT services*, Communications of the ACM **45** (2002), 61–67.

[24] GSM World, *Tapping the potential of roaming*, http://www.gsmworld.com/using/billing/potential.shtml (2006).

[25] D. Hausheer, *PeerMart: Secure decentralized pricing and accounting for peer-to-peer systems*, Dissertation, ETH Zurich, No. 16200, Shaker-Verlag, Aachen, Germany (2006).

[26] D. Hausheer and B. Stiller, *PeerMint: Decentralized and secure accounting for peer-to-peer applications*, 4th International IFIP-TC6 Networking Conference (Networking 2005), Waterloo, ON, Canada, May 2–6, Lecture Notes in Comput. Sci., Vol. 3462, R. Boutaba et al., eds, Springer-Verlag, Berlin (2005), 40–52.

[27] H.-G. Hegering, S. Abeck and B. Neumair, *Integriertes Management vernetzter Systeme – Konzepte, Architekturen und deren betrieblicher Einsatz*, dpunkt-Verlag, Heidelberg (1999), ISBN 3-932588-16-9.

[28] M. Howarth, P. Flegkas, G. Pavlou, N. Wang, P. Trimintzios, D. Griffin, J. Griem, M. Boucadair, P. Morand, H. Asgari and P. Georgatsos, *Provisioning for inter-domain quality of service: The MESCAL approach*, IEEE Communications Magazine **43** (2005), 129–137.

[29] P.C.K. Hung, H. Li and J.J. Jeng, *WS-negotiation: An overview of research issues*, 37th Annual Hawaii International Conference on System Sciences (HICSS'04), Hawaii, USA (2004), 1–10.

[30] *Information Technology – Open Systems Interconnection – Systems Management – Management Functions*, ISO 10164-x, International Organization for Standardization and International Electrotechnical Committee (1991–1997).

[31]  IPDR.org, *IPDR/XML File Encoding Format*, Version 3.5.0.1 (2004).

[32]  R.S. Kaplan and A.A. Atkionson, *Advanced Management Accounting*, 3rd edn, Prentice Hall, Upper Saddle River, USA (1998).

[33]  P. Kurtansky and B. Stiller, *State of the art prepaid charging for IP services*, 4th International Conference on Wired/Wireless Internet Communications (WWIC 2006), May 10–12, Bern, Switzerland, Lecture Notes in Comput. Sci., Vol. 3570, Springer-Verlag, Heidelberg (2006), 143–154.

[34]  C. de Laat, G. Gross and L. Gommans, *Generic AAA Architecture*, Internet Engineering Task Force (IETF) RFC 2903 (2000).

[35]  H. Ludwig, A. Keller, A. Dan, R.P. King and R. Franck, *Web Service Level Agreement (WSLA) Language Specification*, IBM T.J. Watson Research Center, http://www.research.ibm.com/wsla/WSLASpecV1-20030128.pdf (2003).

[36]  B. Moore, E. Ellesson, J. Strassner and A. Westerinen, *Policy core information model – Version 1 specification*, Internet Engineering Task Force (IETF) RFC 3060 (2001).

[37]  P. Morand, M. Boucadair, T. Coadic, P. Levis, R. Egan, H. Asgari, D. Griffin, J. Griem, J. Spencer, M. Howarth, N. Wang, S. Georgoulas, K.H. Ho, P. Flegkas, P. Trimintzios, G. Pavlou, P. Georgatsos, E. Mykoniati, I. Liabotis and T. Damilatis, *Final specification of protocols and algorithms for inter-domain SLS management and traffic engineering for QoS-based IP service delivery*, D1.3 MESCAL Deliverable, http://www.mescal.org/ (2005).

[38]  M3I, *Market Managed Multi-service Internet*, 5th Framework EU Project, IST Program, No. 11429, http://www.m3i.org (2000).

[39]  T.M.T. Nguyen and N. Boukhatem, *Policy-based service level negotiation with COPS-SLS*, Conference on Network Control and Engineering (NetCon'03), Muscat, Oman, October 14–15 (2003), 1–12.

[40]  T. Nolle, *A new business layer for IP networks*, Business Communications Review, http://www.ipsphereforum.org/newsevents/07nollereprint.pdf (2005).

[41]  OASIS, *Automated negotiation of collaboration – Protocol agreements specification*, ebXML Collaboration Protocol Profile and Agreement Technical Committee, http://www.oasis-open.org/committees/ebxml-cppa/negotiation (2003).

[42]  A. Odlyzko, *Internet pricing and the history of communications*, Computer Networks **36** (2001), 493–517.

[43]  A. Osterwalder, *The business model ontology: A proposition in a design science approach*, Ph.D. Thesis, University of Lausanne, HEC (2004).

[44]  R. Párhonyi, *Micro-payment gateways*, University of Twente, CTIT Ph.D. Thesis, Series No. 05-72, Enschede, The Netherlands (2005).

[45]  R. Párhonyi, L.J.M. Nieuwenhuis and A. Pras, *Second generation micro-payment systems: Lessons learned*, 5th IFIP conference on e-Commerce, e-Business, and e-Government (I3E 2005), Poznan, Poland, October 2005 (also published in the Payments and Settlements News No. 24 prepared by the ePSO team at the European Central Bank).

[46]  Paybox Solutions AG, *Mobilizing people – paybox*, http://www.paybox.de/ (2006).

[47]  M.E. Porter, *Competitive Strategy: Techniques for Analyzing Industries and Competitors*, 1st edn., Free Press, New York, USA (1998).

[48]  P. Reichl, P. Flury, J. Gerke and B. Stiller, *How to overcome the feasibility problem for tariffing Internet services: The cumulus pricing scheme*, IEEE International Conference on Communications (ICC 2001), Helsinki, Finland (2001), 2079–2083.

[49]  C. Rigney, *RADIUS accounting*, Internet Engineering Task Force (IETF) RFC 2866 (2000).

[50]  C. Rigney, S. Willens, A. Rubens and W. Simpson, *Remote authentication dial in user service (RADIUS)*, Internet Engineering Task Force (IETF) RFC 2865 (2000).

[51]  Service Level Agreement and SLA Guide, *The SLA Toolkit*, http://www.service-level-agreement.net/ (2006).

[52]  R. Smith and R. Davis, *The contract Net protocol: High level communication and control in a distributed problem solver*, IEEE Transactions on Computers **29** (1980), 1104–1113.

[53]  B. Stiller, *A survey of charging internet services*, Managing IP Networks, Challenges and Opportunities, S. Aidarous and T. Plevyak, eds, Wiley Interscience, Hoboken, NJ (2003).

[54]  B. Stiller, T. Braun, M. Günter and B. Plattner, *The CATI project: Charging and accounting technology for the Internet*, 5th European Conference on Multimedia Applications, Services, and Techniques (EC-

MAST'99), Madrid, Spain, May 26–28, Lecture Notes in Comput. Sci., Vol. 1629, Springer-Verlag, Heidelberg (1999), 281–296.

[55] B. Stiller, P. Reichl and S. Leinen, *Pricing and cost recovery for Internet services: Practical review, classification, and application of relevant models*, Netnomics – Economic Research and Electronic Networking **3** (2001), 149–171.

[56] D.C. Verma, *Policy-based Networking*, Technology Series, New Riders Publishing, Indianapolis, IN (2000).

[57] J. Walker and A. Kulkarni, eds, *Common Open Policy Service (COPS) Over Transport Layer Security (TLS)*, Internet Engineering Task Force (IETF) RFC 4261 (2005).

[58] E. Wustenhoff, *Service level agreement in the data center*, Sun, http://www.sun.com/blueprints/0402/sla.pdf (2002).

[59] Y. Yemini, *The OSI network management model*, IEEE Communications Magazine **31** (1993), 20–29.

# – 7.3 –

# Service Provisioning: Challenges, Process Alignment and Tool Support

Michael Brenner[1], Gabi Dreo Rodosek[2], Andreas Hanemann[3],
Heinz-Gerd Hegering[3], Ralf Koenig[1]

Munich Network Management Team

[1]*Ludwig-Maximilians-Universität München, Oettingenstr. 67, 80538 München, Germany*
*E-mails: {brenner, koenig}@mnm-team.org*
[2]*Universtät der Bundeswehr München, Werner-Heisenberg-Weg 39, 85577 Neubiberg bei München, Germany*
*E-mail: dreo@mnm-team.org*
[3]*Leibniz-Rechenzentrum der Bayerischen Akademie der Wissenschaften, Boltzmannstr. 1,*
*85748 Garching bei München, Germany*
*E-mails: {hanemann, hegering}@mnm-team.org*

## 1. Introduction

The provisioning of services in the field of information and communication technology (ICT) such as network and application services presents many challenges to service providers. Service providers must pay careful attention to operational efficiency and take new approaches to service provisioning in order to compete successfully. They must make better use of existing resource infrastructures, such as networks and servers, as well as reduce the number of operational staff required to deliver services to end users leveraging automation.

Regarding the level of service customization, services span a broad range: from widely deployed commodity services, such as broadband internet access or e-mail services, with millions of equal service instances, to individual services, such as an enterprise extranet or an in-house service like the service architecture of a World Wide Web (WWW) search engine with a single highly customized service instance.

HANDBOOK OF NETWORK AND SYSTEM ADMINISTRATION
Edited by Jan Bergstra and Mark Burgess

The vision of a service provider and also a competitive advantage is the achievement of 'flow-through provisioning', which means that a customer reviews a set of services offered to him by a provider, selects an appropriate service, chooses among the available service quality options, then places an accordingly formed service order, and waits for its almost instantaneous fulfillment. The service provider, who receives such an order, would have automated provisioning systems that activate, monitor and manage the ordered services with minimal additional manual input. So far, this has partly been accomplished for a few types of commodity services, but it remains a vision for most services that require a higher level of customization.

## 1.1. *Terms*

We want to motivate service provisioning and most of the terms used in this chapter with a short example of a commonly known commodity service subscribed by many customers.

EXAMPLE. A customer who wants to order broadband cable access to the Internet typically does so by submitting an order form. He fills in the blanks and chooses among the several available options associated with service parameter values, such as link bandwidth and monthly traffic volume. Finally, after selecting a payment option, he sends the form to the service provider. By executing these actions, a customer has ordered a service.

The form is processed by the service provider and data about the customer and the requested services is registered in some data store. As broadband cable is a shared medium, the internet service provider (ISP) evaluates whether this additional customer can be attached to the particular cable segment that links his home to the cable network, typically based on some (non-disclosed) over-subscription model. If there are too many customers subscribed to a particular segment to keep up with performance guarantees at peak times, the cable operator may have to split the segment. After updating systems for authentication, authorization and accounting, the corresponding network components are configured accordingly to allow traffic from the new customer through the service provider's network to the Internet and vice-versa.

Eventually, a technician is sent to the customer to install and test the interface at which the customer can get access to the ordered service. Depending on the contract, a pre-configured cable modem may be ordered by the service provider and shipped to the customer.

Speaking in more general terms, a *service* is a concept how a certain *functionality* is provided to a *customer* by a *provider*. It abstracts from the specific implementation while all customer-relevant service parameters (e.g., Quality of Service (QoS), cost parameters), such as agreed upon bandwidth or availability in network services, are specified in a *Service Level Agreement (SLA)*.

The service provider is usually free to change the implementation of the service as long as the SLA is not breached. With such an approach, the growing number of shared and dedicated *resources* needed to provide a service as well as the details about the processes needed for its implementation and operation are wrapped for the customer behind a more pleasant interface. This allows the service provider quite a degree of flexibility in the design and management of its resource infrastructure and enables it to leverage various economies of scale and scope, see [21] for details.

The separation of functionality (customer view) and implementation (provider view) is less applicable when customers want specific services being planned and built with regard to their individual requirements, such as existing legacy systems and services that the ordered service has to work together with, thereby limiting the range of implementations for the service provider.

In order to analyze service provisioning in more detail it is necessary to start with the *service life-cycle*. IT services can be associated with a service life-cycle that subsumes the steps from planning to termination of a particular service. A popular simple service life-cycle is called *Plan-Build-Run*, typically re-cycled after a *Change* or *Improvement* step. *Service provisioning* mainly deals with the plan, build, and change parts, providing the necessary input for run-time operations. It is therefore a major part of the service life-cycle. More precisely, service provisioning includes tasks such as planning new services, building the basic infrastructure, SLA negotiation and order processing, identifying adequate resources for service delivery or adapting existing services to specific customer's needs, specifying steps for service implementation and service operation, up to dynamic, near real-time service composition out of service modules based on customer requirements.

In order to offer services at competitive prices, a service provider must work *efficiently* by using as few resources as possible in terms of staff and machinery, while providing services *effectively* by meeting or exceeding the agreed SLA. Therefore, resources are typically shared between multiple customers as long as the SLA does not bind the service provider to use dedicated resources. With new virtualization technologies appearing on the market, the border between physically separated resources and logically shared resources becomes increasingly fuzzy.

EXAMPLE. Imagine the case that five customers need a dedicated server environment with privileged access but they do not need the full computing power of a physical server due to a particular workload profile. To fulfill these service orders, one possible implementation is to configure several virtual servers on one physical server, based on the assumption that the workload profiles overlap only in short periods of time. This way, the service provider can use virtual servers to provide abstract dedicated servers, saving four physical servers including expenses for energy, cooling and rack space. While the service provided to the customer is essentially the same, the implementation is different.

Examples for other shared resources are network links and components shared by multiple virtual private networks or virtual LANs, file servers shared by multiple users, or virtual hosts on a HTTP server. Thus, resource sharing is an optimization method of a provider with regard to resource utilization. Sophisticated mechanisms and tools are necessary to support such sharing to assure that guaranteed service levels are not violated.

When providing *best-effort services,* service providers do not make any service guarantees. The low complexity in service management saves efforts on the provider side and results in lower prices. Best-effort is the typical service quality found in the public internet.

Advanced *quality of service* mechanisms, e.g. reserving a certain portion of end-to-end network bandwidth on demand for video traffic using the Resource Reservation Protocol (RSVP), need to be implemented to provide guaranteed service quality. Many steps in the service provisioning process deal with the assurance of a specified service parameters.

## 1.2. *Differences in provisioning due to the varying level of customization*

Services can be classified regarding a number of properties. For service provisioning, relevant aspects are mainly the number of customers that will subscribe to the service, and the level of customization that these customers require. The number of service subscriptions directly translates to the number of times the order process and the following steps will be executed, while the level of customization translates to the complexity of these processes and therefore the applicability to automated execution.

Regarding the level of customization of service instances, IT services can be roughly grouped into the following three types of services: commodity services, customized services, and individual services (see Table 1 on p. 862). Please note, that the line between these service types is not always sharp and some services may show properties from multiple categories. In the next sections, we will first describe the extremes of this spectrum, namely commodity and individual services, followed by a description of customized services filling the wide range between the former two.

**1.2.1.** *Provisioning of commodity services    Commodity services* are affordable, widely available, well-specified standard services with known functionality provided by many competing service providers. Examples include voice services, broadband internet access, or web hosting.

See Figure 1 for a schematic overview on the provisioning of commodity services. The introduction of new commodity services is typically initiated by service providers sensing market demand and not by specific customer requests.

Commodity services require a basic infrastructure to run services upon, which at first needs to be planned, built, rolled out, and set up by the service provider. Resources are added later on based on market demand analysis and predictive planning.

Resources in this infrastructure and its management are planned and laid out in such a way that service orders from customers can be processed and provisioned in a highly automated fashion, given the fact that the ordering and fulfillment process is run possibly a few thousand or million times. Operations support systems (OSS) and management systems control as many steps as possible in the provisioning workflow, but they need orchestration of devices as well as up-to-date information about the deployed infrastructure.

Commodity services are intended to be subscribed by a large number of customers with a low level of customization, so services are typically comparable even across service providers, mainly for interoperability reasons.

Prices and service parameters are determined by the service provider for pre-packaged service bundles with some service options and are hardly subject to negotiations. Therefore, customers usually have to accept the conditions stated by the service provider and

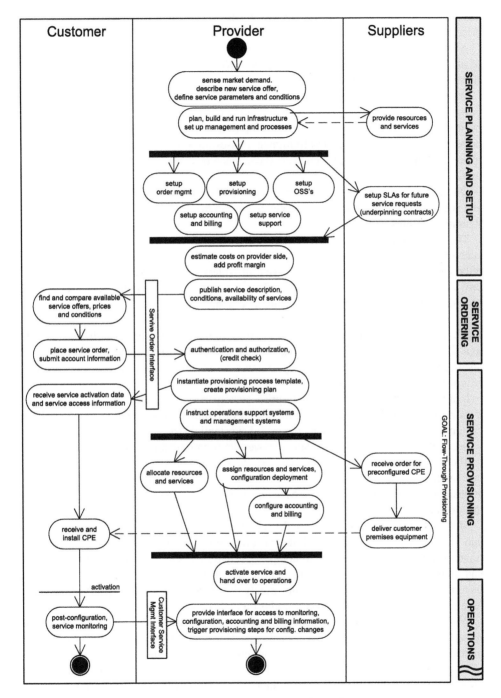

Fig. 1. Provisioning steps for commodity services.

can only search for a service offer that best fits their needs in terms of service features and price.

A large number of essentially equal service orders allows and even requires a high level of automation in the ordering and fulfillment process from service providers in order to successfully compete in the market.

With commodity services, the implementation of a service is typically hidden from the customer, with the service provider being free to change service implementation without prior notice. While service planning and setup of a service infrastructure can take months or years, the service ordering and fulfillment phases for a single service order can be completed within a time range of seconds to days, made possible by the high level of automation.

Due to the high volume of resources in such infrastructures, suppliers offer hardware and software tailored to the requirements of commodity service providers with many configuration options.

The basic infrastructure for many commodity *network* services has been established by state-owned companies in many countries of the world. Market liberation and regulation efforts require established network service providers to open their infrastructures or to provide transit services at certain rates. This led to the establishment of virtual service providers and resellers that use the existing operating infrastructure.

**1.2.2.** *Provisioning of individual services*   Since *individual services* or *solutions* are sold to a single customer, individual negotiations of service parameters and values for service level objectives are an appropriate way to find a compromise between costs and customer requirements.

Individual services for major enterprise customers are typically organized as long-term projects. This is caused by the high level of required customization, resulting from customer-specific requirements documents of substantial size.

See Figure 2 for a simplified, schematic overview on provisioning of individual services.

The provisioning of individual services starts with the customer compiling a service request with detailed specifications of the customer requirements, usually expressed in terms at business process or service level. Many customers use consulting services in compiling this document.

Multiple service providers analyze the service request and refine customer requirements and SLA individually together with the customer. Each provider identifies a set of appropriate resources and subservices to compose or build the solution. After estimating costs and adding a profit margin, bids are sent to the customer. At this point, in most cases there is still a considerable level of uncertainty due to changing or non-specified customer requirements, estimates, and safety margins on both sides.

When the customer decides to accept an offer and places a service order, both sides agree on service functionality, service levels and costs. However, customers may demand specific interfaces, extensions or capabilities not commonly offered by suppliers. In addition, often existing resources or services of the customer may have to be integrated with the new ordered service. Automation in provisioning of such requests is therefore hard to achieve.

Service provisioning for individual services is more project-driven, which involves many people communicating and negotiating with each other. Tools that support the collaboration

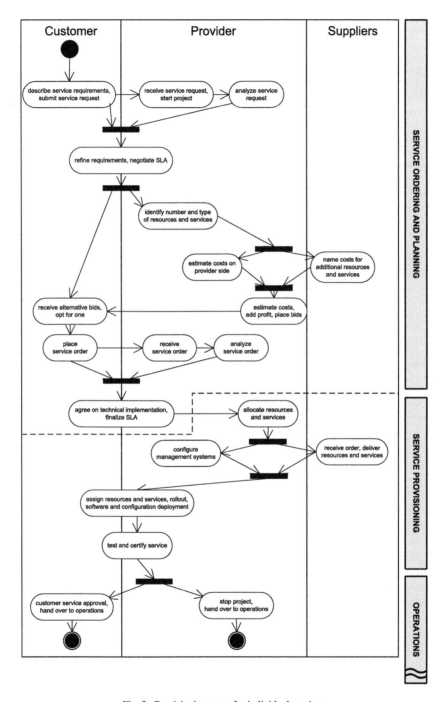

Fig. 2. Provisioning steps for individual services.

Table 1
Types of IT services with respect to the level of customization

|                          | Commodity services | Customized services | Individual services |
|--------------------------|--------------------|---------------------|---------------------|
| Requirements on the customer | customer has to meet certain preconditions | depends on service offer, requirements may be up to negotiations | none, customer requirements are evaluated and implemented |
| Service parameters | are widely standardized across service providers, customer has few service configuration options | customer has multiple options to choose from a pre-defined set of service parameters | the set of service parameters is up to negotiation and can change at run-time |
| Service orders | sold/subscribed at high volumes of equal orders, low price per subscription | low number of equal orders, but large number of orders | one order with substantial budget |
| Pricing | high cost pressure for service providers as services are easy to compare for customers | higher prices may be justified by the level of customization which provides substantial value to the customer | typically expensive to plan and build, individual pricing, customization can pay off in service usage |
| Implementation | hidden from customer, service provider is free to change implementation without prior notice to the customer | typically made by combining and integrating service modules with customer-specific glue code, customer can influence implementation within certain bounds | customer requirements can have substantial influence on implementation, reusability of the result is not a key issue |
| Provisioning tool support | good tool support for provisioning, a high degree of automation | applicability of tools depends on a number of issues (see text) | few tools available, most work done by experienced solutions designers |
| Support and maintenance | costly on-demand support, when not using the least cost intensive way of communication, support is regarded as a cost driver | depends on service offer, service provider may serve as mediator between resource suppliers and customer | typically sold with support and maintenance contracts, support is regarded as a source of income for the provider |

of people such as project management tools, a knowledge base for past solutions and group communication tools provide valuable help and guidance.

Provisioning time – this includes service planning, ordering and fulfillment – is in the range of months to years. Key issues are the fast and thorough analysis of customer requirements and mapping them to needed resources, as well as good communication skills. Experience in building individual services from flexible standard modules by service composition can substantially reduce new developments and therefore provisioning time.

**1.2.3.** *Provisioning of customized services    Customized services* fill the wide range between the former two types of services.

In many cases, customized services are developed by adding more configuration and service options to existing commodity services to better meet the anticipated requirements of a potential set of customers. Other providers develop customized services from individual services. By refactoring the composition of the service into service modules and separating configuration options from service implementation, the former individual service becomes more flexible so that the requirements of more than one customer can be met and development costs per customer decrease.

Customized services typically have a substantial number of orders, but due to the large number of service options and choices for the customer, hardly any two service orders are exactly the same. Depending on the specific service, customized services show features of both commodity and individual services. Customized services are typically built from standard service modules, which allows semi-automatic provisioning. A certain degree of flexibility for the customer is made possible by having the choice between different types of modules and a custom arrangement of the modules to a customized service. This results in a substantial growth of feature combinations with numerous dependencies. Automation would require to model all this complexity, which in turn requires models for all resources and subservices, and thus limits flexibility. The most appropriate level of automation is therefore determined on a case-by-case basis based on many criteria such as cost of human labor, cost and flexibility of automation tools, uniformity of service orders, existing knowledge, supplier relationships and negotiation skills to name just a few.

For comparison: In manufacturing, products built in a similar fashion are called to be 'built to order'. To compete in the market, the sum of costs for development, customization, resource consumption at runtime, and management of customized services must be optimized, which requires advanced skills in supply chain management of a number of specialized suppliers.

## 1.3. *General phases of service provisioning*

As already indicated in the previous discussion, service provisioning consists of the phases (i) service planning, (ii) service ordering, and (iii) service fulfillment (deployment and configuration). Once these steps are completed, the service is activated and handed over to service operations.

The phases were visualized in Figures 1 and 2 for commodity services and individual services. In contrast to the simplified schematics in the two figures, provisioning of all three types of services can be a joint effort of many actors with different roles. Individual services can be planned and built by solutions providers, while service operations is handed to the customer or a third-party service provider. In customized services, service integrators may take over specific sub-tasks to integrate a certain set of resources. Resource suppliers may also offer services to ship resources pre-configured to service provider requirements.

**1.3.1.** *Service planning*    In this phase a service provider plans the provisioning of a new service. Planning includes several steps ranging from a market analysis to determine the business model, the appropriate costs and charges to the verification of technical issues such as the availability of resources. In case of commodity services, that are provided to a large number of customers, a detailed market analysis is a prerequisite for successful service introduction. The trigger to set up and provide such services results mainly from decisions on the provider side. In contrast, if a new service is requested by a customer, which is the typical case with individual services, the planning process is limited to the verification of the available resources and other provider-specific issues, such as policies.

A more detailed observation identifies the following steps:

- Definition of the service and its functionality, service parameters and possible service levels that can be offered to customers.
- Definition of a cost model for internal and external costs.
- Sizing: quantitative estimation of market demand for the service, quantitative mapping of market demand to service demand, quantitative mapping of service demand to resource demand.
- Verification of the available infrastructure (resources) as well as capacity planning with respect to the number of customers and provided service levels.
- Planning processes for resource allocation, resource assignment, service deployment (rollout), and initial configuration.
- Specification of a service ordering process, definition of allowed service changes.
- Planning service operations including service management tools.

**1.3.2.** *Service ordering* During the service ordering phase, a customer selects a service or a service package from the available list of services that a provider offers to customers. By selecting a service, a customer identifies not only the functionality of the service itself but also the requested service quantity and quality, the service cost, the time line when he wants to use the service. The specification of these parameters is done with respect to the agreed values and value ranges as well as thresholds in the respective SLAs. Values, value ranges as well as other issues can be negotiated between a customer and a provider.

**1.3.3.** *Service fulfillment (deployment and configuration)* The service fulfillment phase starts in fact after the submission of the service order. The service provider needs first to check the order with respect to contractual and technical issues and either rejects or grants the service order.

If the service order cannot be granted, it is also possible that the customer and the service provider negotiate about the requested service. If a provider grants the service order, he starts the service fulfillment process, namely resource deployment and configuration as well as the initial configuration of management systems.

Finally, the service provider informs the customer about the successful realization of the service order, and the customer may start to use the service. On the provider side, the service is handed over to the operations department.

**1.3.4.** *Service operations* After service activation, service operations deals with the day-to-day operations to provide services with constant quality meeting or exceeding the SLA.

As a matter of fact, hardware and software resources fail from time to time for a number of reasons and human operators make mistakes. To assure to be notified in case of service failures and degradations of service quality, monitoring the infrastructure is considered necessary by service managers.

Since services are provided upon resources, resource and service management tools must be configured appropriately to gain a service-oriented view that provides the 'right' amount of service-level information to managers from existing resource-level monitoring data. In turn, service management operations must be mapped to operations at resource level.

In addition to providing information to service managers, as well as system and network administrators on the provider side, operations should provide customer-relevant management information via the Customer Service Management (CSM) interface.

## 2. Service provider goals and challenges

Awareness of the main goals of service provisioning is a prerequisite for identifying fundamental challenges of service provisioning that a service provider is confronted with.

While the ultimate goal for a service provider is economic prosperity and providing agreed service levels to his customers, in service provisioning this translates to tuning many parameters to find the right balance between multiple sometimes contradictory goals. Each goal is associated with challenges that have to be faced in terms of meeting the goal. The following discussion will outline the relationship between goals and challenges. It should be noted that the list is not exhaustive although the aim was to provide a broad view; it should give just an idea about the problem area we are talking about.

GOAL 1. *Achievement of business objectives and high service levels.* In today's business environment, services are rolled out in almost real-time. The challenge is of balancing cost-effectiveness with resource utilization needed to meet the dual requirements of ensuring high availability and maximizing SLA performance. Today, 'available' means the right response time, to the right users, right now. So, slow performance, delayed reaction to customer service requests etc. is no longer an option. In fact, the essential goal of a service provider is to move from bits to business values which is also reflected in the goals that a provider has to achieve. However, it differs when talking about individual or commodity services. For commodity services a provider starts a large market analysis to see what the market requests whereas the provision of an individual service is triggered by a specific customer. Business value is in the latter calculated customer-specific.

CHALLENGE 1.1. *How to formalize a SLA?* So far, no generally agreed upon formalized SLAs exist yet. Rather, the inclusion of elements into a SLA (e.g. specification of a service, the service access point (SAP), QoS parameters, cost parameters, accounting models, security levels) depends on a case-by-base basis.

Agreeing on SLAs without a deep knowledge on the area can lower business objectives in a great extent. For example, if specifying only the QoS parameter availability without specifying also the maximal time of an outage, this can lead to a situation where the availability is met, however, the customer is not very satisfied. Beside the challenge to identify all relevant elements of a SLA from the customer and provider side, the resulting automatic service and resource configuration based on SLA parameters is another challenge.

CHALLENGE 1.2. *What service levels can be offered, resp. are required by customers?* This challenge depends on the type of service. In case of commodity services a detailed forecasting and what-if analysis are a precondition to determine adequate service levels that can be offered by a provider. Service levels for individual services are determined on a case-by-case basis. The risk to determine service levels that do not meet business objectives is minimized.

CHALLENGE 1.3. *How can the negotiation process between a customer and a provider be supported?* This challenge refers to the support of the negotiation process between a provider and a customer of the possible service levels. It includes also the contractual and technical verification of customer service requests. Afterwards, the service configuration is started. But service configuration is becoming more difficult all the time, thanks to challenges like personalized service, hybrid networks and end-to-end service guarantees, sending fulfillment costs spiraling.

A negotiation about the service levels between a customer and a provider as well as suppliers can include several iterations until an acceptable solution with respect to cost, quality and other aspects is found.

Before negotiation is started, the verification of contractual and technical issues is performed. The verification of contractual aspects is more or less straightforward. However, the verification of technical issues refers mainly to the resource allocation problem, including the identification of the adequate resources, verification of their availability etc. It is necessary to verify whether other services, respectively customers, may be affected in the provided quality of service if the new service request is granted.

GOAL 2. *Involve customer in service ordering.* As more and more steps in service provisioning are automated with the help of IT systems, all input data must be converted to an electronic form-based format, interpretable by machines.

Traditional points of contact between customers and providers have been branch offices and service hotlines. There, customers expressed service orders or desired changes to existing services, which were then translated by support staff to entries into an internal ordering system. In the same way, output data like service usage statistics and bills, while being generated electronically, was printed and mailed to the customer when internet access had not been widely available to customers.

Disadvantages of this practice on the provider side include expenses for branch offices and staff, printing and mailing. On the customer side, customers could hardly see the full list of available service options and consequences. In addition, call times and opening hours of branch offices are typically limited to about 8 hours per day.

CHALLENGE 2.1. *How to involve the customer in the service provisioning process?* To involve the customer in the service ordering and service change process has two benefits. First, a customer is more satisfied if he is informed about the status of the service provisioning process and he can dynamically change his service orders within the limits of the SLA. Second, by shifting the service ordering and service change process to the customer, a provider can concentrate on service provisioning and can save resources and staff, that would otherwise be needed for receiving service orders and change requests.

A precondition to achieve this goal, however, is to provide a suitable Customer Service Management (CSM) interface where service parameters can be changed or new services can be ordered in a convenient way by the customer in a self-service manner.

CHALLENGE 2.2. *How to design an appropriate interface to the customer that supports all interactions between a service provider and a customer (e.g. service ordering, reporting of trouble reports, accessing performance statistics about the provided service quality etc.)?* A customer may order services or even service packages. In case of the negotiation process it is necessary to think of various types of services as identified already in the introduction. In case of commodity services such a negotiation is almost impossible, in case of individual services it is almost a certainty.

Online customer service management portals, which are typically web-based, more and more replace traditional customer service management interfaces. Via the CSM portal, the service provider receives customer service requests in a known and machine-readable format and provides machine-readable output data like bills and usage statistics, which on the customer side can serve as input for IT systems for electronic payment or accounting.

GOAL 3. *Automation of service fulfillment.* The service provisioning process is quite challenging since the provider has to deal with distribution of services and the inherent heterogeneity of the infrastructure. The heterogeneity arises from the application of many resources from different vendors (technical heterogeneity), the involvement of a variety of different parties within and outside the service provider side (intra-organizational and inter-organizational heterogeneity). Beside of this static complexity, the type and number of changes lead to a dynamic complexity of the service provisioning process.

Depending on conditions like the level of customization of the service (commodity service, customized service, individual service, see Section 1.2 for details) and the organizational structure of the service provider, standard provisioning workflows should be defined for each offered service. Each workflow specifies the collaboration among all involved people and systems. While this often implies features of bureaucracy which appears to limit flexibility, it reduces coordination efforts and works towards a uniform, measurable performance of service provisioning. To achieve higher levels of automation, the provisioning workflow, including steps, roles, and policies must be expressed in machine-interpretable terms.

CHALLENGE 3.1. *What are the necessary service fulfillment and service operational steps? How can these steps be acquired, maintained and improved? How can the acquisition and the execution of these steps be supported by tools or even automated?* In order to realize 'flow-through provisioning' it is necessary to automate service provisioning as far as possible. This means to formalize the necessary provisioning steps and refine them into commands executable by machines, with parameters so that the workflow can be instantiated for similar deployment tasks. These workflow descriptions are collected in a repository of *service provisioning templates* internally within the provider's environment or even externally so that several providers can access it.

Another problem to address is the alignment of service fulfillment steps with the current infrastructure, which refers to the ability to change the steps with respect to changes in the environment either on the service or resource layer. Such an alignment should be done 'almost automatically' by management tools with minimal necessary assistance from operators. In case of an individual service such steps are probably service-dependent and need to be done manually. However, in case of a commodity service such service provisioning steps need to be predefined and executed in an automated way.

The problems associated with the specification of deployment and configuration steps are:

- the granularity of the steps,
- the workflow aspect and
- the alignment of service provisioning to the changing set of resources.

Granularity addresses the refinement of steps to a level of directly executable actions, which ultimately can be performed automatically by existing management tools. An abstract action such as *install_database* needs to be mapped to a more concrete step like *install(oracle9, server1)*, which in turn must be refined to include more details about installation parameters. Given the number of hardware and software resources, resource and service dependencies, distribution of knowledge and the rate of change in all mentioned entities, this presents quite a challenge for complex service provisioning scenarios.

There are two approaches to define provisioning and operational steps:

- to specify all steps with the requested granularity in the initialization phase or
- to specify only a limited set of steps in the initialization phase and improve them afterwards through experience gained during the application of these steps.

The first approach is applicable for environment with simple services and small infrastructure. In case of complex services that depend on other services and large heterogeneous infrastructure the second approach is probably more applicable.

CHALLENGE 3.2. *How fast can a new service be deployed?* Customers may request the provision of new services or changes of the already provisioned ones. In both cases it is necessary to identify the (i) involved resources in terms of resource allocation, (ii) the management tools that are involved in service operation and (iii) the processes that need to be eventually adapted or defined in addition. A precondition for a fast provisioning is an extensive and integrated usage of management tools in order to have the necessary management information about the infrastructure available. Supporting demanding service provisioning means always also to be able to forecast the service usage with respect to e.g. subscribed SLAs, provider policies, measurements and historical data. Furthermore, sophisticated resource planning methods are needed to accomplish this goal.

GOAL 4. *Optimal resource selection and utilization.* While flow-through service provisioning primarily refers to automation of service provisioning processes, optimal resource utilization deals with the question of how the existing infrastructure (i.e. machines and staff) can be used in an optimal way. Talking about the 'optimal' solution raises immediately the question about the objective function of the optimization e.g. whether it is necessary to optimize with respect to service quality, availability, security aspects, costs, specific customer demands or other provider-specific policies. Besides the infrastructure aspect, optimization refers also to the service provisioning process itself. Efficient process definition is quantified with so-called key performance indicators (KPIs).

CHALLENGE 4.1. *How to describe service requests and resource capabilities in an adequate way in order to optimize resource allocation with respect to various provider objec-*

*tives?* This challenge refers to mapping *service requests* to *resource capabilities.* In case of a formalized description of service requests and resource capabilities the mapping is more or less straightforward. However, in addition several provider policies, high availability requirements from customers can influence service provisioning as well, and make the resource allocation problem a hard problem.

CHALLENGE 4.2. *Since several solutions of the service-resource allocation problem are possible, how to support a decision process to select the 'right' one? How can such a decision be supported by management tools?* If several solutions for a resource allocation are possible, it is necessary somehow to determine the 'right' one for the special case. Thus, it is necessary to introduce metrics to evaluate what solution is the most appropriate. In this context, what-if analyses and simulations are very appropriate approaches since a provider can change the values of input parameters and evaluate the influence on the result.

Resource selection becomes even more challenging when considering that resources and subservices can also be bought from suppliers. As provisioning deals with resources and services, how to decide whether a certain service or subservice should be hosted by the service provider himself or ordered from a specialized supplier? From which supplier to buy a particular resource resp. subservice?

CHALLENGE 4.3. *How to assure that a new resource to service assignment will not affect the provided quality of service of other customers or violate provider policies? How to support such what-if analyses?* What-if analyses support the decision process whether to deploy new services upon the existing infrastructure or to extend existing ones or even change the infrastructure of policies. Managers can easily see where demand exceeds supply, where skills need to be enhanced or developed, or if new staff members need to be brought in.

GOAL 5. *Preparation for automated service operations.* A precondition for service operations is to obtain a service-oriented view of the infrastructure which refers to the issue how to configure device-oriented management tools appropriately to gain a service view. This means also to keep track of the relationships and dependencies between services in terms of inter-service dependencies as well as the dependencies between services and resources. Key issues here refer to the (automatic) acquisition of dependencies and the offering of this information to processes, tools and staff that needs this information for various purposes (sales, operations and monitoring, accounting). Furthermore, tracking the distributed provisioned services upon resources is another important aspect as well.

CHALLENGE 5.1. *How to specify the service monitoring process?* This challenge refers to the mapping of service levels to the design of the service monitoring process. If it is for example agreed on resolve times below a few hours, this is necessary to be supported by several processes, for example incident and problem management process. The efficiency of the processes is measured with key performance indicators as described latter on.

CHALLENGE 5.2. *How to configure the device-oriented management tools to monitor the behavior of resources involved in the service provisioning?* Beside the negotiation process itself, operational requirements for various types of services differ as well. Individual services can be handled manually whereas for services offered to hundreds of customers only an automatic approach is a reasonable solution. Furthermore, services can be provided by several providers. In such a case it is necessary to address issues such as what happens if a sub-provider violates the SLA with the provider, and what are than the consequences for the SLAs between the customer and provider.

This challenge deals with the mapping of QoS parameters as specified in a SLA (e.g. availability) to management variables such as node_up or node_down.

GOAL 6. *Adaptivity to changes caused by technology, customers or market demand.* An essential competitive factor among service providers is the ability to realize customer service requests in a fast and convenient way to keep existing customers resp. gain new ones.

CHALLENGE 6.1. *How fast can changes be realized?* Changes can result due to customer requests, as a result of a planning solution or due to changes of provider's policies. Since changes in the infrastructure result in most cases also into changes in the service fulfillment, it is necessary to adapt to these changes in a fast and convenient way. This goal addresses the provider's view of the dynamics of provisioning. It is an important goal as well as a big challenge for a service provider to cope with the dynamics of changes, either of the provided services, the used infrastructure or provider policies. Furthermore, the dynamic nature of SLAs, service and resource configurations, and resource usage needs to be addressed as well.

The ability to adapt to changes means to store service and resources models and keep an up-to-date inventory of all provisioning-related information and policies, and their relationships. Certainly, a periodical verification and update of models as well as inventories, including changes in the infrastructure, is essential.

CHALLENGE 6.2. *How to perform provisioning in a scalable, reliable way, regarding to service/subscription types and order volume?* To ensure reliable provisioning with few errors, especially in the implementation of automated procedures, means to refer to the point of high availability and reliability of the provisioning system, 'carrier-class' as well as to reduce risk in provisioning errors.

Due to the distributed nature of service provisioning, in most cases several persons are involved in the service provisioning process. Their actions need to be specified and coordinated in an appropriate way which brings us to e.g. ITIL, eTOM, as described in the next section.

## 3. Process frameworks and concepts

This section will outline two current concepts of increasing relevance for IT service provisioning: organizational IT Service Management, represented by the two IT business process frameworks ITIL (IT Infrastructure Library) and eTOM (enhanced Telecom Operations Map), and SOA (Service Oriented Architectures), a new paradigm for building service infrastructures. ITIL and eTOM, discussed in the following two sections, present relatively mature frameworks – TOM, the predecessor of eTOM, was first published in 1998, the earliest ITIL publications date back to the late 1980s. SOA by contrast is a currently still evolving idea, that has yet to result in significant standardizations, and will therefore be discussed only briefly in the last section.

### 3.1. *Enhanced Telecom Operations Map (eTOM)*

The eTOM [52] is a process framework that started as an effort for telecommunications provider process standardization, but – influenced by increasing convergence of telecommunications and classic IT services – has evolved into a comprehensive IT service management standard.

**3.1.1.** *Introduction*    The 'e-commerce' or *enhanced Telecom Operations Map* (eTOM) is a business process framework for *Internet and Communications Service Providers* (ICSP). It is owned, published and maintained by the TeleManagement Forum (TM-Forum or TMF, formerly known as the NMF or Network Management Forum). Its predecessor, the Telecom Operations Map (TOM) was first published by the TMF in 1998 and was superseded by the eTOM in 2001. The goal of TOM was the creation of an industry-owned framework of business processes, including the definition of a common enterprise-independent terminology for service management. It was also supposed to serve as a basis for discussing the scope of management information necessary for the execution of the processes. The results of the latter effort have meanwhile spawned their own TMF document family, the *Shared Information and Data Model* (SID) [51]. Together with the guidelines on *Technology Neutral Architecture* (TNA) [50], a life-cycle model [49] and the development of conformity tests [48], eTOM and SID form the pillars of the ambitious *New Generation Operation System and Software* (NGOSS) project [47]. NGOSS aims to define a vendor-independent architecture for management systems, which will permit the construction of complete management solutions from modules with standardized interfaces. Since 2004, eTOM is also an ITU-T Recommendation (M.3050).

**3.1.2.** *Structure and content*    After a brief introduction to the organization of the eTOM document family and the basic structuring principles underlying the herein documented framework, the general content and structure of the *Operations* process area, which is the most relevant to service provisioning, is presented.

*Documents*    Even though the eTOM makes up only a part of the NGOSS project, its description consists of several documents. Various important additions to the main framework document [52] are made in the addenda and application notes. The application note C [54] is a rough draft of a "Public B2B Business Operations Map", addendum D [56] contains "Process Decompositions and Descriptions", addendum F [55] "Process Flow Examples", and finally the "eTOM ITIL Application Note" [53] is a brief outline of the relationship between ITIL and eTOM processes. While the application notes are useful, they are not as essential as the addenda, which in parts contain information critical for the use of eTOM.

*Framework structure*    The eTOM features different views on its processes, defining five *view levels*, from level 0 to level 4. Numerically higher levels offer an increasingly finer view of the service provider processes. In level 0, only three fundamental processes (in level 0 also called 'process areas') are defined, while on level 2, the only level for which process definitions for all level 0 process areas exist, there are already 72 processes. A further refinement into level 3 processes exists currently (eTOM Version 4.0) only for the *Operations* process area, and no level 4 processes have been defined yet.

A level 0 view of eTOM (see Figure 3) shows three fundamental process areas (or processes) of ICS Providers: *Strategy, Infrastructure and Product* (SIP), *Operations* and *Enterprise Management*. Four *functional areas* overlay the SIP and *Operations* process areas. Even if they do not adhere to a clearly defined layering principle, they represent the basic functions involved in service provisioning: Services are build from a combination of enterprise-own resources and sub-services sourced from suppliers and partners, and are then marketed as products to customers.

Also there are five external entities defined, with which a service provider organization needs to interact: *Customers, Suppliers/Partners, Shareholders, Employees* and *Other Stakeholders*. In the context of process management, these entities can be seen as external roles. The definition of roles is not extended in the higher level views of eTOM though, i.e. no enterprise-internal roles, e.g. process owners, are defined. Also – unlike ITIL – it makes no clear distinction between customers and users of a service. Even though it is mentioned in eTOM that the *Customer* entity can be refined to either *Subscriber* or *End User*, this distinction has no apparent impact on any part of the process model. eTOM's role model can be thus described as only rudimentarily developed. Compared to ITIL, the focus of eTOM is therefore more on interaction between operations support systems rather than between human actors.

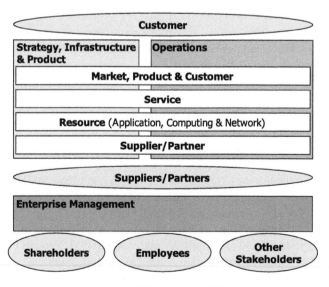

Fig. 3.  eTOM level 0 view [52].

*Operations processes*    Of the process areas defined in level 0 the most relevant to service provisioning is *Operations*, which in many regards represents the core of eTOM. It covers roughly the scope of the original TOM, and contains the most refined processes – level 3 process decompositions, as well as examples of end-to-end process flows are documented only for *Operations* processes.

Figure 4 shows the level 2 view of the *Operations* processes. Each is part of one of the four level 1 processes of *Operations* (aligned to the level 0 functional areas): *Customer Relationship and Management* (CRM), *Service Management and Operations* (SM&O), *Resource Management and Operations* (SM&O) and *Supplier / Partner Relationship Management*. Perpendicular to these are four vertical *end-to-end process groupings*, which yields a process matrix – something that has become a kind of eTOM trademark.

The four end-to-end processes in *Operations* are *Operations Support and Readiness, Fulfillment, Assurance* and *Billing*. The latter three have already been present in TOM and are also often referred to as the 'FAB processes' (Fulfillment, Assurance, Billing). While the functional division of *Operations* into its (horizontal) level 1 processes resembles the organization of an enterprise into departments, the division into vertical end-to-end processes corresponds more closely to the idea of business goal oriented processes. This often causes confusion. Unfortunately there is no precise, universally accepted definition of the term 'business process', but in classic business process management or process reengineering terms, eTOM's end-to-end process groupings would be considered the real

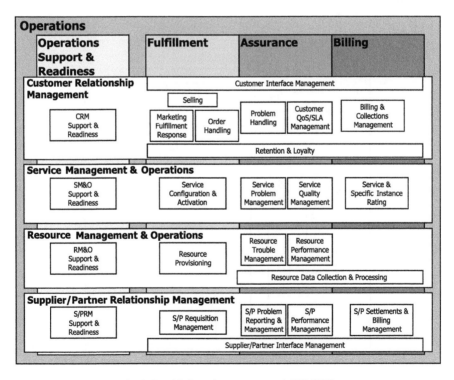

Fig. 4.  Level 2 Operations processes of eTOM [52].

business processes, while most of what eTOM defines as processes would by many people be seen as business activities or business functions.

**3.1.3.** *Service Provisioning in eTOM*   While not being explicitly called 'Service Provisioning', it is the end-to-end process *Fulfillment* that is responsible for providing customers with a requested service in a timely and correct manner. Figure 5 shows an exemplary service provisioning workflow between level 2 process elements:

1. *Customer Interface Management* routes an order for a new service instance to *Selling*.
2. *Selling* requests and receives a priority for the order from *Retention & Loyalty* based on customer data.
3. Prioritized order is handed over to *Order Handling*.
4. *Order Handling* passes a service activation request to *Service Configuration & Activation*.
5. *Service Configuration & Activation* decides which resources are to be supplied internally and issues resource reservation requests, work orders and resource activation requests to *Resource Provisioning* – simultaneously procurement requests are issued to *S/P Requisition Management* for externally supplied resources.
6. *S/P Requisition Management* chooses external sub-providers and issues orders through *S/P Interface Management* and waits for confirmation of external resource activation.

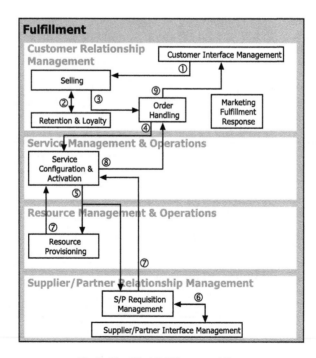

Fig. 5.  Simplified Fulfillment workflow.

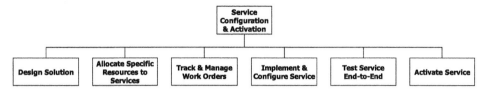

Fig. 6. Level 3 process decomposition of Service Configuration & Activation.

7. *Resource Provisioning* and *S/P Requisition Management* inform *Service Configuration & Activation* when all requested resources have been activated.
8. After final tests, *Service Configuration & Activation* informs *Order Handling* about successful activation of the service.
9. *Order Handling* instructs *Customer Interface Management* to inform the customer about completion of the order.

Of course real service provisioning workflows within eTOM can be much more complex. The above example involves no presales negotiations, no service design activities, nor does it show the necessary transfer of information to initiate *Assurance* and *Billing* of the new process instance. Also, like the workflow examples provided by the TMF in the eTOM addendum F (GB921F), the above description is limited to level 2 processes.

Figure 6 shows the decomposition of *Service Configuration & Activation* into level 3 process elements. Except *Design Solution*, all these sub-processes would take part in the execution of *Service Configuration & Activation* in the above example. A level 3 workflow description of the above service provisioning example would therefore involve well above 20 process elements. Considering the planned decomposition into level 4 elements would multiply this number again.

**3.1.4.** *Summary*   The eTOM defines all generic process elements necessary for service provisioning, for each of which however it describes only the 'what', not the 'how'. Thus, it does not provide ready-to-use internal or cross organizational workflows or interfaces. But the eTOM can lend a common structure to organization-specific workflows, significantly facilitate communications with customers and suppliers, and provide a basis for process automation efforts (see discussion on NGOSS and OSSJ in Section 4.5).

**3.2.** *IT Infrastructure Library*

Despite the similarity of their stated goals, the *IT Infrastructure Library* (ITIL), stemming from a data center and mainframe computer background, and the eTOM, having its roots in a telecommunications services environment, follow quite different approaches.

**3.2.1.** *Introduction*   The origins of ITIL date back to the late 1980s, when the *Central Computer and Telecommunications Agency* (CCTA) started publishing guidance for various aspects of IT operations under the label *IT Infrastructure Library*. Along with other agencies, the CCTA became part of the newly set-up *Office of Government Commerce* (OGC) in 2000, which took over ownership of ITIL. For further ITIL developments, the

OGC is cooperating with the *British Standards Institution* (BSI), the *IT Service Management Forum* (itSMF), an increasingly influential association of ITIL users, and also with the two IT Service Management examination institutes, the Dutch *Exameninstitut voor Informatica* (EXIN) and the British *Information Systems Examination Board* (ISEB). The OGC is still coordinating all official efforts though, and retains the ownership of ITIL.

While other ITIL books cover many other IT-related topics [35], the 'core' of ITIL, its guidance on IT Service Management, is combined in the *Service Support* [34] and *Service Delivery* [36] volumes. Their content is also the scope for tests administered by EXIN and ISEB, to achieve IT Service Management certifications for IT professionals. Since December 2005 there is also an international standard for IT Service Management certifications of IT organizations: ISO/IEC 20000 [27,28], an adaptation of the ITIL-aligned BS 15000 by the BSI.

**3.2.2.** *Structure and content*   ITIL's guidance on IT service management is documented in the books Service Support and Service Delivery, each containing the description of five processes[1] – Service Support additionally covers the business function *Service Desk*, which functions as a single point of contact for the *user*. Unlike eTOM, ITIL distinguishes the roles *user*, i.e. somebody using a service provided by the IT organization, and *customer*, i.e. the contractual partner of the IT organization regarding the conclusion of a *Service Level Agreement* (SLA). For the customer, the Service Level Manager serves as the contact person.

**3.2.3.** *Service Provisioning in ITIL*   There is no explicit 'Service Provisioning' process or another process correspondent to the eTOM *Fulfillment* process grouping defined in ITIL. As a matter of fact, of all the ITIL Service Support and Service Delivery processes, only Incident Management can be more or less clearly mapped to one of eTOM *Operations'* FAB processes, namely to *Assurance*. One of the eTOM application notes, GB921L [53] (see Section 3.1.2), gives some information on ITIL-eTOM mapping, but generally there is no exact one-to-one matching of ITIL service management processes to eTOM processes.

This is due to the fact, that ITIL and eTOM each follow implicit, yet distinct assumptions and philosophies stemming from their origins. Where eTOM, coming from a telecommunications services background, focuses on commodity or customized services offered to a multitude of customers, the ITIL, coming from a data center and mainframe background, concentrates on complex individual services offered to only a few customers (but possibly many users). Consequently, eTOM differentiates clearly between the management of services (as in service-class or product), addressed in the SIP process area and the management of service instances, addressed in the *Operations* process area – a distinction that is not made with this precision in ITIL.

As a result, ITIL processes are not concerned with the efficient, automated provisioning of multiple service instances. Where eTOM's focus is on structuring and automating management activities for more or less standardized services, ITIL's focus is on coordinating the collaboration of IT personnel on not completely pre-determinable tasks. ITIL processes therefore usually cannot be mapped one-to-one on eTOM processes, but have correlations

---

[1] Sometimes *Security Management* is counted as a sixth process of Service Delivery.

to many eTOM processes, sometimes even processes from different eTOM level 0 process areas [53]. ITIL puts most weight on support and coordination of operational activities (and not the activities themselves), an area addressed in eTOM in the processes that form *Operations Support and Readiness* – which overall, even if not acknowledged by the TMFs 'eTOM ITIL application note' [53], have probably the highest correlation to ITILs service management concepts.

In summary, ITIL gives primarily guidance on processes for planning, supporting and controlling operational activities and infrastructure changes. As such, practically all ITIL–ITSM processes have impact on efficiency of service provisioning. The discussion in the following paragraphs will however be limited to presenting the core concepts and terms of the three ITIL processes of special relevance to some of the most difficult issues of service provisioning: *Service Level Management*, central to aligning the services to the customers' needs, *Capacity Management*, ensuring a coherent approach to managing capacity issues and finally *Change Management*, helping to keep the infrastructure in a known and reliable state.

*Service Level Management*    Service Level Management (SLM) is the process responsible within the ITIL framework for the management of interactions between the IT organization and the customer (just like the Service Desk is the single point of contact for the user). The Service Level Manager therefore acts as a middle-man, representing the IT organization in all dealings with the customer, and at the same time being the advocate of customer concerns within the IT provider.

SLM tries to achieve more customer-oriented provisioning of IT services by formulating, concluding and continuously monitoring and adapting a Service Level Agreement between the provider, i.e. the IT organization, and the customer. Figure 7 depicts the interrelationships of the most important concepts and terms central to SLM. An SLA specifies what comprises a service and how it is to be provided. In corporate scenarios (and according to ITIL) an IT service is requested by the customer organization to support one or more of its business processes. The service provider organization can itself however act as a customer, when it needs to procure sub-services for the realization of the offered service. When these sub-services are provided by an internal organization (e.g. another department within the same enterprise), an Operating Level Agreement (OLA) is concluded, an *Underpinning Contract* (UC) when the sub-services are obtained from an external provider. From the viewpoint of these sub-providers however, it is in both cases an SLA. While not part of the ITIL's definition of these terms, OLAs usually do not include clauses about penalties to be paid for failing to meet agreed quality targets, while UCs often do.

The non-functional characteristics that a service needs to have to guarantee satisfactory support of the customer's business process (e.g. service availability and performance) are documented in the *Service Level Requirements*. This can of course not be done by the IT organization alone, but only in collaboration with the customer. In many ways, concluding an SLA is not just subjecting the achievements of the IT to more scrutiny, it often also forces the customer the customer organization to put significant effort in defining its business goals and how these are dependent on IT services more precisely.

To fulfill these requirements, technical and organizational aspects in the IT organization need to be considered. How a service is to be realized on the technical level is docu-

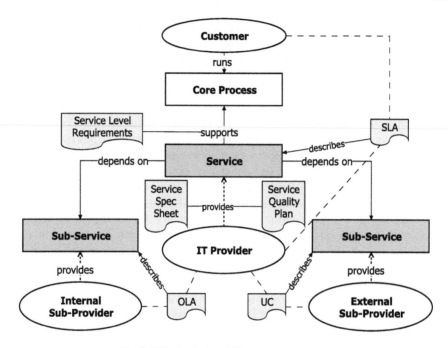

Fig. 7. ITIL Service Level Management concepts.

mented in a *Service Specification Sheet*. This Service Spec Sheet describes what parts of the provider–owned infrastructure, as well as which UCs and OLAs are needed for the provisioning of the service. For the compliance to certain, e.g. availability-related service levels however, not only technical but also organizational aspects need to be managed. The process-oriented performance indicators that need to be met in order to fulfill service level requirements (e.g. incident resolution times, standard change cycle times etc.) are documented in a *Service Quality Plan*.

*Capacity Management*     Capacity Management, like its sibling *Availability Management*, is a planning process primarily concerned with directing infrastructure improvement efforts under a specific quality goal. The goal of capacity management if to make sure that storage, processing and network capacity are provided at in a timely and cost-effective manner.

ITIL divides Capacity Management into three closely interrelated sub-processes:

*Business Capacity Management* Managing the capacity requirements of the business (i.e., customer) – this involves mainly making forecasts of future capacity requirements, based e.g. on trend analysis of service utilization data or communications with the customer.

*Service Capacity Management* Responsible for ensuring that the necessary tools, policies and procedures for the performance management of live, operational IT Services are in place and that collected data is recorded, analyzed and reported.

*Resource Capacity Management* Ensuring effective performance management on the resource level, e.g. by establishing monitoring and reporting of CPU and network load and proactively eliminating bottlenecks before service performance is compromised.

ITIL defines nine principal activities of Capacity Management: *Monitoring, Analysis, Tuning, Implementation, Storage of Capacity Management Data* (into a *Capacity Management Database*), *Demand Management, Modeling, Application Sizing, Production of the Capacity Plan.* Even though Capacity Management is called a process by ITIL, these activities are not arranged in a structured workflow.

The activity in ITIL's Capacity Management that is probably the most neglected by resource-oriented performance management approaches is Demand Management. This refers to supporting the match of capacity requirements and provided capacity, not solely by increasing the latter, but by trying to influence capacity requirements through influencing user behavior. For example, if peak demand can be influenced, e.g. by a lower price for service usage during off-office hours or establishing agreements that non-critical batch jobs should not be started on Fridays when system load is the highest, then peak demand and thus capacity requirements can be reduced.

*Change Management*    ITIL Change Management tries to ensure that changes to the IT infrastructure are executed with the highest possible efficiency and lowest possible impact on service quality. For this purpose, all changes falling into the scope of change management are put under the control of the change management. What falls into change management's scope, i.e. what constitutes a 'change' in the IT organization, needs to be defined. This is usually done aligned to the detail level in which the infrastructure is documented in the *Configuration Management Database* (CMDB) or consequently the Service Specification Sheets. The level of detail required here is usually not very high, as 'change' in an ITIL context relates to structural infrastructure changes (upgrades of hardware, software updates and the like), rather than an altered operational status (e.g. a change in the value of Management Information Base/MIB counters).

Of the three ITIL processes discussed here, Change Management is the only one for which a concrete workflow can be defined [6]. The basic Change Management workflow is illustrated in Figure 8 by an Event-driven Process Chain (EPC) diagram, a graphical notation used in the ARIS framework [2] for modelling workflows. For every change (except for *standard changes*, see below), a *Request For Change* (RFC) must be submitted to change management staff, i.e. in smaller scenarios the change manager. RFCs are screened if they comply to formal guidelines that are laid down by change management (e.g., are name, department and position of the requestor given?). After that, changes are classified according to possible impact and their urgency. At this point, there is a distinction between very regular change procedures and expedited emergency change procedures for very urgent, critical changes (e.g. taking the whole system offline to respond to a major intrusion).

The following describes regular change procedures first. All changes need to be authorized – by whom depends on the classification of the change, e.g. change manager or change management staff for minor impact changes, *Change Advisory Board* (CAB) for changes with significant impact, up to the inclusion of CAB and business management (e.g. CEO) into an executive committee for the authorization for major impact changes. Once they are authorized, they need to be scheduled and executed under the control of change management. Finally every change should be subject to a *Post Implementation Review* (PIR), evaluating if the goals of the change (e.g. availability improvement of a specific ser-

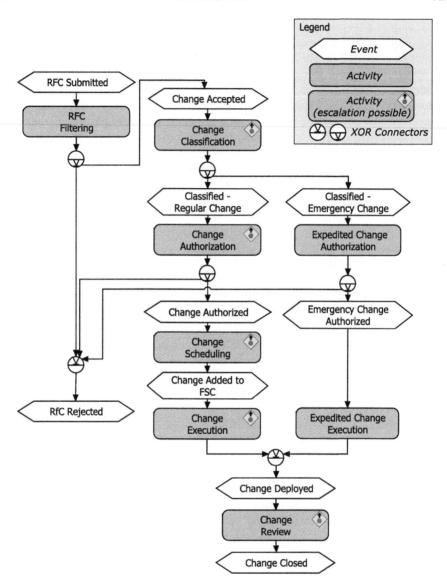

Fig. 8. ITIL Change Management workflow.

vice) were met. Depending on the nature of the change and how achievement of the change goals can be measured, the PIR can happen days, weeks or even months after the change has been executed.

Any of these activities might require a reclassification and consequently a reauthorization of the change. For example, a patch to a server software package might have been classified as a minor change and authorized by the change manager. If then it becomes apparent during testing (a sub-activity of change execution), that not all dependencies had been considered and that for instance the patch necessitates an operating system upgrade or has other unforeseen side-effects, then the change needs to be reconsidered. If, for example, the needed operating system upgrade is considered a significant change, then the CAB must be convened, i.e. the change needs to be escalated.

Such an escalation is not possible for emergency changes. Time is of the utmost importance in emergencies, but emergency change procedures still go through the same process, albeit in an expedited way. An emergency change still needs to be authorized. In practice, emergency changes by their very nature are high risk, thereby of the highest impact classification and warrant the summoning of a CAB meeting. However since there usually is not enough time to hold a regular CAB meeting, the authorization of emergency changes is transferred to the CAB-EC, the *Change Advisory Board Emergency Committee*, constituted of all potential CAB members that can be summoned on short notice. Since the CAB-EC includes all authority that can be assembled within the time constraints, there cannot be any escalation – there simply is no higher authority available at that moment. Also, since action for an emergency change must be immediate, there is no need for change scheduling. Otherwise, the process for emergency changes still includes all change management activities (in an expedited way), as well as a regular Post Implementation Review.

Specific types of changes that are frequent, whose impact is well known and low, and for which detailed procedures can established, can be declared pre-authorized *standard changes*. Typical examples might be the replacement of a hard disk in a RAID-system, replacement of a network interface card or other standard replacement activities. Such standard changes still go formally through the change management process, but usually without much involvement of change management staff. There is no authorization activity, and scheduling, execution and review can often be performed by system administrators according to documented procedures without explicit control by change management staff.

If the provisioning of a service constitutes a change it will in practice be dependent on the type of service and how it is realized: commodity and customized services should be implemented in a way that creation of a new service instance requires only standard changes, at least as long as capacity permits. Individual services however, will often require more complex changes. An institutionalized change management process based on ITIL guidance can improve the coordination of all stakeholders in such a change and make the results much more predictable.

**3.2.4.** *Summary*   While on the surface, ITIL and eTOM share a common goal, their approaches to delivering a framework for IT service management processes are quite different. While eTOM aims for improving efficiency by automation and tries covering the full scope of all provider processes, ITIL sacrifices comprehensiveness for more detailed guidance on 'pain areas' and sees quality improvement as its main goal. For a general discussion of eTOM and ITIL see the chapter by Scheffel and Strassner in this volume [41].

**3.3.** *Service Oriented Architectures*

*Service Oriented Architecture* (SOA) is a concept for designing IT architectures, also often cited as a new approach for addressing *Enterprise Application Integration* (EAI) issues. While it is often referred to in the context of Web Services, it is technology-independent [30]. There are still many different notions about what constitutes a SOA. As yet, there is no standard or even commonly accepted definition of SOA. A very generic definition is provided by the *Organization for the Advancement of Structured Information Standards* (OASIS), an international consortium for e-business standards, describing SOA as "a paradigm for organizing and utilizing distributed capabilities that may be under the control of different ownership domains" [33] – this however describes rather the goal of SOA than the concepts with which to achieve these goals. The OASIS is however also working on a reference model, which will hopefully create a framework to enable better description and specification of SOAs.

Despite the lack of a universally accepted definition, there are some basic ideas and concepts that are commonly attributed to SOA:

*Making functionality network-accessible*    The idea is to make more of the functionality provided by an IT system, previously often dedicated for a specific application, accessible over the network through syntactically standardized interfaces. Compared to objects or software components, services are thought to be usually more 'coarse-grained', i.e. a service should be able to address use cases outside the system in which it is realized. A description of the service and information on how to access and use it is then published in a service repository. This facilitates the reuse of software and system components for purposes that were not anticipated at design time.

*Service composition and loose coupling*    Published, accessible services can be used for building new services through composition. The thinking is that in general, services should be loosely coupled and communicate in an asynchronous manner. This means for instance, that service invocations should be executed in a message-based, rather than a remote procedure call (RPC) style manner, i.e. the requestor will not be blocked until receiving a service response and no communication context needs to be maintained. Also, with no central components and services – in principle not needing to share anything but a network connection and a possibly very simple message exchange mechanism like SOAP over HTTP – cross-organizational service interaction can be realized rather easily.

Compared with previous paradigms for distributed architectures like distributed object models, SOA strives to predetermine only a minimal set of rules for the communication between services (or components). Services could still be realized using object technology, but interaction between services is not dependent on a shared object model, programming paradigm or even type system. SOA is no silver bullet for all problems, as this new flexibility often comes at the cost of lost functionality provided by tight coupling middleware, e.g. reliable messaging or security mechanisms.

The SOA idea holds great promises for the future of provisioning application layer services: Integrating services across different organizational domains, easier load-balancing and failover mechanisms (possibly across organizational boundaries), even automated service discovery and composition – all these are SOA visions.

From a service provisioning perspective, a SOA infrastructure could offer standardized interfaces for service invocation and configuration, thus providing in effect a management interface to the building blocks of more complex, compound services. Though usually request/response software services are described as service examples, services in the SOA sense can also make 'real world' – functionality accessible. So activation and configuration of a network service, e.g. a VPN, could very well be published as a service and integrated into a SOA. In effect the NGOSS architectural guidelines (see also Section 4.3), first published well before the term 'SOA' became popular, aim for such a modular, interface-oriented management system.

There still is a long way to go, but given the substantial industry interest in SOA, it is clear that it addresses important needs and as a concept will probably stay the focus of many efforts in distributed systems – maybe even when the current Web Services protocols have long been superseded.

## 4. Trends in technologies and tools

A variety of tools are already available to support service provisioning processes. In this section state-of-the-art tools are grouped according the service provisioning phases and analyzed with respect to their contribution to the challenges outlined in Section 2. Beside of tools for specific phases, there are workflow tools supporting the whole service provisioning process which are presented in Section 4.5. Due to the amount of tools being offered only a set of selected tools is presented in the following.

### 4.1. *Service planning*

The service planning phase is quite diverse depending on the background of the service to be provided. Usually, a service is described in an abstract manner focusing on its desired functionalities. In addition to performing a market survey to get an estimation about the sales price and the potential number of customers, the provider also has to get an estimation about the costs for the service provisioning using information from sources like inventory management tools, staff salaries, etc. Providers have often developed home-grown solutions for this by retrieving and processing data from SAP-like systems. However, it can be witnessed that the diversity of service planning which is closely tight to the positioning of the provider in the business market makes it difficult to offer standard support tools.

### 4.2. *Service ordering*

In the service ordering phase the details of an order are negotiated with customers. According to the kind of scenario (individual services, customized services, commodity services), this negotiation can be complex or simple. Nevertheless, the service provider has to make sure throughout all provisioning steps that the service order can be fulfilled. Therefore, it has to be verified whether there are currently enough resources or can be made available until the desired fulfillment date keeping in mind the QoS guarantees. In addition, a successful service order has to be registered and transferred to the service implementation. For example, the new service contract has to be added to the accounting process.

Tools which can be applied for this phase comprise web shops (being constructed with tools like Intershop) or home-grown solutions such as web front ends communicating with SAP installations. SAP is frequently used to store customer data and their SLAs.

### 4.3. Service fulfillment

During the service implementation, services and resources have to be ordered from suppliers and configured according to the provisioning requirements. These steps should be automated in order to efficiently deliver the services. Tools should be applied to decide about the sharing of resources and to automatically generate resource configurations (hardware and software). For doing so, it is useful to have a solution database storing resource configurations which have been applied for previous service requests.

**4.3.1.** *Resource virtualization*   During the service implementation the decision about the sharing of resources is very much dependent on the kind of resources in question. For the network it is common practice not to use dedicated hardware, but to create virtual networks using technologies like Multi Protocol Label Switching (MPLS) and/or Virtual Private Networks (VPNs). Hardware components like servers can also be virtualized (e.g. in Web hosting applications) such that these components can be used by customers as if they had dedicated hardware. Another common practice is to share storage space among customers again giving them the impression they are using dedicated resources. Obviously, the implementation of the virtualization has to protect the integrity of each customer. In addition, it is possible to run another operation system on top of the basic operation system which can be helpful depending on the desired software applications. In some cases the virtualization possibilities are already included by vendors. SAP installations can, for example, be shared between different parties and Checkpoint also offers virtual firewalls where a real firewall can filter packets for different target networks.

Customer requirements especially in the area of individual service provisioning have a large influence on virtualization decisions. A virtual firewall, non-dedicated DNS servers, and shared mail servers may be acceptable for customers with small budgets and less strict security requirements, while other requirements may lead to dedicated and consequently much more expensive solutions. In contrast to network sharing and use of tool intrinsic mechanisms, it is currently not usual for service providers to add sharing mechanisms to resources on their own.

**4.3.2.** *Tool support for resource configuration*   Starting from its traditional network management software HP [25] is currently offering a variety of tools to support the service provisioning with respect to ITIL. These tools comprise solutions for service desks and inventory management including an automated configuration detection which is especially useful for Windows Environments. *IBM Tivoli*'s management solutions have gone through a similar evolution as those from HP. IBM also fosters the support of ITIL processes and offers a dedicated tool for service provisioning which is called *IBM Tivoli Provisioning Manager* [26]. It automates the provisioning and configuration of servers, virtual servers, operating systems, middleware, applications, storage, and network devices. This is done by using preconfigured automation packages which are available for a number of products (e.g., *Citrix, MS SQL server, Siebel*). An extension to the *Provisioning Manager* for further automation is the *Intelligent Orchestrator* supporting a scheduling of the automation package installation. Similar to HP and IBM, CA also offers a comparable set of tools.

For software configuration management, Microsoft offers suites for the automated deployment of software in enterprise management scenarios (*Systems Management Server*) and for application management (*Operations Manager*). *Sun N1 software* [44] is designed to support the service life-cycle. Its component *Service Provisioning System* allows for application service provisioning by remotely installing operation systems (Solaris, Linux, Windows), use of XML descriptions to deploy, upgrade, delete, start and stop applications, and record administrator action (allowing for a possible rollback). *Cisco's Configuration Engine* [11] is useful to deploy equipment at customer locations. Its template-based definition of deployment procedures makes it possible to preconfigure equipment so that these configuration steps do not have to be performed by provider staff. In sum, it can be witnessed that tools are already available to ease the configuration of hardware and software. However, the tools are often tied to equipment vendors so that their application may pose additional constraints for the service implementation.

*Cfengine* [8] is a site configuration engine, that is designed to let administrators use configuration files to describe the desired state of a system, rather than describing what should be done to a system to reach that state. The idea of Cfengine is to create a single set of configuration files which describe the setup of all hosts on a network. In a typical setup, Cfengine runs on every host, retrieves configuration data from the master server, and applies changes as needed.

**4.3.3.** *Web service and grid service standards*   A service can be implemented using the Service Oriented Architecture paradigm. This can be useful if the service can be split up into separate tasks being realized by different groups (either internal or partially by external providers). A popular means for the realization of SOAs are *Web services*. These services are specified in an evolving set of standards by *W3C Consortium, DMTF, Open Grid Forum, OASIS* and other standards bodies. Web service specifications deal with the platform and programming language independent specification of autonomous services which can be used to collaborate as distributed applications. They can be described, published, discovered, orchestrated and are designed for the implementation of business processes especially across organizational boundaries.

The most popular Web services standard is *WSDL (Web Services Description Language)* for the specification of Web service interfaces. Other standards deal with the specification of service registry, security, policy, service orchestration, and service agreements. Furthermore, there is an evolving standard for the management of Web services.

For service provisioning in particular an XML-based language called *Service Provisioning Markup Language* [42] is under development aiming at the support for the whole service life-cycle. However, the language is tied to the specification of resources as Web services.

Current problems in using Web services are missing semantics for the service attributes, new challenges for fault management among independent services and a missing concept to deal with the service lifetime.

In the area of grid computing the *Open Grid Service Architecture* (OGSA) [22] aims at automating the provisioning of so called *Grid services*. Grid services can be regarded as an extension of Web services as they are compliant with the WSDL standard, but also define some more features in a standardized XML schema. These features include service states for automated service instantiation and removal.

Figure 9 shows interactions which are defined for the grid service provisioning *(Web Services Resource Framework (WS-RF))*. A service requester would like to create a grid service by contacting the service factory. The service factory composes resources to provide the service and returns an access handle to the service requester. The requester can then communicate with the service instance for the exchange of service data, keep-alives, notifications and for invoking the service. In addition to the service creation on demand, the requester may also contact the service registry where services have been registered by the service factory for reuse by other requesters.

An important extension in WS-RF in comparison to Web services is the explicit creation of services which is part of the infrastructure while this happens in an unspecified manner for standard Web services. This extension has been made to allow for the creation of transient services which may for example be created and only exist to handle a long-term transaction.

The *Configuration Description, Deployment, and Lifecycle Management Specification* (CDDLM) by the Open Grid Forum (former Global Grid Forum) addresses how to describe the configuration of services, deploy them on the Grid, and manage their deployment life-cycle. The document GFD.50 serves as an informational foundation document, while GFD.51 [14], GFD.65 [15] and GFD.69 [16] contain the recommended SmartFrog-based language specification, component model and deployment API, respectively.

In summary, SOAs and Web/Grid services are beginning to play a role in the realization of services in the industry. While the concept of SOAs can be presumed to remain for the longer term, the standards for Web/Grid services are still changing/evolving and may undergo larger revisions. In [1] one of the first realizations of a Web services-based service provisioning in a large scale scenario in the industry is shown.

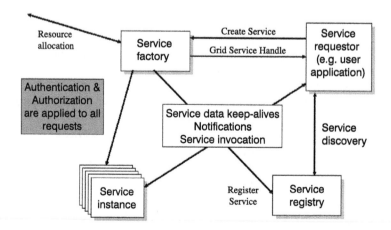

Fig. 9. Grid service provisioning [22].

## 4.4. *Service operations*

In the service operations phase tools are needed to manage the configuration of services and resources, manage the customer SLA data, collect data for SLA reporting and accounting purposes, and to interact with the customers. The latter allow the customers to change the service configuration in a predefined manner as well as to retrieve accounting and SLA reports.

Since the discussion of service operations is out of scope of this article, we are not going into tool details here. However, it can be concluded that a variety of tools exist to support these processes. For the configuration management those tools which have been mentioned for the service implementation support also offer solutions here. The configuration itself can be stored different kinds of databases, which often claim to be compliant to ITILs CMDB, or (more seldom) in LDAP. Vendors like Mercury Interactive offer tools for SLA reporting.

## 4.5. *Service provisioning workflow support*

For the whole process of service provisioning a variety of steps have to be carried out. For following these steps, process management tools can be applied including the definition of conditions which have to be fulfilled for completing a step.

As ITIL has become a very popular standard for service management, many vendors strive to provide ITIL compliant tools as mentioned for HP and IBM. As the ITIL guidelines are written in prose, it is however difficult to define what this compliance actually means. In addition, these tools only address a subset of the ITIL processes.

The Operation Support System through Java (OSSJ) [37] initiative aims at the implementation and deployment of reusable OSS systems using Java technologies. It has provided its own open API specifications, reference implementations and test kits, while commercial products implementing the specifications also became available. In the reference implementation OSSJ uses J2EE to provide strongly-typed, loosely-coupled APIs using XML over Java Messaging Service. Just as important, it applies the Java Community Process for standardization of APIs. In 2004 OSSJ and TeleManagement Forum's *NGOSS* [47] started a collaboration since both initiatives can be regarded as complementary. NGOSS provides technology-neutral business-process and system frameworks, UML information models and resources for building NGOSS-compliant OSS systems and components, while OSSJ provides an OSS implementation in Java.

SAP tools are used by many companies for the support of a variety of business processes. Other tools which are applied in the service provisioning process access SAP to retrieve information needed for the service provisioning. SAP also supports business processes directly by allowing for the modeling of processes using the ARIS business process notation [2]. Other interesting tools in this context comprise Microsoft BizTalk and Oracle's BPEL Engine.

**4.6.** *Summary*

The typical situation in service provisioning can be summarized as follows. A provider usually has a set of home-grown and commercial tools which are applied to service provisioning. These tools are primarily applied to service implementation and operation and each one of them is focused on a specific task. Therefore, the collaboration of these tools is often cumbersome in practice leading to problems like data duplication and inconsistencies. Often, workflows are defined for the service provisioning, but their execution is dependent on the staff and the general service provisioning circumstances.

## 5. Research activities

As the presentation of concepts and frameworks in Section 3 as well as the examination of available tools in the previous section showed, a lot of issues in the context of service provisioning cannot be regarded as satisfactorily solved. Therefore, research being conducted in academia and industry is presented in this section which addresses some of these challenges. The presentation is ordered with respect to the service provisioning phases.

**5.1.** *Service planning*

In some areas of service provisioning such as mobile phone services or context-aware services, a frequent change of the service offers can be witnessed. Therefore, research efforts are aimed at allowing the rapid definition and planning of services, especially to react to changes in the market.

*Service profiles*    Clemm et al. [12] proposed the definition of services profiles which denote a technology-neutral description of services using a template. It includes a set of features that a service may have and for each feature a set of possible parameters. For the later service instantiation, technology-specific provisioning profiles are used to allow for the implementation of the services using different technologies. A service provisioning engine was prototypically implemented and applied for two example scenarios (residential subscriber services, Internet data centers).

**5.2.** *Service ordering*

While the service ordering is quite simple for commodity and customized services, it may require a series of interactions between customer and provider for individual services. The implementation of a CSM as described in the following can be an important step to support the required interactions.

*Customer service management*   Langer and Nerb [31] provided a detailed concept for implementing a *Customer Service Management* (CSM) interface for a given service provisioning scenario. CSM aims at allowing the customer to access provider-internal service management information and to self-manage subscribed services in a predefined manner. With respect to service provisioning the CSM interface allows the customer to order new services or to change parameters of currently subscribed services. The implementation of CSM by e.g. providing a web-based interface allows for the processing of customer change requests since these requests are given in a standardized format.

A central step in service ordering is the negotiation of SLAs. This negotiation requires to know which kinds of SLA mechanisms and guarantees are suitable to allow for stable service operations. Begnum et al. [4] showed that policies using non-linear feedback mechanisms can result in quite complex behavior which may cause the repeated downgrading of service levels when not parameterized with care.

The negotiation phase for service provisioning is quite different in peer-to-peer environments. Here, the negotiation of trust is often a crucial issue since the peers do not have a priori knowledge of each other especially in large-scale deployments. For enabling the service provisioning despite of limited trust, Burgess [10] proposes to accept different trust levels depending on the service's requirements.

In [40], Sarangan and Chen give an overview of protocols for dynamic service negotiation of network services.

### 5.3. *Service fulfillment and operations*

Even though some tools are already in place for service implementation, there are still open issues which need to be addressed in a general manner.

*Service templates and CBR*   In her postdoctoral thesis, Dreo [19] proposes the application of Case-Based Reasoning (CBR) techniques for service provisioning and especially addresses the question how to obtain the requested resp. desired granularity of fulfillment, operational and withdrawal steps. The idea is to describe a service as a *service template*, including service ordering information, the necessary fulfillment and operational steps, the withdrawal steps if the service needs to be removed from the system as well as the policies associated with this service. Having such a service template structure which corresponds to a *case structure*, enables us to apply the CBR paradigm to dynamic service provisioning. The main idea is to think of a *service template as a case*. If a new service is ordered by a customer, then the service template library (case repository) is searched for a similar service order, resp. service template. After a similar service template is found, it is retrieved and adapted to the specific customer requests (specified in the service ordering), and afterwards the adapted steps are executed. The proposed application of the CBR paradigm successfully addresses the following problems: (i) to gain the requested granularity of steps (as precisely as possible) and (ii) to ensure up-to-date provisioning steps since these need to be updated due to changes in the infrastructure, customer requests etc.

*CHAMPS*    Keller and Badonnel [29] describe IBMs research system *CHAMPS (CHAnge Management with Planning and Scheduling)* which is prototypically applied for application service provisioning. It consists of two major components: *Task Graph Builder* and *Planner & Scheduler*. The Task Graph Builder is designed to build a workflow for performing a change in service operations. This workflow is described using *BPEL (Business Process Execution Language)* which is a popular standard for the description of processes in the context of Web services. The Planner & Scheduler component receives the workflows from the Task Graph Builder and generates a schedule for executing the changes under consideration of a set of constraints. These constraints prevent conflicts in the change execution due to resource access conflicts and time conditions as well as non-technical requirements such as SLAs and policies.

*Quantifying complexity*    Brown et al. [7] provide a model to quantify configuration complexity, identifying execution complexity, parameter complexity and memory complexity as the major components of the model. The model is then applied to the CHAMPS change management system mentioned before to show the reduction in human-perceived configuration complexity and installation time.

Diao and Keller [18] address complexity quantification in IT service management. Their goal is to measure the savings in process complexity, when processes are re-engineered or automated. The authors propose a process model based on roles and their participation in tasks producing or consuming business items. They further propose quantifying process complexity along the dimensions execution complexity, coordination complexity, and business item complexity. The authors also describe a tooling concept and a brief evaluation.

*Quartermaster*    HP's *Quartermaster* research project [43] aimed at automating data center management tasks, primarily focused on service fulfillment. The core of the system is a CIM model repository representing four layers: service, application, system, and physical and allowed the definition of policies. A tool chain is arranged around the core providing management functions. Tools include functions of resource allocation and assignment, deployment and run-time control. Automation is achieved by orchestration of tools through workflows. A fundamental concept of Quartermaster is a resource topology. It is a formalized representation (model) of a configured resource environment supporting an application. An editor is offered to design resource topologies. They are stored in the core repository and passed through the tool chain for instantiation. During operation, so-called Service Delivery Controllers, as part of the same tool chain, continuously compare the currently provided service quality with a desired state and re-adjust service operations in a closed control loop.

*Model translation*    Sahai et al. [38] describe the automated translation of resource models, generated by a constraint solver in Quartermaster as MOF models based on CIM, into scripts for software deployment and configuration based on SmartFrog. They address "Built-to-Order" systems, which would well fit in the category of customized services and most individual services. In a follow-up paper [46], automated staging and testing are included in the provisioning workflow. Solution designers can get data about the performance of the deployed solutions via instrumented applications and monitoring tools. In an open-loop like fashion, they can now tweak the input to the constraint solver as well as policies and re-iterate the process. In a deployment of a TPC-W scenario, the performance of several solutions based on servers with different capabilities is compared.

*SunTone*    Sun's *SunTone* program [45] provides an open service model for service delivery especially targeting to avoid 'data silos' (isolated data repositories which lead to data duplication). The model is a three-dimensional cube with the dimensions *tiers – application partitioning*, *layers – hardware/software stack*, and *systemic qualities*. The initiative offers the certification of IT or business processes concerning compliance with the model.

In the service implementation resources have to be allocated to provide the services. As this should be done in an efficient manner, several theoretical approaches have been devised. A promising approach is to model the resource allocation problem as a constraint satisfaction problem [32] for which a set of efficient algorithms exist. This approach can deal with incomplete data and can cope with changes in highly dynamic environments.

For the interaction of resources Burgess [9] proposes to use the *promise theory*. This theory from the area of autonomic computing is based on modeling the resources as autonomous entities. The service fulfillment and operation is therefore a result of the collaboration of independent entities. This theory is in particular interesting for virtualized resources like virtual servers.

### 5.4. *Information modeling*

Since the typical service provisioning that is found within service providers usually implies the distribution of service related information into various data repositories, it is often difficult to get an overview of the service provisioning effectiveness and efficiency. In addition, the data duplication can easily lead to data inconsistencies.

*Service MIB*    Sailer [39] proposes to extend the construction of MIBs, which are usually defined for network and systems management, with respect to services. A *Service MIB* should comprise all information that is needed for the service provisioning including information from all FCAPS (fault, configuration, accounting, performance, security) functional areas. The information modeling for the *Service MIB* is carried out using object-oriented design patterns and by defining a language for the specification of service instances.

For the introduction of services a model-driven approach is proposed in [13] where the composition of components for the service creation is done via a construction GUI. The developed research prototype called *Service Designer* allows for the design of generic service models and the creation of OSS-dependent views. It has been prototypically applied to VoIP and push-to-talk services.

Eilam et al. [20] describe a model-based approach to manage the configuration complexity of distributed applications in data centers. It builds on the principle of separation of concerns. Humans with different roles, such as developers, deployment experts, domain experts, data center operators and business users provide their knowledge as models. These models are taken into account by management components like deployment topology generator, optimizer, planner and provisioning engine. The method is applied to the deployment and configuration of J2EE-based web applications.

## 6. Service provisioning in practice and operations

After service provisioning had been analyzed from the viewpoint of challenges and process alignment in the previous sections, this section is about how service provisioning is currently realized in real provider environments. After giving an overview on current best practices and how they address the challenges listed in Section 2, outlines of interviews with practitioners give some insight into their view on service provisioning. The focus of the questions was on process organization and tool support for service provisioning. We conclude this section with two specific scenarios.

### 6.1. *Best practices in resource allocation and assignment*

Certain patterns can be witnessed in the allocation and assignment of resources with many service providers.

**6.1.1.** *Late assignment of resources to requests*   Service providers can distinguish between resource *allocation*, keeping track of the number of used and spare resources, and resource *assignment*, keeping track of the specific resource instances assigned to the requests. Resources can be allocated, when a new request is received, which needs resources at a future point in time. Specific instances are assigned shortly before the request comes due. This way, the service provider saves assignment efforts, that would have to be redone, when resources fail or requests change. In addition, resources can be allocated, that have not even been delivered by the supplier yet. This best practice works in favor of reducing management efforts and adapting to changes.

EXAMPLE. Imagine that a service request requires one server after three months time for a period of two weeks. Allocation deals with capacities, and therefore would reduce the available capacity in the requested time frame by one server in an allocation calendar. A particular server would be assigned to the request shortly before service activation instead of when receiving the request.

**6.1.2.** *Soft-state assignment of resources to requests*   To adapt to changes and to reclaim resources, resource assignments should regularly be reevaluated. This can be enforced by the service provider by associating a lease time with the assignment of resources to requests. When the lease time is over, the customer is asked to refresh the request.

Soft-state assignment is typically used when providing services that the service provider is not billing for. Due to the lack of obligations, customers can forget about the resources assigned to them, without explicitly expressing that they no longer use the services. Over time this leads to a high percentage of stale entries in data stores that save information on resource assignments, no longer reflecting actual service usage.

EXAMPLE. Service providers use soft-state assignments in their daily operations for many types of user accounts, like for mail, databases, and remote login, and IP addresses. Soft-state assignment is also a typical component of protocols that control resource assignment such as Dynamic Host Configuration Protocol (DHCP) and Resource Reservation Protocol (RSVP).

**6.1.3.** *Configurable resources with remote management support*   All but the simplest resources need configuration to adjust the functionality of the resource to the service provider's individual requirements. The amount of configuration options therefore reflects the flexibility of the resource. In conjunction with resources typically being distributed over several locations, this necessitates devices that can be managed remotely providing an appropriate management interface. The use of management systems, so far mainly for monitoring and remote configuration of devices, requires these management interfaces to comply with well-established standards to ensure interoperability of management systems and managed components. Standards-compliant management interfaces also provide the basis for automating the provisioning process.

Configurable resources with remote management work in favor of fast provisioning, adaptation to changes and scalable provisioning. At the same time they increase the complexity of resource descriptions due to their large number of configuration parameters and options.

EXAMPLE. Managable network components like routers and switches include standard remote management interfaces accessible via Simple Network Management Protocol (SNMP). In addition, they provide command line interfaces (CLI), accessible locally via a serial console port or remotely via telnet or secure shell. While the CLIs are vendor-dependent, vendors in turn work hard to keep a certain common set of command syntax and semantics troughout their product lines.

**6.1.4.** *Resource virtualization*   Resource virtualization techniques have been developed to ease changes in the mapping of logical resources to physical resources without too much increase in visible management complexity. They provide the view of dedicated resources in a shared-resource environment to the users of the virtual resources, aiming at standardization and reducing total resource consumption. A considerable amount of complexity in terms of code and models is put into the virtualization frameworks and therefore hidden from the user and administrator, while some configuration options are exposed via a management interface. See [24] for a more in-depth investigation of the impact of virtualization on management systems.

Resource virtualization can have a mixed impact on provisioning and operations. On the one hand, it can reduce operational complexity by enabling functions that are hard or impossible to achieve on physical resources, e.g., snapshots of virtual servers that can easily be installed on remote servers or adjusting the size of logical volumes at runtime. On the other hand, when things go wrong, virtualization adds a layer of potential reasons for errors, thereby complicating troubleshooting. Finally, the level of influence and security provided by logical separation instead of physical separation in virtualized environments can be hard to assess, leading to mutual influence in shared resource pools.

EXAMPLE. Resource virtualization is used in many ways, such as:

- Virtual server software, like VMWare, Xen and Virtuozzo, create multiple logically separate virtual servers on a single physical server.
- Virtual hosts in an HTTP server allow for virtual web servers.
- Java's Virtual Machine (JVM) decouples applications from the specific processor and operating system architecture.
- Logical volumes in Storage Area Networks (SAN) can be mapped freely to spare hard disks.
- Multi-customer capable systems run service instances for multiple unrelated customers on the same set of physical resources with logical separation.
- Virtual local area networks (VLAN) decouple broadcast domains from the physical network topology, enabling topology changes by remote configuration instead of manually patching cables to different interfaces.
- Techniques, like virtual paths and channels in Asynchronous Transfer Mode (ATM) or label-switched paths in conjunction with the Resource Reservation Protocol (RSVP) in networks utilizing Multi-Protocol Label Switching (MPLS), partition end-to-end link bandwidth to create virtual dedicated network links.
- Virtual private networks (VPN) provide logically separated networks in a shared network topology.

**6.1.5.** *Overprovisioning* When purchasing resources or subscribing to subservices, a precise requirements specification and in-depth performance evaluation is often hard and costly for the service provider mainly due to uncertainties in future resource utilization. Therefore, service providers make educated guesses and add a fair amount of extra headroom to deal with uncertainties as well as to remain flexible to address unpredicted demands or to have room for expansion.

Overprovisioning is therefore a way to keep enough resources for changing or unpredicted demands. It provides flexibility and room for expansion, so that new service requests can be deployed quickly to spare resources.

In addition, overprovisioning reduces the frequency of needed changes. As changes are one of the main reasons of service failures or disruptions, overprovisioning reduces potential service downtimes. Monitoring data such as workload profiles should be regularly analyzed to have a realistic view on resource utilization and the true amount of headroom.

A special form of overprovisioning is assigning dedicated resources to certain services or customers to work around QoS management in shared resource pools even though predicting in advance that the resources will be underutilized. This strategy can make sense for low-cost resources as the win in a reduction of operational complexity, e.g., no need for Quality of Service management, direct mapping of services/customers to particular resources and the other way around in incident and problem management can outweigh the loss resulting from having to provide more resources. With new technology and management systems on the horizon, which may provide resource sharing with only marginal additional management complexity (see 'Resource Virtualization') or better scheduling capabilities, this strategy should be regularly verified by market research, though.

Overprovisioning loosens requirements on choosing the most appropriate set of resources and therefore greatly simplifies the provisioning process at the cost of increased overall resource consumption.

EXAMPLE. Overprovisioning of resources is typically cheap when introducing new hardware resources as the cost of labor and operational disruption is about the same regardless of the type of new hardware resources and the additional price for more capable hardware resources (e.g., fiber optics with a higher nominal bandwidth, more powerful servers, etc.) is typically minor compared to other costs.

**6.1.6.** *Oversubscription*   In many scenarios, customers do not make use of the full share of resources that a peak usage of the subscribed service would require. If planning resource consumption for peak service demand for every customer, resources would be underutilized. In contrast, planning the provisioning for average customer demand allows to provide services to additional customers and therefore increases profit due to increased resource utilization.

Oversubscription works in favor of increasing resource utilization, but leads to challenges of QoS management in shared resource pools. Statistical models as well as a continuous monitoring of the actual resource utilization level are needed in order to avoid SLA breaches.

EXAMPLE. Statistically, not all 500 DSL subscribers connected to a single DSL access multiplexer (DSLAM) download data using the full downstream bandwidth, e.g., 2 Mbit/s, at the same time. While the downstream link from the next router to the DSLAM may only have a bandwidth of 155 Mbit/s (in other words 77x max individual downstream bandwidth), typically all actual concurrent service users may experience a good service quality at full download speed.

**6.2.** *Comments by practitioners in service provisioning*

This section outlines the main findings of three interviews with solution engineers from a large provider of enterprise services. They were asked to comment on the current state of the art in service provisioning in actual daily operations, mainly with respect to processes and tools. The gist of the interviews was compiled into short paragraphs. It is interesting, that the three engineers did not agree in their statements although they work for the same company.

*Comments of the first engineer*   Processes for service provisioning are defined in the industry at such a level of detail that a hardcopy would fill walls. Therefore, people in daily solutions business do not pay too much attention to these processes since they are often hard to understand and follow from the perspective of each employee. An exception are defined synchronization points ('gates'), which terminate certain steps, such as the end of pre-sales negotiations with the customer. By synchronizing at these points, it can be assured that all required steps in the previous phase have been completed and that all necessary information is available to continue. At these synchronization points all affected employees get in contact with the process.

In practice, some concepts from academia have not proved to be useful yet. An example for this is breaking individual services into reusable service components or service modules. It has shown to be quite complicated to identify previous service provisioning projects where a similar solution has been provided. If one takes a firewall as an example, it is often easier for an experienced staff member to customize an off-the-shelf firewall product starting with the default configuration than to search in previous projects whether a similar configuration has been demanded before. An obstacle in using the latter approach is that there is often no suitable kind of directory to search in prior projects for the same or similar configurations. In addition, particularly in individual service provisioning, many previous configurations do not exactly match to the currently given requirements, which requires reconfiguration anyway.

SOAs are not yet relevant in service provisioning for this service provider.

*Comments of the second engineer*    The solution provider uses mostly home-grown tools for service provisioning for historical and political reasons. For service provisioning, a distinction is made between administrative data and technical data. Administrative data, like contracts, supplier addresses, contact information is contained in SAP, which is cumbersome to extend for other related data, while a home-grown application manages technical data stored in a commercial database.

The ordering of new services is done via a web interface in simpler cases, while additional spreadsheets have to be filled in otherwise. Service provisioning is done according to ITIL processes and employees are responsible for the ITIL manager roles (in contrast to the comments of the first engineer). Nowadays, customers sometimes explicitly demand service providers to be organized according to ITIL standards. A solutions database with previous service provisioning cases is available, but the frequency of actual use is unknown.

*Comments of the third engineer*    This engineer was involved in a project where ten thousand branch offices of a customer within a country had to be connected to the common Intranet using standard components. Technically speaking, Cisco SOHO (small office/home office) routers were installed at the branch offices and connected via ISDN to the corporate headquarters where an installation was performed to manage the customer connections including remote management of the routers. While the whole service can be seen as an individual service, the provisioning of equal network access for thousands of branch offices shows properties of a commodity service.

For provisioning, a special rollout system was constructed, which was basically a commercial database combined with a self-made script-based workflow system including interfaces to tools like SAP. The rollout database contained workflow descriptions for the steps to be taken for the equipment installation (schedule appointments with branch offices, carry routers to branch offices, install equipment at sites, configure ISDN connection through provider network, finalize equipment installation in accordance with the branch office staff, collect data for accounting and send customer bills, send orders to suppliers).

All branch offices had to be online at a given date, which requires smooth operation not only from the technical point of view. To name an example of an obstacle which had not been taken into account by either project partner, branch office staff did not accept white network cables as this would not match their furniture. Instead of arguing on technical matters, the solutions provider soon found a supplier for network cables colored in different shades of brown and stocked them in their country-wide warehouses.

The rollout system is not disclosed to nor shared with other providers. Internally, it was used for six or more other related projects, being of substantial value in competition.

Challenges for a network service provider include reliable statements on delivery, functionality and availability of network access, such as: Is DSL available at a number of branch office locations? Which bandwidth is supported in each case? When can it be installed?

This large provider uses a cable management system, that runs in three data centers. It provides an automated mapping of locations to possible network connections. The system has capabilities to reserve certain ports in the network, as well as to configure network components. The actual installation is also dependent on employee availability.

In general, vendors offer pre-configuration services for customer premises equipment. The simplest level of pre-configuration is to configure the address of another configuration server, which is under control of the solutions provider. Configuration services are performed by many individual service providers for large customers including bundling resources from different suppliers.

**6.3.** *Selected scenarios from different articles (case studies)*

Relatively few service providers disclose information about their principles in provisioning, like on technical matters such as resource selection strategies, resource performance comparison or organizational matters such as processes or organizational models. However, two articles have been found which address some of the goals and challenges mentioned above. As these articles deal with commodity service provisioning they are complementary to the interviews for individual service provisioning.

*Provisioning at Google*    In [3], Barroso, Dean and Hölzle describe some details about the Google cluster architecture, including strategies on performance evaluation and resource selection, fault handling, and own developments for a special, highly parallel application. The authors name energy efficiency and price per performance ratio as key design factors.

Google's architecture is based on two principles: First, they believe that the best price per performance trade-off for their applications comes from building a reliable computing infrastructure using clusters of unreliable commodity components. Therefore, reliability is provided in software rather than in high-availability hardware, using more than 15 000 unreliable commodity rack servers to build a highly available computing cluster at a low price. Hardware-based load-balancers are used but no redundant power supplies, RAID systems nor costly fault-tolerant components in the servers. Software-based reliability means that services as well as data are replicated across many different machines, with highly automated management systems in place to prevent, detect and work around failures. GoogleFS [23], a file system designed for and custom-built to the special requirements at Google, has been developed and is now in use in the data centers.

Second, the design is tailored for best aggregate request throughput, not peak server response time, since response times can be managed by parallelizing individual requests. The ratio of price per performance is addressed by purchasing the CPU generation which – at the date of procurement – gives the best performance per unit price, not the CPUs that give the best absolute performance. The authors claim that all software is produced in-house, and system management overhead is minimized through extensive automation and monitoring, which makes hardware and energy costs a significant fraction of the total system operating expenses.

*Provisioning of web hosting services*    Web hosting services are a good example of commodity services, where automation of provisioning is driven to extremes, due to its direct influence on profit. In Germany, two web hosting providers host more than half of the nine million German .de-domains in 2005 [17]. A good description of service provisioning at these providers can be found in [5].

In their web hosting platforms, every provisioning step – from order processing, domain registration to web server configuration – is carried out by thoroughly elaborated and maintained programs written in scripting languages.

Due to the high level of automation, the choice and configuration of resources is crucial to the success of web hosting providers as the following real-world example shows.

- A procured Sun E6500 was one of the biggest servers at a certain period of time, but its strength was more on number crunching than I/O performance.
- A data center was only served by one ISP, resulting in many customers shifting to another provider, when this ISP got into financial trouble.
- Due to budget restrictions, an EMC storage system had not been mirrored to a second one, as had been agreed upon with a contractor.
- Service outages of several days and lost data due to a storage crash caused many customers to transfer their web sites to other hosting providers.

Growing more and more, both companies put substantial efforts in the design of new data centers. Dedicated servers hosted for customers provided a new challenge due to their scale and density, resulting new requirements on available space, energy supply and cooling.


## 7. Summary

Service provisioning has different faces, depending on the number of orders and the level of customization of the service. The high number of equal and relatively simple orders of commodity services allows automation of many steps in the provisioning workflow. Management systems and scripts are set up for this purpose, but they need constant maintenance and orchestrated resources. In contrast, individual services are characterized by a complex set of specific customer requirements and negotiations between service provider and customer on financial and technical matters. Automation is therefore less applicable. In customized services, service providers combine features of individual and commodity services.

As service provisioning includes allocation and configuration of resources, setting the stage for service operations, it has a direct influence on profit. However, many challenges arise from a number of conflicting goals.

Addressing the organizational aspect of IT service management, frameworks of best-practice type processes such as ITIL and eTOM aim at defining terms and providing guidance on how to structure service provisioning and operations. Standards across service providers such as ISO 20000 allow them to be certified to express a certain level of quality to customers. Based on the terms and processes, vendors can develop service support tools for a wide range of service providers. Service Oriented Architectures play only a minor role in the current practice of service provisioning. However, the concepts are supported by major vendors and can therefore be expected to become increasingly relevant.

As mentioned before, tools, ranging from groupware and project management tools in individual services to flow-through provisioning tools in some commodity services, support many steps in service provisioning. Resource virtualization techniques are already applied in many scenarios, but they are expected to bring even new opportunities in the future. Interoperability among commercial provisioning support tools as well as home-grown software and scripts still leaves many open issues.

Service provisioning research is driven by the increasing complexity in terms of scale and heterogeneity of resources and services as well as their interdependencies. Resource allocation algorithms, which have been studied for a long time, are now further elaborated in context with efforts in modeling of organizations and resources including their structure, behavior, capabilities and dependencies. Physical limitations in terms of cooling large data centers, as well as the need for increased energy efficiency in fix and mobile networks increasingly influence decisions in service provisioning, posing open research questions.

Research in self-management techniques initiated especially by IBMs Autonomic Computing Initiative aims at encapsulating complexity in autonomic elements instrumented by business goals and policies. Research in process frameworks tackles complexity from the organizational point of view in single- and multi-domain-scenarios. Implementation of basic infrastructures for grid and utility computing has already started, but still a lot of research is needed to make the visions become reality. Ad-hoc and P2P networks question the traditional centralized approach of provisioning resources under control of a service provider, with service providers such as Skype including these mechanisms in their business models.

In daily operations, provisioning is currently based on experienced network and systems engineers, supported by management systems. Future tools promise to take care of most of the operational procedures, but people will be needed to develop, test, run and maintain these tools.

## 8. Glossary

| | |
|---|---|
| API | Application Programming Interface |
| ARIS | Architecture of Integrated Information Systems |
| ATM | Asynchronous Transfer Mode |
| | |
| B2B | Business to Business |
| BPEL | Business Process Execution Language |
| BS15000 | British Standard 15000 |
| BSI | British Standards Institution |
| | |
| CAB | Change Advisory Board |
| CAB-EC | Change Advisory Board Emergency Committee |
| CBR | Case-Based Reasoning |

| | |
|---|---|
| CCTA | Central Computer and Telecommunications Agency |
| CEO | Chief Executive Officer |
| CHAMPS | CHAnge Management with Planning and Scheduling |
| CIM | Common Information Model |
| CMDB | Configuration Management Database |
| CPU | Central Processing Unit |
| CRM | Customer Relationship Management |
| CSM | Customer Service Management |
| | |
| DNS | Domain Name System |
| DSLAM | DSL Access Multiplexer |
| DSL | Digital Subscriber Line |
| | |
| EPC | Event-Driven Process Chain |
| eTOM | enhanced Telecom Operations Map |
| EXIN | Examination Institute for Information Science |
| | |
| FAB | Fulfillment, Assurance and Billing |
| FCAPS | Fault, Configuration, Accounting, Performance, Security Management |
| | |
| HTTP | Hypertext Transfer Protocol |
| | |
| ICSP | Internet and Communications Service Provider |
| ICT | Information and Communication Technology |
| IEC | International Electrotechnical Commission |
| ISDN | Integrated Services Digital Network |
| ISEB | Information Systems Examinations Board |
| ISO | International Organization for Standardization |
| ISO/IEC 20000 | ISO/IEC 20000: International Standard for IT Service management |
| ISP | Internet Service Provider |
| IT | Information Technology |
| ITIL | IT Infrastructure Library |
| ITSM | IT Service Management |
| itSMF | IT Service Management Forum |
| ITU | International Telecommunication Union |
| | |
| J2EE | Java 2 Enterprise Edition (Sun) |
| | |
| LAN | Local Area Network |
| LDAP | Lightweight Directory Access Protocol |
| | |
| MIB | Management Information Base |
| MNMTeam | Munich Network Management Team |
| MOF | Microsoft Operations Framework |
| MPLS | Multi Protocol Label Switching |

NGOSS    New Generation Operations Systems and Software
NMF      Network Management Forum

OASIS    Organization for the Advancement of Structured Information Standards
OGC      Office of Government Commerce
OGSA     Open Grid Services Architecture
OGSI     Open Grid Services Infrastructure
OLA      Operational Level Agreement
OSS      Operations Support Systems

P2P      Peer-to-Peer
PIR      Post Implementation Review

QoS      Quality of Service

RfC      Request for Change
RPC      Remote Procedure Call
RSVP     Resource Reservation Protocol

SAN      Storage Area Network
SID      Shared Information/Data (model)
SIP      Session Initiation Protocol
SLA      Service Level Agreement
SLM      Service Level Management
SM       Service Management
SOA      Service Oriented Architecture
SOAP     Simple Object Access Protocol
SOHO     Small Office Home Office
SPML     Service Provisioning Markup Language
SQL      Structured Query Language

TMF      TeleManagement Forum
TNA      Technology Neutral Architecture
TOM      Telecom Operations Map
TPC      Transaction Processing Performance Council
TPC-W    Transactional web e-Commerce benchmark by TPC

UC       Underpinning Contract
UML      Unified Modeling Language

VC       Virtual Circuit
VLAN     Virtual Local Area Network
VPN      Virtual Private Network

W3C      World Wide Web Consortium

WSDL   Web Services Description Language
WS-RF   Web Service Resource Framework
WWW   World Wide Web

XML   Extensible Markup Language

## Acknowledgements

The authors wish to thank the members of the Münich Network Management (MNM) Team for helpful discussions and valuable comments on previous versions of this paper. The MNM Team directed by Prof. Dr. Heinz-Gerd Hegering is a group of researchers of the University of Münich, the Münich University of Technology, the University of the Federal Armed Forces Münich, and the Leibniz Supercomputing Center of the Bavarian Academy of Sciences. Its web-server is located at http://www.mnm-team.org.

## References

All URLs were last checked on May 31st 2006.

[1] M. Archarya et al., *SOA in the real world – experiences*, Proceedings of the Third International Conference on Service-Oriented Computing (ICSOC 2005), Lecture Notes in Comput. Sci., Vol. 3826, Amsterdam, The Netherlands (2005), 437–449.

[2] Architecture of Integrated Information Systems (ARIS), http://www.ids-scheer.com/aris.

[3] L.A. Barroso, J. Dean and U. Hölzle, *Web search for a planet: The Google cluster architecture*, IEEE Micro **23** (2) (2003), 22–28.

[4] K. Begnum, M. Burgess, T.M. Jonassen and S. Fagernes, *On the stability of adaptive service level agreements*, eTransactions on Network and System Management **2** (1) (2006), 13–21.

[5] H. Bleich, *Die Masse machts – Wie 1&1 und Strato das Webhosting revolutioniert haben*, c't – Magazin für Computertechnik **19** (2005), 188–192.

[6] M. Brenner, *Classifying ITIL processes – A taxonomy under tool support aspects*, First IEEE/IFIP Workshop on Business-Driven IT Management (BDIM 06), IEEE (2006), http://www.mnm-team.org/pub/Publikationen/bren06/.

[7] A. Brown, A. Keller and J. Hellerstein, *A model of configuration complexity and its application to a change management system*, Proceedings of the 9th IFIP/IEEE International Symposium on Integrated Network Management IM'2005, IEEE (2005), 631–644.

[8] M. Burgess, *Cfengine: A site configuration engine*, Computing Systems **8**, USENIX (1995), 309–337.

[9] M. Burgess, *An approach to policy based on autonomy and voluntary cooperation*, Lecture Notes in Comput. Sci., Vol. 3775 (2005), 97–108.

[10] M. Burgess and K. Begnum, *Voluntary cooperation in a pervasive computing environment*, Proceedings of the Nineteenth Systems Administration Conference (LISA XIX), USENIX Association, Berkeley, CA (2005), 143.

[11] Cisco Configuration Engine, http://cco-rtp-1.cisco.com/en/US/products/sw/netmgtsw/ps4617/index.html.

[12] A. Clemm, F. Shen and V. Lee, *Generic provisioning of services – A close encounter with service profiles*, Computer Networks **43**, Elsevier (2003), 43–57.

[13] M. Cochinwala, H.S. Shim and J.R. Wullert, *A model-driven approach to rapid service introduction*, Proceedings of the 9th IFIP/IEEE International Conference on Integrated Network Management (IM 2005), IFIP/IEEE, Nice, France (2005).

[14] *Configuration description, deployment, and lifecycle management: SmartFrog-based language specification*, Technical Report GFD-R-D.051, Global Grid Forum (2005).

[15] *Configuration description, deployment, and lifecycle management: Component model version 1.0*, Technical Report GFD-R-D.065, Global Grid Forum (2006).

[16] *Configuration description, deployment, and lifecycle management: CDDLM deployment API*, Technical Report GFD-R-D.069, Global Grid Forum (2006).

[17] DENIC eG, *Statistics on .de domains*, http://www.denic.de/de/domains/statistiken/.

[18] Y. Diao and A. Keller, *Quantifying the complexity of it service management processes*, DSOM, Lecture Notes in Comput. Sci., Vol. 4269, R. State, S. van der Meer, D. O'Sullivan and T.Pfeifer, eds, Springer-Verlag (2006), 61–73.

[19] G. Dreo Rodosek, *A framework for IT service management*, Habilitation, University of Munich (LMU) (2002).

[20] T. Eilam, M.H. Kalantar, A.V. Konstantinou, G. Pacifici, J. Pershing and A. Agrawal, *Managing the configuration complexity of distributed applications in internet data centers*, IEEE Communications Magazine **44** (3) (2006), 166–177.

[21] A.L. Factor, *Analyzing Application Service Providers*, Sun Microsystems Press (2002).

[22] I. Foster, C. Kesselman and S. Tuecke, *The Open Grid Services Architecture, The Grid Blueprint for a New Computing Infrastructure*, 2nd edn, I. Foster and C. Kesselman, eds, Morgan Kaufmann (2004).

[23] S. Ghemawat, H. Gobioff and S. Leung, *The Google file system*, SOSP (2003), 29–43.

[24] S. Graupner, R. Koenig, V. Machiraju, J. Pruyne, A. Sahai and A. van Moorsel, *Impact of virtualization on management systems*, Technical Report HPL-2003-125, HP Labs, Palo Alto, CA (2003).

[25] HP OpenView, http://www.managementsoftware.hp.com/.

[26] IBM Tivoli Provisioning Manager, http://www-306.ibm.com/software/tivoli/products/prov-mgr/.

[27] ISO/IEC, *International Standard – ISO/IEC 20000-1:2005 – Information Technology – Service Management – Part 1: Specification* (2005).

[28] ISO/IEC, *International Standard – ISO/IEC 20000-2:2005 – Information Technology – Service Management – Part 2: Code of Practice* (2005).

[29] A. Keller and R. Badonnel, *Automating the provisioning of application services with the BPEL4WS workflow language*, Proceedings of the 15th International Workshop on Distributed Systems: Operations and Management (DSOM 2004), Davis, CA, USA (2004).

[30] D. Krafzig, K. Banke and D. Slama, *Enterprise SOA – Service-Oriented Architecture Best Practices*, Pearson (2005).

[31] M. Langer, S. Loidl and M. Nerb, *Customer service management: A more transparent view to your subscribed services*, Proceedings of the 9th IFIP/IEEE International Workshop on Distributed Systems: Operations & Management (DSOM 98), IFIP/IEEE, Newark, DE, USA (1998), 195–206.

[32] P. Modi, H. Jung, M. Tambe, W. Shen and S. Kulkarni, *A dynamic distributed constraint satisfaction approach to resource allocation*, Principles and Practice of Constraint Programming – CP 2001, Paphos, Cyprus (2001).

[33] OASIS, *Reference model for service oriented architecture*, Committee Draft 1.0 (2006).

[34] OGC, ed., *Service Support*, IT Infrastructure Library, The Stationery Office (2000).

[35] OGC, ed., *Introduction to ITIL*, The Stationery Office (2005).

[36] OGC, ed., *Service Delivery*, IT Infrastructure Library, The Stationery Office (2001).

[37] OSS through Java Initiative, http://java.sun.com/products/oss/.

[38] A. Sahai, C. Pu, G. Jung, Q. Wu, W. Yan and G.S. Swint, *Towards automated deployment of built-to-order systems*, DSOM (2005), 109–120.

[39] M. Sailer, *Towards a service management information base*, Proceedings of the IBM PhD Student Symposium at ICSOC 2005, Amsterdam, The Netherlands (2005).

[40] V. Sarangan and J.-C. Chen, *Comparative study of protocols for dynamic service negotiation in the next-generation internet*, IEEE Communications Magazine **44** (3) (2006), 151–156.

[41] P.F.L. Scheffel and J. Strassner, *Handbook of Network and System Administration*, Elsevier, Amsterdam (2007), Chapter: IT Service Management.

[42] Service provisioning markup language, http://www.openspml.org/.

[43] S. Singhal, S. Graupner, A. Sahai, V. Machiraju and J. Pryne, *Quartermaster: A resource utility system*, Proceedings of the 9th IFIP/IEEE International Conference on Integrated Network Management (IM 2005), IFIP/IEEE, Nice, France (2005).

[44] Sun N1 software, http://www.sun.com/software/n1gridsystem/.

[45] SunTone Service Excellence model, www.suntone.org.

[46] G. Swint, G. Jung, C. Pu and A. Sahai, *Automated staging for built to order IT systems*, NOMS (2006).

[47] Telemanagement Forum, *New generation operations system and software*, http://www.tmforum.org.

[48] Telemanagement-Forum, *NGOSS compliance testing strategy – technical specification*, TMF 050 (2004), Version 4.1.

[49] Telemanagement-Forum, *The NGOSS lifecycle and methodology*, GB927 (2004), Version 1.1.

[50] Telemanagement-Forum, *The NGOSS technology – neutral architecture*, TMF 053 (2004), Version 3.1.

[51] Telemanagement-Forum, *Shared Information/Data (SID) model – concepts, principles, and domains*, GB922 (2003), Version 3.1.

[52] Telemanagement-Forum, *Telecom Operations Map (eTOM) – The Business Process Framework*, GB921 (2004), Version 4.0.

[53] Telemanagement-Forum, *Telecom Operations Map (eTOM) – The business process framework – eTOM-ITIL application note – Using eTOM to model the ITIL processes*, GB921L (2004), Version 4.0.

[54] Telemanagement-Forum, *eTOM – Public B2B Business Operations Map (BOM) application note C – An initial proposal for the scope and structure of ICT business transactions*, GB921C (2004), Version 4.0.

[55] Telemanagement-Forum, *Telecom Operations Map (eTOM) – The business process framework – Addendum F: Process flow examples*, GB921F (2004), Version 4.0.

[56] Telemanagement-Forum, *Telecom Operations Map (eTOM) – The business process framework – Addendum D: Process decompositions and descriptions*, GB921D (2004), Version 4.0.

# – 7.4 –

# IT Service Management

Peter F.L. Scheffel[1], John Strassner[2]

[1]*Verdonck, Klooster & Associates*
*and*
*Waterford Institute of Technology, Waterford, Ireland*
*E-mail: peter.scheffel@vka.nl*
[2]*Autonomic Computing, Motorola Labs, Schaumburg, IL, USA*
*E-mail: john.strassner@motorola.com*

**Abstract**

IT service management is required by both enterprises as well as service providers. There are several approaches that model how enterprise and service providers structure their work in order to define and provide services (such as backup and provisioning) that their users require. This chapter gives a brief overview of two internationally recognized standards for modeling business processes representing IT service management – ITIL (IT Infrastructure Library) and eTOM (enhanced Telecom Operations Map). These are two complementary approaches that each offer value in their own right.

ITIL is a de facto standard for IT/ICT production operations. It provides a comprehensive and consistent set of best practices for IT service management. It is now represented and managed by the itSMF (IT Service Management Forum, www.itsmf.com). The copyright for ITIL is owned by the UK Office of Government Commerce; however, it is a public domain set of best practices. ITIL comprises a fairly simple set of logical processes and a standardized vocabulary, whose main value is to define a common definition of services. The current release of ITIL is v3, which was just released; v2 is described in this chapter.

The eTOM Framework was originally developed in the early 1990s by a set of Service Providers (SPs) that wanted to standardize their business process models, in order to develop interoperable service and resource provisioning and deployment. This helped reduce the dependency that SPs had on individual vendors. The eTOM was first represented as the Telecom Operations Map (TOM), which concentrated on the description of real-time fulfillment, assurance and billing operations processes. This was expanded in the late 1990s to the eTOM, which took a more complete enterprise view of business process modeling. This enables suppliers and other constituencies to better understand SP behavior and requirements. It includes a

HANDBOOK OF NETWORK AND SYSTEM ADMINISTRATION
Edited by Jan Bergstra and Mark Burgess

hierarchy of process definitions that describe how the business should run. The current release of the eTOM is 6.0.

This chapter starts by defining IT services and how we can determine the importance of IT services in relation to their use. This is followed by an introduction to the three pillars of service management: the processes, the organization and the tools. Next, a generic representation for service management implementation is given. This chapter is finalized by a list of common misperceptions in the implementation of service management.

## 1. Terminology

The following is a set of essential definitions used in the IT service management area.

- A process describes a set of sequenced activities that deliver a specified result.
- A business process is a sequence of business activities undertaken by an organization that deliver a specified result.
- A functional process grouping (e.g., customer relationship management) is a set of related processes that involve similar knowledge.
- A hierarchical process decomposition is a systematic approach to modeling processes that define an ordered grouping of processes wherein the specificity of the process (task and description) is more detailed as the hierarchy is traversed downwards. This approach allows processes to be developed more modularly.
- Process elements are fine-grained processes that are used to assemble or aggregate end-to-end business processes. Process elements provide modularity for potential reuse, as well as the ability to model the independent update and/or replacement of a process.
- Resources represent physical and non-physical components used to construct services.
- Services represent units of functionality that are developed by an organization for use by a target constituency (e.g., end-user or administrator). The ITIL associates a service with a business process, while the eTOM defines a service as encapsulated within a product. This definition is neutral, and hence can be related to both standards.

## 2. Introduction to IT services and IT service management

The introduction of IT service management principles within the IT organization is gaining increasing importance in both the enterprise as well as the service provider communities. In particular, the management of the underlying processes that make up IT services enables the organization to represent complex management tasks in terms of simpler hierarchies and sequences of processes. It also enables services to be better understood, and supports the creation of reusable process definitions that can be applied to different services. This latter is very important, as it addresses inherent CAPEX and OPEX principles associated with service management and delivery. Hence, the scope of this chapter is on organizing the operations that an internal IT department performs, which are subsequently transformed into services delivered to its customers. This applies to whether the IT function is outsourced or implemented in house.

## 2.1. *What is an IT service?*

An IT service can take many forms, depending on what type of user the service is intended for. Services such as billing, trouble ticketing, and customer relationship management are used to charge customers for services used as well as to solve problems that customers encounter; functions such as inventory and configuration management are generally performed by the organization to support its internal services; functions such as data backup and recovery are performed both for the needs of the organization as well as on behalf of its customers.

The Shared Information and Data model (SID) is an object-oriented information model that can be used to define the terms used in the ITIL and eTOM service management functions [14]. It takes the following view of services:

- A CustomerFacingService (CFS) is a service that a customer is directly aware of (either through using, ordering, or being billed for the service).
- A ResourceFacingService (RFS) is a service that a customer is not directly aware of; rather, it is internal to the operation of the CFS.

For example, a customer can order different types of VPNs – therefore, a VPN is a CFS. Each type of VPN requires its own internal services to function (e.g., forwarding and routing of traffic). These internal services are in effect invisible to the customer, and hence are classified as RFSs. However, these services are required in order for the CFS to function correctly. This classification is being used by the SID team in order to map eTOM process descriptions into software objects that can be managed.

## 2.2. *Trends in IT service management*

Several trends from the last couple of years have made IT service management grow in importance:

- Organizations' primary processes are becoming more and more dependent on IT services.
- Organizations tend to work more in interdependent chains, concentrating on their own core competences. The current boom in IT outsourcing is an example of this trend. This means that these organizations rely more on information or IT services that are produced in other organizations. This implies a growing need for more reliable IT services.
- There is a higher need for accountability within organizations. Corporate governance laws, like the Sarbanes Oxley act in the United States, are one of the most important factors driving this.
- IT departments still have difficulties in proving their added value to the business side of organizations. They are usually seen as cost centers, and hence are the object of continued cost cutting measures (as opposed to profit centers, which are investment targets).
- IT departments have to cope with constant change in equipment, services demanded, and the constant rapid evolution in technology. These constant changes make it difficult to provide reliable, cost-efficient services.

- The emergence of e-Business (defined as how business partners interact with the aid of ICT technologies) is driving a new dependence on reliable and efficient ICT services.
- The new era of e-Business refers not just to buying and selling, but also customer service, customer satisfaction, and collaboration among business partners.

The above observations show an interesting dichotomy – on the one hand, IT technology is becoming more and more important; on the other hand, IT organizations have to struggle to show their value. One of the instruments for IT managers to bridge the gap between the demand of IT on the business side and their own supply side of IT services is called IT service management.

### 2.3. *What is IT service management*

IT service management has many definitions, depending on the context in which it is used. For the purposes of this chapter, IT service management is the set of methods, processes and techniques that are used to manage a set of services that meet the customer's requirements. Note that a 'customer' can be internal (e.g., an employee of an enterprise) or external (e.g., a subscriber to a service offering provided by a service provider).

Note that this definition of IT service management is rooted in the formal modeling of IT services and their management that are centered on the *customer's* perspective of IT's contribution to the business. As such, it is in direct contrast to technology-centered approaches to IT management and business interaction. This focus on *process* has an impressive lineage, including [1–8,11,12,16]. All of these approaches are independent of technology, and instead abstract technical concerns into a common set of processes.

ITIL and eTOM have enterprise and service provider orientations, and hence define a service differently. In ITIL, a service is defined as "one or more IT systems which enable a business process". In the eTOM, a service is defined as: "Services are developed by a Service Provider for sale within Products. The same service may be included in multiple products, packaged differently, with different pricing, etc. A telecommunication service is a set of independent functions that are an integral part of one or more business processes. This functional set consists of the hardware and software components as well as the underlying communications medium. The Customer sees all of these components as an amalgamated unit".

While these definitions appear to be different, they can be harmonized by realizing that business processes are used to define services. To see this, let us examine the concept of a Service Level Agreement (SLA).

An SLA is a formal negotiated agreement between two parties that defines services that will be provided by the provider to the customer including metrics for measuring services and service quality, tolerances allowed for a service to operate within, liabilities of the provider and customer, actions to be taken in specific scenarios, and penalties if the SLA terms are violated. In this way, the SLA defines what services are to be delivered, within what performance or other tolerances, and what happens if those services are not delivered. In addition, SLAs can cover many aspects of the relationship between the customer and the provider, such as service performance, customer care, billing, provisioning, and reporting.

Given this definition of SLAs, we can equate the ITIL and eTOM definitions as follows:

- ITIL defines the set of business processes that are required to support each service in the SLA.
- eTOM defines the relationship between the provider and the customer for each service in the SLA by defining the business processes·that they interact with.

In ITIL, the SLA forms the starting point for the service management process, in which the customer may change the volume or definition of the necessary services; the supply side engineers its infrastructure and organization to meet the new SLA requirements, as well as delivers and reports on the services; the demand side receives and evaluates the services.

In the eTOM, this is not true. The eTOM considers an SLA no more or less important than other entities, such as customer care, provisioning, and so forth. More specifically, the TeleManagement Forum (TMF) defines an SLA as "a formal negotiated agreement between two parties, sometimes called a Service Level Guarantee. It is a contract (or part of one) that exists between the Service Provider and the Customer, designed to create a common understanding about services, priorities, responsibilities, etc." [13]. Note that the eTOM's use of hierarchical process descriptions enables common tasks and activities to be abstracted into a set of reusable processes; these processes can then be instrumented, realized, and managed in a common way.

The eTOM mapping of an SLA is very different than that of ITIL, mainly because of the way the eTOM is organized and its ultimate goals. The eTOM defines three types of processes: enterprise, operations (e.g., fulfillment, assurance and billing), and strategy, infrastructure, and product. The eTOM sees a service such as email as *more* than a service – it is in fact a packaging of functionality (as a Product) that contains resource (e.g., connectivity) and service (the email itself) dependencies. Thus, e-mail has many different aspects, and those aspects exist concurrently across the enterprise, operations, and strategy-infrastructure-product groupings of the eTOM.

The recommendation of this handbook is to use ITIL best practice specifications to define the overall objectives of an organization, with the more specific eTOM process elements used to fill in the details.

It is evident that a certain service level comes at a certain cost. Upgrading service desk opening hours from business hours to $7 \times 24$ will cause higher personnel costs. Upgrading the storage space per use will cause the need for more storage space, inducing higher costs. In order for a service organization to deliver services in an agile manner, that organization must have a good understanding of how services are delivered and managed. Arguably, the best way to do this is to use a proven model that is extensible, so that it can be mapped to the application-specific needs of an organization. This is what ITIL and eTOM deliver.

**2.4.** *Evolution of IT from technology to services*

From the earlier e-mail service example, it has become clear that customers do not necessarily demand a certain technological solution or specific software application. Rather, they demand certain functionality, and do not care what technology is used as long as it does what it was supposed to do, within the boundaries of the agreed costs.

More importantly, functionality is now being related to added value, defined in terms of the entire customer experience. Customers want personalized services, one-stop shopping, and in general, the empowerment gained from self-service. This requires information-centric business designs to be developed, which in turn drives the development of formalized business process models. This enables a better integration of the supply chain and a unified approach to order entry, fulfillment and delivery, independent of technology and vendors. Formal business process models are required because this integration requires the ability to share and reuse data among suppliers and partners (as well as different parts of an organization).

Put another way, what is important is not technology, but the ability to better manage customer relationships and retain customers. This in turn requires: (1) the ability to provision and activate a broader range of products and services to customers, and (2) the reduction of costs and increase of productivity gains. Both of these require better collaboration between and integration of processes.

## 3. Elements of service management

After defining and investigating the importance and the role of the IT services for an organization, the IT services need to be defined, implemented, tested and managed. ITIL and eTOM are both frameworks that support these and other processes. Both require work in the following areas:

- The management of the infrastructure, delivering the core service. This is outside of the scope of this chapter and is discussed extensively in other parts of this Handbook.
- The management processes for defining, creating, modifying, maintaining and removing services, both for end-users as well as for internal services required for end-users (e.g., a VPN is an end-user service; BGP-based route advertisements are an internal service that the end-user is not directly aware of, but are required for the VPN).
- The set of business processes necessary to provide and maintain a service (e.g., how help is provided to the customer).
- The management processes for defining, creating, modifying and managing the business processes necessary to provide a service.
- The roles and permissions played by the organizational structure (functions, tasks, responsibilities, authorizations) with respect to service management.
- The supporting tools (e.g., administrative, planning and measurement tools).

### 3.1. *Service management: service, business and management processes*

There are several reasons for using standardized models to design service management processes. These include the ability to:

- define and represent dependencies between processes to ensure their correct and optimal deployment,
- design the automation of business processes,

- translate between different constituencies (e.g., a business analyst and a programmer). Models can be built that provide different layers of abstraction, making them good communication tools for this purpose. This requires formal tools and skilled engineers to construct the model. The SID is an example of such a model. Note, however, that such abstractions are derived in the SID from two different sources: (1) the NGOSS 'views' (which defines such constituencies [15]), and (2) classification theory (which governs how the model is built),

- the use of standardized models enables clear communication of features and behavior to vendors, enabling effective selection and implementation of supporting systems. For example, an increasing number of service providers are now issuing request for quotations that explicitly require eTOM and SID compliance.

Since there are many publications and books on ITIL and eTOM, we will therefore not attempt to give detailed descriptions of them. Rather, we will provide an overview of their approach, as well as their similarities and differences. For more information, the reader is kindly referred to the references.

**3.1.1.** *ITIL*   ITIL was initially designed and developed in the 1980s by the British government's Central Computing and Telecommunications Agency (CCTA), now the Office of Government Commerce (OGC). ITIL is a process framework for the delivery and support of IT services.

ITIL is comprised of a set of best practices, bundled as process descriptions. Each process was published in a book – hence the term library. ITIL is under constant revision. The next revision will take place in 2006, totally revising the current processes, with a focus on standardizing process descriptions and bringing them to the same level.

NOTE. This is a common problem with business process descriptions, and the eTOM suffers from it as well. The reason that it is so difficult to standardize is because it is very hard to get agreement from all stakeholders on what a given term, let alone process, means for different organizations. For example, every service provider performs inventory management. However, the means by which inventory management is performed, and the result of the process, varies significantly between vendor and service provider. Furthermore, enterprises tend to focus on different aspects of inventory management. As we will see, there is also a new effort to better synchronize the functionality of ITIL and eTOM, which will be discussed in Section 3.1.6.

ITIL is globally supported by many translations of the basic books and supplementary publications. The IT Service Management Forum (itSMF) is one of the organizations that promote the use of ITIL worldwide. ITIL is a public domain description, meaning that anybody van buy the books and start implement service management according to ITIL, without breaching OGC rights. ITIL is widely supported by training courses, and accreditation schemes. Furthermore, there are several software tools which support the administration of the ITIL processes. On the highest level, the processes are depicted in the following reference model (see Figure 1). The heart of ITIL is formed by the service support set and the service delivery set. These processes can be seen as the most widely used and implemented processes. The service delivery set is shown in more detail in Figure 2.

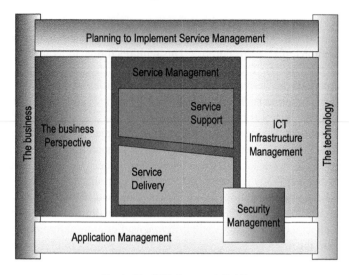

Fig. 1. The ITIL framework [9,10].

The purpose of the Availability Management process is to evaluate the current demands on the availability of the Services (whether or not an SLA is involved), and consequently the ability of the organization to achieve its business goals. It adjusts Services based on availability demands, initiates new designs to achieve availability goals, and maintains, also addresses Service Delivery through the monitoring of the Service, and evaluates whether the introduction of new technology will improve the delivery of the Service.

The Financial Management for IT Services process is focused on the budgeting, accounting and billing of Services, whether these services are internal to the organization and/or provided by external providers.

The Service Continuity Management process is responsible for ensuring that the required Services (including infrastructure, telecommunications, information, applications and support) can be restored in the required time to ensure business continuity. It also includes risk assessment and management, as well as contingency planning.

The service support set is shown in more detail in Figure 3. The Service Desk is not a process, but rather an organizational unit. It provides a single point of contact between the service customer/user and the other processes within the Service Management framework.

The purpose of the Incident Management process is to respond to and manage service requests (e.g., new service requests, and changes to existing service requests) as well as incidents (e.g., outages and disruptions to service). An incident is any event that adversely affects the required service levels as defined by a relevant SLA. The aim of Incident Management is to return service delivery to the required service levels as quickly as possible while minimizing any negative impact upon business operations. Hence, it also is responsible for detecting and categorizing incidents.

The Problem Management process is made up of two complementary parts – proactive and reactive Problem Management. The former addresses the detection and resolution of problems before they escalate into formal incidents. The latter focuses on responding

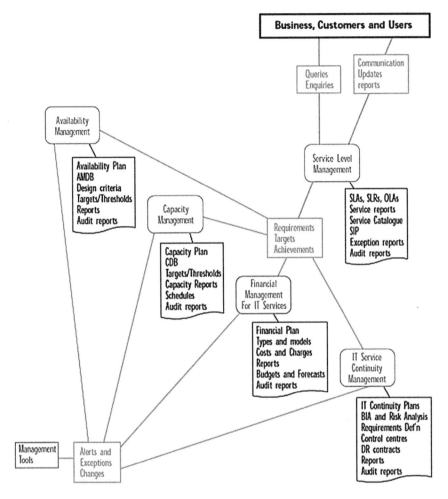

Fig. 2.  ITIL service delivery diagram.

incidents as they occur, again providing problem resolution. The Problem Management process is also responsible for root cause analysis (i.e., the classification and identification of a problem, as well as its associated resolution, which is also called a known Error) for short- as well as long-term issues. Hence, there are two stages to the Problem Management process: (1) Problem Control involves investigating and establishing a root cause of a problem and providing a resolution, and (2) Error Control investigates and eliminates known errors so that they do not reoccur.

The Change Management process is responsible for implementing changes to both Services and Infrastructure in order to provide a benefit to the business. This process provides a structured method for the identification, management, and approval of all changes. This approach allows the resources required to implement a change be directly related to the

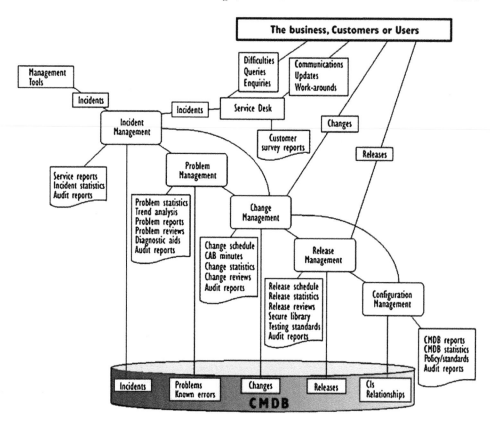

Fig. 3.  ITIL service support diagram.

business impact of a change (i.e., address both benefits and incidents that may arise because of the change).

The Release Management process is an output of the Change Management process, and formalizes the implementation of the change. This process is concerned with all types of service and/or system changes, and takes a wider view of a change, so that all aspects of a change are considered. The Release Management process is also responsible for the Definitive Software Library (DSL) and Definitive Hardware Store (DHS) within the CMDB.

The Configuration Management process is the means by which all the other Service Management Processes are implemented. This process identifies, controls, and reports on all Configuration Items (CIs) within an organization and the relationship between them in the Configuration Management Database (CMDB), and for maintaining the integrity of and updating the CMDB.

**3.1.2.** *eTOM*   The eTOM model is one of the major deliverables of the TeleManagement Forum (TMF). The TMF was founded in 1988 as an international consortium of communications service providers and their suppliers. It aims to help service providers and network operators in automating their business processes effectively. To do this, the TMF uses a business- and customer-driven approach to achieve end-to-end automation, using Commercial Off The Shelf (COTS) software.

The flagship program of the TMF, NGOSS, is shown in Figure 4.

The NGOSS (Next Generation Operational Systems and Software) initiative, is a collection of programs that act as an integrated document suite that spans how NGOSS components and systems are defined, created and built. This consists of a large number of specifications, with associated UML models, and formalizes automation.

As can be seen, the NGOSS is anchored around four key programs – the eTOM, SID, TNA and Compliance programs. The eTOM provides a set of business process descriptions; the SID defines an object-oriented set of class definitions for managed entities in an environment; the TNA defines a distributed, service-oriented architecture for defining, creating, and managing NGOSS components; the Compliance program ensures that vendor contributions comply with the principles of the eTOM, SID and TNA. A high-level overview of the model (showing its level 1 processes) is shown in Figure 5.

NOTE. it is important to understand that the eTOM is not defined in isolation, but rather as a part of an integrated suite of specifications that define how to build NGOSS components that automate business processes. As such, it is no more or less important than the SID (which enables the eTOM process descriptions to be built as software objects) or the TNA (which describes how to build an NGOSS component or system).

The first version of the model was released in 1996 as the SMART model, to be followed by the first publication of the TOM model in 1998. The enhanced TOM, or eTOM model, was first released in 2001. Its current version is 6.0, and it has been standardized by the ITU-T as M.3050. The eTOM model is used by a large number of major service providers worldwide as a reference model.

The eTOM is a reference framework for categorizing the set of business activities that a service provider uses. This is done by first, categorizing the set of business activities that a service provider performs, and second, organizing these processes as a set of activities that facilitates their combination into a set of customizable tasks that an organization can map to their application-specific needs. This is in turn implemented as taxonomy of components of each business process, which are termed process elements. Each process element can be further decomposed to expose appropriate detail, and positioned within a model to describe appropriate relationships (e.g., organizational and functional relationships) that make up process flows of the business.

One of the major goals of the eTOM is to serve as a blueprint for standardizing and categorizing business activities that will in turn be used in the definition and design of Business and Operations Support Systems (BSS and OSS, respectively).

The eTOM itself is divided into three main areas: (1) Operations, (2) Strategy, Infrastructure and Product (SIP), and (3) the Enterprise.

The eTOM splits Service Management into two areas: Service Management and Operations (SM&O) and Service Development and Management (SD&O). SM&O focuses on service delivery and management, and defines Service Management as a grouping of functionality "necessary for the management and operations of communication and information

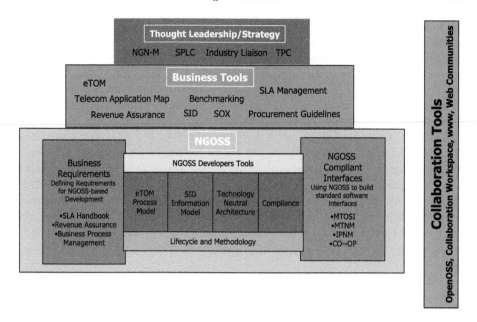

Fig. 4.  An overview of the TMF NGOSS program.

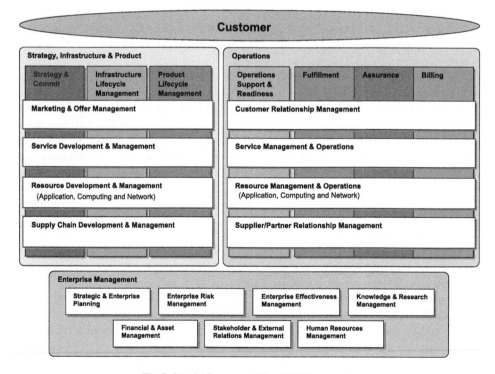

Fig. 5.  Level 1 Processes of the eTOM framework.

services required by or proposed to customers". This includes tactical (e.g., near-term capacity planning, and other functions that involve day-to-day operations and support) and strategic (e.g., planning and developing future services, and lifecycle processes to improve end-user experience). In contrast, SD&O is focused on the management of underlying network services and information technology, including service design and creation.

The Operations process area includes processes that support customer and network operations and management, as well as those that interact directly with the customer. It is made up of four vertical and four horizontal process groupings. The Operations Support and Readiness (OSR) grouping is responsible for providing management, logistics, operational readiness, and administrative support to the Fulfillment, Assurance and Billing (FAB) process groupings. The FAB process groupings focus on Customer Operations processes. The Fulfillment process groupings is responsible for providing customers with their requested products and services. The Assurance process groupings focus on executing proactive and reactive maintenance functions to ensure that products and services provided to customers are available and performing to their SLA requirements. The Billing process grouping is responsible for collecting revenue and resolving billing problems to the customer's satisfaction in a timely manner. The Customer Relationship Management process grouping is focused on customer service and support, retention management, cross-selling, up-selling and direct marketing. The Service Management and Operations process grouping focuses on the delivery and management of communications and information services required by customers. The Resource Management and Operations process grouping focuses on managing application, computing and network infrastructures utilized to deliver and support services required by customers. Finally, the Supplier/Partner Relationship Management process grouping supports the core operational processes of both the customer as well as suppliers and partners that involve FAB operations processes.

The SIP process area focuses on processes that develop strategies to support the development, deployment, and management of infrastructure, products, as well as the development and management of the Supply Chain. In the eTOM framework, infrastructure refers to more than just the network infrastructure that directly supports products and services – significantly, it also includes the operational and organizational infrastructure required to support marketing, sales, service and supply chain processes (e.g., Customer Relationship Management). Hence, these processes direct and enable processes within the Operations process area. The Strategy and Commit process grouping is responsible for supporting the lifecycle processes of infrastructure and products, including the business commitment to support the four horizontal process groupings. The Infrastructure Lifecycle Management process grouping is responsible for the definition, planning and implementation of all necessary application, computing and network infrastructures, and are closely related to the needs of the Product Lifecycle Management vertical end–end processes. The Product Lifecycle Management process groupings are responsible for the definition, planning, design and implementation of all products in the enterprise's portfolio, including customer satisfaction and profit/loss aspects. Note that the SID defines a Product as a grouping of resources and services. The Marketing and Offer Management process groupings focus on the knowledge of running and developing the core business for a service provider, including defining and developing new products, managing existing products and implementing marketing and offering strategies. The Service Development and Management process grouping focuses on planning, developing and delivering services to the Operations domain. It includes processes necessary for defining the strategies for service creation and design, managing existing services, and ensuring that capabilities are in place to meet future service demand. Similarly, the Resource Development and Management process grouping focuses on planning, developing and delivering the resources needed to support services and products to the Operations domain. It includes processes necessary for defining the strategies for development of the network and other physical and non-physical resources, introduction of new technologies and interworking with existing ones, managing existing resources and ensuring that capabilities are in place to meet future service needs. Finally, the Supply Chain Development and Management process grouping focuses on the interactions required with suppliers and partners that are part of the supply chain. These processes include establishing and maintaining all the information flows, managing any mediation required, and financial flows between the provider and supplier.

The Enterprise Management process area includes business processes that are required to run and manage a business. These generic processes focus on both setting and achieving strategic corporate goals and objectives as well as providing those support services that are required throughout an enterprise (e.g., Financial Management, Human Resources Management, and so forth). In general, Enterprise Management processes interface as needed with almost every other process in the Operations and SIP areas.

The eTOM was also developed specifically to address the needs of external interactions with a service provider (e.g., automated B2B and B2C transactions, call centers, and web portals). This is supported by linking the SIP processes to buy and sell operations; the combination of SIP and Operations represent how the eTOM as a whole interacts with the external environment.

**3.1.3.** *E-TOM versus ITIL*   The following differences and similarities can be identified between eTOM and ITIL:

- *Industry Background*. The eTOM originates from the telecom industry, whereas ITIL originates from the IT industry. Although both industries have many similarities (e.g., technology oriented), there are some differences. ITIL originates from the IT department and has a similar, but more limited, scope. The eTOM functions as an overall business framework. Specifically, it describes all business processes required by a Service Provider, and analyzes them at different levels of detail according to their significance and priority for the business. Hence, it acts more as a 'blueprint' for process direction and provides a neutral reference point for internal process reengineering needs, partnerships, alliances, and general working agreements with other enterprises. In contrast, it is important to remember that the eTOM is a part of the overall NGOSS initiative. Hence, its business process descriptions are linked to both a shared information and data model (which was specifically built to support programming) as well as a service-oriented, technology neutral architecture. This is a critical point; for example, the eTOM framework can be used as a tool for analyzing an organization's existing processes and for streamlining them and/or developing new processes. This feeds naturally into the SID (to implement those changes in code) and the TNA (to build a better, improved architecture to support these needs). This is described more in the Section 3.1.6, which considers how to use the best of both approaches.

Table 1

| Comparison item | eTOM | ITIL |
|---|---|---|
| Context | Complete enterprise and interaction with suppliers and partners for Service Providers | ICT Service Management within an Enterprise |
| Cohesiveness | Business view of NGOSS | Best practices |
| Preciseness | Common language for describing business processes; definitions are organized as a hierarchy, but not in the rigid formality of a true specification | Common language for describing service management; definitions are organized as a hierarchy, but not in the rigid formality of a true specification |
| Coverage | The entire enterprise | ITIL maps to eTOM's Operations Assurance, and touches aspects of fulfillment and billing |
| Process flows | Informal examples of process flows are provided; detailed flow diagrams are provided to illustrate end-to-end processes | High-level flow charts are provided |
| Technical content | Very mature; increasing emphasis on application and usage | Relatively mature; major rewrite coming in 2006 |

- *Global Acceptance*. eTOM is a worldwide industry standard applied by telecom service providers (it was these service providers who contributed to the development of the eTOM). The ITU-T has published a previous version of the eTOM as an international standard (M.3050). ITIL has a long history of use in the United Kingdom and the Netherlands but has a fast growing interest in the US, Germany, Japan, Australia and other countries.
- *Scope*. The eTOM provides a business process framework that describes processes and process interactions between different organizational parties at varying levels of abstraction. The eTOM models the buy and sell functions using the Strategy–Infrastructure–Products (SIP) domain, supported by appropriate processes in the Operations and Enterprise domains; it is this combination which enables the eTOM to be easily customized. ITIL, on the other hand, is more applicable for the operational organization and provides less means to optimize the interaction between the service provider and other organizational parts.
- *Approach*. Table 1 summarizes the approaches of eTOM and ITIL.

**3.1.4.** *Bridging the Gap between eTOM and ITIL*   When both eTOM and ITIL processes are drilled down to the lowest level (activities), one-to-one mappings for identical processes/activities can be made. A preliminary mapping of ITIL processes to level 2 eTOM processes has been completed in the TMF. Since the eTOM is a framework, it is much easier to map ITIL onto eTOM than the other way around.

**3.1.5.** *Which model to choose*   The bottom line is that the choice of model to use when planning to implement service management is determined by context and your application-specific needs. Both models that were presented deal with IT services at different levels of abstraction – hence, one is not better than the other. The choice mainly depends on the industry an organization is in, the main objectives of the services and the common language within the organization and between the different stakeholders.

It has proven valuable to introduce eTOM principles into pure IT organizations, because the model has such a strong customer focus. On the other hand, it has proven valuable and practical for telecom operators to start using ITIL processes, since parts of their services tend to look like normal IT services. In general, one can say that the eTOM model will fit organizations whose primary process is IT, like telecom operators or service providers, since the model has more explicit references to customer and marketing processes, which are lacking in most commonly used ITIL processes.

When choosing a model, the most important thing is to use it coherently throughout the organization, to make sure all the benefits of using a standard model are realized.

**3.1.6.** *A third possibility – use both!*    A third approach is, of course, to use the best features of both. The TMF is working on an approach in which the a combined eTOM and ITIL process approach is used. In this approach, ITIL is mapped onto the eTOM process framework to provide additional detail not covered in the original eTOM framework. This can deliver a number of business related advantages:

- Enables different constituencies to describe processes and process flows in a common terminology.
- Enables the customization of services to particular customers, customer segments and organizational business requirements.
- Enables customer satisfaction, loyalty and retention processes to be improved by ensuring that customers, providers, and suppliers/partners are all speaking the same process language.

## 3.2. *Organization*

The service organization is responsible for translating the service demands by its users into services. These services are divided into external (customer-facing) and internal (infrastructure-facing) services. The relationship of the service organization to the rest of the organization controls how these services are used, what priority they are given, and how customers (internal and/or external) are served. There are three basic models to do this:

- A centralized service organization is one in which the service organization is responsible for managing services to all primary business units, customers, or both. There may be equipment which is located at user premises; however, it is still managed by the service organization, and activities on it are coordinated centrally.
- A decentralized service organization is one in which every primary business unit, whether internal to the enterprise or external and supporting customers, is fully self-supporting and self-reliant for services that it produces, and thus is responsible for implementing its own service organization.
- A hybrid service organization is one in which both centralized and decentralized service organizations are combined. There is a centralized service organization that issues high-level guidelines and principles that govern all local service units, and there

are localized service units that translate these and tailor them to the specific needs of their organizations.

It may be clear that the choice for a certain model has different advantages and disadvantages in terms of economies of scale, career opportunities, overhead costs, being able to react to different business priorities, re-use and sharing of best practices and the influence of one individual business unit on IT services and service levels. The choice depends heavily on the size, strategy and business of the organization as a whole, and the importance of information for (parts of) the business. This choice is generally made on corporate level.

After the choice of what type of service organization to implement is determined, the internal organization of the service provider has to take place. A perfect way to organize the service management function does not exist – rather, this organization must be tailored to meet the specific requirements of its business. There are several ways to organize the internal organization:

- By function: this method is structured to reflect different functional roles, and hence is suitable for a more mature organization that performs the same functions. This is because of the rigidity of the organization, which tends to produce 'organizational walls' that make it slow to adapt to new needs. In addition, a service is usually built from elements of all departments, resulting in a lack of overall responsibility for the service. This in turn makes it difficult to prioritize and coordinate inter-department activities.
- By process: this method is structured to group personnel by the types of processes that they support. This provides a quick customer response; however, it is difficult to maintain specialized knowledge, since there has to be functional knowledge of every technical element in each and every department.
- By matrix: this method combines the advantages of both previous types. However, it has the risk of not being able to agree on how to assign resources between the functional and process organizations.

## 3.3. *Tooling*

Service management processes require different types of tooling. Figure 6 shows a high-level overview of some of the different functions and information sets that are used within the service management domain. The white colored elements are different types of on-line infrastructure status reporting tools, directly feeding the incident management process. The orange colored elements are different types of off-line information collection tools, gathering trend information for longer-term tactical processes. These tools form the basis of service level reporting. The green colored elements are different types of project support and direct customer interface functions. This figure is not complete (e.g., ancillary support functions, such as back-up and restore, configuration of services, and software distribution are not shown), but it does cover the major functions required for service management.

The ultimate goal of service management support tools is to deliver an integrated picture of the infrastructure supplying the service and how it connects to the equipment providing the service to the end-user. This provides the service organization insight into the actual status of the services.

Since an IT infrastructure is typically built up of many different types of elements by different vendors with no standardized interfaces, the first problem encountered is defining a common vocabulary, with a common set of meanings (including synonyms!) that enables the data from the different infrastructural elements to be aligned and harmonized with each other. This is best served through the formal introduction of a business process framework, like the eTOM. Once this is done, then the organization can working on getting the right information into the right processes.

There are different solutions on the market that support one or many of the service management functions shown in Figure 6. Tools have usually either grown from the infrastructure up, or from the processes down, depending on the vendor. Both types of tools are rarely truly integrated. This makes the selection of support tools a crucial task in the implementation of service management principles.

The choice of the service management tooling is heavily dependant on the depth of the agreed service levels, the underlying technology and the defined processes. In the end, the tooling dictates what can be detected, what can be reported and what can be registered. In selecting support tools, the answers to the following questions are relevant:

- What are your most important processes and what is the priority of each;

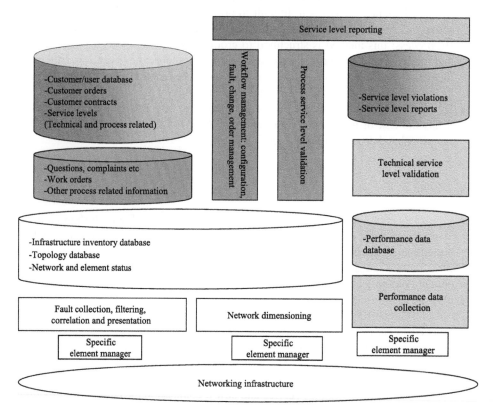

Fig. 6. High-level support system functions.

- Do you want an integrated product suite, or do you want to use best-of-breed solutions;
- Which tasks/information do you need to define, implement and/or automate;
- Who needs a function or specific information, how frequently, why, when, and how reliable should the information be;
- What is your budget.

### 4. Phasing the implementation of service level management

This section describes the main phases of implementing service level management. At the highest level, the approach that is presented is generic and can be applied to any service and/or product development project. The following phases are recognized:

- *Justification phase*, where the service development is analyzed at a global level, the impact on systems, processes and organization is determined, and a high-level GO/NO-GO decision for the next phase is prepared. The aim of the justification phase is to validate whether the *complete* organizational unit (board, business units, service organization) will accept service management principles as the means for interaction and the required policies are well defined or adjusted to facilitate the new way-of-working.
- *Implementation phase*, where all necessary steps are taken within the service provider and customer domain to implement service level management. As such, it includes the selection of Service Level Management (SLM) tools, or other tools that include SLM functions, integration, test; this selection then helps define the set of process design, implementation and test functions for all processes. The process implementation includes the mapping onto the organization and (optional) changes in the organization. The result is a GO/NO-GO decision to enter the operational phase.
- *Operational phase*, where all processes are fully operational and controlled by the service organization. During this phase, the service provider will report on the performance of provided services, such as compliance to all SLAs that define a given service. At the same time, the service provider and customer discuss change requests initiated by the customer for which the service provider needs to perform an impact analysis.

Figure 7 depicts the overall phasing graphically.

### 4.1. *Phase 1: justification phase*

The justification phase has different objectives. The aim of the justification phase is to determine whether the introduction of service level management within the organization can be financially justified. It implies that a complete project plan is developed that guides the introduction of service management within the organization. This includes:

- What are the objectives and scope of the project, and what are the added value and the benefits for the organization?

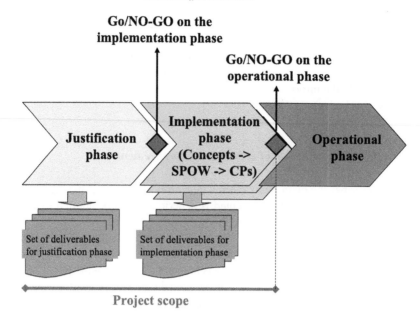

Fig. 7. High-level phasing of service level management.

- What prerequisites does the project impose on the organization as a whole (for example, the implementation of service management usually introduces more formalized procedures, which may cause users to see this as a degradation of services, an annoyance, or both).
- What does the introduction of service level management mean for existing processes, and their mapping onto the organization?
- What systems should be available to support the service level management process?
- Should the above mapping be redesigned to be made more efficient?
- What investments are required and what are the tangible and intangible benefits?
- What is the project organization, which organization is managing the project, what organizational parts are involved and what is the role of each organizational unit?
- How will the project transfer the end results to the service organization (e.g., is there an acceptance procedure)?
- What is the impact on the line organization?
- What is the timeline and sequence of activities (global planning)?
- What are the risks for this project, how could the risks be managed properly?

The introduction of service management should be considered as a highly complex project. This is due to a considerable change in the architecture delivering the services, which in turn makes the processes between customer and service provider change dramatically. While both the service organization as well as the customer *can* benefit from these changes, it must be understood that success is dependent on both embracing the new formal service management processes. In addition, both must realize that benefits may not be immediate. Hence, the underlying crucial issue for this phase is: *does the organization as a whole support the concepts of service management and accept the resulting processes as the means to achieve better service design and delivery to its customers?*

The justification phase should also provide the following deliverables:

- Clear description on the actual services, what is meant with service level management and what is not (defining the exact boundaries of the project).
- Global process design (at least at the organizational level detailing responsibilities for service provider and customer).
- Global systems design including technical feasibility. Are the systems that are required for implementation of service management available, is it proven technology, and can it be integrated in the organizations IT environment?
- How does service level management fit within existing architectures, both for process as well as systems? If existing architecture documents are not sufficient, additions should be described (including options and recommendations).
- Impact analyzes, both for processes and systems, what are the changes in these areas and how could these be implemented (high level).
- Global test plan, again covering both the technical as well as the process aspect.
- Criteria for GO/NO-GO decision.

The previous list of deliverables combined with the detailed project plan should provide a decision making unit the ability to make a conscious GO/NO-GO decision. As such, the justification phase implicitly requests budget allocation for the implementation phase.


**4.2.** *Phase 2: implementation phase*

The implementation phase is guided by the project plan that has been developed during the previous justification phase. As such, it includes the process implementation (design, implementation and test for all service management processes) and system implementation (selection, integration, test). Specifically, it should provide the following high-level deliverables:

- *Agreed service level agreements for all services.* It is recommended to use a roadmap for SLA parameters, as opposed to monitoring and reporting all parameters from the beginning. This allows for a growth scenario between the service provider and its customers and limits the complexity of the process changes that are to be implemented;
- *Agreed and implemented processes for service level management*, both at the service organization as well as in the customer domain. This includes changes in the direct operational processes (e.g., network and service management within a service provider environment, and service level/contract management within the customer domain) and management processes of deliverablesm, such as SLAs. The deliverable includes the transfer to the line organization (i.e., the project will not be involved in the execution of any process);
- *Implemented systems for service level management.* These systems should support the service management processes described in the previous item and be properly integrated into the existing systems architecture for the IT services.

There are several implementation strategies, which can be presented as two different alternatives:

- A 'big bang' scenario, in which the organization adopts a new way of working in a short period of time. This strategy has some practical disadvantages. Because everything has to be perfectly aligned for this type of scenario to succeed, the design phase might take a long time, causing the energy in the organization to decrease, thus losing momentum and focus for the implementation. The change is an organizational change, which is usually not done immediately. The learning curve might be too steep, causing the organization to not achieve the expected service levels. This could lead to customer complaints that will need to be proactively managed.
- A phased approach that, when executed properly, can take away the disadvantages of the big bang approach. In the phased approach, priorities are set according to the most important services. The phased approach uses a specified number of iterations to implement all aspects of the services, over a longer period of time.

The main message of this section is that the implementation of service management is not to be taken lightly. It is a task that can only be successful when all stakeholders are involved and an integrated approach for processes, organization, tools and IT is taken.

### 4.3. *Phase 3: operational phase*

During this phase, all service management processes are fully operational and are controlled by the service organization. The project has been finished. The service provider will report the performance of its services, and the customer will verify the correctness of these reports. At the same time, the service provider and customer will discuss change requests initiated from the customer for which the service provider needs to perform an impact analysis.

The project is finished, but now the game of service management has just started. The current IT environment and customer demands are continuously changing. Speed of response to new demands is of paramount important. The true promise of service management is that it will be able to accommodate continuous improvements and enhancements to its service offerings, whether these are based on cost, performance, or some other set of metrics, in a systematic manner. An organization that has embraced service management is on a journey of continuous improvement of services.

### 4.4. *A final note on ITIL and eTOM co-existence*

Since ITIL is a set of best current practices that are non-prescriptive, the eTOM can be used *in conjunction with* appropriate ITIL business processes to provide additional detail. This is what is being done in the TMF. Note that the SID (Shared Information and Data) [14] model is used to define an object-oriented class hierarchy that maps these business process descriptions into software. In addition, the TNA (Technology Neutral Architecture) [15] defines a distributed architecture that is independent of any one specific language, platform, or protocol to define how different components share and reuse information, as well as communicate with each other. In particular, the concept of a *contract* is very useful to ensure that interoperable communications between components take place. ITIL lacks association to the SID and TNA (or their equivalents); hence, realization of ITIL best practices without links to the eTOM may be more difficult than anticipated.

## 5. Common misperceptions of service management

There are different misperceptions which, when not managed properly, may cause a failure in the implementation of the service management process and thus cause the spillage of valuable time, money and resources:

- Implementing service management is an IT department function. Wrong: without the involvement of the customer, it will not work.
- Send everybody to an ITIL course, then everything will just fall into place. Wrong. ITIL is a set of best common practices. To make it work for you, you have to pick the parts that suit you and apply them specifically to your own situation. In addition, ITIL is not the best approach for service providers.
- OK, send everyone to an eTOM course. Wrong. The eTOM is a comprehensive framework of business process descriptions, but also needs customization to suit the needs of your organization.
- Just write down the processes. Everybody will eventually start working in accordance with them. Wrong: while documenting processes is always good, without any training on how business processes are used, along with how your specific processes can be customized to fit your needs and applications, you will never realize the benefits of service management.
- Why don't we buy one of these new tools, such as incident management? Wrong: without a clear view on what you want to do with it, just buying a tool is useless. This applies to all tools, not just incident management.
- We've bought the ITIL or eTOM compliant tool, so let's just install it and get going. Wrong: Tools are compliant on the processes, but normally need extensive configuration and parameterization (if not just plain programming) to manage your information in an ITIL- and/or eTOM-compliant fashion.
- I've implemented my processes and tools. I have defined my Infrastructure and organization. Now I'm ready. Wrong: You have just started. One of the principles of service management is the process of continuous improvement.
- Implementing service management is hard work, but it pays off: RIGHT.

## References

[1] Business Process Management, http://www.emeraldinsight.com/info/journals/bpmj/bpmj.jsp.
[2] Capability Maturity Model, http://www.sei.cmu.edu/cmmi/, http://www.itmweb.com/f051098.htm.
[3] M.B. Chrissis, M. Konrad and S. Shrum, *CMMI: Guidelines for Process Integration and Product Improvement*, Pearson Education, ISBN 0321154967.
[4] K. Harrison-Broninski, *Human Interactions: The Heart and Soul of Business Process Management*, ISBN 0929652444.
[5] M. Harry and R. Schroeder, *Six Sigma*, Random House (2000), ISBN 0385494378.
[6] M. Havey, *Essential Business Process Modelling*, ISBN 0596008430.

[7]  W. Humphrey, *Managing the Software Process*, Addison–Wesley, ISBN 0201180952.

[8]  M. Murugappan and G. Keeni, *Quality improvement – the six sigma way*, The first Asia-Pacific Conference on Quality Software (2000).

[9]  OGC, *ITIL Best Practice for Service Delivery. Introduction to and guidance on the implementation of ITIL Service Delivery Processes*, OGC, London (2000).

[10] OGC, *ITIL Best Practice for Service Support. Introduction to and guidance on the implementation of ITIL Service Support Processes*, OGC, London (2000).

[11] Six  Sigma,  http://www.motorola.com/content.jsp?globalObjectId=3079,  http://www.motorola.com/content.jsp?globalObjectId=3088, http://www.ge.com/en/company/companyinfo/quality/whatis.htm.

[12] H. Smith and P. Fingar, *Business Process Management: The Third Wave*, ISBN 0929652339.

[13] TeleManagement Forum, *Enhanced Telecom Operations Map (eTOM), The Business Process Framework for the information and communications services industry, GB921, TMF approved version 6.0* (2005).

[14] TeleManagement Forum, *Shared Information and Data model (SID), GB922 (and addenda), version 6.0* (2005).

[15] TeleManagement Forum, *Technology Neutral Architecture (TNA), TMF053 (and addenda), version 5.0* (2005).

[16] Total Quality Management, http://en.wikipedia.org/wiki/Total_Quality_Management.

# – 7.5 –

# Decision and Control Factors for IT-sourcing

## Guus Delen

*Verdonck, Klooster & Associates*
*and*
*Amsterdam School of ICT,*
*Weesperzijde 190, 1097 DZ Amsterdam, The Netherlands*
*E-mail: guus.delen@vka.nl*

**Abstract**

This chapter addresses the paradox that, however ample experience with IT-outsourcing has been built up during the last ten years, still many of them fail or do not meet expectations. Therefore two topics are investigated:

1. the scientific quest for finding which factors influence success or failure of a sourcing;
2. the consultant's question on how to apply the understanding gained on those factors in order to raise the success of actual sourcing projects.

This chapter starts with an introduction and definitions of the various types of sourcing. The second section gives an elaboration on the development, and economic relevance of IT-sourcing. Thereafter a set of ten sourcing-factors is treated, which acts as the initial hypothesis for this research. This set has been validated in a field survey. Section 4 summarises the result of this validation, which shows that nine of the ten factors are relevant for sourcing success and that this set is rather complete. The validation also resulted in an indication of the relative strength of each specific factor. In the last section the sourcing-factors monitor is introduced which makes use of the apparent strength of the nine relevant factors in order to predict and monitor sourcing success in actual cases.

## 1. Introduction

The notion of outsourcing was introduced in 1992 by Loh and Venkatraman after the first big subcontracting of ICT services in 1989 by Eastman Kodak to IBM and DEC. This

HANDBOOK OF NETWORK AND SYSTEM ADMINISTRATION
Edited by Jan Bergstra and Mark Burgess

contract had a value of 250 million dollar and a term of 10 years. Later on this conception was applied retrospectively to the subcontracting of other business processes ('Business Process Outsourcing'), which was common practice long before IT-sourcing came into existence. This way the conception of outsourcing got a much broader scope.

## 1.1. *Outsourcing, insourcing and other forms of sourcing*

In literature one finds many definitions of outsourcing. I quote two well-known definitions, the original one from Loh and Venkatraman and a more recent one:

DEFINITION 1 (Loh and Venkatraman [13]). A process whereby an organisation (1) decides to contract-out or to sell the firm's IT assets, people and/or activities to a third party supplier, (2) who in exchange provides and manages these assets and services for a fee over an agreed period of time.

DEFINITION 2 (Kern and Willcocks [10]). Outsourcing is the decision taken by an organisation to (1) contract out or sell the organisation's assets, people, processes and/or activities to a third party supplier, (2) who in exchange provides and manages assets and services for monetary returns over an agreed period of time.

These and other definitions complement each other without much contradiction. From the very beginning most definitions contain two components: (1) a once-only transfer of assets to an external supplier, (2) followed by receiving back services from that supplier during a certain period of time. This may all be condensed into the phrase 'sale & lease back'. Kern and Willcocks [10] broaden the scope from IT-services to all kinds of services. From the more recent definitions I have derived the following definition:

DEFINITION 3. *Outsourcing* is: (1) the transfer of certain business processes with the associated assets and employees to an external supplier, followed by
(2) receiving back services from that supplier based upon those processes during a number of years and with result obligation.

In real life assets and employees are not always completely transferred to the supplier. When no assets or personnel at all are transferred, one speaks of outtasking.

DEFINITION 4. *Outtasking* is: (1) the transfer of certain business processes without the associated assets and employees to an external supplier, followed by
(2) receiving back services from that supplier based upon those processes during a number of years and with result obligation.

What is considered as outsourcing by the outsourcing company, is considered as insourcing by the supplier. Therefore insourcing may be defined as the complement of outsourcing:

DEFINITION 5. *Insourcing* is: (1) taking over certain business processes with the associated assets and employees from a certain organisation followed by

(2) service delivery to that organisation based upon those processes during a number of years and with result obligation.

In analogy *Intasking* may be defined as:

DEFINITION 6. *Intasking* is: (1) taking over certain business processes without the associated assets and employees from a certain organisation followed by

(2) service delivery to that organisation based upon those processes during a number of years and with result obligation.

An organisation may also outsource new business processes (i.e. processes it has not yet implemented). In that case there is no transfer of processes, but the new processes are developed on order by a supplier.

DEFINITION 7. *Greenfield outsourcing* is: (1) ordering new business processes with the associated assets and employees from an external supplier, followed by

(2) receiving services from that supplier based upon those processes during a number of years and with result obligation.

The complement of greenfield outsourcing is *greenfield insourcing*:

DEFINITION 8. *Greenfield insourcing* is: (1) implementing certain business processes with the associated assets and employees for a certain organisation, followed by

(2) service delivery to that organisation based upon those processes during a number of years and with result obligation.

Despite the absence of any transfer, there is still much resemblance between greenfield sourcing and 'plain' sourcing: The outsourcing process starts the same way, i.e. with decision making (between in-house implementing and ordering of the services), and provider selection. In the delivery phase there is no difference left at all between plain sourcing and greenfield sourcing.

As indicated in the above definitions an outsourcing contract has a term of a number of years. At termination time, the outsourcer may renew the contract, but he may also ask other suppliers for proposals, with the intention of getting a better offer. In the public sector the last option is mandatory, due to the European tendering rules. This may result in the transfer of service delivery from the current supplier (A) to a new supplier (B). This situation differs from first time outsourcing on two points: (1) supplier B will not deliver services to A, but to the original customer of A; (2) despite European Community rulings for 'secondary tupe' (which give personnel the right to follow their jobs to the new provider in case of follow-up sourcing), suppliers do almost never transfer assets and personnel to each other, after all they remain competitors. Therefore follow-up sourcing may be defined as:

DEFINITION 9. *Follow-up sourcing* is: (1) transferring a certain service delivery from the current supplier to new one, followed by

(2) the continuation of that service delivery by the new supplier to the original outsourcer during a number of years and with result obligation.

In about 10% of all cases it happens that the outsourcer reverts to in house service delivery after termination of the sourcing contract. From the point of view of the original outsourcer this may be called insourcing (and by consequence outsourcing as seen from the suppliers point of view). But in order to prevent confusion I adhere tot Kern and Willcocks [10], who call this *Back-sourcing*. Typically there is no transfer of assets and employees, for why should a rejected supplier cede assets and why would specialised IT-personnel consider it a challenge to return to a supportive role in a 'common' organisation. As a wrap up back-sourcing may be defined as:

DEFINITION 10. *Back-sourcing* is: (1) acquisition of the assets and personnel that are required in order to implement certain business processes, which were hitherto delivered as services by an external supplier, followed by
(2) in-house service delivery based on those processes.

On a higher level of abstraction one can state that the process of outsourcing/insourcing separates the organisation of the service recipient from that of the service supplier, whereas those organisations merge again by the process of back-sourcing.
Figure 1 gives on overview of all these kinds of sourcing.

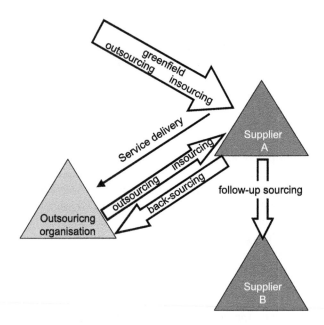

Fig. 1. The different types of sourcing.

## 1.2. *Shared services*

Many benefits from outsourcing, can be achieved within large organisations without the transfer of any assets or personnel to an external party. By merging departments that perform identical processes within an organisation, economies-of-scale and efficiency gains can be achieved. This holds specifically for back-office processes. Such a clustered service department is called a Shared Service Centre (SSC). Strikwerda [17] gives the following definition:

DEFINITION 11. An SSC is a result responsible unit within the internal organisation … which is charged with delivery of specific services to the operational business units based on a agreement and against cost price.

Organising an SSC may be considered as outsourcing *within* the organisation, i.e. *internal outsourcing*. This also offers a good opportunity for introducing best practice procedures and architecture thus achieving a quality improvement. In that case personnel and assets are selectively taken over from the old departments: only those employees that qualify in an assessment and only those assets that fit in the new architecture. After successful implementation of the SSC the organisation still has the option to sell it to an external provider, thus achieving the same the same result as outsourcing, but in a more controlled way. Another option is to allocate the SSC in a joint venture with the provider; this is called *co-sourcing*. This way again almost the same result is achieved as outsourcing, but with a fall back option in case the expectations are not met. A good example of this kind of outsourcing is Atos Euronext, a joint venture of Atos Origin and the European stock market, which has been sold to ATOS in the end [16].

For the record: Sharing of services is as old as big organisations, in the old days this was just called *centralisation*. A good example of an SSC avant la lettre is the Dutch Gemeenschappelijk AdministratieKantoor (Common Administration Office: GAK) of the branch administration offices for social security, which was erected in 1954!

Shared Services has assumed large proportions since 2000, specifically for the supportive Information, Personnel and Finance (IPF) processes of organisations. This holds for the private as well as for the public sector. Strikwerda [17] mentions 100 cases. Examples are:

- DHL Counting House, the European SSC for finance services of DHL.
- Accenture HR Services Ltd, which started when the HR Service Centre of British Telecom (BT) was brought in a joint venture with Accenture. Later on BT sold its shares to Accenture, an the centre now processes the HR administrations of customers all over Europe. This way full outsourcing was reached in three steps.
- The DTO (Defensie Telematica Organisatie) for the Dutch army, navy and air forces.
- The merger of the HRM departments of all Dutch state offices into one Shared Service Centre which is called P-direct (P for Personnel).
- The merger of all IPF functions of all Dutch jail's and re-education institutions of the Utrecht province (started in 2002 and to be rolled out to other provinces in the next phase).
- Bridge, the IT centre of Randstad, the largest temporary workforce intermediary in Holland.

- Achmea active, the IT centre of Achmea, one of the largest insurance companies.
- The concentration of the IT-departments of all 25 regional police corpses into 6 regional centers under the ISC (Information Service Coöperatie) of the police. This operation took several years and was completed in 2003.
- The ISZF (ICT Samenwerkingsverband Zuidwest Fryslân) which was founded on 1-1-2004 by six communities around Bolsward in the Friesland province.

## 2. IT-sourcing

### 2.1. *From systems management to sourcing*

As seen in retrospect, the ongoing development of systems management had around 1990 created the conditions for the breakthrough of sourcing and shared services later on in that decade. The most important developments were:
- Separation between demand and supply in the IT function
- Control based on volume and quality of service
- Centralisation of IT management without consequences for decentralised use of IT.
Next each of these three enablers is treated separately.

*Separation between demand and supply of IT*    The principle of separation of functions in IT management was introduced already with the threefold model of IT management [6]. This model discerns three functions: Functional management, Applications management and Technical management, where the demand side is represented by Functional management, and Applications and Technical management represent the supply side of IT. Such a separation is an important success factor for outsourcing, see the decision factors *Disentanglement* and *Demand management* in Chapter 3.

*Control based on volume and quality of service*    The principle of control on results has been introduced with the ITIL-process *Service Level Management*. In this process all services are specified by compiling a Service catalogue, followed by fixing the service levels in so-called SLAs (Service Level Agreements). This process may be compared with the functional specification of an application according to SDM (Systems Development Methodology). Without a SMART (Specific, Measurable, Acceptable, Result oriented, Time bound) specification there will always remain a gap between the expectations of the principal and the users on one side and the interpretation of the service provider on the other side. Sometimes this is acceptable from the internal IT department, but for an outsourced service specifications are an absolute must. This leads to the Control factor *performance management* in Chapter 3.

*Centralisation of IT management without consequences for decentralised use of IT*    Up and until about 1990 in the so-called 'mainframe era', IT was centrally managed. In the Network era that followed IT management and use was decentralised. Nowadays we have entered the Internet era, where the capacity for electronic communication is so vast that it has become feasible to withdraw all applications from the clients and run them exclusively on the servers. This is called the thin client or the Server Based Computing (SBC) principle. The big advantage of this concentration of applications on the servers is that the IT-managers have only to manage a limited number of versions, so that they are in complete control of the versions and applications that are used. The users do not feel this, because they keep access to the applications from their workstation. Of cause this results in a much higher load on the local network, but thanks to the law of Even More, this hardly needs to bother us any more.

### 2.2. *Domains of IT-sourcing*

A good framework for positioning IT processes is the *Generic model for information management* of Maes [1,15] which is drawn on this page. This model has been derived from the model of Henderson and Venkatraman [8] and has two dimensions: Business versus Technology and Strategy versus Operations. The domains of information management are the middle column and the middle row: the middle row is for the organisational structures that are needed for the translation of strategy into processes and the middle column is for the alignment between business en technology. The alignment goes two ways: the business puts demands on the technology, and the technology has an impact as enabler of the business.

Substituting Technology by Information Technology and filling in the ITIL-processes and operational IT-management processes results in Figure 2, where shades indicate the degree of sourceability.

The IT processes which qualify best for sourcing are the processes that need much IT expertise and little branch expertise, i.e. the processes on the right-hand side of the figure. An operational process will, for its frequency and character, easier develop into a commodity than a tactical or strategic process. Combining these facts results in the following shading of the figure (see Figure 3):

- Management of hardware and software (servers, clients, LAN WAN) is highly standardised and has developed into a commodity. As a consequence of ever more aggressive price competition this is the domain where offshore sourcing is coming up.
- The operational ITIL processes and the management of standard applications packages has a commodity character as well.
- Management of specific applications by a provider offers no additional economy of scale, in this domain off-shore sourcing is the only way to achieve cost savings.
- Application development falls beyond the scope of the figure. An assignment for a project to develop one application is not considered as outsourcing. Only an package assignment to develop a complete application portfolio, can be considered as outsourcing.
- The market for tactical ITIL processes (Service Level management etc.) is still developing. Sometimes these processes form part of an outsourcing deal, but the result falls often behind that of outsourcing operational processes.
- IT strategy and planning are hardly ever outsourced. Providers speak highly of partnerships, but in reality their main drive is for cost efficient service delivery, in order

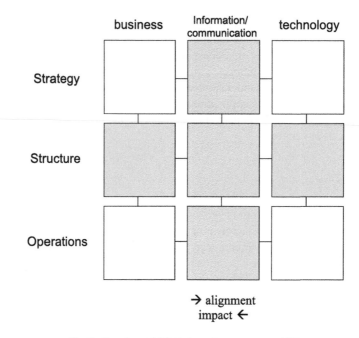

Fig. 2.  Generic model for information management [1].

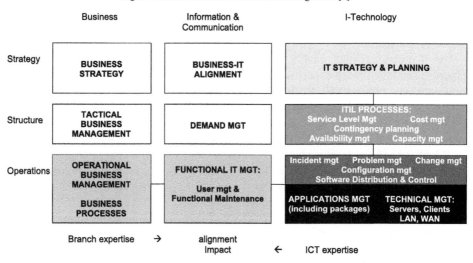

Fig. 3.  Domains of IT-sourcing projected on the generic model for information management of Maes. Legenda:

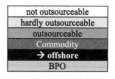

to make profit out of the contract. This way they hardly give attention to proactive tactical management, and even less to IT strategy and planning.

- Sometimes also the operational part of Functional IT management is outsourced. Due to the high degree of branch expertise needed, this is to be considered more as Business Process Outsourcing than as IT-sourcing.

The bulk of outsourced IT services and the most aggressive price competition between providers is encountered in the three commodity domains down to the right of the figure.

### 2.3. *Motives for IT-outsourcing*

Recent research by Corbett [3] has resulted in the following list of motives for IT-sourcing (see Table 1).

These motives have been arranged from operational (efficiency improvement of IT-services) to tactical (effectiveness improvement) to strategic (strengthening the competitive power of the organisation).

*Efficiency-improvement of IT services*　　The oldest motives are cost saving and retention of capital. These motives recur with every economic recession and together they represent 40% of the total. But expectations in this field are not always met. Properly speaking cost savings can be expected only when the internal IT department is not efficient (but in that case saving can be achieved in house as well), or when outsourcing leads to better economies of scale at the provider's site. As a rule of thumb IT departments with less than then 40 employees are less efficient than an outsourced IT service [4].

The exact position of the break even point depends also on the diversity of experience required and the service windows (e.g. $5 \times 8$ or $7 \times 24$ hours).

Table 1
Motives for IT-outsourcing [3]

| Motive | Percentage |
|---|---|
| *Operational*: | |
| cost saving | 35 % |
| retention (of return) of capital | 5 % |
| subtotal | 40 % |
| *Tactical*: | |
| innovation | 2 % |
| quality improvement | 5 % |
| flexible cost structure | 13 % |
| subtotal | 20 % |
| *Strategic*: | |
| revenue growth | 2 % |
| shorter time-to-market | 5 % |
| focus on core competencies | 33 % |
| subtotal | 40 % |
| Total | 100% |

The motive of a capital injection by selling an IT department to an external provider is very attractive in times when organisations have occurred great debts. This motive was explicitly behind the outsourcing around 2002 of the three shared service centres of the Dutch telecom provider KPN to Atos Origin, who payed about 250 million Euro for it.

*Effectiveness improvement of IT services*   Effectiveness is not about a cheaper, but above all a better IT service. External professionals may raise the quality of the IT services with their knowledge and experience, watch trends and signal innovation opportunities in time. Further on outsourcing can make the costs most flexible. The motives of this group make up 20% of the total.

*Strengthening the competitive power of the organisation*   Sourcing is a decision with long term consequences, and should therefore be underpinned with strategic motives. The most strategic motive for outsourcing is strengthening the competitive power of the organisation by focussing it on its core competencies. This way organisations aim at shorter times to market for their products and an increase in revenues. This category grows at the expense of the other motives and nowadays represents 40% of the total [3].

### 2.4. Why IT-sourcing is so special

Outsourcing of catering services, cleaning and housing was very common already before the first great IT-sourcing by Kodak in 1989. After 1989 a giant flood of publications on IT-sourcing has arisen ([2,11,12,18], etc.) with ample attention for sourcings that went wrong. Apparently the impact of IT-sourcing exceeds the impact of outsourcing cleaning or catering services greatly.

I have thought of some explanations:
1. Information management is becoming more and more vital for organisations and in some sectors (e.g. the financial sector) it has developed from a supportive into a primary process. This puts severe demands on its reliability and availability and may cause big damage when anything goes wrong.
2. With the growing of importance the budgets of IT services grow as well. And high budgets always stimulate corporate management to search for cost saving opportunities, if need be even by premature outsourcing.
3. Information management gets more and more interwoven with the processes it supports. This makes it more difficult and risky to place the information management outside the organisation, which stresses the decision factor *careful disentanglement* (see Section 3).
4. As a consequence of the ongoing development and expansion of computer hardware and software, experience with the newest developments is always scarce. The scarcity of that experience alones forces small and middle sized organisations to outsource their IT, but by doing so they make themselves more dependent on external providers as well. Big organisations have more options, but they should at least concentrate their IT experience, e.g. in an SSC.

## 2.5. *Offshore sourcing*

As a direct consequence of outsourcing, the larger part of the IT services will be offered from the location of the provider. In that case one speaks of *remote operations* as opposed to *on-site operations* for the part of the service that is still offered from the client's location. As a consequence of growing bandwidth those remote locations can be farther and farther away, so that ever bigger efficiency gains can be achieved. Gartner [9] make distinction in three grades:

- *On shore sourcing*, where the services are still offered from a location in the clients the same country.
- *Near shore sourcing*, where the services are offered from another country, but still from the clients continent. Near shoring enables bigger economies of scale by concentrating services on a regional scale.
- *Offshore sourcing*, where the services are offered from another continent. Off shoring enables even bigger cost savings by relocating operations to low wage countries.

Offshore sourcing is breaking through for application development and applications management. A more proper term would be *offshore outtasking*, as the intended cost savings can only be achieved by *not* transferring personnel to the low wages country. According to Gartner [7] Indian companies like Wipro, Tata Computing Services (TCS) and Satyam Computer Services do already deliver a volume of five milliard dollar in IT services to Europe and the US, and this amount is growing with 30% a year. Parallel to direct off shoring there is a development of indirect offshore sourcing where providers in Europe or the US like Atos Origin and Capgemini establish daughter companies in India, Malaysia etc. in order to relocate part of their operations offshore. Such companies can offer a mix of on-shore, near-shore and offshore sourcing for the variety of services they offer according to a so-called Global Delivery Model [9].

Beulen [2] rightly indicates that the savings on operational cost by off shoring, should be traded off against a raise of about 50% in the cost of managing the provider. This extra management effort is needed to bridge the difference of time zone, of language but above all of culture.

## 3. Hypothesis: the sourcing factors

In my thesis [5] I have tried to find out which factors influence the success or failure of ICT sourcing. As a starting point, I have identified ten critical factors from literature and from my practical experience. The role of such 'critical factors' varies throughout the sourcing process. During sourcing decision time, the critical factors indicate whether the right decision is taken. When the factors are favourable, it is responsible to continue the sourcing implementation, but during that process the same 'critical factors' must be controlled in order to retain success. Because the factors then change in role and appearance, they are renamed control factors in the implementation stage. So, in practice, decision factors serve to make the right sourcing decision and control factors control the correct implementation of that decision. In time, there comes an end to every sourcing contract, and at that point the outsourcer can decide to renew the contract, switch to another provider, or back-source

Table 2
Critical sourcing factors and their role during the sourcing process

| Sourcing decision: *Decision factors* | Sourcing process: *Control factors* | Follow-up sourcing | Back sourcing |
|---|---|---|---|
| DF1 planned approach | CF1 planned approach | FF1 | BF1 |
| DF2 business case outsourcing | CF2 business case outsourcing | FF2 | BF2 |
| DF3 business case insourcing | CF3 business case insourcing | FF3 | – |
| DF4 disentanglement | CF41 personal transfer | | BF41 |
| | CF42 assets transfer | FF42 | BF42 |
| | CF43 user change support | | BF43 |
| DF5 demand management | CF51 solid contract | FF51 | |
| | CF52 financial agreements | FF52 | – |
| | CF53 performance mgt | FF53 | |
| DF6 sourcing expertise | CF6 retention of expertise | FF6 | BF6 |

to himself. At that time, the critical factors again resume the shape of decision factors, but in a different context. Depending on that context, they are then called follow-up factors or backsource factors. Table 2 shows the evolution of the ten critical sourcing factors during the life cycle of a sourcing deal.

Hereafter the factors are described in some detail:

### 3.1. *Decision factors*

*DF1 Planned approach of the sourcing process*   In order to bring a sourcing process under control, a phasing model is needed. For instance, the system development process only came under control after the introduction of phasing models like SDM (System Development Methodology). Every sourcing should be planned in advance, phase by phase, and that planning may then serve as a frame of reference for the process. A usable model for this is the WCIT sourcing cycle (see Figure 4) with the phases: 1. Sourcing decision. 2. Provider selection. 3. Delivery.

*DF2 A positive business case for outsourcing*   By identifying the benefits and weighing them against the costs, drawbacks and risks of outsourcing, a business case is compiled. The biggest risk is that the outsourcer fails to identify a core competence and outsources it unconsciously. Success in the long run is granted only when the outsourcer continues to use the business case afterwards to evaluate the outsourcing process and the service delivery of the provider.

*DF3 A positive business case for insourcing*   For lasting sourcing success, the sourcing must be profitable for both the outsourcer *and* the insourcer, the so called 'win-win' situation. The business case for insourcing is highly symmetrical to the business case for outsourcing:

- the processes to be transferred should be a core competence for the insourcer;

- the insourcer can offer significant advantage to those processes based on his specific expertise and technologies;
- the insourcer can offer efficiency benefits from his economy-of-scale.

*DF4 Disentanglement of processes*   Before being able to transfer processes and underlying assets to an external organisation, the outsourcer should disentangle those processes from his own organisation. This becomes especially critical when the outsourcer does not want to outsource all his ICT, but only those ICT processes for which he can make the best business cases. An enterprise architecture on all levels, from business processes to ICT components, is indispensable as a frame of reference for this disentanglement.

*DF5 Demand management*   An outsourcer can only outsource his ICT processes when he is sufficiently able to manage the insourcer that takes over those processes; this is called demand management. Which maturity level of demand management is sufficient depends on the degree of outsourcing. For the outsourcing of operational ICT processes, such as network management or server management, the level III, *system oriented* from the EFQM (European Foundation of Quality Management) model is sufficient. However, for the outsourcing of integral ICT departments, the level IV, *chain oriented* is necessary.

*DF6 Sourcing expertise*   The outsourcer and the insourcer should both have experience in ICT outsourcing. However, not all sourcers should be considered equal in terms of experience. Organisations who for the first outsource are *ipso facto* inexperienced, whereas ICT service providers live from insourcing and are therefore extremely experienced. For this reason, first time outsourcers are often dependent on external expertise. They could obtain that expertise from their would-be provider, but there is some difference between outsourcing and insourcing, and it is natural that a provider considers his own interests first. Therefore, outsourcers can better engage an independent sourcing consultant who has previously directed a number of sourcing implementations.

## 3.2. *Control factors*

*CF41 Careful personnel transfer*   With first-time sourcing, personnel is almost always transferred. This transfer is critical to the success of the sourcing, because IT service is a very knowledge intensive branch and the personnel is the main knowledge medium. In the transfer, ample attention should be given to knowledge retention, cultural differences, communication, handling resistance and the workers council.

*CF42 Assets transfer*   The material assets to be transferred when outsourcing are the hardware (WAN, LAN, servers, PC's), the software (systems software and applications), licences and sometimes the premises of an IT department. By transferring these assets, the outsourcer enables an external provider to standardise these assets and integrate them in his own IT infrastructure. This way the provider may achieve efficiency gains in its operations. On the other hand, transferring IT assets to a different environment and a different IT architecture implies a considerable risk, and once the transfer has been made, it may need to be repeated time after time as the provider changes. Therefore, IT assets are, in practice, more often retained than transferred by the outsourcer.

*CF43 User change support*    Outsourcing increases the distance between the users and the IT service provider. Priorities which were self-evident for the domestic IT department, or which were observed when demanded by the users, are exchanged for the formal agreements of a Service Level Agreement (SLA). It is therefore essential that the users are aware in advance of their processes and their critical dependence on IT, and that the outsourcer involves its users in the compilation of the SLA. Of course, every service level has its price, and individual users should understand that they get the service that their company pays for. A good means of improving user awareness is a system of cost charging that confronts the users directly with the expenses of their wishes. To put it shortly, everything revolves here around good expectation management.

*CF51 A solid contract*    A sourcing-contract should offer security as well as flexibility. A contract is valid during the lifetime of the sourcing deal, and in the ideal case it remains in the filing cabinet during that period. When things go well, most of the clauses are never applied and the contract need not be changed or broken up. It can best be compared to a prenuptial marriage agreement, which is inconsequential in the case of a good marriage, but contains some rules to fall back on in case of serious troubles. When such troubles occur, a solid contract can prevent a precipitous escalation to a law suit. Therefore, it is advisable to include a procedure for mediation or arbitration by independent, trusted experts for conflicts that parties cannot resolve themselves. After all, sourcing parties are strongly dependent on each other and have to continue operating without interruption. In a conflict situation, neither party benefits from legal procedures with expensive lawyers and possible damage to their image. And let us not forget that lawsuits can drag on for years without a clear verdict.

Another attention point is the information security. The ISO17799 standard, paragraph 6.2.3 "Addressing security in third party agreements" gives a good checklist for contractual agreements on this issue.

*CF52 Clear financial agreements*    There is a significant difference between the purchasing of products and the establishment of a sourcing relation. When purchasing, one tries to obtain the lowest price for a given product with an established level of quality, but when outsourcing, the parties involved become dependent on each other for a long period of time. When a provider cannot cover his expenses in such a relationship, he may feel forced to limit his efforts, or to expend all of his creativity identifying extra work. In such a situation, anything which is not explicitly included in the contract and the SLA is charged separately, above the standard service prices. Such a situation can easily develop when cost reduction is the only motive for outsourcing. Many IT departments originated from the financial administration of an organisation and do still report to the Chief Financial Officer. In times of economic recession, this may easily lead to a unilateral management of expenses, and this effect worsens when the negotiations are delegated to a purchase department.

A transparent cost price calculation is of paramount importance for a healthy sourcing relationship. The best practice lies in applying the principles of Activity Based Costing (ABC) to the IT service provision.

*CF53 Performance management*    De Looff pointed out in his thesis [14] that "*outsourcing is not justified unless the demands on the service provision can be specified in detail in advance and the compliance to these demands can be measured and enforced*". With this, he identified an important control factor: the ability to specify, measure and monitor the performance of the IT service provider. This cluster of tasks is commonly referred to as *performance management*.

*CF6 Retention of expertise*    Up until the first migration to an external service provider, it can be assumed the outsourcer has the expertise to provide its own IT services. Additionally, during the provider selection phase, the outsourcer acquires the expertise needed to manage a provider. After the transition, during the contract management phase, it becomes important to retain both forms of expertise because, when the outsourcer has lost the expertise to re-establish its own IT service provision from scratch, he loses the option of back-sourcing; that is, he cannot reverse the outsourcing decision. Additionally, if the outsourcer loses the expertise to manage a new provider, he also loses the option of follow-up sourcing and becomes permanently dependent on the one provider that he has.

## 4. Field survey: eighteen cases

The sourcing factors have been evaluated for eighteen cases in practice. From the correlation between success or failure and the compliance to the criteria for the individual sourcing factors, a classification could be derived for the factors, with a weight factor for each class; see Table 3.

## 5. Synthesis: the sourcing-factors monitor

The evaluation of the cases indicates that compliance to the criteria for the most important sourcing factors is necessary for successful sourcing in each phase, and should be established before transition to the next phase is justified. So, the *sourcing decision phase* should result in compliance with the criteria for the most important decision factors before starting the *provider selection phase*.

*Decision and control score*

Just like the first phase transition, the provider selection phase should result in compliance with the criteria for the most important control factors before transition to the transition and contract management phases. Table 3 already indicates how strong or weak the individual factors are. This table can help quantitatively determine the degree to which the sourcing

Table 3
Classification of decision and control factors

| Category | Weight | Decision factors | | Control factors | |
|---|---|---|---|---|---|
| Very strong factors: | 5 points | DF2 | business case for outsourcing | CF2 | business case for outsourcing |
| Non-compliance always implies failure | | | | CF41 | personnel transfer |
| Strong factors: | 3 points | DF3 | business case for insourcing | CF3 | business case for insourcing |
| Non-compliance implies failure or partial success at most. | | DF5 | demand management | CF51 | solid contract |
| | | | | CF52 | financial agreements |
| Common factors: | | | | CF53 | performance management |
| Non-compliance implies failure most times, but | 1 point | DF1 | planned approach | CF1 | planned approach |
| incidental successes | | DF4 | disentanglement | CF43 | user change support |
| occur. | | DF6 | sourcing expertise | CF6 | retention of expertise |
| Weak factors: | | | | | |
| No relation can be proven between compliance and success of sourcing. | 0 | | – | CF42 | assets transfer |
| Maximum score | | $1 \times 5 + 2 \times 3 + 3 = \mathbf{14\ points}$ | | $2 \times 5 + 4 \times 3 + 1 \times 3 = \mathbf{25\ points}$ | |

complies to the complete package of factors. When all decision and control factors are relevant, a maximum of 14 decision points and 25 control points can be scored, as is calculated in Table 3. Then, the decision score (Dsc) is defined as the actual number of decision points scored for the factors that are compliant, divided by the maximum score ($D_{max}$) that can be obtained for all factors that are relevant to a specific sourcing, expressed as a percentage. In the same way, the control score (Csc) is defined as $C_{act}/C_{max}$. After calculating these scores for all 18 cases, it is seen that a successful sourcing corresponds to scores of 85% or more. The partially successful cases correspond to scores between 50% and 85%, and when one score falls below 50%, failure is taken for granted. The fact that such a scoring method can be derived from the ten generic sourcing factors also demonstrates that those ten factors are sufficient to be able to make a reasonable prediction on the success or failure of a specific sourcing.

## Monitoring the sourcing factors

By combining all results to this point, a generic plan can be compiled for monitoring the sourcing factors over the entire lifecycle of a sourcing: the so-called *Sourcing factors monitor*. This monitoring can be considered a kind off early warning system, which flags weaknesses in advance, so that the sourcing process can be kept on the right track. Natural moments for evaluating the scores are the start and the end of each phase of the sourcing

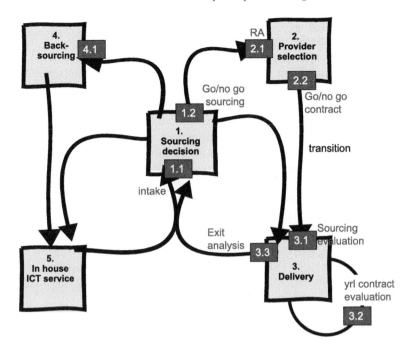

Fig. 4. The sourcing-factors monitor projected on the WCIT sourcing cycle.

cycle. Figure 4 shows a generic schedule for points of measurement in the WCIT sourcing cycle.

The sourcing-factors monitor has been applied to the IT outsourcing of the Dutch Office for Tourism (Nederlands Bureau voor Toerisme en Congressen; NBTC). In this study, it appeared that the factor DF4 *demand management* was weak after the sourcing decision phase. During the provider selection and transition phase, corrective action was taken, which resulted in a rise of the sourcing score from 72 to 100%. This case now has a big chance of success.

## 6. Conclusions

The results of this research may be wrapped up as follows.

1. The success of an IT-sourcing is influenced by nine generic sourcing factors, her placed in order of strength:
   - a positive business case for outsourcing
   - careful personnel transfer
   - a positive business case for outsourcing
   - the factors for demand management:
     - a solid contract
     - transparent financial agreements
     - performance management

- a planned approach
- sourcing expertise
- user change management.

In this success is defined as the degree of willingness of the outsourcer *and* the insourcer to continue the sourcing.

2. The influence of these nine factors together gives sufficient explanation of the success of 18 cases which have chosen arbitrarily from the Dutch sourcing market.

3. The influence of the tenth factor, assets transfer, could not be demonstrated. During the research no other factors have been found, then the ten of the initial hypothesis.

4. A gaugeable decision and control score could be derived by attributing weights to the nine factors, with clear breaking points between success, partial success and failure of a sourcing. Therefore it may be expected that the same scoring method can predict the success of other cases with reasonable probability.

This sourcing factors method may be enhanced further by adding new cases to the database.

## References

[1] T. Abcouwer, R. Maes and J. Truijens, *Contouren van een generiek model voor informatiemanagement*, Prima Vera Working Paper 1997-07, University of Amsterdam (1997).

[2] E.P. Beulen, *Offshore outsourcing van Infrastructure Management: het realiseren van een kostenbesparing op IT-beheerkosten*, IT Service Management Best Practices, Vol. 1, J. van Bon, ed., van Haren Publishing, Zaltbommel (2004), 105–117.

[3] M. Corbett, *The Global Outsourcing Market*, Corbett & Associates, LaGrangeville, NY (2002).

[4] G. Delen, ed., *World Class IT praktijkgids ICT-sourcing*, Tutein Nolthenius, Den Bosch (2003).

[5] G. Delen, *Decision- en Controlfactoren voor IT-sourcing*, van Haren Publishing, Zaltbommel (2005).

[6] G.P.A.J. Delen and M. Looijen, *Beheer van informatievoorziening*, SDM-reeks, Cap Gemini Publishing, Rijswijk (1992).

[7] Gartner, *A Look at India for Offshore Sourcing Options*, Gartner, Stamford, CT (2003).

[8] J.C. Henderson and N. Venkatraman, *Strategic alignment: Leveraging information technology for transforming organizations*, IBM Systems Journal **31** (1) (1993), 4–16.

[9] P. Iyengar and R. Terdiman, *Decision framework for global delivery*, Research Note DF-18-9786, Gartner, Stamford, CT (2003).

[10] T. Kern and L.P. Willcocks, *Exploring information technology outsourcing relationships, theory and practice*, Management Report 61-1999, Erasmus University, Rotterdam (1999).

[11] M.C. Lacity, *Information Systems Outsourcing: Myths, Metaphors and Realities*, Wiley, Chichester (1993).

[12] M.C. Lacity and R.A. Hirschheim, *Beyond the Information Systems Outsourcing Bandwagon*, Wiley, Chichester (1995).

[13] L. Loh and N. Venkatraman, *Diffusion of information technology outsourcing, influence sources and the Kodak effect*, Information Systems Research **4** (1992), 334–358.

[14] L. de Looff, *A model for information systems outsourcing decision making*, Thesis, TU Delft (1996).

[15] R. Maes, *A generic framework for information management*, PrimaVera Working Paper 1999-03, University of Amsterdam (1999).

[16] R. Sanders, *Euronext verkoopt in Nederland zijn belang in Atos Euronext aan Atos Origin*, Computable (17-9-2003).

[17] J. Strikwerda, *Shared Service Centers*, koninklijke van Gorcum, Assen (2003).

[18] L.P. Willcocks, D. Feeny and G. Islei, *Managing IT as a Strategic Resource*, McGraw-Hill, New York (1997).

# – 7.6 –

# How do ICT Professionals Perceive Outsourcing? Factors that Influence the Success of ICT Outsourcing

Diana Hoogeveen

*Verdonck, Klooster & Associates B.V., P.O. Box 7360, 2701 AJ Zoetermeer, The Netherlands*
*E-mail: Diana.Hoogeveen@vka.nl*

**Abstract**

In the past decade, many organizations outsourced their IT departments. Thousands of ICT employees working in these departments were taken over by an ICT service provider. Although some employees may perceive this as a career opportunity, others may feel threatened. Little research was done to evaluate the impact of outsourcing on ICT employees' attitudes and perceptions. We did not find publications about factors influencing the successful transitions of ICT employees to an ICT service provider. We have conducted a close examination of employee attitudes before and after outsourcing to an ICT outsourcing service provider. The propositions that we tested are based on social identity theory and social judgment theory. We have identified a number of important factors that determine the post-outsourcing attitude and that can be applied in the outsourcing practice. We found that it is more desirable to acquire an IT department with satisfied ICT professionals than an IT department with unsatisfied employees. Our findings also indicate that a positive attitude about the professionalism of the new employer can help to ease the transition.

## 1. Introduction

In the past decade, due to ICT outsourcing thousands of ICT employees had to leave their old employer and started working for an ICT service provider. Despite the magnitude of this activity, information systems researchers have done little research on the ICT employees' perspective on this transition. The major focus has been on outsourcing strategy and

---

HANDBOOK OF NETWORK AND SYSTEM ADMINISTRATION
Edited by Jan Bergstra and Mark Burgess

outsourcing risk [3,11,17,18] and recently research interest has shifted towards outsourc-ing relationship management [4,8,12]. Levina and Ross [13] researched the needs of the outsourcing provider seeking positive results, in terms of both cost efficiency and the abil-ity to provide the best possible service to their new customers. Very little has been done to investigate the attitudes and perceptions of the outsourced employees.

It is important to get insight into the attitudes and perceptions of the outsourced employ-ees, for two reasons. First because the future of the ICT employees is at stake. And second because the relationship between the ICT service provider and the transitioned employees influences employee satisfaction and performance. ICT service providers can only provide the best possible service to their customers if their employees are satisfied and perform well.

Employees' responses to outsourcing can vary from very negative to very positive. Will-cocks et al. [18] note that two of the largest Information Systems outsourcing contracts signed during 1994 – British Aerospace and the Inland Revenue – both experienced strikes among the staff affected. Outsourcing may affect job security, job location, management (practices), colleagues, payment and other job conditions of employees. The insecurity about the impact of these changes, the feeling that their former employee does not value the work of the IT department, may negatively influence job motivation which can lead to lower productivity and even sabotage [11]. These problems negatively influence perfor-mance.

Purcell [16] on the other hand, notes that for some employees outsourcing may provide an opportunity for enhanced career and training opportunities if it involves a move to a specialist organization. In some situations outsourcing may provide greater stability and predictability.

In this paper we examine the attitudes and perceptions of outsourced ICT professionals. We try to find factors that influence the response of ICT professionals to the transition to an ICT service provider. Insight in these factors may help ICT service providers to better diagnose individual adjustment problems and prevent long term social and negative performance impact.

We start with a brief review of the research about the employee perspective in ICT outsourcing. Then we will discuss the work from [14]. They developed hypotheses about the relationship between prior outsourcing attitudes and post outsourcing attitudes, which are based on identity theory and social judgment theory and tested these hypotheses in the transportation industry. We replicated their study in the ICT industry. Because ICT professional have some unique differences from non-ICT professionals [1] we wanted to explore the recipe for positive post outsourcing attitudes in the ICT industry. We used a sample of 93 out of 212 ICT professionals whose jobs were outsourced to an ICT service provider. Finally, we suggest how management of the ICT service provider may ease the transition of ICT employees for the success of both the employees and the organizations involved.

## 2. ICT outsourcing research and the ICT professional's perspective

In this research we focus on the type of outsourcing where ICT employees and their tasks are taken over by an ICT service provider (i.e. the services continue to be provided by those employees who have done the work in past, but the original company is no longer responsible for their hiring and direct supervision). Outsourcing often includes at least some employees who fit into this type of outsourcing, augmented by additional employees from the service provider. In the ICT industry, it is often advantageous for the service provider to add to their payroll those employees involved in the acquired ICT function. If the knowledge or capabilities of employees are scarce but necessary to perform the tasks successfully, then it is crucial that the service provider hires the right (number of) employees. This is the case when a service provider starts to offer a new type of business process outsourcing and needs the knowledge about the business processes of the employees involved. Although in some cases the service provider may find it more cost effective not to hire all the employees that previously performed the tasks, for example where the ICT services have become commodity services, like data center operations or desktop management.

Working for an ICT service provider can be intimidating or stimulating, depending upon the employee's point of view. Some ICT professionals prefer the traditional style of employment vs. working as a consultant [10]. ICT professionals may be worried about the job security, pay, pensions, travel distance etcetera. Many ICT professionals may feel that their employer no longer wants them and so the most talented and important professionals will look for jobs elsewhere [2,5]. Other ICT professionals may value the career opportunities that the ICT service provider can offer [10]. They may get the opportunity to work for other customers, learn new methods and techniques, interact with a larger variety of IT professionals, or grow in responsibilities. Because IT is the core business of an ICT service provider there may be more professional opportunities than at the previous employer.

We did not find any publications in the IS journals about the factors that influence these different attitudes about ICT outsourcing. In the International Journal of Human Resource Management, however, we found a recent study from [14], where the employee attitudes were studied both before and after outsourcing in the transportation industry. Using identity theory and social judgment theory, they developed hypotheses about which attitudes stimulate adjustment and which might service to inhibit it.

Logan et al. [14] introduced social identity theory to outsourcing research. Social identity refers to that "part of self-concept which reflects the groups to which a person belongs"; it can affect self-esteem, "so people try to join or remain members of the best possible groups" ([7], page 1151). Outsourcing may threaten social identity because of the effect it has upon groups and group membership [14]. "If the new group is considered to be superior, then employees are likely to join the new group and abandon the old group. But if the new group is considered inferior or undesirable the process is likely to fail as group members strive to protect their identity and the attached self-esteem. The greater the level of success of the outsourced group the more difficult could be the change process" ([14], page 151).

## 2.1. *Hypotheses*

We have tested two hypotheses about employee-specific attitudes to outsourcing that Logan et al. developed and tested in the transportation industry. "The hypotheses make a case for the proposition that employees who have the most positive attitudes and the highest commitment to their previous employer, have the most difficulty in adjusting to a new employment setting. These employees have most invested in the status quo and will show the most resistance to change" ([14], page 151).

HYPOTHESIS 1. Those employees who have high levels of job satisfaction, commitment, perceived success and job involvement prior to outsourcing will be resistant to the changes caused by outsourcing, and this resistance will lead to decreased levels of job satisfaction, commitment, and increased intention to quit after outsourcing.

On the other hand, if individuals see outsourcing as contributing positively to their self-concept or self-esteem, they will support the changes and their attitudes should then show signs of improvement.

HYPOTHESIS 2. Those employees who perceive the new company as having a good reputation, a similarity to their old company and an ability to complement their old company, will support the change, they will have more positive attitudes after outsourcing ([14], page 151).

Logan et al. [14] did an exploratory study of employees in the companies' fleet operations that are taken over by a major trucking firm which is acquiring responsibility for the transportation function. In a survey eighty truck drivers were questioned about their pre- and post acquisition attitudes. They found that if one acquires an operation that is already successful and the employees there feel that the outsourcing provider has a good reputation that the transition of ICT professionals may be easier. Those employees who are the most involved in their jobs and the most satisfied might be the hardest to win over. Logan et al. found that those who were highly involved and perceived themselves and their prior employer as successful are more likely to quit.

The objective of our study was to examine the employee characteristics that are favorable or not favorable to successful ICT outsourcing transitions. We chose to use Logan's research model for this examination for three reasons.

The first reason is that Logan et al. introduced a new theory into outsourcing research that highlights factors that are not so easily identified and dealt with in practice. From the service provider's perspective, it is easier to handle worries about pay, pension and location than managing commitment, job satisfaction and involvement.

The second reason is the hypotheses were only partly supported in Logan et al.'s initial exploratory study. Further examination may provide more insight into the explaining power of these theories. We would like to introduce these theories into the IS field.

The third reason is that ICT professionals may have some unique differences from non-ICT professionals [1,15]. Whether IS and non-IS employees are different, is a debate that has continued within the IS literature. Ferrat and Short found that both could be motivated equally using the same underlying constructs. Im and Hartland disputed their methodology while supporting their outcome. Ferratt and Short did admit that their sample was not random and limited to a specialized group of employees in selected insurance companies in the Midwestern United States: thus they are not representative of all industries. Couger and

Zawacki found that ICT professionals have substantially higher growth needs than any of the job categories surveyed by Hackman and Oldham [6] and were concerned about learning new technology. On the other hand they found that ICT professionals have the lowest social needs among professionals. ICT professionals have more loyalty to their profession than to their company. Therefore ICT professionals may react differently to outsourcing than other professionals. It is worthwhile to test Logan's hypotheses in this different environment to investigate whether their conclusions also hold in an ICT outsourcing situation.

## 3. Method

The client organization is one of the semi-public organizations responsible for the operating of the social security laws in the Netherlands. Their ICT organization consisted of three main groups: the application management group, the data services group and the desktop solutions group and staff. In 2000 the client organization outsourced its ICT organization to an ICT service provider. All the staff and ICT professionals who worked for the host firm were absorbed into the ICT service provider providing the outsourcing. In total 380 people were transferred, 257 full time equivalents ICT professionals and 105 full time equivalents staff. For a long time this outsourcing deal belonged to the top three in terms of volume in the Netherlands.

The service provider employed over 3200 ICT professionals and had a turnover of 345 million euros in 2003. By acquiring this ICT organization, the service provider wanted to increase its market share in the social security market and wanted to acquire knowledge about the exploitation and support of large complex ICT systems. Based on these capabilities the service providers wanted to increase its chances in the outsourcing market.

Before sending the survey we conducted interviews with several ICT professionals that were still working for the service provider, with the HR manager of the service provider, the previous manager of host organization, and an ICT professional that had to leave the service provider. It was evident from these interviews that the employees considered the outsourcing to be a major change. At the first day the name of the ICT organization changed and the new name was put on top of the building. Several interviewees independently came up with the example of the air-bridge that was connecting the building where the ICT organization worked and the building where the client organization worked. As soon as the contract was signed the doors were locked and the bridge could not be used any longer. They all mentioned that they had to get used to the fact that those doors were no longer open for them.

The salaries, pensions and all other job conditions were harmonized. The method of computing pay changed completely, although actual level of wages often went up and never down. The service provider only allowed a very limited amount of part-time work, while the client organization was very liberal towards that.

Most things remained the same during the first 6 months, they continued to work at the same location servicing their host organization and retained the same immediate supervisors. After six months, however, supervisors from other departments of the ICT service provider replaced many supervisors. Each staff member and ICT professional was evaluated, some of them continued doing their old job, some decided to do another job within the service provider and some left. The Capability Maturity Model was implemented to professionalize the work processes and the project work. In the old days at the client organization, money was never an issue for project work. The only important thing was that the functionality was available in time so that legislation could be implemented. This might for example result into the development of several versions with different functionalities to make sure that the right functionality could be implemented in time. But after the outsourcing, it became clear that money was a very important issue for the service provider to stay profitable, and therefore this way of working had to change.

The relationship with the client organization changed. It changed from an informal relationship into a contract management relationship based on service level agreements. Several interviewees indicated that this was the most important difference with the old situation and that they found it hard to get used to. In their experience, the service level agreements resulted in a higher work-pressure.

The interviewees mention both positive and negative aspects about the culture at the service provider. Several people mentioned the kind colleagues at the service provider and the feeling that their new employer is not as formal in terms of dress code and standard work practices as most other ICT service providers. On the negative side they mention that at the service provider people are considered to be a production tool. "If you are not productive anymore or you do not fit in anymore they throw you out", one interviewee said and others confirmed this statement. As a consequence their perceived job security became uncertain. Many people got sick or had to leave because they could not adapt to the pressure of the new environment according to the HR director.

Between 2001 and 2003 the international but also the Dutch ICT market is very bad. Two big reorganizations are necessary for the service provider to stay profitable in the future. Several competitors are merging to increase economies of scale. At the end of 2004 212 ICT professionals out of the original 380 ICT professionals are still working for the service provider.

In terms of profitability the take-over was very successful according to senior management.

## 4. Measures

The survey questionnaire was sent to all survivors of the outsourcing transition. In this survey the ICT professionals were asked to think back to when they worked for the host company (their previous employer) and answer the questions based on how they felt at that time.

In this type of research retrospective data collection can provide useful information, as the object is to compare the pre and post attitudes about the two jobs. "In traditional pre-post-tests, the result could be a similar rating on satisfaction for both jobs even though there had been a tremendous change. For example, the employees thought they were satisfied in their previous job, so in the pre-test they rated themselves as a 4 out of 5. Now after experiencing their current job they 'know what true satisfaction is' and again rate themselves as 4 out of 5 in the post-test. They justify this seeming inconsistency by cognitively re-evaluating the previous job as a 1 or 2. The use of a retrospective measure is designed precisely to capture this shift in attitude anchors over time" ([14], page 154).

To make it possible to compare the outcomes of both studies, we chose to use the same measures as Logan et al. [14] did. Therefore we will not discuss the measures in depth below, an overview of the measures can be found in Table 1. We only explain the measures that were adapted to our specific situation. We translated Logan's questionnaire into the Dutch language.

### 4.1. *Independent variables that were adapted*

**4.1.1.** *Prior involvement*   Logan et al. [14] used the graphical scale for the truck drivers. We chose to use the Likert scale as our ICT professionals have a higher education than the truck drivers and may find the graphical version too simplistic.

**4.1.2.** *Prior personal success and prior success of previous employer*   Prior success and prior success of the previous employer were not combined to form a general success measure because coefficient alpha was too low (0.508).

### 4.2. *Dependent variables that were adapted*

**4.2.1.** *Post-commitment*   We used the same scale as was used from prior commitment but the questions were phrased in the present tense. Coefficient alpha was 0.939.

**4.2.2.** *Post-satisfaction*   We started with the satisfaction scales of the Job Diagnostic Survey [6] to measure overall job satisfaction. It included all factors of job satisfaction including satisfaction with security, pay, growth, supervision and co-worker. Anchors ranged from 1, strongly disagree, to 7, strongly agree. Coefficient alpha for this scale was 0.514. We checked whether deleting several factors had a positive impact on coefficient alpha. We found that the combination of satisfaction with job security, satisfaction with growth and supervision gave the highest alpha score. Coefficient alpha for this measure was 0.657.

## 5. Results

In Table 1 the means and standard deviations for all variables are listed. The ICT professionals showed high involvement, and slightly high satisfaction and success levels at their previous employer. The mean levels for commitment and satisfaction reduced after outsourcing. The new employer scores high on ability to complement.

In Table 2 the zero-order correlations for all variables in the study are presented. Coefficient alpha is provided on the diagonal for multi-item scales. The prior attitudes correlate significantly with one another, as do the post-outsourcing attitudes. High correlation was found between prior job satisfaction and prior commitment. We also found high correlation between post-job satisfaction and post-commitment. We used factor analysis to make

Table 1
Descriptive statistics among all study variables

|                       | Scale    | Mean  | Standard deviation |
|-----------------------|----------|-------|--------------------|
| Prior commitment      | [1...7]  | 4.74  | 0.79               |
| Prior involvement     | [1...7]  | 5.92  | 0.86               |
| Prior satisfaction    | [1...7]  | 5.45  | 0.86               |
| Prior personal success| [1...7]  | 5.22  | 0.93               |
| Prior success employer| [1...7]  | 4.49  | 1.02               |
| Reputation            | [1...7]  | 4.26  | 1.01               |
| Similarity            | [0, 1]   | 0.55  | 0.50               |
| Ability to complement | [0, 1]   | 0.73  | 0.45               |
| Post-commitment       | [1...7]  | 3.80  | 1.18               |
| Post-satisfaction     | [1...7]  | 3.69  | 1.33               |
| Post-intention to quit| [1...7]  | 3.72  | 1.73               |

sure that satisfaction, commitment, perceived success and intention to quit are separate constructs.

Studying the univariate correlations we find a mixed pattern of relationships with respect to the hypotheses. In terms of Hypotheses 1 prior success of the employer correlated in a positive direction with post-satisfaction with co-workers. In the hypothesis we predicted that the relationship between prior success and post-satisfaction would be negative. Hypothesis 1 is not supported by the data.

Turning to Hypothesis 2, we found that ability to complement was positively related to post job satisfaction and post-commitment. Neither the reputation measure nor the similarity measure was significantly correlated to any of the post-outsourcing attitudes. Hypothesis 2 was only partly supported.

We computed multiple regressions in which we used the entire set of pre-outsourcing attitudes as predictors for each of the post-outsourcing attitudes. The results of these regressions can be found in Table 3. For post-outsourcing job satisfaction, they show a significant positive correlation between prior job satisfaction and post job satisfaction. The significant coefficient between ability to complement and post-job satisfaction is replicated. The significant coefficient between ability to complement and post-commitment is also replicated.

## 6. Discussion

Logan et al. [14] found that two sets of employee attitudes prior to outsourcing were predictive of post-outsourcing attitudes. They found that one set of attitudes, which reflected positive assessments of the old employer hindered acceptance of the new employment arrangement. Further they found that the other set of attitudes, which reflected positive assessments of the new company, would enhance adjustment to the outsourcing. Our statistical results, however, only indicate support for the second set of attitudes, which reflect positive assessments of the new employer.

The direction of the effect for prior job satisfaction and post-job satisfaction did not match our expectations. This suggests that it is attractive for the new employer to hire

Table 2
Correlations among all study variables

| | 1 | 2 | 3 | 4 | 5 | 6 | 7 | 8 | 9 | 10 | 11 | 12 |
|---|---|---|---|---|---|---|---|---|---|---|---|---|
| 1. Prior commitment | (0.87) | | | | | | | | | | | |
| 2. Prior Involvement | 0.58** | | | | | | | | | | | |
| 3. Prior satisfaction | 0.66** | 0.54** | (0.61) | | | | | | | | | |
| 4. Prior pers. success | 0.42** | 0.54** | 0.41** | 0.34** | | | | | | | | |
| 5. Prior success empl. | 0.35** | 0.30** | 0.378** | 0.34** | | | | | | | | |
| 6. Reputation | 0.11 | 0.10 | 0.13 | 0.20 | 0.16 | | | | | | | |
| 7. Similarity | −0.01 | −0.13 | −0.19 | 0.05 | −0.07 | −0.05 | | | | | | |
| 8. Ability to complement | −0.13 | 0.00 | −0.11 | −0.02 | −0.21* | 0.28** | −0.06 | | | | | |
| 9. Post-commitment | 0.02 | −0.01 | 0.11 | 0.07 | −0.14 | 0.17 | −0.06 | 0.30** | (0.94) | | | |
| 10. Post-satisfaction | −0.03 | 0.06 | 0.14 | 0.10 | −0.07 | 0.15 | −0.02 | 0.24* | 0.65** | (0.66) | | |
| 12. Post-quit intention | 0.12 | 0.07 | 0.03 | 0.05 | 0.00 | 0.04 | 0.18 | −0.05 | −0.66** | −0.53** | 0.18 | (0.80) |

*Reliability coefficients for multi-item scales are on the diagonal $*p < 0.05$; $**p < 0.01$; $N = 93$.

Table 3
Regression results

| | Post-satisfaction | | Post-commitment | | Post-intention to quit | |
|---|---|---|---|---|---|---|
| | R2 | B | R2 | B | R2 | B |
| Full set of variables | 0.12 | | $0.16^a$ | | 0.06 | |
| Prior commitment | | −0.35 | | 0.00 | | 0.29 |
| Prior involvement | | −0.01 | | −0.21 | | 0.13 |
| Prior satisfaction | | 0.46* | | 0.31 | | −0.09 |
| Prior personal success | | 0.12 | | 0.14 | | −0.06 |
| Prior success employer | | −0.13 | | −0.20 | | −0.08 |
| Reputation | | 0.10 | | 0.11 | | 0.11 |
| Similarity | | 0.11 | | −0.07 | | 0.63 |
| Ability to complement | | $0.60^a$ | | 0.70* | | −0.21 |

$^a p < 0.10$; $^* p < 0.05$; $^{**} p < 0.01$; $N = 93$.
For intention to quit we found no significant predictors.
Overall, these results partially support Hypothesis 2. Hypothesis 1 is not supported.

satisfied ICT professionals, as they may be satisfied in the new environment as well. We found no significant relationship between the other attitudes about the old employer and the post-outsourcing attitudes. This might be explained by the fact that ICT professionals do not feel highly committed to an employer but are more committed to their profession.

The direction of the effect of ability to complement matched our expectations. These findings suggest that it is important for the service provider to communicate to the ICT professionals of the client organization how their new employer can add value to their work by offering them access to new technologies, methodologies, training and work practices and specialized and knowledgeable colleagues. This outcome might be explained by the high growth need that Couger and Zawacki [1] found among ICT professionals.

Ability to complement correlated moderately negative with perceived success of the previous employer. This might indicate that those ICT professionals that perceive their previous employer to be highly successful do not expect the new employer to be able to make a high positive contribution. Although we did not find a direct relationship between prior outsourcing success and post-outsourcing attitudes there seems to be some empirical basis for the conclusion that there is a moderate indirect negative relationship between perceived success of the previous employer and post-outsourcing satisfaction and commitment, through ability to complement.

## 7. Conclusions

We consider this research to be a preliminary investigation. Our sample consisted of only 93 ICT professionals. We applied a retrospective measurement strategy to assess pre-outsourcing attitudes [14]. However, identifying ICT professionals and gaining their co-operation before an outsourcing occurs might just be impossible. From the moment the employees know about the outsourcing their perceptions about the current employer will change.

Kessler et al. [9] argued that three things contribute to employee response to outsourcing: 'context', 'pull' and 'landing'. Our findings indicate that for ICT professionals, 'pull' or the attractiveness of the new employer, has most impact. ICT professionals that feel that the outsourcing provider has the ability to complement their work are most likely to show the highest levels of satisfaction and commitment with their new employer. This supports the proposition based on social identity theory that states that if individuals consider outsourcing as contributing positively to self-concept or self-esteem, they will support the changes and their attitudes should show signs of improvement.

Our data suggest that acquiring a department with satisfied ICT professionals may be easier than acquiring a department with unsatisfied employees. This finding does not support the proposition that was based on social identity theory, that argues that ICT professionals who have the most positive attitudes and the highest commitment to their previous employer have the most difficulty in adjusting to a new employment setting. However, we did find indirect support of the social identity theory. ICT professionals that perceive their previous employer to be highly successful, are more likely to expect the service provider's ability to complement their work lower and might be less satisfied and committed with their new employer.

From our findings we could not conclude whether ICT professionals that are highly committed to their previous employer, will be more satisfied after outsourcing, when they can continue doing the same tasks for their previous employer, instead of being sent off to other customers. We advise to explore this further in future research. It is also interesting to pay more attention to the level and type of commitment of the ICT professionals. Are they indeed more committed to their profession than to their employer and how does that influence the transition to an ICT service provider?

In this study we evaluated the perceived success of the ICT professionals. We found that the perceived professionalism of the ICT service provider is an important factor. But does the ICT service provider indeed offer the ICT professionals new career perspectives. Which employees do get more responsibilities, or get the opportunity to specialize in a particular area and which employees are still doing more or less the same job?

Based on these findings of our study we can give several directions to managers that need to manage ICT outsourcing transitions. The results suggest that satisfaction and a positive perception about the service provider's ability to complement their work are desirable employee characteristics. Although managers usually cannot chose between IT departments to be acquired, this result indicates that managers of outsourcing providers should be very clear about the benefits of outsourcing. They should communicate in an early stage that the salary, pension and all other job conditions will be harmonized, so that it cannot become a source of dissatisfaction. Further they need to emphasize the opportunities that an ICT service provider can offer to ICT professionals. The roles and futures of the ICT professionals should be outlined as well as the techniques, working methods, available training etcetera that can make a positive contribution to the ICT professionals' work. Managers need to pay special attention to those employees that perceive their old employer to be very successful, or risk being stuck with unsatisfied and uncommitted employees.

Our study does not support Logan's [13] work that employee perceptions about the previous job can negatively influence job attitudes after outsourcing. From our findings we can conclude that is more desirable to acquire an IT department with satisfied ICT professionals than an IT department with unsatisfied employees. Our findings also indicate that a positive attitude about the professionalism of the new employer can help to ease the transition.

## Acknowledgements

We wish to acknowledge the advices of a few people in conducting this research. We would like to thank dr. Mary Logan for sharing her questionnaire with us. We thank dr. Soon Ang for advising us on the research design and we like to thank dr. Patrick Groenen for advising us on the statistical analysis of the survey data.

## References

[1] J.D. Couger and R.A. Zawacki, *Motivating and Managing Computer Personnel*, Wiley, New York (1980).

[2] R.T. Due, *The real cost of outsourcing*, Information Systems Management (1992), 78–81.

[3] M.J. Earl, *The risks of outsourcing IT*, Sloan Management Review **37** (3) (1996), 26–32.

[4] J. Goo, R. Kishore and H.R. Rao, *Management of information technology relationships: The role of service level agreements*, Proceeding of the Twenty-Fifth Annual International Conference on Information Systems (2004).

[5] U.T. Gupta and A. Gupta, *Outsourcing the IS function: Is it necessary for your organization?*, Information Management Systems (1992), 44–50.

[6] J.R. Hackman and G.R. Oldham, *Development of the job diagnostic survey*, Journal of Applied Psychology **60** (1975), 159–170.

[7] P.R. Haunschild, R.L. Moreland and A.J. Murrell, *Sources of resistance to mergers between groups*, Journal of Applied Psychology **24** (1994), 1150–1178.

[8] R. Huang, S. Miranda and J.-N. Lee, *How many vendors does it take to change a light bulb? Mitigating the risk of resource dependence in information technology outsourcing*, Proceeding of the Twenty-Fifth Annual International Conference on Information Systems (2004).

[9] I. Kessler, J. Coyle-Shapiro and J. Purcell, *Outsourcing and the employee perspective*, Human Resource Management Journal **9** (1999), 5–19.

[10] M. Khosrowpour, G.H. Subramanian, J. Gunderman and A. Saber, *Managing information technology with outsourcing: An assessment of employee perceptions*, Journal of Applied Business Research **12** (3) (1996), 83–96.

[11] M.C. Lacity, L.P. Willcocks and D.F. Feeny, *The value of selective IT sourcing*, Sloan Management Review **37** (1996), 13–26.

[12] J.N. Lee, M.Q. Huynh, R.C.-W. Kwok and S.M. Pi, *IT outsourcing evolution: Its past, present and future*, Communications of the ACM **45** (3) (2003), 84–89.

[13] N. Levina and J.W. Ross, *From the Vendor's perspective: Exploring the value proposition in information technology outsourcing*, Management Information Systems Quarterly **27** (3) (2003), 331–364.

[14] M.S. Logan, K. Faught and D.C. Ganster, *Outsourcing a satisfied and committed workforce: A trucking industry case study*, International Journal of Human Resource Management **15** (1) (2004), 147–162.

[15] B.L. Mak and H. Sockel, *A confirmatory factor analysis of IS employee motivation and retention*, Information & Management **38** (1999), 265–276.

[16] J. Purcell, *Contingent workers and human resource strategy: Rediscovering the core periphery dimension*, Journal of Professional HRM **5** (1996), 16–23.

[17] L.P. Willcocks and M.C. Lacity, *IT outsourcing in insurance services: Risk, creative contracting and business advantage*, Information Systems Journal **9** (1999), 163–180.

[18] L.P. Willcocks, M.C. Lacity and T. Kern, *Risk mitigation in IT outsourcing strategy revisited: Longitudinal case research at LISA*, Journal of Strategic Information Systems **8** (1999), 285–314.

# 8. Professional and Social Issues

## 8.1. *Commentary*

The administration of systems is as much about humans as it is about technology. A network of computers enables society to be bound together with efficiency and speed but this power carries with it certain responsibilities. In many ways a Network and System Administrator profession is analogous to the medical profession, with its ability to look deep inside peoples' most private places and make life or death decisions about the plumbing.

For a couple of decades now, the profession of system administrators has been seeking to consolidate its identity and define its role, even as the technological and social landscape has changed. It is natural that this has led to the formation of organizations and interest groups, both local and global, who discuss these matters at length. These organizations have not been without their controversies. Indeed, the question of whether system administration is a discipline is still posed every year. In this part we set aside any doubts about the validity of the discipline and address two issues in this human realm.

Chalup writes about the professional organizations that have been responsible for the open environment of especially Unix system administration. In addition to these, there are many industry-academic interest groups such as the Internet Engineering Task Force (IETF), the Distributed Management Task Force (DMTF), the Tele-Management Forum (TMF) and the Network Management Research Group (NMRG).

Fagernes and Ribu offer an overview of the human and ethical challenges and dilemmas associated with working with a demanding technology, often in a position of power over individuals. They root their discussion in classical ethical philosophy and branch into the practical matters of practice and law-making connected with confidential information about users and their activities.

There are many topics that could be added to this part, including human resource management, time management [3], international cyber-law and Digital Rights Management, to name a few. More specialist texts will have to cover these. Terms and conditions of employment and its termination are also a topic that receives little coverage [1] but where many IT employees have to sign agreements affecting their rights.

Finally, Internet governance [2] is a topic that affects not individuals but perhaps countries and societies and will play a larger role in the future of the Internet.

## References

[1]  M. Burgess, *Principles of Network and System Administration*, Wiley, Chichester (2000).
[2]  B. Kahin and J.H. Keller, *Coordinating the Internet*, MIT Press (1997).
[3]  T. Limoncelli, *Time Management*, O'Reilly (2005).

# – 8.2 –

# Systems Administration as a Self-Organizing System: The Professionalization of SA via Interest and Advocacy Groups

Strata R. Chalup

*E-mail: strata@virtual.net*

**Abstract**

Lacking a single vendor or institution both motivated and powerful enough to provide a structure for collaboration, early Unix system administrators had to build their own structures. The gradual coalescence of a cohesive SA community out of the mish-mash of academic, commercial, and government Unix environments was among the most fruitful of these boot-strapping efforts. Looking ahead, SA still lacks the formalized ties into traditional professional societies for the knowledge transfer that will be required for full maturity of the practice and profession. The emergence of university programs in system administration may provide the bridging required to link system administration with relevant formal disciplines such as operations research and computer science.

## 1. Introduction

System administration is clearly shown to be a member of the engineering professions by the plethora of sysadmin membership organizations, each focused on particular specialties, credentials, or topics. Surprisingly, there is no overarching organization today that serves the entire system administrator (SA) community. This may be a side-effect of the largely empirical growth of the profession, which emerged from computing facilities less formalized than the mainframe environment.

Most computing specialties find a home under the wide umbrellas of the Association for Computing Machinery (ACM) [1] or the Institute of Electrical and Electronics Engineers

HANDBOOK OF NETWORK AND SYSTEM ADMINISTRATION
Edited by Jan Bergstra and Mark Burgess

(IEE) [8]. The former considers itself a primarily an "educational and scientific computing society", whereas the latter's focus is "the world's leading professional association for the advancement of technology". The numbers tend to back them up, with over 80,000 ACM members and over 365,000 IEEE members worldwide. None of the existing SA-oriented organizations can come within an order of magnitude of these membership figures.

The ACM and IEEE contain a small number of SA-specific Special Interest Groups (SIGs) amongst their numerous offerings. Both communities are aimed at a different demographic, that of research or of applied technology. Neither currently offers a comfortable middle ground for those whose day to day jobs include a mixture of theory, practice, and old-fashioned ingenuity.

This raises the excellent question of "how can we formally involve SAs in the quality improvement process for technology?" but this is outside the scope of this paper. System administration's status as a poor stepchild of CS and EE trained mainframe career paths of the 1950s–1970s needs to be corrected. A multi-pronged approach will be required to enable appropriate technology and knowledge transfer from the SA field to research to design and implementation. Currently, the information flow tends to be poorly bridged and often in one direction only, generally in ad-hoc ways by organizations and individual contributors. The formal training area that could seemingly most benefit hands-on SAs, namely operations research, is largely absent in the field.

## 2. "Small pieces, loosely joined"

David Weinberger's classic description of the World Wide Web, in contrast to the end to end orientation of the Internet, seems uncannily appropriate to describe the number of different, but slightly overlapping, groups to which the modern-day SA is likely to belong.

ACM and IEEE do not serve their traditional roles as an overarching organization. For example, in the United States, USENIX is the closest thing to an ACM equivalent for Unix-oriented system administrators, but is largely ignored by desktop computer admins and network admins. Network and protocol-oriented admins, along with their counterparts from the programming and development world, tend to be involved in the Internet Engineering Task Force [5], a "large, open, and international community of Internet professionals concerned with the evolution of the Internet architecture". Network administrators primarily rally around the IETF and the North American Network Operators Group [13], which despite its name and description has a strong global following. These are primarily working groups and discussion lists, rather than more passive, benefit-oriented, membership organizations.

A symptom of the disjunction between 'traditional' professional societies and SA organizations is that SAs generally do not become actively involved in ACM or IEEE unless they are classically educated in the computing sciences, or until they reach a certain level of seniority. Rather than the usual model of joining a profession/field oriented parent organization as a student, and then professional societies as one advances, SAs are currently largely self-taught. It is common for SAs to be members of several smaller organizations separately addressing training, specialties and career focus without a common 'mesh' organization.

Examples of career-focused groups include the Independent Computer Consultants Association [7] and the Information Systems Audit and Control Association [19]. Some of these groups emerged more out of the ACM/IEEE culture and model than the more technology-oriented groups which dominate the SA field.

## 3. The evolution of system administration organizations

Managing complex computer infrastructures requires a support structure of peers who can serve as mentors, inspiration, and sounding boards for solutions and wild ideas. Early Unix platforms provided some of the same computational timesharing facilities generally associated with mainframes of the day. The mainframe operators enjoyed a combination of standardization within specific vendor communities, coupled with a high level of formal and informal organizational support from the platform providers. The operator community was correctly identified as a valuable resource by vendors of expensive mainframes, and cultivating the community was part of most vendors' official sales strategy. Groups such as Digital Equipment Corporation's DECUS [6] and Control Data Corporation's PLATO provided an early model for structured interaction between systems staff and vendors, and for peer to peer networking between operators from different sites.

Meanwhile, Unix SAs were not 'feeling the love'. In sharp contrast to the active vendor support enjoyed by mainframe SAs, the pioneers managing Unix environments of the 1970s and 1980s were generally ignored or actively discouraged by vendors. Even Digital Equipment Corporation, whose internal Unix Education Group (UEG) had created a Bell Labs V7 port for the VAX and played key roles in the creation of 4.1BSD, did not consider any Unix version supported until its own release of Ultrix in 1984. Early SAs resignedly kept a VMS or RSTS-E boot tape handy in case their systems required hardware support from DEC. Soon DEC's PDPs and VAXes were joined in Unix capabilities by an ever-growing assortment of Motorola 68000 platforms from established players and startups [2,3].

Key players in the tiny but growing field of Unix system administration set out to channel their irritation and need for peer communication in a useful direction. Ultimately the strong research orientation of early users of Bell Labs' most viral software release would find expression in the peer-reviewed paper model. Presentation-oriented meetings would become the USENIX Annual Technical conference in the late 1970s, and spin off the Large Installation System Administrators (LISA) [9] conference in 1987.

## 4. The grain of sand becomes a pearl

The USENIX Association [20] played a key role as a research and technology enabler for early Unix and for SAs before most of the research and business community had ever heard of Unix [22]. USENIX's Special Technical Group SAGE [14], along with geographically strategic regional groups San Francisco Bay Area's BayLISA [4] and New Jersey's $groupname [23] served as focal points for the growing body of best practices in Unix SA during critical points in the evolution of the field. Bay LISA and $groupname diminished as SAGE grew but they have maintained their regional focus and add value as a venue for rising stars and new technologies to debut at their monthly meetings. SAGE-AU quickly became an entity worthy of great respect, hosting its own conferences and advancing the field considerably.

The new but growing organization known as the League of Professional System Administrators [10] has attracted advocates who appreciate the emphasis on professionalization of SA, rather than research. Approaching professionalization from another angle are vendor-oriented SA conferences such as O'Reilly's Open Source Convention (OSCon) and IDG Publishing's IDC event.

Regional groups such as Boston's Back Bay LISA and Dublin's SAGE-IE meet regularly to provide a local face to face connection outside of the general conference circuit. This function is epitomized by SIG-BEER, described by its practitioners as a 'venerable DC-SAGE tradition' where SAs gather at a local bar monthly to bend the elbow and discuss the profession, the machines, and whatever else comes to mind. There is friendly disagreement as to whether the origin of the group-within-a-group is 'SIG-beer', as in 'Special Interest Group – Beer' ala the ACM, or derives from the Unix-centric explanation of kill SIG-BEER, i.e., 'send the BEER signal to that process' [5]. The success of SIG-BEER inspired the creation of SIG-BEER-WEST when several long-time DC residents moved to the San Francisco Bay Area. Retiring to a local pub, hostelry, or restaurant after a meeting is part of many other SA groups traditions as well.

Most SAs also belong to one or more operating-system and/or technology advocacy groups. Technology advocacy groups tend to be virtual online communities centered around an open source project vital to many SAs daily tasks. Examples of these are the Apache Foundation, various SourceForge projects, the CPAN Perl archive, and monitoring tools such as MRTG and RRDTool.

Operating-system advocacy groups are often regional Linux User Groups (LUGs), which have real-world meetings and troubleshooting sessions (installfests). In areas dominated by particular vendors, there may also be vendor groups for SAS administering vendor Unix, such as Sun Microsystems Solaris, Hewlett-Packard's HP-UX, and IBM's AIX/These are increasingly rare. In general, LUGs now serve where sysadmin groups once trod. There is bottom-up organization, too, as small regional LUGs merge to create meta-event aggregators such as SVLUG (Silicon Valley) or BALE (Bay Area Linux Events). Many SA-centric regional groups are affiliated with organizations such as SAGE, including -AU, -IE, and other country-specific variants, or LOPSA. Regional groups that predate the introduction and popularity of Linux tend to persist and continue meeting, but in many cases Linux User Groups, or LUGs, have taken over the function of Unix SA regional groups and groups have quietly dissolved.

## 5. Back to the future? The return of vendor-oriented communities

The past decade has seen a sharp rise in the number of vendor certification programs offered, and many successful attainers tend to congregate in vendor-specific forums and at vendor-sponsored conferences, such as Oracle's annual conference. Certification-centered communities are evolving around Microsoft's operating systems and their Microsoft Certified Professional (MCP) [12] program. Following closely are programs by other vendors, of which Sun, Red Hat, Oracle, and Cisco are among the most prominent.

Training vendors have also embraced user communities as a way of publicizing their own certificate programs. The Linux Professional Institute (LPI) [11] and the SANS (SysAdmin, Audit, Network, Security) Institute [17] are examples of vendors which do not have membership groups per se, but nonetheless provide support frameworks such as mailing lists, conferences and regional meetups as a combination 'virtual user group' and upselling opportunity. While LPI is a non-profit, and SANS is a for-profit, both organizations allocate considerable resources to providing information for the general public, such as SANS' popular 'Internet Storm Center' [18].

## 6. History in action: SAGE and LOPSA

An excellent example of changing dynamics in SA groups is the multi-year process whereby a new SA group formed to address areas which some members felt were inadequately covered by existing groups. As SAGE has been for many years the pre-eminent SA group, it is natural that it also has been involved with a number of these group formations.

The earliest, and in many ways simplest, division of the SAGE group began when SAs from Australia and several European countries approached the USENIX Association about forming their own versions of SAGE [15,16]. Several international organizations were formed, most with independent charters that offer 'associate' (e.g., non-voting) membership reciprocality for US SAGE members [15,16].

The success of the international groups is credited, with some irony, with sowing the seeds of what became an effort to separate SAGE from its parent organization. Australia's SAGE-AU, and the SAGE-WISE organization of Wales, Ireland, Scotland and England, seemed to beg the question of a SAGE-US. To many SAs, the formation of an independent SAGE or SAGE-like entity seemed logical, with the USENIX Association's Special Technical Group becoming a de facto SAGE-US under a new parent organization.

After many years of impassioned debate between prominent members on the future directions of SAGE, several senior professionals formed LOPSA. As one might imagine, the process of creating LOPSA, and the subsequent repudiation of SAGE by some of its founders, had political repercussions throughout the SA community. The international SAGE groups, such as SAGE-AU, were generally not involved in this matter, considered strictly a North American phenomenon. It was clear to many observers of this protracted debate that all involved strongly believed that the SA community and profession would benefit from following their particular course of action.

As LOPSA reaches its first anniversary, tensions amongst the community seem to have receded, with many SAs participating in both groups. The author's personal slogan during the rocky early days was "It's a community, not a contest", and this seems to have been the most prevalent approach. While the two groups share many similarities, as evidenced by their mission statements included below, many see the current orientation of USENIX and SAGE as more like that of the ACM and LOPSA representing the technology and professional advocacy focus of IEEE. It is too soon to tell how the organizations will grow in response to each other. More than one senior professional has offered the opinion that having multiple organizations thriving is a sign of vigor in a professional community.

> LOPSA: "Our mission is to advance the practice of system administration; to support, recognize, educate, and encourage its practitioners; and to serve the public through education and outreach on system administration issues."

> SAGE: "...to serve the system administration community by:
> - Offering conferences and training to enhance the technical and managerial capabilities of members of the profession.
> - Promoting activities that advance the state of the art or the community.
> - Providing tools, information, and services to assist system administrators and their organizations.
> - Establishing standards of professional excellence and recognizing those who attain them".

## 7. Next steps forward

As we have seen, the system administration community encompasses a wide range of professional, social, and advocacy groups. The ad-hoc nature of group formation and membership somewhat echoes the self-taught aspects of pioneers in the profession. This author's opinion is that a crucial step in the maturation of the profession is to link up in some fashion with traditional organizations such as ACM and IEEE. This is likely to be accomplished by a multi-pronged approach of undergraduate education, continued emphasis on development of both research and profession-oriented groups, and active bridge-building on the part of the SA community.

## 8. A brief timeline of SA organization history in North America

We appreciate information provided by the USENIX Association [21] in constructing this timeline.

1987. April. The first LISA Workshop, chaired by Rob Kolstad and Alix Vasilatos, brings in 75 attendees.

1991. BayLISA founding year. Tradition of meetings being on the 3rd Thursday of the month will be followed by other groups such as $groupname; may date from DECUS days.

1992. SAGE, the System Administrators' Guild, started in early 1992 by a group of San Francisco Bay Area sysadmins, many of whom were also BayLISA founders. Founded as a Special Technical Group of USENIX.

1993. $groupname 1993: "$GROUPNAME is an organization for UNIX system administrators in New Jersey formed to facilitate information exchange pertaining to the field of

Unix system administration. $GROUPNAME is not affiliated with a particular hardware or software vendor or company". (Via their website.)

2002. David Albans moves to the SF Bay Area from Washington, DC, and starts SIG-BEER-WEST in February, continuing the 3rd Saturday tradition. Still going strong in 2006.

2005. LOPSA founding year. LOPSA holds first conference in Phoenix in fall 2006.

## References

[1]  Association for Computing Machinery, http://www.acm.org/.
[2]  AT&T System V, http://en.wikipedia.org/wiki/UNIX_System_V.
[3]  AT&T Version 7, http://en.wikipedia.org/wiki/Unix.
[4]  BayLISA, http://www.baylisa.org/.
[5]  IETF, http://www.ietf.org/.
[6]  In 2000, the DECUS U.S. Chapter incorporated as the independent user group Encompass.
[7]  Independent Computer Consultants Association, http://www.icca.org/.
[8]  Institute of Electrical and Electronics Engineers, http://www.ieee.org/.
[9]  LISA, http://www.usenix.org/events/byname/lisa.html.
[10] LOPSA, http://www.lopsa.org/.
[11] LPI, http://www.lpi.org/.
[12] MCP, http://www.microsoft.com/learning/mcp/mcp/.
[13] NANOG, http://www.nanog.org/.
[14] SAGE (formerly the System Administrators Guild), http://www.usenix.org/.
[15] SAGE Internationals, http://www.guug.de/verein/sage.html?view=print&lang=en.
[16] SAGE Internationals, http://sageweb.sage.org/locals/international.html.
[17] SANS, http://www.sans.org/.
[18] SANS Internet Storm Center, http://isc.sans.org/.
[19] The Information Systems Audit and Control Association, http://www.isaca.org/.
[20] USENIX, http://www.usenix.org/.
[21] USENIX Community Services timeline, http://www.usenix.org/about/history/community.html.
[22] USENIX Firsts timeline, http://www.usenix.org/about/history/firsts.html.
[23] $GROUPNAME, http://www.groupname.org/.

# – 8.3 –

# Ethical, Legal and Social Aspects of Systems

Siri Fagernes, Kirsten Ribu

*Faculty of Engineering, University College Oslo, Pb 4 St. Olavs plass, 0130 Oslo, Norway*
*E-mails: {siri.fagernes,kirsten.ribu}@iu.hio.no*

## 1. Introduction

Communication networks have altered many aspects of life, like work and employment, healthcare, transportation and entertainment. Technology has brought many benefits, but is also the cause of many social and ethical concerns: system failures, the destruction of property, the so-called *digital divide*, workplace monitoring, Internet censorship, and the increasing gap between rich and poor, are some examples of important ethical questions that arise when we reflect on how technology has impacted modern life.

System Administrators play an important part in keeping networks secure. They are also responsible for the uninterrupted operation of the computers to take care of the business needs. A system administrator's duties and responsibilities encompass more than technical knowledge. They are given broad access to the resources of computer systems because their job responsibilities require such access. For instance, in troubleshooting email issues, a system administrator may have to go through the employees' email, but must treat any knowledge so gained as private. Many companies have codes of ethics for employees who in the nature of their job have privileged access to company systems. The content of such codes may include statements like: "do not browse through the computer information of system users while using the powers of privileged access unless such browsing: is a specific part of the job description, do not disclose computer information observed while operating with privileged access, do not copy any computer information for any purpose other than those authorized under their defined job responsibilities, report suspicious incidents to management or police" and so on. Such codes of ethics provide guidelines for

HANDBOOK OF NETWORK AND SYSTEM ADMINISTRATION
Edited by Jan Bergstra and Mark Burgess

professional conduct, and are useful tools. However, situations do arise where it may be hard to decide what to do, for instance, how would you react if your boss asks you to go through all the employees' email?

## 2. Social aspects of technology

New technologies have blossomed during the past century, and have had a tremendous impact on our way of living. Automated methods for developing all kinds of products have lowered the costs and hence the prices, which again has lead to increased purchasing power for households. Less manual labor has reduced the physical strain on the human body, which combined with improved methods within the medical profession, has lead to increased life expectancy.

However, the overall effects of new technology are never fully predictable, and it is obvious that there are several side effects to this development. While less manual labor has led to less physical strain, various health problems might emerge through inactivity. While technology has lead to automated production and increased productivity, the people who earlier performed this work manually have become obsolete.

These and other important issues call for increased awareness regarding introduction of new technology, concerning both short-term and long-term effects on our society. In this section we will present several areas that should be studied in more detail.

### 2.1. *Health issues*

When discussing the social implications of technology, the ones perhaps raising the biggest concerns are those connected to human health. Increased access to technology has reduced the need for human physical labor, and hence reduced the severe physical strain on the human body as a result of daily work. Also, the expected age has increased dramatically as a result of progress within the medical sciences, assisted by better technologies to aid laboratory analysis, test-taking, speeding up the process for giving diagnosis to illnesses and so on.

However, the dramatic increase in the use of so-called *ICTs* the last decade, has also meant increased exposure to other negative side-effects. Extensive research has been performed on this issue, but it seems very difficult to come to any clear conclusion.

When discussing the health effects of information and communication technology, one may distinguish between the physical effects and the psychological effects. It might be of no surprise that the physical effects of ICT on the human body are a lot simpler to investigate than the psychological effects. The first are also the best documented of the two [3]. Of the physical side-effects, the following are among the most common:

1. *Weight issues.* The extensive use of ICTs, at home, in schools and at work, has lead to less physical activity, and hence to an increase in the number of cases of serious overweight or obesity.
2. *Repetitive strain injuries.* As a result of extensive use of a computer mouse or joystick, severe pain in arms and shoulders has been a common problem for those using

a computer as a daily routine. Also, spending long hours sitting in front of a computer screen often lead to extensive strain on the upper body including the back region.

3. *Vision.* The development of the eyes/vision of children could have negative effects of the radiation from TVs and CRT-screens.

4. *Artificial light.* Some types of artificial light (television screens and computer CRTs) might be harmful to biological organisms.

5. *Radiation.* Our daily exposure to electromagnetic signals from several types of networked devices might pose potential risks.

On the psychological side, the results are a bit more difficult to interpret, since the *content* of the media studied could be as important as the amount of consumption [3]. However, there are suggestions of certain physiological effects. Research has shown clear correlations between heavy use of media (television, Internet, video games etc.) and signs of depression and anxiety among young people [3]. Information overload, and the fact that youngsters are not mentally prepared for all these impressions are listed as causes.[1]

Other aspects that could be affected are the general intellectual development, attention span and imagination of users. One particular example is the reliance of external stimulation as a result of heavy consumption of ICT-related media. Also, addiction to certain activities as a result of the new technologies might be a problem in the near future.

## 2.2. *Employment and work*

A common perception is that innovations in technology result in fewer available jobs, because automated production makes human labor redundant. Investigations show that several branches have suffered from reduction in the number of available jobs, both within manufacturing and so-called 'white-collar' jobs [19].

On the other hand, there exist several examples of how innovations within technology have lead to an increase in the number of available jobs, because new technology creates a need for new job functions. Also, innovations cause prices of different goods to drop, again leading to increased purchasing power among people in general, which again leads to increased demand for the products, and hence more jobs.

*Teleworking*, as a result of improved communication technology, has resulted in improved working conditions for those dependent on working from home. This makes it possible for single parents without means to pay for child care, people living in rural areas, and others who normally would have difficulties working normal office hours, to take part in the job market. However, common experiences are a lack of contact with colleagues and office environment, difficulties in separating office hours and leisure time, since office and home are at the same location, and sometimes missed opportunities in getting promoted.

## 2.3. *Changing social and group-dynamical patterns*

One of the most influential technological changes during the last decade has been the mobile phone. Teenagers have adopted this new technology and let it change the way they

---

[1]Editorial comment: Omega-3 deficiency has been related to youth apathy and media addiction in trials in the U.K. indicating that media dependency is part of a wider problem.

interact and keep in touch, even using it as a basis for organizing their circle of friends. This is an example of how *not* having a cellular phone could cause one to be excluded from the social community, as it is the main source of communication and exchange of information.

In the Internet community, there are examples of office environments changing their meeting habits, now using chatrooms as a forum for office meetings. When meetings and general discussions take place 'online' instead of in a physical meeting room, changes in group dynamics emerge, such as who dominates the group, as there typically are different skills needed to succeed online. This is an example of how new technology might fundamentally change the way humans interact, and affect which groups that benefit from the interaction form. While typical 'nerds' might be intimidated by physical meetings, they can 'rule' online.

New technologies also have an impact on the family and home environment. Teenagers tend to adopt new technology more easily than their parents, and hence gain more 'power' in the home, as they become the 'gurus' of the family. This has several implications for the dynamics between parents and their children. For instance, how can the parents control the Internet user habits of their children if the children are in charge of the computers in the home. This phenomenon disturbs the normal family pattern where the parents are the role models.

### 2.4. *The digital divide*

As information technology has gained foothold in our society, a phenomenon known as the *digital divide* has emerged. A common definition of the digital divide is *the situation where some people have access to modern information technology while others do not* [19]. Usually one speaks of the *global* and the *social* divide, where the global divide is a measure of the difference in Internet access between developed and less developed countries, while the social divide is a reference to the difference between individuals within the same country [19]. Parameters affecting the level of Internet access within a particular country are, among others, race, income level and age group.

The global divide is illustrated by the numbers of people in the world with Internet access. In June 2006, approximately 1,043 million people worldwide use the Internet, according to Internet World Stats. The United States, Australia and Europe are well above the average, while other parts of the world are below. In Africa, only about 1 out of 200 persons had Internet access in 2000 (2.6 percent). Because of the high numbers of Internet users in the West, one tends to forget that most people in the world actually have little access to not only the Internet, but also telephones, health care, education and books. These problems are caused by poverty, isolation and politics. Also, there is a digital divide in the United States and other western countries between the rich and the poor, people of different age groups, and between people of different ethnical background. The consequences are serious: people who have not acquired basic computer skills will have trouble finding jobs, and will have little chance to escape poverty. Thus, the global digital divide will continue to exist.

When discussing the digital divide, one important concept is *technological diffusion*, which refers to *the rate at which a new technology is assimilated into a society*. There exist two common theories on how this happens:

1. *The normalization model.* This model predicts that *everyone* in a society will eventually adopt to a new technology, but that the wealthiest part of the population adopts first, then the middle range, and finally the least wealthy part. This model is referred to as the 'optimistic' model.
2. *The stratification model.* This model predicts that the income groups will adopt in the same order as the other model, but it contradicts in the sense that it claims that a larger percentage of the wealthier groups will adopt than the less wealthier. This means that a certain part of the population will *never* take to the new technology, and describes a pessimistic model.

The reason why the digital divide is such a concern nowadays, is the relationship between 'digital literacy' and general wealth or success. Some of the criticism against the digital divide lies in the fact that it appears that digital literacy follows from general success, and not the other way round.

## 3. Ethics

### 3.1. *Ethics and morality*

The term 'ethics' defines the philosophical study of morality, whereas the term 'morality' refers to the way we live and act. To simplify this, we can say that if the traffic laws and regulations are the underlying ethics of driving, morality is abiding by these laws and regulations. Speeding and drunk driving is obviously immoral conduct. Some behavior is not regulated by law, but rather agreed on, for instance polite and considerate driving. Traffic bullies are therefore immoral. In this way we see that ethics reach beyond the law, and philosophers have at all times occupied themselves a great deal with the theoretical study of ethics, or moral philosophy. There are many different perspectives on ethics, as we shall see.

Ethical dilemmas arise when we are faced with problems of right or wrong. The difference between right and wrong may be hard to determine, and is not necessarily a question of what is legal. Tying 'good' and 'bad' to expressions of approval or disapproval of human actions may be difficult, because things can be good or bad independently of our personal approval or disapproval of them [22].

Certain ethical judgments can be classified as feelings or emotions, and these are neither true or false. This view is known as emotivism. However, most philosophers refer to our actions and intentions when discussing moral matters, not to our emotions, because this belongs more to the field of Psychology. The challenge of showing that people are better off acting morally than immorally may be difficult. Personal gain and individual convictions play an important part in most people's lives. Although we know that copying proprietary software or downloading music may be illegal, many people are opposed to the laws regulating such activities, and defend their actions by attacking unpopular regulations. So in order to decide whether something is good or bad, right or wrong, one must rely on experience and common sense, and judge by the consequences or intentions of the act. A knowledge of ethical principles may be a help when trying to analyze difficult ethical dilemmas.

Changes in technology require changes in laws and personal attitudes. The use of computers can cause new ethical dilemmas because people are pushed into unforeseen situations. Although many questions that arise from the use of computers are not very different from problems encountered in other aspects of life, the activities on the Internet, the sheer size of the World Wide Web and the amount of available information often create new ethical dilemmas. Computers are a part of our everyday life, both professional and private, and we are dependent on computer networks in all parts of society. A problem with a computer system may negatively impact a whole company, the company's business partners, financial interests as well as hundreds of users. The interest in ethical reflection on the use of computer systems is steadily growing, and the study of ethics is incorporated into the computer science curricula in many universities and colleges.

The growth of the Internet has created several new legal issues. Boundaries are no longer geographical, and this fact can be used to circumvent national laws. For instance, Internet-based casinos are illegal in many countries, therefore, gambling web-sites are run from places like the Caribbean or Malta. Some of the main concerns in computer ethics today are questions of copyright, privacy and censorship. Traditional rules of conduct are not always applicable to a new medium. It is necessary to reflect on questions like the following: If a certain nation wishes to protect freedom of speech on the Internet, whose laws apply? What about censorship on the Internet? A national constitution protecting freedom of speech is in a sense a local law which does not necessarily apply in other countries. How can issues like freedom of speech, protection of intellectual property, invasions of privacy be governed by law when the whole world is involved?

It is important to become aware of the fact that there are not necessarily correct answers to ethical dilemmas. Every computer professional should therefore have some knowledge of international law and best practices, and an awareness of the responsibilities tied to the job, and be acquainted with the codes of ethics that exist for members of the profession. He or she should reflect on how to deal with ethical dilemmas in situations where there may be considerable pressure to do something illegal or ethically wrong.

## 3.2. *What are computer ethics?*

Ethics and values are today discussed in a context that is not limited to a particular geographic region, a specific religion or culture. Much of the discussion on ethics today has to do with the use of modern technology. Computers are everywhere, playing a large part of peoples' lives, both professionally and in the field of entertainment and relaxation. Computer ethics can therefore be understood as that branch of applied ethics which studies and analyzes social and ethical impacts of information technology.

Computer ethics involve the use of computers, networks and the Internet, and therefore also include global information ethics. Computers have become commonplace, and have great impact on society and the transformation of culture. Because computers have become universal and are part of our daily lives, many ethical problems have to do with computer ethics. Although problems involving computing connect with every-day ethical issues in some way, the field of computer ethics is also unique because of the technology itself. Computers can be designed to perform any task we wish. They are universal tools that can be used in so many ways that it is difficult to foresee all the situations that may arise. Therefore, we do not always have traditions or policies to act as ethical guidelines.

Computers tend to create more novel situations and dilemmas than other tools and technologies. The technology provides opportunities for misuse. For instance, newspapers commonly report stories about identity theft on the Internet, body parts being sold, violation of copyright and privacy, illegal monitoring, terrorists and criminals using the Internet for their own ends, and crimes like the spreading of child pornography.

A great deal of literature on computer ethics discusses the laws and regulations concerning such issues and the use and misuse of the Internet, freedom of speech, privacy, intellectual property, and professional conduct. However, ethics reach beyond the law. Modern technology is developing at a tremendous speed, and traditional ethical theory and every day morals do not always help us to make the right decisions when dealing with computer technology. A system administrator comes into contact with privileged information and must protect the confidentiality of all such information. However, a system administrator may be pressured into exposing such information, for instance snooping in employees' e-mail. The knowledge of ethical principles may serve as a help in order to decide how to act morally under pressure.

### 3.3. *Ethical principles*

Moral truths are those truths that are accepted by many people, despite differences in culture and religion. One example is Human rights. Others could be deeds that create peace and harmony between people. Ethical rules like "*do not steal or cheat, keep promises, pay your debts and do not lie*" ensure that we live good lives in interaction with other people. So how do we decide what is good, fair and honest, in short, how do we behave ethically? One way of looking at it is to place focus on what we ought to do, not what we want to do, which means acting in a certain way according to moral rules. For instance, in a professional context, doing good ethically means doing a good job. On the other hand, good qualities do not necessarily lead to good deeds. Intelligence and courage are good qualities, but can be used in harmful ways, for instance for robbing banks, fraud etc.

A useful ethical theory makes it possible for us to examine moral problems, reach conclusions through logical reasoning and defend the conclusions. Philosophers have always discussed moral matters. Formal study of ethics goes back to the Greek philosopher Socrates. Philosophers have proposed many ethical theories. Traditionally, there are several moral perspectives, for instance ethical principles of consequence (utilitarianism), duty and virtue, formulated by the philosophers John Stuart Mill, Immanuel Kant and Aristotle respectively. A few words about these moral perspectives will help to understand different views on ethics, and show that as individuals, people think differently about moral values. One should regard these theories as tools, or as the 'operating system' for our behavior as human beings. The theories exist as moral perspectives underlying our actions, but are not visible in our every-day lives.

976   S. Fagernes, K. Ribu

There are three main categories of practical ethical theories, or normative ethics. *Normative ethics* is the philosophy of practical moral standards that tell us what is right and what is wrong, and how to live moral lives. They focus on the virtues and good habits that we should acquire, the duties that we should follow, or the consequences of our behavior. These three categories are the ethics of duty (deontology), consequence and virtue.

The *theory of duty* (deontology) focuses on the study of right and wrong, obligation, permissions and duty. Theories of duty propose standards of morality, moral codes and moral rules. One rule that is well known is the so-called golden rule: "Do unto others as you would have them do unto you". This rule focuses on good deeds. Many companies today have moral codes for company behavior based on this principle.

Another way of looking at ethics of duty is to consider what it means to do good by acting according to good will. Good will is the only thing that can be called good without qualification. It is not the same as good deeds, but good in itself. Doing the right thing with no thought of the consequences is an act of duty and good will.

*Utilitarianism* or the *ethics of consequence* (from the Latin utilis, useful) is a theory of ethics based on quantitative maximization of good for society or humanity. This good can be viewed as happiness or pleasure. Utilitarianism is a form of consequentialism, meaning that an act should be judged by the consequences of that act. Utilitarianism is sometimes summarized as "*The greatest happiness for the greatest number of people*". Sometimes people say: 'If I had known the consequences, I would not have acted in this way'. Knowing what is good in a consequentialistic perspective may become more clear in hindsight.

*Virtue ethics* may be identified as the one that emphasizes the virtues, or moral character of a person, in contrast to the approach which emphasizes duties or rules. In philosophy, the phrase *virtue ethics* refers to ethical systems that focus primarily on what sort of person one should try to be. According to virtue ethicists the aim of all humans is to lead a good, happy and fulfilling life.

An ethical principle is a tool for considering whether an act is morally right or wrong. In order to illustrate the difference in the three perspectives described above, we can say that in the case of rescuing a drowning person, a *utilitarian* will point to the fact that the consequences of jumping into the water and pulling the person out will maximize everybody's well-being, the rescuer's own included, since good deeds are rewarded, a *duty ethicist* will state that a rescue act is in accordance with the moral rule "*do unto others as you would be done by*", and not think of any personal gain, whereas a *virtue ethicist* simply knows that rescuing the person is a charitable act, and will do so accordingly.

## 4. Professional ethics

Professional ethics differ from general ethics. The professional is an expert in a field that the customer knows little about, for instance computer science or medicine. Customers rely on the knowledge, expertise and honesty of the professional. The products (medicine, computer networks, bridges, investment advice) affect large numbers of people. A computer professional's work can affect the life, health, finances, freedom and future of a client or a group of people. A professional can cause great harm through dishonesty, carelessness or incompetence. People are often unable to protect themselves, not being involved in the choice of tool or product. Therefore, computer professionals have responsibilities to the general public.

Professional ethics concern one's conduct of behavior and practice when carrying out professional work. Any code may be considered to be a formalization of experience into a set of rules. A code is adopted by a community because its members accept the rules, including the restrictions that apply. Apart from codes of ethics, professional ethics also concern matters such as professional indemnity. No two codes of ethics are identical. They vary by cultural group, by profession and by discipline.

It is often assumed that universal ethical principles exist. However, it is a fact that in some cultures, certain behaviors are unacceptable, but in other cultures the opposite may be true, for instance when it comes to the question of software piracy, several viewpoints exist. Richard Stallman of the Free Software Foundation [5] for instance has argued at length about the superiority of freedom over protectionism in software licensing.

In many situations, there are several options that may be ethically acceptable. It is therefore misleading to divide all acts into two categories: ethically right and ethically wrong. One should rather view acts as ethically obligatory, prohibited or acceptable. One way to try to define these categories are by creating codes of ethics.

After episodes like the Enron scandal of 2001, focus is increasingly placed on ethical behavior in business.[2] The Enron Corporation was an American energy company based in Houston, Texas, United States. Enron employed around 21,000 people and was one of the world's leading electricity, natural gas, pulp and paper, and communications companies. It went bankrupt in 2001, when it was revealed that its reported financial condition was sustained mostly by systematic accounting fraud. It was the biggest bankruptcy in US history, costing 4,000 employees their jobs. The scandal has become a popular symbol of willful corporate fraud and corruption [30].

Many large companies today are being pressured into acting morally, and are forced to take ethics seriously as a part of their culture. There is a heightened awareness of ethical issues in many societies. Attitudes change, and hopefully, corporate businesses will place ethics higher than profit. Corruption can then be more easily revealed, and experience will show that good ethics will pay in the long run. One reason is that customers are more likely to trade with a business known to be honest and trustworthy. Also, employees will join and faithfully serve a company that treats its workers with loyalty and respect.

Although assuming that doing the right thing pays, it is still not easy to know the best way to balance duties to shareholders, the rights of employees, service to customers and obligations to society. But a good ethical policy can help develop a more honest corporate culture.

According to Deloitte and Touche, *"companies that follow both the letter and the spirit of the law by taking a 'values-based' approach to ethics and compliance will have a distinct advantage in the marketplace"*. Deloitte has suggested guidelines for writing a code of ethics, emphasizing how an organization should develop its own code to suit the needs of *"its personnel in defining expected behaviors and in addressing the risks, challenges, and customs in the countries in which it operates, as well as to fit their specific industry and situation"* [23]. The code should not be written as a list of 'do nots' but state expected

---

[2]Editorial comment: the Sarbanes–Oxley (SOX) act was introduced as a measure to avoid future debacles in the US.

behavior, and be global in scope. This example demonstrates that serious attempts are being made in the world of business to behave ethically, although there are, unfortunately, numerous examples of the opposite. However, companies risk having their unethical activities exposed by whistleblowers, who are within their rights to do so, and may be protected by whistleblower protection acts.

### 4.1. *Ethical and professional conduct in system administration*

It is useful to have general guidelines for professional and ethical conduct like the codes of ethics referred to above, but ethical dilemmas arise in unexpected situations, and as a systems administrator you may be at loss. What do you do when for instance you are asked by a colleague or your boss to read someone else's e-mail, or you come across information that reveals plans for sabotaging the company, stealing equipment and reselling via online auction, or illegal financial activities in your company. There are numerous such examples, and you may be faced with a choice: Either 'blow the whistle' and risk losing your job, or keep the information to yourself, having to live with the knowledge that you have maybe not acted ethically.

Regrettably, there are no simple answers to these dilemmas. Moral philosophy assumes that people are rational and make free choices since free choice and the use of rational judgment are characteristics of human beings. Ethical principles may help us out in difficult situations. However, it is human to make mistakes. Besides, a person may not be making a free choice when he or she risks losing a job. Difficult dilemmas arise in professional life, for instance, admitting to a customer that your software is faulty, declining a job for which you are not qualified or speaking out when you know something is wrong takes courage, and is not always rewarded.

### *The role of the system administrator*

A system administrator's job is to ensure proper operation, support, and protection of a company's network services and to take all security measures to protect the computers and the data contained on them. System administrators have access to an enormous amount of information, and therefore have great power and responsibility. This responsibility also includes abiding by copyright laws, as companies caught using pirated software have significant legal and financial liabilities. Protecting the integrity of information includes ensuring that unauthorized users do not access or make changes to data not belonging to them. System administrators are often blamed for copyright violations found on their networks.

Generally, smaller companies employ one or two system administrators to look after the organization. The scenario is different in larger organizations, where there is a larger number of system administrators, each responsible for a specific role within the system.

Unlike many professions, there is no single path to becoming a system administrator. In some cases, candidates are required to possess industry certifications before being considered, but since few colleges or universities have specific programs for system administration, many systems administrators are self-taught. This means that a system administrator may be lacking in knowledge of legal matters and ethical and professional conduct. We therefore seek to give an overview of some important topics that may be useful in order to become aware of ethical questions and how to solve dilemmas. We present references to international laws and regulations, ethical codes and online help-centers. It is our aim to enhance reflection and conscious decision-making.

## 4.2. *Codes of ethics*

Codes of ethics are concerned with a range of issues, including academic honesty, adherence to confidentiality agreements, data privacy, handling of human subjects, impartiality in data analysis and professional consulting, professional accountability resolution of conflicts of interest and software piracy.

Many computer professionals are members of organizations like the ACM or the IEEE, and so have agreed to be bound by one of the following:
- The ACM Code of Ethics and Professional Conduct.
- The ACM/IEEE Software Engineering Code of Ethics and Professional Practice.
- The IEEE Code of Ethics.

These codes are rather general in their phrasing. They present lists of do's and don'ts, which are easy to agree with, but do not necessarily help us out when we are faced with serious ethical dilemmas. The ACM Code includes general moral imperatives, such as "*avoid harm to others*" and "*be honest and trustworthy*". The code also included more specific professional responsibilities like "*acquire and maintain professional competence*" and "*know and respect existing laws pertaining to professional work*" [23]. The IEEE Code of Ethics includes such principles as "*avoid real or perceived conflicts of interest whenever possible*" and "*be honest and realistic in stating claims or estimates based on available data*" [24].

The Accreditation Board for Engineering Technologies (ABET) has long required an ethics component in the computer engineering curriculum. And in 1991, the Computer Sciences Accreditation Commission/Computer Sciences Accreditation Board (CSAC/CSAB) also adopted the requirement that a significant component of computer ethics be included in any computer sciences degree granting program that is nationally accredited. Professional organizations in computer science recognize and insist upon standards of professional responsibility for their members.

### *The ACM software engineering code of ethics and professional practice*

The Association for Computing Machinery (ACM) has created a code of professional ethics for software professionals. "*Commitment to ethical professional conduct is expected of every member (voting members, associate members, and student members) of the Association for Computing Machinery (ACM)*". This Code, found at the URL http://www.acm.org/constitution/code.html, "*consists of 24 imperatives formulated as statements of personal responsibility, and identifies the elements of such a commitment. It contains many, but not all, issues professionals are likely to face... It is understood that some words and phrases in a code of ethics are subject to varying interpretations, and that any ethical principle may conflict with other ethical principles in specific situations. Questions related to ethical conflicts can best be answered by thoughtful consideration of fundamental principles, rather than reliance on detailed regulations"* [24].

The code says:

*As an ACM member I will...*
    1.1. Contribute to society and human well-being.
    1.2. Avoid harm to others.
    1.3. Be honest and trustworthy.
    1.4. Be fair and take action not to discriminate.
    1.5. Honor property rights including copyrights and patents.
    1.6. Give proper credit for intellectual property.
    1.7. Respect the privacy of others.
    1.8. Honor Confidentiality.

## ACM/IEEE-CS joint task force on software engineering ethics and professional practices

Software engineers shall commit themselves to making the analysis, specification, design, development, testing and maintenance of software a beneficial and respected profession. In accordance with their commitment to the health, safety and welfare of the public, software engineers shall adhere to the following 'Eight Principles':

PUBLIC – Software engineers shall act consistently with the public interest.
CLIENT AND EMPLOYER – Software engineers shall act in a manner that is in the best interests of their client and employer consistent with the public interest.
PRODUCT – Software engineers shall ensure that their products and related modifications meet the highest professional standards possible.
JUDGMENT – Software engineers shall maintain integrity and independence in their professional judgment.
MANAGEMENT – Software engineering managers and leaders shall subscribe to and promote an ethical approach to the management of software development and maintenance.
PROFESSION – Software engineers shall advance the integrity and reputation of the profession consistent with the public interest.
COLLEAGUES – Software engineers shall be fair to and supportive of their colleagues.
SELF – Software engineers shall participate in lifelong learning regarding the practice of their profession and shall promote an ethical approach to the practice of the profession. See [24].

## IEEE code of ethics

The Institute of Electrical and Electronics Engineers, IEEE, also has a code of ethics, which is somewhat more explicit. It is found in [7].

We, the members of the IEEE, in recognition of the importance of our technologies in affecting the quality of life throughout the world, and in accepting a personal obligation to our profession, its members and the communities we serve, do hereby commit ourselves to the highest ethical and professional conduct and agree:

1. To accept responsibility in making engineering decisions consistent with the safety, health and welfare of the public, and to disclose promptly factors that might endanger the public or the environment;
2. To avoid real or perceived conflicts of interest whenever possible, and to disclose them to affected parties when they do exist;
3. To be honest and realistic in stating claims or estimates based on available data;
4. To reject bribery in all its forms;
5. To improve the understanding of technology, its appropriate application, and potential consequences;
6. To maintain and improve our technical competence and to undertake technological tasks for others only if qualified by training or experience, or after full disclosure of pertinent limitations;
7. To seek, accept, and offer honest criticism of technical work, to acknowledge and correct errors, and to credit properly the contributions of others;
8. To treat fairly all persons regardless of such factors as race, religion, gender, disability, age, or national origin;
9. To avoid injuring others, their property, reputation, or employment by false or malicious action;
10. To assist colleagues and co-workers in their professional development and to support them in following this code of ethics.

A copy of the code can be downloaded at the IEEE web-site.

## SAGE

The System Administrators Guild, SAGE, has its own code of ethics: The System Administrators' Code of Ethics.

The SAGE 'vow' is the following: *We as professional System Administrators do hereby commit ourselves to the highest standards of ethical and professional conduct, and agree to be guided by this code of ethics, and encourage every System Administrator to do the same.*

The code is found at [20], and contains advice to the system administrator in the form of a vow, for instance, "I will be honest in my professional dealings, and forthcoming about my competence and the impact of my mistakes". The code deals with professionalism, personal integrity. laws and policies, eduction, and professional and ethical responsibilities. It is interesting to note that the code places emphasis on self-education, especially concerning laws, regulations and policies, as well as the responsibility for updating technical knowledge and work-related skills. System administrators often lack formal education, and the code makes it clear that every system administrator is personally responsible for acquiring appropriate knowledge of technical, legal and moral matters.

## The Online Ethics center for engineering and science

The Online Ethics center [12] contains advice on such topics as engineering practice, responsible research, computers and software, diverse workplace, and gives examples of engineers and scientists who have behaved wisely in difficult situations. The Center's mission is to *provide engineers, scientists, and science and engineering students with resources for understanding and addressing ethically significant problems that arise in their work, and to serve those who are promoting learning and advancing the understanding of responsible research and practice in science and engineering* [26].

The Online Ethics center also gives guidelines for citation of web-pages, in order to avoid plagiarism: *The American Psychological Association's recommendations for citing electronic sources, according to the 5th edition of the APA Publication Manual (2001). Citation of the entire website rather than a specific document requires only the address of the site (i.e. http://onlineethics.org). Citation of a specific document requires the author of the page (if listed), the title of the page, the year the page was created or last modified (found below the left-hand menubar on the graphics version of each page on our site and on the third line of the text version), the title of the website, the date you accessed the page, and the URL.*

MLA Guidelines for avoiding plagiarism are found at [12].

*The Modern Language Association recommends the following format for citing individual pages in professional sites on the web. A proper MLA citation should at the least include the author of the page (if listed), the title of the page, the title of the website, the date the page was created or last modified (found below the left-hand menubar on the graphics version of each page on our site and on the third line of the text version), the sponsoring organization, the date you accessed the page, and the URL.*

### Ethics Help-Line

The Ethics Help-Line is intended to provide advice for engineers, scientists, and trainees encountering ethical problems in their work. A principal goal of this Help-Line is to assist scientists and engineers in maintaining high ethical standards and in acting wisely when confronted with multiple and potentially conflicting responsibilities, even where this may lead to conflicts with organizational superiors.

The Ethics Help-Line is sponsored by the Online Ethics Center for Engineering and Science and is cosponsored by the Institute of Electrical and Electronics Engineers (IEEE), and the National Institute for Engineering Ethics (NIEE) [14].

A section of the online ethics center deals with computers and software. '*This section of the OEC contains cases, discussions, and ethical guidelines bearing on the professional responsibilities of computer scientists, computer engineers, and software designers and engineers. These range over fields such as computer theory, computer architecture, and systems engineering*'.

One of the scenarios presented explores the ethical issues involved in the electronic monitoring of the secretarial staff's e-mail [21].

### 4.3. *Whistleblowing*

If you discover unethical or illegal actions at work like a violation of law, rule, regulation or a direct threat to public interest, fraud, health, safety violations and corruption, you must make a decision about what to do with this information. *Whistleblowing* is the term used to define an employee's decision to disclose this information to an authority, for instance your boss, the media or a government official.

Whistleblowing takes moral courage. It means acting according to values of right and wrong when dominant values decree otherwise, being brave enough to stand up to people who have power, considering and understanding the possible consequences and still making the choice to take the action. Despite the dilemmas that potential whistleblowers face, there are increasing numbers of people who are prepared to question corrupt or negligent acts in the workplace, challenging authority by speaking out, even where it may mean risking jobs, career or safety to do so.

### Whistleblower rights

Whistleblowers have rights. Courts have awarded whistleblowers their actual wage loss until they find another job, the difference between their wages for a reasonable time in the future, lost benefits, reinstatement in the previous job, attorney's fees, and punishment, damages. The major question is whether and how to blow the whistle. The Public Interest Disclosure Act 1998 protects workers who blow the whistle about wrong doing. It applies where a worker has a reasonable belief that their disclosure tends to show one or more of the following offenses or breaches:

- A criminal offense.
- The breach of a legal obligation.
- A miscarriage of justice.
- A danger to the health and safety of any individual.
- Damage to the environment; or
- Deliberately covering up of information tending to show any of the above.

Whistleblower protection laws are intended to make it safe for employees to disclose company misconduct. Legal protection for whistleblowing varies from country to country. In the US, the Office of Special Counsel enforces the whistleblower protection provision of the Civil Service Reform Act of 1978, Pub. L. No. 95-454, 92 Stat. 1111 (1979), as amended by the Whistleblower Protection Act (WPA) of 1989. The WPA makes it illegal to take or threaten to take a *personnel action* against a federal employee because the employee has made a protected disclosure. *Personnel action* is broadly defined to include virtually any employment-related decision that has an impact on an employee at the worksite.

In the UK, the Public Interest Disclosure Act 1998 (PIDA) [18]:

> ... provides a framework of legal protection for individuals who disclose information about matters of malpractice, and protects whistleblowers from victimisation and dismissal. Under PIDA, whistleblowers must use prescribed channels for making disclosures in order to retain the Act's protection. The Act's preference is that the disclosure be made to the employer itself or an appropriate public authority, rather than the media. In this manner, the Act protects an employee if he or she makes a disclosure in good faith to the employer.

It is not alway easy to determine when whistleblowing may be the right thing to do. Informing on illegal or unethical practices in the workplace is difficult for several reasons. Ethical justifications for whistleblowing are frequently uncertain. An important question to answer is: when is it in the public interest to do so? Whistleblowers should be quite certain of their case before they decide to blow the whistle.

## The cases of Inez Austin and Per Yngve Monsen

Although whistleblowers are protected by law, they may experience great personal strain and professional problems, with the loss of work and reputation as a result. They may become ostracized and unpopular.

The online Ethics Center for Engineers and Scientists contains several examples of moral conduct. One is the case of the whistleblower Inez Austin, who received the Scientific Freedom and Responsibility Award from the American Association for the Advancement of Science (AAAS) *for her courageous and persistent efforts to prevent potential safety hazards involving nuclear waste contamination* as a senior engineer at the Hanford Site in Washington state, a 586-square-mile former plutonium production facility. According to the AAAS,

> ... her stand in the face of harassment and intimidation reflects the paramount professional duty of engineers to protect the public's health and safety and has served as an inspiration to her co-workers.

The Online Ethics Center presents the case of Inez Austin as an example of someone who followed her ethical convictions in the face of overwhelming adversity and refused to sanction a procedure she believed to be unsafe. In June 1990, Inez Austin refused to approve a plan to pump radioactive waste from an aging underground single-shell tank at the Hanford Site to a double-shell tank, because she believed that the process was too dangerous. By blowing the whistle, her life changed, although she came to be regarded highly by environmental and ethics groups, she was subjected to a career-destroying combination of harassment, bureaucratic maneuvering and ostracization. A second whistleblowing incident a few years later led to the loss of her job [13,26].

This is the dilemma that a whistleblower must consider. Whistleblowers may be faced with harassment and legal pursuit. Although they receive support from media and government, they may later have difficulties finding a job, for who will hire a 'troublemaker?' In all probability, many companies have something to hide, and a well-known whistleblower is seen as a risk.

The case of Per Yngve Monsen, the Norwegian employee at Siemens who blew the whistle on the company for corruption and financial fraud, demonstrates the hazards a whistleblower may be exposed to. He was sacked, and even though he won the court case, has been on sick-leave and out of work for nearly two years. In the media he became a hero, and like Inez Austin, an example of good moral conduct. But from his experience with a powerful company, he advises whistleblowers to be careful, and give great consideration to the risks involved. "If I had known what I know today, I would never have acted like I did", is his conclusion, reported by the Norwegian newpaper Aftenposten. The same newspaper reported on November the 6th, 2006, that inquiries show that people on the whole are afraid to report illegal or unethical activities because of the consequences.

This is indeed unfortunate, and demonstrates in full that when the individual is up against a powerful international company, the results can be disastrous for the person involved. If you do discover illegal or unethical behavior, do not hesitate to get in touch with whistle-blower protection organizations and receive legal advice before you blow the whistle.

## 5. Cyber law

The phrase *Cyber law* usually refers to rules and laws governing the use of communications technology and informations systems. The big challenge regarding this topic is the fact that the Internet spans on a world wide basis, while laws typically varies from country to country. Also, laws regulating our physical society, are not necessarily sufficient to cover all important aspects of cyber space.

Issues covered by cyber law are:

- *Intellectual property.*
- *Privacy.*
- *Freedom of expression.*
- *Jurisdiction.*

### 5.1. *Intellectual property*

Intellectual property denotes the specific legal rights which authors, inventors and other IP holders may hold and exercise, and not the intellectual work itself [8].

*Intellectual property* is *a legal entitlement which sometimes attaches to the expressed form of an idea, or to some other intangible subject matter.*

The motivation of intellectual property is to protect inventors, authors etc by giving them the *ownership* of the result of their brain, similarly to the way people own their physical property. The term intellectual property reflects the idea that this subject matter is the product of the mind of the intellect, and that intellectual property rights may be protected by law in the same way as any other form of property.

Electronic media and the WWW creates new challenges for the protection of intellectual property, and new controversies about intellectual property law. Some of the problems posed by the technology itself are storage and copying of information like text, sound and graphics, and the compressed formats like MP3, which makes copying and distribution of music simple. Downloading and copying text material from the Internet is simple, and makes cheating, for instance, in schools and universities easy. Peer-to-peer technology permits easy transfer of files over the Internet, affecting typically the music and film industry [1].

Digital Rights Management (DRM) is a technology used by copyright owners to control access to digital data. The term often is confused with copy protection, however, DRM restricts the actual use of digital content by way of installing the technology as part of the design. Today, illegal copying and distribution of copied material can result in severe penalties.

Digital Rights Management is a controversial topic. Copyright holders claim that it is their right to restrict copying by the use of technology, because copyright is not respected. On the other hand, the Free Software Foundation is a severe critic of DRM, stating that the word 'Rights' is misleading, and that the term 'Digital Restrictions Management' would be more appropriate. The Electronic Frontier Foundation consider some DRM schemes to be anti-competitive.

We have four main categories of intellectual property; copyright, patents, trade marks and designs.

### 5.2. *Copyright*

Copyright is a protection that covers published and unpublished literary, scientific and artistic works, whatever the form of expression, provided such works are fixed in a tangible or material form. Copyright laws grant the creator the exclusive right to reproduce, prepare derivative works, distribute, perform and display the work publicly. Copyright protects the actual piece of work, not the ideas, knowledge or methods used for creating that work.

*Copyright gives the creators of a wide range of material, such as literature, art, music, sound recordings, films and broadcasts, economic rights enabling them to control use of their material in a number of ways, such as by making copies, issuing copies to the public, performing in public, broadcasting and use on-line. It also gives moral rights to be identified as the creator of certain kinds of material, and to object to distortion or mutilation of it* [8]. Copyright laws enables those who create a piece of work to decide if and how the work can be used, reproduced etc. It is possible to let others use the work freely, or use and distribution must be paid for. There is no such thing as an International Copyright. Protection against unauthorized use in a particular country depends on the national laws of that country. However, most countries do offer protection to foreign works under certain conditions, and these conditions have been greatly simplified by international copyright treaties and conventions. For instance, the Berne Convention for the Protection of Literary and Artistic Works defines the nature and use of intellectual property [2].

Article 2 in the convention defines Protected Works as:
- Literary and artistic works.
- Possible requirement of fixation.
- Derivative works.
- Official texts.
- Collections.
- Works of applied art and industrial designs.
- News.

According to the convention, *the expression 'literary and artistic works' shall include every production in the literary, scientific and artistic domain, whatever may be the mode or form of its expression, such as books, pamphlets and other writings; lectures, addresses, sermons and other works of the same nature; dramatic or dramatico-musical works; choreographic works and entertainments in dumb show; musical compositions with or without words; cinematographic works to which are assimilated works expressed by a process analogous to cinematography; works of drawing, painting, architecture, sculpture, engraving and lithography; photographic works to which are assimilated works expressed by a process analogous to photography; works of applied art; illustrations, maps, plans, sketches and three-dimensional works relative to geography, topography, architecture or science* [2].

Copyright laws vary from country to country, but as a rule do not provide less copyright protection than the Berne Convention, as long as the country is a member.

Software is not specially mentioned in the convention, but proprietary software is also protected by copyright law. The Software and Information Industry association (SIIA) and the Business Software Alliance (BSA) estimate that the value of pirated software world-wide has been approximately 11–13 billion US dollars per year for many years. Of course these figures are insecure as the activity is illegal, but the gap between the sales number for software licenses and the calculated actual number of applications installed on computers gives some idea of the range of software piracy, as estimated by the SIIA. In China, for instance, dozens of factories are said to produce pirated CDs for export to other countries, even though laws exist to protect intellectual property rights, particularly for foreign work, but the laws are not enforced [1].

## Fair use

There is much discussion about the ethics of copying. A difficult area of copyright law is the so-called 'Fair Use'. The actual specifics of what is acceptable will be governed by national laws, and although similar, laws will vary from country to country. Cases dealing with fair use can be complex, as decisions are based on individual circumstances and judgments. To avoid problems, if in any doubt, it is advised to always get the permission of the owner, prior to use.

The UK Copyright Service provides copyright registration for original works by writers, musicians, artists, designers, software providers, authors, companies, organizations and individuals. Known as Copyright Witness internationally, and the UK Copyright Service in the UK, the service supports international copyright protection by securing independent evidence that will help prove originality and ownership in any future claims or disputes [27].

The UK copyright service defines fair use in the following way: *In copyright law, there is a concept of fair use, also known as; free use, fair dealing, or fair practice. Fair use sets out certain actions that may be carried out, but would not normally be regarded as an infringement of the work. The idea behind this is that if copyright laws are too restrictive, it may stifle free speech, news reporting, or result in disproportionate penalties for inconsequential or accidental inclusion.*

Under fair use rules, it may be possible to use quotations or excerpts, where the work has been made available to the public (i.e. published), provided that:

- The use is deemed acceptable under the terms of fair dealing.
- That the quoted material is justified, and no more than is necessary is included.
- That the source of the quoted material is mentioned, along with the name of the author.

Typical free uses of work include:

- Inclusion for the purpose of news reporting.
- Incidental inclusion.
- National laws typically allow limited private and educational use.

'Incidental inclusion' is where part of a work is unintentionally included. A typical example of this would be a case where holiday movie inadvertently captured part of a copyright work, such as some background music, or a poster that just happened to be on a wall in the background.

### Copyright and the Internet

The Internet and 'public domain' are not synonymous. Work published on the Internet is not automatically placed it in the public domain. In the US, material found on the web may be copied freely only if the information is created by the federal government, if the copyright has expired or the copyright has been abandoned by the holder.

The following example illustrates the problem: Google is planning to scan books and publish them through their service Google Print. This undoubtedly violates copyright protection. An author has two rights: The right to create a piece of work, and the right to publish that work. Scanning a book means creating copies of that book, and publishing the book on the Internet means making copies available for the public. The author must agree to this approach, there is no such thing as a silent agreement. Google's arguments are that this must be regarded as fair use, but in most European countries, there is no rule which makes the copying of books fair use. In all probability, it is not acceptable under US law either. Some countries have specific laws defining when it is legal to use parts of published work without the author's agreement, for instance for private and educational purposes. But making a piece of intellectual property public on the Internet is not covered by any of these regulations [28].

### 5.3. *Patenting*

A patent is related to inventions that are supposed innovative (or new), useful and *non-obvious*. An inventor may apply for a patent, and after the patent has been issued, the patent holder has an *exclusive right to commercially exploit the invention for a certain period of time (typically twenty years from the filing date of a patent application)*.

A patent gives an inventor the right over the invention for a limited period of time, restricting others from copying, selling or using the invention. Most patents today are for improvements in known technology.

Specific conditions must be fulfilled to get a patent. Major ones are that the invention must [8]

- Be new or at least undocumented. The invention must not form part of the 'state of the art', meaning everything that has been made available to the public before the date of applying for the patent. This includes published documents and articles, but also use, display, description.
- Involve an inventive step. As well as being new, the invention must not be obvious from the state of the art.
- Be industrially applicable. This condition requires that the invention can be made or used in any kind of industry.

**5.4.** *Trade marks*

A trade mark is *a distinctive sign which is used to distinguish the products or services of one business from those of another business (Wikipedia).*

A trade mark is *any sign which can distinguish the goods and services of one trader from those of another. A sign includes words, logos, colours, slogans, three-dimensional shapes and sometimes sounds and gestures. A trade mark is therefore a 'badge' of trade origin. It is used as a marketing tool so that customers can recognise the product of a particular trader. To be registrable in the UK it must also be capable of being represented graphically, that is, in words and/or pictures* [8]. Copying a logo or slogan is a violation of copyright law.

**5.5.** *Designs*

Design is defined as product appearance, *of the whole or a part of a product resulting from the features of, in particular, the lines, contours, colours, shape, texture or materials of the product itself or its ornamentation.* Design is usually protected by three legal rights: registered designs, unregistered design right and artistic copyright.

Design registration gives the owner a monopoly on their product design, i.e. the right for a limited period to stop others from making, using or selling a product to which the design has been applied, or in which it has been incorporated, without their permission and is additional to any design right or copyright protection that may exist automatically in the design [8].

**5.6.** *Privacy*

The Internet with its millions of users, represents several threats to personal security and privacy. Sensitive information is published openly, including telephone conversations, electronic mail, trade secrets and health records. The technological development makes everybody vulnerable to unwanted snooping by governments, business competitors, terrorists, hackers and thieves, in addition to identity theft and fraud. *Privacy International (PI)* is a human rights group formed in 1990 as a watchdog on surveillance and privacy invasions by governments and corporations. PI is based in London, England, and has an office in Washington, DC. PI has conducted campaigns and research throughout the world on issues ranging from wiretapping and national security, to ID cards, video surveillance, data matching, police information systems, medical privacy, and freedom of information and expression [17].

The protection of privacy is strong in international law. Privacy is a fundamental human right and one of the most important human rights of the modern age. It is protected in the Universal Declaration of Human Rights, the International Covenant on Civil and Political Rights, and in many other international and regional human rights treaties. Nearly every country in the world includes a right of privacy in its constitution. Modern constitutions include specific rights to access and control one's personal information. International agreements that recognize privacy rights are for instance the International Covenant on Civil and

Political Rights or the European Convention on Human Rights (full name: The Convention for the Protection of Human Rights and Fundamental Freedoms, Rome 4 November 1950). Article 8 in this convention states that:

> Everyone has the right to respect for his private and family life, his home and his correspondence. There shall be no interference by a public authority with the exercise of this right except such as is in accordance with the law and is necessary in a democratic society in the interests of national security, public safety or the economic well-being of the country, for the prevention of disorder or crime, for the protection of health or morals, or for the protection of the rights and freedoms of others.

Privacy International (PI) is a human rights group formed in 1990 as a watchdog on surveillance by governments and corporations. PI is based in London, England, and has an office in Washington, DC PI has conducted campaigns throughout the world on issues ranging from wiretapping and national security activities, to ID cards, video surveillance, data matching, police information systems, and medical privacy.

Privacy protection is also controlled by individual users. Internet users employ various programs and systems that provide privacy and security of communications. These include encryption, anonymous remailers, proxy servers and digital cash. Not all tools effectively protect privacy [17].

### *Phil Zimmermann and PGP – Pretty Good Privacy*

Philip R. Zimmermann is the creator of Pretty Good Privacy, an email encryption software package. Zimmermann, a software engineer with more than 20 years of experience in cryptography and data security originally designed PGP as designed as a human rights tool, and published it for free on the Internet in 1991. As a consequence, Zimmermann was subjected to three years of criminal investigation, because the US government maintained that US export restrictions for cryptographic software were violated when PGP spread worldwide. However, the government dropped the case in 1991, and PGP has since become the most popular email encryption software in the world. Zimmermann founded PGP Inc. in 1991, a company that was acquired by Network Associates Inc (NAI) in December 1997. In 2002 PGP was acquired from NAI by a new company called PGP Corporation. Zimmermann is now a consultant for companies and industry on cryptography. His home page is found at http://www.philzimmermann.com/.

### **5.7.** *Privacy and Human Rights*

The Universal Declaration of Human Rights from 1948 article 12 states that "*No one shall be subjected to arbitrary interference with his privacy, family, home or correspondence, nor to attacks upon his honour and reputation. Everyone has the right to the protection of the law against such interference or attacks*".

According to the International Covenant on Civil and Political Rights – 1966 article 17:

1. *No one shall be subjected to arbitrary or unlawful interference with his privacy, family, home or correspondence, nor to unlawful attacks on his honour and reputation.*

2. *Everyone has the right to the protection of the law against such interference or attacks.*

Public information is defined as information knowable by all, for instance information you have provided to an organization that has a right to share it with other organizations. An example of such information is the telephone directory. On the other hand, personal information is not part of a public record, for example your religious and political beliefs. If you personally disclose this information to an organization with the right to inform other organizations, it becomes public information.

### 5.8. *Freedom of speech on the Internet: the role of Freenet*

> *The Internet is a powerful and positive forum for free expression. Internet users, online publishers, library and academic groups and free speech and journalistic organizations share a common interest in opposing the adoption of techniques and standards that could limit the vibrance and openness of the Internet as a communications medium. Indeed, content 'filtering' techniques already have been implemented in ways inconsistent with free speech principles, impeding the ability of Internet users to publish and receive constitutionally protected expression* [9].

Article 19 of the Universal Declaration of Human Rights and Article 10 of the European Convention on Human Rights are international human rights proclamations that support the freedom of speech, although implementation of this fundamental human right is unfortunately lacking in many countries. However, the right to freedom of speech is not unlimited, many governments may prohibit certain forms of expression. According to international law, restrictions on free speech must be provided by law, pursue an aim recognized as legitimate, and be necessary for the accomplishment of that aim. Some of the aims that are considered legitimate are protection of the rights and reputations of others and the protection of national security and public order, health and morals. There are differences of opinion among people of different nations and cultures as to when restriction of free speech meets these criteria [31].

The Internet has opened new possibilities for achieving freedom of speech, giving anybody the opportunity to publish whatever they want independent of legal measures. Pseudonymity and data havens (such as Freenet [6]) allow unlimited free speech, because the technology guarantees that material cannot be removed or censored.

> Freenet is free software which lets you publish and obtain information on the Internet without fear of censorship. To achieve this freedom, the network is entirely decentralized and publishers and consumers of information are anonymous. Without anonymity there can never be true freedom of speech, and without decentralization the network will be vulnerable to attack.

Communications by Freenet nodes are encrypted and are routed through other nodes to make it extremely difficult to determine who is requesting the information and what its content is. Users contribute to the network by giving bandwidth and a portion of their hard drive (called the 'data store') for storing files. Unlike other peer-to-peer file sharing networks, Freenet does not let the user control what is stored in the data store. Instead, files are kept or deleted depending on how popular they are, with the least popular being discarded to make way for newer or more popular content. Files in the data store are encrypted to reduce the likelihood of prosecution by persons wishing to censor content.

Websites which are censored by national governments are often re-hosted on a server in a country with no censorship restrictions. The United States has in many ways the world's least restrictive governmental policies on freedom of speech. Many websites re-host their content on an American server to escape censorship. This is especially the case with neo-nazi and other sites promoting racial hatred, since these are illegal in many European countries.

Another organization dedicated to protecting freedom of speech on the Internet is The Electronic Frontier Foundation (EFF), found at www.eff.org. The '*EFF is a donor-funded nonprofit and depends on your support to continue successfully defending your digital rights. Litigation is particularly expensive; because two-thirds of our budget comes from individual donors, every contribution is critical to helping EFF fight, and win, more cases*'.

### 5.9. *Google and China*

The Chinese government has imposed Internet censorship in order to control or eliminate access to information on controversial and sensitive topics such as the Tiananmen Square protests of 1989, Falun Gong, Tibet, Taiwan, pornography or democracy. They have also enlisted the help of some American companies like MSN, who have subsequently been criticized by proponents of freedom of speech.

Google launched a China-based search engine that will be self-censored to avoid posting results that antagonize China's communist government.

Google.cn uses the Chinese Web suffix *.cn* in addition to the existing Chinese-language Google website hosted on servers in the USA.

> In order to operate from China, we have removed some content from the search results available on Google.cn, in response to local law, regulation or policy. Removing search results is inconsistent with Google's mission, but providing no information is more inconsistent with our mission.
>
> Senior policy counsel Andrew McLaughlin, Google.

The Google mirror *elGoog* makes it possible to search for words like 'democracy' from inside China. http://elgoog.rb-hosting.de/index.cgi.

### 5.10. *Transborder data flows*

Data protection laws can be circumvented by transferring personal information to third countries, where the national law of certain countries do not apply. The data can then be processed in these countries, called 'data havens'. This is why most data protection laws include restrictions on the transfer of information to third countries unless the information is protected in the destination country. For example, Article 12 of the Council of Europe's 1981 Convention places restrictions on the transborder flows of personal data [4]. Similarly, Article 25 of the European Directive imposes an obligation on member States to ensure that any personal information relating to European citizens is protected by law when it is exported to countries outside Europe. It states:

> The Member States shall provide that the transfer to a third country of personal data which are
> undergoing processing or are intended for processing after transfer may take place only if the third
> country in question ensures an adequate level of protection.

Therefore, there is currently growing pressure outside Europe for the passage of strong data protection laws. Those countries that refuse to adopt privacy laws may be unable to conduct certain types of information flows with Europe, particularly if they involve sensitive data. The European Commission determines a third country's system for protecting privacy, where the main principle in this determination process is that the level of protection in the receiving country must be *adequate* rather than *equivalent*. In this way, privacy protection is guaranteed from the third party, although the precise dictates of the Directive are not strictly followed.

## 6. Educational aspects

### 6.1. *Teaching ethics*

In order for computer professionals to incorporate an *ethical responsibility*, it has been stressed that it is important to have this topic introduced to them already during their educational years.

As system administration is not a common educational programme yet, most of the topics on this issue have been inspired by the more general field of Computer Science and Information Systems. However, most of these issues are also relevant for the system administration education.

There has been significant controversy in the field of teaching ethics to computer science graduates. There have been debates on all imaginable aspects of the issue, on what kind of *subjects* should be covered by the curriculum, *how* the material should be integrated (separate course or modules in other courses), and *who* should actually teach these subjects:

However, *how* to incorporate social and ethical aspects into the technical programmes has not proven trivial. Common problems have been:

- *What* should the students learn?
- *Who* should teach these untraditional subjects?
- *How* should these issues be integrated in the traditional curriculum?

*Course content and learning objectives*    The *ACM/IEEE Computing Curriculum 1991*, demanded an inclusion of ethical and social aspects in the Computer Science curriculum, therefore, the project *ImpactCS* was funded [25]. The objective of this project was to define the core content and develop a framework for integrating social and ethical topics into the computer science curriculum [29].

The project delivered three reports [11], where each of them covered different aspects of the process of including ethics in the CS curriculum. Among the final results was a very broad list of topics, which included the following *learning objectives*:

- *Responsibility of the Computer Professional for Computer Science.* This topic again contains the subtopics:
  - History of the development and impact of computer technology.
  - Why be ethical?

 – Major ethical models.
 – Definition of computing as a profession.
- *Basic Elements of Ethical Analysis for Computer Science.* The elements covered by this section are:
 – Ethical claims can and should be discussed rationally.
 – Ethical choices cannot be avoided.
 – Easy ethical approaches are questionable.
- *Basic Skills of Ethical Analysis for Computer Science*:
 – Arguing from example, analogy, and counter-example.
 – Identifying stakeholders in concrete situations.
 – Identifying ethical issues in concrete situations.
 – Applying ethical codes to concrete situations.
 – Identifying and evaluating alternative courses of action.
- *Basic elements of Social Analysis for Computer Science*:
 – Social context influences the development and use of technology.
 – Power relations are central in all social interactions.
 – Technology embodies the values of the developers.
 – Populations are always diverse.
 – Empirical data are crucial to design and development processes.
- *Basic Skills of Social Analysis for Computer Science*:
 – Identifying and interpreting the social context of a particular system.
 – Identifying assumptions and values embedded in a particular system.
 – Using empirical data to evaluate a particular implementation of a technology.

As the list shows, the project ImpactCS provides a very broad range of subjects, not supported by all members of this community. David Preston claims this list is too long, and that most of the subjects are out of scope for an ethics course to be interesting for the Computer Science students [16]. A more narrow approach is recommended, to assure better relevance and hence interest from the students. In particular, they recommend spending considerable time focusing on the students' *own* experiences, as these tend to arouse more interest and participation among the students.

### 6.2. *How should professional ethics be taught?*

After establishing the learning objectives, there is still the issue of how to determine the teaching form for this kind of material. Using *case studies* is a very common approach, as this is claimed to improve the understanding of certain ethical dilemmas and increase interest and hence participation among the students. Some authors object to this, with the argument that case studies do not bring sufficient realism to the issues, and that the students should be encouraged to bring their *own* experiences to class [16].

The form of the lectures is also heavily debated, although it seems to be agreed upon that a mixture of *group discussions*, *guest lectures*, student-prepared *panel debates* and so on, is to be preferred over more traditional teacher-led lectures. One particular issue is whether the communication between teacher and students should be *one-way*, or if it should be more negotiation-based. Regarding so-called *open sessions*, where external speakers give talks with interaction with the students, there seem to be very difficult to predict which topics appeal to the students.

There is also significant emphasis on using a variety of technical resources in the classes, for instance local, national or international newspapers, local leaflets, film, music, student-generated material, fiction etc.

*Who should teach ethics?*   One of the problems with integrating an ethics and social aspects course in the computer science or system administration curriculum, is the lack of knowledge of the teachers and faculty. The discussion has been about who should teach these subjects: faculty of engineering or teachers of philosophy or social science.

The clear advantage of having teachers from the computer science community teach computer ethics is that they are familiar with the technological aspects of the use of computers, and the work situations in which ethical dilemmas arise. The use of case studies from industry is a recommended approach, and computer professionals have first-hand knowledge of the issues that are described in well-known cases and examples. Teachers of Philosophy may tend to present ethical theory and cases in a too theoretical and academic manner, which may have little relevance for computer professionals.

### 6.3. *Teaching students to hack: a necessity or security risk?*

The emergence of diverse security threats has lead to increased focus on information security courses in the computer science education. A significant part of such courses has been the *security labs*, where students can achieve hands-on experience with important tools and concepts within information security, both regarding potential threats and how to protect the systems from diverse attacks [10].

Typical student activities in these labs are:
1. Hacking (accessing systems or networks where the students do not have access).
2. Find vulnerable systems and penetrate them.
3. Remove evidence of penetration.

The argument for teaching the students such skills have been that the system administrators must possess the same skills as the attackers in order to be able to protect the system from the attacks (know your enemy). The reason for this is mainly the need for so-called *ethical hackers*.

The term *ethical hacking* comes from the practice of skilled 'hackers' trying to break into computer systems with the goal of discovering potential vulnerabilities in the system, in order to prevent attacks. More precisely, the ethical hackers aim at answering the following questions [15]:
- What part of the system is visible to an intruder?
- What can an intruder do with that information?
- Does anyone at the target notice the intruder's activities?

Arguments in favor of this practice has been [10]:
- "that hacking skills are equivalent to audit skills as both are designed to discover flaws in the protection of data and secure operation of a system. Just as auditors test systems for security or operational flaws, hackers 'test' systems through attack;

- Knowledge of hacking skills and practice in attacking secured systems improves security by informing network administrators of how an exploit can be executed; and
- To provide the best security defense, a systems administrator must possess the same skills as the attacker".

This kind of work obviously requires people with significant skills and experience, which has lead to several universities and colleges including 'hacking' exercises as part of their computer security curriculum. Examples of such exercises have been writing port scanners, propagating viruses and exploit programs, in addition to activities like packet sniffing and injecting packets.

However, recently there has been increased focus on the downsides of this kind of organized student activity. There obviously is a significant risk of actually *educating* hackers or the 'bad guys'. Further, hacking skills does not necessarily make system administrators able to prevent such attacks.

To avoid malicious behavior from students, some universities offer security labs only at the graduate level, with the motivation that only the most mature, and hence more responsible students will have access to this kind of information. Other universities have tried avoiding hands-on content or keeping the labs away from student access.

These methods have proved to have weaknesses, so as alternative methods for educating the students, the following has been suggested [10]:

- Teaching recovery activities.
- Intensive vulnerability assessments of a simulated corporate network.
- Network forensic investigations.

The following questions are asked by the authors of [10]:

- How should students be taught to implement network and desktop security?
- Is hacking and virus-writing a pre-requisite for developing strong technical skills in detecting malicious activity?
- What course content (if any) should be off-limits?
- Will hacking skills for network administrators necessarily improve security and their employability?

## 7. Summary

The use of computers and computer networks poses ethical questions that often differ from our every-day ethical problems, mainly because of the nature of the technology itself. Computers have become universal tools, computers and networks are everywhere, shaping our lives, work and leisure time. With the rapid development of computer technology, new situations constantly arise and create policy vacuums, because the answers to ethical dilemmas are not clear. The law will not always be a help, because laws are designed to fit the technologies and problems of the day, and because technology is developing at such a tremendous speed. It takes time for the legal issues to catch up with the technical possibilities.

System administrators are often faced with ethical dilemmas in situations where they must make decisions, and take proper action. There are many paths to becoming a system administrator, and many lack formal education. Decisions may therefore often be founded on intuition, or be taken under pressure. The consequences can be serious, like the risk of losing a job, either through negligence, or because of being pressured into some illegal or unethical activity. For a system administrator it is therefore of the greatest importance to be acquainted with the rules, laws and policies regarding the performance of the work duties.

A system administrator must be able to make sound decisions based on reflection, knowledge, common sense and professional responsibility, and is, according to the sage code of Ethics, also obliged to educate himself/herself on technical, legal and ethical issues. In this chapter we have sought to outline some of the most important laws, ethical principles and guidelines that may assist the system administrator in fulfilling his or her professional duties.

## References

[1] S. Baase, *A Gift of Fire*, Pearson Education (2003).
[2] Berne Convention for the Protection of Literary and Artistic Works, http://www.wipo.int/treaties/en/ip/berne/trtdocs_wo001.html.
[3] P.N. Edwards, C.A. Lee and B.S. Williams, *Evaluating informations and communications technology: Perspectives for a balanced approach* (2001).
[4] EU Data Protection Directive, http://www.opsi.gov.uk/ACTS/acts1998/19980023.htm.
[5] Free Software Foundation (FSF), http://www.fsf.org/.
[6] Freenet project, http://freenetproject.org/ (2006).
[7] IEEE, *Code of ethics*, http://www.ieee.org/portal/pages/about/whatis/code.html.
[8] Intellectual Property, http://www.intellectual-property.gov.uk/index.html.
[9] Internet Free Expression Alliance, http://www.ifea.net.
[10] P.Y. Logan and A. Clarkson, *Teaching students to hack: Curriculum issues in information security*, SIGCSE, ACM (2005).
[11] C.D. Martin and E. Yale Weltz, *From awareness to action: Integrating ethics and social responsibility into the computer science curriculum*, Computers and Society (1999).
[12] Online Ethics Centre, http://onlineethics.org/cite-link.html (2006).
[13] Online Ethics Centre, http://onlineethics.org/moral/austin/index.html (2006).
[14] Online Ethics Helpline, http://onlineethics.org/helpline/index.html (2006).
[15] C.C. Palmer, *Ethical hacking*, http://www.research.ibm.com/journal/sj/403/palmer.html (2001).
[16] D. Preston, *What makes professionals so difficult: An investigation into professional ethics teaching*, Proceedings of POLICY'98: Ethics and Social Impact, ACM Press (1998).
[17] Privacy International, http://www.privacyinternational.org.
[18] Protection for Whistleblowers, http://www.ignet.gov/randp/f01c10.pdf.
[19] M.J. Quinn, *Ethics for the Information Age*, Pearson Education (2006).
[20] SAGE, *Code of ethics*, http://www.sage.org/ethics.
[21] Scenarios about computers and software, http://onlineethics.org/com/scenarios.html.
[22] J.P. Sterba, *Ethics: The Big Questions*, Blackwell Publishers (2002).
[23] Suggested Guidelines for Writing a Code of Ethics/Conduct, http://www.deloitte.com/dtt/cda/doc/content/Suggested%20Guidelines%20for%20writing%20a%20code%20of%20ethics%2C%20conduct.pdf.
[24] The ACM Software Engineering Code of Ethics and Professional Practice, http://www.acm.org/constitution/code.html.
[25] The ImpactCS Home Page, http://www.seas.gwu.edu/ impactcs/.
[26] The Online Ethics Center for Engineering and Science, http://www.onlineethics.org.
[27] The UK Copyright Service, http://www.copyrightservice.co.uk/.
[28] O. Torvund, *Copyright: An introduction* – (Norwegian), http://www.torvund.net/artikler/art-opphav.asp.
[29] L.H. Werth, *Getting started with computer ethics*. Proceedings of the Twenty-Eighth SIGCSE Technical Symposium on Computer Science Education, ACM Press (1997), 1–5.
[30] Wikipedia, 13.08.2006. http://www.wikipedia.org.
[31] Wikipedia, http://en.wikipedia.org/wiki/Freedom_of_speech.

# Subject Index

Printed and bound by CPI Group (UK) Ltd, Croydon, CR0 4YY

03/10/2024

01040333-0010